NANOMATERIALS FOR ENVIRONMENTAL PROTECTION

NANOMATERIALS FOR ENVIRONMENTAL PROTECTION

Edited by

BORIS I. KHARISOV
OXANA V. KHARISSOVA
H. V. RASIKA DIAS

WILEY

Cover Images: iStockphoto© theasis/PaulFleet/PLAINVIEW/shaunl

Published by John Wiley & Sons, Inc., Hoboken, New Jersey
Published simultaneously in Canada

Library of Congress Cataloging-in-Publication Data:

Nanomaterials for environmental protection / edited by Boris I. Kharisov, Oxana V. Kharissova, H. V. Rasika Dias.
1 online resource.
Includes bibliographical references and index.
Description based on print version record and CIP data provided by publisher; resource not viewed.
ISBN 978-1-118-84535-6 (pdf) – ISBN 978-1-118-84554-7 (epub) – ISBN 978-1-118-49697-8 (hardback) 1. Sanitary engineering–Equipment and supplies. 2. Environmental protection–Equipment and supplies. 3. Water–Purification–Materials. 4. Nanostructured materials. 5. Nanofiltration. I. Kharisov, Boris I. II. Kharissova, Oxana Vasilievna. III. Dias, H. V. Rasika.
TD192
628.028′4–dc23

2014026672

Printed in the United States of America

10 9 8 7 6 5 4 3 2 1

CONTENTS

PREFACE

Air, water, and ground pollutants are undeniably a major problem for the environment. Adverse effects caused by auto emissions, unsafe disposal of toxic waste, and harmful by-products of industrial processes on the environment and human health are well established. Generation and accumulation of contaminants is a continuous process, and therefore is an increasing concern, which is especially noticeable near large cities and industrial centers. Classic purification techniques are frequently insufficient or too expensive and cannot be relied upon as the only ways of removing pollutants from air, groundwater, and soils. It is exciting to see that nanotechnology has been used increasingly to address these environmental pollution control and remediation needs. During the past decade, most of the nanotechnology applications related to the environment have focused on the water sector: removal of toxic organic compounds with special emphasis on chlorinated aromatics, remediation of heavy metals such as chromium(VI) or radionuclides, disinfection of drinking water, and desalination of seawater. To achieve these goals, nanostructurized metals (such as silver and iron), photocatalysis and nanofiltration methods, nanoceramics, nanoadsorbents, and nanoparticle-containing membranes have been employed. Several techniques have reached pilot plant and larger scale applications, such as the use of Fe^0 nanoparticles, for remediation of organic pollutants in soils. Other uses of nanomaterials to address environmental issues are growing but are still at the research level.

In this book, we have presented a comprehensive view of applications of nanomaterials for environmental protection and remediation. The chapters have been written by well-known specialists in corresponding fields worldwide. On behalf of the Editorial Office and John Wiley & Sons, we would like to thank all participating authors for their hard work in creation of this book.

The Editors

LIST OF CONTRIBUTORS

Aracely Angulo-Molina, Departamento de Ciencias de la Salud, Universidad de las Américas Puebla, Puebla, Mexico

Marco Antonio Quiroz Alfaro, Departamento de Ciencias Químico-Biológicas, Universidad de las Américas Puebla, Puebla, México

Swechha Anand, Centre of Experimental Medicine and Surgery, Institute of Medical Sciences, Banaras Hindu University, Varanasi, India

Kevin John Pulikotil Anthony, GS Institute of Bio and Nanotechnology, Coimbatore, India

David D.J. Antia, DCA Consultants Ltd., Scotland, UK

Xin Ba, Institute of Nanoscience and Nanotechnology, College of Physical Science and Technology, Central China Normal University, Wuhan, China

Amro El Badawy, Pegasus Technical Services Inc., Cincinnati, OH, USA

Jamuna Bai A., Department of Studies in Microbiology, University of Mysore, Mysore, India

Erick R. Bandala, Departmento de Ingeniería Química, Alimentos y Ambiental, Universidad de Las Américas Puebla, Puebla, Mexico

Vipul Bansal, NanoBiotechnology Research Laboratory, Centre for Advanced Materials and Industrial Chemistry, School of Applied Sciences, RMIT University, Melbourne, Australia

Bhalchandra M. Bhanage, Department of Chemistry, Institute of Chemical Technology, Mumbai, India

Suresh K. Bhargava, NanoBiotechnology Research Laboratory, Centre for Advanced Materials and Industrial Chemistry, School of Applied Sciences, RMIT University, Melbourne, Australia

Amit Bhatnagar, Department of Biology and Environmental Science, Faculty of Health and Life Sciences, Linnaeus University, Kalmar, Sweden

Mónica Cerro-López, Departmento de Ciencias Químico-Biológicas, Universidad de las Américas Puebla, Puebla, México

Levan Chkhartishvili, Department of Engineering Physics, Georgian Technical University and Laboratory for Boron and Powdered Composite Materials, F. Tavadze Institute of Metallurgy and Materials Science, Tbilisi, Georgia

Rute F. Domingos, Centro de Química Estrutural, Instituto Superior Técnico, Universidade de Lisboa, Lisbon, Portugal

Mayra Z. Figueroa-Torres, Departamento de Ecomateriales y Energía, Facultad de Ingeniería Civil-Universidad Autónoma de Nuevo León, Avenida Universidad y Avenida Fidel Velázquez S/N, Ciudad Universitaria, San Nicolás de los Garza, Nuevo León, México

Mayank Gangwar, Department of Microbiology and Infectious Diseases and Department of Pharmacology, Institute of Medical Sciences, Banaras Hindu University, Varanasi, India

Olga Gapurova, Institute of Nuclear Physics, Uzbekistan Academy of Sciences, Tashkent, Uzbekistan

Gennady Gerasimov, Institute of Mechanics, Moscow State University, Moscow, Russia

Elsa Nadia Aguilera González, Corporación Mexicana de Investigación en Materiales, S.A. de C.V., Calle Ciencia y Tecnología No. 790, Saltillo 400, Saltillo, Coahuila, Mexico

Hongbo Gu, Integrated Composites Laboratory (ICL), Dan F. Smith Department of Chemical Engineering, Lamar University, Beaumont, TX, USA

Zhanhu Guo, Integrated Composites Laboratory (ICL), Dan F. Smith Department of Chemical Engineering, Lamar University, Beaumont, TX, USA

Sangiliyandi Gurunathan, GS Institute of Bio and Nanotechnology, Coimbatore, India

Richard G. Haire, Oak Ridge National Laboratory, Oak Ridge, TN, USA

Kirk Hatfield, Engineering School of Sustainable Infrastructure & Environment, University of Florida, Gainesville, FL, USA

Teresa de Jesús Palacios-Hernández, Departamento de Ciencias Biológicas, Universidad Popular Autónoma del Estado de Puebla, Puebla, Mexico

Yudong Huang, School of Chemical Engineering and Technology, Harbin Institute of Technology, Harbin, Heilongjiang, China

Yunpeng Huang, State Key Laboratory for Modification of Chemical Fibers and Polymer Materials and College of Chemistry, Chemical Engineering and Biotechnology, Donghua University, Shanghai, China

Chaudhery Mustansar Hussain, Department of Chemistry and Environmental Science, New Jersey Institute of Technology, Newark, NJ, USA

Muniyandi Jeyaraj, GS Institute of Bio and Nanotechnology, Coimbatore, India

Stepan N. Kalmykov, Chemistry Department, Lomonosov Moscow State University, Moscow, Russia

Boris I. Kharisov, Facultad de Ciencias Químicas and CIIDIT, Universidad Autónoma de Nuevo León, Monterrey, Mexico

Oxana V. Kharissova, CIIDIT and Facultad de Ciencias Físico-Matemáticas, Universidad Autónoma de Nuevo León, Monterrey, Mexico

Rashid A. Khaydarov, Institute of Nuclear Physics, Uzbekistan Academy of Sciences, Tashkent, Uzbekistan

Renat R. Khaydarov, Institute of Nuclear Physics, Uzbekistan Academy of Sciences, Tashkent, Uzbekistan

Vladimir A. Kholodov, Chemistry Department, Lomonosov Moscow State University, and Dokuchaev Soil Science Institute of RAAS, Moscow, Russia

Marta I. Litter, Gerencia Química, Comisión Nacional de Energía Atómica and Consejo Nacional de Investigaciones Científicas y Técnicas and Instituto de Investigación e Ingeniería Ambiental, Universidad de Gral. San Martín, Buenos Aires, Argentina

Peng Liu, State Key Laboratory of Applied Organic Chemistry and Key Laboratory of Nonferrous Metal Chemistry and Resources Utilization of Gansu Province, College of Chemistry and Chemical Engineering, Lanzhou University, Lanzhou, China

Ubaldo Ortiz Méndez, CIIDIT, Universidad Autónoma de Nuevo León, Monterrey, Mexico

Miguel A. Méndez-Rojas, Departmento de Ciencias Químico-Biológicas, Universidad de las Américas Puebla, Puebla, México

Jose Ruben Morones-Ramirez, Chemical Engineering Department, School of Chemistry, Universidad Autónoma de Nuevo León, Pedro de Alba, S/N. Ciudad Universitaria, San Nicolás de los Garza, Mexico

Muralidharan Murugan, GS Institute of Bio and Nanotechnology, Coimbatore, India

Mohammad Mahdi Najafpour, Department of Chemistry, Center for Research in Climate Change and Global Warming, and Institute for Advanced Studies in Basic Sciences (IASBS), Zanjan, Iran

Alexander P. Novikov, Vernadsky Institute of Geochemistry and Analytical Chemistry of RAS, Moscow, Russia

Yit Thai Ong, School of Chemical Engineering, Engineering Campus, Universiti Sains Malaysia, Pulau Pinang, Malaysia

Sukanchan Palit, Department of Chemical Engineering, University of Petroleum and Energy Studies, Dehradun, India

Aniruddha B. Patil, Department of Chemistry, Institute of Chemical Technology, Mumbai, India

Irina V. Perminova, Chemistry Department, Lomonosov Moscow State University, Moscow, Russia

José P. Pinheiro, Centro de Biomedicina Molecular Estrutural (CBME)/IBB, LA, DQF/FCT, Universidade do Algarve, Faro, Portugal

Sergey A. Ponomarenko, Chemistry Department, Lomonosov Moscow State University, and Enikolopov Institute of Synthetic Polymeric Materials of RAS, Moscow, Russia

Rajesh Ramanathan, NanoBiotechnology Research Laboratory, Centre for Advanced Materials and Industrial Chemistry, School of Applied Sciences, RMIT University, Melbourne, Australia

Isaías Juárez-Ramírez, Departamento de Ecomateriales y Energía, Facultad de Ingeniería Civil-Universidad Autónoma de Nuevo León, Avenida Universidad y Avenida Fidel Velázquez S/N, Ciudad Universitaria, San Nicolás de los Garza, Nuevo León, México

Yit Thai Ong, School of Chemical Engineering, Engineering Campus, Universiti Sains Malaysia, Pulau Pinang, Malaysia

Aniruddha B. Patil, Department of Chemistry, Institute of Chemical Technology, Mumbai, India

Natalia Quici, Gerencia Química, Comisión Nacional de Energía Atómica, and Consejo Nacional de Investigaciones Científicas y Técnicas, Buenos Aires, Argentina

Ravishankar Rai V., Department of Studies in Microbiology, University of Mysore, Mysore, India

David G. Rickerby, European Commission Joint Research Centre, Institute for Environment and Sustainability, Ispra, Italy

Luis Ángel Garza Rodríguez, Facultad de Ingeniería y Arquitectura, Universidad Regiomontana A.C., Monterrey, Mexico

Sodeh Sadjadi, Nuclear Science and Technology Research Institute, Tehran, Iran

Hafiz H.M. Salih, Illinois Sustainable Technology Center, University of Illinois at Urbana-Champaign, Champaign, IL, USA

José Luis Sánchez-Salas, Departamento de Ciencias Químico-Biológicas, Universidad de las Américas Puebla, Puebla, Mexico

Natalia S. Shcherbina, Chemistry Department, Lomonosov Moscow State University, Moscow, Russia, and Paul Scherrer Institute (PSI), Villigen, Switzerland

Mika Sillanpää, Laboratory of Green Chemistry, Department of Energy and Environmental Technology, Faculty of Technology, Lappeenranta University of Technology, Mikkeli, Finland

Mingwu Shen, College of Chemistry, Chemical Engineering and Biotechnology, Donghua University, Shanghai, China

Santosh K. Singh, Centre of Experimental Medicine and Surgery, Institute of Medical Sciences, Banaras Hindu University, Varanasi, India

Xiangyang Shi, State Key Laboratory for Modification of Chemical Fibers and Polymer Materials; College of Chemistry, Chemical Engineering and Biotechnology, Donghua University, Shanghai, China; and CQM-Centro de Química da Madeira, Universidade da Madeira, Campus da Penteada, Portugal

Ravi Shukla, NanoBiotechnology Research Laboratory, Centre for Advanced Materials and Industrial Chemistry, School of Applied Sciences, RMIT University, Melbourne, Australia

Soon Huat Tan, School of Chemical Engineering, Engineering Campus, Universiti Sains Malaysia, Pulau Pinang, Malaysia

Leticia M. Torres-Martínez, Departamento de Ecomateriales y Energía, Facultad de Ingeniería Civil-Universidad Autónoma de Nuevo León, Avenida Universidad y Avenida Fidel Velázquez S/N, Ciudad Universitaria, San Nicolás de los Garza, Nuevo León, México

Vel Murugan Vadivelu, School of Chemical Engineering, Engineering Campus, Universiti Sains Malaysia, Pulau Pinang, Malaysia

Vijay C. Verma, Department of Microbiology and Infectious Diseases, Institute of Medical Sciences, Banaras Hindu University, Varanasi, India

Shige Wang, State Key Laboratory for Modification of Chemical Fibers and Polymer Materials, Donghua University, Shanghai, China

Huige Wei, Integrated Composites Laboratory (ICL), Dan F. Smith Department of Chemical Engineering, Lamar University, Beaumont, TX, USA

Suying Wei, Department of Chemistry and Biochemistry, Lamar University, Beaumont, TX, USA

Liangbin Xiong, School of Physics and Electronic-Information Engineering, Hubei Engineering University, Xiaogan, China

Xingru Yan, Integrated Composites Laboratory (ICL), Dan F. Smith Department of Chemical Engineering, Lamar University, Beaumont, TX, USA

Qian Wen Yeang, School of Chemical Engineering, Engineering Campus, Universiti Sains Malaysia, Pulau Pinang, Malaysia

Kian Fei Yee, School of Chemical Engineering, Engineering Campus, Universiti Sains Malaysia, Pulau Pinang, Malaysia

Huaqing Yu, School of Physics and Electronic-Information Engineering, Hubei Engineering University, Xiaogan, China

Ying Yu, Institute of Nanoscience and Nanotechnoloogy, College of Physical Science and Technology, Central China Normal University, Wuhan, China

Sharif Hussein Sharif Zein, School of Chemical Engineering, Engineering Campus, Universiti Sains Malaysia, Pulau Pinang, Malaysia and Current address: Faculty of Science and Engineering, School of Engineering, University of Hull, Hull, UK

Wenpei Zhang, Institute of Nanoscience and Nanotechnology, College of Physical Science and Technology, Central China Normal University, Wuhan, China

Jiahua Zhu, Department of Chemistry and Biochemistry, and Integrated Composites Laboratory (ICL), Dan F. Smith Department of Chemical Engineering, Lamar University, Beaumont, TX, USA

LIST OF ABBREVIATIONS

2,4-DCP	2,4-dichloro phenol
2-CP	2-chlorophenol
2D	two-dimensional
4-CP	4-chlorophenol
4NCB	4-nitrochlorobenzene
AA	acrylic acid
AAS	atomic absorption spectroscopy
AASp	atomic absorption spectrophotometry
A-ATP	aminopropyl attapulgite
AB	ammonia borane
AC	activated carbon
AEAPTMS	*N* β-aminoethyl-γ-aminopropyl trimethoxysilane
AEPE	2-(3 (2-aminoethylthio)propylthio)ethanamine
AET	1-(*p*-acetylphenyl)-3-(*o*-ethoxyphenyl)triazene
AF	acid fuchsin
AFC	alkaline fuel cell
AFM	atomic force microscopy
AFS	atomic fluorescence spectrometry
Ag NP	silver nanoparticle
AGNES	absence of gradients and Nernstian equilibrium stripping
ALP	alkaline phosphatase
ALT	alanine aminotransferase
AM	acrylamide
amine-CNT	amine-modified carbon nanotube
Am-PAA-B	amine-modified polyacrylamide/bentonite
AN	acrylonitrile
ANOVA	analysis of variance
AO	acrylamide oxime
AO7	acid orange 7
AOC	assimilable organic carbon
AOM	advanced oxidation method
AOP	advanced oxidation process
APDC	ammonium pyrrolidinedithiocarbamate
APMDS	3-(2-aminoethylamino) propyl-methyldimethoxysilane

ApqZ	aquaporin Z
APTS	aminopropyltriethoxysilane
AQDS	anthraquinone disulfonate
AR	alizarin red
ARB	Acid Red B
ARP	advanced reductive process
ASV	anodic stripping voltammetry
ATP	attapulgite
Au NP	gold nanoparticle
BALF	bronchoalveolar lavage fluid
BAT	best available technology
BB41	Basic Blue 41
BBB	blood–brain barrier
bbc	body-centered-cubic
BE	benzethonium
BET	Brunauer–Emmett–Teller
bis-MPA	2,2-bis (hydroxymethyl) propionic acid
BLM	biotic ligand model
BM	biomacromolecule
BMSAPD	bis(4-methoxysalicylaldehyde)-1,2-phenylenediamine
BNT	bentonite
BPA	bisphenol A
BQ	1,4-benzoquinone
BR46	Basic Red 46
BSA	bovine serum albumin
BTEX	benzene, toluene, ethylbenzene, and xylenes
BUN	blood urea nitrogen
BW	brackish water
BY28	Basic Yellow 28
CA	cathodic arc
CAc	cellulose acetate
CAD	catalytic arc discharge
Ca-Mt	calcium-saturated montmorillonite
CB	conduction band
CBDA	cetylbenzyldimethylammonium
CCD	charged coupled device
CCS	carbon capture and storage
CD	cyclodextrin
CDTA	*trans*-1,2-diaminocyclohexane-*N*,*N*,*N′*,*N′*-tetraacetic acid
CDTEA	carboxydecyltriethylammonium
CEC	cation exchange capacity
CFU	colony-forming unit
CK	creatine kinase
CLE-AdSV	competitive ligand exchange adsorptive cathodic stripping voltammetry
CLM	confocal laser microscopy
CM	chloromethane
CMC	carboxymethyl cellulose
CNF	carbon nanofiber
CNIM	carbon nanotube-immobilized membrane
CNM	carbon nanomaterial
CNT	carbon nanotube
CNT–MMM	CNT–mixed matrix membrane
CPBr	cetylpyridinium bromide
CR	Congo red

CS	impregnated chitosan
CSTR	continuously stirred tank reactor
CT	catechol
CTAB	cetyltrimethyl ammonium bromide
CuO NP	copper oxide nanoparticle
CuO	copper oxide
CuTSPc	copper tetrasulfonated phthalocyanine
CVD	chemical vapor deposition
CVG	chemical vapor generation
Cys	cysteine
DBP	disinfection by-product
DC	dip coating
DCA	dichloroacetate
DCB	dichlorobenzene
DCE	dichloroethylene
DCF	diclofenac
DCM	dichloromethane
DDT	dichloro-diphenyl-trichloroethane
DEA	diethanolamine
DFT	density functional theory
DGT	diffusive gradients in thin film
DHE	dihydroethidium
DIZ	diameter of the inhibition zone
DLM	Diffuse Layer Model
DLS	dynamic light scattering
DLVO	Derjaguin–Landau–Verwey–Overbeek
DM	direct mixing method
DMA	dimethylarsinic acid
DMAPS	*N*-dimethylaminopropyltrimethoxysilane
DMF	dimethylformamide
DMFC	direct-methanol fuel cell
DMT	Donnan membrane technique
DNAPL	dense nonaqueous-phase liquid
DNT	dinitrotoluene
DOC	dissolved organic carbon
DP	alkyl-1,3-diaminopropane-diaminopropane
DR23	Direct Red 23
DSC	differential scanning calorimetry
DSPME	dispersive solid-phase microextraction
DTC	dithiocarbamate
DTMA	decyltrimethylammonium
DTPA	diethylenetriamine pentacetic acid
EDA	ethylene diamine
EDB	ethylene dibromide
EDSp	energy-dispersive spectrometry
EDS	*N*-(3-dimethylaminopropyl)-*N*′-ethylcarbodiimide hydrochloride
EDTA	ethylenediaminetetraacetic acid
EDX	energy-dispersive X-ray
EE	extraction efficiency
EF	enrichment factor
EFOM	effluent organic matter
en	ethylenediamine
ENM	engineered nanomaterial
EOR	enhanced oil recovery

EPA	Environmental Protection Agency
EPS	extracellular polymeric substance
ERGO	electrochemically reduced graphene oxide
ETAAS	electrothermal atomic absorption spectrometry
EURL ECVAM	European Union Reference Laboratory for Alternatives to Animal Testing
EZVI	emulsified zero-valent iron
FAAS	flame atomic absorption spectrometry
fcc	face-centered cubic
FCS	fluorescence correlation spectroscopy
FDA	Food and Drug Administration
FeGP	Fe-guanylyl cofactor
FE-SEM	field-emission scanning election microscopy
FET	field-effect transistor
FFF	field flow fractionation
FIAM	free ion activity model
FS	faradic supercapacitor
FTIR	Fouricr transform infrared
FTO	fluorine-doped tin oxide
FUL	fullerene
G'	storage modulus
G''	loss modulus
GA	glutaraldehyde
GAC	granular activated carbon
GC	gas chromatography
GC/FID	gas chromatography/flame ionization detector
GC/MS	gas chromatography/mass spectrometry
GMNO	graphene–Mn_3O_4
GMR	giant magnetoresistance
GO	graphene oxide
GO_x	glucose oxidase
GPTMS	3-glycidoxypropyltrimethoxysilane
GRD	glass-rod dosimeter
GSH	glutathione
HA	humic acid
HAAC	*N*-2-hydroxypropyl trimethyl ammonium chloride chitosan
HAP	hydroxyapatite
HAPE	hyperbranched aliphatic polyester
HAPE-ATP	hyperbranched aliphatic polyester-grafted attapulgite
Hb	hemoglobin
HBA	horse blood agar
h-BN	hexagonal boron nitride
HCB	hexachlorobenzene
HDC	hydrodynamic chromatography
HDPy	hexadecylpyridinium
HDTMAB	hexadecyltrimethylammonium bromide
Hect.	hectorite
HEDTA	hydroxyethylenediaminetetraacetic acid
HEG	hydrogen-exfoliated graphene
HEK	human epidermal keratinocyte
HF	Hartree–Fock
HGMS	high-gradient magnetic separation
HMI	heavy metal ion
HMSC	human mesenchymal stem cell
HMX	tetramethylenetetranitramine

HNT	halloysite nanotube
HOMO	highest-occupied molecular orbital
HOPG	highly ordered pyrolytic graphite
HP	heterogeneous photocatalysis
HPC	heterogeneous photocatalysis
HPLC	high-pressure liquid chromatography
HPOM	heteropolyoxometallate
HPRB	humic permeable reactive barrier
HQ	hydroquinone
HRTEM	high-resolution transmission electron microscopy
HS	humic substance
HS/GC/MS	headspace/gas chromatography/mass spectrometry
HTB	hexagonal tungsten bronze
HTG	hydrothermal-reduced graphene
ICP-AES	inductively coupled plasma atomic emission spectroscopy
ICP-MS	inductively coupled plasma mass spectrometry
ICP–OES	inductively coupled plasma optical emission spectrometry
IE	ion exchange
IL	ionic liquid
IOT	iron ore tailing
IPCC	Intergovernmental Panel on Climate Change
ISE	ion-selective electrode
JCPDF	Joint Committee on Power Diffraction File
LA	laser ablation
LB	Luria–Bertani
LCST	lower critical solubility temperature
LDH	layered double hydroxide
LED	light emitting diode
L–H	Langmuir–Hinshelwood
LMCT	ligand-to-metal charge transfer
LSPR	localized surface plasmon resonance
LUMO	lowest unoccupied molecular orbital
MA	maleic anhydride
MAS	magic angle spinning
M-ATP	modified ATP
MB	methylene blue
MBC	minimum bactericidal concentration
MBI	2-mercaptobenzimidazole
MCHC	mean corpuscular hemoglobin concentration
MCM	mobile composition of matter
MCNC	magnetic carbon nanocomposite
MCV	mean corpuscular volume
MD	membrane distillation
MDin	molecular dynamic
MDA	malondialdehyde
MEA	monoethanolamine
MEAm	2-mercaptoethylammonium
MED	multieffect distillation
MEEE	2-(2-(2-mercaptoethoxy)ethoxy)ethanol
MeO	methyl orange
MES	2-mercaptoethanesulfonic acid
MFFF	magnetic field flow fractionation
MG	malachite green
MGNC	magnetic graphene nanoplatelet composite

MgO	magnesium oxide
MHC	mean corpuscular hemoglobin
MIC	minimum inhibitory concentration
MIR	mid-infrared
MMA	monomethylarsonic acid
MMP	mitochondrial membrane potential
MMT	montmorillonite
MMWCNT	magnetic multiwalled carbon nanotube
MNDO	medial neglecting by differential overlapping
MNP	magnetic nanoparticle
MO	methyl orange
MOCVD	metal organic chemical vapor deposition
MOF	metal–organic framework
MP	methyl parathion
MPDET	*N*-[3-(trimethoxysilyl)propyl]diethylenetriamine
MPS	3-mercaptopropyltrimethoxysilane
MR	magnetoresistance
MRr	membrane reactor
MS	mass spectrometry
MSF	multistage flash
MSM	metal–semiconductor–metal
MSP	monomer-stabilized polymerization
MSW	municipal solid waste
MT	metallothionein
MTBE	methyl *tert*-butyl ether
MW	molecular weight
MWCNT	multiwalled carbon nanotube
MWNT	multiwalled nanotube
NCC	nanocarbon colloids
NCMC	nanocarbon–metal compositions
NDMA	nitrosodimethylamine
NF	nanofiber
NFil	nanofiltration
NF-κB	nuclear factor-κB
NGCC	natural gas combined cycle
NHE	normal hydrogen electrode
NICA	nonideal competitive adsorption
NM	nanomaterial
NMR	nuclear magnetic resonance
NOAA	National Oceanic and Atmospheric Administration
NOM	natural organic matter
NP	nanoparticle
NPC	nanophotocatalysts
NS	naphthalenesulfonate
NSTF	nanostructured thin film
NTA	nanoparticle tracking analysis
NTAc	nitriloacetic acid
NTP	National Toxicology Program
NW	nanowire
nZVI	nanoscale zero-valent iron
OD	organic dye
ODTMA	2-oxyhydrazino-*N*-(2-methylen-yl-hydroxyphenyl)pyridinium
OECD	Organization for Economic Co-operation and Development
OREC	organic modified rectorite

ORR	oxygen reduction reaction
P2AT	poly(2-aminothiphenol)
PA	palmitic acid
PAm	polyamide
PAA	polyacrylic acid
CFU	colony-forming unit
PAAM	poly(acrylamide)
PAA–nanoclay	polyacrylic acid–nanoclay
PAB	polyaminoborane
PAG	polycyclic aromatic hydrocarbon
PAGE	polyacrylamide gel electrophoresis
PAM	polyacrylamide
PAMAM	polyamidoamine
PAM-ATP	polyacrylamide-grafted attapulgite
PANa	1-(2-pyridylazo)-2-naphthol
PAN	polyacrylonitrile
BSA	bovine serum albumin
PANI	polyaniline
PANI-NF	PANI nanofiber
PC	photocatalytic
PCh	phytochelatin
PCA	principle component analysis
PCB	polychlorinated biphenyl
PCE	perchloroethylene
PCE	tetrachloroethylene
PCNA	proliferating cell nuclear antigen
PDADMAC	poly(diallyldimethylammonium chloride)
PDC	poly-α,β-DL-aspartyl-L-cysteine
PDMA	poly(*N*,*N*-dimethylacrylamide)
PDT	photodynamic therapy
PE	polyelectrolyte
PECh	photoelectrochemical
PEC	photoelectrocatalytic
PECC	photoelectrochemical cell
PECVD	plasma-enhanced chemical vapor deposition
PEDOT	poly(3,4-ethylenedioxythiophene)
PEF	photoelectro–Fenton
PEFC	proton exchange fuel cell
PEG	polyethylene glycol
PEI	polyethyleneimine
PEO	polyethylene oxide
PES	polyethersulfone
PGA	poly-γ-glutamic acid
PIB	polyiminoborane
PNC	polymer nanocomposite
PNIPAAm/PNIPAM	poly(*N*-isopropylacrylamide)
POM	polyoxometalate
PP	polypropylene
PPCP	pharmaceutical and personal care product
PPTA	poly(paraphenylene terephthalamide)
PPy	polypyrrole
PRB	permeable reactive barrier
PSA	prostate-specific antigen
PSD	particle size distribution

PSf	polysulfone
PSS	polystyrene sulfonate
PTFE	polytetrafluoroethylene
p-TSA	*p*-toluenesulfonic acid
PV	photovoltaic
PVA	poly(vinyl alcohol)
PVC	polyvinyl chloride
PVDF	poly(vinylidene) difluoride
PVP	poly(vinylpyrrolidone)
QCM	quartz crystal microbalance
QD	quantum dot
RB	rhodamine B
RB5	Reactive Black 5
RBC	Red blood cell count
RCl	reduced chlorophenol
REC	natural rectorite
Redox	reduction–oxidation
RESS	rapid expansion of supercritical solution
RF	radiofrequency
RH	relative humidity
RhB	rhodamine B
RNA	ribonucleic acid
RNIP	reactive nanoscale iron product
RO	reverse osmosis
ROS	reactive oxygen species
RRG	relative root growth
RSD	relative standard deviation
RT-CES	real-time cell electronic sensing
RWQCB	Regional Water Quality Control Board
SAW	surface acoustic wave
SBME	stir bar microextraction
SC	semiconductor
SCF	supercritical fluid
SCW	supercritical water
SDH	sorbitol dehydrogenase
SDS	sodium dodecyl sulfate
SEC	size exclusion chromatography
SEM	scanning electron microscopy
SEMic	scanning electron microscope
SEN	single enzyme nanoparticle
SERS	surface-enhanced Raman scattering
SI-ATRP	surface-initiated atom transfer radical polymerization
SIP	surface-initiated polymerization
Sm	samarium
SOA	small molecular weight organic acid
SOD	superoxide dimutase
SPC	silica polyamine composite
SPE	solid-phase extraction
SPES	sulfonated poly(ether) sulfone
SPIMHN	2-((3-silylpropylimino)methyl)-2-hydroxy-1-naphthol
SPION	superparamagnetic iron oxide nanoparticle
SPME	solid-phase microextraction
SPR	surface plasmon resonance
SSCP	scanned stripping chronopotentiometry

STXM	X-ray microscopy
SUVA	specific UV absorbance
SVOC	semivolatile organic compound
SW	seawater
SWCNT	single-walled carbon nanotube
SWNT	single-walled nanotube
SWRO	seawater reverse osmosis
TAA	tri(2-aminoethyl)amine
TA-AC	tannin-immobilized activated clay
TATP	triacetone triperoxide
TBA	tetrabutylammonium bromide
t-BN	turbostratic boron nitride
TCE	trichloroethylene
TCM	trichloromethane
TCP	trichlorophenol
TCr	total Cr
TDN	titanium dioxide nanotube
TEG	tetraethylene glycol
TEM	transmission electron microscopy
TEMic	transmission electron microscope
TEPA	tetraethylenepentamine
TFC	thin-film composite
TGA	thermal gravimetric analysis
TGAn	thermal gravimetric analyzer
TGA	thermogravimetric analysis
THF	tetrahydrofuran
THM	trihalomethane
TiO_2 NP	titanium dioxide nanoparticle
TJ	tight junction
TLD	thermo-luminescent dosimeters
TM	transition metal
TMAT	trimethylammoniumethanethiol
TMFE	thin mercury film electrode
TMP	transmembrane pressure
TNT	trinitrotoluene
TOC	total organic carbon
TOF	turnover frequency
TP	tea polyphenol
TRLIF	time-resolved laser-induced fluorescence
TRTA	triethylene tetraamine
UFP	ultrafine particle
UN	United Nations
UN-MDGs	United Nations Millennium Development Goals
UV	ultraviolet
UV/vis/NIR	ultraviolet–visible–near-infrared
VB	valence band
VC	vinyl chloride
VL	visible light
VOC	volatile organic compound
VRH	variable range hopping
VTSA	vacuum temperature swing adsorption
w/o	water in oil
WHO	World Health Organization
WOC	water-oxidizing complex

WPMN	Working Party on Manufactured Nanomaterials
XPS	X-ray photoelectron spectroscopy
XRD	X-ray diffraction
ZnONP	zinc oxide nanoparticle
ZnO	zinc oxide
ZrP	zirconium phosphate
ZVI NP	zero-valent iron NP
ZVI	zero-valent iron
ZVM	zero valent metal
μ-SPE	microscale solid-phase extraction

PART I

REMEDIATION WITH USE OF METALS, METAL OXIDES, COMPLEXES AND COMPOSITES

1

GROUNDWATER WATER REMEDIATION BY STATIC DIFFUSION USING NANO-ZERO VALENT METALS (Fe^0, Cu^0, Al^0), n-FeH^{n+}, n-$Fe(OH)_X$, n-FeOOH, n-Fe-$[O_XH_Y]^{(n+/-)}$

DAVID D.J. ANTIA
DCA Consultants Ltd., Scotland, UK

1.1 INTRODUCTION

Zero valent metals (ZVM) and ZVM combinations, including magnetic ZVM [1–4] and reduced ZVM [5–7], are highly efficient water treatment agents [8–12]. They will remove ions, chemicals, compounds, and biota from water. The principal ZVM used in pilot and commercial water treatment are Fe^0 and Fe^0+Cu^0 [13–17]. Prices for ZVM powders are a function of commodity prices, particle size, particle shape (e.g., dendritic, spherical, platy, irregular, etc), manufacturing method, source, and quantity ordered. Current (June 2014) prices (k$ FOB t^{-1}) for powders are (i) 13,000–400,000 nm (Fe = 0.6–1.6; Cu = 1–7; Al = 5–20), (ii) <3,000 to >8,000 nm (Fe = 5–15; Cu = 9–18; Al = 9–850), (iii) 10–1000 nm (Fe = 30–1100; Cu = 9–1000; Al = 9–900). n-ZVM powders are either used to treat water in a reactor [10], or are injected into an aquifer [10, 17], or are placed in a permeable reactive barrier (PRB) within the aquifer [10, 18].

Use of n-ZVM in a fixed (packed) bed reactor (where all the feed water flows through the n-ZVM) is impractical, as n-Fe^0 rapidly corrodes and expands to form hydrated, low-density hydroxides ($Fe(OH)_2$, $Fe(OH)_3$) and peroxides (FeOOH). This results [10] in a decrease in porosity, decrease in permeability, increase in the proportion of dead end pores, and a decrease in pore throat size. Associated gas bubble formation (O_2, H_2 [10]) results in permeability reduction [18, 19] due to gas occlusion switching water flow from viscous flow to Knudsen diffusion [20–22]. The net effect is a major reduction in permeability (and water flow rate) over a short time period (Fig. 1.1a) in the reactor, or aquifer [10]. Fluid flow (Q_f, m^3 m^{-2} s^{-1}) = k_p D_f [20–22]. A list of abbreviations is provided in Appendix 1.A. The dominant fluid flow mechanism switches over a period of 2–6 weeks from viscous flow to Knudsen diffusion due to the generation and presence of trapped nano/micron-sized gas bubbles [10, 20–22].

Changing the reactor type to a diabatic diffusion reactor (where a body of water overlies a static body of ZVM, and all water enters and leaves the reactor through the water body), mimics the situation that occurs in an aquifer, during remediation, following n-ZVM injection. A policy of groundwater abstraction, treatment, and reinjection allows the remediation to be undertaken in a short time period in a controlled environment without creating long-term damage to the aquifer [10]. In an unconfined, diabatic, diffusion environment (e.g., shallow contaminated aquifer or soil), the principal controls on Eh, pH, EC (electrical conductivity), and remediation are [10]: (i) flowing water space velocity ($SV = Q_{fr}/W_{zvm}$); (ii) the stored water to ZVM ratio, [$SWZ = S_w/W_{zvm}$]; (iii) the water composition; (iv) atmospheric/groundwater temperature fluctuations; (v) atmosphere composition variation (principally humidity); (vi) atmospheric pressure fluctuation; (vii) vertical infiltration recharge (associated with storm events) into the aquifer/soil; (viii) water losses from the aquifer/soil, due to evaporation, leaching, and the interaction of the n-ZVM (and n-ZVM products) with minerals and biota; and (ix) porosity occlusion resulting from the movement of displaced air as the water levels rise and fall (during and following infiltration recharge) [20–23].

Nanomaterials for Environmental Protection, First Edition. Edited by Boris I. Kharisov, Oxana V. Kharissova, and H. V. Rasika Dias.

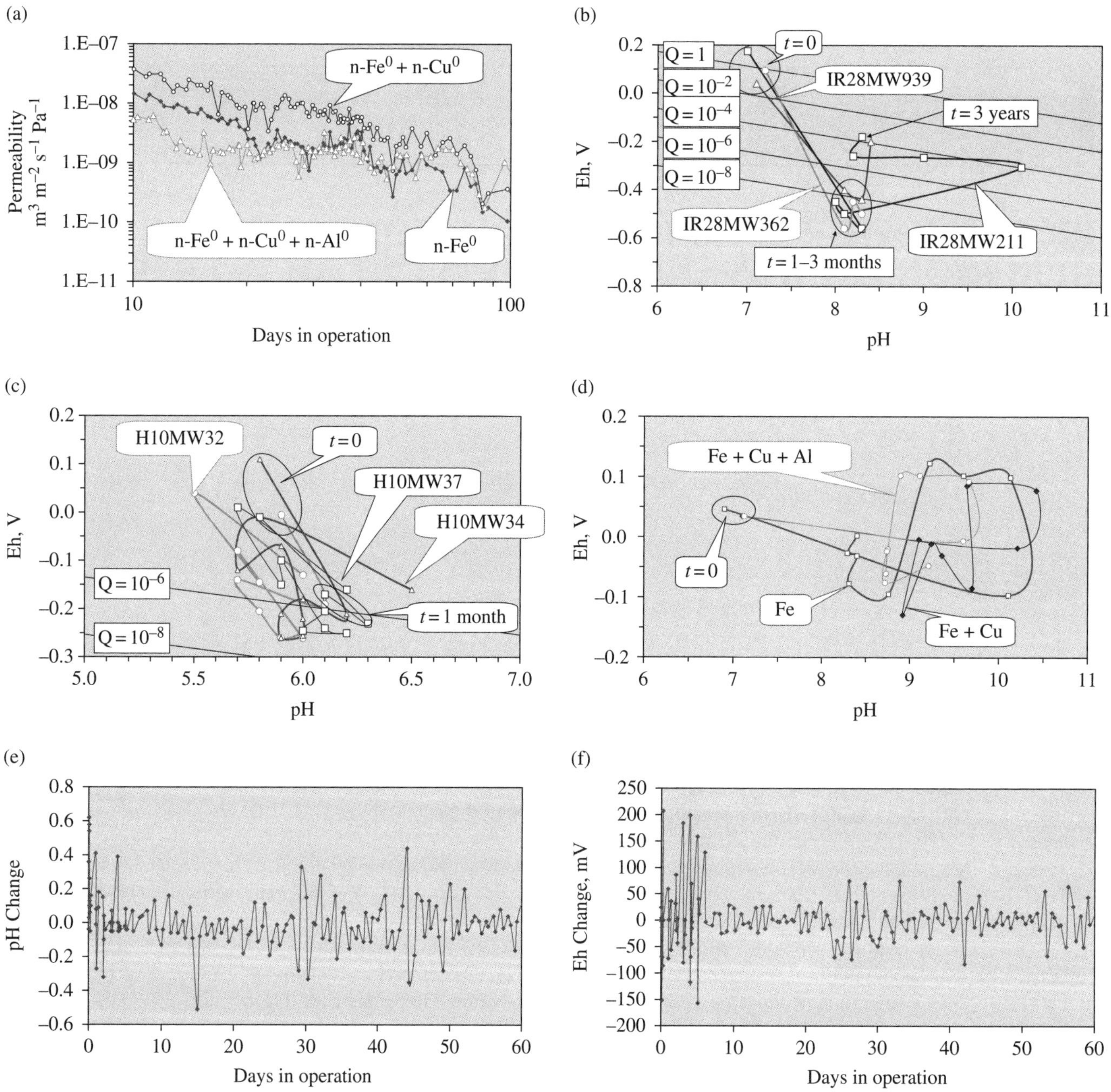

FIGURE 1.1 Nano-Material Behavior During Remediation. (a) n-ZVM permeability decline in a fixed (packed) bed multi-tubular reactor [10]. (b) Impact of 7.2 t, pneumatically injected n-Fe^0 (>1000 nm) into 1839 m^3 soil, on Eh and pH with time. Hunters Point, San Francisco, USA) [17]. Q = equilibrium Mol l^{-1} TCE/DCE; (c) Impact of 7.63 kg infiltrated n-Fe^0 (50–300 nm) (containing 0.15% Pd) into 808 m^3 soil, on Eh and pH with time (12 months): NAS Jacksonville, Florida, USA [17]. (d) n-ZVM: pH vs. Eh over time in a static diffusion reactor showing typical trajectories. Data points taken at 0, 1, 10, 20, 30, 40, 50, 60 day intervals [135]. (e) pH oscillation with time, n-Fe^0[135]. Oscillation value = change in pH from previous measurement. (f) Eh oscillation with time, n-Fe^0: Data: [135]. Oscillation value = change in Eh from previous measurement.

This review considers (i) the contaminants that can be removed by n-Fe^0; (ii) the factors and mechanisms that impact on the remediation rates; (iii) the interaction between n-Fe^0, water, ZVM corrosion/remediation products; and (iv) Eh, pH oscillations, and trajectories and their impact on remediation.

1.2 CONTAMINANTS REMOVED BY n-Fe^0, n-Cu^0, AND n-Al^0

Contaminants removed from water in a diffusion environment at temperatures in the range [<0 to >70°C] by n-Fe^0, n-Cu^0, and n-Al^0, include

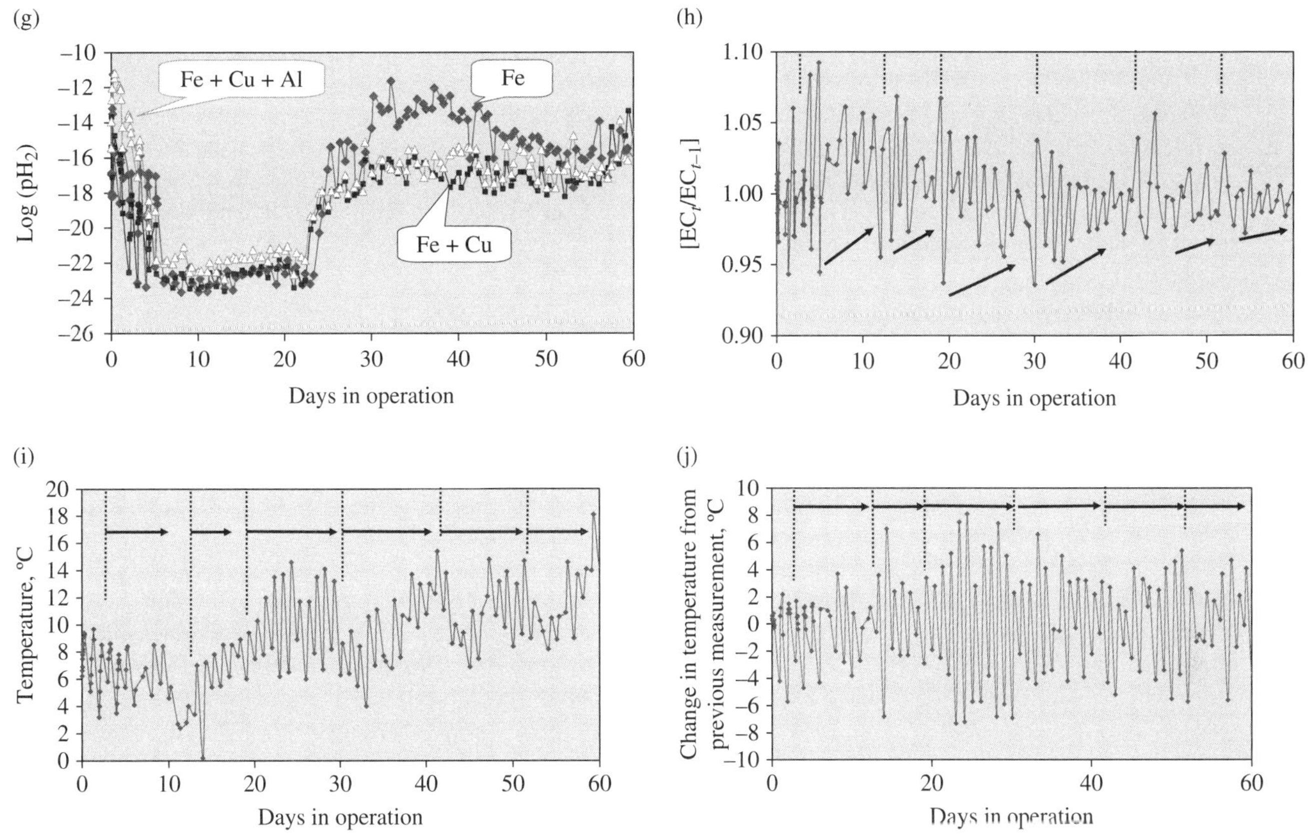

FIGURE 1.1 (*Continued*) (g) pH_2 vs. time in a static diffusion reactor for nano-ZVM (Fig. 1.1e and f). (h) Typical declining EC oscillations with time. n-Fe^0 [135]. (i) Variation in temperature with time (Figs. 1.1e–h) (j) Temperature oscillations with time.

1. Pollutants of the form MO_x, where M = S, N, C, P, Cl (e.g., nitrates [24–30], nitrites [30 33], perchlorates [34–37], carbonic acids [38–45], phosphates [46], sulfates [47–50]). Pollutants are removed within green rust, or ZVM-hydroxide/peroxide structures, or polyoxometallate (POM) structures, by cation/anion substitution, or by layer accretion [38–50]. Green rusts are highly reactive structures comprising [38–50] alternating positively and negatively charged hydroxide/peroxide layers and hydrated anion layers with the general composition $[\mathrm{Cation}_a{}^{I}\mathrm{Cation}_c{}^{II}\ \mathrm{Cation}_b{}^{III}\ (OH)_m\ (OOH)_d]^{x+}$ $[(A_{y/nv}\ yiH_2O]^{x-}$. Cations and anions can be substituted [38–50]. *A* is an anion (e.g., Cl^-, SO_4^{2-}, CO_3^{2-}, Br^-, I^-, NO_3^-, ClO_4^-, SO_3^{2-}, SeO_4^{2-}, PO_4^{2-}, OH^-, OOH^-, O_x^{y-}, etc.); nv = valency; yi = the inter-layer water and is typically between 2 and 4. A typical green rust forms as plates 5–2000 nm in diameter and about 40 nm thick, for example, [50]. Green rusts (ZVM degradation products) are highly efficient anion and cation scavengers and may be as reactive, or more reactive, than Fe^0 [9, 49]. During scavenging operations, the "green rusts" can incorporate cation layers of the form $[\mathrm{Cation}_e{}^{I}(OH)_m\ (OOH)_d]^{x+}$ and $[\mathrm{Cation}_f{}^{IV}(OH)_m\ (OOH)_d]^{x+}$ and higher valent cation hydroxides/peroxides.
2. Gases, including H_2S [51], O_2 [52], CO_2 [53], CO [53], H_2 [53].
3. Halogenated ions of the form $[\mathrm{halogen}]_xO_y$ (e.g., chlorates, bromates, perchlorates, etc.), and $C_x[\mathrm{Halogen}]_yO_z$ [34–37] and halogenated organic compounds of the general form $C_xH_y[\mathrm{Halogen}]_z$, where *y* can be 0. The halogen is one or more of Cl, Br, I, F. [54–57], for example, chloromethane (CM), trichloromethane (TCM), dichloromethane (DCM), tetrachloromethane; perchloroethylene (PCE), trichloroethylene (TCE), dichloroethylene (DCE); vinyl chloride (VC); hexachloroethane, tetrachloroethane, trichloroethane, dichloroethane, chloropropane (etc.), chlorobutane (etc.), chlorobenzene, (etc.), ethylene dibromide (EDB), perchlorate, polychlorinated biphenyls (PCB's). The end degradation products take the generic form C_xH_y (e.g., methane, ethyne, ethene, ethane, propane, butane, pentane, hexane, heptane, octane). These may be further altered to form products of the form: $H_xC_yO_z$ or ring structures.
4. Organic peroxides (e.g., triacetone triperoxide (TATP))[58].
5. Organic nitrogenous compounds, including azo dyes [59–61], atrazine [62, 63], cyclonite/hexogen (RDX) [14, 64], dinitrotoluene (DNT) [65, 66], nitrosodimethylamine (NDMA) [67, 68], nitrocellulose [69], tetramethylenetetranitramine (HMX) [70–72], trinitrotoluene (TNT) [73–75], disinfection by-products (DBPs) [76, 77], fertilizers [78, 79], pesticides [80–83], herbicides [84, 85], fungicides [86].

6. Organic compounds including Methyl *tert*-butyl Ether (MTBE) [87], aromatics (e.g., BTEX) [88–91]), hydrocarbons [21, 92–98], hormonal pollutants [99]. Metal carbonyl pollutants (e.g., $Fe(CO)_5$, $Fe(CO)_4$) can be reduced to n-Fe^0 (5–15 nm) by thermolysis in the presence of functional polymers [100].
7. Most metals, metalloids, and nonmetals, including their oxides, hydrides, hydroxides, peroxides, nitrates, nitrites, sulfides, sulfates, halides, carbonates, bicarbonates, and phosphates. ZVM is used to adjust the Eh and pH. This shifts the water redox environment into a redox environment, which will allow either direct precipitation, or precipitation by substitution of Fe in a precipitated Fe corrosion product [10, 101–113]. Examples of contaminant ions and the associated precipitated products, which can be formed by the presence of ZVM, are summarized in Appendix 1.B.
8. Microbiota [10, 114–128] including *Escherichia coli* [115–118, 123, 124, 128], colliforms (e.g., *Enterococcus faecium*, *Enterococcus faecilis*) [128], *Klebsiella pneumoniae* [125], *Salmonella typhimurium* [10], *Salmonella enterica* [124], *Salmonella paratyphi* [125], *Shigella* spp. [125], *Salmonella* spp.[124], *Staphylococcus aureus* [117, 118], *Streptococci* spp. [126], *Bacillus cereuis* [118], *Bacillus subtilis var. niger* [116, 119, 123], *Dehalococcoides* spp. [123], *Pseudomonas spp.* [118], *Pseudomonas fluorescens* [116, 118, 123], *Pseudomonas aeruginosa* [125], *Vibrio parahaemolyticus* [118], *Vibrio cholerae* [126], phiX174/FX174 [120, 128], T1 [121], Aichi virus [120], adenovirus 41 [120], MS-2 [116, 120, 128], Hepatitis A [122], norovirus [122], rotavirus [122], f2 virus [128], *Alcaligenes eutrophus* [123], *Aspergillus versicolor* [116, 119, 122], *Cryptosporidium* spp. [126], *Naeglaeria* spp. [126], *Naeglaeria fowleri* [128], *Giardia* spp. [126], *Hartmannella veriformis* [128], *Tetrahymena pyriformis* [128], *Daphnia magna* [116], *Pseudokirchneriella subcapitata* [116], *Dunaliella tertiolecta* [116], *Thalassiorsria pseudonana* [116], *Isochrysis galbbana* [116], fungi [127], prions [127], viruses [127], protozoa [127], bacteria [127], algae [127], etc. n-Fe^0 (20–30 nm) rapidly inactivates microorganisms by coating them with Fe(OOH) [119]. Inactivation is by one or more of Eh:pH changes and the interaction of Fe corrosion products (oxides, hydroxides, and peroxides), for example, [114, 119].
9. Macrobiota. n-Fe^0 in soil (0.1 to >1 g n-Fe^0 kg^{-1} soil) adversely affect worms (e.g., *Eisenia fetida* and *Lumbricus rubellus*) and springtails (e.g., *Folsomnia candida*) [123].
10. Plants. Concentrations of n-Fe^0 in excess of 250 mg kg^{-1} soil have been found to stunt the growth of rye grass and clover [123].

1.3 REMEDIATION MECHANISMS

The mechanisms associated with ZVM remediation are the subject of conflicting, overlapping, and competing hypotheses, and more than one mechanism applies in each remediation environment. The principal hypotheses are

1. *Catalyst Model:* ZVM acts as a Langmuir-Hinshelwood catalyst (e.g., [55, 95], that is, adsorption of reactants on ZVM surface and desorption of products [55, 130, 131]), or Eley-Rideal catalyst (e.g., [95, 129], i.e., adsorption of one or more reactants on the ZVM surface with reaction of the adsorbed species with one or more fluid-phase reactants that are not adsorbed on the ZVM surface to produce a product [129–131]), or acid catalyst ($Fe\text{-}H^{n+}$) [10, 21, 96–98].
2. *Redox Model:* ZVM changes the water Eh and pH, thereby forcing remediation by changing both K and ΔG for the remediation reaction [10]. Under this model, n-ZVM reactions are essentially fluid phase electrochemical reactions, or contact surface reactions [10].
3. *Galvanic Model:* ZVM ionization (Appendix 1.B, Appendix 1.C) results in n-ZVM acting as self-charging galvanic cells (Fig. 1.2) that adjust the water pH and Eh. This adjustment forces a change in the cation:anion equilibrium state within the water [10]. The change in equilibrium state forces the reduction/oxidation of specific cations and anions, and a change in the Gibbs Free Energy associated with the remediation reaction [103, 104]. The presence of ZVM (and ZVM-ion adducts) in water creates (in a diabatic environment) a perpetual oscillation between higher and lower Eh and higher and lower pH [10] (Fig. 1.1b–j). This oscillation, which can be interpreted as alternating charging and discharging of the galvanic cells (Fig. 1.2): (i) creates, discharges, and adsorbs H^+ (protons, H_3O^+, $H_5O_2^+$, $H_7O_3^+$, $H_9O_4^+$, FeH^{2+}, FeH_2^+), e^- (H^-, H_2O^-, electrons), O^-, O^{2-}, O_2^-, H_2O_2, OH, OH^-, O_2H, and O_2H^-; (ii) creates a unique (ZVM specific) trajectory of Eh/pH change with both residence time and space velocity [10]. This galvanic discharge–recharge mechanism results in substantial water consumption (>0.18 *t* H_2O t^{-1} n-Fe^0), but drives fluid phase (and ZVM/ion surface) Fenton Reactions, electron shuttle reactions, proton shuttle reactions, and oxide ($H_xO_y^{(c+/-)}$) shuttle reactions within water [10, 96]. These reactions undertake the reduction/oxidation of pollutants,

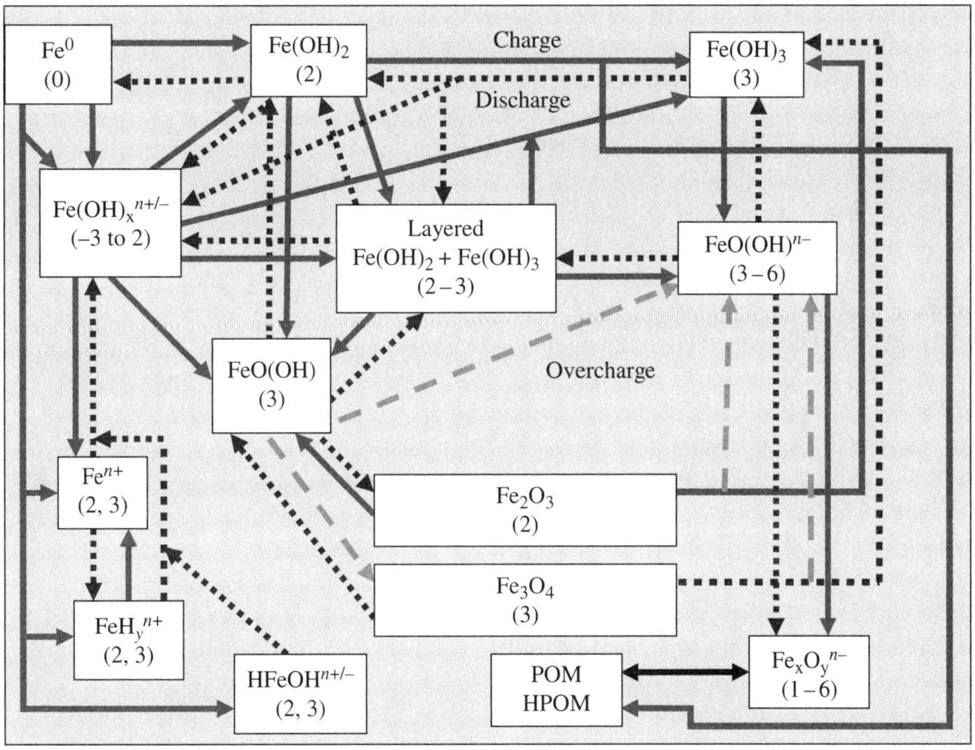

FIGURE 1.2 Fe–Hydrogen Redox Cell: Simplified relationship between n-Fe^0, Fe^0 products, oxidation number (brackets), and stored charge in the various ZVM components.

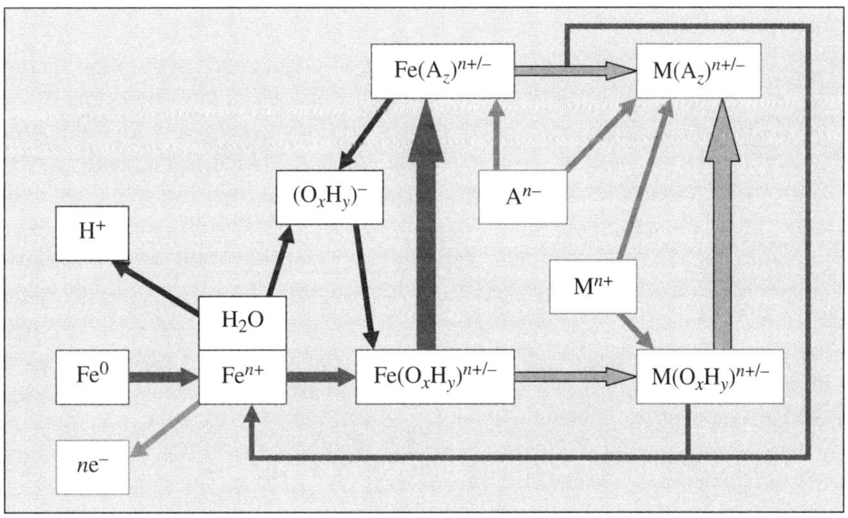

FIGURE 1.3 Fe-water Redox cell, simplified sequence of anion and cation exchange.

removal by incorporation into hydroxide/peroxide precipitates of anions and cations, and the reformulation of organic pollutants into simple alkanes and alkenes [10, 21, 95–98].

4. *Adsorption Model*: ion substitution (Fig. 1.3) of ZVM corrosion products and nano–molecular growth in self-assembly molecules nucleating around ZVM corrosion products results in the removal (by substitution/adsorption) of pollutant ions [38–50, 101, 102]. This model is treated in this study as a subset (Figs. 1.2 and 1.3) of the Galvanic Model.

The generic remediation reaction takes the form $aA+bB=dD+eE$. If $C_{t=0}$ = the contaminant at time, $t=0$, then the observed rate of reaction ($k_{observed}$), between $t=0$, and $t=m$, can be determined [130, 131] as

$$k_{observed} = \frac{\left(C_{t=m} / C_{t=0}\right)}{(t = m(s))} \tag{1.1}$$

$$\mathrm{Ln}\left(\frac{C_{t=0}}{C_{t=m}}\right) = \left[k_{observed}\right]t \tag{1.2}$$

Equations 1.1, 1.2 apply to each remediation model.

1.3.1 Catalyst Model

The hypothesis [48] that ZVM acts as a catalyst will result in decreasing particle size, increasing particle surface area, and/or increasing the quantity of ZVM, automatically increasing the observed rate of reaction ($k_{observed}$) [48, 55, 129–131]:

$$k_{sa} = \frac{k_{observed}}{a_{s(t=n)}p_m} \text{[Normalised Reaction Rate]} \tag{1.3}$$

$$k_{sa} = A_{sa} \exp\left(\frac{-E_{a(sa)}}{RT}\right) \tag{1.4}$$

$$k_{observed} = A_{(observed)} \exp\left(\frac{-E_{a(observed)}}{RT}\right) \tag{1.5}$$

$$\text{Reaction rate}, v, \left(\text{mol l}^{-1}\text{s}^{-1}\right) = k[\text{A}]^m [\text{B}]^n [\text{C}_a]^p \tag{1.6}$$

It is commonly assumed that if a plot of ln(k_{sa} or $k_{observed}$) vs. time and pollutant concentration can be interpreted as a negative, or positive, zero-, first-, second- or third-order reaction [130, 131], then the ZVM must be acting as a catalyst. However, the primary interaction of the ZVM is with water (e.g., n-$Fe^0+H_2O=HFeOH^{2+}+2e^-$), and this interaction generates e^- [103] (Appendix 1.C). e^- is a powerful catalyst (used in electron shuttle reactions) [130]. It is therefore possible that much of the catalytic activity attributed to n-Fe^0 (and other ZVM) has been misattributed, and the actual catalytic activity/remediation reactions are undertaken by e^- [10] (as the availability of e^- is directly linked to the corrosion of ZVM (Appendix 1.B, 1.C)).

The catalytic model assumes that the remediation reactions may take the form, A+ZVM={A[ZVM]}=products, or A+ZVM hydride, oxide, hydroxide, peroxide={A[ZVM hydride, oxide, hydroxide, peroxide]}=products, The associated reaction rates are [130]: k_d=A+ZVM={A[ZVM]}; k_{-d}={A[ZVM]}=A+ZVM; k_r={A[ZVM]}=products. The overall rate of reaction (v)=k_r[{A[ZVM]}]=k_d k_r [A{ZVM}]/(k_{-d}+k_r) [130] and the overall rate coefficient $k_{observed}$=v/{A[ZVM]}=$k_d k_r$/(k_{-d}+k_r) [130]. The equilibrium constant ($K_{\{A[ZVM]\}}$) for the encounter pair {A[ZVM]} is k_d/k_{-d} and $k_{observed}=k_r K_{\{A[ZVM]\}}$[130].

In groundwater, the ZVM diffusion environment results in $k_{observed}$ (m^3 s^{-1})=$4\pi r_{\{A[ZVM]\}} D_{\{A[ZVM]\}}$[130]. Transition state theory (absolute rate theory) [130] defines: $k_{observed}=k_B T/h \exp(-\Delta G^{‡}/RT)$. The concentration of dissolved ions in the water impacts directly on the reaction rate (k), that is, $k_{observed}=(k_B T/h)\, K^{‡}\, (\gamma_A\, \gamma_{ZVM}/\gamma_{\{A[ZVM]\}})$ [130]. These interactions are rarely accounted for in studies that suggest that ZVM acts as a remediation catalyst.

1.3.2 Redox Model

In groundwater [103, 104, 131]:

$$\Delta E(\text{Eh}) = \Delta E° - \frac{RT}{nF}\ln[\text{Q}] = \frac{\Delta G}{(nF)} = -\Delta E° - \frac{RT}{nF}\ln[\text{K}] \tag{1.7}$$

$$\Delta E° = \frac{RT}{nF}\ln[\text{K}] = \frac{\Delta G°}{(nF)} \tag{1.8}$$

A non-catalytic redox remediation reaction can be expressed as $a[A]+mH^+ + ne^- = b[B]+c[H_2O]$. This generic equation allows the relationship between Eh, pH, E, $\Delta G°$, and ($[B]^b/[A]^a$) associated with each remediation reaction to be summarized [103, 131] as

1. when A is an aqueous ion (oxide, hydroxide, peroxide) and B is a precipitate as $Eh=\Delta G°/(nF)+(0.0591/n)\log([B]^b/[A]^a)+([-0.0591m/n\,[pH]])$; where $\Delta G°/(nF)=\Delta E° = -RT\ln[K]$: $pH=-\log H^+$; $(H_2O)^c=1$
2. when [A] and [B] are dissolved substances ($M\,L^{-1}$), and $m>0$ and $n=0$, then, $\log([B]^b/[A]^a)=\Delta E° + m[pH]$
3. when [A] and [B] are dissolved substances ($M\,L^{-1}$), and $m=0$ and $n>0$, then, $Eh=\Delta E° + (0.0591/n)\log([B]^b/[A]^a)$
4. when [A] and [B] are dissolved substances ($M\,L^{-1}$), and $m=>0$ and $n>0$, then, $Eh=\Delta E° + (0.0591/n)\log([B]^b/[A]^a)+([-0.0591m/n\,[pH]]$
5. when [A] and [B] are solid substances, and $m>0$ and $n>0$, then, $Eh=\Delta E° + ([-0.0591m/n\,[pH]])$
6. when [B] is a solid substance and [A] is a dissolved substance ($M\,l^{-1}$), and $m>0$ and $n=0$, then, $\log([A])=\Delta E° + m\,[pH]$
7. when [B] is a solid substance and [A] is a dissolved substance ($M\,l^{-1}$), and $m>0$ and $n>0$, then, $Eh=\Delta G°/(nF)+(0.0591/n)\log([A])+([-0.0591m/n\,[pH]])$.

The partial pressure of the gaseous reactants/ions (e.g., H, O, CO, CO_2, C_xH_y, etc.) alters $k_{observed}$, as $k_{observed}=k(P_p)^{xm}$ and $K_p=K(RP)^{cp}$, [21, 131].

The interactions between Eh, pH, partial pressure of (pH_2), and partial pressure of $[O_2]$ (pO_2) are defined by the relationships [103]: (i) Hydrogen: $Eh=0.00–0.0591\,pH–0.0295\log(pH_2)$ $[2H^+ + 2e^- = H_2(g, aq)]$, (ii) Oxygen: $Eh=1.228–0591\,pH+0.0147\log(pO_2)$ $[2H_2O=O_2(g, aq)+4H^+ + 4e^-]$.

These relationships imply [10, 103, 104] that if ZVM is able to alter the Eh and pH of water, that the resultant remediation (e.g., Appendix 1.B) is both non-catalytic, and a natural consequence of an Eh, pH modification of pore water chemistry. This model assumes that the primary role of ZVM during the remediation process is to alter the water Eh and pH [10].

1.3.3 Galvanic Model

The presence of ZVM creates two primary products in water [103, 104]. They are e^- and H^+. Secondary products include H, H_2, O, O_2, O_2^-, O^{2-}, OH, OH^-, O_2H, O_2H^-, H_2O_2 [103, 104, 132]. The ZVM gradually degrades to produce ZVM ions [Fe^{n+}, Al^{3+}, Cu^{n+}] and associated ion adducts [103, 104, 132].

1.3.3.1 Diabatic Environment Remediation by ZVM injection into soil, or groundwater (<25 m depth), takes place in a diabatic environment where the temperature, T, is a function of atmospheric temperature [133, 134]. T varies during the day and seasonally over the year [133, 134]. Daily variations in T decrease with increasing depth [133]; daily variations of T are within the range <1 to >15°C; annual variations are within the range <1 to >50°C. Changing T will change the partial pressures of H_2 (and O_2) and one or more of pH, Eh, K, Q, $k_{observed}$ [103, 104, 131, 132]. Where the remediation reaction is reversible, and $E_a>0$, decreases in temperature may result in $k_d<k_{-d}$ and reversal of the remediation reaction (and vice versa). When Eh and pH are largely unaffected by changes in T, and ion removal is by precipitation (Appendix 1.B) then, a change in T of 1°C changes $\log([B]^b/[A]^a)$ by $(R\ln[K])/(0.0591/n)$ [103]. Consideration of temperature variation is therefore a major variable when predicting the effectiveness of a groundwater remediation program.

1.3.3.1.1 Redox Trajectory Placement of n-ZVM in a diabatic groundwater environment results in a gradual change in Eh, pH over time [10, 17, 135] as the oxidation state of the Fe^0 increases (Fig. 1.1b–d). The redox trajectory is a function of Fe^0 particle size [10, 17, 135] (Fig. 1.1b and c), Fe^0:water ratio [10, 17] (Fig. 1.1b and c) and ZVM composition [10, 135] (Fig. 1.1d) [10, 135]. Daily variations in temperature [134] force an oscillation in both Eh and pH [10, 96, 135] (Fig. 1.1e and f), while maintaining a relatively constant hydrogen partial pressure (pH_2) (Fig. 1.1g). pH_2 can be independent of ZVM composition (Fig. 1.1g).

The general redox oscillation (Fig. 1.1e and f) is accompanied by a cyclic oscillation in EC [10, 135] (Fig. 1.1h), which reflects adjusting $Fe(OH)_x$, FeOOH, Fe_xO_y composition [10, 96, 135]. Each oscillation cycle commences with a large swing in EC, which dampens with time (Fig. 1.1h). These EC oscillations (Fig. 1.1h) reflect oscillations in Eh, pH, K, $\log(B]^b/[A]^a)$ and are directly linked to cyclic changes in temperature (Fig. 1.1i and j).

1.3.3.2 Remediation Types Fe^0 remediation reactions fall into two basic groups: (i) irreversible, ZVM, or e^-, catalyzed reactions, or reaction sequences (Type A) (e.g., nitrate, PCE removal [10]), and (ii) reversible redox, or ZVM (oxide, hydroxide, peroxide) substitution reactions (Type B) [10, 101–104]. Type A reactions are described by the Catalyst model, Redox model, and Galvanic model, while Type B reactions are described by the Redox model, Galvanic model, and Adsorption model.

1.3.3.2.1 Galvanic Type A Reactions Type A reactions require e^-, or H^+, or O_2 (e.g., electron shuttle and Fenton Reactions [10, 135]) to produce a product. They are favored by changes to the redox (Eh:pH) environment as [103]

$$\log\left(\frac{[B]^b}{[A]^a}\right) = Eh - \Delta E° + \frac{[-0.0591m/n[pH]]}{0.0591/n} \quad (1.9)$$

Type A reactions are theoretically reversible, but in practice many are effectively irreversible (e.g., nitrate removal [10, 24–33, 129], TCE removal [10, 17, 54–57, 136]). For example [57], PCE (C_2Cl_4) degrades to C_2Cl_2 and TCE (C_2CH_3H). TCE degrades to DCE ($C_2Cl_2H_2$) and C_2ClH. DCE degrades to VC (C_2ClH_3), C_2H_2, and C_2H_4. VC degrades to C_2H_4. C_2Cl_2 degrades to C_2ClH, which then degrades to C_2H_2, which is then hydrogenated to C_2H_4. C_2H_2 and C_2H_4 are hydrogenated to C_2H_6 and C_xH_y [94, 96]. E_a for nitrate removal is in the range 21–46 kJ mol^{-1}[6, 30]. E_a for PCE/TCE/chlorinated hydrocarbon removal is in the range 9.8–80 kJ mol^{-1}[11, 136]. Since $\Delta E°=-RT\ln[K]$ and $\ln[K]=\Delta E°/RT$ [103, 104], it follows that increasing temperature, while maintaining a constant Eh and pH (when $\Delta E°>0$), will decrease the equilibrium ratio ($[B]^b/[A]^a$). It will also increase the reaction rate ($k_{observed}$) (Eq. 1.4).

From Equation 1.3, it follows that the principal controls on a Type A remediation program are ZVM particle size, particle type, mass ratio of pollutant:injected ZVM, and the injected ZVM:water/gas slurry concentration (g l) [137]. A relatively small reduction in particle size (from >1000 to 50–300 nm) can allow a major reduction in the amount of ZVM required to remove greater than 99% of the TCE in the groundwater (Fig. 1.1b and c).

From Equation 1.9, it follows that remediation is enhanced by increasing the availability of e^- by increasing the O_2 saturation of the pore water [138, 2, 139–141], while maintaining a constant, or decreasing, pH, and/or decreasing the aquifer pH by injection of CO_2 [94, 96, 2, 139–141] or addition of acidic components, for example, $FeCl_y$, while maintaining a constant or decreasing Eh [[10], [21], [95], [103], [142]]. It also follows (from Eqs. 1.3–1.5) that increasing the groundwater temperature by water injection, steam injection, or gas injection may reduce the time and amount of n-ZVM required to achieve a specific level of remediation from, for example, 100 days, to between <1 day and >50 days.

1.3.3.2.1.1 GALVANIC TYPE A REACTIONS: IMPACT OF OXYGENATION In oxygenated water, n-Fe^0 behaves as an iron–oxygen redox cell [138], where the overall reaction is $Fe^0+0.5O_2+H_2O=Fe(OH)_2$ [Cathode {+} reaction : $0.5O_2+H_2O+2e^-=2OH^-$; Anode [−] reaction: $Fe^0+2OH^-=Fe(OH)_2+2e^-$; pH=<10.53]. $Fe^0=Fe^{2+}+2e^-$; $Fe^{2+}+2OH^-=Fe(OH)_2$ when the Fe^{2+} concentration is greater than (log (Fe^{2+})=13.29–2pH [103]. At a pH>10.53, $FeOOH^-+H^+=Fe(OH)_2$ when the $FeOOH^-$ concentration is greater than (log ($FeOOH^-$)=−18.30–pH [103]. The relative stability of the Fe^{2+} and $FeOOH^-$ ions is provided by the molar relationship log[$FeOOH^-/Fe^{2+}$]=−31.58+3pH [103]. The addition of oxygen into the iron–air cell modifies the standard redox cell used to produce $Fe(OH)_2$ from: (i) $Fe+2H_2O=Fe(OH)_2+2H^++2e^-$ (Eh for phase boundary is [103]: Eh=−0.047–0.0591 pH) to; (ii) $Fe^0+0.5O_2+H_2O=Fe(OH)_2$ (Eh for phase boundary is [103]: Eh=−1.29–0.0591 pH). The net effect is an increase in the availability of e^-, and an increase in the associated remediation rates. At any given time, the concentration of e^- in the water is [103]: e^- [M l^{-1}]=$10^{((Eh\ (water)+1.125)/0.0295)-pH\ (water))}$. Magnetised n-$Fe^0$ will preferentially attract O_2 (e.g., $Fe^0+O_2+2H^+=Fe^{2+}+H_2O_2$; $Fe^{2+}+H_2O_2=Fe^{3+}+2HO+e^-$) [2]. Chlorinated organics are removed from oxygenated water by an electron shuttle mechanism using Fe^0 or Al^0. A simple shuttle mechanism, where e^- acts as a catalyst [130], is provided as $H_zC_xCl_y+e^-+H=[H_{z+1}C_xCl_{y-1}]+Cl+e^-$. The electron shuttle model predicts that increasing the availability of e^- by oxygenation, or another mechanism, will increase the remediation rate. Experiments have established that oxygenation increases the rate of remediation reaction (for As removal) by greater than 4 fold (over a 60-min period) but does not necessarily reduce Eh [139–141], through the reversible equilibrium reactions $Fe^0+2H_2O=Fe^{2+}+H_2+2OH^-$; $Fe^{2+}+H+e^-=FeH^+$; $FeH^++O=FeOH^+$; $Fe^0+2H_2O+O_2=2Fe^{2+}+4OH^-$; $2Fe^{2+}+nOH^-=Fe(OH)_n$, etc. (Fig. 1.2). Effective anion removal (e.g., As) is enhanced in an acidic environment [101–104, 2, 139–141]. This can be achieved by acidifying the water by CO_2 injection [139–141] or acid injection [2, 139–141], prior to n-Fe^0 injection, and oxidation [139–141]. e^- generation through a strategy of cyclic n-Fe^0 oxidation and reduction appears to be effective over greater than 4000 redox cycles [138].

1.3.3.2.2 Galvanic Type B Reactions Type B remediation reactions occur when (i) the interaction of T, Eh, pH changes resulting from the presence of ZVM, results in a change in K, which allows pollutant ions to be precipitated as oxides, peroxides, hydroxides, sulfides, carbonates, etc (e.g., Appendix 1.A), and (ii) when the Fe^0 corrodes to one or more of n-FeH^{n+}, n-$Fe(OH)_x$, n-FeOOH, n-Fe-$[O_xH_y]^{(n+/-)}$ (Fig. 1.2). Subsequent Fe ion substitution/adsorption (or Fe ion adduct formation) of cations and anions (Fig. 1.3) results in pollutant cation/anion removal from the water, with precipitation within a n-Fe-$[O_xH_y]$ structure–Adsorption Model [101, 102].

At any specific (constant) pH (e.g., pH = 7), both cations and anions are removed [143]. The total amount of cations removed may increase, or decrease, with changing temperature [144], and may be a function of both pollutant concentration, and the concentration of other anions (e.g., humic acids) within the groundwater [144]. Anion and cation removal increases with time [145], and the ratio of cations:anions removed by incorporation/substitution varies with the pollutant ion adduct:Fe^{n+} ratio in the water [144]. Fe(II) ions (and other cations) hydrolize on the surface of FeOOH particles [146].

1.3.3.2.2.1 GALVANIC TYPE B REACTIONS: POM AND HPOM PRODUCTS Polyoxometalates (POMs) (Fig. 1.2) are self-assembly accretionary molecules that take the form of a sandwich composed of a central core fragment $Fe^{II}{}_n Fe^{III}{}_m O_z.(H_2O)_y$ surrounded by external fragments of $Fe^{II}{}_n Fe^{III}{}_m O_z.H_2O_y$ linked by two distinct edge sharing dimeric clusters of $(Fe(OH)_2$ [147, 148]. The formation of POMs greatly increases the rate of n-ZVM water remediation by serving as an electron shuttle and ion chelating agent [149]. A POM (Fig. 1.2) may potentially remove (Fig. 1.3) greater than 10 g Cation g^{-1} n-Fe^0. The associated by-product Type A reactions, involving e^- catalysis, may remove greater than 1 g pollutant g^{-1} n-Fe^0.

Heteropolyoxometallates (HPOM) are derived from metal cages of the form $(MO_n)_x$, which incorporate anion templates of the form (AO_x^{n-}) [150]. However, their pentagonal building blocks form around a pentagonal bipyramidal core (MO_n), which can be hydrated [150]. A typical HPOM nucleates around a cluster of 2 Fe ions (oxidation state 2^+ or 3^+ or 4^+). They seed a linkage, which allows clusters of pentagonal, or another structural form, of $M(1)O_n$ to accrete [150]. In saline water, the monomer may take the form $[K_{8+x}Na_{9+y}H_{29+z}[H_{34}M(1)_{119}M(2)_8Fe_2O_{420}(H_2O)_{34+n}]]^{(8-x-y-z)-}$; the diamer may take the form $[K_{16+x}Na_{1\,9+y}H_{57+z}[H_{34}M(1)_{119}M(2)_8Fe_2O_{420}(H_2O)_{74+n}]]^{(16-x-y-z)-}$ [150]. M(1) and M(2) are different metal cations incorporated in the HPOM from the water. An individual HPOM molecule may have a size of less than 3 nm [150]. HPOM formation is slow and conversion of 4% of the n-Fe^0 to HPOM may take greater than 4 weeks [150]. However, they are highly effective remediation agents [149] with a potential absorption capacity of greater than 100 g pollutant cation g^{-1} n-Fe [150]. Injection of 100 kg n-Fe^0 into groundwater can potentially result in greater than 400 kg of pollutant cations being removed in HPOM structures over a 4-week period.

1.3.3.2.2.2 GALVANIC TYPE B REACTIONS: IMPACT OF HYDROGEN In poorly oxygenated water, the n-Fe^0 gradually corrodes (Appendix 1.C, Fig. 1.2) to form a corrosion zone of Fe-hydroxides and peroxides at the n-Fe^0–water interface [10]. The interface acts as a hydrogen electrode (cathode) and the Fe^0 acts as a current electrode (anode). During remediation, Fe/Cu acts as a cathode to an Al anode. The Cu acts as a cathode to a Fe anode [151–154]. The basic process involves charge transfer (and OH ion formation) at the ZVM–water interface and includes electron transfer via conduction, electron insertion into active sites, and conduction by hopping through electron deficient lattice sites within the active material [151].

In a diabatic environment, the perpetual oscillation and change in temperature (Fig. 1.1i and j), results in a perpetual oscillation between forward and backward reactions (Fig. 1.2). This oscillation allows the hydrides/hydroxides/peroxides/oxides (Fig. 1.2) to be used as stores of protons (H^+) and electrons (e^-) [151–154], which can be accessed for Type A remediation reactions. All changes that increase the oxidation number of the ZVM ion adducts, effectively result in electron storage (charging) occurring and vice versa [151] (Fig. 1.2). Ions (aqueous or solid) that contain an oxidation number greater than the stoichiometric charge are overcharged [151] (Fig. 1.2).

1.3.3.2.2.3 GALVANIC TYPE B REACTIONS: DISCHARGE During discharge, electrons flow from the current electrode (Fe^0 particles $[Fe^0 = Fe^{n+} + ne^-]$ and other ZVM and ZVM adducts (Figs. 1.2 and 1.3, Appendix 1.C)), through the hydroxides, peroxides (Fig. 1.2)–charge transfer sites [e.g., $FeOOH + H_2O + e^- = Fe(OH)_3 + e^-$; $FeOOH + OH^- + H^+ + e^- = Fe(OH)_3 + e^-$] to the hydrogen electrode (Cu^0 particles) [151]. Hydrogen generation occurs at the particles, which act as a hydrogen electrode [$2H_2O + 2e^- = 2H^+ + 2OH^- + 2e^-$; $2H_2O + 2e^- = H_2 + 2OH^-$].

1.3.3.2.2.4 GALVANIC TYPE B REACTIONS: RECHARGE During recharge, the electron flow is reversed [151] and oxygen forms at the cathode as a by-product of electron generation [cathode–electron generation: $4OH^- = O_2\ (g) + 2H_2O + 4e^-$; hydroxide reduction to peroxide in the charge transfer sites: $Fe(OH)_3 = FeOOH + H_2O + e^-$; Fe ion reduction to Fe^0] [151].

1.3.3.2.2.5 GALVANIC TYPE B REACTIONS: GAS EVOLUTION During recharge, oxygen accumulates in the charge transfer sites [151]. During discharge, hydrogen accumulates in the charge transfer sites [151]. Both gases show very different morphologies at the ZVM–water interface [10]. Oxygen bubbles tend to form in, and are commonly encased by, the cathodic particles (e.g., Cu) [10], and form rapidly after a ZVM mixture (Fe+Cu, Fe+Cu+Al (Fig. 1.4a–d)) is placed in the reactor. The initial corrosion reactions are recharge reactions forming FeOOH. The FeOOH forms active charge sites. The formation of hydrogen gases initially results in the adsorption of the O_2 gas bubbles, with no hydrogen gas discharge (i.e., $2H_2 + O_2 = 2H_2O + heat$) [151].

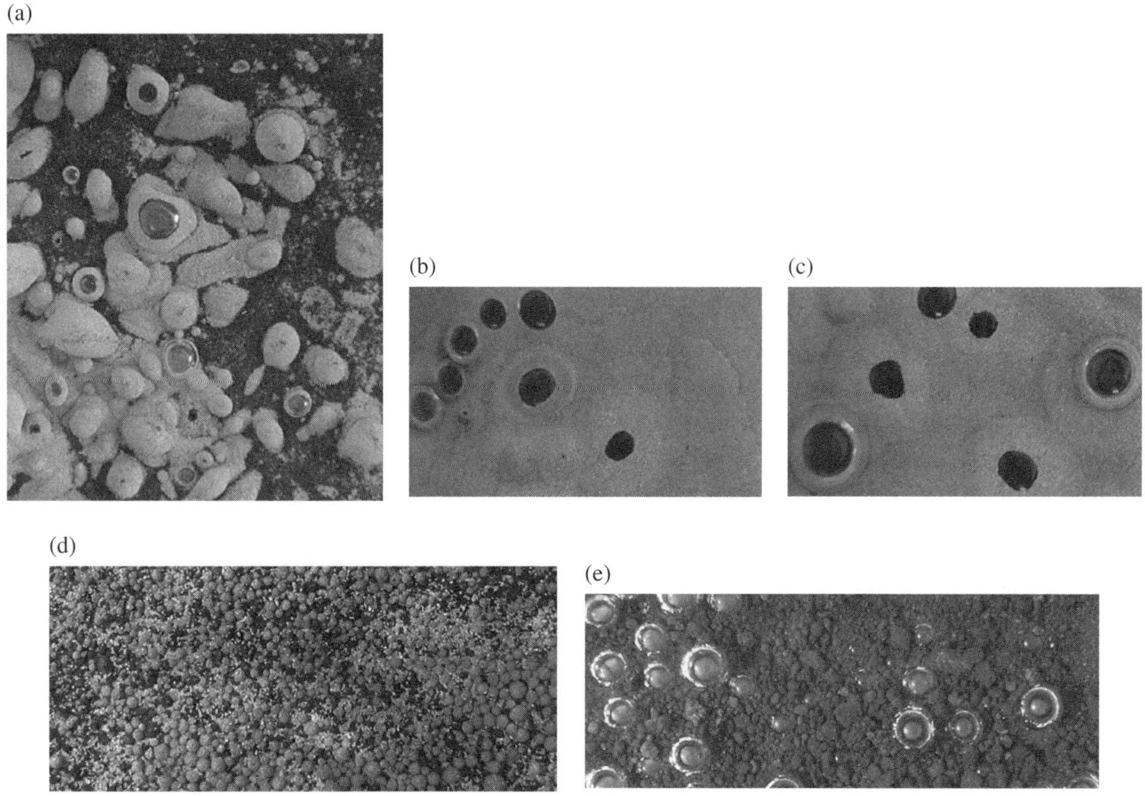

FIGURE 1.4 Morphology of common gas bubbles associated with n-Fe^0 (Nanofer Star (supplied by nanoiron s.r.o.; www.nanoiron.cz), 50 nm, BET = 20 m^2 g^{-1}; mixed with n-Al^0 and n-Cu^0) (a) Oxygen bubbles encased by Cu^0 on the surface of n-Fe^0 [5 g n-Fe^0 + 5 g n-Cu^0 + 0.25 l saline H_2O [Eh = 0.095 V; pH = 7.01; EC = 1.993 mS cm^{-1}; T = 12.8 C – gas composition checked using TCD GC]]. (b) O_2 gas venting where n-Al^0 rests on top of n-Fe^0. The O_2 gas bubbles are encased by n-Cu^0. [5 g n-Fe^0 + 5 g n-Cu^0 + 5 g n-Al^0 + 0.25 l saline H_2O [Eh = 0.073 V; pH = 7.00; EC = 1.981 mS cm^{-1}; T = 12.9 C– gas composition checked using TCD GC]]. (c) O_2 gas venting where n-Al^0 rests on top of n-Fe^0. [5 g n-Fe^0 + 5 g n-Cu^0 + 5 g n-Al^0 + 0.25 l saline H_2O [Eh = 0.073 V; pH = 7.00; EC = 1.981 mS cm^{-1}; T = 12.9 C– gas composition checked using TCD GC]]. (d) O_2 filled spheres of n-Cu^0 developing on the n-Fe^0 – water interface, 5 min after loading into a reactor. [40% n-Fe^0 + 20% n-Cu^0 + 40% n-Al^0]. (e) H_2 gas bubbles developing on the ZVM-water interface (Fig. 1.4d), 3 weeks after loading [H_2 composition verified by TCD GC]. Part of the n-Fe^0 has been corroded to form agglomerated FeOOH and Fe_3O_4 nodules or clods (0.5–4 mm in diameter). Some of the nodules are coated with n-Cu^0. Each nodule forms an accreting galvanic cell (Fig. 1.2) with an anodic core (e.g., n-Fe^0, n-Al^0, $Fe(OH)_2$) and a cathodic exterior (e.g., n-Cu^0, n-FeOOH, n-Fe_3O_4). Individual gas bubbles are 3–6 mm in diameter.

After the FeOOH corrosion products (Fig. 1.2) reach a critical mass, the ZVM switches from operating in a net recharge mode, to operation in a net discharge mode. During this phase, distinctive hydrogen gas bubbles form on the ZVM/FeOOH surface (Fig. 1.4e). Unlike the O_2 bubbles, H_2 bubbles are not associated with a specific cathodic ZVM, but instead form on the surface (and in) active charge transfer sites (e.g., FeOOH, Fe_3O_4 (Fig. 1.4e)).

1.3.3.2.2.6 GALVANIC TYPE B REACTIONS: HYDROGEN EVOLUTION The amount of hydrogen generated is a function of ZVM composition, water composition, and operating conditions (pressure, temperature) [155–157]. The maximum hydrogen production occurs when the n-ZVM is reduced to the ZVM oxide (Fig. 1.2). For example, xZVM + $y H_2O$ = ZVM_xO_y + $y H_2$. For the reaction $3Fe + 4H_2O = Fe_3O_4 + 4H_2$ (Figs. 1.2 and 1.4e), 167 g n-Fe^0 (50 nm) + 72 g H_2O = Fe_3O_4 + 8 g H_2 (89.64 l) [158]. This process can be undertaken over a short time period using n-Fe^0 (50 nm). Increasing the temperature of a water:n-Fe^0 mixture from <20 to 350°C over a 90-min period, in a sealed diffusion reactor, will result in a H_2 yield of about 450–540 m^3 H_2 t^{-1} n-Fe^0, and a gas pressure of greater than 5 MPa [159]. Cooling the reactor to 20°C provides a deliverable H_2 gas at less than 3 MPa [159]. Reduction of the Fe_3O_4 to Fe^0 allows the cycle to be repeated (e.g., $Fe_3O_4 + 4CO = 3Fe^0 + 4CO_2$; $Fe_3O_4 + 4H_2 = 3Fe^0 + 4H_2O$) [159, 160]. In a confined diffusion reactor, the general reactions (Fig. 1.2), result (at T = <50°C) in low levels of pressurized H_2 gas evolution as the Fe^0 oscillates between charged (Fe^{III}) and discharged (Fe^{II}) states [155–157].

The oscillating combination of H^+ and e^- generation from the cathodic sites during recharge and discharge [151, 153] creates the driving force for chlorinated hydrocarbon (and other Type A) remediation [161].

1.4 REMEDIATION MARKET

Contaminated sites (soils and groundwater) vary in size from <100 m^2 to >10 km^2. The number of contaminated sites, which could benefit from n-ZVM treatment, is estimated at 350,000–400,000 in Europe, 235,000–355,000 in the United States [136, 162]. There are probably a similar number of contaminated sites in Canada, S. America, China, Russia, India, The Middle East, Asia, Australia, and Africa. To date, only a few sites have been treated using n-ZVM.

1.4.1 Remediation Costs

A typical PCE/TCE/DCE groundwater remediation costs around \$200–\$700 kg n-Fe^0 used [137, 162], and utilizes less than 1–280 *t* n-Fe^0 for each t PCE/TCE/DCE in the soil/aquifer [17, 137]. The cost comprises a n-Fe^0 cost (e.g., \$30–\$100 kg^{-1}) + injection/infiltration + monitoring costs. Since the radius of influence of an injection well/infiltration point source is typically less than 40 m [22, 23], reducing the n-Fe^0 cost will not necessarily reduce the costs associated with injection/infiltration and monitoring. Remediation adds value by either allowing the land to be rehabilitated, for industrial, domestic, or agricultural applications, or by allowing the water to be used for municipal, industrial or agricultural purposes. The sustainable remediation cost is a function of the overall value added by the remediation.

1.4.1.1 Reduction of Type A Remediation Costs Type A remediation costs are reduced by (i) reducing both particle size and the amount of ZVM injected (Eq. 1.3). Compare with Figures 1.1b, c, which both achieved greater than 99% removal of 25–88 mg TCE l^{-1} H_2O [17]; (ii) increasing groundwater temperature [130] (Eq. 1.4, and/or oxygen levels [138, 2, 139–141], and/or increasing groundwater acidity [10, 103, 104, 135, 142] (Eqs. 1.4 and 1.9), in order to both accelerate the remediation and reduce the overall amount of n-Fe^0 required. The catalytic model assumes that 1 mol n-Fe^0 can only generate 2 or 3 mol e^- (Appendix 1.C) and that increasing particle size will increase the active life of the n-Fe^0 [163]. The galvanic model assumes that the perpetual oscillations [10] within the groundwater will allow a substantially greater amount of e^- and H^+ to be formed using a cyclic process. That is,

1. $H_2O^- + Fe^{2+} = FeH^{2+} + OH^-$; $H + Fe^{2+} = FeH^{2+}$;
2. $H_3O^+ + FeH^{2+} = H_2O + H_2 + Fe^{3+}$; $H_2 = 2H^+ + 2e^-$; $H^+ + e^- = H$
3. $H + Fe^{3+} = H^+ + Fe^{2+}$ [132]

Fresh oxygen contained in recharge water entering the remediation zone will be initially removed [132] as $O_2 + Fe^{2+} = O_2^- + Fe^{3+}$; $O_2 + FeOH^+ = O_2^- + FeOH^{2+}$; $O_2 + Fe(OH)_2 = O_2^- + Fe(OH)_2^+$, $O_2 + Fe(OH)_3^- = O_2^- + Fe(OH)_3$, etc. The O_2^- interacts with $FeO_xH_y^{n+/-}$, H_2O, O_2H, OH and H to form O^-, O^{2-}, O_2 O_2H, OH, H_2O_2 and $FeO_xH_y^{n+/-}$ [132]. This allows recharge of oxygenated water (from surface precipitation and subsurface flow) to provide a natural drive for the galvanic cell.

1.4.1.1.1 Reduction of Type A Remediation Costs: Catalytic Model The cathodic model focuses on reducing particle size and increasing temperature to increase remediation rates and reduce the amount of ZVM required. Figures 1.1b, c demonstrate that the same degree of TCE remediation can be achieved using 3.9 kg n-Fe^0 (>1000 nm) m^3 soil and 0.009 kg n-Fe^0 (50–300 nm) m^3 soil. The total n-Fe^0 surface area in Figure 1.1b is about 20 times greater than the n-Fe^0 surface area in Figure 1.1c. Brownfield development land may economically sustain a remediation cost of \$3–\$6 MM/acre (i.e., \$75–\$1500 m^3 soil/aquifer), depending on location and final use. Comparative costs for surface reactor treatment of industrial water and agricultural water to remove chlorinated hydrocarbons and nitrates using ZVM in fixed/packed bed reactors are in the order of \$0.03 m^{-3} H_2O for greater than 90% removal [10, 13].

1.4.1.1.2 Reduction of Type A Remediation Costs: Galvanic Model The galvanic model indicates that the concentration of Fe^{2+}, $FeO_xH_y^{n+/-}$ ions and the presence of a controlled instability in the groundwater following ZVM injection (e.g., temperature variation, oxygen variation, acidification) controls the rate of Type A remediation. These factors facilitate remediation through electron shuttle reactions [164–166]. A galvanic cell of this type (Fig. 1.2) can be sustained through greater than 200,000 cycles/oscillations [151]. Application of this model to brownfield site remediation will (i) reduce the amount of n-Fe^0 required to achieve a specific level of remediation within a specific timeframe; and (ii) reduce the remediation time required using a specific amount of n-Fe^0. Remediation time frames for TCE removal can be potentially reduced from >1 year to <1 week.

1.4.1.2 Reduction of Type B Remediation Costs Type B remediation (Appendix 1.B, Fig. 1.3) can be undertaken using n-ZVM or ZVM corrosion products (Fig. 1.2). The remediation occurs over a long timeframe (days to years), which is controlled by the ZVM concentration, Eh, pH, and ion type. The amount of contaminant removed increases with time, and is typically in the range less than 0.01–0.3 g contaminant g^{-1} n-Fe^0. The galvanic model allows the timeframe required to remove specific pollutants (Appendix 1.B), and the total amount of pollutant removed to be reduced by controlling the Eh, pH environment, and ZVM composition with time. Remediation rates during active galvanic management can potentially exceed 1 g pollutant g^{-1} n-Fe^0. Active subsurface Eh:pH management using the galvanic model may be able to reduce the treatment costs to less than \$2 MM/acre (i.e., \$5–\$150 m^3 soil/aquifer).

1.5 CONCLUSIONS

Groundwater remediation (Type A and B) using ZVM is typically undertaken by ZVM infiltration, or pneumatic injection of ZVM [17, 137] using a passive process of injection followed by monitoring over a number of years. This approach, which assumes that the catalytic model applies, provides little, or no, effective day to day control over the rate of remediation. The observation that bimetallic ZVM (e.g., n-Fe^0 + one or more metals where $E° < E°\ Fe^{II}$ (Appendix 1.C)) shows increased reactivity (and delayed rates of Fe^0 oxidation) when compared with n-Fe^0 [179] is consistent with the galvanic model. The close proximity of the cathodic and anodic species coupled with diabatic oscillations results in continual oscillating reduction and oxidation of the bi-metal species. In mono n-Fe^0 the initial oxidation (formation of Fe-$(OH)_2$) (associated with galvanic oscillation between Fe^{II} and Fe^{III} (Appendix 1.C)) results initially in exponential particle growth [179]. This switches to logarithmic particle growth as the cathodic species $Fe(OH)_3$, FeOOH, and Fe_xO_y start to form [179]. The associated by-products, which react [10] to remove contaminants, are [e.g., 103] e^-, H, H^+, OH, OH^-, O_2H, O_2H^-, H_2O_2, O, O^-, O_2^-, and O^{2-}. Particle growth and agglomeration is rapid with 50 nm particles forming agglomerations of greater than 1 mm within 21 days (e.g., Fig. 1.4e, see also [10]). n-FeO_xH_y expulsion with (H_2, O_2) gas bubbles results in a rapid and effective dispersion of colloidal Fe^{II}–Fe^{III} galvanic cells throughout the water column. These grow with time (Figs. 1.2 and 1.3) to form colloidal particles greater than 1 mm in diameter, which settle on the ZVM–water interface [10]. The n-colloid clouds in the water within the diffusion environment tend to be mono-specific, color coordinated (e.g., white − $Fe(OH)_2$; yellow/orange = $Fe(OH)_3$; blue-green = green rust; dark-red brown/black = FeOOH; oxygenated blood red = Fe_2O_3; grey/black = Fe_3O_4), and indicate the galvanic charge status within the reaction environment (Fig. 1.2). The dominant colloid species changes with Eh, pH, and charge status of the water. The accreting growing colloidal particles, which can grow from 50 nm to greater than 5 mm, obtain buoyancy from H and O, which are present on the active sites.

An understanding of the corrosion of n-Fe^0 in the remediation environment and the controls that allow the net reaction directions (Fig. 1.1e–h) to be switched between recharge (formation of Fe^{III} ion adducts) and discharge (formation of Fe^{II} ion adducts) (Figs. 1.2 and 1.3) is an essential prerequisite to understanding how to reduce the cost and increase the efficiency of the remediation program.

The galvanic model requires active post-injection management of the groundwater Eh, pH temperature, and oxygenation levels. It has the potential to allow 15–100 nm Fe^0, Cu^0, Al^0 (typically spherical/blocky) particles with a surface area of 10–80 $m^2\ g^{-1}$, and costing \$20,000–\$850,000 t^{-1}, to be restructured and replaced by specific galvanic components (5–80 nm) with a layered structure [(e.g., $Fe(OH)_x$, FeOOH, etc. (Fig. 1.2)) and a surface area of <100 to >30,000 $m^2\ g^{-1}$ Fe^0, costing around \$300–\$15,000 t^{-1}]. The net effect of this restructuring is to reduce the amount of n-Fe^0 required, the rate of remediation, the time frame for remediation, and the overall cost of the remediation while increasing the amount of pollutant removed g^{-1} n-Fe^0.

APPENDIX 1.A LIST OF ABBREVIATIONS AND EQUATION SYMBOLS

1. *a*, *b*, *d*, *e*, *m*, *n*, and *p* are constants which are determined experimentally. In a simple non-catalytic example where $b = 0$, m = the reaction order [130]. The reaction order is calculated as m + n + p [130].
2. A' = a constant (0.509 $dm^{1.5}$ $mol^{-0.5}$ at 298K);
3. A_f = pre-exponential factor [$E_{a(sa)}$ & $A_{(sa)}$ = normalized for p_m and a_s; $E_{a(observed)}$ & $A_{(observed)}$ = E_a and A_f calculated without correction or normalization for p_m and a_s.
4. $a_{s(t=n)}$ = ZVM surface area(m^2 gm) at time t. $a_{s(t=n)}$ decreases with increasing time as the ZVM surfaces become oxidized;
5. B' = a constant; a = radius of the ion;
6. C_a = catalyst (e.g., ZVM);
7. $C_{t=n}$ = contaminant concentration at time, t = n (seconds) [mg l^{-1}, M l^{-1}];

8. $C_{t=0}$ = initial contaminant concentration at $t = 0$ [mg l^{-1}, M l^{-1}];
9. c_i = concentration (mol dm^{-3}; mol l^{-1}) of the ith ion of charge, z_i.
10. *cp* = moles gaseous reactants-moles gaseous products;
11. $D_{\{A[ZVM]\}}$ = diffusion coefficient = $k_B T/\ \pi\beta\eta\ r_s$;
12. D_f = Driving force, Pa (1 m head = 10,000 Pa);
13. E_a = Activation energy (kJ Mol^{-1});
14. *F* = Faraday constant;
15. *H* = Planck's constant;
16. I = ionic strength of the water = $0.5\ \Sigma\ c_i Z_i^2$;
17. $\log(\gamma_A) = -A' z_A^{\,2}\ I^{0.5}$ (lower ionic strengths);
18. $\log(\gamma_{A)} = -A' z_A^{\,2}\ I^{0.5}/(1+B' a\ I^{0.5})$ (higher ionic strengths);
19. $\log(k_{observed}/k_o) = 1.018\ z_A\ z_B\ I^{0.5}$;
20. k_B = Boltzmann's constant;
21. k_o = the rate coefficient at zero ionic strength = $k_r K_{AB}$;
22. $k_{observed}$ = observed contaminant removal rate constant [proportion removed s^{-1}]; $k_{observed}$ can also be expressed as mg removed s^{-1}, or [moles removed mole reactant^{-1} s^{-1}], or another suitable set of units;
23. k_p = permeability, $m^3\ m^{-2}\ s^{-1}\ Pa^{-1}$;
24. k_{sa} = reaction rate constant which has been normalized for ZVM surface area and ZVM concentration in the water;
25. K, K_{AB} = Equilibrium constant;
26. K_p = equilibrium constant adjusted for pressure;
27. $K^{\ddagger}(\gamma_A\ \gamma_{ZVM}/\gamma_{\{A[ZVM]\}}) = \exp(-\Delta G^{\ddagger}/RT)$ and $RT(\mathrm{Ln}(K^{\ddagger}(\gamma_A\ \gamma_{ZVM}/\gamma_{\{A[ZVM]\}})) = -\Delta G^{\ddagger}$;
28. *n* = number of electrons transferred;
29. *P* = pressure;
30. P_p = reactant partial pressure;
31. p_m = mass concentration of ZVM at *t* (g l^{-1}). p_m decreases as ZVM ions are removed with the product water and as ZVM is replaced with ZVM-hydroxides, peroxides, oxides, carbonates, sulphates, sulphides, etc.;
32. *Q* = reaction quotient;
33. Q_{fr} = flow rate, m^3 hr;
34. *R* = gas constant;
35. $r_{\{A[ZVM]\}}$ = encounter radius (nm) of the reactant [A]: ZVM interaction;
36. r_s = hydrodynamic radius (nm) of the diffusing species;
37. SV = space velocity, $m^3\ hr^{-1}\ t^{-1}$ ZVM;
38. SWZ = stored water to ZVM ratio in the reaction environment, m^3 H_2O t^{-1} ZVM;
39. S_w = Volume of water contained in the reaction environment, m^3;
40. T = Temperature, K;
41. *t* (time) can be expressed in seconds, minutes, hours, days;
42. W_{zvm} = weight (*t*) of ZVM in the reaction environment;
43. *Xm* = number of moles of the gaseous reactant;
44. z_A = charge number of ion species;
45. z_B = charge number of ion species, B;
46. ZVM = zero valent metal;
47. H = viscosity;
48. B = a constant (continuum solvent, $\beta = 6$; molecular diffusion, $\beta = 4$);
49. K = transmission coefficient (e.g., 1.0);
50. ΔE^o = standard potential for the reaction; $\Delta E = Eh$;
51. $\Delta G^{\ddagger}$ = overall Gibbs free energy of activation = $\Delta G^o_{\{A[ZVM]\}} + \Delta G^*$;
52. ΔG^o = standard Gibbs free energy for a reaction;

53. $\Delta G^{o}_{\{A[ZVM]\}}$ = free energy change on forming the encounter pair;
54. ΔG^{*} = free energy of activation from the encounter pair;
55. $\Delta G = \Delta H - T\,\Delta S$;
56. ΔH = heat of reaction;
57. ΔS = entropy;

APPENDIX 1.B IONS (OXIDES, HYDRIDES, PEROXIDES, AND HYDROXIDES) REMOVED BY PRECIPITATION DUE TO THE ALTERATION OF Eh AND pH IN GROUNDWATER BY ZVM

Data Sources: [10, 103, 104, 167–175]

In the simplest case, n-ZVM addition leaves pH effectively unaltered (e.g., Fig. 1.1c).

Eh prior to addition of n-ZVM = Eh [103, 104, 131] = $\Delta E^{o} + (0.0591/n)\log([B]^{b}/[A]^{a})_{t=0}$. For an example contaminant removal reaction,

$$Cd^{2+} + H_2 = Cd(s) + 2H^{+}$$

(the half reactions are $Cd^{2+} + 2e^{-} = Cd^{0}$ and $H_2 = 2H^{+} + 2e^{-}$; see Appendix 1.C); $K = Q = [H^{+}]^2/([Cd^{2+}]P_{H2}) = B^{b}/A^{a}$ [131]. After n-ZVM addition, at time $t = m$, the Eh changes (Fig. 1.1b–d) result in a new equilibria, where the new $\log([B]^{b}/[A]^{a})_{t=m} = (Eh - \Delta E^{o})/(0.0591/n)$; ΔE^{o} is corrected to the actual groundwater temperature. In this example, if the groundwater at t = 0 contains a 0.001 M $Cd^{2+}\,l^{-1}$ and an Eh of 0.13 V (Fig. 1.1c), then Eh = 0.13 = ΔE^{o} (−0.4 V—Appendix 1.B) + 0.0591/2 log Q; that is, log Q = 18; if −log (H^{+}) = pH [103, 131], then for pH = 6.5, at t = 0, $P_{H2} = 10^{-22}$. Changing the Eh to −0.2 V (Fig. 1.1c) after 1 month, while maintaining a pH of 6.5, changes log Q to 6.7. The Cd^{2+} concentration in the water at time, t = 1 month, is therefore a function of P_{H2} in the groundwater resulting from the presence of n-Fe^{0} (Fig. 1.4e). Increasing P_{H2} to 10^{-10} could achieve the observed Eh (−0.2 V) while leaving the Cd^{2+} concentration unchanged. Increasing P_{H2} to 10^{-8} reduces the Cd^{2+} concentration in water to 0.00001 M $Cd^{2+}\,l^{-1}$ from 0.001 M $Cd^{2+}\,l^{-1}$; that is, the effectiveness of the n-Fe^{0} treatment program for any specific Eh and pH, where the product is a zero valent metal (Appendix 1.B), is maximized by increasing the H_2 partial pressure. The alternative remediation strategy of using O_2 injection to oxidize cations (e.g., $Cd^{2+} + 0.5O_2 + H_2O = Cd(OH)_2$, where $0.5O_2 + H_2O + 2e^{-} = 2OH^{-}$; $Cd^{2+} + 2OH^{-} = Cd(OH)_2$), and $H_2 = 2H^{+} + 2e^{-}$) effectively changes Q to $Q = [H^{+}]^2/([Cd^{2+}]\,P_{H2}P_{O2})$, and ΔE^{o} to 0.4 V [177]. This alternative strategy uses the n-Fe^{0} to control the groundwater pH (i.e., H^{+} and P_{H2}) and the P_{O2} associated with O_2 injection to control the rate and degree of remediation [139–141]. For example, if at t = 0, Eh = 0.13 V, pH = 6.5, and the water contains 0.001 M $Cd^{2+}\,l^{-1}$ and $P_{H2} = 10^{-22}$, $P_{O2} = 0$, then instigation of an oxygen injection scheme following n-Fe^{0} injection into the groundwater, while maintaining a constant Eh and pH, will result in both P_{H2} and P_{O2} increasing [e.g., [139–141]]. Once P_{H2} and P_{O2} have exceeded a critical level (e.g., 10^{-11}), any subsequent increases in partial pressure will be compensated for by either decreases in Eh, or the removal of Cd^{2+} as $Cd(OH)_2$. Increasing P_{H2} and P_{O2} to 10^{-9}, will reduce the molar concentration of Cd^{2+} to 0.0000001 M $Cd^{2+}\,l^{-1}$ (i.e., 0.146 g $Cd(OH)_2\,l^{-1}\,H_2O$ will have been precipitated into the ZVM bed). This simple example has been used to demonstrate how a traditional ZVM remediation program [e.g., [17]] can be modified using the galvanic model [138, 2, 139–141] to both accelerate and control the rate of remediation. Once the bulk of the cations have been converted to oxides/hydroxides/peroxides, the diabatic galvanic model (Figs. 1.2 and 1.3) controls the rate of remediation.

Contaminant Ion/Ion Adduct	Potentially precipitated by ZVM as
Ac^{3+}, $AcOH^{2+}$, $Ac(OH)_2^{+}$	$Ac(OH)_3$, AcOOH
Ag^{n+}, AgO^{+}, AgO^{-}, AgOH, $AgOH_2^{-}$, $AgCl_2^{-}$	Ag, AgCl, AgOH, Ag_2O, Ag_2O_2, Ag_2O_3
Al^{n+}, $HAlO_2$, AlO_2^{-}, $AlOH^{2+}$, AlOH3, $Al(OH)_2^{+}$, $Al(OH)_4^{-}$	$Al(OH)_3$, AlOOH, Al_2O_3
Am^{n+}, $AmOH^{2+}$, AmO_2^{+}, $Am(OH)_2^{+}$	$Am(OH)_3$, $Am(OH)_4$, AmO_2
AsH_3, $HAsO_2$, AsO^{+}, H_3AsO_4, $H_2AsO_4^{-}$, $HAsO_4^{2-}$, AsO_2^{-}, AsO_4^{3-}	As, AsO_3
Au^{n+}, H_3AuO_3, $H_2AuO_3^{-}$, $HAuO_2^{2-}$	Au, $Au(OH)_3$, AuOOH, AuO_2
Ba^{2+}, $BaOH^{+}$	$Ba(OH)_2$, BaO_2
Be^{2+}, $Be_2O_2^{-}$	$Be(OH)_2$, BeO, $Be_2O(OH)_2$
Bi^{3+}, $BiOH^{2+}$, BiO^{+}, BiO_2^{-}, BiO_3^{-}	Bi, $Bi(OH)_3$, BiOOH, Bi_2O_3, Bi_2O_5, Bi_4O_7, Bi_2O_4
Ca^{2+}, $CaOH^{+}$	$Ca(OH)_2$, CaO_2, $CaCO_3$, $CaSO_4$
Cd^{2+}, $CdOH^{+}$, $HCdO_2^{-}$	Cd, $Cd(OH)_2$
Ce^{3+}, CeO^{+}, $Ce(OH)^{3+}$, $Ce(OH)_2^{2+}$	$Ce(OH)_3$, CeOOH, $Ce_2(CO_3)_3$, CeO_2
Cm^{3+}, $CmOH^{2+}$, $Cm(OH)_2^{+}$	$Cm(OH)_3$, CmOOH

Contaminant Ion/Ion Adduct	Potentially precipitated by ZVM as
Co^{2+}, $HCoO_2^-$	Co, CoO_2, $Co(OH)_2$, $Co(OH)_3$, CoOOH, CoS
Cr^{n+}, $Cr_2O_7^-$, $Cr_2O_4^{2-}$, CrO_4^{2-}, $HCrO_4^-$, CrO_2^-, CrO_3^{3-}, $CrOH^{2+}$, $Cr(OH)_2^+$, $Cr(OH)_4^-$	Cr, $Cr(OH)_2$, $Cr(OH)_3$, $Cr(OH)_4$, CrOOH, Cr_2O_3
Cs^+	CsO_2
Cu^{n+}, $Cu(OH)^+$, $HCuO_2^-$, CuO_2^{2-}, CuCl2-, $Cu(OH)_2^-$, $Cu(OH)_4^{2-}$	Cu, $Cu(OH)_2$, CuO, Cu_2O, $CuCl_2.3Cu(OH)_2$
Dy^{3+}, $DyOH^{2+}$, DyO^+, DyO_2^-	$Dy(OH)_3$, DyOOH, Dy_2O_3, $Dy_2(CO_3)_3$
Er^{3+}, $ErOH^{2+}$, ErO^+, ErO_2^-	$Er(OH)_3$, ErOOH, Er_2O_3
Eu^{n+}, $EuOH^{2+}$, EuO^+, EuO_2^-	$Eu(OH)_3$, EuOOH, EuO_3H_3, $Eu_2(CO_3)_3$
Fe^{n+}, $FeO_xH_y^{n+/-}$, $FeSO_4^+$, $FeSO_4$, $Fe(SO_4)_2^-$,	Fe_2O_3, Fe_3O_4, $Fe(OH)_x$, FeOOH, $Fe_8O_8(OH)_6SO_4$, $Fe_6(OH)_{12}SO_4$, $FeCO_3$, FeS_2, FeS, $Fe(HS)_2$
Ga^{3+}, $GaOH^{2+}$, GaO^+, GaO_2^-, $HGaO_3^{2-}$, GaO_3^{3}	$Ga(OH)_3$, GaOOH, Ga_2O_3
Gd^{n+}, $GdOH^{2+}$, GdO^+, GdO_2H, GdO_2^-	$Gd(OH)_3$, GdOOH, Gd_2O_3, $Gd_2(CO_3)_3$
Ge^{2+}, H_2GeO_3, $HGeO_3^-$, GeO_3^{2-}	Ge, $Ge(OH)_2$, GeO, GeO_2
Hf^{4+}, HfO^{2+}, $HHfO_2^-$, $HHfO_3^-$, HfO^{2-}	$Hf(OH)_n$, $HfO(OH)_2$, HfO_2
Hg^{2+}, $HHgO_2^-$, $Hg(OH)_2$	Hg, HgO, HgO_2
Ho^{3+}, $HoOH^{2+}$, HoO^+, HoO_2H, HoO_2^-	$Ho(OH)_3$, HoOOH, Ho_2O_3
In^{3+}, In^+, $InOH^{2+}$, $In(OH)_2^+$, $HInO_2$, InO^+, InO_2^-	In, $In(OH)_3$, InOOH, In_2O_3
Ir^{3+}, IrO_4^{2-}	Ir, IrO_2
La^{3+}, $LaOH^{2+}$, LaO^+, LaO_2H, LaO_2^-	$La(OH)_3$, LaOOH, $La_2(CO_3)_3$, La_2O_3
Lu^{3+}, $LuOH^{2+}$, LuO^+, LuO_2^-	$Lu(OH)_3$, LuOOH, Lu_2O_3
Mg^{n+}, $MgOH^+$	MgO_2, $Mg(OH)_2$, $MgCO_3$
Mn^{2+}, $HMnO_2^-$, MnO_4^-, MnO_4^{2-}, $MnOH^+$, $Mn(OH)_3^-$, $Mn(OH)_4^{2-}$, MnO, MnOH, MnO_2^{2-}	$Mn(OH)_2$, MnO_2, Mn_2O_3, Mn_3O_4, MnS, $MnHCO_3$, $MnCO_3$
Mo^{3+}, $HMoO_4^-$, MoO_4^{2-}, MoO_2^+, H_2MoO_4, MoO_2OH^+	Mo, $Mo(OH)_2$, $Mo(OH)_3$, $Mo(OH)_4$, $MoO(OH)_2$, $MoO(OH)_3$, MoOOH, MoS_2, H_2MoO_4, MoO_2, MoO_3
Nb^{n+}, NbO_2^{n+} $Nb(OH)_5$, NbO_3^-, $HNbO_3$, $Nb(OH)_6^-$	Nb, NbO, NbO_2, Nb_2O_5
Nd^{3+}, $NdOH^{2+}$, NdO^+, NdO_2H, NdO_2^-	$Nd(OH)_3$, Nd_2O_3, NdOOH, Nd_nO_m, $Nd_2(CO_3)_3$
Ni^{2+}, $HNiO_2^-$, NiO_2^{2-}, $Ni(OH)_3^-$, $Ni(OH)_4^{2-}$, $NiOH^+$	Ni, NiO_2, $Ni(OH)_2$, $Ni(OH)_3$, NiS, Ni_3O_4, Ni_2O_3, Ni_2H, Ni_2O_4
Np^{n+}, NpO_2OH, NpO_2OH^+, $NpOH^{3+}$, $Np(OH)_5^-$,$NpO_2OH_2^-$, NpO_2^+, NpO_2^{2+}, NpO_2F^+, NpF_2^{2+}, $NPO_2CO_3^-$, $NpO_2(CO_3)_2^{3-}$, $NpO_2(CO_3)_3^{5-}$	NpO_2, NpO_3, NpO_5, $Np(OH)_4$
OsO_4^{2-}, H_2OsO_5, $HOsO_5^-$, OsO_5^{2-}	Os, OsO_2, OsO_4
Pa^{4+}, $PaOH^{3+}$, $Pa(OH)_2^{2+}$, $Pa(OH)_3^+$, $PaO(OH)_3$, $PaO(OH)_2$, $PaOOH^{2+}$, PaO_2^+	Pa_xO_y, $Pa(OH)_x$
Pb^{2+}, $HPbO_2^-$, PbO_3^{2-}, $PbOH^+$, $HPbO_2^-$, $Pb(OH)_3^-$	Pb, PbO, $Pb(OH)_2$, PbO_2, Pb_2O_3
Pd^{n+}, $PdOH^+$, PdO_2^{2-}, PdO_3^{2-}, $PdCl_4^{2-}$	Pd, Pd_2H, PdO_3, $Pd(OH)_2$, $Pd(OH)_4$
Pm^{3+}	$Pm(OH)_3$, PmOOH, Pm_nO_m
Po^{2+}, $Po(OH)_2^{4+}$, PoO_3^{2-}	Po, PoO_2, PoO_3, H_2PoO_3, $Po(OH)_4$, $PoO(OH)_2$
Pr^{3+}, $Pr(OH)_2^+$, $PrOH^{2+}$, PrO^+, PrO_2H, PrO_2^-	$Pr(OH)_3$, PrOOH, PrO_2, $Pr_2(CO_3)_3$, Pr_2O_3
Pt^{2+}, PtO, $Pt(OH)^{n+}$, PtO_2^{2-}, PtO_4^{2-}	Pt, $Pt(OH)_2$, PtO_2, PtO_3
Pu^{3+}, $PuOH^{2+}$, $PuO_2(OH)_2$, PuO_2^+, PuO^{2+}, PuF_3^+, $PuO_2F_3^-$, $PuO_2(OH)_2H$ CO_3^-,PuO_2OH^+, $Pu(OH)_5^-$	PuO_2, PuO_3, Pu_2O_3, $Pu(OH)_3$, $Pu(OH)_4$
Re^-, Re^+, ReO_4^-	Re, ReO_2, ReO_3, Re_2O_3
Rh^{n+}, $RhOH^+$, RhO, RhO^+, RhO_4^{2-}	Rh, RhO_2, Rh_2O_3
Ru^{3+}, RuO^+, H_2RuO_5, $HRuO_5^-$, $Ru(OH)_2^+$, $Ru(OH)_2^{2+}$, H_2RuO_5, $HRuO_5^-$, RuO_4^{2-}, RuO_4^-, RuO_4^{2-}	Ru, RuO_2, RuO_4, RuOOH, $Ru(OH)_3$
SbO_2^+, SbO^+, SbO_3^-, SbO_2^-, $HSbO_2$	Sb, Sb_2O_5, Sb_2O_3, $HSb(OH)_6$
Sc^{3+}, ScO^+, $HScO_2$, ScO_2^-, $ScOH^{2+}$	$Sc(OH)_3$, ScOOH, Sc_2O_3
H_2Se, HSe^-, SeO_4^{2-}, H_2SeO_3, $HSeO_4^-$, $HSeO_3^-$, SeO_3^{2-}	Se
Sm^{n+}, $SmOH^{2+}$, $Sm(OH)_2^+$, SmO^+, SmO_2^-	Sm_2O_3, $Sm(OH)_3$, SmOOH, $Sm_2(CO_3)_3$
Sn^{2+}, Sn^{4+}, $SnOH^+$, $SnOOH^+$, $Sn(OH)_2^{2+}$, $Sn(OH)_3^+$, $Sn(OH)_5^-$, $Sn(OH)_5^{2-}$, SnO, $HSnO_2^-$, SnO_3^{2-}	Sn, $Sn(OH)_2$, $Sn(OH)_4$, SnO_2
Sr^{2+}, $SrOH^+$	$Sr(OH)_2$, SrO_2
Tb^{3+}, $TbOH^{2+}$, TbO^+, TbO_2H, TbO_2^-	$Tb(OH)_3$, TbOOH, Tb_2O_3, Tb_7O_{12}, Tb_6O_{11}, TbO_2, $Tb_2(CO_3)_3$
Tc^{2+}, TcO_4^-, $TcOOH^+$, $TcOOH_3^-$, $TcO(OH)_2$, $TcO(OH)_3^-$, TcO^{2+}	Tc, Tc_3O_4, TcOH, Tc_4O_7, TcO_2, $Tc(OH)_2$
Te^{2-}, Te^{4+}, H_2TeO_4, $HTeO_4^-$, $Te(OH)_3^+$, TeO_4^{2-}, $HTeO_2^+$, $HTeO_3^-$, TeO_3^-	Te, TeO_2, $TeO_3.3H_2O$, H_2TeO_4
Th^{4+}	$Th(OH)_n$, ThO_2

(*Continued*)

Contaminant Ion/Ion Adduct	Potentially precipitated by ZVM as
Ti^{n+}, $HTiO_3^-$	$Ti(OH)_3$, TiOOH, TiO_2
Tl^{3+}, Tl^+, TlOH, TlO_2^-, $HTlO_2$	Tl, $Tl(OH)_3$, TlOOH, TlOH, Tl_2O_4
Tm^{n+}, $TmOH^{2+}$, TmO^+, TmO_2H, TmO_2^-	$Tm(OH)_3$, TmOOH, Tm_2O_3
U^{n+}, UO_2F^+, UO_2^+, $U(OH)_x^{n+/-}$, $H_2O_{13}U_3^-$, UO_2^{2+}, UO_4^{2-}, HUO_4^-, $H_5O_5U^-$, UO_2^+, $UO_2(OH)_2$, $UO_2(OH)_3^-$, $UO_2(CO_3)_2^{2-}$, $UO_2(CO_3)_3^{4-}$	$U(OH)_4$, UO_2, UO_3, UO_4, U_3O_8, U_4O_9, UF_4, UO_2CO_3
V^{n+}, VO_2^+, VO^{2+}, VOH^{2+}, VO^+, $H_3V_2O_7^-$, $H_2VO_4^-$, HVO_4^{2-}, VO_4^{2-}, $VOOH^+$, $(VO)_2(OH)_5^-$, $V(OH)_m^{n+/-}$, $HV_2O_5^-$	$V(OH)_n$, V_2O_5, V_2O_4, V_2O_3, V_3O_5, V_2O_2
WO_4^{2-}, WO_5^{2-}	W, WO_2, W_2O_5, WO_3
Y^{3+}, YOH^{2+}, YO^+, HYO_2, YO_2^-	$Y(OH)_3$, YO_3H_3, YOOH, $Y_2(CO_3)_3$, Y_2O_3
Yb^{n+}, $YbOH^{2+}$, YbO_2, YbO_2^-	$Yb(OH)_3$, YbOOH
Zn^{2+}, $HZnO_2^-$, $ZnOH^+$, $Zn(OH)_3^-$, $Zn(OH)_4^{2-}$, ZnO_2^{2-}	Zn, ZnS, $Zn(OH)_2$, ZnO_2
Zr^{n+}, ZrO^{2+}, $HZrO_2^+$, $HZrO_3^-$	$Zr(OH)_n$, ZrO_2

APPENDIX 1.C HALF REACTIONS AND REDOX POTENTIALS ASSOCIATED WITH ZVM

All ZVM, when present in the water form redox half cells with the potentials indicated. These form separate hydrogen cells, which interact with the ZVM and ZVM corrosion products (Figs. 1.2 and 1.3). Species that act as cathodes to n-Fe^0 have $E^o < E^o$ Fe^0. Fe^0 can act as both a cathode (Fe^{III}) and an anode (Fe^{II}). Species that act as anodes to n-Fe0 have $E^o > E^o$ Fe. Each ZVM species acts as a cathode or anode to one or more ZVM species present in the water. Figure 1.2 illustrates the dominant corrosion species types associated with the anodic ZVM species. Similar relationships exist between ZVM, hydroxides, oxides, and peroxides (where appropriate) for each species listed in Appendix 1.B and 1.C in the diffusion remediation environment.

Data Source: [131, 176–178]

Half Reaction	E°(V)	Half reaction	E°(V)	Half reaction	E°(V)
$Sr^0 = Sr^+ + e^-$	4.10	$Hf^0 = Hf^{4+} + 4e^-$	1.55	$Ge^0 = Ge^{4+} + 4e^-$	−0.12
$Ca^0 = Ca^+ + e^-$	3.80	$Zr^0 = Zr^{4+} + 4e^-$	1.45	$Ge^0 = Ge^{2+} + 2e^{--}$	−0.24
$Li^0 = Li^+ + e^-$	3.04	$Mn^0 = Mn^{2+} + 2e^-$	1.19	$Re^0 = Re^{3+} + 3e^-$	−0.30
$Cs^0 = Cs^+ + e^-$	3.03	$V^0 = V^{2+} + 2e^-$	1.18	$Bi^0 = Bi^{3+} + 3e^-$	−0.31
$Rb^0 = Rb^+ + e^-$	2.98	$Nb^0 = Nb^{3+} + 3e^-$	1.10	$Cu^0 = Cu^{2+} + 2e^-$	−0.34
$K^0 = K^+ + e^-$	2.93	$Cr^0 = Cr^{2+} + 2e^-$	0.91	$Tc^0 = Tc^{2+} + 2e^-$	−0.40
$Ba^0 = Ba^{2+} + 2e^-$	2.91	$Zn^0 = Zn^{2+} + 2e^-$	0.76	$Ru^0 = Ru^{2+} + 2e^-$	−0.46
$Sr^0 = Sr^{2+} + 2e^-$	2.90	$Cr^0 = Cr^{3+} + 3e^-$	0.74	$Bi^0 = Bi^+ + e^-$	−0.5
$Ca^0 = Ca^{2+} + 2e^-$	2.87	$Ta^0 = Ta^{3+} + 3e^-$	0.60	$Cu^0 = Cu^+ + e^-$	−0.52
$Na^0 = Na^+ + e^-$	2.71	$Ga^0 = Ga^{3+} + 3e^-$	0.55	$Te^0 = Te^{4+} + 4e^-$	−0.57
$Mg^0 = Mg^+ + e^-$	2.70	$Fe^0 = Fe^{2+} + 2e^-$	0.44	$Rh^0 = Rh^{2+} + 2e^-$	−0.60
$La^0 = La^{3+} + 3e^-$	2.38	$Cd^0 = Cd^{2+} + 2e^-$	0.40	$Rh^0 = Rh^+ + e^-$	−0.60
$La^0 = La^{3+} + 3e^-$	2.38	$In^0 = In^{3+} + 3e^-$	0.34	$Tl^0 = Tl^{3+} + 3e^-$	−0.74
$Mg^0 = Mg^{2+} + 2e^-$	2.37	$Tl^0 = Tl^+ + e^-$	0.34	$Rh^0 = Rh^{3+} + 3e^-$	−0.76
$Ce^0 = Ce^{3+} + 3e^-$	2.33	$Co^0 = Co^{2+} + 2e^-$	0.28	$Po^0 = Po^{4+} + 4e^-$	−0.76
$Th^0 = Th^{4+} + 4e^-$	1.90	$Ni^0 = Ni^{2+} + 2e^-$	0.26	$Ag^0 = Ag^+ + e^-$	−0.80
$Be^0 = Be^{2+} + 2e^-$	1.85	$Ga^0 = Ga^+ + e^-$	0.20	$2Hg^0 = Hg_2^{2+} + 2e^-$	−0.80
$U^0 = U^{3+} + 3e^-$	1.80	$Mo^0 = Mo^{3+} + 3e^-$	0.20	$Pd^0 = Pd^{2+} + 2e^-$	−0.95
$Al^0 = Al^{3+} + 3e^-$	1.66	$In^0 = In^+ + e^-$	0.14	$Ir^0 = Ir^{3+} + 3e^-$	−1.16
$Md^0 = Md^{3+} + 3e^-$	1.65	$Fe^0 = Fe^{3+} + 3e^-$	0.04	$Pt^0 = Pt^{2+} + 2e^-$	−1.18
$Ti^0 = Ti^{2+} + 2e^-$	1.63	$H_2 = 2H^+ + 2e^-$	0.00	$Au^0 = Au^{3+} + 3e^-$	−1.50
$Hf^0 = Hf^{4+} + 4e^-$	1.55	$2H^+ + 2e^- = H_2$	0.00	$Au^0 = Au^+ + e^-$	−1.69
$Zr^0 = Zr^{4+} + 4e^-$	1.45	$W^0 = W^{3+} + 3e^-$	−0.10		

REFERENCES

[1] Rosicka D, Sembera J. Assessment of influence of magnetic forces on aggregation of zero-valent iron nanoparticles. Nanoscale Res Lett 2010;6 (10):6.

[2] Kim DH, Kim J, Choi W. Effect of magnetic field on the zero valent iron induced oxidation reaction. J Hazard Mater 2011;192:928–931.

[3] Dalla VE, Coisson M, Appino C, Vinai F, Sethi R. Magnetic characterisation and interaction modelling of zerovalent iron nanoparticles for the remediation of contaminated aquifers. J Nanosci Nanotechnol 2009;9:3210–3218.

[4] Wang Q, Kanei SR, Park H, Ryu A, Choi H. Controllable synthesis, characterisation, and magnetic properties of nano-scale zerovalent iron with high Brunauer-Emmett-Teller surface area. J Nanopart Res 2009;11:749–755.

[5] Huang CP, Wang HW, Chiu PC. Nitrate reduction by metallic iron. Water Res 1998;32:2257–2264.

[6] Liou YH, Lo S-L, Lin C-J, Kuan WH, Weng SC. Effects of iron surface pre-treatment on kinetics of aqueous nitrate reduction. J Hazard Mater 2005;B126:189–195.

[7] Solovyov S, Goldman A. *Mass Transport and Reactive Barriers in Packaging: Theory, Applications & Design*. Lancaster, PA: DES Tech Publications Inc.; 2007. p 558.

[8] Crane RA, Scott TB. Nanoscale zero-valent iron: future prospects for an emerging water treatment technology. J Hazard Mater 2012;211–212:112–125.

[9] Noubactep C. Aqueous contaminant removal by metallic iron: is the paradigm shifting? Water SA 2011;37:419–426.

[10] Antia DDJ. Sustainable zero-valent metal (ZVM) water treatment associated with diffusion, infiltration, abstraction and recirculation. Sustainability 2010;2:2988–3073.

[11] Junyapoon S. Use of zero-valent iron for waste water treatment. KMITL Sci Technol J 2005;5:587–595.

[12] Zhang W-X. Nanoscale iron particles for environmental remediation: an overview. J Nanopart Res 2003;5:323–332.

[13] Ma L, Zhang W-X. Enhanced biological treatment of industrial waste water with bi-metallic zero-valent iron. Environ Sci Technol 2008;42:5384–5389.

[14] Comfort SD, Shea PJ, Machacek TA, Satapanajaru T. Pilot-scale treatment of RDX-contaminated soil with zerovalent iron. J Environ Qual 2003;32:1717–1725.

[15] Loraine G, Burris D, Li L, Schoolfield J. Mass transfer effects on kinetics of dibromoethane reduction by zero valent iron in packed-bed reactors. J Environ Eng 2002;128:85–93.

[16] Zang Y, Jing Y, Quan X, Liu Y, Onu P. A built-in zerovalent iron anaerobic reactor to enhance treatment of azo dye waste water. J Water Sci Technol 2011;63:741–746.

[17] Gavaskar A, Tatar L, Condit W. Cost and performance report: nanoscale zero-valent iron technologies for source remediation. NAVFAC Naval Facilities Engineering Command. Contract Report CR-05-007-ENV. Port Hueneme: Engineering Service Center; 2005. p 44.

[18] Henderson AD, Demond AH. Impact of solids formation and gas production on the permeability of ZVI PRBs. J Environ Sci 2011;137:689–696.

[19] Jeen SW, Amos RT, Blowes DW. Modelling gas formation and mineral precipitation in a granular iron column. Environ Sci Technol 2012;46:6742–6749.

[20] Antia DDJ. Formation and control of self-sealing high permeability groundwater mounds in impermeable sediment: implications for SUDS and sustainable pressure mound management. Sustainability 2009;1:855–923.

[21] Antia DDJ. Polymerisation theory—formation of hydrocarbons in sedimentary strata (hydrates, clays, sandstones, carbonates, evaporates, volcanoclastics) from CH_4 and CO_2: parts I to IV. Indian J Petrol Geol 2009;17(1):49–86; 2009;17(2):11–70; 2010;18(1):1–50; 2011;18(2):1–45.

[22] Antia DDJ. Interpretation of overland flow associated with infiltration devices placed in boulder clay and construction fill. In: Wong TSW, editor. *Overland flow and surface runoff*. New York: Nova Science Publishers; 2012. p 211–285.

[23] Antia DDJ. Interacting infiltration devices (field analysis, experimental observation and numerical modeling): prediction of seepage (overland flow) locations, mechanisms and volumes—implications for SUDS, groundwater raising projects and carbon sequestration projects. In: Hirsch G, Kappel B, editors. *Hydraulic Engineering: Structural Applications, Numerical Modeling and Environmental Impacts*. New York: Nova Science Publishers; 2011. p 85–156.

[24] Westerhof P, James J. Nitrate removal in zero-valent iron packed columns. Water Res 2003;37:1818–1830.

[25] Chen YM, Li CW, Chen SS. Fluidized zero valent iron bed for nitrate removal. Chemosphere 2005;59:753–759.

[26] Hsu J-C, Liao C-H, Wei Y-L. Nitrate removal by synthetic nanoscale zero-valent iron in aqueous recirculated reactor. Sustain Environ Res 2011;21:353–359.

[27] Ji MK, Park WB, Khan MA, Abou-Shanab RA, Kim Y, Cho Y, Choi J, Song H, Jeon BH. Nitrate and ammonium ions removal from groundwater by a hybrid system of zero-valent iron combined with absorbents. J Environ Monit 2012;14:1153–1158.

[28] Anotai J, Liao C-H, Ruangchanikom C. Nitrate removal by Fe^0/CO_2 process using an innovative continuous flow reactor. J Environ Eng Manage 2010;20:77–84.

[29] Cho D-W, Abou-Shnab RAI, Kim Y, Jeon B-H, Song H. Enhanced reduction of nitrate in groundwater by zero-valent iron with activated red mud. Geosyst Eng 2011;14:65–70.

[30] Ginner JL, Alvarez PJJ, Smith SL, Scherer MM. Nitrate and nitrite reduction by Fe^0: influence of mass transport, temperature and denitrifying microbes. Environ Eng Sci 2004;21:219–229.

[31] Liang F, Fan J, Guo Y, Fan M, Wang J, Yang H. Reduction of nitrite by ultrasound-dispersed nanoscale zero-valent iron particles. Ind Eng Chem Res 2008;47:8550–8554.

[32] Zhang Z, Hao Z, Yang Y, Zhang J, Wang Q, Xu X. Reductive denitrification kinetics of nitrate by zero-valent iron. Desalination 2010;257:158–162.

[33] Huang YH, Zhang TC. Nitrite reduction and formation of corrosion coatings in zerovalent iron systems. Chemosphere 2006;64:937–943.

[34] Xu J, Gao N, Tang Y, Deng Y, Sui M. Perchlorate removal using granular activated carbon supported iron compounds: synthesis, characterisation, and reactivity. J Environ Sci 2010;23:1807–1813.

[35] Huang H, Sorial GA. Perchlorate remediation in aquatic systems by zero valent iron. Environ Eng Sci 2007;24:917–926.

[36] Lee C, Batchelor B, Park SH, Han DS, Abdel-Wahab A, Kramer TA. Reduction of perchlorate using zero-valent titanium (ZVT) anode: reaction mechanism. Adv Environ Res 2012;1:37–55.

[37] Lien H-L, Yi CC, Lee Y-C. Perchlorate removal by acidified zero-valent aluminium and aluminium hydroxide. Chemosphere 2010;80:888–893.

[38] Murad E, Taylor RM. The Mossbauer spectra of hydroxycarbonate green rusts. Clay Min 1984;19:77–83.

[39] Ahmed IAM, Shaw S, Benning LG. Formation of hydroxysulphate and hydroxycarbonate green rusts in the presence of zinc using time-resolved in situ small and wide angle X-ray scattering. Min Mag 2008;72:159–162.

[40] Su C, Puls RW. Significance of Iron (II,III) hydroxycarbonate green rust in arsenic remediation using zerovalent iron in laboratory column tests. Environ Sci Technol 2004;38:5224–5231.

[41] O'Loughlin EJ, Burris DR. Reduction of halogenated ethanes by green rust. Environ Toxicol Chem 2004;23:41–48.

[42] Schwertmann U, Fechter H. The formation of green rust and its transformation to lepidocrocite. Clay Min 1994;29:87–92.

[43] Hansen HCB, Taylor RM. Formation of synthetic analogues of double metal-hydroxy carbonate minerals under controlled pH conditions: I. synthesis of pyroaurite and reevesitie. Clay Min 1990;25:161–179.

[44] Hansen HCB, Taylor RM. Formation of synthetic analogues of double metal-hydroxy carbonate minerals under controlled pH conditions: II. synthesis of desautelsite. Clay Min 1991;26:507–525.

[45] Ruby C, Abdelmoula M, Naille S, Renard A, Khare V, Ona-Nguema G, Morin G, Genin J-MR. Oxidation modes and thermodynamics of Fe^{II-III} oxyhydroxycarbonate green rust: dissolution-precipitation versus in situ deprotonation; about fougerite mineral. Geochim Cosmochim Acta 2009;74:953–966.

[46] Hansen HCB, Poulsen IF. Interaction of synthetic sulphate green rust with phosphate and the crystallization of vivianite. Clays Clay Min 1999;47:312–318.

[47] Ahmed IAM, Benning LG, Kakonyi G, Sumoondur AD, Terrill NJ, Shaw S. Formation of green rust sulphate: a combined in situ time-resolved X-ray scattering and electrochemical study. Langmuir 2010;26:6593–6603.

[48] Wilkin R, McNeil MS. Laboratory evaluation of zero-valent iron to treat water impacted by acid mine drainage. Chemosphere 2003;53:715–725.

[49] Wander MCF Environmental redox reactions of iron [PhD thesis]. New York: Stoney Brook University; 2007.

[50] Davesne E, Dideriksen K, Christiansen BC, Sonne M, Ayala-Luis KB, Koch CB, Hansen HCB, Stipp SLS. Free energy of formation for green rust sodium sulphate ($NaFe^{II}_6 Fe^{III}_3 (OH)_{18} (SO_4)_{2(s)}$). Geochim Cosmochim Acta 2010;74:6451–6467.

[51] Yang C-W. Removal of hydrogen sulphide with nanoscale zero-valent iron from piggery waste water [MSc thesis]. Taiwan: National University Kaoshiung; 2011.

[52] Keenan CR, Sedlak DL. Factors affecting the yield of oxidants from the reaction of nanoparticulate zero-valent iron and oxygen. Environ Sci Technol 2008;42:1262–1267.

[53] Vesselli E, Rizzi M, de Rogatis L, Ding X, Baraldi A, Cornelli G, Savio L, Vattuone L, Rocca M, Fornasiero P, Baldereschi A, Perissi M. Hydrogen assisted transformation of CO_2 on Nickel; the role of formate and carbon monoxide. J Phys Chem Lett 2010;1:402–416.

[54] Gillham RW, O'Hannesian SF. Enhanced degradation of halogenated aliphatics by zero valent iron. Ground Water 1994;32:958–967.

[55] Janda V, Vasek P, Bizova J, Belohlav Z. Kinetic models for volatile chlorinated hydrocarbons removal by zero-valent iron. Chemosphere 2004;54:917–925.

[56] Liu Y, Majetich SA, Tilton RD, Sholl DS, Lowry GV. TCE dechlorination rates, pathways, and efficiency of nanoscale iron particles with different properties. Environ Sci Technol 2005;39:1338–1345.

[57] Ulsamer S. A model to characterize the kinetics of dechlorination of tetrachloroethylene and trichloroethylene by a zero valent iron permeable reactive barrier [MSc thesis]. USA: Worcester Polytechnic Institute; 2011.

[58] Clausen III C, Geiger CL, Sigman M, Fidler R. Safe, in situ methodologies for the destruction of triacetone triperoxide and other explosive peroxides. US Patent 8062442 B1. November 22, 2011.

[59] Nam S, Tratnyek PG. Reduction of azo dyes with zero valent iron. Water Res 2000;34:1837–1845.

[60] Shih Y-H, Tso C-P, Tung L-Y. Rapid degradation of methyl orange with nanoscale zerovalent iron particles. J Environ Eng Manage 2010;20:137–143.

[61] Chen ZX, Jin XY, Chen Z, Megharai M, Naidu R. Removal of methyl orange from aqueous solution using bentonite-supported nanoscale zero-valent iron. J Colloid Interface Sci 2011;15:601–607.

[62] Dombek T, Dolan E, Schultz J, Klarup D. Rapid reductive dechlorination of atrazine by zero valent iron under acidic conditions. Environ Pollut 2001;111:21–27.

[63] Kim G, Jeong W, Choe S. Dechlorination of atrazine using zero valent iron (Fe^0) under neutral pH conditions. J Hazard Mater 2008;155:502–506.

[64] Oh SY, Cha DK, Chiu PC, Kim BJ. Zero-valent iron treatment of RDX-containing and perchlorate containing wastewaters from an ammunition-manufacturing plant elevated temperatures. Water Sci Technol 2006;54:47–53.

[65] Thomas JM, Hernandez R, Kuo CH. Single-step treatment of 2,4-dinitrotoluene via zero-valent metal reduction and chemical oxidation. J Hazard Mater 2008;155:193–198.

[66] Oh SY, Kang SG, Chiu PC. Degradation of 2,4-dinitrotoluene by persulphate activated with zero-valent iron. Sci Total Environ 2010;408:3464–3468.

[67] Schaefer CE, Topoleski C, Fuller ME. Effectiveness of zerovalent iron and nickel catalysts for degrading chlorinated solvents and n-nitrosodimethylamine in natural groundwater. Water Environ Res 2007;79:57–62.

[68] Mitch WA, Sharp JO, Trussell RR, Valentine RL, Alvarez-Cohen L, Sedlak DL. n-nitrosodimethylamine (NDMA) as a drinking water contaminant: a review. Environ Eng Sci 2003;20:389–404.

[69] Clausen C, Geiger C, Sigman M, Fidler R. Degradation of TATP, TNT, and RDX using mechanically alloyed metals. US Patent 8092622 B1. January 10, 2012.

[70] Oh S-Y, Chiu PC, Kim BJ, Cha DK. Enhancing Fenton oxidation of TNT and RDX through pretreatment with zero-valent iron. Water Res 2003;37:4275–4283.

[71] Fuller ME, Schafer CE, Lowey JM. Degradation of explosives-related compounds using nickel catalysts. Chemosphere 2007;67:419–427.

[72] Oh SY, Cha DK, Kim BJ, Chin PC. Transformation of hexahydro 1,3, 5-trinitro-1,3,5-triazine (RDX), octahydroxo-1,3,5,7-tetranitro-1,3,5,7-tetrazocine (HMX), and methylenedinitramine (MDNA) with elemental iron. Environ Toxicol Chem 2005;24:2812–2819.

[73] Echols E. Environmental remediation of TNT using nanoscale zero-valent iron [MSc thesis]. Florida: University of South Florida; 2009.

[74] Cho C, Bae S, Lee W. Enhanced degradation of TNT and RDX by bio-reduced iron bearing soil minerals. Adv Environ Res 2012;1:1–14.

[75] Barreto-Rodrigues M, Silva FT, Paiva TCB. Combined zero-valent iron and Fenton processes for the treatment of Brazilian TNT industry wastewater. J Hazard Mater 2009;165:1224–1228.

[76] Jin Y, Chiu P. Mitigation of irrigation water using zero-valent iron treatment. Patent WO/2011/163346. June 22, 2011.

[77] Chun CL, Hozalski RM, Arnold WA. Degradation of disinfection byproducts by carbonate green rust. Environ Sci Technol 2007;41:1615–1621.

[78] Lin K-S, Chang N-B, Chuang T-D. Fine structure characterisation of zero-valent iron nano-particles for decontamination of nitrites and nitrates in waste water and ground water. Sci Technol Adv Mater 2008;9:025015 9.

[79] Rocca CD, Belgiorno V, Meric S. Overview of in-situ applicable nitrate removal processes. Desalination 2007;204:46–62.

[80] Boussahel R, Harik D, Mammar M, Lamara-Mohamed S. Degradation of obsolete DDT by Fenton oxidation with zero-valent iron. Desalination 2007;206:369–372.

[81] Singh R, Singh A, Misra V, Singh RP. Degradation of lindane contaminated soil using zero-valent iron, nanoparticles. J Biomed Nanotechnol 2011;7:175–176.

[82] Cong X, Xue N, Wang S, Li K, Li F. Reductive dechlorination of organochlorine pesticides in soils from an abandoned manufacturing facility by zero-valent iron. Sci Total Environ 2010;408:3418–3423.

[83] Ghauch A. Rapid removal of flutriafol in water by zero-valent iron powder. Chemosphere 2008;71:816–826.

[84] Thompson JM, Chislom BJ, Bezbaruah AN. Reductive dechlorination of chloroacetanilide herbicide (alachor) using zero-valent iron nanoparticles. Environ Eng Sci 2010;27:227–232.

[85] Suntornchot P, Satapanajaru T, Comfort SD. Application of nano-zerovalent iron for treating metolachor in aqueous solution. World Acad Sci Eng Technol 2010;72:625–628.

[86] Ghauch A. Degradation of benomyl, picloram, and dicamba in a conical apparatus by zero-valent iron powder. Chemosphere 2001;43:1109–1117.

[87] Berendahl JA, Thies TP. Fenton's oxidation of MTBE with zero valent iron. Water Res 2004;38:327–334.

[88] Chang M-C, Shu H-Y, Hsieh W-P, Wang M-C. Using nanoscale zero-valent iron for the remediation of polycyclic aromatic hydrocarbons contaminated soil. J Air Waste Manage Assoc 2005;55:1200–1207.

[89] Chang M-C, Shu H-Y, Hsieh W-P, Wang M-C. Remediation of soil contaminated with pyrene using ground nanoscale zero-valent iron. J Air Waste Manage Assoc 2007;57:221–227.

[90] Kim Y-H, Shin WS, Ko S-O, Kim M-C. Reduction of aromatic hydrocarbons by zero-valent iron and palladium catalyst. In: Zachry T, editor, *Environmental and Waste Management Symposium*; March 28–April 1, 2004; American Chemical Society; 2004. 5 pp. Available at http://ersdprojects.science.doe.gov/ersd/workshop_pdfs/california_2004/p132.pdf. Accessed June 2, 2014.

[91] Sanchez I, Stuber F, Font J, Fortuny A, Fabregat A, Bengoa C. Elimination of phenol and aromatic compounds by zero valent iron and EDTA at low temperature and atmospheric pressure. Chemosphere 2007;68:338–344.

[92] Hori Y, Murata A, Takahashi R. Formation of hydrocarbons in the electrochemical reduction of carbon dioxide at a copper electrode in aqueous solution. J Chem Soc Faraday Trans 1 1985;85:2309–2326.

[93] Hardy LI, Gilham RW. Formation of Hydrocarbons from the reduction of aqueous CO_2 by zero valent iron. Environ Sci Technol 1995;30:57–65.

[94] Deng B, Cambell TJ, Burris DR. Hydrocarbon formation in metallic iron/water systems. Environ Sci Technol 1997;31:1185–1190.

[95] Antia DDJ. Oil polymerisation and fluid expulsion from low temperature, low maturity, over pressured sediments. J Petrol Geol 2008;31:263–282.

[96] Antia DDJ. Hydrocarbon formation in immature sediments. Adv Petrol Expl Dev 2011;1:1–13.

[97] Antia DDJ. Polymerisation theory for a low temperature catalytic formation of petroleum hydrocarbons involving carbon dioxide, methane and hydrogen in sedimentary rocks. J Appl Geochem 2011;13:142–148.

[98] Antia DDJ. Oil reserves attributable to low temperature and high pressure catalytic processes. Indian J Petrol Geol 2011;19 (1):1–44.

[99] Cai K, Phillips DH, Elliott C, Van der Heiden E, Scippo M-L, Muller M, Connolly L. Removal of androgens and estrogens from water by reactive materials. J Water Res Protect 2010;2:990–993.

[100] Smith T, Wychick D. Colloidal iron dispersions prepared via the polymer catalysed decomposition of iron pentacarbonyl. J Phys Chem 1980;84:1621–1629.

[101] Dold B. Basic concepts of environmental geochemistry of sulphide mine-waste. Mineralogia, geoquimica y geomicrobiologia para el manejo ambiental de desechos mineros. XXIV Curso Latino de Metalogina, August22–September 2, 2005, Lima, Peru Colorado: UNESCO-SEG, SEG; 2005. 36 pp.

[102] Dold B. Basic concepts in environmental geochemistry of sulfidic mine-waste management. In: Kumar ES, editor. *Waste Management*. Croatia: Intech; 2010. p 173–198.

[103] Pourbaix M. *Atlas of electrochemical equilibria in aqueous solutions*. Houston: NACE International; 1974.

[104] Verink ED. Simplified procedure for constructing Porbaix diagrams. In: Revie RW, editor. *Uhlig's Corrosion Handbook*. 3rd ed. Hoboken: John Wiley & Sons Inc; 2011.

[105] Thirunavukkarasu OS, Viraraghavan T, Subramanian KS. Arsenic removal from drinking water using granular ferric hydroxide. Water SA 2003;29:161–170.

[106] Giles DE, Mohapatra M, Issa TB, Anad S, Singh P. Iron and aluminium based adsorption strategies for removing arsenic from water. J Environ Manage 2011;92:3011–3022.

[107] Mamindy-Pajany Y, Hurel C, Marmier N, Romeo M. Arsenic (V) adsorption from aqueous solution onto goethite, hematite, magnetite and zero-valent iron: effects of pH, concentration and reversibility. Desalination 2011;281:93–99.

[108] Tanbooncuy V, Grisdanurak N, Liao CH. Background species effect on aqueous arsenic removal by nano-zero-valent iron using fractional factorial design. J Hazard Mater 2012;205–206:40–46.

[109] Duarte AALS, Cardoso SJA, Alcada AJ. Emerging and innovative techniques for arsenic removal applied to a small water supply system. Sustainability 2009;1:1288–1304.

[110] Nikolaidis NP, Dobbs GM, Lackovic JA. Arsenic removal by zero-valent iron: field, laboratory and modelling studies. Water Res 2003;37:1417–1425.

[111] Jain CK, Singh RD. Arsenic removal using adsorptive media treatment process. India Water Week. Water Energy and Food Security: Call for Solutions, April 10–14, 2012 New Delhi: Ministry of Water Resources; 2012.

[112] Gottinger AM, Wild DJ, McMartin D, Moldovan B, Wang D. Development of an iron-amended biofilter for removal of arsenic from rural Canadian prairie potable water. NRCC-53267. Ottawa: National Research Council Canada; 2010. 14 pp.

[113] Noubactep C. Characterizing the reactivity of metallic iron in Fe^0/As-rock/H_2O systems by long-term column experiments. Water SA 2012;38:511–518.

[114] Noubactep C. On the mechanism of microbe inactivation by metallic iron. J Hazard Mater 2011;198:383–386.

[115] Lee C, Kim JY, Lee WII, Nelsom KL, Yoon J, Sedlak DL. Bactericidal effect of zero-valent iron nanoparticles on Escherichia coli. Environ Eng Sci 2008;42:4927–4933.

[116] Keller AA, Garner K, Miller RJ, Leniham HS. Toxicity of nano-zero valent iron to freshwater and marine species. PLoS One 2012;7 (8):e43983.

[117] Mahdy SA, Raheed QJ, Kalaichelvan PT. Antimicrobial activity of zero-valent iron nanoparticles. Int J Mod Eng Res 2012;2:578–581.

[118] Prema P, Selvarani M. Inactivation of bacteria using chemically fabricated zerovalent iron nanoparticles. Int Res J Pharm Sci 2012;3 (1):37–41.

[119] Diao M, Yao M. Use of zero-valent iron nano-particles in inactivating microbes. Water Res 2009;43:5243–5251.

[120] Chunjian S, Jie W, Yan J, Kniel KE, Chiu PC. Removal of viruses and bacteriophages from drinking water using zero-valent iron. Sep Purit Technol 2012;84:72–78.

[121] Jin Y, Chiu P. Filtration media coated with zero-valent metals, their process of making and use. Patent US 2011/0139726 A1. June 16, 2011.

[122] Devi NL, Shihua YICQ. Recent status of arsenic contamination in groundwater of Northeastern India—a review. Rep Opin 2009;1:22–32.

[123] Sevcu A, El-Temsah YS, Joner EJ, Cernik M. Oxidative stress induced microorganisms by zero valent iron nanoparticles. Microbes Environ 2011;26:271–281.

[124] Jin Y, ChiuP. Mitigation of irrigation water using zero valent iron treatment. Patent US2011/0309021 A1. December 22, 2011.

[125] Mahapatra O, Ramaswamy S, Nune SVK, Yadavalli T, Gopalakrishnan C. Corn flake-like morphology of iron nanoparticles and its antibacterial property. J Gen Appl Microbiol 2011;57:59–62.

[126] Rima J, Li QX, Aouuezova L. Generation of free radicals, analytical methods, bacterial disinfections, and oxidative destruction of organic chemicals using zero valent iron and other metals. Patent US 8048317 B2. November 1, 2011.

[127] Jin Y, Chiu P. Removal of microorganisms and disinfection byproduct precursors using elemental iron or aluminium. Patent US8114279. February 14, 2012.

[128] Miranda-Rios M, Luna-Pabello VM, de Velasquez MTO, Barrera-Godinez JA. Removal of Escherichia coli from biological effluents using natural and mineral aggregates. Water SA 2011;37:213–220.

[129] Rodriquez-Maroto JM, Garcia-Herruzo F, Garcia-Rubio A, Gomez-Lahoz C, Vereda-Alonso C. Kinetics of the chemical reduction of nitrate by zero-valent iron. Chemosphere 2009;74:804–809.

[130] Pilling MJ, Seakins PW. *Reaction Kinetics*. Oxford: Oxford University Press; 1995.

[131] Ebbing DD, Gammon SD. *General Chemistry*. Boston: Houghton Mifflin Co.; 2005 (8th ed), 2008 (9th ed), 2012 (10th ed).

[132] Bouniol P. Influence of iron on water radiolysis in cement based materials. J Nucl Mater 2010;403:167–183.

[133] Kasenow M. *Applied Ground-Water Hydrology and Well Hydraulics*. 2nd ed. Colorado: Water Resources Publications, LLC, 2001.

[134] Zheng D, Hunt ER Jr, Running SW. A daily soil temperature model based on air temperature and precipitation for continental applications. Climate Res. 1993;2:183–191.

[135] Antia DDJ. Modification of aquifer pore-water by static diffusion using nano-zero valent metals. Water 2011;3:79–112.

[136] Lee C-L, Jou C-JG. Integrating suspended copper/iron bimetal nanoparticles and microwave irradiation for treating chlorobenzene in aqueous solution. Environ Pollut 2012;1:160–168.

[137] Mueller NC, Braun J, Bruns J, Cernik M, Rissing P, Rickerby D, Nowack B. Application of nanoscale zero valent iron (NZVI) for groundwater remediation in Europe. Environ Sci Pollut Res 2012;19:550–558.

[138] Narayan RS, Suraya Prakash GK, Kindler A. Iron-air rechargeable battery. Patent US 2012/0187918 A1. July 26, 2012.

[139] Tanbonchuy V, Hsu JC, Grisdanurak N, Liao CH. Arsenate removal by nano zero-valent iron in the gas bubbling system. World Acad Sci Eng Technol 2010;41:263–265.

[140] Tanbonchuy V, Hsu JC, Grisdanurak N, Liao CH. Gas-bubbled nano-zero-valent iron process for high concentration arsenate removal. J Hazard Mater 2011;186:2123–2128.

[141] Tanbonchuy V, Liao CH, Grisdanurak N. Arsenic removal by nanoiron in the gas-bubbled aqueous solution. 2011 International Conference on Environment Science and Engineering. Volume 8; IPCBEE; 2011. pp 237–241. Available at www.ipcbee.com/vol8/53-S20011.pdf. Accessed June 2, 2014.

[142] Dong J, Zhao Y, Zhao R, Zhou R. Effects of pH and particle size on kinetics of nitrobenzene reduction by zero-valent iron. J Environ Sci 2010;22:1741–1747.

[143] Taylor RM. Influence of chloride on the formation of iron oxides from Fe(II) chloride. II effect of [Cl] on the formation of Lepidocrocite and its crystallinity. Clays Clay Min 1984;32:175–180.

[144] Olu-Owolabi BI, Ajayi SO. Cation adsorption on goethite-humic acid complex. Sci Iranica 2003;10:329–333.

[145] Hansel CM, Benner SG, Neiss J, Dohnalkova A, Kukkadapu RK, Fendorf S. Secondary mineralization pathways induced by dissimilatory iron reduction of ferrihydrite under advective flow. Geochim Cosmochim Acta 2003;67:2977–2992.

[146] Christiansen BC, Balic-Zunic T, Petit P-O, Frandsen C, Merup S, Geckeis H, Katerinpoulou A, Stipp SLS. Composition and structure of iron bearing, layered double hydroxide (LDH)—green rust sodium sulphate. Geochim Cosmochem Acta 2009;73:3579–3592.

[147] Dong X, Xhang Y, Liu B, Zhen Y, Hu H, Xue G. Double sandwich polyoxometalate and its Fe(II) substituted derivative $[As_2Fe_5Mo_{21}O_{82}]^{17-}$ and $[As_2Fe_6Mo_{20}O_{80}(H_2O)_2]^{16-}$. Inorg Chem 2012;51:2318–2324.

[148] Yan J, Gao J, Long D-L, Miras HN, Cronin L. Self-assembly of a nanosized, saddle-shaped, solution-stable polyoxometallate anion built from pentagonal building blocks: $[H_{34}W_{119}Se_8Fe_2O_{420}]^{54-}$. J Am Chem Soc 2010;132:11410–11411.

[149] Lee C, Keenan CR, Sedlak DL. Polyoxometalate-enhanced oxidation of organic compounds by nanoparticulate zero-valent iron and ferrous iron in the presence of oxygen. Environ Sci Technol 2008;42:4921–4926.

[150] Vila-Nadal L, Mitchell SG, Rodriguez-Fortea A, Miras HN, Cronin L, Poblet JM. Connecting theory with experiment to understand the initial nucleation steps of heteropolyoxometalate clusters. Phys Chem Chem Phys 2011;13:20136–20145.

[151] Zimmerman, AH. *Nickel-Hydrogen Batteries: Principles and Practice*. Aerospace Press, AIAA, El Segundo, 2011.

[152] Lyons MEG, Rebouillat S. Paving thc way to the integration of smart nanostructures: part I: nanotethering and nanowiring via material nanotechnology and electrochemical identification. Int J Electrochem Sci 2009;4:481–515.

[153] Rebouillat S, Lyons MEG, Brandon MP, Doyle RL. Paving the way to the integration of smart nanostructures: part II: nanostructured microdispersed hydrated metal oxides for electrochemical energy conversion and applications. Int J Electrochem Sci 2011; 6:5830–5917.

[154] Thaller LH, Zimmerman AH. Overview of the design, development, and application of nickel-hydrogen batteries. NASA/TP-2003-211905. Hanover: NASA Center for Aerospace Information; 2003.

[155] Chen K-F, Li S, Zhang W. Renewable hydrogen generation by bimetallic zero valent iron particles. Chem Eng J 2011;170:562–567.

[156] Reardon EJ. Styles of corrosion and inorganic control on hydrogen pressure buildup. Environ Eng Sci 2005;39:7311–7317.

[157] Zhao C, Reardon EJ. H2 gas charging of zero-valent iron and TCE degradation. J Environ Protect 2012;3:272–279.

[158] Soukup K, Rogut J, Grabowski J, Wiatowski M, Ludwik-Pardala M, Schneider P, Solcova O. Porous iron and ferric oxide pellets for hydrogen storage: texture and transport characteristics. Advances in Control, Chemical Engineering and Mechanical Engineering. WSEAS Press, 2010. pp 99–103. Available at http://www.wseas.us/e-library/conferences/2010/Tenerife/MECHECICON/MECHECICON-19.pdf. Accessed June 2, 2014.

[159] Rogut, J. The potential of nanoscience and nanotechnology in the development of innovative thermochemical processes of separation, purification and compression of hydrogen and carbon dioxide in emerging technologies. *Nanotech Europe* Institute for Energy; 2009. 30 pp. ftp://data.cc-nanochem.dc/NanotechEurope2009/167.pdf

[160] Singh A, Al-Raqom F, Klausner J, Petrasch J. Hydrogen production via iron/iron oxide looping cycle. Proceedings of the ASME 2011 5th International Conference on Energy Sustainability; 2011, ES2011-54499, 1–10. Available at http://itme000.louisiana.edu/assign/Solar%20Thermal%20Project/Literature/ASME%20ES%202011/ES2011-54499.pdf. Accessed June 2, 2014.

[161] Gunawardana B, Singhal N, Swedlund P. Degradation of chlorinated phenols by zero valent iron and bimetals of iron: a review. Environ Eng Res 2011;16:187–203.

[162] Mueller NC, Nowack B. Nano zero valent iron—THE solution for water and soil remediation? Report of the Observatory NANO. EMPA; 2010. 34 pp. Available at www.observatorynano.eu

[163] Noubactep C, Care S, Crane R. Nanoscale metallic iron for environmental remediation: prospects and limitations. Water Air Soil Pollut 2012;223:1363–1382.

[164] Chrysochoou M, McGuire M, Dahal G. Transport characteristics of tea green nano-scale zero valent iron as a function of soil mineralogy. Chem Eng Trans 2012;28:6.

[165] Chrysochoou M, Johnston C, Dahal G. A comparative evaluation of hexavalent chromium treatment in contaminated soil by calcium polysulfide and green-tea nanoscale zero valent iron. J Hazard Mater 2012;201–202:33–42.

[166] Bokare AD, Chi W. Zero-valent aluminium for oxidative degradation of aqueous organic pollutants. Environ Sci Technol 2009;43:7130–7135.

[167] Taylor RM, Robbins RG. Treatment of Berkley Pitlake water using the green precipitate process. Proceedings of the 1998 Conference on Hazardous Waste Research*58* Utah, May 18–21, 1998. 1–14. Available at www.engg.ksu.edu/HSRC/98Proceed/

[168] Zang W, Fyfe WS, Barnett RL. A silver-palladium alloy from the Bahia lateritic gold deposit, Carajas, Brazil. Mineral Mag 1992;56:47–51.

[169] Beverskog B, Puigdomenech I. Pourbaix diagrams for the system copper-chlorine at 5–100°C. SKI Rapport 98:19. Swedish Nuclear Power Inspectorate, 1998.

[170] Hu QH, Zavarin M, Rose TP. Effect of reducing groundwater on the retardation of redox-sensitive radionuclides. Geochem Trans 2008;9:12. DOI: 10.1186/1467-4866-9-12.

[171] Davoodi A, Pakshir M, Babaiee M, Ebrahimi GR. A comparative H2S corrosion study of 304L and 316L stainless steels in acidic media. Corrosion Sci 2011;53:399–408.

[172] Hem JD. Chemical equilibria affecting the behaviour of manganese in natural water. Hydrolog Sci J 1963;8:30–37.

[173] Brookins DG. Eh-pH diagrams for the rare earth elements at 25°C and one bar pressure. Geochem J 1983;17:223–229.

[174] Henderson T. Geochemical reduction of hexavalent chromium in Trinity sand aquifer. Ground Water 1994;32:477–486.

[175] Takena, N. Atlas of Eh-pH Diagrams. Geological Survey of Japan Open File Report No 419, 2005. Available at http://www.fssm.ucam.ac.ma/biblioadmin/opac_css/chimie/Atlas_Eh-pH_diagrams.pdf. Accessed June 2, 2014.

[176] Ellis H. *Book of Data*. London: Longman; 1993.

[177] Lide DR. *CRC Handbook of Chemistry and Physics*. 89th ed. Boca Raton: CRC Press; 2008.

[178] Stark JG, Wallace HG. *Chemistry Data Book*. London: Murray; 1982.

[179] Greenlee LF, Torrey JD, Amaro RL, Shaw JM. Kinetics of zero valent iron nanoparticle oxidation in oxygenated water. Environ Sci Technol 2012;46:12913–12920. DOI: 10.1021/es303037k Web publication date 7/11/2012.

2

NANOSTRUCTURED METAL OXIDES FOR WASTEWATER DISINFECTION

ERICK R. BANDALA[1], MARCO ANTONIO QUIROZ ALFARO[2], MÓNICA CERRO-LÓPEZ[2], AND MIGUEL A. MÉNDEZ-ROJAS[2]

[1]*Departmento de Ingeniería Química, Alimentos y Ambiental, Universidad de las Américas Puebla, Puebla, Mexico*

[2]*Departamento de Ciencias Químico-Biológicas, Universidad de las Américas Puebla, Puebla, Mexico*

2.1 INTRODUCTION

Wastewater generation and release into the environment is a major concern because of its impact on surface and underground water quality. This is particularly important in developing countries where, in many of the cases, domestic and industrial wastewater is discharged into natural water bodies with improper or no treatment. As an example, in Mexico, the annual production of both domestic and industrial wastewater is estimated to be as high as 13.5 km^3/year. Only 30% (about 3.9 km^3/year) of this wastewater is treated. The estimated amount of pollutants generated is 9 million tons of BOD_5 per year, with only 20% (1.94 million tons of BOD_5) being removed [1].

There may be many different undesirable consequences associated with the release of such significant amounts of contaminants into natural water courses. Among them, probably the most disturbing is the dissemination of pathogenic microorganisms causing waterborne diseases through the ingestion of wastewater effluents in the water source downstream. The prevalence of such pathogenic microorganisms in water is probably among the most important human health problems. It is known that unsafe drinking water is related to millions of deaths yearly (around 1,400,000 yearly worldwide) as a result of waterborne diseases [2, 3]. One-sixth of the world's population has been estimated to lack access to water and sanitation services [4] usually associated with poverty [5]. Waterborne diseases affect the poorest sectors of society and are recognized by the United Nations to be the cause of infections for around one billion people in developing countries [6]. Removal of pathogenic microorganisms capable of causing waterborne diseases has become one of the most important scientific and technological tasks, and many different sanitary engineering processes have been developed for achieving this goal [7, 8]. Chlorination is probably the most widely used cost-effective water disinfecting methodology around the world; nevertheless, in some cases chlorination is not capable of inactivating these microorganisms, and it has been associated with unpleasant taste and several concerns on the toxicity of its reaction by-products [9].

Among the alternative methodologies developed to generate safe drinking water, advanced oxidation processes (AOPs) have been identified as emerging technologies for inactivating pathogens in water. Destruction of parasites by applying ozone, ultraviolet (UV) radiation, and homogeneous and heterogeneous photocatalysis (HP) has been discussed in the past for bacteria, fungi, viruses, and other highly resistant microorganisms [10–13]. Among all these interesting alternatives, HP is the technology most extensively discussed in the literature [14] for inactivating microorganisms in water.

Nanomaterials for Environmental Protection, First Edition. Edited by Boris I. Kharisov, Oxana V. Kharissova, and H. V. Rasika Dias.

2.2 PHOTOACTIVE METAL OXIDES

Photocatalysis is defined as the acceleration of a reaction by using a photocatalyst [15]. The initial step in HP is the absorption of a quantum of energy by the catalyst (usually a semiconductor). Once energy absorption occurs, the catalyzer (C) generates charge carriers (electrons, e^-, and holes, h^+), and excited electrons are transferred to the oxidant (Ox_1). Simultaneously, the catalyzer accepts electrons from the reducer (Red_2) to fill the holes generated at its valence band. Electron flux in both directions is null and the catalyzer remains unaltered as proposed in reaction sequence (2.1) [16]:

$$\begin{aligned} &C \xrightarrow{h\nu} C(e^- + h^+) \\ &h^+ + Red_2 \rightarrow Ox_2 \\ &e^- + Ox_1 \rightarrow Red_1 \end{aligned} \tag{2.1}$$

HP uses a dispersed solid semiconductor (TiO_2, ZnO, etc.) to generate a stable colloidal suspension under radiation for stimulating a reaction at the solid/liquid interface. At the interphase between the semiconductor and a solution containing a redox pair, charge transference occurs to balance chemical potentials between the two faces. Metallic oxides and sulfurs are among the most commonly used semiconductor materials available for photocatalytic (PC) purposes. Nowadays, titanium dioxide (TiO_2) is frequently used as a semiconductor for HP processes and is reported to be among the most active [17]. Table 2.1 gives some of the most common semiconductor materials used in PC reactions, their band-gap energy, and the maximal wavelength values required for catalyst activation.

Degradation of organic pollutants by HP is among the most successful applications of AOPs for eliminating toxic substances in water [18] as suggested by the wide variety of research groups, facilities, scientific reports, and patents around the use of this technology. Use of HP water disinfection has given rise to the design and synthesis of new nanostructurated photocatalysts with interesting properties for more efficient environmental applications.

2.3 KINETICS AND REACTION MECHANISMS

For a long time, several groups working on the analysis of HP mechanisms have hypothesized on the photogeneration of holes (h^+) and surface-trapped hydroxyl radicals ($HO^\bullet$) [19]. Initial steps involved in band-gap irradiation of TiO_2 particles (or any other semiconductor) have been studied in detail by laser-flash photolysis measurements [20, 21]. It is well established that TiO_2 illumination with radiation of the proper wavelength ($\geq E_g$) generates an electron/hole pair that can recombine or dissociate (both reactions are in competition) to produce, in the latter case, a conduction band electron and a valence band hole that are able to migrate to the particle surface. Once on the surface, both charge carriers are able to interact with adsorbed electron acceptors and oxidize electron donors. In the aqueous phase, oxygen is often present as an

TABLE 2.1 Band-gap energy and activation wavelength for some semiconductors [16]

Material	Band-gap energy (eV)	Activation wavelength (nm)
$BaTiO_3$	3.3	375
CdO	2.1	590
CdS	2.5	497
CdSe	1.7	730
Fe_2O_3	2.2	565
GaAs	1.4	887
GaP	2.3	540
SnO_2	3.9	318
$SrTiO_3$	3.4	365
TiO_2	3.2	387
WO_3	2.8	443
ZnO	3.2	390
ZnS	3.7	336

electron acceptor and HO^- and H_2O are available as electron donors to yield hydroxyl radicals. It is well documented that these trapping reactions occur in a lapse time shorter than 30 ps [22].

Considering the importance of mass transference in the process, initial practical approaches for a quantitative description of HP kinetics have been commonly carried out using a Langmuir–Hinshelwood (L–H) kinetics model [23]. This mathematical model assumes that the reaction occurs on the catalyst surface. According to the L–H model, the reaction rate (r) is proportional to the fraction of particle surface covered by the pollutant (θ_x). Mathematically,

$$r = -\frac{dC}{dt} = k_r\theta_x = \frac{k_r KC}{1 + KC + K_s C_s} \tag{2.2}$$

where k_r is the reaction rate constant, K is the pollutant adsorption constant, C is the pollutant concentration at any time, K_s is the solvent adsorption constant, and C_s is the concentration. During the 1980s, many authors presented their data using the L–H kinetic approach. Nevertheless, despite fitting well with the experimental data, the L–H approach does not consider the role of the radiation field on the mechanism [24, 25].

Other kinetic studies on HP suggest that the reaction rate increases with catalyst concentration to get a maximum value concentration depending on the compound and the reactor used. For these concentrations, the reaction rate remains unchanged or decreases with further increments of catalyst concentration [25]. An interesting problem is the relation between catalyst concentration, reaction rate, radiation absorption, and process improvement. Considering this, many different models have been proposed. Studies have suggested relationships aiming to estimate the radiation absorbed by the catalyst [24, 25]. From these results, several models, most of them based on complex mathematical or statistical computational approaches, have been developed. These models are able to predict radiation absorption and scattering as a function of catalyst concentration, optical path and catalyst type, and its relation to pseudokinetic constants obtained experimentally [24–26]. Based on the radiation absorbed by the catalyst, Alfano's group as well as other authors have focused on the a priori design of photochemical reactors, the improvement of HP reactions, and the generation of intrinsic reaction kinetic that may lead to process scaling-up [27–29].

Besides reactor design, heterogeneous PC degradation reaction can be enhanced by the use of higher active catalyst or inorganic oxidizing species. In the first case, activation of TiO_2 under visible light is a desirable technological approach. In order to utilize visible light for TiO_2 excitation, dye-sensitized and ion-doped TiO_2 has been developed in recent years and has yielded promising results for the photocatalyzed degradation of different substrates [30, 31].

2.4 VISIBLE LIGHT ABSORBING SEMICONDUCTORS

According to reaction sequence (2.1), the production of charge carriers is a fundamental step in the degradation processes using HP. Once generated, these species may lead to hydroxyl radical generation (and the subsequent organic matter degradation) or can recombine to generate the initial state and energy emission. This latter reaction, known as recombination, is a practical problem when using a TiO_2 catalyst, and it is extremely efficient (reaction rate = 10^{-9} s) when no proper electron acceptor is present in the reaction media [32]. This side process is energy-wasting and limits the achievement of high quantum yields (i.e., number of primary chemical reactions per photon absorbed). In most of the cases, dissolved oxygen is used as the electron scavenger, and several works have dealt with its efficiency as an oxidant agent to complete organic matter mineralization [33]. Nevertheless, it has been demonstrated that only low mineralization is reached when dissolved oxygen is used as an oxidant agent [34].

The recombination of charge carriers seriously affects the actual photonic yield of the photocatalyst; however, this is not the only variable related with the effectiveness of the process. One of the main disadvantages in its use is related to the wavelength required for activating TiO_2. The HP process using titanium dioxide occurs only when it is irradiated with ultraviolet (UV, $\lambda < 400$ nm) radiation and photon energy is absorbed by the crystal structure of the semiconductor, transferring electrons from the valence to the conducting band. This specific characteristic limits the photocatalyst sensitivity when solar radiation is intended to be used since only a small part of the solar spectrum (about 5%) falls within the UV radiation [2, 35]. In order to avoid this limitation, several modifications have been attempted to the TiO_2 PC structure in order to make it active under visible spectral irradiation, improve its photosensitivity and quantum yield, as well as reduce its band-gap energy requirements for PC activation. These modifications follow different approaches: (i) dye sensitization, (ii) TiO_2 coupling with other semiconducting materials with appropriate band-gap energy, (iii) surface deposition of metal clusters, and (iv) doping the crystal lattice with metallic and nonmetallic foreign atoms; see Figure 2.1 [36, 37].

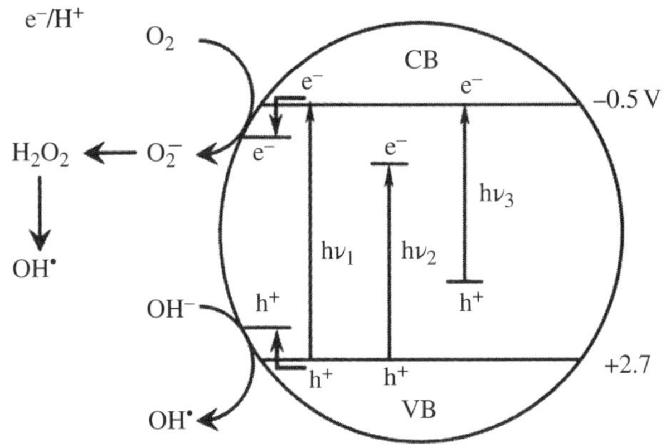

FIGURE 2.1 TiO_2 photocatalyst's band-gap modification for undoped ($h\nu_1$), metal doping ($h\nu_2$), and nonmetal doping ($h\nu_3$).

Several years ago, Sato discovered that the addition of NH_4OH in a titania sol–gel formulation followed by calcination of the precipitated powder at temperatures of 300 °C and higher resulted in a material that exhibited visible light sensitization [38]. It was not until a few years ago that a first explanation of this phenomenon was provided by Asahi and coworkers who reported that nitrogen-doped TiO_2 could absorb in the visible light. It was a yellowish material prepared by sputtering a TiO_2 target under a N_2(40%)/Ar gas mixture followed by annealing under N_2 at 550 °C for 4 h, and it was photocatalytically active for the degradation of methylene blue and acetaldehyde under visible light [39–42].

Nitrogen is one of the most widely reported dopants for shifting TiO_2 spectral absorption into the visible range. Substitution of N in the titania lattice has been proposed to contribute to a narrow band gap by merging nitrogen and oxygen 2p states on the top of the valence band as nitride (Ti–N) or oxynitride (Ti–O–N). Another theory is the formation of oxyanion species at interstitial lattice sites, where N is bonded to one or more oxygen atoms creating localized intergap states. Both arrangements are considered to change the energy photothreshold responsible for the red shift of the optical absorption toward the visible spectra region [43].

Several recent studies in this emerging field of N-doped TiO_2 catalyst activated by visible light deal with the fundamental physics to explain the electronic properties of N-doped TiO_2 [44–46] and provide solid evidence for the promising properties of this environmentally "green" material [47, 48]. However, some other reports have suggested that anion-doped TiO_2 photocatalysts have low efficiency due to charge carrier recombination from anion doping [49] and that metal ion modification influences light absorption and photoreactivity of TiO_2 acting as electron/hole traps [50].

Silver-, vanadium-, iron-, and palladium-modified nitrogen-doped TiO_2 have been tested to enhance titania's PC performance in inactivating Gram-negative bacteria, such as *Escherichia coli*, *Pseudomonas aeruginosa*, and *Prevotella intermedia* [50–55], as well as some Gram-positive bacteria, such as Methicillin-resistant *Staphylococcus aureus* (MRSA), *Staphylococcus epidermidis*, *Staphylococcus saprophyticus*, *Streptococcus pyogenes*, and *Saccharomyces cerevisiae* [56–59], and a few protozoa, such as *Tetraselmis suecica*, *Amphidinium carterae*, *Chlorella vulgaris* [60, 61], in water.

Nonmetallic ion doping has also been shown to be effective to induce modifications of the electronic structure of TiO_2 by creating surface oxygen vacancies due to charge compensation between the nonmetallic ion and Ti^{4+} but without producing a significant change in the optical absorption of the photocatalyst [37, 62]. Moreover, codoping of TiO_2 with nitrogen and fluorine has demonstrated high PC activity in the visible region with beneficial effects induced by both dopants [63–65]. Huang et al. confirmed strong visible light absorption and high PC activity of N–FTiO_2 for *p*-chlorophenol and rhodamine B degradation under visible light irradiation [64]. Xie et al. effectively decomposed methyl orange with a visible light–activated N–F–TiO_2 photocatalyst [65]. Both attributed their findings to the synergistic effect of nitrogen and fluorine doping. More recently, Dionysiou's group has shown the effectiveness of N–TiO_2 in removing cyanobacterial toxins in water by using visible radiation [37, 66]. Castillo-Ledezma et al. have recently demonstrated the capability of N–F–TiO_2 in inactivating *E. coli* using solar radiation under several different reaction conditions, including pH value and radiation wavelengths (visible and UV + visible), and they showed that the N–F–TiO_2 process possesses higher efficiency than the regular TiO_2 and solar disinfection processes [2]. Similar results were found by Wong et al. for nitrogen-doped TiO_2 in inactivating *E. coli* and several other pathogenic microorganisms [67]. They also found that proteins and light-absorbing contaminants reduce the bacterial activity of the photocatalyst as a result of their light-shielding effects. Liu et al. found that extracellular polymeric substances (EPS) generated by some heterotrophic bacteria play an important role in controlling the kinetics of the solar-induced PC process by generating a protective layer against the presence of photogenerated reactive oxygen species (ROS) during the photocatalyst activation [68].

In a very interesting synergistic approach, Wu et al. developed and tested a metal-/nonmetal-doped TiO_2 catalyst for the inactivation of Gram-negative (*E. coli* and *P. aeruginosa*) and Gram-positive (*S. aureus*) cells [69]. They found that by combining both metal and nonmetal dopants, PC activity was much higher than using the dopants separately under visible light illumination.

2.5 SLURRIES OR IMMOBILIZED PHOTOCATALYST

The generation of catalyst sludges as by-products is one of the main disadvantages of HP processes in water treatment. At the pilot-plant level, HP uses suspended TiO_2 in photoreactors where the semiconductor is recovered after the treatment. According to various lab-scale research reports [14], the use of TiO_2 in suspensions is more efficient than their use in immobilized forms. Nevertheless, this latter form possesses specific advantages, such as cost reductions, a decrease in material losses, and the likelihood of skipping recovery steps in the process, which make it possible to produce immobilized titania photocatalysts with higher efficiency as compared with those reported to date [70, 71].

Several supporting materials, from sand to quartz optical fiber, have been reported so far for TiO_2 immobilization. During the past few years, the in situ generation of catalysts seems to be the most promising technology for catalyst immobilization [71]. Other authors [72] have demonstrated that by using this method, the generated fixed form of titanium dioxide presents equal efficiency as Degussa P-25 (considered the most efficient form of titanium dioxide) suspended catalyst for bacteria inactivation. However, more research is needed before this can be considered for the future design of efficient PC processes.

2.6 TiO_2 PARTICLES AND NANOTUBES

Nanostructured TiO_2 with different sizes, shapes, and morphologies (nanoparticles, nanofibers, nanowires, nanorods, nanoporous materials, and nanotubes) have been prepared and their unique properties tested for several applications [73]. The preparation method plays a key role on the variety of structures formed. Some well-known methods to prepare TiO_2 nanostructures are sol–gel, microemulsion, hydrothermal, solvothermal, vapor deposition, and anodization [74, 75]. The main idea behind this molecular design is to improve the physical and chemical properties of TiO_2, including their PC activity and PV response [76].

One of the most promising TiO_2 nanostructures for environmental applications is titanium dioxide nanotubes (TDN). Literature on TDN is the second most abundant only after that on carbon nanotubes. TDNs prepared from the anodization of metal plates have attracted interest because they can improve performance on well-known titania applications such as water photoelectrolysis, photocatalysis, heterojunction solar cells, gas sensing, and environmental purification [77, 78] since they provide a highly ordered TiO_2 nanotubular array with a large specific surface area. Two important structural characteristics of these TDN arrays are their inner diameter of approximately 100 nm and an average length of 3 μm (Fig. 2.2).

The highly ordered structure can improve the transport of photogenerated electrons in the TiO_2 film by providing a unidirectional electric channel and reducing grain boundaries in photocatalysis and photoelectrochemical applications, while the large surface area can allow more efficient light harvesting [79, 80].

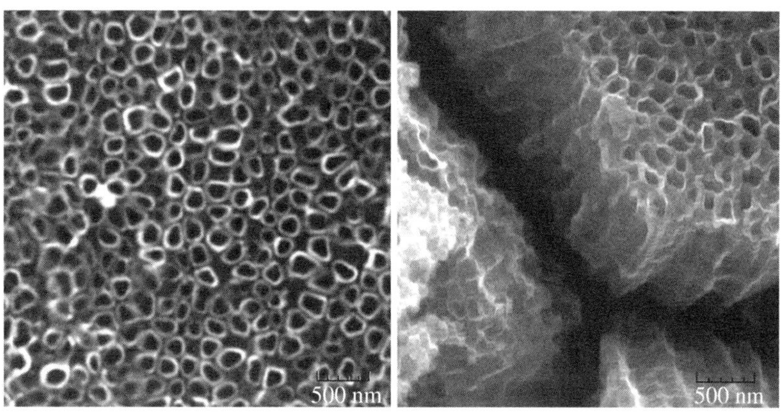

FIGURE 2.2 Standard electron microscope (SEM) images of TDN obtained by anodization. Left top view; right cross-sectional image.

As a material for electrocatalytic oxidation of azo dyes, TDNs show null or very low capability to oxidize them [80]. However, when electrochemical oxidation is combined with UV irradiation, degradation becomes more effective as in the case of methyl orange degradation reported by Zhang, who achieved a degradation efficiency of 56.3% for an irradiation time of 90 min with an initial dye concentration of 2×10^{-4} M [80].

Also, in electrocatalysis, TiO_2 nanotubes have been used as a substrate to support electroactive materials like noble metals (Ag, Au, Pd, Pt, and Ru) [81–83] or metal oxides. As mentioned earlier, TiO_2 itself is a widely used material in electrochemistry due to its semiconductive properties, insertion host capacity, and long-term stability [81]. These characteristics make TDNs very suitable to hold other materials providing a large surface area and a chemically stable structure that can be repeatedly used [83]. Gold nanoparticles (3 nm) loaded on TDN layers have proven to be more effective for the reduction of oxygen than gold nanoparticles dispersed on a compact flat TiO_2 film. Gold as well as other noble metals like silver, palladium, platinum, and ruthenium have been deposited on TDNs in order to enhance TiO_2 PC properties [81].

2.7 PHOTOCATALYSIS ON TiO_2 NANOTUBES

In recent years colloidal and particulated TiO_2 have been used to photodegrade pollutants in both liquid and gas phases. However, these suspended systems must face three inherent technical problems: (i) separation or filtration after the reaction, (ii) particle aggregation, and (iii) the problems associated with continuous flow systems. Methods to prepare TiO_2 on a solid support have been reported, but the efficiency of the immobilized system is much lower than that of suspended ones because of the reduction of surface active area on the immobilized catalyst. Therefore, TiO_2 nanotubes might overcome this drawback due to its high specific surface area [84]. Toward this end, the PC activity of TDNs toward methyl orange discoloration was evaluated and compared with that of a TiO_2 nanoparticle film. In this study, the authors reported a more efficient degradation activity on the nanotubes than on the nanoparticle film. This higher efficiency is attributed to a more effective separation of the photogenerated electron–hole pairs and the higher inner surface area of the nanotube structure [84].

In general, in order to enhance TiO_2 PC activity, a deposit of noble metal particles increases electron–hole separation and promotes electron transfer processes. It has been stated that nanotubes loaded with gold and silver nanoparticles form local Schottky junctions that present a higher potential gradient than at the TiO_2/electrolyte interface [82]. Photocatalysis can be improved if two metals are simultaneously used to modify TiO_2 as it was investigated for a Au–Pd-comodified TiO_2 nanotube film (Au–Pd–TiO_2). The activities of naked TiO_2 and Au–Pd–TiO_2 toward malathion degradation were compared. The study showed that 73.8% of malathion was removed in the presence of naked TiO_2, but 98.2% malathion removal was achieved when Au–Pd–TiO_2 was used instead of naked TiO_2 [83].

As we can see, one of the major trends in the environmental use of TiO_2 nanotubes is combining them with another material, metal, or semiconductor in order to enhance their PC properties or to obtain a dual material that has the properties of both parent materials. In some cases, the TiO_2 nanotube array plays the role of a substrate, which has a large surface area, good mechanical resistance, and greater adsorption capacity to support a highly electroactive metal oxide such as PbO_2 (Fig. 2.3a) or SnO_2 doped with Sb (Fig. 2.3b) [85–87].

PbO_2 and SnO_2 are known to have a high oxygen evolution potential, which makes them more efficient for electro-oxidation of organic pollutants. These active oxides loaded on ceramics, Ti, and other support materials are suitable for environmental

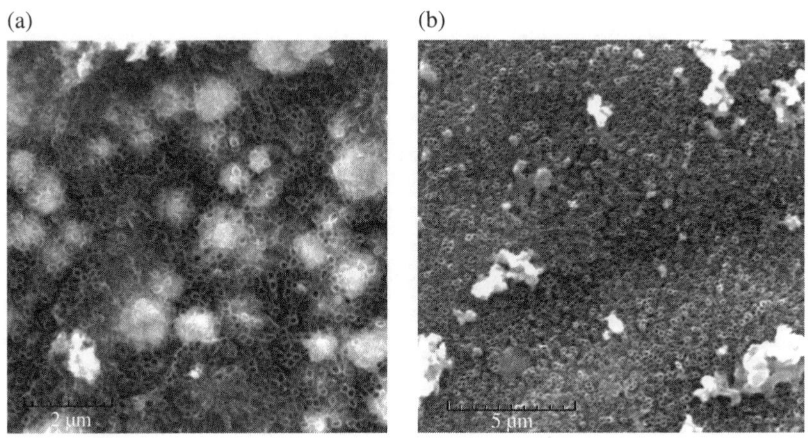

FIGURE 2.3 (a) PbO_2 deposited on TDN; (b) Sb–SnO_2 deposited on TDN.

applications due to their low cost, good corrosion resistance, and high conductivity, but mechanical instability of their coatings is a major disadvantage that has to be overcome. To achieve that goal, several efforts are being made, including the use of TiO_2 nanotubes prepared by the anodization of Ti plates, which provides a stake structure where the electroactive materials can penetrate and combine more firmly, thereby increasing the stability of electrodes [85, 87].

Relatively few works dealing with TDN tests for water disinfection have been reported. Ng et al. synthesized TDNs via electrochemical anodization followed by calcination and used these materials for *E. coli* inactivation in water [88]. They found that over 95% of the initial *E. coli* viable counts were inactivated in only 50 min under UV-A radiation. Presence of ionic species and organic compounds in water did not produce any observable effect in the inactivation rate.

2.8 PHOTOELECTROCATALYSIS ON TDN

The idea of combining TiO_2 with other oxides in order to obtain an electrode that presents both PC and electrocatalytic properties is not new [89]. However, the possibility of using TDNs, prepared anodically on Ti plates, as both support and photoactive material, makes it easier to grow or implant electroactive oxides to create an electrode material with photoelectrocatalytic (PEC) properties. Thus, great attention has been paid recently to prepare electrodes of this type.

The use of naked TDN as photoelectrocatalysts has been reported for azo decolorization, for example, in the cases of methyl orange (MeO) or acid orange 7. In the case of methyl orange, PEC and PC activities of the nanotube electrode were compared for decolorization experiments where a 0.5-V potential was applied and a UV illumination was utilized. A 99% removal was achieved under PEC conditions, while only 21.5% of the MeO was removed by the PC process for a total time of 90 min [80]. For acid orange 7, total removal (discoloration) was achieved after 45 min of PC decomposition under an applied potential of 1.0 V versus the Ag/AgCl reference electrode [90]. According to this author, applying an electrochemical potential helps to control the band bending, which can result in more efficient charge carrier separation.

One way to enhance the PEC properties of TDNs is to couple them with other materials like Sb-doped SnO_2, SiO_2, Bi_2O_3, Fe_2O_3, ZnO, and CdS [87, 91, 92]. Thus, for example, a photocatalyst like Bi_2O_3 that is capable of oxidizing water under a visible light irradiation could be loaded onto a TiO_2 nanotube array electrode, and the composite electrode achieved higher catalytic activities toward 2,4-dichlorophenol degradation than the individual Bi_2O_3 or TiO_2 electrode materials did. An effective photocatalyst for the elimination of environmental pollutants should present high PC activity both in the UV and in the visible light regions. This can be accomplished by a composite electrode like Bi_2O_3/TiO_2, where Bi_2O_3 is photoactive under visible light and TiO_2 is active under UV irradiation [87].

PEC using TDN has also been briefly reported for application as a promising and powerful tool for bacteria inactivation [93]. Current work deals with the development of TDN through electrochemical anodization in aqueous solution. The nanotube electrode built by this procedure was compared with a mesoporous TiO_2 electrode for *E. coli* inactivation. High surface TiO_2 nanotubes resulted in high photocurrent and an extremely rapid *E. coli* inactivation rate (~10^6 CFU mL^{-1} killed in less than 10 min). Recently, a $Ag/AgBr/TiO_2$ nanotube array with enhanced visible light activity was synthesized and its PEC activity tested for inactivation of *E. coli* under visible radiation ($\lambda > 400$ nm) resulting in complete sterilization highly superior than with other reference photocatalysts [94]. However, other authors disagree with these results, and there is a controversy on the real advantages of PEC for bacteria inactivation [95].

2.9 OTHER NANOSTRUCTURED METAL OXIDES

Metals have been used in water treatment for long. For example, silver has been in use for the treatment of infections and wounds as well as for water disinfection due to its antimicrobial activity since Roman times. More recently, other metal derivatives, such as metal oxides but also metal chalcogens, have drawn attention because of their potential applications in inhibiting microorganism growth due to their high surface areas, unusual crystal morphologies, and high catalytic activity. The use of oxide nanomaterials for environmental remediation has been reviewed recently [96, 97] and is a hot research field. They can be used for fast and cost-effective cleaning procedures for contaminants in comparison to traditional methods [98, 99]. In particular, the use of nanomaterials for wastewater treatment has attracted the attention of some groups, as there are several reports on the specific biocide action of some nanomaterials against different types of microorganisms that show low general toxic profiles and good stability [100, 101]. This property of nanomaterials may be exploited for use as effective disinfectant agents. It remains a complex issue to understand the several factors involved in toxicity, such as chemical composition, surface reactivity, size, distribution, cell type, experimental setup, as it requires an interdisciplinary view. However, before their large-scale use for water disinfection, it is necessary to understand the real environmental impact that engineered nanomaterials have [98].

Magnesium oxide (MgO) nanoparticles showed biocidal activity against certain Gram-positive and Gram-negative bacteria, as well as spores, in particular against *E. coli*, *B. megaterium*, *Streptococcus pneumoniae*, and *Bacillus subtilis*. [102, 103]. Alumina (aluminum oxide, Al_2O_3) nanoparticles have also shown mild antibacterial activity against *E. coli*, *S. pneumoniae*, *P. aeruginosa*, and *B. subtilis* cultures [104–106]. Bulk versus nanoalumina comparison showed that the nanosized Al_2O_3 had higher toxicity at the same concentration [107, 108].

Copper (II) oxide (CuO) and copper (I) oxide (Cu_2O) are both semiconducting compounds that are cytotoxic to several types of cells [109, 110]. In fact, they are considered among the most toxic nanomaterials based on several studies. In the case of environmental effects, it has been reported than CuO nanoparticles with sizes lesser than 50 nm are toxic when tested against *Nitellopsis obtusa*, *Chlorella*, shrimp *Thamnocephalus platyurus*, and rotifer *Brachionus calyciflorus* [111]. In that sense, copper oxides may be useful for applications in water disinfection. For example, CuO nanoparticles were effective in killing several pathogenic bacteria (*P. aeruginosa*, *Proteus* spp., *B. subtilis*) when used in high concentrations [112]. Several reports relate the toxicity to copper ions more than to the presence of nanoparticles, but sometimes the nature of the specific toxic agent is unclear. Coordinative interactions with chemical moieties with great affinity to copper ions such as carboxylic or amine groups, as well as the probability of intercalation into DNA structures to form complexes with nucleotides, may cause disruption of biochemical process and thus be the reason for their toxicity [113]. Cu_2O nanoparticles supported onto natural clinoptilolite showed 100% antiprotozoan activity against *Paramecium caudatum* and *Eurytemora affinis* after 1 h of contact and inhibitory growth effects on *E. coli* and *S. aureus* in tests specifically designed to measure the effectiveness of these materials for disinfection of secondary effluent water and removal of pathogenic microorganisms in the tertiary stage of wastewater treatment [114].

Zinc oxide (ZnO), on the other hand, has been used in several commercial applications such as in sunscreens, cosmetics, pharmaceutical, and the plastic and food industry due to its semiconducting properties, stability, and antibacterial activity. It has been tested against bacteria such as *B. subtilis*, *E. coli*, *Pseudomonas fluorescens*, *Listeria monocytogenes*, *Salmonella enteritidis*, *Salmonella typhimurium*, and *S. aureus*, showing from mild to strong microbicidal activity, although the studies were more oriented toward finding novel food and agricultural applications than toward finding a utility for wastewater treatment [107, 115–118]. The mechanism of antibacterial activity is suggested to proceed through generation of hydrogen peroxide on the ZnO surface. It has also been suggested that the release of Zn(II) ions may damage the cell membrane, and they may interact with specific biomolecules [119, 120]. Cobalt-doped ZnO antibacterial activity was tested against *E. coli*, *Klebsiella pneumonia*, *Shigella dysenteriae*, *Salmonella typhi*, *P. aeruginosa*, *B. subtilis*, *S. aureus*, and the fungi *Fusarium* sp., showing significant results [121, 122]. The use of thin films of ZnO is also being considered to avoid the growth of biofilms, which may be very useful for wastewater treatment technologies involving large active surfaces or membranes [123].

Cerium oxide (CeO_2) nanoparticles have shown antimicrobial activity by electrostatically binding the cell membrane of Gram-negative cells [124]. They have been detected by monitoring the bacterial growth on a solid medium of *S. aureus*, *P. aeruginosa*, *Klebsiella* sp., *S. pneumoniae*, and *S. aureus* [104, 125]. When compared with bulk and nano CeO_2, bulk CeO_2 showed less toxicity than their nanosized counterparts [126].

Nitrogen-doped ZrO_2 has been reported to effectively inhibit the growth of *E. coli* and biofilm heterotrophic bacteria under solar light illumination [68]. Iron oxides (FeO, α-Fe_2O_3, β-Fe_2O_3, γ-Fe_2O_3, Fe_3O_4) are also of chemical interest as they are biocompatible and some of them are magnetically active. They are able to participate in light-driven or pH-induced redox changes that are responsible for the generation of ROS and other highly reactive chemical species (such as primarily hydroxyl radicals), following Fenton-like mechanisms. Just as in the case of TiO_2, the use of such reactive species for water disinfection is attractive as they are produced by the illumination of a suspension of contaminated water and nanoparticles for a short time, and solar light may be used when the shifting of the optical response from UV to visible light is to be achieved. In that sense, the growth of *S. aureus* cultures was inhibited significantly compared to control samples when 3 mg/ml of iron oxide dose was used [127]. It has also been reported than Fe_3O_4 nanoparticles have an inhibitory effect on *E. coli* in a concentration-dependent manner; although no correlation to light exposition was made, the authors suggested than ROS may be responsible for growth inhibition [128]. However, a recent study warns that superparamagnetic iron oxide nanoparticles (SPIONs) may have a negative effect for wastewater remediation as they induce an increase of *P. aeruginosa* biofilm biomass [129].

2.10 CONCLUSIONS

The specific biocide activity of some nanomaterials against microorganisms present in wastewater is a very attractive property for their eventual incorporation in large-scale methods for water disinfection. However, several important challenges remain with respect to their commercial use. First, it is necessary to remove particles from water as they may negatively interfere in secondary treatment processes or, due to their inherent toxicity, may present a risk for the environment (biomass) or water consumers. Second, electrostatic or van der Waals interactions among nanoparticles to form larger aggregates may be useful to obtain particle sizes easily trapped by traditional filtration methods or to induce spontaneous precipitation. But then, they may be incorporated into activated sludge, creating another type of problem—affecting microorganisms present there. Their size may also increase due to interactions with organic matter dispersed in solution (humic and fulvic acids, carbonaceous materials, among others). Core–shell nanoparticles involving a ferromagnetic layer is another attractive option as the latter may be used to remove nanoparticles by simply applying a strong external magnetic field. Functionalization of their surfaces to maintain their solubility, increase their stability against aggregation, decrease their chemical reactivity at a wide range of pH values, or avoid interactions with chemical substances present at solution is also key for their eventual consideration in wastewater treatment procedures. Biological interactions with other organisms (not only microbes, but also other phyla such as invertebrates, vertebrates, and plants) may prove to be another challenge, as the nanoparticles may biodegrade, suffering biologically induced chemical and physical transformations, which affects their stability and chemical reactivity, or even changes their biological activity from nontoxic to highly toxic or vice versa.

All these problems make the field more attractive for further research in order to overcome the difficulties and better exploit the unique properties of nanomaterials for wastewater treatment.

REFERENCES

[1] Comisión Nacional del Agua (CNA). 2011 Water statistics in Mexico. Available at. http://www.conagua.gob.mx/ocb/ConsultaTemasInteres.aspx?n0=f16ed22d-197b-4b43-bc12-4af596c47ba9 (Accessed March 15, 2013).

[2] Castillo-Ledezma JH, Sánchez-Salas JL, López-Malo A, Bandala ER. Effect of pH, solar irradiation and semiconductor concentration on the photocatalytic disinfection of *Escherichia coli* in water using Nitrogen-doped TiO_2. Eur. Food Res. Tech. 2011;233:825–834.

[3] Shannon MA, Bohn PW, Elimelech M, Georgiadis JG, Mariñas BJ, Mayes AM. Science and technology for water purification in the coming decades. Nature 2008;452:302–310.

[4] Montgomery MA, Elimelech M. Water and sanitation in developing countries: Including health in the equation. Environ. Sci. Technol. 2007;41:17–24.

[5] Gelover S, Gomezo L, Reyes K, Leal T. A practical demonstration of water disinfection using TiO_2 films and sunlight. Water Res. 2006;40:3274–3280.

[6] UN. 2012. The millennium development goals report, Available at http://www.un.org/en/development/desa/publications/mdg-report-2012.html (Accessed March 11, 2013).

[7] Bandala, E.R. Castillo-Ledezma, J. Gonzalez, L. Sanchez-Salas, J.L. Solar driven advanced oxidation processes for inactivation of pathogenic microorganisms in water. In *Recent Research Development in Photochemistry and Photobiology*. Volume 8, India: Transworld Research Networks; 2011. p. 1–16.

[8] Riahi K, Ben Mamou A, Ben Thayer B. Date-palms fiber media filters as a potential technology for tertiary domestic wastewater treatment. J. Hazard. Mat. 2009;161:608–613.

[9] Sichel C, Tello J, Cara M, Fernandez-Ibañez P. Effect of UV solar intensity and dose on the photocatalytic disinfection of bacteria and fungi. Catal. Today 2007;129:152–160.

[10] Orta MT, Martinez J, Monje I, Rojas MN. Destruction of helminto (*Ascaris suum*) eggs by ozone. Ozone-Sci. Eng. 2004;26:359–366.

[11] Alouini Z, Jemli M. Destruction of helminth eggs by photosensitized porphyrin. J. Environ. Monitor. 2004;3:548–551.

[12] Bandala ER, Corona-Vasquez B, Guisar R, Uscanga M. Inactivation of highly resistant microorganisms in water using solar driven phoocatalytic processes. Int. J. Chem. React. Eng. 2009;7:A7.

[13] Guisar R, Herrera MI, Bandala ER, García J, Corona-Vasquez B. Inactivation of waterborne pathogens using solar photocatalysis. J. Adv. Oxid. Technol. 2007;10:453–438.

[14] McGuigan KG, Conroy RM, Mosler H, Du Preez M, Ubomba-Jaswa E, Fernandez-Ibañez P. Solar water disinfection (SODIS): A review from bench-top to roof-top. J. Hazard Mater 2012;235/236:29–46.

[15] Kish H. What is photocatalysis? In: Serpone N, Pelizzetti E, editors. *Photocatalysis Fundamentals and Applications*. New York: John Wiley and Sons; 1989.

[16] Malato S. *Solar Photocatalytic Decomposition of Pentachlorophenol Dissolved in Water*. Madrid, Spain: Editorial CIEMAT; 1999.

[17] Blanco J, Fernandez P, Malato S. Solar photocatalytic detoxification and disinfection of water: Recent overview. J Solar Energy Eng Trans ASME, 2007;129:4–15.

[18] Bandala ER, Estrada C. Comparison of solar collection geometries for application to photocatalytic degradation of organic contaminants. J. Sol. Energ.-T. ASME 2007;129:22–26.

[19] Romero M, Blanco J, Sánchez B, Vidal A, Malato S, Cardona AI, Garcia E. Solar photocatalytic degradation of water and air pollutants: challenges and perspectives. Sol. Energy 1999;66:169–182.

[20] Bahnemann DW, Hilgendorff M, Memming R. Charge carriers dynamic at TiO_2 particles: Reactivity of free and trapped holes. J. Phys. Chem. B 1997;101:4265–4275.

[21] Serpone N. Relative photonic efficiencies and quantum yields in heterogeneous photocatalysis. EPA Newsletter 1996;58:54–81.

[22] Serpone N, Lawless D, Khairutdinov R, Pelizzetti E. Subnanosecond relaxation dynamics in TiO_2 colloidal sols (Particle size = 1.0-13.4 nm). Relevance to heterogeneous photocatalysis. J. Phys. Chem. 1995;99:16655–16661.

[23] a) Al-Ekabi H, Serpone N. Kinetic studies in heterogenoeous photocatalysis. Photocatalytic degrdation of chlorinated phenols in aereated aqueous solutions over TiO_2 suported on a glass matrix. J. Phys. Chem. 1988;92:5726–5731.

b) Al-Ekabi H, Serpone N, Pelizzetti E, Minero C, Fox MA, Barton R. Kinetics studies in heterogeneous photocatalysis. TiO_2-mediated degradation of 4-chlorophenol alone and in three-component mixture of 4-chlorophenol, 2,4-dichlorophenol and 2,4,5-trichlorphenol in air-equilibrated aqueous media. Langmuir 1989;5:250–255.

[24] Bandala ER, Arancibia C, Orozco S, Estrada C. Solar photoreactors comparison based on oxalic acid photocatalytic degradation. Sol. Energy 2004;77:509–512.

[25] Arancibia C, Bandala ER, Estrada C. Radiation absorption and rate constants for carbaryl photocatalytic degradation in a solar collector. Catal. Today 2002;76:149–159.

[26] Curco D, Gimenez J, Addardak A, Cervera S, Esplugas S. Effects of radiation absorption and catalyst concentration on the photocatalytic degradation of pollutants. Catal. Today 2002;75:177–188.

[27] Alfano OM, Bahnemann D, Cassano AE, Dillert R, Goslich R. Photocatalysis in water environments using artificial and solar light. Catal. Today 2000;58:199–230.

[28] Cassano AE, Alfano OM. Reaction engineering of suspended solid heterogeneous photocatalytic reactors. Catal. Today 2000;58:167–197.

[29] Brandi R, Alfano OM, Cassano AE. Evaluation of radiation absorption in slurry photocatalytic reactors 1. Assessment of methods in use and new proposals. Environ. Sci. Technol. 2000;34:2631–2639.

[30] Bae E, Choi W. Highly enhanced photoreductive degradation of perchlorinated compounds on dye-synthetized metal/TiO_2 under visible light. Environ. Sci. Technol. 2003;37:147–152.

[31] Lin YM, Tseng YH, Huang JH, Chao CC, Chen CC, Wang I. Photoctalytic activity for degradation of nitrogen oxides over visible light responsive titania-based photocatalyst. Environ. Sci. Technol. 2006;40:1616–1621.

[32] Hoffmann MR, Martin ST, Choi W, Bahnemann DW. Environmental applications of semiconductor photocatalysis. Chem. Rev. 1995;95:69–96.

[33] Martin ST, Lee AT, Hoffmann MR. Photocatalytic degradation of pesticide-acaricides in aqueous suspensions of TiO_2. Environ. Sci. Technol. 1995;29:2567–2573.

[34] Minero C, Pelizzetti E, Malato S, Blanco J. Large solar plant photocatalytic water decontamination: Effect of operational parameters. Sol. Energy 1996;56:421–428.

[35] Srinivasan C, Somasundaram N. Bactericidal and detoxification effects of irradiated semiconductor catalyst, TiO_2. Curr. Sci. India 2003;85:1431–1438.

[36] Popa M, Diamandescu L, Vasiliu F, Teodorescu CM, Cosoceanu V, Baia M, Feder M, Baia L, Danciu V. Synthesis, structural characterization and photocatalytic properties of iron-doped TiO_2 aerogels. J. Mater. Sci. 2009;44:358–364.

[37] Pelaez MA, de la Cruz AA, Stathatos E, Falaras P, Dionysiou DD. Visible light activated N-F-codoped TiO_2 nanoparticles for the photocatalytic degradation of microcystin-LR in water. Catal. Today 2009;144:19–25.

[38] Sato S. Photocatalytic Activity of NOx-Doped TiO_2 in the Visible Light region. Chem. Phys. Lett. 1986;123:126–128.

[39] Asah R, Morikawa T, Ohwaki T, Aoki L, Taga Y. Visible-light photocatalyst in nitrogen-doped titanium dioxides. Science 2001;293:269–271.

[40] Sato S, Nakamura R, Abe S. Visible-light Sensitization of TiO_2 Photocatalysts by Wet-method N doping. App. Catal. A-Gen. 2005;284:131–137.

[41] Morikawa T, Asahi R, Ohwaki T, Aoki K, Taga Y. Band-gap Narrowing of Titanium Dioxide by Nitrogen Doping. Jpn. J. Appl. Phys. 2, 2001;40:L561–L563.

[42] Premkumar J. Development of Super-Hydrophilicity on Nitrogen-Doped TiO_2 Thin Film Surface by Photoelectrochemical Method under Visible Light. Chem. Mater. 2004;16:3980–3981.

[43] Emeline, A.V.; Kuznetsov, V.N.; Rybchuk, V.K.; Serpone, N. Visible light active titania photocatalyst: The case of N-doped TiO_2s-properties and some fundamental issues. Int. J. Photoenergy, 2008, 1-19.

[44] Wang ZP, Cai WM, Hong XT, Zhao XL, Xu F, Cai CG. Photocatalytic Degradation of Phenol in Aqueous Nitrogen-doped TiO_2 Suspensions with Various Light Sources. Appl. Catal. B-Environ. 2005;57:223–231.

[45] Miyauchi M, Ikezawa A, Tobimatsu H, Irie H, Hashimoto K. Zeta Potential and Photocatalytic Activity of Nitrogen Doped TiO_2 Thin Films. Phys. Chem. Chem. Phys. 2004;6:865–870.

[46] Burda C, Lou YB, Chen XB, Samia ACS, Stout J, Gole JL. Enhanced Nitrogen Doping in TiO_2 Nanoparticles. Nano Letters 2003;3:1049–1051.

[47] Gole JL, Stout JD, Burda C, Lou YB, Chen XB. Highly Efficient Formation of Visible Light Tunable $TiO_{2-x}N_x$ Photocatalysts and their Transformation at the Nanoscale. J. Phys. Chem. B 2004;108:1230–1240.

[48] Kisch H, Macyk W. Visible-light Photocatalysis by Modified Titania. ChemPhysChem 2002;3:399–400.

[49] Torres GR, Lindgren T, Lu J, Granqvist CG, Lindquist SE. Photoelectrochemical study of nitrogen-doped titanium dioxide for water oxidation. J. Phys. Chem. B 2004;108:5995–6003.

[50] Li Q, Wu P, Xie R, Shang JK. Enhanced photocatalytic disinfection of microorganisms by transition metal ion modification of nitrogen doped titanium oxide. J. Mater. Res. 2010;25:167–176.

[51] a) Wu P, Xie R, Shang JK. Enhanced visible-light photocatalytic disinfection of bacteria spores by palladium-modified nitrogen-doped titanium oxide. J. Am. Ceram. Soc. 2008;91:2957–2962.

b) Wu P, Xie R, Imlay K, Shang JK. Visible light induced bactericidal activity of titanium dioxide co-doped with nitrogen and silver. Environ. Sci. Technol. 2010;44:6992–6997.

[52] Sun C, Li Q, Gao S, Cao L, Shang JK. Enhanced photocatalytic disinfection of *Escherichia coli* bacteria by silver modification of nitrogen-doped titanium oxide nanoparticle photocatalyst under visible-light illumination. J. Am. Ceram. Soc. 2010; 93:3880–3885.

[53] Ubonchonlakate K, Sikong L, Saito F. Photocatalytic disinfection of *P. aeruginosa* bacterial Ag-doped TiO_2 film. Procedia Eng. 2012;32:656–662.

[54] Nair RG, Roy J, Samdarshi SK, Mukherjee AK. Mixed phase V doped titania shows high photoactivity for disinfection of *Escherichia coli* and detoxification of phenol. Sol Energ. Mat. Sol. C. 2012;105:103–108.

[55] Mo AC, Xu W, Xian S, Li Y, Bai S. Antibacterial activity of silver-hydroxypatite/titania nanocomposite coating on titanium against oral bacteria. Key Eng. Mat. 2007;330/332:455–458.

[56] Sheel, D.W.; Brook, L.A.; Ditta, I.B.; Evans, P.; Foster, H.A.; Steele, A.; Yates, H.M. Biocidal silver and silver/titania composite films grown by chemical vapor deposition. *Int. J. Photoenergy*, 2008, *article ID 168185*, 11p.

[57] Chen WJ, Tsai PJ, Chen YC. Functional Fe_3O_4/TiO_2 core/shell magnetic nanoparticles as photokilling agents for pathogenic bacteria. Small 2008;4:485–491.

[58] Lu JW, Li FB, Guo T, Lin LW, Hou MF, Liu TX. TiO_2 photocatalytic antifungal technique for crop diseases control. J. Environ. Sci.-China 2006;18:397–401.

[59] Erkam A, Bakir U, Karakas G. Photocatalytic microbial inactivation over Pd doped SnO_2 and TiO_2 thin films. J. Photochem. Photobiol. A 2006;184:313–321.

[60] Rodriguez-Gonzalez V, Alfaro SO, Torrez-Martinez LM, Cho SH, Lee SW. Silver-TiO_2 nanocomposites: Synthesis and harmful algae bloom UV photoelimination. Appl. Catal. B-Environ. 2010;98:229–234.

[61] Foster HA, Ditta IB, Varghese S, Steele A. Photocatalytic disinfection using titanium dioxide: spectrum and mechanism of antimicrobial activity. Appl. Microbiol. Biot. 2011;90:1847–1868.

[62] Reginfo JA, Pulgarin C. Photocatalytic activity of N, S co-doped and N-doped commercial anatase TiO_2 powders towards phenol oxidation and E. coli inactivation under simulated solar light irradiation. Sol. Energy 2010;84:37–43.

[63] Li D, Ohashi N, Hishita S, Kolodiazhnyi T, Haneda H. Origin of visible-light-driven photocatalysis: A comparative study on N/F doped and N-F codoped TiO_2 powders by means of experimental characterizations and theoretical calculations. J. Solid. State Chem. 2005;178:3293–3302.

[64] Huang DG, Liao SJ, Liu JM, Dang J, Petrik L. Preparation of visible-light responsive N-F codoped TiO_2 photocatalyst by a sol-gel-solvothermal method. J. Photoch. Photobio. A 2006;184:282–288.

[65] Xie Y, Li Y, Zhao X. Low temperature preparation and visible light induced catalytic activity of anatase N-F codoped TiO_2. J. Mol. Catal. A-Chem. 2007;277:119–126.

[66] Choi H, Antoniou MG, Pelaez M, de la Cruz AA, Shoemaker JA, Dionysiou DD. Mesoporous Nitrogen-doped TiO_2 for the Photocatalytic Destruction of the Cyanobacterial Toxin Microcystin-LR under Visible Light. Environ. Sci. Technolo. 2007;41:7530–7535.

[67] Wong MS, Chu WC, Sun DS, Huang HS, Chen JH, Tsai PJ, Lin NT, Yu MS, Hsu SF, Wang SL, Chang HH. Visible-light-induced bactericidal activity of a nitrogen-doped titanium photocatalyst against human pathogens. Appl. Environ. Microb. 2006;72:611–6116.

[68] Liu Y, Li J, Qiu X, Burda C. Bactericidal activity of nitrogen-doped metal oxide nanocatalysts and the influence of bacterial extracellular polymeric substances (EPS). J. Photoch. Photobio. A 2007;190:94–100.

[69] Wu P, Xie R, Imlay JA, Shang JK. Visible light induced photocatalytic inactivation of bacteria by composite photocatalysts of palladium oxide and nitrogen-doped titanium oxide. Appl. Catal. B 2009;88:576–581.

[70] Balasubramanian G, Dionysiou DD, Suidan MT, Baudin I, Laine JM. Evaluating the activities of immobilized TiO_2 powder films for the photocatalytic degradation of organic contaminants in water. Appl. Catal. B-Environ. 2004;47:73–84.

[71] Gelover S, Mondragon P, Jimenez A. Titanium dioxide sol-gel deposited over glass and its applications as photocatalyst for water decontamination. J. Photoch. Photobio. A 2004;165:241–246.

[72] Guillard C, Disdier J, Monnet C, Dussaud J, Malato S, Blanco J, Maldonado MI, Herrmann JM. Solar efficiency of a new deposited titania catalyst: chlorophenol, pesticide and dye removal applications. Appl. Catal. B-Environ. 2003;46:319–332.

[73] Chuangchote S, Jitputti J, Sagawa T, Yoshikawa S. Photocatalytic Activity for Hydrogen Evolution of Electrospun TiO_2 Nanofibers. ACS Appl. Mater Interfaces 2009;5:1140–1143.

[74] Wang J, Zhao L, Lin V, Lin Z. Formation of various TiO_2 nanostructures form electrochemically anodized titanium. J. Mater. Chem. 2009;19:3682–3687.

[75] Hahn R, Macak JM, Schmuki P. Rapid anodic growth of TiO_2 and WO_3 nanotubes in fluoride free electrolytes. Electrochem. Commun. 2007;9:947–952.

[76] Kamat, P. TiO_2 Nanostructures: Recent Physical Chemistry Advances. The J. Phys. Chem. C, 2012, 116, 11849–11851.

[77] Zhang H, Quan X, Chen S, Zhao H. Fabrication and Characterization of Silica/Titania Nanotubes Composite Membrane with Photocatalytic Capability. Environ. Sci. Technol. 2006;40:6104–6109.

[78] Sun W, Yu Y, Pan H, Gao X, Chen Q, Peng L. Cds Quantum Dots Sensitized TiO_2 Nanotube-Array Photoelectrodes. J. Am. Chem. Soc. 2008;130:1124–1125.

[79] Liu Z, Zhang X, Nishimoto S, Jing M, Tryk D, Murakami T, Fujishima A. Highly Ordered TiO_2 Nanotube Arrays with Controllable Length for Photoelectrocatalytic Degradation of Phenol. J. Phys. Chem. C 2008;112:253–259.

[80] Zhang Z, Yuan Y, Shi G, Fang Y, Liang L, Ding H, Jin L. Photoelectrocatalityc Activity of Highly Ordered TiO_2 Nanotube Arrays Electrode for Azo Dye Degradation. Environ. Sci. Technol. 2007;41:6259–6263.

[81] Macak J, Schmidt-Stein F, Schmuki P. Efficient oxygen reduction on layers of ordered TiO_2 nanotubes loaded with Au nanoparticles. Electrochem. Commun. 2007;9:1783–1787.

[82] Paramasivam I, Macak J, Schmuki P. Photocatalytic activity of TiO_2 nanotube layers loaded with Ag and Au nanoparticles. Electrochem. Commun. 2008;10:71–75.

[83] Yu H, Wang X, Sun H, Huo M. Photocatalytic degradation of malathion in aqueous solutions using an Au-Pd-TiO_2 nanotube film. J. Hazard. Mater. 2010;184:753–758.

[84] Zhuang H, Lin C, Lai Y, Sun L, Li J. Some Critical Structure Factors of Titaniun Oxide Nanotube Array in its Photocatalytic Activity. Environ. Sci. Technol. 2007;41:4735–4740.

[85] Zhao G, Cui X, Liu M, Li P, Zhang Y, Cao T, Li H, Lei Y, Liu L, Li D. Electrochemical Degradation of Refractory Pollutant Using a Novel Microstrured TiO_2 Nanotubes/Sb-Doped SnO_2 electrode. Environ. Sci. Technol. 2009;43:1480–1486.

[86] Li P, Zhao G, Cui X, Zhang Y, Tang Y. Constructing Stake Structured TiO_2-Nts/Sb-doped SnO_2 Electrode Simultaneously with High Electrocatalytic and Photocatalytic Performance for Complete Mineralization of Refractory Aromatic Acid. J. Phys. Chem. C 2009;113:2375–2383.

[87] Zhao X, Liu H, Qu J. Photoelectrocatalytic degradation of organic contaminants at Bi_2O_3/TiO_2 nanotube array electrode. Applied Surface Science. 2011;257:4621–4624.

[88] Ng J, Zhang X, Zhang T, Pou J, Du JA, Sun D. Construction of self-organized free-standing TiO_2 nanotube arrays for effective disinfection of drinking water. J. Chem. Technol. Biotechnol. 2010;85:1061–1066.

[89] Li J, Zhen L, Li L, Shi G, Xian Y, Jin L. Photoelectro-Synergistic Catalysis at Ti/TiO_2/PbO_2 Electrode and Its Applications on Determination of Chemical Oxygen Demand. Electroanal. 2006;18:2251–2256.

[90] Zlamal M, Macak J, Schmuki P, Krýsa J. Electrochemically assisted photocatalysis on self-organized TiO_2 nanotubes. Electrochem. Commun. 2007;9:2822–2826.

[91] Jeon T, Choi W, Park H. Photoelectrochemical and Photocatalytic Behaviors of Hematite-Decorated Titania Nanotube Arrays: Energy Level Mismatch versus Surface Specific Reactivity. J. Phys. Chem. C 2011;115:7134–7142.

[92] Zhang Y, Zhao G, Lei Y, Li P, Li M, Jin Y, Lv B. CdS-Encapsulated TiO_2 Nanotube Arrays Lidded with ZnO Nanorod Layers and Their Photoelectrocatalytic Applications. ChemPhysChem. 2010;11:3491–3498.

[93] Baram N, Starosvetsky D, Starosvetsky J, Epshtein M, Armon R, Ein-El Y. Enhanced inactivation of *E. coli* bacteria using immobilized porous TiO_2 photoelectrocatalysis. Electrochim. Acta 2009;54:3381–3386.

[94] Hou Y, Li H, Zhao Q, Chen G, Raston CL. Role of hydroxyl radicals and mechanisms of *Escherichia coli* inactivation on Ag/AgBr/TiO_2 nanotube array electrode under visible light irradiation. Environ. Sci. Technol. 2012;46:4042–4050.

[95] Egerton TA. Does PEC by TiO_2 work? J. Chem. Technol. Biotechnol. 2011;86:1024–1031.

[96] Tiwari DK, Behari J, Sen P. Application of nanoparticles in waste water treatment. World. Appl. Sci. J. 2008;3:417–433.

[97] Khin MM, Nair AS, Babu VJ, Murugan R, Ramakrishna S. A review on nanomaterials for environmental remediation. Energy Environ. Sci. 2012;5:8075–8109.

[98] Brar SK, Verma M, Tyagi RD, Surampalli RY. Engineered nanoparticles in wastewater and wastewater sludge – Evidence and impacts. Waste Manage. 2010;30:504–520.

[99] Hajipour MJ, Fromm KM, Ashkarran AA, de Aberastruri DJ, de Larramendi IR, Rojo T, Serpooshan V, Parak WJ, Mahmoudi M. Antibacterial properties of nanoparticles. Trends Biotechnol. 2012;30:499–511.

[100] Li Q, Mahendra S, Lyon DY, Brunet L, Liga MV, Li D, Alvarez PJJ. Antimicrobial nanomaterials for water disinfection and microbial control: Potential applications and implications. Water Res. 2008;42:4591–4602.

[101] Taylor E, Webster TJ. Reducing infections through nanotechnology and nanoparticles. Int. J. Nanomedicine 2011;6:1463–1473.

[102] Koper O, Klabunde J, Marchin G, Klabunde KJ, Stoimenov P, Bohra L. Nanoscale Powders and Formulations with Biocidal Activity Toward Spores and Vegetative Cells of Bacillus Species, Viruses, and Toxins. Curr. Microbiol. 2002;44:49–55.

[103] Ravishankar RV, Jamuna BA. Nanoparticles and their potential application as antimicrobials. In: Mendez-Vilas A, editor. *Science against microbial pathogens: communicating current research and technological advances*. Extremadura, Spain: Formatex Research Center; 2011. p 197–209.

[104] Gokulakrishnam R, Ravikumar S, Raj JA. *In vitro* antibacterial potential of metal oxide nanoparticles against antibiotic resistant bacterial pathogens. Asian Pac. J. Trop. Med. 2012;2:411–413.

[105] Sadiq M, Chowdhury B, Chandrasekaran N, Mukherjee A. Antimicrobial sensitivity of *Escherichia coli* to alumina nanoparticles. Nanomed. Nanotechnol. 2009;5:282–286.

[106] Mukherjee A, Mohammed SI, Prathna TC, Chandrasekaran N. Antimicrobial activity of aluminum oxide nanoparticles for potential clinical applications. In: Mendez-Vilas A, editor. *Science against microbial pathogens: communicating current research and technological advances*. Extremadura, Spain: Formatex Research Center; 2011. p 245–251.

[107] Jiang W, Mashayekhi H, Xing B. Bacterial toxicity comparison between nano- and micro-scaled oxide particles. Environ. Pollut. 2009;157:1619–1625.

[108] Balasubramanyam A, Sailaja N, Mahboob M, Rahman MF, Hussain SM, Grover P. In vitro mutagenicity assessment of aluminium oxide nanomaterials using the *Salmonella*/microsome assay. Toxicol. in Vitro 2010;24:1871–1876.

[109] Fahmy B, Cormier SA. Copper oxide nanoparticles induce oxidative stress and cytotoxicity in airway epithelial cells. Toxicol. in Vitro 2009;23:1365–1371.

[110] Karlsson HL, Cronholm P, Gustafsson J, Möller L. Copper oxide nanoparticles are highly toxic: a comparison between metal oxide nanoparticles and carbon nanotubes. Chem. Res. Toxicol. 2008;21:1726–1732.

[111] Manusadzianas L, Caillet C, Fachetti L, Gylyte B, Grigutyte R, Jurkoniene S, Kartonas R, Sadauskas K, Thomas F, Vitkus R, Férard J-F. Toxicity of copper oxide nanoparticle suspensions to aquatic biota. Environ. Toxicol. Chem. 2012;31:108–114.

[112] Ren G, Hu D, Cheng EWC, Vargas-Reus MA, Reip P, Allaker RP. Characterization of copper oxide nanoparticles for antimicrobial applications. Int. J. Antimicrob. Ag. 2009;33:587–590.

[113] Rupareli JP, Chatterjee AK, Duttagupta SP, Mukherji S. Strain specificity in antimicrobial activity of silver and copper nanoparticles. Acta Biomater 2008;4:707–771.

[114] Hrenovic J, Milenkovic J, Daneu N, Kepcija RM, Rajic N. Antimicrobial activity of metal oxide nanoparticles supported onto natural clinoptilolite. Chemosphere 2012;88:1103–1107.

[115] Franklin NM, Rogers NJ, Apte SC, Batley GE, Gadd GE, Casey PS. Comparative toxicity of nanoparticulate ZnO, bulk ZnO, and $ZnCl_2$ to a freshwater microalga (*Pseudokirchneriella subcapitata*): the importance of particle solubility. Environ. Sci. Technolo. 2007;41:8484–8490.

[116] Gordon T, Perlstein B, Houbara O, Felner I, Banin E, Margel S. Synthesis and characterization of zinc/iron oxide composite nanoparticles and their antibacterial properties. Colloid. Surface A 2011;1/8:374–379.

[117] Liu Y, He L, Mustapha A, Li H, Hu ZQ, Lin M. Antibacterial activities of zinc oxide nanoparticles against *Escherichia coli* O157:H7. J. Appl. Microbiol. 2009;107:1193–1201.

[118] Jin T, Sun D, Su Y, Zhang H, Sue HJ. Antimicrobial efficacy of zinc oxide quantum dots against *Listeria monocytogenes*, *Salmonella enteritidis* and Escherichia coli O157:H7. J. Food. Sci. 2009;74:46–52.

[119] Brayner R, Ferrari-Iliou R, Brivois N, Djediat S, Benedetti MF, Fievet F. Toxicological impact studies based on Escherichia coli bacteria in ultrafine ZnO nanoparticles colloidal medium. Nano Letters 2006;6:866–870.

[120] Huang Z, Zheng X, Yan D, Yin G, Liao X, Kang Y, Yao Y, Huang D, Hao B. Toxicological effect of ZnO nanoparticles based on bacteria. Langmuir 2008;24:4140–4144.

[121] Sharma D, Rajput J, Kaith BS, Kaur M, Sharma S. Synthesis of ZnO nanoparticles and study of their antibacterial and antifungal properties. Thin Solid Films 2010;519:1224–1229.

[122] Nair MG, Nirmala M, Rekha K, Anukaliani A. Structural, optical, photo catalytic and antibacterial activity of ZnO and Co doped ZnO nanoparticles. Mater. Lett. 2011;65:1797–1800.

[123] Jones N, Ray B, Ranjit KT, Manna AC. Antibacterial activity of ZnO nanoparticle suspensions on a broad spectrum of microorganisms. FEMS Microbiol. Lett. 2008;279:71–76.

[124] Thill A, Zeyons O, Spalla O, Chauvat F, Rose J, Auffan M, Flank AM. Cytotoxicity of CeO_2 nanoparticles for *Escherichia coli*: Physico-chemical insight of the toxicity mechanism. Environ. Sci. Technol. 2006;40:6151–6156.

[125] Negahdary M, Mohseni G, Fazilati M, Parsania S, Rahimi G, Rad S, Rezaei-Zarchi S. The antibacterial effect of cerium oxide nanoparticles on *Staphylococcus aureus* bacteria. Ann. Biol. Res. 2012;3:3671–3678.

[126] Kuang Y, He X, Zhang Z, Li Y, Zhang H, Ma Y, Wu Z, Chai Z. Comparison study on the antibacterial activity of nano- or bulk-cerium oxide. J. Nanosci. Nanotechnol. 2011;11:4103–4108.

[127] Tran N, Mir A, Mallik D, Sinha A, Nayar S, Webster TJ. Bactericidal effect of iron oxide nanoparticles on *Staphylococcus aureus*. Int. J. Nanomedicine 2010;5:277–283.

[128] Chatterjee S, Bandyopadhyay A, Sarkar K. Effect of iron oxide and gold nanoparticles on bacterial growth leading towards biological application. J. Nanobiotechnol. 2011;9:1–7.

[129] Haney C, Rowe JJ, Robinson JB. SPIONs increase biofilm formation by *Pseudomonas aeruginosa*. J. Biomater. Nanobiotechnol. 2012;3:508–518.

3

Cu_2O-BASED NANOCOMPOSITES FOR ENVIRONMENTAL PROTECTION: RELATIONSHIP BETWEEN STRUCTURE AND PHOTOCATALYTIC ACTIVITY, APPLICATION, AND MECHANISM

LIANGBIN XIONG[1], HUAQING YU[1], XIN BA[2], WENPEI ZHANG[2], AND YING YU[2]

[1] *School of Physics and Electronic-Information Engineering, Hubei Engineering University, Xiaogan, China*

[2] *Institute of Nanoscience and Nanotechnology, College of Physical Science and Technology, Central China Normal University, Wuhan, China*

3.1 INTRODUCTION

Photocatalytic organics degradation and H_2 evolution by using semiconductor photocatalysts have attracted considerable interest since the pioneering work of Fujishima and Honda [1], who proved that water can be photoelectrochemically decomposed into hydrogen and oxygen by using a semiconductor (TiO_2) electrode under ultraviolet (UV) irradiation. Photocatalysis for water purification and wastewater treatment was also extensively studied, and it was proven to be a cost-effective, safe, and promising alternative since Matsunaga et al. [2] first reported the efficiency of photocatalytic oxidation of *Saccharomyces cerevisiae* (yeast), *Lactobacillus acidophilus*, *Escherichia coli*, and *Chlorella vulgaris* (green algae) in water using a Pt–TiO_2 photocatalyst upon illumination with near-UV light in 1985. In the past few decades, improvements in photocatalytic efficiency have focused on the following two aspects: (1) increasing the absorption range of the solar spectrum and enhancing quantum efficiency by modifying TiO_2, for example, with noble metals [3–10], by doping, [11–14] and by coating with other semiconductors (Fe_2O_3 [15, 16], WO_3 [17–19], V_2O_5 [20], Cu_2O [21], CdS [22]), which were done in the late 1980s to develop second-generation TiO_2 photocatalysts that could absorb both UV (290–400 nm) and visible (400–700 nm) light and thereby enhance the overall efficiency; (2) screening other substitute materials of photocatalytic semiconductors, for example, CdS [23], Cr_2O_3 [24], BiOX (X = Cl, Br, and I) [25], $SrTiO_3$ [26], ZnS [27], Fe_2O_3 [28], ZnO [29], and Cu_2O [30].

Copper (I) oxide (Cu_2O, cuprous oxide) was the first substance known to behave as a semiconductor, together with selenium. In 1916, photoconductivity of cuprous oxide was observed. After that, its semiconductor properties were investigated. Historically, the first real solar cell with Cu_2O was fabricated by the end of 1920 [31, 32]. Rectifier diodes based on this material were used industrially as early as 1924, and most theories on semiconductors were developed using the data on Cu_2O-based devices. Before the application of silicon, the use of Cu_2O in the transistor industry was reported. In the early 1950s, doping of silicon and germanium was discovered; these elements became the standard, and the attention on Cu_2O declined.

As a matter of fact, during the oil crisis of the 1970s, significant research effort was devoted to the study of alternative energy supplies, and Cu_2O, a very potential candidate material for solar energy conversion, came back into the limelight after several decades. In 1998, Hara et al. [30] reported that illuminated Cu_2O particles acted as photocatalysts in the evolution of H_2 and O_2 from water and that the particles displayed long-term (up to 1900 h) stability. Since then, extensive investigations on the photocatalytic

Nanomaterials for Environmental Protection, First Edition. Edited by Boris I. Kharisov, Oxana V. Kharissova, and H. V. Rasika Dias.

activity of Cu_2O have been carried out. The reasons Cu_2O is considered a potential solar-driven photocatalyst are as follows: (i) it is nontoxic; (ii) it is composed of two very abundant elements; (iii) it has a good absorption coefficient for light with wavelength above 580 nm [33] and high theoretical solar cell conversion efficiency (up to 18%) [34]; and (iv) it is easy to prepare. However, it is well known that Cu_2O can easily be deactivated by photocorrosion [35], although it was reported that photocatalytic splitting of water on Cu_2O powder proceeded without any noticeable deactivation for more than 1900 h [30]. Stability of Cu_2O has been studied by some researchers [36–41]. It is very clear that Cu_2O can be oxidized easily if photogenerated holes cannot be scavenged efficiently because it is thermodynamically favorable for photogenerated holes to oxidize Cu_2O to CuO. In order to improve stability and photocatalytic activity of Cu_2O, numerous efforts have been undertaken on Cu_2O modification for its application in the degradation of organic pollution, photocatalytic disinfection, and photosplitting of water. In this chapter, based on the work done by other groups and in our lab, Cu_2O-based nanocomposites for environmental protection have been reviewed for disinfection and degradation of organic pollutants, with a focus on the structural characteristics of Cu_2O and its modification, the relationship between structure and photocatalytic activity, and photocatalytic mechanism.

3.2 STRUCTURAL FEATURE AND Cu_2O MODIFICATION

3.2.1 Structural Feature of Cu_2O

Cu_2O has a very simple structure that crystallizes in a high-symmetry cubic lattice, as shown in Figure 3.1 [42]. The crystal structure of Cu_2O is called the cuprite structure [43], which can be viewed as two sublattices, a body-centered cubic (bcc) sublattice of oxygen anions (Fig. 3.1a), and a face-centered cubic (fcc) sublattice of copper cations (Fig. 3.1b). The oxygen atoms occupy tetrahedral interstitial positions relative to the copper sublattice, a simple cubic Bravais lattice with the symmetry of the 224th space group $(O_h^4, Pn\bar{3}m)$, so that oxygen is tetrahedrally coordinated by copper. However, copper is linearly coordinated by two neighboring oxygen atoms. These low coordination numbers are very unusual for metal oxides and only two other substances—Ag_2O and Pb_2O—have this crystal structure. Also, a short copper–oxygen bond length in Cu_2O is unusual, which is strongly suggestive of metal–metal (Cu–Cu) bonding.

In order to gain a deeper insight into the chemical bonding of Cu_2O, Gordienko et al. [44] proposed the mechanism responsible for the formation of Cu–O bonds and possible direct Cu–Cu bonds. The d states of the metal hybridize not only with the p states of the anion but also with the s and p orbitals of the metal itself. The elucidation of the mechanism responsible for s–d and p–d hybridizations is of particular importance in explaining the mechanism of chemical bonding between atoms with filled d shells. As mentioned earlier, Cu_2O has a cuprite structure, in which oxygen is tetrahedrally coordinated by copper, whereas copper is linearly coordinated by two neighboring oxygen atoms. The linear coordination of copper to oxygen and the short interatomic distances in crystals cannot be explained by simple models of interaction between O^{2-} and Cu^+ ions with a completely filled 4d shell but can be interpreted only if it is assumed that copper atoms form a direct Cu–Cu bond. Several experimental and theoretical works have been dedicated to the study of the electronic structure of Cu_2O [45–48]. As shown in Figure 3.2, the valence band (VB) of Cu_2O is mainly formed by Cu 3d levels, together with a small contribution from 4s orbitals, and the hybridization of Cu 3d levels and empty Cu 4s levels forms the conduction band (CB) [49]. The same parity of the bands suggests that Cu_2O is a semiconductor with a direct gap at

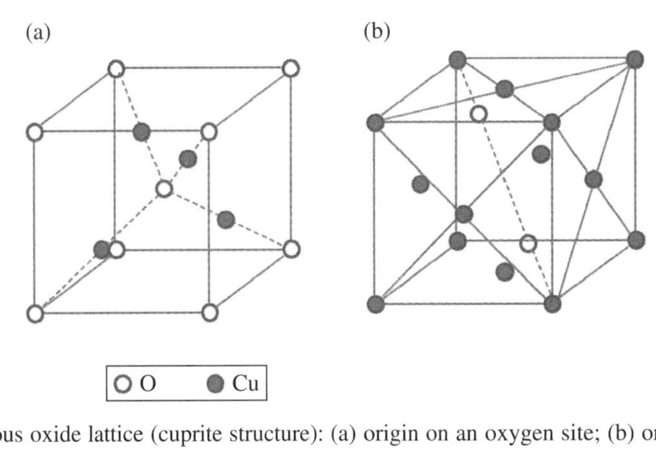

FIGURE 3.1 Unit cell of cuprous oxide lattice (cuprite structure): (a) origin on an oxygen site; (b) origin on a copper site. Dashed lines represent the bonds whereas solid lines are only a guide for the eyes. This structure contains two Cu_2O formula per unit cell [42].

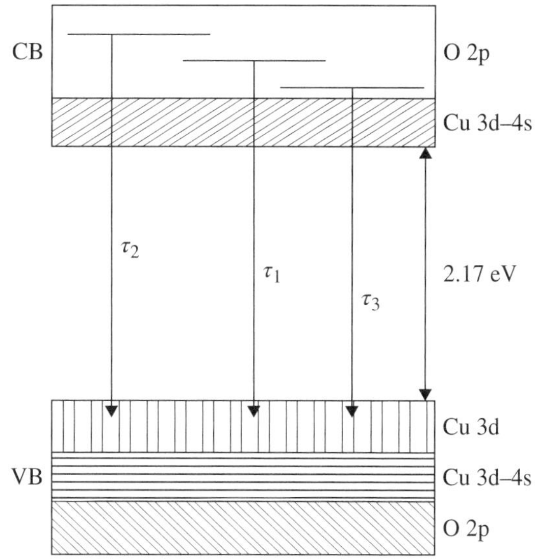

FIGURE 3.2 Schematic diagram of the energy band of bulk Cu_2O semiconductor. Reproduced by permission from Ref. [45]. © 2010, American Institute of Physics.

the center of the Brillouin zone (Γ point) [42]. The experimental band gap is about 2.0 eV [33, 50]. Theoretical calculations using local density functional theory (LDFT) yielded values between 0.45 and 0.8 eV [51], while Hartree–Fock (H–F) calculations reported a value of 9.7 eV [48]. It is well known that LDFT underestimate band gaps while H–F calculations overestimate them.

The special crystal structure of Cu_2O results in an energy band configuration different from other common metal oxide semiconductors. The top of the VB of common metal oxide semiconductors is usually composed of an O2p orbit with a local area distribution charge so that its band gap can only be narrowed by doping with other elements. As for Cu_2O, the Cu^+ ion electronic structure ends up with $3d^{10}4s^0$, in which the energy for 4s orbitals is only slightly higher than that for the 3d levels. So, the band gap of Cu_2O can be widened by reducing the interaction of Cu d–d of the upper VB, and the band gap of Cu_2O can also be adjusted by doping with different elements to change Cu–Cu interaction [52].

3.2.2 Modification

Though Cu_2O has been reported to be used as a stable photocatalyst with exposing (111) facets [53], Cu(I) in Cu_2O is unstable because it is easily oxidized to Cu(II) by photogenerated holes or another oxidant, or it might be reduced to Cu(0) by photogenerated electrons. The thin-layer CuO on the surface of Cu_2O can prevent further photocorrosion of Cu_2O to form a stable Cu_2O/CuO core/shell structure [36], but it also reduces the photocatalytic activity of Cu_2O because CuO has a less narrow band gap (1.4 eV [46]), which suggests a weak redox ability and high combination rate of photogenerated electrons and holes. Additionally, though Cu(I) in Cu_2O may be reduced to Cu(0) by photogenerated electrons, it cannot easily happen since photogenerated electrons are dominantly captured by oxygen in the process of photocatalysis [40]. In the contrary, Cu(0) is usually oxidized to Cu(I) to form a passivation layer of Cu_2O if Cu(0) is exposed to air [54]. Thus, in order to improve stability of Cu_2O, it is very important to prevent Cu_2O from photocorrosion by holes. Doping [55] and hybridization [56] have been proven to be effective in the improvement of photocatalytic activity and stability of narrow semiconductors. Our theoretical calculations and experimental results show that Zn-doped CdS [55] indeed had much higher photocatalytic activity and stability than pristine CdS. So, in order to improve photocatalytic activity and stability for Cu_2O, it is promising to modify Cu_2O by doping it with other elements, coating with other semiconductors, and morphology control.

3.2.2.1 Doping It is known that impurities in semiconductors play a fundamental role in determining the electronic quality of a semiconductor. Even a very small concentration of impurities can significantly alter the properties of a semiconductor, and therefore a systematic study of the most common impurities and their effects as a function of concentration is necessary. The existence of Cu–Cu interactions in the cubic Cu_2O structure plays a key role in the unusual behavior of Cu_2O. Through doping, these interactions can be disrupted and the band gap for Cu_2O may be changed.

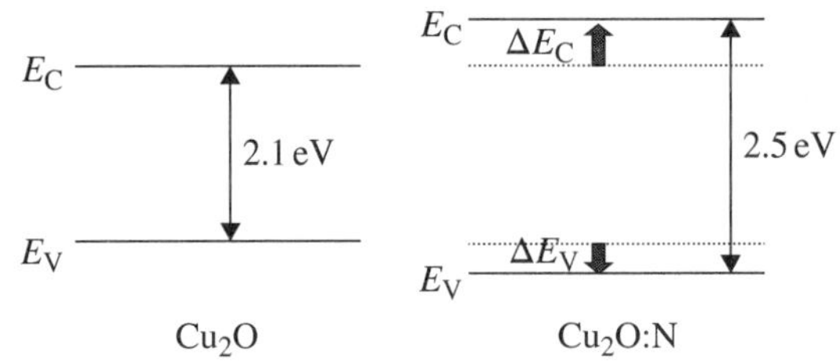

FIGURE 3.3 Energy level diagram of Cu_2O reference film and Cu_2O:N film with N-doping concentration of 2.9%. Reproduced by permission from Ref. [57]. © 2009, American Institute of Physics.

Band structure and electric properties of doped Cu_2O are extensively investigated based on the first principles of DFT. Nolan et al. [52] proposed two conditions related to the dopant that are key for increasing the band gap. The first condition is that dopants with ionic radii (La^{3+}, Sr^{2+}, and Ba^{2+} dopants) larger than Cu^+ will result in a larger band gap over stoichiometric Cu_2O, while other dopants with ionic radii smaller than Cu^+ will show no enhancement of the band gap. The second condition is avoiding misalignment of dopant electronic states with Cu_2O bands, which turns out to be the most important aspect of doping. If dopant electronic states interact with the VB or CB of Cu_2O, for example, In^{3+} or Cd^{2+}, or produce defect states in the Cu_2O band gap, for example, Ce^{4+}, the band gap is reduced. Nolan et al. [49] also reported that substitutional cation doping with Al and Au/Ag decreases the band gap of Cu_2O. Martínez-Ruiz [48] have drawn the conclusion that doping with Ag atoms decreases the band gap, while doping with Ni results in a p-type semiconductor with impurity levels above the allowed VB maximum and doping with Zn results in an n-type semiconductor with impurity levels above the CB minimum.

N [57–59] and Cl [60–62] elements are mostly experimentally studied as dopants for Cu_2O. Nitrogen is an impurity that can be easily incorporated in Cu_2O at high concentrations [57] and is a very effective p-type dopant [58], in agreement with the hypothesis that nitrogen acts as a substitutional impurity for oxygen atoms. Nakano et al. [57] investigated the effect of N doping for Cu_2O films deposited by reactive magnetron sputtering. N-doped Cu_2O films have a positive shift in the position of the VB edge. Here, the CB shift ΔE_C of the Cu_2O:N films is estimated from the optical band gap ΔE_g and VB shift ΔE_V. More interestingly, the CB edge is found to be negatively shifted, with a value a little more than the ΔE_V. Band-gap diagrams of Cu_2O and Cu_2O:N with N-doping concentration of 2.9% are typically shown in Figure 3.3. In this case, the ΔE_V and ΔE_C are estimated to be 0.17 and 0.23 eV, respectively. Both band edge shifts tend to enlarge with increasing N-doping concentration up to 3%. However, Malerba et al. [63] excluded the band-gap widening observed by Nakano et al. [57]. They studied the optical properties of sputtered Cu_2O thin films doped with nitrogen concentrations between 1 and 2.5 at.%. All of the doped samples exhibited two clearly defined absorption bands at energies below the gap, with an intensity well correlated with the N concentration. This result show that both the two-subband absorption processes are related to optical transitions toward two impurity levels introduced by nitrogen or by more complex defects formed by nitrogen. The current main reason for the studies on Cl-doped Cu_2O is that Cu_2O has potential for application in solar cells [60, 61]. Han et al. [61, 62] found a remarkable improvement in electric conductivity and no significant change in the band gap for Cu_2O. Other nonmetal-doped (such as Si [64, 65]) and transition metal (Cd [66], Ni [67], Co [68], Mo [69])-doped Cu_2O were also investigated. Additionally, Tseng et al. [70] reported that the absorption edges shift to the right (red shift) with an increase in Ag content, which implies shortening of optical band gap due to Ag doping.

3.2.2.2 Hybridization The semiconductor hybridization approach has been shown to be another effective method for improving photocatalytic activity through good photogenerated charge separation with the formation of heterojunction structure. In this configuration, several advantages can be obtained: (i) an improvement of charge separation; (ii) an increase of the charge carrier lifetime; and (iii) an enhancement of the interfacial charge transfer efficiency to adsorbed substance on the surface. As is well known, Cu_2O used as a photocatalyst has low energy-conversion efficiency (<1% [71, 72]), which is due to the fact that the light-generated charge carriers in Cu_2O cannot be efficiently transferred to the surface and are lost during recombination [73]. Moreover, semiconductor materials with band gaps suitable for capturing a significant fraction of the incident solar spectrum energy (≈0.8–2.4 eV) typically suffer from UV-induced photocorrosion [74], as does Cu_2O. Combining Cu_2O composites with other semiconductors or conductive materials may be a good solution to these problems. Cu_2O-based cocatalysts such as Cu_2O/TiO_2 [75, 76], Cu_2O/ZnO [77], Cu_2O/chitosan [78], Cu_2O/multiwalled carbon nanotubes [79], and Au/Cu_2O [80, 81] show enhanced charge separation efficiency. Cu_2O-based heterojunction devices such as Cu_2O/ZnO [82] and Cu_2O/NiO [83] show better electric and photo properties or higher quantum efficiency than pure Cu_2O. Among these composites, the Cu_2O/TiO_2 system has attracted the most extensive research. The band-gap energies of Cu_2O and TiO_2 are around 2.0 and 3.2 eV, respectively. As shown in Figure 3.4, the CB minimum and the VB maximum of Cu_2O lie above those of TiO_2. Therefore, the

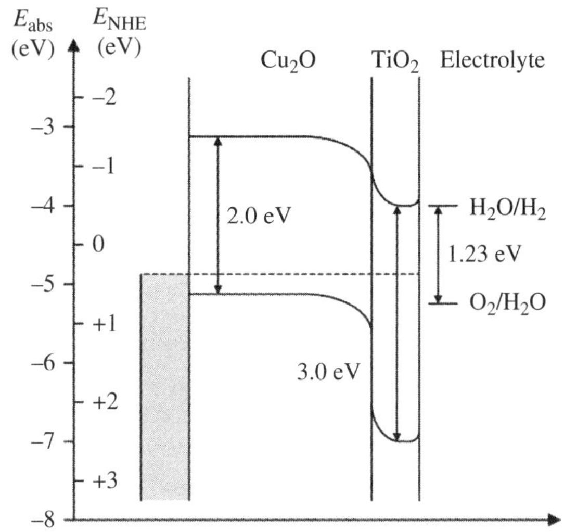

FIGURE 3.4 Energy band diagram for the Cu_2O/TiO_2 heterostructure. The TiO_2 layer is very thin and considered fully depleted with negligible band bending at the electrolyte interface compared to the Cu_2O/TiO_2 interface. Reproduced by permission from Ref. [56]. © 2002, Elsevier Science B.V.

electrons excited to the CB of Cu_2O can transfer to TiO_2, whereas the holes generated in the VB in TiO_2 prefer opposite transfer to Cu_2O. Charge carriers separated in different semiconductors effectively reduce the chance of electron–hole pair recombination and also prolong their lifetime, thereby increasing quantum efficiencies. In addition, the working wavelength range extends to a visible region due to the absorption of visible light (VL) by Cu_2O, which further enhances the efficiency of solar energy transition. Synergy between these effects enables the Cu_2O/TiO_2 system to exhibit great potential for solar cell and photocatalysis applications. Additionally, the stability of Cu_2O and heterojunctions thereof are significantly affected by the material that Cu_2O is coupled with. For example, Cu_2O heterojunctions with In_2O_3, SnO_2, and ZnO were reported to be unstable, while Cu_2O–CdO junctions were relatively stable [84, 85]. And a thin layer of TiO_2 deposited on p-type Cu_2O successfully protects the Cu_2O film against photocorrosion, suggesting that TiO_2 may be one of the best candidates to couple with CuO and/or Cu_2O to enhance stability [56].

3.2.2.3 ***Morphology Control*** Because of the anisotropy of crystals, people found different crystal faces usually exhibit different properties, such as stability and activity. In the past few decades, great achievements have been made by investigating the chemical properties of definite crystal faces by employing bulk single crystals. Many photocatalysts have been observed to show different photoactivities on different faces; for example, the {001} facets of anatase TiO_2 show higher activity than the {101} facets [86], the {001} facets of ZnO display high activity than that of other facets [87], the {110} facets of Ag_3PO_4 exhibit higher surface energy than that of {100} planes [88], and the tetrahexahedral Pt and Au nanocrystals enclosed by 24 {037} or {122} facets possess excellent electro-oxidation activity [89]. The study of the properties of definite crystal facets not only helps us to precisely control their structures and shapes, but also offers an opportunity for discovering multifunctional materials with potentially exciting and unique properties. Hence, it is necessary to develop effective and facile methods for the controllable synthesis of Cu_2O as well as to investigate corresponding growth mechanisms. Cu_2O nanocrystals are relatively easy to prepare and low in cost because of abundant copper sources, simple preparation process, and low energy consumption. They can also form a wide variety of morphologies. Various interesting Cu_2O nanostructures such as rods [90], wires [91, 92], spheres [93–95], flowers [96], towers [97], flower leaves [98], dendritic structures [99], cacti [100], whiskers [101], bipyramids [102], eight-pod cubes, six-armed star-like structures [103], different polyhedral structures (such as cubes [104–106], octahedra [107–109], 50-facet microcrystals [110], truncated octahedrons [111], triangular nanoplates [112]), and different hollow structures (such as spheres [113–116], cubes, [117] truncated rhombic dodecahedral Cu_2O nanocages and nanoframes [118]) have been synthesized.

Here, we pay more attention to the synthesis of Cu_2O nanostructures with well-defined morphologies with sharp facets. These nanostructures possess well-defined facets and sharp edges and are ideal for facet-dependent property studies. Hollow Cu_2O nanostructures are also reviewed here. These are usually synthesized directly without the use of templates or through the dissolution of preformed solid Cu_2O nanocrystals. For this reason, it is possible that they have well-defined morphologies such as cubic, octahedral, and other complex but symmetrical shapes.

3.2.2.3.1 Controllable Synthesis of Cu_2O with Different Polyhedral Structures A number of studies have described the synthesis of Cu_2O nanocubes. Guo and Murphy [119] prepared Cu_2O nanocubes with edge lengths of approximately 450 nm by mixing a solution of $CuSO_4$, cetyltrimethylammonium bromide (CTAB) surfactant, sodium ascorbate, and NaOH at a reaction temperature of 55°C. Wang et al. [104] synthesized uniform crystalline Cu_2O cube in high yields by reducing the copper–citrate complex solution with glucose. A series of shape evolutions for Cu_2O particles from the transient species such as multipod and star-shaped particles to cubic crystals have been captured based on transmission electron microscope (TEMic) and scanning electron microscope (SEMic) observations. It was concluded that a higher growth rate along the [111] direction induces shrinking of the eight {111} facets, while six {100} facets remain to form Cu_2O cubes because of their lower growth rate. Kim et al. [120] heated a solution of ethylene glycol at 140°C and added NaCl, $Cu(NO_3)_2$, and poly(vinyl pyrrolidone) (PVP) to synthesize Cu_2O nanocubes with an average edge length of 410 nm. Chloride ions were found to play a pivotal role in the formation of nanocubes, and only polycrystalline spheres were generated in the absence of chloride ions.

Reports primarily on the synthesis of octahedral Cu_2O nanocrystals are also available. Zhang et al. [107] successfully prepared monodispersed submicron-sized Cu_2O octahedra in large quantities assisted by the capping reagent PVP-K30 (MW = 58,000). The mechanism of Cu_2O octahedra formation can be explained as the cooperative effect coming from the growth units of the anion coordinative polyhedra theoretical model and polymer selective adsorption. He et al. [108] prepared monodispersed cuprous oxide octahedron nanocrystals with sizes smaller than 100 nm under controlled conditions. The method is based on the reduction of copper nitrate in Triton X-100 water-in-oil (w/o) microemulsions by γ-irradiation. The average edge length of the octahedron-shaped nanocrystals varies from 45 to 95 nm as a function of the dose rate. Guo et al. [109] described a simple electrochemical route for the controlled synthesis of a Cu_2O microcrystal from perfect octahedra to monodispersed colloid spheres via the adjustment of electrodeposition potential without the introduction of any template or surfactant. High yields (~100%) of perfect Cu_2O octahedra and monodispersed colloid spheres were obtained. In their electrochemical synthesis system, $Cu(OH)_4^{2-}$ exists in the following equilibrium condition 3.1:

$$Cu(OH)_4^{2-} \leftrightarrow Cu^{2+} + 4OH^- \tag{3.1}$$

Cu^{2+} can be slowly reduced to Cu_2O in an alkaline solution under a certain potential range (from −0.5 to −0.6 V). Also, the OH^- ions might be selectively adsorbed on (111) facets of Cu_2O crystals, and it may slow the growth rates along the [111] direction, which results in the final morphology of octahedra with (111) facets. As a comparison, when the electrodeposition potential was changed to −0.4 V for 15 min, no particles could be obtained. It can be concluded that the reduction potential of $Cu(OH)_4^{2-}/Cu_2O$ can be estimated at −0.4 V versus Ag/AgCl (saturated KCl), which is slightly higher than the potential range employed (from −0.5 to −0.6 V). Thus, $Cu(OH)_4^{2-}$ will gradually reduce in this case. However, changing the electrodeposition potential to −0.7 V will improve the reaction kinetics and result in an increase in the reaction rates. Thus, it is hard to adjust the growth of certain facets of Cu_2O via OH^- adsorption. Therefore, monodispersed Cu_2O colloid spheres are finally produced. In addition, the potential-dependent adsorption of OH^- on the preformed Cu_2O facets {100} and {111} is probably a key factor. When the potential is moved to a more negative one, fewer OH^- ions are adsorbed on the Cu_2O surface because the surface concentration of OH^- decreases, leading to the formation of spherical particles without preferential growth on certain facets.

The synthesis of polyhedral Cu_2O with high-index facets is of great interest since it may exhibit higher chemical activities in practical applications. Leng et al. [110] reported a solution-based approach for the preparation of unusual polyhedral 50-facet Cu_2O microcrystals, whose surfaces are enclosed by high-index {311} facets in addition to low-index {100}, {110}, and {111} facets. The formation of the 50 facets can be geometrically viewed as the truncation of all 24 vertices of a small rhombicuboctahedron with 26 facets (Fig. 3.5). During growth in solutions, the anisotropic growth rates along the [100], [110], and [111] directions might be responsible for the formation of this morphology. The Miller index of the 24 nearly isosceles trapezoids could be assigned to the {311} planes based on geometrical analysis, and this was verified by simulated models using the WinXmorph software and supported by TEMic and electron diffraction (ED) observations. Sun et al. [121] described the synthesis of highly symmetric multifaceted polyhedral 50-facet polyhedral Cu_2O crystals with high-index {522} facets and polyhedral architectures Cu_2O crystals with high-index {544} and {211} planes via a template-free complex–precursor solution route. Later, they [122] presented another route for the synthesis of highly symmetric polyhedron-aggregated multifaceted Cu_2O homogeneous structures composed of a cubic core and many abridged polyhedral building blocks with different index facets (including high-index {522} facets and low-index {111}, {100}, and {110} facets). Wang et al. [123] developed a facile wet chemical method to synthesize truncated concave octahedral Cu_2O microcrystals mainly enclosed by {*hhl*} high-index facets using an aqueous solution of $Cu(CH_3COO)_2$, sodium dodecyl sulfate (SDS) surfactant, NaOH, and D-(+)-glucose reductant. In this system, SDS was found to be crucial for the formation of the concave Cu_2O microcrystals, and the crystal growth rate also affected the morphology of Cu_2O microcrystals. Wang et al. [124] described a facile strategy for crystal engineering of Cu_2O polyhedrons with high-index facets developed in the system of $CuSO_4$/NaOH/ascorbic acid. Novel micron-sized Cu_2O 74-facet polyhedrons with the exposure of

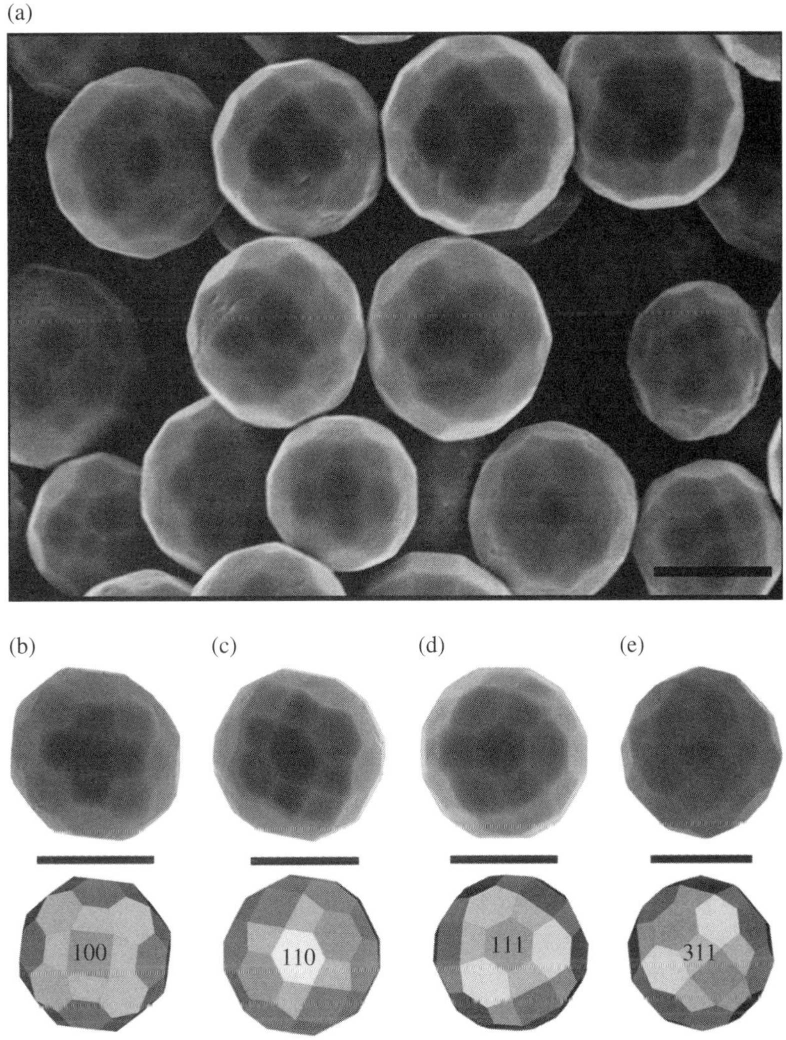

FIGURE 3.5 Typical FESEM images of as-synthesized polyhedral 50-facet Cu_2O microcrystals (a) and observed and simulated (bottom) individual polyhedra with different polygons facing up (b)–(e). Scale bar: 2.0 μm. Reproduced by permission from Ref. [110]. © 2010, American Chemical Society.

high-index {744} and {211} facets (Fig. 3.6) were first achieved in the process of systematic shape evolution. According to the facile strategy, the morphologies of the products could be precisely tuned from cubic, 18-facet, and 26-facet polyhedrons exposing low-index facets to the 50- and 74-facet polyhedrons exposing high indexes of {211}, {522}, and {744}, by simply altering the NaOH concentration. Based on the control experiments, it was found that high NaOH concentration could promote crystal growth along the [100] direction and increase the ratio (R) of the growth rates along the [100] and [111] directions, which caused the shape evolution of the Cu_2O polyhedrons with different index facets. Figure 3.7 shows the SEMic images and the corresponding simulated structures of a series of Cu_2O polyhedrons by changing the concentration of NaOH from 1.5 to 9.5 M in the aqueous solution of $CuSO_4$ and ascorbic acid. Figures 3.7a and b shows the SEMic images of Cu_2O cubes when the NaOH concentration was set as 1.5 M. Figure 3.7c presents the corresponding simulated structure of the Cu_2O cube with exposed {100} facets, which are square facets. As the NaOH concentration increased to 2.5 M, the 18-facet polyhedron was obtained (Fig. 3.7d and e). Compared with the Cu_2O cube (Fig. 3.7c), the emerging new facets in the 18-facet Cu_2O crystal are {110} facets hexagon facets, which grow between any two adjacent {100} facets. The corresponding simulated {110} facets are hexagon facets, which are shown in Figure 3.7f. When the NaOH concentration was elevated to 4 M, the 26-facet polyhedron evolved along with the {111} facets (Fig. 3.7g and h). The {111} facets in Figure 3.7i are presented by triangles, grown among the relatively newer facets {110} shown in Figure 3.7f. As the NaOH concentration reached 6.5 M, a 50-facet Cu_2O crystal exposing {522} facets was formed (Fig. 3.7j and k), and every four {522} facets, given by hexagons adjacent to squares facets, grows around every single (100) facet

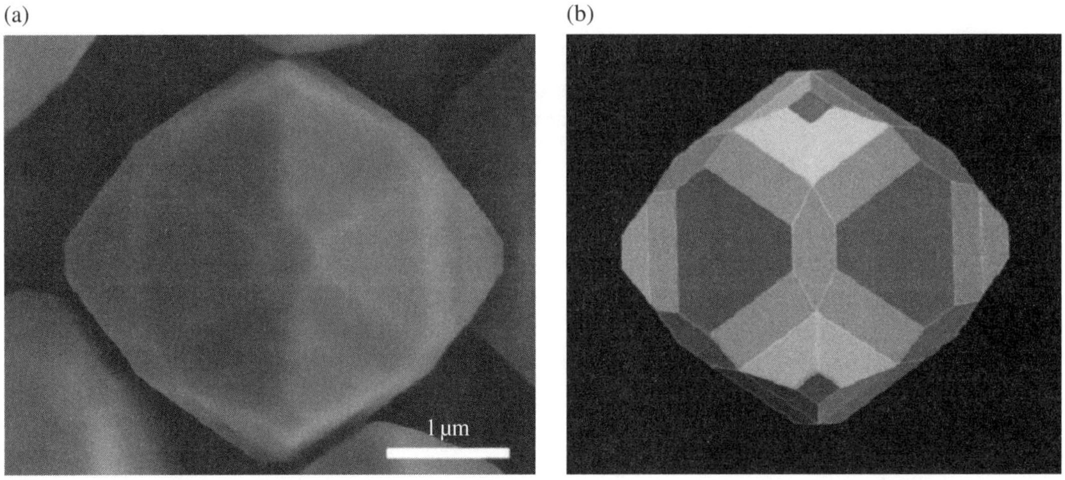

FIGURE 3.6 SEMic image (a) and the corresponding simulated structure (b) of a single74-facet Cu_2O crystal: squares for {100}, hexagons for {110}, truncated triangles for {111}, long quadrangles for {211} and isosceles trapezoids for {744}. Reproduced by permission from Ref. [124]. © 2013, The Royal Society of Chemistry.

(Fig. 3.7l). Because there are 6 (100) facets in a 50-facet Cu_2O crystal, 24 {522} facets are observed in a 50-facet Cu_2O crystal. When the NaOH concentration further increased to 8 M, another type of 50-facet Cu_2O crystal exposing {211} facets was obtained (Fig. 3.7m and n). The new exposing {211} facets shown as long hexagons adjacent to squares facets also grow around every single (100) facet (Fig. 3.7o). Finally, when NaOH concentration reached 9.5 M, a well-defined 74-facet Cu_2O polyhedron with exposing {744} facets was produced (Fig. 3.7p and q). The newest {744} facets given by isosceles trapezoids grew between a single (100) facet and 4 newer {211} facets (Fig. 3.7r), so there are 24 {744} facets and 24 {211} facets in a single 74-facet Cu_2O crystal.

3.2.2.3.2 Controllable Synthesis of Cu_2O with Hollow Structures The general method for chemical synthesis of hollow nanostructures is template strategy, using either a hard or a soft template. In addition, templateless synthesis through self-assembly of nanoparticles (NPs) is also achievable for forming hollow materials [125]. The use of templates that serve as cores is for the subsequent growth of shells. The cores need to be removed by dissolution, etching, or thermal treatment. Templateless synthesis by self-construction, which is based on the ability of spherical nanocrystallites to stabilize certain crystallographic planes as they undergo two-dimensional (2D) and then 3D aggregation, represents another mechanism for fabricating hollow nanostructures [118]. Zhang et al. [113] reported a simple liquid reduction approach to prepare a Cu_2O hollow microsphere film and hollow nanosphere powder with $Cu(OH)_2$ nanorods as the precursor and ascorbic acid as the reductant at 60°C. When a $Cu(OH)_2$ nanorod array film grown on a copper foil was used as a precursor, a Cu_2O thin film made up of hollow microspheres with average diameter of 1.2 mm was successfully prepared. When the $Cu(OH)_2$ nanorods were scraped from the copper foil and then used as the precursor, Cu_2O hollow nanosphere powder with average diameter of 270 nm was obtained. Xu et al. [115] synthesized the nanosized Cu_2O/poly (ethylene glycol 400) (PEG-400) composite hollow spheres (HSs, 50–80 nm in diameter) with mesoporous shells of approximately 15–20 nm. In the hollow nanostructures, the polymer content is about 18.1 wt.%, and the mean sizes of the component nanocrystals and the pore diameter are about 5 and 3.8 nm, respectively. In the fabrication process, PEG-400 molecules self-assemble to form micelles, which act as templates for the formation of hollow structures. PEG also acts as a reducing agent, solvent, and complexing agent. The formation of mesoporous structures is due to the oriented aggregation of composite NPs. Xu et al. [116] successfully prepared multishelled structures as well as a single-crystalline shell wall Cu_2O by two stages. In the first stage, Cu_2O NPs (2–5 nm) underwent a 2D-oriented attachment under the direction of CTAB multilamellar vesicles to form the primary well-crystallized porous-shelled hollow spheres, which serve as the base framework for the single-crystalline multishelled hollow spheres (MHSs). In the second stage, densification of this porous shell wall via Ostwald ripening took place to achieve compact multishelled hollow spheres with a single-crystalline shell wall. Remarkably, the formation of the porous shell wall is important for the formation of the multishelled hollow spheres for providing exchange channels for reactants. Pang and Zeng [93] successfully fabricated monodispersed Cu_2O spheres with diameters of 130–135 nm. Using the Cu_2O spheres thus prepared as solid precursors, uniform hollow CuS and CuSe derivatives have also been synthesized. More importantly, a range of 2D and 3D superlattices of Cu_2O, CuS, and CuSe solid/hollow spheres have been assembled for the first time. Xu et al. [117] assembled single-crystalline hollow Cu_2O nanocubes with well-defined shape and tunable size (50–200 nm) (Fig. 3.8) by a PEG-200-assisted precursor hydrolysis process in aqueous solution at room

FIGURE 3.7 And the corresponding simulated structures of Cu_2O crystals were expressed with different polygons in Fig. 1c, f, i, l, o and r: square facets for {100} facets, hexagons for {110} facets, triangles for {111} facets, hexagons adjacent to squares facets for {522} facets, long hexagons adjacent to squares facets for {211} facets, and isosceles trapezoids for {744} facets. Reproduced by permission from Ref. [124].

FIGURE 3.8 TEMic images and SAED patterns (a), (b), HR-TEMic image (c), and FE-SEMic images (d)–(f) of as-prepared products. Reprinted with permission from Ref. [117]. © 2010, American Chemical Society.

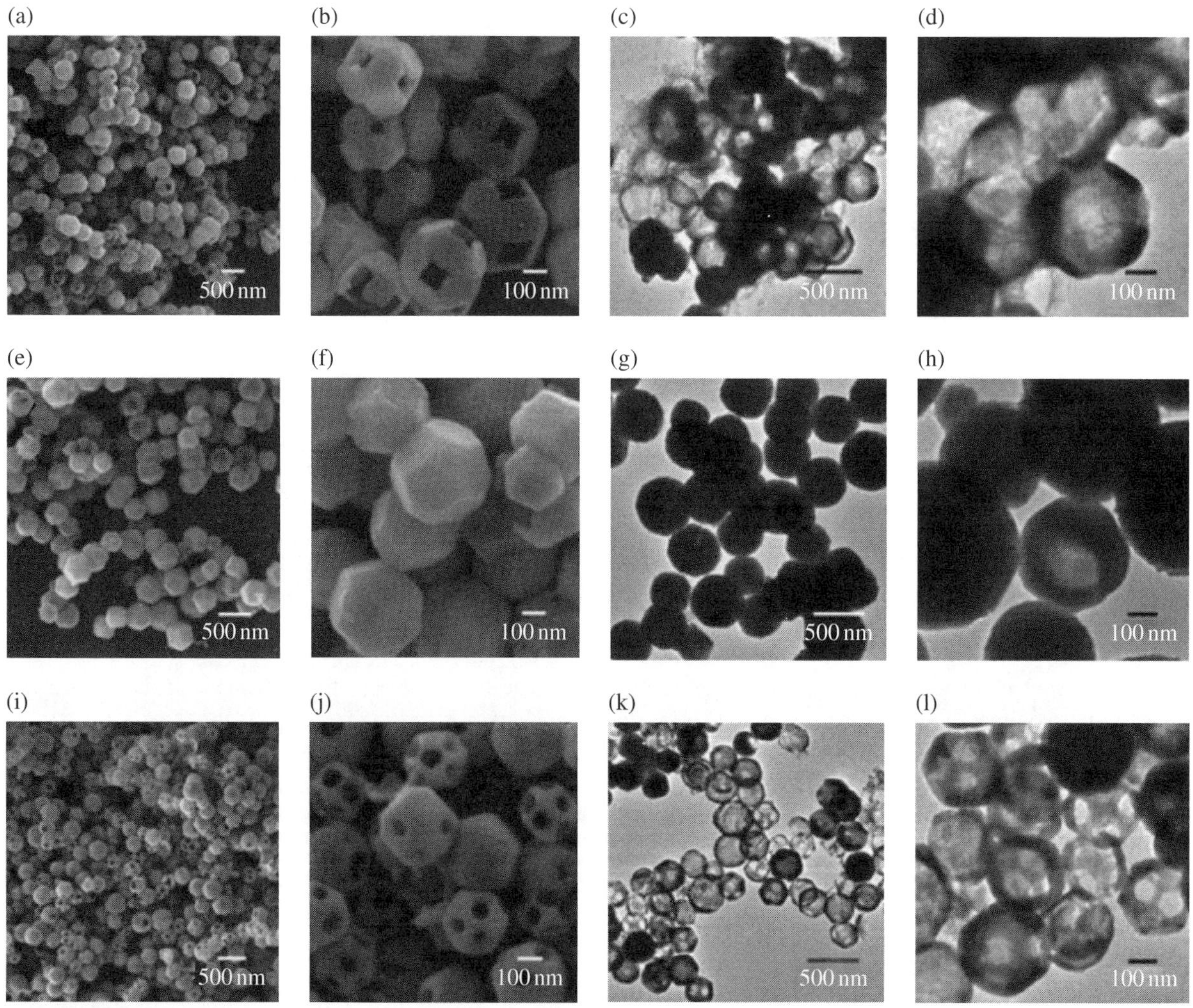

FIGURE 3.9 SEMic and TEMic images of the truncated rhombic dodecahedral Cu_2O nanoparticles: (a)–(d) type-I nanoframes, (e)–(h) nanocages, and (i)–(l) type-II nanoframes. The magnified images clearly show the hollow structures of the nanoparticles. Reprinted with permission from Ref. [118]. © 2008, American Chemical Society.

temperature. Figure 3.9 displays SEMic and TEMic images of the Cu_2O nanoframes and nanocages. Each truncated rhombic dodecahedral particle contains 12 hexagonal {110} faces and 6 {100} faces. Type I nanoframes are constructed from the hexagonal {110} skeleton. They are 300–350 nm in diameter. The {100} faces are formed in the nanocages so that they have a truncated rhombic dodecahedral morphology. Nanocages have larger diameters (350–400 nm) and thicker walls than type I nanoframes. The added HCl promotes the etching process. When ethanol is added and sonication of the solution is carried out, the adsorption of SDS molecules on the nanocage surfaces may be temporarily disrupted. Removal of SDS facilitates the reaction of Cu_2O and HCl to form $HCuCl_2$. The faster etching rate on the {110} faces than on the {100} faces transforms the nanocages into type II nanoframes with thinner walls and smaller particle sizes. Type II nanoframes are 200–250 nm in diameter. The nanocages are not stable under acidic conditions and collapse after the solution is aged for 6 h.

3.3 Cu_2O-BASED NANOCOMPOSITES FOR ENVIRONMENTAL PROTECTION

Semiconductor photocatalysis has been one of potential solutions to energy shortage and environmental pollution worldwide in recent years. The ability of this advanced oxidation technology has been widely demonstrated to remove persistent organic compounds and microorganisms from water, and produce hydrogen via water splitting, which is not only low cost but environmentally friendly. In this section, we will first discuss Cu_2O-based nanocomposites for environmental protection through the degradation of organic pollutants, focusing on the relationship between structure and photocatalytic activity and photocatalytic mechanism. Then, Cu_2O-based nanocomposites for disinfection and hydrogen evolution will be introduced.

3.3.1 Cu_2O-Based Nanocomposites for Organics Degradation

This section begins with an introduction to facet-dependent photocatalytic activity of Cu_2O crystals. Then, some examples of organics degradation over Cu_2O without sharp crystals is described. Finally, Cu_2O-based nanocomposites for organics degradation is presented. Here, the Cu_2O/TiO_2 nanocomposite, the most extensively investigated Cu_2O-based cocatalyst, will be highlighted by introducing related mechanisms for organics degradation. Particularly, the mechanism for Cu_2O working as a photo-oxygen cathode will be described in detail for photocatalysis due to the formation of photogenerated Fenton reagents in the presence of Fe^{2+}.

3.3.1.1 Facet-Dependent Photocatalytic activity of Cu_2O Crystals As we know, on the one hand, crystallography of a semiconductor photocatalyst determines the semiconductor's electric and energy band structure, which is crucial to its physical and chemical properties. On the other hand, the morphology of the photocatalyst is also important for its photocatalytic properties. Over the past few decades, a lot of work has been done on the chemical properties of definite crystal planes. It is generally considered that the chemical reactivity of crystals can be significantly affected by their shapes, due to surface atom arrangement, bonding, surface energy, and so on. Inorganic single crystals with highly reactive surfaces are therefore desired. Unfortunately, surfaces with high reactivity usually possess high surface energy, and such surfaces tend to shrink rapidly during the crystal growth process to minimize total surface energy. Thus, crystals with a dominant high reactivity surface are of great interest, yet challenging. The facet-dependent photocatalytic activities of Cu_2O nanocrystals have been studied recently. The special O–Cu–O 180° linear coordination makes the crystalline surfaces of {111} and {100} possess distinctive chemical activities.

Xu et al. [126] used the synthesized Cu_2O nanocrystals with well-defined structures and sharp faces to investigate their comparative photocatalytic activity. They found that the octahedral Cu_2O with exposed {111} crystal surfaces show improved ability on adsorption and photodegradation of methyl orange than Cu_2O cubes with exposed {100} crystal surfaces. Two years later, Kuo et al. [127] reported similar results that octahedral Cu_2O crystals with entirely {111} facets are photocatalytically more active than truncated cubic crystals with mostly {100} facets. The {111} planes of Cu_2O are also found to show higher catalytic activity (two times at 30°C) than that of {100} facets on the N-arylation reaction of iodobenzene with imidazole [128]. A crystal model analysis shows that the (100) planes contain oxygen atoms as they do in the unit cell. However, a cut of the unit cell over one of the (111) planes reveals the presence of surface Cu atoms with dangling bonds. This simple comparison indicates that the {111} faces have higher surface energy and are expected to be more catalytically active than the {100} facets. Furthermore, Cu_2O crystals bounded by the {111} facets contain positively charged copper atoms at the surfaces, whereas those bounded by the {100} facets such as the cubes are electrically neutral. This observation suggests that Cu_2O bounded by the {111} facets should interact more strongly with negatively charged molecules, and the corresponding photodegradation of these molecules is more effective. That is why methyl orange, a negatively charged molecule, was first chosen for the photodecomposition experiments. So, Cu_2O has a poor performance with regard to the degradation of a positively charged molecule. Kuo and Huang [129] mentioned in their review paper that when methylene blue, a positively charged molecule, was used for photodegradation experiments, neither the cubes nor the octahedra caused any photodegradation. On the other hand, Cu_2O crystals bounded by the {100} are less sensitive to the charge of the adsorbed molecules and are not photocatalytically active. Pang et al. [130] reported that the efficiency and selectivity of bacteriostatic effects vary with Cu_2O morphology. Octahedral Cu_2O with {111} crystal surfaces exhibits high selectivity and better bacteriostatic effect for five different bacteria than cubic Cu_2O with {100} crystal surfaces. These results are further confirmed by Ren et al. [131], that is, they observed that Cu_2O octahedral crystals exhibited higher activity in inactivating *E. coli* than Cu_2O cubic crystals. They suggested the difference in the antibacterial activity of Cu_2O crystals can be ascribed to the atomic arrangements of different exposed surfaces.

Cu_2O with high-index facets may exhibit higher chemical activities that are of great importance in practical applications. In 2010, Leng et al. [110] reported the synthesis of 50-facet Cu_2O polyhedral microparticles partially enclosed by {311} high-index planes in a water–ethanol mixed alkaline solvent using D-(+)-glucose as a reductant at 60°C. The microcrystals prepared in this manner show a higher specific catalytic rate toward CO oxidation. The observed enhanced specific catalytic rate toward CO oxidation over the polyhedral 50-facet Cu_2O microcrystals is ascribed to the presence of these high-index facets. In order to study the shape-dependent catalytic activity of Cu_2O with high-index planes, Leng et al. [132] selected five of the typical morphologies of Cu_2O, including cubes, octahedra, 18-facet polyhedra, small rhombicuboctahedral, and a 50-facet polyhedral. They found that among the three kinds of particles enclosed by low-index planes, for example, (111), (110), or (100) planes, the octahedral particles show the highest specific catalytic activity followed by rhombic dodecahedral and then cubic particles. The much higher specific catalytic activity observed for the 50-facet polyhedral particles (big 50-facet polyhedra) prepared using D-(+)-glucose as the reductant can be attributed to the presence of (311) high-index planes possessing surface defects or steps, and kink atoms with low coordination numbers. Sun et al. [133] also systematically investigated the crystal-facet-dependent effect of polyhedral Cu_2O microcrystals on photodegradation of MO. The sequence of photodegradation of the MO aqueous solution under UV irradiation in the presence of different polyhedral Cu_2O microcrystals is as follows: 50-facet polyhedral Cu_2O with high-index {522} facets greater than 50-facet polyhedral Cu_2O with high-index {211} facets greater than 26-facet Cu_2O cubes greater than 26-facet polyhedral Cu_2O octahedra. The photocatalytic superiority of the novel 50-facet polyhedral architectures can be attributed to the introduction of highly active components of high-index surfaces, which can offer a higher number of unsaturated Cu dangling bonds and surface oxygen vacancies, and accelerate the formation of highly oxidative •OH radicals, leading to enhanced decomposition of MO dye. This study not only provides convincing evidence that the performance of catalysts can be enhanced by crystal-facet tailoring, but also promotes the synthesis and application of other polyhedral crystals with high-index surfaces in the catalytic field.

Additionally, the catalytic activity of Cu_2O with high-index surfaces was also investigated by Wang et al. [123]. They synthesized truncated concave octahedral Cu_2O microcrystals mainly enclosed by {*hhl*} high-index facets using an aqueous solution of $Cu(CH_3COO)_2$, SDS surfactant, NaOH, and D-(+)-glucose reductant. In this system, SDS was found to be crucial for the formation of concave Cu_2O microcrystals, and the crystal growth rate also affected the morphology of Cu_2O microcrystals. In the catalytic oxidation of CO, truncated concave octahedral Cu_2O enclosed mainly by {332} high-index facets exhibited enhanced catalytic activity in comparison with that of low-index {111} and {100} facets, which can be due to the presence of high-density steps on {332} high-index facets. The order of CO catalytic activities of the crystal facets was found to follow the sequence {332} > {111} > {100}. These results indicate that we can make use of surface-engineering strategies to improve the catalytic activity of materials due to the different activities of different crystal facets.

3.3.1.2 Cu_2O without Sharp Crystals for Organics Degradation Cu_2O without sharp crystals is also important for application in photocatalysis. The photocatalytic activity of materials is closely correlated not only to the crystal plane structure, but also to the surface state and microstructure. It is demonstrated that the Cu_2O nanowhiskers composed of quantum dots have good photocatalytic activity under VL [134]. Tang et al. [135] reported that the highest decolorization ratio of 90% was observed for the Cu_2O sample prepared by the addition of 0.05 g/l CTAB. Zheng et al. [53] compared the efficiency of MO degradation between Cu_2O microcrystals and N-doped TiO_2 (1 g/l). Although the Cu_2O microcrystals prepared thus have lower surface areas, they show much higher photocatalytic activities for MO bleaching. After being irradiated under VL for 60 min, the MO dye was completely bleached over Cu_2O in the first run, while only 40% of MO dye was bleached over N-doped TiO_2. However, as the irradiation time increases, the photocatalytic activity of Cu_2O gradually decreases. After 16 runs of MO bleaching, Cu_2O microcrystals completely transformed into nanosheets, and the photocatalytic activity of nanosheets was only half of that from microcrystals in the first run. However, it still shows higher activity than N-doped TiO_2. Shoeib et al. [136] reported that the decolorization efficiencies of MO reached 98.7% in 30 min without any further decrease in photocatalytic efficiency with increase in the irradiation time to 120 min. Zhang et al. [137] reported that the photocatalytic degradation ratio of methyl orange over Cu_2O microcubes reached 98.1% under natural solar light with constant stirring for 3 h at room temperature. Besides MO, other organics or toxic gases have also been reported as models for the studies on Cu_2O photocatalytic activity. Leng et al. [110]

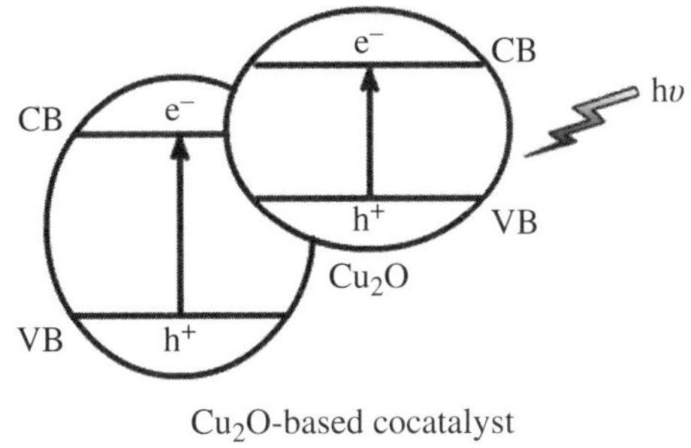

FIGURE 3.10 Scheme of band structure Cu_2O-based cocatalysts.

reported a solution-based approach to prepare unusual polyhedral 50-facet Cu_2O microcrystals with a morphological yield higher than 70%. And the microcrystals prepared in this manner showed a higher specific catalytic rate toward CO oxidation. Active brilliant red X-3B and *p*-chloronitrobenzene were used as models for the studies on Cu_2O photocatalytic activity by our group [134, 138]. The degradation of methylene blue can reach a maximum of 89% after 150-min irradiation due to the synergistic effect of Cu_2O and H_2O_2 [139].

3.3.1.3 Cu_2O-Based Nanocomposites for Organics Degradation Cu_2O-based cocatalysts (such as Cu_2O/TiO_2 [35, 140–145], Cu_2O/ZnO [77, 145, 146], WO_3/Cu_2O [147], Cu_2O/CeO_2 [148], $Cu_2O/BiVO_4$ [149], Cu_2O/Bi_2O_3 [77], Cu_2O/SiC [150], Au/Cu_2O [80, 81, 151], Ag/Cu_2O [152], $Ag/POM/Cu_2O$ [153], $Fe_3O_4/C/Cu_2O$ [154], Cu_2O/polyaniline (PANI) [155], Cu_2O nanocube/polycarbazol [81], Cu_2O/carbon [156], Cu_2O nanocubes and multiwall carbon nanotubes with poly tetrafluoroethylene (PTFE) (Cu_2O/CNTs/PTFE) [157]) have been extensively investigated for photocatalytic organics degradation. Among these Cu_2O-based cocatalysts, bi-semiconductor systems (Cu_2O/TiO_2, Cu_2O/ZnO, WO_3/Cu_2O, Cu_2O/CeO_2, $Cu_2O/BiVO_4$, Cu_2O/Bi_2O_3, and Cu_2O/SiC) have similar band matching structure (Fig. 3.10). The CB of Cu_2O is higher than that of other semiconductors, while the VB of Cu_2O is lower. Thus, photogenerated carriers experience similar transfer process when these bi-semiconductor systems undergo proper irradiation. The generated electrons in Cu_2O and holes in counterpart semiconductors migrate to the CB of counterpart semiconductors and the VB of Cu_2O, respectively. Thus, the photogenerated electron–hole pairs are separated more effectively in the Cu_2O-based cocatalysts. As for the noble metal/Cu_2O composite structure (such as Au/Cu_2O, Ag/Cu_2O), the noble metal/semiconductor architecture usually exhibits improved carrier separation and plasmonic enhancement of VL photocatalytic activity. Three-body composite systems and Cu_2O/organic composite systems are also found to have an improved photocatalytic activity for the synergistic effect at work in these compounds. However, we are not going to enumerate the photocatalytic activity for all Cu_2O-based co-catalysts because of space constraints. Here, we just want to focus on some typical Cu_2O-based cocatalysts: (i) Cu_2O/wide-gap semiconductor (such as Cu_2O/TiO_2, Cu_2O/WO_3), (ii) noble metal/Cu_2O (such as Ag/Cu_2O, Au/Cu_2O), and (iii) Cu_2O/polymer (such as Cu_2O/PANI).

3.3.1.3.1 Cu_2O/Wide-Gap Semiconductor for Organics Degradation The combination of Cu_2O and TiO_2 is most extensively investigated for photocatalytic organics degradation. Our group synthesized Cu_2O/TiO_2 particle composites and Cu_2O/TiO_2 composite films for organic pollutant degradation under UV or VL. In 2004, we [75] synthesized nanosized Cu_2O/TiO_2 particle composites by an electrochemical method and found that the photocatalytic activities of the composites prepared by this method exceed that of Degussa TiO_2 (P25). In 2007, we [141] prepared Cu_2O/TiO_2 composites by a simple electrochemical method, and we coated them on a glass matrix through a spraying method. It was found that the photocatalytic activity of the Cu_2O/TiO_2 composite film in the presence of $FeSO_4$ and EDTA was much higher than that for a similar system with only TiO_2 and Cu_2O film, respectively. Without the presence of $FeSO_4$ and ethylene diamine tetraacetic acid (EDTA), there is no degradation of methylene blue. As can be observed from Figure 3.11, the Cu_2O/TiO_2 composite film with the ratio of $TiO_2:Cu_2O=1:1$ (w/w) has much higher activity for photodegradation of methylene blue under VL than that of the TiO_2, Cu_2O control film, and the composite film with only 3 wt.% Cu_2O, respectively. Furthermore, the photocatalytic activity of the composite film with only 3 wt.% Cu_2O is slightly better than that of the TiO_2 and Cu_2O film. Thus, there should be a synergistic effect in the composite film. Additionally, the ratio of TiO_2 and Cu_2O for the composite is critical for the activity of the composite film. The mechanism for the great improvement of photocatalytic activity of the TiO_2/Cu_2O composite film is proposed by the VB theory. Photodegradation of organic pollutants by a semiconductor is usually completed by photoexcited electrons and holes together. Thus, it is important to improve quantum efficiency by effective electron and hole separation.

The band edge positions of TiO_2 and Cu_2O are presented in Figure 3.4. The internal energy scale is shown on the right for comparison to normal hydrogen electrode (NHE). It is well known that the band gap of TiO_2 is about 3.2 eV and the CB of TiO_2

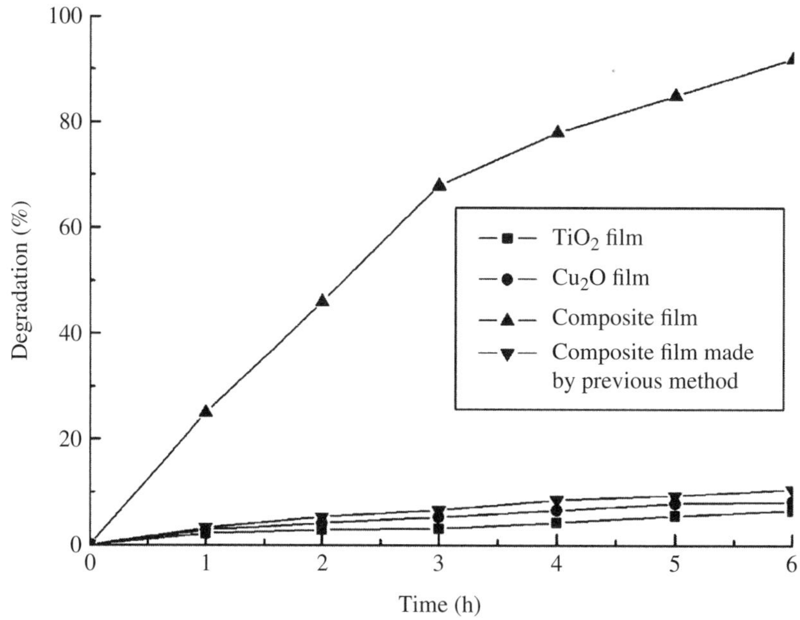

FIGURE 3.11 Photodegradation of methylene blue by TiO_2/Cu_2O composite film (with the ratio of TiO_2 and Cu_2O 1:1 (w/w)), the composite film made by previous method (with only 3 wt.% Cu_2O), TiO_2 film and Cu_2O film, respectively, in the same condition. pH 4, [$FeSO_4$] 0.10 g/l, [EDTA] 0.05 g/l. Reproduced by permission from Ref. [141]. © 2007, American Chemical Society.

is about −0.2 eV [158]. The difference between rutile TiO_2 and anatase TiO_2 is not under consideration here. Cu_2O is one of the semiconductors that have the highest CB. The band gap of Cu_2O is about 2.0 eV, and the CB of Cu_2O is −1.4 eV [159]. Coupling of the two kinds of semiconductors allows the vectorial displacement of electrons from Cu_2O to TiO_2. Thus, under VL excitation, photogenerated electrons resulting from Cu_2O accumulate in the lower-lying CB of TiO_2 to form the Ti^{3+} electron center, whereas holes can accumulate in the VB Cu_2O to form the hole center, which can be consumed during weak oxidation. Accumulated electrons in the CB of TiO_2 can be transferred to oxygen adsorbed on the surface to form O^{2-} or O_2^{2-}, which combines with H^+ to form H_2O_2. Therefore, the probability of photogenerated electrons in TiO_2 to recombine with the holes in the Cu_2O VB is very low, which leads to the enhancement of the photocatalytic activity of Cu_2O. The photocatalytic reaction paths for the TiO_2/Cu_2O composite system are shown in Equations 3.2–3.7. Highly mobile electrons first transfer from the CB of Cu_2O to that of TiO_2. The hole stays in the Cu_2O VB for weak oxidation reaction. Finally, the hydrocarbon resulting from the photodegradation of a pollutant (methylene blue) and Fe^{2+} (which is a kind of effective hole scavenger) causes hole consumption.

$$Cu_2O + h\upsilon \rightarrow (h_{vb}^{+} + e_{cb}^{-}), \tag{3.2}$$

$$e_{cb}^{-} + {>}Ti^{IV}OH \rightarrow {>}Ti^{III}OH \tag{3.3}$$

$$h_{vb}^{+} + \text{hole scavenger} \rightarrow \text{weak oxidation} \tag{3.4}$$

Ti^{3+} acts like e_{cb}^{-} with the reaction of O_2.

$$Ti^{III}OH + O_2 \rightarrow O_2^{\bullet-} + {>}Ti^{IV}OH \tag{3.5}$$

As H_2O_2 is evolved from $O_2^{\bullet-}$, with the existence of an electronic donor, the following equations can be driven.

$$O_2^{\bullet-} + H^{+} \rightarrow HO_2^{\bullet} \tag{3.6}$$

$$HO_2^{\bullet} + H^{+} + e_{cb}^{-} \rightarrow H_2O_2 \tag{3.7}$$

Because the VB of Cu_2O is not high enough, the oxidation ability of holes from Cu_2O under VL to form the peroxide radical is limited. When Fe^{2+} is present in the solution, Fenton reagent with H_2O_2, which evolves according to Equations 3.5, 3.6, and 3.7, is formed, and the advanced oxidation process can start. This is why the TiO_2/Cu_2O composite reported earlier [76] has good photocatalytic activity only under UV light without the presence of Fe^{2+} and EDTA. This photocatalytic reaction path works

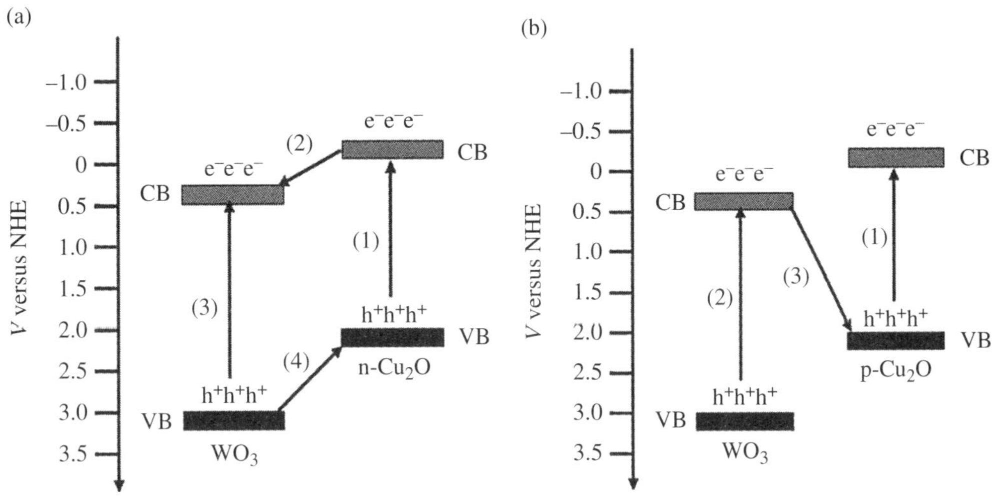

FIGURE 3.12 Schematic diagram for the charge-transfer process in WO_3/n-Cu_2O composite film (a) and in WO_3/p-Cu_2O composite film (b). Reproduced by permission from Ref. [147]. © 2012, Elsevier B.V.

well in the decomposition of organic contaminants. The efficiency for this kind of photocatalytic system is related to the ratio of TiO_2 and Cu_2O in the composite. When the amount of TiO_2 in the composite is very close to that of Cu_2O, almost all of the electrons generated from Cu_2O have place to transfer. If the amount of TiO_2 is much higher than that of Cu_2O, a significant number of the electrons generated from Cu_2O under VL cannot be transferred, leading to low efficiency for dye degradation, or the inability to produce enough H_2O_2.

Wei et al. [147] fabricated WO_3/n-Cu_2O and WO_3/p-Cu_2O composite films on titanium (Ti) substrates with a consecutive cathodic electrodeposition route. They found that the photocatalytic activity of WO_3/n-Cu_2O was higher than that of WO_3 but lower than that of n-Cu_2O. On the contrary, WO_3/p-Cu_2O exhibited higher photocatalytic activity than both WO_3 and p-Cu_2O alone. The suppression of photocatalytic activity for n-Cu_2O when it is combined with WO_3 can be interpreted using a schematic diagram of energy band structure of the WO_3/n-Cu_2O film shown in Figure 3.12a. Under simulated natural light illumination, the electrons in the VB of Cu_2O and WO_3 are excited to their corresponding CB according to processes (1) and (3) shown in Figure 3.12a, respectively. The CB edge of Cu_2O (−0.28 V versus NHE) is higher than that of WO_3 (+0.4 V versus NHE); the VB edges of Cu_2O and WO_3 are situated at +1.92 and +3.1 V versus NHE, respectively [160, 161]. From the thermodynamic viewpoint, the photogenerated electrons can transfer from CB of Cu_2O to that of WO_3 (process (2)), while the photogenerated holes can migrate in the opposite direction from VB of WO_3 to that of Cu_2O (process (4)). Because it is impossible for the photogenerated electrons in the CB of WO_3 to be consumed by the adsorbed oxygen through one-electron reduction, such a charge transfer weakens the stronger reduction power of photogenerated electrons in the CB of Cu_2O. As a result, the WO_3/n-Cu_2O composite film shows photocatalytic activity lower than that of a single n-Cu_2O film. This mechanism can also be used to explain similar results obtained by others for ZnO/Cu_2O and TiO_2/Cu_2O composite photocatalysts [162, 163]. For the WO_3/p-Cu_2O composite film, it is more likely that the WO_3/p-Cu_2O composite film shows a different mechanism for photogenerated charge transfer. This can be explained based on the model of the p–n photochemical diode shown in Figure 3.12b rather than the charge separation model (shown in Fig. 3.12a). In this p–n heterojunction, the majority of electrons in WO_3 and the majority of holes in p-Cu_2O combine by transfer through the interface between the two semiconductors (process (3)), while the recombination of photogenerated charges in the respective semiconductors is suppressed. Consequently, the photogenerated electrons with strong reduction power in the CB of Cu_2O and the photogenerated holes with strong oxidation power in the VB of WO_3 are retained. In such a case, two photons must be absorbed to generate one net electron–hole pair for redox reactions at the photocatalyst surface.

3.3.1.3.2 Noble Metal/Cu_2O for Organics Degradation Semiconductor/noble metal nanocomposites as photocatalysts have been extensively studied [164–167]. It is believed that in these semiconductors/noble metals, photoexcited electrons in the CB can be transferred to noble metals, which act as electron sinks due to the Schottky barrier at the metal/semiconductor interface, while the holes can remain on the semiconductor surface [168]. The recombination of electrons and holes can therefore be prolonged and the photocatalytic efficiency thereby improved [169]. In addition, the surface plasmon resonance of noble metal NPs is expected to enhance the absorption of incident photons, which enhance the photocatalytic efficiency of semiconductors [170].

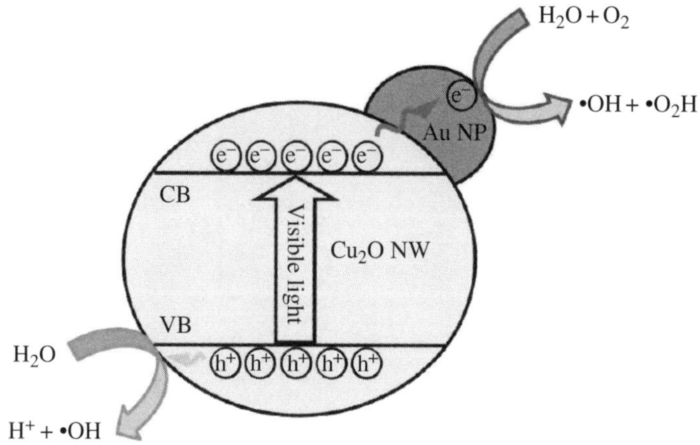

FIGURE 3.13 Scheme of proposed •OH generation process on the surface of Cu_2O-Au nanocomposites. Reproduced by permission from Ref. [151]. © 2012, American Chemical Society.

Pan et al. [151] have developed a facile strategy for coupling Au NPs onto the surface of Cu_2O nanowires. The presence of Au NPs enhances the photodegradation efficiency of Cu_2O nanowires. These Cu_2O/Au nanocomposites display tunable optical properties, and their photocatalytic properties are dependent on the coverage density of Au NPs. The enhanced photocatalytic efficiency is believed to be due to the enhanced light absorption by surface plasmon resonance and the electron sink effect of Au NPs. Charge separation prevents the recombination of electrons and holes (as shown in Fig. 3.13) and thus enhances the photocatalytic activity of Cu_2O nanowires. At the same time, the absorption of VL for semiconductors can be enhanced by coupling with plasmonic metal nanostructures through a local field enhancement effect, leading to an enhanced photocatalytic property.

Ag@Cu_2O core–shell NPs exhibit photocatalytic activity over an extended wavelength range because of the presence of localized surface plasmon resonance in the Ag core in comparison to pristine Cu_2O NPs [151]. The photocatalysis action spectra and transient absorption measurements show that the plasmonic energy is transferred from the metal to the semiconductor via plasmon-induced resonant energy transfer and direct electron transfer simultaneously, which can generate electron–hole pairs in the semiconductor. The localized surface plasmon resonance band of the Ag@Cu_2O core–shell shows a red shift with an increase in the Cu_2O shell thickness, extending the light absorption of Ag@Cu_2O heterostructures to longer wavelengths. As a result, the photocatalytic activity of the Ag@Cu_2O core–shell NPs is varied by the modulation of the shell thickness on the nanometer scale. The Ag@Cu_2O core–shell heterostructure is an efficient VL plasmonic photocatalyst, which allows for tunable light absorption over the entire VL region by tailoring the shell thickness.

3.3.1.3.3 Cu_2O/Polymer for Organics Degradation Besides inorganic quantum dots [171], organic dye molecules [172] have also been used as "photosensitizers." Wang et al. [155] successfully synthesized Cu_2O/(PANI) nanocomposites by a simple, one-step hydrothermal method. The Cu_2O NPs with a diameter of 40 nm disperse in the PANI. The nanocomposites, for the first time, are used as a photocatalyst for the degradation of organic pollutants. The combination of Cu_2O NPs with PANI leads to high photocatalytic activity for the degradation of dyes under VL irradiation, which is much higher than that of TiO_2 and pure Cu_2O. A possible photocatalytic mechanism is proposed to explain the improvement of the photocatalytic activity (Fig. 3.14). The significant enhancement of photocatalytic performance is attributed to the synergistic effect between PANI and Cu_2O. The band gap of Cu_2O is 2.17 eV, and PANI has a band gap of 2.8 eV, which indicates that both Cu_2O and PANI can be excited by VL. After the Cu_2O/PANI composites are synthesized, the two types of materials closely combine together and form interfaces. The Cu_2O in the photocatalyst absorbs photons and excites electron and hole pairs when the system is irradiated with VL. The PANI also absorbs photons to induce π–π^* transition, transporting the excited electrons to the π^*-orbital. Thus, the excited-state electrons produced by PANI are injected into the CB of Cu_2O. Subsequently, simultaneous holes on the VB of Cu_2O migrate to the π-orbital of PANI because of the enjoined electric fields of the two materials. The photoexcited electrons are effectively collected by Cu_2O, and the holes by PANI. The recombination process of the electron–hole pairs is hindered, and charge separation and stabilization are achieved. Therefore, efficient electron–hole separation leads to significant enhancement of photocatalytic dye degradation in the Cu_2O/PANI composites. Based on this understanding, the role played by PANI can be illustrated by injecting electrons into Cu_2O CB under VL illumination and triggering the formation of a very reactive radical superoxide radial

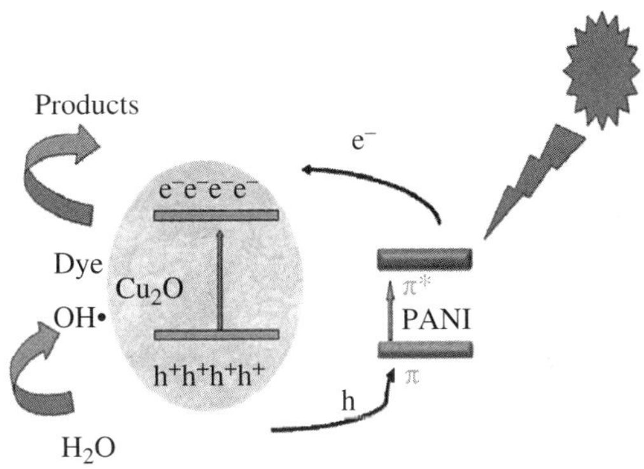

FIGURE 3.14 Scheme of the separation and transfer of photogenerated charge carriers in the Cu_2O/PANI system under visible light irradiation. Reproduced by permission from Ref. [155]. © 2012, Elsevier B.V.

ion O^{2-} and a hydroxyl radical •OH, which are responsible for the degradation of the organic compound. The possible photocatalytic reactions are proposed as follows (Eqs. 3.8–3.12):

$$Cu_2O / PANI + h\upsilon \rightarrow Cu_2O / PANI^+ + e^- \quad (3.8)$$

$$e^- + O_2 \rightarrow O^{2-} \quad (3.9)$$

$$O^{2-} + 2e^- + 2H^+ \rightarrow OH + OH^- \quad (3.10)$$

$$h^+ + (H_2O \leftrightarrow H^+ + OH^-) \rightarrow H^+ + \bullet OH \quad (3.11)$$

$$Dye + \bullet OH \rightarrow \text{oxidation products.} \quad (3.12)$$

3.3.2 Cu_2O-Based Nanocomposites for Water Disinfection

Water disinfection means the removal, deactivation, or killing of pathogenic microorganisms including a variety of helminthes, protozoa, fungi, bacteria, rickettsiae, viruses, and prions. The conventional method for water disinfection is commonly accomplished by the addition of a strong oxidant such as free chlorine. However, the treatment process results in the formation of toxic disinfection by-products (DBPs) such as trihalomethanes and haloacetic acids. Moreover, free chlorine is ineffective in controlling some pathogens like *Cryptosporidium parvum* and *Mycobacterium avium*. UV/chlorine and ozone/chlorine combinations are effective in killing the *C. parvum* oocyst. However, viruses are resistant to both UV and combined chlorine disinfection. Additionally, ozonation and chlorination of water containing bromide can form other unregulated DBPs like haloacetonitriles and iodoacetic acid, which are more toxic and carcinogenic than those associated with free chlorine. Thus, disinfection regulations now require minimization of such by-products, and there is a strong demand for alternative technologies [173].

A promising disinfection method is to enhance the utilization of photons by integrating engineered photocatalytic nanostructures in the treatment system. In 1985, Matsunaga et al. [2] reported the microbicidal effects of TiO_2. Their excellent research work gave rise to extensive research in the field of oxidative photocatalysis, mainly dealing with TiO_2 suspensions and targeting bacteria, viruses, fungi, algae, and protozoa. However, this approach has its own limitations: (i) The efficiency of photocatalytic materials is limited by the recombination of photogenerated charges, and the bacteria need to contact with the catalyst; (ii) the adsorption–desorption problems in the reaction systems have not been fully addressed; (iii) it is still quite challenging to fabricate stable photocatalytic materials with high efficiency in the VL spectral range. The discovery of new photocatalytic materials will be a major driving force for future developments in the field of photocatalyzed disinfection. Novel photocatalysts such as titanium-free materials, coupled binary nanostructures, and even complex ternary composites are of great interest. Copper has a long history as an antimicrobial agent, with its earliest recorded use in 2600 BC [174]. This metal in the form of copper ions [175], metals and alloys [176–178], and oxides (Cu_2O, CuO) [179] has been shown to exhibit excellent antimicrobial activity against a number of microorganisms including bacteria and viruses, yet it is relatively safe for humans [180]. Cu_2O is commercially used in paints in the health care sector and as antifouling surfaces on a ship hull [181]. Crystallography facet-dependent antibacterial activities of Cu_2O were reported by Ren's [131] and Pang' groups [130]. Ren et al. [131] synthesized Cu_2O cubic and octahedral crystal samples and evaluated their bactericidal activity by the inactivation of *E. coli*. Owing to the

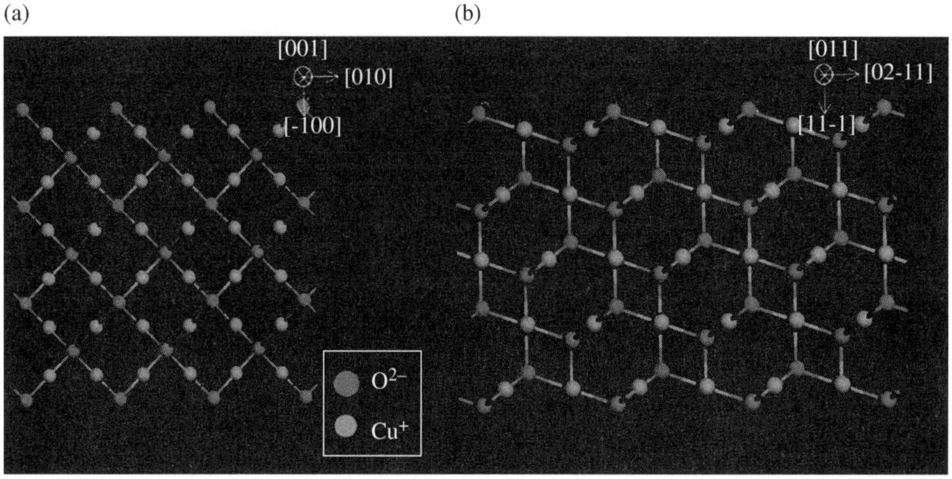

FIGURE 3.15 The atomic arrangement in (100) (a) and (111) (b) planes of the Cu_2O structure, respectively. Reproduced by permission from Ref. [131]. © 2011, American Chemical Society.

different atomic arrangement of the exposed surfaces, Cu_2O octahedral crystals bounded by {111} facets exhibited higher activity in killing *E. coli* than cubic ones bounded by {100} facets. There is a difference in the inactivation activities between Cu_2O cubic crystals and octahedral crystals because of the atomic arrangement of the exposed crystallographic facets. Cu_2O has a cuprite crystal structure, which is composed of a body-centered cubic packing of oxygen atoms with copper atoms occupying half of the tetrahedral sites. As shown in Figure 3.15, the structural arrangement perpendicular to the [100] direction alternates between copper- and oxygen-containing planes. The two layers repeat alternatively to maintain stoichiometry and charge neutrality. An ideal {100} plane should be either copper- or oxygen-terminated. However, considering that the synthesis process was carried out in an aqueous medium, a copper-terminated plane would be rather unstable due to the active interaction with the hydroxyl group. Here, the oxygen-terminated {100} plane is expected for Cu_2O cubic crystals. As regards the [111] direction, an ideal {111} plane possesses hexagonal symmetry. Every Cu-containing plane is sandwiched between two O-containing planes. The three-plane unit repeats to satisfy stoichiometry and charge neutrality. Thus, the ideal {111} plane should be terminated by an outer layer of oxygen anions, with a second atomic layer of Cu^+ cations and a third atomic layer of oxygen anions. Every two Cu atoms have a dangling bond perpendicular to the {111} planes. Thus, the {111} plane is expected to possess a higher energy status than the {100} plane that is 100% oxygen terminated. On the basis of the analysis of the relationship between the morphology of different exposed surfaces and chemical activities, it is predicted that the octahedral crystals bounded by the {111} facet are more active than the cubic crystals exposed with the {100} plane. Experimental results agree well with this analysis.

Though antibacterial activities of Cu_2O were extensively investigated, very little attention has been paid to photocatalytic antibacterial activities of Cu_2O. To the best of our knowledge, the photocatalytic activity of disinfection for pristine Cu_2O has been reported only by our group [182] until now. We prepared different morphologies of Cu_2O films by copper anode oxidation by controlling the reaction conditions (current density, temperature, and the reaction time). With the increase in current density, nanonet, nanosheet, and nanorod arrays of Cu_2O can be obtained. The Cu_2O nanorod shows the best photocatalytic activity for disinfection, and the disinfection efficiency can reach 90.85% in 1 h and 100% in 2 h. As we know, Cu_2O is unstable in aqueous solution under irradiation, and the oxidation product Cu^{2+} ion or CuO and the reduction product Cu can inactivate bacterium directly. Thus, compared with traditional TiO_2 photocatalysts for water disinfection, Cu_2O can not only harvest solar light due to its narrow band gap and inactivate bacteria under irradiation, but also kill bacterium in the dark.

Cu_2O-based NPs were extensively investigated for photocatalytic water disinfection. Wang et al. [183] prepared Cu_2O NPs-sensitized ZnO nanorod arrays on the indium tin oxide coated glass substrate via a facile electrodeposition process. Under simulated sunlight illumination, the Cu_2O NPs-sensitized ZnO nanorod arrays show a significant photocatalytic antibacterial performance. Wong P.K.'s group [184] synthesized a VL-driven photocatalyst Cu_2O–CuO/$Sr_3BiO_{5.4}$ by the precipitation method; 5.4 log of *E. coli* can be inactivated within 3 h in a Cu_2O–CuO/$Sr_3BiO_{5.4}$ suspension under VL irradiation provided by the fluorescent lamps. Compared with Cu_2O–CuO/$Sr_3BiO_{5.4}$ suspension, a Cu_2O/TiO_2 nanotube (TNT) [185] film

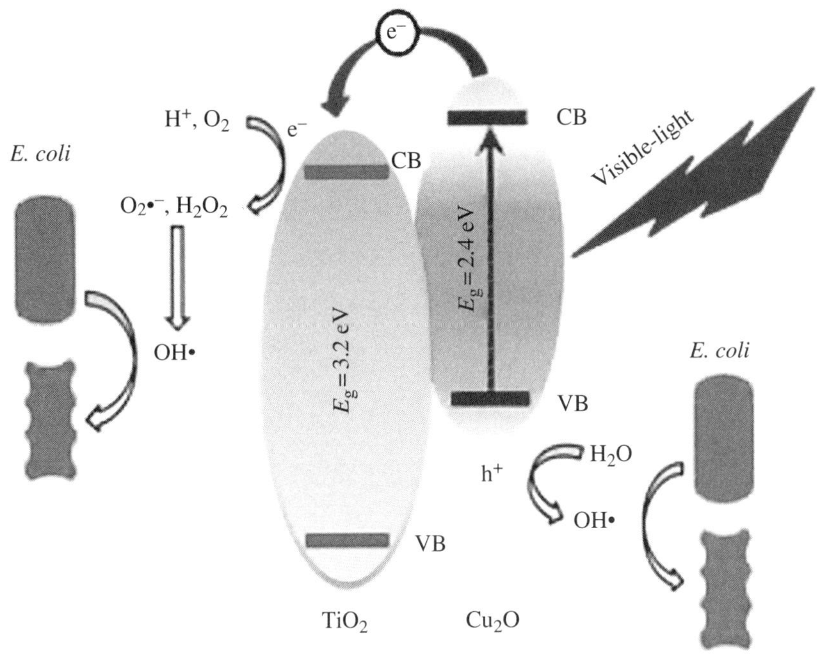

FIGURE 3.16 Scheme of inactivation mechanisms of *E. coli* by Cu_2O/TNTs under visible-light irradiation. Reproduced by permission from Ref. [185]. © 2012, Springer-Verlag.

photocatalyst shows much higher activity in the inactivation of *E. coli*. It is capable of complete inactivation of *E. coli* in 5×10^7 colony-forming units/ml within a record short disinfection time of 20 min under VL irradiation. The average bactericidal percentages of the Cu_2O/TNTs for *E. coli* are 20 times and 6.6 times higher than those of TNTs under the same conditions and Cu_2O/TNTs without light, respectively. This superior bactericidal performance is mainly attributed to the high ability to produce •OH radicals by both the photogenerated electron and the hole of the prepared photocatalyst under VL. The mechanism of Cu_2O/TNTs for photocatalytic disinfection is proposed based on Figure 3.16, which schematically illustrates photocatalytic generation of •OH at the Cu_2O/TNTs film. Cu_2O is a p-type semiconductor having a direct band gap and its CB edge potential is approximately 1.2 eV more negative than that of TiO_2, which makes it suitable as a VL sensitizer for TiO_2 [76]. The narrow band gap of Cu_2O makes it capable of absorbing VL to generate photoelectrons and holes at the CB and VB, respectively. The CB edge potential difference between Cu_2O and TiO_2 serves as a motive force driving the photoelectrons from the CB of Cu_2O to the CB of TiO_2. The synergetic effect results from the matched electron structure of Cu_2O and TiO_2 facilitates the charge separation to minimize the recombination, which in turn prolongs the lifetime of holes to enhance the production of •OH at the VB of Cu_2O via an oxidative process. On the other hand, the electrons transferred to the CB of TiO_2 react with adsorbed oxygen to form active oxygen species such as O^{2-}, O^{2-}, and H_2O_2 via a reductive process; •OH can then be produced through further chemical reactions of the active oxygen species. The ability to produce •OH by both a photogenerated electron and a hole is believed to be an important advantage for the high concentration of •OH and thereafter the high bacterial performance.

It is worth mentioning that Fenton-like reactions play an important role in photocatalytic disinfection due to Cu^+/Cu^{2+} reaction [186]. Paschoalinoa et al. [186] proposed the mechanism of Fenton-like reactions. They described CuO powders with different specific surface areas that inactivate *E. coli* in aqueous solution under VL irradiation >360 nm. The CuO in contact with the bacterial suspension shows a change in its surface oxidation state from Cu^{2+} to Cu^+. The outermost layer of the catalyst (5–7 nm) becomes mainly Cu_2O (80%) and CuO (20%), as observed by X-ray photoelectron spectroscopy (XPS). The suggested mechanism of reaction at the CuO under light irradiation is as follows:

$$CuO + h\upsilon \rightarrow CuO\,(e_{cb}^-,\ h_{vb}^+) \tag{3.13}$$

The e_{cb}^- in Equation 3.13 is produced from the CuO (p type) with a band-gap energy of 1.7 eV, a flat-band potential of −0.3 V saturated calomel electrode (SCE) (pH 7), and a VB at +1.4 V SCE [75]. The electron–hole pair is formed when photon energies exceed the band gap of CuO. The excited electron can either react directly with O_2 for $O_2^{\bullet-}$ formation (reaction 3.14) or reduce the Cu^{2+} lattice to Cu^+, leading to reactions (3.15)–(3.16) with the formation of $O_2^{\bullet-}$ radicals

$$CuO\ (e_{cb}^-) + O_2 \rightarrow CuO + O_2^{\bullet-} \tag{3.14}$$

$$CuO\ \left(e_{cb}^-\right) \rightarrow CuO\ (Cu^+) \tag{3.15}$$

$$CuO(Cu^+) + O_2 \rightarrow CuO(Cu^{2+}) + O_2^{\bullet-} \tag{3.16}$$

Equation 3.17 shows that the equilibrium between H^+ and $O_2^{\bullet-}$ leads to the formation of the $HO_2^{\cdot}$ radical. The $HO_2^{\bullet}$ in Equation 3.17 leads to the production of H_2O_2 in Equation 3.18:

$$H^+ + O_2^- = HO_2^{\bullet}\ (\text{partial}\ O_2^{\bullet-}) \tag{3.17}$$

$$CuO\ (Cu^+) + HO_2^{\bullet}\ (\text{partial}\ O_2^{\bullet-}) + H^+ \rightarrow CuO\ (Cu^{2+}) + H_2O_2 \tag{3.18}$$

or the alternative reaction

$$2Cu(I) + O_2 + 2H^+ \rightarrow 2Cu(II) + H_2O_2 \tag{3.19}$$

Equation 3.18 is analogous to the behavior of Fe^{2+} ions (Fenton reaction) in the Haber–Weiss cycle [187]. Indeed, the production of H_2O_2 as expected based on Equation 3.19 was evaluated experimentally in both the dark and illuminated culture medium in the presence and absence of bacterial cells, respectively. Production was observed to be higher than that for the sample kept in the dark at the beginning of the experiment, but showed similar values after a 4-h reaction. This behavior was observed both in the presence and in the absence of *E. coli*, although higher levels of H_2O_2 were measured in the culture medium containing *E. coli*. Paschoalinoa et al. suggested that the Cu(I) ions on the CuO is the active catalytic species in accordance with Equation 3.20:

$$CuO(Cu^{1+}) \rightarrow CuO_{\text{vacancy}} + Cu^{1+} \tag{3.20}$$

The oxidative side of the redox cycle would involve the h^+ in Equation 3.14 and convert H_2O into •OH, allowing for the hole transfer and finally the formation of H_2O_2 in Equation 3.22.

$$h^+ + H_2O \rightarrow H^+ + \bullet OH \tag{3.21}$$

$$\bullet OH + \bullet OH \rightarrow H_2O_2 \tag{3.22}$$

This mechanism for copper toxicity implies the capacity of CuO to generate reactive oxygen species (ROS) involving Cu ions via Fenton-like reactions through the Haber–Weiss cycle. The inhibition of *E. coli* enzymatic activity is due to Cu ion binding with the *E. coli* cell walls and thiol proteins [188] since Cu^+ has the ability to bind indiscriminately to cell walls [189]. It is worth noting that Fenton-like Cu^+/Cu^{2+} reactions can also take place in the absence of irradiation [190, 191]. This reaction without irradiation can be concisely described as copper-catalyzed Haber–Weiss reactions. Cu(II) is initially reduced by the reductant X_{red} (Eq. 3.23), followed by reoxidation with hydrogen peroxide (Eq. 3.24), resulting in a net production of the hydroxyl radical [165].

$$X_{red} + Cu(II) \rightarrow X_{ox} + Cu(I) \tag{3.23}$$

$$Cu(I) + H_2O_2 \rightarrow Cu(II) + OH^- + \bullet OH \tag{3.24}$$

3.3.3 Cu_2O-Based Nanocomposites for H_2 Evolution

TiO_2 nanomaterials represent the most important semiconductor catalysts for hydrogen production from splitting water. The principle of the TiO_2 photocatalytic system for water in Fujishima and Honda's pioneering work is depicted in Figure 3.17 [56]. A photocatalyst absorbs UV and/or VL irradiation from sunlight or an illuminated light source. The electrons in the VB of the photocatalyst are excited to the CB, while the holes are left in the VB. This, therefore, creates the negative-electron (e^-) and positive-hole (h^+) pairs. This stage is referred to as the semiconductor's "photoexcited" state. After photoexcitation, the excited electrons and holes separate and migrate to the surface of the photocatalyst. Here, in the photocatalytic water-splitting reaction, they act as both a reducing agent and an

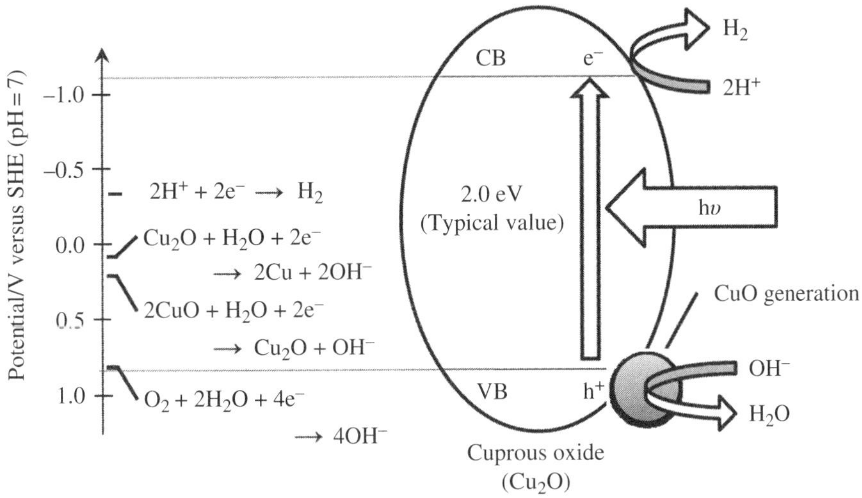

FIGURE 3.17 Fundamental principle of semiconductor-based photocatalytic water splitting for hydrogen generation. Reproduced by permission from Ref. [199]. © 2008, The Electrochemical Society.

oxidizing agent to produce H_2 and O_2, respectively. Water splitting into H_2 and O_2 is an uphill reaction. It needs the standard Gibbs free energy change ΔG^0 of 237 kJ/mol, corresponding to 1.23 eV, as shown in Equation 3.25:

$$H_2O \rightarrow O_2 + H_2,\ \Delta G = +237\ \text{kJ/mol} \tag{3.25}$$

Therefore, the band-gap energy (E_g) of the photocatalyst should be greater than 1.23 eV to achieve water splitting. Furthermore, the width of the band gap and the potential of CB and VB are critical for efficient water splitting. The bottom level of the CB has to be more negative than the reduction potential of H^+/H_2 (0 V versus NHE), while the top level of the VB has to be more positive than the oxidation potential of O_2/H_2O (1.23 V).

Since the band-gap energy (E_g) of the photocatalyst should be greater than 1.23 eV to achieve water splitting, the band gap of a semiconductor should be less than 3.0 eV (>400 nm) to use VL. However, TiO_2 is a wide band-gap semiconductor (E_g = 3.2 eV), which limits the absorption of sunlight to the major portion of the solar spectrum. The approach to solving the water-splitting problem has focused on exploring new materials for both anodic and cathodic processes and integrating configurations that utilize photovoltaic cell junctions. The ultimate goal of these efforts is an efficient photoelectrolysis cell design that can simultaneously drive, in an unassisted fashion, both hydrogen evolution and water oxidation reactions. Water-splitting cells require semiconductor materials that are able to support rapid charge transfer at a semiconductor/aqueous interface, exhibit long-term stability, and efficiently harvest a large portion of the solar spectrum. In 1998, Hara et al. [30] reported that illuminated Cu_2O particles were needed for the evolution of H_2 and O_2 from water. Since then, Cu_2O- and Cu_2O-based semiconductors were extensively investigated for photocatalytic water splitting. In Hara et al.'s experiments, H_2 and O_2 were evolved over a period of 1900 h, without a noticeable decrease in activity. In later studies, they also observed that water splitting continued under stirring conditions for many hours after the light was switched off [192]. This dark reaction was observed for a variety of oxides including Cu_2O, NiO, Co_3O_4, and Fe_3O_4 with a reaction rate about one-third that of the photochemical experiments with Cu_2O. To explain their results, Ikeda et al. [193] proposed that the mechanical energy supplied by the stirring was converted into chemical energy with the metal oxide acting as a catalyst. However, de Jongh et al. [194, 195] presented a negative result that the presence of Cu_2O electrodes for reduction of water is kinetically unfavorable, although high photocurrent efficiency was observed in the presence of oxygen or electron scavengers. Vacuum studies on the decomposition of D_2O by Cu_2O using UV light, on the other hand, showed no evidence of photosplitting [196]. Most recently, photocatalysis by Cu_2O powders was investigated in distilled water, and it was found that hydrogen evolution was possible only with freshly prepared Cu_2O powder, while oxygen evolution could not be observed [197]. With a cathodic bias of −0.3 V versus Ag/AgCl, Nian et al. [198] directly evidenced that the photoinduced charges in the CB of p-type Cu_2O are capable of reducing water to H_2. Kakuta and Abea [199] evaluated the photocatalytic potential of Cu_2O powder by studying its ability to stoichiometrically decompose water into O_2 and H_2. Although thus far it has been indicated that Cu_2O can function as a photocatalyst for overall water splitting, H_2 evolution is found to take place solely when there is no generation of O_2. Moreover, the evolution rate of H_2 gradually decreases with prolonged irradiation, and, consequently, ceases. From the structural analyses of deactivated Cu_2O, it appears that H_2 evolution commences with the self-oxidation of Cu_2O to CuO. From the band structure of Cu_2O in Figure 3.17,

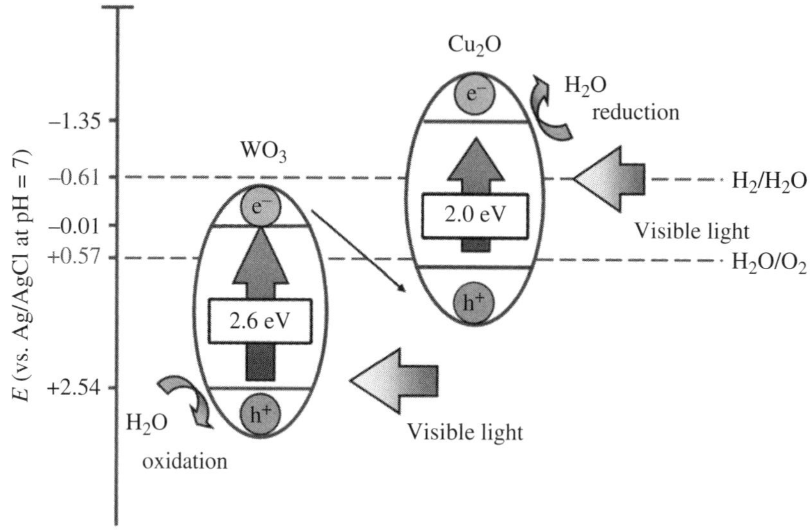

FIGURE 3.18 Reaction scheme of photoinduced water splitting over coupled p-type Cu_2O and n-type WO_3 photocatalysts in a suspension system. Reprinted with permission from Ref. [202]. © 2008, Elsevier B.V.

it can be concluded that the redox potentials for H_2 and O_2 evolution and the oxidation of Cu_2O to CuO are all within the band gap, and, therefore, all of these processes are possible in principle. However, the driving force for water oxidation is minimal, while the oxidation of Cu_2O is thermodynamically favorable. In the presence of light, morphological instabilities due to the oxidation and reduction of Cu_2O induced by photogenerated electrons have been observed, and the photocorrosion of Cu_2O has been recognized as a reason for the decay of its photocatalytic property.

In order to be a viable material to be used in direct H_2 production, particularly under solar energy, Cu_2O was used to couple with other semiconductors such as TiO_2 [21, 200, 201], WO_3 [202], and Cu/TiO_2 [203]. Our previous studies found a Cu_2O/TiO_2 bilayer film [200] and Cu_2O/TiO_2 [201] nanocomposites as photocatalysts for water splitting, while Cu_2O film and Cu_2O NPs prepared by the chemical deposition method were completely nonactive. Similar results were also reported by Yasomanee' group [201]. In our study, neither Cu_2O nor TiO_2 could produce H_2 from water splitting, but the combined systems could produce H_2 from water splitting. The mechanism of photocatalytic reaction is proposed based on energy band theory and experimental results. The photogenerated electrons from Cu_2O were captured by Ti^{4+} ions in TiO_2, and Ti^{4+} ions were further reduced to Ti^{3+} ions. Thus, the photogenerated electrons were stored in Ti^{3+} ions as a form of energy. These electrons trapped in Ti^{3+} can be released if a suitable electron acceptor is present, whereas holes stay in the VB of Cu_2O to form hole centers. This hampers the recombination of the photogenerated electrons with the holes. Thus, electrons trapped in Ti^{3+} ions may have a long lifetime and can be transferred to the interface between the composite and solution. These electrons can be directly combined with H+ to produce H_2. Additionally, holes are accumulated in the VB of Cu_2O to form hole centers, which can be consumed by weak oxidation. Senevirathna and Pitigala [21] deposited quantum-sized Cu_2O on TiO_2 NPs to produce the catalyst TiO_2/Cu_2O and found that this composite system can photogenerate hydrogen from water under sacrificial conditions.

Hu et al. [202] coupled Cu_2O with WO_3 to form a heterocontact for the improvement of photoinduced charge separation in Cu_2O. Prior to the experiments of the Cu_2O/WO_3 suspension system, pristine Cu_2O was detected and no trace of H_2 was evolved from Cu_2O/pure water suspensions even in the presence of methanol (a sacrificial hole scavenger). The results indicate that the photoinduced carriers within Cu_2O are lost to back reactions due to inefficient charge transport. This is generally encountered for narrow-gap semiconductors. For the Cu_2O/WO_3 suspension system, charge transfer between p- and n-type semiconductors through a mechanism denoted as "interparticle electron transfer" is essential for improving charge separation. The n-type semiconductor employed should have a VB edge more positive than that of the p-type Cu_2O. The working scheme is demonstrated in Figure 3.18. In this p–n coupling, two photons have to be absorbed to produce one net electron–hole pair for redox reactions at the solid/liquid interface. This electron–hole pair consists of the minority hole and minority electron from the n- and p-type semiconductors, respectively. On the other hand, the majority electrons and holes of the respective semiconductors combine during the instant particle contact. Through this coupling, the number of photoinduced charges lost to back reactions in Cu_2O can be reduced. In this p–n coupling, WO_3 was chosen as the n-type semiconductor to couple with the p-type Cu_2O since it has a VB edge (about +2.54 V versus Ag/AgCl at pH = 7), much more positive than the water oxidation potential and its CB edge. Additionally, its CB edge is not negative enough to reduce H_2O to H_2 (Fig. 3.18).

Li et al. [203] fabricated an ohmic nanojunction photocatalyst by assembling Cu@Cu_2O core–shell NPs on TNT arrays. The prepared photocatalysts show great enhancement in the evolved H_2 rate due to the efficient transfer of photoexcited electrons and holes. This approach puts forward a facile, low-cost method to improve photocatalytic efficiency and is helpful for exploring and developing new photocatalysts based on the heterojunction working principle.

3.4 CONCLUSIONS AND OUTLOOK

Structural characteristics, modification, photocatalytic mechanism, the relationship between structure and photocatalytic activity, and research progress of Cu_2O-based nanocomposites for environmental protection are summarized. As a nontoxic, abundant, high absorption coefficient and theoretical solar cell conversion efficiency semiconductor, Cu_2O is a potential candidate in solar energy applications. However, the low efficiency and weak oxidation ability of Cu_2O hinder its application in photocatalysis particularly for the application of solar hydrogen production from water. Moreover, Cu_2O is unstable since it typically suffers from photocorrosion under irradiation. Thus, we may need to continue this study focusing on these two aspects. On the one hand, controllable synthesis of Cu_2O nanocrystals shows that there are still significant challenges to tune the band structure and morphology for the enhancement of photocatalytic efficiency. On the other hand, based on solid-state electronics, electrochemistry/photoelectrochemistry, different modification approaches such as doping, coupled or mixed with other materials, need further screening and investigation for the improvement of stability under irradiation. In addition, the studies related to mechanisms of photocatalytic activity of Cu_2O-based nanocomposites are still insufficient. Fenton-like Cu^+/Cu^{2+} reactions are proposed and mainly used for interpreting Cu_2O disinfection in the absence of irradiation. We need to delve into what happens during Fenton-like Cu^+/Cu^{2+} reactions in photocatalytic organics degradation and H_2 evolution under irradiation. Besides, since H_2O_2 proves to be one of the ROS produced both in pristine Cu_2O and in Cu_2O/TiO_2 nanocomposites under VL, it is of interest to discover the synergistic effect in detail for Cu_2O-based nanocomposites in the presence of Fe^{2+}. If so, it is possible to generate low-cost and high-efficiency in situ Fenton reagents from Cu_2O-based nanocomposites, which do not require the supply of H_2O_2 in comparison with traditional Fenton reagents.

ACKNOWLEDGMENTS

Dr. Xiong acknowledges the financial support from the Project of Hubei Provincial Department of Education (XD201010406), the Doctor Foundation of Huanggang Normal University (2011cd233), the Project of Industry Academy Research Cooperation of Huanggang Normal University (2012026203), the Project of Hubei Silver Dragon Pipes Industry Ltd. (2012013111), and the National Natural Science Foundation of China (11104101). Prof. Yu thanks support from the National Natural Science Foundation of China (No. 21377044), Wuhan Planning Project of Science and Technology (No. 2014010101010023) and self-determined research funds of CCNU from the colleges' basic research and operation of MOE (No. CCNU13F027).

REFERENCES

[1] Fujishima A, Honda K. Electrochemical photolysis of water at a semiconductor electrode. Nature 1972;238 (5358):37.

[2] Matsunaga T, Tomodam R, Nakajima T, Wake H. Photoelectrochemical sterilization of microbial cells by semiconductor powders. FEMS Microbiol Lett 1985;29 (1–2):211.

[3] Sato S, White JM. Photodecomposition of water over Pt/TiO_2 catalysts. Chem Phys Lett 1980;72 (1):83.

[4] Sadeghi M, Liu W, Zhong TG, Stavropoulos P, Levy B. Role of photoinduced charge Carrier separation distance in heterogeneous photocatalysis oxidative degradation of CH_3OH vapour in contact with Pt-TiO_2 and confirmed TiO_2-Fe_2O_3. J Phys Chem 1996;100 (50):19466.

[5] Sakata T, Kawai T, Hashimoto K. Photochemical diode model of Pt/TiO_2 particle and its photocatalytic activity. Chem Phys Lett 1982;88 (1):50.

[6] Abrahams J, Davidson RS, Morrison C. Optimization of the photocatalytic properties of titaniumdioxide. J Photochem Photobiol A 1985;29 (3–4):353.

[7] Suri RPS, Liu J, Hand DW, Crittenden JC, Perram DL, Mullins ME. Heterogeneous photocatalytic oxidation of hazardous organic contaminants in water. Water Environ Res 1993;65 (5):665.

[8] Iliev V, Tomova D, Bilyarska L, Eliyas A, Petrov L. Photocatalytic properties of TiO_2 modified with platinum and silver nanoparticles in the degradation of oxalic acid in aqueous solution. Appl Catal B 2006;63 (3–4):266.

[9] Sakthivel S, Shankar MV, Palanichamy M, Arabindoo B, Bahnemann DW, Murugesan V. Enhancement of photocatalytic activity by metal deposition: characterisation and photonic efficiency of Pt, Au and Pd deposited on TiO_2 catalyst. Water Res 2004;38 (13):3001.

[10] Yang YF, Sangeetha P, Chen YW. Au/TiO_2 catalysts prepared by photo-deposition method for selective CO oxidation in H_2 stream. Int J Hydrogen Energy 2009;34 (21):8912.

[11] Sato S. Photocatalytic activity of nitrogen oxide (NOx)-doped titanium dioxide in the visible region. Chem Phys Lett 1986;123 (1–2):126.

[12] Asahi R, Morikawa T, Ohwaki T, Aoki K, Taga Y. Visible-light photocatalysis in nitrogen-doped titanium oxides. Science 2001;293 (5528):269.

[13] Fujishima A, Zhang XT, Tryk DA. TiO_2 photocatalysis and related surface phenomena. Surf Sci Rep 2008;63 (12):515.

[14] Ryu JG, Choi WY. Photocatalytic oxidation of arsenite on TiO_2: understanding the controversial oxidation mechanism involving superoxides and the effect of alternative electron acceptors. Environ Sci Technol 2006;40 (22):7034.

[15] Long RQ, Yang RT. Selective catalytic oxidation of ammonia to nitrogen over Fe_2O_3-TiO_2 prepared with a sol-gel method. J Catal 2002;207 (2):158.

[16] Celik E, Yildiz AY, Ak Azema NF, Tanoglu M, Toparli M, Emrullahoglu OF, Ozdemir I. Preparation and characterization of Fe_2O_3-TiO_2 thin films on glass substrate for photocatalytic applications. Mater Sci Eng B 2006;129 (1–3):193.

[17] Ke DN, Liu HJ, Peng TY, Liu X, Dai K. Preparation and photocatalytic activity of WO_3/TiO_2 nanocomposite particles. Mater Lett 2008;62 (3):447.

[18] Yang XL, Daia WL, Guo CG, Chen H, Cao Y, Li HX, He HY, Fan KN. Synthesis of novel core-shell structured WO_3/TiO_2 spheroids and its application in the catalytic oxidation of cyclopentene to glutaraldehyde by aqueous H_2O_2. J Catal 2005;234 (2):438.

[19] Kwon YT, Song KY, Lee WI, Choi GJ, Do YR. Photocatalytic behavior of WO_3-loaded TiO_2 in an oxidation reaction. J Catal 2000;191 (1):192.

[20] Rau MA, Bukallah SB, Hamadi A, Sulaiman A, Hammadi F. The effect of operational parameters on the photoinduced decoloration of dyes using a hybrid catalyst V_2O_5/TiO_2. Chem Eng J 2007;129 (1–3):167.

[21] Senevirathna MKI, Pitigala PKDDP, Tennakone K. Water photoreduction with Cu_2O quantum dots on TiO_2 nano-particles. J Photochem Photobiol A 2005;171 (3):257.

[22] Jangaum JS, Kimb HG, Joshia UA, Janga JW, Lee JS. Fabrication of CdS nanowires decorated with TiO_2 nanoparticles for photocatalytic hydrogen production under visible light irradiation. Int J Hydrogen Energy 2008;33 (21):5975.

[23] Jangaum JS, Joshia UA, Lee JS. Solvothermal synthesis of CdS nanowires for photocatalytic hydrogen and electricity production. J Phys Chem C 2007;111 (35):13280.

[24] Maeda K, Teramura K, Lu D, Saito N, Inoue Y, Domen K. Noble-metal/Cr_2O_3 core/shell nanoparticles as a cocatalyst for photocatalytic overall water splitting. Angew Chem Int Ed Engl 2006;118 (46):7970–7973.

[25] An HZ, Du Y, Wang TM, Wang C, Hao WC, Zhang JY. Photocatalytic properties of BiOX (X=Cl, Br, and I). Rare Metals 2008;27 (3):243.

[26] Kato H, Kudo A. Visible-light-response and photocatalytic activities of TiO_2 and $SrTiO_3$ photocatalysts codoped with antimony and chromium. J Phys Chem B 2002;106 (19):5029.

[27] Kudo A, Sekizawa M. Photocatalytic H_2 evolution under visible light irradiation on Ni-doped ZnS photocatalyst. Chem Commun 2000;15:1371.

[28] Li LL, Chu Y, Liu Y, Dong LH. Template-free synthesis and photocatalytic properties of novel Fe_2O_3 hollow spheres. J Phys Chem C 2007;111 (5):2123.

[29] Chakrabarti S, Dutta BK. Photocatalytic degradation of model textile dyes in wastewater using ZnO as semiconductor catalyst. J Hazard Mater 2004;112 (3):269.

[30] Hara M, Kondo T, Komoda M, Ikeda S, Shinohara K, Tanaka A, Kondo JN, Domen K. Cu_2O as a photocatalyst for overall water splitting under visible light irradiation. Chem Commun 1998;3:357.

[31] Grondahl LO, Geiger PH. A new electronic rectifier. AIEE. 1927;46:357.

[32] Tucker CW. A study of photovoltaic cells. J Phys Chem 1927;31 (9):1357.

[33] Agekyan VT. Spectroscopic properties of semiconductor crystals with direct forbidden energy gap. Phys Status Solidi A 1977;43 (1):11.

[34] Musa AO, Akomolafe T, Carter MJ. Production of cuprous oxide, a solar cell material, by thermal oxidation and a study of its physical and electrical properties. Sol Energy Mater Sol Cells 1998;51 (3–4):305.

[35] Bessekhouad Y, Robert D, Weber JV. Photocatalytic activity of Cu_2O/TiO_2, Bi_2O_3/TiO_2 and $ZnMn_2O_4$/TiO_2 heterojunctions. Catal Today 2005;101 (3–4):315.

[36] Neogi P. Interfacial stability of electrodeposition of cuprous oxide films. J Chem Phys 2004;121 (19):9630.

[37] Sowers KL, Fillinger A. Crystal face dependence of p-Cu_2O stability as photocathode. J Electrochem Soc 2009;156 (5):F80.

[38] Tennakone K, Kumarsinghe AR, Sirimanne PM. The effect of methyl viologen ion on photocurrent and stability of Cu_2O photocathodes in I^-/I^{3-} redox electrolyte. J Photochem Photobiol A 1995;88 (1):39.

[39] Tachibana Y, Muramoto R, Matsumoto H, Kuwabata S. Photoelectrochemistry of p-type Cu_2O semiconductor electrode in ionic liquid. Res Chem Intermed 2006;32 (5):575.

[40] Khan KA. Stability of a Cu_2O photoelectrode in an electrochemical cell and the performances of the photoelectrode coated with Au and SiO thin films. Appl Energy 2009;65 (1–4):59.

[41] Huang L, Peng F, Yu H, Wang HJ. Preparation of cuprous oxides with different sizes and their behaviors of adsorption, visible-light driven photocatalysis and photocorrosion. Solid State Sci 2009;11 (1):129.

[42] Biccari F. Defects and doping in Cu_2O [Ph.D. thesis]. Sapienza: University of Rome; March 15, 2010. Available at http://biccari.altervista.org. Accessed April 21, 2014.

[43] Önsten A, Göthelid M, Karlsson UO. Atomic structure of Cu_2O (111). Surf Sci 2009;603 (2):257.

[44] Gordienko AB, Zhuravlev YN, Fedorov DG. Band structure and chemical bonding in Cu_2O and Ag_2O Oxides. Phys Solid State 2007;49 (2):223.

[45] Das K, Sharma SN, Kumar M, De SK. Luminescence properties of the solvothermally synthesized blue light emitting Mn doped Cu_2O nanoparticles. J Appl Phys 2010;107 (2):024316.

[46] Ghijsen J, Tjeng LH, Elp JV, Eskes H, Westerink J, Sawatzky GA. Electronic structure of Cu_2O and CuO. Phys Rev B 1988;38 (16):11322.

[47] Ruiz AM, Moreno MG, Takeuchi N. First principles calculations of the electronic properties of bulk Cu_2O, clean and doped with Ag, Ni, and Zn. Solid State Sci 2003;5 (2):291.

[48] Ruiz E, Alvarez S, Alemany P, Evarestov RA. Electronic structure and properties of Cu_2O. Phys Rev B 1997;56 (12):7189.

[49] Nolan M, Elliott SD. Tuning the electronic structure of the transparent conducting oxide Cu_2O. Thin Solid Film 2008;516 (7):1468.

[50] Uihlein C, Fröhlich D, Kenklies R. Investigation of exciton fine structure in Cu_2O. Phys Rev B 1981;23 (6):2731.

[51] French M, Schwartz R, Stolz H, Redmer R. Electronic band structure of Cu_2O by spin density functional theory. J Phys Condens Matter 2009;21 (1):015502.

[52] Nolan M, Elliott SD. Tuning the transparency of Cu_2O with substitutional cation doping. Chem Mater 2008;20 (17):5522.

[53] Zheng ZK, Huang BB, Wang Z, Guo M, Qiu X, Zhang X, Wang P, Dai Y. Crystal faces of Cu_2O and their stabilities in photocatalytic reactions. J Phys Chem C 2009;113 (32):14448.

[54] Vesaghi MA, Shafiekhani A, Baghizadeh A, Lameii M, Ghodselahi T. XPS study of the Cu@Cu_2O core-shell nanoparticles. Appl Surf Sci 2008;255 (5):2730.

[55] Yang F, Yan N, Huang S, Sun Q, Zhang L, Yu Y. Zn-Doped CdS nanoarchitectures prepared by hydrothermal synthesis: mechanism for enhanced photocatalytic activity and stability under visible light. J Phys Chem C 2012;116 (16):9078.

[56] Siripala W, Ivanovskaya A, Jaramillo TF, Baeck SH, McFarland EW. A Cu_2O/TiO_2 heterojunction thin film cathode for photoelectrocatalysis. Sol Energy Mater Sol Cells 2003;77 (3):229.

[57] Nakano Y, Sacki S, Morikawa T. Optical bandgap widening of p type Cu_2O films by nitrogen doping. Appl Phys Lett 2009;94 (2):022111.

[58] Ishizuka S, Akimoto K. Control of the grow orientation and electrical properties of polycrystalline Cu_2O thin films by group-IV elements doping. Appl Phys Lett 2004;85 (21):4920.

[59] Ishizuka S, Kato S, Maruyama T, Akimoto K. Nitrogen doping into Cu_2O thin films deposited by reactive radio-frequency magnetron Bibliography 205 sputtering. Jpn J Appl Phys 2001;40 (4B):2765.

[60] Olsen LC, Addis FW, Miller W. Experimental and theoretical studies of Cu_2O. Sol Cells 1982;7 (3):247.

[61] Han XF, Han K, Tao M. N-type Cu_2O by electrochemical doping with Cl. Solid-State Lett 2009;12 (4):H89.

[62] Han XF, Han K, Tao M. Characterization of Cl-doped n-type Cu_2O prepared by electrodeposition. Thin Solid Film 2010;518 (19):5363.

[63] Malerba C, Ricardo CLA, D'Incau M, Biccari F. Nitrogen doped Cu_2O: a possible material for intermediate band solar cells? Sol Energy Mater Sol Cells 2012;105:192.

[64] Ishizuka S, Kato S, Okamoto Y, Akimoto K. Control of hole carrier density of polycrystalline Cu_2O thin films by Si doping. Appl Phys Lett 2002;80 (6):950.

[65] Nemoto T, Nakano T. The effect of silicon in cuprous oxide. Jpn J Appl Phys 1967;6:543.

[66] Tapiero M, Noguet C, Zielinger JP, Schwab C, Pierrat D. Conversion photovoltaïque dans Cu_2O. Rev Phys Appl 1979;14 (1):231.

[67] Kikuchi N, Tonooka K. Electrical and structural properties of Ni–doped Cu_2O films prepared by pulsed laser deposition. Thin Solid Film 2005;486 (1–2):33.

[68] Kale S, Ogale S, Shinde S, Sahasrabuddhe M, Kulkarni VN, Greene RL, Venkatesan T. Magnetism in cobalt-doped Cu_2O thin films without and with Al, V, or Zn codopants. Appl Phys Lett 2003;82 (13):2100.

[69] Chang GS, Kurmaev EZ, Boukhvalov DW, Moewes A, Finkelstein LD, Wei M, MacManus-Driscoll JL. Defect-induced ferromagnetism in Mn-doped Cu_2O. J Phys Condens Matter 2008;20 (21):215216.

[70] Tseng CC, Hsieh JH, Liu SJ, Wu W. Effects of Ag contents and deposition emperatures on the electrical and optical behaviors of Ag-doped Cu_2O thin film. Thin Solid Film 2009;518 (5):1407.

[71] Rothenberger G, Moser J, Graetzel M, Serpone N, Sharma DK. Charge carrier trapping and recombination dynamics in small semiconductor particles. J Am Chem Soc 1985;107 (26):8054.

[72] Tang YW, Chen ZG, Jia ZJ, Zhang LS, Li JL. Electrodeposition and characterization of nanocrystalline cuprous oxide thin films on TiO_2 films. Mater Lett 2005;59 (4):434.

[73] Wang SJ, Yin SF, Li L, Xu BQ, Ng CF, Au CT. Investigation on modification of Ru/CNTs catalyst for the generation of COx-free hydrogen from ammonia. Appl Catal B 2004;52 (4):287.

[74] Hardee KL, Bard AJ. Semiconductor electrodes X. Photoelectrochemical behavior of several polycrystalline metal oxide electrodes in aqueous solutions. J Electrochem Soc 1977;124 (2):215–224.

[75] Li JL, Liu L, Yu Y, Tang YW, Li HL, Du FP. Preparation of highly photocatalytic active nano-size TiO_2-Cu_2O particle composites with a novel electrochemical method. Electrochem Commun 2004;6 (9):940.

[76] Hou Y, Li XY, Zhao QD, Quan X, Chen GH. Fabrication of Cu_2O/TiO_2 nanotube heterojunction arrays and investigation of its photoelectrochemical behavior. Appl Phys Lett 2009;95 (9):093108.

[77] Jiang TF, Xie TF, Chen LP, Fu ZW, Wang DJ. Carrier concentration-dependent electron transfer in Cu_2O/ZnO nanorod arrays and their photocatalytic performance. Nanoscale 2013;5 (7):2938.

[78] Chen JY, Zhou PJ, Li JL, Li SQ. Depositing Cu_2O of different morphology on chitosan nanoparticles by an electrochemical method. Carbohydr Polym 2007;67 (4):623.

[79] Brittl PJ, Santhanam KSV, Rubio A, Alonso JA, Ajayan PM. Improved charge transfer at carbon nanotube electrodes. Adv Mater 1999;11 (2):154.

[80] Kong LN, Chen W, Ma DK, Yang Y, Liu SS, Huang SM. Size control of Au@Cu_2O octahedra for excellent photocatalytic performance. J Mater Chem 2012;22 (2):719.

[81] Wang WC, Lyu LM, Huang MH. Investigation of the effects of polyhedral gold nanocrystal morphology and facets on the formation of Au@Cu_2O core-shell heterostructures. Chem Mater 2011;23 (10):2677.

[82] Katayama J, Ito K, Matsuok M, Tamali J. Performance of Cu_2O/ZnO solar cell prepared by two-step electrodeposition. J Appl Electrochem 2004;34 (7):687.

[83] Miyata T, Tanaka H, Sato H, Minami T. P-type semiconducting Cu_2O–NiO thin films prepared by magnetron sputtering. J Mater Sci 2006;41 (17):5531.

[84] Herion J, Niekisch EA, Scharl G. Investigation of metal oxide/cuprous oxide heterojunction solar cells. Sol Energy Mater 1980;4 (1):101.

[85] Papadimitriou L, Economou NA, Trivich D. Heterojunction solar cells on cuprous oxide. Sol Cells 1981;3 (1):73.

[86] Yang HG, Sun HC, Qiao SZ, Zou J, Liu G, Smith SC, Cheng HM, Lu GQ. Anatase TiO_2 single crystals with a large percentage of reactive facets. Nature 2008;453 (7195):638.

[87] Lu HB, Wang SM, Zhao L, Li JC, Dong BH, Xu ZX. Hierarchical ZnO microarchitectures assembled by ultrathin nanosheets: hydrothermal synthesis and enhanced photocatalytic activity. J Mater Chem 2011;21 (12):4228.

[88] Bi YP, Ouyang SX, Umezawa N, Cao JY, Ye JH. Facet effect of single-crystalline Ag_3PO_4 sub-microcrystals on photocatalytic properties. J Am Chem Soc 2011;133 (17):6490.

[89] Tian N, Zhou ZY, Sun SG, Ding Y, Wang ZL. Synthesis of tetrahexahedral platinum nanocrystals with high-index facets and high electro-oxidation activity. Science 2007;316 (5825):732.

[90] Wei MZ, Huo JZ. Preparation of Cu_2O nanorods by a simple solvothermal method. Mater Chem Phys 2010;121 (1–2):291.

[91] Tan YW, Xue XY, Peng Q, Zhao H, Wang TH, Li YD. Controllable fabrication and electrical performance of single crystalline Cu_2O nanowires with high aspect ratios. Nano Lett 2007;7 (12):3723.

[92] Orel ZC, Anlovar A, Drai G, Igon M. Cuprous oxide nanowires prepared by an additive-free polyol process. Cryst Growth Des 2007;7 (2):453.

[93] Pang ML, Zeng HC. Highly ordered self-assemblies of submicrometer Cu_2O spheres and their hollow chalcogenide derivatives. Langmuir 2010;26 (8):5963.

[94] Li X, Tao FF, Jiang Y, Xu Z. 3-D ordered macroporous cuprous oxide: fabrication, optical, and photoelectrochemical properties. J Colloid Interface Sci 2007;308 (2):460.

[95] Lee WR, Piao LH, Park CH, Limc YS, Dod YR, Yoond SH, Kimb SH. Facile synthesis and size control of spherical aggregates composed of Cu_2O nanoparticles. J Colloid Interface Sci 2010;342 (1):198.

[96] Luo YS, Li SQ, Ren QF, Liu JP, Xing LL, Wang Y, Yu Y, Jia ZJ, Li JL. Facile synthesis of flowerlike Cu_2O nanoarchitectures by a solution phase route. Cryst Growth Des 2007;7 (1):87.

[97] Zhang HW, Zhang X, Li HY, Qu ZK, Fan S, Ji MY. Hierarchical growth of Cu_2O double tower-tip-like nanostructures in water/oil microemulsion. Cryst Growth Des 2007;7 (4):820.

[98] Sun F, Guo YP, Song WB, Zhao JZ, Tang LQ, Wang ZC. Morphological control of Cu_2O micro-nanostructure film by electrodeposition. J Cryst Growth 2007;304 (2):425.

[99] McShane CM, Choi KS. Photocurrent enhancement of n-type Cu_2O electrodes achieved by controlling dendritic branching growth. J Am Chem Soc 2009;131 (7):2561.

[100] Luo YS, Yu BH, Tu YC, Liang Y, Zhang YG, Liu JP, Li JL, Jia ZJ. Self-assembly synthesis of cactuslike Cu_2O 3D nanoarchitectures via a low-temperature solution approach. Mater Res Bull 2008;43 (8–9):2166.

[101] Qu YL, Li XY, Chen GH, Zhang HJ, Chen YY. Synthesis of Cu_2O nano-whiskers by a novel wet-chemical route. Mater Lett 2008;62 (6–7):886.

[102] Jimenez-Cadena G, Comini E, Ferroni M, Sberveglieri G. Synthesis of Cu_2O bi-pyramids by reduction of $Cu(OH)_2$ in solution. Mater Lett 2010;64 (3):469.

[103] Liu HR, Miao WF, Yang S, Zhang ZM, Chen JF. Controlled synthesis of different shapes of Cu_2O via γ-irradiation. Cryst Growth Des 2009;9 (4):1733.

[104] Wang DB, Mo MS, Yu DB, Xu LQ, Li FQ, Qian YT. Large-scale growth and shape evolution of Cu_2O cubes. Cryst Growth Des 2003;3 (5):717.

[105] Wang YQ, Liang WS, Satti A, Nikitin K. Fabrication and microstructure of Cu_2O nanocubes. J Cryst Growth 2010;312 (9):1605.

[106] Yao KX, Yin XM, Wang TH, Zeng HC. Synthesis, self-assembly, disassembly, and reassembly of two types of Cu_2O nanocrystals unifaceted with {001} or {110} planes. J Am Chem Soc 2010;132 (17):6131.

[107] Zhang X, Xie Y, Xu F, Liu XH, Xu D. Shape-controlled synthesis of submicro-sized cuprous oxide octahedral. Inorg Chem Commun 2003;6 (11):1390.

[108] He P, Shen XH, Gao HC. Size-controlled preparation of Cu_2O octahedron nanocrystals and studies on their optical absorption. J Colloid Interface Sci 2005;284 (2):510.

[109] Guo SJ, Fang YX, Dong SJ, Wang EK. Templateless, surfactantless, electrochemical route to a cuprous oxide microcrystal: from octahedra to monodisperse colloid spheres. Inorg Chem 2007;46 (23):9537.

[110] Leng M, Liu MZ, Zhang YB, Wang ZQ, Yu C, Yang XG, Zhang HJ, Wang C. Polyhedral 50-facet Cu_2O microcrystals partially enclosed by {311} high-index planes: synthesis and enhanced catalytic CO oxidation activity. J Am Chem Soc 2010;132 (48):17084.

[111] Wang HH, Tian F, Li XP, Liu FL, Shen Q. Preparation and shape evolution of cuprous oxide in the solution phases of copper (II) dodecyl sulfate. Powder Technol 2010;197 (3):298.

[112] Ng CHB, Fan WY. Shape evolution of Cu_2O nanostructures via kinetic and hermodynamic controlled growth. J Phys Chem B 2006;110 (42):20801.

[113] Zhang WX, Luan CY, Yang ZH, Liu XT, Zhang DP, Yang SH. Preparation and optical properties of Cu_2O hollow microsphere film and hollow nanosphere powder via a simple liquid reduction approach. Appl Surf Sci 2007;253 (14):6063.

[114] Yu HG, Yu JG, Liu SW, Mann S. Template-free hydrothermal synthesis of CuO/Cu_2O composite hollow microspheres. Chem Mater 2007;19 (17):4327.

[115] Xu YY, Chen DR, Jiao XL, Xue KY. Nanosized Cu_2O/PEG400 composite hollow spheres with mesoporous shells. J Phys Chem C 2007;111 (44):16284.

[116] Xu HL, Wang WH, Zhou L. A growth model of single crystalline hollow spheres: oriented attachment of Cu_2O nanoparticles to the single crystalline shell wall. Cryst Growth Des 2008;8 (10):3486.

[117] Xu YY, Jiao XL, Chen DR. PEG-assisted preparation of single-crystalline Cu_2O hollow nanocubes. J Phys Chem C 2008;112 (43):16769.

[118] Kuo CH, Huang MH. Fabrication of truncated rhombic dodecahedral Cu_2O nanocages and nanoframes by particle aggregation and acidic etching. J Am Chem Soc 2008;130 (38):12815.

[119] Gou LF, Murphy CJ. Solution-phase synthesis of Cu_2O nanocubes. Nano Lett 2003;3 (2):23.

[120] Kim MH, Lim B, Lee EP, Xia YN. Polyol synthesis of Cu_2O nanoparticles: use of chloride to promote the formation of a cubic morphology. J Mater Chem 2008;18 (34):4069.

[121] Sun SD, Kong CC, Yang SC, Wang LQ, Song XP, Ding BG, Yang ZM. Highly symmetric polyhedral Cu_2O crystals with controllable-index planes. CrystEngComm 2011;3 (7):2217.

[122] Sun SD, Zhang H, Song XP, Liang SH, Kong CC, Yang ZM. Polyhedron-aggregated multi-facet Cu_2O homogeneous structures. CrystEngComm 2011;13 (20):6040.

[123] Wang X, Liu C, Zheng BJ, Jiang YQ, Zhang L, Xie ZX, Zheng LS. Controlled synthesis of concave Cu_2O microcrystals enclosed by {hhl} high-index facets and enhanced catalytic activity. J Mater Chem A 2013;1 (2):282.

[124] Wang XP, Jiao SH, Wu DP, Li Q, Zhou JG, Jiang K, Xu DS. A facile strategy for crystal engineering of Cu_2O polyhedrons with high-index facets. CrystEngComm 2013;315 (10):1849.

[125] Su XD, Zhao JZ, Bala H, Zhu YC, Gao Y, Ma SS, Wang ZC. Fast synthesis of stable cubic copper nanocages in the aqueous phase. J Phys Chem C 2007;111 (40):14689.

[126] Xu H, Wang W, Zhu W. Shape evolution and size-controllable synthesis of Cu_2O octahedra and their morphology-dependent photocatalytic properties. J Phys Chem B 2006;110 (28):13829.

[127] Kuo CH, Huang MH. Facile synthesis of Cu_2O nanocrystals with systematic shape evolution from cubic to octahedral structures. J Phys Chem C 2008;112 (47):18355.

[128] Xu Y, Wang H, Yu YF, Tian L, Zhao WW, Zhang B. Cu_2O nanocrystals: surfactant-free room-temperature morphology-modulated synthesis and shape-dependent heterogeneous organic catalytic activities. J Phys Chem C 2011;115 (31):15288.

[129] Kuo CH, Huang MH. Morphologically controlled synthesis of Cu_2O nanocrystals and their properties. Nanotoday 2010;5 (2):106.

[130] Pang H, Gao F, Lu QY. Morphology effect on antibacterial activity of cuprous oxide. Chem Commun 2009;9:1076.

[131] Ren J, Wang WZ, Sun SM, Zhang L, Wang L, Chang J. Crystallography facet-dependent antibacterial activity: the case of Cu_2O. Ind Eng Chem Res 2011;50 (17):10366.

[132] Leng M, Yu C, Wang C. Polyhedral Cu_2O particles: shape evolution and catalytic activity on cross coupling reaction of iodobenzene and phenol. CrystEngComm 2012;14 (24):8454.

[133] Sun SD, Song XP, Sun YX, Deng DC, Yang ZM. The crystal-facet-dependent effect of polyhedral Cu_2O microcrystals on photocatalytic activity. Catal Sci Technol 2012;2 (5):925.

[134] Yu Y, Huang WY, Du FP, Ma LL. Synthesis and characteristic of cuprous oxide nano-whiskers with photocatalytic activity under visible light. Mater Sci Forum 2005;475–479:3531.

[135] Tang AD, Xiao Y, Ouyang Y, Nie S. Preparation, photo-catalytic activity of cuprous oxide nano-crystallites with different sizes. J Alloys Compd 2008;457 (1–2):447.

[136] Shoeib MA, Abdelsalam OE, Khafagi MG, Hammam RE. Synthesis of Cu_2O nanocrystallites and their adsorption and photocatalysis behavior. Adv Powder Technol 2012;23 (3):298.

[137] Zhang XX, Song JM, Jiao J, Mei XF. Preparation and photocatalytic activity of cuprous oxides. Solid State Sci 2010;12 (7):1215.

[138] Ma LL, Li JL, Sun HZ, Qiu MQ, Wang JB, Chen JY, Yu Y. Self-assembled Cu_2O flowerlike architecture: polyol synthesis, photocatalytic activity and stability under simulated solar light. Mater Res Bull 2010;45 (8):961.

[139] Xu L, Xu HY, Wu SB, Zhang XY. Synergy effect over electrodeposited submicron Cu_2O films in photocatalytic degradation of methylene blue. Appl Surf Sci 2012;258 (11):4934.

[140] Hou Y, Li XY, Zou XJ, Quan X, Chen GH. Photoeletrocatalytic activity of a Cu_2O-loaded self-organized highly oriented TiO_2 nanotube array electrode for 4-Chlorophenol degradation. Environ Sci Technol 2009;43 (3):858.

[141] Zhang YG, Ma LL, Li JL, Yu Y. In situ Fenton reagent generated from TiO_2/Cu_2O composite film: a new way to utilize TiO_2 under visible light irradiation. Environ Sci Technol 2007;41 (17):6264.

[142] Xu YH, Liang DH, Liu ML, Liu DZ. Preparation and characterization of Cu_2O-TiO_2: efficient photocatalytic degradation of methylene blue. Mater Res Bull 2008;43 (12):3474.

[143] Zhang SS, Zhang SQ, Peng F, Zhang HM, Liu HW, Zhao HJ. Electrodeposition of polyhedral Cu_2O on TiO_2 nanotube arrays for enhancing visible light photocatalytic performance. Electrochem Commun 2011;13 (8):861.

[144] Wang C, Wu JC, Wang PF, Ao YH, Hou J, Qian J. Photoelectrocatalytic determination of chemical oxygen demand under visible light using Cu_2O-loaded TiO_2 nanotube arrays electrode. Sens Actuators B 2013;181:1.

[145] Liu H, Wang J, Fan XM, Zhang FZ, Liu HR, Dai J, Xiang FM. Synthesis of Cu_2O/T-ZnO_w nanocompound and characterization of its photocatalytic activity and stability property under UV irradiation. Mater Sci Eng B 2013;178 (2):158.

[146] Zhu QW, Zhang YH, Zhou FS, Lv FZ, Ye ZF, Fanc FD, Chu PK. Preparation and characterization of Cu_2O-ZnO immobilized on diatomite for photocatalytic treatment of red water produced from manufacturing of TNT. Chem Eng J 2011;171 (1):61.

[147] Wei SQ, Ma YY, Chen YY, Liu L, Liu Y, Shao ZC. Fabrication of WO_3/Cu_2O composite films and their photocatalytic activity. J Hazard Mater 2011;194:243.

[148] Hu SC, Zhou F, Wang LZ, Zhang JL. Preparation of Cu_2O/CeO_2 heterojunction photocatalyst for the degradation of Acid Orange 7 under visible light irradiation. Catal Commun 2011;12 (9):794.

[149] Wang WZ, Huang XW, Wu S, Zhou YX, Wang LJ, Shi HL, Liang YJ, Zou B. Preparation of p–n junction $Cu_2O/BiVO_4$ heterogeneous nanostructures with enhanced visible-light photocatalytic activity. Appl Catal B 2013;134–135:293.

[150] Li HL, Lei YG, Huang Y, Fang YP, Xu YH, Zhu L, Li X. Photocatalytic reduction of carbon dioxide to methanol by Cu_2O/SiC nanocrystallite under visible light irradiation. J Nat Gas Chem 2011;20 (2):145.

[151] Pan YL, Deng SZ, Polavarapu L, Gao NY, Yuan PY, Sow CH, Xu QH. Plasmon-enhanced photocatalytic properties of Cu_2O nanowire-Au nanoparticle assemblies. Langmuir 2012;28 (33):12304.

[152] Li JT, Cushing SK, Bright J, Meng FK, Senty TR, Zheng P, Bristow AD, Wu NQ. Ag@Cu_2O core-shell nanoparticles as visible-light plasmonic photocatalysts. ACS Catal 2013;3 (1):47.

[153] Zhou WH, Li N, Caon MH, Hu CW. Three-dimensional Ag/POM/Cu_2O tricomponent nanohybrids with enhanced visible-light photocatalytic activity. Mater Lett 2013;99:68.

[154] Li SK, Huang FZ, Wang Y, Shen YH, Qiu LG, Xie AJ, Xu SJ. Magnetic Fe_3O_4@C@Cu_2O composites with bean-like core/shell nanostructures: synthesis, properties and application in recyclable photocatalytic degradation of dye pollutants. J Mater Chem 2011;2 (20):7459.

[155] Wang XF, Chen GM, Zhang J. Synthesis and characterization of novel Cu_2O/PANI composite photocatalysts with enhanced photocatalytic activity and stability. Catal Commun 2013;31:57.

[156] Zhou KQ, Shi YQ, Jiang SH, Hu Y, Gui Z. Facile preparation of Cu_2O/carbon sphere heterostructure with high photocatalytic activity. Mater Lett 2013;98:213.

[157] Ai ZH, Xiao HY, Mei T, Liu J, Zhang LZ, Deng KJ, Qiu JR. Electro-Fenton degradation of rhodamine B based on a composite cathode of CuO nanocubes and carbon nanotubes. J Phys Chem C 2008;112 (31):11929.

[158] Grätzel M. Photoelectrochemical cells. Nature 2001;414 (6861):338.

[159] Zhang L, Li J, Chen Z, Tang Y, Yu Y. Preparation of Fenton reagent with H_2O_2 generated by solar light-illuminated nano-Cu_2O/MWNT scomposites. Appl Catal A 2005;299:292.

[160] Xu Y, Schoonen MAA. The absolute energy positions of conduction and valence bands of selected semiconducting minerals. Am Mineral 2000;85 (3–4):543.

[161] Miyauchi M. Photocatalysis and photoinduced hydrophilicity of WO_3 thin films with underlying Pt nanoparticles. Phys Chem Chem Phys 2008;10 (41):6258.

[162] Xu C, Cao L, Su G, Liu W, Liu H, Yu Y, Qu X. Preparation of ZnO/Cu_2O compound photocatalyst and application in treating organic dyes. J Hazard Mater 2010;176:807.

[163] Huang L, Peng F, Wang H, Yu H, Li Z. Preparation and characterization of Cu_2O/TiO_2 nano–nano heterostructure photocatalysts. Catal Commun 2009;10:1839.

[164] Georgekutty R, Seery MK, Pillai SC. A highly efficient Ag-ZnO photocatalyst: synthesis, properties, and mechanism. J Phys Chem C 2008;112 (35):13563.

[165] Wang P, Huang BB, Qi XY, Zhang XY, Dai Y, Wei JY, Whangbo MH. Ag@AgCl: a highly efficient and stable photocatalyst active under visible light. Angew Chem Int Ed Engl 2008;47 (41):7931.

[166] Zhang L, Blom DA, Wang H. Au-Cu_2O core-shell nanoparticles: a hybrid metal-semiconductor heteronanostructure with geometrically tunable optical properties. Chem Mater 2011;23 (20):4587.

[167] Li XZ, Li FB. Study of Au/Au^{3+}-TiO_2 photocatalysts toward visible photooxidation for water and wastewater treatment. Environ Sci Technol 2001;35 (11):2381.

[168] Hoffmann MR, Martin ST, Choi WY, Bahnemann DW. Environmental applications of semiconductor photocatalysis. Chem Rev 1995;95 (1):69.

[169] Wang ZH, Zhao SP, Zhu SY, Sun YL, Fang M. Photocatalytic synthesis of M/Cu_2O (M = Ag, Au) heterogeneous nanocrystals and their photocatalytic properties. CrystEngComm 2011;13 (7):2262.

[170] Schaadt DM, Feng B, Yu ET. Enhanced semiconductor optical absorption via surface plasmon excitation in metal nano-particles. Appl Phys Lett 2005;86 (6):063106.

[171] Wang D, Zhao H, Wu NQ, El Khakani A, Ma D. Tuning the charge-transfer property of PbS-quantum dot/TiO_2 nanobelt nanohybrids via quantum confinement. J Phys Chem Lett 2010;1 (7):1030.

[172] Grätzel M. Dye-sensitized solar cells. J Photochem Photobiol C 2003;4 (2):145.

[173] Zhang DQ, Li GS, Yu JM. Inorganic materials for photocatalytic water disinfection. J Mater Chem 2010;20 (22):4529.

[174] Dollweth H, Sorenson J. Historic uses of copper compounds in medicine. Trace Elem Med 1985;2 (2):80.

[175] Sagripanti JL, Routson LB, Bonifacino AC, Lytle CD. Mechanism of copper-mediated inactivation of herpes simplex virus. Antimicrob Agents Chemother 1997;41 (4):812.

[176] Santo CE, Taudte N, Nies DH, Grass G. Contribution of copper ion resistance to survival of *Escherichia coli* on metallic copper surfaces. Appl Environ Microbiol 2008;74 (4):977.

[177] Faundez G, Troncoso M, Navarrete P, Figueroa G. Antimicrobial activity of copper surfaces against suspension of *Salmonella enterica* and *Campylobacter jejuni*. BMC Microbiol 2004;4:19.

[178] Noyce JO, Michels H, Keevil CW. Inactivation of influenza A virus on copper versus stainless steel surfaces. Appl Environ Microbiol 2007;73 (8):2748.

[179] Sunada K, Minoshima M, Hashimoto K. Highly efficient antiviral and antibacterial activities of solid-state cuprous compounds. J Hazard Mater 2012;235–236:265.

[180] Gadi B, Jeffrey G. Copper as a biocidal tool. Curr Med Chem 2005;12 (18):2163.

[181] Cooney JJ, Tang RJ. Quantifying effects of antifouling paints on microbial biofilm formation. Methods Enzymol 1999;310:637.

[182] Yan LL, Wang Y, Xiong LB, Li JL, Yin YH, Wong PK, Ying Y. Preparation and photocatalytic sterilization property of Cu_2O nanostructure with copper anode oxidation method. Chin J Inorg Chem 2009;25 (11):1960.

[183] Wang Y, She GW, Xu HT, Liu YY, Mu LX, Shi WS. Cu_2O nanoparticles sensitized ZnO nanorod arrays: electrochemical synthesis and photocatalytic properties. Mater Lett 2012;67:110.

[184] Zhao W, Wong KH, Hu C, Yu JC, Chan CY, Qi T, Wong PK. Synthesize of Cu_2O-CuO/$Sr_3BiO_{5.4}$ and its photocatalytic activity. Appl Surf Sci 2012;258 (16):5955.

[185] Zhang SS, Liu C, Liu XL, Zhang H, Liu PR, Zhang SQ, Peng F, Zhao HJ. Nanocrystal Cu_2O-loaded TiO_2 nanotube array films as high-performance visible-light bactericidal photocatalyst. Appl Microbiol Biotechnol 2012;96 (5):1201.

[186] Paschoalino M, Guedes NC, Jardim W, Mielczarski E, Mielczarski JA, Bowen P, Kiwi J. Inactivation of *E. coli* mediated by high surface area Cu_2O accelerated by light irradiation >360 nm. J Photochem Photobiol A 2008;199 (1):105.

[187] Petasnem RG, Zika RG. Hydrogen peroxide lifetimes in south Florida coastal and offshore waters. Marine Chem 1997;56 (3–4):215.

[188] Letelier ME, Lepe AM, Faúndez M, Salazar J, Marín R, Aracen P, Speisky H. Possible mechanisms underlying copper-induced damage in biological membranes leading to cellular toxicity. Chem Biol Interact 2005;151 (2):71.

[189] Nishikawa T, Lee ISM, Shiraishi N, Ishikawa T, Ohta Y, Nishikimi M. Identification of S100b protein as copper-binding protein and its suppression of copper-induced cell damage. J Biol Chem 1997;272:23037.

[190] Park HJ, Nguyen TTM, Yoon JY, Lee CH. Role of reactive oxygen species in *Escherichia coli* inactivation by cupric ion. Environ Sci Technol 2012;46 (20):11299.

[191] Bao M, Wang D, Liu SY, Kuang LL, Sun J, Wang F, Wen YX. Electrodeposition and electrocatalytic activity of Cu_2O film on stainless steel substrate. Appl Surf Sci 2012;258 (20):8008.

[192] Hara M, Kodama M, Hasei H, Yashima M, Ikeda S, Tanaka T, Kondo JN, Domen K. A study of mechano-catalysts for overall water splitting. J Phys Chem B 2000;104 (4):780.

[193] Ikeda S, Takata T, Kondo T, Hitoki G, Hara M, Kondo JN, Domen K, Hosono H, Kawazoe H, Tanak A. Mechano-catalytic overall water splitting. Chem Commun 1998;20:2158.

[194] de Jongh PE, Vanmaekelbergh D, Kelly JJ. Cu_2O: a catalyst for the photochemical decomposition of water? Chem Commun 1999;12:1069.

[195] de Jongh PE, Vanmaekelbergh D, Kelly JJ. Photoelectrochemistry of electrodeposited Cu_2O. J Electrochem Soc 2000;147 (2):486.

[196] Walker AV, Jr Yates JT. Does cuprous oxide photosplit water? J Phys Chem B 2000;104 (38):9038.

[197] Kakuta SJ, Abe T. A novel example of molecular hydrogen generation from formic acid at visible-light-responsive photocatalyst. Appl Mater Interfaces 2009;1 (12):2707.

[198] Nian JN, Hu CC, Teng HS. Electrodeposited p-type Cu_2O for H_2 evolution from photoelectrolysis of water under visible light illumination. Int J Hydrogen Energy 2008;33 (12):2897.

[199] Kakuta SJ, Abea T. Structural characterization of Cu_2O after the evolution of H_2 under visible light irradiation. Electrochem Solid-State Lett 2009;12 (3):1.

[200] Xiong LB, Ouyang ML, Yan LL, Li LJ, Qiu MQ, Yu Y. Visible-light energy storage by Ti^{3+} in TiO_2/Cu_2O bilayer film. Chem Lett 2009;38 (12):1154.

[201] Yasomanee JP, Bandara J. Multi-electron storage of photoenergy using Cu_2O-TiO_2 thin film photocatalyst. Sol Energy Mater Sol Cells 2008;92 (3):348.

[202] Hu CC, Nian JJ, Teng HS. Electrodeposited p-type Cu_2O as photocatalyst for H_2 evolution from water reduction in the presence of WO_3. Sol Energy Mater Sol Cells 2008;92 (9):1071.

[203] Li ZH, Liu JW, Wang DJ, Gao Y, Shen J. $Cu_2O/Cu/TiO_2$ nanotube Ohmic heterojunction arrays with enhanced photocatalytic hydrogen production activity. Int J Hydrogen Energy 2012;37 (8):6431.

4

MULTIFUNCTIONAL NANOCOMPOSITES FOR ENVIRONMENTAL REMEDIATION

Suying Wei[1], Jiahua Zhu[1,2], Hongbo Gu[2], Huige Wei[2], Xingru Yan[2], Yudong Huang[3], and Zhanhu Guo[2]

[1] *Department of Chemistry and Biochemistry, Lamar University, Beaumont, TX, USA*
[2] *Integrated Composites Laboratory (ICL), Dan F. Smith Department of Chemical Engineering, Lamar University, Beaumont, TX, USA*
[3] *School of Chemical Engineering and Technology, Harbin Institute of Technology, Harbin, Heilongjiang, China*

4.1 INTRODUCTION

Nanocomposites are materials that are composed of more than two components with at least one in the nanoscale (1–100 nm) range. The central theme of making nanocomposites is to integrate nanomaterials with others into one entity, thus generating so-called synergistic properties and/or performances. Multifunctional nanocomposites are an excellent illustration of targeting nanocomposites, in which each of the components is structure- and property-integrated uniquely to produce hybrid materials that possess multifunctionalities in both structure and properties. Reflecting the authors' research interests and experiences, in this chapter we mainly focus on polymer-based and carbon-based multifunctional nanocomposites. Polymers have been considered as excellent host matrices for nanocomposite materials. They have been extensively investigated for their potential wide applications because of their easy processability, low-cost manufacturing, good adhesion to substrates, and unique physicochemical properties. Polymer nanocomposites (PNCs) are polymer composites using nanostructured materials as fillers. Depending on the type of filler material, different properties can be achieved for PNCs. The types of polymer matrices, insulating or conducting, also render the nanocomposites with significantly different synergy. Carbon-based structures as another type of matrix for nanocomposites may include carbon nanotubes, carbon nanofibers (NFs), carbon nanoplatelets, and graphenes. The unique and interesting chemical and electronic structures of carbon have rendered carbon-based nanocomposites a variety of noteworthy applications, including anticorrosion in electronic devices and sensors, environmental remediation for heavy metals and other toxic species, and magnetic data storage and magnetic imaging.

Owing to the inherent synergistic properties of multifunctional nanocomposites, including electronic, optical, magnetic, and mechanical, they have seen wide applications spanning the broad areas of electronics, energy, space, and biomedicine. For example, magnetic PNCs that generally consist of magnetic nanofillers have shown great potential in high-density magnetic recording, magnetic sensors, magnetic carriers, magnetic storage, and color imaging [1–4]. Multifunctional nanocomposites have also found applications in electrochemical energy storage especially for developing Faradic supercapacitors (FS), which currently suffer from low power density due to the slow faradic process during charge/discharge and poor cycling stability as a result of the redox reaction that takes place on the electrode [5–7]. Multifunctional nanocomposites may comprise pseudoactive materials (large faradic capacitance) such as metal oxide [8, 9] or conductive polymer nanostructures [10, 11] in a carbon matrix (high conductivity and large specific surface area); thus it can potentially enhance the capacitance of the resulting composite electrodes. For example, flexible polyaniline (PANI) film/graphene composite paper was produced via an in situ anodic electropolymerization process, and this material generated a capacitance of 233 F/g [12]. Furthermore, PANI nanowire arrays [13]

Nanomaterials for Environmental Protection, First Edition. Edited by Boris I. Kharisov, Oxana V. Kharissova, and H. V. Rasika Dias.

and PANI nanofibers (PANI-NFs) [14] were successfully grafted on a graphite oxide (GO) sheet using a rapid mixing reaction method; both composite materials reached high capacitances of 555 and 480 F/g, respectively.

Moreover, multifunctional nanocomposites have also found great potential in biomedical applications, which can be attributed to the complex structure and functionalities inherent in this type of synergistic materials. Compared to the widely accepted traditional metallic and ceramic biomaterials, multifunctional nanocomposites are able to offer high biocompatibility, light weight, anticorrosion, relatively easy fabrication and processing, and low cost. In addition, it has been realized that the mismatch in the mechanical elastic modulus and tensile strength between the metal and ceramic biomaterials and the targeting hard tissues is due to the negative effects in tissue remodeling. Composites, however, can offer greater potential to structurally match the biological environment, by adjusting the reinforcing material types and morphology, volume fraction, and interaction and distribution in the matrix [15]. Furthermore, a broad range of functionalities and properties are able to be tailored into the composite entity to suit various intended biological functions.

In this chapter, we mainly review and report the environmental applications of polymer-based and carbon-based multifunctional nanocomposites, using our own research as well as published results by other groups. The targeting remediation is centered on nanocomposite solid wastes, wastewater, heavy metals, arsenic, and organic dyes. In the following, we will also introduce the major techniques used for developing these multifunctional nanocomposites in our lab, as well as the major corresponding characterization and property analysis methods. Finally, we will give our opinion on the perspectives of multifunctional nanocomposites, especially their potential for environmental remediation and sustainability. As a result, this chapter does not extensively cover every aspect of multifunctional nanocomposites, but rather selectively reports what we have achieved to date.

4.2 MULTIFUNCTIONAL NANOCOMPOSITES DEVELOPMENT: FROM FABRICATION TO PROCESSING

In order to produce synergistic entity nanocomposites with multiple functionalities, the method of dispersion of the nanofiller in the matrix and subsequent processing are crucial for their targeted properties and applications. Since nanofillers have diverse chemical compositions and physical forms, it has been one of the most challenging and targeted topic of discussion in the nanocomposites regime. The traditional direct mixing method is not able to attain uniform dispersion and retain the strong connection between the filler and the matrix, thus failing to achieve the desired properties. Scientists and researchers have explored alternative means to integrate multiple components through either ex situ or in situ techniques. The key step very often lies at the surface functionalization stage of one or more components simultaneously to the blending process. In the following, we will elaborate what we have achieved in polymer-based nanocomposites as well as carbon-based nanocomposites.

4.2.1 Fabrication and Processing of Multifunctional Polymer-Based Nanocomposites

Over the years, we have developed several effective methods to fabricate and process polymer-based multifunctional nanocomposites, namely, surface-initiated polymerization (SIP) and monomer-stabilized polymerization (MSP), which have been proven to enhance the interaction between the polymer matrix and the nanofillers, improve the dispersion of the nanofillers in the polymer matrix, and greatly contribute to the respective mechanical, magnetic, and conducting properties [16, 17]. In the SIP method, briefly speaking, it typically starts with mixing the monomers and the nanofillers and then interconnection is initiated when the preadsorbed catalyst or initiator starts polymerization of the monomers from the nanofiller surface. For example, highly flexible polyurethane nanocomposites with high loading of iron nanoparticles (NPs) were successfully prepared using the SIP method, and the resulting nanocomposites displayed favorable mechanical strength and magnetic properties. The SIP scheme is demonstrated in Figure 4.1.

The SIP method has also been widely applied to other PNC systems, including conducting polymers PANI-, polypyrrole (PPy)-, and thermoplastic polymers polypropylene (PP)-and polyethylene (PE)-based. Nanofillers are generally nanostructured ceramic and ceramic oxides, as well as the various carbon nanostructures. For example, magnetic polyaniline PNCs [18] were successfully prepared using the SIP method and showed negative permittivity in both the neat PANI and the PANI–magnetite nanocomposites. More interestingly, large-room-temperature magnetoresistance (MR) was observed in both the neat PANI and the PANI–magnetite nanocomposites, with the latter being much larger (MR ~95%) than the former (MR ~53%). Furthermore, temperature-dependent resistivity study indicated a variable range hopping (VRH) mechanism for the electron conduction in all the materials in the aforementioned PANI PNCs.

In the PANI–barium titanate system synthesized from the SIP method, an interesting correlation was observed between the particle size, loading level, and mixing methods of the ferroelectric barium titanate in the PANI matrix, and the corresponding

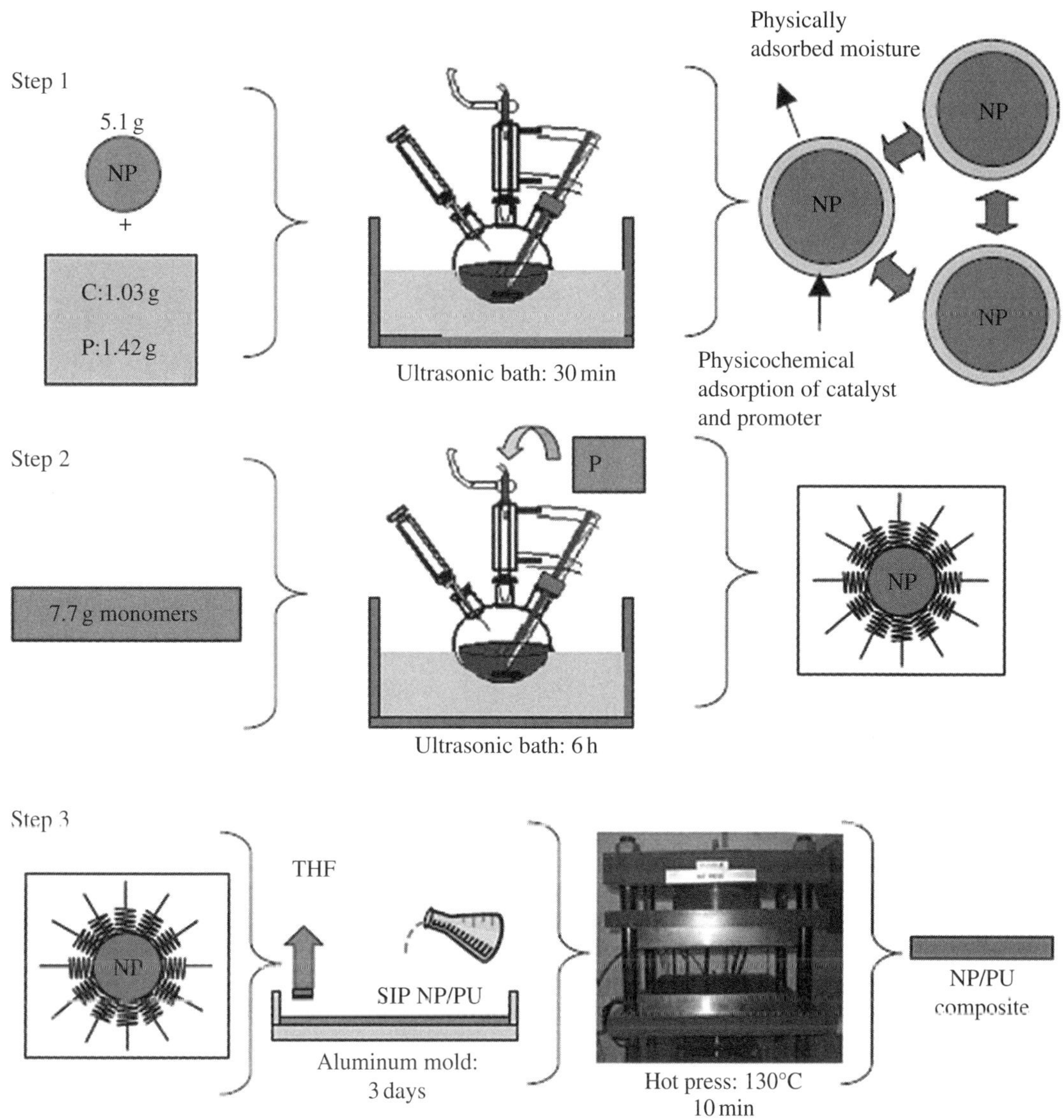

FIGURE 4.1 Schematic of the surface-initiated polymerization (SIP) process. Adapted with permission from Ref. [16]. © IOP.

magnetic and dielectric properties. The PNC samples synthesized using the SIP method showed negative permittivity in contrast to the positive real permittivity in the PNCs prepared by physical mixing. Moreover, positive MR was observed in all the PNCs, which was analyzed by the wave function shrinkage model. The external magnetic field contracted the electronic wave function at the impurity center, while the introduction of barium titanate reduced the contraction effect and thus PNCs showed lower magnetic resistance than the neat PANI.

The MSP method usually avoids the use of additional surfactants, but instead the monomer itself serves as the coupling agent with one end bonding to the NP surface and the other continuing polymerization with other monomers, thus forming a well-dispersed and uniform PNC network. For example, through this unique MSP method, a robust entity of Fe/vinyl ester resin nanocomposites was fabricated, which demonstrated enhanced mechanical strength [17]. Figure 4.2 shows a scheme in which iron NPs are stabilized by vinyl ester monomer and facilitate further polymerization with other monomers, thus forming the interconnecting system. In addition, we have also successfully prepared PP- and PE-based PNCs by in situ thermal decomposition of the organometallic iron precursor in the polymer solution at high temperatures [19–22] in which size- and phase-controlled iron NPs were synthesized in situ and uniformly dispersed in the polymer matrix to form PP or PE magnetic nanocomposites.

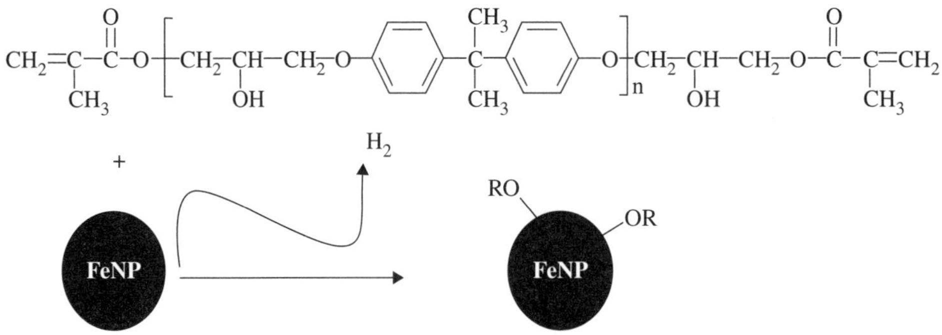

FIGURE 4.2 Scheme of the vinyl ester stabilization of iron NPs; OR represents the hydroxyl group in the vinyl ester. Adapted with permission from Ref. [17]. © Elsevier.

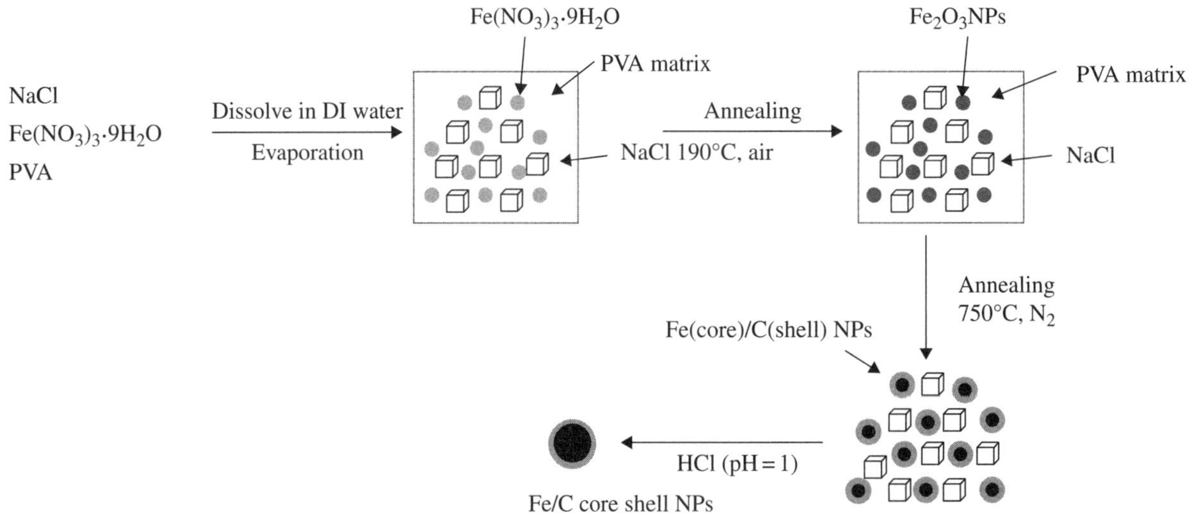

FIGURE 4.3 Schematic of the formation of Fe@C NPs. Adapted with permission from Ref. [23]. © RSC.

4.2.2 Fabrication and Processing of Multifunctional Carbon-Based Nanocomposites

In general, carbon-based nanocomposites may be prepared via two pathways. One is by directly using carbon-based chemical structures (carbon nanotubes, carbon NFs, carbon nanoplatelets, graphene, and/or graphite) as the composite matrix and doping other fillers into it. The limited doping level and the chemical mechanical compatibility between the carbon matrix and the nanofillers have witnessed a great change in this research field. The other is by starting with PNCs, and, through carbonization and/or graphitization, forming carbon-based nanocomposites. As PNCs have been developed over a much longer period, various techniques and/or potential to make well-compatible and synergistic composites are available. The carbon-based nanocomposites in our research were generally derived using the second pathway, and it typically resulted in complex carbon structures; the nanofillers incorporated initially often facilitate this carbonization process, resulting in various preferred carbon nanostructures.

Carbon-stabilized iron NPs have been prepared using poly(vinyl alcohol) (PVA) as the carbon source, and carbonization was achieved by annealing at a high temperature of 750°C. Briefly speaking, iron nitrate, sodium chloride, and PVA were dissolved in deionized water at an iron nitrate to sodium chloride ratio of 1 : 20. The solution was then heated to 70°C to remove deionized water. The solid sample was further heat-treated through two necessary steps. First, it was heated at 190°C in a tube furnace for about 2 h under air atmosphere to produce $Fe(NO_3)_3$, and then it was decomposed to form Fe_2O_3 NPs. In the second step, PVA was carbonized and Fe_2O_3 was reduced by the carbon decomposed from PVA to Fe NPs at 750°C in a nitrogen atmosphere, and the resulting carbon served as a protection shell against oxidation of the iron in air. Salt, NaCl, was initially added to serve as a spacer to prevent iron NP aggregation during the formation process. When the core–shell C@Fe NPs were formed, NaCl was removed by washing with deionized water multiple times. Hydrochloric acid at a pH of 1.0 was used to remove uncoated Fe NPs as well as to introduce carboxylic acid on the NP surfaces. Finally, the sample was dried in a vacuum oven at 40°C for 24 h to remove any residual water and acid. Finally, the sample was dried in a vacuum oven at 40°C for 24 h to remove any residual water and acid. A schematic of the carbonization of PVA and the formation of core–shell C@Fe is shown in Figure 4.3 [23].

4.3 CHARACTERIZATION AND PROPERTY ANALYSIS OF MULTIFUNCTIONAL NANOCOMPOSITES

As their nature indicates, multifunctional nanocomposites generally demonstrate a variety of unique properties associated with the integration of multicomponents in the entity and because of the synergy induced during material fabrication and processing. In this section, we mainly focus on the thermal, mechanical, magnetic, and rheological properties relevant to our research interests.

4.3.1 Thermal Properties Analysis with TGA and DSC

Thermal stability of multifunctional nanocomposites can be analyzed by thermal gravimetric analysis (TGA) either in air or of inert gases such as nitrogen. Through TGA, changes in the mass of nanocomposites over a wide range of temperatures will reflect the thermal behavior of integrated components within the composites as well as those adsorbed on the surface, such as moisture and residual solvents. It typically involves both thermal degradation and oxidative decomposition of the polymer if analyzed in air, but merely thermal degradation in inert gases. The residual mass is usually metal or ceramic species that will not decompose and be released into the gas phase at the final high temperature. Incorporation of nanofillers has, in general, been proven to enhance thermal stability; that is, the polymer or carbon matrix can only be degraded at higher temperatures. Differential scanning calorimetry (DSC) is another important thermal analysis technique that is widely used to determine glass transition temperature (T_g), melting temperature (T_m), as well as crystalline properties of nanocomposite samples. The inherent characteristic temperatures T_g and T_m reflect the atomic, molecular, and stereomicrostructures of the respective polymer and/or carbon matrix in multifunctional composites, and nanofiller components may affect both temperatures by interaction with the matrix, thereby affecting the corresponding rigidity as well as the crystallinity of the polymer and/or carbon.

In the PP/iron core–shell nanocomposite system, thermal behavior was affected not only by the molecular weight of the PP, but also by the relative density of the grafting maleic anhydride (MA) functional groups on the polymer chain. In Figure 4.4, both neat low molecular weight (LM)–PP and high molecular weight (HM)–PP show different thermal decomposition profiles in air and in nitrogen; neat LM–PP demonstrated slower degradation kinetics than HM–PP in both air and nitrogen, which can be attributed to the relatively higher density of MA groups on LM–PP compared to HM–PP, and thus much higher intermolecular interactions (mostly notably through hydrogen bonding) between polymer chains for LM–PP. The peaks of melting temperature recrystallization in the DSC graph indicate that HM–PP is highly crystalline while LM–PP is amorphous.

4.3.2 Mechanical Properties Analysis: Young's Modulus and Tensile Strength

Mechanical properties are critical evaluation aspects for multifunctional nanocomposites, no matter what type of application they are targeting, which may include electronic devices, magnetic sensors, drug delivery vehicles, and so on. Generally speaking, the mechanical properties in consideration mainly focus on Young's modulus, which is the ratio of the tensile stress (MPa) and the tensile strain (dimensionless) in the elastic portion of the stress–strain curve. Figure 4.5 shows the tensile stress–strain curves for polyurethane/iron nanocomposites formed through either the direct method (DM) or the SIP method. It demonstrates that the product derived from the former method showed large cracks, while that derived from the latter was much more flexible and showed no cracks. Figure 4.5 presents a quantitative measurement of the mechanical behaviors via the tensile stress–tensile strain curves of the Fe_2O_3/polyurethane (PU) composites fabricated from the DM and SIP methods, respectively. Young's moduli and tensile strengths are similar in both composites. However, the elongation of the SIP composite is about four times that of the DM composite. The strong chemical bonding between Fe_2O_3 NPs and polyurethane as well as the uniform particle distribution within the polymer matrix contribute toward the mechanical behavior of the SIP composites.

4.3.3 Magnetic Properties, MR, and Giant Magnetoresistance in Multifunctional Nanocomposites

Multifunctional magnetic composites have found wide applications in electronics, magnetic sensing, and environmental remediation. Compared to their metallic counterpart, polymer- and carbon-based multifunctional nanocomposites have the advantages of light weight, flexibility, processability, and anticorrosion. The high specific area of these structured materials offers enhanced adsorption capacity, while their magnetic property enables recycling of the new family of adsorbents. Figure 4.6 shows the magnetic properties of magnetic graphene nanoplatelet composites (MGNCs); a large coercivity of 496 Oe was observed in core–double shell MGNCs, which indicates that they are ferromagnetic and harder than bare Fe NPs (5.0 Oe) of comparable size [17, 24] at room temperature after they were placed on the graphene sheet. MGNCs are tested to be stable after a 4-h immersion in 1 M HCl [25]. This further confirms the formation of a protective carbon shell around the magnetic core. This unique magnetic property enables easy recycling of MGNC adsorbents for environmental remediation, leaving no secondary pollution, and simultaneously improving the economic value of MGNCs.

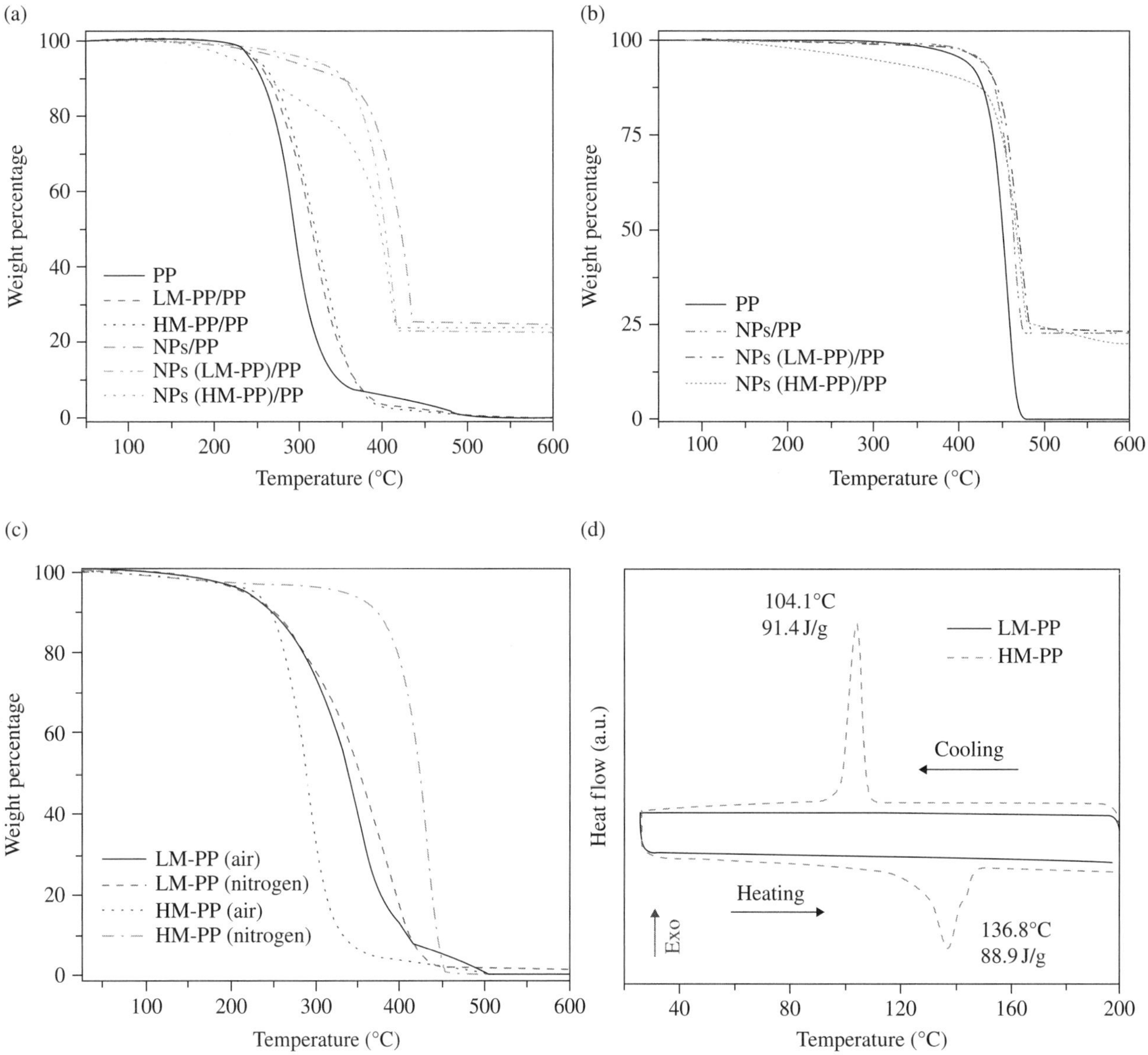

FIGURE 4.4 TGA thermograms of PP, M-PP/PP blends, and PNCs in (a) air and (b) nitrogen, (c) TGA of LM and HM in both air and nitrogen, and (d) DSC of LM–PP and HM–PP. Adapted with permission from Ref. [20]. © RSC.

MR is a phenomenon where resistance of a material changes when an external magnetic field is applied. Giant magnetoresistance (GMR) is the large resistance change that occurs when the relative orientation of the magnetic domains in adjacent layers is adjusted from antiparallel to parallel under an applied magnetic field. It is defined as the ratio $(R_{AP}-R_P)/R_P$, where R_P and R_{AP} are the resistances of materials for parallel and antiparallel alignments, respectively. The GMR phenomenon was first discovered in multilayered structural materials, in which ferromagnetic Fe metal layers were separated by a nonmagnetic Cr layer in 1988. Large GMR values can be obtained typically at very low temperatures, while obtaining large-room-temperature GMR is still a challenge. In addition, similar GMR was reported in granular-structured materials, in which a granular structure with ferromagnetic nanograins is dispersed in a nonmagnetic metal. The first reported granular GMR was in Cu/Co composites where Co granules were embedded in a Cu matrix [26]. Since then, other metal composites have been found to exhibit GMR with a reported maximum signal of 60% at 4.2 K [27].

Most recently, molecular and composite systems have exhibited interesting MR phenomena in which both positive and negative resistance changes were observed upon applying an external magnetic field. This is intriguing, in particular, for conducting polymer–based multifunctional nanocomposites, as the conducting polymer is a conjugated structure with electrons being exchanged and transported along the backbone in the case of oxidative remediation of heavy metal species. Figure 4.7 shows

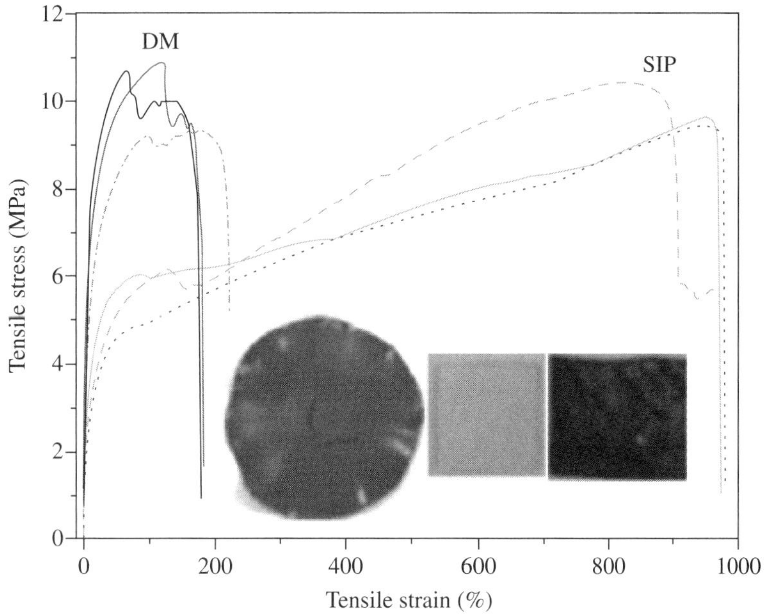

FIGURE 4.5 Tensile stress–strain curves of 65 wt% nanocomposites fabricated by the DM and SIP methods; insets show the optical micrographs of composites obtained using DM (left), pure polyurethane (middle), SIP (right). Adapted with permission from Ref. [16]. © IOP.

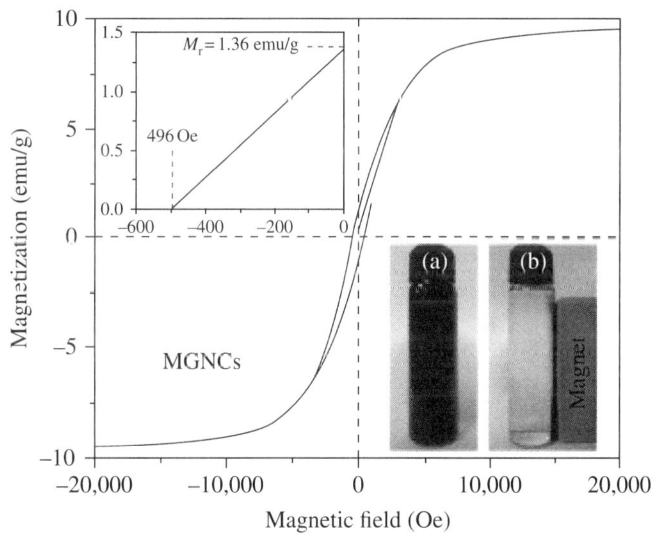

FIGURE 4.6 Room-temperature hysteresis loop of MGNCs. Top inset shows the enlarged partial curve and bottom inset shows MGNCs dispersed in acid and magnetic separation. Adapted with permission from Ref. [25]. © ACS.

the MR properties of carbon nanostructure–derivatized PANI-based nanocomposites at two different temperatures; at the lower temperature of 130 K, all samples demonstrated small, negative MR, while at the higher temperature of 290 K all samples showed large, positive MR. This phenomenon indicated that at different temperatures, electron transport is affected differently by the external magnetic field, which further indicates that at different temperatures, electron transport takes place via different mechanisms.

4.3.4 Rheological Analysis of Multifunctional Nanocomposites

Dispersion of nanofillers in a polymer or carbon matrix may have a significant impact on the rheological behavior, which subsequently affects the processing and properties of multifunctional nanocomposites. The difference in rheological behavior can in turn provide insights into the structural variations of nanofillers in host polymer or carbon matrix. The characteristic

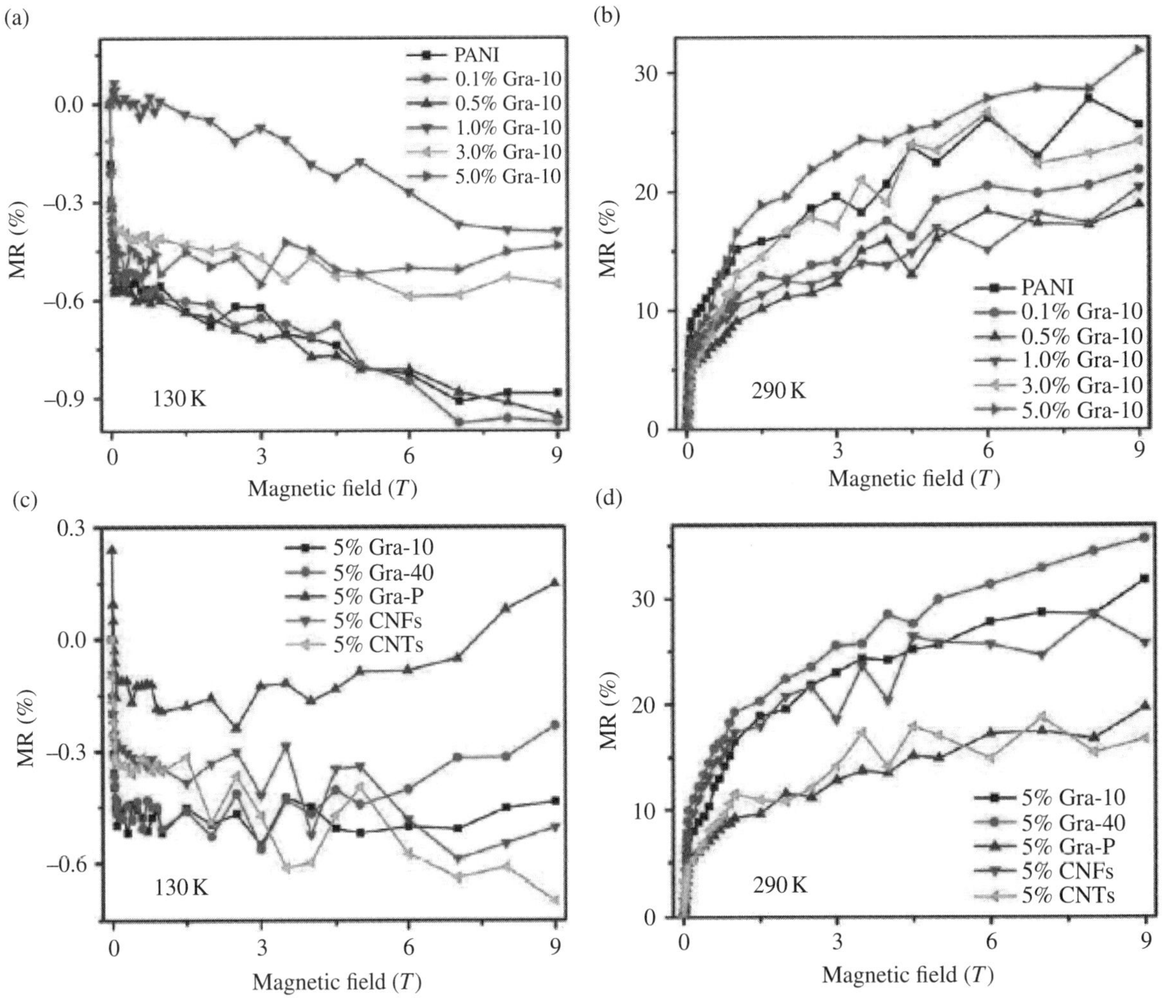

FIGURE 4.7 MR behavior of pure PANI and its PNCs as a function of Gra-10 loading at (a) 130 K and (b) 290 K; MR behavior of the PNCs filled with 5% weight loading of different carbon nanostructures at (c) 130 K and (d) 290 K. Adapted with permission from Ref. [28]. © ACS.

parameters involved in the rheological analysis versus the oscillation frequency sweeping include complex viscosity (η^*), storage modulus (G'), loss modulus (G''), and damping property (tan δ). There are generally two major types of fluid behaviors: the Newtonian frequency–independent fluid and the shear thinning fluid. Figure 4.8 shows an example of a PP-based nanocomposite where the PP matrix is doped with a small percentage of MA-grafted PP, with in situ synthesized cobalt NPs as nanofillers. As is noted from the complex viscosity graph, the PP/PP-g–MA demonstrated Newtonian behavior at a lower frequency range (below 1 Hz), while it showed apparent shear thinning behavior at a higher frequency range (between 1 and 100 Hz). PNCs with relatively low-level nanofillers show similar behavior as the blend polymer matrix, while with high-level nanofillers, the shear thinning phenomenon occurs throughout the whole frequency range.

4.4 ENVIRONMENTAL REMEDIATION THROUGH MULTIFUNCTIONAL NANOCOMPOSITES

Rapid industrialization has led to an increased discharge of wastes into the environment, posing potential harm to society and human health. Among these are the heavy metal species found in underground and surface water, as well as the solid plastic wastes accumulating in landfills and other locations.

Among these heavy metal species, Cr(VI) is a commonly identified contaminant because of its high toxicity and mobility [30]. The maximum permissible limit of Cr in drinking water as recommended by the U.S. Environmental Protection Agency (EPA) is 100 μg/l [31]. Arsenic is also known for its toxicity and carcinogenicity to human beings [32–34]; drinking water contaminated with arsenic is becoming an increasingly significant concern throughout the world. Long-term exposure to arsenic

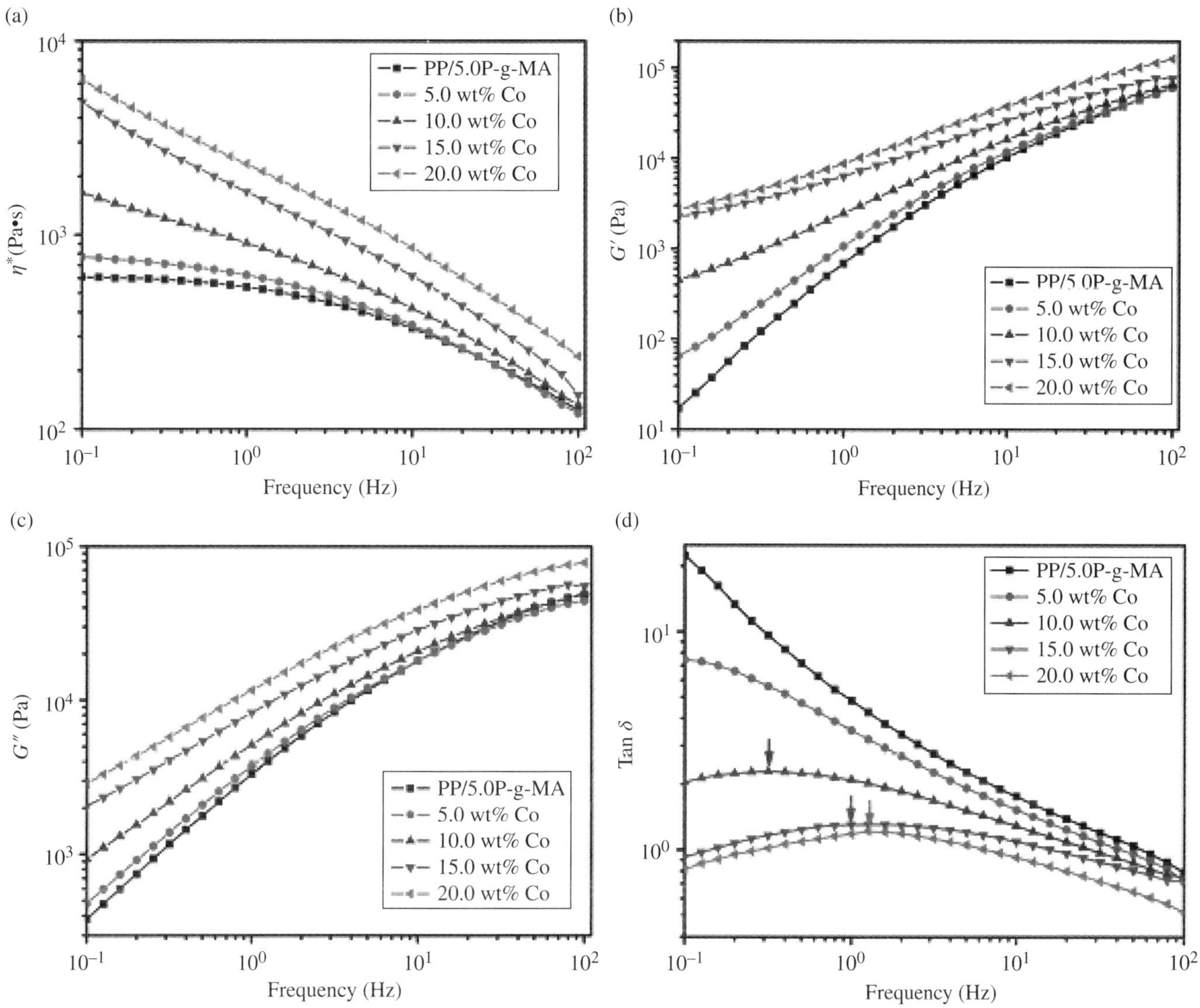

FIGURE 4.8 Melt rheological properties of PP-based nanocomposites: (a) complex viscosity, (b) storage modulus, (c) loss modulus, and (d) damping property of PP and its nanocomposites with different particle loadings. Adapted with permission from Ref. [29]. © ACS.

can cause cancers of the bladder, lungs, skin, kidney, liver, and prostate [35]. Many countries have set strict regulations to limit arsenic level in drinking water, which adds to the technical challenges as well as places financial burden on water facilities to reach the 10 ppb requirements set by the U.S. EPA. Thus there is urgent need for efficient and economic methods to remove heavy metal species and arsenic.

In the following sections, we will discuss in more detail our research on chromium removal, arsenic remediation, and sustainable treatment of plastic solid wastes. Both polymer- and carbon-based multifunctional nanocomposites have been explored for application in environmental remediation especially for heavy metal removal, targeting trace-level metal concentration as well as fast kinetics and comprehensive utilization of composite materials.

4.4.1 Solid Wastes Treatment—Comprehensive and Sustainable Recycling

Solid wastes treatment has become a challenge with the advancement of industrialization and the expanded use of plastics. There are generally three strategies to deal with solid wastes: incineration, landfilling, and recycling. Incineration may generate toxic gases causing secondary pollution, and landfilling is also not practical in the long term considering the limited land resources and possible leaching of hazardous substances from landfill sites. Therefore, comprehensive recycling has become the most favorable method for the treatment of solid wastes. Here we take PS/Ni@NiO PNCs recycling as an example [36]

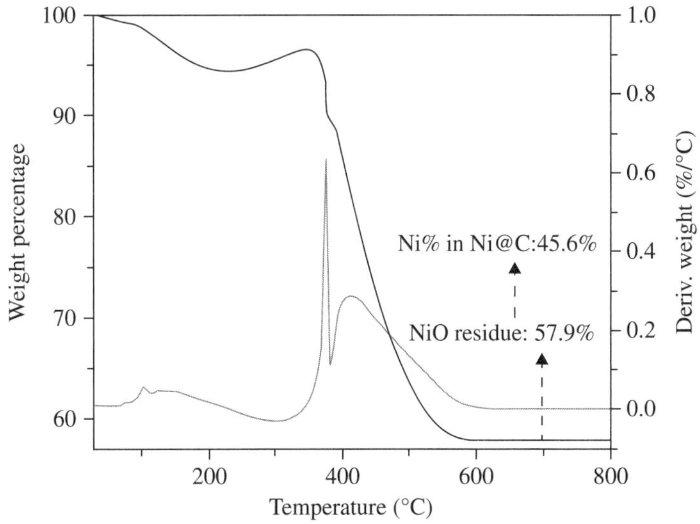

FIGURE 4.9 Thermogravimetric curves of the core@shell structured Ni@C NPs. Adapted with permission from Ref. [36]. © RSC.

where magnetic core@shell structured Ni@C NPs as well as liquid fuel products were generated. The former is an excellent adsorbent that can be used for environmental remediation due to the high specific surface area of the carbon shell and the high magnetization of the Ni core. Liquid fuels produced from the PNCs contain more saturated components than those from PS, which is attributed to catalytic hydrogenation by the Ni nanofiller during the pyrolysis process (Fig. 4.9).

4.4.2 Wastewater Treatment—Adsorption and/or Redox Remediation

A number of technologies for wastewater treatment and remediation, such as cyanide treatment [37], electrochemical precipitation [38], reverse osmosis (RO) [39, 40], ion exchange (IE) [41, 42], and adsorption [23, 43–49], have been developed. Among these technologies, however, chlorination of cyanides can result in highly toxic intermediates and other toxic organochlorines. These compounds together with residual chlorine create additional environmental problems. Precipitation is considered to be the most applicable and economical approach, but this technique may generate a large amount of precipitate sludge that requires additional processes for further treatment. Though RO can effectively reduce metal ions, its applications are limited due to high costs and low pH value [50]. There are very limited studies on the removal of Cr(III) by IE [41, 42, 51], and its operation cost is also higher compared to other techniques [39]. In addition, Cr(VI) reduction by bulk zero-valence Fe [52, 53], Fe(II) [54, 55], hydrogen [56, 57], dissolved organic compounds [58], and sulfur compounds [59] was also reported to have reached satisfactory removal capacity over an extended treatment time.

More recently, nanostructured iron and iron oxides have proven highly efficient for heavy metal removal through either reduction or adsorption [60–62]. However, using these nanomaterials still has two major challenges; one is easy oxidation/dissolution of pure Fe NPs especially in acidic solutions and the other is the recovery of minute NPs, especially in a continuous flow system. To prevent oxidation/dissolution of NPs, a chemically inert shell structure is usually introduced around the NP core. The shell material may include silica [63–66], polymer [67, 68], carbon [23], and noble metals [69–71]. Among these alternatives, carbon shell has proven to be favorable due to its low cost and high specific surface area [72, 73]. Furthermore, among the various carbon materials, graphene, an ultrathin and high surface area carbon structure, stands out as the best choice as an adsorbent owing to its unique electronic and chemical structures. Magnetic graphene nanocomposites, with added magnetic property as well as enhanced specific surface areas, have been prepared by our group, and have successfully been applied as efficient and fast adsorbents in heavy metal remediation. Figure 4.10 demonstrates chromium removal efficiency by pristine graphene and magnetic graphene nanocomposites; the latter demonstrated much higher efficiency than the former, and within 5 min of treatment, Cr(VI) can be removed almost completely at an adsorbent loading level of 3.0 g/l.

4.4.3 Wastewater Treatment—Arsenic Removal from Wastewater

Arsenic is another species that poses serious health and environmental concerns. In general, arsenic in inorganic form is more toxic than in organic form, and arsenite [As(III)] is more mobile and toxic than arsenate [As(V)] [74]. There have been various efforts on remediation of arsenic from the disposal scene, including ion exchange [75], RO [76], and adsorption. Compared to

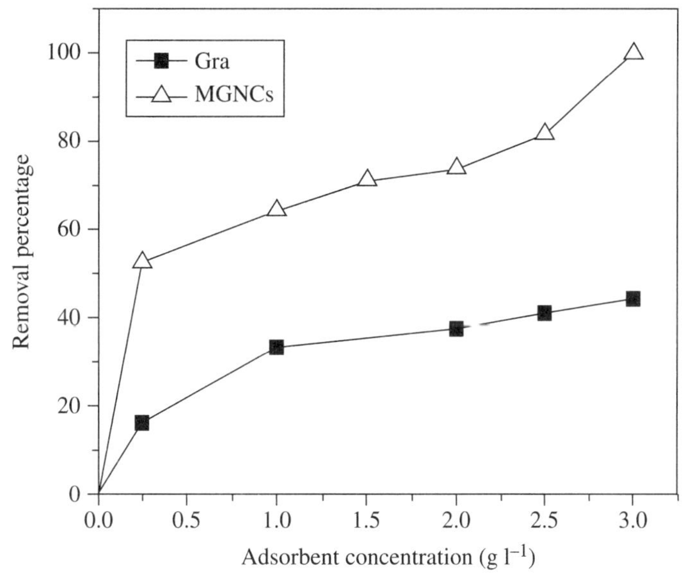

FIGURE 4.10 Cr(VI) removal percentage based on different loadings of Gra and MGNCs. ([Cr(VI)] = 1000 μg/l, pH 7, treatment time: 5 min). Adapted with permission from Ref. [25]. © ACS.

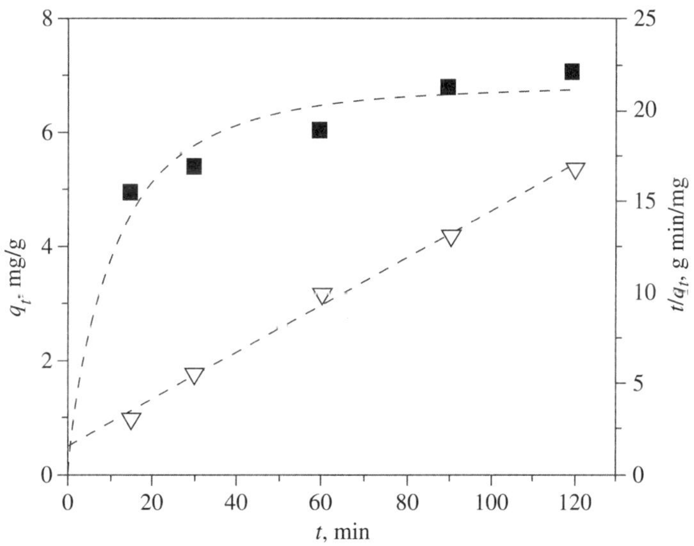

FIGURE 4.11 Adsorption kinetic behavior of As(III) by MGNCs: As(III) removal rate q_t versus time t (curve with squares) and the transformed rate plot t/q_t versus t (curve with triangles). (MGNC concentration: 0.2 g/l [As(III)] = 4 ppm, pH = 7, room temperature). Adapted with permission from Ref. [77]. © ECS.

others methods, adsorption has the advantages of low cost and low technology demand for operation and maintenance. As for chromium removal, MGNCs with core–shell Fe–Fe_2O_3 NPs also showed great potential in the remediation of arsenic species [77]. As can be seen in Figure 4.11, the removal rate of arsenic reached 7 mg/g after a 2-h interaction at room temperature when the starting concentration of As(III) was 4 ppm and the pH 7.

4.4.4 Wastewater Treatment—Residual Dye Removal

Dyes are commonly used in many industries such as paper, textile, pulp, pharmaceutical, and bleaching. Residual dyes discharged into natural water resources consist of serious organic pollutants, posing health and environmental concerns. The textile industry is a major source of dye pollutants in significant amounts. As the degradation products of those dyes, such as benzidine, naphthalene, and other aromatic compounds, are potentially carcinogenic or mutagenic to life forms, treatment of dye-polluted water is of urgent importance. Because of the superior performance of multifunctional nanocomposites in the remediation of heavy metal species and arsenic, they could be potential candidates for organic dye removal from wastewater, as the dyes are basically organic molecules, which are anionic, nonionic, or cationic. Thus by adjusting the pH, porosity, and integrated functionalities of multifunctional nanocomposite adsorbents, organic dyes can be removed, and their comprehensive reuse is expected in the near future.

4.5 SUMMARY

In this chapter, we mainly focused on what we have achieved in environmental remediation with multifunctional polymer- and carbon-based magnetic nanocomposites. The preparation, characterization, and property analysis of these materials are detailed. Moreover, the superior performance of these synergistic materials in the remediation of waste species such as for chromium and arsenic removal, especially when compared to the widely accepted activated carbon, indicates their great potential for application in other types of pollutants, including, but not limited to, organic dyes that are discharged into the environment.

REFERENCES

[1] Sun XC, Nava N, Reyes GJ. Microstructure and magnetic properties of Fe(C) and Fe(O) nanoparticles. *MRS Proc*; 2001;704:W9.5.1.

[2] Diaz L, Santos M, Ballesteros C, Marysko M, Pola J. IR laser-induced chemical vapor deposition of carbon-coated iron nanoparticles embedded in polymer. J Mater Chem 2005;15:4311–4317.

[3] Cao H, Huang G, Xuan S, Wu Q, Gu F, Li C. Synthesis and characterization of carbon-coated iron core/shell nanostructures. J Alloys Compd 2008;448:272–276.

[4] Fan N, Ma X, Ju Z, Li J. Formation, characterization and magnetic properties of carbon-encapsulated iron carbide nanoparticles. Mater Res Bull 2008;43:1549–1554.

[5] Wang G, Zhang L, Zhang J. A review of electrode materials for electrochemical supercapacitors. Chem Soc Rev 2012;41 (2):797–828.

[6] Li Y, Tan B, Wu Y. Mesoporous Co_3O_4 nanowire arrays for lithium ion batteries with high capacity and rate capacity. Nano Lett 2008;8 (1):265–270.

[7] Zhang LL, Zhao XS. Carbon-based materials as supercapacitor electrodes. Chem Soc Rev 2009;38 (9):2520–2531.

[8] Yu G, Hu L, Vosgueritchian M, Wang H, Xie X, McDonough JR, Cui X, Cui Y, Bao Z. Solution-processed graphene/MnO_2 nanostructured textiles for high-performance electrochemical capacitors. Nano Lett 2011;11 (7):2905–2911.

[9] Lu Q, Lattanzi MW, Chen Y, Kou X, Li W, Fan X, Unruh KM, Chen JG, Xiao JQ. Supercapacitor electrodes with high-energy and power densities prepared from monolithic NiO/Ni nanocomposites. Angew Chem Int Ed 2011;50 (30):6847–6850.

[10] Hughes M, Chen GZ, Shaffer MSP, Fray DJ, Windle AH. Electrochemical capacitance of a nanoporous composite of carbon nanotubes and polypyrrole. Chem Mater 2002;14 (4):1610–1613.

[11] Zhu J, Chen M, Qu H, Zhang X, Wei H, Luo Z, Colorado HA, Wei S, Guo Z. Interfacial polymerized polyaniline/graphite oxide composites toward electrochemical energy storage. Polymer 2012;53 (25):5953–5964.

[12] Wang DW, Li F, Zhao J, Ren W, Chen ZG, Tan J, Wu ZS, Gentle I, Lu GQ, Cheng HM. Fabrication of graphene/polyaniline composite paper via in situ anodic electropolymerization for high-performance flexible electrode. ACS Nano 2009;3 (7):1745–1752.

[13] Xu J, Wang K, Zu SZ, Han BH, Wei Z. Hierarchical nanocomposites of polyaniline nanowire arrays on graphene oxide sheets with synergistic effect for energy storage. ACS Nano 2010;4 (9):5019–5026.

[14] Zhang K, Zhang LL, Zhao XS, Wu J. Graphene/polyaniline nanofiber composites as supercapacitor electrodes. Chem Mater 2010;22 (4):1392–1401.

[15] Ramakrishna S, Huang Z-M, Kumar GV, Batchelor AW, Mayer J. *An Introduction to Biocomposites. Vol. 1 Series on Biomaterials and Bioengineering*. London: Imperial College Press; 2004.

[16] Guo Z, Park S, Wei S, Pereira T, Moldovan M, Karki AB, Young DP, Hahn HT. Flexible high-loading particle reinforced polyurethane magnetic nanocomposite fabrication through particle surface initiated polymerization. Nanotechnology 2007;18:335704.

[17] Guo Z, Lin H, Karki AB, Wei S, Young DP, Park S, Willis J, Hahn HT. Facile monomer stabilization approach to fabricate lion/vinyl ester resin nanocomposites. Compos Sci Technol 2008;68:2551–2556.

[18] Gu H, Huang Y, Zhang X, Wang Q, Zhu J, Shao L, Haldolaarachchige N, Young DP, Wei S, Guo Z. Magnetoresistive polyaniline-magnetite nanocomposites with negative dielectrical properties. Polymer 2012;53:801–809.

[19] He Q, Yuan T, Zhu J, Luo Z, Haldolaarachchige N, Sun L, Khasanov A, Li Y, Young DP, Wei S, Guo Z. Magnetic high density polyethylene nanocomposites reinforced with in-situ synthesized Fe@FeO core-shell nanoparticles. Polymer 2012;53:3642–3652.

[20] Zhu J, He Q, Luo Z, Khasanov A, Li Y, Sun L, Wang Q, Wei S, Guo Z. Property manipulated polypropylene-iron nanocomposites with maleic anhydride polypropylene. J Mater Chem 2012;22:15928–15938.

[21] Zhu J, Wei S, Li Y, Sun L, Haldolaarachchige N, Young D, Southworth C, Khasanov A, Luo Z, Guo Z. Surfactant-free synthesized magnetic polypropylene nanocomposites: rheological, electrical, magnetic and thermal properties. Macromolecules 2011;44: 4382–4391.

[22] He Q, Yuan T, Wei S, Haldolaarachchige N, Luo Z, Young DP, Khasanov A, Guo Z. Morphology and phase controlled iron oxide nanoparticles stabilized with maleic anhydride grafted polypropylene. Angew Chem Int Ed 2012;51:8842–8845.

[23] Zhang D, Wei S, Kaila C, Su X, Wu J, Karki AB, Young DP, Guo Z. Carbon-stabilized iron nanoparticles for environmental remediation. Nanoscale 2010;2:917–919.

[24] Guo Z, Park S, Hahn HT, Wei S, Moldovan M, Karki AB, Young DP. Giant magnetoresistance behavior of an iron/carbonized polyurethane nanocomposite. Appl Phys Lett 2007;90:053111.

[25] Zhu J, Wei S, Gu H, Rapole SB, Wang Q, Luo Z, Haldolaarachchige N, Young DP, Guo Z. One-pot synthesis of magnetic graphene nanocomposites decorated with core@double-shell nanoparticles for fast chromium removal. Environ Sci Technol 2012;46:977–985.

[26] Berkowitz AE, Mitchell JR, Carey MJ, Young AP, Zhang S, Spada FE, Parker FT, Hutten A, Thomas G. Giant magnetoresistance in heterogeneous Cu-Co alloys. Phys Rev Lett 1992;68 (25):3745–3748.

[27] Xiong P, Xiao G, Wang JQ, Xiao JQ, Jiang JS, Chien CL. Extraordinary hall effect and giant magnetoresistance in the granular Co-Ag system. Phys Rev Lett 1992;69 (22):3220–3223.

[28] Zhu J, Gu H, Luo Z, Haldolaarachchige N, Young DP, Wei S, Guo Z. Carbon Nanostructures derived polyaniline metacomposites: electrical dielectric and giant magnetoresistive properties. Langmuir 2012;28 (27):10246–10255.

[29] He Q, Yuan T, Zhang X, Luo Z, Haldolaarachchige N, Sun L, Young DP, Wei S, Guo Z. Magnetically soft and hard polypropylene/cobalt nanocomposites: role of maleic anhydride grafted polypropylene. Macromolecules 2013;46:2357–2368.

[30] Hsu LC, Wang SL, Lin YC, Wang MK, Chiang PN, Liu JC, Kuan WH, Chen CC, Tzou YM. Cr(VI) removal on fungal biomass of Neurospora crassa—the importance of dissolved organic carbons derived from the biomass to Cr(VI) reduction. Environ Sci Technol 2010;44:6202–6208.

[31] Xu Y, Zhao D. Reductive immobilization of chromate in water and soil using stabilized iron nanoparticles. Water Res 2007;4:2101–2108.

[32] Kitchin KT, Conolly R. Arsenic-induced carcinogenesis-oxidative stress as a possible mode of action and future research needs for more biologically based risk assessment. Chem Res Toxicol 2009;23:327–335.

[33] Chandra V, Park J, Chun Y, Lee JW, Hwang IC, Kim KS. Water-dispersible magnetite-reduced graphene oxide composites for arsenic removal. ACS Nano 2010;4:3979–3986.

[34] Nata IF, Sureshkumar M, Lee CK. One-pot preparation of amine-rich magnetite/bacterial cellulose nanocomposite and its application for arsenate removal. RSC Adv 2011;1:625–631.

[35] Navas-Acien A, Sharrett AR, Silbergeld EK, Schwartz BS, Nachman KE, Burke TA, Guallar E. Arsenic exposure and cardiovascular disease: a systematic review of the epidemiologic evidence. Am J Epidemiol 2005;162:1037–1049.

[36] Zhu J, Wei S, Li Y, Pallavkar S, Lin H, Haldolaarachchige N, Luo Z, Young DP, Guo Z. Comprehensive and sustainable recycling of polymer nanocomposites. J Mater Chem 2011;21:16239–16246.

[37] Monser L, Adhoum N. Modified activated carbon for the removal of copper, zinc, chromium and cyanide from wastewater. Sep Purif Technol 2002;26:137–146.

[38] Kongsricharoern N, Polprasert C. Chromium removal by a bipolar electro-chemical precipitation process. Water Sci Technol 1996;34:109–116.

[39] Hafez A, El-Mariharawy S. Design and performance of the two-stage/two-pass RO membrane system for chromium removal from tannery wastewater. Part. 3. Desalination 2004;165:141–151.

[40] Modrzejewska Z, Kaminski W. Separation of Cr(VI) on chitosan membranes. Ind Eng Chem Res 1999;38:4946–4950.

[41] Rengaraj S, Joo CK, Kim Y, Yi J. Kinetics of removal of chromium from water and electronic process wastewater by ion exchange resins 1200H, 1500H andIRN97H. J Hazard Mater 2003;102:257–275.

[42] Rengaraj S, Yeon KH, Moon SH. Removal of chromium from water and wastewaters by ion exchange resins. J Hazard Mater 2001;87:273–287.

[43] Mohan D, Singh KP, Singh VK. Removal of hexavalent chromium from aqueous solution using low-cost activated carbons derived from agricultural waste materials and activated carbon fabric cloth. Ind Eng Chem Res 2005;44:1027–1042.

[44] Sharma YC, Singh B, Agrawal A, Weng CH. Removal of chromium by riverbed sand from water and wastewater effect of important parameters. J Hazard Mater 2008;151:789–793.

[45] Hu J, Chen G, Lo IMC. Removal and recovery of Cr(VI) from wastewater by maghemite nanoparticles. Water Res 2005;39:4528–4536.

[46] AydIn YA, Aksoy ND. Adsorption of chromium on Chitosan: optimization, kinetics and thermodynamics. Chem Eng J 2009;151:188–194.

[47] Zhao Y, Peralta-Videa JR, Lopez-Moreno ML, Ren M, Saupe G, Gardea-Torresdey JL. Kinetin increases chromium absorption, modulates its distribution, and changes the activity of catalase and ascorbate peroxidase in Mexican Palo Verde. Environ Sci Technol 2011;45:1082–1087.

[48] Sawalha MF, Gardea-Torresdey JL, Parsons JG, Saupe G, Peralta-Videa JR. Determination of adsorption and speciation of chromium species by saltbush (atriplex canescens) biomass using a combination of XAS and ECP-OES. Microchem J 2005;81:122–132.

[49] Parsons JG, Hejazi M, Tiemann KJ, Henning J, Gardea-Torresdey JL. An XAS study of the binding of copper(II), zinc(II), chromium(III) and chromium(VI) to hopbiomass. Microchem J 2002;71:211–219.

[50] Modrzejewska Z, Kaminski W. Separation of Cr (VI) on chitosan membranes. Ind Eng Chem Res 1999;38:4946–4950.

[51] Pansini M, Colella C, Gennaro M. Chromium removal from water by ion exchange using zeolite. Desalination 1991;83:145–157.

[52] Melitas N, Chuffe-Moscoso O, Farrell J. Kinetics of soluble chromium removal from contaminated water by zerovalent iron media: corrosion inhibition and passive oxide effects. Environ Sci Technol 2001;35:3948–3953.

[53] Wei S, Wang Q, Zhu J, Sun L, Lin H, Guo Z. Multifunctional composite core-shell nanoparticles. Nanoscale 2011;3:4474–4502.

[54] Deng B, Stone AT. Surface-catalyzed chromium(VI) reduction: reactivity comparisons of different organic reductants and different oxide surfaces. Environ Sci Technol 1996;30:2484–2494.

[55] Jardine PM, Fendorf SE, Mayes MA, Larsen IL, Brooks SC, Bailey WB. Fate and transport of hexavalent chromium in undisturbed soil. Environ Sci Technol 1999;33:2939–2944.

[56] Powell RM, Puls RW, Hightower SK, Sabatini DA. Coupled iron corrosion and chromate reduction: mechanisms for subsurface remediation. Environ Sci Technol 1995;29:1913–1922.

[57] Loyaux-Lawniczak S, Refait P, Ehrhardt J-J, Lecomte P, Génin JMR. Trapping of Cr by formation of ferrihydrite during the reduction of chromate ions by Fe(II)-Fe(III) hydroxysalt green rusts. Environ Sci Technol 1999;34:438–443.

[58] Reardon EJ. Anaerobic corrosion of granular iron measurement and interpretation of hydrogen evolution rates. Environ Sci Technol 1995;29:2936–2945.

[59] Gangoly N, Markey DC, Thodos G. *Water-reuse: water's interface with energy, air, and solids*. CONF-750530. New York: American Institute of Chemical Engineers; March 19–20, 1975.

[60] Zhang W. Nanoscale iron particles for environmental remediation: an overview. J Nanopart Res 2003;5:323–332.

[61] Ponder SM, Darab JG, Mallouk TE. Remediation of Cr(VI) and Pb(II) aqueous solutions using supported, nanoscale zero-valent iron. Environ Sci Technol 2000;34:2564–2569.

[62] Hu J-S, Zhong L-S, Song W-G, Wan L-J. Synthesis of hierarchically structured metal oxides and their application in heavy metal ion removal. Adv Mater 2008;20:2977–2982.

[63] Lu Y, Yin Y, Mayers BT, Xia Y. Modifying the surface properties of superparamagnetic iron oxide nanoparticles through a sol-gel approach. Nano Lett 2002;2:183–186.

[64] Wang G, Harrison A. Preparation of iron particles coated with silica. J Colloid Interface Sci 1999;217:203–207.

[65] Zhu J, Wei S, Haldolaarachchige N, Young DP, Guo Z. Electromagnetic field shielding polyurethane nanocomposites reinforced with core-shell Fe silica nanoparticles. J Phys Chem C 2011;115:15304–15310.

[66] Zhu J, Wei S, Lee IY, Park S, Willis J, Haldolaarachchige N, Young DP, Luo Z, Guo Z. Silica stabilized iron particles toward anti-corrosion magnetic polyurethane nanocomposites. RSC Adv 2012;2:1136–1143.

[67] Pellegrino T, Manna L, Kudera S, Liedl T, Koktysh D, Rogach AL, Keller S, Rädler J, Natile G, Parak WJ. Hydrophobic nanocrystals coated with an amphiphilic polymer shell: a general route to water soluble nanocrystals. Nano Lett 2004;4:703–707.

[68] Lin C-AJ, Sperling RA, Li JK, Yang T-Y, Li P-Y, Zanella M, Chang WH, Parak WJ. Design of amphiphilic polymer for nanoparticle coating and functionalization. Small 2008;4:334–341.

[69] Cho S-J, Idrobo J-C, Olamit J, Liu K, Browning ND, Kauzlarich SM. Growth mechanisms and oxidation-resistance of Au-coated Fe nanoparticles. Chem Mater 2005;17:3181–3186.

[70] Lu Z, Prouty MD, Guo Z, Golub VO, Kumar CSSR, Lvov YM. Magnetic switch of permeability for polyelectrolyte microcapsules embedded with Co@Au nanoparticles. Langmuir 2005;21:2042–2050.

[71] Guo Z, Kumar C, Henry L, Doomes E, Hormes J, Podlaha EJ. Displacement synthesis of Cu shells surrounding Co nanoparticles. J Electrochem Soc 2005;152:D1–D5.

[72] Oliveira LCA, Rios RVRA, Fabris JD, Garg V, Sapag K, Lago RM. Activated carbon/iron oxide magnetic composites for the adsorption of contaminants in water. Carbon 2002;40:2177–2183.

[73] Vaughan RL Jr, Reed BE. Modeling As(V) removal by a iron oxide impregnated activated carbon using the surface complexation approach. Water Res 2005;39:1005–1014.

[74] Schnoor JL, editor. *Environmental Modeling: Fate and Transport of Pollutants in Water, Air, and Soil*. New York: John Wiley & Sons; 1996.

[75] Kim J, Benjamin MM. Modeling a novel ion exchange process for arsenic and nitrate removal. Water Res 2004;38:2053.

[76] Kang M, Kawasaki M, Tamada S, Kamei T, Magara Y. Effect of pH on the removal of arsenic and antimony using reverse osmosis membranes. Desalination 2000;131:293.

[77] Zhu J, Sadu R, Wei S, Chen D, Haldolaarachchige N, Luo Z, Gomes J, Young DP, Guo Z. Magnetic graphene nanoplatelet composites toward arsenic removal. ECS J Solid State Sci Technol 2012;1 (1):M1–M5.

5

NANOMATERIALS FOR THE REMOVAL OF VOLATILE ORGANIC COMPOUNDS FROM AQUEOUS SOLUTIONS

Amro El Badawy[1] and Hafiz H.M. Salih[2]

[1] *Pegasus Technical Services Inc., Cincinnati, OH, USA*

[2] *Illinois Sustainable Technology Center, University of Illinois at Urbana-Champaign, Champaign, IL, USA*

5.1 INTRODUCTION

Volatile organic compounds (VOCs) are naturally occurring or man-made hydrocarbons with high saturation vapor pressures (>102 kPa) at room temperature conditions (25°C) [1]. Examples of VOCs include BTEX (i.e., benzene, toluene, ethylbenzene, and *p*-xylene), chlorobenzenes, and chlorinated alkenes. Extensive use of VOCs in various industrial processes results in the discharge of large annual amounts of these compounds into the aqueous environment [2]. VOCs are highly reactive and are among the most toxic chemicals for human health and the environment [1]. Most VOCs are known as carcinogens (i.e., benzene), while some (toluene, xylenes) are associated with acute effects [3]. Persistent release of toxic VOCs into the aquatic environment is a growing concern. Therefore, there is great demand to develop novel, cost-effective, and environmentally friendly methods for their destruction [3].

Many remediation processes such as sorption, redox reactions, and photocatalytic transformations are currently being improved or initiated by using nanomaterials (NMs) for the removal of organic pollutants from wastewater and drinking water [4]. For instance, sorption into NMs has proven to be an effective method for the removal of VOCs from aqueous solution. Carbon nanotubes (CNTs) as well as metal and metal oxide nanoparticles (NPs) have unique properties that make them more efficient sorbents for the removal of VOCs as compared to traditional sorbent materials. Two key properties that make NMs attractive sorbents are (i) on a mass basis, NMs have significantly higher surface areas as compared to the bulk species and (ii) NMs can be functionalized with various types of functional groups to increase their affinity toward target compounds [5]. Photocatalysis and redox reactions mediated by NPs have also been tested for the decomposition of VOCs present in aqueous solutions [6, 7]. The current chapter aims to give an overview of the potential uses of NMs for the removal of VOCs from aqueous solutions.

5.2 NMs FOR BTEX REMOVAL

The BTEX VOCs benzene (B), toluene (T), ethylbenzene (E), and *p*-xylene (X) are important solvents commonly used in industrial operations [2]. A large amount of BTEX wastewater is discharged annually into the aqueous environment, posing public health and environmental risks [2]. Research studies reported the use of various types of NMs such as CNTs and gold NP–poly(dimethylsiloxane) (Au–PDMS) nanocomposites for the removal of BTEX from contaminated water [2, 8].

Nanomaterials for Environmental Protection, First Edition. Edited by Boris I. Kharisov, Oxana V. Kharissova, and H. V. Rasika Dias.

5.2.1 CNTs for BTEX Adsorption

Since their discovery in 1991 by Sumio Iijima, CNTs, a new member of the carbon family, have gained widespread attention because of their outstanding physicochemical properties [9, 10]. Nanotubes are categorized as (i) multiwalled carbon nanotubes (MWCNTs) that consist of up to several tens of graphitic shells with adjacent shell separation of ~0.34 nm, diameters of ~1 nm, and a large length to diameter ratio and (ii) single-walled carbon nanotubes (SWCNTs) that usually have a diameter close to 1 nm, with a tube length that can be many millions of times longer [9, 11].

Because CNTs provide chemically inert surfaces, large specific surface areas, hollow structure, and light mass density, they have been tested as adsorbents for various types of hazardous pollutants such as heavy metals, radionuclides, and organic compounds [12–14]. Furthermore, CNTs demonstrated superior adsorption capabilities than conventional adsorbents (e.g., activated carbon) because their structure at the atomic scale is far more well defined and uniform [9, 15]. The adsorption properties of CNTs depend on the adsorption sites, surface area, surface functional groups, purity, and porosity [16].

CNTs have four possible adsorption sites: internal sites, interstitial channels, external groove sites, and external surfaces. On the external sites (grooves and outer surfaces), the adsorption reaches equilibrium much faster than on the internal sites (interstitial and inside the tube) under same pressure and temperature conditions [9]. This is attributed to the direct exposure of the external sites to the adsorbing materials as compared to the adsorption process on the internal sites, which has to be initiated at the end of the pores, followed by diffusion to the internal sites [17, 18].

Su et al. [2] employed MWCNTs fabricated by the catalytic chemical vapor deposition method and oxidized by sodium hypochlorite (NaOCl) solution as adsorbent for BTEX in aqueous solution. The CNTs (NaOCl) exhibited significantly higher BTEX adsorption capacity than raw CNTs. The oxidation process of the CNTs using NaOCl resulted in a decrease in the surface area of CNTs (NaOCl) in all pore size ranges (1.7, 1.7–5, and 5–100 nm). The authors attributed this decrease in surface area to the formation of functional groups on the external and internal surface of CNTs (NaOCl) that are a direct product of oxidation. The introduction of surface oxygen functional groups (e.g., carboxylic groups) into CNTs provides various chemical sites for BTEX adsorption. Additionally, X-ray diffraction (XRD) data showed that the intensity of the CNT peak become considerably stronger after CNTs are oxidized. This indicates an increase in the purity of CNTs after oxidation.

The affinity of BTEX to CNTs (NaOCl) was in the order of X > E > T > B. The increase in molecular weight (B < T < E, X), the decrease in solubility (B > T > E > X), and the increase in boiling point (B < T < E, X) may explain the observed order of BTEX adsorption on CNTs (NaOCl). Changes in solution pH [3–11] did not significantly impact the BTEX adsorption on CNTs (NaOCl), which imply that BTEX are in molecular form during the adsorption process and that ion exchange does not play a part in the BTEX solution. Similar to pH, changes in solution ionic strength (0–0.2 M NaCl) had no significant impact on BTEX adsorption, reflecting high stability of CNTs (NaOCl) as BTEX adsorbents in a wide range of ionic strengths for solutions.

In another study by Su et al. [19] NaOCl-oxidized CNTs were compared with CNTs oxidized by other chemical agents including HCl, H_2SO_4, HNO_3, or NaOCl solutions for enhancing BTEX adsorption in aqueous solution. The results showed that NaOCl-oxidized CNTs showed the greatest enhancement followed by HNO_3-oxidized CNTs, and then H_2SO_4-oxidized CNTs.

In the same study, Su et al. [19] tested BTEX adsorption on granular activated carbon (GAC) and GAC (NaOCl) for comparison with CNTs (NaOCl). The GAC (NaOCl) achieved lower BTEX adsorption than GAC as a result of the collapse of the pore structure after oxidation. The NaOCl-oxidized CNTs showed the best adsorption performance of BTEX followed by GAC and then CNTs. This adsorption affinity trend may be explained with two possible reasons: (1) the surface carboxylic groups concentration followed the order of CNT (NaOCl) (1.039 mmol/g) > GAC (0.252 mmol/g) > CNTs (0.161 mmol/g) and (2) the surface total acidity that can enhance electrostatic interactions between BTEX molecules and adsorbent surface followed the order of CNTs (NaOCl) (1.17 mmol/g) > GAC (0.7 mmol/g) > CNTs (0.22 mmol/g). Furthermore, the order of BTEX adsorption was inconsistent with the order of surface area, pore volume, and average pore diameter. This clearly indicates that BTEX adsorption is dependent on the surface chemistry of the adsorbents rather than on its characteristics.

The mechanism responsible for the adsorption of BTEX on CNT (NaOCl) was identified as π–π electron–donor–acceptor mechanism involving the carboxylic oxygen atom of the CNTs (NaOCl) acting as the electron donor and the aromatic ring of BTEX acting as the electron acceptor, as presented in Figure 5.1 [19]. This mechanism is similar to the adsorption mechanism of BTEX on powdered activated carbon.

Carrillo-Carrion et al. [20] exploited the strong adsorption affinity of CNTs toward organic compounds for the development of analytical methodologies. In their study, BTEX were liquid–liquid extracted from olive oil samples using aqueous dispersion of sodium dodecyl sulfate (SDS)-coated MWCNTs as the extracting medium. Then, headspace/gas chromatography/mass spectrometry (HS/GC/MS) was used for the analysis of the aqueous phase. Liquid–liquid extraction was the key step in the entire analytical process for measuring BTEX concentrations in the samples as it is the source for sensitivity and selectivity enhancement. Selectivity and sensitivity determination is enhanced by the presence of SDS-coated MWCNTs in the aqueous phase resulting in detection limits at least 10 times lower than the direct HS method. Coating the MWCNTs with surfactants such as SDS enhances their adsorption properties because it

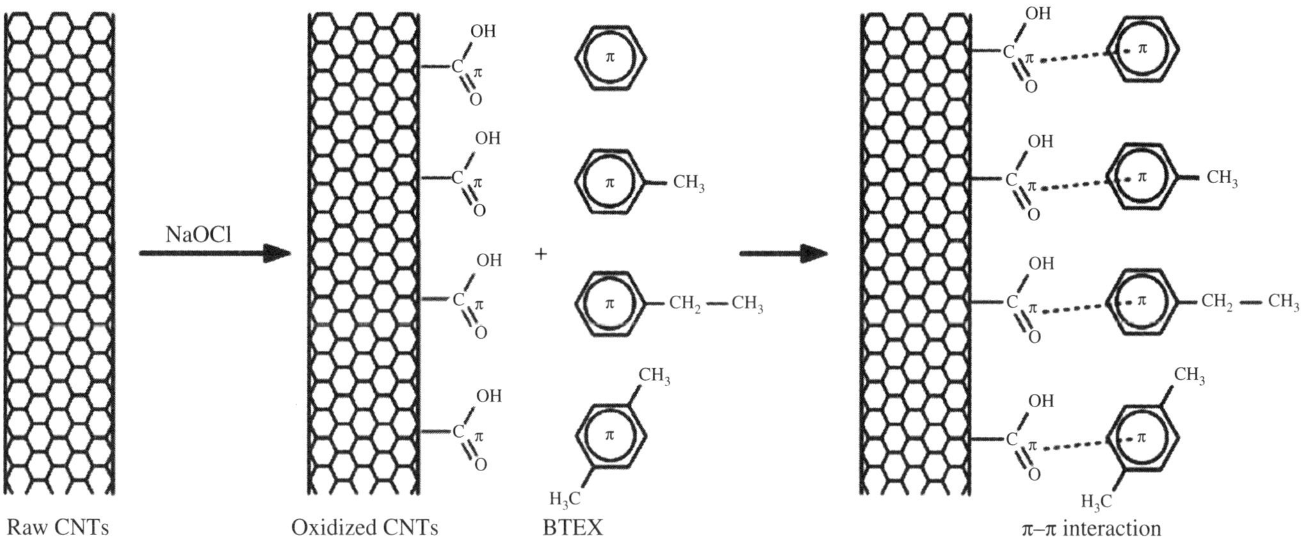

FIGURE 5.1 Diagram of the mechanism of BTEX adsorption on NaOCl-oxidized CNTs. Adapted with permission from Ref. [19], © Elsevier.

reduces the aggregation potential of MWCNTs and consequently results in maintaining the active surface area of the materials, which is responsible for the material's effectiveness.

5.2.2 Nanocomposites for BTEX Removal

Gold NP–poly(dimethylsiloxane) (Au–PDMS) nanocomposite was employed by Gupta and Kulkarni [8] for the removal of BTEX compounds from water. The Au–PDMS nanocomposite was prepared by incorporating the highly compressible porous foam PDMS with Au NPs (10–15 nm). Au–PDMS exhibits high swelling ability (~600%) against BTEX. Swelling ability is a property that has been utilized to remove oil spills from water. The use of Au NPs in nanocomposites results in a highly expanded porous PDMS structure compared to normal PDMS. Furthermore, an increase in the percentage of Au in the nanocomposite results in an increase in the degree of compressibility of the PDMS foam. The sorbent capacity of the Au–PDMS nanocomposite for BTEX in aqueous solution was nearly six times higher than that of PDMS foam without the Au NPs. An advantage of using the Au–PDMS nanocomposite for adsorbing BTEX from water compared with sorbents in the form of particulates is its capacity to be molded into any shape during or after synthesis and to fit into water bodies for practical applications.

5.3 NANOMATERIALS FOR CHLOROBENZENE REMOVAL

Large quantities of chlorobenzenes are widely employed for many industrial processes [21]. Because of their extensive use over several decades, chlorobenzenes are common in natural and engineered environmental systems. Chlorobenzenes contaminate groundwater, wastewater, and the marine environment. Because of their acute toxicity, chlorobenzenes are of grave health and environmental concern. The U.S. Environmental Protection Agency has listed dichlorobenzene (DCB), 1,2,4-trichlorobenzene (1,2,4-TCB), 1,2,4,5-tetrachlorobenzene (1,2,4,5-TeCB), and hexachlorobenzene (HCB) as priority pollutants [22]. Adsorption is the traditional method for the removal of chlorobenzenes. Besides adsorption, advanced oxidation processes (AOPs) are successfully employed to destroy chlorobenzenes. Various types of NMs have recently been utilized in the removal of chlorobenzenes from aqueous solutions through adsorption and AOPs.

5.3.1 Metal Oxides for Chlorobenzene Degradation

Selli et al. [23] compared the efficiency of three different advanced oxidation techniques, that is, photolysis, photocatalysis using TiO_2 NPs, and sonolysis, for the degradation and mineralization of 1,4-DCB. The results showed that photocatalysis on TiO_2 achieved faster removal of 1,4-DCB as compared to sonolysis and direct photolysis. The highest degradation and mineralization rate of 1,4-DCB was attained under sonophotocatalytic conditions (i.e., the combined use of photocatalysis and sonolysis).

Generally, photocatalysis includes (i) photogeneration of electrons in the conduction band trapped by a recipient (e.g., superoxide/hydroxyl radicals) and (ii) photogenerated holes in the valence band consumed by the donors, that is, organic pollutants [24]. Similar to the photocatalytic process, sonolysis of aqueous solution is known to produce hydroxyl radicals and other radical species that can rapidly initiate the degradation of 1,4-DCB through the direct attack on the aromatic ring [23]. Additionally, under ultrasound, VOCs such as 1,4-DCB may undergo degradation via another path, that is, by direct pyrolysis in the vapor phase of pulsating or collapsing cavitation bubbles within the hot interfacial region between the vapor and the surrounding liquid phases.

5.3.2 CNTs for Chlorobenzene Sorption

In a study by Liu et al. [25], MWCNTs were used as solid-phase extraction (SPE) sorbents for chlorobenzenes. Chromatographic techniques are the most widely used methods for analyzing chlorobenzenes, but preliminary separation and enrichment of trace chlorobenzenes is usually necessary in order to increase chlorobenzene concentration and to eliminate matrix effects [25]. Therefore, Liu et al. [25] utilized MNCNTs as a solid phase for extracting 1,2-chlorobenzene from aqueous solution and compared its adsorption characteristics with commonly used SPE sorbents such as C_{18} silica and activated carbon. The adsorption capacities of MWCNTs, C_{18} silica, and activated carbon for 1,2-chlorobenzene at an equilibrium concentration of 90 μg/ml were 237, 189, and 150 mg/g, respectively. These results show that MWCNTs can be efficiently applied for the determination of chlorobenzenes and other VOCs in natural and polluted water. Peng et al. [26] investigated the use of CNTs as an adsorbent for 1,2-DCB in aqueous solution. Their results showed that it took only 40 min for the 1,2-DCB sorption onto CNTs to reach equilibrium with a maximum sorption capacity of 30.8 mg/g.

5.4 NMs FOR CHLORINATED ALKENES REMOVAL

Hazardous chlorinated compounds, especially chlorinated alkenes, contaminate groundwater and continue to be a significant environmental problem [27]. Examples of volatile chlorinated alkenes include trichloroethylene (TCE), tetrachloroethylene (PCE), and 1,1,1-trichloroethane (1,1,1-TCA). More attention is given to TCE as it is a common contaminant in soils and groundwater. TCE can also remain in the subsurface as dense nonaqueous-phase liquids (DNAPL) and remain a continuous long-term threat to the environment [28, 29]. TCE is a possible carcinogen at low concentrations. The Safe Drinking Water Act issued by the U.S. Environmental Protection Agency determined the maximum contaminant level of TCE at 5 μg/l [7]. Conventional treatment methods for the removal of chlorinated compounds include activated carbon adsorption, air stripping, and catalysis. While adsorption and air stripping successfully remove these compounds, they only replace the contaminant to another phase rather than convert it to nonhazardous products. Catalysis is a better approach to completely remove chlorinated compounds from the environment since it converts them into safer, nonchlorinated compounds [27]. NMs have also been employed as new adsorbents or catalysts to remove chlorinated alkenes from aqueous solution.

5.4.1 Metallic NPs for the Dechlorination of Chlorinated Alkenes

Laboratory studies have shown that nano zero-valent iron (nZVI) is a powerful reductant for targeting chlorinated ethylenes (e.g., PCE and TCE) (sometimes referred to as chlorinated ethenes) [30]. The electrons transferred from the metallic nZVI particles to the chlorinated ethenes convert them to environmentally benign chloride and nonchlorinated organic compounds. Typically, nZVI particles have ~30 times higher surface area than their larger-sized granular iron counterparts and can degrade chlorinated ethylenes at rates orders of magnitude faster than granular ZVI.

Despite the advantages of using nZVI particles for environmental remediation of chlorinated ethylenes, bare nZVI forms larger aggregates of particles shortly after production. This aggregation limits the ability of nZVI particles to migrate in aquifers and therefore cannot treat a large volume of the subsurface in situ by direct injection. To overcome this drawback, carboxymethyl cellulose (CMC) was effectively used as a stabilizing agent to generate highly dispersed nZVI particles [31]. Bennett et al. [30] investigated the transport of CMC-stabilized nZVI particles (CMC-ZVI) in saturated sediments and their reactivity toward chlorinated ethenes in a series of single-well push–pull tests. The results indicated that CMC-nZVI particles were mobile in the aquifer but appeared to lose mobility with time as a result of the interactions of the particles and the aquifer sediments. The total mass of destroyed ethenes from groundwater was low because the injected solutions "pushed" the dissolved chlorinated ethenes away from the injection well. These results indicate that for in situ remediation programs using metallic NPs, high advective groundwater velocities are needed to deliver the NP substantial distances from the point of injection. For example, groundwater recirculation would likely be an efficient method for enhancing NP migration by maintaining high postinjection groundwater velocities.

Another issue with the utilization of pure nZVI for the dehalogenation of chloroethylenes is that nZVI particles are pyrophoric and react spontaneously with atmospheric oxygen [32]. In order to prevent nZVI from oxidation, Tiehm et al. [32] synthesized air-stable nZVI particles (20–25 nm) by applying ultrasound to a solution of $Fe(CO)_5$ in edible oil (corn oil) and dispersing the resulting nZVI particles in a carbon matrix. This synthesis approach resulted in the formation of air-stable nZVI particles with a protecting layer composed of graphitic and disordered carbon. The authors utilized the air-stable nZVI for the dehalogenation of chloroethylenes in synthetic aqueous medium as well as polluted groundwater. The results indicated faster mass-normalized degradation kinetics of chloroethylenes (trichlororethylene, *cis*-dichloroethylene, vinyl chloride, ethane, and ethane) using the air-stable nZVI, in both synthetic media and groundwater samples, as compared to conventional nZVI.

Choi and Lee [7] utilized the nZVI Fenton system with Cu(II) (Cu/nZVI) for the degradation of TCE. Among the available methods for TCE degradation, Fenton reaction has attracted attention because of its strong oxidative capacity for organic contaminants. In the conventional Fenton process, the decomposition of H_2O_2 to hydroxyl radicals (OH·) in the presence of Fe(II) is the main mechanism for producing a representative oxidant. The main drawback of this conventional process is that the pH should be kept in the acidic range to avoid the precipitation of $Fe(OH)_{3(s)}$ [33]. When using the Cu(II)/nZVI Fenton system, the degradation of TCE was not influenced by the initial pH [7]. The Cu(II)/nZVI Fenton system degraded 95% of the TCE in 10 min with 20 mM Cu(II) at an initial pH of 3.0. This was significantly higher than the TCE degradation (25%) using the conventional nZVI Fenton process without Cu(II) under the same experimental conditions.

The authors [7] suggested that oxidative TCE degradation in Cu(II)/nZVI Fenton systems involves the following four steps: (1) Fe(II) is released from nZVI surfaces via oxidative substitution of Fe(0) by Cu(II); (2) Cu species are then reduced by Fe(0) on the nZVI surface; (3) the OH· are generated by aqueous Fe(II) oxidation with H_2O_2, which results in the degradation of TCE by the OH· in Cu(II)/nZVI suspension; and (4) the Fe(II) are released by direct oxidation with H_2O_2 on the nZVI surface, and the full degradation of TCE occurs because of the continuous production of OH·. The results of this study imply that the Cu(II)/nZVI Fenton system can be applied to systems contaminated with TCE at circum-neutral pH.

5.4.2 Biometallic NPs for the Dechlorination of Chlorinated Alkenes

As discussed earlier, several approaches utilizing nZVI NPs are tested for the dechlorination of chlorinated alkenes. Nonetheless, these approaches have some disadvantages such as the formation of more toxic by-products (e.g., vinyl chloride), slow reaction rates, and the difficulty to treat several halogenated organic compounds together [34, 35]. As a result of its superior catalytic capacity in hydrodehalogination reactions, palladium can overcome these shortcomings. Hennebel et al. [29] utilized microbially produced palladium NPs (bio-Pd) for the remediation of TCE-contaminated groundwater. The authors [29] utilized a method to produce a nano bio-Pd catalyst by precipitating palladium on *Shewanella oneidensis* in order to form NPs. Bacteria, in the presence of a hydrogen donor, can reduce Pd(II) and subsequently precipitate it as Pd nanocrystals on their cell wall and in their periplasmatic space. The bacteria in this case serve as a biological carrier for the Pd NPs, keeping them in suspension and preventing their aggregation. The bacteria can also interact with the contaminant and contribute toward more efficient removal.

The use of *S. oneidensis* in this study [29] has a major advantage over other bacteria because it the only known bacteria that can optimally reduce Pd when it is grown under controlled conditions of oxygen, carbon source, and hydrogen donor. In their experiment, Hennebel et al. [29] used hydrogen gas, formate, and formic acid as hydrogen donors. The results showed that TCE was efficiently removed and ethane was the only organic degradation product and that no intermediate chlorinated reaction products were detected. Hydrogen gas was the most appropriate hydrogen donor used in the process. The authors also impregnated the bio-Pd in a continuous plate membrane reactor (MRr) for the treatment of TCE-contaminated water. The MRr system achieved removal rates of up to 2515 mg TCE/day/g Pd using H_2 gas as the hydrogen donor. The authors concluded that biocatalysis with bio-Pd resulted in complete, efficient, and rapid removal of TCE.

In another study, Xiu et al. [28] investigated the effect of nZVI particles on a mixed culture dechlorinating TCE. Biological degradation by dechlorinating bacteria is a promising technology for DNAPL remediation. Examples of anaerobic bacteria used for the reductive dechlorination of TCE and PCE include *Desulfuromonas*, *Sulfurospirillum multivorans*, and *Dehalobacter*. However, several field studies identified many challenges facing bioremediation within DNAPL source zones such as low dechlorination rates, insufficient supply of suitable electron donors such as H_2 and acetate to the dechlorinating bacteria, and toxicity because of high concentrations of PCE or TCE [36].

Because corrosion of nZVI produces H_2, which is a highly favorable electron donor for anaerobic dechlorinating bacteria, Xiu et al. [28] investigated the effect of synergetic use of nZVI particles with microbial dechlorinating ability of bacteria on the dechlorination efficiency of TCE. The results showed that the presence of nZVI initially inhibited the dechlorinating organisms. However, dechlorination activity showed recovery after a lag period. During the active period of the dechlorinating bacteria, the H_2 resulting from the cathodic corrosion of nZVI was utilized as an electron donor by the bacteria, which resulted in the recovery of the bacterial activity following partial oxidation and most likely passivation of the nZVI. This study showed that the overall dechlorination efficiency of TCE was enhanced by the concurrent or subsequent addition of nZVI particles to the media. The authors [28] also suggested that, in environmental systems, nZVI can be used to abiotically degrade a large fraction of DNAPL mass in the source zone directly, while subsequent biological dechlorination can serve as a polishing step to remove residual chloroethylenes.

5.4.3 Bimetallic NPs for the Dechlorination of Chlorinated Alkenes

The oxidation of nZVI by molecular oxygen and the formation of a metal hydroxide surface layer results in the decrease of the surface reactivity of the NPs over time [4]. By using bimetallic NPs such as Pd/Fe and Ni/Fe, this problem can be avoided. Cho and Choi [37] employed nanosized palladium–iron (Pd/Fe) bimetallic particles for the degradation of TCE, PCE, and 1,1,1-TCA (1,1,1-trichloroethane) from aqueous solution. The results showed that the reactivity of (nZVI) particles was enhanced by modifying its surface using Pd and CMC to form CMC-stabilized Pd/Fe bimetallic NPs (CMC-Pd/Fe). The bimetallic NPs were spherical in shape with an average diameter of 98.5 nm. CMC prevented individual NPs from forming aggregates and thus enhanced the remediation efficiency of chlorinated organic compounds.

The CMC-Pd/Fe NPs achieved a significant increase in the removal of TCE (~85%) as compared to the relatively low removal (~15%) when conventional nZVI was used. Removal efficiencies of approximately 80 and 56% were achieved for PCE and 1,1,1-TCA, respectively, when CMC-Pd/Fe was used. The authors [37] identified reductive dechlorination as the main degradation mechanism for the investigated chlorinated compounds. The results of their study also suggested that the position of the chlorine in chlorinated organic compounds has a significant effect on the degradation kinetics.

Bimetallic nickel–iron (Ni/Fe) NPs (3- to 30-nm diameter) have been tested as a reagent for the dechlorination of TCE in aqueous solution [38]; 0.1 g of Ni/Fe NPs reduced the concentration of TCE from a 40 ml saturated aqueous solution (24 mg/l) to <6 mg/l in 20 min. The use of Ni/Fe NPs resulted in dechlorination surface area–normalized reaction rate that was 50 times faster than that using nZVI. This is because of the nickel-catalyzed hydrodechlorination of the TCE. In this process, as the iron corrodes, the cathodically protected nickel chemisorbs hydrogen ions, and thus TCE adsorbed on the nickel surface is hydrogenated. Toxic dechlorination by-products such as vinyl chloride are formed in trace amounts and do not persist. This study suggested that bimetallic NPs containing good catalysts (e.g., Ni and Pd) for hydrogenation can be used for the dehalogenation of chlorinated organics at faster rates and with the absence of toxic by-products.

The authors [38] noted that even though Ni/Fe NPs did not leach nickel in a 33-day time period, there is still a possibility that nickel will leach once the iron has been exhausted after long burial times. This suggests that palladium may be a better choice as a catalyst for most applications because it is more stable than nickel. However, when a more active catalyst such as Pd is used, the initial corrosion rate of the iron increases because Pd catalyzes both the hydrogenation reaction and the evolution of molecular hydrogen. As the iron surface becomes passivated, the difference between catalytic metals becomes less distinct. Moreover, because of the fast corrosion rate, the useful time of nZVI is more limited than that of lower surface area iron, and thus Ni/Fe and Pd/Fe NPs are suitable for shorter-term remediation applications such as injection into contaminated groundwater than for long-term reactive barriers. More research is needed to synthesize bimetallic NPs with slower corrosion rates while retaining the appropriate degree of reactivity for remediation applications [38].

Other studies in the literature also showed that bimetallic NPs exhibit high dechlorination efficiency for chlorinated alkenes. Shih et al. [39] reported that Pd/Fe bimetallic NPs achieved a more rapid dechlorination rate for HCB than Fe NPs. Meyer et al. [40] have successfully prepared reactive membranes by incorporating bimetallic Fe/Pt NPs (24 nm) into cellulose acetate films. The prepared membranes successfully reduced TCE with ethane as the only observed by-product. Zhao and Nagy [41] synthesized a nanosorbent for the removal of TCE and PCE by incorporating SDS into magnesium–aluminum layered double hydroxide (LDHs), and their results showed higher adsorption capacity of the SDS–Mg/Al LDHs toward both TCE and PCE in aqueous solution as compared to organoclays.

5.5 NMs FOR PHENOL REMOVAL

5.5.1 Metal Oxide NPs for Phenol Degradation

Phenol is among the phenolic compounds that are considered as priority pollutants since they are harmful to organisms even at low concentrations, and many of them have been classified as hazardous pollutants because of their potential harm to human health [24]. NMs were also tested for the removal of phenol from aqueous solutions, which is an important challenge facing environmental protection. Chitose et al. [6] investigated the radiolysis of an aqueous phenol solution containing TiO_2 NPs (30-nm diameter). Titanium dioxide NPs with high surface area and band-gap energy of 3.0 eV were used. This makes TiO_2 NPs an effective photocatalyst. The tested irradiation techniques were UV, γ-ray, and electron beam irradiation. The results showed that phenol was decomposed completely using three irradiation techniques in the presence of NPs. However, in the case of ionizing radiation, phenol decayed more efficiently than in the case of UV photocatalysis. Also, the absorbed energy of the ionizing radiation (γ-ray and electron beam) needed for phenol decomposition was much lower than that for UV photocatalysis.

The addition of TiO_2 in the case of ionizing radiation had a significant effect on phenol decomposition since phenol was also efficiently decomposed in the absence of TiO_2 NPs [6]. However, the presence of TiO_2 NPs drastically increased the removal of total organic carbon by ionizing radiation. Ionizing radiation processes induced the formation of radical products from water molecules such e^-_{aq}, H atoms, and OH radicals. These water decomposition products were responsible for phenol decomposition by ionizing radiation.

5.5.2 Nanocomposites for Phenol Degradation

Zinc oxide (ZnO) NPs are gaining significant attention as a photocatalyst for the degradation of organic pollutants because of the generation of a more negative potential in the electrons derived from them as compared to those derived from TiO_2 [42, 43]. Meshram et al. [24] investigated the removal of phenol using ZnO–bentonite nanocomposite as a photocatalyst under UV irradiation in a batch as well as a continuously stirred tank reactor (CSTR). Clay is widely available and inexpensive and is an attractive substrate for the immobilization of various photocatalysts. Ion exchange intercalation process was utilized to intercalate ZnO NPs (20–30 nm) in the lattice structure of the bentonite matrix. The advantages of intercalation are (i) proper dispersion of nanosized photocatalysts in a solid support and thus the generation of distinct reaction sites; (ii) participation of solid support in the adsorption of the contaminants, which can increase the rate of photocatalytic degradation; and (iii) less amount of photocatalyst required for the degradation of the organic pollutant [24].

The results from the batch experiments showed that the removal of phenol followed first-order reaction kinetics with Langmuir type of adsorption characteristics [24]. The CSTR experiment showed that phenol can be continuously removed from aqueous stream using ZnO–bentonite nanocomposite as a photocatalyst under UV irradiation. The utilized ZnO–bentonite nanocomposite achieved ~70% removal of phenol from the effluent at a low flow rate and under basic pH conditions (12.0), with the generation of phenoxide ions. The photocatalytic degradation of phenol using ZnO–bentonite nanocomposite resulted in the formation of adipic acid and 2,4,6-triphenoxy phenol through a radical mechanism.

5.6 THE IMPACT OF NMs ON VOC REMOVAL BY OTHER PROCESSES

Salih et al. [44] investigated the impact of the presence of three commercially available NPs in aqueous stream on TCE adsorption by GAC. The presence of background materials in natural water can highly impact the adsorption efficiency in removing VOCs by activated carbon. Since NMs are emerging as new contaminants in water, how their presence impacts the efficiency of VOC removal by GAC needs investigation. Using GAC as adsorbent, Salih et al. [44] conducted TCE adsorption isotherms and column breakthrough experiments in the presence and absence of silicon dioxide (SiO_2 NPs), titanium dioxide (TiO_2 NPs), and iron oxide (Fe_2O_3 NPs) NPs. The results of the TCE adsorption isotherms showed that the presence of neither of the three NPs had an impact on TCE adsorption by GAC. The authors attributed this lack of effect to the fast adsorption kinetics of TCE on GAC. On the other hand, during the column breakthrough studies, the presence of any of the three NPs decreased the amount of TCE adsorbed on GAC. This is a result of the preloading pore blockage phenomenon. The presence of Fe_2O_3 NPs, which had the highest particle size distribution (PSD) among the investigated NPs, resulted in the fastest TCE breakthrough followed by TiO_2 NPs, while SiO_2 with the smallest PSD showed the least impact.

5.7 CHALLENGES IN THE USE OF NMs FOR VOC REMEDIATION

Although NPs proved efficient for the removal of VOCs from aqueous solution, these investigations were conducted only in a laboratory scale setup and thus the scale-up of these processes to pilot and full scale may or may not be practical or economically feasible. There are many challenges facing the use of NMs for environmental remediation such as (i) the availability of suppliers that can provide large quantities of NMs at economically viable prices; (ii) the integration of NMs into existing water treatment plants; and (iii) the fate and toxicity of NMs, which are critical issues that have to be taken into account when selecting NMs for water remediation [5]. However, as research progresses, the properties of NMs can be enhanced, for example, by synthesizing more stable NMs, finding environmentally friendly techniques to introduce functional groups on NMs, and finding energy-efficient techniques to separate NMs from aqueous solution at the end of the remediation process. Therefore, more research is still needed before a conclusion can be made with regard to the use of NMs for the remediation of VOCs in aqueous solutions.

REFERENCES

[1] Berenjian A, Chan N, Malmiri HJ. Volatile organic compounds removal methods: a review. Am J Biochem Biotechnol 2012;8:220–229.

[2] Lu C, Su F, Hu S. Surface modification of carbon nanotubes for enhancing BTEX adsorption from aqueous solutions. Appl Surf Sci 2008;254:7035–7041.

[3] Tsoukleris DS, Maggos T, Vassilakos C, Falaras P. Photocatalytic degradation of volatile organics on TiO_2 embedded glass spherules. Catal Today 2007;129:96–101.

[4] Prasse C, Ternes T. Removal of organic and inorganic pollutants and pathogens from wastewater and drinking water using nanoparticles—a review. In: Frimmel FH, editor. *Nanoparticles in the Water Cycle: Properties, Analysis and Environmental Relevance*. Berlin: Springer; 2010. p 55–79.

[5] Savage N, Diallo MS. Nanomaterials and water purification: opportunities and challenges. J Nanopart Res 2005;7:331–342.

[6] Chitose N, Ueta S, Seino S, Yamamoto TA. Radiolysis of aqueous phenol solutions with nanoparticles. 1. Phenol degradation and TOC removal in solutions containing TiO_2 induced by UV, γ-ray and electron beams. Chemosphere 2003;50:1007–1013.

[7] Choi K, Lee W. Enhanced degradation of trichloroethylene in nano-scale zero-valent iron Fenton system with Cu (II). J Hazard Mater 2012;211–212:146–153.

[8] Gupta R, Kulkarni GU. Removal of organic compounds from water by using a gold nanoparticle–poly (dimethylsiloxane) nanocomposite foam. ChemSusChem 2011;4:737–743.

[9] Ren X, Chen C, Nagatsu M, Wang X. Carbon nanotubes as adsorbents in environmental pollution management: a review. Chem Eng J 2011;170:395–410.

[10] Iijima S. Helical microtubules of graphitic carbon. Nature 1991;354:56–58.

[11] Popov VN. Carbon nanotubes: properties and applications. Mater Sci Eng R 2004;43:61–102.

[12] Hsieh SH, Horng JJ, Tsai CK. Growth of carbon nanotube on micro-sized Al_2O_3 particle and its application to adsorption of metal ions. J Mater Res 2006;21:1269–1273.

[13] Wang X, Chen C, Hu W, Ding A, Di X, Zhou X. Sorption of 243Am (III) to multiwall carbon nanotubes. Environ Sci Technol 2005;39:2856–2860.

[14] Tournus F, Charlier JC. Ab initio study of benzene adsorption on carbon nanotubes. Phys Rev B 2005;71:165421.

[15] Lu C, Chiu H. Adsorption of zinc(II) from water with purified carbon nanotubes. Chem Eng Sci 2006;61:1138–1145.

[16] Agnihotri S, Mota JP, Rostam-Abadi M, Rood MJ. Theoretical and experimental investigation of morphology and temperature effects on adsorption of organic vapors in single-walled carbon nanotubes. J Phys Chem B 2006;110:7640–7647.

[17] Burde JT, Calbi MM. Physisorption kinetics in carbon nanotube bundles. J Phys Chem C 2007;111:5057–5063.

[18] Rawat DS, Calbi MM, Migone AD. Equilibration time: kinetics of gas adsorption on closed-and open-ended single-walled carbon nanotubes. Phys Chem C 2007;111:12980–12986.

[19] Su F, Lu C, Hu S. Adsorption of benzene, toluene, ethylbenzene and *p*-xylene by NaOCl-oxidized carbon nanotubes. Colloid Surf A 2010;353:83–91.

[20] Carrillo-Carrión C, Lucena R, Cárdenas S, Valcárcel M. Liquid–liquid extraction/headspace/gas chromatographic/mass spectrometric determination of benzene, toluene, ethylbenzene, (*o*-, *m*- and *p*-)xylene and styrene in olive oil using surfactant-coated carbon nanotubes as extractant. J Chromatogr A 2007;1171:1–7.

[21] Beurskens JE, Dekker CG, Van Den Heuvel H, Swart M, de Wolf J, Dolfing J. Dechlorination of chlorinated benzenes by an anaerobic microbial consortium that selectively mediates the thermodynamic most favorable reactions. Environ Sci Technol 1994;28:701–706.

[22] Sarrion MN, Santos FJ, Galceran MT. Strategies for the analysis of chlorobenzenes in soils using solid-phase microextraction coupled with gas chromatography–ion trap mass spectrometry. J Chromatogr A 1998;819:197–209.

[23] Selli E, Bianchi CL, Pirola C, Cappelletti G, Ragaini V. Efficiency of 1,4-dichlorobenzene degradation in water under photolysis, photocatalysis on TiO_2 and sonolysis. J Hazard Mater 2008;153:1136–1141.

[24] Meshram S, Limaye R, Ghodke S, Nigam S, Sonawane S, Chikate R. Continuous flow photocatalytic reactor using ZnO–bentonite nanocomposite for degradation of phenol. Chem Eng J 2011;172:1008–1015.

[25] Liu G, Wang J, Zhu Y, Zhang X. Application of multiwalled carbon nanotubes as a solid-phase extraction sorbent for chlorobenzenes. Anal Lett 2004;37:3085–3104.

[26] Peng X, Li Y, Luan Z, Di Z, Wang H, Tian B, Jia Z. Adsorption of 1,2-dichlorobenzene from water to carbon nanotubes. Chem Phys Lett 2003;376:154–158.

[27] Wong MS, Alvarez PJ, Fang YL, Akçin N, Nutt MO, Miller JT, Heck KN. Cleaner water using bimetallic nanoparticle catalysts. J Chem Technol Biotechnol 2009;84:158–166.

[28] Xiu ZM, Jin ZH, Li TL, Mahendra S, Lowry GV, Alvarez PJ. Effects of nano-scale zero-valent iron particles on a mixed culture dechlorinating trichloroethylene. Bioresource Technol 2010;101:1141–1146.

[29] Hennebel T, Simoen H, De Windt W, Verloo M, Boon N, Verstraete W. Biocatalytic dechlorination of trichloroethylene with bio-palladium in a pilot-scale membrane reactor. Biotechnol Bioeng 2009;102:995–1002.

[30] Bennett P, He F, Zhao D, Aiken B, Feldman L. In situ testing of metallic iron nanoparticle mobility and reactivity in a shallow granular aquifer. J Contam Hydrol 2010;116:35–46.

[31] He F, Zhao D. Preparation and characterization of a new class of starch-stabilized bimetallic nanoparticles for degradation of chlorinated hydrocarbons in water. Environ Sci Technol 2005;39:3314–3320.

[32] Tiehm A, Kraßnitzer S, Koltypin Y, Gedanken A. Chloroethene dehalogenation with ultrasonically produced air-stable nano iron. Ultrason Sonochem 2009;16:617–621.

[33] Pignatello JJ, Oliveros E, MacKay A. Advanced oxidation processes for organic contaminant destruction based on the Fenton reaction and related chemistry. Crit Rev Environ Sci Technol 2006;36:1–84.

[34] Lin CJ, Lo SL, Liou YH. Dechlorination of trichloroethylene in aqueous solution by noble metal-modified iron. J Hazard Mater 2004;116:219–228.

[35] Wang CB, Zhang WX. Synthesizing nanoscale iron particles for rapid and complete dechlorination of TCE and PCBs. Environ Sci Technol 1997;31:2154–2156.

[36] Amos BK, Daprato RC, Hughes JB, Pennell KD, Löffler FE. Effects of the nonionic surfactant Tween 80 on microbial reductive dechlorination of chlorinated ethenes. Environ Sci Technol 2007;41:1710–1716.

[37] Cho Y, Choi SI. Degradation of PCE, TCE and 1,1,1-TCA by nanosized FePd bimetallic particles under various experimental conditions. Chemosphere 2010;81:940–945.

[38] Schrick B, Blough JL, Jones AD, Mallouk TE. Hydrodechlorination of trichloroethylene to hydrocarbons using bimetallic nickel–iron nanoparticles. Chem Mater 2002;14:5140–5147.

[39] Shih YH, Chen YC, Chen MY, Tai YT, Tso CP. Dechlorination of hexachlorobenzene by using nanoscale Fe and nanoscale Pd/Fe bimetallic particles. Colloid Surf A 2009;332:84–89.

[40] Meyer DE, Wood K, Bachas LG, Bhattacharyya D. Degradation of chlorinated organics by membrane-immobilized nanosized metals. Environ Prog 2004;23:232–242.

[41] Zhao H, Nagy KL. Dodecyl sulfate–hydrotalcite nanocomposites for trapping chlorinated organic pollutants in water. J Colloid Interface Sci 2004;274:613–624.

[42] Wu JJ, Tseng CH. Photocatalytic properties of nc-Au/ZnO nanorod composites. Appl Catal B 2006;66:51–57.

[43] Richard C, Bosquet F, Pilichowski JF. Photocatalytic transformation of aromatic compounds in aqueous zinc oxide suspensions: effect of substrate concentration on the distribution of products. J Photochem Photobiol A 1997;108:45–49.

[44] Salih HH, Patterson CL, Sorial GA. Comparative study on the implication of three nanoparticles on the removal of trichloroethylene by adsorption–pilot and rapid small-scale column tests. Wat Air Soil Pollut 2013;224:1–11.

6

HYBRID METAL NANOPARTICLE-CONTAINING POLYMER NANOFIBERS FOR ENVIRONMENTAL APPLICATIONS

YUNPENG HUANG[1,2], SHIGE WANG[1], MINGWU SHEN[2], AND XIANGYANG SHI[1,2,3]

[1] *State Key Laboratory for Modification of Chemical Fibers and Polymer Materials, Donghua University, Shanghai, China*

[2] *College of Chemistry, Chemical Engineering and Biotechnology, Donghua University, Shanghai, China*

[3] *CQM-Centro de Química da Madeira, Universidade da Madeira, Campus da Penteada, Portugal*

6.1 INTRODUCTION

In recent years, nanomaterials have found applications in various fields such as pharmaceutical technology [1–4], electronic industry [5–7], textiles [8, 9], mechanical industry [10, 11], and environmental remediation [12–15]. Among numerous kinds of nanomaterials, metal nanoparticles (NPs) have been widely studied in environmental applications due to their unique surface features, reactivity, and catalytic properties [16, 17]. However, the usage of metal NPs in environmental technology encountered some obstacles due to the difficulty in recycling colloid metal NPs or the possibility of introducing secondary contamination. In order to overcome this bottleneck, several categories of nanomaterials have been employed as supporting materials to incorporate metal NPs, forming multifunctional composite or hybrid materials [13–15, 18–20]. Compared with other nanomaterials, one-dimensional nanofibers possess superior properties such as high aspect ratio, large specific surface area, high porosity, and stability in liquid media [13, 21]. Therefore, nanofibers have turned out to be one of the best candidates for the formation of hybrid nanomaterials for environmental applications [13, 22, 23].

Electrospinning has been actively exploited as an intriguing technology for generating long polymer fibers with diameters ranging from tens of nanometers to several micrometers [24, 25]. Various composite polymeric nanofibers and nanostructured materials with a high aspect ratio and a specific surface area [26, 27] have been fabricated for various applications including, but not limited to, solar cells [28], filtration [29], environmental remediation [12–15, 30], protective clothing [31], biosensors [32], and tissue engineering scaffolds [33–38]. Electrospun composite or hybrid nanofibers are promising due to the coexisting physicochemical properties possessed by both the host and guest materials.

This chapter reviews recent advances in the fabrication of metal NP-doped electrospun nanofibers, and their applications in environmental remediation. Following a brief introduction, the challenges of environmental nanotechnology and the basic knowledge of electrospinning technology are introduced. Then, the fabrication of different metal NP-immobilized nanofibers and their applications in the remediation of different environmental contaminants in water resources are discussed.

Nanomaterials for Environmental Protection, First Edition. Edited by Boris I. Kharisov, Oxana V. Kharissova, and H. V. Rasika Dias.

6.2 CHALLENGES OF ENVIRONMENTAL NANOTECHNOLOGY

With industrialization, many issues of water, air, and soil pollution have arisen gradually. An alarming situation is that, by the end of 2025, more than 50% of the countries in the world will face freshwater shortage [39]. Water pollution will increase the worldwide pressure on freshwater resources.

Water contaminants are generally divided into both organic and inorganic substances. Most of the contaminants are contributed by industrial plants like mining, textile, leather, and paper production. Until now, a wide range of conventional treatment techniques including chemical coagulation [40], activated sludge [41], biosorption [42], and plant accumulation [43] have been extensively used to remediate these water pollutants. For example, Annadurai et al. [44] used chitosan as an adsorbent material for reactive dye removal and demonstrated that a high absorption capacity of 91.47–130.0 mg/g toward the removal of Remazol black 13 dye could be achieved. Chen et al. [45] reported that the carboxymethylated bacterial cellulose could efficiently remove copper and lead ions from aqueous solution. In another work, Fillipi et al. [46] introduced a ligand-modified ultrafiltration system to selectively remove copper ions.

With superior advantages such as large surface area, reactive surface features, and quantum size–related properties, various NPs have been used for environmental remediation applications [47]. For instance, innovative granular activated carbon composites doped with iron/palladium (Fe/Pd) bimetallic NPs can be used to physically absorb and chemically dechlorinate polychlorinated biphenyls. Zirconia- and silica-supported zero-valent iron NPs' (ZVI NPs) ability to separate and reduce pertechnetate ions from complex waste mixtures has been demonstrated [48]. However, one critical problem is that used colloidal NPs such as noble metals (e.g., Au, Ag, and Pd) may lead to further environmental contamination in aqueous media or soil. Therefore, the development of various supporting materials that can incorporate or immobilize reactive metal NPs and can be easily separated from the environment is crucial in not only improving the longevity of the particles, but also in avoiding secondary contamination of the environment.

6.3 ELECTROSPINNING TECHNOLOGY

Since a series of patents on electrospinning process were issued to Formhals [49] in 1934, electrospinning technology has been identified as one of the powerful techniques to produce fibers with diameters ranging from tens of nanometers to a few micrometers. Figure 6.1 shows a schematic of a basic setup of electrospinning technology, which consists of three major modules: a high-voltage power generator, a syringe pump, and a collector. The needle connected to a syringe is used as a spinneret. With the assistance of the pump, the polymer solution hosted in the syringe can be fed through the spinneret continuously and gently. The mutual electrostatic repulsion between polymer chains causes a force directly counterbalancing the surface tension under the applied voltage. Then, the hemispherical-shaped liquid can be distorted to form a conical shape known as the Taylor cone [50].

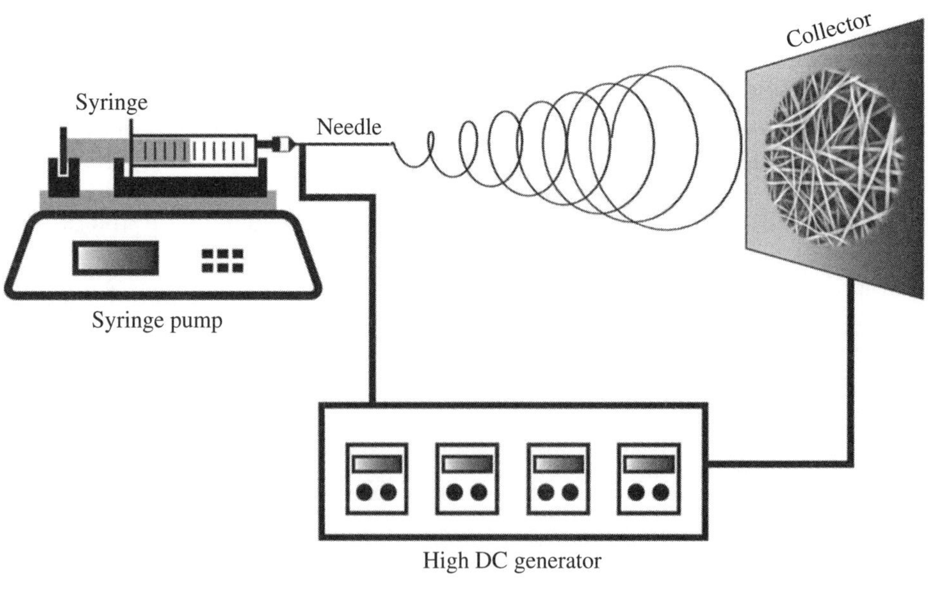

FIGURE 6.1 Schematic illustration of a basic setup of the electrospinning technology.

Once the intensity of the electric field exceeds a critical value, a jet flow of charged solution is ejected from the tip of the Taylor cone continuously, which undergoes stretching and elongation. Finally, the solvent evaporates, and the formed nanofibers deposit on the collector in a random manner [51].

A variety of materials such as naturally occurring polymers, synthetic polymers, ceramics, metals, or their mixture have been electrospun to produce ultrafine nanofibers. The sizes, composition, and morphology of the electrospun nanofibers are able to be well-controlled through the regulation of the electrospinning parameters [52–55]. A number of electrospinning methods have emerged. For instance, Chu et al. [56] designed an array of multiple needles for large-scale production of electrospun nanofibers. In some other approaches, core–shell electrospinning, auxiliary electric fields, rotating tube, rotating disk, needleless electrospinning, patterned electrodes, or magnetic electrospinning were also utilized to produce different nanofibers [57–63]. The objectives of these explorations were to fabricate nanofibers with controlled alignment or orientation or to render nanofibers with special functionalities.

The electrospinning process is influenced by many parameters, including (i) solution properties such as polymer molecular weight, viscosity, conductivity, elasticity, and surface tension; (ii) processing parameters such as applied voltage, flow rate, distance between needle tip and collector, and needle inner diameter; and (iii) ambient conditions such as temperature, humidity, and air velocity in the electrospinning chamber [64]. For example, the polymer solution should have an appropriate concentration to maintain polymer entanglement. The distance between the needle and the collector can influence the morphology of nanofibers, and incorrect distance may result in a rough or beaded fibrous mat, or significantly decreased porosity of the fibers. Since the applied voltage can produce ionized solution and impact the elongation and splitting of the jet flow, smooth and uniform fiber structures can be formed only at a proper applied voltage. Extremely high voltage often results in unstable electrospinning [65].

6.4 FABRICATION OF HYBRID METAL NP-CONTAINING POLYMER NANOFIBERS

6.4.1 Zero-Valent Iron Nanoparticles

ZVI NPs have been used as a cost-effective and environmentally friendly material for environmental remediation [66]. Much effort has been devoted to the synthesis and application of ZVI NPs. To improve stability and mitigate aggregation, ZVI NPs have been immobilized or incorporated within or onto different supporting materials for environmental applications [12–14, 19, 20, 30].

In our previous work, polyelectrolyte (PE) multilayers assembled onto electrospun cellulose acetate (CAc) nanofibrous mats were used as nanoreactors to immobilize ZVI NPs [20, 67]. Figure 6.2 illustrates the whole experimental process. In brief, smooth and continuous CAc nanofibers with a mean diameter of 295 ± 145 nm were first produced through electrospinning. Then, polyacrylic acid (PAA)/poly(diallyldimethylammonium chloride) (PDADMAC) multilayers were layer-by-layer (LbL)-assembled onto the prepared electrospun CAc nanofibers via electrostatic interaction. SEM (scanning electron microscopy) morphology observation of the CAc nanofibers deposited with three, four, six, and nine bilayers of PAA/PDADMAC shows that when the number of bilayers is below six, the porous structure of the nanofibrous mats can be well maintained. Therefore, six-bilayer-assembled CAc nanofibrous mats were selected to immobilize ZVI NPs. The six-bilayer-assembled CAc nanofibrous mats were then immersed into an aqueous solution of $FeCl_2$ to allow Fe^{2+} ions to complex with the free carboxyl groups of PAA through ionic exchange, followed by in situ reduction with $NaBH_4$ to generate ZVI NPs-immobilized polymer nanofibrous mats. Transmission electron microscopy (TEM) images (Figure 6.3) show that the formed ZVI NPs are evenly distributed onto the CAc nanofibers with an average size of 1.4 ± 0.28 nm. We show that the loading capacity of ZVI NPs can be tuned by changing the number of PE layers and the cycles in the binding/reduction process, and increasing the number of binding/reduction cycles leads to slightly bigger ZVI NPs.

To generate hybrid nanofibers incorporated with ZVI NPs both on the fiber surface and inside the nanofibers, we utilized PAA/poly(vinyl alcohol) (PVA) nanofibers as a nanoreactor for reductive formation of ZVI NPs [12, 14]. First, a mixture solution of PAA and PVA was electrospun to form nanofibers. To render them with water stability, freshly prepared PAA/PVA nanofibers were cross-linked at 145°C for 30 min. Then, water-stable electrospun PAA/PVA nanofibrous mats were soaked in an aqueous solution of ferric trichloride to allow Fe^{3+} to complex with available free carboxyl groups of PAA through ionic exchange, followed by $NaBH_4$ reduction to form ZVI NPs. This process enables the generation of ZVI NPs with a diameter of 1.6 nm, uniformly distributed both within the nanofibers and on the fiber surface. It was possible to control the loading percentage and size of the ZVI NPs simply by varying the number of ferric ion binding/reduction cycles [68].

To enhance the mechanical durability of the ZVI NP-containing PAA/PVA nanofibers, we fabricated multiwalled carbon nanotube (MWCNT)-reinforced PAA/PVA composite nanofibrous mats by electrospinning a PAA/PVA mixture solution containing well-dispersed MWCNTs. The so-formed MWCNT-containing PAA/PVA nanofibers were processed to immobilize

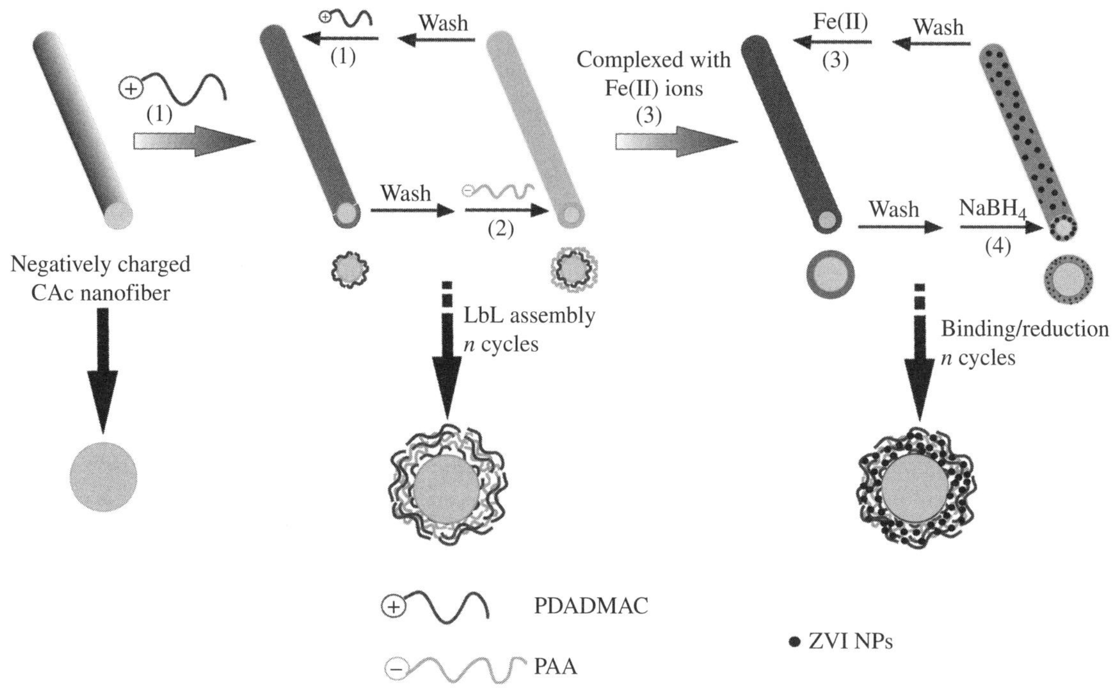

FIGURE 6.2 Schematic illustration of immobilizing ZVI NPs onto PE multilayer-assembled CAc nanofibers. Reprinted with permission from Ref. [20]. © 2009, American Chemical Society.

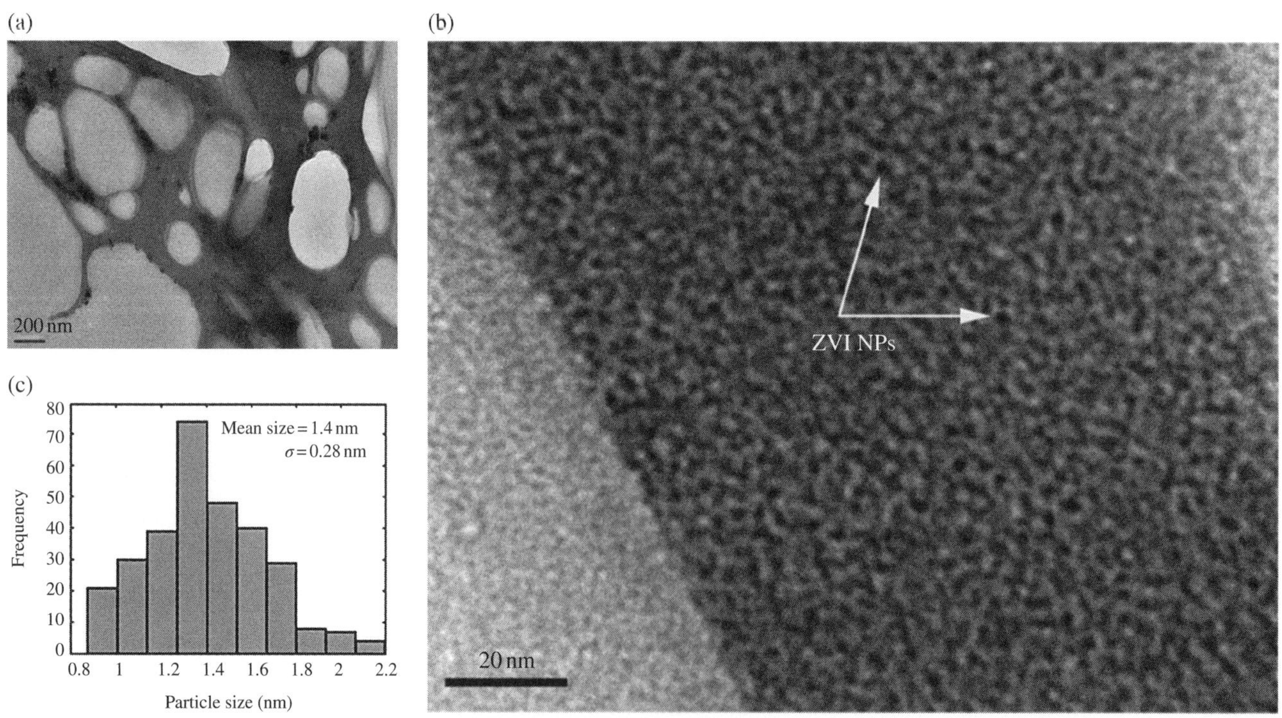

FIGURE 6.3 Cross-sectional TEM images of ZVI NP-immobilized CAc nanofibrous mats: (a) overview of the TEM image; (b) magnified TEM image showing individual ZVI NPs; (c) size distribution histogram of the formed ZVI NPs. Reprinted with permission from Ref. [20]. © 2009, American Chemical Society.

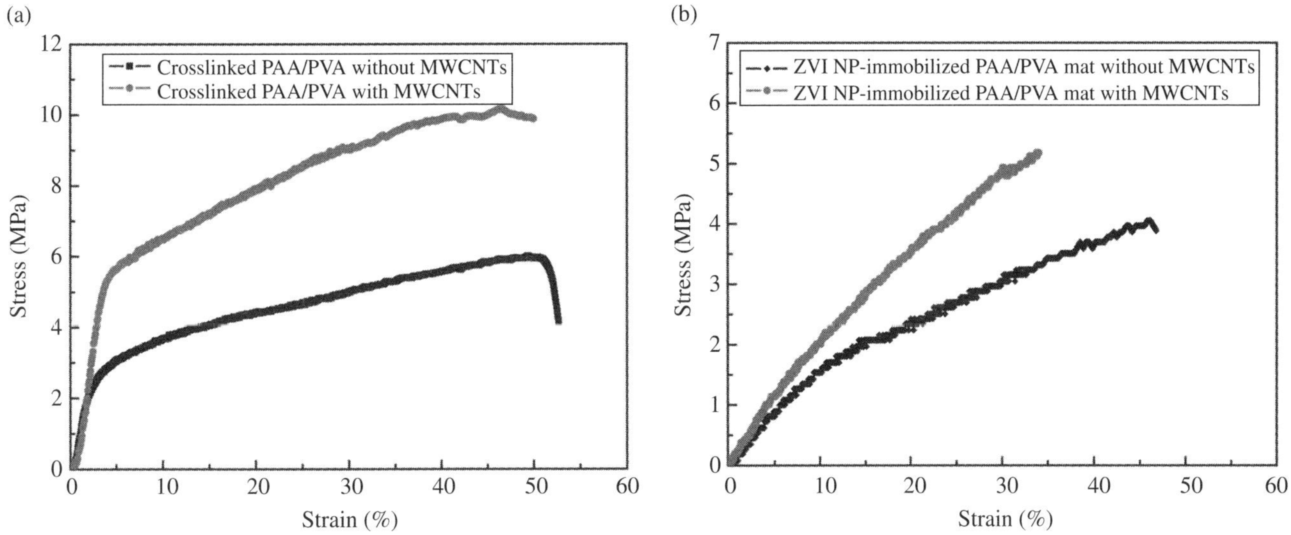

FIGURE 6.4 Typical stress–strain curves of nanofibrous mats. (a) Cross-linked PAA/PVA nanofibrous mats with and without MWCNTs; (b) ZVI NP-immobilized nanofibrous mats with and without MWCNTs. Reproduced with permission from Ref. [13]. © Royal Society of Chemistry.

ZVI NPs according to the procedures described earlier [14]. Typical stress–strain curves (Figure 6.4) show that the mechanical properties of the electrospun nanofibrous mats with and without ZVI NP immobilization were significantly enhanced even with only 1.0 wt% MWCNTs incorporated within the nanofibers [13, 19].

6.4.2 Bimetallic Iron-Based NPs

Due to the metastability and easy oxidation of ZVI NPs, their capability to remediate environmental pollutants is limited. Therefore, synthesis of iron-based bimetallic NPs could be an effective way to maintain the reactivity of ZVI NPs. In our recent work, we investigated the feasibility of using electrospun PAA/PVA nanofibers as a nanoreactor to prepare bimetallic Fe/Pd NP-immobilized composite nanofibers [69]. In our approach, the cross-linked PAA/PVA nanofibrous mats were dipped in an aqueous ferric trichloride solution to allow Fe^{3+} to complex with available free carboxyl groups of PAA through ionic exchange, followed by reduction via $NaBH_4$ to form ZVI NPs. These ZVI NP-containing nanofibrous mats were then immersed into a solution of palladium chloride. Then, bimetallic Fe/Pd NPs were finally formed by a replacement reaction due to the high metal activity of Fe than that of Pd. The mean diameter of the Fe/Pd NPs was estimated to be 2.8 ± 0.92 nm. Elemental analysis of the Fe/Pd NP-immobilized nanofibers demonstrated that the Fe and Pd loading percentages were 19.44 and 2.44 wt%, respectively.

6.4.3 Other Metal NPs

Besides the immobilization of metal iron-based NPs within and/or on the surface of polymer nanofibers, other noble metal NPs are also capable of being immobilized. Recently, we reported the formation and immobilization of Au NPs into electrospun nanofibers through in situ reduction [70, 71]. In brief, freshly prepared polyethyleneimine (PEI)/PVA nanofibrous mats were cross-linked under glutaraldehyde (GA) vapor to render them water-stable. Then, the cross-linked PEI/PVA nanofibrous mats were immersed into an aqueous solution of $HAuCl_4$ to allow $AuCl_4^-$ ions to complex with the free amine groups of PEI via ionic exchange, and the Au NPs were finally formed by reducing the complexed $AuCl_4^-$ with $NaBH_4$. SEM morphology observation shows that the Au NP-immobilized nanofibrous mats retain a uniform and porous structure, similar to the cross-linked electrospun PEI/PVA nanofibers without Au NPs. Cross-sectional TEM images show that the Au NP-immobilized PEI/PVA nanofibers have round-shaped patterns of Au NPs that are uniformly distributed within the cross section of the fibers with a mean diameter of 11.8 ± 3.3 nm. Energy-dispersive spectroscopy and Fourier transform infrared spectroscopy characterization further demonstrated the successful immobilization of Au NPs within the nanofibers.

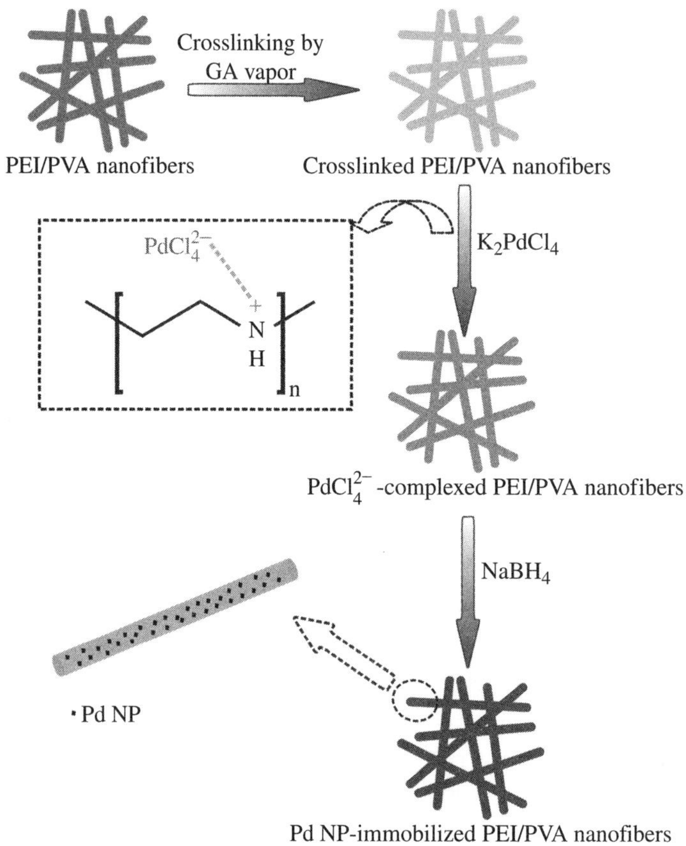

FIGURE 6.5 Schematic illustration of the immobilization of Pd NPs within electrospun PEI/PVA nanofibers. Reprinted with permission from Ref. [72]. © 2012, American Chemical Society.

On the basis of work related to the immobilization of Au NPs, Huang et al. [72] used the same electrospun PEI/PVA nanofibrous mats as nanoreactors to immobilize Pd NPs. The complete fabrication process is shown in Figure 6.5. In brief, cross-linked PEI/PVA nanofibrous mats were soaked in a K_2PdCl_4 aqueous solution to allow $PdCl_4^{2-}$ ions to complex with available PEI amine groups through ionic exchange. Then, $NaBH_4$ solution was added to reduce $PdCl_4^{2-}$ to form Pd NPs. Similar to the immobilization of ZVI or Au NPs within nanofibers [13, 14, 69, 70], the immobilization of Pd NPs does not appreciably change the morphology of PEI/PVA nanofibers (Figure 6.6). EDS analysis confirmed the existence of Pd element within the nanofibers. TEM images show that these Pd NPs with a mean diameter of 2.6 nm are quite uniformly distributed within the fibers with a small portion of particles having a denser distribution at the outer surface of the fibers. Furthermore, thermogravimetric analysis shows that the loading capacity of Pd NPs within the nanofibrous mat is about 13.1%. And this considerable amount of Pd NPs loaded endows the nanofibrous material with the potential to tackle tough environmental problems.

6.5 ENVIRONMENTAL APPLICATIONS OF HYBRID METAL NP-CONTAINING POLYMER NANOFIBERS

6.5.1 Decoloration

The major advantage of NP-immobilized polymer nanofibers is that the fibrous mats can be easily separated from the contaminated water, which is an ideal property enabling their use as reusable and recyclable filtration or sorption materials, and simultaneously to avoid possible secondary contamination. The prepared ZVI NPs immobilized onto the $(PAA/PDADMAC)_6$-coated CAc nanofibrous mats have been used to decolorize dye-contaminated water [20]. In that study, Xiao et al. used ZVI NP-immobilized $(PAA/PDADMAC)_6$-coated CAc nanofibrous mats for the decoloration of acid fuchsin (AF), a common organic dye in textile industry. The results show that the decoloration efficiency of the ZVI NPs with a size of 1.4 nm formed via two

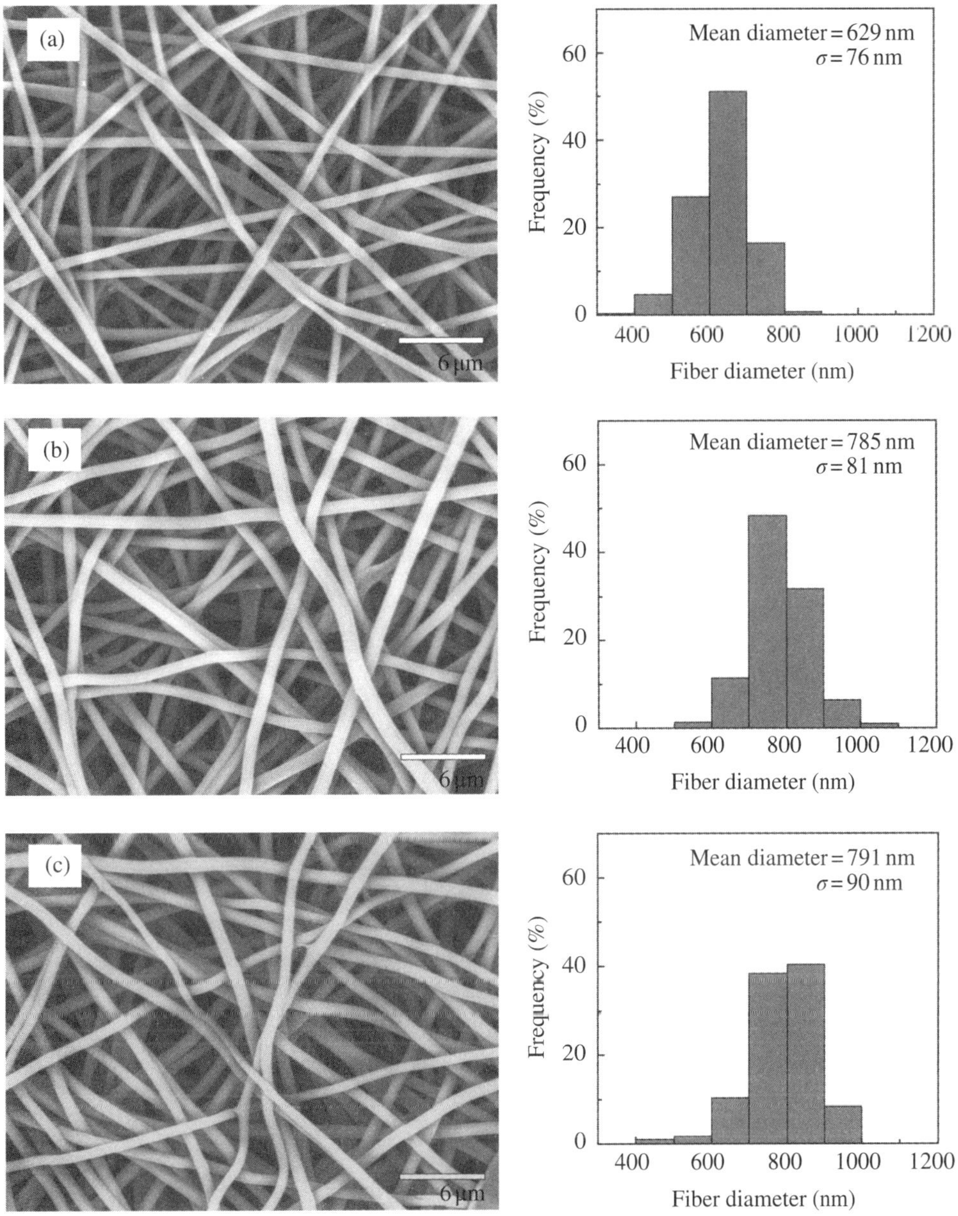

FIGURE 6.6 SEM images and diameter distribution histograms of (a) the noncrosslinked PEI/PVA nanofibers, (b) the cross-linked PEI/PVA nanofibers, and (c) the Pd NP-containing PEI/PVA nanofibers. Reprinted with permission from Ref. [72]. © 2012, American Chemical Society.

cycles of Fe(II) binding and reduction is much higher than that with a size of 2.3 nm formed via six cycles of Fe(II) binding and reduction (Figure 6.7). This could be due to the smaller size and thus the greater specific surface area of the ZVI NPs, thereby leading to a greater reactivity toward dye decoloration.

Similarly, ZVI NP-immobilized PAA/PVA nanofibrous mats synthesized by Xiao et al. were also able to decolorize AF with a decoloration percentage approaching 95.8% within 40 min [14]. When compared to the synthesis of ZVI NP powder using the method in the literature [73], the prepared ZVI NPs (1.6 nm) uniformly distributed in the polymer nanofibers have much higher reactivity to decolorize AF. For practical environmental applications, it is essential to fabricate nanofibrous materials with enhanced mechanical durability. In another study of ours, MWCNT-reinforced PAA/PVA nanofibrous mats containing ZVI NPs were also used to decolorize dyes such as methyl blue, acridine orange, and AF [13]. The efficiencies of MWCNT-reinforced PAA/PVA nanofibrous mats containing ZVI NPs to decolorize the three dyes are listed in Table 6.1. After an exposure time of 40 min, the percentage of all dyes remaining was lower than 10%.

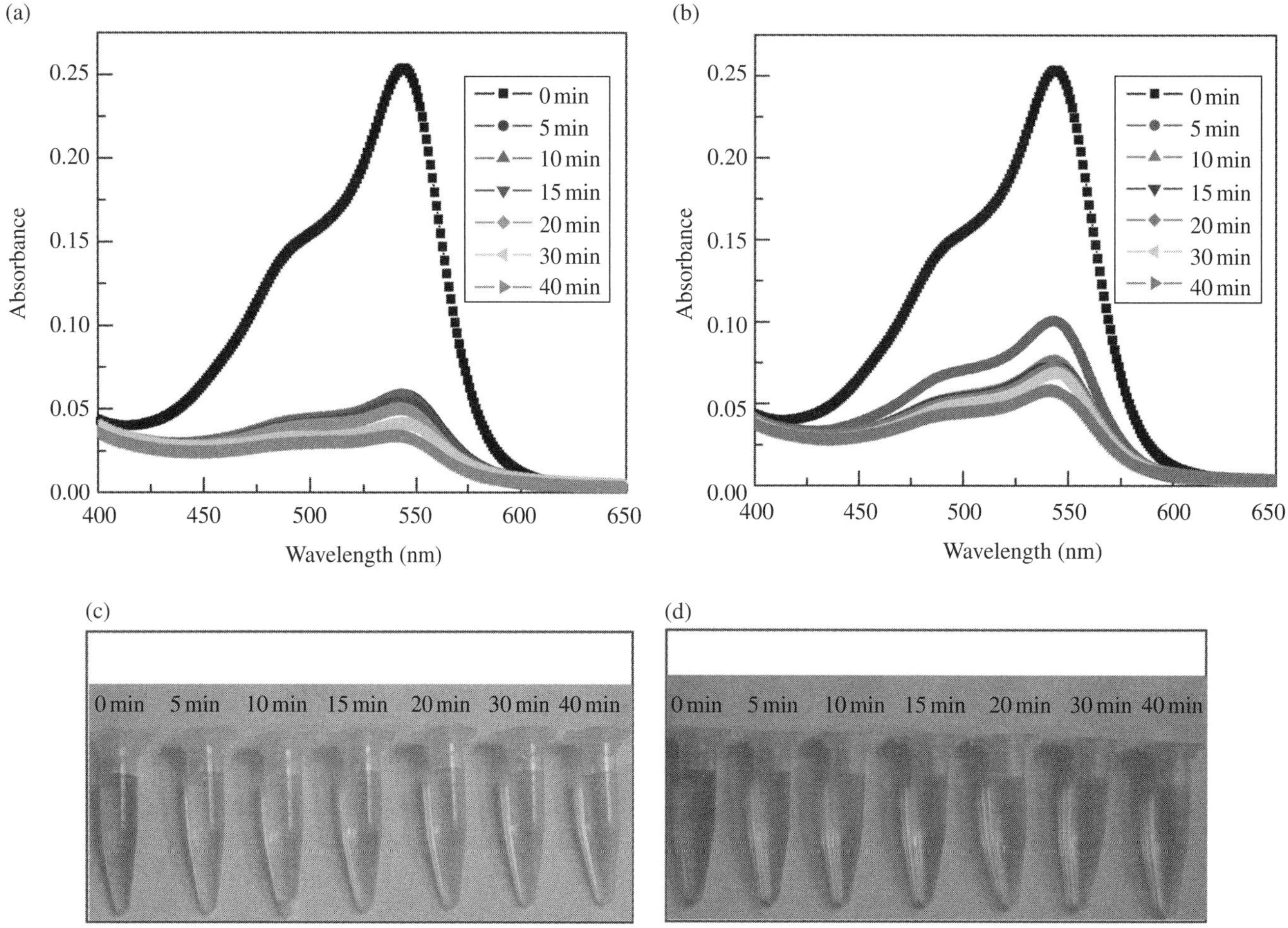

FIGURE 6.7 UV-vis spectra of a solution of acid fuchsin (AF) (60 ml/l, 30 ml) in the presence of ZVI NP-immobilized hybrid nanofibrous mats synthesized with two (a) and six (b) cycles of binding/reduction at a time interval of 0, 5, 10, 15, 20, 30, and 40 min, respectively, and photos of the AF solution treated with ZVI NP-immobilized hybrid nanofibrous mats synthesized with two (c) and six (d) cycles of binding/reduction at different time intervals. Reprinted with permission from Ref. [20]. © 2009, American Chemical Society.

TABLE 6.1 Decoloration efficiencies of dyes treated with MWCNT-incorporated nanofibrous mats containing ZVI NPs at 5 and 40 min

Sample	Initial concentration (mg/l)	pH value of dye solution	Maximum absorption wavelength (nm)	Reaction time (min)	Remaining percentage (%)
Methyl blue	60	5.68	600	5	8.91
				40	3.84
Acridine orange	60	5.6	493	5	6.93
				40	0.626
Acid fuchsin	60	6.8	544	5	37.4
				40	9.75

Reproduced with permission from Ref. [13]. © Royal Society of Chemistry.

Different from physical absorption of dyes using nonmetal incorporated nanofibers [71], the decoloration using ZVI NP-immobilized polymer nanofibers is predominantly based on the chemical reactivity of the immobilized ZVI NPs. Although complete details of the mechanism are still unclear, it is believed that the chromophore of the selected dyes can be destroyed after the iron reduction process. The reaction between Fe^0 and H_2O or H^+ can generate atomic H, destroying the chromophore group of the dyes [74]. Moreover, the intermediate products of Fe^0 such as Fe^{2+}, Fe^{3+}, $Fe(OH)_y^{2-y}$, and $Fe(OH)_x^{3-x}$ were thermodynamically unstable and active [75].

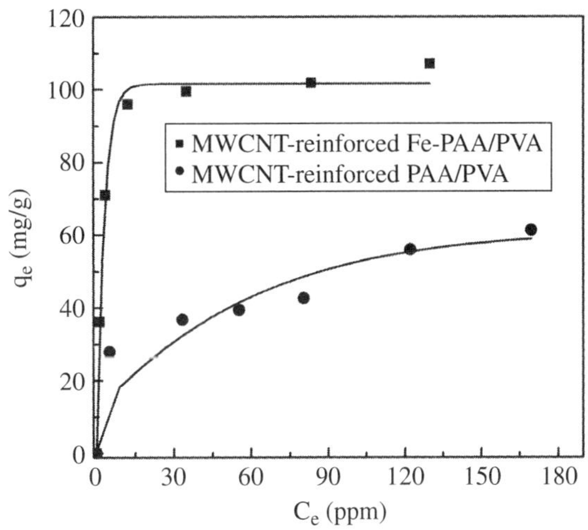

FIGURE 6.8 The Cu(II) ions sorption isotherms (contact time: 60 min; [mat] = 0.5 mg/ml). Reprinted with permission from Ref. [19], pp. 48–54. © 2011, Elsevier.

6.5.2 Heavy Metal Ion Remediation

For toxic heavy metal ion remediation, electrospun PAA/PVA nanofibrous mats have been used to remove copper (II) ions from aqueous solution [15]. The PAA/PVA nanofibrous mats were able to remove about 98% of Cu(II) in the initial 30 min at room temperature. The Cu(II) removal is attributed to the strong complexation effect of free carboxyl group of PAA with Cu(II) ions [76, 77]. In addition, the water-stable PAA/PVA nanofibrous mats displayed excellent selectivity in the absorption of Cu(II) in the presence of both Cu(II) and Ca(II) ions with the same concentration. Furthermore, it was shown that under the selected ionic strengths ([NaCl] = 0, 0.1, and 0.5 M), the absorption capability of the cross linked PAA/PVA nanofibrous mats was not significantly impacted. When compared with other materials used for heavy metal ion removal, the high surface area to volume ratio of the porous PAA/PVA fiber material allows for complete interaction with the ions to be removed, thereby significantly enhancing the capability of this material for environmental remediation.

By incorporating ZVI NPs within PAA/PVA nanofibers, heavy metal ions could be remediated via both physical absorption and chemical absorption processes. In our recent study [19], MWCNT-reinforced PAA/PVA nanofibers immobilized with ZVI NPs were also used to remediate Cu(II) in aqueous solution. Cu(II) ions from aqueous solution were removed using hybrid ZVI NP-immobilized nanofibers under an array of varying conditions, including solution pH and contact time. We showed that Cu(II) removal by hybrid mats containing ZVI NPs was not affected by pH variation and the removal percentage remained at around 91%. In contrast, the sorption ability of PAA/PVA nanofibrous mats without ZVI NP immobilization was affected by the pH change: The Cu(II) removal percentage increased markedly with pH, presumably due to the fact that more carboxylic groups in the PAA/PVA nanofibers became deprotonated, leading to favorable ion sorption via electrostatic interaction [76, 78]. Furthermore, MWCNT-reinforced nanofibrous mats containing ZVI NPs were able to significantly remove Cu(II) within the first 30 min, and then reached an equilibrium (75.3 mg/g) within 60 min. By comparison, the MWCNT-reinforced nanofibrous mats without ZVI NPs exhibited a much lower level of Cu(II) sorption. Although the incorporation of MWCNTs within the nanofibers did not enhance the Cu(II) removal ability of PAA/PVA nanofibrous mats, our results clearly suggest that the immobilized ZVI NPs improved Cu(II) removal via a chemical reduction reaction. Figure 6.8 shows the results of isotherm experiments. In the case of the ZVI NP-immobilized hybrid mats, sorption of Cu(II) increased sharply with the aqueous concentration below 12 mg/l at equilibrium and then reached a plateau near 100 mg/g when the aqueous Cu(II) concentration was above 12 mg/l, indicating saturation of the binding sites. For the mats that were not ZVI NP-immobilized, sorption of Cu(II) ions exhibited a gradual increase and reached 61.4 mg/g when the aqueous concentration at equilibrium was 169.3 mg/l. Obviously, the sorption capacity of the hybrid mats was greatly enhanced in the presence of immobilized ZVI NPs.

6.5.3 Dechlorination

Trichloroethylene (TCE) is a widely used chlorinated solvent, which is one of the common pollutants in soil and groundwater. Dechlorination of TCE is able to significantly reduce its toxicity. MWCNT-reinforced ZVI NP-immobilized electrospun PAA/PVA nanofibrous mats fabricated by Xiao et al. have proven to be highly effective in the dechlorination of TCE [13]. The results show that both ZVI NP-containing nanofibrous mats with and without MWCNTs exhibit excellent performance in the

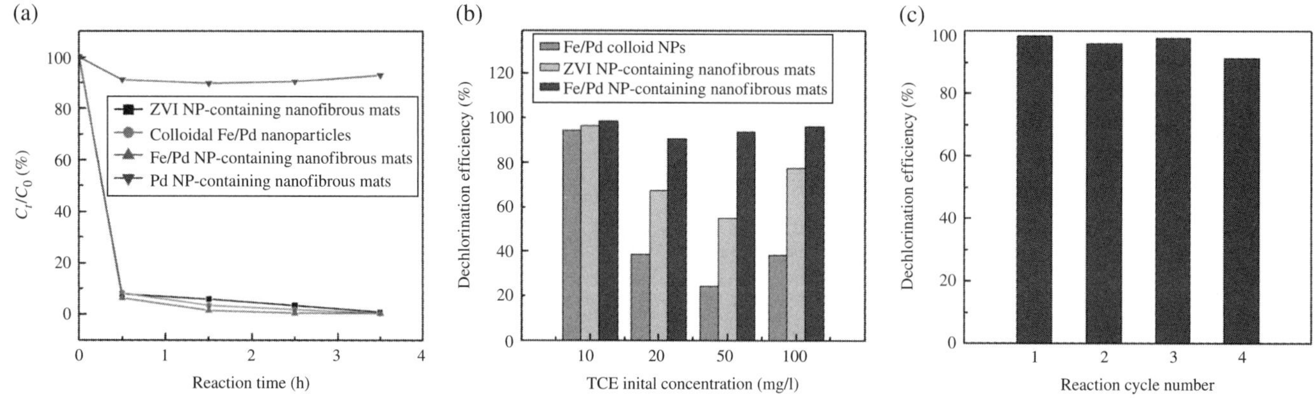

FIGURE 6.9 (a) Remaining fraction of TCE as a function of time after treatment with Fe/Pd colloid NPs, ZVI NP-containing fibrous mats, and Fe/Pd NP-containing nanofibrous mats. C_0 and C represent the initial and final concentration of TCE, respectively. The initial concentration of TCE was 10 mg/l. (b) Removal efficiency of TCE by Fe/Pd colloid NPs and ZVI NP- and Fe/Pd NP-containing nanofibrous mats with different initial TCE concentrations. The reaction time was 1.5 h. (c) Removal efficiency of TCE after treatment with the same Fe/Pd NP-containing nanofibrous mats for the first, second, third, and fourth time. The reaction time was 1.5 h and the initial TCE concentration was 10 mg/l. Reprinted with permission Ref. [69], pp. 349–356. © 2012, Elsevier.

degradation of TCE. During the first 30 min, the degradation process reached a dynamic equilibrium, and the remaining fraction of TCE treated with ZVI NP-immobilized nanofibrous mats with and without MWCNTs was estimated to be 0.034 and 0.086, respectively. It should be noted that the major advantage of the MWCNTs within the nanofibers is to improve the mechanical durability of the nanofibers. The presence of MWCNTs does not compromise the excellent capability of the immobilized ZVI NPs to dechlorinate TCE. Moreover, it may be helpful to concentrate the hydrophobic chlorinated organic contaminants for efficient remediation.

In another study, we investigated the feasibility of using bimetallic Fe/Pd NP-immobilized PAA/PVA electrospun nanofibers to dechlorinate TCE [69]. Freshly synthesized Fe/Pd colloidal NPs and single-metal ZVI NP- and Pd NP-immobilized nanofibrous mats were used as controls. As shown in Figure 6.9a, except Pd NP-immobilized nanofibrous mats, the other three materials were able to degrade TCE, indicating that immobilized Pd NPs do not contribute to the dechlorination effect of TCE. The slightly decreased remaining fraction of TCE may be ascribed to the adsorption of TCE onto the polymer nanofibers with a high specific surface area. It is noticeable that the remaining fraction of TCE treated with Fe/Pd colloid NPs and ZVI NP- and Fe/Pd NP-immobilized nanofibrous mats were quite close at the initial TCE concentration of 10 mg/l. In order to further investigate the enhanced dechlorination ability of Fe/Pd NP-immobilized nanofibrous mats, the initial TCE concentration was increased to 20, 50, and 100 mg/l, respectively. It is clear that in all the cases, the efficiency of TCE dechlorination using Fe/Pd NP-containing nanofibrous mats is above 90.6% (Figure 6.9b), much higher than that using ZVI NP-containing nanofibrous mats and the Fe/Pd colloid NPs. Reusability and recyclability assessment results showed that after exposure to sodium borohydride aqueous solution for 10 min, nanofibrous mats can be regenerated. The regenerated Fe/Pd NP-containing nanofibrous mats exhibited similar performance in the second, third, and fourth round of TCE dechlorination, comparable to that of freshly prepared mats (Figure 6.9c).

6.5.4 Environmental Catalysis

The generated hybrid metal NP-containing polymer nanofibers can be used as an efficient catalyst to remediate heavy metal ions. In our recent report, catalytic active Pd NP-immobilized electrospun PEI/PVA nanofibrous mats were fabricated for catalytic transformation of hexavalent chromium (Cr(VI)) to trivalent Cr (Cr(III)) [72]. It is known that Cr(VI) has acute mutagenicity and carcinogenicity, while Cr(III) is far less toxic. In the presence of formic acid used as a reducing agent, a piece of Pd NP-immobilized PEI/PVA nanofibrous mat was immersed in a $K_2Cr_2O_7$ solution as a catalyst. Meanwhile, in order to evaluate the reusability of this catalyst, more reaction cycles were performed using the same mat. As shown in Figure 6.10, the intensity of the characteristic absorption peak at 350 nm for $Cr_2O_7^{2-}$ decreased with time. At 12 min, the absorption peak completely vanished, indicating the complete transformation of Cr(VI). By comparison, in the presence of PEI/PVA nanofibrous mats without Pd NPs (control group), the characteristic peak at 350 nm did not change considerably (Figure 6.10b). In addition, the Cr(VI) solution treated by formic acid without any catalysts did not show any significant change in the absorption peak at 350 nm within 25 min (Figure 6.10c). These results demonstrate that the excellent catalytic transformation of Cr(VI) to Cr(III) can

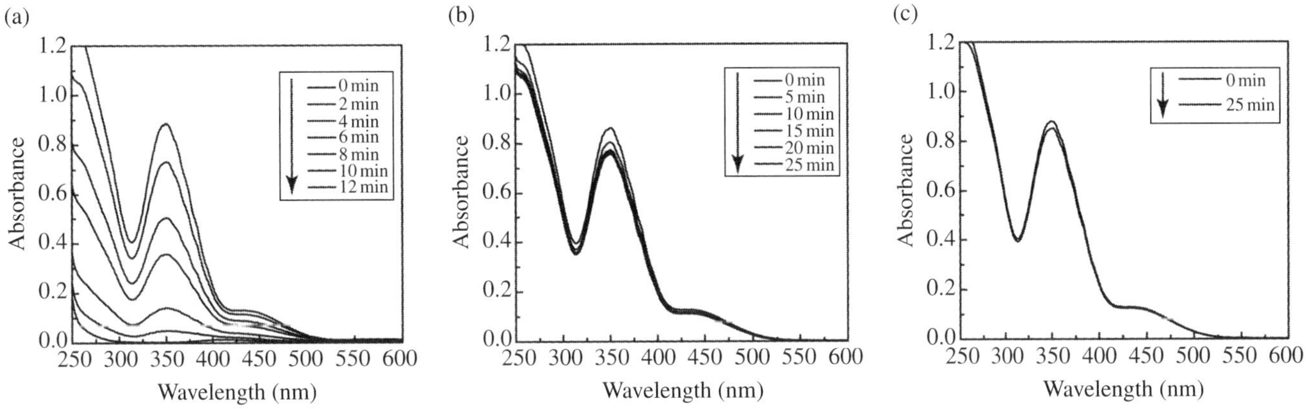

FIGURE 6.10 UV–vis spectra of Cr(VI) aqueous solution (a) treated with the Pd NP-immobilized PEI/PVA nanofibrous mat in the presence of formic acid, (b) treated with the PEI/PVA nanofibrous mat without Pd NPs in the presence of formic acid, and (c) treated with only formic acid. Reprinted with permission from Ref. [72]. © 2012, American Chemical Society.

be solely attributed to the immobilized Pd NPs. Reusability results show that more than 99% of Cr(VI) can be transformed to Cr(III) after 12, 15, and 15 min in the first, second, and third reaction cycles, respectively, confirming the superior reusability of the Pd NP-immobilized PEI/PVA nanofibrous mats.

6.6 CONCLUSIONS AND OUTLOOK

In summary, this chapter has described the fabrication of different metal NP-immobilized polymer nanofibers for environmental remediation applications. In particular, ZVI NP-immobilized PAA/PVA nanofibers have been fabricated via different approaches. To further improve mechanical durability, MWCNTs have been incorporated within the nanofibers prior to the in situ formation of ZVI NPs. Furthermore, to enhance the reactivity of immobilized ZVI NPs, bimetallic Fe/Pd NPs were formed within PAA/PVA nanofibers. These ZVI NP-based nanofibers are able to decolorize dying contaminants, to remove toxic heavy metal ions, and to dechlorinate chlorinated organic solvents (e.g., TCE). For the catalytic remediation of heavy metal ions, Pd NP-immobilized polymer nanofibers have been fabricated, and the hybrid fibers are able to efficiently transform Cr(VI) to Cr(III) in aqueous solution. The formed metal NP-containing polymer nanofibers are stable, and no leakage of the metal NPs are observed during storage and remediation processes.

There is no doubt that electrospun polymer nanofibrous mats have exhibited great advantages over other conventional media as an ideal substrate to immobilize active metal NPs for environmental remediation applications. The electrospinning technology has received increasing attention in the fabrication of various metal NP-immobilized polymer nanofibers, owing to their high porosity, huge specific surface area, and flexibility in surface functionalization. However, there still remain some challenges for practical environmental applications. For example, the lack of mass production of high-quality electrospun nanofibers quite restricts its application in practice. The selection of suitable polymer materials with improved mechanical durability and the surface modification of preformed nanofibers with new functionalities for selective remediation of environmental pollutants are additional issues to be addressed. These challenges will drive the development of a variety of metal NP-incorporated hybrid polymer nanofibers for various environmental applications.

REFERENCES

[1] Kreuter J. Nanoparticlebased drug delivery systems. J Control Release 1991;16:169.

[2] Paciotti GF, Myer L, Weinreich D, Goia D, Pavel N, McLaughlin RE, Tamarkin L. Colloidal gold: a novel nanoparticle vector for tumor directed drug delivery. Drug Deliv 2004;11:169.

[3] Shi X, Wang S, Meshinchi S, Van Antwerp ME, Bi X, Lee I, Baker JR. Dendrimer-entrapped gold nanoparticles as a platform for cancer cell targeting and imaging. Small 2007;3:1245.

[4] Trewyn BG, Giri S, Slowing II, Lin VSY. Mesoporous silica nanoparticle based controlled release, drug delivery, and biosensor systems. Chem Commun 2007:3236.

[5] Govorov AO, Bryant GW, Zhang W, Skeini T, Lee J, Kotov NA, Slocik JM, Naik RR. Exciton-plasmon interaction and hybrid excitons in semiconductor-metal nanoparticle assemblies. Nano Lett 2006;6:984.

[6] Kotov NA, Dekany I, Fendler JH. Layer-by-layer self-assembly of polyelectrolyte-semiconductor nanoparticle composite films layer-by-layer self-assembly of polyelectrolyte-semiconductor nanoparticle composite films. J Phys Chem 1995;99:13065.

[7] Wang Y, Lee JY, Deivaraj TC. Tin nanoparticle loaded graphite anodes for Li-ion battery applications. J Electrochem Soc 2004;151:A1804.

[8] Jeon HJ, Yi SC, Oh SG. Preparation and antibacterial effects of Ag-SiO_2 thin films by sol-gel method. Biomaterials 2003;24:4921.

[9] Parikh DV, Fink T, Rajasekharan K, Sachinvala ND, Sawhney APS, Calamari TA, Parikh AD. Antimicrobial silver/sodium carboxymethyl cotton dressings for burn wounds. Text Res J 2005;75:134.

[10] Canter N. Trends in extreme pressure additives. Tribol Lubric Technol 2007;63:10.

[11] Lee J, Cho S, Hwang Y, Lee C, Kim S. Enhancement of lubrication properties of nano-oil by controlling the amount of fullerene nanoparticle additives. Tribol Lett 2007;28:203.

[12] Fang X, Xiao S-L, Shen M-W, Guo R, Wang S-Y, Shi X-Y. Super dye removal capability of zero-valent iron nanoparticle-immobilized electrospun polymer nanofibrous membranes. J Mod Text Sci Eng 2010;1:1.

[13] Xiao S, Shen M, Guo R, Huang Q, Wang S, Shi X. Fabrication of multiwalled carbon nanotube-reinforced electrospun polymer nanofibers containing zero-valent iron nanoparticles for environmental applications. J Mater Chem 2010;20:5700.

[14] Xiao S, Shen M, Guo R, Wang S, Shi X. Immobilization of zerovalent iron nanoparticles into electrospun polymer nanofibers: synthesis, characterization, and potential environmental applications. J Phys Chem C 2009;113:18062.

[15] Xiao S, Shen M, Ma H, Guo R, Zhu M, Wang S, Shi X. Fabrication of water-stable electrospun polyacrylic acid-based nanofibrous mats for removal of copper (II) ions in aqueous solution. J Appl Polym Sci 2010;116:2409.

[16] Dandapat A, Jana D, De G. Pd nanoparticles supported mesoporous γ-Al_2O_3 film as a reusable catalyst for reduction of toxic Cr^{VI} to Cr^{III} in aqueous solution. Appl Catal Gen 2011;396:34.

[17] Omole MA, K'Owino IO, Sadik OA. Palladium nanoparticles for catalytic reduction of Cr(VI) using formic acid. Appl Catal Environ 2007;76:158.

[18] Huang Q, Shi X, Pinto RA, Petersen EJ, Weber WJ Jr. Tunable synthesis and immobilization of zero-valent iron nanoparticles for environmental applications. Environ Sci Technol 2008;42:8884.

[19] Xiao S, Ma H, Shen M, Wang S, Huang Q, Shi X. Excellent copper(II) removal using zero-valent iron nanoparticle-immobilized hybrid electrospun polymer nanofibrous mats. Colloid Surf A Physicochem Eng Asp 2011;381:48.

[20] Xiao S, Wu S, Shen M, Guo R, Huang Q, Wang S, Shi X. Polyelectrolyte multilayer-assisted immobilization of zero valent iron nanoparticles onto polymer nanofibers for potential environmental applications. ACS Appl Mater Interfaces 2009;1:2848.

[21] Xia Y, Yang P, Sun Y, Wu Y, Mayers B, Gates B, Yin Y, Kim F, Yan H. One-dimensional nanostructures: synthesis, characterization, and applications. Adv Mater 2003;15:353.

[22] Vohra V, Calzaferri G, Destri S, Pasini M, Porzio W, Botta C. Toward white light emission through efficient two-step energy transfer in Hybrid nanofibers. ACS Nano 2010;4:1409.

[23] Qi RL, Liu HJ. Electrospun MWCNTs/Poly (lactic-co-glycolic acid) composite nanofibrous drug delivery system. Adv Mater Res 2012;424:1220.

[24] Hou H, Reneker DH. Carbon nanotubes on carbon nanofibers: a novel structurebased on electrospun polymer nanofibers. Adv Mater 2004;16:69.

[25] Li D, Xia Y. Electrospinning of nanofibers: reinventing the wheel? Adv Mater 2004;16:1151.

[26] Reneker DH, Yarin AL. Electrospinning jets and polymer nanofibers. Polymer 2008;49:2387.

[27] Yoon K, Hsiao BS, Chu B. Functional nanofibers for environmental applications. J Mater Chem 2008;18:5326.

[28] Fujihara K, Kumar A, Jose R, Ramakrishna S, Uchida S. Spray deposition of electrospun TiO_2 nanorods for dye-sensitized solar cell. Nanotechnology 2007;18:365709.

[29] Lebrun L, Vallée F, Alexandre B, Nguyen QT. Preparation of chelating membranes to remove metal cations from aqueous solutions. Desalination 2007;207:9.

[30] Xiao S, Wu S, Shen M, Guo R, Wang S, Shi X. Polyelectrolyte multilayer film-assisted formation of zerovalent iron nanoparticles onto polymer nanofibrous mats. J Phys Conf Ser 2009;188:012015/1.

[31] Schreuder-Gibson H, Gibson P, Senecal K, Sennett M, Walker J, Yeomans W, Ziegler D, Tsai PP. Protective textile materials based on electrospun nanofibers. J Adv Mater 2002;34:44.

[32] Wang X, Kim Y-G, Drew C, Ku B-C, Kumar J, Samuelson LA. Electrostatic assembly of conjugated polymer thin layers on electrospun nanofibrous membranes for biosensors. Nano Lett 2004;4:331.

[33] He W, Ma ZW, Yong T, Teo WE, Ramakrishna S. Fabrication of collagen-coated biodegradable polymer nanofiber mesh and its potential for endothelial cells growth. Biomaterials 2005;26:7606.

[34] Hong KH. Preparation and properties of electrospun poly (vinyl alcohol)/silverfibre web as wound dressings. Polym Eng Sci 2007;47:43.

[35] Liu FJ, Guo R, Shen MW, Wang SY, Shi XY. Effect of processing variables on the morphology of electrospun poly[(lactic acid)-co-(glycolic acid)] nanofibers. Macromol Mater Eng 2009;294:666.

[36] Liu F, Guo R, Shen M, Cao X, Mo X, Wang S, Shi X. Effect of the porous microstructures of poly (lactic-co-glycolic acid)/carbon nanotube composites on the growth of fibroblast cells. Soft Mater 2010;8:239.

[37] Liao H, Qi R, Shen M, Cao X, Guo R, Zhang Y, Shi X. Improved cellular response on multiwalled carbon nanotube-incorporated electrospun polyvinyl alcohol/chitosan nanofibrous scaffolds. Colloid Surf B Biointerfaces 2011;84:528.

[38] Wang S, Cao X, Shen M, Guo R, Bányai I, Shi X. Fabrication and morphology control of electrospun poly(gamma-glutamic acid) nanofibers for biomedical applications. Colloid Surf B Biointerfaces 2012;89:254.

[39] Veleirinho B, Rei MF, Lopes-Da-Silva JA. Solvent and concentration effects on the properties of electrospun poly(ethylene terephthalate) nanofiber mats. J Polym Sci B 2008;46:460.

[40] Lin SH, Lin CM. Treatment of textile waste effluents by ozonation and chemical coagulation. Water Res 1993;27:1743.

[41] Ganesh R, Boardman GD, Michelsen D. Fate of azodyes in sludges. Water Res 1994;28:1367.

[42] Volesky B, Holan ZR. Biosorption of heavy metals. Biotechnol Prog 1995;11:235.

[43] Kumar P, Dushenkov V, Motto H, Raskin I. Phytoextraction: the use of plants to remove heavy metals from soils. Environ Sci Technol 1995;29:1232.

[44] Annadurai G, Ling LY, Lee JF. Adsorption of reactive dye from an aqueous solution by chitosan: isotherm, kinetic and thermodynamic analysis. J Hazard Mater 2008;152:337.

[45] Chen S, Zou Y, Yan Z, Shen W, Shi S, Zhang X, Wang H. Carboxymethylated-bacterial cellulose for copper and lead ion removal. J Hazard Mater 2009;161:1355.

[46] Fillipi BR, Scamehorn JF, Christian SD, Taylor RW. A comparative economic-analysis of copper removal from water by ligand-modified micellar-enhanced ultrafiltration and by conventional solvent-extraction. J Membr Sci 1998;145:27.

[47] Choi H, Al-Abed SR, Agarwal S, Dionysiou DD. Synthesis of reactive nano Fe/Pd bimetallic system-impregnated activated carbon for the simultaneous adsorption and dechlorination of PCBs. Chem Mater 2008;20:3649.

[48] Darab JG, Amonette AB, Burke DSD, Orr RD, Ponder SM, Schrick B, Mallouk TE, Lukens WW, Caulder DL, Shuh DK. Removal of pertechnetate from simulated nuclear waste streams using supported zero valent iron. Chem Mater 2007;19:5703.

[49] Formhals A. Process and apparatus for preparing artificial threads. US patent US1975504. 1934; Method and apparatus for spinning. US patent US2160962. 1939; Artificial thread and method of producing same. US patent US2187306. 1940; Production of artificial fibers from fiber forming liquids. US patent US2323025. 1943; Method and apparatus for spinning. US patent US2349950. 1944.

[50] Wang S, Zhao Y, Shen M, Shi X. Electrospun hybrid nanofibers doped with nanoparticles or nanotubes for biomedical applications. Ther Deliv 2012;3:1155.

[51] Bhardwaj N, Kundu SC. Electrospinning: a fascinating fiber fabrication technique. Biotechnol Adv 2010;28:325.

[52] Li D, Xia Y. Fabrication of titania nanofibers by electrospinning. Nano Lett 2003;3:555.

[53] Li MY, Mondrinos MJ, Gandhi MR, Ko FK, Weiss AS, Lelkes PI. Electrospun protein fibers as matrices for tissue engineering. Biomaterials 2005;26:5999.

[54] Luu YK, Kim K, Hsiao BS, Chu B, Hadjiargyrou M. Development of a nanostructured DNA delivery scaffold via electrospinning of PLGA and PLA-PEG block copolymers. J Control Release 2003;89:341.

[55] Wu H, Zhang R, Liu X, Lin D, Pan W. Electrospinning of Fe, Co, and Ni nanofibers: synthesis, assembly, and magnetic properties. Chem Mater 2007;19:3506.

[56] Fang D, Hsiao BS, Chu B. Multiple-jet electrospinning of non-woven nanofiber articles. Polym Preprs 2003;44:59.

[57] Carnell LS, Siochi EJ, Wincheski RA, Holloway NM, Clark RL. Electric field effects on fiber alignment using an auxiliary electrode during electrospinning. Scr Mater 2009;60:359.

[58] Theron A, Zussman E, Yarin AL. Electrostatic field-assisted alignment of electrospun nanofibres. Nanotechnology 2001;12:384.

[59] Yang D, Lu B, Zhao Y, Jiang X. Fabrication of aligned fibrous arrays by magnetic electrospinning. Adv Mater 2007;19:3702.

[60] Wang X, Niu H, Lin T. Needleless electrospinning of nanofibers with a conical wire coil. Polym Eng Sci 2009;49:1582.

[61] Teo WE, Kotaki M, Mo XM, Ramakrishna S. Porous tubular structures with controlled fibre orientation using a modified electrospinning method. Nanotechnology 2005;16:918.

[62] Zhang YZ, Huang ZM, Xu XJ, Lim CT, Ramakrishna S. Preparation of core-shell structured PCL-r-gelatin bi-component nanofibers by coaxial electrospinning. Chem Mater 2004;16:3406.

[63] Wu Y, Dong Z, Wilson S, Clark RL. Template-assisted assembly of electrospun fibers. Polymer 2010;51:3244.

[64] Doshi J, Reneker DH. Electrospinning process and applications of electrospun fibers. J Electrostat 1995;35:151.

[65] Zuo W, Zhu M, Yang W, Yu H, Chen Y, Zhang Y. Experimental study on relationship between jet instability and formation of beaded fibers during electrospinning. Polym Eng Sci 2005;45:704.

[66] Petersen EJ, Pinto RA, Shi X, Huang Q. Impact of size and sorption on degradation of trichloroethylene and polychlorinated biphenyls by nano-scale zerovalent iron. J Hazard Mater 2012;243:73.

[67] Ding B, Kim J, Kimura E, Shiratori S. Layer-by-layer structured films of TiO_2 nanoparticles and poly (acrylic acid) on electrospun nanofibres. Nanotechnology 2004;15:913.

[68] Xiao S, Shen M, Ma H, Fang X, Huang Q, Weber WJ Jr, Shi X. Manipulation of the loading and size of zero-valent iron nanoparticles immobilized in electrospun polymer nanofibers. J Nanosci Nanotechnol 2011;11:5089.

[69] Ma H, Huang Y, Shen M, Guo R, Cao X, Shi X. Enhanced dechlorination of trichloroethylene using electrospun polymer nanofibrous mats immobilized with iron/palladium bimetallic nanoparticles. J Hazard Mater 2012;211–212:349.

[70] Fang X, Ma H, Xiao S, Shen M, Guo R, Cao X, Shi X. Facile immobilization of gold nanoparticles into electrospun polyethyleneimine/polyvinyl alcohol nanofibers for catalytic application. J Mater Chem 2011;21:4493.

[71] Fang X, Xiao S, Shen M, Guo R, Wang S, Shi X. Fabrication and characterization of water-stable electrospun polyethyleneimine/polyvinyl alcohol nanofibers with super dye sorption capability. New J Chem 2011;35:360.

[72] Huang Y, Ma H, Wang S, Shen M, Guo R, Cao X, Zhu M, Shi X. Efficient catalytic reduction of hexavalent chromium using palladium nanoparticle-immobilized electrospun polymer nanofibers. ACS Appl Mater Interfaces 2012;4:3054.

[73] Wang C-B, Zhang W-X. Synthesizing nanoscale iron particles for rapid and complete dechlorination of TCE and PCBs. Environ Sci Technol 1997;31:2154.

[74] Fan J, Guo Y, Wang J, Fan M. Rapid decolorization of azo dye methyl orange inaqueous solution by nanoscale zerovalent iron particles. J Hazard Mater 2009;166:904.

[75] Sohn K, Kang SW, Ahn S, Woo M, Yang SK. Fe(0) nanoparticles for nitratereduction: Stability, reactivity, and transformation. Environ Sci Technol 2006;40:5514.

[76] Feng N, Guo X, Liang S. Adsorption of copper(II) by chemicallymodified orange peel. J Hazard Mater 2009;164:1286.

[77] Rivas BL, Schiappacasse LN, Pereira UE, Moreno-Villoslada I. Interactions of polyelectrolytes bearing carboxylate and/or sulfonate groups with Cu(II) and Ni(II). Polymer 2004;45:1771.

[78] Wang TC, Rubner MF, Cohen RE. Polyelectrolyte multilayer nanoreactors for preparing silver nanoparticle composites: controlling metal concentration and nanoparticle size. Langmuir 2002;18:3370.

7

NANOMATERIALS ON THE BASIS OF CHELATING AGENTS, METAL COMPLEXES, AND ORGANOMETALLICS FOR ENVIRONMENTAL PURPOSES

BORIS I. KHARISOV[1,2], OXANA V. KHARISSOVA[2,3], AND UBALDO ORTIZ MÉNDEZ[2]

[1]*Facultad de Ciencias Químicas, Universidad Autonoma de Nuevo León, Monterrey, Mexico*

[2]*CIIDIT, Universidad Autónoma de Nuevo León, Monterrey, Mexico*

[3]*Facultad de Ciencias Físico-Matemáticas, Universidad Autónoma de Nuevo León, Monterrey, Mexico*

7.1 INTRODUCTION

Chelating agents and their metal complexes are at the heart of modern coordination chemistry and in their original non-nano forms were used for environmental protection, for instance, for increasing the solubility of heavy metals in soil and therefore in enhancing phytoextraction [1]. In the past 20 years, in accordance with general developmental trends in nanoscience and nanotechnology, a variety of nanomaterials and nanocomposites have been created. These materials have been used, among other things, for environmental purposes, such as remediation/recovery/selective collection of metals from soils, water, or seawater [2–6].

In this chapter, we will discuss available nanomaterials and nanocomposites containing metal complexes and a few organometallic compounds used for the improvement of the environment. In addition to removal of heavy metals, they can be applied in catalysis. Moreover, they have antibacterial, sorbent, and sensor properties and can be precursors for the creation of other nanomaterials, utilized for environmental protection. N-, N,O-, S-, and P-containing ligands are the basis of these nanomaterials/nanocomposites, including well-known materials in nanotechnology such as polyaniline (PANi) and chitosan.

7.2 ELEMENTAL METALS FUNCTIONALIZED WITH CHELATING LIGANDS

A few functionalized (ligand-capped) or supported metal nanoparticles (NPs) (generally Au, Ag, Fe, and bimetallics formed from them), containing chelating ligands, have been applied for remediation of toxic metals by their chelation, as sensors for metal cations.

Multifunctional biocompatible nanochelators present a class of chelating agents that have the potential to compensate for the drawbacks of traditional ligands used in chelation therapy. The nanoscale dimensions of these nanochelators allow for a high surface to volume ratio to accommodate multiple ligands on the surface and have good solubility for easy absorption across the blood–brain barrier. A series of such nanochelators on the basis of gold and iron NPs with chelating ligands that are analogues of natural phytochelatin (PCh, Fig. 7.1) ligands found in plants were prepared [7, 8]. These natural PCh peptides are cysteine-rich ligands that are known to take up metals with high affinity and specificity. Ligand-capped Au NPs (~4-nm metal core, ~340-bipy

Nanomaterials for Environmental Protection, First Edition. Edited by Boris I. Kharisov, Oxana V. Kharissova, and H. V. Rasika Dias.

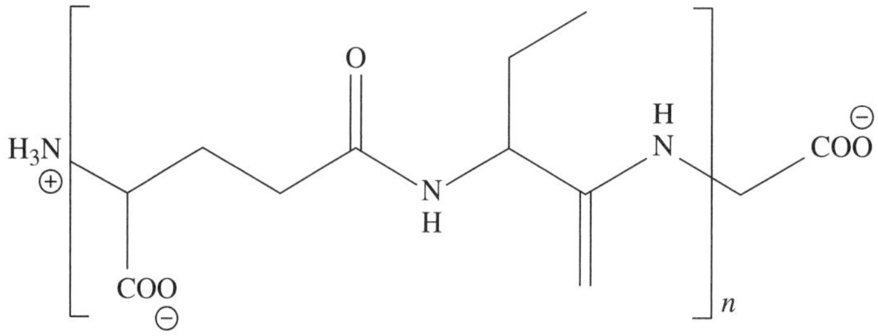

FIGURE 7.1 Natural phytochelatin.

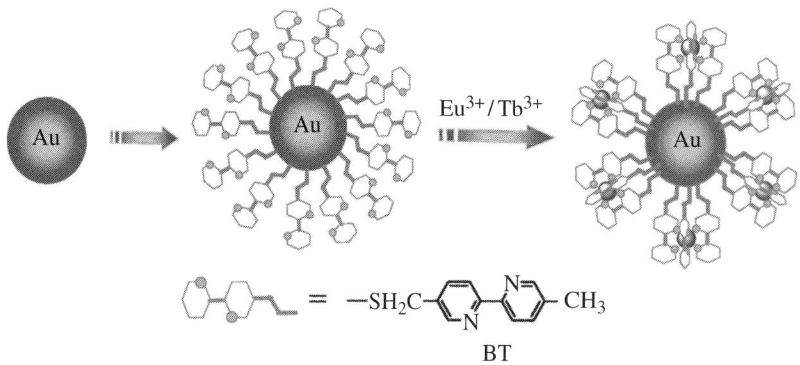

FIGURE 7.2 Design strategy for luminescent nanomaterials. Reproduced with permission from Ref. [9]. © American Chemical Society.

ligand shell) were synthesized (Fig. 7.2) [9] by capping monothiol derivatives of 2,2′-dipyridyl onto the surface of Au NPs (Au–BT). The high local concentration of the chelating ligands (~5 M) around the Au NP makes these particles excellent ion sponges, and their complexation (1:3 complexation between Eu^{III}/Tb^{III} ions and bipyridines) with Eu^{III}/Tb^{III} ions yields phosphorescent nanomaterials. The red-emitting Au–BT:Eu^{III} complex exhibited a long lifetime of 0.36 ms with six line-like emission peaks, whereas the green-emitting Au–BT:Tb^{III} complex exhibited a lifetime of 0.7 ms with four line-like emission peaks. These phosphorescent nanomaterials, designed by linking BT:Eu^{III} complexes to Au NPs, were further used as sensors for metal cations. Using tea polyphenols (TPs) as a reductant, Ag NPs supported on halloysite (1:1 aluminosilicate clay mineral with the empirical formula $Al_2Si_2O_5(OH)_4$) nanotubes (HNTs) were greenly synthesized [10] for photocatalytic decomposition of methylene blue (MB). HNTs were initially functionalized by *N*-β-aminoethyl-γ-aminopropyl trimethoxysilane (AEAPTMS) to introduce amino groups to form N-HNTs to fasten the Ag NPs; then the Ag NPs were synthesized and "anchored" on the surface of the HNTs. The chelating interaction between the Ag NPs and N atoms together with the TP molecules was revealed. The photocatalytic activity measurements of the prepared AgNPs@N-HNTs catalyst, evaluated by the decomposition of MB, exhibited excellent catalytic activity and high adsorption capability to MB.

Nano zero-valent iron (nZVI) technology is becoming an increasingly popular choice for environmental remediation and remediation of contaminated sites as iron is inexpensive, nontoxic, and environmentally compatible [11]. Nanoparticles are attractive for remediation of various contaminants because of their unique physicochemical properties, especially its high surface area over iron filings. At the same time, it is difficult to obtain pure nZVI without impurities of 15 other possible compounds (oxides and hydroxides). It could be reached using chelating agents. Thus, synthesis of air-stable nZVI is undertaken in the presence of ethylene diamine tetraacetic acid (EDTA), diethylenetriamine pentacetic acid (DTPA), nitriloacetic acid (NTAc), *trans*-1,2-diaminocyclohexane-*N,N,N′,N′*-tetraacetic acid (CDTA), hydroxyethylenediaminetetraacetic acid (HEDTA), triethylene tetraamine (TRTA), and *N*-cetyl-*N,N,N*-trimethyl ammonium bromide (CTAB) chelating agents [12]. The chelating effect was found to be the best for EDTA, NTAc, and HEDTA, but the least for CDTA and CTAB. Hydroxyl groups, lone pair electrons on the nitrogen atom, and steric effects of the cyclohexane ring and bulky surroundings played an important role in providing air stability toward synthesized nZVI. Chitosan–Fe^0 NPs (chitosan–Fe^0), prepared using nontoxic and biodegradable chitosan (see section N-Containing Ligands) as a stabilizer [13, 14], were used for Cr(VI) removal from water. The overall disappearance of Cr(VI) may include both physical adsorption of Cr(VI) onto the chitosan–Fe^0 surface and the subsequent reduction of Cr(VI) to Cr(III). It was revealed that after the reaction, relative to Cr(VI) and Fe(0), Cr(III) and Fe(III) were the predominant species on the surface of chitosan–Fe^0. Chitosan has also been found to inhibit Fe(III)–Cr(III) precipitation due to its high efficiency in chelating Fe(III) ions. In addition, immobilization of bimetallic NPs (Fe/Ni and Fe/Pd NP systems (<40 nm)) in polymer membrane (such as cellulose acetate, polyvinylidene fluoride (PVDF), polysulfone, chitosan) media led to the creation of materials for remediation of organic contaminants [15]. The second dopant metal (such as Ni, Pd) plays a very important role in terms of catalytic property (hydrodechlorination) and a significant reduction in the formation of intermediates. In addition to the rapid degradation (by Fe/Ni) of TCE to ethane, a complete dechlorination of selected PCBs (trichloroethylene (TCE) and selected polychlorinated biphenyls (PCBs)) using milligram quantities of immobilized Fe/Pd NPs in the membrane domain was achieved.

7.3 N-CONTAINING LIGANDS

Among simple N-containing ligands, we note ethylenediamine (en), whose complexes have been used as precursors of compounds having catalytic activities. Thus, mesoporous Co-doped ZnS and ZnO nanoplates were fabricated [16] by calcination of a $Zn_{0.95}Co_{0.05}S(en)_{0.5}$ complex (en = ethylenediamine), which was hydrothermally synthesized using ethylenediamine as a single solvent and chelating agent. Photocatalytic performance of the prepared materials was studied for decomposition of the azo dye (acid red 14). The $Zn_{0.95}Co_{0.05}S$ calcined at 500°C exhibited the highest photocatalytic activity under ultraviolet (UV) irradiation and also showed photocatalytic performance under visible light irradiation. Composite ruthenium-containing silica nanomaterials ($[RuO_2]@SiO_2$) from amine-stabilized ruthenium NPs as elemental bricks were found to possess a high specific surface area making them attractive materials for catalysis [17]. Bifunctional $H_2N–(CH_2)_x–Si(OEt)_3$ amines were used as stabilizing ligands (x = 3, 11) for the synthesis of ruthenium NPs, from [Ru(COD)(COT)] (COD = 1,3-cyclooctadiene, COT = 1,3,5-cyclooctatriene) as the metal precursor. The functionalization [18] of Fe_3O_4 NPs with carboxyl (succinic acid), amine (ethylenediamine), and thiol (2,3-dimercaptosuccinic acid) led to the formation of nanoadsorbents showing superparamagnetic behavior at room temperature with strong field-dependent magnetic responsivity. These products were found to be useful in enhancing the efficiency of these NPs for the removal of toxic metals (Cr^{3+}, Co^{2+}, Ni^{2+}, Cu^{2+}, Cd^{2+}, Pb^{2+}, and As^{3+}) and bacterial pathogens (*Escherichia coli*) from water. Depending on the surface functionality (COOH, NH_2, or SH), magnetic nanoadsorbents capture metals either by forming chelate complexes or by the ion exchange process or electrostatic interaction. It was observed that the capture efficiency of bacteria strongly depends on the concentration of nanoadsorbents and their inoculation time.

Pyridine-containing derivatives are also common chelators for trace amounts of metals. Thus, silica-coated Fe_3O_4 NPs were modified with 2,6-diaminopyridine and used for selective magnetic solid-phase extraction of trace amounts of metal ions (Cu(II) and Zn(II)) [19]. Their quantitative extraction from mixed-ion solutions was accomplished at an optimal pH value of 6 within less than 10 min. Magnetic NPs (MNPs) prepared from Fe_3O_4 and functionalized with pyridine were applied as an adsorbent for the solid-phase extraction of trace quantities of Pd(II) [20]. The pyridine group was immobilized on the surface of the MNPs by covalent bonding of isonicotinamide. The modified MNPs can be readily separated from an aqueous solution by applying an external magnetic field. The detection limit and preconcentration factor were found to be 0.15 μg/l and 196, respectively. In addition, the technique for the solid-phase extraction of gold using various kinds of pyridine-functionalized nanoporous silica prior to its determination in various samples using flame atomic absorption spectroscopy (FAAS) was developed [21].

Multiwalled carbon nanotubes (MWCNTs) were dispersed and loaded with 1-butyl-3-methylimidazolium hexafluorophosphate ($[BMIM]PF_6$), supported on sawdust and used as a new adsorbent for preconcentration of trace amount of Bi [22]. Bi(III) ions were retained by the adsorbent MWCNT–$[BMIM]PF_6$ in a column after the formation of anionic complex BiI_4^- with iodide BiI_4^- complexes through electrostatic interactions with positively charged imidazolium ion. The method was applied for detecting Bi(III) in river water, tap water, and drug samples. Chelating polymer sorbents (oligomer containing a bis(pyrazole-1-yl)methane fragment, Fig. 7.3) can be used in analytical chemistry and the environmental protection sphere to recover, separate, and concentrate heavy and rare metals from natural and industrial waters [23]. Modified magnetite NPs functionalized with triazene groups (Fig. 7.4) were designed and prepared for extraction/preconcentration of sub-ppb levels of mercury ions in water and fish samples prior to their determination with inductively coupled plasma optical emission spectrometry (ICP–OES) [24]. In the separation process, an aqueous solution of Hg^{2+} ions was mixed with 150 mg of Fe_3O_4 magnetite NPs modified with 1-(*p*-acetylphenyl)-3-(*o*-ethoxyphenyl)triazene (AET) and then an external magnetic field was applied for isolation of magnetite NPs containing mercury ions. The sorption capacity of functionalized Fe_3O_4 NPs under optimum conditions was found to be 10.26 mg of Hg^{2+} per gram at pH 7 with a preconcentration factor of 500 (2 ml of elution for a 1000 ml sample volume).

PANi (Fig. 7.5) is known as a classic material and supporting agent in nanotechnology. The ability of organometallic titanium–PANi hybrid materials to function as gas sensors at room temperature was investigated [25]. To form these hybrid materials, commercially available PANi powders were directly added into organometallic titanium sols, synthesized using

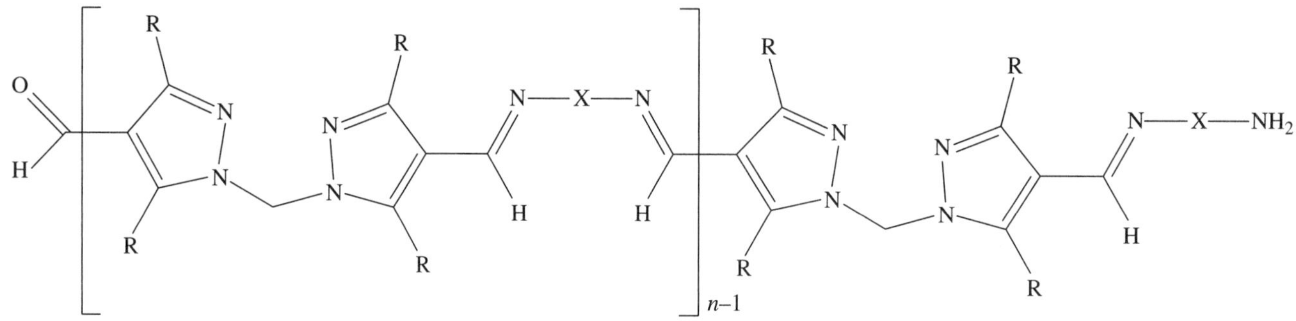

FIGURE 7.3 Bis(pyrazole-1-yl)methane fragment.

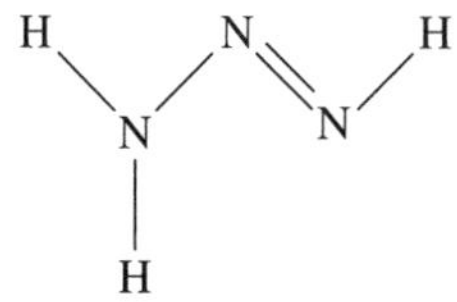

FIGURE 7.4 Triazene.

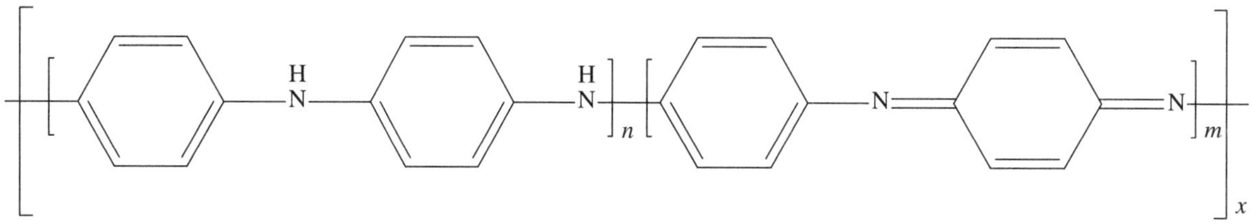

FIGURE 7.5 Polyaniline (PANi).

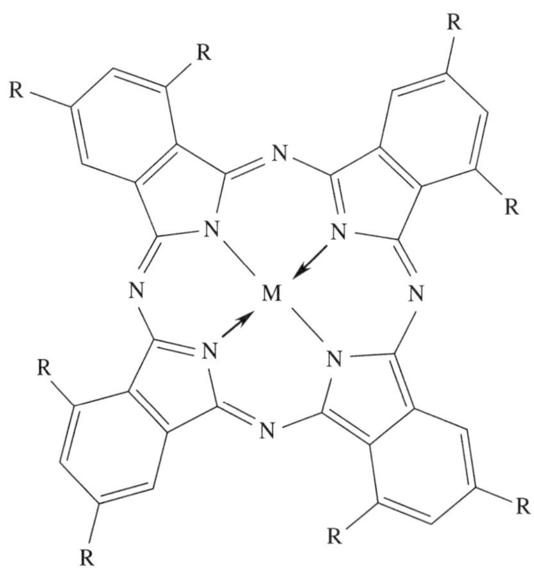

FIGURE 7.6 General formula of metal phthalocyanines.

the sol–gel method. For gas sensing tests, the products were exposed to ethanol vapor, revealing that the composite sensors required an appropriate ratio to exhibit optimum sensing properties.

The phthalocyanines (Fig. 7.6), known for more than 100 years as pigments, among a series of other applications, have been used for purposes of environmental protection too. Thus, iron(II) phthalocyanine (nano FePc) could serve as a powerful functional material for the development of sensors and electrocatalytic devices [26]. The nano FePc exhibited enhanced

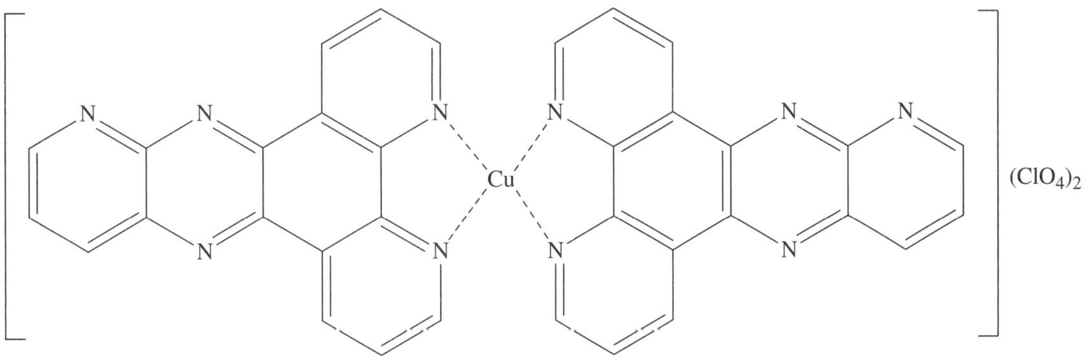

FIGURE 7.7 $[Cu(L)_2](ClO_4)_2$ of Schiff base ligand (L = 4,5,9,13,14-pentaaza-benzo[b]triphenylene).

electrocatalytic properties for the detection of thiocyanate and nitrite in aqueous solutions compared to the bulk FePc counterpart. Polycopper tetraaminophthalocyanine (CuTAPc) nanowires and nanotubes, fabricated [27] on porous alumina templates by electropolymerization, showed that their lengths could be controlled by the number of cycles applied and the monomer concentrations, while their diameters are confined by the pore size of the template. The product of electropolymerization (whether nanowires or nanotubes) was revealed to be a function of the monomer concentrations. These nanostructures may have applications in microelectronics, chemical sensing, and catalysis. Copper–tetrasulfonated phthalocyanine (CuTSPc)–sensitized mesoporous SnO_2, namely, CuTSPc/SnO_2 (with specific surface area and pore size of 0.1 mol.% CuTSPc/SnO_2 236 m^2/g and 2.6 nm, respectively), prepared by a template-free hydrothermal method, contained the sulfonated groups of CuTSPc bonded to the tin ion in a chelating bidentate mode. Its photocatalytic activity and circulating degradation rate (with 0.1 mol.% CuTSPc/SnO_2) under low-power visible light (15 W), detected using rhodamine B (RhB) as the objective decomposition substance, were revealed up to 87% during 180 min of reaction [28].

Among other N-containing ligands that are nanomaterials, we note a solid nanocomposite material on the basis of NPs of a metal coordination polymer with CN ligands including M^{n+} cations (Ni^{2+}, Co^{2+}, Mo^{5+}, Fe^{2+}, or Fe^{3+}) {M is a transition metal and n is 2 or 3} and $[M¢(CN)m]^{x-}$ anions ($[Fe(CN)_6]^{3-}$ or $[Fe(CN)_6]^{4-}$, $[Mo(CN)_8]^{3-}$, $[Co(CN)_6]^{3-}$) {where M¢ is a transition metal, x is 3 or 4, and m is 6 or 8} [29]. These compounds were patented as a method for fixing an inorganic pollutant, such as radioactive cesium. Schiff bases, classic objects in coordination chemistry, also revealed activities in nanostructured form. Thus, the Cu(II) Schiff base complex (Fig. 7.7), $[Cu(L)_2](ClO_4)_2$ of Schiff base ligand (L = 4,5,9,13,14-pentaaza-benzo[*b*]triphenylene) was synthesized, and its nanostructured compound was prepared [30] by sonochemical and solvothermal methods. The free Schiff base and its metal complex were screened for antibacterial activities; the metal complex was observed to be more active than its free Schiff base ligand. Very high adsorption capacity of trace amounts of Pb(II), Cd(II), and Cu(II) in environmental and biological samples was revealed for iron oxide silica magnetic particles with a Schiff base (Fe_3O_4/SiO_2/L) [31]. According to the authors, this magnetic solid phase is an ideal support because it has a large surface area and good selectivity and can be easily retrieved from large volumes of aqueous solutions. The detection limits were 0.14, 0.19, and 0.12 μg/l for Pb(II), Cd(II), and Cu(II) ions, respectively.

7.4 O-CONTAINING LIGANDS

Pure O-containing ligands as part of nanomaterials are represented with fewer examples. Thus, the preconcentration of trace heavy metal ions in environmental samples, based on the sorption of Cu(II), Cd(II), Ni(II), and Cr(III) ions with salicylic acid (Fig. 7.8) as the chelate on silica-coated magnetite NPs (Fig. 7.9), was used for evaluating these trace and toxic metals in various waters, foods, and other samples [32]. These magnetic NPs carrying the target metals could be easily separated from the aqueous solution simply by applying an external magnetic field; no filtration or centrifugation was necessary. An efficient method was undertaken [33] for synthesizing highly electrocatalytic Pt NPs on a carbon nanofiber (CNF) material, based on absorption of $Pt(acac)_2$ molecules on the functionalized CNFs and their further reduction to Pt NPs (2.9 ± 0.4 nm and 100% loading yield) by diffusion-limited sublimation in a confined space. The MeOH oxidation current density per mg Pt of Pt-loaded CNF sample was found to be approximately 60 times as high as in the commercially available sample and superior to other existing samples. In addition, 3,4-dihydroxy-9,10-dioxo-2-anthracenesulfonate (Alizarin red, AR, Fig. 7.10) chelates TiO_2 NPs through the catechol moiety [34]. Reduction of Cr(VI) to Cr(V) in the coupled AR@TiO_2 system utilizes a high fraction of the photogenerated electrons and induces degradation of the complex. Addition of the chelating polymer polyacrylic acid (PAA,

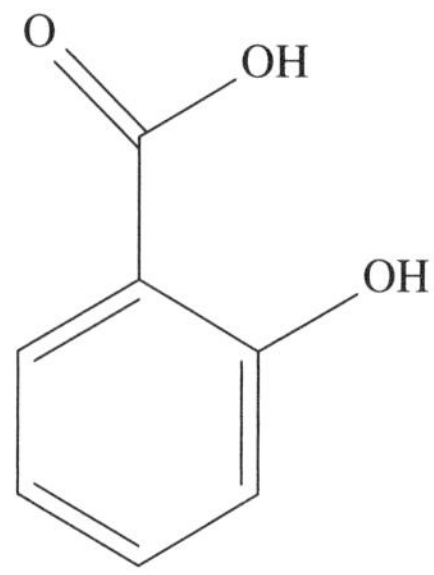

FIGURE 7.8 Salicylic acid.

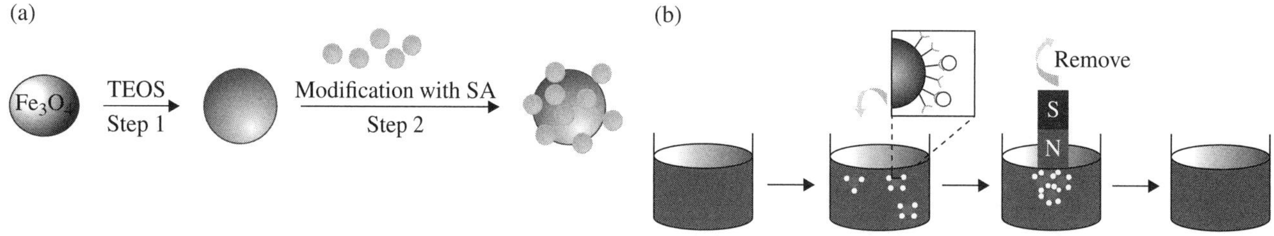

FIGURE 7.9 Preparation of silica-coated magnetite nanoparticles (NPs) modified with salicylic acid: preparation of adsorbent (a) and solid-phase extraction of the analytes (b). Reproduced with permission from Ref. [32]. © Chemistry Central.

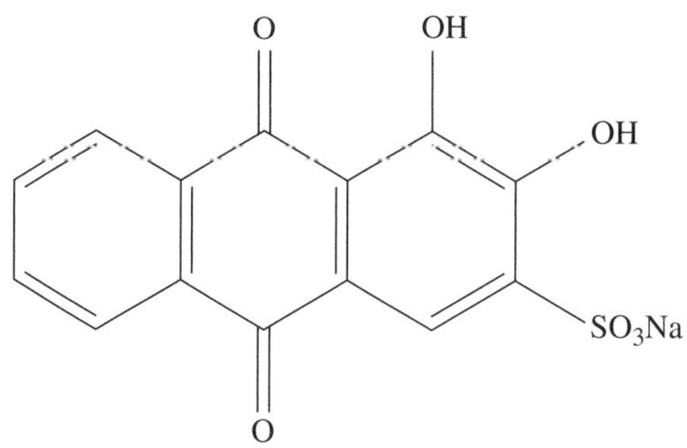

FIGURE 7.10 Alizarin red (AR), sodium 3,4-dihydroxy-9,10-dioxo-9,10-dihydro-2-anthracenesulfonate.

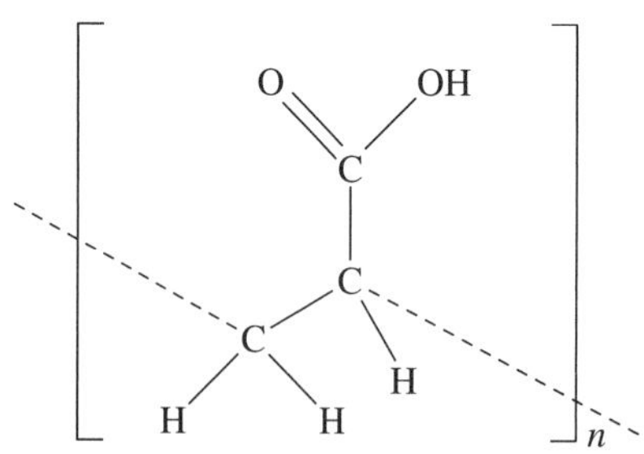

FIGURE 7.11 Polyacrylic acid. Reproduced with permission from Ref. [43]. © 2005, Elsevier Science.

Fig. 7.11) to assist in Mn removal from groundwater by membranes was examined [35] using membranes with different pore sizes under various operating conditions. Negligible Mn removal was achieved by the authors with ultrafiltration and nanofiltration membranes at acidic pH, but removal greater than 90% could be achieved at elevated pH (pH 9), presumably due to the formation of manganese hydroxides. Mn removal increased substantially when PAA was added to the feed solution, due to Mn chelation by PAA and rejection of chelates by the membranes. The chelate could be broken at acidic pH, by releasing free PAA, which could then be separated from Mn ions and reused.

7.5 N,O-CONTAINING LIGANDS

7.5.1 EDTA

A series of reports are dedicated to EDTA-based (Fig. 7.12) nanochelators due to commercial availability of this classic ligand. Thus, separation of cerium ions from dilute cerium feed solution was carried [36] out by nanofiltration. Cerium rejection by nanofiltration assisted by complexation with EDTA as chelating agent was markedly influenced by pH (2–10), where a higher degree of cerium removal was achieved for the basic medium than the acidic medium. The maximum observed rejection of Ce(III) was found to be 94.37 and 90.03% for an initial feed concentration of 10 and 80 mg/l, respectively. Attachment of EDTA-like chelators to carbon-coated metal nanomagnets resulted in a magnetic reagent for the rapid removal of heavy metals (Cd, Pb, Cu) from solutions or contaminated water by three orders of magnitude to concentrations as low as μg/l [37]. In addition, EDTA was used as a precursor for metal complex formation and further to control the growth of NPs. Thus, monoclinic scheelite-type $BiVO_4$ NPs with large surface area, synthesized using $Bi(NO_3)_3$ and NH_4VO_3 as raw materials, through a hydrothermal process in the presence of EDTA [38], exhibited a good photocatalytic performance for degrading phenol solution as a model organic pollutant under visible illumination. The key feature of this method was found to be the chelating role of the EDTA group in the synthetic process, which can greatly control the concentration of Bi^{3+}, leading to growth inhibition of $BiVO_4$ crystallites.

7.5.2 Chitosan-Based Nanomaterials

Chitosan (Fig. 7.13) has attracted considerable interest because of its unique combination of properties, such as biocompatibility, biodegradability, metal complexation, and antibacterial activity. Therefore, chitosan has a variety of current and potential applications in various fields, for example, biotechnology, pharmaceutics, wastewater treatment, cosmetics, and food science.

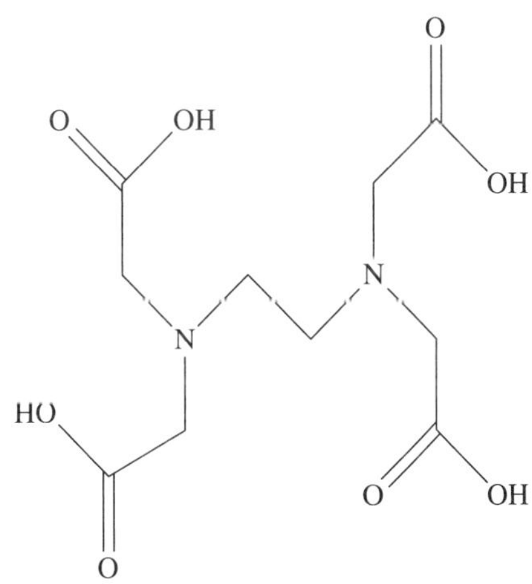

FIGURE 7.12 EDTA.

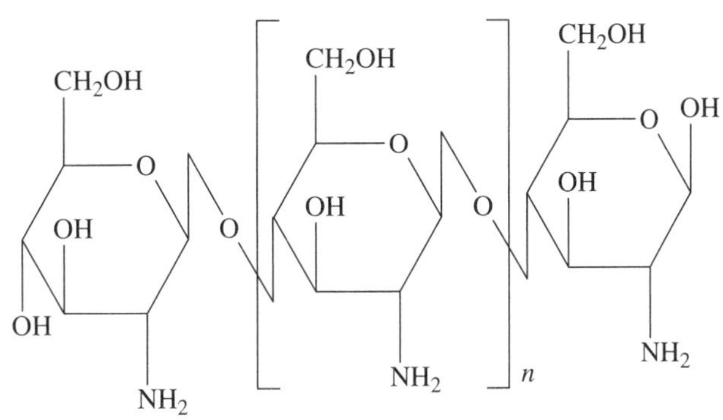

FIGURE 7.13 Chitosan.

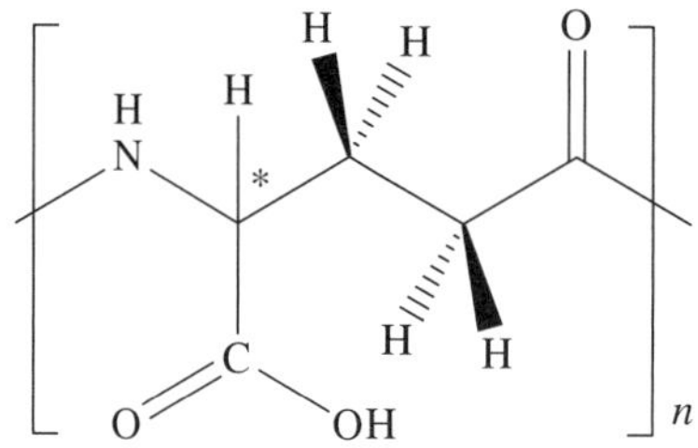

FIGURE 7.14 Polyglutamic acid.

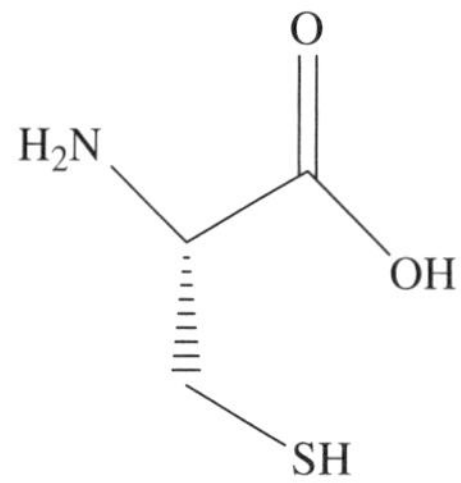

FIGURE 7.15 Cysteine. Reproduced with permission from Ref. [43]. © 2005, Elsevier Science.

Thus, the sorting of Pb^{2+} ions in a bloodstream flow was carried out [39] by mixing the ions in a chitosan nanobead aquasolution and then using a chelating mechanism to selectively sponge them up. To obtain an efficient chelating reaction and chelating nanobeads that can be attracted and separated from the bloodstream flow, a microfluidic device that had a microchannel with electrodes was designed and fabricated by a microelectromechanical process. In this device, the microchannel with electrodes provided a local dielectrophoretic field that was strong enough to manipulate and separate the chelating nanobeads in the continuous bloodstream flow. Crystalline Cu_2O NPs were synthesized via the templating method by taking advantage of the chelation of chitosan with copper ions [40]. The resulting Cu_2O-embedded film exhibited higher photocatalytic activity toward methyl orange degradation under visible light irradiation. Chitosan NPs were prepared based on ionic gelation between chitosan and sodium tripolyphosphate and then Ag^+, Cu^{2+}, Zn^{2+}, Mn^{2+}, or Fe^{2+} was individually loaded onto chitosan NPs [41]. Their antibacterial activities were evaluated by the determination of minimum inhibitory concentration (MIC) and minimum bactericidal concentration (MBC) against *E. coli* 25922, *Salmonella choleraesuis* ATCC 50020, and *Staphylococcus aureus* 25923 *in vitro*, showing that antibacterial activity was significantly enhanced by the metal ions loaded, except for Fe^{2+}.

7.5.3 Amino Acids

Amino acids and poly(amino acids) are natural chelating agents for various metal ions. In particular, zinc ions were encapsulated [42] in situ in a conductive polypyrrole film using polyglutamic acid (Fig. 7.14) as a localized complexing agent within the film. The subsequent electrochemical reduction of the metal ions to zero-valent metal led to the formation of NPs. An important advantage of this electrochemical approach is facile regeneration of the particles, as well as prevention of the aggregation of NPs in the conductive polymeric film. The formed NPs were composed of zinc and were 18 ± 7 nm in diameter. In addition, the NP/polymer composite was used to reduce halogenated organics, indicating its potential usefulness in remediation applications.

A series of reports are dedicated to cysteine-containing chelating nanomaterials. Thus, an adsorbent for the capture of mercuric chloride vapor from flue gases on the basis of a chelating ligand on cysteine basis (Figs. 7.15, 7.16, and 7.17) with an ionizing surface nano-layer on a mesoporous substrate was reported [43]. The maximum theoretical (equilibrium) capacity for mercury removal was estimated to be 33 mg Hg/g, and thermal stability tests indicated stable operation up to 135°C. In addition, poly-α,β-DL-aspartic acid is well known as the green chelant of various metal ions. To provide a nanochelant for treating Pb(II) poisoning, poly-α,β-DL-aspartic acid was modified [44] with L-Cys to form poly-α,β-DL-aspartyl-L-cysteine (PDC). DL-Asp was converted into polysuccinimide through thermal polycondensation, and the amidation of polysuccinimide with L-Cys provided PDC. In water, PDC formed various porous nanospecies, which benefited the removal of Pb(II). PDC did not remove the essential metals, including Cu^{2+}, Fe^{2+}, Mn^{2+}, Zn^{2+}, and Ca^{2+}, in treated mice.

Nanomaterials on the basis of carbon allotropes can also be functionalized with amino acids and used for absorption of pollutants such as cadmium. Thus, graphene oxide (GO) nanosheets were decorated with a cysteine-rich metal-binding

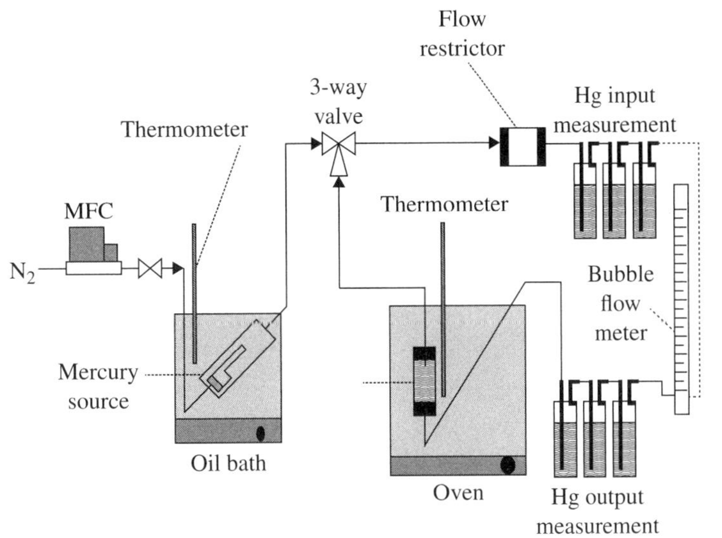

FIGURE 7.16 Fixed-bed apparatus for the determination of mercuric chloride capture characteristics. Reproduced with permission from Ref. [43]. © Elsevier Science.

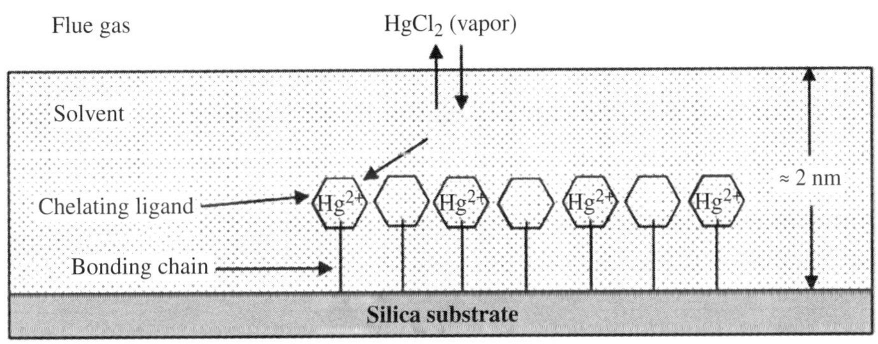

FIGURE 7.17 Schematic representation of chelating adsorbent for vapor-phase mercury. Reproduced with permission from Ref. [43]. © Elsevier Science.

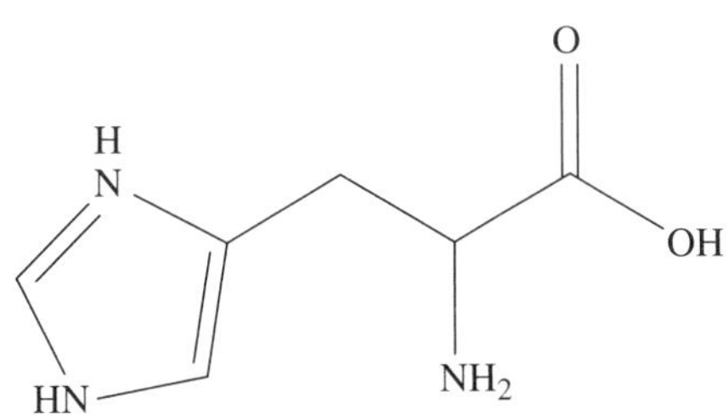

FIGURE 7.18 Histidine.

protein, cyanobacterium metallothionein (SmtA) [45]. The formed SmtA–GO composites were found to exhibit ultrahigh selectivity toward the adsorption of Cd and were then assembled onto the surface of cytopore microbeads and used for highly selective adsorption and preconcentration of ultratrace Cd. In comparison with bare GO (carboxyl-rich GO)-loaded cytopore (GO@cytopore), SmtA–GO–loaded cytopore (SmtA–GO@cytopore) shows a 3.3-fold improvement over the binding capacity of Cd, that is, 7.70 mg/g for SmtA–GO@cytopore compared to 2.34 mg/g for that by GO@cytopore. Another example is the functionalization of MWCNTs with histidine (Fig. 7.18), which was shown as a simple, rapid, and reliable approach for improving the selectivity and adsorption capacity of MWCNTs toward Cd^{2+} [46]. An additional advantage is

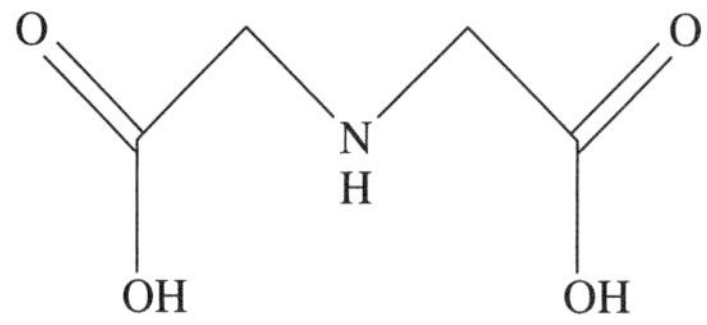

FIGURE 7.19 Iminodiacetic acid.

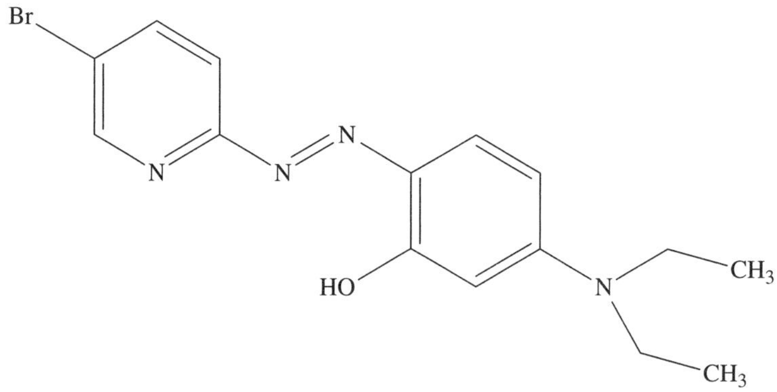

FIGURE 7.20 2-(5-Bromo-2-pyridylazo)-5-diethylamino-phenol.

that the use of histidine as substrate, associated with the inherent features of MWCNTs, avoids the incompatibility problem between MWCNTs and aqueous medium in flow injection analysis. Efficient functionalization with histidine was confirmed from 100 preconcentration/elution steps without loss of adsorption capacity toward Cd^{2+}. The same nanocomposite was also applied for selective preconcentration of V(V) and its online ultrasensitive determination in a variety of biological and environmental samples [47]. With the consumption of 5.0 ml sample solution and a preconcentration time of 1 min, an enhancement factor of 35 with a detection limit (3 s) of 9 ng/l or 0.2 nM was achieved. In addition, an amperometric glassy carbon biosensing electrode for glucose, based on the immobilization of a highly sensitive glucose oxidase (GO_x) by affinity interaction on carbon nanotubes (CNTs) functionalized with iminodiacetic acid (Fig. 7.19) and metal chelates, was fabricated [48]. This technique for immobilization exploits the affinity of Co(II) ions to the histidine and cysteine moieties on the surface of GO_x. The resulting biosensor was found to be capable of detecting glucose at levels as low as 0.01 mM and had excellent operational stability.

7.5.4 Other N,O Ligands

Poly[aniline-*co*-5-sulfo-2-anisidine] nanograins with a rough and porous structure demonstrated ultrastrong adsorption and highly efficient recovery of silver ions [49]. The 50/50 copolymer nanograins exhibited much stronger Ag^+ adsorption than PANi and all other reported sorbents. The maximum Ag^+ sorption capacity of less than or equal to 2034 mg/g is the highest found thus far and also much higher than the maximum Hg ion sorption capacity. It was revealed that Ag^+ sorption occurs mainly due to the redox mechanism involving reduction of Ag^+ to separable silver nanocrystals, chelation between Ag^+ and -NH-/-N=/-NH_2/-SO_3H/-OCH_3, and ion exchange between Ag^+ and H^+ on -SO_3-H. Copolymer NPs bearing many functional groups on their rough and porous surface can be directly used to recover and separate silver nanocrystals from practical Ag^+ wastewaters containing Fe, Al, K, and Na ions. A sorbent extraction procedure for Pb(II), Cu(II), Ni(II), and Fe(III) ions on single-walled carbon nanotube disks was established [50]. Analyte ions were converted to 2-(5-bromo-2-pyridylazo)-5-diethylamino-phenol (Fig. 7.20) chelates, then adsorbed on the disk, and further desorbed by 10 ml 2 M HNO_3.

High performance was reported for NP γ-alumina coated with sodium dodecyl sulfate and 4-(2-pyridylazo)-resorcinol (SDS–PAR, Fig. 7.21) in sorbent solid-phase extraction [51]. A sorbent with adjusted nanometer-sized alumina was used to preconcentrate and separate Cu in plants belonging to the legume family and natural water samples. The method was applied to detect Cu ions at trace levels in substantial samples such as cucumber, eggplant, mint, tomato, potato, parsley, spinach, apple and apple core, mangosteen, kiwi, banana, macaroni, pea, wheat flour, red beans, lentil, barley, tap water, river water, and sea

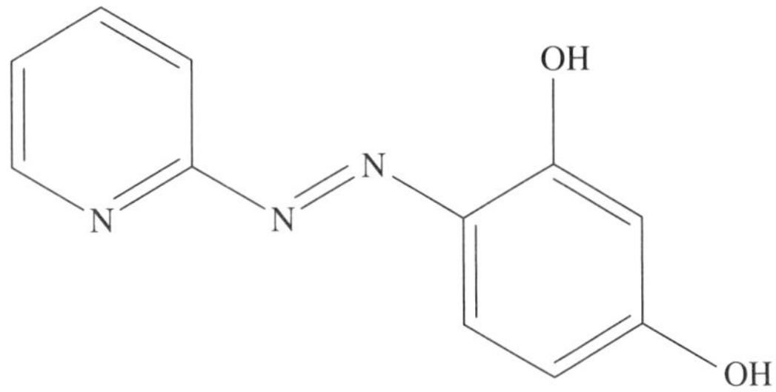

FIGURE 7.21 4-(2-Pyridylazo)-resorcinol.

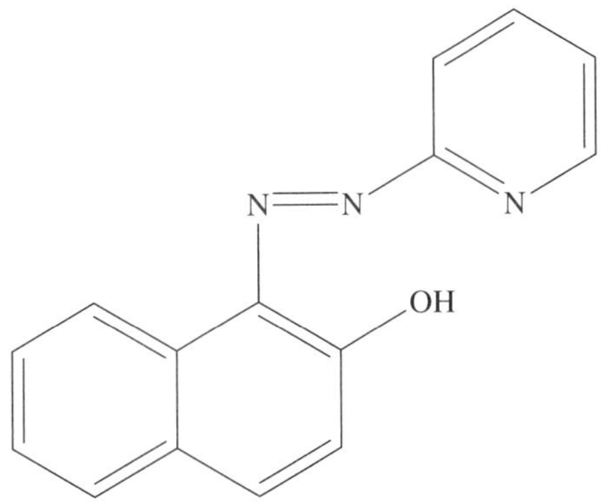

FIGURE 7.22 1-(2-Pyridylazo)-2-naphthol.

water. Also, 1-(2-pyridylazo)-2-naphthol (PANa, Fig. 7.22)–modified MNPs were used for the separation and preconcentration of Mn^{2+} ions from aqueous samples [52]. The detection limit of the proposed method was 0.11 $\mu g \cdot l^{-1}$. A selective solid-phase extractor (stable in 6 mol/l HCl, common organic solvents, and buffer solutions at pH 2.0–8.0) was prepared based on the modification of silica gel with 2-((3-silylpropylimino)methyl)-2-hydroxy-1-naphthol (SPIMHN) [53]. This nanosorbent was applied for the enrichment of trace amounts of Fe^{3+}, Pb^{2+}, Cu^{2+}, Ni^{2+}, Co^{2+}, and Zn^{2+} subsequent to their determination by flame atomic absorption spectrometry. Sorbent based on gold NPs loaded in activated carbon (AuNP–AC) was modified by bis(4-methoxysalicylaldehyde)-1,2-phenylenediamine (BMSAPD) [54]. This sorbent (AuNP–AC–BMSAPD) was applied for the enrichment and preconcentration of trace amounts of Co^{2+}, Cu^{2+}, Ni^{2+}, Fe^{2+}, Pb^{2+}, and Zn^{2+} ions in real samples (recoveries in the range of 95–99.6% and a relative standard deviations (RSD) <4.0%). Following the optimum conditions, a preconcentration factor of 200 was obtained for all the metal ions under study with detection limits of 1.4–2.6 ng/ml.

7.6 S-CONTAINING LIGANDS

S-containing ligands are commonly used as sensors; thus, the chelate sensor method for detecting mercury, based on the high affinity of thiol-containing ligands for the mercuric ion, has been known for a long time [55]. The S-containing ligands, used for environmental applications, are mainly represented by dithiocarbamates (DTCs). Thus, a dispersive solid-phase microextraction (DSPME) with MWCNTs as solid sorbent and ammonium pyrrolidinedithiocarbamate (APDC) as chelating agent was developed for the determination of selenium in the presence of heavy metal ions and alkali metals [56]. The Se(IV)–APDC complex was adsorbed on MWCNTs dispersed in aqueous samples, which were then filtered, and MWCNTs with selenium chelate were collected in a filter. Under optimized conditions, Se can be determined with very good recovery (97 ± 3%), precision (RSD = 3.2%), and detection limits (from 0.06 to 0.2 ng/ml). The high efficiency of nanometer-sized γ-alumina coated with sodium dodecyl sulfate–pyrrolidine dithiocarbamate (SDS–4-BPDC) for sorbent solid-phase extraction was also reported [57]. Sorbent with modified nanometer-sized alumina was used to preconcentrate and separate cobalt in food and environmental water samples. The proposed method was applied to determine cobalt at trace levels in real samples such as rice, tobacco, orange, green pepper, black tea, honey, potato, spinach, mangosteen, tomato, strawberry, carrot, apple, kiwi, mushroom, cucumber, lettuce, milk, wheat sprout, tap water, river water, and sea water with satisfactory

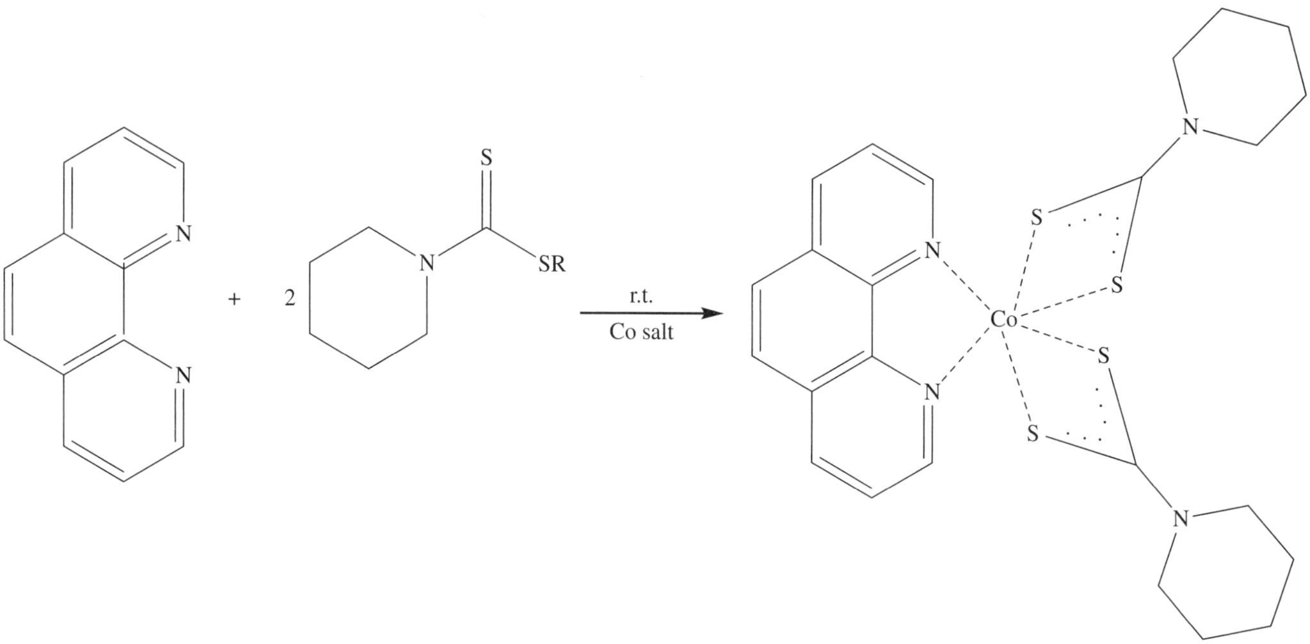

FIGURE 7.23 Synthesis of nano dithiocarbamate (DTC) complex.

results. In addition, a nano DTC complex was prepared (Fig. 7.23) by reaction between DTC and a metal salt under ultrasound irradiation [58]. The antibacterial activity of its NP derivatives was tested against microorganisms (*E. coli*, *Klebsiella pneumoniae*, *S. aureus*, and *Bacillus subtilis*) using the disk diffusion method and compared with non-nano conditions. The compounds showed significant antibacterial activity against bacterial strains within the zone of inhibition, 23 mm at MIC of 30.0 μg/disk.

7.7 P-CONTAINING LIGANDS

In addition to nanocomposites with sulfur-containing ligands, use of phosphorus-containing ligands in environmental applications as nanocomposites or their precursors is rare. Thus, the nanostructured conjugated polymer poly[1,1′-bis(ethynyl)-4,4′-biphenyl-(bis-tributylphosphine)Pd(II)], known as Pd–diethynylbiphenyl (DEBP), and the hybrid system Pd/Pd–DEBP, obtained by the dispersion of Pd(0) nanoclusters into the organometallic conjugated polymer Pd–DEBP, was studied as an active membrane for the development of surface acoustic wave (SAW)-based chemical sensors [59]. The sensors based on Pd–DEBP and Pd/Pd–DEBP membranes showed an RH (relative humidity) sensitivity of 600 and 320 Hz/RH%, and a H_2 sensitivity of 650 and 670 Hz/ppm, respectively. These values, compared with literature data, showed an enhancement of sensitivity in the low RH range of 0–30%. There is a strong chelating ability between phosphates and metal ions that can be used for generating biosensors. Thus, a porous nanomaterial on the basis of zirconium phytate (derivative of phytic acid, Fig. 7.24) was synthesized [60] by direct precipitation, and its nanoporous film was employed as a substrate for making an horseradish peroxidase (HRP)-based biosensor by the drop-coating method and investigating the electrochemical behavior of enzyme. It was shown that the absorbed HRP retained its bioactivity and realized direct electron transfer due to improved biocompatibility of zirconium phytate. Moreover, the biosensor displayed good bioelectrocatalytic ability toward the reduction of H_2O_2 with a linear response to H_2O_2 over a concentration range from 6.67×10^{-7} to 6×10^{-6} mol/l, and a detection limit of 5.3×10^{-7} mol/l at a signal to noise ratio (S/N) = 3. This biosensor exhibited a low detection limit, high enzymic activity, and good reproducibility and stability. Silica polyamine composites (SPC) made from silanized amorphous nanoporous silica gel and poly(allylamine) (BP-1) were functionalized [61] with phosphorus acid using the Mannich reaction, resulting in a phosphonic acid–modified composite (BPAP). Zr(IV) was immobilized on BPAP with a loading of 1.12 mmol/g. Zr loading was analyzed by mass gain, ICP/atomic emission spectrometry (AES), and scanning electron microscopy (SEM)/dispersive X-ray spectroscopy (EDX). Arsenate anions were found to be strongly adsorbed on the ZrBPAP composite at pH 2–8, while arsenite only adsorbed well at pH 10. The sorption mechanism (a chelation between arsenate or arsenite and the Zr(IV)–phosphonic acid complex of BPAP) is discussed.

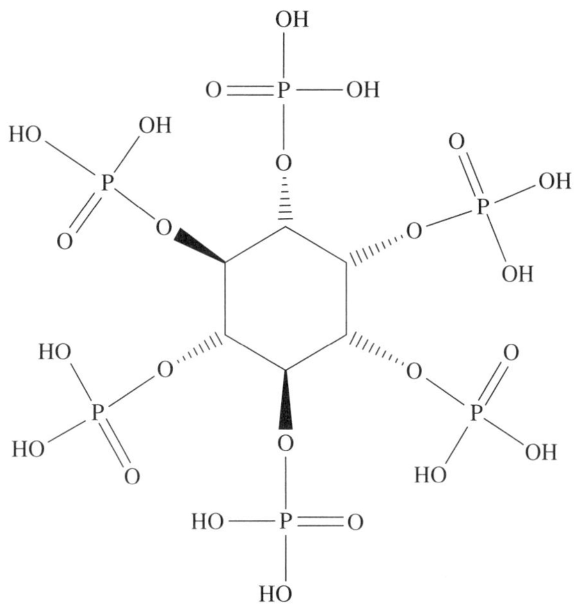

FIGURE 7.24 Phytic acid.

7.8 NANOCOMPOSITES ON THE BASIS OF ORGANOMETALLIC COMPOUNDS

A few organometallics or, much more frequently, products of their decomposition have been used for environmental purposes, mainly in sensing procedures [62], for instance, $[Au_2Ag_2(C_6F_5)_4(NH_3)_2]_n$ [63] or ferrocene–peptido conjugates [64]. Using another cene derivative, on titanium basis, hierarchical anatase TiO_2 materials (assembled from very thin TiO_2 nanosheets, which are composed of numerous highly crystallized anatase nanocrystals) with flower-like morphologies were prepared [65] via a one-step template-free hydrothermal method starting from titanocene dichloride as precursor and 1,2-ethylenediamine (EDA) as chelating agent in aqueous solution. These hierarchical TiO_2 materials showed very good photocatalytic performance when applied to photodegradation of MB, which can be related to the unique features of hierarchical structures, large specific surface areas, and high crystallization degree of the TiO_2 materials obtained. Organometallic titanium–PANi hybrid materials (titanium:PANi = from 1 to 5 wt.%) can act as gas sensors (ethanol vapor) at room temperature [25]. To obtain these hybrid materials, commercially available PANi powder was directly added to organometallic titanium sols, which were synthesized using the sol–gel method. The sensing mode was based on the variation in the electric conductivity based on the interaction between the gas molecules and the film. The composite sensors required appropriate ratio to exhibit optimum sensing properties. Nanocrystalline tin dioxide (SnO_2) ultrathin films (30–35 nm thick; mean crystallite size ranging from 4 to 7 nm) were obtained employing a straightforward solution-based route that involved the calcination of bridged polystannoxane films processed by the sol–gel process from bis(triprop-1-ynylstannyl)alkylene and -arylene precursors [66]. In the presence of H_2 and CO gases, these layers led to highly sensitive, reversible, and reproducible responses.

During the past few decades, organometallic methodologies have generated a number of highly effective electrocatalyst systems based on mono- and bimetallic NPs having controlled size, composition, and structure, used in fuel cells [67]. As an example, iron, cobalt, and CNF (FeCo–CNF) composite electrocatalysts for the oxygen reduction reaction (ORR) in alkaline fuel cells (AFCs) were fabricated via electrospinning and the subsequent pyrolysis of a mixture of a nitrogen-containing polymer and organometallic compounds [68]. The resultant FeCo–CNFs catalysts demonstrated comparable electrocatalytic activity and stability to commercial carbon-supported platinum (Pt/C) for ORR, a direct four-electron reduction pathway, and better ethanol tolerance than Pt/C in an alkaline electrolyte.

7.9 CONCLUSIONS

A series of nanomaterials on the basis of coordination compounds and few organometallics are currently used for environmental properties. Among others, classic ligands in metal complex chemistry, such as dipyridyl, chitosan, ethylenediamine, 2,6-diaminopyridine, triazene, PANi, phthalocyanines, Schiff bases, salicylic acid, EDTA, amino acids, DTCs, and their derivatives have been reported. Some of them were used as precursors for inorganic nanocomposites. The discussed nanomaterials have been synthesized by classic wet chemistry routes, as well as by sonochemical and solvothermal processes, electropolymerization, the sol–gel method, and so on.

The obtained nanocomposites possess properties frequently better in comparison with their nonnano analogues and have numerous environmental applications, in particular the removal, recovery, concentration, or separation of heavy/toxic (Cr^{3+}, Co^{2+}, Ni^{2+}, Cu^{2+}, Cd^{2+}, Pb^{2+} and As^{3+}) and rare metals from natural and industrial waters and the elimination of bacterial pathogens (*E. coli*) from water, also being sensors for metal cations, nanosorbents, and composing fuel cells. They can also serve as catalysts (hydrodechlorination) or photocatalysts (degradation of phenol or MB). Despite these applications, coordination and organometallic nanomaterials have a limited number of applications in the field of environmental protection.

REFERENCES

[1] Evangelou MWH, Ebel M, Schaeffer A. Chelate assisted phytoextraction of heavy metals from soil. Effect, mechanism, toxicity, and fate of chelating agents. Chemosphere 2007;68 (6):989–1003.

[2] Lestan D, Luo C-L, Li X-D. The use of chelating agents in the remediation of metal-contaminated soils: a review. Environ Pollut 2008;153 (1):3–13.

[3] Kim J, Tsouris C, Mayes RT, Oyola Y, Saito T, Janke CJ, Dai S, Schneider E, Sachde D. Recovery of uranium from seawater: a review of current status and future research needs. Sep Sci Technol 2013;48 (3):367–387.

[4] Fontenot SA, Carter TG, Johnson DW, Addleman RS, Warner MG, Yantasee W, Warner CL, Fryxell GE, Bays JT. Nanostructured materials for selective collection of trace level metals from aqueous systems. In: Pierce DT, Zhao JX, editors. *Trace Analysis with Nanomaterials*. Weinheim: Wiley-VCH Verlag GmbH & Co. KGaA; 2010. p 191–221.

[5] Liu R, Zhou Q, Ma Q. Applications of nanomaterials in remediation of contaminated water and soil: a review. Shengtaixue Zazhi 2010;29 (9):1852–1859.

[6] Rosenberg E. Silica polyamine composites: advanced materials for metal ion recovery and remediation. In: Abd-El-Aziz, AS, Carraher, CE, Pittman, CU, Zeldin, M, editors. Macromolecules Containing Metal and Metal-Like Elements: Group IVA Polymers. Volume 4. Hoboken, NJ: John Wiley & Sons, Inc.; 2005. p 51–78.

[7] Wu RH, Gowrishankar M, Azimi S, Mackiewicz MR. Biocompatible gold nanomaterials for chelation of toxic metals [abstracts]. In: *66th Northwest Regional Meeting of the American Chemical Society*, Portland; June 26–29 2011, NORM-28.

[8] Mackiewicz MR, Wu RH, Mau T. Multifunctional biocompatible iron and gold nanomaterials for metal chelation applications [abstract]. In: *241st ACS National Meeting & Exposition*, Anaheim; March 27–31, 2011, INOR-673.

[9] Ipe BI, Yoosaf K, Thomas KG. Functionalized gold nanoparticles as phosphorescent nanomaterials and sensors. J Am Chem Soc 2006;128 (6):1907–1913.

[10] Zou ML, Du ML, Zhu H, Xu CS, Fu YQ. Green synthesis of halloysite nanotubes supported Ag nanoparticles for photocatalytic decomposition of methylene blue. J Phys D Appl Phys 2012;45 (32):325302/1–325302/7.

[11] Kharisov BI, Dias R, Jimenez-Perez VM, Kharissova OV, Olvera Perez B, Muñoz Flores B. Iron-containing nanomaterials: synthesis, properties, and environmental applications. RSC Adv 2012;2 (25):9325–9358.

[12] Allabaksh MB, Mandal BK, Kesarla MK, Kumar KS, Reddy PS. Preparation of stable zero valent iron nanoparticles using different chelating agents. J Chem Pharmaceut Res 2010;2 (5):67–74.

[13] Geng B, Jin Z, Li T, Qi X. Kinetics of hexavalent chromium removal from water by chitosan-Fe^0 nanoparticles. Chemosphere 2009; 75 (6):825–830.

[14] Geng B, Jin Z, Li T, Qi X. Preparation of chitosan-stabilized Fe^0 nanoparticles for removal of hexavalent chromium in water. Sci Total Environ 2009;407 (18):4994–5000.

[15] Xu J, Bachas L, Bhattacharyya D. Synthesis of nanostructured bimetallic particles in polyligand-functionalized membranes for remediation applications. In: Savage, N, editor. *Nanotechnology Applications for Clean Water*. Norwich, NY: William Andrew; 2009. p 311–335.

[16] Jang JS, Kim ES, Choi SH, Kim DH, Kim HG, Lee JS. Synthesis, electronic property and photocatalytic applications of mesoporous cobalt-doped ZnS and ZnO nanoplates. Appl Catal A Gen 2012;427–428:106–113.

[17] Tristany M, Philippot K, Guari Y, Colliere V, Lecante P, Chaudret B. Synthesis of composite ruthenium-containing silica nanomaterials from amine-stabilized ruthenium nanoparticles as elemental bricks. J Mater Chem 2010;20 (42):9523–9530.

[18] Singh S, Barick KC, Bahadur D. Surface engineered magnetic nanoparticles for removal of toxic metal ions and bacterial pathogens. J Hazard Mater 2011;192 (3):1539–1547.

[19] Zhai Y, He Q, Han Q, Duan S-E. Solid-phase extraction of trace metal ions with magnetic nanoparticles modified with 2,6-diaminopyridine. Microchim Acta 2012;178 (3–4):405–412.

[20] Bagheri A, Behbahani M, Amini MM, Sadeghi O, Tootoonchi A, Dahaghin Z. Preconcentration and separation of ultra-trace palladium ion using pyridine-functionalized magnetic nanoparticles. Microchim Acta 2012;178 (3–4):261–268.

[21] Moazzen E, Daei N, Hosseini SM, Ebrahimzadeh H, Monfared A, Amini MM, Sadeghi O. Comparison of the performance of pyridine-functionalized nanoporous silica particles for the extraction of gold(III) from natural samples. Microchim Acta 2012;178 (3–4):367–372.

[22] Pourreza N, Sheikhnajdi K. Multi-walled carbon nanotube modified with 1-butyl-3-methylimidazolium hexafluorophosphate supported on sawdust as a selective adsorbent for solid phase extraction of Bi(III). Talanta 2012;99:507–511.

[23] Potapov AS, Khlebnikov AI. Oligomers containing bis(pyrazole-1-yl)methane fragments as sorbents of lanthanide and heavy metal ions. 2008, Patent RU 2322289 C1 20080420.

[24] Rofouei MK, Rezaei A, Masteri-Farahani M, Khani H. Selective extraction and preconcentration of ultra-trace level of mercury ions in water and fish samples using Fe_3O_4-magnetite-nanoparticles functionalized by triazene compound prior to its determination by inductively coupled plasma-optical emission spectrometry. Anal Method 2012;4 (4):959–966.

[25] Izzuddin I, Jumali MHH, Yahaya M, Salleh MM. New hybridization approach of titanium organometallic: PANi thin films as room temperature gas sensors. Sains Malays 2012;41 (8):1017–1021.

[26] Mamuru SA, Ozoemena KI. Impedimetric and electrocatalytic properties of nanostructured iron(II) phthalocyanine at pyrolytic graphite electrode. Mater Chem Phys 2009;114 (1):113–119.

[27] Gu F, Xu GQ, Ang SG. Fabrication of CuTAPc polymer nanowires and nanotubes by electropolymerization. Nanotechnology 2008;19 (14):145606/1–145606/7.

[28] Guo L, Pan H, Shen S, Huang J. Preparation of nanosized mesoporous CuTSPc/SnO_2 and its photocatalytic activity under visible light irradiation. Cuihua Xuebao 2009;30 (1):53–59.

[29] Grandjean A, Barre Y, Guari Y, Larionova J, Guerin C. Solid hexa- and octacyanometallate nanocomposite material for fixing radioactive cesium nanoparticles. PCT International Application. WO 2010133689. 2010.

[30] Saghatforoush LA, Mehdizadeha R, Chalabian F. Synthesis of CuO nanoparticles from a new nano-structured copper(II) Schiff base complex by ultrasonic and solvothermal methods: structural, thermal and antibacterial studies. J Chem Pharm Res 2011; 3 (2):691–702.

[31] Bagheri H, Afkhami A, Saber-Tehrani M, Khoshsafar H. Preparation and characterization of magnetic nanocomposite of Schiff base/silica/magnetite as a preconcentration phase for the trace determination of heavy metal ions in water, food and biological samples using atomic absorption spectrometry. Talanta 2012;97:87–95.

[32] Shishehbore MR, Afkhami A, Bagheri H. Salicylic acid functionalized silica-coated magnetite nanoparticles for solid phase extraction and preconcentration of some heavy metal ions from various real samples. Chem Cent J 2011;5:41.

[33] Nguyen TX, Sanchez DPV, Kim TH, Park HI, Yun M, Lee YH. Diffusion-limited reduction of organometallic compound on carbon nanofiber mat for catalytic applications. J Mater Chem 2010;20 (26):5468–5473.

[34] Di Iorio Y, San Roman E, Litter MI, Grela MA. Photoinduced reactivity of strongly coupled TiO_2 ligands under visible irradiation: an examination of an Alizarin Red@TiO_2 nanoparticulate system. J Phys Chem C 2008;112 (42):16532–16538.

[35] Han S-C, Choo K-H, Choi S-J, Benjamin MM. Removal of manganese from water using combined chelation/membrane separation systems. Water Sci Technol 2005;51 (6–7):349–355.

[36] Murthy ZVP, Choudhary A. Separation of cerium from feed solution by nanofiltration. Desalination 2011;279 (1–3):428–432.

[37] Koehler FM, Rossier M, Waelle M, Athanassiou EK, Limbach LK, Grass RN, Guenther D, Stark WJ. Magnetic EDTA: coupling heavy metal chelators to metal nanomagnets for rapid removal of cadmium, lead and copper from contaminated water. Chem Commun 2009;32:4862–4864.

[38] Sun W, Xie M, Jing L, Luan Y, Fu H. Synthesis of large surface area nano-sized $BiVO_4$ by an EDTA-modified hydrothermal process and its enhanced visible photocatalytic activity. J Solid State Chem 2011;184 (11):3050–3054.

[39] Wang M-W. Using chelating chitosan nanobeads and a microfluidic-microelectric trap to sort lead(II) in a continuous bloodstream flow. J Electrochem Soc 2011;158 (3):D166–D171.

[40] Zhang A, Zhang N, Hong S, Zhang M. Nanocrystalline cuprous oxide in chitosan membrane: preparation, characterization, and photocatalytic properties. Kinet Catal 2009;50 (5):748–751.

[41] Du W-L, Niu S-S, Xu Y-L, Xu Z-R, Fan C-L. Antibacterial activity of chitosan tripolyphosphate nanoparticles loaded with various metal ions. Carbohydr Polym 2009;75:385–389.

[42] Venkatachalam K, Gavalas VG, Xu S, de Leon AC, Bhattacharyya D, Bachas LG. Poly(amino acid)-facilitated electrochemical growth of metal nanoparticles. J Nanosci Nanotechnol 2006;6 (8):2408–2412.

[43] Abu-Daabes MA, Neville G. Synthesis and characterization of a nano-structured sorbent for the direct removal of mercury vapor from flue gases by chelation. Chem Eng Sci 2005;60 (7):1901–1910.

[44] Li L, Wu J, Zhao M, Wang Y, Zhang H, Zhang X, Gui L, Liu J, Mair N, Peng S. Poly-α,β-DL-Aspartyl-L-Cysteine: a novel nanomaterial having a porous structure, special complexation capability for Pb(II), and selectivity of removing Pb(II). Chem Res Toxicol 2012; 25 (9):1948–1954.

[45] Yang T, Liu L-h, Liu J-w, Chen M-L, Wang J-H. Cyanobacterium metallothionein decorated graphene oxide nanosheets for highly selective adsorption of ultra-trace cadmium. J Mater Chem 2012;22 (41):21909–21916.

[46] Lima G, Midori de Oliveira F, Ohara M, Segatelli MG, Tarley CRT. Evaluation of histidine functionalized multiwalled carbon nanotubes for improvement in the sensitivity of cadmium ions determination in flow analysis. In: Bianco, S, editor. *Carbon Nanotubes*. Rijeka: INTECH 2011. p 67–80.

[47] Liu Y, Li Y, Yang L. Histidine functionalized multi-walled carbon nanotubes as sorbent for flow injection-electrothermal atomic absorption spectrometric ultrasensitive determination of trace vanadium (V) in biological and environmental samples. Microchem J 2012;104:56–61.

[48] Tu X, Zhao Y, Luo S, Luo X, Feng L. Direct electrochemical sensing of glucose using glucose oxidase immobilized on functionalized carbon nanotubes via a novel metal chelate-based affinity method. Microchim Acta 2012;177 (1–2):159–166.

[49] Li X-G, Feng H, Huang M-R. Redox sorption and recovery of silver ions as silver nanocrystals on poly(aniline-co-5-sulfo-2-anisidine) nanosorbents. Chem A Eur J 2010;16 (33):10113–10123.

[50] Yilmaz E, Alothman ZA, Sumayli HMT, Ibrahim M, Soylak M. Sorbent extraction of Pb(II), Cu(II), Ni(II), and Fe(III) ions as 2-(5-bromo-2-pyridylazo)-5-diethylamino-phenol chelates on single-walled carbon nanotube disks prior to their flame atomic absorption spectrometric determinations in animal feeds and natural water samples. J AOAC Int 2012;95 (4):1205–1210.

[51] Mirabi A, Hosseini SN. Use of modified nano sorbent materials for extraction and determination of trace amounts of copper ions in food and natural water samples. Trends Appl Sci Res 2012;7 (7):541–549.

[52] Khajeh M, Sanchooli E. Preconcentration of trace amounts of manganese in water samples based on (1-(2-pyridylazo)-2-naphthol) modified magnetic nanoparticles. Int J Env Anal Chem 2012;92 (5):613–619.

[53] Ghaedi M, Sharifpour E. Chemically modified nano silica gel with 2-((3-silylpropylimino) methyl)-2-hydroxy-1-naphthol (SPIMHN) as good and efficient adsorbent for solid phase extraction. Desalin Water Treat 2012;41 (1–3):315–324.

[54] Karimipour G, Ghaedi M, Sahraei R, Daneshfar A, Biyareh MN. Modification of gold nanoparticle loaded on activated carbon with bis(4-methoxysalicylaldehyde)-1,2-phenylenediamine as new sorbent for enrichment of some metal ions. Biol Trace Elem Res 2012;145 (1):109–117.

[55] Kido H, Szurdoki F, Gustin MS, Hammock BD. Chelate sensor method for mercury. In: Sarkar B, editor. Metals and Genetics, *Proceedings of the International Symposium on Metals and Genetics*, Toronto, ON, Canada, May 26–29, 1998. 2nd ed. New York: Kluwer; 1999. p 101–116.

[56] Skorek R, Turek E, Zawisza B, Margui E, Queralt I, Stempin M, Kucharski P, Sitko R. Determination of selenium by X-ray fluorescence spectrometry using dispersive solid-phase microextraction with multiwalled carbon nanotubes as solid sorbent. J Anal Atom Spectrom 2012;27 (10):1688–1693.

[57] Mirabi A, Fasamanesh A. Application of a new modified nano alumina as a sorbent for preconcentration and determination of cobalt ions in food and environmental water samples prior to flame atomic absorption spectroscopy. Asian J Chem 2012;24 (8):3420–3424.

[58] Nabipour H. Synthesis of a new dithiocarbamate cobalt complex and its nanoparticles with the study of their biological properties. Int J Nano Dim 2011;1 (3):225–232.

[59] Caliendo C, Contini G, Fratoddi I, Irrera S, Pertici P, Scavia G, Russo MV. Nanostructured organometallic polymer and palladium/polymer hybrid: surface investigation and sensitivity to relative humidity and hydrogen in surface acoustic wave sensors. Nanotechnology 2007;18 (12):125504/1–125504/7.

[60] Dong J, Wen Y, Miao Y, Xie Z, Zhang Z, Yang H. A nanoporous zirconium phytate film for immobilization of redox protein and the direct electrochemical biosensor. Sensor Actuator B Chem 2010;B150 (1):141–147.

[61] Kailasam V, Rosenberg E, Nielsen D. Characterization of surface-bound Zr(IV) and its application to removal of As(V) and As(III) from aqueous systems using phosphonic acid modified nanoporous silica polyamine composites. Ind Eng Chem Res 2009;48 (8):3991–4001.

[62] Jiménez-Cadena G, Riu J, Rius FX. Gas sensors based on nanostructured materials. Analyst 2007;132:1083–1099.

[63] Elosua C, Arregui FJ, Zamarreno CR, Bariain C, Luquin A, Laguna M, Matias IR. Volatile organic compounds optical fiber sensor based on lossy mode resonances. Sensor Actuator B Chem 2012;173:523–529.

[64] Martic S, Labib M, Shipman PO, Kraatz H-B. Ferrocene-peptido conjugates: from synthesis to sensory applications. Dalton Trans 2011;40 (28):7264–7290.

[65] Zhu T, Li J, Wu Q. Construction of TiO_2 hierarchical nanostructures from nanocrystals and their photocatalytic properties. ACS Appl Mater Interfaces 2011;3 (9):3448–3453.

[66] Renard L, Babot O, Saadaoui H, Fuess H, Broetz J, Gurlo A, Arveux E, Klein A, Toupance T. Nanoscaled tin dioxide films processed from organotin-based hybrid materials: an organometallic route toward metal oxide gas sensors. Nanoscale 2012;4 (21):6806–6813.

[67] Boennemann H, Khelashvili G. Efficient fuel cell catalysts emerging from organometallic chemistry. Appl Organomet Chem 2010; 24 (4):257–268.

[68] Jeong B, Uhm S, Lee J. Iron-cobalt modified electrospun carbon nanofibers as oxygen reduction catalysts in alkaline fuel cells. ECS Trans 2010;33 (1):1757–1767.

PART II

REMEDIATION WITH USE OF CARBON NANOTUBES

8

CARBON NANOTUBES: NEXT-GENERATION NANOMATERIALS FOR CLEAN WATER TECHNOLOGIES

Yit Thai Ong, Kian Fei Yee, Qian Wen Yeang, Sharif Hussein Sharif Zein[1,*], and Soon Huat Tan

[1] *School of Chemical Engineering, Engineering Campus, Universiti Sains Malaysia, Pulau Pinang, Malaysia*

* *Current address: Faculty of Science and Engineering, School of Engineering, University of Hull, Hull, UK*

8.1 INTRODUCTION

Water is an important resource for daily human activities. The growing population in the past decade has led to increased demand for water supply. Although 70% of the Earth's surface is covered with water, only 2% is fresh water. Furthermore, 90% of the Earth's fresh water supply is frozen in glaciers [1], which indicates limited availability of fresh water. This limited fresh water supply has worsened with anthropogenically caused water pollution. Statistical data from the World Water Assessment Programme show that 2 million tons of human waste is disposed into water bodies each day, and 70% of industrial waste in developing countries is directly discharged with no proper treatment [2]. The demand for water has been projected to exceed the supply by 56% by the year 2025 [3]. This indicates the urgent need for improving water quality through water treatment processes to secure an adequate water supply.

Advances in nanoscale science and engineering can be used to find a solution to water quality problems using nanomaterials. Carbon nanotubes (CNTs) are a type of nanomaterial that have been extensively studied and viewed as a promising candidate for improving water quality because of their unique physicochemical properties. Research on the novel application of CNTs in water treatment has shown the superior performance of CNTs in producing clean water. In fact, the innovative application of CNTs has been anticipated as a new trend in clean water technologies. Our discussion on the potential of CNTs begins with a brief introduction to CNTs and their modifiable surface properties. The applications of CNTs in water treatment regarding nanoparticles and membranes are then discussed. At the end of this chapter, a lifecycle assessment on CNTs is also performed to evaluate their potential environmental impact.

8.2 MODIFICATION OF CNTs

The discovery of CNTs by Iijima et al. [4] and Bethune et al. [5] has revolutionized the future of nanotechnology. CNTs are allotropes of carbon that appear as cylindrical-shaped macromolecules with diameters in the nanometer range and lengths of up to several micrometers. The walls of CNTs are made of sp^2-hybridized carbon atoms arranged in a hexagonal lattice, analogous to the atomic planar of graphite, and capped with a spherical fullerene-like molecule [6]. CNTs may exist in two different

Nanomaterials for Environmental Protection, First Edition. Edited by Boris I. Kharisov, Oxana V. Kharissova, and H. V. Rasika Dias.

conformations: single-walled CNTs (SWCNTs) or multiwalled CNTs (MWCNTs). SWCNTs comprise a single-layer cylindrical graphite sheet, whereas MWCNTs comprise two or more concentric cylindrical shells of graphite coaxially arranged around a central hollow area with a space between layers [7, 8]. Both SWCNTs and MWCNTs can generally be synthesized by subjecting a precursor, such as graphite, to an arc discharge [9] or laser irradiation [10] in the presence of a catalyst metal particle such as iron (Fe), cobalt (Co), or nickel (Ni). An alternative synthesis route is chemical vapor deposition (CVD) [11], in which gaseous hydrocarbon sources are catalytically decomposed with a metal nanoparticle supported on a substrate.

The distinctive properties of the sp^2-hybridized bonding in CNTs provide unusual mechanical stability, high electrical and thermal conductivity, and strong chemical reactivity [12]. However, CNTs tend to agglomerate because of strong van der Waals interactions, causing these CNTs to be insoluble and hard to disperse in almost all solvents. This phenomenon critically affects the performance of CNTs in water treatments and the quality of their hybrid materials. Modification of CNTs by functionalization is indispensable to solving the problem. Functionalization is an effective strategy for improving the surface characteristics of CNTs by introducing suitable functional moieties into CNT walls [13]. Figure 8.1 shows that the various extensively developed functionalization approaches can be categorized into covalent and noncovalent sidewall functionalization.

The local strain arising from the pyramidalization and misalignment of the π orbitals of the sp^2-hybridized carbon atom causes CNTs to become more reactive and more likely to be covalently attached onto other chemical species [14]. The most popular process in covalent sidewall functionalization can be achieved by defect functionalization. In this process, acidic solutions (such as hydrochloric acid (HCl), nitric acid (HNO_3), or sulfuric acid (H_2SO_4)) or oxidizing agents (such as potassium permanganate ($KMnO_4$) or hydrogen peroxide (H_2O_2)) are used to create a defect site in CNTs by opening their end caps and attaching the caps to hydroxyl, carbonyl, and carboxyl groups [15]. The presence of these oxygen-containing functional groups exfoliates CNTs from aggregation and facilitates the dispersion of CNTs. These functional groups also serve as anchor sites for amidation, esterification, thiolation, and silanization reactions. Various functional molecules including biomolecules [16, 17] and polymers [18, 19] are then enabled to be covalently bonded with CNTs.

CNTs are also known for their reactivity with increased curvature [20]. Thus, the treatment of CNTs with highly reactive species such as radical aryls, nitrenes, carbines, or halogens [21, 22] can induce a change in hybridization from sp^2 to sp^3 in the carbon atom and a simultaneous loss of the π conjugation system on the graphene layer [23]. These changes lead to the occurrence of direct sidewall functionalization. Direct coupling of functional groups such as fluorine onto the π-conjugated carbon framework can be further modified by nucleophilic substitution reactions and replaced by hydroxyl, amino, or alkyl groups [6, 24]. This direct coupling is similar to defect functionalization. This approach reduces the formation of defect sites that may inflict damage to the CNT structures and affect their intrinsic properties. An example of direct sidewall functionalization is through the N_2 plasma technique. In this approach, active species –C* or –C–N* are produced from the reaction between CNTs and active nitrogen species (N*) and serve as active sites for grafting functional molecules onto CNT frameworks [25]. This method has a great advantage of eliminating the use of large amounts of chemicals.

Noncovalent functionalization is particularly attractive because it involves physical interactions between CNTs and functional molecules without disturbing the bonding structure of CNTs. Van der Waals interaction and π–π stacking are the physical interactions responsible for noncovalent bonding. Surfactant adsorption is one of the typical methods of noncovalent functionalization. Surfactants are well known for their amphiphilic properties because of the hydrophilic region composed of a polar head group and a hydrophobic region composed of an alkyl chain tail in the structure [26] of surfactants. Thus, dispersing CNTs in surfactants such as sodium dodecyl sulfate (SDS), sodium dodecylbenzene sulfonate, triton X, and siloxane polyether copolymer [27–29] facilitates the interaction of CNTs with the hydrophobic region

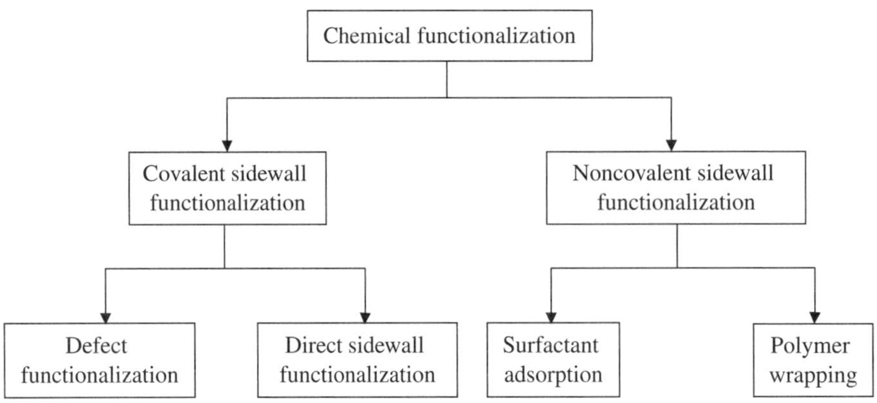

FIGURE 8.1 Functionalization of CNTs.

of the surfactants. The interaction leads to adsorption of the surfactants onto the surface of CNTs. Adsorption is believed to become stronger if the hydrophobic region of the surfactant molecules contains aromatic groups, which increase their π–π stacking with the graphitic CNT sidewalls [13, 26]. CNTs can also be noncovalently functionalized with a polymer by allowing polymeric molecules to wrap around the sidewall of the CNTs based on the aforementioned physical interactions. Polymer wrapping can be easily implemented by mixing the CNTs in a solution containing a polymer and forming a supramolecular complex. Polymers are believed to have the tendency to wrap around CNTs in a helical geometry to minimize the strain in their conformations [30].

The modifiable surface of CNTs demonstrates the adaptation of their tunable surface properties with viable applications and the optimization of their surface characteristic for enhancing the performance of CNTs.

8.3 ADSORPTION OF HEAVY METALS

Heavy metals naturally exist in the ecosystem. However, the introduction of heavy metals to water by anthropogenic sources such as industrial and agricultural waste disposal has created a serious pollution issue that has drawn major global concern. The persistence of heavy metals in water resources threatens human health even at relatively minor levels of exposure because they are not degradable and possess toxicity. Therefore, removing heavy metals from the ecosystem has become an indispensable task. Adsorption has been the most effective and economical method for removing heavy metals from an aqueous environment over the past decades [31]. In addition to easy handling and flexibility in design, the reversibility of the adsorption process makes the adsorbent reusable after regeneration through desorption, making the process cost-effective.

CNTs are recognized as a highly efficient adsorbent for removing heavy metals from water because of their large surface active sites and controlled pore size distribution. Surface-modified CNTs are more popular for adsorption studies than pristine CNTs because of the enhanced adsorption capability of surface-modified CNTs. One of the most common methods for modifying the CNT surface is chemical oxidation. CNTs are treated with strong oxidizing agents to make possible the attachment of oxygen-containing functional groups, which increase adsorption affinity toward heavy metal ions including lead (Pb(II)) [32–34], zinc (Zn(II)) [35–37], nickel (Ni(II)) [38–41], cadmium (Cd(II)) [42], and copper (Cu(II)) [43]. Another CNT modification method is functionalization with metal oxides [44–46], polymers [18, 19, 47, 48], and organic substances [49–52] to form a new composite called CNT-based nanocomposites. These functional components serve as efficient anchors for heavy metals by taking advantage of the large surface area of CNTs. The high adsorption of CNT/tri(2-aminoethyl)amine (TAA) toward Pb(II) was reported by Cui et al. [49], who attributed the adsorption to the coordination between the nitrogen group in the TAA and the metal ions. Shao et al. [48] also reported that the amide functional groups in MWCNT/poly(acrylamide) and MWCNT/poly(*N*,*N*-dimethylacrylamide) serve as efficient adsorption sites for Pb(II). The adsorption performances of various types of CNTs are listed in Table 8.1 [18, 33, 35, 37, 42, 43, 48, 49, 53, 54].

Adsorption of heavy metals onto CNTs involves complementary steps that may be attributed to physical adsorption, electrostatic attraction, precipitation, and chemical interaction [55]. Among these, the major step is the chemical interaction between heavy metals and the surface functional groups of CNTs, where adsorption occurs by sharing the lone electron pair from the functional group with the heavy metal [25]. This finding is consistent with the study of Wang et al. [56], who reported that 75.3% of the total Pb(II) adsorbed onto CNTs depends on the chemical interaction with an oxygen-containing functional group to form a chemical complex. Such a dominant mechanism allows adsorption onto CNTs to reach equilibrium in a shorter time than conventional porous-structure adsorbents where the adsorbate needs to diffuse from the exterior into the inner surface [39]. The adsorption behavior of heavy metals on the CNT surface mainly follows the Langmuir isotherm, where the heavy metals are adsorbed onto the CNT surface in a monolayer and are localized at the adsorption site without any interaction with other heavy metal ions [45, 51, 57].

The surface properties of the CNTs are prone to changes depending on pH, which affects the adsorption capacity. The amount of heavy metal adsorbed can be correlated to pH called the "point of zero charge" (pH_{PZC}), in which a net charge of the adsorbent is equal to zero [55]. The adsorption phenomena in CNTs are favored at pH higher than pH_{PZC} because the surface of CNTs is more negatively charged and induces an electrostatic attraction, which promotes the adsorption of heavy metals that usually appear as cation species. However, decreasing pH to less than pH_{PZC} reduces heavy metal adsorption because the CNT surfaces become positively charged and compete with H^+ for adsorption sites [35, 38, 58]. Previous reports [43, 59–61] have indicated that the pH_{PZC} of pristine CNTs lies between pH 4 and 6, whereas pH_{PZC} decreases with modified CNTs. The modified CNTs demonstrate better adsorption properties because the adsorption of CNTs is effective over a broader range of pH. In most cases, low pH is used for the desorption of heavy metals from CNTs with acidic solutions such as HNO_3 and HCl [33, 51, 53] instead of adsorption. Over 90% of the adsorption sites on CNTs can be recovered and reused [53, 62]. The adsorption performance of CNTs is found to remain stable for over 10 cycles of adsorption and desorption [50, 53, 63–65].

TABLE 8.1 Performance of various types of CNTs in the adsorption of heavy metal ions

Adsorbents	Adsorbed materials	Maximum adsorption capacity, q_m (mg/g)	Details	Reference
MWCNTs (HNO_3)	Pb(II)	85	The high adsorption rate of Pb(II) is attributed to the oxygenous functional groups formed on the surface of MWCNTs that react with Pb(II) to form salt or complex deposited onto the surface of MWCNTs.	[33]
SWCNTs (NaClO)	Zn(II)	43.66	The attachment of oxygen-containing functional groups increase CNTs hydrophilicity and the ion exchange capacity for Ni(II) adsorption.	[35]
MWCNTs (NaClO)		32.68		
MWCNTs	Zn(II)	10.21	The adsorption of Zn(II) depends on the chemical interaction between Zn(II) and surface functional groups on CNTs rather than surface area and pore volume.	[37]
MWCNTs (HNO_3/H_2SO_4)		18.14		
MWCNTs (HNO_3)		27.20		
MWCNTs ($KMnO_4$)		28.01		
CNTs	Cd(II)	1.1	The oxidation ability of HNO_3 is the highest.	[42]
CNTs (H_2O_2)		2.6	The highest adsorption in $KMnO_4$ can be attributed to the adsorbed residual MnO_2 particles on CNTs.	
CNTs ($KMnO_4$)		11.0		
CNTs (HNO_3)		5.1		
MWCNTs	Cu(II)	8.25	Modification with NaOCl and HNO_3 increases the area of active adsorption sites of CNTs and the proportion of available adsorption sites.	[43]
MWCNTs (HNO_3)		13.87	The oxidizing ability of NaOCl exceed that of HNO_3	
MWCNTs (NaClO)		47.39		
MWCNTs (NaClO)	Ni(II)	38.46	The adsorption performance of both MWCNTs and SWCNTs remain stable after 10 cycles of sorption/desorption process.	[53]
SWCNTs (NaClO)		47.85		
MWCNT/Poly(2-aminothiphenol) (P2AT)	Cd(II) Pb(II)	178.7 186.4	Metal ions are adsorbed by sharing an electron pair of =N and —S— in P2AT with Cd(II) and Pb(II).	[18]
MWCNT/poly(acrylamide) (PAAM)	Pb(II)	35.7	The amide functional groups in PAAM and PDMA act as efficient anchors for Pb(II).	[48]
MWCNT/Poly(*N*,*N*-dimethylacrylamide) (PDMA)		25.8	The sorption of MWCNT/PAAM is higher because of the higher amide group content.	
MWCNT/tri(2-aminoethyl)amine (TAA)	Pb(II)	38	The high adsorption of Pb(II) can be attributed to the coordination interaction between the nitrogen group in TAA and metal ions	[49]
MWCNT/2-vinylpyridine (VP)	Pb(II)	37.0	The pyridyl group in VP has a strong affinity to interact with Pb(II)	[54]

Thermodynamic studies provide an overview of the macroscopic properties of heavy metal adsorption onto CNTs and are conducted based on the thermodynamic parameters enthalpy change ($\Delta H°$), Gibbs free energy change ($\Delta G°$), and entropy change ($\Delta S°$). The adsorption of heavy metals onto CNTs generally involves a positive enthalpy change ($\Delta H°$), indicating that the adsorption process is endothermic, and a negative Gibbs free energy change ($\Delta G°$), indicating that the sorption of the heavy metal on CNTs occurs spontaneously. $\Delta G°$ becomes more negative with increased temperature because of the endothermic characteristics. A more negative $\Delta G°$ indicates more efficient adsorption at a higher temperature. $\Delta S°$ of the adsorption usually has a positive value, which means an increase in the degree of freedom at the solid–liquid interface, and increases the probabilities of heavy metal adsorption onto CNTs [32, 38, 40, 43, 46, 48].

8.4 ADSORPTION OF BACTERIAL PATHOGENS

The exceptional adsorption ability of CNTs is not limited to heavy metal ions. Srivastava et al. [66] revealed the potential of CNTs to adsorb and remove bacterial pathogens from contaminated water without any surface modification. Numerous studies [67–70] have also demonstrated that CNTs possess high binding affinity toward bacteria such as *Bacillus subtilis*, *Escherichia coli*, and *Staphylococcus aureus* because of their high aspect ratio and large surface area. The interaction between CNTs and bacteria basically relies on physical adsorption, wherein bacteria are spontaneously adsorbed in the mesopores and macropores of the fibrous structure of CNTs [68, 70]. The forces responsible for bacterial adsorption can be van der Waals forces, electrostatic interaction, and hydrophobic interaction [68, 71–73]. The bacterial cells that have direct contact with CNTs upon adsorption have a strong tendency to form biofilms or aggregates by connecting individual cells. The adsorption of bacteria onto CNTs may occur through the following steps: (i) diffusion of bacteria from a solution to the outer surface of CNTs; (ii) diffusion of bacteria into the macropores of CNTs; (iii) adsorption of bacteria onto the internal surface of CNTs; and (iv) formation of biofilm or aggregates on the CNT surface [67].

The main reason for using CNTs as adsorbents apart from providing adsorption sites to capture bacteria is the antimicrobial property of CNTs, which is a unique feature. The antimicrobial property enables CNTs to inhibit bacterial cell activity, causing cell inactivation. The antimicrobial property of CNTs can be attributed to the impairment of cell function through the destruction of major constituents (for instance, the cell wall), interference with cell metabolic processes, and inhibition of cell growth by blocking the key cellular constituent (for instance, DNA, coenzymes, and cell wall proteins) [74]. Bacterial cells, especially those in direct contact with CNTs, may immediately lose their cellular integrity after exposure to CNTs. The antimicrobial activity of CNTs is significantly influenced by time. A longer exposure time is known to increase CNT antibacterial activity and increase cell inactivation. The increased activity is due to the increased probability of contact between CNTs and bacterial cells, which diminishes the resistance of bacteria after a long exposure time [75, 76]. Two types of bacteria exist based on their cell wall structure: Gram-positive and Gram-negative. The cell wall of Gram-positive bacteria is usually composed of a thicker peptidoglycan layer that is susceptible to attack by CNTs. The cell wall of Gram-negative bacteria comprises a thin peptidoglycan layer and an additional outer layer composed of phospholipids and lipopolysaccharides. CNTs reportedly possess stronger antimicrobial activity in Gram-positive bacteria than in Gram-negative bacteria. Therefore, the difference in cell wall structure can affect the antimicrobial activity of CNTs [77].

The nanoparticle size is also another factor that affects the antimicrobial activity of CNTs. Kang et al. [78] compared the antimicrobial activities of SWCNTs and MWCNTs toward *E. coli* and found that bacterial cell damage was much more pronounced in SWCNTs than in MWCNTs. Scanning electron microscopy (SEM) images of *E. coli* after exposure to SWCNTs and MWCNTs are shown in Figure 8.2. The result implies that the smaller diameter and shorter length of SWCNTs benefit antimicrobial activity by providing a larger surface area to increase the likelihood of CNTs making contact with bacterial cells. The smaller diameter of SWCNTs also facilitates the partitioning and partial penetration of CNTs into the cell wall. Arias et al. [77] reported stronger interactions in SWCNTs and observed that SWCNTs have tight contact with bacterial cells. These researchers also believed that the needlelike appearances of SWCNTs surrounding cell aggregates are more likely to inflict severe cell damage. Bacterial cells make only loose contact with MWCNTs by settling on the MWCNT surface, which may not effectively damage cell walls, unlike SWCNTs.

SWCNTs demonstrate higher affinity for interacting with bacterial cells than MWCNTs, but SWCNTs involve high synthesis costs. Therefore, MWCNTs are modified to enhance their performance. Yuan et al. [79] grafted MWCNTs with polyamidoamine (PAMAM) and deposited silver nanoparticles. The presence of PAMAM debundles MWCNTs, enhancing dispersion. Silver nanoparticles significantly improve the antimicrobial efficiency of MWCNTs against *S. aureus*. Zardini et al. [80] functionalized

(a) (b)

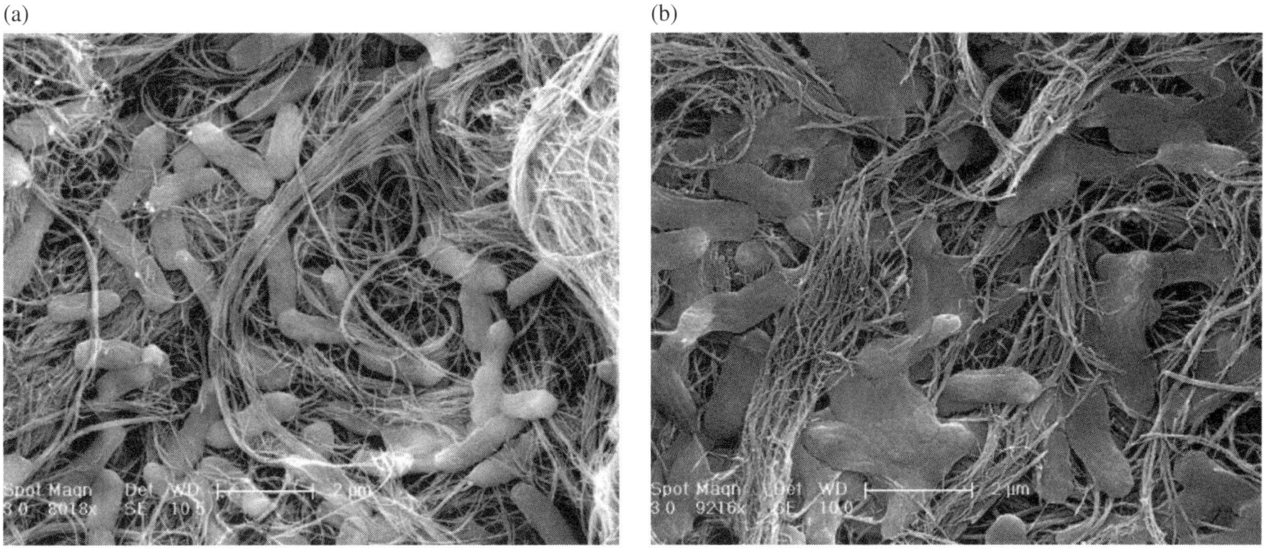

FIGURE 8.2 Scanning electron microscopy (SEM) images of *E. coli* after exposure to (a) MWCNTs and (b) SWCNTs [78].

MWCNTs with two different types of amino acids: lysine and arginine. MWCNTs functionalized with both amino acids were found to enhance the interaction of MWCNTs with *Salmonella typhimurium*, *E. coli*, and *S. aureus*. The positively charged arginine and lysine groups improved the adsorption of bacterial cells in MWCNTs, increasing the effectiveness of antimicrobial activity. Qi et al. [81] performed another modification by immobilizing MWCNTs with cefalexin and investigated the contact of MWCNTs with Gram-positive bacteria (*S. aureus* and *B. subtilis*) and Gram-negative bacteria (*E. coli* and *Pseudomonas aeruginosa*). The result shows that the antimicrobial activity of the produced nanocomposite is more effective in Gram-positive bacteria because cefalexin mainly inhibits peptidoglycon synthesis.

8.5 CNT-BASED MEMBRANES

The excellent physicochemical properties of CNTs are attracting attention because of their possible application in membrane technologies. A recent discovery on fast water transportation in the inner hollow cavity of CNTs also increases the potential of CNTs as a material for water treatment. Extremely rapid water transport is about two to five times higher than the value predicted through the Hagen–Poiseuille equation [82, 83]. Theoretical studies using molecular dynamic simulation have shown that water can fill the empty inner cavity of CNTs within a few tens of picoseconds. The weak interaction of water molecules with the hydrophobic walls of CNTs enables frictionless transport and high water flux [84]. Detailed studies on the transport mechanism have been conducted because of the rapid water transport to the nanoscale confinement in the CNT channel. Confining water inside the CNT channel narrows the interaction energy distribution, eventually lowering the chemical potential and free energy [85]. The confinement also induces stronger attraction between water molecules than the interaction with CNT walls [86], which promotes a higher water flux.

The beneficial effect of CNTs in water transport can be exploited in the development of membrane-separation technologies. Their presence is a massive breakthrough in membrane technologies and provides a synergistic effect for water treatments. CNT-based membranes are versatile materials for various membrane-separation applications including ultrafiltration [87–89], nanofiltration [90–93], reverse osmosis [94–96], and membrane distillation (MD) [97–101]. Studies have shown that CNT-based membranes can overcome the limitation of conventional membranes. The use of CNT-based membranes can also address cost-effectiveness issues.

CNTs may have different macroscopic structures and arrangements in membranes depending on the synthesis method. The most common strategy is the incorporation of CNTs as fillers in the polymer matrix by blending them with polymer solution to form so-called CNT–mixed matrix membranes (CNT–MMMs) [102]. CNTs may also appear in bundle forms arranged like paper sheets, which are known as buckypapers. They may also be uniformly aligned perpendicular to the membrane surface, which forms vertically aligned CNT membranes.

8.5.1 CNT–MMMs

Figure 8.3 shows that CNTs are randomly oriented inside the polymer matrix of CNT–MMMs. The presence of a small amount of CNTs can improve the properties and generate new characteristics in the resultant membranes. Shaw et al. [94] synthesized CNT–MMMs composed of MWCNTs and polyamide for use in water treatment. CNTs are well recognized for their ability to sustain high energy loading. Thus, CNTs serve as reinforcing agents that can improve the Young's modulus, toughness, and tensile strength of CNT–MMMs. However, ensuring adequate interfacial bonding between the polymer matrix and CNTs is necessary to elicit the maximum reinforcing effect of CNTs. Hence, CNTs are usually modified with functional molecules prior to blending with polymer.

CNT–MMMs also enhance the chemical resistance of polymeric membranes particularly against chlorine, which is commonly found in water [95, 104]. Polymeric reverse osmosis membranes such as polyamides are susceptible to degradation upon contact with chlorine, which in turn causes deteriorated membrane-separation performance. Polyamide degradation is mainly attributed to the chlorination of N–H in amide bonds upon exposure to chlorine, followed by oxidation to the quinoid structure through hydrolysis reaction with water. The deformation of an amide network causes serious defects in the membrane structure, thereby decreasing salt rejection [104]. The presence of CNTs effectively increases the chemical resistance of polyamide membranes against chlorine. Experimental results showed that salt rejection in CNT–MMMs surpasses that in conventional polyamide membranes after exposure to high-concentration chlorine solution. For polyamide membranes, salt rejection decreased by 21.8% after chlorination. However, addition of 0.1, 0.5, and 1% (w/v) CNT–MMMs decreases salt rejection by 15.8, 11.2, and 10.1%, respectively. The interaction between carboxylic groups in acid-modified CNTs and the amide bond in polyamide is believed to inhibit the chlorination of amide bond, which enhances the chlorine resistance of CNT-based membranes [95].

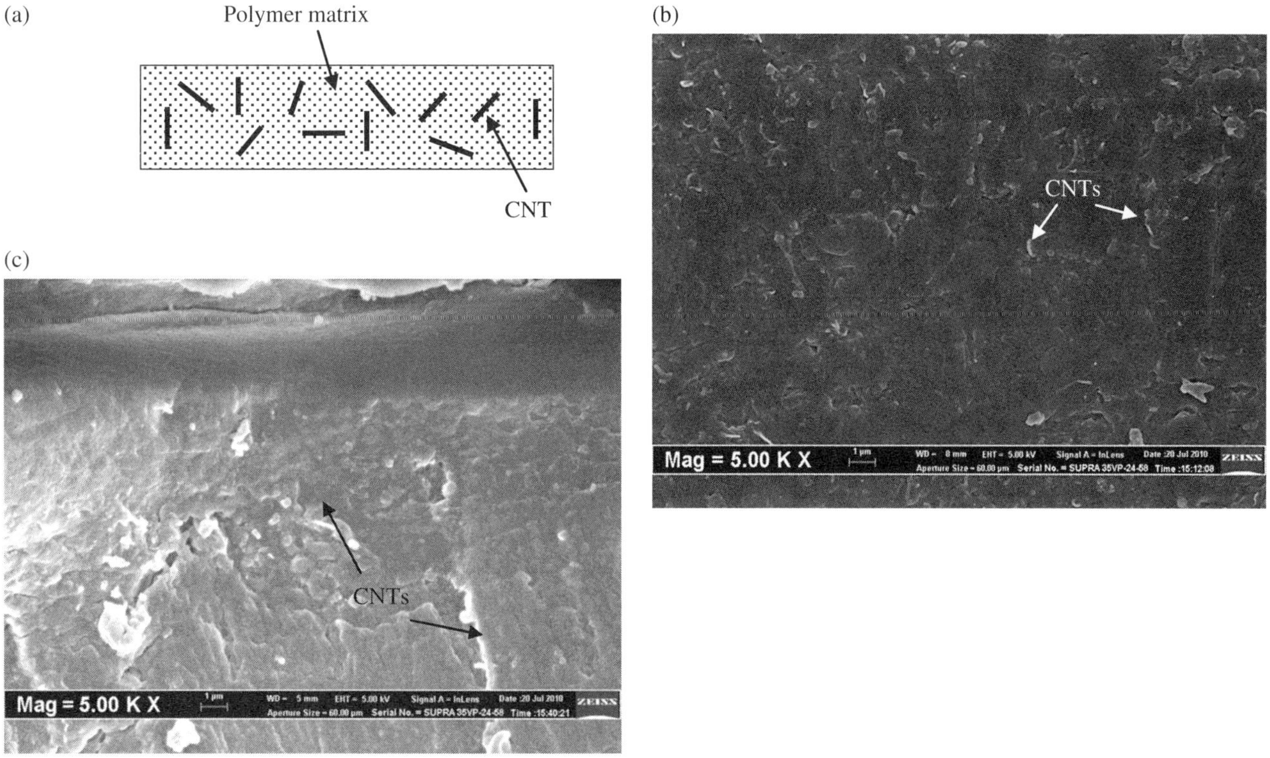

FIGURE 8.3 (a) Schematic of CNT–MMMs, (b) SEM images of the surface of CNT–MMMs, and (c) SEM images of the cross section of CNT–MMMs [103].

Furthermore, considerable attention has been paid to the fabrication of porous-structured CNT–MMMs because of their application in nanofiltration and ultrafiltration. The results show that CNT–MMMs enhance the water permeability and antifouling properties of the polymer host matrix. Generally, porous-structured CNT–MMMs are prepared by the phase-inversion method. The polymer is dissolved into a suitable solvent, casted, and then immersed in a nonsolvent, which results in precipitation, eventually leading to the formation of porous-structure membranes. Interestingly, CNTs may alter the membrane structure during membrane formation. The morphology and porous structure of CNT–MMMs change depending on the amount of CNTs added to membranes [87]. The pore size and surface roughness increase with increased concentration of CNTs and reach a maximum value at a critical concentration. Previous reports have indicated that the critical concentration ranges from 0.2 to 5 wt.%. Beyond the critical concentration, the high density of CNTs increases the solution viscosity and delays the exchange between solvent and nonsolvent, thereby leading to reduced pore sizes and smoother surfaces [87–89, 92, 105]. The pore size of CNT–MMMs is important in controlling the water flux and solute rejection. Larger pore sizes increase water flux but compromise solute rejection. Hence, high water flux and solute rejection can be achieved by adding the appropriate amount of CNTs [92]. CNTs modified with oxygen-containing functional groups reportedly spontaneously migrate onto the surface of the membrane during phase inversion because of their hydrophilic effect [88, 93]. This phenomenon promotes water permeability by attracting water molecules inside the membrane matrix, which facilitates their passage through the membrane. The improved surface hydrophilicity also enables CNT–MMMs to reduce the occurrence of fouling. A major problem in membrane separation is membrane fouling, wherein solutes are deposited onto the surface or in the pores of the membrane, thereby decreasing membrane performance. The tendency of the CNT–MMM surface to interact with a water layer in an aqueous solution reduces possible interactions with the solute. Antifouling properties of CNT–MMMs were studied by Celik et al. [87], who observed a significant decline in the flux of polyethersulfone (PES) membranes with respect to operating time; nevertheless, the decrement is minimal and reaches a plateau with increased operating time. Vatanpour et al. [92] compared the antifouling performances of pristine and CNT/PES membranes and found that membrane fouling resulting from bovine serum albumin solution can be reduced in the presence of CNTs. These results also confirm the importance of surface roughness in the antifouling resistance of CNT-based membranes. Increased surface roughness may enhance membrane fouling because the foulant can easily be entrapped in the valleys of membranes with a coarser surface, thereby clogging the valleys. However, the charges supplied by the functional moieties on the surface of CNTs also contribute to the antifouling property by inducing strong electrostatic repulsion forces to repel charged solutes. The surface of CNT-based membranes is thus shielded, and blockages caused by the solute deposition are prevented.

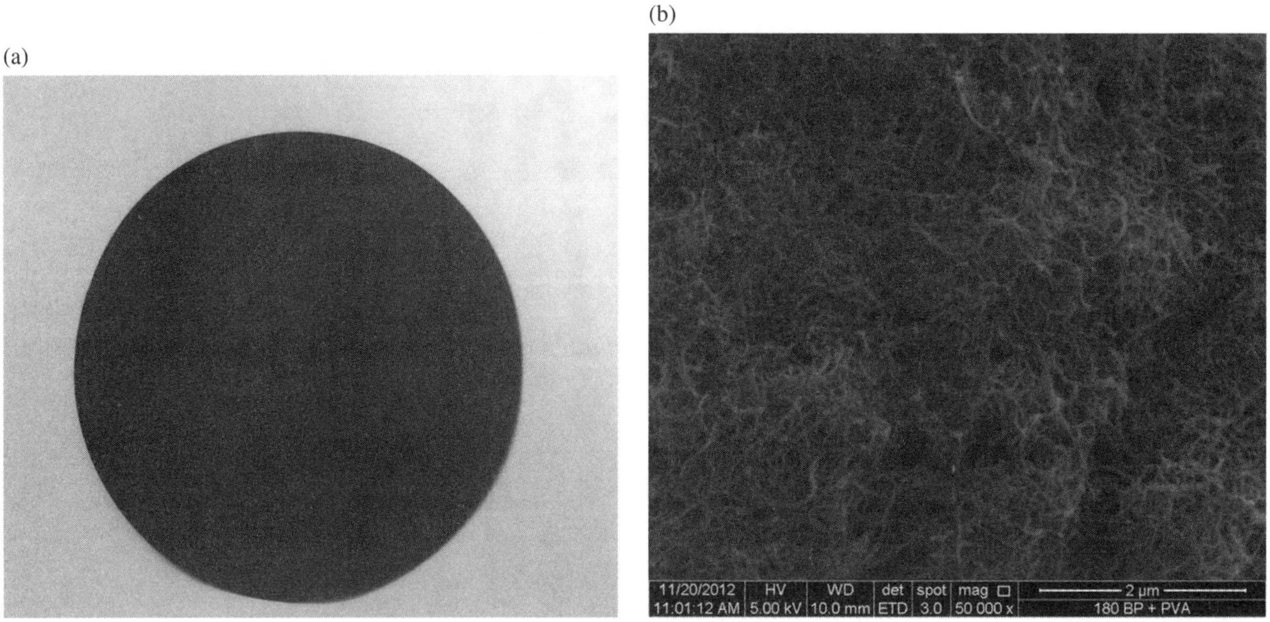

FIGURE 8.4 (a) Photograph of buckypapers and (b) scanning electron microscopy (SEM) image of the surface of buckypapers.

8.5.2 Buckypapers

Buckypapers consist solely of packed CNT bundles entangled with one another, held together, and arranged as a planar film through van der Waals forces, as shown in Figure 8.4. Normally, self-supporting CNT-based membranes are synthesized by dispersing CNTs in a suitable solvent such as ethanol [106], propan-2-ol [107], and dimethylformamide [108], followed by vacuum filtration using a porous membrane filter. In some cases, surfactants such as Triton X-100 [109] or SDS [110] are also used as the dispersion medium. During the drying process, the solvent evaporates and the capillary forces of the solvent bring CNTs together to a closely knit structure [111]. Buckypapers form a highly porous network structure, whereas CNTs are randomly oriented, although they are predominantly arranged in a plane parallel to the surface of the buckypapers [112]. The porous characteristics of buckypapers can be controlled by the length distribution of CNTs [113, 114].

The application of buckypapers in water treatment has gained interest in the past few years. Dumee et al. [99] experimentally discovered the potential application of buckypapers in MD. MD is a thermally driven process in which only vapor molecules are in contact with membranes and transported across membranes through membrane pores. MD is anticipated to be an alternative to reverse osmosis and desalination. Essentially, buckypapers meet the requirement of serving as membranes for MD. Given their hydrophobic nature, buckypapers prepared from pristine CNTs exhibit high contact angle with water droplets, which indicates their weak surface interaction with water, thereby preventing leakage of the aqueous solution onto the membrane surface. The presence of CNTs in buckypapers also enables better thermal stability that restricts the thermal degradation of membranes at high temperatures. This observation is critical because MD preferably operates at high temperatures because the vapor pressure exponentially increases with increased temperature according to Antoine's equation [115]. Moreover, buckypapers can reduce pore clogging caused by water condensation during vapor permeation. CNTs have a high thermal conductivity that enables a relatively uniform temperature distribution in buckypapers, which reduces the possibilities of liquid condensation in membrane pores caused by a temperature gradient [101]. In addition, the high porosity of buckypapers provides an alternative pathway for efficient vapor transport. Dumee et al. [99] showed the advantages of buckypapers in MD in terms of the membranes demonstrating 99% salt rejection and 12 $kg/m^2 \cdot h$ flux rate. The outcome is comparable with that using conventional MD membranes, such as polytetrafluoroethylene (PTFE). However, buckypapers endure a flux decline with prolonged operating time as a result of crack formation caused by mechanical aging. Several approaches have been proposed to solve this problem, including the preparation of buckypapers from alkoxysilane-functionalized CNTs [97] or coating of thin PTFE layers on buckypapers [116].

Apart from MD, the porous structure of buckypapers makes them appropriate for filtration-related applications. In fact, instead of buckypapers, a number of researchers have successfully used the concept of bundle CNTs in filtering processes to remove nanoparticles, dye molecules, and viruses from water [117, 118]. In line with these results, Roy et al. [90] produced buckypapers with 1.3 g/cm^3 density by a vacuum filtration technique. The resultant buckypapers can intercept and remove nanoparticles with average sizes of about 14–15 nm, with a flux of approximately 1000 $l/h \cdot m^2 \cdot bar$. Despite the controlled pore size, the potential demonstrated by the buckypapers for filtering nanoparticles can be attributed to the high tortuous paths of buckypapers.

8.5.3 Vertically Aligned CNT Membranes

Vertically aligned CNT membranes are composed of highly organized CNTs arranged in a vertical orientation. Among CNT-based membranes, those that are vertically aligned fully exploit the advantages of the inner cavity of CNTs as flow channels. However, obtaining high-quality CNTs in vertical arrays poses a significant challenge in the production of vertically aligned CNT membranes. To date, most vertically aligned CNTs are produced by CVD, where forests of aligned CNTs are grown on either a silicon or a quartz substrate [83]. Normally, these vertically aligned CNTs are highly porous because of the interstitial space between individual CNTs and their lumina. Interstitial spaces between CNT arrays are commonly filled with an impermeable material by infiltration or spin coating that forms a continuous matrix. Subsequently, the excess matrix material and substrate are removed, and the CNT tips are uncapped through plasma chemistry [112]. The resultant CNT-based membrane has a near-unity tortuosity factor, which suggests the presence of straight pores [119]. The procedure of producing vertically aligned CNT membranes is illustrated in Figure 8.5.

Since the introduction of the first prototype of vertically aligned CNT membranes by Hind et al. [120], a number of researchers have envisioned their potential in water technology development. The measured water flow rates across vertically

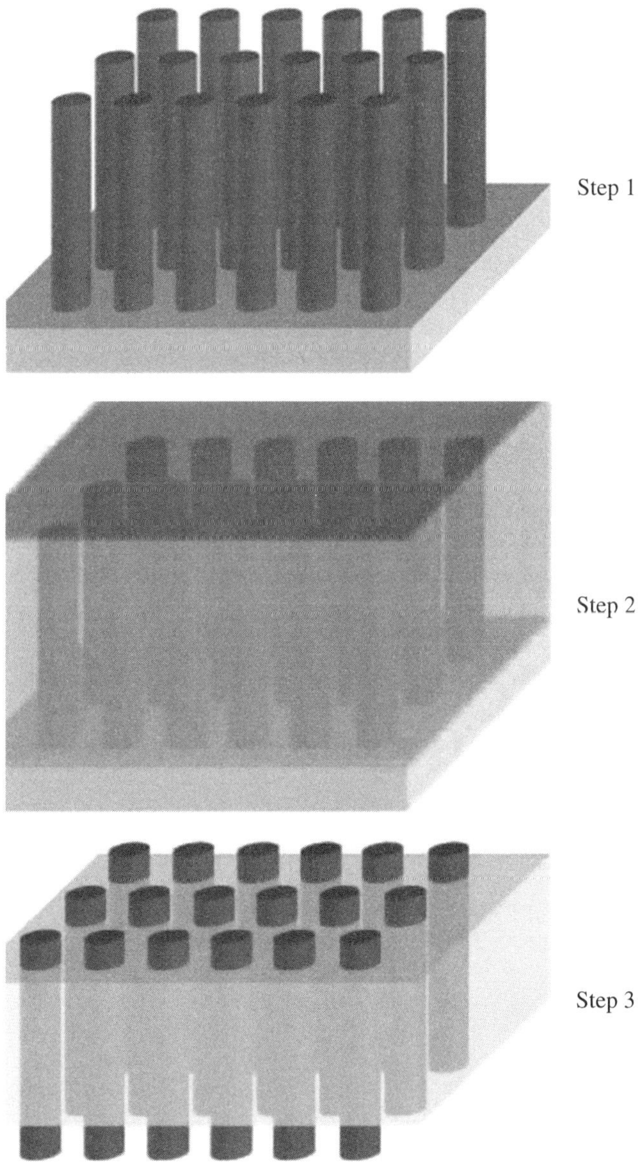

FIGURE 8.5 Schematic of the general approach to producing vertically aligned carbon nanotube (CNT) membranes [112]. Step 1: Forest vertical arrays of CNTs are grown on a substrate. Step 2: The interstitial spaces between CNTs are filled matrix material. Step 3: The substrate is removed and the CNT termini are uncapped.

aligned CNT membranes composed of less than 2-nm-diameter CNTs demonstrate an enhancement within three orders of magnitude that are faster than the values calculated from continuum hydrodynamic models. These findings verify the fast water permeation from the nearly frictionless graphitic wall of CNTs because the interstitial space of the arrays is filled with silicon nitride [121]. Apart from frictionless transport, the vertically aligned CNT structure that exposes the termini of CNTs to the outer surface also enables selective mass transport. Several experiments on pressure-driven flow have demonstrated the effectiveness of vertically aligned CNT membranes in eliminating particles with sizes ranging from 1 to 10 nm. Furthermore, recent simulation studies have revealed that the nanometer diameters of CNTs also provide a platform for ion removal from water [122]. The ability of CNTs to allow water permeation and prevent ions from passing through is critically influenced by the CNT diameter [123]. Raw CNTs offer a low energy barrier that enables rapid water permeation but may not adequately prevent small ionic species such as Na^+ and Cl^- from passing through. Meanwhile, functionalized CNTs may increase the ability to gate molecule transport through their inner cavities. The attachment of a functional group may impose a charged effect and cause steric blockage at the CNT terminals. This phenomenon reduces water flux but improves ion rejection [124]. Fornasiero et al. [125] investigated ion transport through sub-2-nm vertically aligned CNT membranes. Negatively charged groups are introduced to CNT terminals by plasma treatment, and the result shows that the ion rejection of membrane reaches as high as 98% when the aqueous electrolyte solutions pass through pores. The exclusion of ions from the CNTs with negatively charged functional groups is mainly attributed to the Donnan-type rejection mechanism, where electrostatic interactions of the charged group tend to exclude ions of the same charge while being freely permeable to ions of opposite charge. Therefore, the combination of fast transport and high ion rejection of vertically aligned CNT membranes may lead to efficient water desalination.

8.6 LIFECYCLE ASSESSMENT OF CNTs

CNTs have demonstrated their valuable potential in water treatment technologies. However, concerns on the environmental impact originating from their large-scale application prompt the need to assess the lifecycle of CNTs from the “cradle” (raw material acquisition) to the “grave” (disposal or recycling) [126]. The lifecycle assessment begins with the identification of the required raw materials, including carbon feedstock (graphite or CO), metal catalyst particles (Ni, Co, and Fe), and acid reagents (HNO_3, H_2SO_4, and HCl). Carbon feedstock and metal catalysts are consumed for CNT production, whereas acid reagents are used to treat the produced CNTs. CNTs are often produced by arc discharge, laser ablation, or CVD methods. Regardless of the production method, the energy needed to produce CNTs is higher than that needed to produce traditional materials such as aluminum, steel, and polypropylene [127]. This high energy consumption has a significant environmental impact because energy generation is always associated with greenhouse gas emission. The production process may also lead to the potential release of airborne pollutants such as polycyclic aromatic hydrocarbons and volatile organic compounds [128]. In addition, subsequent treatment with acid reagent produces acidic wastewater, which may cause water pollution if discharged into bodies of water without adequate treatment. However, among the production methods, CVD is considered to exert the least environment impact because of its low energy consumption and increased yield of high-purity CNTs, which reduce the need for rigorous postproduction treatment [129]. Upon application, the risk of releasing CNTs into the environment is present. Nevertheless, the exceptional abilities of CNTs for removing hazardous pollutants from water reduce its environmental burden. Furthermore, the application of CNTs in membranes enables a low-pressure-driven separation process that reduces energy consumption. At the end of the lifecycle, CNTs may be disposed by landfilling or incineration. Nonetheless, CNTs significantly possess superior durability, which in turn increases their lifespan during their application. Moreover, the superior durability of CNTs also allows recovery and reuse for the processing of new materials. The reusability permits CNTs to embrace the 3R (reduce, reuse, and recycle) concept [74]. These findings aid in reducing waste disposal and the subsequent environmental impact caused by landfilling and incineration. The environmental impacts from the CNT lifecycle are summarized in Table 8.2.

TABLE 8.2 Environment impact analysis of CNTs

	Positive impact	Negative impact
Raw material acquisition		
CNT production		High-energy consumption Release of airborne pollutants Generation of acidic wastewater
CNT application	Promotion of energy efficient operation Reduction of hazardous pollutant concentration	Potential release of CNTs into the environment
End of life	Superior durability that increases their lifespan Capability to be recovered and reused	

8.7 CONCLUSION

CNTs offer tremendous opportunity and possibilities for resolving water quality–related problems. Their modifiable surface properties through covalent and noncovalent functionalization enable the attachment of suitable functional moieties. Thus, the tunable surface properties of CNTs meet application requirements. Given their high aspect ratio and large surface area, CNT particles have high sorption capacity toward heavy metals and microorganisms and exhibit superior reusability by maintaining a stable adsorption performance for over 10 cycles of sorption and desorption. Their contact with microorganisms also reveal their antimicrobial property, which allows them to inhibit cell activities that eventually lead to cell inactivation. CNTs are viewed as novel membrane materials that induce immense improvement in membrane-separation processes. CNT–MMMs formed by embedding them into a polymer matrix increase the mechanical properties of the membranes and enhance their chemical resistance against chlorine. In the fabrication of porous-structured CNT–MMMs by the phase-inversion method, the presence of CNTs facilitates alterations in the membrane structure, which increases membrane hydrophilicity. Consequently, high water flux and solute rejection are enabled, and the antifouling properties of the membrane are improved. CNTs can also bundle up as a planar film known as buckypapers. Buckypapers are favorable for MD because of their exceptionally high thermal stability and efficient vapor transport. Apart from these observations, the highly porous structures possessed by buckypapers also make them applicable in the filtration and removal of nanoparticles. Vertically aligned CNT membranes are another form of CNT-based membranes, where CNTs are uniformly arranged vertically. Although their practical application in water treatment is scarcely reported, simulation studies have shown that the vertically aligned CNT structure can fully utilize the advantages of the inner cavities of CNTs as flow channels. The termini of CNTs can serve as gatekeepers to control mass transport selectively. The environmental impacts of CNTs based on lifecycle assessment reveal the potential release of hazardous materials to the environment during CNT manufacture. However, their highly effective applications potentially reduce the environmental burden and outweigh their negative impact.

ACKNOWLEDGMENT

The authors acknowledge the fellowship support of the MyPhD program of the Ministry of Higher Education Malaysia, Universiti Sains Malaysia (USM) Research University Grant, USM Membrane Cluster Grant, USM fellowship, and the Fundamental of Research Grant Scheme.

REFERENCES

[1] Sheikholeslami R. Strategies for future research and development in desalination—challenges ahead. Desalination 2009;248 (1–3): 218–224.

[2] Penza M, Rossi R, Alvisi M, Signore MA, Cassano G, Dimaio D, Pentassuglia R, Piscopiello E, Serra E, Falconieri M. Characterization of metal-modified and vertically-aligned carbon nanotube films for functionally enhanced gas sensor applications. Thin Solid Films 2009;517 (22):6211–6216.

[3] Di Francia G, Alfano B, La Ferrara V. Conductometric gas nanosensors. J Sensors 2009;2009:659275–659275.

[4] Iijima S. Helical microtubules of graphitic carbon. Nature 1991;354 (6348):56–58.

[5] Bethune DS, Kiang CH, De Vries MS, Gorman G, Savoy R, Vazquez J, Beyers R. Cobalt-catalysed growth of carbon nanotubes with single-atomic-layer walls. Nature 1993;363 (6430):605–607.

[6] Balasubramanian K, Burghard M. Chemically functionalized carbon nanotubes. Small 2005;1 (2):180–192.

[7] Ajayan PM. Nanotubes from carbon. Chem Rev 1999;99 (7):1787–1799.

[8] Daniel S, Rao TP, Rao KS, Rani SU, Naidu GRK, Lee HY, Kawai T. A review of DNA functionalized/grafted carbon nanotubes and their characterization. Sens Actuators B 2007;122 (2):672–682.

[9] Journet C, Maser WK, Bernier P, Loiseau A, Lamy de la Chapelle M, Lefrant S, Deniard P, Lee R, Fischer JE. Large-scale production of single-walled carbon nanotubes by the electric-arc technique. Nature 1997;388 (6644):756–758.

[10] Guo T, Nikolaev P, Thess A, Colbert DT, Smalley RE. Catalytic growth of single-walled nanotubes by laser vaporization. Chem Phys Lett 1995;243 (1–2):49–54.

[11] Dai H, Rinzler AG, Nikolaev P, Thess A, Colbert DT, Smalley RE. Single-wall nanotubes produced by metal-catalyzed disproportionation of carbon monoxide. Chem Phys Lett 1996;260 (3–4):471–475.

[12] Jie H. *Structures and Properties of Carbon Nanotubes*. New York: CRC Press; 2005.

[13] Hirsch A. Functionalization of single-walled carbon nanotubes. Angew Chem Int Ed 2002;41 (11):1853–1859.

[14] Banerjee S, Hemraj-Benny T, Wong SS. Covalent surface chemistry of single-walled carbon nanotubes. Adv Mater 2005;17 (1):17–29.

[15] Musameh MM, Hickey M, Kyratzis IL. Carbon nanotube-based extraction and electrochemical detection of heavy metals. Res Chem Intermediat 2011;37 (7):675–689.

[16] Liu Y, Li Y, Yan X-P. Preparation, characterization, and application of L-cysteine functionalized multiwalled carbon nanotubes as a selective sorbent for separation and preconcentration of heavy metals. Adv Funct Mater 2008;18 (10):1536–1543.

[17] Parodi B, Savio M, Martinez LD, Gil RA, Smichowski P. Study of carbon nanotubes and functionalized-carbon nanotubes as substrates for flow injection solid phase extraction associated to inductively coupled plasma with ultrasonic nebulization: application to Cd monitoring in solid environmental samples. Microchem J 2011;98 (2):225–230.

[18] Nabid MR, Sedghi R, Bagheri A, Behbahani M, Taghizadeh M, Abdi Oskooie H, Heravi MM. Preparation and application of poly(2-amino thiophenol)/MWCNTs nanocomposite for adsorption and separation of cadmium and lead ions via solid phase extraction. J Hazard Mater 2012;203–204:93–100.

[19] Nabid MR, Sedghi R, Hajimirza R, Oskooie HA, Heravi MM. A nanocomposite made from conducting organic polymers and multi-walled carbon nanotubes for the adsorption and separation of gold(III) ions. Microchim Acta 2011;175 (3–4):315–322.

[20] Bahr JL, Tour JM. Covalent chemistry of single-wall carbon nanotubes. J Mater Chem 2002;12 (7):1952–1958.

[21] Holzinger M, Vostrowsky O, Hirsch A, Hennrich F, Kappes M, Weiss R, Jellen F. Sidewall functionalization of carbon nanotubes. Angew Chem Int Ed 2001;40 (21):4002–4005.

[22] Tasis D, Tagmatarchis N, Bianco A, Prato M. Chemistry of carbon nanotubes. Chem Rev 2006;106 (3):1105–1136.

[23] Ma PC, Siddiqui NA, Marom G, Kim JK. Dispersion and functionalization of carbon nanotubes for polymer-based nanocomposites: a review. Compos A Appl Sci Manuf 2010;41 (10):1345–1367.

[24] Herrero Latorre C, Álvarez Méndez J, Barciela García J, García Martín S, Peña Crecente RM. Carbon nanotubes as solid-phase extraction sorbents prior to atomic spectrometric determination of metal species: a review. Anal Chim Acta 2012;749:16–35.

[25] Sitko R, Zawisza B, Malicka E. Modification of carbon nanotubes for preconcentration, separation and determination of trace-metal ions. Trends Analyt Chem 2012;37:22–31.

[26] Hu CY, Xu YJ, Duo SW, Zhang RF, Li MS. Non-covalent functionalization of carbon nanotubes with surfactants and polymers. J Chin Chem Soc 2009;56 (2):234–239.

[27] Britz DA, Khlobystov AN. Noncovalent interactions of molecules with single walled carbon nanotubes. Chem Soc Rev 2006;35 (7):637–659.

[28] Hu C, Liao H, Li F, Xiang J, Li W, Duo S, Li M. Noncovalent functionalization of multi-walled carbon nanotubes with siloxane polyether copolymer. Mater Lett 2008;62 (17–18):2585–2588.

[29] Priya BR, Byrne HJ. Investigation of sodium dodecyl benzene sulfonate assisted dispersion and debundling of single-wall carbon nanotubes. J Phys Chem C 2008;112 (2):332–337.

[30] O'Connell MJ, Boul P, Ericson LM, Huffman C, Wang Y, Haroz E, Kuper C, Tour J, Ausman KD, Smalley RE. Reversible water-solubilization of single-walled carbon nanotubes by polymer wrapping. Chem Phys Lett 2001;342 (3–4):265–271.

[31] Rao GP, Lu C, Su F. Sorption of divalent metal ions from aqueous solution by carbon nanotubes: a review. Sep Purif Technol 2007;58 (1):224–231.

[32] Li YH, Di Z, Ding J, Wu D, Luan Z, Zhu Y. Adsorption thermodynamic, kinetic and desorption studies of Pb^{2+} on carbon nanotubes. Water Res 2005;39 (4):605–609.

[33] Wang HJ, Zhou AL, Peng F, Yu H, Chen LF. Adsorption characteristic of acidified carbon nanotubes for heavy metal Pb(II) in aqueous solution. Mater Sci Eng A 2007;466 (1–2):201–206.

[34] Xu D, Tan X, Chen C, Wang X. Removal of Pb(II) from aqueous solution by oxidized multiwalled carbon nanotubes. J Hazard Mater 2008;154 (1–3):407–416.

[35] Lu C, Chiu H. Adsorption of zinc(II) from water with purified carbon nanotubes. Chem Eng Sci 2006;61 (4):1138–1145.

[36] Lu C, Chiu H, Liu C. Removal of zinc(II) from aqueous solution by purified carbon nanotubes: kinetics and equilibrium studies. Ind Eng Chem Res 2006;45 (8):2850–2855.

[37] Lu C, Chiu H. Chemical modification of multiwalled carbon nanotubes for sorption of Zn^{2+} from aqueous solution. Chem Eng J 2008;139 (3):462–468.

[38] Chen C, Wang X. Adsorption of Ni(II) from aqueous solution using Oxidized multiwall carbon nanotubes. Ind Eng Chem Res 2006;45 (26):9144–9149.

[39] Lu C, Liu C. Removal of nickel(II) from aqueous solution by carbon nanotubes. J Chem Technol Biotechnol 2006;81 (12):1932–1940.

[40] Lu C, Liu C, Su F. Sorption kinetics, thermodynamics and competition of Ni^{2+} from aqueous solutions onto surface oxidized carbon nanotubes. Desalination 2009;249 (1):18–23.

[41] Yang S, Li J, Shao D, Hu J, Wang X. Adsorption of Ni(II) on oxidized multi-walled carbon nanotubes: effect of contact time, pH, foreign ions and PAA. J Hazard Mater 2009;166 (1):109–116.

[42] Li YH, Wang S, Luan Z, Ding J, Xu C, Wu D. Adsorption of cadmium(II) from aqueous solution by surface oxidized carbon nanotubes. Carbon 2003;41 (5):1057–1062.

[43] Wu CH. Studies of the equilibrium and thermodynamics of the adsorption of Cu^{2+} onto as-produced and modified carbon nanotubes. J Colloid Interface Sci 2007;311 (2):338–346.

[44] Di ZC, Ding J, Peng XJ, Li YH, Luan ZK, Liang J. Chromium adsorption by aligned carbon nanotubes supported ceria nanoparticles. Chemosphere 2006;62 (5):861–865.

[45] Hsieh SH, Horng JJ, Tsai CK. Growth of carbon nanotube on micro-sized Al_2O_3 particle and its application to adsorption of metal ions. J Mater Res 2006;21 (5):1269–1273.

[46] Wang SG, Gong WX, Liu XW, Yao YW, Gao BY, Yue QY. Removal of lead(II) from aqueous solution by adsorption onto manganese oxide-coated carbon nanotubes. Sep Purif Technol 2007;58 (1):17–23.

[47] Salam MA, Makki MSI, Abdelaal MYA. Preparation and characterization of multi-walled carbon nanotubes/chitosan nanocomposite and its application for the removal of heavy metals from aqueous solution. J Alloys Compd 2011;509 (5):2582–2587.

[48] Shao D, Ren X, Hu J, Chen Y, Wang X. Preconcentration of Pb^{2+} from aqueous solution using poly(acrylamide) and poly(N,N-dimethylacrylamide) grafted multiwalled carbon nanotubes. Colloids Surf A Physicochem Eng Asp 2010;360(1–3):74–84.

[49] Cui Y, Liu S, Hu ZJ, Liu XH, Gao HW. Solid-phase extraction of lead(II) ions using multiwalled carbon nanotubes grafted with tris(2-aminoethyl)amine. Microchim Acta 2011;174 (1):107–113.

[50] Li R, Chang X, Li Z, Zang Z, Hu Z, Li D, Tu Z. Multiwalled carbon nanotubes modified with 2-aminobenzothiazole modified for uniquely selective solid-phase extraction and determination of Pb(II) ion in water samples. Microchim Acta 2011;172 (3–4):269–276.

[51] Wang J, Ma X, Fang G, Pan M, Ye X, Wang S. Preparation of iminodiacetic acid functionalized multi-walled carbon nanotubes and its application as sorbent for separation and preconcentration of heavy metal ions. J Hazard Mater 2011;186 (2-3):1985–1992.

[52] Zang Z, Hu Z, Li Z, He Q, Chang X. Synthesis, characterization and application of ethylenediamine-modified multiwalled carbon nanotubes for selective solid-phase extraction and preconcentration of metal ions. J Hazard Mater 2009;172 (2-3):958–963.

[53] Lu C, Liu C, Rao GP. Comparisons of sorbent cost for the removal of Ni^{2+} from aqueous solution by carbon nanotubes and granular activated carbon. J Hazard Mater 2008;151 (1):239–246.

[54] Ren X, Shao D, Zhao G, Sheng G, Hu J, Yang S, Wang X. Plasma induced multiwalled carbon nanotube grafted with 2-vinylpyridine for preconcentration of Pb(II) from aqueous solutions. Plasma Process Polym 2011;8 (7):589–598.

[55] Ren X, Chen C, Nagatsu M, Wang X. Carbon nanotubes as adsorbents in environmental pollution management: a review. Chem Eng J 2011;170 (2–3):395–410.

[56] Wang H, Zhou A, Peng F, Yu H, Yang J. Mechanism study on adsorption of acidified multiwalled carbon nanotubes to Pb(II). J Colloid Interface Sci 2007;316 (2):277–283.

[57] Li YH, Ding J, Luan Z, Di Z, Zhu Y, Xu C, Wu D, Wei B. Competitive adsorption of Pb^{2+}, Cu^{2+} and Cd^{2+} ions from aqueous solutions by multiwalled carbon nanotubes. Carbon 2003;41 (14):2787–2792.

[58] Weng C-H, Huang CP. Adsorption characteristics of Zn(II) from dilute aqueous solution by fly ash. Colloids Surf A Physicochem Eng Asp 2004;247 (1–3):137–143.

[59] Gao Z, Bandosz TJ, Zhao Z, Han M, Qiu J. Investigation of factors affecting adsorption of transition metals on oxidized carbon nanotubes. J Hazard Mater 2009;167 (1–3):357–365.

[60] Pillay K, Cukrowska EM, Coville NJ. Multi-walled carbon nanotubes as adsorbents for the removal of parts per billion levels of hexavalent chromium from aqueous solution. J Hazard Mater 2009;166 (2–3):1067–1075.

[61] Stafiej A, Pyrzynska K. Adsorption of heavy metal ions with carbon nanotubes. Sep Purif Technol 2007;58 (1):49–52.

[62] Afzali D, Mostafavi A. Potential of modified multiwalled carbon nanotubes with 1-(2-pyridylazo)-2-naphtol as a new solid sorbent for the preconcentration of trace amounts of cobalt(II) ion. Anal Sci 2008;24 (9):1135–1139.

[63] Afzali D, Mostafavi A, Etemadi F, Ghazizadeh A. Application of modified multiwalled carbon nanotubes as solid sorbent for separation and preconcentration of trace amounts of manganese ions. Arab J Chem 2012;5 (2):187–191.

[64] Tajik S, Taher MA. A new sorbent of modified MWCNTs for column preconcentration of ultra trace amounts of zinc in biological and water samples. Desalination 2011;278 (1–3):57–64.

[65] Vellaichamy S, Palanivelu K. Preconcentration and separation of copper, nickel and zinc in aqueous samples by flame atomic absorption spectrometry after column solid-phase extraction onto MWCNTs impregnated with D2EHPA-TOPO mixture. J Hazard Mater 2011;185 (2–3):1131–1139.

[66] Srivastava A, Srivastava ON, Talapatra S, Vajtai R, Ajayan PM. Carbon nanotube filters. Nat Mater 2004;3 (9):610–614.

[67] Deng S, Upadhyayula VKK, Smith GB, Mitchell MC. Adsorption equilibrium and kinetics of microorganisms on single-wall carbon nanotubes. IEEE Sens J 2008;8 (6):954–962.

[68] Moon HM, Kim JW. Carbon nanotube clusters as universal bacterial adsorbents and magnetic separation agents. Biotechnol Prog 2010;26 (1):179–185.

[69] Upadhyayula VKK, Deng S, Mitchell MC, Smith GB, Nair VK, Ghoshroy S. Adsorption kinetics of *Escherichia coli* and *Staphylococcus aureus* on single-walled carbon nanotube aggregates. Water Sci Technol 2008;58:179–184.

[70] Upadhyayula VKK, Deng S, Smith GB, Mitchell MC. Adsorption of *Bacillus subtilis* on single-walled carbon nanotube aggregates, activated carbon and NanoCeram™. Water Res 2009;43 (1):148–156.

[71] Mozes N, Marchal F, Hermesse MP, Van Haecht JL, Reuliaux L, Leonard AJ, Rouxhet PG. Immobilization of microorganisms by adhesion: interplay of electrostatic and nonelectrostatic interaction. Biotechnol Bioeng 1987;30 (3):439–450.

[72] Stenström TA. Bacterial hydrophobicity, an overall parameter for the measurement of adhesion potential to soil particles. Appl Environ Microbiol 1989;55 (1):142–147.

[73] Van Loosdrecht MC, Lyklema J, Norde W, Schraa G, Zehnder AJ. The role of bacterial cell wall hydrophobicity in adhesion. Appl Environ Microbiol 1987;53 (8):1893–1897.

[74] Ong YT, Ahmad AL, Zein SHS, Tan SH. A review on carbon nanotubes in an environmental protection and green engineering perspective. Braz J Chem Eng 2010;27 (2):227–242.

[75] Kang S, Mauter MS, Elimelech M. Microbial cytotoxicity of carbon-based nanomaterials: implications for river water and wastewater effluent. Environ Sci Technol 2009;43 (7):2648–2653.

[76] Kang S, Pinault M, Pfefferle LD, Elimelech M. Single-walled carbon nanotubes exhibit strong antimicrobial activity. Langmuir 2007;23 (17):8670–8673.

[77] Arias LR, Yang L. Inactivation of bacterial pathogens by carbon nanotubes in suspensions. Langmuir 2009;25 (5):3003–3012.

[78] Kang S, Herzberg M, Rodrigues DF, Elimelech M. Antibacterial effects of carbon nanotubes: size does matter!. Langmuir 2008;24 (13):6409–6413.

[79] Yuan W, Jiang G, Che J, Qi X, Xu R, Chang MW, Chen Y, Lim SY, Dai J, Chan-Park MB. Deposition of silver nanoparticles on multi-walled carbon nanotubes grafted with hyperbranched poly(amidoamine) and their antimicrobial effects. J Phys Chem C 2008;112 (48):18754–18759.

[80] Zardini HZ, Amiri A, Shanbedi M, Maghrebi M, Baniadam M. Enhanced antibacterial activity of amino acids-functionalized multi walled carbon nanotubes by a simple method. Colloids Surf B Biointerfaces 2012;92:196–202.

[81] Qi X, Gunawan P, Xu R, Chang MW. Cefalexin-immobilized multi-walled carbon nanotubes show strong antimicrobial and anti-adhesion properties. Chem Eng Sci 2012;84:552–556.

[82] Ahadian S, Kawazoe Y. An artificial intelligence approach for modeling and prediction of water diffusion inside a carbon nanotube. Nanoscale Res Lett 2009;4 (9):1054–1058.

[83] Pendergast MM, Hoek EMV. A review of water treatment membrane nanotechnologies. Energy Environ Sci 2011;4 (6):1946–1971.

[84] Noy A, Park HG, Fornasiero F, Holt JK, Grigoropoulos CP, Bakajin O. Nanofluidics in carbon nanotubes. Nano Today 2007;2 (6):22–29.

[85] Hummer G, Rasaiah JC, Noworyta JP. Water conduction through the hydrophobic channel of a carbon nanotube. Nature 2001;414 (6860):188–190.

[86] Kolesnikov AI, Zanotti JM, Loong CK, Thiyagarajan P, Moravsky AP, Loutfy RO, Burnham CJ. Anomalously soft dynamics of water in a nanotube: a revelation of nanoscale confinement. Phys Rev Lett 2004;93 (3):035503.

[87] Celik E, Park H, Choi H. Carbon nanotube blended polyethersulfone membranes for fouling control in water treatment. Water Res 2011;45 (1):274–282.

[88] Qiu S, Wu L, Pan X, Zhang L, Chen H, Gao C. Preparation and properties of functionalized carbon nanotube/PSF blend ultrafiltration membranes. J Membr Sci 2009;342 (1–2):165–172.

[89] Wu H, Tang B, Wu P. Novel ultrafiltration membranes prepared from a multi-walled carbon nanotubes/polymer composite. J Membr Sci 2010;362 (1–2):374–383.

[90] Roy S, Jain V, Bajpai R, Ghosh P, Pente AS, Singh BP, Misra DS. Formation of carbon nanotube bucky paper and feasibility study for filtration at the nano and molecular scale. J Phys Chem C 2012;116 (35):19025–19031.

[91] Roy S, Ntim SA, Mitra S, Sirkar KK. Facile fabrication of superior nanofiltration membranes from interfacially polymerized CNT-polymer composites. J Membr Sci 2011;375 (1–2):81–87.

[92] Vatanpour V, Madaeni SS, Moradian R, Zinadini S, Astinchap B. Fabrication and characterization of novel antifouling nanofiltration membrane prepared from oxidized multiwalled carbon nanotube/polyethersulfone nanocomposite. J Membr Sci 2011;375 (1–2):284–294.

[93] Vatanpour V, Madaeni SS, Moradian R, Zinadini S, Astinchap B. Novel antibifouling nanofiltration polyethersulfone membrane fabricated from embedding TiO2 coated multiwalled carbon nanotubes. Sep Purif Technol 2012;90:69–82.

[94] Shawky HA, Chae SR, Lin S, Wiesner MR. Synthesis and characterization of a carbon nanotube/polymer nanocomposite membrane for water treatment. Desalination 2011;272 (1–3):46–50.

[95] Park J, Choi W, Kim SH, Chun BH, Bang J, Lee KB. Enhancement of chlorine resistance in carbon nanotube-based nanocomposite reverse osmosis membranes. Desalin Water Treat 2010;15 (1–3):198–204.

[96] Zhang L, Shia GZ, Qiua S, Cheng LH, Chen HL. Preparation of high-flux thin film nanocomposite reverse osmosis membranes by incorporating functionalized multi-walled carbon nanotubes. Desalin Water Treat 2011;34 (1–3):19–24.

[97] Dumée L, Germain V, Sears K, Schütz J, Finn N, Duke M, Cerneaux S, Cornu D, Gray S. Enhanced durability and hydrophobicity of carbon nanotube bucky paper membranes in membrane distillation. J Membr Sci 2011;376 (1–2):241–246.

[98] Dumée L, Sears K, Schütz J, Finn N, Duke M, Gray S. Carbon nanotube based composite membranes for water desalination by membrane distillation. Desalin Water Treat 2010;17 (1–3):72–79.

[99] Dumée LF, Sears K, Schütz J, Finn N, Huynh C, Hawkins S, Duke M, Gray S. Characterization and evaluation of carbon nanotube Bucky-Paper membranes for direct contact membrane distillation. J Membr Sci 2010;351 (1–2):36–43.

[100] Gethard K, Sae-Khow O, Mitra S. Water desalination using carbon-nanotube-enhanced membrane distillation. ACS Appl Mater Interfaces 2010;3 (2):110–114.

[101] Gethard K, Sae-Khow O, Mitra S. Carbon nanotube enhanced membrane distillation for simultaneous generation of pure water and concentrating pharmaceutical waste. Sep Purif Technol 2012;90:239–245.

[102] Ismail AF, Goh PS, Sanip SM, Aziz M. Transport and separation properties of carbon nanotube-mixed matrix membrane. Sep Purif Technol 2009;70 (1):12–26.

[103] Ong YT, Ahmad AL, Zein SHS, Sudesh K, Tan SH. Poly(3-hydroxybutyrate)-functionalised multi-walled carbon nanotubes/chitosan green nanocomposite membranes and their application in pervaporation. Sep Purif Technol 2011;76 (3):419–427.

[104] Park J, Choi W, Cho J, Chun BH, Kim SH, Lee KB, Bang J. Carbon nanotube-based nanocomposite desalination membranes from layer-by-layer assembly. Desalin Water Treat 2010;15 (1–3):76–83.

[105] Choi JH, Jegal J, Kim WN. Fabrication and characterization of multi-walled carbon nanotubes/polymer blend membranes. J Membr Sci 2006;284 (1–2):406–415.

[106] Kim YA, Muramatsu H, Hayashi T, Endo M, Terrones M, Dresselhaus MS. Fabrication of high-purity, double-walled carbon nanotube buckypaper. Chem Vap Deposition 2006;12 (6):327–330.

[107] Gou J, Liang Z, Wang B. Experimental design and optimization of dispersion process for single-walled carbon nanotube bucky paper. Int J Nanosci 2004;3 (3):293–307.

[108] Ogino SI, Sato Y, Yamamoto G, Sasamori K, Kimura H, Hashida T, Motomiya K, Jeyadevan B, Tohji K. Relation of the number of cross-links and mechanical properties of multi-walled carbon nanotube films formed by a dehydration condensation reaction. J Phys Chem B 2006;110 (46):23159–23163.

[109] Boge J, Sweetman LJ, In Het Panhuis M, Ralph SF. The effect of preparation conditions and biopolymer dispersants on the properties of SWNT buckypapers. J Mater Chem 2009;19 (48):9131–9140.

[110] Chapartegui M, Barcena J, Irastorza X, Elizetxea C, Fernandez M, Santamaria A. Analysis of the conditions to manufacture a MWCNT buckypaper/benzoxazine nanocomposite. Compos Sci Technol 2012;72 (4):489–497.

[111] Whitten PG, Spinks GM, Wallace GG. Mechanical properties of carbon nanotube paper in ionic liquid and aqueous electrolytes. Carbon 2005;43 (9):1891–1896.

[112] Sears K, Dumee L, Schuetz J, She M, Huynh C, Hawkins S, Duke M, Gray S. Recent developments in carbon nanotube membranes for water purification and gas separation. Materials 2010;3 (1):127–149.

[113] Kukovecz A, Smajda R, Kónya Z, Kiricsi I. Controlling the pore diameter distribution of multi-wall carbon nanotube buckypapers. Carbon 2007;45 (8):1696–1698.

[114] Smajda R, Kukovecz A, Kónya Z, Kiricsi I. Structure and gas permeability of multi-wall carbon nanotube buckypapers. Carbon 2007;45 (6):1176–1184.

[115] Al-Obaidani S, Curcio E, Macedonio F, Di Profio G, Al-Hinai H, Drioli E. Potential of membrane distillation in seawater desalination: thermal efficiency, sensitivity study and cost estimation. J Membr Sci 2008;323 (1):85–98.

[116] Dumée L, Campbell JL, Sears K, Schütz J, Finn N, Duke M, Gray S. The impact of hydrophobic coating on the performance of carbon nanotube bucky-paper membranes in membrane distillation. Desalination 2011;283:64–67.

[117] Li H, Gui X, Zhang L, Wang S, Ji C, Wei J, Wang K, Zhu H, Wu D, Cao A. Carbon nanotube sponge filters for trapping nanoparticles and dye molecules from water. Chem Commun 2010;46 (42):7966–7968.

[118] Mostafavi ST, Mehrnia MR, Rashidi AM. Preparation of nanofilter from carbon nanotubes for application in virus removal from water. Desalination 2009;238 (1–3):271–280.

[119] López-Lorente AI, Simonet BM, Valcárcel M. The potential of carbon nanotube membranes for analytical separations. Anal Chem 2010;82 (13):5399–5407.

[120] Hinds BJ, Chopra N, Rantell T, Andrews R, Gavalas V, Bachas LG. Aligned multiwalled carbon nanotube membranes. Science 2004;303 (5654):62–65.

[121] Holt JK, Park HG, Wang Y, Stadermann M, Artyukhin AB, Grigoropoulos CP, Noy A, Bakajin O. Fast mass transport through sub-2-nanometer carbon nanotubes. Science 2006;312 (5776):1034–1037.

[122] Bakajin O, Noy A, Fornasiero F, Grigoropoulos CP; Holt JK, In JB; Kim S, Park HG. Nanofluidic carbon nanotube membranes: applications for water purification and desalination. In: Nora S, Mamadou D, Jeremiah D, Anita S, Richard S, editors. *Nanotechnology Applications for Clean Water*. Boston: William Andrew Publishing; 2009. p 77–93.

[123] Corry B. Designing carbon nanotube membranes for efficient water desalination. J Phys Chem B 2007;112 (5):1427–1434.

[124] Corry B. Water and ion transport through functionalised carbon nanotubes: implications for desalination technology. Energy Environ Sci 2011;4 (3):751–759.

[125] Fornasiero F, Hyung GP, Holt JK, Stadermann M, Grigoropoulos CP, Noy A, Bakajin O. Ion exclusion by sub-2-nm carbon nanotube pores. Proc Natl Acad Sci U S A 2008;105 (45):17250–17255.

[126] Murphy R. *Chapter 3—Life Cycle Assessment*. Cambridge: Woodhead Publishing Limited; 2004.

[127] Khanna V, Bakshi BR, Lee LJ. Carbon nanofiber production: life cycle energy consumption and environmental impact. J Ind Ecol 2008;12 (3):394–410.

[128] Plata DL, Hart AJ, Reddy CM, Gschwend PM. Early evaluation of potential environmental impacts of carbon nanotube synthesis by chemical vapor deposition. Environ Sci Technol 2009;43 (21):8367–8373.

[129] Upadhyayula VKK, Meyer DE, Curran MA, Gonzalez MA. Life cycle assessment as a tool to enhance the environmental performance of carbon nanotube products: a review. J Clean Prod 2012;26:37–47.

PART III

PHOTOCATALYTICAL REMEDIATION

9

NEW ADVANCES IN HETEROGENEOUS PHOTOCATALYSIS FOR TREATMENT OF TOXIC METALS AND ARSENIC

Marta I. Litter[1,2,3] and Natalia Quici[1,2]

[1]Gerencia Química, Comisión Nacional de Energía Atómica, Buenos Aires, Argentina

[2]Consejo Nacional de Investigaciones Científicas y Técnicas, Buenos Aires, Argentina

[3]Instituto de Investigación e Ingeniería Ambiental, Universidad de Gral. San Martín, Buenos Aires, Argentina

9.1 INTRODUCTION

Conventional methods for water treatment such as biological methods, ozonation, and activated carbon are not completely suitable to reduce the concentration of certain pollutants to regulation levels, especially in the case of resilient compounds; furthermore, even when successful, these processes can be slow and expensive [1, 2]. A relatively new alternative to traditional water treatments are the so-called advanced oxidation or reductive processes (AOPs/ARPs), which appear to be attractive options for the removal of difficult contaminants. These processes are based on the formation of very active oxidative and reductive species, mainly the hydroxyl radical, HO• [3–9]. The most investigated technologies from the fundamental point of view and their application in real systems have been AOPs. However, some pollutants can be transformed only via a reductive pathway, for which ARPs can be good choices [3–11].

One of the most extensively investigated AOPs/AORs is heterogeneous photocatalysis (HP), where a broadband semiconductor (SC) is irradiated by ultraviolet (UV) or visible light leading to oxidation–reduction reactions and to the transformation of pollutants present in aqueous or gaseous systems. HP leads to the eventual mineralization of organic contaminants and to the transformation of metallic ions and metalloids to less toxic species or species easier to be separated from the system in a subsequent treatment step [4, 8, 12].

The main aspect that will be covered in this chapter is the utilization of reductive and oxidative HP for the treatment of heavy metals (hexavalent chromium, uranyl, mercury (II) and lead (II)) and arsenic (As(III)/(V)), all of them known for their high toxicity and environmental damage.

9.2 BASIC PRINCIPLES OF HP

In HP, the SC is excited in a primary process by absorption of photons of energy equal to or higher than the SC band gap (E_g), leading to the promotion of an electron from the valence band (VB) to the conduction band (CB), with the creation of a hole:

$$SC + h\nu \rightarrow e^-_{CB} + h^+_{VB} \tag{9.1}$$

Nanomaterials for Environmental Protection, First Edition. Edited by Boris I. Kharisov, Oxana V. Kharissova, and H. V. Rasika Dias.

The SC can de-energize in different ways. A very likely but unfavorable consequence is the electron–hole recombination, a detrimental unproductive process. Another possibility is electron transfer reactions between electrons or holes and organic or inorganic species adsorbed on the SC surface or close to it (and not necessarily adsorbed). The SC can donate electrons, reducing electron acceptors (A), or a hole can receive an electron from a donor species (D), which suffers oxidation. The probability of these transfer processes depends on the relative redox positions of the CB and VB in the SC in relation to the redox potential of the adsorbed species and of the rate of competitive processes.

TiO_2 is by far the most useful SC material for photocatalytic purposes because of its exceptional optical and electronic properties, chemical stability, nontoxicity, and low cost. The energy band gaps of the most used photocatalytic forms of TiO_2, anatase and rutile, are 3.23 eV (corresponding to 384 nm) and 3.02 eV (411 nm), respectively [9]. The most popular commercial TiO_2 material for HP was originally produced by the German company Degussa (now Evonik) with the denomination P-25. Some other commercial products, such as Hombikat UV100, and different products of Cristal Global (PC50, PC100, PC500), Fluka, and so on, have been tested in several photocatalytic systems [13–15]. Although the data are divergent in the literature, the redox levels of h^+_{VB} and e^-_{CB} photogenerated by HP in P-25 that are taken into consideration in this chapter are +2.9 and −0.3 V,[1] respectively [16].

It has been proposed that in HP carried out under normal illumination (e.g., black light or xenon lamps), the redox processes take place through monoelectronic step reactions, due to the low frequency of photon absorption. In the case of acceptors:

$$A + e^-_{CB} \rightarrow A^- \tag{9.2}$$

The holes can directly attack the D species adsorbed on TiO_2, or adsorbed water molecules/surface hydroxyl groups, generating $HO^\bullet$. These radicals are highly oxidizing species with a reduction potential of +2.8 V [16]:

$$h^+_{VB} + D_{ads} \rightarrow D^{\bullet +}_{ads} \tag{9.3}$$

$$h^+_{VB} + \left(HO^-_{surf}\right) H_2O_{ads} \rightarrow HO^\bullet \ (+H^+) \tag{9.4}$$

The flatband position of a SC[2] follows Nernstian law and is reduced in 59 mV per unit of pH [17]; thus, the driving force of the heterogeneous redox reactions can be varied by controlling the pH.

HP oxidative reactions are usually performed in the presence of molecular oxygen, with the objective of enhancing the mineralization of organic compounds; e^-_{CB} reduce O_2 with hydroperoxyl or superoxide radical $\left(HO_2^\bullet / O_2^{\bullet -}\right)$ formation, decreasing the probability of electron–hole recombination. The value for the reduction potential for the hydroperoxyl/superoxide radical has been reported as $E^0\left(O_2/HO_2^\bullet\right) = -0.05\,V$ [18]:

$$O_2 + e^-_{CB} (+H^+) \rightarrow O_2^{\bullet -}\left(HO_2^\bullet\right) \tag{9.5}$$

$$2HO_2^\bullet \rightarrow H_2O_2 + O_2 \tag{9.6}$$

$$H_2O_2 + O_2^{\bullet -} \rightarrow HO^\bullet + HO^- + O_2 \tag{9.7}$$

$$H_2O_2 + e^-_{CB} \rightarrow HO^\bullet + HO^- \tag{9.8}$$

This indicates that reduction of oxygen by e^-_{CB} would be thermodynamically feasible at low pH. At higher pH values, a low driving force would make this reaction more difficult.

When a metal ion is present in a HP aqueous system solution, it can undergo reduction or oxidation reactions. Figure 9.1 shows the reduction potential of different couples together with the energy levels for e^-_{CB} and h^+_{VB}; the comparison of the redox levels leads to a first approximation of the thermodynamic feasibility of transformation of the species. All couples with redox potentials more positive than the level of e^-_{CB} can be directly reduced and all couples with redox potentials less positive than the level of h^+_{VB} can be, in principle, oxidized.

[1] All reduction potentials in this work are standard values versus NHE; therefore, the values correspond to pH 0 unless a different condition is specified.
[2] Potential at which no excess charge exists in the SC and there is no electric field and no space charge region so that the (conduction and valence) bands are not bent.

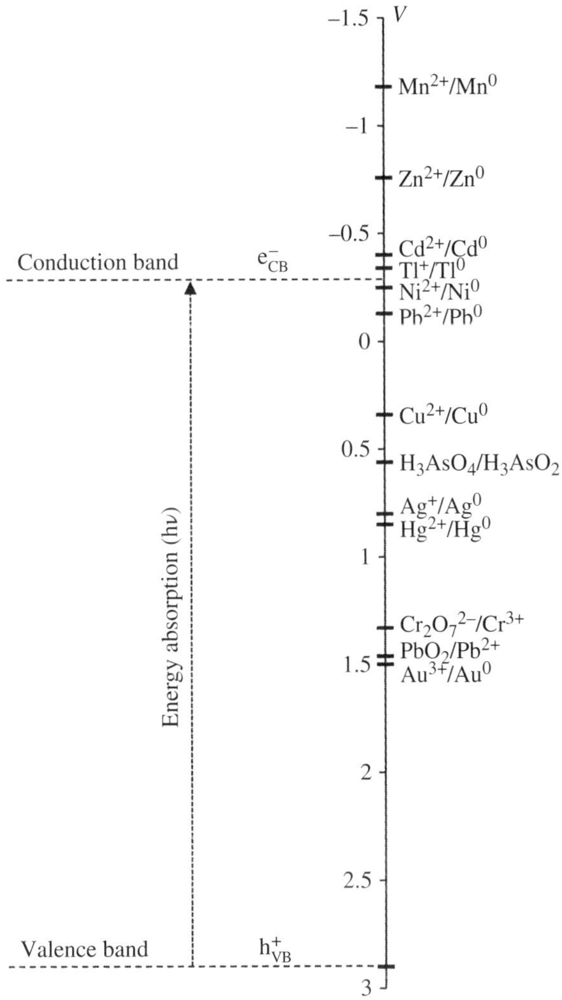

FIGURE 9.1 Position of the reduction potentials of various metallic couples related to the energy levels of the CB and VA of P-25. Adapted from Ref. [10].

Figure 9.1 shows global (multielectronic) reactions, and these can be favorable (exergonic) in principle. However, if one-electron steps are considered, only thermodynamically allowed reactions would occur. Accordingly, the ion could be reduced by e^-_{CB} in a direct reduction step, a reaction that requires the e^-_{CB} reduction potential to be more negative than the one corresponding to the $M^{n+}/M^{(n-1)+}$ pair:

$$M^{n+} + e^-_{CB} \rightarrow M^{(n-1)+} \tag{9.9}$$

Alternatively, oxidation of the metal ion can occur by reaction with holes or $HO^{\bullet}$, reaching a higher oxidation state:

$$M^{n+} + h^+_{VB}/HO^{\bullet} \rightarrow M^{(n+1)+} \tag{9.10}$$

Some metallic species (such as Cr(VI), Hg(II), or U(VI)) cannot be transformed to a higher oxidation state, but they can be directly reduced by e^-_{CB} [11]. This process can be accelerated and even produced in thermodynamically not feasible conditions, that is, when the redox potential of the couple to be reduced is more negative than the level of e^-_{CB}, by the addition of sacrificial donors to the solution. This donor-mediated reduction of the metal or metalloid constitutes an indirect reduction process. The donors can be categorized into two different groups: direct h^+_{VB} acceptors (as in Eq. 9.3), for example, ethylenediaminetetraacetic acid (EDTA), citric acid, salicylic acid, or indirect acceptors via HO• reaction, such as methanol, acetic acid, or formaldehyde (Eq. 9.11).

$$ROH + HO^{\bullet} \rightarrow {}^{\bullet}ROH \quad (9.11)$$

$$M^{z+} \xrightarrow{\bullet ROH} \xrightarrow{\bullet ROH} M^{0} \quad (9.12)$$

The reaction with h^{+}_{VB} or HO• reduces the electron–hole recombination and generates very strong reducing radicals [7, 10, 11], which allow the reduction of species that are not likely to be reduced directly by e^{-}_{CB}, as is the case for Pb(II) and As(V) [10, 19–21]. A synergetic effect generally exists between oxidative and reductive processes [7, 10], especially in the presence of organic compounds that can undergo oxidation.

Photocatalytic-based treatments can take advantage of these transformations in order to obtain the following:

1. A metallic solid form deposited on the photocatalyst surface; in this way, the species is removed from the solution. This procedure can also be applied for metal recovery (copper, platinum, silver, gold).
2. A less toxic soluble species (Cr(VI) to Cr(III), As(III) to As(V)), also easier to be removed by an ulterior treatment.

9.3 ARSENIC

Arsenic contamination in water can be anthropogenic (mining, use of biocides, wood preservers), but it mainly comes from natural sources due to dissolution of minerals in surface water or groundwater, or to volcanic processes [22, 23]. Chronic ingestion of arsenic for prolonged periods of time results in arsenicosis, a hydric disease, which causes severe skin lesions including pigmentation changes, palmoplantar keratosis, and other syndromes, ending generally in cancer [24]. Arsenic in drinking water constitutes a serious problem, affecting several million people all over the world. The World Health Organization (WHO) [24] recommends 10 $\mu g\,l^{-1}$ as the maximum allowable As concentration in drinking water, a value that most national regulatory agencies accept as the norm.

Arsenic in water can be found in its tri- and pentavalent oxidation states; the predominant As(III) form is arsenite (H_3AsO_3), while As(V) frequently occurs as arsenates ($H_2AsO_4^-$ or $HAsO_4^{2-}$). Arsenite removal from water is difficult; it is more mobile and with a toxicity 20 times higher than that of As(V).

Conventional water treatments used to remove As are oxidation/coagulation/adsorption processes on iron or aluminum hydroxides, ionic interchange, activated alumina, lime softening, and reverse electrodialysis and osmosis [25, 26]. Transformation to As(V) makes the application of conventional technologies such as ion exchange and adsorption easier. However, new emerging techniques should be investigated for low-cost solutions to the arsenic problem, especially for low-income populations, as mentioned in some references [22, 27, 28].

Photocatalysis of arsenic systems was described in our previous review [10] where examples of oxidative TiO_2 HP [29–34] and mechanisms for As(III) oxidation [30, 35] were reported in studies that spanned concentrations from the micro- to the millimolar range, showing in every case very fast oxidation in 10–100 min. On the other hand, reduction of As(V) to As(III) or As(0) is thermodynamically favored even with mild reductants, as can be inferred from the Latimer diagram for As (Fig. 9.2). Further reduction to arsine (As(−3)) is more difficult. All the reductive steps are less favored at higher pH values.

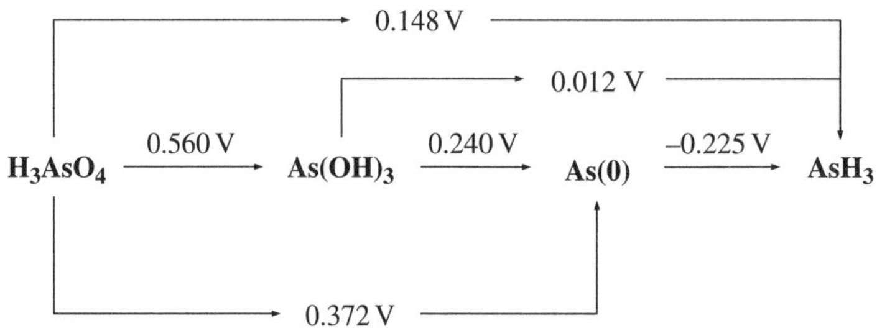

FIGURE 9.2 Latimer diagram connecting the different As species [36].

Oxidative TiO_2 HP has been shown to be a good alternative for As(III) oxidation. In an effort to explain the mechanism of oxidation, several authors have suggested monoelectronic steps with formation of As(IV) [30, 33, 35] that could involve the action of $HO^\bullet$ (Eq. 9.13), reaction with VB holes (Eq. 9.14), or reaction with $O_2^{\bullet-}$ (Eq. 9.15):

$$\mathrm{As(III)} + \mathrm{HO}^\bullet_{\mathrm{ads}} \left(\mathrm{or\ HO}^\bullet_{\mathrm{free}}\right) \rightarrow \mathrm{As(IV)} + \mathrm{OH}^- \tag{9.13}$$

$$\mathrm{As(III)} + h^+_{\mathrm{VB}} \rightarrow \mathrm{As(IV)} \tag{9.14}$$

$$\mathrm{As(III)} + \mathrm{HO}_2^\bullet / \mathrm{O}_2^{\bullet-} \rightarrow \mathrm{As(IV)} + \mathrm{HO}_2^- / \mathrm{O}_2^- \tag{9.15}$$

The reduction potential of the As(IV)/As(III) couple has been reported as $E^0 \approx +2.4\,\mathrm{V}$ [37]. Therefore, the formation of As(IV) by the attack on holes, $HO^\bullet$, or $HO_2^\bullet/O_2^{\bullet-}$ is thermodynamically possible. Regardless of the first oxidation step, As(IV) transforms into As(V), the stable form:

$$\mathrm{As(IV)} + \mathrm{O}_2 / \mathrm{O}_2^{\bullet-} / h^+_{\mathrm{VB}} \rightarrow \mathrm{As(V)} \tag{9.16}$$

In recent years, some examples of further work on oxidative As(III) removal using HP under UV light were reported, in addition to those described in our previous review [10], and are detailed in the following paragraphs.

The mechanism of photocatalytic oxidation of As(III) has been a matter of discussion for many years, the controversy being centered on determining the major oxidant in the system, $O_2^{\bullet-}$ [32, 33, 38, 39], $HO^\bullet$, or h^+_{VB} [30, 31, 35, 40–43]. However, whatever the oxidant, it is not possible to deny the efficiency of the HP process to transform As(III) into As(V). The real problem, however, is to remove the later dissolved As(V).

In this line, an interesting review on As HP removal has been published recently, and most of the articles therein are also reviewed in the following texts [44]. The authors focused on the application of TiO_2 and TiO_2-based materials for removing inorganic and organic arsenic, highlighting that TiO_2-based arsenic removal methods developed to date focus on the photocatalytic oxidation of arsenite to arsenate and on the adsorption of inorganic and organic arsenic (monomethylarsonic acid (MMA) and dimethylarsinic acid (DMA)) on TiO_2. They observed that improvements in the photocatalytic process should focus on the combination with adsorbents (e.g., slag–iron oxide–TiO_2, meso-TiO_2/α-Fe_2O_3 composites) to achieve higher photocatalytic abilities and adsorption capacity and also to immobilize TiO_2 to facilitate the ulterior separation of the solid from the aqueous phase. Another important point highlighted in the chapter, which is also a problem in real groundwater and wastewaters, is the influence of the coexisting solutes (silicate, phosphate, carbonate oxyanions, and humic acid (HA)) on HP of arsenic systems because of the competition of these species with As for adsorption on TiO_2. In this context, Tsimas et al. [45] studied the simultaneous HP oxidation of As(III) and HA—usually found together in groundwater—in a ternary As(III)/HA/TiO_2 system. It was shown that the efficiency of both As(III) and HA oxidation decreased in the ternary system compared with that of the corresponding binary systems (As(III)/TiO_2 and HA/TiO_2), indicating that one species inhibits the oxidation of the other. In another work [46], a hybrid system combining TiO_2 with nanoscale zero-valent iron (nZVI) in a tubular photoreactor with recirculation was used for As(III) photooxidation. The addition of nZVI significantly enhanced arsenite removal, reducing the required mass of TiO_2 by five times and achieving As values below the WHO-recommended concentrations (from $500\,\mu g\,l^{-1}$ to $<10\,\mu g\,l^{-1}$). Xu and Meng [47] studied the effect of the crystalline size (6.6 and 30.1 nm) of TiO_2 on As(III) and As(V) adsorption and photocatalytic oxidation employing single-phase anatase nanoparticles. The adsorption capacity of TiO_2 for As(III) and As(V) increased linearly with the BET surface area of the particles, and no significant difference in the rate of As(III) photooxidation was found between the nanoparticles of 6.6 and 14.8 nm diameter, but a clear decrease was obtained with the 30.1 nm particles, indicating the importance of the size of the nanoparticles to be used.

Regarding low-cost applications, several papers describe the use of very simple photocatalytic systems to remove As from water at household levels. Several works in our laboratory [22, 27, 28, 48] report results of an application to remove arsenic from natural waters in rural areas; to perform HP experiments, walls of PET plastic bottles were impregnated with TiO_2 by a very simple technique [49]. As(III) solutions ($[\mathrm{As(III)}]_0 = 1000\,\mu g\,l^{-1}$, pH 7.8) were added in these bottles, which were put in a horizontal position and irradiated for 6 h by UV light (366 nm, $800\,\mu W\,cm^{-2}$). After irradiation, nongalvanized packing iron wire pieces (usually employed in agricultural applications) were added. After 24 h settlement in the dark, As removals ranging 80–86% were obtained. The same bottle could be reused at least three times without loss of efficiency. The HP procedure was proposed to remove As (500–1800 $\mu g\,l^{-1}$, unidentified speciation) from well water samples of Las Hermanas (Santiago del Estero Province, Argentina). In this case, a $FeCl_3$ solution was added at the end of solar irradiation, and more than 94% As removal took place, attaining concentrations below the WHO limits. Fostier et al. [50] proposed a similar low-cost application for As(III) oxidation using TiO_2 immobilized on PET bottles combined with Fe(II) addition to the solution before exposure to sunlight.

Photocatalytic arsenite oxidation could also be carried out under visible irradiation. There, the mechanism is different from the one that takes place under UV light, involving charge injection from an excited dye molecule into the CB of the SC as the primary mechanism for the production of an oxidized dye radical ($Dye^{\bullet+}$) [51]:

$$\text{Dye} + h\nu\ (\text{visible}) \rightarrow \text{Dye}^* \tag{9.17}$$

$$\text{Dye}^* + \text{TiO}_2\left(e^-_{CB}\right) \rightarrow \text{Dye}^{\bullet+} + \text{TiO}_2\left(e^-_{CB}\right) \tag{9.18}$$

$$\text{Dye}^{\bullet+} + \text{As(III)} \rightarrow \text{Dye} + \text{As(IV)} \tag{9.19}$$

In the presence of oxygen, superoxide is produced through the cathodic pathway (Eqs. 9.5–9.8), which can contribute to As(III) oxidation (Eq. 9.15).

Park et al. [52] used titania nanoparticles sensitized with an organic dye (OD, (E)-3-(5-(5-(4-(bis(4-((2-(2-methoxyethoxy) ethoxy)methyl)phenyl)amino)phenyl)-thiophen-2-yl)thiophen-2-yl)-2-cyanoacrylic acid))) for the oxidation of arsenite and compared the efficiency with those of bare TiO_2 and TiO_2 sensitized with a ruthenium bipyridyl complex ($Ru^{II}(4,4\text{-bpy}(COOH)_2)_3$, RuL_3). The results under visible light ($\lambda > 420$ nm) showed better activities for OD/TiO_2 in comparison with other photocatalysts, with a constant photoactivity over a wide pH range. The superior stability of OD/TiO_2 over RuL_3/TiO_2 was related to the good chemical anchoring between the dye and the surface; in addition, this type of dye is better than Ru-containing dyes, which are expensive and toxic.

In another work, a high degree of As(III) removal under visible light illumination was obtained using palladium-modified nitrogen-doped titanium oxide (TiON/PdO) nanoparticles [53]. The strong adsorption (for both As(III) and As(V)) and photo-oxidation by TiON/PdO led to efficient As removal. The system was good enough to reach As concentrations in agreement with the WHO regulation limits at initial As(III) concentrations lower than 0.4 mg l^{-1}. The enhanced photocatalytic activity under visible light illumination was explained by a strong optoelectronic coupling between PdO and TiON.

In contrast, reductive photocatalytic systems for arsenic transformation are still scarce. Reduction of As(III) by TiO_2 e^-_{CB} is possible either in the presence or in the absence of an electron donor in the aqueous system, but, in the case of As(V), direct reduction does not seem to be feasible because the reported reduction potential of the As(V)/As(IV) couple is highly negative ($E^0 \approx -1.2$ V [37]). Therefore, the indirect pathway by using sacrificial electron donors like alcohols or carboxylic acids (e.g., methanol, formic acid) is indispensable [21]. The indirect reductive mechanism for As(V) photocatalytic reduction was first proven in anoxic conditions in the presence of methanol by Yang et al. [34]. Later, our group [21] studied the reaction starting from As(V) and also from As(III) in the presence of methanol; As(III) reaction in the absence of the organic compound was also analyzed. A mechanism based on the formation of a hydroxymethyl radical produced by hole/$HO^\bullet$ attack on methanol was proposed. This radical can donate electrons to the CB or itself be effective as an As(V) reductant, with formaldehyde generation in both cases:

$$\text{CH}_3\text{OH} + h^+_{VB}\{\text{HO}^\bullet\} \rightarrow {}^\bullet\text{CH}_2\text{OH} + \text{H}^+\{\text{H}_2\text{O}\} \quad E^0_{({}^\bullet\text{CH}_2\text{OH}/\text{CH}_3\text{OH})} = 1.45\text{V} \tag{9.20}$$

$${}^\bullet\text{CH}_2\text{OH} \rightarrow \text{CH}_2\text{O} + \text{H}^+ + e^-_{CB} \quad E^0_{({}^\bullet\text{CH}_2\text{OH}/\text{CH}_2\text{O})} \approx -0.9 \text{ to } -1.18\text{V} \tag{9.21}$$

$${}^\bullet\text{CH}_2\text{OH} + \text{As(V)} \rightarrow \text{CH}_2\text{O} + \text{As(IV)} + \text{H}^+ \tag{9.22}$$

CH_2O can be transformed to formic acid and finally mineralized to CO_2 with the generation of $CO_2^{\bullet-}$ $\left(E^0_{(CO_2/CO_2^{\bullet-})} \approx -2.0\text{V}\right)$ [19], another strong radical that can contribute to the reducing process:

$$\text{CO}_2^{\bullet-} + \text{As(V)} \rightarrow \text{CO}_2 + \text{As(IV)} \tag{9.23}$$

In a further step, As(IV) is reduced by CB or trapped electrons, or by $^\bullet CH_2OH$ or $CO_2^{\bullet-}$ to As(III). As mentioned earlier, unlike the case of As(V), Levy et al. [21] observed direct photocatalytic As(III) reduction by e^-_{CB} in the absence of MeOH; this indicates that although the value for the monoelectronic couple As(III)/As(II) is not known, it should be below the CB level:

$$\text{As(III)} + e^-_{CB} \rightarrow \text{As(II)} \tag{9.24}$$

Therefore, after As(III) formation, consecutive monoelectronic steps would lead to the formation of stable products such as As(0) and AsH_3, which were unambiguously identified by the authors in all cases through X-ray photoelectron spectroscopy (XPS) and X-ray absorption near edge structure (XANES) analyses of the solid residues formed on the TiO_2 surface. In this way, reduction to solid As(0) was achieved, which is a very convenient method for As removal from water. However, careful attention must be paid during these kinds of reactions as they can result in the formation of AsH_3, one of the most toxic forms of As. Proposed strategies for AsH_3 treatment include adsorption on suitable materials or oxidative treatment by a coupled gas-phase photocatalytic system, which our research group is currently developing.

The photoconversion of As(III) and As(V) (arsenate) over mesoporous TiO_2 electrode was investigated in a photoelectrochemical cell (PECC) [54] for a wide range of concentrations from micro- to nanomolar. It was shown that As(III) can be oxidized and As(V) can be reduced in anoxic conditions under UV irradiation using the photoelectrochemical setup. The reduction or increment of the photooxidation current depended on the As(III) concentration range. This abnormal concentration-dependent behavior was ascribed to As(IV) species, which can be either oxidized or reduced depending on the experimental conditions. Because As(V) is an electron acceptor, its addition consistently lowered the photocurrent in the entire concentration range.

Although, as mentioned earlier, inorganic As(III) and As(V) are the predominant contaminant species in water, the removal or transformation of organoarsenic species cannot be neglected. MMA and DMA are used as herbicides in agriculture and employed in industrial activities [41, 55] and are also introduced to the environment to some extent through As(V) methylation carried out by a variety of organisms [56]. Also, complex organoarsenic species (e.g., with substituted phenyl rings in their structure) are used to control intestinal parasites in poultry and swine [57, 58]. It has been shown that TiO_2 photocatalysis is effective for the oxidation of some organic arsenic species.

Xu et al. [41] studied the TiO_2 photocatalytic degradation of MMA and DMA. MMA seems to be the primary DMA degradation product to be degraded later and the arsenic moiety oxidized to As(V). Results obtained in the presence of a radical scavenger (terbutanol) indicated that the hydroxyl radical was the primary oxidant in the reaction. Further work [59] undertaken with other radical scavengers (superoxide dimutase (SOD), sodium bicarbonate, and sodium azide) also supported $HO^\bullet$ as the primary oxidant and showed that the methyl groups in MMA and DMA are transformed into formic acid and possibly methanol. Other authors [55] compared the photocatalytic oxidation on MMA and DMA by different AOPs, that is, UV/H_2O_2, UV/TiO_2, and UV/$S_2O_8^{2-}$, finding the last one the most effective.

Zheng et al. [58] studied the TiO_2 photocatalytic degradation of phenylarsonic acid (used as feed additive in the poultry industry), where $HO^\bullet$ was found to play a major role in oxidation, as was the case for DMA and MMA.

Due to the extensive, good-quality work done on photocatalytic systems for As removal and the importance of the problem of As in the environment, it is the vision of the authors that research on practical applications of these photocatalytic systems should be strengthened in the coming years in order to provide cost-effective systems.

9.4 CHROMIUM

Chromium is a metal with multiple industrial and technological applications including metallurgy, electroplating, the textile industry, leather tanning, and wood preservation. Because of its extensive use, Cr is a frequent contaminant in wastewater, mainly in the form of Cr(III) and Cr(VI). Cr(VI) presents the highest environmental threat due to its toxicity for biological organisms together with its high solubility and mobility. The WHO [60] has established the maximum contaminant level of Cr(VI) in drinking water to be $0.05\,mg\,l^{-1}$, while total Cr to be discharged should be below $0.1\,mg\,l^{-1}$ according to the U.S. Environmental Protection Agency (EPA) [61]. Cr(III) is considered to be nontoxic or of very low toxicity [62, 63], and its mobility is lower than that of Cr(VI). The conventional treatment for the removal of Cr(VI) dissolved in water involves its reduction to Cr(III) with the use of sodium thiosulfate, ferrous sulfate [64, 65], sodium metabisulfite, sulfur dioxide, or other chemicals, with subsequent economical costs and generation of residues [66].

Figure 9.3 shows the redox potentials of chromium in acid conditions taken from reference [67].

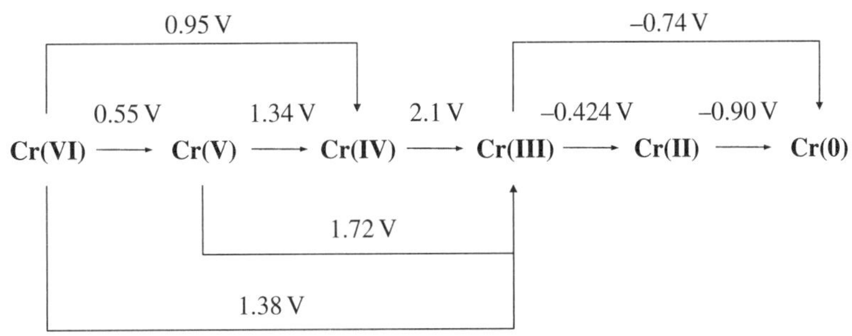

FIGURE 9.3 Latimer diagram connecting the different Cr species [67].

Cr(VI) reduction is enhanced in acid solutions, and the net reaction can be given as:

$$2Cr_2O_7^{2-} + 16H^+ \rightarrow 4Cr^{3+} + 8H_2O + 3O_2 \quad \Delta G^0_{298} = -115.8\,kJ \tag{9.25}$$

At neutral pH:

$$4Cr_2O_7^{2-} + 20H^+ \rightarrow 4Cr^{3+} + 10H_2O + 3O_2 \tag{9.26}$$

Although the process is thermodynamically feasible, $Cr_2O_7^{2-}$ is stable at room temperature in acidic solutions due to the high overpotential of the water oxidation conjugate reaction. Therefore, strategies for improving the reduction reaction should be implemented.

A considerable number of works have been published on the photocatalytic reduction of Cr(VI) employing TiO_2, Fe/TiO_2, Pt/TiO_2, and other SCs such as ZnO, CdS, ZnS, and WO_3 [68–74]. Many examples have already been described in our previous reviews [7, 10] including reactors for technological applications, microparticles used in slurries or conveniently supported, nanoparticles, and nanotubes [69, 71–112]. The major conclusions that emerge from past publications can be summarized in the following points:

1. Cr(VI) HP reduction depends upon pH. Low pH favors the net reaction (Eq. 9.19); however, neutral or alkaline conditions favor precipitation and immobilization of Cr(III) as the hydroxide, facilitating further separation.
2. Cr(VI) reduction can be accelerated by the addition of organic compounds acting as holes or $HO^{\bullet}$ scavengers.
3. Molecular oxygen does not affect Cr(VI) HP reduction, at least at acidic pH, as will be explained in detail later.

From a thermodynamic point of view, e^-_{CB} have the appropriate potential to directly reduce Cr(VI), Cr(V), and Cr(IV) (see Figs. 9.1 and 9.3), and the HP mechanism for the reduction of Cr(VI) to Cr(III) is proposed to occur in three monoelectronic steps [113–115]:

$$Cr(VI) + e^-_{CB} \rightarrow Cr(V) \tag{9.27}$$

$$Cr(V) + e^-_{CB} \rightarrow Cr(IV) \tag{9.28}$$

$$Cr(IV) + e^-_{CB} \rightarrow Cr(III) \tag{9.29}$$

This was proven by our group through electron paramagnetic resonance (EPR) spectroscopy, which allowed the identification of paramagnetic Cr(V) species [113–115].

Detrimental reoxidation of reduced Cr species by holes or hydroxyl radicals, however, is possible:

$$Cr(V/IV/III) + h^+_{VB}(HO^{\bullet}) \rightarrow Cr(V/IV/III) \tag{9.30}$$

An enhancement of the reaction takes place if suitable electron donors are present either by reducing the probability of recombination or through an indirect reduction driven by radicals formed by hole/$HO^{\bullet}$ attack (Eqs. 9.3 and 9.11), which, at high concentrations, hinder reaction (9.30) and themselves contribute to Cr species reduction.

The role of O_2 in photocatalytic Cr(VI) systems is noteworthy due to its unique behavior compared with other metal cations, because it appears to be independent of the presence of O_2, at least at acidic pH. This process was explained by a very strong association between Cr(VI) and TiO_2 through the formation of a charge transfer complex, identified by an absorption band at 380 nm [117]. Due to the fast capture of electrons by Cr(VI) as a result of the formation of this complex, no competition for O_2 by e^-_{CB} (Eq. 9.5) is present, and, consequently, no inhibition by O_2 is observed either in the absence of organic donors or in the presence of EDTA and oxalic or citric acid [113–115, 117, 118]. Experiments on platinized photocatalysts supported this independence, the same temporal evolution of Cr(VI) concentration being obtained using Pt/TiO_2 or bare TiO_2 either in the presence or in the absence of O_2 [116–120]. If O_2 had exerted any effect, Pt would have enhanced the reaction by decreasing the overpotential for water oxidation.

Cr(VI) photocatalytic reduction continues to be a very rich and interesting study system, and extensive work has been carried out in recent years, either with the aim of water treatment or as a practical system for the evaluation of the photocatalytic activity of several SC samples in the aqueous phase [121–132]. Most of the recent papers use the HP Cr(VI) reduction reaction as a model system for testing new photocatalysts; this is the case of immobilized nanotubular TiO_2 electrodes [130], WO_3-doped long TiO_2 nanotube arrays [133], and Au/TiO_2 heterojunction nanotube composites [126], among other materials. Other papers shed more light on some specific features of the Cr(VI) photocatalytic system. For example, Mu et al. [127] explored the HP Cr(VI) reduction using TiO_2 nanofibers prepared via an alkaline hydrothermal reaction and TiO_2 nanoparticles with varied structural properties. The results showed that Cr(VI) reduction is affected mostly by the specific surface area of the catalyst rather than by the crystalline-phase composition. This agrees with the hypothesis of the formation of the strong complex between Cr(VI) and TiO_2, mentioned earlier, which would be more favorable at a higher surface area of the photocatalyst. Pandikumar et al. [131] evaluated TiO_2–Au nanocomposites for UV light Cr(VI) reduction and simultaneous methylene blue oxidation; the authors observed a synergy between these two processes due to the presence of Au nanoparticles on the TiO_2 surface. In another work, TiO_2 coupled to nonfunctionalized carbon nanotubes (CNTs) was found more active than OH-functionalized CNTs because of the higher Cr(VI) adsorption on the first samples [128]. Similarly, Au/TiO_2 heterojunction composite nanotube arrays showed a higher efficiency than unmodified TiO_2 nanotube arrays for the simultaneous transformation of Cr(VI) and acid orange 7 (AO7) [126]. In another paper, Cr(VI) reduction in the presence of EDTA [124] was employed to evaluate the activity of anatase TiO_2 films deposited by cathodic arc (CA) on glass substrates and compared with the activity of P-25 films obtained by dip-coating (DC) immersion. Despite DC films showing higher photoactivity for Cr(VI) reduction, its adhesion properties are very poor compared to CA films, making this last type of films a promising material for photocatalytic applications that require immobilized catalysts.

Photocatalytic Cr(VI) reduction is also possible under visible irradiation. Kim and Choi [123] investigated the formation of surface complexes between TiO_2 and electron donors such as methanol, formic acid, acetic acid, triethanolamine, or EDTA, which can absorb visible light through a ligand-to-metal charge transfer (LMCT) mechanism; the TiO_2–EDTA complex, especially, showed an outstanding visible light activity for Cr(VI) reduction. A similar approach was presented by Wang et al. [129] using low molecular weight organic acids (SOAs) such as tartaric, citric, and lactic acids with TiO_2 under light irradiation above 420 nm; the authors support a charge transfer complex mechanism in which the photogenerated electrons are transferred from SOAs to TiO_2 and then accepted by the adsorbed Cr(VI). Another example is simultaneous Cr(VI) removal and naphthalenesulfonate (NS) oxidation in textile wastewater by the combination of visible light photocatalysis and ion exchange membrane processes (electrodialysis) [122]. While NS acts as a hole scavenger attenuating electron–hole recombination, the ion exchange membrane prevents Cr(III) reoxidation. A $CuAl_2O_4$/TiO_2 heterosystem irradiated with visible light was used to achieve Cr(VI) reduction, the reaction being enhanced in the presence of salicylic acid as a sacrificial agent [134]. Also, Cr(VI) was effectively reduced under visible light irradiation employing TiO_2-coated CdS nanowires (NWs) [125]. Other modified TiO_2 materials showed activity for visible light–driven photocatalytic Cr(VI) reduction such as SnS_2/TiO_2 nanocomposites or N-doped and N–F codoped TiO_2 [121, 132].

Dye-photosensitized TiO_2 samples were proven to reduce Cr(VI) under visible light; some examples follow. Di Iorio et al. [116] studied Alizarin red chelated to TiO_2. Results indicate a high efficiency for Cr(VI) reduction, almost independent of the photon flux and of the irradiation wavelength, and slightly dependent on Cr(VI) concentration in the explored range (40–200 μM), with similar results in air and nitrogen atmosphere. EPR experiments confirmed Cr(V) formation. In another work [135], TiO_2 coated with hydroxylaluminumtricarboxymonoamide phthalocyanine (AlTCPc) irradiated under visible light in the presence of 4-chlorophenol (4-CP) as the sacrificial donor showed a very rapid Cr(VI) reduction to Cr(III). As discussed in the section on Arsenic, in the presence of dyes, the photocatalytic mechanism is different from that taking place under UV light, according to the scheme shown in Figure 9.4: after excitation of the dye, electron injection into the CB promotes Cr(VI) reduction [51]. A very good protective effect of 4-CP preventing AlTCPc photobleaching was observed, as reactive oxygen species (ROS, e.g., $O_2^{\bullet-}$, $HO_2^{\bullet}$), which could oxidize the dye, could be formed due to the fast reaction of e_{CB}^- with Cr(VI). In another paper, Park et al. [52] tested the visible light activity of the OD-sensitized TiO_2 catalyst described earlier (Section 9.3) for the reduction of Cr(VI). As in the case of As(III), Cr(VI) reduction with OD/TiO_2 showed better results than with both RuL_3/TiO_2 and bare TiO_2. The addition of an electron acceptor such as Fe^{3+} significantly retarded Cr(VI) reduction because of competition for e_{CB}^-, confirming that Cr(VI) reduction proceeds by accepting electrons from the dye-sensitized TiO_2. In a very recent work, Park et al. continued studies on this system [136] and tested the conversion of Cr(VI) to Cr(III) using a very similar dye, that is, (E)-3-(5-(5-(4-(bis(4-((2-(2-methoxyethoxy)methyl)phenyl)amino)phenyl)thiophen-2-yl)thiophen-2-yl)-2-cyanoacrylic acid), agglomerated with bare TiO_2 nanoparticles; the authors indicate an enhancement in the reduction rate of Cr(VI) attributed to the retardation of the charge recombination between the oxidized dye and the injected electron by the agglomeration with bare TiO_2.

A very unusual mode of titania photosensitization was presented by Kuncewicz et al. [120]. The authors proposed that as the broad LMCT band of CrO_4^{2-} in aqueous solutions (with a maximum at 373 nm at pH 7) extends to visible light, excitation of the CrO_4^{2-} species adsorbed on the titania surface at 440 nm (2.8 eV) should result in electron transfer from O^{-II} to Cr(VI)

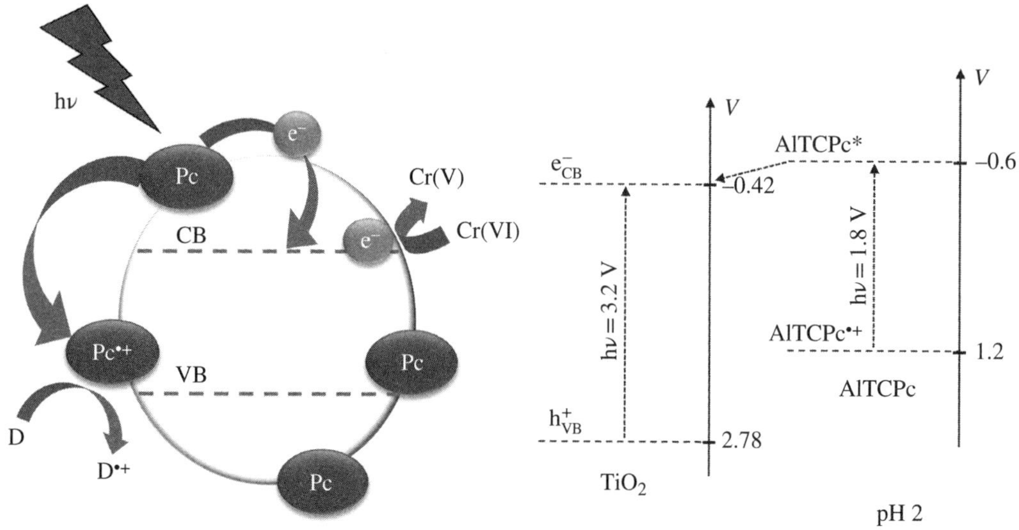

FIGURE 9.4 Visible light HP Cr(VI) reduction with TiO_2 modified with AlTCPc at pH 2.

with the generation of Cr(V) and O^{-I}. According to the authors, CrO_4^{2-} can act as a photosensitizer promoting hole injection to the TiO_2 VB. In this system, the generated O^{-I} would oxidize TiO_2, while Cr(V) would reduce O_2 as an external electron acceptor.

Not many works have explored the photocatalytic reduction of Cr(VI) beyond the laboratory scale. In a very recent work, HP technology for Cr(VI) and HA was applied in a pilot reactor [137], starting from the laboratory and scaling up the system to a pilot reactor five times the volume and height of the laboratory reactor by maintaining geometric parameters and nominal powers. The authors obtained very good results: after almost 4 h of reaction in a 15 l pilot reactor ($[Cr(VI)]_0 = 10\,mg\,l^{-1}$, $[HA]_0 = 15\,mg\,l^{-1}$, $[TiO_2] = 1.5\,g\,l^{-1}$), removal of Cr(VI) and HA was around 98 and 87%, respectively.

The scientific community has been successful in the generation of knowledge about fundamental mechanisms, mechanism pathways, the role of O_2 and electron donors, and so on, related to a very rich reductive photocatalytic Cr(VI) system. More studies and efforts should be undertaken to incorporate the knowledge of these promising technologies into practical applications.

9.5 URANIUM

Uranium is commonly found in its hexavalent form, U(VI), but it also occurs naturally in the +2, +3, +4, and +5 valence states. The ^{238}U isotope constitutes more than 99% of naturally occurring uranium, with ^{235}U and ^{234}U isotopes comprising minor percentages. Natural uranium can be found in granites and various other mineral deposits, and it is present in the environment as a result of leaching from natural deposits, release in mill tailings, emissions from the nuclear industry, the combustion of coal and other fuels, and the use of uranium-containing phosphate fertilizers [138]. The most common form of U(VI) in water is uranyl ion, UO_2^{2+}.

The health impact associated with high levels of naturally occurring uranium in drinking water is considered to result mainly from its chemical toxicity rather than the risk from exposure to radioactivity. The primary chemically induced effect of uranium in humans is nephritis; high levels of uranium have been associated with high blood pressure, bone dysfunction, and likely reproductive impairment in human populations [139]. For human beings, a guideline value of $0.015\,mg\,l^{-1}$ in drinking water has been indicated, which serves only as a provisional guideline value due to limited information on the health effects [138].

Several methods are available for the removal of uranium from drinking water, such as ionic exchange, ultrafiltration [140], adsorption on granular ferric hydroxide, iron oxides (ferrihydrite, hematite, magnetite, and goethite) [141–143], pine sawdust [144], zeolites [145], activated carbon [146] or TiO_2 [147, 148], bioreduction [149], and zero-valent iron [150–153]. Normally, the treatment of wastewater containing radioactive uranium includes the processes of concentration and solidification. Reductive photocatalysis can be proposed for U(VI) reduction because two processes can be combined into one step, by reducing and depositing uranium from aqueous solutions [154].

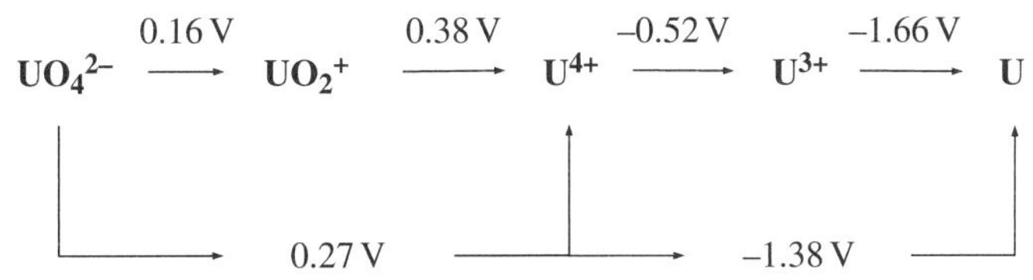

FIGURE 9.5 Latimer diagram connecting the different U species [152].

Reports concerning HP for uranium treatment in water are still scarce. In our previous reviews, a few cases were reported [154–156] but, due to their importance and limited diffusion, they are included again in this work. Most of the investigations focus on uranyl ion reduction, with final formation of a deposit of uranium oxide (IV) (UO_2).

If standard reduction potentials are taken into account (Fig. 9.5), U(VI) can be photocatalytically reduced by TiO_2 e^-_{CB} to U(V) and then to U(IV).

In addition, U(V) rapidly disproportionates to U(VI) and U(IV) in the presence of water or hydrolytic compounds [157] with a very complex chemistry; U(V) is very sensitive to atmospheric oxygen with rapid reoxidation to U(VI) [158].

In a pioneering work, Amadelli [155] reported photoreduction of uranyl solutions (initial concentration between 1.2 and 12.4 mM) on UV-illuminated TiO_2 suspensions and electrodes in the presence of excess of hole scavengers (2-propanol, acetate, or formate); working with TiO_2 suspensions, the final product was uranium oxide of stoichiometry close to U_3O_8 found as a dark gray solid on the SC surface. Later, Chen et al. [154] presented results of U(VI) HP photocatalytic removal from aqueous solutions in the presence of EDTA (initial U(VI) concentration of 50 mg l^{-1} and variable U(VI): EDTA ratio) and absence of oxygen; the authors remarked on the importance of the presence of adequate donors to sustain appreciable reaction rates. Further exposure to air reoxidized and redissolved the uranium species. This method was proposed for recovering uranyl from wastewaters of nuclear power plants, where chelating agents (normally EDTA, hydroxyethylenediaminetetraacetic acid (HEDTA) and oxalic and citric acid) and their uranium (VI) complexes are usually present.

In another work [159], using methanol as hole scavenger, U(VI) from phosphate-containing waste was photocatalytically reduced in a 98% from an initial concentration of 40 mg l^{-1}. Complexation of uranyl with the hole scavenger was found to play an essential role in the photoredox process, and, in addition, this indicates the predominance of the indirect reductive pathway (see Section 9.1) [155]. Eliet and Bidoglio [160] studied the U(VI) photocatalytic reduction in TiO_2 aqueous suspensions by time-resolved laser-induced fluorescence (TRLIF), proposing a reaction mechanism based mainly on the role of adsorbed uranium species. Later, Selli et al. [156] observed that the U(VI) photoreduction in TiO_2 aqueous suspensions was enhanced when complexed with HA. The authors proposed a kinetic model considering that the photoreduction takes place in both the aqueous phase and the photocatalyst solution interface. Boxall et al. [161] explored the use of HP in actinide valence state control as a possible application of waste minimization in nuclear fuel processing and with the separation of Np, Pu, and U. Other authors [162] studied the U(VI) photocatalytic reduction on the surface of TiO_2 anatase nanotubes with and without UV light irradiation. XPS measurements of the catalyst after the experiments showed adsorbed U(VI) and U(IV) species, with a predominant amount of the reduced form both in the dark and in illuminated experiments; a higher amount of U(IV) was obtained when light was used.

While mechanistic studies on the photocatalytic removal of uranium-related species are scarce and merit further research, additional investigation is mandatory to clarify major reaction engineering issues following this approach.

9.6 LEAD

Lead has many industrial uses such as in the production of lead acid batteries, solder, alloys, cable sheathing, pigments, rust inhibitors, ammunition, glazes, and plastic stabilizers. Consequently, lead pollution is mainly anthropogenic, coming from industrial effluents. Lead has been extensively used in plumbing fittings and as solder in water distribution systems, and lead pipes still subsist in old structures. Polyvinyl chloride (PVC) pipes now in use also contain lead compounds that can be leached resulting in high lead concentrations in drinking water. Lead effects on the central nervous system can be serious, with pregnant women, fetuses, infants, and children up to 6 years of age the most likely to suffer the adverse health effects of its cumulative poison [163]. A guideline value of 0.01 mg l^{-1} is maintained but designated as provisional by the WHO [163].

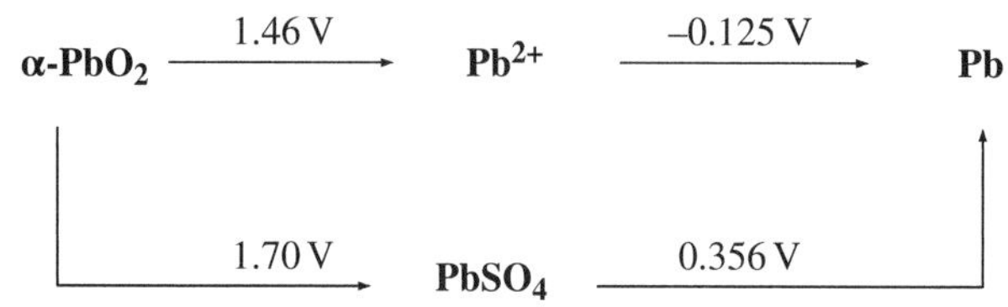

FIGURE 9.6 Latimer diagram connecting the different Pb species [164].

Figure 9.6 shows the Latimer diagram for Pb.

HP of lead systems has received scarce attention. In our previous reviews [7, 10] we cited a number of early papers [75, 78, 165–180], but information continues to be rather scant. The mechanisms of transformation of lead (II) in water by UV–TiO_2 are especially attractive because they depend very much on the reaction conditions, which are related to the nature of the photocatalyst, the effect of oxygen, and the presence of electron donors.

Our group and others have investigated both the oxidative and the reductive route. In the first case, Pb(IV) formation through hole or $HO^\bullet$ attack takes place through two consecutive monoelectronic steps. The one-electron reduction potential for the Pb(III)/Pb(II) couple is not reported, but it is not unreasonable to assume that Pb(II) can be easily oxidized to this unstable intermediate.

$$\mathrm{Pb(II)} + \mathrm{h}^+_{\mathrm{VB}} / \mathrm{HO}^\bullet \rightarrow \mathrm{Pb(III)} \qquad (9.31)$$

Pb(III) transforms into Pb(IV) by simple oxidation by O_2 or by stronger oxidants present in the system (h^+_{VB}, ROS, etc.); disproportionation is also possible:

$$\mathrm{Pb(III)} + \mathrm{h}^+_{\mathrm{VB}} / \mathrm{HO}^\bullet / \mathrm{ROS} \rightarrow \mathrm{Pb(IV)} \qquad (9.32)$$

$$2\mathrm{Pb(III)} \rightarrow \mathrm{Pb(II)} + \mathrm{Pb(IV)} \qquad (9.33)$$

In the presence of O_2, a dark brown PbO_2 deposit on TiO_2 was observed [19]. Equation 9.5, that is, reduction of O_2 to superoxide, is normally a very slow reaction due to its high overpotential. In this context, Pb(II) removal by oxidation is poor. The use of platinized samples (Pt/TiO_2) enhanced the reaction, because Pt decreases this overpotential, accelerating the reaction in oxic systems [10, 19].

Ozone addition promotes HP oxidation of Pb(II), because of the photochemical formation of H_2O_2 and other ROS, including the ozonide radical $\left(HO_3^\bullet / O_3^{\bullet-}\right)$ [178, 181–183]; this was proven by X-ray diffraction (XRD), where the patterns revealed deposition of PbO_2 and $PbO_{1.37}$ on the photocatalyst, the unstoichiometric oxide being formed by the action of the ozonide:

$$\mathrm{Pb}^{2+} \xrightarrow{\mathrm{O}_3^{\bullet-}} \xrightarrow{\mathrm{O}_3^{\bullet-}} \mathrm{PbO}_{1.37} \qquad (9.34)$$

On the other hand, the one-electron reduction potential of the Pb(II)/Pb(I) couple is very negative ($E^0 = -1.0$V [184]), preventing a direct reductive route by e^-_{CB}, as observed in experiments performed employing Pt/TiO_2 under N_2 [19, 165, 172]. In fact, direct reduction of Pb(II) to Pb(0) by a bielectronic process has been reported under laser irradiation, where, due to the high photonic frequency, the authors propose that accumulation of electrons may allow multielectronic injection [176, 179].

However, the photocatalytic reduction of Pb(II) is possible under regular illumination via the indirect route in the presence of electron donors (Eqs. 9.3 and 9.11). Different organic compounds have been used in the past to promote reductive lead HP under N_2 and over pure TiO_2 [19, 20, 79, 114, 171].

Murruni et al. [20] obtained a high efficiency in Pb(II) reduction when using 2-propanol and formic acid as sacrificial donors, the latter being a more environmentally friendly alternative as its degradation products are not toxic. Two monoelectronic reduction steps were proposed:

$$\mathrm{Pb(II)} + \rightarrow \mathrm{Pb(I)} \qquad (9.35)$$

$$\mathrm{Pb(I)} \rightarrow \mathrm{Pb(0)} \qquad (9.36)$$

It was reported that Pb(0) produced via indirect reduction forms colloidal zero-valent Pb that eventually deposits on the lamp used as an irradiation source when it is immersed in the suspension.

Recently, Li et al. [185] explored the photocatalytic Pb(II) reduction over TiO_2 and Ag/TiO_2 catalysts in the presence of formic acid. They showed that the initial rate of Pb(II) reduction over Ag/TiO_2 is dependent on the concentrations of both Pb(II) and formic acid. Solution pH impacts on Pb(II) reduction with 98.6% removal at pH 3.5 and just 11.8 at pH 0.8; the result can be explained by a stronger adsorption of formate at the higher pH on the photocatalyst surface favoring the formation of $CO_2^{\bullet-}$. Previously, other authors [186] employed a quartz crystal microbalance (QCM) technique to study the kinetics of photocatalytic reduction of Pb(II) on nanocrystalline TiO_2 coatings. The photocatalyst was used as an ultrathin coating deposited on the surface of quartz crystal, and Pb(II) removal could be monitored by measuring the change in frequency of the quartz crystal resonator. Again, it was found that the indirect reduction of Pb(II) depends greatly on the organic additives, with the following order of efficiency: $HCOOH > H_2C_2O_4 > CH_3OH > C_2H_5OH$.

Titania–silica (TiO_2–SiO_2) photocatalyst [187] (silica essentially being used as support to increase the specific surface area) was synthesized and employed in a study that focused on the simultaneous oxidation of cyanide and the removal of heavy metal ions, Pb(II) among them, under mild conditions in aqueous solutions. The photocatalytic performance was markedly dependent on the catalyst, target concentrations, and reaction time.

In conclusion, the photocatalytic treatment of Pb(II) has many potential advantages, including the following:

1. Mixtures of lead (II) and organic scavengers may be present in industrial wastes, which could give rise to economical methods for its removal.
2. The photocatalytic treatments of Pb(II) generate its transformation and immobilization as lead oxides, metallic deposits, or colloidal zero-valent lead, making the ulterior separation and treatment of solid residues easier.

Although the Pb photocatalytic system has been extensively studied, significant effort must be undertaken to transfer the knowledge, in the search for technological solutions.

9.7 MERCURY

Mercury is a metallic element that can be found naturally in the environment. The solubility of mercury in water and therefore its presence in natural or residual waters depends on its chemical state: while Hg(0) and HgS are insoluble, Hg(II) is easily soluble and Hg(I) presents slow solubility [188]. Organic mercury species such as methyl or phenylmercury salts have a higher toxicity than inorganic species, with organic mercury being chemically stable due to the carbon–mercury bond. The toxicity of mercury (II) is a matter of concern; the toxic effects of inorganic mercury compounds are mainly renal, whereas methyl- and ethylmercury salts are responsible for neurological disturbances. The guideline value for inorganic mercury is $0.006\,mg\,l^{-1}$ in drinking water [188].

The use of mercury in industrial processes is significant, and it is used in the electrolytic production of chloride and caustic soda, in electrical appliances, and as raw material for various mercury compounds. Mercury compounds are used as agricultural pesticides, fungicides, antiseptics, preservatives, pharmaceuticals, electrodes, and reagents [188].

Figure 9.7 shows the Latimer diagram for Hg.

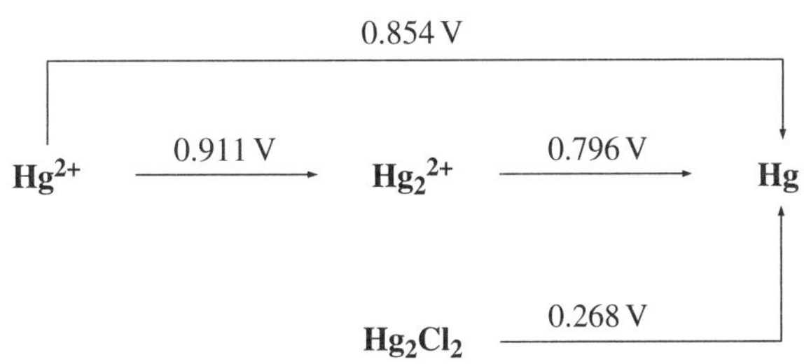

FIGURE 9.7 Latimer diagram connecting the different Hg species [192].

HP appears as a suitable alternative for Hg reduction, and many authors have studied this technology with TiO_2, ZnO, and WO_3, exploring the use of UV, visible, and solar irradiation [10, 177, 190–207]. The main conclusions drawn from published works can be condensed in the following points:

1. Mercury photocatalytic reduction strongly depends on pH conditions. Higher conversion rates can be obtained at basic pH. The reason for this behavior seems to be a larger driving force for Hg(II) reduction produced by the shift in the potential of the CB electrons with the increase in the pH: ECB decreases 0.059 V per unit of pH (at 25 °C) [206]. In acidic conditions, Hg(0) photooxidation can occur inhibiting Hg(II) reduction, as found by Serpone et al. [200] in experiments at pH 0. In a very recent work, López-Muñoz et al. [208] studied the influence of pH and the addition of methanol, formic acid, and oxalic acid as sacrificial agents for the photocatalytic removal of Hg(II). At pH 10, an efficient removal of Hg(II) was achieved even in the absence of organic additives, attaining final mercury concentrations in the solution at trace levels ($\mu g\,l^{-1}$). In acidic conditions, the addition of sacrificial organic molecules significantly increased the rate and extent of aqueous Hg(II) removal.
2. The presence of oxygen inhibits mercury reduction in acidic or neutral pH conditions, due to the competition between Equations 9.5, 9.37, and 9.38 for a direct reduction mechanism:

$$Hg^{2+} + e^-_{CB} \rightarrow Hg(I) \tag{9.37}$$

$$Hg(I) + e^-_{CB} \rightarrow Hg(0) \tag{9.38}$$

   A mechanism involving monoelectronic steps has been proposed by our group [212] for direct Hg(II) HP reduction, particularly in the case of inorganic salts. Evidence from experiments performed with $HgCl_2$ yielding calomel Hg_2Cl_2 (together with Hg(0)) support this hypothesis.
3. The addition of electron donors to the system favors reductive Hg(II) HP indicating the parallel occurrence of an indirect mechanism. These donors can be added to the system with or without the formation of complexes with Hg(II), or as part of an organic mercury species, with the organic moiety acting as electron donor. Electron donors added to the system can avoid reoxidation of mercury species by reacting with h^+_{VB} / $HO^•$ and preventing the following reaction from taking place [209, 210]:

$$Hg(0)/(I) + h^+_{VB}/HO^• \rightarrow Hg(I)/Hg(II) \tag{9.39}$$

The nature and distribution of mercury products deposited on the catalyst depend on the reaction conditions. Starting from inorganic mercury salts, $Hg(NO_3)_2$, $HgCl_2$ and $Hg(ClO_4)_2$, Hg(0), HgO, or Hg_2Cl_2 are formed on the photocatalyst surface [209]. In the absence of additives, Hg_2Cl_2 and Hg(0) were respectively identified in acidic and neutral/alkaline media as main reduced species on the titania surface [208]. The addition of organic additives enhances the photocatalytic reduction to Hg(0). On the other hand, a direct correlation between Hg(II) dark adsorption on the TiO_2 [208] surface and the efficiency of Hg(II) photoreduction could not be established [206, 208].

Recently, Lenzi et al. [211] studied Hg(II) photoreduction using $HgCl_2$ in the presence of formic acid at pH 4 on TiO_2 and Ag/TiO_2 prepared by sol-gel and impregnation methods. The Ag/TiO_2 sample prepared by impregnation proved to be better than the sample prepared by the sol-gel method, because, due to the high dispersion of Ag in the last sample, Ag may act as an electron–hole recombination center, leading to decreased Hg(II) photoreduction. In contrast, in the sample prepared by impregnation, Ag addition seems to prevent electron–hole recombination, with better photoactivity.

As was pointed out before, organic mercury species are much more toxic than inorganic ones. Metallic mercury has been reported as the product of HP treatment of methylmercury in the presence of methanol and in the absence of oxygen [200, 212]. Later, Miranda et al. [198] found optimal conditions at alkaline pH for this reaction. Although reduction of Hg(II) was found both in the presence and in the absence of oxygen, different products were identified from the organic moiety of the organometallic compounds: methane was found in oxic conditions while methanol formed under a nitrogen atmosphere.

Two monoelectronic successive steps are proposed where the organic moiety acts as an electron donor. Our group studied the UV/TiO_2 photocatalysis of phenylmercury salts in aqueous solutions starting from acetate ($C_6H_5HgCH_3CO_2$, phenylmercury acetate (PMA)) and chloride (C_6H_5HgCl, phenylmercury chloride (PMC)) salts [195]. The results were very promising as it was possible to remove mercury to a large extent with the simultaneous mineralization of the organic portion of the compound.

Metallic Hg was found as a solid deposit when PMA was used and Hg_2Cl_2 in the case of PMA. The best condition for Hg(II) removal and mineralization was at pH 11 in the absence of O_2. Phenol was detected as an intermediate in the process in both cases and with no formation of dangerous alkylmercury species. Calomel formation from PMC under nitrogen supported the two successive one-electron transfer reactions hypothesis, as in the case of inorganic salts. A very interesting application for the TiO_2–Hg system is the use of the photocatalytic reduction of mercury in the field of analytical chemistry as a cost-effective, fast, and sensitive method for mercury determination in environmental and biological samples. Chemical vapor generation (CVG) is used as a sample introduction method for specific trace element determination using atomic spectrometry [213]. A photocatalytic TiO_2 system appears as an attractive alternative to achieve the transformation Hg(II) → Hg(0) without the need for the use of concentrated chemical reductants. This application of the HP system for Hg(II) reduction has been studied by some authors in the past few years [213–215]. Yin et al. [213] studied Hg vapor generation by direct photocatalysis employing UV–TiO_2 nanoparticles with online atomic fluorescence spectrometry (AFS) determination for the first time. The authors employed formic acid and sodium formate as the hole scavenger mixture. Han et al. [214] studied a similar system using low molecular weight alcohols, aldehydes, or carboxylic acids such as glycol, 1,2-propanediol, glycerol, oxalic acid, and malonic acid. The authors obtained good reduction of Hg(II) with all the tested compounds using only UV light irradiation initially (high-pressure Hg vapor UV lamp); the addition of TiO_2 to the system led to improved efficiency, with very low detection limits (0.02–0.04 $\mu g\, l^{-1}$). Other authors [215] studied Hg concentration, pH, formic acid concentration, the effect of oxygen, and UV irradiation time for this process. In their work, the authors investigated the determination of low Hg(II) concentrations (3 $\mu g\, l^{-1}$) and compared the photocatalytic activity of unmodified TiO_2, Ag–TiO_2, and ZnO for reduction of mercury. They obtained a recovery in the range of 95–99% when it was catalyzed by nano Ag–TiO_2 in optimized conditions with a detection limit of 0.13 $\mu g\, l^{-1}$.

Despite the multiple investigations conducted to study the photocatalytic reduction of Hg(II), many of the processes involved are still unknown due to the complexity of Hg(II) chemistry, which involves various species in solution and in the solid phase. More fundamental research studies should be undertaken in order to clarify the mechanisms, especially for extremely toxic organomercurial species. On the other hand, it is important to remember that many mercury species present high tendency toward volatilization, and the coupling of liquid and gaseous HP setups can be the solution to the problems associated to these issues.

9.8 CONCLUSIONS

Heterogeneous photocatalytic treatment of metals and metalloids can be a valuable option for the removal of these species from water, which does not require expensive reagents or equipment. UV lamps are economical and the possibility of using costless solar light is open, with much more research in the past few years concerning the efficiency of photocatalysts with activity in the visible range. However, an overview of the literature on the subject for the species described here—arsenic, chromium, uranium, mercury and lead—indicates that important fundamental and applied research is still missing on several aspects.

The application of HP to arsenic species is very promising taking into account that both oxidative and reductive mechanisms may lead to less toxic species (As(V) compared with As(III)) or the solid phases (As(0)). Although the oxidative system has been studied in detail, the reductive pathway is a challenging alternative that should be taken into account and improved.

The mechanistic features of the photocatalytic reduction of chromium (VI) to the less toxic Cr(III) form have been thoroughly studied. The impressive synergistic effect of organic donors present simultaneously with Cr(VI) in wastewaters as a result of different industrial processes reveals that this system is very appropriate to be scaled for real-world applications. In order to achieve this, research on adequate photoreactor design and industrial innovation is a priority. Cr(VI) photocatalytic reduction is unique because it is not inhibited by oxygen, at least at acidic pH, and this represents an additional advantage for implementation.

Research on photocatalytic removal of uranium salts from water is still scarce. Photoreduction is possible only in the presence of hole scavengers in the absence of oxygen with the formation of uranium oxides; however, rapid reoxidation and redissolution takes place after exposure to air. These are aspects that should be considered and improved to fully demonstrate the potential of the technology.

Mechanistic studies on Pb(II) HP are almost complete and ready for real-world applications. The best method is the reductive indirect pathway, driven by species formed in the presence of alcohols or carboxylates, which, as in the case of Cr(VI), can be present together with Pb(II) in wastewaters, rendering the technology appropriate for Pb(II) removal.

Mercury (II) photocatalysis also has not been studied much, although it is known that reaction occurs easily through direct reduction driven by CB electrons. Interesting and encouraging results have been found in the case of the extremely toxic organomercurial species, and further investigation on possible applications deserves attention.

To sum up, fundamental research is still needed in order to improve TiO_2 photocatalysis of arsenic and heavy metal systems, such as photocatalyst performance, improvement in photon efficiencies, achievement of higher overall rates, and decrease in the conversion time. Effective immobilization of the photocatalyst is a major challenge, which will have a significant effect on photocatalytic reactor design and costs, suppressing additional operations of separating catalysts from treated water.

However, it is important to take into account that these treatments have to be applied only at small and medium scales. Combination with other AOPs/ARPs or conventional treatments should also be considered to improve the efficiency of the removal of contaminants or mixtures of contaminants. Application in real waters should also take into account the presence of interferences such as phosphates, silicates, and carbonates/bicarbonates, which could influence removal. In this sense, arsenic removal is very sensitive to these species.

REFERENCES

[1] Degremont G. *Water Treatment Handbook*. Paris: Lavoisier Publishing; 1991.

[2] Tchobanoglous G, Burton FL, Stensel HD. *Wastewater Engineering Treatment and Reuse*. New York: McGraw-Hill; 2003.

[3] Domènech X, Jardim WF, Litter MI. Tecnologias avanzadas de oxidacion para la eliminacion de contaminantes. In: Blesa MA, Sánchez Cabrero B, editors. Eliminación de contaminantes por fotocatálisis heterogénea. Texto Colectivo elaborado por la Red CYTED VIII-Ged. Ediciones CIEMAT. Madrid, España; 2004. p 7–34.

[4] Hoffmann MR, Martin ST, Choi W, Bahnemann DW. Environmental applications of semiconductor photocatalysis. Chem Rev 1995;95:69–96.

[5] Huang CP, Dong C, Tang Z. Advanced chemical oxidation: its present role and potential future in hazardous waste treatment. Waste Manage 1993;13:361–377.

[6] Legrini O, Oliveros E, Braun AM. Photochemical processes for water treatment. Chem Rev 1993;93:671–698.

[7] Litter MI. Heterogeneous photocatalysis: transition metal ions in photocatalytic systems. Appl Catal B 1999;23:89–114.

[8] Litter MI. Introduction to photochemical advanced oxidation processes for water treatment. In: Boule P, Bahnemann DW, Robertson PKJ, editors. *Environmental Photochemistry Part II (The Handbook of Environmental Chemistry*, Vol. 2, Part M). Berlin, Heidelberg: Springer-Verlag; 2005. p 325–366.

[9] Rajeshwar K. Photoelectrochemistry and the environment. J Appl Electrochem 1995;25:1067–1082.

[10] Litter MI. Treatment of chromium, mercury, lead, uranium, and arsenic in water by heterogeneous photocatalysis. Adv Chem Eng 2009;36:37–67.

[11] Litter MI, Domènech X, Mansilla HD, Blesa MA, Sánchez Cabrero B. Remoción de contaminantes metálicos. In: Blesa MA, Sánchez Cabrero B, editors. Eliminación de contaminantes por fotocatálisis heterogénea. Texto Colectivo elaborado por la Red CYTED VIII-Ged. Ediciones CIEMAT. Madrid, España; 2004. p 163–187.

[12] Grela MA, Loeb B, Restrepo GM, Lagorio MG, San Román E. Los mecanismos dedestrucción decontaminantes orgánicos. In: Blesa MA, Sánchez Cabrero B, editors. Eliminación de contaminantes por fotocatálisis heterogénea. Texto Colectivo elaborado por la Red CYTED VIII-Ged. Ediciones CIEMAT. Madrid, España; 2004. p 125–162.

[13] Liu I, Lawton LA, Bahnemann DW, Liu L, Proft B, Robertson PKJ. The photocatalytic decomposition of microcystin-LR using selected titanium dioxide materials. Chemosphere 2009;76:549–553.

[14] Hufschmidt D, Bahnemann D, Testa JJ, Emilio CA, Litter MI. Enhancement of the photocatalytic activity of various TiO_2 materials by platinisation. J Photochem Photobiol A 2002;148:223–231.

[15] Emilio CA, Litter MI, Kunst M, Bouchard M, Colbeau-Justin C. Phenol photodegradation on platinized-TiO_2 photocatalysts related to charge-carrier dynamics. Langmuir 2006;22:3606–3613.

[16] Martin ST, Herrmann H, Hoffmann MR. Time-resolved microwave conductivity. Part 2.—Quantum-sized TiO_2 and the effect of adsorbates and light intensity on charge-carrier dynamics. J Chem Soc Faraday Trans 1994;90:3323–3330.

[17] Fujishima A, Zhang X, Tryk DA. TiO_2 photocatalysis and related surface phenomena. Surf Sci Rep 2008;63:515–582.

[18] Wardman P. Reduction potentials of one-electron couples involving free radicals in aqueous solution. J Phys Chem Ref Data 1989;18:1637–1755.

[19] Murruni L, Leyva G, Litter MI. Photocatalytic removal of Pb(II) over TiO_2 and Pt–TiO_2 powders. Catal Today 2007;129:127–135.

[20] Murruni L, Conde F, Leyva G, Litter MI. Photocatalytic reduction of Pb(II) over TiO_2: new insights on the effect of different electron donors. Appl Catal B 2008;84:563–569.

[21] Levy IK, Mizrahi M, Ruano G, Zampieri G, Requejo FG, Litter MI. TiO_2-photocatalytic reduction of pentavalent and trivalent arsenic: production of elemental arsenic and arsine. Environ Sci Technol 2012;46:2299–2308.

[22] Litter MI, Morgada ME, Bundschuh J. Possible treatments for arsenic removal in Latin American waters for human consumption. Environ Pollut 2010;158:1105–1118.

[23] Bundschuh J, Litter MI, Ciminelli V, Morgada ME, Cornejo L, Garrido Hoyos S, Hoinkis J, Alarcón-Herrera MT, Armienta MA, Bhattacharya P. Emerging mitigation needs and sustainable options for solving the arsenic problems of rural and isolated urban areas in Latin America–a critical analysis. Water Res 2010;44:5828–5845.

[24] WHO. Arsenic in drinking-water, background document for development of WHO—Guidelines for Drinking-water Quality. Geneva: WHO; 2011. Report nr WHO/SDE/WSH/03.04/75/Rev/1.

[25] Ingallinella AM, Fernándes R, Stecca L. Proceso ARCIS-UNR para la remoción de arsénico y flúor en aguas subterráneas: una experiencia de aplicación. Revista Ingeniería Sanitaria y Ambiental (AIDIS Argentina); 2003. Parte 1 en No.66, p 36–39; Parte 2 en No.67. p 61–65.

[26] Emett MT, Khoe GH. Photochemical oxidation of arsenic by oxygen and iron in acidic solutions. Water Res 2001;35:649–656.

[27] Litter MI, Fernández RG, Cáceres RE, Cobián DG, Cicerone D, Cirelli AF. Low-cost technologies for arsenic removal at low- and mediun scale. Revista Ingeniería Sanitaria y Ambiental 2008;100:41–50.

[28] Morgada ME, Mateu M, Bundschuh J, Litter MI, e-Terra. 2008. Arsenic in the Iberoamerican Region. The Iberoarsen Network and a possible economic solution for arsenic removal in isolated rural zones. Available at http://"mailto:editores.e-terra@geopor.pt" e-terra@geopor.pt. Accessed May 31, 2014.

[29] Bissen M, Vieillard-Baron MM, Schindelin AJ, Frimmel FH. TiOg—catalyzed photooxidation of arsenite to arsenate in aqueous samples. Chemosphere 2001;44:751–757.

[30] Dutta PK, Pehkonen SO, Sharma VK, Ray AK. Photocatalytic oxidation of arsenic(III): evidence of hydroxyl radicals. Environ Sci Technol 2005;39:1827–1834.

[31] Jayaweera PM, Godakumbra PI, Pathiartne KA. Photocatalytic oxidation of As(III) to As(V) in aqueous solutions: a low cost pre-oxidative treatment for total removal of arsenic from water. Curr Sci 2003;84:541–543.

[32] Lee H, Choi W. Photocatalytic oxidation of arsenite in TiO_2 suspensions: kinetics and mechanisms. Environ Sci Technol 2002;36:3872–3878.

[33] Ryu J, Choi W. Effects of TiO_2 surface modifications on photocatalytic oxidation of arsenite. The role of superoxides. Environ Sci Technol 2004;38:2928–2933.

[34] Yang H, Lin WY, Rajeshwar K. Homogeneous and heterogeneous photocatalytic reactions involving As(III) and As(V) species in aqueous media. J Photochem Photobiol A 1999;123:137–143.

[35] Ferguson MA, Hoffmann MR, Hering JG. TiO_2-photcatalyzed As(III) oxidation in aqueous suspensions: reaction kinetics and effects of adsorption. Environ Sci Technol 2005;39:1880–1886.

[36] Santhanam KS, Sundaresan NS. Arsenic. In: Bard JA, Parsons R, editors. *Standard Potentials in Aqueous Solutions*. New York: Marcel Dekker, Inc.; 1985. p 162–171.

[37] Kläning UK, Bielski BHJ, Sehested K. Arsenic(IV): a pulse-radiolysis study. Inorg Chem 1989;28:2717–2724.

[38] Choi W, Yeo J, Ryu J, Tachikawa T, Majima T. Photo catalytic oxidation mechanism of As(III) on TiO_2: unique role of As(III) as a charge recombinant species. Environ Sci Technol 2010;44:9099–9104.

[39] Ryu J, Choi W. Photocatalytic oxidation of arsenite on TiO_2: understanding the controversial oxidation mechanism involving superoxides and the effect of alternative electron acceptors. Environ Sci Technol 2006;40:7034–7039.

[40] Fei H, Leng W, Li X, Cheng X, Xu Y, Zhang J, Cao C. Photocatalytic oxidation of arsenite over TiO_2: is superoxide the main oxidant in normal air-saturated aqueous solutions? Environ Sci Technol 2011;45:4532–4539.

[41] Xu T, Cai Y, O'Shea KE. Adsorption and photocatalyzed oxidation of methylated arsenic species in TiO_2 suspensions. Environ Sci Technol 2007;41:5471–5477.

[42] Yoon SH, Lee JH. Oxidation mechanism of As(III) in the UV/TiO_2 system: evidence for a direct hole oxidation mechanism. Environ Sci Technol 2005;39:9695–9701.

[43] Yoon SH, Oh SE, Yang JE, Lee JH, Lee M, Yu S, Pak D. TiO_2 photocatalytic oxidation mechanism of As(III). Environ Sci Technol 2009;43:864–869.

[44] Guan X, Du J, Meng X, Sun Y, Sun B, Hu Q. Application of titanium dioxide in arsenic removal from water: a review. J Hazard Mater 2012;215–216:1–16.

[45] Tsimas ES, Tyrovola K, Xekoukoulotakis NP, Nikolaidis NP, Diamadopoulos E, Mantzavinos D. Simultaneous photocatalytic oxidation of As(III) and humic acid in aqueous TiO_2 suspensions. J Hazard Mater 2009;169:376–385.

[46] Nguyen TV, Vigneswaran S, Ngo HH, Kandasamy J, Choi HC. Arsenic removal by photo-catalysis hybrid system. Sep Purif Technol 2008;61:44–50.

[47] Xu Z, Meng X. Size effects of nanocrystalline TiO_2 on As(V) and As(III) adsorption and As(III) photooxidation. J Hazard Mater 2009;168:747–752.

[48] Morgada ME, Levy K, Mateu M, López G, Meichtry M, Bahnemann D, Dillert R, Bhattacharya P, Bundschuh J, Litter M. Low-cost solar technologies for arsenic in drinking water. In: Kabay N, Bundschuh J, Hendry B, Bryjak M, Yoshizuka K, Bhattacharya P, Anac S,

editors. *The Global Arsenic Problem: Challenges for Safe Water Production*. London: CRC Press/Balkema, Taylor & Francis Group; 2010. p 209–218.

[49] Meichtry JM, Lin HJ, de la Fuente L, Levy IK, Gautier EA, Blesa MA, Litter MI. Low-cost TiO_2 photocatalytic technol-ogy for water potabilization in plastic bottles for isolated regions. Photocatalyst fixation. J Sol Energy-T ASME 2007;129:119–126.

[50] Fostier AH, Pereira Mdo SS, Rath S, Guimarães JR. Arsenic removal from water employing heterogeneous photocatalysis with TiO_2 immobilized in PET bottles. Chemosphere 2008;72:319–324.

[51] Vinodgopal K, Wynkoop DE, Kamat PV. Environmental photochemistry on semiconductor surfaces: photosensitized degradation of a textile azo dye, Acid Orange 7, on TiO_2 particles using visible light. Environ Sci Technol 1996;30:1660–1666.

[52] Park Y, Lee SH, Kang SO, Choi W. Organic dye-sensitized TiO_2 for the redox conversion of water pollutants under visible light. Chem Commun 2010;46:2477–2479.

[53] Li Q, Easter NJ, Shang JK. As(III) removal by palladium-modified nitrogen-doped titanium oxide nanoparticle photocatalyst. Environ Sci Technol 2009;43:1534–1539.

[54] Monllor-Satoca D, Gómez R, Choi W. Concentration-dependent photoredox conversion of As(III)/As(V) on illuminated titanium dioxide electrodes. Environ Sci Technol 2012;46:5519–5527.

[55] Yoon S-H, Lee S, Kim T-H, Lee M, Yu S. Oxidation of methylated arsenic species by $UV/S_2O_8^{2-}$. Chem Eng J 2011;173:290–295.

[56] Cullen WR, Reimer KJ. Arsenic speciation in the environment. Chem Rev 1989;89:713–764.

[57] Bednar AJ, Garbarino JR, Ferrer I, Rutherford DW, Wershaw RL, Ranville JF, Wildeman TR. Photodegradation of roxarsone in poultry litter leachates. Sci Total Environ 2003;302:237–245.

[58] Zheng S, Cai Y, O'Shea KE. TiO_2 photocatalytic degradation of phenylarsonic acid. J Photochem Photobiol A 2010;210:61–68.

[59] Xu Z, Jing C, Li F, Meng X. Mechanisms of photocatalytical degradation of monomethylarsonic and dimethylarsinic acids using nanocrystalline titanium dioxide. Environ Sci Technol 2008;42:2349–2354.

[60] WHO. Guidelines for drinking-water quality, third edition incorporating the first and second addenda. Volume 1. Recommendations. 3rd ed. Geneva: WHO; 2008.

[61] Baral A, Engelken RD. Chromium-based regulations and greening in metal finishing industries in the USA. Environ Sci Policy 2002;5:121–133.

[62] Gauglhofer J, Bianchi V. *Metals and Their Compounds in the Environment*. Weinheim: VCH; 1991.

[63] Hamilton JW, Whcterhan KE. *Handbook of Toxicity of Inorganic Compounds*. New York: Marcel Dekker, Inc.; 1988.

[64] Fendorf SE, Li G. Kinetics of chromate reduction by ferrous iron. Environ Sci Technol 1996;30:1614–1617.

[65] Seaman JC, Bertsch PM, Schwallie L. In situ Cr(VI) reduction within coarse-textured, oxide-coated soil and aquifer systems using Fe(II) containing solutions. Environ Sci Technol 1999;33:938–944.

[66] Beukes JP, Pienaar JJ, Lachmann G, Giesekke EW. The reduction of hexavalent chromium by sulphite in wastewaters. Water SA 1999;25:363–370.

[67] Niki K. Chromium. In: Bard JA, Parsons R, Jordan J, editors. *Standard Potentials in Aqueous Solution*. New York: Marcel Dekker, Inc.; 1985. p 453–461.

[68] Distribución del arsénico en la región Ibérica e Iberoamericana. Distribución del arsénico en la región Ibérica e Iberoamericana In: Bundschuh J, Pérez Carrera A, Litter MI, editors. Editorial Programa Iberoamericano de Ciencia y Tecnología para el Desarrollo, octubre de; Argentina. Available at http://www.cnea.gov.ar/pdfs/actividades_ambientales/LIBROdistribucionarsenico.pdf. Accessed May 31, 2014. Edición CYTED. Argentina; 2008.

[69] Domènech X, Muñoz J. Photochemical elimination of Cr(VI) from neutral-alkaline solutions. J Chem Technol Biotechnol 1990;47:100–107.

[70] Dozzi MV, Saccomanni A, Selli E. Cr(VI) photocatalytic reduction: effects of simultaneous organics oxidation and of gold nanoparticles photodeposition on TiO_2. J Hazard Mater 2012;211–212:188–195.

[71] Prairie MR, Evans LR, Martinez SL. Chemical oxidation: technology for the nineties. Proceedings of the second international symposium. Volume 2. Nashville, TN; 1992. p 13.

[72] Prairie MR, Evans LR, Stange BM, Martinez SL. An investigation of titanium dioxide photocatalysis for the treatment of water contaminated with metals and organic chemicals. Environ Sci Technol 1993;27:1776–1782.

[73] Prairie MR, Stange BM. Photocatalysis for the treatment of waste: Applications involving the removal of metals. AIChe Symp Ser 295 1993;89:460–465.

[74] Yoneyama H, Yamashita Y, Tamura H. Heterogeneous photocatalytic reduction of dichromate on n-type semiconductor catalysts. Nature 1979;282:817–818.

[75] Aarthi T, Madras G. Photocatalytic reduction of metals in presence of combustion synthesized nano-TiO_2. Catal Commun 2008;9:630–634.

[76] Aguado MA, Giménez J, Cervera-March S. Continuous photocatalytic treatment of Cr(VI) effluents with semiconductor powders. Chem Eng Commun 1991;104:71–85.

[77] Cappelletti G, Bianchi CL, Ardizzone S. Nano-titania assisted photoreduction of Cr(VI). The role of the different TiO_2 polymorphs. Appl Catal B 2008;78:193–201.

[78] Chen D, Ray AK. Removal of toxic metal ions from wastewater by semiconductor photo-catalysis. Chem Eng Sci 2001;56:1561–1570.

[79] Chenthamarakshan CR, Rajeshwar K, Wolfrum EJ. Heterogeneous photocatalytic reduction of Cr(VI) in UV-irradiated titania suspensions: effect of protons, ammonium ions, and other interfacial aspects. Langmuir 2000;16:2715–2721.

[80] Colón G, Hidalgo MC, Navío JA. Influence of carboxylic acid on the photocatalytic reduction of Cr(VI) using commercial TiO_2. Langmuir 2001;17:7174–7177.

[81] Colón G, Hidalgo MC, Navío JA. Photocatalytic deactivation of commercial TiO_2 samples during simultaneousphotoreduction of Cr(VI) and photooxidation of salicylic acid. J Photochem Photobiol A 2001;138:79–85.

[82] Das DP, Parida K, De BR. Photocatalytic reduction of hexavalent chromium in aqueous solution over titania pillared zirconium phosphate and titanium phosphate under solar radiation. J Mol Catal A Chem 2006;245:217–224.

[83] Domènech J, Andrés M, Muñoz J. Utilización de la energía solar en descontaminación de aguas; aplicación en la eliminación de Hg(II) y Cr(VI). Tecnología del Agua 1986;6:35–42.

[84] Domènech J, Muñoz J. Photocatalytical reduction of Cr(VI) over ZnO powder. Electrochim Acta 1987;32:1383–1386.

[85] Fu H, Lu G, Li S. Adsorption and photo-induced reduction of Cr(VI) ion in Cr(VI)-4CP(4-chlorophenol) aqueous system in the presence of TiO_2 as photocatalyst. J Photochem Photobiol A 1998;114:81–88.

[86] Giménez J, Aguado MA, Cervera-March S. Photocatalytic reduction of chromium(VI) with titania powders in a flow system. Kinetics andcatalyst activity. J Mol Catal A Chem 1996;105:67–78.

[87] Hidalgo MC, Colón G, Navio JA, Macias M, Kriventsov VV, Kochubey DI, Tsodikov MV. EXAFS study and photocatalytic properties of un-doped and iron-doped ZrO_2-TiO_2 (photo-)catalysts. Catal Today 2007;128:245–250.

[88] Horváth O, Bodnár E, Hegyi J. Photoassisted oxidative degradation of surfactants and simultaneous reduction of metals in titanium dioxide dispersions. Colloids Surf A 2005;265:135–140.

[89] Jiang F, Zheng Z, Xu Z, Zheng S, Guo Z, Chen L. Aqueous Cr(VI) photo-reduction catalyzed by TiO_2 and sulfated TiO_2. J Hazard Mater 2006;134:94–103.

[90] Kajitvichyanukul P, Ananpattarachai J, Pongpom S. Sol–gel preparation and properties study of TiO_2 thin film for photocatalytic reduction of chromium (VI) in photocatalysis process. Sci Tech Adv Mater 2005;6:352–358.

[91] Kanki T, Yoneda H, Sano N, Toyoda A, Nagai C. Photo-catalytic reduction and deposition of metallic ions in aqueous phase. Chem Eng J 2004;97:77–81.

[92] Khalil LB, Mourad WE, Rophael MW. Photocatalytic reduction of environmental pollutant Cr(VI) over some semiconductors under UV/visible light illumination. Appl Catal B 1998;17:267–273.

[93] Ku Y, Jung IL. Photocatalytic reduction of Cr(VI) in aqueous solutions by UV irradiation with the presence of titanium dioxide. Water Res 2001;35:135–142.

[94] Lin W-Y, Wei C, Rajeshwar K. Photocatalytic reduction and immobilization of hexavalent chromium at titanium dioxide in aqueous basic media. J Electrochem Soc 1993;140:2477–2482.

[95] Miyake M, Yoneyama H, Tamura H. An electrochemical study on reactions of quinones in methanol on an illuminated rutile catalyst. Electrochim Acta 1977;22:319–324.

[96] Mohapatra P, Samantaray SK, Parida K. Photocatalytic reduction of hexavalent chromium in aqueous solution over sulphate modified titania. J Photochem Photobiol A 2005;170:189–194.

[97] Muñoz J, Domènech X. TiO_2 catalysed reduction of Cr-(VI) in aqueous solutions under ultraviolet illumination. J Appl Electrochem 1990;20:518–521.

[98] Papadam T, Xekoukoulotakis NP, Poulios I, Mantzavinos D. Photocatalytic transformation of acid orange 20 and Cr(VI) in aqueous TiO_2 suspensions. J Photochem Photobiol A 2007;186:308–315.

[99] Rengaraj S, Venkataraj S, Yeon JW, Kim Y, Li XZ, Pang GKH. Preparation, characterization and application of Nd–TiO_2 photocatalyst for the reduction of Cr(VI) under UV light illumination. Appl Catal B 2007;77:157–165.

[100] Ryu J, Choi W. Substrate-specific photocatalytic activities of TiO_2 and multi-activity test for water treatment application. Environ Sci Technol 2008;42:294–300.

[101] Sabate J, Anderson MA, Aguado MA, Giménez J, Cervera-March S, Hill CG Jr. Comparison of TiO_2 powder suspensions and TiO_2 ceramic membranes supported on glass as photocatalytic systems in the reduction of chromium(VI). J Mol Catal 1992;71:57–68.

[102] Schrank SG, José HJ, Moreira RFPM. Simultaneous photocatalytic Cr(VI) reduction and dye oxidation in a TiO_2 slurry reactor. J Photochem Photobiol A 2002;147:71–76.

[103] Selli E, De Giorgi A, Bidoglio G. Humic acid-sensitized photoreduction of Cr(VI) on ZnO particles. Environ Sci Technol 1996;30:598–604.

[104] Shim E, Park Y, Bae S, Yoon J, Joo H. Photocurrent by anodized TiO_2 photoelectrode for enzymatic hydrogen production and chromium(VI) reduction. Int J Hydrogen Energy 2008;33:5193–5198.

[105] Tuprakay S, Liengcharernsit W. Liengcharernsit, lifetime and regeneration of immobilized titania for photocatalytic removal of aqueous hexavalent chromium. J Hazard Mater 2005;124:53–58.

[106] Tzou YM, Wang SL, Wang MK. Fluorescent light induced Cr(VI) reduction by citrate in the presence of TiO_2 and Ferric ions. Colloids Surf A 2005;253:15–22.

[107] Wang L, Wang N, Zhu L, Yu H, Tang H. Photocatalytic reduction of Cr(VI) over different TiO_2 photocatalysts and the effects of dissolved organic species. J Hazard Mater 2008;152:93–99.

[108] Wang S, Wang Z, Zhuang Q. Photocatalytic reaction of environmental pollutants Cr(VI) over CdS powder. Appl Catal B 1992;1:257–270.

[109] Xu XR, Li HB, Gu JD. Simultaneous decontamination of hexavalent chromium and methyl tert-butyl ether by UV/TiO_2 process. Chemosphere 2006;63:254–260.

[110] Xu Y, Chen X. Photocatalytic reduction of dichromate over semiconductor catalysts. Chem Ind 1990(15):497–498.

[111] Yang JK, Lee SM. Removal of Cr(VI) and humic acid by using TiO_2 photocatalysis. Chemosphere 2006;63:1677–1684.

[112] Zheng S, Xu Z, Wang Y, Wei Z, Wang B. On the enhanced catalytic activity of TiO_2-supported layered compounds for Cr(VI) photo-reduction. J Photochem Photobiol A 2000;137:185–189.

[113] Meichtry JM, Brusa M, Mailhot G, Grela MA, Litter MI. Heterogeneous photocatalysis of Cr(VI) in the presence of citric acid over TiO_2 particles: relevance of Cr(V)–citrate complexes. Appl Catal B 2007;71:101–107.

[114] Testa JJ, Grela MA, Litter MI. Experimental evidence in favor of an initial one-electron transfer process in the heterogeneous photocatalytic reduction of chromium (VI) over TiO_2. Langmuir 2001;17:3515–3517.

[115] Testa JJ, Grela MA, Litter MI. Heterogeneous photocatalytic reduction of chromium (VI) over TiO_2 particles in the presence of oxalate. Involvement of Cr(V) species. Environ Sci Technol 2004;38:1589–1594.

[116] Di Iorio Y, Román ES, Litter MI, Grela MA. Photoinduced reactivity of strongly coupled TiO_2 ligands under visible irradiation. An examination of Alizarin Red @TiO_2 nanoparticulate system. J Phys Chem C 2008;112:16532–16538.

[117] Meichtry JM. Tratamiento de Cr(VI) por Fotocatálisis Heterogénea con TiO_2 [Doctoral Thesis]. Buenos Aires: Universidad de Buenos Aires; 2011.

[118] Testa JJ. Estudios fotocatalíticos de reducción y oxidación con TiO_2, Fe/TiO_2 y Pt/TiO_2 [Doctoral Thesis]. Buenos Aires: Universidad de Buenos Aires; 2012.

[119] Alam M, Henderson MA, Kaviratna PD, Herman GS, Peden CHF. Chromyl chloride chemistry on the TiO_2(110) surface. J Phys Chem B 1998;102:111–122.

[120] Kuncewicz J, Ząbek P, Kruczała K, Szaciłowski K, Macyk W. Photocatalysis involving a visible light-induced hole injection in a chromate(VI)-TiO_2 system. J Phys Chem C 2012;116:21762–21770.

[121] Giannakas AE, Seristatidou E, Deligiannakis Y, Konstantinou I. Photocatalytic activity of N-doped and N–F co-doped TiO_2 and reduction of chromium(VI) in aqueous solution: an EPR study. Appl Catal B 2013;132–133:460–468.

[122] Hsu HT, Chen SS, Chen YS. Removal of chromium(VI) and naphtha-lenesulfonate from textile wastewater by photocatalysis combining ionic ex-change membrane processes. Sep Purif Technol 2011;80:663–669.

[123] Kim G, Choi W. Charge-transfer surface complex of EDTA-TiO_2 and Its effect on photocatalysis under visible light. Appl Catal B 2010;100:77–83.

[124] Kleiman A, Márquez A, Vera ML, Meichtry JM, Litter MI. Photocatalytic activity of TiO_2 thin films deposited by cathodic arc. Appl Catal B 2011;101:676–681.

[125] Liu S, Zhang N, Tang Z-R, Xu Y-J. Synthesis of one-dimensional CdS@TiO_2 core-shell nanocomposites photocatalyst for selective redox: the dual role of TiO_2 shell. ACS Appl Mater Interfaces 2012;4:6378–6385.

[126] Luo S, Xiao Y, Yang L, Liu C, Su F, Li Y, Cai Q, Zeng G. Simultaneous detoxification of hexavalent chromium and acid orange 7 by a novel Au/TiO_2 heterojunction composite nanotube arrays. Sep Purif Technol 2011;79:85–91.

[127] Mu R, Xu Z, Li L, Shao Y, Wan H, Zheng S. On the photocatalytic properties of elongated TiO_2 nanoparticles for phenoldegradation and Cr(VI) reduction. J Hazard Mater 2010;176:495–502.

[128] Shaham Waldmann N, Paz Y. Photocatalytic reduction of Cr(VI) by titanium dioxide coupled to functionalized CNTs: an example of counterproductive charge separation. J Phys Chem C 2010;114:18946–18952.

[129] Wang N, Zhu L, Deng K, She Y, Yu Y, Tang H. Visible light photocatalytic reduction of Cr(VI) on TiO_2 in situ modified with small molecular weight organic acids. Appl Catal B 2010;95:400–407.

[130] Yoon J, Shim E, Bae S, Joo H. Application of immobilized nanotubular TiO_2 electrode for photocatalytic hydrogen evolution: reduction of hexavalent chromium (Cr(VI)) in water. J Hazard Mater 2009;161:1069–1074.

[131] Pandikumar A, Ramaraj R. Titanium dioxide–gold nanocomposite materials embedded in silicate sol–gel film catalyst for simultaneous photodegradation of hexavalent chromium and methylene blue. J Hazard Mater 2012;203–204:244–250.

[132] Zhang YC, Li J, Xu HY. One-step in situ solvothermal synthesis of SnS_2/TiO_2 nanocomposites with high performance in visible light-driven photocatalytic reduction of aqueous Cr(VI). Appl Catal B 2012;123–124:18–26.

[133] Yang L, Xiao Y, Liu S, Li Y, Cai Q, Luo S, Zeng G. Photocatalytic reduction of Cr(VI) on WO_3 doped long TiO_2 nanotube arrays in the presence of citric acid. Appl Catal B 2010;94:142–149.

[134] Gherbi R, Nasrallah N, Amrane A, Maachi R, Trari M. Photocatalytic reduction of Cr(VI) on the new hetero-system $CuAl_2O_4/TiO_2$. J Hazard Mater 2011;186:1124–1130.

[135] Meichtry JM, Rivera V, Di Iorio Y, Rodríguez HB, Román ES, Grela MA, Litter MI. Photoreduction of Cr(VI) using hydroxoaluminiumtricarboxymonoamide phthalocyanine adsorbed on TiO_2. Photochem Photobiol Sci 2009;8:604–612.

[136] Park Y, Kim W, Monllor-Satoca D, Tachikawa T, Majima T, Choi W. Role of interparticle charge transfers in agglomerated photocatalyst nanoparticles: demonstration in aqueous suspension of dye-sensitized TiO_2. J Phys Chem Lett 2013;4:189–194.

[137] Cid LC, Grande MC, Acosta EO, Ginzberg B. Removal of Cr(VI) and humic acid by heterogeneous photocatalysis in a laboratory reactor and a pilot reactor. Ind Eng Chem Res 2012;51:9468–9474.

[138] WHO. Uranium in drinking-water, background document for development of WHO—Guidelines for Drinking-water Quality. Geneva: WHO; 2004. Report nr WHO/SDE/WSH/03.04/118.

[139] Nriagu J, Nam DH, Ayanwola TA, Dinh H, Erdenechimeg E, Ochir C, Bolormaa TA. High levels of uranium in groundwater of Ulaanbaatar, Mongolia. Sci Total Environ 2012;414:722–726.

[140] Kryvoruchko AP, Yurlova LY, Atamanenko ID, Kornilovich BY. Ultrafiltration removal of U(VI) from contaminated water. Desalination 2004;162:229–236.

[141] Behrends T, Van Cappellen P. Competition between enzymatic and abiotic reduction of uranium(VI) under iron reducing conditions. Chem Geol 2005;220:315–327.

[142] Liger E, Charlet L, Van Cappellen P. Surface catalysis of uranium(VI) reduction by iron(II). Geochim Cosmochim Acta 1999;63:2939–2955.

[143] Zeng H, Giammar DE. U(VI) reduction by Fe(II) on hematite nanoparticles. J Nanopart Res 2011;13:3741–3754.

[144] Zou W, Zhao L. Removal of uranium(VI) from aqueous solution using citric acid modified pine sawdust: batch and column studies. J Radioanal Nucl Chem 2012;291:585–595.

[145] Camacho LM, Deng S, Parra RR. Uranium removal from groundwater by natural clinoptilolite zeolite: effects of pH and initial feed concentration. J Hazard Mater 2010;175:393–398.

[146] Mellah A, Chegrouche S, Barkat M. The removal of uranium(VI) from aqueous solutions onto activated carbon: kinetic and thermodynamic investigations. J Colloid Interface Sci 2006;296:434–441.

[147] Pal S, Tewari PK. Composite sorbent for attrition minimization. Desalin Water Treat 2011;28:183–188.

[148] Wazne M, Meng X, Korfiatis GP, Christodoulatos C. Carbonate effects on hexavalent uranium removal from water by nanocrystalline titanium dioxide. J Hazard Mater 2006;136:47–52.

[149] Martins M, Faleiro ML, da Costa AMR, Chaves S, Tenreiro R, Matos AP, Costa MC. Mechanism of uranium (VI) removal by two anaerobic bacterial communities. J Hazard Mater 2010;184:89–96.

[150] Crane RA, Dickinson M, Popescu IC, Scott TB. Magnetite and zero-valent iron nanoparticles for the remediation of uranium contaminated environmental water. Water Res 2011;45:2931–2942.

[151] Crane RA, Scott TB. Nanoscale zero-valent iron: future prospects for an emerging water treatment technology. J Hazard Mater 2012;211–212:112–125.

[152] Klimkova S, Cernik M, Lacinova L, Filip J, Jancik D, Zboril R. Zero-valent iron nanoparticles in treatment of acid mine water from in situ uranium leaching. Chemosphere 2011;82:1178–1184.

[153] Yan S, Hua B, Bao Z, Yang J, Liu C, Deng B. Uranium(VI) reduction by nanoscale zerovalent iron under anoxic conditions. Environ Sci Technol 2010;44:7783–7789.

[154] Chen J, Ollis DF, Rulkens WH, Bruning H. Photocatalyzed deposition and concentration of soluble uranium(VI) from TiO_2 suspensions. Colloids Surf A 1999;151:339–349.

[155] Amadelli R, Maldotti A, Sostero S, Carassiti V. Photodeposition of uranium oxides onto TiO_2 from aqueous uranyl solutions. J Chem Soc Faraday Trans 1991;87:3267–3273.

[156] Selli E, Eliet V, Spini MR, Bidoglio G. Effects of humic acids on the photoinduced reduction of U(VI) in the presence of semiconducting TiO_2 particles. Environ Sci Technol 2000;34:3742–3748.

[157] Craft ES, Abu-Qare AW, Flaherty MM, Garofolo MC, Rincavage HL, Abou-Donia MB. Depleted and natural uranium: chemistry and toxicological effects. J Toxicol Environ Health B 2004;7:297–317.

[158] Selbin J, Ortego JD. Chemistry of uranium (V). Chem Rev 1969;69:657–671.

[159] Evans CJ, Nicholson GP, Faith DA, Kan MJ. Photochemical removal of uranium from a phosphate waste solution. Green Chem 2004;6:196–197.

[160] Eliet V, Bidoglio G. Kinetics of the laser-induced photoreduction of U(VI) in aqueous suspensions of TiO_2 particles. Environ Sci Technol 1998;32:3155–3161.

[161] Boxall C, Le Gurun G, Taylor RJ, Xiao S. The applications of photocatalytic waste minimisation in nuclear fuel processing. In: Boule P, Bahnemann DW, Robertson PKJ, editors. *Environmental Photochemistry Part II (The Handbook of Environmental Chemistry*, Vol. 2, Part M). Berlin, Heidelberg: Springer-Verlag; 2005. p 451–481.

[162] Bonato M, Allen G, Scott T. Reduction of U(VI) to U(IV) on the surface of TiO_2 anatase nanotubes. Micro Nano Lett 2008;3:57–61.

[163] WHO. Lead in drinking-water, background document for development of WHO—Guidelines for Drinking-water Quality. Geneva: WHO; 2011. Report nr WHO/SDE/WSH/03.04/09/Rev/1.

[164] Galus Z. Carbon, silicon, germanium, tin, and lead. In: Bard JA, Parsons R, Jordan J, editors. *Standard Potentials in Aqueous Solution*. New York: Marcel Dekker, Inc.; 1985. p 189–236.

[165] Chenthamarakshan CR, Yang H, Savage CR, Rajeshwar K. Photocatalytic reactions of divalent lead ions in UV-irradiated titania suspensions. Res Chem Intermed 1999;25:861–876.

[166] Inoue T, Fujishima A, Honda K. Photoelectrochemical imaging processes using semiconductor electrodes. Chem Lett 1978;7:1197–1200.

[167] Inoue T, Fujishima A, Honda K. Photoelectrochemical characteristics of photo- electrochemical imaging system.... J Electrochem Soc 1980;127:1582–1588.

[168] Kabra K, Chaudhary R, Sawhney RL. Effect of pH on solar photo catalytic reduction and deposition of Cu(II), Ni(II), Pb(II), and Zn(II): speciation modeling and reaction kinetics. J Hazard Mater 2007;149:680–685.

[169] Kabra K, Chaudhary R, Sawhney RL. Solar photocatalytic removal of Cu(II), Ni(II), Zn(II), and Pb(II): speciation modeling of metal–citric acid complexes. J Hazard Mater 2008;155:424–432.

[170] Kobayashi T, Taniguchi Y, Yoneyama H, Tamura H. Effective surfaces of semiconductor catalysts for light-induced heterogeneous reactions evaluated by simultaneous photodeposition of both oxidation and reduction products. J Phys Chem 1983;87:768–778.

[171] Lawless D, Res A, Harris R, Serpone N, Minero C, Pelizzetti E, Hidaka H. Removal of toxic metal from solutions by photocatalysis using irradiated platinized titanium dioxide: removalof lead. Chim Ind 1990;72:139–146.

[172] Maillard-Dupuy C, Guillard C, Pichat P. The degradation of nitrobenzene in water by photocatalysis over TiO_2—kinetics and products—simultaneous elimination of benzamide or phenol or Pb2+ cations. New J Chem 1994;18:941–948.

[173] Mishra T, Hait J, Aman N, Jana RK, Chakravarty S. Effect of UV and visible light on photocatalytic reduction of lead and cadmium over titania based binary oxide materials. J Colloid Interface Sci 2007;316:80–84.

[174] Rajeshwar K, Chenthamarakshan CR, Ming Y, Sun W. Cathodic photoprocesses on titania films and in aqueous suspensions. J Electroanal Chem 2002;538–539:173–182.

[175] Rajh T, Ostafin AE, Micic OI, Tiede DM, Thurnauer MC. Surface modification of small particle TiO_2 colloids with cysteine for enhanced photochemical reduction: an EPR study. J Phys Chem 1996;100:4538–4545.

[176] Rajh T, Tiede DM, Thurnauer MC. Surface modification of TiO_2 nanoparticles with bidentate ligands studied by EPR spectroscopy. J Non-Cryst Solids 1996;205–207:815–820.

[177] Tennakone K. Photoreduction of carbonic acid by mercury coated n-titanium oxide. Sol Energy Mater 1984;10:235–238.

[178] Tennakone K, Wijayantha KGU. Heavy-metal extraction from aqueous medium with an immobilized TiO_2 photocatalyst and a solid sacrificial agent. J Photochem Photobiol A 1998;113:89–92.

[179] Thurnauer MC, Rajh T, Tiede DM. Surface modification of TiO_2: correlation between structure, charge separation and reduction properties. Acta Chem Scand 1997;51:610–618.

[180] Torres J, Cervera-March S. Kinetics of the photoassisted catalytic oxidation of Pb(II) in TiO_2 suspensions. Chem Eng Sci 1992;47:3857–3862.

[181] Brillas E, Calpe JC, Cabot PL. Degradation of the herbicide 2,4-dichlorophenoxyacetic acid by ozonation catalyzed with Fe2+ and UVA light. Appl Catal B 2003;46:381–391.

[182] Kasprzyk-Hordern B, Ziółek M, Nawrocki J. Catalytic ozonation and methods of enhancing molecular ozone reactions in water treatment. Appl Catal B 2003;46:639–669.

[183] Tanaka K, Abe K, Hisanaga T. Photocatalytic water treatment on immobilized TiO_2 combined with ozonation. J Photochem Photobiol A 1996;101:85–87.

[184] Breitenkamp M, Henglein A, Lilie J. Mechanism of the reduction of lead ions in aqueous solution (a pulse radiolysis study), Ber. Bunsenges. Phys Chem Chem Phys 1976;80:973–979.

[185] Li L, Jiang F, Liu J, Wan H, Wan Y, Zheng S. Enhanced photocatalytic reduction of aqueous Pb(II) over Ag loaded TiO_2 with formic acid as hole scavenger. J Environ Sci Health A 2012;47:327–336.

[186] Yang Z-P, Zhang C-J. Kinetics of photocatalytic reduction of Pb(II) on nanocrystalline TiO_2 coatings: a quartz crystal microbalance study. Thin Solid Films 2010;518:6006–6009.

[187] Harraz FA, Abdel-Salam OE, Mostafa AA, Mohamed RM, Hanafy M. Rapid synthesis of titania–silica nanoparticles photocatalyst by a modified sol–gel method for cyanide degradation and heavy metals removal. J Alloys Compd 2013;551:1–7.

[188] WHO. Mercury in drinking-water, background document for development of WHO—Guidelines for Drinking-water Quality. Geneva: WHO; 2005. Report nr WHO/SDE/WSH/05.08/10.

[189] Balej J. Mercury. In: Bard JA, Parsons R, Jordan J, editors. *Standard Potentials in Aqueous Solution*. New York: Marcel Dekker, Inc.; 1985. p 265–286.

[190] Aguado MA, Cervera-March S, Giménez J. Continuous photocatalytic treatment of mercury(II) on titania powders. Kinetics and catalyst activity. Chem Eng Sci 1995;50:1561–1569.

[191] Chen LX, Rajh T, Wang Z, Thurnauer MC. EXAFS studies of surface structures of TiO_2 nano-particles. J Phys Chem B 1997;101:10688–10697.

[192] de la Fournière EM, Leyva AG, Gautier EA, Litter MI. Treatment of phenylmercury salts by heterogeneous photocatalysis over TiO_2. Chemosphere 2007;69:682–688.

[193] Hegyi J, Horváth O. Photocatalytic reduction of mercury(II) and simultaneous oxidative degradation of surfactants in titanium dioxide suspensions. Prog Colloid Polym Sci 2004;125:10–16.

[194] Horváth O, Hegyi J. Light-induced reduction of heavy-metal ions on titanium dioxide dispersions. Prog Colloid Polym Sci 2002;117:211–216.

[195] Kaluza U, Boehm HP. Titanium dioxide catalyzed photooxidation of mercury. J Catal 1971;22:347–358.

[196] Khalil LB, Rophael MW, Mourad WE. The removal of the toxic Hg(II) salts from water by photocatalysis. Appl Catal B 2002;36:125–130.

[197] Lau LD, Rodriguez R, Henery S, Manuel D, Schwendiman L. Photoreduction of mercuric salt solutions at high pH. Environ Sci Technol 1998;32:670–675.

[198] Miranda C, Yáñez J, Contreras D, Garcia R, Jardim WF, Mansilla HD. Photocatalytic removal of methylmercury assisted by UV-A irradiation. Appl Catal B 2009;90:115–119.

[199] Scott Rader W, Solujic L, Milosavljevic EB, Hendrix JL, Nelson JH. Photochemistry of aqueous solutions of dicyanomercury(II) and potassium tetracyanomercurate(II). J Sol Energy-T ASME 1994;116:125–129.

[200] Serpone N, Ah-You YK, Tran TP, Harris R, Pelizzetti E, Hidaka H. AM1 simulated sunlight photoreduction and elimination of Hg(II) and CH_3Hg(II) chloride salts from aqueous suspensions of titanium dioxide. Sol Energy 1987;39:491–498.

[201] Skubal LR, Meshkov NK. Reduction and removal of mercury from water using arginine-modified TiO_2. J Photochem Photobiol A 2002;148:211–214.

[202] Tanaka K, Harada K, Murata S. Photocatalytic deposition of metal ions onto TiO_2 powder. Sol Energy 1986;36:159–161.

[203] Tennakone K, Ketipearachchi US. Photocatalytic method for removal of mercury from contaminated water. Appl Catal B 1995;5:343–349.

[204] Tennakone K, Thaminimulle CTK, Senadeera S, Kumarasinghe AR. TiO_2-catalysed oxidative photodegradation of mercurochrome: an example of an organo-mercury compound. J Photochem Photobiol A 1993,70.193–195.

[205] Tennakone K, Wickramanayake S. Cyclic photocleavage of water with the intermediate redox couple mercurous oxide/mercury. J Phys Chem 1986;90:1219–1222.

[206] Wang X, Pehkonen SO, Ray AK. Photocatalytic reduction of Hg(II) on two commercial TiO_2 catalysts. Electrochim Acta 2004;49:1435–1444.

[207] Wang ZH, Zhuang QX. Photocatalytic reduction of pollutant Hg(II) on doped WO_3 dispersion. J Photochem Photobiol A 1993;75:105–111.

[208] López-Muñoz MJ, Aguado J, Arencibia A, Pascual R. Mercury removal from aqueous solutions of $HgCl_2$ by heterogeneous photocatalysis with TiO_2. Appl Catal B 2011;104:220–228.

[209] Botta SG, Rodríguez DJ, Leyva AG, Litter MI. Features of the transformation of HgII by heterogeneous photocatalysis over TiO_2. Catal Today 2002;76:247–258.

[210] Custo G, Litter MI, Rodríguez D, Vázquez C. Total reflection X-ray fluorescence trace mercury determination by trapping complexation: application in advanced oxidation technologies. Spectrochim Acta A 2006;61:1119–1123.

[211] Lenzi GG, Fávero CVB, Colpini LMS, Bernabe H, Baesso ML, Specchia S, Santos OAA. Photocatalytic reduction of Hg(II) on TiO_2 and Ag/TiO_2 prepared by the sol–gel and impregnation methods. Desalination 2011;270:241–247.

[212] Serpone N, Borgarello E, Pelizzetti E. Photoreduction and photodegradation of inorganic pollutants: I. Cyanides. In: Schiavello M, editor. *Photocatalysis and Environment*. Amsterdam: Kluwer Academic Publishers; 1988. p 499–526.

[213] Yin Y, Liang J, Yang L, Wang Q. Vapour generation at a UV/TiO_2 photocatalysis reaction device for determination and speciation of mercury by AFS and HPLC-AFS. J Anal At Spectrom 2007;22:330–334.

[214] Han C, Zheng C, Wang J, Cheng G, Lv Y, Hou X. Photo-induced cold vapor generation with low molecular weight alcohol, aldehyde, or carboxylic acid for atomic fluorescence spectrometric determination of mercury. Anal Bioanal Chem 2007;388:825–830.

[215] Matusiewicz H, Stanisz E. Evaluation of the catalyzed photo-cold vapour generation for determination of mercury by AAS. J Brazil Chem Soc 2012;23:247–257.

10

NANOSTRUCTURED TITANIUM DIOXIDE FOR PHOTOCATALYTIC WATER TREATMENT

David G. Rickerby
European Commission Joint Research Centre, Institute for Environment and Sustainability, Ispra, Italy

10.1 INTRODUCTION

Heterogeneous photocatalysis has attracted interest as a method for water treatment because it is very effective in both degrading organic and inorganic chemical pollutants and destroying viruses and bacteria [1]. Additional applications include air purification, self-cleaning and sterilizing surfaces, and water photolysis [2]. Anatase titanium dioxide is the most commonly employed photocatalytic material but its efficiency is limited by an absorption spectrum confined to the ultraviolet (UV) range and a high charge carrier recombination rate. These disadvantages have stimulated investigation of alternative photocatalyst materials such as other transition metal oxides, main group element oxides, rare earth oxides and binary sulfides, as well as ternary and quaternary compounds [3]. Until now, however, these efforts have met with relatively limited success.

Titanium dioxide exists in three naturally occurring phases: tetragonal rutile, tetragonal anatase, and orthorhombic brookite (rutile is the most stable phase except in the case of very small crystal sizes where anatase and brookite become more stable). The anatase phase has a band gap, E_g, of 3.2 eV and is thus able to absorb light only of wavelengths shorter than 388 nm. Its (001) surface is highly photoreactive due to a high degree of coordinative unsaturation: only fivefold coordinated Ti and twofold coordinated O atoms are present in the first layer [4]. However, less than 10% of the surface area of typical anatase crystals consists of {001} facets, while a major part of the surface area is made up of stable {101} facets.

The application of surface science techniques has provided molecular-level insights into the basic mechanisms of photocatalytic phenomena. Among the key topics of interest are photon absorption, charge carrier transport, trapping and recombination, electron transfer dynamics, adsorption and the adsorbed state, photocatalytic reaction mechanisms, effects of inhibitors and promoters, and the phase and form of the photocatalyst [5]. Studies of the relationships between crystal and surface structure, surface chemistry, and electronic, electrochemical, and photoelectrochemical properties have advanced our understanding of the behavior of photocatalytic materials, thereby enabling improvements in their performance [6]. Bulk and surface defects, such as oxygen vacancies, exert a significant influence on the electronic properties and catalytic activity [7].

Absorption by the photocatalyst of photons of energy greater than that of the band gap leads to the formation of electron–hole pairs that can initiate redox reactions at the semiconductor surface. However, this process is relatively inefficient because only a small fraction of the incident photons take part in photocatalytic reactions [8]. The use of nanocrystalline materials has been widely recognized as one way of improving the efficiency of the photocatalytic reaction [9]. Not only is the reactivity increased due to the higher specific surface area of the nanomaterial but quantum confinement effects can also occur when the Bohr radius approaches or becomes larger than the crystal dimensions. Quantization of the energy levels in the conduction and valence bands and modification of the band gap result in an increased charge transfer rate, which may lead to improved photocatalytic efficiency when the reaction rate is charge transfer limited [10].

Nanomaterials for Environmental Protection, First Edition. Edited by Boris I. Kharisov, Oxana V. Kharissova, and H. V. Rasika Dias.

The main focus of current work on methods for increasing photocatalytic activity is on the modification of the surface chemistry and controlled growth of TiO_2 nanocrystals. There has been much recent progress on the synthesis of nanocrystals with a higher proportion of reactive facets, and theoretical modeling has assisted by providing an understanding of the influence of surface energy effects in determining morphology as the crystal size is reduced [11]. Attempts have also been made to increase the reaction rate by surface loading with noble metal clusters [12, 13], to enhance photocatalytic activity by doping with metals to inhibit electron–hole recombination by charge trappping [14, 15], and to achieve visible light photocatalysis by doping with various metals and nonmetals [16, 17].

There are many existing and potential applications of photocatalysis in drinking water purification and wastewater treatment, and it is often claimed that the advantage of photocatalytic treatment with respect to chlorination is that it is able to degrade organic pollutants without the production of toxic by-products. However, this is not entirely accurate as problems may still be experienced due to the formation of long-lived toxic intermediates by complex photodegradation processes [18, 19]. It is therefore essential to develop a more comprehensive knowledge of the reaction pathways in order to ensure the complete elimination of any toxic compounds produced.

10.2 PHOTOCATALYTIC REACTION

The basic mechanism of photocatalysis is illustrated schematically in Figure 10.1. Incident photons with energies greater than that of the semiconductor band gap create electron–hole pairs (e^-, h^+) by excitation of electrons from the valence band to the conduction band. These electron–hole pairs take part in redox reactions with acceptor and donor species on the surface of the photocatalyst [20, 21]. Reactions between valence band holes and adsorbed water molecules or surface hydroxide ions lead to the formation of hydroxyl radicals according to the equations:

$$h^+ + H_2O \rightarrow HO^\bullet + H^+ \tag{10.1}$$

$$h^+ + OH^- \rightarrow HO^\bullet \tag{10.2}$$

Dissolved oxygen molecules trap conduction band electrons to form superoxide ions that can react with hydrogen ions to form hydroperoxyl radicals:

$$e^- + O_2 + H^+ \rightarrow O_2^{\bullet -} + H^+ \rightarrow HO_2^\bullet \tag{10.3}$$

Hydrogen peroxide may be formed by reaction of hydroperoxyl radicals with hydrogen ions and may act as an oxidant or dissociate into hydroxide ions and hydroxyl radicals:

$$e^- + HO_2^\bullet + H^+ \rightarrow H_2O_2 \rightarrow OH^- + HO^\bullet \tag{10.4}$$

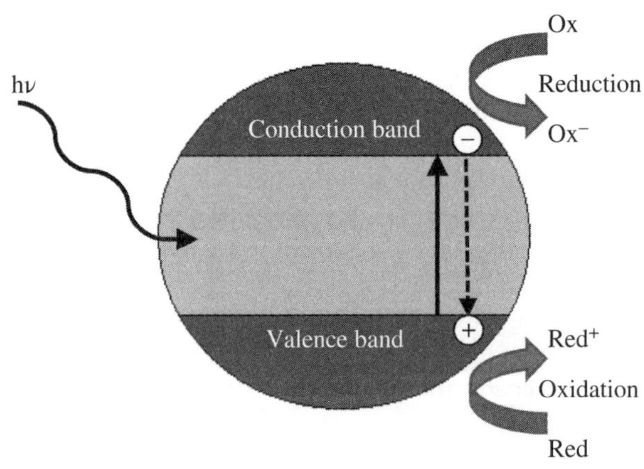

FIGURE 10.1 Processes involved in the photocatalytic reaction. Reproduced with permission from Ref. [83]. © 2011, Nano Science and Technology Institute.

Photoreduction and photooxidation reactions occurring on or close to the photocatalyst surface are able to degrade organic and inorganic pollutants and destroy microorganisms. A general expression describing photocatalytic oxidation of a chlorinated hydrocarbon compound may be written [8, 10]:

$$C_xH_yCl_z + \left[\frac{x+(y-z)}{4}\right]O_2 \rightarrow xCO_2 + zH^+ + zCl^- + \left[\frac{(y-z)}{2}\right]H_2O \quad (10.5)$$

The photonic efficiency ζ is used to quantify photocatalyst performance. It can be determined from the ratio of the initial rate of photocatalytic degradation to the rate of incident photons [22] and should not be confused with the quantum yield Φ, which relates the amount of reactant consumed to the number of photons absorbed by the catalyst at a given wavelength λ. The photonic efficiency has an inverse dependence on the incident photon rate [8]. This is due to the limitation of the reaction rate because of electron–hole recombination, which competes with the reaction of electrons and holes with acceptor and donor molecules and results in a nonlinear dependence of the photonic efficiency on the photon reaction rate. Generation of electrons and holes is not sufficient by itself to ensure that a photocatalytic reaction occurs. The electron–hole recombination rate must, in addition, be low enough to allow an adequate number of carriers to reach the surface of the photocatalyst and react with adsorbed molecules.

10.3 SYNTHESIS TECHNIQUES

A wide range of methods are available for the preparation of nanostructured TiO_2. These include sol–gel techniques, micelle and inverse micelle methods, nonhydrolic sol processes, hydrothermal and solvothermal methods, direct oxidation, chemical vapor deposition, physical vapor deposition, electrodeposition, sonochemical methods, and microwave techniques [23]. Many different types of low-dimensional and three-dimensional nanostructures can be fabricated using these synthesis methods, such as nanoparticles, nanorods, nanowires, nanotubes, and nanosheets. Size-related changes in mechanical, chemical, electronic, and optical properties are observed in these nanomaterials; phase transitions may also be affected.

It is possible to control the particle size and crystal structure of TiO_2 nanomaterials by varying the synthesis conditions. The proportions of rutile and anatase in mixed-phase thin films deposited by pulsed laser deposition are dependent on the irradiation energy and the oxygen working pressure [24] and can additionally be modified by high-temperature annealing [25]. The crystal structure of TiO_2 thin films produced by radiofrequency magnetron sputtering is strongly influenced by the deposition temperature, with higher temperatures favoring formation of the rutile phase [26]. Pure anatase films can be obtained by increasing the substrate to target distance [27]. This results in increased scattering of the sputtered particles, leading to higher energy losses before they impinge on the substrate and a consequent reduction in the thermal energy transmitted to the growing film. In the case of films produced by chemical vapor deposition, both the crystal structure and the grain size are highly dependent on the synthesis temperature [28].

The physical characteristics of nanopowders are related to the preparation method and the synthesis parameters. For the sol–gel process the chemistry of the precursor is critical in determining both the particle size and the phase content [29], while the diameter of single-phase anatase nanoparticles prepared by heating amorphous TiO_2 in air increases with increasing temperature [30]. The proportions of anatase and rutile phases present in TiO_2 nanoparticles obtained by thermal plasma synthesis can be controlled by varying the working pressure. Rutile is the more abundant phase for low-pressure synthesis, while anatase predominates at higher pressures [31]. Temperature and pressure play similarly important roles in determining particle size and crystallinity in nanopowders synthesized by direct current reactive magnetron sputtering [32]. The properties of anatase nanoparticles prepared by the solvothermal technique, on the other hand, can be controlled by varying the temperature, precursor concentration, and hydrolysis ratio [33].

Nanoparticle size is a critical parameter influencing both the thermal stability [34] and the adsorption of molecules on the surface [35], which are governed by size-dependent effects on the surface enthalpy and surface free energy. Due to the influence of particle size on the surface free energy, adsorption increases much more rapidly with decreasing particle size than would be expected on the basis of the increase in surface area alone. The photocatalytic activity of nanoparticles prepared by the sol–gel method is dependent on the calcination temperature due to its influence on the particle size, physical properties, and crystallinity [36]. Particle shape is the main factor affecting photocatalytic activity in nanoparticles produced by the flame hydrolysis technique [37–39].

Surface defect chemistry has a considerable influence on the photocatalytic reaction [40]. The principal defects of interest on the TiO_2 surface are Ti^{3+} point defects and oxygen vacancies, which act as traps for charge carriers [41]. Ti^{3+} surface defects have an additional important function in photocatalytic reactions because they act as preferential binding sites for adsorbed species. The semiconducting behavior of TiO_2 is closely related to the defect chemistry due to its effect on the electron and hole concentrations and charge carrier mobility. Control of the defect structures during processing of semiconductor materials thus allows significant modification of their photocatalytic characteristics [42]. Hydrogen thermal treatment is effective in generating oxygen vacancies and Ti^{3+} point defects in commercially available TiO_2 nanopowders [41]. The surface density of the Ti^{3+} defects may be increased by increasing crystallite size in anatase nanoparticles produced by solvothermal synthesis [43, 44]. Plasma treatment can create new oxygen vacancy states in nanocrystalline TiO_2 that change its light absorption properties [45].

10.4 STRATEGIES FOR INCREASING PHOTOREACTIVITY

Because of the relatively poor efficiency of the photocatalytic reaction, a great deal of the current research in this field is directed toward development of improved photocatalysts by chemical or structural modification [46]. A photocatalyst immobilized on a substrate has a smaller effective surface area compared to nanoparticles in aqueous suspension. The specific surface area is greatly increased by fabricating highly ordered arrays of TiO_2 nanotubes by anodic oxidation of a titanium substrate. These nanotube arrays display enhanced photocatalytic activities compared to thin films [47]. The effect of varying the nanotube diameter is comparatively modest, however, because the increase in the active surface area of the photocatalyst as the diameter is decreased is offset by a reduction in light transmission.

Doping with low concentrations of transition metal ions can increase the photocatalytic activity of TiO_2 by creating charge carrier traps that prevent recombination [14, 15]. For relatively large particles, volume recombination is the primary mechanism, while for extremely small particles the reduction of the transport distance of electrons and holes to the surface increases the charge carrier transfer rate. Below a critical diameter, surface recombination becomes dominant, reducing the photocatalytic efficiency [48]. A systematic investigation of TiO_2 doped with Fe^{3+}, V^{4+}, Re^{5+}, Mo^{5+}, Ru^{3+}, Mn^{3+}, and Rh^{3+} ions [14] indicated that the photoreactivity with chloroform and carbon tetrachloride increased exponentially until the solubility limit for the metal was exceeded, after which it decreased again, due to the effect of the excess dopant present on the surface. Another study [15] found that doping with Nd^{3+} and Pd^{2+} ions was effective in increasing the degradation rate for 2-chlorophenol, while Pt^{4+} resulted in only a marginal improvement and Fe^{3+} was detrimental. The position of the dopant in the crystal lattice is determined by its size relative to the Ti^{4+} ion; interstitial dopants are more effective carrier traps than substitutional ones because they cause greater perturbation of the localized energy levels. Experimental evidence [49] indicates that the effect on photocatalytic activity is greatest when the ionic radius of the dopant is approximately 0.8 nm.

The photoreactivity of metal ion-doped TiO_2 depends in a complex manner on the concentration and distribution of the dopants, their energy levels within the lattice, and the d electron configuration [14]. Powders doped with different transition metal ions by a wet impregnation method generally showed higher photocatalytic activity for degradation of 4-nitrophenol than the undoped photocatalyst [50]; however, no simple correlation could be made between photoreactivity and the physicochemical properties of the photocatalysts. The effect of nontransition metals as dopants has also been investigated. Sb-doped nanocrystalline TiO_2 synthesized by a coprecipitation method was found to be more effective in degrading methylene blue than similar material in the undoped state [51]. TiO_2 nanoparticles prepared by the sol–gel process and codoped with Fe^{3+} and Eu^{3+} ions that functioned as traps for holes and electrons, respectively, showed significantly enhanced photodegradation efficiency for chloroform as a result of the synergistic effect of the two different dopants [52].

Surface loading with noble metal clusters can increase the photoreactivity of TiO_2 by shifting the energy levels in the band gap [12] or by injecting electrons into the conduction band [13]. For very small clusters of atoms, the physical, chemical, and electronic properties may additionally be modified by quantum confinement effects. These result in a metal to insulator transition in Au, Pd, and Ag when the particle diameter is below approximately 3 nm [53, 54]. The peak catalytic activity occurs in the same size range, indicating a strong correlation with the electronic properties of the metal cluster and in particular the band gap [55]. Figure 10.2 shows a conceptual model of the evolution of the morphology with cluster size and the resultant effect on the photocatalytic activity. For a diameter of approximately 3 nm, the cluster has a bilayer configuration and the turnover frequency (TOF) is a maximum.

Comparison may be made between this idealized structure and the appearance of an approximately 5-nm-diameter Au cluster, seen in Figure 10.3, deposited on the surface of mixed-phase TiO_2 from an aqueous solution of $HAuCl_4$ by a precipitation method [56]. Loading TiO_2 with Au in this way has been shown to increase the photodegradation efficiency for 4-chlorophenol [56] and methyl tertiary-butyl ether [57]. Corresponding improvements have also been reported in the photocatalytic activity toward methyl orange using Ag-modified TiO_2 films prepared by radiofrequency (RF) magnetron sputtering [58]. The presence of nanosized Pt deposits on the surface of TiO_2 has furthermore been shown to affect the kinetics and mechanisms of the photocatalytic degradation of trichloroacetate [59].

UV light constitutes only about 5% of the solar radiation spectrum compared to approximately 46% for the visible wavelengths. This has motivated attempts to realize visible light photocatalysts by doping TiO_2 with anions or cations to narrow the

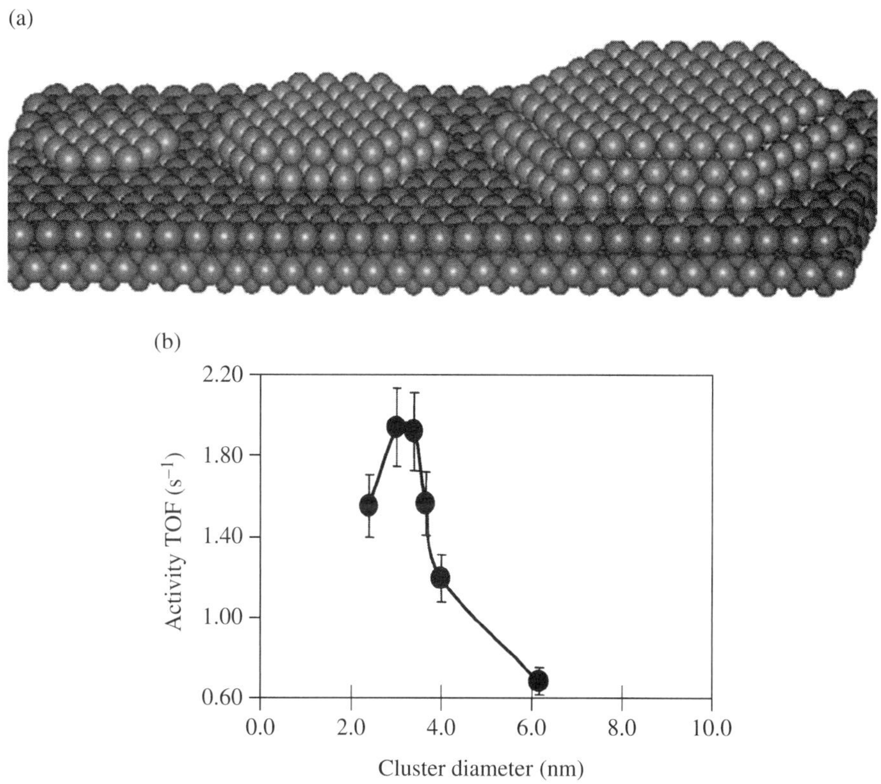

FIGURE 10.2 (a) Size dependence of the morphology of gold clusters on the TiO_2 (110) surface; (b) relationship of the cluster diameter to photocatalytic activity. Reproduced with permission from Ref. [55]. © 2002, Academic Press.

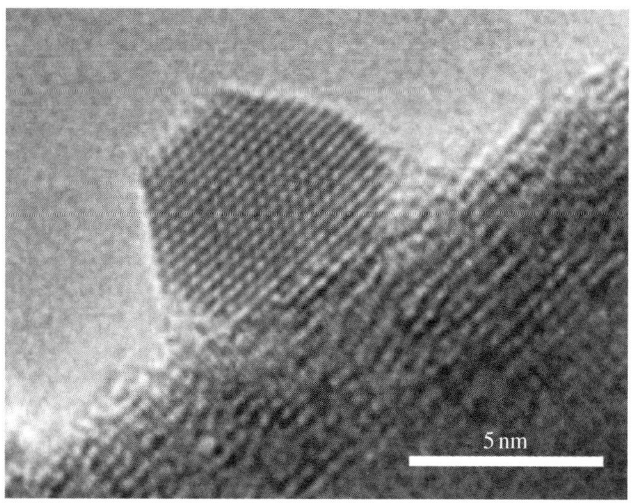

FIGURE 10.3 High-resolution transmission electron microscopy image of a gold cluster on a TiO_2 surface. Reproduced with permission from Ref. [56]. © 2004, Springer.

band gap and thereby expand the spectral range of absorbed wavelengths. Doping may be carried out either by ion implantation or by chemical processes such as sol–gel. Extensive research activity followed initial success in using ion implantation to inject small amounts of Cr and V that resulted in increased photocatalytic activity [60, 61]. Cations occupy substitutional Ti lattice sites and cause a red shift of the absorption band by an amount dependent on the implantation energy, which determines the depth distribution of the implanted ions. Doping with other metals, such as Fe, Ni, Rh, Mn, and Nb, produces similar effects to varying degrees. Interestingly, no effect is observed for implantation with Ar, Mg, or Ti, suggesting that the absorption shift arises from interaction with the implanted metal ions rather than local structural modification due to high-energy implantation. Recent work has indicated that doping with Ag, Rb, Y, and La is likewise ineffective in shifting the absorption spectrum [62], evidently because the incompatible ionic radii prevent occupation of Ti lattice sites.

Doping TiO_2 with anions likewise increases light absorption and photocatalytic activity, but in this case the dopants occupy substitutional O sites in the lattice [63]. Thin films deposited by sputtering in a low-pressure N_2/Ar atmosphere and powders heat-treated in NH_3/Ar were more photoreactive than the same materials in the undoped condition [64]. Nanocrystalline powders doped with N have been prepared by different synthesis routes from a wide variety of precursors [16, 17]. Proposed methods include hydrolysis of tetra-butyl titanate in an aqueous NH_3 solution [65] and the synthesis of nanopowders with controlled levels of N by sequential reaction of $TiCl_4$ in toluene with H_2O and NH_3 followed by final calcination [66]. Visible light photocatalysis was demonstrated in B-doped and B-/N-codoped nanopowders prepared by a sequential reaction method [67] and by the sol–gel process [68]. Synthesis of C-doped anatase nanoparticles with visible light photoreactivity has been carried out by a modified sol–gel technique [69]. The sol–gel process has also been used to produce N-/F-codoped nanopowders [70, 71] (with NH_4F and NH_4Cl as precursors). N-/C-codoped TiO_2 photocatalysts (with NH_3 vapor as the N precursor and alcohols as the C precursor) have been prepared by a hydrothermal method [72].

Crystal shape is a major determining factor in the photocatalytic behavior of anatase TiO_2 because of the influence of the highly reactive {001} facets. On the basis of the standard Wulff construction, less than 6% of the total surface area of a typical anatase crystal is composed of these facets, while more stable {101} surfaces constitute the predominant fraction [73]. Substantial research efforts are therefore currently dedicated to the methods for controlled growth of TiO_2 nanocrystals containing an elevated percentage of high-energy {001} facets with the aim of increasing their photocatalytic activity. Before discussing methods of tailoring crystal morphology it is first necessary to consider the underlying thermodynamic principles that govern crystal shape due to the preferential growth of low-energy facets. It should also be remarked that it has recently been discovered [74] that, at variance with the generally accepted explanation, clean {001} surfaces are under certain conditions less reactive than {101} in photooxidation reactions that produce hydroxide radicals and photoreduction reactions that generate hydrogen.

10.5 PARTICLE SHAPE CALCULATION

A thermodynamic model has been developed that allows calculation of the particle shape and phase stability of faceted TiO_2 nanocrystals [75–77]. On the basis of this model it has been established that there is a size dependence of the shape as a result of the increasing surface contribution to the total free energy of the crystal with decreasing diameter. The model takes into account the variation of the free energy with crystal size and shape due to the increasing contribution of surface effects as the crystal size is decreased.

For a nanoparticle of a specific phase, the total free energy G^0 can be derived from the sum of the bulk and surface components:

$$G^0 = G^{\text{bulk}} + G^{\text{surf}} \tag{10.6}$$

The free energy in the formation of a nanocrystal is determined by summing the weighted values of the individual surfaces:

$$G^0 = \Delta_f G^0 + M(1-e)\frac{\left[q\sum_i f_i \gamma_i\right]}{\rho} \tag{10.7}$$

where $\Delta_f G^0$ is the free energy of formation of bulk material, M is the molar mass, e is the volume dilation due to surface tension, q is the surface to volume ratio, f_i is a weighting factor, γ_i is the surface energy of surface i, and ρ is the density. The size dependence of the free energy arises due to the increase in the surface to volume ratio and the volume dilation with decreasing crystal size. The shape dependence results from the changes in the surface to volume ratio and the surface free energy term related to the fraction of a given surface present in the crystal.

The average surface tension σ can be calculated from the weighted sum of the individual values σ_i of each surface of the crystal:

$$\sigma = \sum_i f_i \sigma_i \tag{10.8}$$

The effective pressure P_{eff} due to surface tension in a particle of radius R is estimated by applying the Laplace–Young equation:

$$P_{eff} = \frac{2\sigma}{R} \tag{10.9}$$

The volume dilation e due to the effect of surface tension is given by:

$$e = \frac{2\beta\sigma}{R} \tag{10.10}$$

where β is the compressibility.

A numerical minimization procedure is employed to determine the crystal shape with the lowest free energy for a given nanoparticle size using values of γ_i and σ_i calculated by density functional theory for clean and hydrated low index surfaces of anatase and rutile [75]. Anatase is the thermodynamically stable phase for particle sizes below approximately 10 nm in vacuum and approximately 15 nm in water, although metastable anatase can persist even for particle diameters greater than this. The phase transition corresponds to the point of intersection of the theoretically calculated free energies. For very small crystals (<3-nm diameter) the contribution to the total surface energy from the edge and corner components may in addition become significant [78]. Since γ and σ are sensitive to the surface state, the shape and phase stability of TiO_2 nanocrystals depend on the degree of hydrogenation [75, 76]. It is well known that pH influences both the particle size and the shape in sol–gel preparation of nanoscale titanium dioxide [79]. This allows the possibility of controlling the morphology and phase transition by modifying the surface chemistry; environmentally sensitive phase maps for TiO_2 nanocrystals can be constructed using the phase boundaries derived from the thermodynamic model [80].

Natural anatase crystals typically exhibit a truncated tetragonal bipyramidal shape, which has also been observed in the nanocrystalline form [81]. This morphology is characterized by A, the side of the base of the pyramid, and B, the side of the square {001} facets at the opposing ends of the crystal [79]. The ratio B/A defines the degree of truncation. As a result of the size dependence of the total free energy, this ratio is smaller for nanocrystals than for macroscopic crystals. Simulated images of anatase nanocrystals [82, 83] generated by the thermodynamic model and the standard Wulff construction are contrasted in Figure 10.4; only the Ti atoms are shown for clarity and to facilitate comparison with transmission electron micrographs. The shape predicted by the thermodynamic theory corresponds closely to that of anatase nanoparticles synthesized under near-equilibrium conditions by the sol–gel [84] or hydrothermal routes [85], as confirmed by the high-resolution transmission electron micrograph in Figure 10.5. Conversely, the shape of nanocrystals synthesized under nonequilibrium conditions by gas-phase crystallization can deviate considerably from the theoretical equilibrium morphology [86].

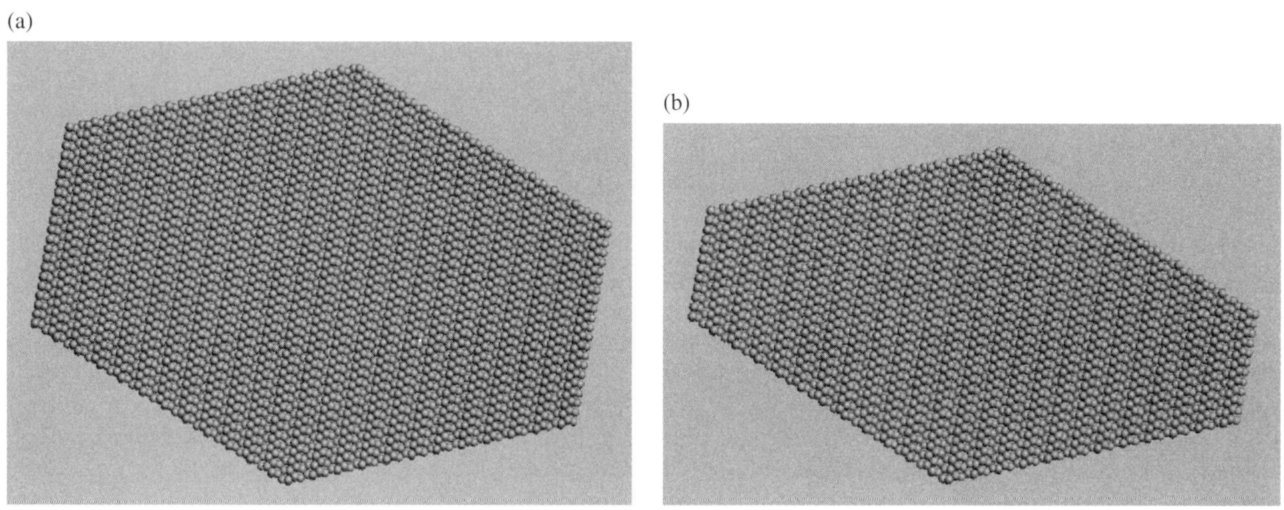

FIGURE 10.4 Simulated anatase TiO_2 nanocrystals in the [100] projection based on: (a) the Wulff construction; (b) the thermodynamic model. Reproduced with permission from Ref. [83]. © 2011, Nano Science and Technology Institute.

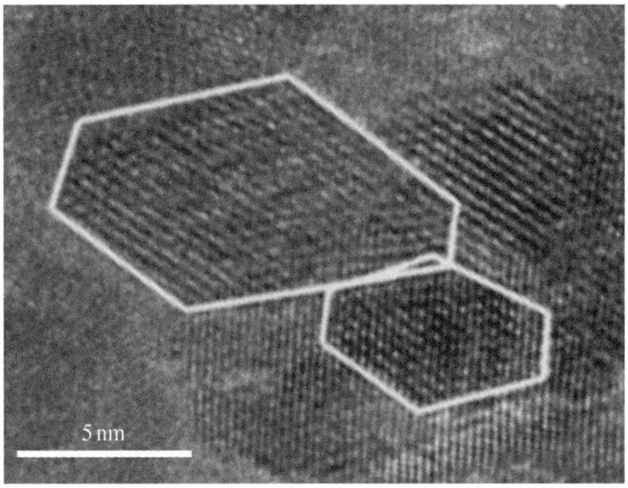

FIGURE 10.5 High-resolution transmission electron microscopy image showing faceted TiO_2 nanocrystals in the [100] projection. Reproduced with permission from Ref. [84]. © 2008, Academic Press.

10.6 CONTROLLED CRYSTAL GROWTH

Growth of stable {101} facets is generally favored with respect to reactive {001} facets in anatase because it results in a minimum energy configuration. While reducing the particle size increases the specific surface area, it also decreases the total fraction of photoreactive {001} facets. For this reason, reducing the particle size alone may not be the most effective way of improving photocatalytic activity. Nevertheless, anatase nanocrystals with well-defined, truncated bipyramidal shapes were found to have superior photocatalytic properties than commercially produced nanoparticles [87]. This can be explained by high-resolution transmission electron microscopy studies of commercial TiO_2 nanoparticles [88] that revealed a wide range of shapes defined by other low index facets in addition to {001} and {101}. Some of the recent progress in crystal shape engineering by selective growth of {001} facets using novel synthesis methods will be described next.

Macroscopic (~1-μm-diameter) anatase crystals with a higher percentage area of reactive facets were synthesized by a hydrothermal method using an aqueous solution of TiF_4 as the precursor and HF as a crystallographic control agent [89]. Due to its low bonding energy, the presence of F strongly bound to Ti lowers the surface energy and thereby promotes preferential growth of {001} facets. The degree of truncation of the crystals could be controlled by varying the pH of the solution and B/A ratios of up to 0.84, equivalent to a surface area consisting of 47% {001} facets, was achieved. An alternative solvothermal method was subsequently developed [90], using a 2-propanol and HF mixture as the capping agent, which allowed synthesis of anatase TiO_2 nanosheets with 64% of {001} facets. Measurements of the rate of hydroxyl radical formation under UV irradiation indicated an increased photoreactivity per unit area compared with conventional TiO_2; however, the practical photocatalytic efficiency was restricted due to the large crystal size. Improved results were obtained with nanosized {001} faceted TiO_2 crystals synthesized from $Ti(SO_4)_2$ in HF by means of a hydrothermal technique [91].

Anatase nanocrystals with a truncated bipyramidal form were prepared by dissolving TiO_2 fibers in an aqueous acetic acid solution followed by a hydrothermal treatment [92]. Their shape was dependent on the acidity of the solution: 9.6% of the overall surface area consisted of {001} facets at a pH of 1.6, while lower pH values tended to suppress their formation, resulting in a sharp tetragonal shape. An ionic liquid-assisted hydrothermal synthesis method, using a $TiCl_4$ precursor in HF or H_2SO_4, was effective in controlling the crystal shape and inhibiting phase transformation [93]. The nanoparticles produced with HF varied from square plates to highly truncated bipyramids, while those prepared using H_2SO_4 had a zigzag morphology, elongated in the [001] direction, that consisted of alternating truncated pyramidal and parallelepiped sections connected by common {001} interfaces. TiO_2 nanosheets with elevated percentages of {001} facets have also been produced by a different hydrothermal method using tetrabutyl titanate in a 47% aqueous solution of HF as the precursor [94].

While hydrothermal methods are the most commonly employed, various other synthesis methods for nanoparticles with controlled shapes have been reported. Decahedral anatase nanocrystals with high photocatalytic activities were prepared by gas-phase reaction of $TiCl_4$ with O_2 using a rapid heating and quenching technique [95, 96]. Nanocrystals with a large percentage of exposed high-energy {001} and {010} facets, synthesized by a nonaqueous method, using titanium isopropoxide as the precursor [97], showed significantly increased photocatalytic activity in comparison to commercial TiO_2 nanopowder. A solvothermal technique has been developed to enable production of various nanocrystal morphologies, including rhombic, truncated

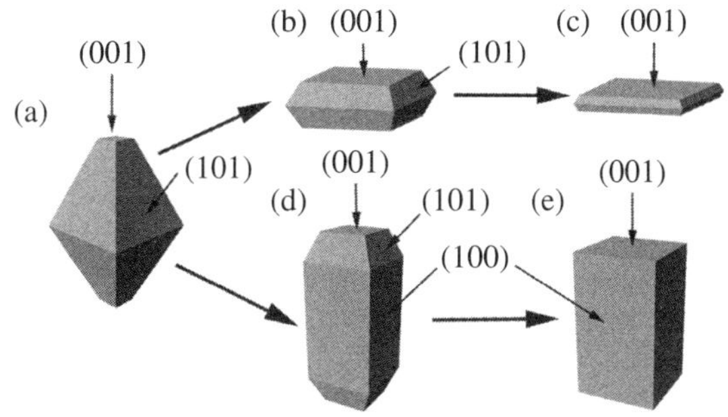

FIGURE 10.6 Anatase nanocrystal shapes: (a) truncated tetragonal bipyramid; (b) truncated tetragonal bipyramid with a large percentage area of {001} facets; (c) square sheets with dominant {001} facets; (d) elongated truncated tetragonal bipyramid with a large percentage of lateral {100} facets; (e) tetragonal cuboid enclosed by {001} and {100} facets. Reproduced with permission from Ref. [100]. © 2011, Wiley—VCH Verlag GmbH & Co.

pyramidal, elongated rhombic, spherical, dog-bone, and bar-shaped, from a titanium butoxide precursor [98, 99]. Water vapor was utilized as the hydrolysis agent and oleic acid and oleylamine as capping surfactants to control crystal growth. The concentrations of the surfactants played a crucial role in determining the nanocrystal shape: higher amounts of oleic acid acted to suppress growth in the [011] direction by selectively binding to the (001) surface.

A great variety of nanocrystal shapes can be obtained using these synthesis methods [74, 100]. Some of the possible modifications of the parent equilibrium crystal form are illustrated in Figure 10.6. Theoretical modeling provides insights on the influence of surface chemistry on the preferential growth of low-energy facets and the crystal morphology. An increase in the photocatalytic efficiency requires both reduction in particle size and control of crystal shape. However, the percentage of high-energy facets that can be achieved in nanosized crystals is ultimately limited by thermodynamic considerations. One way of overcoming this obstacle is through the use of hierarchical structures with a high surface area of reactive facets. A low-temperature hydrothermal process has been described [101] that uses Ti powders in an aqueous HF solution to produce flower-like nanostructures consisting of highly truncated tetragonal pyramidal anatase with {001} exposed facets and {101} sides.

Additional improvements in the photoreactivity of TiO_2 may be possible by combining methods for controlled nanocrystal growth with anion or cation doping. N-doped anatase sheets with predominant {001} facets, synthesized from TiN and HF by a hydrothermal technique, strongly absorb visible light [102]. Nanocrystals with a high percentage of {001} facets, prepared from a TiF_4 precursor by a solvothermal process and codoped with N and S by adding CH_4N_2S prior to calcination, show visible light photocatalytic activity [103]. Selective deposition of low amounts of Pt nanoparticles on the {101} facets of anatase nano crystals improves the efficiency of photoreduction and photooxidation processes by specific surface-induced separation of electrons and holes [104]. The ratio of {001} to {101} facets, and hence the crystal shape, is critical to achieving the optimal balance between recombination and redox reaction rates.

10.7 APPLICATION OF PHOTOCATALYSIS IN WATER TREATMENT

Photocatalysis is effective for the destruction of a wide range of organic and inorganic pollutants and the deactivation of bacteria and viruses. The classes of organic compounds that can be treated include organic acids, aromatic and chlorinated hydrocarbons, alkanes, haloalkanes, alcohols, surfactants, herbicides, pesticides, and dyes [9, 10, 21]. Hazardous inorganic substances comprise bromate, chlorate, azide, halides, nitric oxide, platinum, palladium and rhodium species, silver and sulfur species, metal ions and salts, cyanide, organometallics, thiocyanate, ammonia, nitrates, and nitrites [10, 21]. Pathogens susceptible to photocatalytic disinfection include *Escherichia coli* [105–107], *Salmonella enterica* and *Pseudomonas aeruginoas* [107], and *Clostridium perifringens* [108]. Photocatalysis is a suitable alternative to chemical disinfection of drinking water and is also effective against chlorine-resistant organisms.

The chemical reactions governing photocatalytic degradation of specific compounds are well known [10]. In other cases the degradation pathways are less well understood and there is a potential risk of generation of toxic intermediates [18]. Chlorine and bromine species may be formed by hydrolization, and ozone may be produced as a result of the oxidation of adsorbed hydroxide ions, giving rise to oxygen species that can react with molecular oxygen to form ozone [19]. While the presence of these compounds may assist disinfection processes, reactions between chlorine, bromine, ozone, and hydroxyl radicals can lead to the formation of chlorinated or brominated by-products. Photocatalytic degradation of pesticides and herbicides may result in the production of other toxic compounds [109–111] that, particularly during the initial phase of treatment phase, can be more toxic than the original substance. Thorough evaluation of transformation products and their potential toxicity is therefore essential, together with correct process control, to ensure a sufficient degree of mineralization and transformation to innocuous by-products.

The design of the photocatalytic reactor is critical to providing high light intensity and a large surface area of the catalyst coating [112, 113]. Small-scale laboratory studies have shown the effectiveness of a stirred tank reactor, using nanoparticle TiO_2 films immobilized on glass sheets, in degrading atrazine [114] and inactivating *E. coli* [115]. The feasibility of the use of photocatalytic degradation for the treatment of water contaminated with the pesticides diuron, imidacloppriad, formetanate, and methomyl was demonstrated at the pilot scale with a solar plant employing specially designed compound parabolic collectors to intensify the radiation [116]. Several alternative types of solar reactor design have been proposed such as the parabolic trough reactor, thin-film fixed bed reactor, and double-skin sheet reactor [117]. Use of immobilized photocatalysts avoids the difficulty of separating the nanoparticles after treatment. Not all the potential applications of photocatalysis involve large-scale installations; a miniaturized photocatalytic reactor for purification of drinking water has been designed by applying microfluidic technology [118]. Interest in photocatalytic water treatment has so far has been relatively limited, despite a number of successful field trials on contaminated groundwater and industrial wastewater. However, the development of more efficient, visible light photocatalysts can be expected to increase its competitiveness with conventional water treatment systems.

REFERENCES

[1] Mills A, Le Hunte S. An overview of semiconductor photocatalysis. J Photochem Photobiol A 1997;108:1–35.

[2] Hashimoto K, Irie H, Fujishima A. TiO_2 photocatalysis: a historical overview and future prospects. Jpn J Appl Phys 2005;44: 8269–8285.

[3] Di Paola A, García-López E, Marcì G, Palmisano L. A survey of photocatalytic materials for environmental remediation. J Hazard Mater 2012;211–212:3–29.

[4] Vittadini A, Casarin M, Selloni A. Chemistry of and on TiO_2-anatase surfaces by DFT calculations: a partial review. Theor Chem Acc 2007;117:663–671.

[5] Henderson MA. A surface science perspective on TiO_2 photocatalysis. Surf Sci Rep 2011;66:185–297.

[6] Fujishima A, Zhang X, Tryk DA. TiO_2 photocatalysis and related surface phenomena. Surf Sci Rep 2008;63:515–582.

[7] Ganduglia-Pirovano MV, Hofmann A, Sauer J. Oxygen vacancies in transition metal and rare earth oxides: current state of understanding and remaining challenges. Surf Sci Rep 2007;62:219–270.

[8] Bahnemann DW. Current challenges in photocatalysis: improved photocatalysts and appropriate photoreactor engineering. Res Chem Intermed 2000;26:207–220.

[9] Gnaser H, Huber B, Ziegler C. In: Nalwa HS, editor. *Encyclopedia of Nano-Science and Nanotechnology*. Volume 6, Stevenson Ranch: American Scientific Publishers; 2004. p 505–535.

[10] Hoffmann MR, Martin ST, Choi W, Bahnemann DW. Environmental applications of semiconductor photocatalysis. Chem Rev 1995; 95:69–96.

[11] Liu G, Yu JC, Lu GQ, Cheng H-M. Crystal facet engineering of semiconductor photocatalysts: motivations, advances and unique properties. Chem Commun 2011;47:6763–6783.

[12] Sakthivel S, Shankar MV, Palanichamy M, Arabindoo B, Bahnemann DW, Murugesan V. Enhancement of photocatalytic activity by metal deposition: characterisation and photonic efficiency of Pt, Au and Pd deposited on TiO_2 catalyst. Water Res 2004;38:3001–3008.

[13] Primo A, Corma A, García H. Titania supported gold nanoparticles as photocatalyst. Phys Chem Chem Phys 2011;13:886–910.

[14] Choi W, Termin A, Hoffmann MR. The role of metal ion dopants in quantum-sized TiO_2: correlation between photoreactivity and charge carrier recombination dynamics. J Phys Chem 1994;98:13669–13679.

[15] Shah SI, Li W, Huang C-P, Jung O, Ni C. Study of Nd3+, Pd2+, Pt4+ and Fe3+ dopant effect on photoreactivity of TiO_2 nanoparticles. Proc Natl Acad Sci 2002;99 (Suppl 2):6482–6486.

[16] Chatterjee D, Dasgupta S. Visible light induced photocatalytic degradation of organic pollutants. J Photochem Photobiol C 2005;6: 186–205.

[17] Pelaez M, Nolan NT, Pillai SC, Seery MK, Falaras P, Kontos AG, Dunlop PSM, Hamilton JWJ, Byrne JA, O'Shea K, Entezari MH, Dionysiou DD. A review on the visible light active titanium dioxide photocatalysts for environmental applications. Appl Catal B 2012;125:331–349.

[18] Konstantinou IK, Albanis TA. Photocatalytic transformation of pesticides in aqueous titanium dioxide suspensions using artificial and solar light: intermediates and degradation pathways. Appl Catal B Environ 2003;42:319–335.

[19] Selcuk H, Dunlop PSM, Byrne JA, Anderson MA. In: Anastasia N, Rizzo L, Selcuk H, editors. *Control of Disinfection By-Products in Drinking Water Systems*. Hauppauge: Nova Science Publishers; 2004. p 447–456.

[20] Pirkanniemi K, Sillanpää M. Heterogeneous water phase catalysis as an environmental application: a review. Chemosphere 2002; 48:1047–1060.

[21] Al-Rasheed RA. Water treatment by heterogeneous photocatalysis: an overview. Proceedings of the 4th SWCC Acquired Experience Symposium Jeddah; 2005.

[22] Serpone N, Salinaro A. Terminology, relative photonic efficiencies and quantum yields in heterogeneous photocatalysis. Part I: suggested protocol. Pure Appl Chem 1999;71:303–320.

[23] Chen X, Mao SS. Titanium dioxide nanomaterials: synthesis, properties, modifications and applications. Chem Rev 2007;107:2891–2959.

[24] Kitazawa S, Choi Y, Yamamoto S, Yamaki T. Rutile and anatase mixed crystal TiO_2 thin films prepared by pulsed laser deposition. Thin Solid Films 2006;515:1901–1904.

[25] Zhao L, Han M, Lian J. Photocatalytic activity of TiO_2 films with mixed anatase and rutile structures prepared by pulsed laser deposition. Thin Solid Films 2008;516:3394–3398.

[26] Heo CH, Lee S-B, Boo J-H. Deposition of TiO_2 thin films using RF magnetron sputtering method and study of their surface characteristics. Thin Solid Films 2005;475:183–188.

[27] Ogawa H, Higuchi T, Nakamura A, Tokita S, Miyazaki D, Hattori T, Tsukamoto T. Growth of TiO_2 thin film by reactive RF magnetron sputtering using oxygen radical. J Alloy Compd 2008;449:375–378.

[28] Nolan MG, Pemble ME, Sheel DW, Yates HM. One step process for chemical vapour deposition of titanium dioxide thin films incorporating controlled structure nanoparticles. Thin Solid Films 2006;515:1956–1962.

[29] Azizi R, Rasouli S, Ahmadi NP, Kolaei AJJ, Azizi M. A systematic investigation of experimental conditions on the particle size and structure of TiO_2 nanoparticles synthesized by a sol-gel method. J Ceram Process Res 2012;13:164–169.

[30] Zhang H, Finnegan M, Banfield JF. Preparing single-phase nanocrystalline anatase from amorphous titania with particle sizes tailored by temperature. Nano Lett 2001;1:81–85.

[31] Banerjee I, Karmaker S, Kulkarni NV, Nawale AB, Mathe VL, Das AK, Bhoraskar SV. Effect of ambient pressure on the crystalline phase of nano TiO_2 particles synthesized by a dc thermal plasma reactor. J Nanopart Res 2010;12:581–590.

[32] Dreesen L, Colomer J-F, Limage H, Giguère A, Lucas S. Synthesis of titanium dioxide nanoparticles by reactive DC magnetron sputtering. Thin Solid Films 2009;518:112–115.

[33] Wahi RK, Liu Y, Falkner JC, Colvin VL. Solvothermal synthesis and characterization of anatase TiO_2 nanocrystals with ultrahigh surface area. J Colloid Interface Sci 2006;302:530–536.

[34] Li W, Ni C, Lin H, Huang CP, Shah SI. Size dependence of thermal stability of TiO_2 nanoparticles. J Appl Phys 2004;96:6663–6668.

[35] Zhang H, Penn RL, Hamers RJ, Banfield JF. Enhanced adsorption on surfaces of nanocrystalline materials. J Phys Chem B 1999;103:4656–4662.

[36] Han Y, Kim H-S, Kim H. Relationship between synthesis conditions and photocatalytic activity of nanocrystalline TiO_2. J Nanomater 2012;2012:10.

[37] Balázs N, Mogyorósi K, Srankó DF, Pallagi A, Alapi T, Oszkó A, Dombi A, Sipos P. The effect of particle shape on the activity of nanocrystalline TiO_2 photocatalysts in phenol decomposition. Appl Catal B 2008;84:356–362.

[38] Balázs N, Srankó DF, Dombi A, Sipos P, Mogyorósi K. The effect of particle shape on the activity of nanocrystalline TiO_2 photocatalysts in phenol decomposition. Part 2: The key synthesis parametersinfluencing the particle shape and activity. Appl Catal B 2010;96:569–576.

[39] Mogyorósi K, Balázs N, Srankó DF, Tombácz E, Dékány I, Oszkó A, Sipos P, Dombi A. The effect of particle shape on the activity of nanocrystalline TiO_2 photocatalysts in phenol decomposition. Part 3: The importance of surface quality. Appl Catal B 2010;96:577–585.

[40] Xiong L-B, Li J-L, Yang B, Yu Y. Ti^{3+} in the surface of titanium dioxide: generation, properties and photocatalytic application. J Nanomater 2012;2012:13.

[41] Liu H, Ma HT, Li XZ, Li WZ, Wu M, Bao XH. The enhancement of TiO_2 photocatalytic activity by hydrogen thermal treatment. Chemosphere 2003;50:39–46.

[42] Nowotny MK, Sheppard LR, Bak T, Nowotny J. Defect chemistry of titanium dioxide. Application of defect engineering in processing of TiO_2-based photocatalysts. J Phys Chem C 2008;112:5275–5300.

[43] Sirisuk A, Klansorn E, Praserthdam P. Effects of reaction medium and crystallite size on Ti^{3+} surface defects in titanium dioxide nanoparticles prepared by solvothermal method. Catal Commun 2008;9:1810–1814.

[44] Kongsuebchart W, Praserthdam P, Panpranot J, Sirisuk A, Satayaprasert C, Supphasrirongjaroen P. Effect of crystallite size on the surface defect of nano-TiO_2 prepared via solvothermal synthesis. J Cryst Growth 2006;297:234–238.

[45] Nakamura I, Negishi N, Kutsuna S, Ihara T, Sugihara S, Takeuchi K. Role of oxygen vacancy in the plasma-treated TiO_2 photocatalyst with visible light activity for NO removal. J Mol Catal A 2000;161:205–212.

[46] Zhang D, Li G, Li H, Lu Y. The development of better photocatalysts through composition- and structure-engineering. Chem Asian J 2013;8:26–40.

[47] Zhuang H-F, Lin C-J, Lai Y-K, Sun L, Li J. Some critical structure factors of titanium oxide nanotube array in its photocatalytic activity. Environ Sci Technol 2007;41:4735–4740.

[48] Beydoun D, Amal R, Low G, McEvoy SJ. Role of nanoparticles in photocatalysis. Nanopart Res 1999;1:439–458.

[49] Chen J, Yao M, Wang XJ. Investigation of transition metal ion doping behaviours on TiO_2 nanoparticles. Nanopart Res 2008;10: 163–171.

[50] Di Paola A, Marcì G, Palmisano L, Schiavello M, Uosaki K, Ikeda S, Ohtani B. Preparation of polycrystalline TiO_2 photocatalysts impregnated with various transition metal ions: characterization and photocatalytic activity for the degradation of 4-Nitrophenol. J Phys Chem B 2002;106:637–645.

[51] Moon J, Takagi H, Fujishiro Y, Awano M. Preparation and characterization of the Sb-doped TiO_2 photocatalysts. J Mater Sci 2001;36:949–955.

[52] Yang P, Lu C, Hua N, Du Y. Titanium dioxide nanoparticles co-doped with Fe^{3+} and Eu^{3+} ions for photocatalysis. Mater Lett 2002;57:794–801.

[53] Chen MS, Goodman DW. Catalytically active gold: from nano-particles to ultra-thin films. Acc Chem Res 2006;39:739–746.

[54] Chen MS, Goodman DW. Structure-activity relationships in supported Au catalysts. Catal Today 2006;111:22–33.

[55] Goodman DW. Model catalysts: from imagining to imaging a working surface. J Catal 2003;216:213–222.

[56] Orlov A, Jefferson DA, Macleod N, Lambert RM. Photocatalytic properties of TiO_2 modified with gold nanoparticles in the degradation of 4-chlorophenol in aqueous solution. Catal Lett 2004;92:41–47.

[57] Orlov A, Jefferson DA, Tikhov M, Lambert RM. Enhancement of MTBE photocatalytic degradation by modification of TiO_2. Catal Commun 2007;8:821–824.

[58] Meng F, Sun Z. Enhanced photocatalytic activity of silver nanoparticles modified TiO_2 thin films prepared by RF magnetron sputtering. Mater Chem Phys 2009;118:349–353.

[59] Kim S, Choi W. Dual photocatalytic pathways of trichloroacetate degradation on TiO_2: effects of nano-sized Pt deposits on kinetics and mechanism. J Phys Chem B 2002;106:13311–13317.

[60] Anpo M. Use of visible light. Second-generation titanium oxide photocatalysts prepared by the application of an advanced metal ion-implantation method. Pure Appl Chem 2000;72:1787–1792.

[61] Anpo M, Kishiguchi S, Ichihashi Y, Takeuchi M, Yamashita H, Ikeue K, Morin B, Davidson A, Che M. The design and development of second-generation titanium oxide photocatalysts able to operate under visible light irradiation by applying a metal ion-implantation method. Res Chem Intermed 2001;27:459–467.

[62] Choi J, Park H, Hoffmann MR. Effects of single metal-ion doping on the visible light photoreactivity of TiO_2. J Phys Chem C 2010;114:783–792.

[63] Di Valentin C, Pacchioni G, Selloni A. Origin of the different photoactivity of N-doped anatase and rutile TiO_2. Phys Rev B 2004;70:085116.

[64] Asahi R, Morikawa T, Ohwaki T, Aoki K, Taga Y. Visible-light photocatalysis in nitrogen-doped titanium oxides. Science 2001;293:269–271.

[65] Wang Z, Cai W, Hong X, Zhao X, Xu F, Cai C. Photocatalytic degradation of phenol in aqueous nitrogen-doped TiO_2 suspensions with various light sources. Appl Catal B Environ 2005;57:223–231.

[66] In S, Orlov A, García F, Tikhov M, Wright DS, Lambert RM. Efficient visible light-active N-doped TiO_2 photocatalysts by a reproducible and controllable synthetic route. Chem Commun 2006;40:4236–4238.

[67] In S, Orlov A, Berg R, García F, Pedrosa-Jimenez S, Tikhov MS, Wright DS, Lambert RM. Effective visible light-active B-doped and B,N-codoped TiO_2 photocatalysts. J Am Chem Soc 2007;129:13790–13791.

[68] Gombac V, De Rogatis L, Gasparotto A, Vicario G, Montini T, Barreca D, Balducci G, Fornasiero P, Tondello E, Graziani M. TiO_2 nanopowders doped with boron and nitrogen for photocatalytic applications. Chem Phys 2007;339:111–123.

[69] Lui G, Han C, Pelaez M, Zhu D, Liao S, Likodimos V, Ioannidis N, Kontos AG, Falaras P, Dunlop PSM, Byrne JA, Dionysiou DD. Synthesis, characterization and photocatalytic evaluation of visible light activated C-doped TiO_2 nanoparticles. Nanotechnology 2012;23:294003.

[70] Livraghi S, Elghniji K, Czoska AM, Paganini MC, Giamello E, Ksibi M. Nitrogen-doped and nitrogen-fluorine-codoped titanium dioxide. Nature and concentration of the photoactive species and their role in determining the photocatalytic activity under visible light. J Photochem Photobiol A 2009;205:93–97.

[71] Li X, Zhang H, Zheng X, Yin Z, Wei L. Visible light responsive N-F codoped TiO_2 photocatalysts for the degradation of 4-chlorophenol. J Environ Sci 2011;23:1919–1924.

[72] Dolat D, Quici N, Kusiak-Nejman E, Morawski AW, Li Puma G. One-step hydrothermal synthesis of nitrogen, carbon co-doped titanium dioxide (N,C,TiO_2) photocatalysts. Effect of alcohol degree and chain length as carbon dopant precursors on photocatalytic activity and catalyst deactivation. Appl Catal B 2012;115–116:81–89.

[73] Lazzeri M, Vittadini A, Selloni A. Structure and energetics of stoichiometric TiO_2 anatase surfaces. Phys Rev B 2001;63:155409.

[74] Pan J, Liu G, Lu GQ, Cheng H-M. On the true photoreactivity order of {001}, {010}, and {101} facets of anatase TiO_2 crystals. Angew Chem Int Ed 2011;50:2133–2137.

[75] Barnard AS, Zapol P. Predicting the energetics, phase stability, and morphology evolution of faceted and spherical anatase nanocrystals. J Phys Chem B 2004;108:18435–18440.

[76] Barnard AS, Zapol P, Curtiss LA. Modeling the morphology and phase stability of TiO_2 nanocrystals in water. J Chem Theory Comp 2005;1:107–116.

[77] Barnard AS, Curtiss LA. Computational nano-morphology: modeling shape as well as size. Rev Adv Mater Sci 2005;10:105–109.

[78] Hummer DR, Kubicki JD, Kent PRC, Post JE, Heaney PJ. The origin of nanoscale stability reversals in titanium oxide polymorphs. J Phys Chem C 2009;113:4240–4245.

[79] Barnard AS, Curtiss LA. Prediction of TiO_2 nanoparticle phase and shape transitions controlled by surface chemistry. Nano Lett 2005;5:1261–1266.

[80] Barnard AS, Xu H. An environmentally sensitive phase map of titania nanocrystals. ACS Nano 2008;2:2237–2242.

[81] Shklover V, Haibach T, Bolliger B, Hochstrasser M, Erbudak M, Nissen H-U, Zakeeruddin SM, Nazeeruddin MK, Grätzel M. Crystal structures of Ru complex sensitizers of TiO_2-anatase nanopowders. J Sol State Chem 1997;132:60–72.

[82] Rickerby DG. Nanotechnological medical devices and nanopharmaceuticals: the European regulatory framework and research needs. J Nanosci Nanotechnol 2007;7:1–8.

[83] Rickerby DG. Increasing the efficiency of photocatalytic water treatment with titanium dioxide nanoparticles by control of crystal shape. Proceedings of Nanotech 2011. Volume 3; Boston, Danville: Nano Science and Technology Institute; 2011. pp 585–588.

[84] Djerdj I, Arčon D, Jagličić Z, Niederberger M. Nonaqueous synthesis of metal oxide nanoparticles: short review and doped titanium dioxide as case study for the preparation of transition metal-doped oxide nanoparticles. J Solid State Chem 2008;181: 1571–1581.

[85] Yeredla RR, Xu H. An investigation of the nanostructured rutile and anatase plates for improving the photosplitting of water. Nanotechnology 2008;19:055706.

[86] Ahonen PP, Moisala A, Tapper U, Brown DP, Jokiniemi JK, Kauppinen EI. Gas-phase crystallization of titanium dioxide nanoparticles. J Nanopart Res 2002;4:43–52.

[87] Čaplovičová M, Billik P, Čaplovič L, Brezová V, Turáni T, Plesch G, Fejdi P. On the true morphology of highly photoactive anatase TiO_2 nanocrystals. Appl Catal B 2012;117–118:224–235.

[88] Feldhoff A, Mendive C, Bredow T, Bahnemann D. Direct measurement of size, three-dimensional shape, and specific surface area of anatase nanocrystals. ChemPhysChem 2007;8:805–809.

[89] Yang HG, Sun CH, Qiao SZ, Zou J, Liu G, Smith SC, Cheng HM, Lu GQ. Anatase TiO_2 single crystals with a large percentage of reactive facets. Nature 2008;453:638–641.

[90] Yang HG, Liu G, Qiao SZ, Sun CH, Jin YG, Smith SC, Zou J, Cheng HM, Lu GQ. Solvothermal synthesis and photoreactivity of anatase TiO_2 nanosheets with dominant {001} facets. J Am Chem Soc 2009;131:4078–4083.

[91] Liu G, Sun CH, Yang HG, Smith SC, Wang L, Lu GQ, Cheng HM. Nanosized anatase TiO_2 single crystals for enhanced photocatalytic activity. Chem Commun 2010;46:755–757.

[92] Dai Y, Cobley CM, Zeng J, Sun Y, Xia Y. Synthesis of anatase TiO_2 nanocrystals with exposed {001} facets. Nano Lett 2009;9:2455–2459.

[93] Ding K, Miao Z, Hu B, An G, Sun Z, Han B, Liu Z. Shape and size controlled synthesis of anatase nanocrystals with the assistance of ionic liquid. Langmuir 2010;26:5129–5134.

[94] Han X, Kuang Q, Jin M, Xie Z, Zheng L. Synthesis of titania nanosheets with a large percentage of exposed (001) facets and related photocatalytic properties. J Am Chem Soc 2009;131:3152–3153.

[95] Amano F, Yasumoto T, Prieto-Maheney OO, Uchida S, Shibayama T, Terada Y, Ohtani B. Highly active titania photocatalyst particles of controlled crystal phase, size, and polyhedral shape. Top Catal 2010;53:455–461.

[96] Sugishita N, Kuroda Y, Ohtani B. Preparation of decahedral anatase titania particles with high-level photocatalytic activity. Catal Today 2011;164:391–394.

[97] Wu B, Guo C, Zheng N, Xie Z, Stucky GD. Nonaqueous production of nanostructured anatase with high-energy facets. J Am Chem Soc 2008;130:17563–17567.

[98] Dinh C-T, Nguyen T-D, Kleitz F, Do T-O. Shape-controlled synthesis of highly crystalline titania nanocrystals. ACS Nano 2009;3:3737–3743.

[99] D'Arenzio M, Carbajo J, Bahamonde A, Crippa M, Polizzi S, Scotti R, Wahba L, Morazzoni F. Photogenerated defects in shape-controlled TiO_2 anatase nanocrystals: a probe to evaluate the role of crystal facets in photocatalytic processes. J Am Chem Soc 2011;133:17652–17661.

[100] Zhao X, Jin W, Cai J, Ye J, Li Z, Ma Y, Xie J, Qi L. Shape- and size-controlled synthesis of uniform anatase TiO_2 nanocuboids enclosed by active {100} and {001} facets. Adv Funct Mater 2011;21:3554–3563.

[101] Liu M, Piao L, Lu W, Ju S, Zhao L, Zhou C, Li H, Wang W. Flower-like TiO_2 nanostructures with exposed {001} facets: facile synthesis and enhanced photocatalysis. Nanoscale 2010;2:1115–1117.

[102] Liu G, Yang HG, Wang X, Cheng L, Pan J, Lu GQ, Cheng H-M. Visible light responsive nitrogen doped anatase TiO_2 sheets with dominant {001} facets derived from TiN. J Am Chem Soc 2009;131:12868–12869.

[103] Shi W, Yang W, Li Q, Gao S, Shang P, Shang JK. The synthesis of nitrogen/sulfur co-doped TiO_2 nanocrystals with a high specific surface area and a high percentage of {001} facets and their enhanced visible-light photocatalytic performance. Nanoscale Res Lett 2012;7:590.

[104] Liu C, Han X, Xie S, Kuang Q, Wang X, Jin M, Xie Z, Zheng L. Enhancing the photocatalytic activity of anatase TiO_2 by improving the specific facet-induced spontaneous separation of photogenerated electrons and holes. Chem Asian J 2013;8:282–289.

[105] Dunlop PSM, Byrne JA, Manga N, Eggins BR. The photocatalytic removal of bacterial pollutants from drinking water. J Photochem Photobiol A 2002;148:355–363.

[106] Rincón A-G, Pulgarin C. Bactericidal action of illuminated TiO_2 on pure *Escherichia coli* and natural bacterial consortia: post-irradiation events in the dark and assessment of the effective disinfection time. Appl Catal B 2004;49:99–112.

[107] Robertson JMC, Robertson PKJ, Lawton LA. A comparison of the effectiveness of TiO_2 photocatalysis and UVA photolysis for the destruction of three pathogenic micro-organisms. J Photochem Photobiol A 2005;175:51–56.

[108] Dunlop PSM, McMurray TA, Hamilton JWJ, Byrne JA. Photocatalytic inactivation of Clostridium perfringens spores on TiO_2 electrodes. J Photochem Photobiol A 2008;196:113–119.

[109] Manilal VB, Haridas A, Alexander R, Surender GD. Photocatalytic treatment of toxic organics in wastewater: toxicity of photodegradation products. Water Res 1992;26:1035–1038.

[110] Herrmann JM, Guillard C, Arguello M, Agüera A, Tejedor A, Piedra L, Fernández-Alba A. Photocatalytic degradation of pesticide pirimiphos-methyl: determination of the reaction pathway and identification of intermediate products by various analytical methods. Catal Today 1999;54:353–367.

[111] Malato S, Cáceres J, Fernández-Alba AR, Piedra L, Hernando MD, Agüera A, Vial J. Photocatalytic treatment of diuron by solar photocatalysis: evaluation of main intermediates and toxicity. Environ Sci Technol 2003;37:2516–2524.

[112] Ray AK. Design, modeling and experimentation of a new large-scale photocatalytic reactor for water treatment. Chem Eng Sci 1999;54:3113–3125.

[113] Mukherjee PS, Ray AK. Major challenges in the design of a large-scale photocatalytic reactor for water treatment. Chem Eng Technol 1999;22:253–260.

[114] McMurray TA, Dunlop PSM, Byrne JA. The photocatalytic degradation of atrazine on nanoparticulate TiO_2 films. J Photochem Photobiol A 2006;182:43–51.

[115] Alrousan DMA, Dunlop PSM, McMurray TA, Byrne JA. Photocatalytic inactivation of *E. coli* in surface water using immobilised nanoparticle TiO_2 films. Water Res 2009;43:47–54.

[116] Malato S, Blanco J, Vidal A, Alarcón D, Maldonado MI, Cáceres J, Gernjak W. Applied studies in solar photocatalytic detoxification: an overview. Solar Energy 2003;75:329–336.

[117] Bahnemann D. Photocatalytic water treatment: solar energy applications. Solar Energy 2004;77:445–459.

[118] Lindstrom H, Wooton R, Iles A. High surface area titania photocatalytic microfluidic reactors. AIChE J 2007;53:695–702.

11

DESTRUCTION OF CHLOROORGANIC COMPOUNDS WITH NANOPHOTOCATALYSTS

Rashid A. Khaydarov, Renat R. Khaydarov, and Olga Gapurova

Institute of Nuclear Physics, Uzbekistan Academy of Sciences, Tashkent, Uzbekistan

11.1 INTRODUCTION

Photocatalysis based on the application of nanocatalysts is a very promising method for the treatment of contaminated water, air, and land. Photocatalytic systems equipped with artificial ultraviolet (UV) light systems can be applied at ambient temperature to degrade various chemical and microbiological pollutants: organic materials, organic acids, estrogens, pesticides, dyes, crude oil, microbes (including viruses and chlorine-resistant organisms), and inorganic substances.

There are several potential photocatalysts, but the most widespread substance is nano titanium dioxide. Nanosized particles are preferred to bulk TiO_2 because they are significantly more reactive than larger particles due to their larger surface area. TiO_2 is chemically stable and has a high ability to break molecular bonds leading to degradation. To avoid free nanoparticles in water, nano TiO_2 particles are usually immobilized on a substrate or integrated into thin films and other materials. On the other hand, the lifetime of nanoparticles before their agglomeration can be easily regulated [1], and the nanoparticles can be transformed into large particles that are absolutely safe for the environment. This method opens an opportunity to increase the efficiency of catalysts significantly by using colloidal nanophotocatalysts (NPC) homogeneously dispersed on the surface of land to be treated. At present, the development of new types of NPCs with enhanced catalytic properties is a challenge, as they have many applications in science and technology.

This chapter deals with a new method for the destruction of chloroorganic compounds that is based on the use of colloidal nanocarbon–metal compositions (NCMC) as NPC dispersed on polluted land, soil, and water. Dichlorodiphenyltrichloroethane (DDT), aldrin, lindane, and polychlorinated biphenyls (PCBs) have been chosen as the chloroorganic compounds to test NPC efficiency.

The production and agricultural use of DDT was banned in the early 1970s. The environmental distribution and effects of DDT arise from its unusual chemical stability and hence its persistence. DDT and some of the breakdown products of DDT, principally dichlorodiphenyldichloroethylene (DDE), are highly persistent in soil, sediment, and biota as they are relatively resistant to breakdown by the enzymes and higher organisms found in the soil. Thus contamination by these substances can last long after DDT application.

Aldrin is an organochlorine insecticide that was widely used until the 1970s, when it was banned in most countries. Before the ban, it was heavily used as a pesticide to treat seed and soil. In soil, on plant surfaces, or in the digestive tracts of insects, aldrin oxidizes to epoxide dieldrin, which is a stronger insecticide. Like related polychlorinated pesticides, aldrin is highly lipophilic. Its solubility in water is only 0.027 mg/l, which exacerbates its persistence in the environment. It was banned by the Stockholm Convention on Persistent Organic Pollutants.

Nanomaterials for Environmental Protection, First Edition. Edited by Boris I. Kharisov, Oxana V. Kharissova, and H. V. Rasika Dias.

The World Health Organization classifies lindane as "moderately hazardous," and its international trade is restricted and regulated under the Rotterdam Convention on Prior Informed Consent. In 2009, the production and agricultural use of lindane was banned under the Stockholm Convention on Persistent Organic Pollutants. Lindane is currently permitted only for application as a second-line pharmaceutical treatment for lice and scabies skin conditions. Lindane is a persistent organic pollutant: it has a relatively long life in the environment, it can be transported for a long distance by natural processes like global distillation, and it can be bioaccumulated in food chains, though it is rapidly eliminated when exposure is discontinued. Lindane is stable to light, air, heat, carbon dioxide, and strong acids. Trichlorobenzenes and hydrochloric acid are formed during dechlorination processes in the presence of alkali or after prolonged exposure to heat.

PCBs were often used as a dielectric fluid in transformers and capacitors. PCBs are very stable compounds, and they do not decompose promptly. This is due to their chemical inability to oxidize and reduce in the natural environment. Furthermore, PCBs have a long half-life (8–10 years) and are insoluble in water, which contributes to their stability. Their destruction by chemical, thermal, and biochemical processes is extremely difficult and presents the risk of generating extremely toxic dibenzodioxins and dibenzofurans through partial oxidation. Intentional degradation as a treatment of unwanted PCBs generally requires high heat or catalysis. In the atmosphere, PCBs can be degraded by hydroxyl radicals. Starting in the early 1970s, production and new uses of PCBs have been banned due to concerns about the accumulation of PCBs and toxicity of their by-products. Today PCBs can still be released into the environment from poorly maintained hazardous waste sites that contain PCBs, illegal or improper dumping of PCB wastes, leaks or releases from electrical transformers containing PCBs, and disposal of PCB-containing consumer products into municipal or other landfills not designed to handle hazardous waste. PCBs may also be released into the environment by the burning of some wastes in municipal and industrial incinerators. At present, most concentrations of PCBs are detected (up to 280 μg/kg) in bottom sediments.

11.2 PRINCIPLE OF THE METHOD

The principle of the method is based on the destruction of organic substances and bacteria by NPC dispersed on polluted open land or water. The NPC under natural UV radiation form OH radicals in the presence of water molecules, and the radicals destroy organic substances and bacteria. In brief, photocatalytic reactions of an aqueous NPC suspension system can be described as follows [2]:

$$\mathrm{NPC} + h\nu \rightarrow \mathrm{NPC} + e^-_{CB} + h^+_{VB} \tag{11.1}$$

$$h^+_{VB} + \mathrm{OH}^- (\mathrm{surface}) \rightarrow \mathrm{OH}^\bullet \tag{11.2}$$

$$h^+_{VB} + H_2O \rightarrow \mathrm{OH}^\bullet + H^+ \tag{11.3}$$

$$e^-_{CB} + O_2 \rightarrow O_2^- \tag{11.4}$$

$$e^-_{CB} + h^+_{VB} \rightarrow \mathrm{heat} \tag{11.5}$$

where $h\nu$ is the UV irradiation, h^+_{VB} is the valence band holes, and e^-_{CB} is the conduction band electrons. It is known that active oxygen and radical species existing in the presence of oxygen and water take part in the oxidation–reduction reaction and destroy organic molecules and bacteria.

The NPC must be harmless and their concentration in water must be lower than the permissible level. Moreover, nanoparticles must form agglomerates within the stipulated time with further coagulation and precipitation; that is, they must form safe ordinary particles. Many semiconductive metal compounds are used as photocatalysts and the most well known among them is TiO_2. Usually photocatalysts are applied on a carrier [3–9], have sizes greater than 1 μm, and cannot be used for dispersion on a large area. That is why it is necessary to develop a new type of NPC that can meet all the requirements described earlier.

As is well known, nanocomposites combine the properties of two or more different materials with the possibility of novel mechanical, physical, or chemical behavior arising [10]. Nanocomposites of conjugated materials and metal nanoparticles are prepared from different metals, different types of conjugated polymers, and oligomer linkers [11–18].

Another type of nanoscale materials are nanocomposites of carbon nanoparticles and polymers. Indeed, for example, electrolytically generated nanocarbon colloids (NCC) have functional groups such as carbonyl, hydroxyl, and carboxyl groups formed on the surface of carbon nanoparticles [19–22]. These nanocomposites can be modified by attaching different cations. On the other hand, most polymers can react with different ions and molecules and also participate in the modification of nanocomposites, for example, nanocarbon–polymer nanocomposites prepared via electrolytically generated NCC and polyethylenimine [1, 23]. A similar method can be used to prepare nanocarbon–metal nanocomposites (NCMC) as the NPC to control cyanobacteria [24].

11.3 MATERIALS AND EQUIPMENT

NCC was prepared by the electrochemical method described in our previous works [1, 21, 23]. The process was based on the use of an inexpensive two-electrode device in which an anode and a cathode made from high-density isotropic graphites OEG4 (Russia) (65 mm × 30 mm × 15 mm) were transformed into carbon colloidal particles. The anode and the cathode were immersed in a plastic electrolytic cell (120 mm × 140 mm × 105 mm) filled with distilled or deionized water as the electrolyte. The distance between the electrodes was varied from 10 to 120 mm in the current density range 0.1–3 mA/cm^2 at a constant voltage of 60 V. The electrolytic cell was installed on the magnetic stirrer in order to provide water flowing between the electrodes. That allows saturating the electrolyte with carbon colloids and discharging the gas generated on the surface of electrodes due to the electrolysis of water. The process of device operation involved two repeatable consecutive steps: (1) electrolysis for 10 min (2) and electrolyte stirring for 60 s. The process was executed automatically using ST-T2 twin timer controller (RIKO Opto-electronics Technology Co., Ltd., Taiwan).

The NCC preparation process was executed in two stages: anode activation and carbon nanoparticles generation. At the first stage, the electrolyte has low conductivity, electric current density is small, about 0.1–0.2 mA/cm^2, and the oxidation reaction is slow. The duration of this stage is about 50 h and depends on the quality (density) of graphite. At this stage, the voltage between the electrodes is high, about 60–100 V. As the reaction proceeds, the conductivity of the electrolyte is abruptly increased, the current density goes beyond 10 mA/cm^2, and the oxidization reaction sets in. As a result, the carbon electrode is finely split with follow-up covering by the carboxyl group.

At the second stage, the electric current density between the electrodes is about 3–4 mA/cm^2. The NCC is stable for at least 150 days. If the current density values are greater than 8–10 mA/cm^2, the rate of oxygen diffusion through the electrode is so high that the pressure inside the electrode causes its disintegration. In this case, the synthesized NCC is not stable, and its precipitation is observed in 2–3 weeks.

The size and shape of nanoparticles were determined with transmission electron microscopy (TEM) (LEO-912-OMEGA, Carl Zeiss, Germany). The concentration of Ti in the solutions was determined by neutron activation analysis by irradiating water samples in the nuclear reactor of the Institute of Nuclear Physics (Tashkent, Uzbekistan).

The Ge(Li) detector with a resolution of about 1.9 keV at 1.33 MeV and a 4096-channel analyzer was used for the detection of γ-ray quanta. The area under the γ-peak of radionuclide ^{51}Ti (half-life $T_{1/2} - 5.8$ min, energy of the γ peak $E\gamma = 0.319$ MeV) was measured to determine the concentration of Ti.

PCBs from the transformer ("transformer oil") were used. The analysis of PCBs was performed by gas chromatography–mass spectrometry (GC/MS) (Agilent Technologies 6890N network gas chromatography system with 5973 inert mass selective detection (MSD)) operating in the SIM mode. The PCBs were separated using a DB-5 (60 m, 0.25 mm, 0.25 lm) chromatographic column.

The concentrations of DDT, $C_{12}H_8Cl_6$, and $C_6H_6Cl_6$ were determined in accordance with the Standard Methods (Russia) for Testing Chemical Procedures (method for determining residual quantities of pesticides) for the evaluation of disinfection [25].

The photocatalytic destruction of DDT, $C_{12}H_8Cl_6$, and $C_6H_6Cl_6$ in the NCMC (Ti) suspension under UV illumination was investigated in order to evaluate the photocatalytic activity of the nanoparticles. Aldrin and lindane water solutions, aldrin, linden and DDT powders, and transformer oils containing PCBs were filled or put into Petri dishes. A 60-W UV lamp (DB-60, Russia) fixed at a distance of 25 cm above the solution surface was used as the UV light source and provided radiation power of 1 W/m^2 within the 220- to 320-nm region.

11.4 RESULTS AND DISCUSSIONS

11.4.1 Synthesis of NPC

Experiments have shown that the yield of NCMC (Ti) in the electrolysis process depends on the voltage V between the electrodes and the pH of the solution. Figure 11.1 shows the yield dC_{Ti}/dt of Ti 6 min after the start of the electrolysis process, where C_{Ti} is the concentration of Ti in the electrolyte. The yield dC_{Ti}/dt increases with increasing voltage between the electrodes up to 12–13 V but then it decreases slowly. This behavior is explained by the formation of trivalent titanium on the surface of the Ti

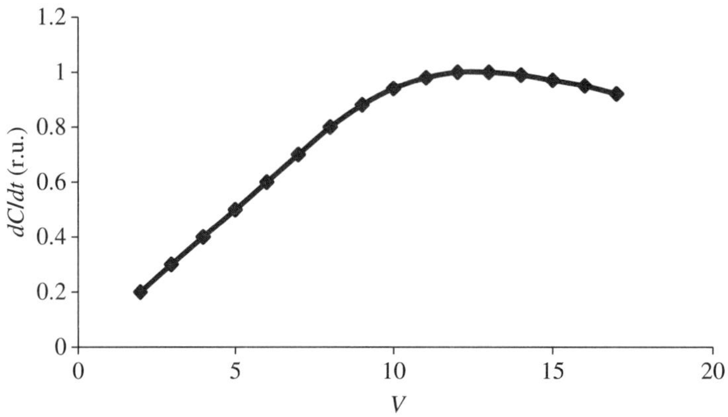

FIGURE 11.1 Dependence of the yield of titanium 6 min after the start of the electrolysis process versus electrolysis voltage.

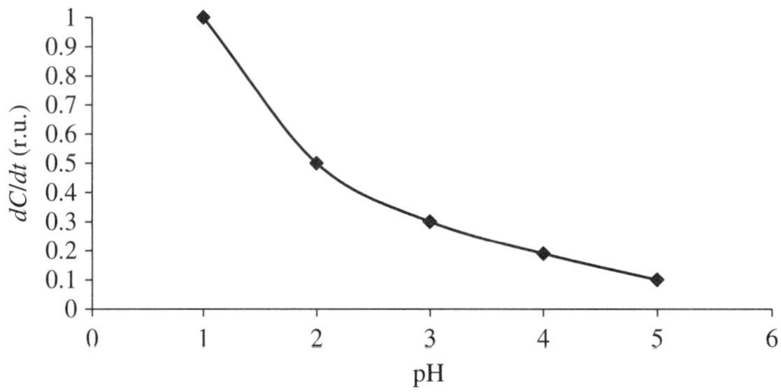

FIGURE 11.2 Dependence of the yield of titanium 6 min after the start of the electrolysis process versus pH of the solution.

electrode, which becomes blue. Figure 11.2 demonstrates the dependence of dC_{Ti}/dt against the concentration of H_2SO_4 in the solution 6 min after the start of the electrolysis process. When the pH of the electrolyte comes close to a neutral value, trivalent titanium is formed on the surface of the Ti electrode.

At the first stage of the electrolysis process when the Ti electrode is the anode, electric current between the electrodes is about 3–4 mA/cm^2. In the second stage, after changing the polarity, electric current increases up to 180–200 mA/cm^2 in about 0.1–0.2 s. During the first stage, the oxygen is released on the titanium anode, titanium oxides and sulfates are formed on the surface, and titanium ions leaving the anode are oxidized by oxygen near the surface of the anode in the solution or react with NCC that have carboxyl groups as the active groups. The thin semiconductor layer that forms on the surface of the titanium electrode has high resistance, and the electric current between the electrodes is small, about 3–4 mA/cm^2. At the same time, the negatively charged carbon nanoparticles go away from the graphite cathode and functional groups such as carbonyl (>C=O), hydroxyl (–OH), and carboxyl (–COOH) groups are formed on the surface of carbon particles.

During the second stage, the oxidation process occurs at the carbon anode. The magnitude of repulsion forces between the stacked layers of graphite becomes larger than that of van der Waals attraction forces between the layers, leading to the formation of carbon nanoparticles, which changes the polarity of electrodes. The oxides are cleaned from the surface of the titanium cathode, and the electric current between the electrodes increases up to 180–200 mA/cm^2. Titanium ions and charged particles from the titanium oxide interact with carbon nanoparticles and form NCMC (Ti). The oxygen adsorbed on the surface of the particles forms -Ti(OH)-O-Ti(OH)-, which can help the photogenerated holes h^+ to change into a $OH^•$ free radical. Otherwise, the oxidization activity of $OH^•$ is the strongest in aqueous solution. A typical TEM micrograph of NCMC (Ti) is given in Figure 11.3 and shows that nanoparticles have a spherical morphology. TEM measurements of particles have shown that the average size of nanoparticles is 6 ± 2 nm.

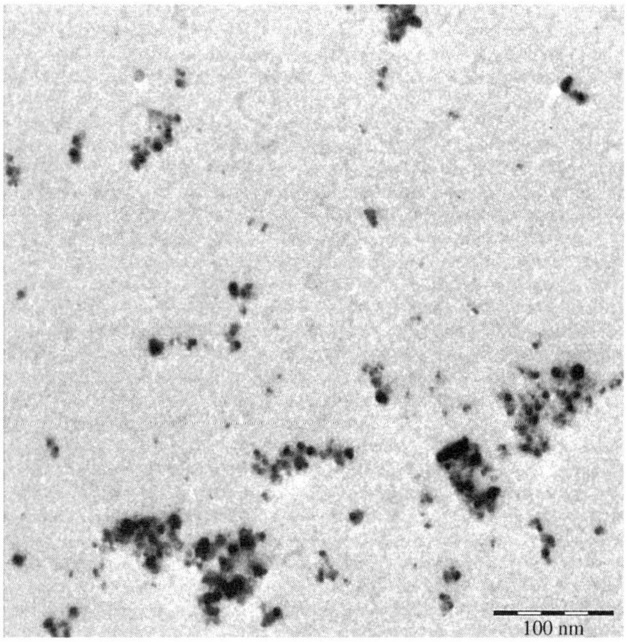

FIGURE 11.3 Typical transmission electron microscopy (TEM) micrograph of NCMC (Ti).

11.4.2 Laboratory Tests

The photodegradation ability of aldrin, lindane, DDT, and PCBs by NCMC (Ti) was investigated in laboratory conditions.

Colloidal solutions of NCMC (Ti) were added to two Petri dishes containing water with aldrin. The first sample was irradiated by UV lamp. The second Petri dish was irradiated by sunlight for 2 days from 11 a.m. to 4 p.m. The third sample did not contain NCMC (Ti) and was irradiated by a UV lamp. Concentrations of NCMC (Ti) and aldrin in solutions were 5 and 10 μg/l, respectively. The temperature of solutions during irradiation under UV lamp was 25°C and under sunlight was 26–30°C. Dependences of aldrin concentrations in water samples against exposure time are given in Figure 11.4. The results show that aldrin in water is degraded in 5–6 h. We did not observe any degradation of aldrin without NCMC (Ti). In another experiment, emulsion of a fine powder of aldrin with particle size of about 10 μm was dispersed to reach a surface density 10 μg/cm^2 on the bottom of two Petri dishes containing water and NCMC (Ti). The concentration of NCMC (Ti) was 10 μg/l, and the water layer thickness was 5 mm. Dependences of aldrin surface density in samples irradiated by UV lamp and sunlight against exposure time are given in Figure 11.5. The results show that aldrin powder in water is degraded in 8–10 h.

For the investigating photodegradation of lindane, colloidal solutions of NCMC (Ti) were added to two Petri dishes containing water with lindane. The first sample was irradiated by UV lamp. The second Petri dish was irradiated by sunlight for 2 days from 11 a.m. to 4 p.m. The third sample did not contain NCMC (Ti) and was irradiated by UV lamp. Concentrations of NCMC (Ti) and lindane in solutions were 1 and 8 mg/l, respectively. The temperature of solutions during irradiation under UV lamp was 25°C and under the sunlight it was 26–30°C. The dependencies of lindane concentrations in water samples against exposure time are given in Figure 11.6. The results show that lindane in water is degraded in 6–7 h. We did not observe any degradation of lindane without NCMC (Ti). Trichlorobenzenes are formed as reaction products during the first 2 h of irradiation, but they are destroyed 3 h after the start of the irradiation process. In another experiment, a fine powder of lindane with particle size of about 10 μm was dispersed to reach a surface density of about 8 mg/cm^2 on the bottom of two Petri dishes containing water and NCMC (Ti). The concentration of NCMC (Ti) was 1 mg/l, and the water layer thickness was 5 mm. Dependences of lindane surface density in samples irradiated by UV lamp and sunlight against the exposure time are given in Figure 11.7. The results show that lindane powder in water is degraded in 8–10 h.

We investigated the photodegradation of DDT powder with particle size of about 10 μm under UV lamp and sunlight. The powder was dispersed on the bottom of two Petri dishes containing water and NCMC (Ti). The concentration of NCMC (Ti) was 1 mg/l, the thickness of water was 5 mm, and the surface density of the powder was about 5 mg/cm^2. Dependences of DDT surface density in samples irradiated by UV lamp and sunlight against exposure time are given in Figure 11.8. The results show that the surface density of DDT powder in water decreases by two times in 8–10 h.

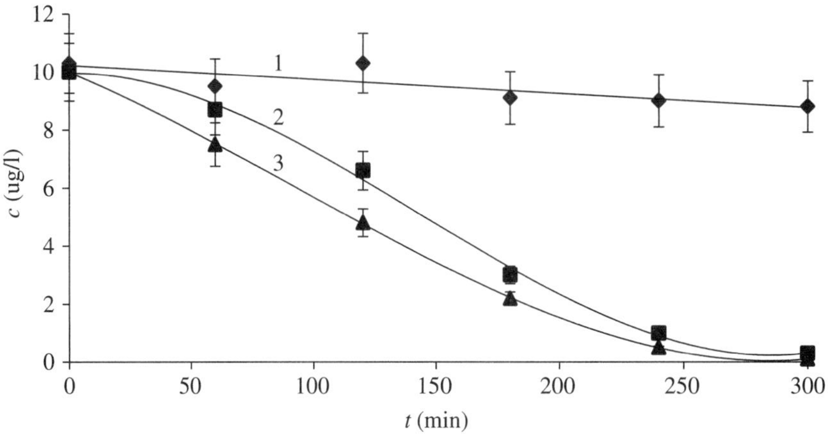

FIGURE 11.4 Photodegradation of aldrin in water without NCMC (Ti) (1) and with NCMC (Ti) under sunlight (2) and UV lamp (3).

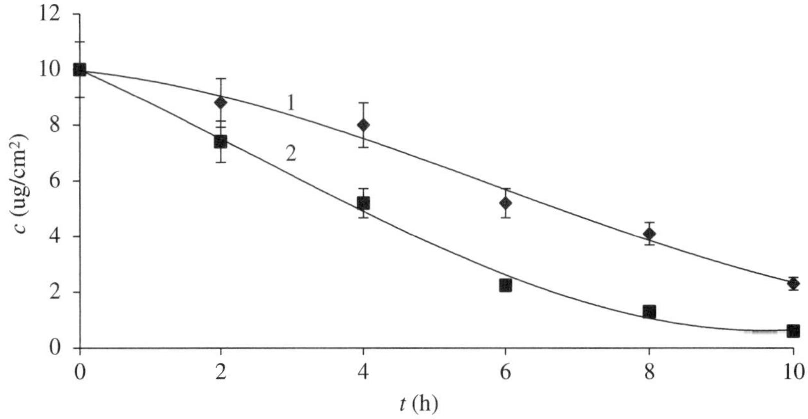

FIGURE 11.5 Photodegradation of aldrin powder in water under sunlight (1) and UV lamp (2).

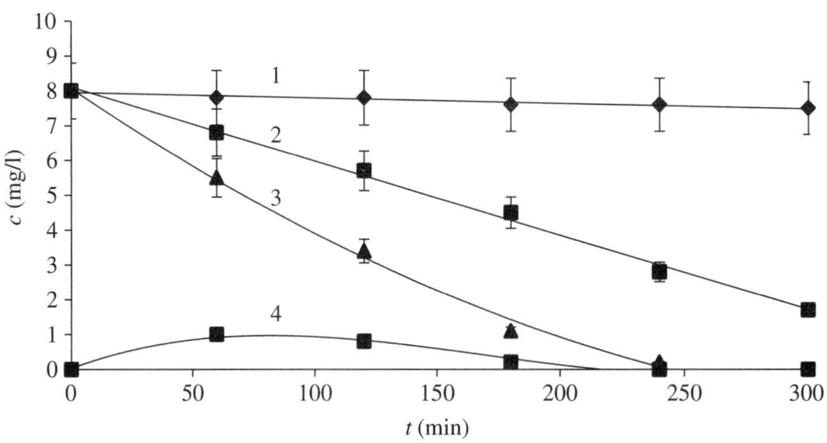

FIGURE 11.6 Photodegradation of lindane in water without NCMC (Ti) (1) and with NCMC (Ti) under sunlight (2) and UV lamp (3). Behavior of trichlorobenzenes formed during the irradiation of lindane.

The photodegradation of PCBs by NCMC (Ti) was studied in laboratory conditions. A colloidal solution of NCMC (Ti) was added to Petri dishes containing water with transformer oil with concentration of 800 μg/l and PCBs with concentration of 80 μg/l. Petri dishes were placed under the sunlight at 11 a.m. and tested for 6 h. The concentration of NCMC (Ti) in water was 1 mg/l. Dependences of PCB concentrations in the samples irradiated by sunlight and UV lamp against exposure time are given in Figure 11.9. The results show that PCBs are degraded for 5–6 h.

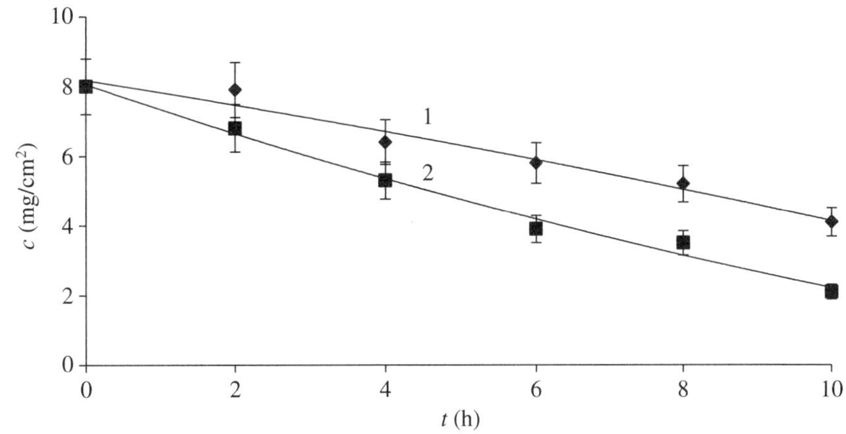

FIGURE 11.7 Photodegradation of lindane powder in water under sunlight (1) and UV lamp (2).

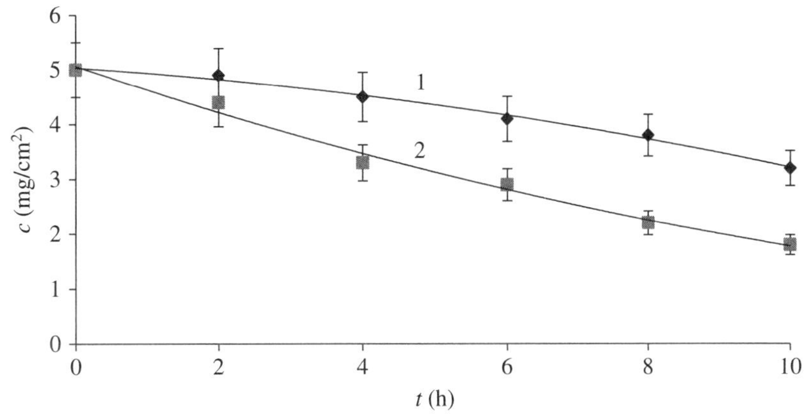

FIGURE 11.8 Photodegradation of DDT powder in water under sunlight (1) and UV lamp (2).

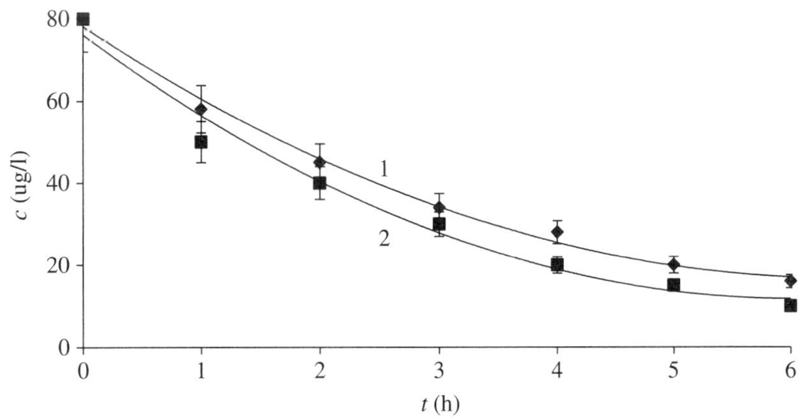

FIGURE 11.9 Photodegradation of PCBs under solar (1) and UV lamp (2) irradiation by NCMC (Ti) at a concentration of 1 mg/l. Temperature of solutions is 30°C.

11.4.3 Field Tests

Field tests of the technology have been conducted with DDT on the surfaces of concrete slabs. A water emulsion of DDT was sprayed to get a concentration of about 1–2 g/m^2 on the concrete slab. Then the surfaces of the concrete slabs were pulverized by water solution of NCMC (Ti) at a concentration of 2 mg/l to get a surface density of about 0.01 g/m^2 for nanocompositions.

Air humidity in the experiments was about 40% at a temperature of 32°C. Test results have shown that the concentration of DDT on the surface of the slab decreased to 0.1–0.2 g/m^2 in 3 days.

11.5 CONCLUSIONS

Colloidal solutions of electrochemically synthesized NCMC (Ti) have shown a high efficiency as photocatalysts for the destruction of chloroorganic substances such as aldrin, lindane, DDT, and PCBs. Photodegradation occurs within the NCMC (Ti) concentration range of 1 μg/l to 10 mg/l. The time for degradation of aldrin, lindane, and PCBs dissolved in water under the UV lamp and sunlight is about 5–7 h. Trichlorobenzenes are formed as reaction products during the first 2 h of irradiation of lindane, but are completely destroyed 3 h after the start of the irradiation process. The concentrations of organic reaction products in the photodegradation process of aldrin, DDT, and PCBs were lower than the detection limit. The destruction time of powders of aldrin, lindane, and DDT is about 8–10 h. The concentration of DDT sprayed on the concrete slab surface decreased from 1–2 to 0.1–0.2 g/m^2 in 3 days. In the near future, for decontamination of soil, we are going to develop an apparatus that continuously stirs the soil samples and irradiates them by sunlight or UV.

REFERENCES

[1] Khaydarov RA, Khaydarov RR, Gapurova O. Water purification from metal ions using carbon nanoparticle-conjugated polymer nanocomposites. Water Research 2010;44:1927–1933.

[2] Tang WZ, An H. UV/TiO_2 photocatalytic oxidation of commercial dyes in aqueous solutions. Chemosphere 1995;31:4157–4171.

[3] Kim SB, Hwang HT, Hong SC. Photocatalytic degradation of volatile organic compounds at the gas–solid interface of a TiO_2 photocatalyst. Chemosphere 2002;48 (4):437–444.

[4] Kim SB, Cha WS, Hong SC. Photocatalytic degradation of gas-phase methanol and toluene using thin-film TiO_2 photocatalyst. II. kinetic study for the effect of initial concentration and photon flux. Journal of Industrial and Engineering Chemistry 2002;8 (2):162–167.

[5] Abu Bakar WAW, Othman R, Ali MY. Photocatalytic degradation and reaction pathway studies of chlorinated hydrocarbons in gaseous phase. Transactions C: Chemistry and Chemical Engineering 2010;17 (1):1–14.

[6] Jiang D, Zhao H, Zhang S, John R. Comparition of photocatalytic degradation kinetic characteristics of different organic compounds at anatase TiO_2nanoporous film electrodes. Journal of Photochemistry and Photobiology A Chemistry 2006;177 (2–3):253–260.

[7] Sopyan I. An efficient TiO_2 thin-film photocatalyst: photocatalytic properties in gas-phase acetaldehyde degradation. Journal of Photochemistry and Photobiology A Chemistry 1996;98 (1–2):79–86.

[8] Yu H, Zhang K, Rossi C. Theoretical study on photocatalytic oxidation of VOCs using nano-TiO_2 photocatalyst. Journal of Photochemistry and Photobiology A: Chemistry 2007;188:65–73.

[9] Oki K, Tsuchida S, Nishikiori H, Tanaka N, Fujii T. Photocatalytic degradation of chlorinated ethenes. International Journal of Photoenergy 2003;5:11–15.

[10] Bryan CS, Michael OW. Metal nanoparticle—conjugated polymer nanocomposites. Chemical Communications 2005;27:3375–3384.

[11] Daniel M-C, Astruc D. Gold nanoparticles: assembly, supramolecular chemistry, quantum-size-related properties, and applications toward biology, catalysis, and nanotechnology. Chemical Reviews 2004;104:293–346.

[12] Rao CNR, Cheetham AK. Science and technology of nanomaterials: current status and future prospects. Journal of Materials Chemistry 2001;11:2887–2894.

[13] Schmid G, Simon U. Gold nanoparticles: assembly and electrical properties in 1-3 dimensions. Chemical Communications 2005;6:697–710.

[14] Tan Y, Li Y, Zhu D. Synthesis of poly (2-methoxyaniline)/Au nanoparticles in aqueous solution with chlorauric acid as the oxidant. Synthetic Metals 2003;847:135–136.

[15] Mulvaney P. Not all that's gold does glitter. MRS Bulletin 2001;26:1009–1014.

[16] Murphy CJ, Jana NR. Controlling the aspect ratio of inorganic nanorods and nanowires. Advanced Materials 2002;14:80–82.

[17] Khlebtsov NG, Trachuk LA, Mel'nikov AG. Sensitivity of metal nanoparticle surface plasmon resonance to the dielectric environment. Optika i Spectrosk 2005;98:77–86.

[18] Mock JJ, Smith DR, Schultz S. Interparticle coupling effects on plasmon resonances of nanogold particles. Nano Letters 2003;3:1087–1090.

[19] Hsu WK, Terrones M, Hare JP, Terrones H, Kroto HW, Walton DRM. Electrolytic formation of carbon nanostructures. Chemical Physics Letters 1996;262:161–166.

[20] Kim D, Hwang Y, Cheong SI, Lee JK, Hong D, Moon S, Lee JE, Kim SH. Production and characterization of carbon nano colloid via one-step electrochemical method. Journal of Nanoparticle Research 2005;10 (7):1121–1128.

[21] Khaydarov RR, Khaydarov RA, Gapurova O. Remediation of contaminated groundwater using nano-carbon colloids. In: Linkov I, editor. *Nanomaterials: Risk and Benefits, NATO Science for Peace and Security Series C: Environmental Security*. the Netherlands: Springer; 2009. p 219–224.

[22] Peckett JW, Trens P, Gougeon RD, Poppl A, Harris RK, Hudson MJ. Electrochemically oxidised graphite. Characterisation and some ion exchange properties. Carbon 2000;38:345–353.

[23] Khaydarov RR, Khaydarov RA, Gapurova O. Nanocarbon–polymer nanocomposites. International Journal of Nanoparticles 2012;5 (1):88–98.

[24] Khaydarov RA, Khaydarov RR, Gapurova O. Regulation of cyanobacteria in large open water reservoirs. In: Baba A, editor. *Climate Change and its Effects on Water Resources. NATO Science for Peace and Security Series C. Environmental Security*. Volume 3, the Netherlands: Springer; 2011. p 147–156.

[25] GOSSTANDARD. Method for determining residual quantities of pesticides. GOST 13496.20-87 of Russian Federation, 2008.

12

SEMICONDUCTOR NANOMATERIALS FOR ORGANIC DYE DEGRADATION AND HYDROGEN PRODUCTION VIA PHOTOCATALYSIS

LETICIA M. TORRES-MARTÍNEZ, ISAÍAS JUÁREZ-RAMÍREZ, AND MAYRA Z. FIGUEROA-TORRES

Departamento de Ecomateriales y Energía, Facultad de Ingeniería Civil-Universidad Autónoma de Nuevo León, Avenida Universidad y Avenida Fidel Velázquez S/N, Ciudad Universitaria, San Nicolás de los Garza, Nuevo León, México

12.1 INTRODUCTION

The combination of nanotechnology and photocatalysis for clean environment and for providing a renewable energy source has been one of the most important research subjects in recent years. In particular, photocatalytic processes using semiconductor materials in nanoparticle size have been demonstrated to be effective technologies for organic compound degradation and hydrogen production from water splitting. The interest is because in these photocatalytic processes, solar energy can be used as a primary energy source to activate the semiconductor material used as catalyst. It is well known that the photocatalytic process is carried out through the activation of a semiconductor material with ultraviolet (UV) or visible light irradiation [1, 2].

The photocatalytic process involves the generation of electron–hole pairs in the valence and conduction bands of the semiconductor. However, the recombination reaction of electrons and holes commonly tends to occur quickly, diminishing the semiconductor activity [3–9]. Therefore, an electron acceptor added to the reaction mixture is adequate. Hence, it is necessary to carry out modifications of electronic structure, crystal structure, or physicochemical properties in order to enhance the activity of semiconductor materials [10, 11]. In recent years, several chemical methods such as hydrothermal, solvo-combustion, sol–gel, and coprecipitation have been employed to synthesize new and better semiconductor materials with strict control of composition, homogeneity, size and particle shape, as well as use low temperature for their synthesis.

Particularly, during the past decade, several semiconductor nanomaterials with specific structural, physicochemical, and electronic characteristics have been designed and prepared in our laboratory for use as active materials in photoinduced processes. These catalysts could present better photocatalytic performance than TiO_2, due to the different methods of preparation like sol–gel, hydrothermal, colloidal, coprecipitation, and mechanical milling. Additionally, the presence of metal transition cations as dopant agents modifies the band gap value in order to activate the catalysts under UV and visible light irradiation. Therefore, in this chapter several ceramic semiconductor nanomaterials corresponding to the following binary and ternary oxides families are presented: (i) perovskite-type structure, for example, $NaTaO_3$, La:$NaTaO_3$, and Sm:$NaTaO_3$; (ii) pyrochlore-type structure, for example, Sm_2MTaO_7 (M=Fe, In, Ga) and Bi_2MTa0_7 (M=Fe, In, Ga); (iii) rectangular tunnel–type structure, for example, $Na_2Ti_6O_{13}$; and (iv) SiC–TiO_2 compounds. These materials have been tested as catalysts for the photodegradation of methylene blue, rhodamine B, and indigo carmine under UV or visible light source, as well as a catalyst for hydrogen production from water splitting under UV light [12–23]. The main research activities of our group focus on both the performance–structure aspects and the effects of the synthesis route.

Nanomaterials for Environmental Protection, First Edition. Edited by Boris I. Kharisov, Oxana V. Kharissova, and H. V. Rasika Dias.

12.1.1 Perovskite-Type Structure Compounds: $NaTaO_3$, La : $NaTaO_3$ and Sm : $NaTaO_3$

One of the semiconductors mostly used in photocatalysis is the perovskite-type oxide $NaTaO_3$ because it demonstrated excellent results as a catalyst for hydrogen production from water splitting reaction and for degradation of organic compounds [4, 7, 11]. This material has been considered as a catalyst due to its ability to carry out photoabsorption and charge separation, its mobility, as well as its use in reduction–oxidation (redox) reactions in the photocatalytic system. In general, $NaTaO_3$ presents a cubic framework where Ta–O–Ta bonds form an angle close to 180° (Fig. 12.1).

In our research group, $NaTaO_3$, La : $NaTaO_3$, and Sm : $NaTaO_3$ semiconductors have been synthesized by the sol–gel method [12, 13]. The incorporation of La and Sm into $NaTaO_3$ reduces the mobility of electrons and holes to avoid the recombination and retards the crystallization of $NaTaO_3$, provoking the formation of $NaTaO_3$ nanoparticles with high specific surface area. Figure 12.2 shows the X-ray diffraction (XRD) patterns of $NaTaO_3$ doped with La. This material was prepared by the sol–gel method. On the basis of observations, its patterns agreed well with the file pattern reported in the Joint Committee on Power Diffraction File (JCPDF) database.

The material prepared by the sol–gel method was obtained in nanoparticle size, and as can be seen in the TEM micrographs, particles tend to agglomerate each other, increasing their size to greater than 200 nm (Fig. 12.3).

The specific surface area and the energy band gap of $NaTaO_3$ perovskite-type compounds revealed interesting photocatalytic applications (Table 12.1).

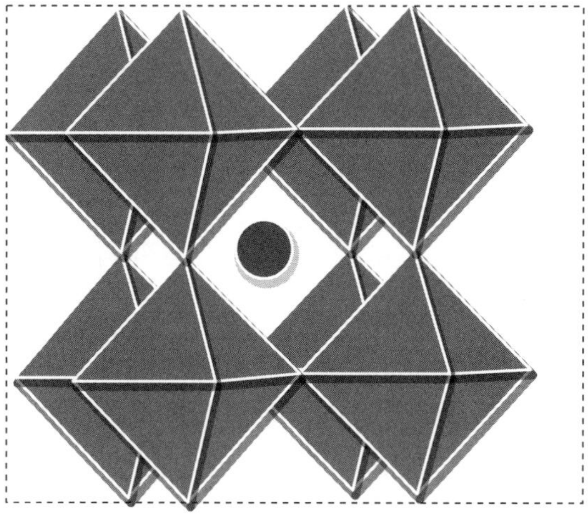

FIGURE 12.1 Crystal structure of $NaTaO_3$.

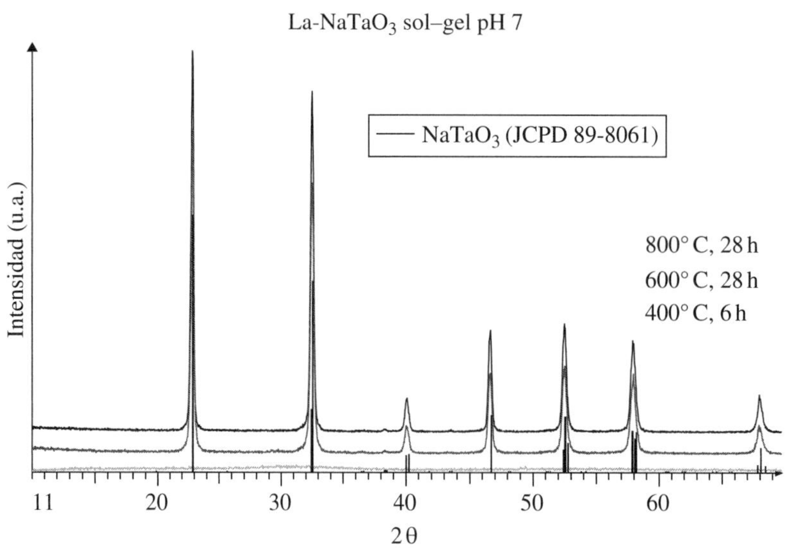

FIGURE 12.2 X-ray diffraction (XRD) patterns of La : $NaTaO_3$ synthesized by the sol–gel method.

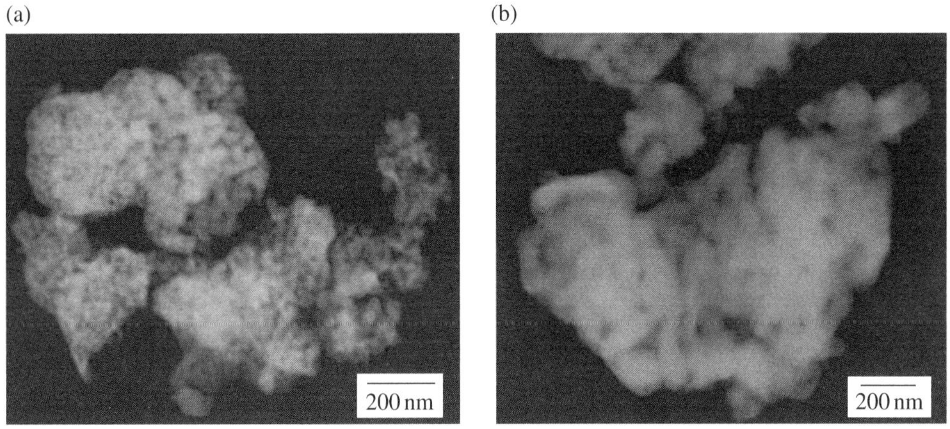

FIGURE 12.3 Transmission electron microscopy (TEM) micrographs of the nanometric size of (a) La : $NaTaO_3$ and (b) Sm : $NaTaO_3$ prepared via the sol–gel method.

TABLE 12.1 Specific surface area and the energy band gap of $NaTaO_3$ perovskite-type compounds prepared by the sol–gel method and compared with solid-state compounds

Synthesis method	Material	Surface area ($m^2 \cdot g^{-1}$)	Band gap energy (eV)
Solid state 850°C	$NaTaO_3$	5	4.0
	La : $NaTaO_3$	5	4.0
	Sm : $NaTaO_3$	5	3.9
Sol-gel 600°C	$NaTaO_3$	14	4.0
	La : $NaTaO_3$	14	4.0
	Sm : $NaTaO_3$	22	4.0

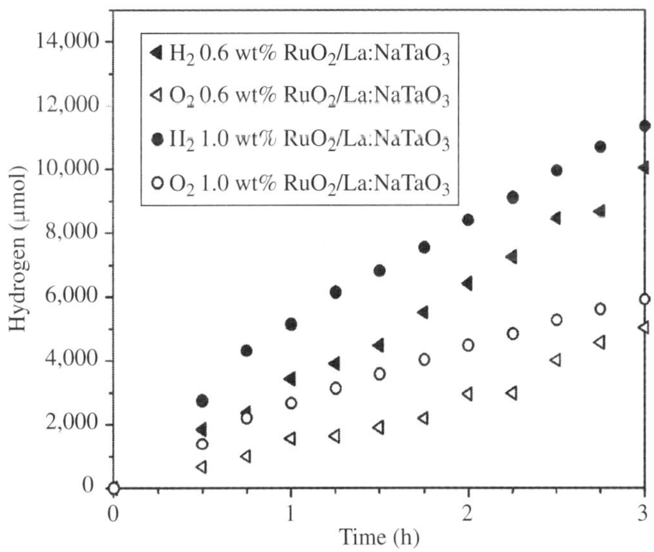

FIGURE 12.4 Hydrogen production using La : $NaTaO_3$ as a catalyst prepared via the sol–gel method. Reproduced with permission from Ref. [13]. © 2010, Elsevier Limited.

Synthesized samples were tested as a photocatalyst for the water splitting reaction. It was shown that by impregnating La : $NaTaO_3$ with Ru, the activity for the water splitting increased; the water splitting activity was almost 25 times greater than that obtained with the La : $NaTaO_3$ semiconductor (Fig. 12.4). RuO_2 acts as an electron trap for the excited electrons induced by UV irradiation, generating very active semiconductors for hydrogen production for water splitting [13].

Results revealed that after 3 h of irradiation, hydrogen production reaches almost 11,500 micromols (μmol). The hydrogen and oxygen production from the water splitting reaction is shown in Table 12.2. Evaluation was carried out using pure water and a xenon UV lamp of 400 W.

TABLE 12.2 Hydrogen production using $NaTaO_3$ as catalyst synthesized by several methods

Compound	Cocatalyst (wt.%)	Hydrogen ($\mu \cdot h^{-1}$)	Oxygen ($\mu \cdot h^{-1}$)
$Sm:NaTaO_3$	None	140	65
$La:NaTaO_3$	None	160	80
$La:NaTaO_3$	RuO_2 0.2	660	377
$La:NaTaO_3$	RuO_2 0.6	3,246	1,623
$La:NaTaO_3$	RuO_2 1.0	4,108	2,160
$La:NaTaO_3$	RuO_2 1.0	11,300 (after 3 h of irradiation)	10,050 (after 3 h of irradiation)

TABLE 12.3 Kinetic parameters obtained from the photocatalytic degradation of methylene blue under UV light and using $NaTaO_3$ perovskite-type compounds

Semiconductor	Sol–gel synthesis	Degradation (%)	k (min^{-1})	$t_{1/2}$ (min)
$NaTaO_3$	SG-600	95	0.0078	89
$La:NaTaO_3$	SG-600	66	0.0084	83
$Sm:NaTaO_3$	SG-600	91	0.0106	65

Therefore, RuO_2 enhances the activity of $La:NaTaO_3$ samples for hydrogen production because it traps the electrons and avoids easy recombination of the electron–hole pair during the reaction.

On the other hand, $NaTaO_3$ perovskite-type compounds have also been tested as a photocatalyst for degradation of organic dyes [12]. When $NaTaO_3$ compounds were used as a photocatalyst for methylene blue degradation, excellent activity was observed. In this case, sample doped with Sm showed the best half-life for this reaction (Table 12.3).

In comparison with TiO_2, $NaTaO_3$ perovskite-type compounds are more effective for methylene blue degradation [12].

12.1.2 Pyrochlore-Type Structure: Sm_2MTaO_7 (M = Fe, In, Ga), and Bi_2MTa0_7 (M = Fe, In, Ga)

Several investigations have been conducted that focus on pyrochlore-type compounds with the formula A2BB′O7 where A and B sites can be substituted by metal ions in order to develop photocatalysts with a narrow band gap. This is because it has been reported that some metal oxides containing cations with d^{10} electronic configuration such as Ga^{3+}, In^{3+}, Ge^{4+}, Sn^{4+}, and Sb^{5+} are attractive for the synthesis of pyrochlore-type compounds [24–27]. Furthermore, the conduction band of d^{10} metal oxides presents larger dispersion than that of d^0 transition metals, which allows high mobility of the photoexcited electrons [28]. Therefore, semiconductor oxides combining cations with $4f–d^{10}–d^0$ electronic configurations are considered interesting for water splitting reaction and organic dye degradation. In particular, pyrochlore-type structure compounds prepared by means of a solid state reaction have shown excellent activity in photoinduced processes. This activity is due to the distortion of their octahedral, which causes an increase in the concentration of holes, thus having an important effect on the charge mobility.

Particularly in our group, novel pyrochlore-type compounds, for example, Sm_2MTaO_7 (M = Fe, In, Ga), and Bi_2MTa0_7 (M = Fe, In, Ga), have been synthesized via the sol–gel method with attractive characteristics as potential photocatalyst materials [14–19]. The prepared compounds had improved photophysical properties and demonstrated excellent results for the inactivation of microorganisms as well as for organic dye degradation and hydrogen production.

Sm_2GaTaO_7 was prepared for the first time using this process, and its structural features are presented in Figure 12.5, which indicate that it is a compound with monoclinic structure and consisting of irregular Ga/Ta octahedra linked at the corners and interconnected into a hexagonal tungsten bronze (HTB)-type network forming two dimensional (2D) HTB blocks, similar to the one reported for Sm_2FeTaO_7 [14]. Therefore, Sm_2GaTaO_7 can be considered as a pyrochlore-related compound.

Sm_2GaTaO_7 was evaluated as a photocatalyst for the water splitting reaction, where it was found that it is able to produce hydrogen without loading the cocatalyst; however, when RuO_2 was added, hydrogen evolution was enhanced (Fig. 12.6). The material $0.2RuO_2/Sm_2GaTaO_7$ showed the highest hydrogen evolution; however, when the amount of RuO_2 increased the hydrogen evolution decreased considerably and it was even less than for Sm_2GaTaO_7. These results revealed that RuO_2 as a cocatalyst provides efficient active sites for hydrogen evolution; however, an excess of RuO_2 has a negative effect on the reaction. It is assumed that the overload of RuO_2 limited the light absorption by Sm_2GaTaO_7, reducing the generation of electron–hole pairs and also acting as recombination centers.

The results revealed that the crystal structure and the constitutive elements play an important role in the photocatalytic activity and the presence of RuO_2 is necessary because it acts as an effective cocatalyst to enhance hydrogen evolution activity of Sm_2GaTaO_7 [15].

On the other hand, Bi_2MTaO_7 pyrochlore-type compounds have been tested as photocatalysis for organic dye degradation [16–19]. According to the results, activity is higher in samples prepared via sol–gel than in samples prepared via the solid state.

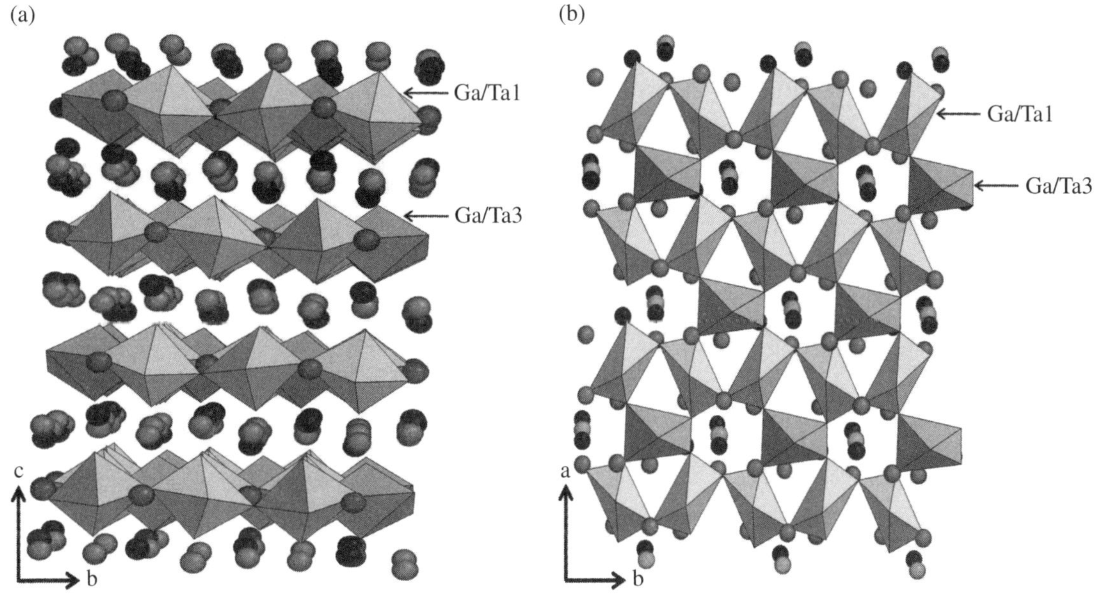

FIGURE 12.5 Structural features of monoclinic Sm_2GaTaO_7. (a) Sm_2GaTaO_7 with hexagonal tungsten bronze (HTB) arrangement and (b) Sm_2GaTaO_7 forming 2D HTB blocks structure. Reproduced with permission from Ref. [29]. © IJHE.

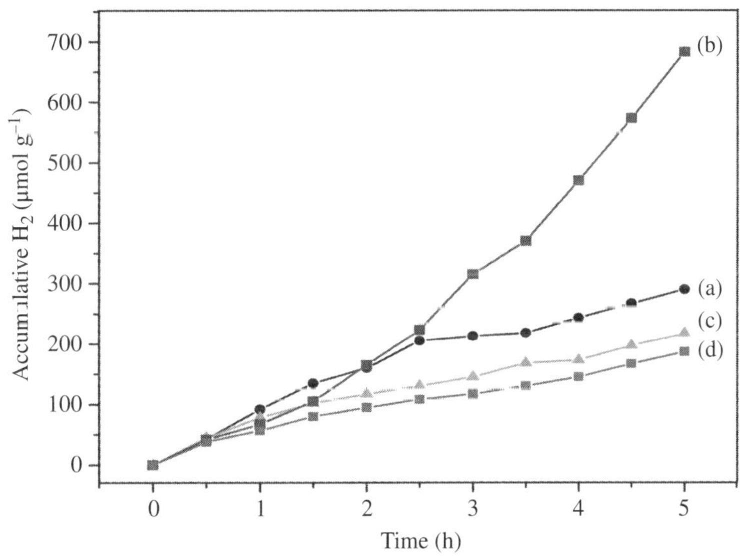

FIGURE 12.6 Photocatalytic hydrogen production of Sm_2GaTaO_7 materials: (a) Sm_2GaTaO_7, (b) 0.2 RuO_2/Sm_2GaTaO_7, (c) 0.9 RuO_2/Sm_2GaTaO_7, and (d) 1.5 RuO_2/Sm_2GaTaO_7. Reproduced with permission from Ref. [29]. © IJHE.

TABLE 12.4 Photocatalytic activity of Bi_2MTaO_7 pyrochlore-type compounds for Alizarin red degradation

	Bi_2AlTaO_7		Bi_2FeTaO_7		Bi_2GaTaO_7		Bi_2InTaO_7	
Synthesis T (°C)	% D	$t_{1/2}$ (h)	% D	$t_{1/2}$ (h)	% D	$t_{1/2}$ (h)	% D	$t_{1/2}$ (h)
SG-400	30	6.5	90	1.0	50	4.0	90	1.0
SG-600	70	2.0	80	1.5	30	8.0	~100	0.5
SG-800	80	1.5	75	2.5	15	16.5	80	1.5
Solid state	60	2.5	40	6.5	20	10.5	~100	1.0

As given in Table 12.4, it is observed that Bi_2InTaO_7 shows the highest activity.

Bi_2InTaO_7 was also tested as a photocatalyst for red tide and green tide degradation. Figure 12.7 shows the photocatalytic behavior of these reactions [19].

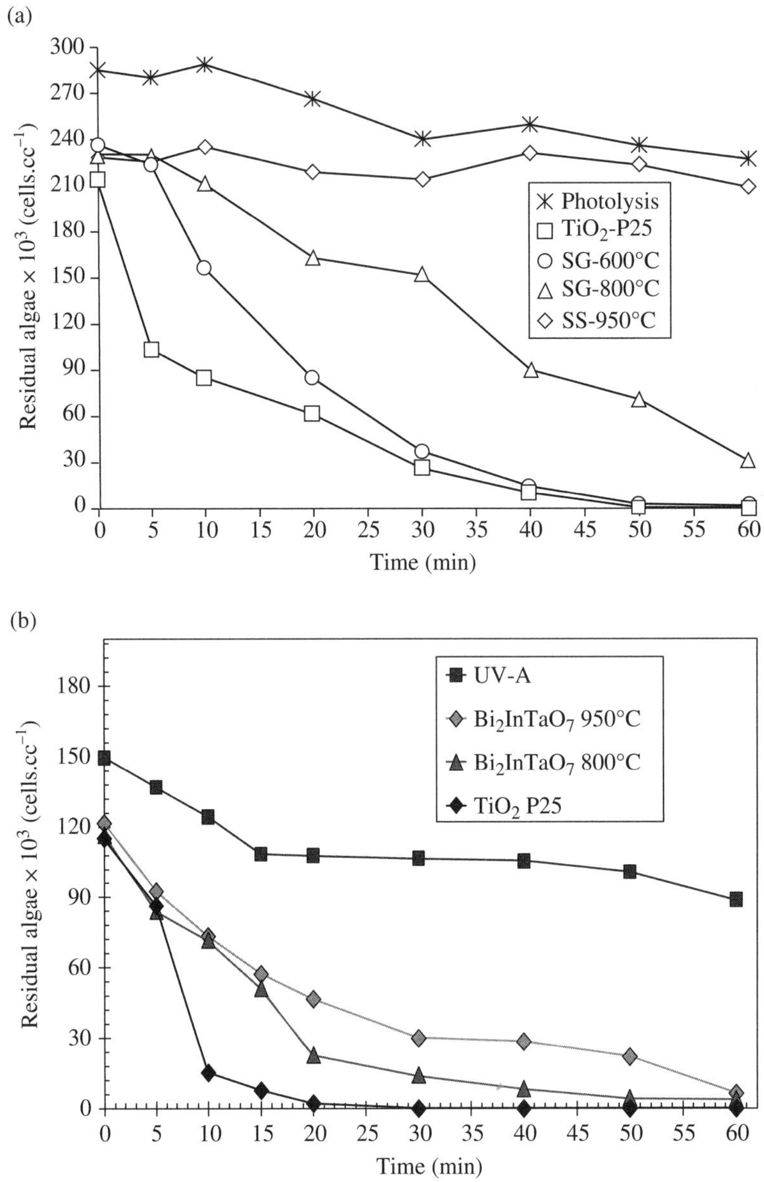

FIGURE 12.7 Photocatalytic removal of Marine Plankton, Algae-tide (a) and microorganism-tide (b) using Bi_2InTaO_7 as a photocatalyst under ultraviolet (UV) light. Reproduced with permission from Ref. [19]. © 2011, JCPR.

12.1.3 Rectangular Tunnel-Type Structure: $Na_2Ti_6O_{13}$

It is well known that the photocatalytic process can be increased by surface modification of the photocatalysts by loading a cocatalyst. The cocatalyst can be a noble metal (Pt and Rh) or a transition metal oxide (NiO and RuO_2), which acts as an electron trap decreasing the electron–hole recombination process and also generates active sites for redox reactions.

$Na_2Ti_6O_{13}$ was obtained for the first time using the sol–gel method and tested as a photocatalyst in methylene blue degradation [20, 21].The crystal structure of $Na_2Ti_6O_{13}$ is represented in Figure 12.8 where the presence of a tunnel structure is observed.

Through scanning electron microscopy (SEM) micrographs the presence of the $Na_2Ti_6O_{13}$ compound was corroborated. This material was obtained in the form of microrods (Fig. 12.9).

The photocatalytic evaluation of $Na_2Ti_6O_{13}$ on methylene blue revealed that the best photocatalytic performance on the degradation of methylene blue under UV light was obtained using a photocatalyst sample heated at 400°C. However, total crystallization of $Na_2Ti_6O_{13}$ occurred above 600°C; therefore, in this case, the activity reported for a sample at 400°C is influenced by the presence of the amorphous material (Table 12.5).

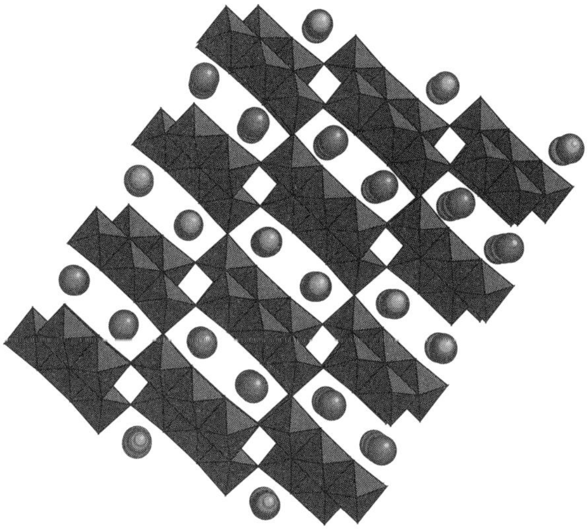

FIGURE 12.8 Crystal structure of $Na_2Ti_6O_{13}$. Reproduced with permission from Ref. [20]. © 2008, Springer.

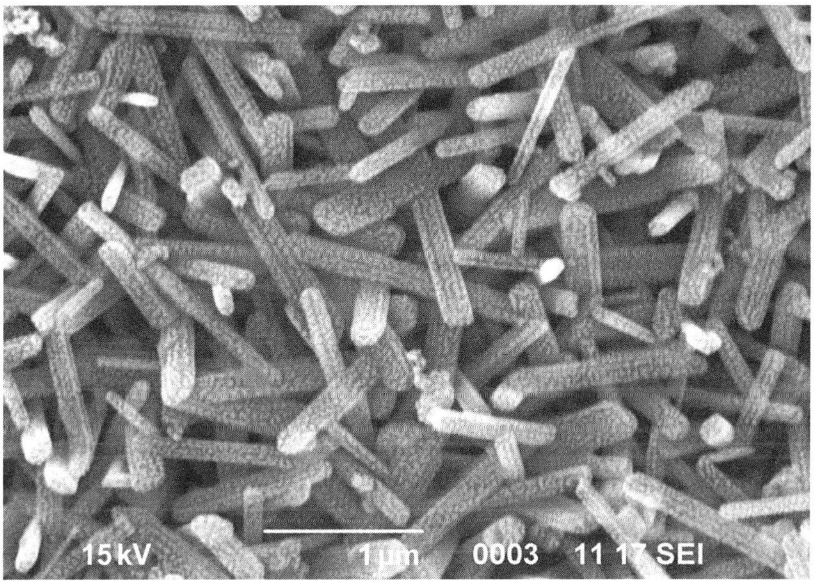

FIGURE 12.9 Scanning electron microscopy (SEM) micrographs of $Na_2Ti_6O_{13}$ where the presence of microrods can be observed. Reproduced with permission from Ref. [20]. © 2008, Springer.

TABLE 12.5 Kinetic parameters of methylene blue degradation using $Na_2Ti_6O_{13}$ as photocatalyst

	Sol–gel	
$Na_2Ti_6O_{13}$	k (min^{-1})	$t_{1/2}$ (min)
400°C	0.0149	46
600°C	0.0011	630

12.1.4 SiC–TiO_2 Compounds

SiC has been considered as a support due to its properties like high thermal conductivity, high mechanical strength, as well as low chemical reactivity. In this work, TiO_2 was deposited on SiC by the sol–gel method, but in order to increase the efficiency of the photocatalytic activity, time of deposition was decreased considerably. SiC–TiO_2 has been tested as a catalyst for the degradation of indigo carmine and methylene blue dyes. In this case, SiC–TiO_2 was prepared via sol–gel method [22, 23].

According to the results, SiC did not undergo any reaction after the deposition of TiO_2 (Fig. 12.10).

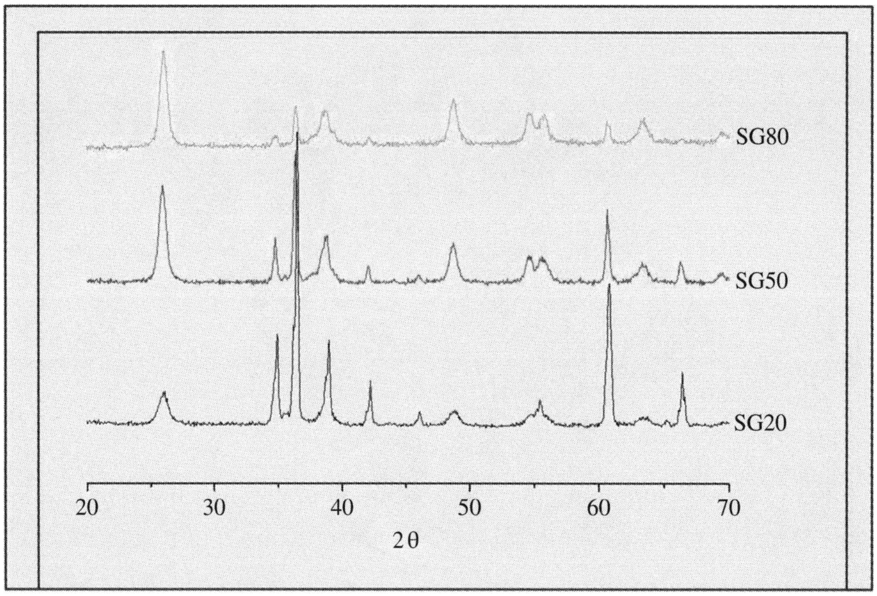

FIGURE 12.10 X-ray diffraction (XRD) patterns of SiC–TiO_2 samples prepared via the sol–gel method.

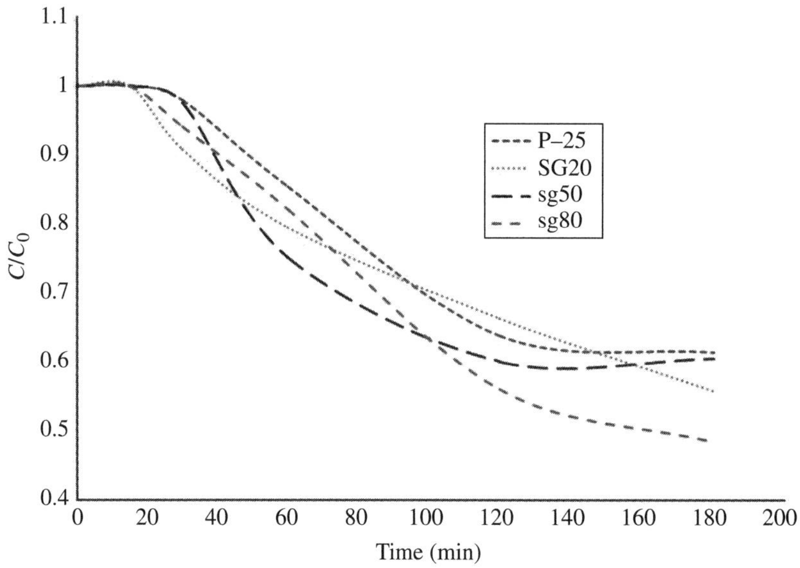

FIGURE 12.11 Methylene blue degradation using SiC–TiO_2 compounds prepared via the sol–gel method.

In this case, it was observed that this condition induces a synergistic effect between the two components of the catalyst that reduces the recombination of the electron–hole pairs , favoring the photocatalytic activity.

In Figure 12.11, the photocatalytic curves for methylene blue degradation are shown; it is observed that SiC–TiO_2 samples have better degradation capacity than TiO_2.

The kinetic parameters for this reaction are showed in Table 12.6. Although half-life reaches almost 200 min, all SiC–TiO_2 compounds are more active than TiO_2.

The degradation of indigo carmine occurred faster than that of methylene blue. This is because the nature of the dye allows easy degradation (Fig. 12.12).

Table 12.7 shows the kinetic parameters for indigo carmine photodegradation. The half-life in this case is lower than the half-life for methylene blue. In fact, total degradation of indigo carmine is achieved in less than 1 h.

As mentioned, the catalysts discussed here were designed and prepared with specific structural, physicochemical, and electronic characteristics in order to be considered as active materials in photoinduced processes.

TABLE 12.6 Kinetic parameters of methylene blue degradation using SiC–TiO_2 as photocatalyst

Material	k (min^{-1})	$t_{1/2}$ (min)
TiO_2 P-25	0.0024	289
SG20	0.0027	257
SG50	0.0025	277
SG80	0.0031	224

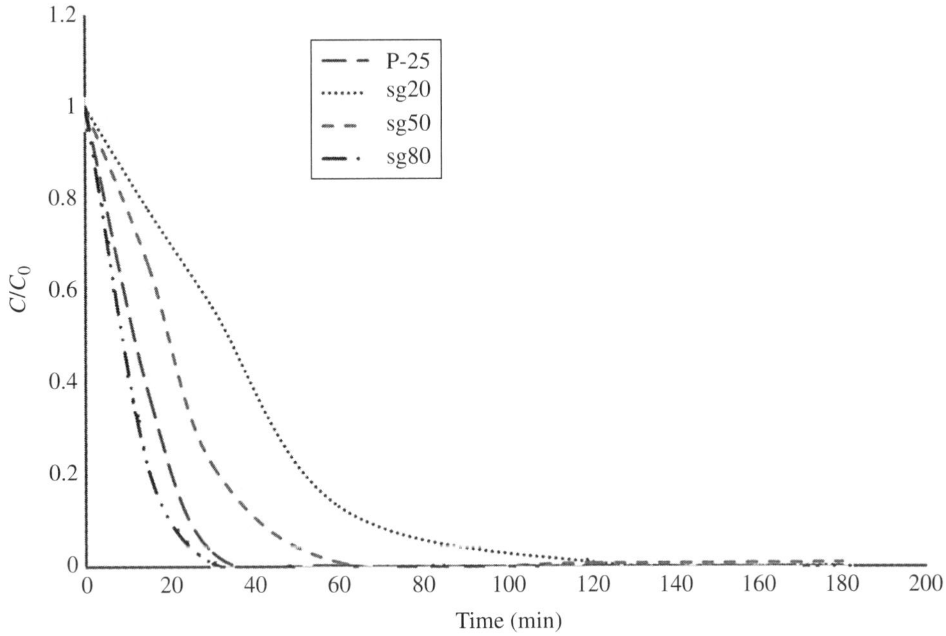

FIGURE 12.12 Indigo carmine degradation using SiC–TiO_2 compounds prepared via the sol–gel method.

TABLE 12.7 Kinetic parameters of indigo carmine degradation using SiC–TiO_2 as photocatalyst

Material	k (min^{-1})	$t_{1/2}$ (min)
TiO_2 P-25	0.0154	45
SG20	0.0145	48
SG50	0.0165	42
SG80	0.0331	21

In some cases, these catalysts presented photocatalytic performance better than or similar to TiO_2, particularly due to the different methods of preparation like sol–gel, which allows the control of morphology, surface area, and porosity enhancement. Additionally, the presence of metal transition cations as dopant agents modified the band gap value in order to activate the catalysts under UV and visible light irradiation. In this sense a relationship was established between the performance–structure and the synthesis route of several catalyst materials.

ACKNOWLEDGMENTS

The authors thank CONACYT for the financial support obtained through projects CB168730-2011, CB177099-2012, INFR03-2011-173625 and INFRA-2012-01-187552, CNPq Mexico-Brasil 174247, and FOINS/75/2012. Thanks to PAICYT-UANL-2012 projects and PIFI-2011. Additionally, the authors thank Christian Gomez-Solis, Ph.D., and Miguel A. Ruiz-Gomez, Ph.D., for their invaluable collaboration.

REFERENCES

[1] Maeda K. Photocatalytic water splitting using semiconductor particles: history and recent developments. J Photochem Photobiol C 2011;12 (4):237–268.

[2] Jing D, Guo L, Zhao L, Zhang X, Liu H, Li M, Shen S, Liu G, Hu X, Zhang X, Zhang K, Ma L, Guo P. Efficient solar hydrogen production by photocatalytic water splitting: from fundamental study to pilot demonstration. Int J Hydrogen Energy 2010;35 (13): 7087–7097.

[3] Zhu J, Zach M. Nanostructured materials for photocatalytic hydrogen production. Curr Opin Colloid Interface Sci 2009;14 (4): 260–269.

[4] Kudo A, Miseki Y. Heterogeneous photocatalyst materials for water splitting. Chem Soc Rev 2009;38:253–278.

[5] Maeda K, Abe R, Domen K. Role and function of ruthenium species as promoters with TaON-based photocatalysts for oxygen evolution in two-step water splitting under visible light. J Phys Chem C 2011;115 (7):3057–3064.

[6] Chen Y, Yang H, Liu X, Guo L. Effects of cocatalysts on photocatalytic properties of la doped Cd_2TaGaO_6 photocatalysts for hydrogen evolution from ethanol aqueous solution. Int J Hydrogen Energy 2010;35 (13):7029–7035.

[7] Kudo A. Recent progress in the development of visible light-driven powdered photocatalysts for water splitting. Int J Hydrogen Energy 2007;32 (14):2673–2678.

[8] Hernandez-Alonso MD, Fresno F, Suarez S, Coronado JM. Development of alternative photocatalysts to TiO_2: challenges and opportunities. Energy Environ Sci 2009;2:1231–1257.

[9] Domen K, Kudo A, Onishi T. Mechanism of photocatalytic decomposition of water into H_2 and O_2 over NiO-$SrTiO_3$. J Catal 1986;102 (1): 92–98.

[10] Cho IS, Bae ST, Kim DH, Hong KS. Effects of crystal and electronic structures of ANb_2O_6 (A = Ca, Sr, Ba) metaniobate compounds on their photocatalytic H_2 evolution from pure water. Int J Hydrogen Energy 2010;35 (23):12954–12960.

[11] Kato H, Asakura K, Kudo A. Highly efficient water splitting into H_2 and O_2 over lanthanum-doped $NaTaO_3$ photocatalysts with high crystallinity and surface nanostructure. J Am Chem Soc 2003;125 (10):3082–3089.

[12] Torres-Martinez LM, Cruz-Lopez A, Juarez-Ramirez I, Meza-de la Rosa ME. Methylene blue degradation by $NaTaO_3$ sol–gel doped with Sm and La. J Hazard Mater 2009;165 (1–3):774–779.

[13] Torres-Martinez LM, Gomez R, Vazquez-Cuchillo O, Juarez-Ramirez I, Cruz-Lopez A, Alejandre-Sandoval FJ. Enhanced photocatalytic water splitting hydrogen production on RuO_2/La:$NaTaO_3$ prepared by sol-gel method. Catal Commun 2010;12 (4):268–272.

[14] Torres-Martinez LM, Ruiz-Gomez MA, Figueroa-Torres MZ, Juarez-Ramirez I, Moctezuma E, Lopez-Cuellar E. Synthesis by two methods and crystal structure determination of a new pyrochlored-related compound Sm_2FeTaO_7. Mater Chem Phys 2012;133: 839–844.

[15] Ruiz-Gomez MA, Torres-Martinez LM, Juarez-Ramirez I, Moctezuma E, Figueroa-Torres MZ. Sm_2FeTaO_7 photocatalyst for degradation of indigo carmine dye under solar light irradiation. Int J Photoenergy 2012;2012:7.

[16] Torres-Martinez LM, Ruiz-Gomez MA, Moctezuma E, Figueroa-Torres MZ, Juarez-Ramirez I. Sol–gel synthesis of Sm_2InTaO_7 and its photocatalytic activity on degradation of crystal violet dye and reduction of Cr(VI) ions. Res Chem Intermed 2013;39 (4):1533–1544.

[17] Torres-Martinez LM, Juarez-Ramirez I, Ramos-Garza JS, Vazquez-Acosta F, Lee SW. Bi_2MTaO_7 (M = Al, Fe, Ga, In) photocatalyst for organic compounds degradation under UV and visible light. WSEAS Trans Environ Dev 2010;6:286–295.

[18] Torres-Martinez LM, Juarez-Ramirez I, Ramos-Garza JS, Vazquez-Acosta F, Lee SW. Sol-gel preparation of Bi_2InTaO_7 and its photocatalytic behavior for organic compounds degradation. Mater Sci Forum 2010;658:491–494.

[19] Torres-Martinez LM, Juarez-Ramirez I, Ramos-Garza JS, Vazquez-Acosta F, Gomez R, Lee SW. Bi_2InTaO_7 compounds as promising photocatalysts for the marine plankton removal. J Ceram Process Res 2011;12:1–4.

[20] Torres-Martinez LM, Juarez-Ramirez I, Del Angel-Sanchez K, Garza-Tovar L, Cruz-Lopez A, Del Angel G. Rietveld refinement of $Na_2Ti_6O_{13}$ prepared by sol-gel, and its photocatalytic performance on the degradation of methylene blue under UV light. J Solgel Sci Technol 2008;47 (2):158–164.

[21] Torres-Martinez LM, Cruz-Lopez A, Garza-Tovar L, Del Angel K, Juarez-Ramirez I. Synthesis of sol gel $Na_2Zr_xTi6_{-x}O_{13}$ $(0 < x < 1)$ materials and their photocatalytic performance on organic dyes degradation. Res Chem Intermed 2008;34 (4):403–416.

[22] Juarez-Ramirez I, Moctezuma E, Torres-Martinez LM, Gomez-Solis C. Short time deposition of TiO_2 nanoparticles on SiC as photocatalysts for the degradation of organic dyes. Res Chem Intermed 2013;39 (4):1523–1531.

[23] Gomez-Solis C, Juarez-Ramirez I, Moctezuma E, Torres-Martinez LM. Photodegradation of indigo carmine and methylene blue dyes in aqueous solution by SiC–TiO_2 catalysts prepared by sol-gel. J Hazard Mater 2012;217–218:194–199.

[24] Sato J, Saito N, Nishiyama H, Inoue Y. New photocatalyst group for water decomposition of RuO_2-loaded p-block metal (In, Sn, and Sb) oxides with d^{10} configuration. J Phys Chem B 2001;105 (26):6061–6063.

[25] Sato J, Kobayashi H, Ikarashi K, Saito N, Nishiyama H, Inoue Y. Photocatalytic activity for water decomposition of RuO_2-dispersed Zn_2GeO_4 with d^{10} configuration. J Phys Chem B 2004;108 (14):4369–4375.

[26] Arai N, Saito N, Nishiyama H, Shimodaira Y, Kobayashi H, Inoue Y, Sato K. Photocatalytic activity for overall water splitting of RuO_2-loaded $Y_xIn_{2-x}O_3$ (x = 0.9–1.5). J Phys Chem C 2008;112 (13):5000–5005.

[27] Tang XD, Ye HQ, Liu H, Ma CX, Zhao Z. A novel visible-light-driven photocatalyst Sm_2InNbO_7 for H_2 or O_2 evolution. Chem Phys Lett 2009;484 (1–3):48–53.

[28] Tang X, Ye H, Liu H, Ma C, Zhao Z. Photocatalytic splitting of water under visible-light irradiation over the NiO_x-loaded Sm_2InTaO_7 with 4f-d^{10}-d^0 configuration. J Solid State Chem 2010;183 (1):192–197.

[29] Ruiz-Gómez MA,, Torres-Martínez LM, Figueroa-Torres MZ, Moctezuma E, Juárez-Ramírez I. Hydrogen evolution from pure water over a new advanced photocatalyst Sm_2GaTaO_7 Original Research Article. Int J Hydrogen 38(28):12554–12561.

PART IV

NANOADSORBENTS AND NANOFILTRATION

13

ADVANCED OXIDATION PROCESSES, NANOFILTRATION, AND APPLICATION OF BUBBLE COLUMN REACTOR

SUKANCHAN PALIT
Department of Chemical Engineering, University of Petroleum and Energy Studies, Dehradun, India

13.1 INTRODUCTION

Environmental engineering is moving toward the next generation of science and technology. The advanced oxidation process (AOP) is a major water treatment and waste treatment procedure. Due to strict and stringent environmental restrictions and regulations, scientists are keen on discovering newer and novel environmental engineering procedures. So, AOPs or membrane separation procedures are opening up new windows of innovation in this age of green technology. The world of imagination has brought the scientific age to a new reality. Integration of various AOPs for wastewater treatment has opened up exciting and exhilarating insights for the scientific community. Our aim in this chapter is to outline in a far-reaching review and experimental validation the importance of various AOPs, the potential of nanofiltration, and the applications of a visionary tool—the bubble column reactor. Our basic aim is to outline the importance of AOPs and nanofiltration of dye wastewater and integrate them with applications of a bubble column reactor.

13.2 VISION OF AOP

The vision of a scientist is versatile and inspiring. Research, scientific vision, and scientific steadfastness has compelled scientists to strive further. The world of chemical engineering and environmental engineering is ushering in a new global era, and a new world of scientific research is emerging. The scientific vision and immense scientific understanding of advanced oxidation processes (AOPs), nanofiltration, and bubble column reactors is awesome, ever-growing, and far-reaching. The basic advantages of a bubble column reactor are that it has feasible hydrodynamics and excellent heat and mass transfer characteristics. Extensive research work is being undertaken by scientists, researchers, and the learned scientific community. The world of the unknown is opening up to a new age and a new vision. Environmental restrictions and stringent regulations are the forerunners of this scientific vision.

AOPs comprise a promising and challenging technology for the treatment of wastewaters containing organic compounds that are not easily removable. All AOPs are designed to produce hydroxyl radicals, which act with high efficiency to destroy organic compounds. AOPs combine ozone(O_3), ultraviolet (UV), hydrogen peroxide (H_2O_2), and/or a catalyst to offer a powerful water treatment solution for the reduction and/or removal of residual organic compounds as measured by chemical oxygen demand (COD), biological oxygen demand (BOD), or total organic carbon (TOC). This chapter delineates in detail a general review of

Nanomaterials for Environmental Protection, First Edition. Edited by Boris I. Kharisov, Oxana V. Kharissova, and H. V. Rasika Dias.

efficient AOPs developed to decolorize and/or degrade organic pollutants for environment protection. The fundamentals involved in the main applications of typical methods such as Fenton, electro-Fenton, photo-Fenton, ozonation, and UV methods are deliberated and discussed in great detail. Various combinations of these processes and their industrial applications are outlined in this study.

13.3 VISION OF NANOFILTRATION AND THE WIDE DOMAIN OF MEMBRANE SCIENCE AND ITS IMPACT ON SCIENTIFIC RESEARCH PURSUIT

It is urgent and imperative to review and discuss the past to provide a compelling vision for the future. The research domain of membrane science and nanofiltration is undergoing rapid and spontaneous changes. The scientist's vision has widened and the emerging scientific imagination is compelling. The purpose of this chapter is to review the history of development of membranes and membrane processes particularly nanofiltration for water production in general and seawater desalination in particular. We want to highlight some new trends in the following areas: membrane development, membrane characterization, membrane transport, and membrane system design. The future prospects in the four areas are discussed in minute detail. Membrane development deals with recent progress in the development of reverse osmosis (RO) membranes used for desalination. There are two different approaches, both based on in situ polycondensation. One is to develop membranes for desalination of brackish water operable at ultralow pressure and the other is to develop membranes operable at high pressures to achieve high pure water recovery in seawater desalination. In the membrane characterization section, atomic force microscopy (AFM) is featured as a new tool to investigate the nature of membrane surfaces. The effects of surface roughness, which can be measured by AFM, on membrane productivity and membrane fouling are outlined.

An in-depth insight into the drawbacks of nanofiltration is presented by Van der Bruggen et al. [1]. According to their definition, nanofiltration was defined as "a process intermediate between reverse osmosis and ultrafiltration that rejects molecules which have a size in the order of one nanometer." They have reviewed every aspect of the subject of nanofiltration. Nanofiltration was introduced in the late 1980s, mainly aiming at combined softening and organics removal. Since then, the applications of nanofiltration have extended tremendously. An insight into the branch of nanofiltration showed the giant steps science has taken for the well-being of mankind.

The review delineates six challenging areas for nanofiltration where solutions and remedies are still scarce: (i) avoiding membrane fouling, and the possibility of remediation; (ii) improving the separation between solutes that can be achieved; (iii) further treatment of concentrates and an increase in the efficiency of separation; (iv) chemical resistance and limited/short life span of membranes; (v) insufficient and low rejection of pollutants in water treatment; and (iv) the urgent need for modeling and simulation tools. This chapter gives a holistic idea of the state of the art in this field and what the scientific fraternity should aim at as well as its vision. All six thrust areas or domains are interlinked and could possibly reach out in a greater way to bring about remedies and solutions.

Nanofiltration has new possibilities such as in drinking water production, arsenic removal, the removal of pesticides, the production of endocrine disruptors and chemicals, and partial desalination.

Matsuura [2] dealt with progress in membrane science and technology for seawater desalination in a phenomenal and visionary review paper. This review outlines some new trends in the following four areas: membrane development, membrane characterization, membrane transport, and membrane system design. Future targets, vision, and prospects in these four areas are delineated. The review deals with membrane development highlighting recent progress in the development of RO membranes used for desalination. There are two different approaches, both based on in situ polycondensation. One is to develop and devise membranes used for desalination of brackish water operable at ultralow pressures, and the other is to develop and devise membranes operable at high pressures to achieve high pure water recovery in seawater desalination. In the membrane characterization section, the application of AFM in investigating and discovering the science behind membrane surfaces is discussed. Also, the review paper by Matsuura (2001) deals with transport models made primarily for charged membranes. Membrane transport deals with transport models made primarily for charged membranes. Hybrid systems for seawater desalination in which membrane processes are incorporated are discussed. According to the Conclusion section in this review paper, there is an enormous and sizable potential to reduce desalination costs by combining membrane processes with novel separation techniques/unit operations.

Sidek et al. [3] reviewed a phenomenal paper on the factors governing the nanofiltration membrane separation process. The main objectives of this paper are to review the performance of nanofiltration membranes in removing unwanted particles from a solution by evaluating the factors, such as Donnan and steric interaction and transmembrane pressure (TMP), that influence rejection by the membrane. The right combination of membrane pore size (steric effect) and its effective charge density (Donnan effect) leads to optimum separation performance. However, the effect of TMP on nanofiltration rejection is not consistent. At high TMP, rejection can be either increased or decreased, depending on other operating parameters such as pH, ionic strength, the presence of salt. pH and feed concentration (ionic strength) play a significant role in nanofiltration membrane separation.

They outlined membrane characteristics governing factors. Trans membrane pressure drop (TMP) is outlined in major detail. The delineated features are effects of TMP on flow rate and rejection. They described in detail the physics behind the effects of TMP on flux rate. The effects of the TMP on the permeate flux at different solute concentrations can be observed by keeping the operating temperature and pH constant. The permeate flux at the steady state increases, with the applied pressure at all concentrations. An increase in flux was noted with an increase in the operating pressure. Since nanofiltration is basically a pressure-driven filtration process, flux is supposed to increase with pressure. The phenomenon can be mathematically explained.

Hong et al. [4] delineated the chemical and physical aspects of natural organic matter (NOM) fouling of nanofiltration membranes. The role of chemical and physical interactions in NOM fouling of nanofiltration membranes is systematically investigated. Results of fouling experiments with three basic acids demonstrate that membrane fouling increases with increasing electrolyte (NaCl) concentration, decreasing solution pH, and addition of divalent cations (Ca^{2+}). At fixed solution ionic strength, the presence of calcium ions, at concentrations typical of those found in natural waters, has a marked effect on membrane fouling. Divalent cations interact specifically with carboxyl functional groups and thus substantially reduce charge and the electrostatic repulsion between humic macromolecules.

In recent years, membrane filtration has emerged as a viable treatment alternative to comply with existing and pending water quality regulations. Of particular interest is the use of nanofiltration as a treatment alternative for the removal of NOM, a precursor of disinfection by-products, in anticipation of more stringent regulations. Nanofiltration technology also offers a versatile approach to meeting multiple water quality objectives, such as the control of organic, inorganic, and microbial contaminants.

Successful application of nanofiltration technology, however, requires efficient control of membrane fouling. Fouling, often associated with the accumulation of substances on the membrane surface or within the membrane pore structure, worsens membrane performance and ultimately shortens membrane life.

Hilal et al. [5] reviewed research work on using AFM toward improvement in nanofiltration membrane properties for desalination pretreatment.

Seawater is characterized by having a high degree of hardness, varying turbidity and bacterial contacts, and high total dissolved solids (TDS). These properties give rise to major problems such as scaling, fouling, high energy requirements, and the requirement of high-quality construction materials. To solve seawater desalination problems and to minimize their effect on productivity and water cost of conventional plants, a new approach using nanofiltration as pretreatment to both RO and thermal processes has been shown to enhance the production of desalted water and reduce the cost, yet it is an environmentally friendly process.

The following areas were covered:

a. Development of high-performance nanofiltration membranes
b. Development of accurate and practical characterization methods
c. Development of a good predictive modeling technique

The use of AFM in membrane studies was also outlined in detail.

Hilal et al. [6] attempted a comprehensive review of nanofiltration membranes and dealt with its treatment, pretreatment, modeling, and AFM. This review addresses the application of AFM in studying the morphology of membrane surfaces as part of nanofiltration membrane characterization.

A comprehensive review of nanofiltration in water treatments is presented including a review of the applications of nanofiltration in treating water as well as in the pretreatment process for desalination; the mechanism as well as minimization of nanofiltration membrane fouling problems; and the theories for modeling and transport of salt and charged and noncharged organic compounds in nanofiltration membranes.

Ashaghi et al. [7] dealt with nanofiltration in detail in their research review on ceramic ultrafiltration and nanofiltration membranes for oilfield-produced water treatment. Produced water is any fossil water that is brought to the surface along with crude oil or natural gas. By far, produced water is the largest by-product or waste stream by volume associated with oil and gas production. The volume of produced water is dependent upon the state of maturation of the field. There is an urgent need for new technologies for produced water treatment due to increased focus on water conservation and environmental regulation. Ceramic ultrafiltration and nanofiltration membranes represent a relatively new class of materials available for the treatment of produced water. According to their research, the issues needing to be addressed are the prevention of membrane fouling during operation and the provision of an expedient, cost-effective, and nonhazardous means of cleaning fouled membranes. The researchers embarked on the present study because there are not enough existing studies related to the treatment of oilfield-produced water using ceramic membranes. Ceramic membrane systems under nanofiltration and ultrafiltration conditions have proven to be economically attractive for the treatment of produced waters with elevated concentrations of oils and low to medium diameters of the particles.

No research review is complete unless the mechanisms of membrane science or nanofiltration are not taken into account.

Wijmans et al. [8] dealt with the solution–diffusion model in their review. The solution–diffusion model has emerged over the past 20 years as the most widely accepted explanation for transport in dialysis, RO, gas permeation, and pervaporation. In this review paper they dealt with the phenomenological equations for transport in these processes using the solution–diffusion model and derived the equations starting from the fundamental statement that flux is proportional to a gradient in chemical potential. The direct and indirect evidence for the model's validity is presented, together with a brief discussion of the transition between a solution–diffusion membrane and a pore-flow membrane seen in nanofiltration membranes and some gas-permeation membranes. The principal property of membranes used in separation applications is the ability to control the permeation of different species. Two models are used to describe this permeation process. The first is the solution–diffusion model, in which permeants dissolve in the membrane material and then diffuse through the membrane down a concentration gradient. A separation is achieved between different permeants because of differences in the amount of material that dissolves in the membrane and the rate at which the material diffuses through the membrane. The second is the pore flow model in which permeants are separated by pressure-driven convective flow through tiny pores.

Nghiem et al. [9] delineated the removal of natural hormones by nanofiltration membranes with measuring, modeling, and mechanisms. The removal of four natural steroid hormones—estradiol, estrone, testosterone, and progesterone—by nanofiltration membranes was investigated. Two nanofiltration membranes with quite different salt retention characteristics were utilized.

Renou et al. [10] reviewed the opportunities in landfill leachate treatment. In most countries, sanitary landfilling is nowadays the most common way to eliminate municipal solid wastes (MSW). In spite of many advantages, the generation of heavily polluted leachates, with significant variations in both volumetric flow and chemical composition, constitutes a major drawback. Year after year, the recognition of landfill leachate effect on the environment has forced authorities to fix more and more stringent requirements for pollution control. This paper is a review of landfill leachate treatments. The advantages and drawbacks of the various treatments are discussed under the following topics: (i) leachate transfer, (ii) biodegradation, (iii) chemical and physical methods, and (iv) membrane processes.

Childress et al. [11] related nanofiltration membrane performance to membrane charge (electrokinetic) characteristics. The performance (i.e., water flux and solute rejection) of a thin-film composite (TFC) aromatic polyamide nanofiltration membrane and its relation to membrane surface charge (electrokinetic) characteristics was studied.

Properties of nanofiltration membranes, model development, and industrial applications have been delineated in a dissertation by Johannes Martinus Koen Timmer [12]. The dissertation deals with industrial membrane processes and aspects of nanofiltration. It also deals with the transport of lactic acid through RO and nanofiltration membranes. A model for mass transport is described. In the pursuit of excellence, they delineated and described the entire pressure-driven membrane process. Their research encompassed the dairy industry and the applications of nanofiltration.

Water engineering or environmental engineering is a boost to this research pursuit when we consider the work done by Xia et al. [13] for arsenic removal by nanofiltration and its application in China. According to their study, arsenic contamination of groundwater and the associated health risks have been reported in many parts of China. Nanofiltration is a promising technology for arsenic removal since it requires less energy than traditional RO membranes. In this study, the removal of arsenic from synthetic waters by nanofiltration membranes was investigated. Arsenic rejection experiments included variation of arsenic feed concentration, pH, and existence of other ionic compounds. The possible influence of NOM on As(V) rejection by nanofiltration membranes was also explored.

Orecki and Tomaszewska [14] did a fundamental research on oily wastewater treatment using nanofiltration process. The authors lucidly and intensely dealt with a positive objective on the domain of oily wastewater treatment using the formidable intellectual challenge of the application area of nanofiltration. The nanofiltration studies were carried out with a permeate obtained from ultrafiltration (UF) (used for the treatment of the oily wastewater from metal industry). The influence of transmembrane pressure on a permeate flux, the degree of rejection of oil and inorganic compounds were investigated with great precision. The studies on the nanofiltration treatment of oil wastewater demonstrated a high effectiveness of the rejection of oil and inorganic compounds. The permeate obtained from the treatment was free of oil. The nanofiltration process was carried out in a pilot plant equipped with a tubular module with the AFC 30 membrane (PCI)—(the working area equal to 0.9 m^2) and a spiral wound module (with the nanofiltration 270-2540 membrane (Film Tec—the working area equal to 2.6 m^2). The studies earlier predicted and performed showed that the membranes differed in molecular weight cut-off (MWCO). The MWCO of the membranes were found out to be equal to 250 g/mol for nanofiltration 270-2540 and nanofiltration AFC30, respectively. The results and discussion showed a remarkable pattern. The raw oily wastewater used in these studies was collected from metal treatment industry. Apart from oil, the wastewater contained a lot of other contaminants, including solid state, lubricants, metal fines and sometimes dissolved metals. Although the UF membranes rejected oil in 90%, the permeate still contained different solutes. The authors performed an integrated ultrafiltration and nanofiltration technique. The conclusions in this study were affirmative. As a result of nanofiltration used as a second stage of oily waste water treatment the removal of organic compounds (TOC) for the studied membranes (nanofiltration 270-2540 and nanofiltration AFC 30) exceeded 65%. Moreover the cations are rejected by 75% and the sulfates were rejected by 95%. The permeate did not contain oil. Orecki and Tomaszewska [14] established a theory of the effectiveness of integrated ultrafiltration—nanofiltration process. Hilal et al. [15] dealt with nanofiltration of highly concentrated salt solutions approaching seawater salinity. Nanofiltration membranes have been employed in pretreatment unit operations in both thermal and membrane seawater desalination processes. This has resulted in reduction of chemicals used in pretreatment processes as well as lowering the energy consumption and water production cost and therefore has led to a more environmentally friendly process. In order to predict nanofiltration membrane performance, a systematic study on the filtration performance of selected commercial nanofiltration membranes against brackish water and seawater is required. In this study, three commercial nanofiltration membranes (NF90, NF270, NF30) have been used to treat highly concentrated (NaCl) salt solutions up to 25,000 ppm, a salinity level similar to that of seawater.

A phenomenal review on cotton textile processing, its waste generation, and effluent treatment was undertaken by Babu et al. [16]. This review discusses cotton textile processing and methods of treating effluents in the textile industry. This area of research combines a discussion of waste production from textile processes, such as desizing, mercerizing, bleaching, dyeing, finishing, and printing with a discussion of advanced methods of effluent treatment, such as electrooxidation, biotreatment, and photochemical and membrane processes.

Ning [17] undertook research work on arsenic removal by RO. A short review on the RO process and applications was dealt with by Garud et al. [18]. The short review discusses the applicability of an RO system for treating effluents from the beverage industry, distillery spent wash, groundwater treatment, the recovery of phenol compounds and the reclamation of wastewater, and seawater RO (SWRO) treatment indicating the efficiency and applicability of RO technology.

13.4 FURTHER RESEARCH ENDEAVOR IN THE FIELD OF NANOFILTRATION

Nanofiltration presents vast and challenging opportunities for knowledge advancement in the field of desalination technology. The world of environmental engineering will usher in a new era in the field of membrane science with immense challenges and vision. Our endeavor encompasses research on AOPs and the application of a bubble column reactor. The challenges that lie ahead are significant and far-reaching. Desalination technology and nanofiltration have a close umbilical relationship. These two branches of engineering will definitely solve the intricate problems of water technology. Man's scientific vision will be enhanced and emphasized if the hurdles are overcome and frontiers scaled. The world of environmental engineering and nanofiltration will surely be a challenge to the drinking water problems of the suffering millions.

13.5 RESEARCH THRUST AREAS IN THE FIELD OF AOPs AND THE VISION TOWARD EFFECTIVE OZONATION PROCEDURES

Palit et al. [19] dealt with membrane separation processes and RO in a detailed review. The application area and thrust was on wastewater treatment. The author has widely described the scope of RO and its potential applications.

Stasinakis [20] in an insightful review dealt with immense depth on the use of selected advanced oxidation processes for wastewater treatment. The purpose and the aim of the study was to review the use of titanium dioxide/UV light process, hydrogen peroxide/UV light process and Fenton's reactions in wastewater treatment. The main reactions and the operating parameters (initial concentration of the target compounds, amount of oxidation agents and catalysts, nature of the wastewater etc) affecting these processes are reported, while several recent applications to wastewater treatment are presented.

Koch et al. [21] dealt with immense details on ozonation of hydrolyzed azo dye reactive yellow 84(CI). According to scientific innovation and immense scientific understanding, the combination of chemical and biological water treatment processes is a promising technique to reduce recalcitrant wastewater loads. Ozonation has been applied to many fields in water and wastewater treatment. Especially for textile mill effluents ozonation can achieve high color removal, enhance biodegradability, destroy phenols and reduce chemical oxygen demand (COD). This work unfolds the reaction intermediates and products formed during ozonation. The work mainly deals with the degradation of hydrolyzed Reactive Yellow 84 (Color Index), a widely used azo dye in textile finishing processes with two monochlorotriazine anchor groups. The authors have investigated the formation of intermediate products and the reaction kinetics of the entire procedure of ozonation. A general review on AOPs for wastewater treatment was done by Sharma et al. [22]. Efficient AOPs developed to decolorize and/or degrade organic pollutants for environmental protection were covered. The fundamentals and main applications of typical methods such as Fenton, electro-Fenton, photo-Fenton, ozonation, and UV radiation were discussed. Various combinations of these processes and their industrial applications are the visionary aspects of this study.

Research on AOPs for the treatment of textile and dye wastewater was attempted by Kalra et al. [23]. Their paper reviews different AOPs like ozonation, hydrogen peroxide, UV radiation, and their combination for comparison of treatment efficiencies for remediation of textile wastewater. This paper reveals that the treatment efficiencies depend on the characteristics of wastewater to be treated.

The use of AOP in an ozone+ hydrogen peroxide system for the removal of cyanide from water was dealt with by Kepa et al. [24]. The results of laboratory tests are presented in this paper, which indicate that AOPs can be used for the removal of cyanide from water. A comparative analysis was carried out for the processes of ozonation, oxidation with hydrogen peroxide, and advanced oxidation in the O_3+H_2O_2 system.

Zhou et al. [25] wrote a review paper on advanced technologies in water and wastewater treatment. They dealt with three emerging treatment technologies including membrane filtration, AOPs, and UV irradiation that hold great promise to provide alternatives for greater protection of human health and environment.

Kdasi et al. [26] provided a clear picture of the treatment of textile wastewater by AOPs in a review. An overview of basic treatment efficiency for different AOPs is considered and presented based on specific features. The review covers a lucid introduction, textile wastewater characteristics, a description of AOPs, application areas of a UV lamp, ozone, O_3/UV, H_2O_2, O_3/H_2O_2 (peroxane), O_3/H_2O_2/UV, and a broad conclusion.

Palit et al. [27] touched upon membrane separation processes and AOPs for dyes in a bubble column reactor in a keen and far-reaching overview. Topics covered included the dependence of rate constant, order of reaction, and subsequently rate of reaction on the pH and oxidation–reduction potential associated with the ozonation of dye.

Gogate et al. [28] dealt with imperative technologies for wastewater treatment with emphasis on oxidation technologies at ambient conditions in a review. This work highlights the basis of the different oxidation processes including the operation parameters for the reactor design with a complete overview of the various applications for wastewater treatment in the recent years.

Chiron et al. [29] explained in a review the state of the art of pesticide chemical oxidation. This review reveals a general lack of data on kinetics of formation and disappearance of the major by-products. The efficiency of AOPs has scarcely been investigated at the industrial scale, that is, in presence of a mixture of active ingredients together with their formulating agents and at concentration levels above 10 mg/l.

Kos et al. [30] dealt with the subject of decoloration of real textile wastewater with AOPs. The efficiency rates of AOPs for the decoloration of different types of textile wastewater from textile plants in Lodz, Poland, were compared on the basis of the results obtained. AOPs with the use of ozone, gamma radiation, hydrogen peroxide, and UV radiation gave good decoloration results.

Suty et al. [31] described the applications of AOPs with emphasis on the present and future aspects. The use of AOPs to remove pollutants in various water treatment applications has been the subject of study for around 30 years. Most of the available AOPs have been investigated in depth, and a considerable body of knowledge has been built up about the reactivity of many pollutants. Nevertheless, it is difficult to obtain an accurate picture of the use of AOPs, and their applications for a range of water treatment processes have not been determined to date. The purpose of this overview is to discuss these processes and provide an indication of future trends and prospects.

Abdelmalek et al. [32] dealt in the area of removal of pharmaceutical and personal care products (PPCPs) from RO retentate using AOPs. Studies focusing on pharmaceutical and PPCPs have raised questions concerning their concentrations in the RO retentate. AOPs are alternatives for destroying these compounds in retentate that contains high concentration of effluent organic matter (EFOM) and other inorganic constituents.

Huber et al. [33] dealt with oxidation of pharmaceuticals during ozonation and AOP applications in drinking water treatment. It was shown that the second-order rate constants determined in pure aqueous solutions could be applied to predict the behavior of pharmaceuticals dissolved in natural water. Overall, it can be concluded that ozonation and AOPs are promising processes for efficient removal of pharmaceuticals in drinking water.

Huber et al. [34] explained decolorization of process waters in deinking mills and similar applications in a review. Process waters in deinking mills often feature a strong coloration, due to dyes and pigments released from the recovered paper. This can usually be remediated by pulp bleaching treatment with appropriate chemicals. In this review, the available technologies for process water decolorization are discussed (chemical methods, physicochemical methods, and biological treatments).

13.6 MAN'S SCIENTIFIC MIND TOWARD NOVEL ENVIRONMENTAL ENGINEERING PROCEDURES

Man's scientific vision is targeted toward improving the lot of the suffering millions and toward tackling drinking water issues. The hidden scientific truth and scientific vision is targeted in every step towards progress of purposeful and definitive research pursuit. The aim of this chapter is to delineate the intricacies of wastewater treatment and bring before the scientific community rigorous scientific inquiry. The greatness of the past and present will frame the future. So the aim and mission towards a better and congenial human society is ushering in hope and insightful destiny.

13.7 APPLICATION OF A BUBBLE COLUMN REACTOR AND VISION OF INNOVATIVE IDEAS

The bubble column reactor is a visionary tool for its application in AOPs. Challenges lie ahead in the integration of nanofiltration and the bubble column reactor. Excellent heat and mass transfer characteristics urges the scientific community to declare it as a challenging and visionary tool. The world of environmental engineering and the domain of wastewater treatment will usher in a new era with the application of bubble column reactor.

Bubble column reactors are intensively used as multiphase reactors and contactors in the chemical, biological, and pharmaceutical domains. They provide several advantages during operation and maintenance such as high heat and mass transfer characteristics, compactness, and low operating and maintenance costs. Three-phase bubble column reactors are widely used in chemical reaction engineering, that is, in the presence of catalysts, and also in biochemical applications where microorganisms are utilized as solid suspensions in order to manufacture industrially valuable bioproducts. Investigation of design parameters characterizing the operation and transport phenomena of bubble columns has led to a better understanding of hydrodynamic properties, heat and mass transfer characteristics, and flow regime mechanisms at work during the operation. The review [35] also targets and focuses on bubble column reactors, their description, design and operation, application areas, fluid dynamics, and regime analysis encountered.

The application of bubble column reactors has had a tremendous impact in recent years in the field of environmental engineering. The world of unknown in the areas of AOPs and nanofiltration will open up new windows of innovation and intuition in the years to come. Bubble column reactors have new applications with regard to AOPs. The vision is clear-cut with the frontiers of environmental science and engineering expanding to unprecedented levels. The grit and determination of the scientist's vision is inspiring and immense. The common man's problems in the field of desalination will mostly be solved by the science of nanofiltration. Research areas will slowly grow to visionary proportions.

13.8 DOCTRINE OF HEAT AND MASS TRANSFER CHARACTERISTICS OF A BUBBLE COLUMN REACTOR

Heat and mass transfer predictabilities are the major targets of the operation of bubble column reactor. Theoretically, a scientist's vision is toward better operation and maintenance of this visionary tool. The aim is to achieve better hydrodynamics as well as better reactivity, heat, and mass transfer.

13.9 DOCTRINE OF HYDRODYNAMICS OF BUBBLE COLUMN REACTOR AND MULTIPHASE FLOW

The importance of hydrodynamics of bubble column and the associated regime of multiphase flow is targeted toward efficient reactivity and linked to heat and mass transfer characteristics.

Bubble column reactors belong to the general class of multiphase reactors that include mainly three categories, that is, trickle bed reactors (fixed or packed bed), fluidized bed reactor, and bubble column reactor [36–39].

13.10 HIDDEN TRUTH IN THE DOMAIN OF ENVIRONMENTAL ENGINEERING AND NANOFILTRATION

Hurdles and unsurpassed barriers shapes the future and the scientific rigor is a vibrant witness. The greatness of environmental engineering tools are unraveled and the human society aims at a sustainable future. The Loeb–Sourirajan model has revolutionized the entire field of membrane science. Principles of membrane science have been established through scientific research. Newer and novel methods of water treatment and desalination are shaping and reshaping scientific intuition. Environmental engineering will usher in a new dawn in science and technology. The backbone of this unending scientific endeavor is strict and stringent environmental regulations. Developed and developing nations are gearing up toward new challenges to embrace the far-reaching environmental engineering frontier.

13.11 FUTURE VISION AND FUTURE FLOW OF THOUGHTS

Nanofiltration has been found to be extremely effective and visionary and will be very much in demand in the years to come [36–39]. Vision, opportunity, and diligent truth will propel the scientist's vision toward a new scientific generation. The greatness and the faults of the technology are envisioned in its application. Bubble column reactors and AOPs are also pillars of the ongoing review. Their applications in the field of environmental engineering are extremely purposeful, intricate, and far-reaching. Man's scientific vision and scientific endeavor is powered by a determination—the determination to serve the suffering millions. The problem of drinking water is immense, and stringent hurdles need to be overcome in both developed and developing countries. So the ultimate target and vision is to make technology accessible to the common masses and venture out to heal the wounds of science and technology. Nanofiltration, AOPs, and the application of bubble column reactors is just a beginning. The application areas of nanofiltration are absolutely far-reaching and visionary. The world will face new challenges in these scientific domains in years to come.

ACKNOWLEDGMENT

The author wishes to acknowledge Dr. Bhaskar Sengupta of Queen's University, Belfast, United Kingdom, under whose guidance he did research work on ozonation of dyes in a bubble column reactor. The author gratefully acknowledges the contributions of past and present teachers of Jadavpur University, Kolkata, India. Also, the contribution of staff, students, teachers, and management of University of Petroleum and Energy Studies, Dehradun, India, needs to be mentioned.

REFERENCES

[1] Bruggen V, Manttari M, Nystrom M. Drawbacks of applying nanofiltration and how to avoid them: a review. Sep Purif Technol 2008;63:251–263.

[2] Matsuura T. Progress in membrane science and technology for seawater desalination—a review. Desalination 2001;134:47–54.

[3] Sidek NM, Ali N, Fauzi SAA. The governing factors of nanofiltration membrane separation process performance: a review, UMTAS, Empowering Science Technology and Innovation towards a Better Tomorrow; 2011, 241–248.

[4] Hong S, Elimelech M. Chemical and physical aspects of natural organic matter (NOM) fouling of nanofiltration membranes. J Membr Sci 1997;132:159–181.

[5] Hilal N, Mohammad AW, Atkin B, Darwish NA. Using atomic force microscopy towards improvement in nanofiltration membranes properties for desalination pretreatment—a review. Desalination 2003;157:137–144.

[6] Hilal N, Al-Zoubi H, Darwish NA, Mohammad AW, Abu Arabi M. A comprehensive review of nanofiltration membranes: treatment, pretreatment, modeling and atomic force microscopy. Desalination 2004;170:281–308.

[7] Ashaghi KS, Ebrahimi M, Czermak P. Ceramic ultrafiltration and nanofiltration membranes for oilfield produced water treatment: a mini review. Open Environ J 2007;1:1–8.

[8] Wijmanns JG, Baker RW. The solution–diffusion model—a review. J Membr Sci 1995;107:1–21.

[9] Nghiem LD, Schafer AI, Elimelech M. Removal of natural hormones by nanofiltration membranes: measurement, modeling and mechanisms. Environ Sci Technol 2004;38:1888–1896.

[10] Renou S, Givaudan JG, Poulain S, Dirassouyan F, Moulin P. Landfill leachate treatment: review and opportunity. J Hazard Mater 2008;150:468–493.

[11] Childress A, Elimelech M. Relating nanofiltration membrane performance to membrane charge (electrokinetic) characteristics. Environ Sci Technol 2000;34:3710–3716.

[12] Timmer JMK. Properties of nanofiltration membranes; model development and industrial application [PhD thesis]. Eindhoven: Technische Universiteit; 2001.

[13] Xia S, Dong B, Zhang Q, Xu B, Gao N, Causseranda C. Study of arsenic removal by nanofiltration and its application in China. Desalination 2007;204:374–379.

[14] Orecki A, Tomaszewska M. The oily wastewater treatment using the nanofiltration process. Pol J Chem Technol 2007; 9 (4):40–42.

[15] Hilal N, Al-Zoubi H, Mohammad AW, Darwish N. Nanofiltration of highly concentrated salt solutions up to seawater salinity. Desalination 2005;184:315–326.

[16] Babu RB, Parande AK, Raghu S, Prem Kumar T. Cotton textile processing: waste generation and effluent treatment. J Cotton Sci 2007;11:141–153.

[17] Ning RY. Arsenic removal by reverse osmosis. Desalination 2002;143:237–241.

[18] Garud RM, Kore SV, Kore VS, Kulkarni GS. A short review on process and applications of reverse osmosis. Univ J Environ Res Technol 2011;1 (3):233–238.

[19] Palit S. A short review of applications of reverse osmosis and other membrane separation procedures. Int J Chem Sci Appl 2012;3 (2):302–305.

[20] Stasinakis AS. Use of selected advanced oxidation processes (AOPs) for wastewater treatment: a mini review. Global NEST J. 2008;10 (3): 376–385.

[21] Koch M. Yediler A, Lienert D, Insel G, Kettrup A. Ozonation of hydrolyzed azo dye reactive yellow 84(CI). Chemosphere 2002;46:109–113.

[22] Sharma S, Ruparelia JP, Patel ML. A general review on advanced oxidation processes for waste water treatment. International Conference on Current Trends in Technology (NUiCONE-2011). Institute of Technology, Nirma University, Ahmedabad, India; December 8–10, 2011.

[23] Kalra SS, Mohan S, Sinha A, Singh G. Advanced oxidation processes for treatment of textile and dye wastewater: a review. 2nd International conference on environmental science and development IPCBEE 2011, Vol. 4. Singapore: IACSIT Press.

[24] Kepa U, Stanczyk-Mazanek E, Stepniak L. The use of advanced oxidation process in the ozone+ hydrogen peroxide system for the removal of cyanide from water. Desalination 2008;223:187–193.

[25] Zhou H, Smith DW. Advanced technologies in water and wastewater treatment. J Environ Eng Sci 2002;1:247–264.

[26] Al-Kdasi A, Idris A, Saed K, Guan CT. Treatment of textile wastewater by advanced oxidation processes: a review. Global Nest Int J 2004;6 (3):222–230.

[27] Palit S. Membrane separation processes and advanced oxidation processes of dyes in bubble column reactor-a keen and far reaching overview. Int J ChemTech Res 2012;4 (3):862–866.

[28] Gogate PR, Pandit AB. A review of imperative technologies for wastewater treatment I: oxidation technologies at ambient conditions. Adv Environ Res 2004;8:500–551.

[29] Chiron S, Fernandez-Alba A, Rodriguez A, Garcia-Calvo E. Pesticide chemical oxidation: state of the art. Water Res 2000;34 (2):366–377.

[30] Kos L, Perkowski J. Decolouration of real textile wastewater with advanced oxidation processes. Fibres Text East Eur 2003;11 (4):43.

[31] Suty H, De Traversay C, Cost M. Applications of advanced oxidation processes: present and future. Water Sci Technol 2004;49 (4):227–233.

[32] Abdelmelek SB, Greaves J, Ishida KP, Cooper WJ, Song W. Removal of pharmaceutical and personal care products from reverse osmosis retentate using advanced oxidation processes. Environ Sci Technol 2011;45:3665–3671.

[33] Huber MM, Canonica S, Park GY, Gunten UV. Oxidation of pharmaceuticals during ozonation and advanced oxidation processes. Environ Sci Technol 2003;37:1016–1024.

[34] Huber P, Carre B. Decolorization of process waters in deinking mills and similar applications: a review. BioResources 2012;7 (1):1366.

[35] Kantarci N, Borak F, Ulgen KO. Bubble column reactors. Process Biochem 2005;40:2263–2283.

[36] Sukanchan P. Progress in membrane separation processes, ozonation and other advanced oxidation processes-a review. Int J Chem Anal Sci 2012;3 (1):1290–1292.

[37] Sukanchan P. Ozone treatment as an effective advanced oxidation process for the degradation of textile dye-effluents. Int J Chem Anal Sci 2012;3 (1):1293–1295.

[38] Palit S. Ozonation of direct red—23 dye in a fixed bed batch bubble column reactor. Ind J of Sci Technol 2009;2 (10).

[39] Palit S. Ozonation associated with nanofiltration as an effective procedure in treating dye effluents from textile industries with the help of a bubble column: a review. Int J Chem Chem Eng 2011;1 (1):53–60.

14

CARBON NANOMATERIALS AS ADSORBENTS FOR ENVIRONMENTAL ANALYSIS

CHAUDHERY MUSTANSAR HUSSAIN

Department of Chemistry and Environmental Science, New Jersey Institute of Technology, Newark, NJ, USA

14.1 INTRODUCTION

Nanotechnology can be defined as the science and engineering involved in the design, synthesis, characterization and application, organization, and devices of nanomaterials. In the past few years, nanotechnology has grown by leaps and bounds, and this multidisciplinary scientific field is undergoing explosive development. It can prove to be a boon for human health care because of the huge potential benefits in areas as diverse as drug development, water decontamination, information and communication technologies, and the production of stronger, lighter materials. A complete list of the potential applications of nanotechnology is too vast and diverse to discuss in detail, but without doubt, one of the greatest contributions of nanotechnology is in the development of analytical sciences [1, 2].

Nanomaterials, with bodily structures less than 100 nm in one or more dimensions, have attracted significant attention from scientists in recent years, mainly due to their unique, attractive, thermal, mechanical, electronic, and biological properties. Their high surface to volume ratio, the possibility of surface functionalization, and favorable thermal features provide the flexibility needed for a broad range of analytical applications. The physicochemical properties of nanomaterials can be controlled through structural design, incorporation of suitable components, or modification of their surfaces [3]. Widespread applicability and outstanding performance of nanomaterials have not only accelerated the development of materials science, but also provided many opportunities in related disciplines. Research on application of nanomaterials in analytical chemistry has also experienced impressive growth in terms of the number of papers published in scientific literature. This chapter highlights the most important recent advances in the use of carbon nanomaterials (CNMs) for environmental analysis. Finally, an outlook in which challenges and opportunities are identified is given.

14.2 CNMs FOR ENVIRONMENTAL ANALYSIS

Environmental analysis is the use of analytical chemistry techniques to measure and detect pollutants in the environment. It involves the determination of natural and harmful concentrations of chemical constituents in the environment. CNMs are increasingly being considered one of the most promising materials for future applications. They can be combined with other types of nanomaterials to form nanocomposites, thus incorporating different properties in a single new material. Moreover, their unique electrical, optical, and mechanical properties make them useful for developing the next generation of miniaturized, low-power, ubiquitous sensors. Additionally, CNMs can be used as nanoadsorbents for liquid- as well as gas-phase adsorption of environmental pollutants because of their special nano-scale adsorbent properties [4, 5]. Therefore, CNMs are poised to overtake other well-established adsorbent materials; however, their commercial exploitation is still a long way off. Among the

Nanomaterials for Environmental Protection, First Edition. Edited by Boris I. Kharisov, Oxana V. Kharissova, and H. V. Rasika Dias.

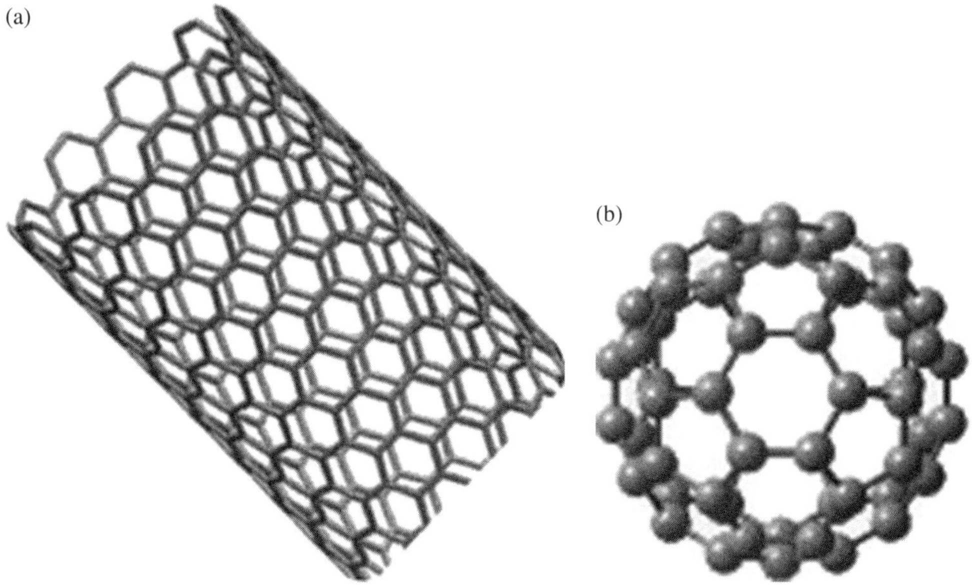

FIGURE 14.1 Models of (a) CNT, (b) C60.

CNMs, carbon nanotubes (CNTs) and fullerenes (FULs) receive the maximum attention because of their unique properties, which includes thermal conductivity, stability, tensile strength, and their ability to act as conductors and semiconductors depending on the chemical and physical treatments. Models of a CNT and C60 are shown in Figure 14.1. A general comparison of their role as nanoadsorbents in environmental analysis on the basis of published data and their application in separation sciences are shown in Figure 14.2.

14.2.1 Carbon Nanotubes

CNTs, which were first noticed by Ijima in 1991, have been used for different purposes in the analytical sciences due to their mechanical, electric, optical, and magnetic properties as well as their extremely large surface area. CNTs are hollow graphitic material composed of one or multiple layers of graphene sheets (single-walled CNTs, SWNTs, and multiwalled CNTs, MWNTs, respectively. The synthesis of CNTs can be carried out by means of three main techniques: chemical vapor deposition (CVD), laser ablation (LA), and catalytic arc discharge (CAD). CVD seems to be the most appropriate procedure for the synthesis of CNTs for analytical applications due to the high purity and desirable tuning obtained at low temperature. However, for all the synthesis methods, the presence of different undesired by-products (such as carbonaceous residues, amorphous carbon, metal impurities) makes it necessary to purify CNTs. To purify synthesized nanotubes, different strategies including chemical oxidation, physical separation, or combinations of both chemical and physical techniques have been employed. Chemical oxidation is a purification system based on the fact that carbonaceous impurity residues are oxidized sooner than CNTs. The main advantage is their ease, but it should be noted that the oxidation process affects the structure of the nanotube introducing functional groups (hydroxyl, carbonyl, and carboxyl) and the defects in the side walls. Physical purification procedures are based on the different physical properties (such as size, ratio, weight, and electrical and magnetic characteristics) between impurities and CNTs. Filtration, centrifugation, chromatography, and electrophoresis are the commonly employed techniques. The disadvantages of these procedures are that (i) the elimination of certain impurities is inefficient; (ii) a high dispersion of CNTs is required; and (iii) only a low quantity of CNT can be purified [6]. The effect of purification can be seen in Figure 14.3.

The sorption sites on CNTs are on the wall and in the interstitial spaces between tubes. These sites are easily accessible for both adsorption and rapid desorption. The impurities on CNTs reduce their availability because the sorbate has to diffuse through these impurities to reach the CNTs. Moreover, the porous structure of impurities introduces mass transfer limitations, slowing both adsorption and desorption. Understanding these characteristics is important for their application in separation media. The excellent features of CNTs, along with their nano-scale features, make them ideal candidates for micro-scale devices for gas- as well as liquid-phase analysis. In gas-phase analysis, one can employ a microconcentrator or a microsorbent trap, which have been used in a variety of online chromatography and sensing applications. The purpose of such devices is usually to act as a fast preconcentrator or to modulate the concentration of a stream for real-time monitoring. They have been

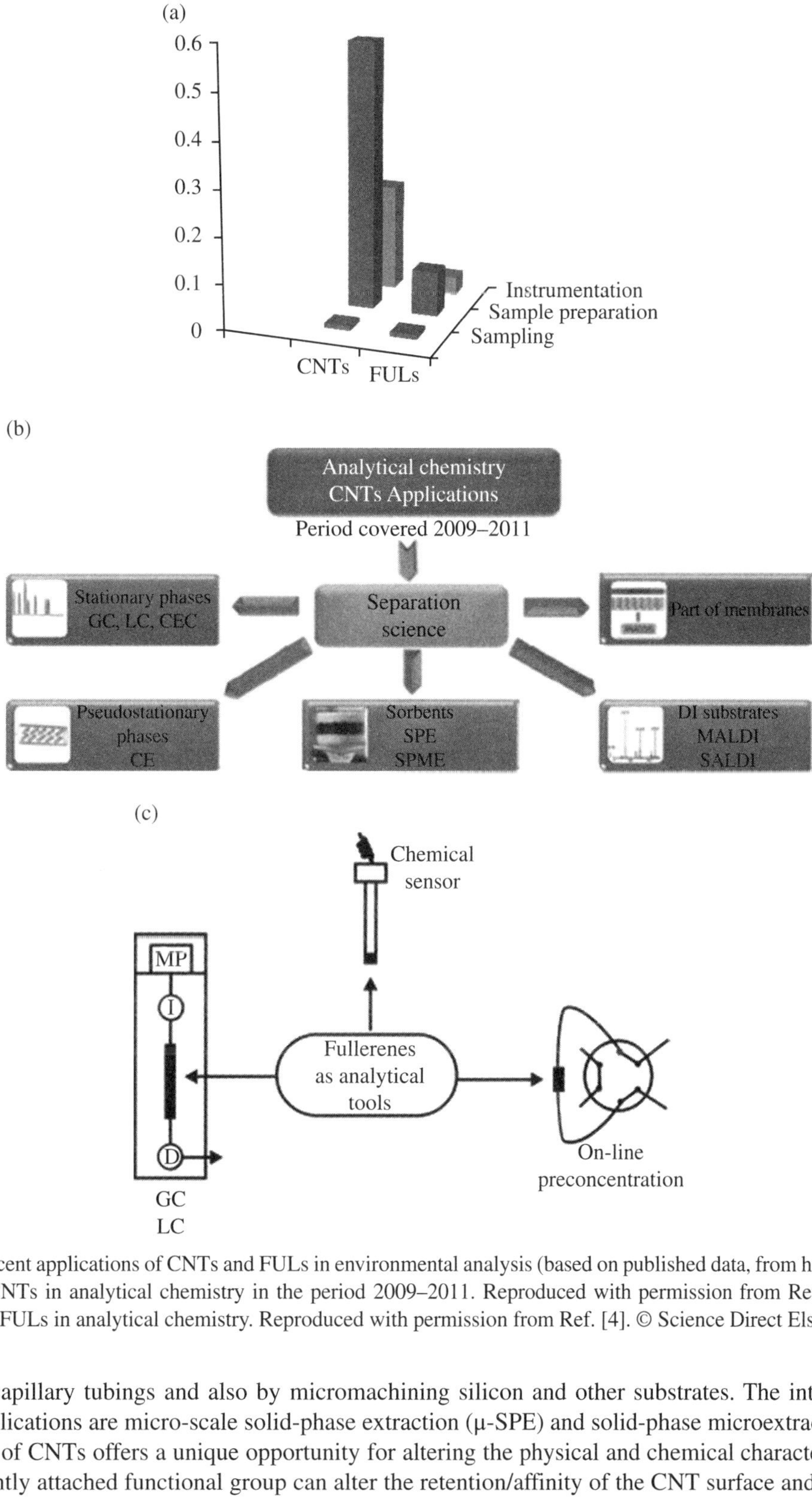

FIGURE 14.2 (a) Recent applications of CNTs and FULs in environmental analysis (based on published data, from http://www.sciencedirect.com). (b) The use of CNTs in analytical chemistry in the period 2009–2011. Reproduced with permission from Ref. [3]. © Science Direct Elsevier. (c) The use of FULs in analytical chemistry. Reproduced with permission from Ref. [4]. © Science Direct Elsevier.

fabricated in small capillary tubings and also by micromachining silicon and other substrates. The interesting liquid-phase preconcentrating applications are micro-scale solid-phase extraction (μ-SPE) and solid-phase microextraction (SPME) [4–6].

Functionalization of CNTs offers a unique opportunity for altering the physical and chemical characteristics of CNTs. The presence of a covalently attached functional group can alter the retention/affinity of the CNT surface and important properties such as polarity, hydrophilicity, and other specific interactions. The functional groups may also alter diffusional resistance and reduce the accessibility and affinity of CNT surfaces for certain analytes. Functionalization also enhances interaction with polymers and other materials, thus facilitating the formation of composites that can be used as preconcentrating substrates. These include polymers and sol–gel–type immobilization [6]. Covalent and noncovalent functionalization are shown in Figure 14.4.

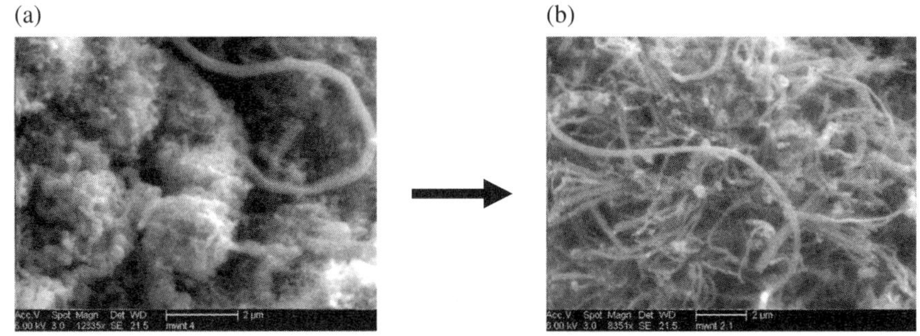

FIGURE 14.3 CNTs: (a) Impure, (b) Pure. Reproduced with permission from Ref. [6]. © Springer.

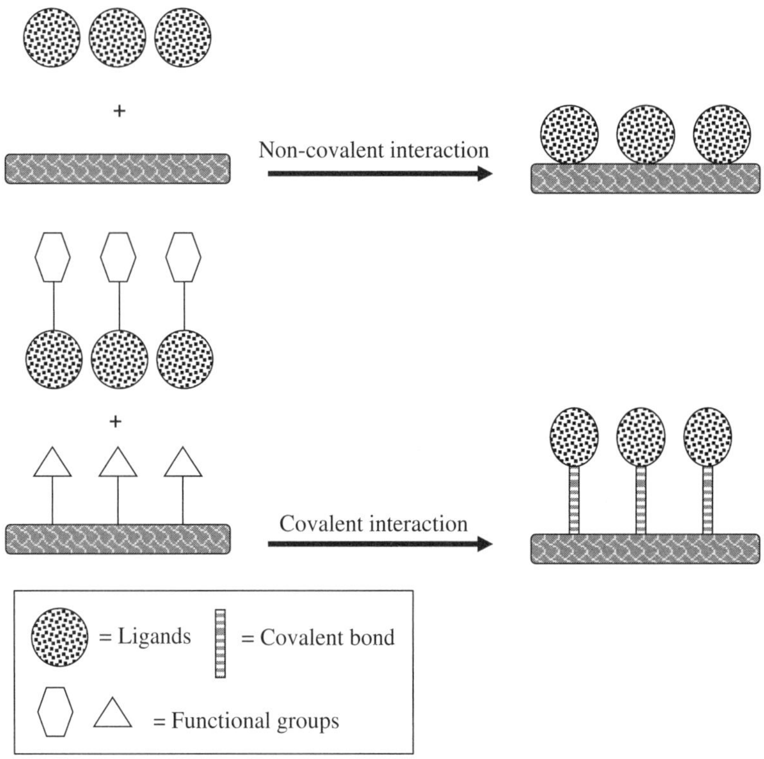

FIGURE 14.4 Sketch of covalent and noncovalent functionalization on a CNT.

14.2.2 Fullerenes

FULs have attracted considerable attention in different fields of science since their discovery in 1985. Physical, chemical, and biological properties of FULs indicate that they have promising applications. It is inferred that size, hydrophobicity, three-dimensionality, and electronic configurations make them an appealing subject in chemistry. FULs are often accompanied with large amounts of soots; therefore, purification is necessary before their real applications are realized. Purification of FULs normally involves extraction from soots and then separation by liquid chromatography. Additionally, other purification techniques, like elution on molecular sieves or activated charcoal, gel permeation, and supercritical fluids purification, have also been used [4].

FULs are practically insoluble in water and generally tend to form aggregates in aqueous media. To disperse them in water, prolonged stirring for days is needed. The addition of additives, for example, surfactants, cyclodextrins (CDs), or sugar polymers, aids their efficient dispersion in water. FULs can also be dispersed in water by first dissolving them in an organic solvent and then adding the solution to water and evaporating off the organic solvent.

Incorporation of FULs into polymeric materials to improve their processability has become a common practice in applications such as optoelectronics, photovoltaics, fuel cell membranes, and the life sciences. The unique properties of FULs, such as their spherical shape, conjugated three-dimensional π-electronic system, and the ability to exhibit donor–acceptor or π–π interactions, make them attractive candidates for analytical applications. As a result, these have been used as chromatographic stationary phases in a wide temperature range (80–360°C), with good thermal stability and high selectivity for aromatic compounds. Chemical functionalization of FULs is often carried out to alter their chemical properties and/or improve miscibility with host polymers [7–9].

14.3 ADSORPTION ON CNMs

Adsorption of pollutants on CNMs is controlled by hydrophobic, dispersion, and weak dipolar forces. The higher equilibrium rates of CNMs over conventional carbons are attributed to π–π electron donor–acceptor interactions with sorbates, reduced heterogeneity of adsorption energies, and the absence of pore diffusion as an intermediate mechanism in adsorption. Additionally, CNMs demonstrate a virtual absence of hysteresis between adsorption and desorption isotherms for liquids and gases under atmospheric pressure. Various carbon-based sorbents have been utilized for the preconcentration of analytes. These include activated carbon, carbon molecular sieves, graphitized carbon black, and porous carbon. These differ in their physicochemical characteristics, such as pore size/shape, surface area, pore volume, and surface functionality. Kinetic and thermodynamic properties of carbon-based sorbents including breakthrough volumes, adsorption isotherms, intermolecular interaction mechanisms at the adsorbate/carbon sorbent interface strongly influence preconcentration [6]. Table 14.1 provides the general characteristics of some popular adsorbents and CNTs. The adsorbent can affect adsorption by providing a large number of available sorption sites and by facilitating specific interactions such as hydrogen bonding.

The sorption mechanism between CNTs and other conventional sorbents (Table 14.1) is shown in Figure 14.5. The diffusion into the porous structure leads to slow mass transfer, and consequentially the quantitative release of large molecules becomes a limiting factor. On the other hand, in CNTs, sorption occurs mainly on the outer surface and in the hollow spaces of the open tubes (provided the tubes are uncapped). Therefore, their release by both thermal processes and solvents are relatively simpler and are not limited by diffusion. As such, a wide range of compounds from small molecules to large semivolatile compounds can be easily concentrated and desorbed from CNTs [6, 7].

There are other factors that come into play. The exceptionally high aspect ratio (in millions) of CNTs provides a special confinement effect, which leads to completely different physical behavior when compared to that of more conventional sorbents. Moreover, outer tube surfaces have a large number of intertubular spaces that provide specific adsorption sites. As a result, the sorption capacity is higher than what one would expect based on Brunauer–Emmett–Teller (BET) surface area measurement. Various approaches have been used to describe the adsorption of CNTs. The high sorption capacity of CNTs has been explained by the presence of high-energy adsorption sites, such as CNT defects and interstitial and groove regions between CNT bundles, and the phenomenon of multilayer adsorption during microsorption on CNTs. Capillary forces in nanotubes can also be strong, which may draw molecules from vapor or liquid phases by van der Waals attractive forces and dipole-induced dipole interactions. Hydrophobicity and capillarity can provide ordering and orientation of sorbents on the sorbate. Adsorption studies detail rapid equilibrium rates, high adsorption capacity, low sensitivity to pH range, minimal hysteresis in dispersed CNTs, and consistency with traditional Langmuir, BET, or Freundlich isotherms. Moreover, increase in dispersion energy and the overlapping force of adjacent carbon walls expand sorbent–sorbate and sorbent–sorbent interactions, resulting in condensation within the nanotube. The filling of nanotubes may explain why some models describe adsorption capacity above the physical surface area of a CNT [8–10].

TABLE 14.1 Properties of porous adsorbents and CNTs

Sorbent	Strength	Surface area (m^2/g)	Maximum T (°C)	Analytes	Drawback
Tenax TA	Weak	35	350	Nonpolar compounds and less volatile polar compounds	Undergo chemical decomposition in highly oxidizing atmospheres (i.e., in the presence of reactive gases such as 0_3 and NO_2),
Carbotrap	Medium-weak	100	>400	VOCs including ketones, alcohols, aldehydes and perfluorocarbon tracer gases	No sorption for polar compounds
Carbopack	Medium	240	>400	Hydrocarbons, BTEX	No sorption for polar compounds
Carbosieve	Very strong	800	>400	C2, C3, and C4 hydrocarbons	Not suited for higher organics
Carboxen 1000	Very strong	>1200	>400	Ultravolatile hydrocarbons	Not suited for higher organics
CNTs	Very strong	150–1500	>400	Very volatile to semivolatile organics (Methane to PAHs)	Need purification, relatively expensive

Reproduced with permission from Ref. [6]. © Springer.

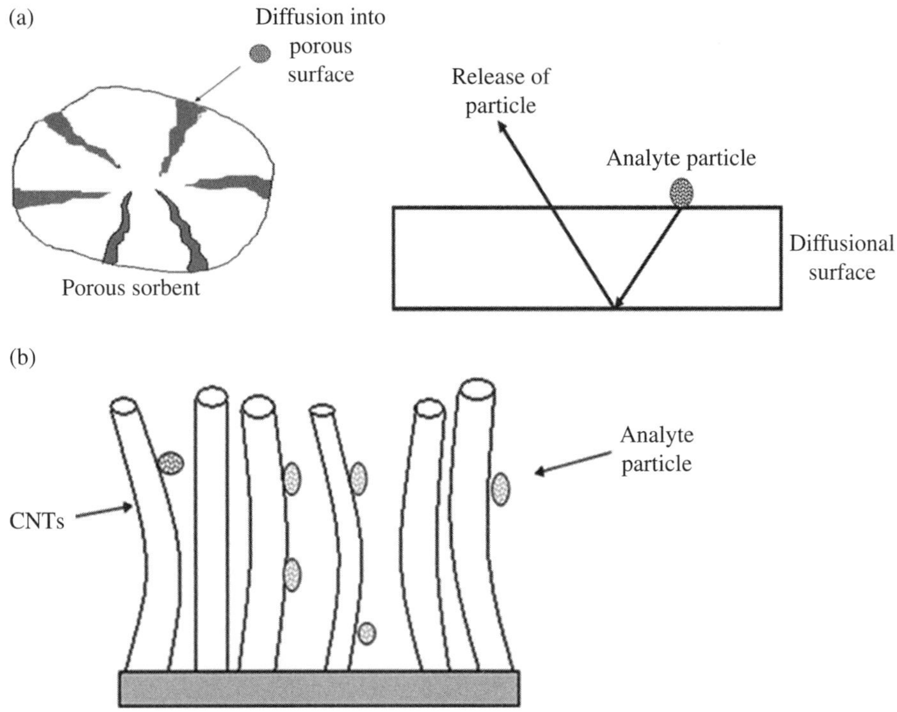

FIGURE 14.5 (a) Sketch of a porous carbon sorbent and diffusion into pores, (b) sorption on CNTs for analytes. Reproduced with permission from Refs. [6, 7]. Ref. [6] © Springer and Ref. [7] © Science Direct Elsevier.

Adsorption on FULs is a subject of increasing experimental and theoretical interest. A few research groups have explored the adsorption properties of FULs and generally focused on the adsorption of simple gases or vapors of organic compounds on C60 solid. Both types of mechanisms—physisorption and chemisorption—have been realized on FULs both theoretically and experimentally. Some studies have reported interfacial interactions of organic molecules with C60 in aqueous media. FULs exhibit strong hydrophobicity, and various methods have been used to disperse FULs in aqueous media. These include sonication, interaction with polymers, redox reactions, and extended mixing [11].

The occurrence of porosity on the FUL surface has been studied via adsorption. Some research groups by using Kr, N_2, O_2, and CO_2 adsorption and others by using N_2 and O_2 adsorption proved the existence of micropores in C60. Moreover, research also pointed out that surface area values normally depend on sample preparation and purification, as well as sample age. More recent work indicates that FUL powders with higher degrees of purity consist of nonporous particles. Overall, the porous texture of FULs is very sensitive to the synthesis and purification procedures followed in their preparation [7, 12].

The desorption behavior of pyrene, phenanthrene, and naphthalene on FULs and CNTs was examined wherein FULs showed significant desorption in contrast to CNTs. The reason was the difference in their distinct geometries. Spherical FULs resulted in aggregates providing two space structures, that is, the external surface of aggregates, where adsorption was reversible, and closed interstitial spaces in small aggregates. Interstitial spaces within small aggregates were unavailable for adsorption, but these contributed to pore deformation. Rearrangement of small aggregates and penetration of adsorbate molecules into closed interstitial spaces between small aggregates during adsorption led to molecular entrapment and hysteresis. However, CNTs did not form closed interstitial spaces in their aggregates due to their length; hence, no adsorption–desorption hysteresis was observed. Table 14.2 provides applications of FULs as an adsorbent for a variety of analytes [12, 13].

14.4 CNMs FOR PRECONCENTRATION OF ENVIRONMENTAL POLLUTANTS

14.4.1 Solid-Phase Extraction

Solid-phase extraction (SPE) is commonly employed in trace analysis of environmental pollutants because it was demonstrated to be one of the most suitable methods to preconcentrate the environmental analyte and to eliminate potential interferences. The advantages of SPE are its high enrichment factor, good recovery, rapidity, the use of small quantities of organic solvents, as well as the possibility of automatization of the whole process.

TABLE 14.2 Applications of FULs in as adsorbent

Analyte	Sample	Technique[a]	Detection limit (ng/ml)	R.S.D.(%)	Comments
Pb	Waters	FAAS	5	2.1	Study of adsorption isotherms
Cd	Biological	FAAS	0.3–2.0	1.9–2.3	Comparison of APDC and 8-hydroxyquinoline as chelating reagents
Cu	Waters	FAAS	0.3–3.0	1.7–3.1	Comparison of C_{60} and C_{70} fullerene. Sorption of neutral chelates and ion pairs
Cd	Waters	ETAAS	0.002		Tungsten coil atomizer
Pb			0.023		
Ni			0.075		
Co	Wheat flour	ETAAS	0.008	4.0	Comparative study of C_{60} fullerene and RP-C_{18}. Autosampler as interface
Lead species	Waters	FAAS	0.5–4	3.5	Precipitation of inorganic lead. Selectivity related to column conditioning
Organic and organometallic compounds	Waters	GC/FID, FAAS	5–15	2.4–3.0	Systematic study of retention of organic and organometallic compounds on C_{60} fullerene
Alkyl lead species	Waters	FAAS, GC/MS	0.5 0.001–0.004	6.0	Screening by FAAS and speciation by GC. Derivatization with Grignard reagent
Alkyl lead species	Waters	GC/MS	0.004–0.012	4.5	In situ derivatization with $NaBPr_4$
Dithiocarbamates of Zn, Mn, and Fe	Grain	FAAS	1–5	2.5	Group speciation according to the metal present in the fungicide

FAAS, flame atomic absorption spectrometry; ETAAS, electrothermal atomic absorption spectrometry; GC/FID, gas chromatography with flame ionization detector; GC/MS, gas chromatography/mass spectrometry; R.S.D., relative standard deviation, $n = 11$.

14.4.1.1 CNTs as SPE Material CNTs have been extensively studied as sorbent material for conventional SPE. In-depth detailed reviews have been published in the literature where CNTs as SPE sorbent have been successfully used for diverse organic and inorganic analytes from water, seawater, food, and pharmaceutical samples [3, 14–17].

As already mentioned, in general, CNTs tend to demonstrate higher capacity than conventional carbon-based adsorbents. CNTs have been used in conventional SPE preconcentration of a variety of polar and nonpolar analytes including endocrine disrupters such as phenolic compounds, highly polar sulfonylurea herbicides, and molecules of pharmaceutical origin. They have been compared to other conventional sorbents such as C18 and XAD. They demonstrated advantages such as superior breakthrough volumes, better retention of polar species, better efficiency with seawater, and higher extraction efficiencies [3, 17].

Covalent functionalization that usually incorporates hydroxyl, carboxyl, or carbonyl groups onto side walls of CNTs can selectively alter their selectivity. CNTs are also covalently immobilized onto solid supports such as steel or silica, and thus changes in the geometry of the SPE device can take place. Alternatively, noncovalent side-wall functionalization of CNTs also provides a change in selectivity. This type of functionalization includes ionic interactions (e.g., dipole–dipole), π–π stacking, hydrogen bonds, electrostatic forces, van der Waals forces, dative bonds, and hydrophobic interactions. The combination of two or more similar or different interactions increases the stability and the selectivity of the system. Nonpolar organic analytes generally show high capacity sorption with unfunctionalized CNTs as hexagonal arrays of carbon atoms in graphene sheets interact strongly with aromatic rings, due to van der Waals interactions, whereas functionalized CNTs showed higher retention of polar organic analytes. SPE using CNTs has been carried out in cartridges and disks. The disk format provides a larger surface area leading to good mass transfer and fast flow rates. MWNTs are assembled into a disk by means of a filtration process, firmly fixed on a piece of polytetrafluoroethylene (PTFE) membrane filter, and then used as a preconcentrator. Enrichment factors as high as 4000 with recoveries in the range of 87–110% have been reported [3, 7, 16].

14.4.1.2 FULs as SPE Materials A limited number of applications of FULs for SPE have been reported. Ballesteros et al. studied the adsorption behavior of FULs C60 using organic compounds and organometallic compounds as model analytes. The adsorption capacity of FULs was found to be significantly higher for organometallic compounds. Extraction of benzene, toluene, ethylbenzene, and xylenes (BTEX) from water samples on a home-made C60-packed minicolumn was also reported, where FULs were found to be better than conventional C18 and Tenax TA in terms of sensitivity and precision. Adsorption of

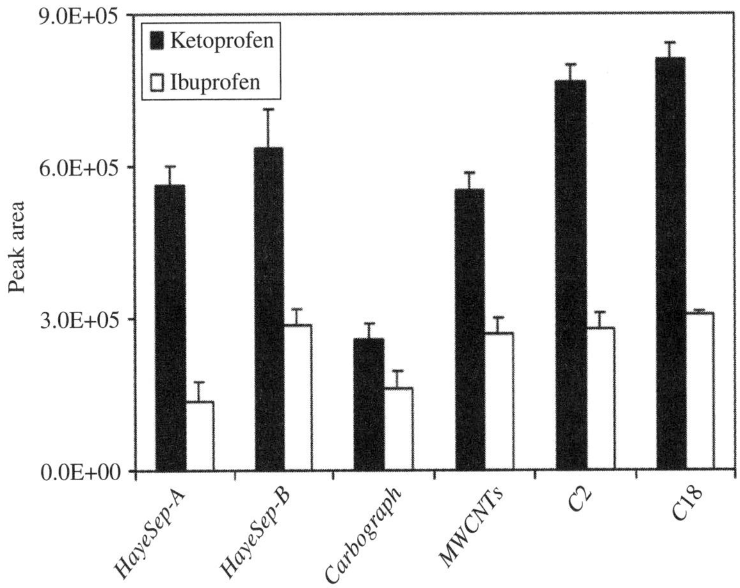

FIGURE 14.6 Effect of sorbent type on extraction efficiency (EE) of μ-SPE. Reproduced with permission from Ref. [18]. © ACS.

metallic compounds on FULs has shown higher extraction efficiencies for lead in water, cobalt in wheat flour, as well as inorganic and organic cadmium and lead during metal speciation [4, 7–9].

14.4.2 Micro Solid-Phase Extraction

To achieve high enrichment while reducing the sorbent and solvent amount, μ-SPE has been developed recently. The objective was to approach the size and convenience of SPME but retain the lower detection limits offered by SPE. These techniques attempt to miniaturize conventional SPE to smaller beds and solvent volumes and offer the convenience of easier analyte recovery and shorter sample preparation time. The small sorbent beds contain a small amount of sorbent, and high-performance sorbent materials are required to obtain satisfactory extraction efficiencies.

A membrane-assisted μ-SPE device containing C-18 and CNTs has been reported [18]. Here the sorbent is held within a polypropylene membrane envelope. A square (15×15 mm), heat-sealed polypropylene sheet membrane envelope was packed with 6 mg of MWCNT. This μ-SPE device may be dropped into and stirred in aqueous samples. The approach is somewhat similar to stir bar microextraction (SBME). The analytes from the sample diffuse through the membrane and are adsorbed on the MWNT filling. After extraction, the device is removed, rinsed, dried, and dipped in organic solvents under ultrasound to desorb the extracted species. MWNT was found to be an efficient sorbent for this new μ-SPE device and was compared to conventional sorbents like C18, C2, Carbograph, HayeSep A, and HayeSep B. In Figure 14.6, C18 showed the highest extraction efficiency (EE) for both compounds. CNTs were better than Carbograph and comparable with HayeSep A. However, as compared to C18, CNTs showed lower EE. The reason was insufficient electrostatic interaction between MWNTs and target analytes [18].

A μ-SPE device consisting of a syringe where CNTs were packed and self-assembled within an attached 0.53-mm ID needle has been reported. The device is shown in Figure 14.7. The enrichment factor (EF) and EE for self-assembled CNTs, SWNTs, and MWNTs were compared to C-18. Table 14.3 lists these values for 2-nitrophenol, 2,6-dichloroaniline, and naphthalene. The results show that 300 μg of C-18 was not sufficient to accomplish EF higher than one. However, CNTs were quite effective as a μ-SPE sorbent with higher EE and EF. This is in agreement with previous reports where the CNTs were more effective than C-18 as SPE devices [19].

14.4.3 Solid-Phase Microextraction

SPME was first proposed in 1990 by Arthur and Pawliszyn [20]. It can be considered as a miniaturized sample pretreatment technique whose advantages over other well-established methodologies have been clearly identified. One of the advantages is its easy online coupling which chromatographic and electrophoretic techniques. Operationally, this technique is based on the partition of the analyte between the sample and the stationary phase coated on a solid support. CNMs have emerged in the past decade

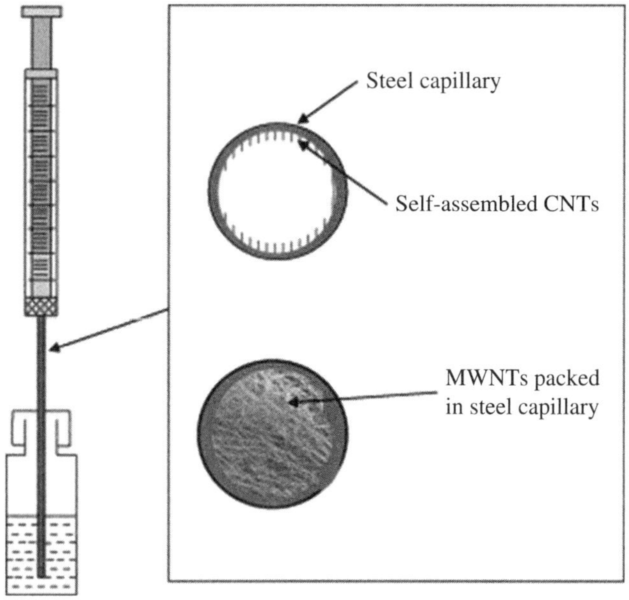

FIGURE 14.7 Schematic diagram of the μ-SPE system. Reproduced with permission from Ref. [19]. © Science Direct Elsevier.

TABLE 14.3 Comparison of breakthrough times (min) of different organic compounds on functionalized CNTs, unfunctionalized CNTs, and Carbopack

	2-Nitrophenol		2,6-Dichloroaniline		Naphthalene		Breakthrough
Sorbents	EF	EE (%)	EF	EE (%)	EF	EE (%)	Volume (ml)
C-18	0.3	1.3	0.6	2.5	0.6	2.5	5
Self-assembled CNTs	0.6	2.3	0.7	2.8	1.9	7.5	7.5
SWNTs	2.8	11.1	4.1	16.2	6.2	24.7	17.5
MWNTs	3.4	13.7	3.7	14.8	6.8	27.1	17.5

Reproduced with permission from Ref. [19]. © Science Direct Elsevier.

as a strong alternative to commercial SPME coatings because of their remarkable electronic properties and their excellent mechanical, electrical, and chemical properties.

14.4.3.1 CNTs as SPME Material CNTs have been employed as a sorbent for SPME fibers. Since SPME fibers require immobilization of the CNTs on a wire of micron-scale diameter, significant effort has gone into investigating various approaches to implementing CNTs on a fiber. Some general approaches have been dip coating in a CNT suspension, making a polymer composite, immobilization by the sol–gel approach, and electrodeposition. Generally, CNT-based SPME coatings have acceptable reproducibility; high solvent, mechanical, and thermal stabilities; reduced fiber affinity to water molecules in the headspace (HS), and improved EE. A detailed review was recently published on CNT-based SPME coatings covering most of these aspects [21].

Immobilization by sol–gel technology has been the most dominant SPME fibers built-technique for CNTs coating on fiber solution (FS) fibers, which exhibited higher EE and thermal stability and longer life span compared with commercial SPME fibers. The peak areas of chromatographic responses of CNT fibers for analytes were more than those of poly(dimethylsiloxane) (PDMS) and polyethylene glycol (PEG) fibers. This is mainly due to the highly porous structure obtained from sol–gel technology, which leads to larger surface area and better accessibility for analytes toward inner sorbents [22].

Fragile FS fibers were replaced with stainless steel (SS) wires and introduced a repeatable coating procedure to enhance the reproducibility of sol–gel CNT fibers. In this method, the SS wire was drawn at a constant rate during sol–gel layer coating. The unbreakable sol–gel CNT SPME metal fiber showed excellent thermal stability, solvent durability, and good EE because of inherent features of MWNTs and the innovative repeatable sol–gel method. The sol–gel technique, although having many advantages, was replaced by other methods because of its inadequate reproducibility and because the fragile FS fiber was mostly used as a support in this method. This deficiency in the reproducibility was the outcome of a great number of optimization factors in the sol–gel coating technique. The chemical and electrochemical methods of coating preparation emerged to overcome this obstacle [23].

Recently, electrodeposition of CNTs and conductive polymers on metal wires was carried out. Aniline dissolved CNTs through the donor–acceptor complex, and the polyaniline (PANI/MWNT) coating was electrochemically prepared. This fiber showed higher EE than the PANI fiber because of its porous structure and the π–π interaction of CNTs with aromatic compounds. The MWNT/PANI fiber also showed high thermal stability and excellent reusability due to the chemical bonding between the Pt substrate and coating and the interaction of PANI with MWCNTs. The MWCNT/polypyrrole (PPy) SPME-coated fiber showed higher EE for the pyrethroids than the PPy-coated fiber and selected commercial fibers. The mechanical strength of the MWCNT/PPy-coated fiber and the resistance of the coating to different organic solvents and strong acidic and basic aqueous solutions illustrated the high chemical stability of the fiber [24].

Nafion was also used as a binder to immobilize MWNTs on a SS wire to prepare the MWNT/Nafion fiber. This fiber was used as a working electrode for the EE-SPME of basic drugs in urine samples and organic ionic compounds. The EE-SPME of anionic (deprotonated carboxylic acids) and cationic (protonated amines) compounds in aqueous solutions showed more effective and selective extraction in comparison with DI-SPME, due to electrophoresis and complementary charge interaction [25].

14.4.3.2 FULs as SPME Material FULs have a large surface to volume ratio and a thermal stability that make them ideal for SPME application. The main problem in using FULs as an SPME coating is their poor solubility in solvents, which is crucial for preparing SPME coatings; so, to use FULs in SPME, they have to be attached to a polymeric matrix. Xiao et al. used a polysiloxane–FUL composite for SPME. Synthesis of the coating involved the reaction between a polyazidosilicone backbone and C60. They then used epoxy resin glue to coat the treated silica capillary with the resultant polymeric FUL [26].

Sol–gel technology was used to develop the FUL–polysiloxane surface-bonded porous coating to the FS fiber. The sol–gel hydroxyfullerene coating exhibited high thermal stability, excellent solvent (organic and inorganic) resistance, and long lifespan because of the unique properties of FUL and the chemical bonding between the coating and fiber surface. The charge transfer and structural similarity between the analytes, polychlorinated biphenyls (PCBs), and hydroxyfullerene caused the high extraction capabilities of coating for the planar PCB congeners [27].

14.4.4 Gas-Phase Micropreconcentration on CNT

The gas adsorption capacity of CNTs has been studied by computational methods as well as by experimental measurements. Adsorption of CO and CO_2 and of Ar, N_2, and CH_4 has shown that CNTs are a favorable sorbent for lighter gases. Limited studies have been carried out with organics, where adsorption isotherms for benzene, methanol, and methane as well as affinity of dioxins have been studied. Adsorption kinetics of ethanol, isopropanol, cyclohexane, benzene, and hexane and sorption capacities of a variety of analytes on SWNTs and MWNTs have also been studied. CNTs were used for the preconcentration of organic vapors on SWNTs and MWNTs, and retention was found to be stronger than for Tenax [6, 7, 28–30].

The high capacity of CNTs makes them excellent sorbents for gas-phase microconcentration, where the amount of sorbent is limited. Typical methods of release for the trapped analyte is thermal desorption, and the ease of desorption of large molecules from the CNT surface makes them high-performance sorbents. The microconcentration device can be made by two methods. The first approach is the packing of CNTs into the capillary tubing. The packing of a nano-scale sorbent requires special efforts to ensure uniform packing and low pressure drop. Powdered CNTs as well as nanotube paper have been used as a packing material in a metal preconcentrator tube for a vapor preconcentrator. The second approach is a self-assembled format, where the CNTs are synthesized in situ within preconcentration devices like a steel tubing or a microfabricated device by CVD. The self-assembled format of CNTs provided a consistent ordered structure along the tubing that facilitated a compatible flow within a device, especially for a serpentine microfabricated structure that could be difficult to pack. The self-assembly of CNTs in 250 and 500-μm i.d. capillaries has been reported using CVD of CO, C_2H_4, and ethanol. The surface of stainless-steel tubing was found to be effective because it could be made catalytically active for CNT growth. The self-assembled layer was then heated in a stream of oxygen for 5 h at 300°C to oxidize any impurities formed during the growth process [6, 7, 31–33]. An online monitoring CNT microtrap and chromatogram generated by it with a microfabricated SWNT device is shown in Figure 14.8.

The trapping efficiency of a sorbent depends upon its surface area as well as physisorption/chemisorption on the surface. As mentioned earlier, the sorption capacity in a flow system is usually estimated by studying the breakthrough time, which is defined as the time required by an analyte to elute through, or the time for which the solute is retained on the sorbent. Breakthrough time is known to be a function of the capacity factor, length, and flow rate. CNTs have been used for the preconcentration of volatile organic compounds (VOCs) as well semivolatile organics (SVOCs). Breakthrough times of selected VOCs obtained on a microtrap using different types of CNTs are presented in Table 14.4. Comparative data using as-prepared MWNTs with NTCs, purified MWNTs, and SWNTs are presented. The breakthrough volume on impure MWNT was similar to that of

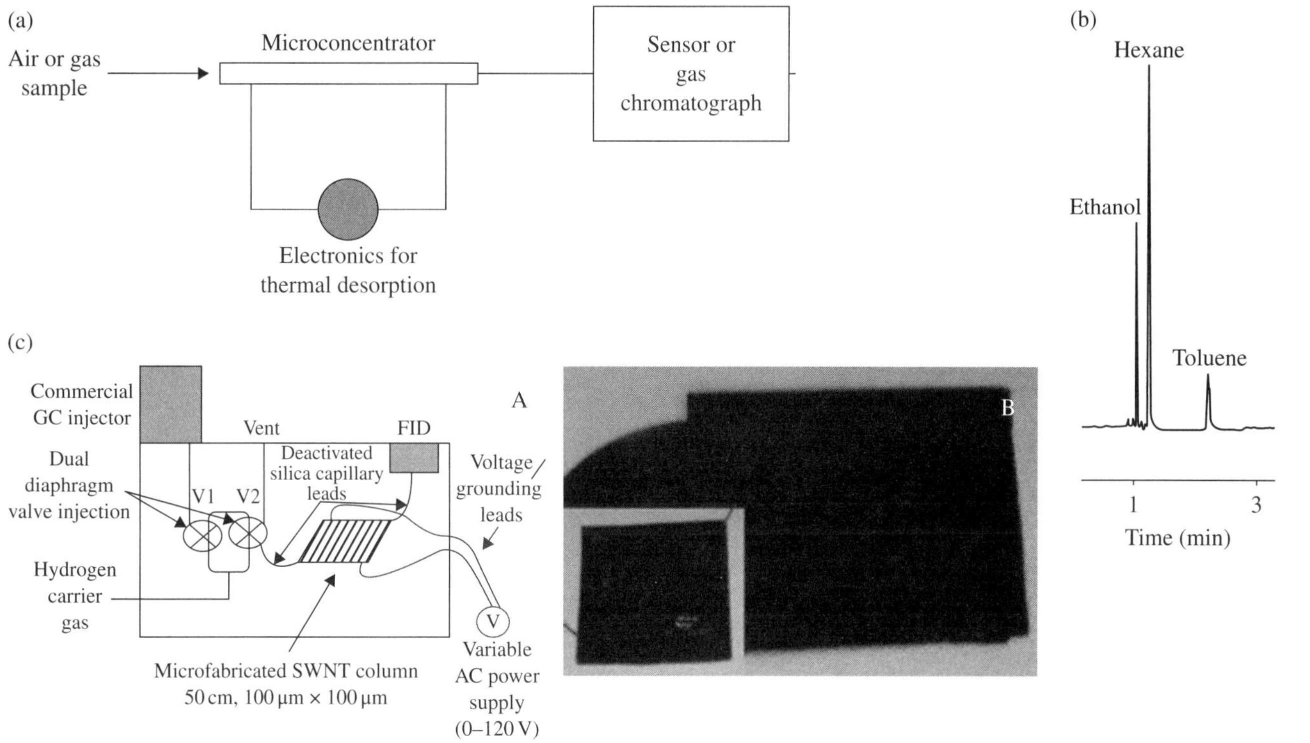

FIGURE 14.8 (a) Online monitoring with a CNT microtrap; (b) chromatogram generated by a CNT microconcentrator; (c) microfabricated SWNT column. Reproduced with permission from Refs. [31–33]. Ref. [31] © ACS Publications, Ref. [32] © Science direct Elsevier, and Ref. [33] © RSC Publication.

TABLE 14.4 Breakthrough times (min) for VOCs and SVOCs with different types of CNTs, where the concentration of the analytes was in the range of 9–12 ppm at a flow rate of 3 ml/min, and the amount of sorbent in the microtrap was 20 mg

Type of microtrap	Dichloromethane	Hexane	Toluene	O nitro phenol	Naphthalene
MWNT-1C[a]	5	15	40	50.0	120.0
SWNT[b]	10	20	50	30.0	90.0
MWNT-1[c]	2	5	6	5.0	15.0
Carbopack	2	5	6	5.0	12.0

Reproduced with permission from Refs. [32–34]. © Science Direct Elsevier.
[a]Cleaned MWNTs (Fig. 14.3b).
[b]Commercially available purified SWNTs.
[c]Synthesized MWNTs with significant amount of impurities (Fig. 14.3a).

Carbopack. This is mainly because of the nonnanotubular nature of adsorption. Once the impurities were removed, the breakthrough volume increased quite dramatically. SWNTs showed the highest adsorption capacity in terms of breakthrough volume. This was attributed to the higher aspect ratio due to the smaller diameter, which led to a higher specific area. For example, the breakthrough time of dichloromethane (DCM) on SWNTs was five times higher than on Carbopack or unpurified MWNTs. CNTs were also excellent for sampling small molecules such as CH_4 to large SVOC molecules [32, 33].

The effect of CNT functionalization on gas-phase preconcentration was also studied. Carboxylated CNT showed significantly different retention characteristics. The breakthrough time for polar compounds increased dramatically—nearly 300% for ethanol, 200% for DCM, and 200% for propanol. This was because the presence of polar functional groups led to strong interactions with polar compounds. Therefore, CNTs aid preconcentration for a wide range of polar and nonpolar compounds, and mixed sorbent beds may be used for a wide range of compounds [34].

The phenomenon of desorption is complementary to adsorption. Quantitative desorption is important for accurate quantitation as well as for the reuse of expensive CNTs. Desorption efficiency often determines the range of compounds, especially the larger molecules that can be used with a microconcentrator. Rapid desorption from a microtrap is also essential for producing a sharp concentration pulse to serve as an injection for gas chromatography (GC) or GC/mass spectrometry (MS), and for

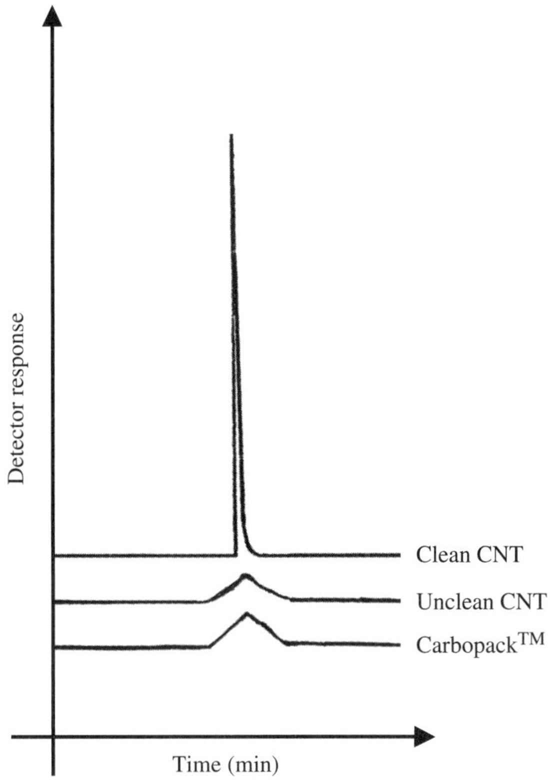

FIGURE 14.9 Comparison of desorption profile from different CNTs and Carbopack. Reproduced with permission from Refs. [33, 34]. © RSC Publication.

producing a high signal to noise ratio for sensors. Figure 14.9 shows the desorption profile of naphthalene from CNT, unpurified CNT, and Carbopack™. Under the same conditions, the desorption band from SWNTs was much narrower than that from either Carbopack™ or unpurified MWNTs. The purified MWNT was similar to the SWNT and is not presented here [33].

14.5 CHROMATOGRAPHIC APPLICATIONS OF CNMs

Most recent work deals with CNMs used as separation media in GC, and in particular in the past 15 years various research groups have been investigating the gas chromatographic properties of FULs and CNTs. Essentially two general approaches have been practiced, that is, their usage in packed and in capillary columns. Despite the lower efficiency of the former, simplicity in preparation made packed columns attractive for easily studying the chromatographic behavior of the selected stationary phase. Different techniques have been considered to prepare open tubular columns based on CNMs, mostly entailing CVD or chemicals methods, the latter widely employed especially for FULs bonded stationary phases [3, 4, 13].

14.5.1 Chromatographic Stationary Phases with CNTs

Li and Yuan were the first who tested purified MWNTs in glass columns for different classes of compounds for the GC stationary phase. Good separation was achieved compared to graphitized carbon especially in the case of polar analytes, due to the low content of polar groups in CNTs. The role of sorbent morphology was elucidated by an analysis of selected probes: it was observed that for the same surface area and amount of material packed in the columns, CNTs exhibited higher adsorption than graphitized carbon (Carbopack B) due to their higher number of sorption sites in comparison with a planar carbon surface in spite of the higher number of theoretical plates of the graphitized packing; on the other hand, porous material such as activated charcoal demonstrated strong retention toward aromatics, making any separation unfeasible. Such findings were also substantiated in another work where pristine and purified MWNTs were compared with activated charcoal for the separation of light alkanes and aromatic hydrocarbons. The pristine material was HCl-treated to remove metallic impurities, as these influenced surface properties such as acceptor/donor characteristics and the dispersive component of the surface energy, without altering their original structure due to the nonoxidative nature of HCl. Results clearly showed the outstanding chromatographic properties of MWNTs in comparison with activated charcoal. Linear/branched alkanes and aromatics could be analyzed on CNTs, contrary to charcoal where very strong adsorption and lack of selectivity were evident, and not even temperature over 300°C could desorb alkyl benzenes in a reasonable timeframe [35, 36].

Saridara and Mitra reported the immobilization of CNTs on the inner wall of capillary columns by using CVD to prepare self-assembled CNTs in capillary tubes with ethylene as the carbon source. Radial-aligned MWNTs were grown in situ on the wall of a stainless-steel tube, which was properly treated to generate a catalytically active surface. Importantly, the thickness of the CNT coating, responsible for different retention capabilities, could be varied by tuning experimental conditions; cogeneration of amorphous and nontubular carbon was overcome by selective annealing after assembling, due to the higher thermal stability of CNTs. The column showed good thermal stability and satisfactory efficiency in the separation of C2–C6 alkenes. It has to be underlined that in situ self-assembly enables preservation of CNTs. Nano characteristics may be lost due to agglomeration when nanotubes are packed as a powder. Such synthetic route was studied further by the same research group. SWNTs were CVD-grown on an open tubular GC column by using ethanol and metallic cobalt as the carbon source and catalyst, respectively. In contrast to a previous report, noodle-like structures of CNTs were observed. C2–C14 alkanes and polycyclic aromatic hydrocarbons (PAHs) were separated, although the peak shape was far from ideal. Temperature up to 425°C could be reached for eluting the highest molecular weight compounds, with no bleeding. Further optimization of CNT self-assembly via CVD was possible by a one-step process based on ethanol as the carbon source, yielding highly pure CNTs (Fig. 14.10). Despite its relatively short length (1.5 m), this column showed good chromatographic properties in terms of resolution and efficiency, enabling analysis of compounds whose volatility is very different (Table 14.5) [6, 7, 37–39].

Safavi et al. synthesized packed capillary columns by the online CVD method. CNTs showed selectivity for separation of gases. Ar, H_2, and CO_2 were easily resolved on SWNTs, yielding better results than MWNTs, whereas no separation was possible on activated carbon and carbon nanofibers. Completely different procedures, based on chemical methods (i.e., oxidation, acylation and nucleophilic substitution), were adopted to anchor CNTs on the inner surface of capillary columns. SWNTs–COCl, obtained by treating SWNTs–COOH with thionyl chloride, were covalently grafted on the column wall exposing free amino groups previously inserted by amino silanization of the fused-silica capillary tube with 3-aminopropyl-trietoxysilane. Interestingly, bonded SWNTs were end-to-end-linked forming a CNT network, as a result of the reaction between SWNTs–COCl and SWNT–COOH possibly present in the reaction medium. The column, tested for separation of light hydrocarbons, alcohols, and aromatics, showed selective adsorption in terms of retention times, but resolution was far from optimal, with marked band broadening and peak tailing [40].

Stadermann et al. fabricated the SWNTs-based stationary phase of the microchannels consisting of a 100 m by 100 m^2 and a 50-cm-long microcolumn placed in a functional chip with an integrated resistive heater and interfaced with a flame ionization detector (Fig. 14.8c). The device was initially patterned by photolithography and then the CNT film was grown in situ by CVD, with variable thickness being obtained depending on the growth time. The layer was quite uniform, despite the presence of voids. In less than 1 min four C6–C12 alkanes, as well as a functional group test mixture (methanol, 2-pentanone, anisole, decane), were separated at flow rates of 12–14 ml min^{-1}. Due to the small thermal mass, a fast temperature gradient of 60°C s^{-1} was applicable, with short equilibration time (90 s) between each thermal gradient run. The micro-GC system was further improved, in particular a smaller device was implemented (50 by 50 m^2 and a 30-cm-long channel) with a different chip bonding technique and a new growth 183 recipe, obtaining a homogeneous CNT mat, important for minimizing band broadening. Such improvements led to better resolution and peak shape, as apparent from the separation of five alkanes in a frame time of about 5 s. Control tests performed on the bare chip highlighted the chromatographic role of SWNTs [31, 41] (Fig. 14.11).

14.5.2 Chromatographic Stationary Phases with FULs

Chemical linking of C60 to polysiloxane by bond formation was undertaken to prepare a capillary column for separation of PCBs. Starting from functionalized FULs, the synthesis consists of a single simple reaction step involving the coupling of polyaminopropyl(methyl)siloxane with the reactive ester cyclohexylaminocarbonyl-(N)-succinimyloxycarbonyl-1,2-dihydro-[60] FULs. Increase in analyte retention was seen in going from tetra-ortho- to mono-ortho-substituted PCBs, as the biphenyls with low number of chlorine atoms in the ortho position are liable to assume planar conformations; this promotes charge transfer/dipole interactions between the solute and the sorbent. High-resolution separation of other high-boiling-point compounds such as PAHs and phthalic esters was achieved on a polysiloxane capillary column modified with FULs. In this case the polymeric phase was firstly chemically modified by electrophilic addition to allyl bromide, followed by reaction with sodium azide and finally by 1,3-dipolar addiction with C60. C12–C32 n-alkanes and C1–C10 phthalic esters, as well as standard mixtures of PAHs, which eluted in the order of their increasing dispersion force, were nicely separated [42, 43].

As expected, stacking was one of the main interactions governing adsorption, as demonstrated by higher retention times of methyl esters of unsaturated fatty acids in comparison with homologous saturated analytes. Also, alcoholic and aromatic positional isomers were resolved, thus evidencing structural selectivity. FULs loading was an influencing parameter with regard

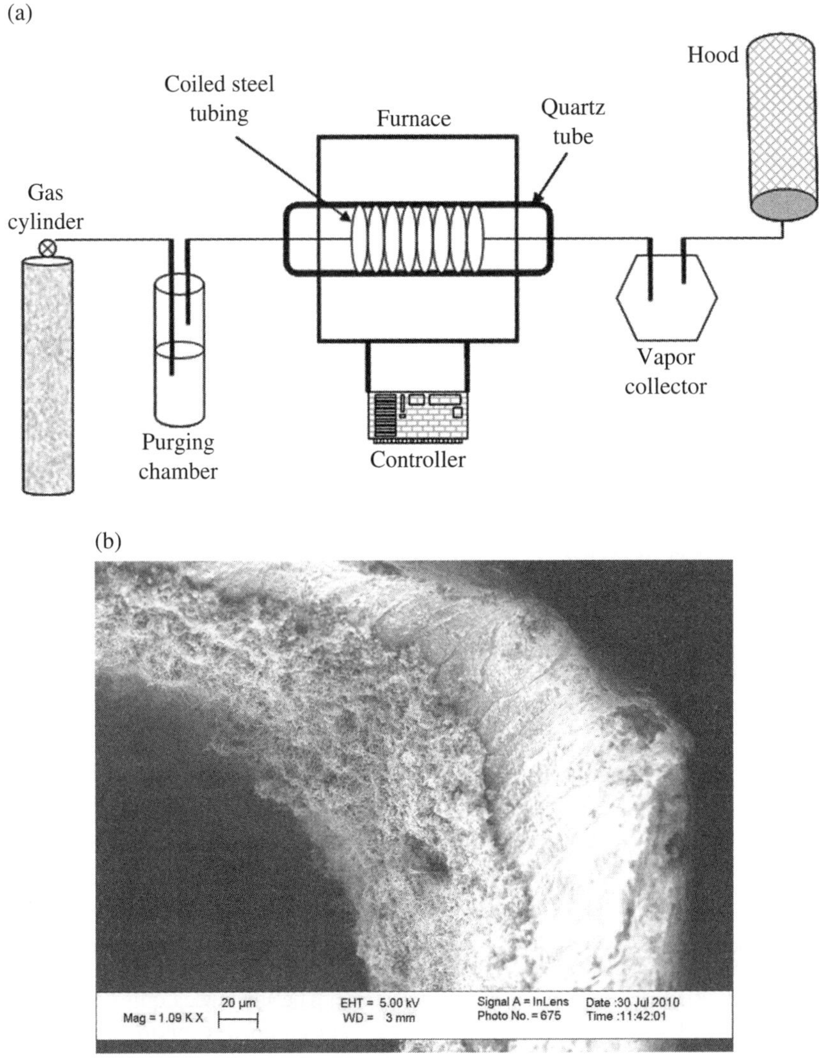

FIGURE 14.10 (a) CVD for self-assembly of CNTs and (b) SEM image of self-assembled CNTs inside a steel tubing. Reproduced with permission from Ref. [39]. © ACS Publications.

TABLE 14.5 Chromatographic data of a CNT-coated GC column with select compounds, where, k' is the capacity factor and N is the number of theoretical plates

Compound	Boiling point (°C)	k (CNT coating)	N
Methanol	64.7	1.02	1820
Hexane	69	0.82	1710
Ethanol	78	2.05	1920
Carbon tetrachloride	76.76	1.21	1620
Benzene	80	1.59	1724
Acetonitrile	81.6	1.67	1780
Propanol	97	3.62	1684
Toluene	111	2.59	1450
1-Butanol	117	5.21	1675
Xylene	139	5.02	1684
Phenol	182	5.97	1520
Naphthalene	217.9	7.16	1350

Reproduced with permission from Ref. [39]. © Science Direct Elsevier.

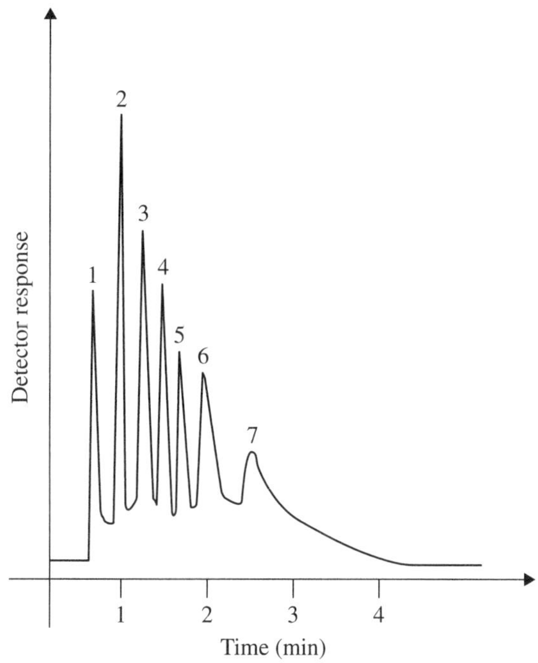

FIGURE 14.11 Chromatographic separations using a CNT-coating column: methanol (1), ethanol (2), carbon tetrachloride (3), acetonitrile (4), xylene (5), phenol (6) and naphthalene (7). Reproduced with permission from Ref. [39]. © ACS Publications.

to column performance; specifically, by increasing CNMs (from 3 to 6 wt%), retention and selectivity were enhanced, although efficiency did not improve. Bonding of C60 imparted a certain polarity to the siloxane phase, resulting in a column with higher polarity than commercial SE-30 and similar to that of OV-3. To enhance FULs selectivity for separation of both polar and apolar analytes, different stationary phases based on FULs but with mixed composition have been prepared. Crown ethers and CDs were selected as "model modifiers" with hydrophilic and hydrophobic cavities, respectively. A synergistic effect of FULs and crown ethers was considered to explain the complete separation of structural isomers of butanol on the C60-(dibenzo-24-C8) sorbent, not achieved on the nonmixed phases. Also, increased retention of heterocyclic compounds containing nitrogen or sulfur was observed. FULs coupled with CDs were prepared to separate aromatic hydrocarbons, based on the ability of CDs to yield inclusion complexes, due to hydrophobic and stacking interactions [42, 43].

14.5.3 High-Pressure Liquid Chromatography

The use of CNMs for designing novel stationary phases for high-pressure liquid chromatography (HPLC) is a current trend in analytical chemistry. The synthesis of hybrid materials is very promising, in which CNTs or FULs play the role of active modifiers responsible for retention and selectivity, while the bulk material in which they are incorporated just acts as the support (i.e., silica), not taking part in the separation process. This is an approach especially adopted for packed-column liquid chromatography. Also, CNT embedment in polymeric monoliths has been undertaken by chemical and physical methods [44, 45].

However, one of the main troubles encountered in this application is the preparation of sorbent phases with suitable chemical stability—mandatory for long-term high-quality performance in particular with regard to the chemical route used for grafting the nanomaterials. An alternative approach is the preparation of nanostructured stationary phases. Overall, it is has to be considered that the chromatographic behavior of the sorbent would be heavily affected by some major factors, such as support particle size, porosity, homogeneity of functionalization, and nanomaterials [46].

14.6 MEMBRANES APPLICATIONS OF CNMs

The natural characteristics of CNMs such as high porosity and surface area make them attractive for the preparation of membranes. CNTs-based membranes have been used to distinguish between molecules based on their size; additionally, functionalized CNTs have been used to influence molecular transport. Because of this, the fabrication of CNTs-based membranes has dramatically increased in recent years, particularly for their use in gas separation [3]. Several physicochemical studies have demonstrated that fluids can flow through the cores of CNTs (e.g., hydrogen, hydrocarbon gases, water, and methane). Dense arrays of aligned MWNTs have been described in literature, which could potentially be used for solute transport. Atomistic simulations have predicted CNTs have exceptional flux and selectivity properties compared to other known nanomaterials and the transport of gases in CNTs with diameter 1 nm was predicted to be orders of magnitude faster than in zeolites. These exceptionally high transport rates have been attributed to the inherent molecular smoothness of the nanotubes. Some of these theoretical predictions have been verified experimentally with larger CNTs. Holt et al. have constructed nanotube–Si_3N_4 composite membranes using CVD [47]. In related work, Hinds and coworkers constructed polymer–nanotube composite membranes using MWNTs having large diameters (6–7 nm) [48]. They verified that transport of liquids (alkanes, water) is orders of magnitude faster than can be accounted for by conventional hydrodynamic flow [49].

In the analytical field, Kamilah et al. and O. Sae-Khow et. al. Prepared a CNT-immobilized membrane (CNIM) where functionalized CNTs were immobilized. One membrane image is shown in Figure 14.10 [50, 51] (Fig. 14.12).

Both SWNTs and MWNTs were used to fabricate the CNIM hollow fiber. Immobilization was carried out such that the CNT surface was fully accessible to adsorption/desorption. It was found that incorporation was quite rugged, and the membrane did not lose the CNTs in spite of several washes with water and solvent. Several organic compounds including trichloroacetic acid and tribromoacetic acid, two important disinfection by-products in water treatment, were selected as the model solutes for CNIM hollow fibers. CNT-mediated membrane extraction provided higher enrichment for all compounds, with improvements of up to 240% over the polypropylene membrane. This is also attributed to the enhanced partitioning of the uncharged acids in the CNTs prior to preconcentration into the basic acceptor (Fig. 14.10).

14.7 OTHER APPLICATIONS OF CNMs

In addition to sorbent properties, CNMs present excellent electric, magnetic, and thermal properties. In general, analytical tools based on carbon nanostructures exploit two or more properties. In the previous sections, we described applications in which sorption was the most important property. However, there are applications in which sorption plays a secondary role, especially the participation of CNTs in developing biosensors and sensors based on field-effect transistors (FETs). Due to their excellent electrical properties, CNTs can be used to develop electrical sensors in which CNTs mediate the electron-transfer reaction with electroactive species [52].

The sorption properties of CNTs facilitate immobilization of biomolecules (e.g., antibodies, cofactors, or enzymes) and the redox process by immobilizing the analyte on the electrode surface. Although CNTs are inert structures, their electrical properties very much depend on the effects of charge transfer and chemical doping. CNT-FETs have been used to detect gases, so the electronic structures of target molecules near semiconducting nanotubes produce changes in the electrical conductivity of CNTs. CNT-FETs can be applied to detect NH_3, CO, and CO_2. However, we need to make it clear that the applicability of CNT-FETs is limited when the analytes have low adsorption energy or poor charge transfer to CNTs [53, 54].

14.8 CONCLUSIONS AND FUTURE VISION

CNMs, especially in the form of CNTs and FULs, have a tremendous potential in environmental analysis as nanoadsorbents in a wide variety of environmental analytical applications. However, this is only the starting point because there are large numbers of other CNMs with exceptional properties that have not yet been explored from the environmental analytical point of view. One of the main challenges and research is to find cost-effective, scalable production methods that retain the essential properties of CNMs. Additionally, the combination of CNMs with other new materials (e.g., quantum dots or ionic liquids) can enhance sorption properties of these, which will help in the development of new analytical tools that will simplify the environmental analytical process. Moreover, to bring CNM development a step forward, the sorption kinetics on CNMs need to be studied in more detail for understanding the interaction of gases and liquids. Although CNMs have demonstrated their potential because of their unique chemical and physical properties, the applications of CNMs in environmental analysis are still limited and at an early stage.

In conclusion, the applications of CNMs in the field of environmental analysis are very interesting and endless; however, the commercial production of CNMs is still some way off, and there is a need for important breakthroughs. Along with the growth of interest in CNMs, the potential effects on human health and the environment, both adverse and beneficial, need to be considered. Their short- and long-term effects on the human body, immunotoxicity, and phototoxicity will also require detailed exploration. Increasingly, a study of their fate and environmental impact is becoming important due to the discharges already occurring to the environment. The likely further increase in CNM discharges along with the dramatic industry growth, and the immense knowledge gaps in risk assessment and management, would necessitate further studies in this area.

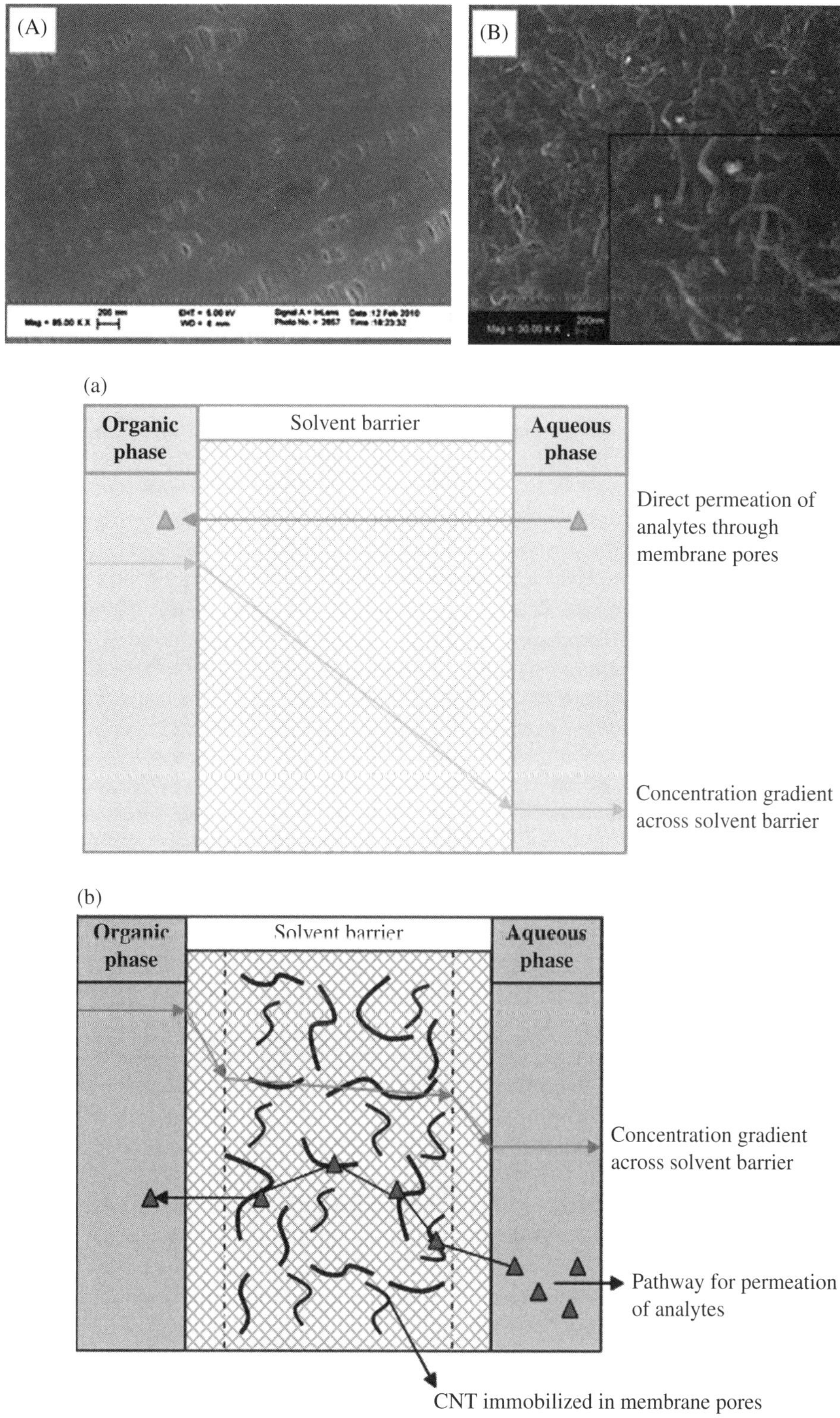

FIGURE 14.12 SEM images of (A) unmodified composite membrane and (B) membrane immobilizing with CNTs. Schematic representation of solvent retention and solute transport in (a) the plain membrane and (b) the CNIM. The triangles represent the path of the analyte molecules, and the dashed lines mark the solvent barriers. Reproduced with permission from Refs. [50, 51]. Ref. [50] © Science direct Elsevier and Ref. [51] © ACS Publications.

REFERENCES

[1] Malshe Ajay P, Rajurkar KP. Environmental applications: environmentally benign nanomanufacturing. In: Karn B, Masciangioli T, Zhang W-x, Colvin V, Alivisatos P, editors. *Nanotechnology and the Environment*. Volume 890, American Chemical Society; 2004. p 304–312.

[2] Rose-Pehrsson Susan L, Pehrsson Pehr E. Environmental applications: sensors and sensor systems: overview. In: Karn B, Masciangioli T, Zhang W-x, Colvin V, Alivisatos P, editors. In: *Nanotechnology and the Environment*. Volume 890, American Chemical Society; 2004. p 154–156.

[3] Herrera-Herrera AV, González-Curbelo MÁ, Hernández-Borges J, Rodríguez-Delgado MÁ. Carbon nanotubes applications in separation science: a review. Anal Chim Acta 2012;734:1–30.

[4] Baena JR, Gallego M, Valcárcel M. Fullerenes in the analytical sciences. TrAC Trend Anal Chem 2002;21 (3):187–198.

[5] Scida K, Stege PW, Haby G, Messina GA, García CD. Recent applications of carbon-based nanomaterials in analytical chemistry: critical review. Anal Chim Acta 2011;691 (1–2):6–17.

[6] Hussain C, Mitra S. Micropreconcentration units based on carbon nanotubes (CNT) *Anal Bioanal Chem* 2011;399 (1):75–89.

[7] Hussain CM, Mitra S. 2.20—Nanomaterials for sample preparation. In: Janusz P, editor. *Comprehensive Sampling and Sample Preparation*. Oxford: Academic Press; 2012. p 389–418.

[8] Filella M. 1.07 - Nanomaterials. In: Janusz P, editor. *Comprehensive Sampling and Sample Preparation*. Oxford: Academic Press; 2012. p 109–124.

[9] Sanchís J, Farré M, Barceló D. Chapter 4—Analysis and fate of organic nanomaterials in environmental samples. In: Marinella F, Damià B, editors. *Comprehensive Analytical Chemistry*. Volume 59, Amsterdam: Elsevier; 2012. p 131–168.

[10] Ren X, Chen C, Nagatsu M, Wang X. Carbon nanotubes in environmental pollution management, a review. Chem Eng J 2011;170 (2–3):395–410.

[11] Dresselhaus MS, Eklund PC, Dresselhaus G. Chapter 2—Fullerenes and nanotubes. In: Timothy DB, editor. *Carbon Materials for Advanced Technologies*. Oxford: Elsevier Science Ltd; 1999. p 35–94.

[12] Lucena R, Simonet BM, Cárdenas S, Valcárcel M. Potential of nanoparticles in sample preparation. J Chromatogr A 2010;1218 (4):620–637.

[13] Valcárcel M, Cárdenas S, Simonet BM, Moliner-Martínez Y, Lucena R. Carbon nanostructures as sorbent materials in analytical processes. TrAC Trend Anal Chem 2008;27 (1):34–43.

[14] Ravelo-Pérez LM, Herrera-Herrera AV, Hernández-Borges J, Rodríguez-Delgado MÁ. Carbon nanotubes: solid-phase extraction. J Chromatogr A 2010;1217 (16):2618–2641.

[15] Pyrzynska K. Carbon nanotubes as sorbents in the analysis of pesticides. Chemosphere 2011;83 (11):1407–1413.

[16] Herrero Latorre C, Álvarez Méndez J, Barciela García J, García Martín S, Peña Crecente RM. Carbon nanotubes as solid-phase extraction sorbents prior to atomic spectrometric determination of metal species: a review. Anal Chim Acta 2012;749:16–35.

[17] Sitko R, Zawisza B, Malicka E. Modification of carbon nanotubes for preconcentration, separation and determination of trace-metal ions. TrAC Trend Anal Chem 2012;37:22–31.

[18] Basheer C, Alnedhary AA, Rao BSM, Valiyaveettil S, Lee HK. Development and application of porous membrane-protected carbon nanotube micro-solid-phase extraction (m-SPE) combined with gas chromatography/mass spectrometry. Anal Chem 2006;78 (8): 2853–2858.

[19] Sae-Khow O, Mitra S. Carbon nanotubes as the sorbent for integrating μ-SPE within the needle of a syringe. J Chromatogr A 2009;1216 (12):2270–2274.

[20] Lord H, Pawliszyn J. Microextraction of drugs. J Chromatogr A 2000;902 (1):17–63.

[21] Mehdinia A, Aziz-Zanjani MO. Recent advances in nanomaterials utilized in fiber coatings for solid-phase microextraction. TrAC Trend Anal Chem 2013;42:205–215.

[22] Jiang R, Zhu F, Luan T, Tong Y, Liu H, Ouyang G, Pawliszyn J. Carbon nanotube-coated solid-phase microextraction metal fiber based on sol-gel technique. J Chromatogr A 2009;1216 (22):4641–4647.

[23] Chen L, Chen W, Ma C, Du D, Chen X. Electropolymerized multiwalled carbon nanotubes/polypyrrole fiber for solid-phase microextraction and its applications in the determination of pyrethroids. Talanta 2011;84 (1):104–108.

[24] Li Q, Wang X, Yuan D. Preparation of solid phase micro-extraction fibre coated with single-walled carbon nanotubes by electrophoretic deposition and its application in extracting phenols from aqueous samples. J Chromatogr A 2009;1216 (9):1305–1311.

[25] Chen W, Zeng J, Chen J, Huang X, Jiang Y, Wang Y, Chen X. High extraction efficiency for polar aromatic compounds in natural water samples using multiwalled carbon nanotubes/Nafion solid-phase microextraction coating. J Chromatogr A 2009;1216 (52):9143–9148.

[26] Xiao C, Han S, Wang Z, Xing J, Wu C. Application of the polysilicone fullerene coating for solid-phase microextraction in the determination of semi-volatile compounds. J Chromatogr A 2001;927 (1–2):121–130.

[27] Yu J, Dong L, Wu C, Wu L, Xing J. Hydroxyfullerene as a novel coating for solid-phase microextraction fiber with sol-gel technology. J Chromatogr A 2002;978 (1–2):37–48.

[28] Bondavalli P, Legagneux P, Pribat D. Carbon nanotubes based transistors as gas sensors: state of the art and critical review. Sens Actuators B Chem 2009;140 (1):304–318.

[29] Su F, Lu C, Cnen W, Bai H, Hwang JF. Capture of CO_2 from flue gas via multiwalled carbon nanotubes. Sci Total Environ 2009;407 (8):3017–3023.

[30] Llobet E. Gas sensors using carbon nanomaterials: a review. Sens Actuators B Chem 2013;179:32–45.

[31] Stadermann M, McBrady AD, Dick B, Reid VR, Noy A, Synovec RE, Bakajin O. Ultrafast gas chromatography on single-wall carbon nanotube stationary phases in microfabricated channels. Anal Chem 2006;78 (16):5639–5644.

[32] Hussain CM, Saridara C, Mitra S. Microtrapping characteristics of single and multi-walled carbon nanotubes. J Chromatogr A 2008;1185 (2):161–166.

[33] Hussain CM, Saridara C, Mitra S. Carbon nanotubes as sorbents for the gas phase pre-concentration of semivolatile organics in a microtrap. Analyst 2008;133 (8):1076–1082.

[34] Hussain CM, Saridara C, Mitra S. Modifying the sorption properties of multi-walled carbon nanotubes via covalent functionalization. Analyst 2009;134 (9):1928–1933.

[35] Li Q, Yuan D. Evaluation of multi-walled carbon nanotubes as gas chromatographic column packing. J Chromatogr A 2003;1003 (1–2):203–209.

[36] Speltini A, Merli D, Quartarone E, Profumo A. Separation of alkanes and aromatic compounds by packed column gas chromatography using functionalized multi-walled carbon nanotubes as stationary phases. J Chromatogr A 2010;1217 (17):2918–2924.

[37] Saridara C, Mitra S. Chromatography on self-assembled carbon nanotubes. Anal Chem 2005;77 (21):7094–7097.

[38] Karwa M, Mitra S. Chromatography of self-assembled singe walled carbon nanotubes. Anal Chem 2006;78 (6):2064–2070.

[39] Hussain CM, Saridara C, Mitra S. Self-assembly of carbonnanotubes via ethanol chemical vapor deposition for thesynthesis of gas chromatography columns. Anal Chem 2010;82 (12):5184–5188.

[40] Safavi A, Maleki N, Doroodmand MM. Single-walled carbon nanotubes as stationary phase in gas chromatographic separation and determination of argon, carbon dioxide and hydrogen. Chim Acta 2010;675 (2):207–212.

[41] Reid VR, Stadermann M, Bakajin O, Synovec RE. Talanta 2009;77 (4):1420–1425.

[42] Glausch A, Hirsch A, Lamparth I, Schurig V. Retention behaviour of polychlorinated biphenyls on polysiloxane-anchored C60 in gas chromatography. J Chromatogr A 1998;809 (1–2):252–257.

[43] Li Y, Chen Y, Xiang R, Ciuparu D, Pfefferle LD, Horváth C, Wilkins JA. Incorporation of single-wall carbon nanotubes into an organic polymer monolithic stationary phase for mu-HPLC and capillary electrochromatography. Anal Chem 2005;77:1398–1406.

[44] Chambers SD, Holcombe TW, Svec F, Fréchet JMJ. Effect of multi-walled carbon nanotubes incorporation into benzyl methacrylate monolithic columns in capillary liquid chromatography. Porous polymer monoliths functionalized through copolymerization of a C_{60} fullerene-containing methacrylate monomer for highly efficient separations of small molecules. Anal Chem 2011;83 (24): 9478–9484.

[45] Chambers SD, Svec F, Fréchet JMJ. Incorporation of carbon nanotubes in porous polymer monolithic capillary columns to enhance the chromatographic separation of small molecules. J Chromatogr A 2011;1218 (18):2546–2552.

[46] Aqel A, Yusuf K, Al-Othman ZA, Badjah-Hadj-Ahmed AY, Alwarthan AA. Analyst 2012;137 (18):4309–4317.

[47] Hinds BJ, Chopra N, Rantell T, Andrews R, Gavalas V, Bachas LG. Aligned multiwalled carbon nanotube membranes. Science 2004;303 (5654):62–65.

[48] Holt JK, Park HG, Wang Y, Stadermann M, Artyukhin AB, Grigoropoulos CP, Noy A, Bakajin O. Fast mass transport through sub-2-nanometer carbon nanotubes. Science 2006;312 (5776):1034–1037.

[49] Noy A, Park HG, Fornasiero F, Holt JK, Grigoropoulos CP, Bakajin O. Nanofluidics in carbon nanotubes. Nano Today 2007;2 (6):22–29.

[50] Hylton K, Chen Y, Mitra S. Carbon nanotube mediated microscale membrane extraction. J Chromatogr A 2008;1211 (1–2):43–48.

[51] Sae-Khow O, Mitra S. Carbon nanotube immobilized composite hollow fiber membranes for pervaporative removal of volatile organics from water. J Phys Chem C 2010;114 (39):16351–16356.

[52] Vashist SK, Zheng D, Al-Rubeaan K, Luong JHT, Sheu F-S. Advances in carbon nanotube based electrochemical sensors for bioanalytical applications. Biotechnol Adv 2011;29 (2):169–188.

[53] Jacobs CB, Peairs MJ, Venton BJ. Review: carbon nanotube based electrochemical sensors for biomolecules. Anal Chim Acta 2010;662 (2):105–127.

[54] Chikkadi K, Roman C, Durrer L, Süss T, Pohle R, Hierold C. Scalable fabrication of individual SWNT chem-FETs for gas sensing. Proc Eng 2012;47:1374–1377.

SUGGESTED WEBSITES

http://www.epa.gov/
books.google.com/books?isbn=1420053647
http://pubs.acs.org/journal/ancac3
http://www.nature.com/nnano/index.html
http://www.sciencedirect.com/science/article/pii/S0165993614000594
http://www.springer.com
http://www.Azonano.com
http://www.Nanovip.com
http://www.Nanotechweb.org
http://www.Nano.org.uk
http://www.Nano.gov
http://www.Nanotechwire.com
http://www.Nanotechnology.com

15

APPLICATION OF NANOADSORBENTS IN WATER TREATMENT

AMIT BHATNAGAR[1] AND MIKA SILLANPÄÄ[2]

[1]*Department of Biology and Environmental Science, Faculty of Health and Life Sciences, Linnaeus University, Kalmar, Sweden*

[2]*Laboratory of Green Chemistry, Department of Energy and Environmental Technology, Faculty of Technology, Lappeenranta University of Technology, Mikkeli, Finland*

15.1 INTRODUCTION

Water is a scarce resource, and with the pressures of climate change, rapid industrialization, and population growth, water has become even scarcer, especially in developing regions. It is estimated that about one-sixth of the population living in developing countries have no access to clean drinking water, while about 2.6 billion people do not have access to basic sanitation facilities [1]. This situation becomes worse with contamination of water by a wide variety of toxic pollutants from natural, domestic, or industrial sources. Despite their presence at low concentration ranges, environmental pollutants pose serious threats to freshwater supply, living organisms, and public health. Water pollution is one of the critical problems of the twenty-first century that humanity is facing. The quest to ensure that all people have access to clean drinking water is now enshrined in the United Nations' (UN) Millennium Development Goals, which aim to halve the proportion of people without sustainable access to safe drinking water by 2015. Robust and technological improvements are urgently required to treat polluted water economically and efficiently. In recent years, various technologies have been developed to address this problem [2, 3]. Among them, adsorption has been found very promising to remove diverse types of pollutants from water and wastewater [4]. One of the most prominent examples of a conventional adsorbent for environmental application is activated carbon [5]. This conventional material is cost-effective in application and thus widespread in the European drinking water industry for water purification. However, in recent years, nanotechnology has emerged as a powerful platform, among other twenty-first-century technologies, and various nanoadsorbents have been developed and tested toward water treatment applications [6]. Nanostructured adsorbents offer the opportunity of an even greater adsorption capacity and might be designed to target specific contaminants. Nanotechnology, related to the manipulation of materials and systems at the nanoscale (<100 nm), is considered one of the most important achievements in science and technology. At the nanoscale, materials show unique characteristics and, because of their small size, they possess a large surface area and "surface area to volume" ratio [7]. In this regard, promoting nanoadsorbents presents opportunities to develop local and practical solutions for tackling global water pollution. This chapter presents a brief overview of the technical applicability of different nanoadsorbents for removing various aquatic pollutants. Present challenges, future research directions, and potential implications to public health, resulting from their applications in water treatment, are also discussed.

Nanomaterials for Environmental Protection, First Edition. Edited by Boris I. Kharisov, Oxana V. Kharissova, and H. V. Rasika Dias.

15.2 NANOADSORBENTS IN WATER TREATMENT

Nanoadsorbents are nanoscale particles from organic or inorganic materials that have a high affinity to adsorb substances. Because of their high porosity, small size, and active surface, nanoadsorbents not only are capable of sequestering contaminants with varying molecular size, hydrophobicity, and speciation behavior, but they also enable manufacturing processes to consume raw materials efficiently without releasing a toxic payload [8]. Nanoadsorbents not only work rapidly, but also have considerable metal-binding capacities. They can also be chemically regenerated after being exhausted [9]. For these reasons, recently, scholarly interest in nanotechnology has been growing rapidly worldwide.

15.2.1 Carbon Nanotubes

Due to the high specific area and large micropore volume, carbon nanotubes (CNTs) have been widely utilized for the adsorption of various aquatic contaminants such as organic compounds and inorganic ions in the past decade. CNTs include single-walled carbon nanotubes (SWCNTs) and multiwalled carbon nanotubes (MWCNTs), and both can be used for the adsorptive removal of various pollutants from water. The unique physicochemical and electrical properties of CNTs surpass those of its prominent counterpart, activated carbon. Some of the studies describing the adsorptive removal of various contaminants by CNTs are discussed herein: Adsorptive removal of zinc (Zn^{2+}) using purified CNTs (MWCNTs and SWCNTs) has been examined by Lu and Chiu [10]. The CNTs were subjected to thermal treatment and treated with sodium hypochlorite. As reported in other studies [11], purification resulted in the removal of metal catalysts and amorphous carbon from CNTs, and large quantities of nanotube bundles with a hollow inner tube diameter were observed in the purified CNTs. The Brunauer–Emmett–Teller (BET) analysis showed that SWCNTs possess higher surface area and pore volume than MWCNTs, but average pore diameter of SWCNTs was lower than that of MWCNTs. The zeta potentials of purified CNTs were more negative than that of raw CNTs due to the presence of negative functional groups on the surface of purified CNTs. Adsorption of Zn^{2+} increased in the pH range of 1–8. But low Zn^{2+} adsorption occurred at low pH due to the competition between H^+ and Zn^{2+} for the adsorption sites. After an increase in Zn^{2+} adsorption up to pH 8.0, adsorption remained constant in the pH range of 8.0–11.0. The Zn^{2+} species prevails until pH 8.0, and the process governing Zn^{2+} removal is adsorption. The kinetics of Zn^{2+} adsorption onto the CNTs was fast (60 min). The kinetics and isotherm study confirmed that the adsorption capacity of Zn^{2+} onto purified SWCNTs was higher than that onto MWCNTs, which was in agreement with the BET results. The maximum adsorption capacities of Zn^{2+} calculated by the Langmuir model were reported as 43.66, and 32.68 mg/g with SWCNTs and MWCNTs, respectively, at an initial Zn^{2+} concentration range of 10–80 mg/l. The isotherm data was better explained by the Langmuir model than by the Freundlich model. Gupta et al. [12] studied Cr(III) removal by combining the magnetic properties of iron oxide with adsorption properties of MWCNTs. Acid treatment (HNO_3) of MWCNTs was performed to modify the surface of the CNTs with carbonyl and hydroxyl groups. Furthermore, the MWCNTs/nano iron oxide composites were prepared using ferric chloride hexahydrate and ferrous chloride. Scanning electron microscopy (SEM) images of the prepared composite exhibited an entangled network of oxidized MWCNTs with clusters of iron oxides attached to them. The presence of maghemite (γ-Fe_2O_3) and magnetite (Fe_3O_4) was confirmed by the X-ray diffraction (XRD) analysis of the composites. The additional adsorbing sites provided by the oxygen atoms of iron oxide nanoparticles (NPs) on the surface of MWCNTs became available for the electrostatic interaction with Cr(III) and, thus, the prepared composite showed higher Cr(III) capacity. The composite was effective for Cr(III) adsorption in the pH range above pH 3 and below pH 7 due to the presence of $Cr(OH)^{2+}$ and $Cr(OH)_2^+$ species of chromium. Also, at pH values of 4.0–7.0, the net negative surface charge on the former allowed increased adsorption of chromium species on MWCNTs. The results of the fixed bed experiments revealed that lower flow rate favored Cr(III) removal due to the increased contact time between Cr(III) and adsorbent. An increase in fixed bed layers also revealed an increase in Cr(III) uptake, which was attributed to the availability of more adsorption sites. Several researchers have also modified CNTs to evaluate the efficiency of the former with the unmodified CNTs for the removal of various contaminants. Konicki et al. [13] evaluated the performance of magnetic MWCNT–Fe_3C nanocomposites (MMWCNTs–ICN) for the adsorption of the anionic dye Direct Red 23 (DR23), which is present in various industrial effluents. Chemical vapor deposition (CVD) was used to obtain the nanocomposites. Thermogravimetric analysis (TGA) confirmed that the composites were composed of graphite and iron carbide (Fe_3C) phases. Adsorption equilibrium was attained in 160–340 min depending on the initial dye concentration. The availability of a large number of vacant sites resulted in faster adsorption in the initial stage. The driving force of concentration gradient increased with an increase in the initial concentration of the dye and resulted in increased dye adsorption capacity. With an increase in pH (from 3.7 to 11.1), the adsorption capacity decreased due to the electrostatic interaction between the anionic dye and the partially negatively charged nanocomposites. At lower pH, the positive surface charge was dominant due to the protonation of oxygen-containing functional groups. Temperature also favored the adsorption of DR23 onto the nanocomposites. The kinetic data and experimental isotherm data fitted well with the pseudo-second-order model and Freundlich isotherm model, respectively.

Chatterjee et al. [14] impregnated chitosan (CS) hydrogel beads with MWCNTs and studied the adsorption efficiency of the modified adsorbent for Congo red (CR) dye removal. CNT impregnation rendered the beads materially denser and the porosity of the CS/CNT beads also increased due to the addition of CNT. The field-emission SEM (FE-SEM) image of the modified adsorbent revealed that CNT was present both uniformly throughout the CS matrix and as small aggregates in the CS/CNT beads. The energy-dispersive X-ray (EDX) analysis of the modified adsorbent after CR adsorption revealed higher S (wt.%) in the CS/CNT beads (4.62%) than in the CS beads (1.83%), indicating that CNT impregnation increased the adsorption capacity of CS/CNT beads. A negative effect of CNT concentration in CS beads was observed. An increase from 0.01 to 0.05 wt.% CNT showed a decrease in adsorption capacity. It was attributed to the formation of larger aggregates of CNTs, which hindered the access of CR to the adsorption site of CS and cetyltrimethylammonium bromide (CTAB). CTAB is a dispersant present in CS/CNT beads that interacts with the hydrophobic moieties of CR and partly has ionic interaction with the anionic charge on CR. The adsorption of CR onto CS/CNT was highly pH dependent, and maximum adsorption by the CS/CNT beads (423.1 mg/g) occurred at pH 4.0. Since beads formation was carried out in a sodium hydroxide precipitation bath, the amine groups of CS molecules were not protonated. An increase in initial pH 4.0–5.6 and from pH 5.0 to 6.0 during adsorption indicated a decrease in H^+ concentration in the solution. However, such behavior was absent when the initial pH of the CR solution was neutral or basic as protonation of amine groups did not occur at pH 7.0 and 9.0 due to absence of free H^+ in solution. The Langmuir isotherm model showed a better fit with the experimental isotherm data than the Freundlich isotherm model. The maximum adsorption capacity of CS/CNT beads for CR was reported as 450.4 mg/g. The values obtained from pseudo-first-order, pseudo-second order kinetic models and the intraparticle diffusion model showed that the mass transfer rate was fairly increased by the addition of CNT. Bazrafshan et al. [15] studied the removal of reactive red 120 textile dye (RR-120) using SWCNTs. The SWCNTs were synthesized by the catalytic CVD method. They were further functionalized by dispersing the former in nitric acid. The adsorptive removal of RR-120 was influenced by pH. The optimum pH for the process was 5.0 as below and above this pH adsorptive removal of the dye was found to decrease. Maximum adsorption capacity of SWCNTs for the dye was reported as 426.49 mg/g, with 85.3% dye removal taking place at pH 5.0 with initial dye concentration of 50 mg/l. The protonation of electron Π–rich regions on the surface of SWCNTs at lower pH values forms positive surface charge, increasing the uptake of negative charged dye. However, with an increase in pH, the carboxylic groups ionized and caused an increase in negative surface charge density resulting in reduced removal of RR-120. The increased adsorbent dosage resulted in an increase in the surface area and adsorption sites that helped in achieving maximum dye adsorption (88.28%, 882.83 mg/g). Adsorption equilibrium was achieved in 180 min contact time. The BET model fitted well with the isotherm data.

In another study, resorcinol was taken as a model phenolic derivative and the efficiency of MWCNTs was assessed for the adsorption of the former [16]. Nitric acid treatment was conducted to oxidize the MWCNTs. The kinetics of resorcinol onto MWCNTs was fast. The experiments were not performed at pH values greater than 8.0 as in basic conditions resorcinol is oxidized. Resorcinol uptake increased with the decrease in solution pH since its solubility decreases with the lowering of pH. A comparison of the adsorption properties of treated and nontreated MWCNTs was performed. A reduction in the uptake capacity of resorcinol was observed due to the electrostatic repulsion between negatively charged resorcinol and acid-treated MWCNTs. Also, the carboxylic groups, those that are present on the surface of MWCNTs, act as electron withdrawing groups localizing electrons from the Π system of MWCNTs and interfering and weakening the forces between the aromatic ring of resorcinol and the graphite structure of MWCNTs. The hydroxyl group (located in meta-position) present in the structure of resorcinol also positively influenced its adsorption. The potential of MWCNTs for herbicide adsorption has also been tested by researchers [17]. Diuron and dichlobenil were the two representative herbicides studied, and the effect of lead on their adsorption onto MWCNTs was observed. The MWCNTs with outer diameters of less than 8 nm (MWCNTs8) exhibited maximum adsorption capacity for both herbicides due to their large surface area and pore volume, and the experimental isotherm data was in agreement with the Polanyi–Manes model. It was also observed that an increase in oxygen content of MWCNTs decreased the adsorption of the herbicides due to deprotonation of carboxylic groups at pH 6.0. Since MWCNTs with outer diameters of 10–20 nm and oxygen content of 1.52% (MWCNTs-O (1.52%)), and MWCNTs-O (2.66%) and MWCNTs-O (7.58%), had quite similar diameters and surface areas, but different oxygen contents, they were chosen to study the effect of lead on the sorption of herbicides. In case of Pb(II), in a single-solute system, MWCNTs-O (2.66%) and MWCNTs-O (7.58%) showed Pb(II) adsorption capacities of 17.75 and 91.67 mg/g for the same Pb^{2+} equilibrium concentration implying that the surface O-containing groups improved the ion exchange capabilities of MWCNTs and, therefore, resulted in higher Pb^{2+} adsorption. The adsorption of diuron or dichlobenil on MWCNTs-O (7.58%) was suppressed significantly (~10–30 and ~15–30%, respectively) in the presence of Pb^{2+}. But less effect of Pb^{2+} on herbicide removal was noticed when MWCNTs-O (2.66%) and MWCNTs-O (1.52%) were used as adsorbent. It was concluded that complexation of Pb^{2+} with carboxylic groups was responsible for the reduced herbicide removal. Also, hydrated lead cations occupied partial surface of MWCNTs-O and shielded the hydrophobic and hydrophilic sites of MWCNTs, leading to inhibition in herbicides adsorption around the metal-complexed moieties.

Natural organic matter (NOM) removal by CNTs has also been studied by researchers, and the results were found to be promising. Lu and Su [18] investigated the role of thermally treated MWCNTs (at 400°C for 60 min) in the removal of NOM in terms of dissolved organic carbon (DOC) and assimilable organic carbon (AOC). A comparative study of CNT and granular activated carbon (GAC) adsorption of NOM was also conducted. The NOM had anionic character across the pH range 3.0–9.0. The thermally treated CNTs had greater surface area than their counterparts. Boehm titration results confirmed that the acidic and basic site concentrations were 0.4 and 0.25 mmol/g, respectively for raw (untreated) CNTs and 0.3 and 0.2 mmol/g, respectively for heated CNTs. The kinetics of DOC removal was faster than that of AOC onto CNTs due to higher initial DOC concentration that creates a greater diffusion driving force on the CNT surface. The experimental isotherm model fitted well with the Langmuir model, and the equilibrium adsorption of DOC and AOC obtained by the treated adsorbent was found to be more than that obtained by raw CNTs. The increase in solution pH indicated that the hydroxyl ions were released from the CNT surface when NOM adsorption occurred. But on increasing the initial pH from 4.0 to 5.0, DOC and AOC removal decreased due to the competition between NOM and OH^- on the same CNT site. GAC exhibited DOC and AOC adsorption capacities as 14.711 and 0.341 mg/g, respectively. The raw and modified CNTs exhibited 22.003 and 26.138 mg/g adsorption capacity for DOC and 0.485 and 0.548 mg/g for AOC. Joseph et al. [19] investigated NOM removal from a variety of potential drinking water sources through a combination of coagulation and adsorption with CNTs. Both SWCNTs and MWCNTs were examined for their removal efficiencies of NOM parallel to the coagulation process in the presence of alum and ferric chloride metal coagulants. Synthetic seawater (SW) and brackish water (BW) (contained humic acid as its representative NOM), synthetic (old and young) landfill leachates (OL, YL) (contained glucose as its representative NOM), and broad river water (BRW) in Columbia, South Carolina were chosen to obtain a detailed knowledge of NOM and its behavior with regard to adsorption and coagulation. Humic acid used in the study contained approximately 40% DOC. The isotherm data of BW, SW, and BRW fitted well with the Freundlich model, and SWCNTs consistently had a higher adsorption capacity than MWCNTs for NOM. The higher specific surface area of SWCNTs (407 m^2/g) as compared to that of MWCNTs (60 m^2/g) was assumed to be responsible for the greater NOM uptake capacity of the former.

Adsorption of trihalomethanes (THMs) from water using CNTs has also been investigated [11]. MWCNTs were fabricated by catalytic decomposition of the CH_4/H_2 mixture at 700°C using Ni particles as catalyst. The as-prepared CNTs were treated with concentrated acids for 24 h to remove metal catalysts and then washed by deionized water. After cleaning, the CNTs were again dispersed into concentrated acid solutions and refluxed at 80°C for 2 h to remove amorphous carbon. Finally, the CNTs-containing solution was filtered to obtain purified CNTs. The THM solution consisted of equivalent concentrations of $CHCl_3$, $CHCl_2Br$, $CHClBr_2$, and $CHBr_3$. The SEM images confirmed the cylindrical shape of the CNTs, and the transmission electron microscopy (TEM) image exhibited that CNTs possessed multiple walls with hollow inner diameter. The purity of CNTs increased from 95.05 to 98.87% after acid treatment as revealed by TGA. The kinetic studies showed that the diffusion mechanism controlled the THMs adsorption onto CNTs and the smallest molecule, $CHCl_3$, was preferred over other THMs ($CHCl_2Br$, $CHClBr_2$, and $CHBr_3$) because $CHCl_3$ could easily enter the inner cavities through the pores. The adsorption capacity of $CHCl_3$ was observed to be the highest (2.41 mg/g), followed by $CHBrCl_2$ (1.23 mg/g), $CHBr_2Cl$ (1.08 mg/g), and then $CHBr_3$ (0.92 mg/g). As the surface tension of $CHCl_3$ (27.14 dyn/cm) is much lower than that of $CHBr_3$ (46.2 dyn/cm), adsorption of the former onto CNTs was much easier. Also, the polarity of the different THM compounds influenced the adsorption process. Since the polarity of $CHCl_3$ is the highest, maximum adsorption was noticed in the case of $CHCl_3$. THMs adsorption fluctuated only a little in the pH range of 3.0–7.0. This was attributed to the purification of CNTs by an acid solution that made the CNTs resistant to the acidic environment. But at higher pH (>7.0), THMs adsorption decreased due to the ionization of oxygen-containing groups on CNTs surface resulting in the blocking of access of THM molecules to adsorption sites. The comparative adsorption study of THMs by powdered activated carbon (PAC) and CNTs showed that the adsorption capacity of $CHCl_3$ on CNTs was twice that on PAC due to the presence of functional groups on CNTs surface. Several other reports are also available where carbon nanomaterials have been used for the removal of different pollutants [20–32].

15.2.2 Metal Oxide-Based NPs

Several adsorption studies have also been conducted to evaluate the performance of iron oxide–based nanomaterials. Many studies have successfully shown the use of iron oxide as composite materials with host materials to fabricate magnetic sorbents. Agrawal and Bajpai [33] synthesized iron oxide–based gelatin NPs and evaluated their use for the removal of Cr(VI) from aqueous solutions. An emulsion cross-linking method was followed to synthesize the gelatin-based NPs. The cross-linking occurred between gelatin and glutaraldehyde. The gelatin NPs were allowed to swell in Fe^{2+}/Fe^{3+} solution and further treated with NH_4OH (an alkali) to precipitate the Fe ions into iron oxide within the NP matrix. The SEM characterization revealed the smooth, homogeneous, and relatively spherical particle surface of the gelatin NPs. The TEM images confirmed that after Cr(VI) adsorption onto the nanomaterial surface, the average size of the adsorbent increased. The XRD patterns of the FeOGel nanocomposites showed the impregnation of iron oxide into gelatin NPs. The percent crystallinity of the FeOGel was also higher than of the bare gelatin NPs confirming the impregnation of iron oxide in the gelatin NPs. The zeta potential measurements indicated that in basic media the adsorbent surface exhibited a negative potential due to the formation of FeO^-, while in acidic media, positive surface charge prevailed due to the formation of $FeOH^{2+}$. Maximum Cr(VI) removal was observed at pH 2.0–3.0. Such adsorptive behavior was confirmed by the zeta potential measurements. Electrostatic attraction between the positively charged surface hydroxyl groups $FeOH^{2+}$ of FeOGel and negatively charged chromium species ($HCrO_4^-$) was responsible for the maximum uptake of Cr(VI) at pH 2.0. At pH less than 1.0, the H_2CrO_4 species is predominant and results in less adsorption of Cr(VI) onto the nanomaterial. At higher pH, FeO species occur primarily, and the adsorbent surface mainly carries net negative charge thus repelling the metal anions (CrO_4^{2-}). The kinetics of Cr(VI) uptake on FeOGel was fast, and equilibrium was achieved in 90 min. The increase in FeOGel dosage resulted in the increase of chromium uptake. The adsorption data fit well with the Langmuir isotherm model. The maximum adsorption capacity of FeOGel for Cr(VI) was determined to be 120 mg/g.

Iron (Fe^{+3}) oxide/hydroxide NPs-based agglomerates suspension as adsorbent was evaluated by Zelmanov and Semiat [34] for the removal and recovery of Cr(VI) from water. Iron (Fe^{3+}) oxide/hydroxide–based NPs (NanoFe) sol and iron (Fe^{+3}) oxide/hydroxide NP agglomerates (AggFe) suspension were prepared by hydrolysis and by adjusting solution pH, respectively. In case of Cr(VI) removal by NanoFe and AggFe separately, a significant influence of nanoparticle concentration on residual chromium concentration was reported. The effect of pH on Cr(VI) removal by NanoFe corroborated the observations obtained by Agrawal and Bajpai [33] for the removal of Cr(VI) by FeOGel nanocomposites. A sharp rise in residual Cr(VI) concentration was observed when the solution pH was increased. At basic pH (8.3–8.5), Cr(VI) adsorption onto NPs was not observed due to the presence of hydroxyl groups. The effect of pH was similar in case of Cr(VI) removal by AggFe. The kinetics of Cr(VI) onto NanoFe and AggFe was extremely fast (<1 min). The large external surface adsorption area of AggFe was found to be responsible for the rapid adsorption kinetics of Cr(VI). Precipitation of chromium as $BaCrO_4$ crystals was found to be possible due to which 99% chromium could be recovered. Thermal hydrolysis method was employed by Nieto-Delgado and Rangel-Mendez [35] to generate iron oxide NPs inside the activated carbon pores considering that other techniques employ voluminous chemicals such as surfactants, chelating agents, and organometallic compounds that have a poor diffusion into the activated carbon pores. The efficiency of synthesized iron-modified activated carbon was further tested for arsenic (As(V)) removal. Three different activated carbons viz. (i) Filtrasorb 400–activated carbon (F400), (ii) a bagasse-based activated carbon produced by $ZnCl_2$ (ACZ), and (iii) carbon produced by H_3PO_4 (ACP) were used in the study. The results were analyzed based on response surface methodology. The iron content in ACZ-Fe (0.92–4.63%) and ACP-Fe (2.18–5.27%) was found to be higher than in F400-Fe (0.73–2.2%) activated carbons. The characterization results revealed that the F400 carbon was basic and had less oxygenated surface groups. On the contrary, ACZ and ACP were found to be acidic containing larger amount of oxygenated surface groups. The results also suggested that the oxygenated surface groups promote the anchorage of iron to the adsorbent surface. In the study, adsorption capacity of 1.4 mg As/g C was reported for F400, whereas it was almost null for ACP and ACZ pristine carbons. SEM analysis revealed that F400-Fe had the smallest iron hydro(oxides) particle size and a higher surface area and hence the highest iron efficiency to remove As(V). In contrast, ACZ-Fe, which is mainly microporous, had a reduced surface area after modification. However, ACP-Fe, which is mesoporous in nature, had a moderate surface area on modification. Along with the particle size of iron hydro(oxides) loaded onto activate carbon, the carbon surface charge was also crucial. ACP-Fe and ACZ-Fe had low pH_{pzc} (3.1–3.6), whereas pH_{pzc} of F400-Fe was 8.6, which favored the adsorption of negative species of arsenic. It was confirmed via XRD analyses that iron existed as hematite crystals in the adsorbent; thus the adsorption capacity of modified carbon was unaltered with time. For the three modified carbons, the adsorption capacities for As(V) were reported as 3.25, 1.45, and 1.65 mg As/g for F400-Fe, ACZ-Fe, and ACP-Fe, respectively, at pH 7.0 (with 1.0 ppm arsenic concentration).

Iron oxide magnetic NPs (MNPs) have also been applied as immobilization carriers for a biomass strain (*Phanerochaete chrysosporium*), and the prepared adsorbent has been tested for lead (Pb(II)) removal [36]. Iron oxide MNPs (Fe_3O_4 MNPs) and Ca alginate were encapsulated in the *P. chrysosporium* pellets to prepare the adsorbent. The adsorption equilibrium of Pb(II) onto the prepared adsorbent was achieved within 8 h. The pseudo-second-order model fitted well with the experimental kinetics data. Maximum Pb(II) adsorption capacity was reported as 176.33 mg/g at initial Pb(II) concentration of 500 mg/l. The Langmuir isotherm model fitted well with the experimental isotherm data. The interfacial adsorption and intraparticle diffusion were found to influence Pb(II) adsorption onto the adsorbent. The electrostatic attraction between MNPs and metal ions also influenced the sorption process. It was proven from the Fourier transform infrared (FTIR) spectra of the adsorbent, before and after Pb(II) adsorption, that the functional groups present on the surface of *P. chrysosporium* also played a critical role in Pb(II) removal. The adsorption efficiency of the adsorbent for five adsorption–desorption cycles of Pb(II) was almost constant (90%). Synthesis of magnetite nanoparticles using a low-cost approach was also undertaken [37]. Iron ore tailings (IOT) as the primary source of MNPs were taken into consideration during the study. Methylene blue (MB) and CR dyes were taken as the model pollutants. Complete recovery of iron from IOT was made with HCl. Finally, ferric (hydro)oxide (FHO) (primarily constituting of goethite) was obtained after treatment with concentrated ammonia. The adsorbent pH was raised with NaOH and stabilized using sodium dodecyl sulfate (SDS). The chemical analysis of synthesized MNPs showed that the total iron content was close to the theoretical value (72.36%). The pH_{pzc} of MNPs was found to be 7.1. The results of adsorption studies of MB and CR dyes exhibited that the kinetics of the process was rapid, showing that adsorption occurred only on the surface of MNPs. The behavior of MNPs toward varying pH was explained on the basis of pHpzc (pH_{pzc}: ~7.1), and the adsorption of cationic MB was found to be above pH_{pzc} and anionic CR adsorption was higher in the lower pH range. The optimized pH for MB and CR dye adsorption was approximately 9.2 and 6.2, respectively. The aggregation of MNPs with increase in the MNPs dose led to a non-linear adsorption of dyes. An increase in dye concentration led to an increase in dye uptake by MNPs, but the percentage of adsorption decreased. The monolayer adsorption capacities of MNPs (70.4 and 172.4 mg/g for MB and CR, respectively) were obtained. Desorption studies of MB and CR dyes were also performed, and 95% CR was found to desorb at pH greater than 8.0. However, MB desorption was more effective (85% MB desorption) when methanolic solution of acetic acid (4% (v/v) acetic acid in methanol) was used as an eluent.

In an attempt to modify the iron oxide NPs for rendering them effective for dye removal, Inbaraj and Chen [38] synthesized poly(γ-glutamic acid) (PGA)–coated iron oxide NPs. PGA coating was found to improve the stability of the MNPs, which was confirmed by the leaching experiments in different aqueous matrices (deionized water, tap water, river water, acidic solutions, basic solutions). The MB dye experimental isotherm data fitted well with the Redlich–Peterson and Langmuir models. The maximum adsorption capacity of MB dye onto PGA MNPs obtained from the Langmuir model was 78.67 mg/g. The adsorption kinetics of MB onto synthesized nanomaterials was fast (equilibrium was achieved in 5 min). In the presence of coexisting ions, MB adsorption was lowered. A declining trend in the presence of $CaCl_2$ was more significant than in the presence of NaCl due to the valence effect. Acidic pH (1.0–3.0) was not favorable for the adsorption of MB dye onto PGA MNPs due to the competition between dye cations and hydrogen ions in the solution for the α-carboxylate anions present on the surface of the adsorbent. The ion exchange mechanism that dominated at higher pH favored the MB adsorption onto PGA MNPs. On the other hand, desorption was favorable in acidic media (ionized water). Salih et al. [39] investigated the fate and transport of Fe_2O_3 NPs in a GAC adsorber bed and its impact on trichloroethylene (TCE) adsorption. The TEM image revealed that Fe_2O_3 NPs formed chains and branches due to intermolecular attraction. The presence of Fe_2O_3 NPs enhanced TCE removal (70%) from the aqueous phase as compared to kinetics onto GAC alone. But, desorption was similarly fast due to the lack of a suitable pore size distribution for permanent adsorption of TCE. Characterization results revealed that only 3% of Fe_2O_3 NP total surface area was microporous resulting in their easy desorption into bulk solution. It was concluded from the values obtained from the kinetic models that batch systems might not be optimal to study TCE adsorption. Column breakthrough studies were also undertaken in which it was observed that TCE breakthrough occurred in a shorter time in the presence of Fe_2O_3 NPs than in their absence. This was attributed to the attachment of Fe_2O_3 onto GAC. Also, the impact of Fe_2O_3 NPs on TCE breakthrough was found to be concentration-dependent.

Jegadeesan et al. [40] reported the sorption of As(III) and As(V) species on nanosized amorphous and crystalline TiO_2. The isotherms indicated TiO_2 sorption capacities as dependent on the site density, surface area, and crystalline structure. The adsorbent surface remained almost constant for particles between 5 and 20 nm. But As(V) surface coverage increased with the degree of crystallinity, which was confirmed by X-ray absorption spectroscopic analysis. The data indicated binuclear bidentate inner sphere complexation of As(III) and As(V) on amorphous TiO_2 at neutral pH. Belessi et al. [41] synthesized TiO_2 NPs and used them for the removal of reactive red 195 azo dye. The effects of pH, concentration of dye, and adsorbent dose have been studied on the removal of dye. The equilibrium data fitted well with Langmuir and pseudo-second-order kinetic models. At pH 3.0 and 30°C, the maximum adsorption capacity was 87.0 mg/g. The kinetic studies indicated a rapid sorption of dye in the first 30 min with equilibrium at 1 h.

Few other efforts have been made to use other nano metal oxides for the removal of various aquatic pollutants. Arsenic (As(III) and As(V)) removal using nano copper oxide has been tested [42, 43]. The kinetics of arsenic onto CuO NPs was fast, but As(III) took a comparatively longer time than As(V) to adsorb [42]. As(III) adsorption onto CuO NPs was maximum at pH 9.3; on the contrary, As(V) adsorption was relatively independent of pH. Sulfate and silicate are the main competing ions in case of As(III), whereas only phosphate was found to affect As(V) adsorption. Hydrous cerium oxide (HCO) NPs have also been reported for the removal of arsenic [44]. The kinetics of As(III) and As(V) was observed to be rapid onto HCO NPs. The increase in pH from 3.0 to 7.0 resulted in the decrease of percent removal of As(V) from 99.7 to 54.7%. On the contrary, only a slight increase in As (III) removal percent was noticed when the pH was increased from 3.0 to 7.0. Unlike the case of other metal oxyhydroxides, the increase in solution pH affected As(III) removal only slightly, suggesting that a strong affinity exists between the As(III) and HCO NPs. The repulsion effect from the increase in solution pH was easily overcome by the adsorption energy. Mainly $H_2PO_4^-$ had an adverse effect on As(III) and As(V) removal efficiency, which was similar to the results obtained in previous studies on arsenic adsorption. The FTIR results of HCO NPs after adsorption process exhibited the formation of Ce-O-As and As-OCe bonds in case of As(III)–HCO and As(V)–HCO NPs, respectively, indicating the formation of inner-sphere complexes.

15.2.3 Miscellaneous Nanoadsorbents

CS–Fe(0) NPs were prepared using biodegradable CS as a stabilizer, and batch experiments were conducted to evaluate the influences of initial Cr(VI) concentration and other factors on Cr(VI) reduction on the surface of CS–Fe(0) [45]. The authors suggested that the overall disappearance of Cr(VI) might include both physical adsorption of Cr(VI) onto the CS–Fe(0) surface and subsequent reduction of Cr(VI) to Cr(III). Characterization with high-resolution X-ray photoelectron spectroscopy revealed that after the reaction, relative to Cr(VI) and Fe(0), Cr(III) and Fe(III) were the predominant species on the surface of CS–Fe(0). CS also inhibited the formation of Fe(III)–Cr(III) precipitation due to its high efficiency in chelating Fe(III) ions. The adsorption of eosin Y, as a model anionic dye, from aqueous solution using CS NPs prepared by the ionic gelation between CS and tripolyphosphate was examined by Du et al. [46] The adsorption capacity was found to be 3.33 g/g. The adsorption process was endothermic in nature with an enthalpy change (ΔH) of 16.7 kJ/mol at 20–50°C. The optimum pH value for eosin Y removal was found to be 2–6. The dye was desorbed from the CS NPs by increasing the pH of the solution. The desorption percentage was about 60% within 60 min at pH 11.0, whereas 98.5% of the dye could be eluted at pH 12 in 150 min. The potential of nanochitosan to remove acid dyes from aqueous solution was also explored by the researchers [47]. The monolayer adsorption capacities were determined to be 1.77, 4.33, 1.37, and 2.13 mmol/g nanochitosan for acid orange 10, acid orange 12, acid red 18, and acid red 73 dyes, respectively. The differences in capacities might be due to the differences in the particle size of dye molecules and the number of sulfonate groups on each dye molecule. The results have demonstrated that monovalent and smaller dye molecular sizes have superior capacities due to the increase in dye/CS surface ratio in the system and deeper penetration of dye molecules into the internal pore structure of nanochitosan. The mechanism of the adsorption process of acid dye on nanochitosan was proposed to be the ionic interactions of the colored dye ions with the amino groups on the CS. By encapsulating zirconium phosphate (ZrP) NPs into three macroporous polystyrene resins with various surface groups, that is, $-CH_2Cl$, $-SO_3^-$, and $-CH_2N^+(CH_3)_3$, three nanocomposite adsorbents (denoted as ZrP–Cl, ZrP–S, and ZrP–N, respectively) were fabricated for lead removal from water [32]. Effect of the functional groups on nano ZrP dispersion and effect of ZrP immobilization on the mechanical strength of the resulting nanocomposites were investigated. Charged functional groups ($-SO_3^-$ and $-CH_2N^+(CH_3)_3$) are more favorable than the neutral $-CH_2Cl$ group to improve nano ZrP

TABLE 15.1 List of some nanoadsorbents used in the removal of different aquatic pollutants

Nanoadsorbent	Adsorbate	Uptake capacity	Reference
SWCNTs	Zn(II)	43.66 mg/g	Lu and Chiu [10]
MWCNTs	Zn(II)	32.68 mg/g	Lu and Chiu [10]
CS/CNT beads	Congo red	423.1 mg/g	Chatterjee et al. [14]
SWCNTs	Reactive red 120	426.49 mg/g	Bazrafshan et al. [15]
FeOGel	Cr(VI)	120 mg/g	Agrawal and Bajpai [33]
Magnetite nanoparticles	Methylene blue	70.4 mg/g	Giri et al. [37]
Magnetite nanoparticles	Congo red	172.4 mg/g	Giri et al. [37]
Poly(γ-glutamic acid)-coated iron oxide nanoparticles	Methylene blue	78.67 mg/g	Stephen and Chen [38]
TiO_2 nanoparticles	Reactive red 195	87.0 mg/g	Belessi et al. [41]
Chitosan nanoparticles	Eosin Y	3.33 g/g	Du et al. [46]
Nanochitosan	Acid orange 10	1.77 mmol/g	Cheung et al. [47]
Nanochitosan	Acid orange 12	4.33 mmol/g	Cheung et al. [47]
Nanochitosan	Acid red 18	1.37 mmol/g	Cheung et al. [47]
Nanochitosan	Acid red 73	2.13 mmol/g	Cheung et al. [47]
P(A-O)/AT nano-adsorbent	Pb(II)	109.9 mg/g	Jin et al. [48]
Fe–Al–Ce nanoadsorbent (coated granules)	Fluoride	2.77 mg/g	Chen et al. [49]
Aligned carbon nanotubes	Fluoride	4.5 mg/g	Li et al. [61]
Nanoscale aluminum oxide hydroxide	Fluoride	3259 mg/kg	Wang et al. [62]
CaO nanoparticles	Fluoride	163.3 mg/g	Patel et al. [63]
Fe_3O_4@$Al(OH)_3$ NPs	Fluoride	88.48 mg/g	Zhao et al. [64]
Nanoalumina	Fluoride	14.0 mg/g	Kumar et al. [66]
Nanoalumina	Nitrate	4.0 mg/g	Bhatnagar et al. [65]
Nanoalumina	Bromate	6 mg/g	Bhatnagar and Sillanpää [67]
Mg-doped nano ferrihydrite	Fluoride	64 mg/g	Mohapatra et al. [57]
Hierarchical porous CeO_2 nanospheres	Congo red	942.7 mg/g	Ouyang et al. [56]

dispersion (i.e., to achieve smaller ZrP NPs). ZrP–N and ZrP–S had higher capacity than ZrP–Cl for lead removal. As compared to ZrP–N, ZrP–S exhibits higher preference toward lead ion at high calcium levels as a result of the potential Donnan membrane effect. On the other hand, nano ZrP immobilization would simultaneously reinforce both the compressive strength and the wear performance of the resulting nanocomposites with the ZrP loadings up to 5 wt%.

The organic–inorganic hybrid of poly(acrylic acid–acrylonitrile)/attapulgite, P(A-N)/AT nanocomposites, were prepared by in situ polymerization and composition of acrylic acid (AA) and acrylonitrile (AN) onto modified AT NPs [48]. The resulting P(A-N)/AT nanocomposites were transformed into a novel nanoadsorbent of poly(acrylic acid-acryloamidoxime)/AT by further functionalization, that is, P(A-O)/AT nanoadsorbent. The adsorption properties of P(A-O)/AT toward metal ions were determined, and the results indicated that the adsorbents with nanocomposite structure held a good of selectivity to Pb(II) among numerous metal ions. The maximum removal capacity of Pb(II) was up to 109.9 mg/g, and it was notable to see that the adsorption removal of P(A-O)/AT nanoadsorbent for Pb(II) was more than 96.6% when the initial concentration of Pb(II) was 120.0 mg/l. Besides the nanoadsorbents mentioned, several other materials have also been examined as nanoadsorbents for the removal of different aquatic pollutants [49–75]. Table 15.1 lists some of the nanoadsorbents used to for the removal of different aquatic pollutants.

15.3 CONCLUSIONS

In the past few decades, various types of nanoadsorbents have been developed and tested for the removal of different types of aquatic pollutants. Nanoadsorbents present a potent alternative to conventional treatment methods due to increased adsorption capacity and substance specificity. However, most applications are not yet ready for the market due to technical challenges (e.g., scale-up, system setup), environmental concerns, and cost-effectiveness, and only a few commercial products are available in the market. Another important aspect that needs to be considered is the selection of the most suitable nanoadsorbent in removing target pollutants depending on the characteristics of effluents to be treated, technical applicability, discharge standards, cost-effectiveness, regulatory requirements, and long-term environmental impacts. Although the amount of available literature data for the application of nanoadsorbents in water and wastewater treatment is increasing at a tremendous pace, there are still several gaps that need more attention, such as investigation of these materials with real industrial effluents, regeneration studies, and continuous flow studies. Despite various drawbacks and challenges that currently exist, widespread and significant progress in this area can be expected in the future.

REFERENCES

[1] WHO. *Guidelines for drinking-water quality*. Geneva, Switzerland: WHO Press; 2004.

[2] Cheremisinoff NP. *Handbook of Water and Wastewater Treatment Technologies*. Burlington: Elsevier; 2001. p 576.

[3] Zhou H, Smith DW. Advanced technologies in water and wastewater treatment. J Environ Eng Sci 2002;1:247–264.

[4] Faust SD, Aly OM. *Adsorption Process for Water Treatment*. Stoneham: Butterworths Publishers; 1987.

[5] Bansal RC, Goyal M. *Activated Carbon Adsorption*. CRC Press; 2005.

[6] Kurniawan TA, Sillanpää MET, Sillanpää M. Nanoadsorbents for remediation of aquatic environment: local and practical solutions for global water pollution problems. Crit Rev Environ Sci Technol 2011;42 (12):1233–1295.

[7] Yonzon CR, Stuart DA, Zhang X, McFarland AD, Haynes CL, Van Duyne RP. Towards advanced chemical and biological nanosensors—An overview. Talanta 2005;67 (3):438–448.

[8] Pacheco S, Medina M, Valencia F, Tapia J. Removal of inorganic mercury from polluted water using structured nanoparticles. J Environ Eng 2006;132 (3):342–349.

[9] Yang K, Xing B. Desorption of polycyclic aromatic hydrocarbons from carbon nanomaterials in water. Environ Pollut 2007;145 (2):529–537.

[10] Lu C, Chiu H. Adsorption of zinc(II) from water with purified carbon nanotubes. Chem Eng Sci 2006;61 (4):1138–1145.

[11] Lu C, Chung Y-L, Chang K-F. Adsorption of trihalomethanes from water with carbon nanotubes. Water Res 2005;39 (6):1183–1189.

[12] Gupta VK, Agarwal S, Saleh TA. Chromium removal by combining the magnetic properties of iron oxide with adsorption properties of carbon nanotubes. Water Res 2011;45 (6):2207–2212.

[13] Konicki W, Pełech I, Mijowska E, Jasińska I. Adsorption of anionic dye direct red 23 onto magnetic multi-walled carbon nanotubes-Fe_3C nanocomposite: Kinetics, equilibrium and thermodynamics. Chem Eng J 2012;210:87–95.

[14] Chatterjee S, Lee MW, Woo SH. Adsorption of congo red by chitosan hydrogel beads impregnated with carbon nanotubes. Bioresour Technol 2010;101 (6):1800–1806.

[15] Bazrafshan E, Mostafapour FK, Hosseini AR, Raksh Khorshid A, Mahvi AH. Decolorisation of reactive red 120 dye by using single-walled carbon nanotubes in aqueous solutions. J Chem 2013;2013:8.

[16] Liao Q, Sun J, Gao L. The adsorption of resorcinol from water using multi-walled carbon nanotubes. Colloids Surf A Physicochem Eng Asp 2008;312 (2–3):160–165.

[17] Chen G-C, Shan X-Q, Pei Z-G, Wang H, Zheng L-R, Zhang J, Xie Y-N. Adsorption of diuron and dichlobenil on multiwalled carbon nanotubes as affected by lead. J Hazard Mater 2011;188 (1–3):156–163.

[18] Lu C, Su F. Adsorption of natural organic matter by carbon nanotubes. Sep Purif Technol 2007;58 (1):113–121.

[19] Joseph L, Flora JRV, Park Y-G, Badawy M, Saleh H, Yoon Y. Removal of natural organic matter from potential drinking water sources by combined coagulation and adsorption using carbon nanomaterials. Sep Purif Technol 2012;95:64–72.

[20] Pan B, Lin D, Mashayekhi H, Xing B. Adsorption and hysteresis of bisphenol A and 17α-ethinyl estradiol on carbon nanomaterials. Environ Sci Technol 2008;42 (15):5480–5485.

[21] Yang K, Xing B. Adsorption of organic compounds by carbon nanomaterials in aqueous phase: Polanyi theory and its application. Chem Rev 2010;110 (10):5989–6008.

[22] Chen W, Duan L, Zhu D. Adsorption of polar and nonpolar organic chemicals to carbon nanotubes. Environ Sci Technol 2007;41 (24):8295–8300.

[23] Lu C, Su F, Hu S. Surface modification of carbon nanotubes for enhancing BTEX adsorption from aqueous solutions. Appl Surf Sci 2008;254 (21):7035–7041.

[24] Pan B, Xing B. Adsorption mechanisms of organic chemicals on carbon nanotubes. Environ Sci Technol 2008;42 (24):9005–9013.

[25] Upadhyayula VKK, Deng S, Mitchell MC, Smith GB. Application of carbon nanotube technology for removal of contaminants in drinking water: A review. Sci Total Environ 2009;408 (1):1–13.

[26] Yan XM, Shi BY, Lu JJ, Feng CH, Wang DS, Tang HX. Adsorption and desorption of atrazine on carbon nanotubes. J Colloid Interface Sci 2008;321 (1):30–38.

[27] Ji L, Chen W, Bi J, Zheng S, Xu Z, Zhu D, Alvarez PJ. Adsorption of tetracycline on single-walled and multi-walled carbon nanotubes as affected by aqueous solution chemistry. Environ Toxicol Chem 2010;29 (12):2713–2719.

[28] Hu J, Tong Z, Hu Z, Chen G, Chen T. Adsorption of roxarsone from aqueous solution by multi-walled carbon nanotubes. J Colloid Interface Sci 2012;377 (1):355–361.

[29] Upadhyayula VKK, Ruparelia JP, Agrawal A. Use of carbon nanotubes in water treatment. In: Mukhopadhyay SM, editor. *Nanoscale Multifunctional Materials*. Chichester: John Wiley & Sons, Inc.; 2011. p 321–368.

[30] Bina B, Amin M, Rashidi A, Pourzamani H. Benzene and toluene removal by carbon nanotubes from aqueous solution. Arch Environ Protect 2012;38:3.

[31] Bahgat M, Farghali AA, El Rouby W, Khedr M, Mohassab-Ahmed MY. *Adsorption of methyl green dye onto multi-walled carbon nanotubes decorated with Ni nanoferrite*. Nanoscience: *Applied*; 2012. p 1–11.

[32] Zhang Q, Pan B, Zhang S, Wang J, Zhang W, Lv L. New insights into nanocomposite adsorbents for water treatment: a case study of polystyrene-supported zirconium phosphate nanoparticles for lead removal. J Nanopart Res 2011;13 (10):5355–5364.

[33] Agrawal P, Bajpai AK. Synthesis of iron oxide based gelatin nanocomposites and their applications in removal of Cr(VI) ions from aqueous solutions. J Macromol Sci A 2010;48 (1):47–56.

[34] Zelmanov G, Semiat R. Iron (Fe^{+3}) oxide/hydroxide nanoparticles-based agglomerates suspension as adsorbent for chromium (Cr^{+6}) removal from water and recovery. Sep Purif Technol 2011;80 (2):330–337.

[35] Nieto-Delgado C, Rangel-Mendez JR. Anchorage of iron hydro(oxide) nanoparticles onto activated carbon to remove As(V) from water. Water Res 2012;46 (9):2973–2982.

[36] Xu P, Zeng GM, Huang DL, Lai C, Zhao MH, Wei Z, Li NJ, Huang C, Xie GX. Adsorption of Pb(II) by iron oxide nanoparticles immobilized Phanerochaete chrysosporium: equilibrium, kinetic, thermodynamic and mechanisms analysis. Chem Eng J 2012;203:423–431.

[37] Giri SK, Das NN, Pradhan GC. Synthesis and characterization of magnetite nanoparticles using waste iron ore tailings for adsorptive removal of dyes from aqueous solution. Colloids Surf A Physicochem Eng Asp 2011;389 (1–3):43–49.

[38] Stephen IB, Chen BH. Dye adsorption characteristics of magnetite nanoparticles coated with a biopolymer poly(γ-glutamic acid). Bioresour Technol 2011;102 (19):8868–8876.

[39] Salih HH, Patterson CL, Sorial GA, Sinha R, Krishnan R. The implication of iron oxide nanoparticles on the removal of trichloroethylene by adsorption. Chem Eng J 2012;193–194:422–428.

[40] Jegadeesan G, Al-Abed SR, Sundaram V, Choi H, Scheckel KG, Dionysiou DD. Arsenic sorption on TiO_2 nanoparticles: size and crystallinity effects. Water Res 2010;44 (3):965–973.

[41] Belessi V, Romanos G, Boukos N, Lambropoulou D, Trapalis C. Removal of reactive red 195 from aqueous solutions by adsorption on the surface of TiO_2 nanoparticles. J Hazard Mater 2009;170 (2–3):836–844.

[42] Martinson CA, Reddy KJ. Adsorption of arsenic(III) and arsenic(V) by cupric oxide nanoparticles. J Colloid Interface Sci 2009;336 (2):406–411.

[43] Goswami A, Raul PK, Purkait MK. Arsenic adsorption using copper(II) oxide nanoparticles. Chem Eng Res Des 2012;90 (9):1387–1396.

[44] Li R, Li Q, Gao S, Shang JK. Exceptional arsenic adsorption performance of hydrous cerium oxide nanoparticles: Part A. Adsorption capacity and mechanism. Chem Eng J 2012;185–186:127–135.

[45] Geng B, Jin Z, Li T, Qi X. Kinetics of hexavalent chromium removal from water by chitosan-Fe^0 nanoparticles. Chemosphere 2009;75 (6):825–830.

[46] Du WL, Xu ZR, Han XY, Xu YL, Miao ZG. Preparation, characterization and adsorption properties of chitosan nanoparticles for eosin Y as a model anionic dye. J Hazard Mater 2008;153 (1–2):152–156.

[47] Cheung WH, Szeto YS, McKay G. Enhancing the adsorption capacities of acid dyes by chitosan nano particles. Bioresour Technol 2009;100 (3):1143–1148.

[48] Jin X, Yu C, Li Y, Qi Y, Yang L, Zhao G, Hu H. Preparation of novel nano-adsorbent based on organic–inorganic hybrid and their adsorption for heavy metals and organic pollutants presented in water environment. J Hazard Mater 2011;186 (2–3):1672–1680.

[49] Chen L, Wang T-J, Wu H-X, Jin Y, Zhang Y, Dou X-M. Optimization of a Fe–Al–Ce nano-adsorbent granulation process that used spray coating in a fluidized bed for fluoride removal from drinking water. Powder Technol 2011;206 (3):291–296.

[50] Hashem FS. Adsorption of methylene blue from aqueous solutions using Fe_3O_4/bentonite nanocomposite. Open Access Sci Rep 2012;1:1–5.

[51] Yang Y, Han S, Fan Q, Ugbolue SC. Nanoclay and modified nanoclay as sorbents for anionic, cationic and nonionic dyes. Textil Res J 2005;75 (8):622–627.

[52] Liu P, Zhang L. Adsorption of dyes from aqueous solutions or suspensions with clay nano-adsorbents. Sep Purif Technol 2007;58 (1):32–39.

[53] Yaroshenko NA, Sazonova VF, Perlova OV, Perlova NA. Sorption of uranium compounds by zirconium-silica nanosorbents. Russ J Appl Chem 2012;85 (6):849–855.

[54] Li X-G, Feng H, Huang M-R. Strong adsorbability of mercury ions on aniline/sulfoanisidine copolymer nanosorbents. Chemistry—A. Euro J 2009;15 (18):4573–4581.

[55] Park HS, Lee Y-C, Choi BG, Hong WH, Yang J-W. Clean and facile solution synthesis of Iron(III)-entrapped γ-alumina nanosorbents for arsenic removal. ChemSusChem 2008;1 (4):356–362.

[56] Ouyang X, Li W, Xie S, Zhai T, Yu M, Gan J, Lu X. Hierarchical CeO_2 nanospheres as highly-efficient adsorbents for dye removal. New J Chem 2013;37 (3):585–588.

[57] Mohapatra M, Hariprasad D, Mohapatra L, Anand S, Mishra BK. Mg-doped nano ferrihydrite—A new adsorbent for fluoride removal from aqueous solutions. Appl Surf Sci 2012;258 (10):4228–4236.

[58] Nassar NN. Kinetics, mechanistic, equilibrium, and thermodynamic studies on the adsorption of acid red dye from wastewater by γ-Fe_2O_3 nanoadsorbents. Sep Sci Technol 2010;45 (8):1092–1103.

[59] Sharma YC, Srivastava V, Singh VK, Kaul SN, Weng CH. Nano-adsorbents for the removal of metallic pollutants from water and wastewater. Environ Technol 2009;30 (6):583–609.

[60] Li Y-H, Wang S, Cao A, Zhao D, Zhang X, Xu C, Luan Z, Ruan D, Liang J, Wu D, Wei B. Adsorption of fluoride from water by amorphous alumina supported on carbon nanotubes. Chem Phys Lett 2001;350 (5–6):412–416.

[61] Li Y-H, Wang S, Zhang X, Wei J, Xu C, Luan Z, Wu D. Adsorption of fluoride from water by aligned carbon nanotubes. Mater Res Bull 2003;38 (3):469–476.

[62] Wang S-G, Ma Y, Shi Y-J, Gong W-X. Defluoridation performance and mechanism of nano-scale aluminum oxide hydroxide in aqueous solution. J Chem Technol Biotechnol 2009;84 (7):1043–1050.

[63] Patel G, Pal U, Menon S. Removal of fluoride from aqueous solution by CaO nanoparticles. Sep Sci Technol 2009;44 (12):2806–2826.

[64] Zhao X, Wang J, Wu F, Wang T, Cai Y, Shi Y, Jiang G. Removal of fluoride from aqueous media by Fe_3O_4@$Al(OH)_3$ magnetic nanoparticles. J Hazard Mater 2010;173 (1–3):102–109.

[65] Bhatnagar A, Kumar E, Sillanpää M. Nitrate removal from water by nano-alumina: Characterization and sorption studies. Chem Eng J 2010;163 (3):317–323.

[66] Kumar E, Bhatnagar A, Kumar U, Sillanpää M. Defluoridation from aqueous solutions by nano-alumina: characterization and sorption studies. J Hazard Mater 2011;186 (2–3):1042–1049.

[67] Bhatnagar A, Sillanpää M. Sorption studies of bromate removal from water by nano–Al_2O_3. Sep Sci Technol 2011;47 (1):89–95.

[68] Tuutijärvi T, Lu J, Sillanpää M, Chen G. As(V) adsorption on maghemite nanoparticles. J Hazard Mater 2009;166 (2–3):1415–1420.

[69] Tuutijärvi T, Lu J, Sillanpää M, Chen G. Adsorption mechanism of arsenate on crystal nanoparticles. J Environ Eng 2010;136 (9):897–905.

[70] Tuutijärvi T, Repo E, Vahala R, Sillanpää M, Chen G. Effect of competing anions on arsenate adsorption onto maghemite nanoparticles. Chin J Chem Eng 2012;20 (3):505–514.

[71] Afkhami A, Saber-Tehrani M, Bagheri H. Modified maghemite nanoparticles as an efficient adsorbent for removing some cationic dyes from aqueous solution. Desalination 2010;263 (1–3):240–248.

[72] Afkhami A, Moosavi R. Adsorptive removal of Congo red, a carcinogenic textile dye, from aqueous solutions by maghemite nanoparticles. J Hazard Mater 2010;174 (1–3):398–403.

[73] Yuan G. Natural and modified nanomaterials as sorbents of environmental contaminants. J Environ Sci Health A 2004;39 (10):2661–2670.

[74] Camtakan Z, Erenturk S, Yusan S. Magnesium oxide nanoparticles: Preparation, characterization, and uranium sorption properties. Environ Progr Sustain Energy 2012;31 (4):536–543.

[75] Giraldo L, Erto A, Moreno-Piraján J. Magnetite nanoparticles for removal of heavy metals from aqueous solutions: synthesis and characterization. Adsorption 2013:1–10.

16

ORGANO-CLAY NANOHYBRID ADSORBENTS IN THE REMOVAL OF TOXIC METAL IONS

PENG LIU

State Key Laboratory of Applied Organic Chemistry and Key Laboratory of Nonferrous Metal Chemistry and Resources Utilization of Gansu Province, College of Chemistry and Chemical Engineering, Lanzhou University, Lanzhou, China

16.1 INTRODUCTION

Toxic heavy metals in air, soil, and water lead to global ecological problems that are a growing threat to humanity [1]. Contamination of water by toxic heavy metals is a major environmental problem now. Heavy metals discharged in water bodies through wastes also affect aquatic life and destroy their self-purification power [2]. In view of these facts, it is important to prevent water pollution caused by heavy metals. Research is now focused on developing suitable clean and green technologies either to prevent heavy metal pollution or to reduce it to very low levels. This can be achieved either by decreasing the afflux of heavy metals to the receiving bodies (rivers, sewer and lake, etc.) or by their removal from contaminated media. The heavy metal pollution arising from anthropogenic activities can be prevented, while that of natural origin is unavoidable.

Various treatment technologies have been developed for the removal of heavy metals from water and wastewater. The most widely used conventional methods for removing heavy metals from wastewater include ion exchange, chemical precipitation, reverse osmosis, evaporation, and membrane filtration. The type of treatment technique required for a particular industry depends upon the nature, composition, and flow rate of the effluent together with the quality control needed to be achieved. The effectiveness of the treatment plant can be optimized by adopting any one of the techniques discussed or the appropriate combination of two or more techniques. But most of these methods suffer from some drawbacks, such as high capital and operational cost or the disposal of the residual metal sludge, and are not suitable for small-scale industries [3]. Besides high capital and operational cost, most of the conventional methods are often ineffective or uneconomical when the heavy metal concentration is in the range of 10–100 mg/l [4].

Adsorption has been found to be an efficient and economic process to remove heavy metals from industrial wastewater. The process is suitable even when the metal ions are present in concentration as low as 1 mg/l [5]. Activated carbon (powdered or granular) [6] and polymer ion exchangers and adsorbents [7] are the most widely used adsorbents in the removal of heavy metals in water. However, the cost of the adsorbent becomes relatively high; therefore there is an increasing trend for substituting these sorbents with nonconventional adsorbents [8] (such as natural by-products [9], fly ash [10], sand [11], and so on) in order to make the process economically feasible.

Clays are hydrous aluminosilicates broadly defined as those minerals that make up the colloid fraction (<2 μm) of soils, sediments, rocks, and water [12] and may be composed of mixtures of fine-grained clay minerals and clay-sized crystals of other minerals such as quartz, carbonate, and metal oxides. Clays play an important role in the environment by acting as a natural scavenger of pollutants by taking up cations and anions either through ion exchange or through adsorption or both. Thus, clays invariably contain exchangeable cations and anions held to the surface. Large specific surface area, chemical and mechanical

Nanomaterials for Environmental Protection, First Edition. Edited by Boris I. Kharisov, Oxana V. Kharissova, and H. V. Rasika Dias.

stability, layered structure, high cation exchange capacity (CEC), and so on, have made clays excellent adsorbent materials toward heavy metal ions (HMI) [13], dyes [14], pesticides [15], and other organic matters [16].

The prominent cations and anions found on the clay surface are Ca^{2+}, Mg^{2+}, H^+, K^+, NH_4^+, Na^+, and SO_4^{2-}, Cl^-, PO_4^{3-}, and NO_3^-. These ions can be exchanged with other ions relatively easily without affecting the clay mineral structure. The edges and the faces of clay particles can adsorb anions, cations, and nonionic and polar contaminants from natural water. The contaminants accumulate on the clay surface leading to their immobilization through the processes of ion exchange, coordination, or ion–dipole interactions. Sometimes the pollutants can be held through H bonding, van der Waals interactions, or hydrophobic bonding arising from either strong or weak interactions. The strength of the interactions is determined by various structural and other features of the clay mineral [13].

There have been various attempts to improve the quality and characteristics of clays by modifying them using different techniques, such as acid activation [17], which can improve their exchangeability, or surface modifications with inorganic materials such as magnetic nanoparticles [18–20] or organic molecules [21–26]. However, the adsorption selectivity of clay adsorbents seems to be low due to the ion-exchanging interaction between pristine clays, acid-treated clays, or the magnetic clay nanocomposites and HMI. On the contrary, the functional groups introduced via surface modifications with organic molecules can obviously improve their adsorption selectivity toward HMI [13, 25]; therefore, the surface organo-modification of clay minerals has become increasingly important for improving the practical applications of clays and clay minerals.

In the present chapter, recent advances in the preparation of the organo-clay nanohybrids via surface modification with small organic molecules or polymers, as well as their adsorption performance toward the toxic HMI, are reviewed. The surface modification techniques and the adsorption selectivity of the resultant organo-clay nanohybrid adsorbents are emphasized. And the practical application prospects of the organo-clay nanohybrid adsorbents are also prospected.

16.2 ORGANO-MODIFICATION WITH SMALL MOLECULES

16.2.1 Electrostatic Interactions or Intermolecular Forces

Due to high CEC and exchangeable cations, ion exchange with organic cations has been used to modify the surface properties of clay minerals. Organic cation intercalation or modification plays an important role in clay/polymer nanocomposite formation by providing a hydrophobic environment for the intragallery adsorption of the polymer precursor [27].

16.2.1.1 Quaternary Ammonium Salts or Amines Wang et al. developed a simple two-step approach to design several modified clays for selective removal and recovery of heavy metals via ion exchange and hydrophobic anchoring of several surfactants such as long-chain alkyldiamines (cetylbenzyldimethylammonium, CBDA), long-chain dialkylamines (alkyl-1,3-diaminopropane, DT), and long-chain carboxylic acids (palmitic acid, PA) onto clay matrices (hectorite (Hect.) or montmorillonite (MMT)) [28]. The adsorption capacities and affinity constants of the organo-modified clays were found to approach those of the commercial chelating resin (Chelex 100, Bio-Rad). Adsorption had been shown to be mainly through chemical complexation rather than ion exchange. The maximum adsorption capacity of the modified clay toward Cd(II) ion was found to be 42 ± 0.8 mg/g clay with affinity constant of 3.0 ± 0.1 mg/l. The adsorption of the metal ions was pH-dependent, so pH can act as a molecular switch to regenerate the modified clay complex adsorbents.

Bhattacharyya et al. prepared the organo-clay adsorbent for Cd(II) [29] and Cu(II) ions [30] via the immobilization of tetrabutylammonium bromide (TBA) onto the surfaces of kaolin and MMT. After calcination of the TBA-modified MMT (TBA–MMT), it also had a higher adsorption capacity than the corresponding clay. The calcined TBA–MMT had only 44, 41, and 42% of the adsorption capacity of the parent MMT with respect to Fe(III), Co(II), and Ni(II), indicating an actual decrease in the number of adsorption sites for taking up metal ions [31].

2-Oxyhydrazino-*N*-(2-methylen-yl-hydroxyphenyl)pyridinium (OMHP) ion was immobilized onto Na–MMT clay, and the modified clay (OMHP–MMT) was used in the removal of Cu(II) [32]. It showed good removal efficiency and selectivity toward Cu(II) at pH 3.0–8.0 and stirring time 10 min with a removal capacity of 119 mEq/100 g. Most common ions did not interfere with the removal process except Fe^{3+}. And the adsorbed Cu(II) could be quantitatively recovered by 10 ml 1% thiourea in 0.1 mol l^{-1} HCl with 100-fold preconcentration factor. The same group had also modified bentonite (BNT) with methylene blue (MB) [33]. The modified clay (MB–BNT) showed good selectivity toward Hg^{2+} with an extraction capacity of 37 mequiv./100 g in the presence of I^-, giving rise to a ratio of MB/Hg^{2+}/I^- 1:1:3 in the clay phase. And the adsorbed Hg^{2+} could be quantitatively recovered by ammonia buffer (0.05 M NH_4Cl/NH_4OH) at pH 8.5.

The unmodified clay minerals show no affinity for chromate due to their negatively charged surfaces. Prakash et al. had modified the surfaces of the clay mineral such as kaolinite (kaolin), MMT, and pillared MMT with hexadecyltrimethylammonium bromide (HDTMAB) for the removal of the chromate [34]. It was found that MMT could adsorb a quantity of HDTMAB equivalent to its CEC. The amount of the modified MMT adsorbed was maximum at pH ≤ 1 and decreased to almost half the quantity at pH 2, and remained constant up to pH = 6. Above pH = 8, the amount adsorbed was negligible. And the pH dependence of chromate adsorption was attributed to the pH-dependent equilibria $HCr_2O_7^- \leftrightharpoons Cr_2O_7^{2-} \leftrightharpoons CrO_4^{2-}$. HDTMA was also used for the modification of various minerals for the removal of chromate. The optimum pH for Cr(VI) adsorption was 4 in all the examined cases. The maximum adsorption capacity followed the order vermiculite (27 mg/g) > BNT (24 mg/g) > attapulgite (ATP) (15 mg/g) > zeolite (13 mg/g) [35].

With regard to the Cr(VI) ion, several works were reported to compare the adsorption property of the clay minerals after different treatments. It was found that the adsorption capacity of organo-modified clay was higher than that of acid- and heat-treated ones [36, 37]. It indicated that organo-modified clay could be used as the adsorbent for Cr(VI) removal.

Natural rectorite (REC) was modified with the surfactant of dodecyl benzyl dimethyl ammonium chloride, HDTMAB, and octadecyl trimethyl ammonium bromide via the cation exchange reaction. The three kinds of organic modified rectorites (OREC) were used as adsorbents for Cr(VI) removal in aqueous solution [38]. They showed better adsorption of Cr(VI) at pH range of 2–6, with maximum adsorption near 100% with the octadecyl trimethyl ammonium bromide–modified REC within 40 min. And its maximum adsorption capacity (q_m) was calculated to be 3.57 mg/g.

Stearyl trimethylammonium chloride (STAC) was used for the modification of REC via a similar procedure [39]. The adsorption capacity of the modified REC (STAC–REC) reached as high as 21 g/kg for Cr(VI), or 400 mmol/kg of chromate. Higher Cr(VI) adsorption on the STAC–REC occurred in acidic solutions (pH 4), and the amount of adsorbed Cr(VI) decreased rapidly with increasing pH. So the formation of the Cr(VI)–STAC complex was inhibited during adsorption.

The halloysite nanotubes (HNTs) were modified with the surfactant of HDTMAB to form a new adsorbent. The modified HNTs exhibited a rapid adsorption rate for chromates and approached 90% of the maximum adsorption capacity within 5 min [40]. The regeneration of modified HNTs could be realized by an eluent (mixed solution of $NaNO_3$ and NaOH), and the recovered adsorbent could be used again for Cr(VI) removal.

Naidu et al. modified BNT with a commercially available, low-cost alkyl ammonium surfactant Arquad® 2HT-75 (Aq) or adsorbing Cr (VI) from aqueous solution [41]. It was found that more ordered solid-like conformation of surfactant molecules was obtained in the organo-clay as the surfactant loading increased. And higher surfactant loadings provide better adsorption efficiency.

Two hydrous phyllosilicates, a MMT, and a vermiculite were organo-modified with hexadecylpyridinium (HDPy), HDTMA, and benzethonium (BE), and the Cr(VI) adsorption properties were investigated [42]. After the organic cation treatment, the surface charge of the clays shifted from negative to positive values up to 269 $mmol_c$/kg; it was highly affected by the specific surface area, layer charge, and pH. Depending on the kinds of the organic cations used, the adsorption capacity of Cr(VI) was shown to have a comparable effect of 0.38 and 0.30 mol/kg for organo-vermiculites HDPy and HDTMA and 0.37 and 0.26 mol/kg for the corresponding BNTs at pH 4.5, respectively. For all organo-clays, the adsorbed amount of Cr(VI) decreased with increasing pH. The effect of competing anions, Cl^-, NO_3^-, and SO_4^{2-}, on Cr(VI) adsorption appeared to be small.

Atia modified BNT with cetylpyridinium bromide (CPBr) for the removal of the oxyanions of Cr(VI) and Mo(VI) from aqueous solutions [43]. It was found that the surfactant was adsorbed on MMT in a bilayer structure, and the adsorption was an ion exchange mechanism.

The cationic surfactant octadecyl benzyl dimethyl ammonium was used for the modification of BNT to remove As(V) and As(III) from aqueous solution [44]. Its adsorption capacities were 0.288 mg/g for As(V) and 0.102 mg/g for As(III), which were much higher than 0.043 and 0.036 mg/g for the unmodified BNT. The adsorption of As(V) and As(III) was strongly dependent on solution pH. Addition of anions did not impact As(III) adsorption, while they clearly suppressed adsorption of As(V).

Majdan et al. modified BNT with octadecyltrimethylammonium bromide (ODTMA) for the sorption of U(VI) [45]. It was reported that ODTMA–BNT exhibited high sorption affinity for U(VI) ions in a pH range of 6–10, and the most probable anionic complex was the $(UO_2)_3(OH)_7^-$ ions.

Besides the quaternary ammonium salts mentioned earlier, alkylamine, diamines, and the amphoteric or other surfactants containing an amino group have also been used for the organo-modification of clay minerals as adsorbents for HMI. Guerra et al. modified the BNT via an intercalation process with polar n-alkylamine molecules of general formula $H_3C(CH_2)_n$–NH_2 (n = 1–4) for mercury removal from water [46]. It was found that the Hg(II)–nitrogen interactions was favorable for the adsorption of Hg(II).

Dodecylamine-intercalated Na–MMT was prepared as a novel adsorbent for the removal of chromium from tannery wastewater [47]. The adsorption involved the electrostatic interaction of hydrogentetraoxochromate(VI) anion with the protonated dodecylamine and the surface hydroxyl groups of the clay material with a adsorption capacity of 23.69 mg/g. The adsorbent could be regenerated using NaOH solution.

In order to increase the adsorbed amounts of Zn^{2+} and Cu^{2+}, MMT was previously intercalated with ethylenediamine (EDA-MMT) [48]. EDA caused an increase in the adsorbed copper ions of about 28% due to the formation of $[Cu(EDA)_2]^{2+}$ in the interlayer. However, the adsorbed amounts of the Zn^{2+} ions on the EDA-MMT was three- to fourfold higher than CEC, but no complex with EDA was observed. It was resulted from the expansion of a space between the MMT layers, which facilitates the intercalation of the zinc ions and nitrate compensating the excessive positive charge.

Amphoteric surfactants-modified clay exhibited the potential for simultaneously adsorbing heavy metals and organic pollutants. Boyd et al. used the carboxylic group–bearing carboxydecyltriethylammonium $[(HOOC)C_{10}H_{20}N(C_2H_5)_3]^+$ (CDTEA) cations to substitute exchangeable inorganic cations of the MMT [49]. The dual sorptive properties of the CDTEA–MMT toward Pb^{2+} and chlorobenzene were compared with those of its nonfunctionalized analogue, decyltrimethylammonium (DTMA)–MMT, and with Na–MMT. The results indicated that CDTEA–MMT possessed the dual sorptive properties for both heavy metals and organic contaminants with the noncompetitive nature of the sorption processes.

An amphoteric biologically based ligand, cysteine (Cys), was also used for the modification of the BNT (N-Ben). The modified sorbent (Cys-Ben) demonstrated affinity for soft and moderately soft HMI, such as Cd(II) and Pb(II), probably as a result of the soft basic character of the thiol ligand side chains [50]. Cys-Ben was found to be effective for metal binding with capacities of 0.503 and 0.525 mmol/g for Pb(II) and Cd(II), respectively, much higher than that of N-Ben.

2-(3-(2-aminoethylthio)propylthio)ethanamine (AEPE) was used as the ligand for Hect. and MMT to increase the chelating affinity toward Hg(II) ions, due to the presence of both N and S donor atoms in the molecule and the ease of synthesis [51]. The X-ray diffraction (XRD) patterns indicated that AEPE was mainly grafted on the external surface of MMT, while AEPE was grafted on both the external and interlayer surfaces of Hect. The AEPE-modified clay minerals were good chelating materials for Hg(II) ions, compared to the unmodified clay minerals. The adsorption capacity of AEPE–MMT and AEPE–Hect. for Hg(II) was 46.1 and 54.7 mg/g, respectively, for solution containing 140 mg/l Hg(II) ions (pH 4).

16.2.1.2 Other N-Containing Molecules The organic chelating agents for HMI have also been immobilized onto the clay minerals as adsorbents. De Leon et al. revealed the potential of pillared clay as an adsorbent for the removal of metal ions from effluents [52]. They pillared a Brazilian BNT with 1.10-phenanthroline via the chemical adsorption process for the removal of Cu^{2+} ions. The pillared clay showed a maximum uptake of Cu^{2+} ions at approximately 110 mg/g of adsorbent, which represented more than 10 times the uptake capacity of the untreated BNT. This Cu^{2+} ion adsorption was independent of medium pH, and it was very high when compared with other known adsorbents. However, the adsorbed Cu^{2+} ions could not be desorbed from the pillared clay in either acidic or basic solutions.

2,2′-Dipyridyl was used as a complexation agent to be immobilized onto BNT via ion exchange as an adsorbent in the removal of copper(II) ions [53]. The results indicated that the maximum adsorption capacity was 54.07 mg/g from the Langmuir isotherm model at 50°C. The thermodynamic parameters indicated that the adsorption process is spontaneous, endothermic, and chemical in nature.

8-Hydroxy quinolin was modified onto BNT by intercalation of 8-hydroxy quinolinium ion for the removal of Pb(II) [54], Cd(II) [55], and Cu(II) [56] ions. The maximum adsorption capacity was 142.94, 61.35, and 56.55 mg/g, respectively, for the three HMI from the Langmuir isotherm model.

In order to enhance the adsorption selectivity toward Hg(II) ions, some sulfur-containing molecules such as 2-mercaptobenzothiazole [57] and 2-mercaptobenzimidazole (MBI) [58] were impregnated onto the clay surface for the removal of some HMI from water samples. The adsorption of Hg(II) increased with increasing pH and reached a plateau in the pH range 4.0–8.0. The removal of Hg(II) was found to be >99% at an initial concentration of 50 mg/l. The adsorbents could also be used to remove Cd(II) and Pb(II) from wastewaters.

16.2.1.3 Anionic Surfactants Ion exchange with cationic surfactants has been widely used to alter the surface properties of the clay minerals in order to improve sorption ability. Relatively few studies have been made for sorption of heavy metal cations on clays modified with anionic surfactants such as sodium dodecyl sulfate (SDS). The sorption of metal ions by naturally available clays did not receive much attention, perhaps due to the weak binding strength between them. Since BNT clays have a tendency to swell, large anionic species such as SDS can easily enter and become fixed strongly in the interlayer region of, for example, MMT [59]. In order to adsorb metal cations and form complexes, the clay surface must possess negatively charged sites or there should be a replacement of weakly held counterions in the solution. Sorption of metal cations on anionic surfactant-modified clays is due to the formation of a surface cation complex [60]. Lin and Juang modified the naturally available MMT by the anionic surfactant SDS, which penetrated into the interlamellar region of the clay by the expansion of clay sheets in the c-axis, for the removal of Cu^{2+} and Zn^{2+} ions from aqueous solutions [61].

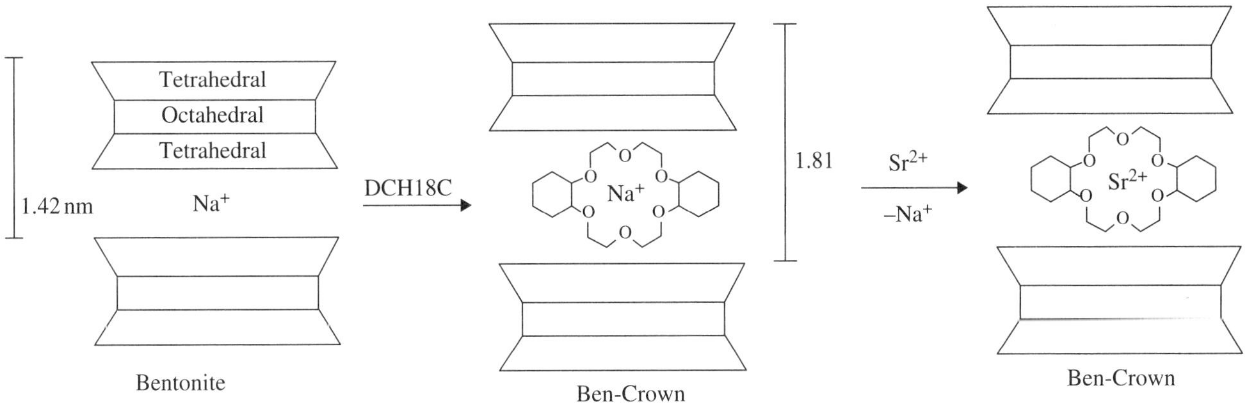

FIGURE 16.1 Intercalation of crown ether in bentonite (BNT), leading to propping of the layers and the ion exchange behavior of Ben-Crown.

Humic acid–modified Ca–MMT was prepared to adsorb copper (II), cadmium (II), and chromium (III) ions from aqueous solutions [62]. Batch experiments were carried out to investigate the possible adsorption mechanisms of the metal ions on humic acid–modified Ca–MMT. The results showed that the adsorption capacities of the modified clay for the metal ions were improved significantly as compared to that of the raw clay. The maximum adsorption capacities followed the order of Cr(III) > Cu(II) > Cd(II) for both materials.

16.2.1.4 Neutral Molecules Crown ether is known to complex with numerous metals and is widely used for the adsorption of HMI. They can also be intercalated between the layers of 2:1 BNT saturated with alkaline or alkaline earth cations. Dicyclohexano-18-crown-6 was modified to BNT for the adsorption of cesium and strontium (Fig. 16.1) [63]. The extraction data were fitted by the Langmuir adsorption model, and the apparent experimental exchange capacity obtained by linear regression analysis was in good agreement with the amount of crown ether (0.22 mmol/g) intercalated in BNT.

Liu et al. modified BNT with 4′-methylbenzo-15-crown-5 for the adsorption of HMI (Cu^{2+} and Pb^{2+}) [64]. For BNT modified with MB15C5 (MB15C5-BNT) and natural BNT (N-BNT), the equilibrium data closely fitted the Langmuir model and showed the affinity order $Pb^{2+} > Cu^{2+}$, and the adsorption capacity of MB15C5-BNT was higher than that of N-BNT for Pb^{2+} and Cu^{2+}. The adsorption of Cu^{2+} and Pb^{2+} increases with increasing pH, and the adsorption of Cu^{2+} and Pb^{2+} reached a maximum at pH 3.5–6.

16.2.2 Covalent Modifications

The most widely used surface covalent organo-modification method is to modify the clay with functional silanes. Celis et al. functionalized sepiolite by covalently grafting 3-mercaptopropyltrimethoxysilane (MPS) to the surface silanol groups of the clay, whereas MMT was functionalized by replacement of the interlayer inorganic cation (Na^+) by 2-mercaptoethylammonium (MEAm) cations [65]. The functionalized clays adsorbed most of the Hg(II) ions present in solution up to saturation and were also good adsorbents of Pb(II) at low metal ion concentrations (i.e., <0.02 mM). They were, however, less effective toward Pb(II) and Zn(II) at high metal ion concentrations. The presence of $NaNO_3$ or $Ca(NO_3)_2$ as background electrolytes at concentrations ranging from 0.001 to 0.1 M did not alter the great adsorption capacity of functionalized sepiolite for Hg(II).

The BNT from Campina Grande (Paraíba, PB), Brazil, was modified by acid treatment followed by immobilization of ligands containing thiol (–SH) groups by covalent grafting with surface and interlayer silanol groups with MPS under anhydrous conditions (Fig. 16.2) [66]. The unmodified BNT adsorbed silver ions in negligible amounts (0.06–0.08 mmol/g of freeze-dried BNT) due to the cation exchange mechanism. The functionalized samples revealed a high affinity toward Ag^+ (0.85–1.03 mmol/g of clay), about 10 times higher than the unmodified one. The results suggested that the mechanism of adsorption involves primarily silver ion complexation by the thiol groups instead of cation exchange.

Natural and acid-activated sepiolite samples were functionalized with MPS, and the functionalization of sepiolites by grafting silane reagents occurs mainly on the surface, whereby their crystalline structure remains unchanged [67]. The efficiency of the adsorbents in Cr(VI) removal from aqueous solutions follows the order: functionalized acid-activated sepiolite > functionalized natural sepiolite > acid-activated sepiolite > natural sepiolite. The adsorption capacity was strongly dependent on the pH of the solution from which the adsorption occurred. Maximum Cr(VI) removal was ca. 8.0 mg Cr(VI) per gram of functionalized

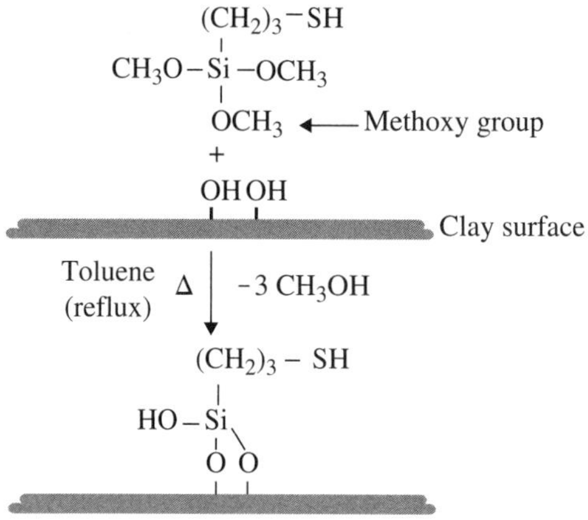

FIGURE 16.2 Schematic illustration of the functionalization with functional silane.

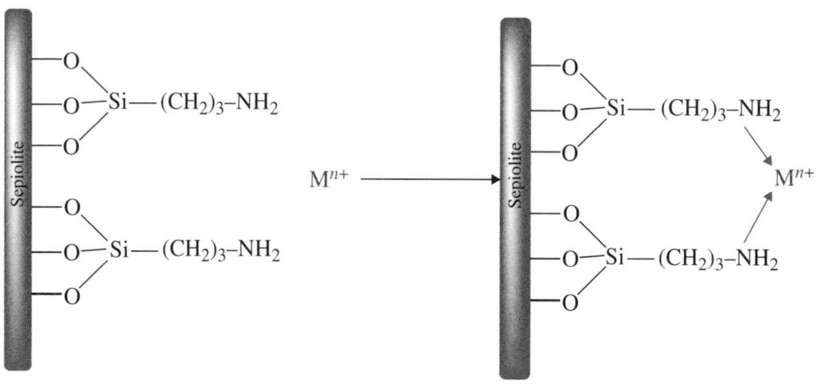

FIGURE 16.3 Reaction mechanism between the modified sepiolite and the metal ions.

acid-activated sepiolite at initial pH 3, when the pH of solution reached 4.7, and ca. 2.7 mg Cr(VI) per gram of functionalized natural sepiolite at initial pH 2, when the pH of the solution reached 2.5. Cr(VI) removal by the functionalized sepiolites was mainly attributed to electrostatic attraction between the protonated mercapto groups and the negatively charged Cr(VI) species.

Besides the functional silane-containing thiol (–SH) groups, those containing amino ($-NH_2$) groups were widely investigated as surface modifiers. 3-Aminopropyltriethoxysilane (APTS) was used for the surface modification of sepiolite as an adsorbent in the removal of various HMI such as Fe, Mn, Co, Zn, Cu, Cd, and Ni from aqueous solutions (Fig. 16.3) [68]. The results showed that the amount adsorbed increases with solution pH in the pH range of 1.5 and 7.0, indicating that the modified sepiolite adsorbed Fe and Mn ions more than other metal ions such as Co, Zn, Cu, Cd, and Ni. It was also found that temperature had an important effect on metal ion adsorption by the modified sepiolite.

Smectite clay was used for pillaring and the organo-functionalization process in the present study. The pillaring ions were obtained through chemical reaction of $AlCl_3{\cdot}6H_2O$/NaOH, $Ti(OC_2H_5)_4$/HCl, $Cr_3(OH)_4^{4+}$/NaOH and $ZrOCl_2{\cdot}8H_2O$ solutions. The resulting materials were organo-functionalized with APTS to obtain the $S_{Al/APS}$, $S_{Zr/APS}$, $S_{Ti/APS}$, and $S_{Cr/APS}$ samples [69]. The adsorption affinity of the divalent metal Pb^{2+} ions onto modified clays edge sites exhibited the following order:

$$S_{Zr/ASP} > S_{Al/APS} > S_{Cr/APS} > S_{Ti/APS} > S_{nat}$$

MMT clay was intercalated with pyridine/APTS and dimethyl sulfoxide/3-aminopropyltri-ethoxysilane for mercury cation adsorption from aqueous solutions [70]. The specific area of the natural MMT, 41.4 m^2/g, increased after the immobilization process to give 198.5 m^2/g. Therefore, the adsorption capacity for mercury removal increased after the chemical modification.

Three organo-MMT, APTS–MMT, SDS–MMT, and HDTMAB–MMT, were prepared by grafting with 3-aminopropyltriethoxysilane (APTS), or surface modification with HDTMAB and SDS from the calcium-saturated MMT (Ca–MMT), respectively [71]. The organo-MMT were used for the adsorption of Sr(II). The adsorption capacity of APTS–MMT was 65.6 mg/g, much larger than those of Ca–MMT, SDS–MMT, and HDTMAB–MMT (13.23, 26.85, and 3.91 mg/g, respectively) under the chosen conditions. Sr(II) removal by Ca–MMT was mainly by ion exchange; nevertheless, removal by SDS–MT and APTS–MMT

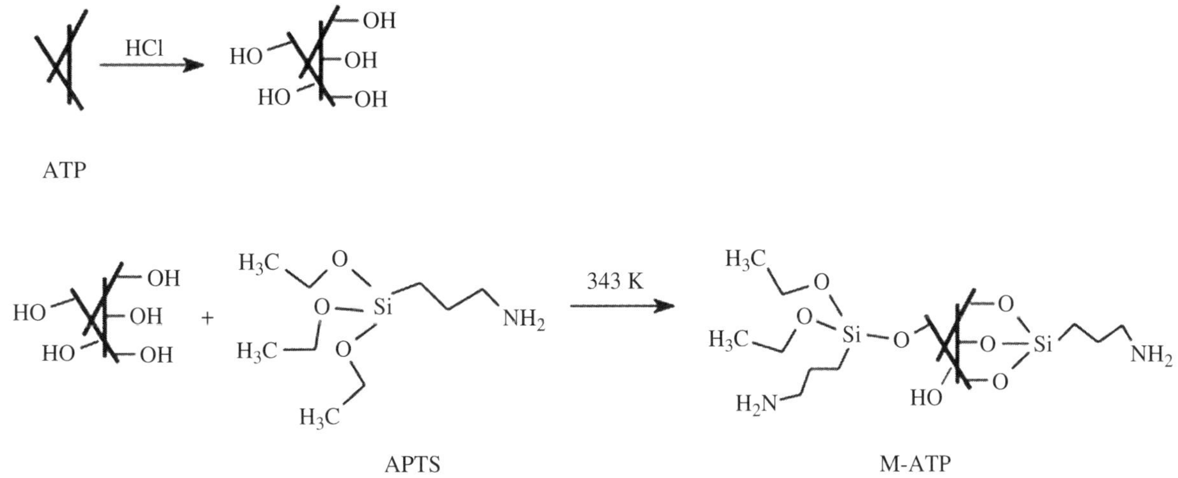

FIGURE 16.4 Schematic illustration of the surface modification of attapulgite (ATP) by 3-aminopropyltriethoxysilane (APTS).

were through surface adsorption and ligand adsorption, respectively. The organo-Mts were used for the adsorption of Sr(II). The adsorption capacity of APTS–MT was 65.6 mg/g, much larger than those of Ca–MT, SDS–MT, and HDTMAB–MT (13.23, 26.85, and 3.91 mg/g, respectively) under the chosen conditions. Sr(II) removal by Ca–MT was mainly by ion exchange; nevertheless, removal by SDS–MT and APTS–MT were through surface adsorption and ligand adsorption, respectively.

APTS was also used for the surface modification of the natural fibrous crystalloid hydrous magnesium aluminum silicate mineral ATP after being activated with acid, in order to develop an effective adsorbent for aqueous Hg(II) removal (Fig. 16.4) [72]. The adsorption capacities significantly increased from 5 (raw ATP) to 90 mg/g modified ATP (M-ATP), due to the complexation between mercury and the amine functional groups on the M-ATP. Batch adsorption results showed that the adsorption process was rapid and over 90% of aqueous Hg(II) was removed within 1 h. The efficiency of the adsorbent was found to remain almost constant over a wide pH range (3–11). Ionic strength and coexisting ions had a slight influence on the adsorption capacity. Hg(II) adsorbed onto M-ATP could be effectively desorbed in 1:1 (m/m) chlorohydric acid/thiourea solution.

The surface of pyrophyllite mineral was modified by coating with a 3-(2-aminoethylamino) propyl-methyldimethoxysilane (APMDS) coupling agent and the utility of the APMDS-modified pyrophyllite was investigated as an adsorbent for removal of Pb(II) ions from aqueous solutions [73]. The APMDS modified pyrophyllite adsorbed approximately 93% of Pb(II) ions at an initial concentration of 20 mg/l, while natural pyrophyllite did only 35% under the same conditions. The adsorption of Pb(II) increased with increasing pH and reached a maximum value in the pH range 6.5–7 and at an initial concentration of 20 mg/l. The presence of some metal ions such as Na^+, K^+, Ca^{2+}, Mg^{2+}, Fe^{3+}, Cu^{2+}, Mn^{2+}, and Zn^{2+} and the ligand type of lead complex added to the solution (e.g., $Pb(OOC\text{-}CH_3)_2$) negatively affected the adsorption of Pb(II) from solutions under the optimized conditions. It was suggested that the functional amino groups on the surface of APMDS-modified pyrophyllite are the main available interaction sites for Pb(II) ions.

The surface modification of sepiolite with [3-(2-aminoethylamino)propyl]trimethoxysilane has been employed for the adsorption of metal ions [74]. Adsorption of metal ions onto the modified sepiolite varies with the type of metal cations. The available basic nitrogen centers covalently bonded to the sepiolite skeleton were studied for Co(II), Cu(II), Mn(II), Zn(II), Fe(III), and Cd(II) adsorption from aqueous solutions. It was found that the amount of metal ion adsorbed onto the modified sepiolite increases with an increase in solution equilibrium pH and temperature, whereas it generally decreases with the ionic strength. The experimental data correlated reasonably well with the Langmuir adsorption isotherm. The ability to adsorb the cations gave a capacity order of Zn(II) > Cu(II) similar to Co(II) > Fe(III) > Mn(II) > Cd(II) with affinities of 2.167×10^{-4}, 1.870×10^{-4}, 1.865×10^{-4}, 1.193×10^{-4}, 0.979×10^{-4}, and 0.445×10^{-4} mol/g, respectively.

Triethoxy-3-(2-imidazolin-1-yl)propylsilane was modified onto sepiolite clay for the adsorption of HMI (Fig. 16.5) [75]. The adsorption of metal ions onto modified sepiolite as an adsorbent increased with increasing pH and temperature, but ionic strength exhibited no significant effect. The affinity of metal ions for the modified sepiolite surface is a result of both chemical and electrostatic affinity. The adsorption followed the sequence Co(II) > Cu(II) > Cd(II) > Mn(II) > Fe(III) > Zn(II).

The compound N-[3-(trimethoxysilyl)propyl]diethylenetriamine (MPDET) was anchored onto the Amazon kaolinite surface (KLT) by the heterogeneous route. The adsorption of uranyl on natural (KLT) and modified (KLT(MPDET)) kaolinite clays was investigated as a function of the solution pH, metal concentration, temperature, and ionic strength [76]. The ability of these materials to remove U(VI) from aqueous solution was followed by a series of adsorption isotherms adjusted to a Sips equation at room temperature and pH 4.0. The maximum number of moles adsorbed was determined to be 8.37×10^{-3} and 13.87×10^{-3} mmol/g for KLT and KLT(MPDET) at 298 K, respectively.

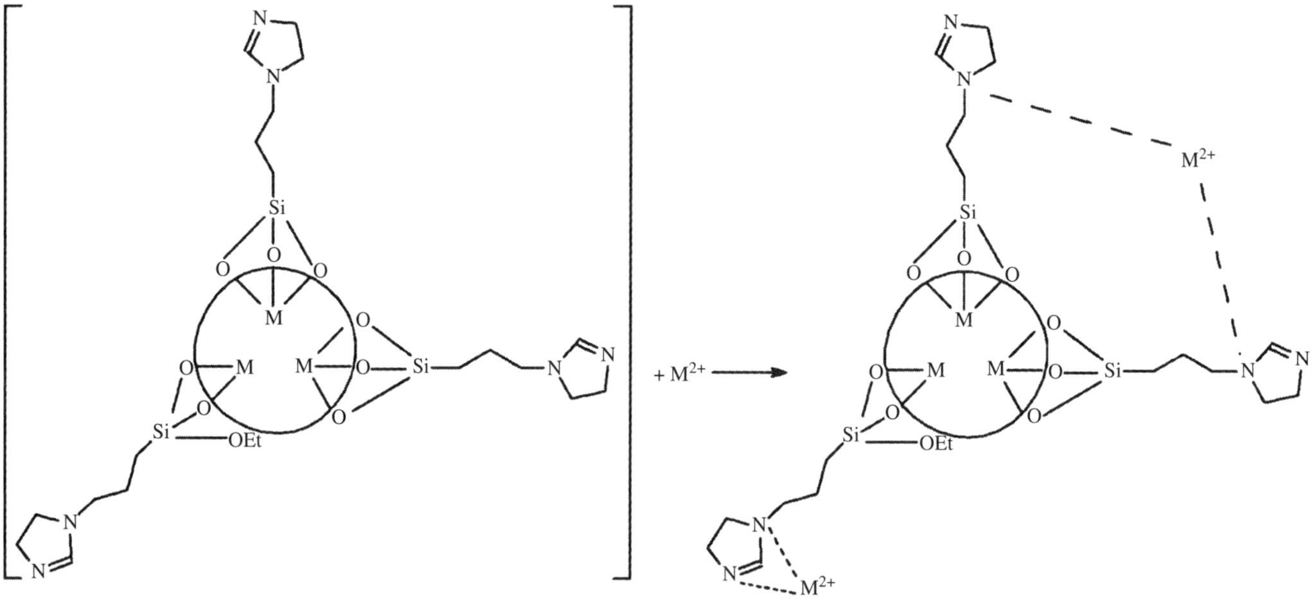

FIGURE 16.5 Reaction between the modified sepiolite and the metal ions.

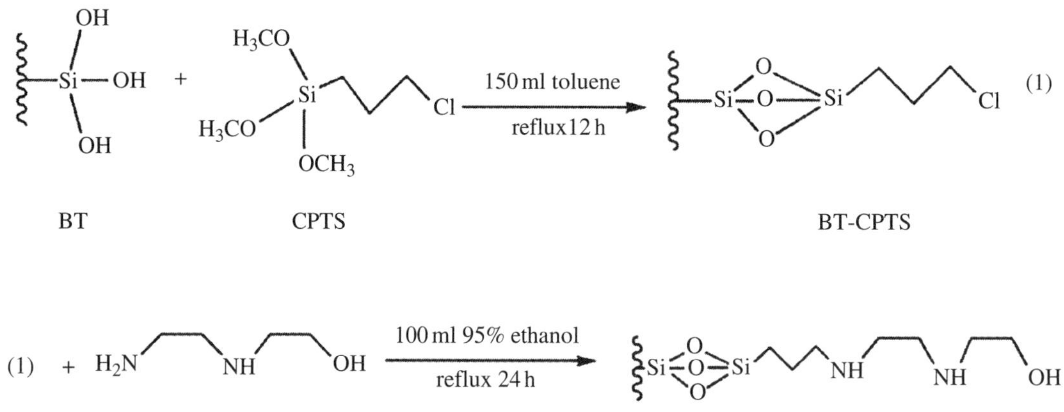

FIGURE 16.6 Synthesis route of the BT-NHED adsorbent.

A new material has been synthesized using dry process to activate bentonite followed by *N*-(2-hydroxyethyl) ethylenediamine (NHED) connecting chlorosilane coupling agent (Fig. 16.6) [77]. The most interesting trait of the new material was its selective adsorption of rare earth elements. Samarium (Sm) was quantitatively adsorbed at pH 4 and shaking time of 2 min onto the new material. Under these conditions, the maximum static adsorption capacity of Sm(III) was found to be 17.7 mg/g. The adsorbed Sm(III) ions were quantitatively eluted by 2.0 ml, 0.1 mol/l HCl, and 5% CS $(NH_2)_2$ solution.

An efficient adsorbent was synthesized by anchoring N-methylimidazole onto activated palygorskite (Fig. 16.7) [78]. The N-methylimidazole-modified palygorskite was examined as a strongly basic anion exchange adsorbent for the adsorption of Pb(II) ions from wastewater. The maximum adsorption capacity was found to be 714.29 mg/g at pH 4 and temperature 283 K. The maximum adsorption capacity of functional palygorskite obtained for Pb(II) ions in this study was found to be comparable and the highest for all of corresponding clay-related adsorbents reported in the literature.

Hectorite clay has been modified with MBI using homogeneous and heterogeneous routes (Fig. 16.8) [79]. Due to the increase in basic centers attached to the pendant chains, the metal adsorption capability of the final chelating materials was found to be higher than for its precursor. The adsorption of metal ions from aqueous solution follows the order $Eu^{3+} > U^{6+} > Th^{4+}$ for all systems. The maximum number of moles adsorbed was determined to be 11.63, 12.85, and 14.01 mmol/g (H_{HET}) for Th^{4+}, U^{6+}, and Eu^{3+}, respectively. The competitive sorption behavior, with variation of pH, was favorable for the separation of thorium(IV) from binary mixtures with uranium(VI) and europium(III).

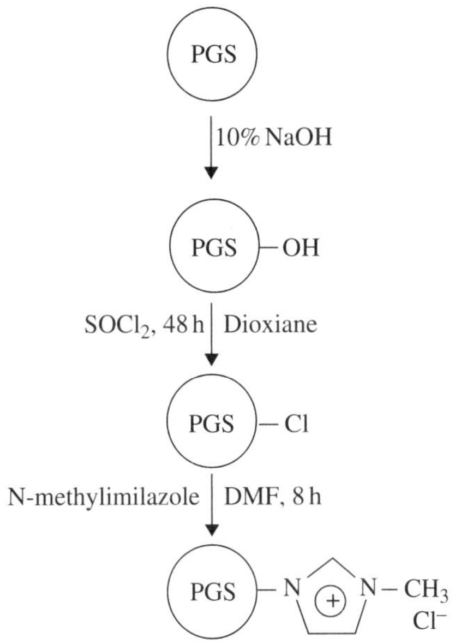

FIGURE 16.7 Preparation of the N-methylimidazole-modified palygorskite.

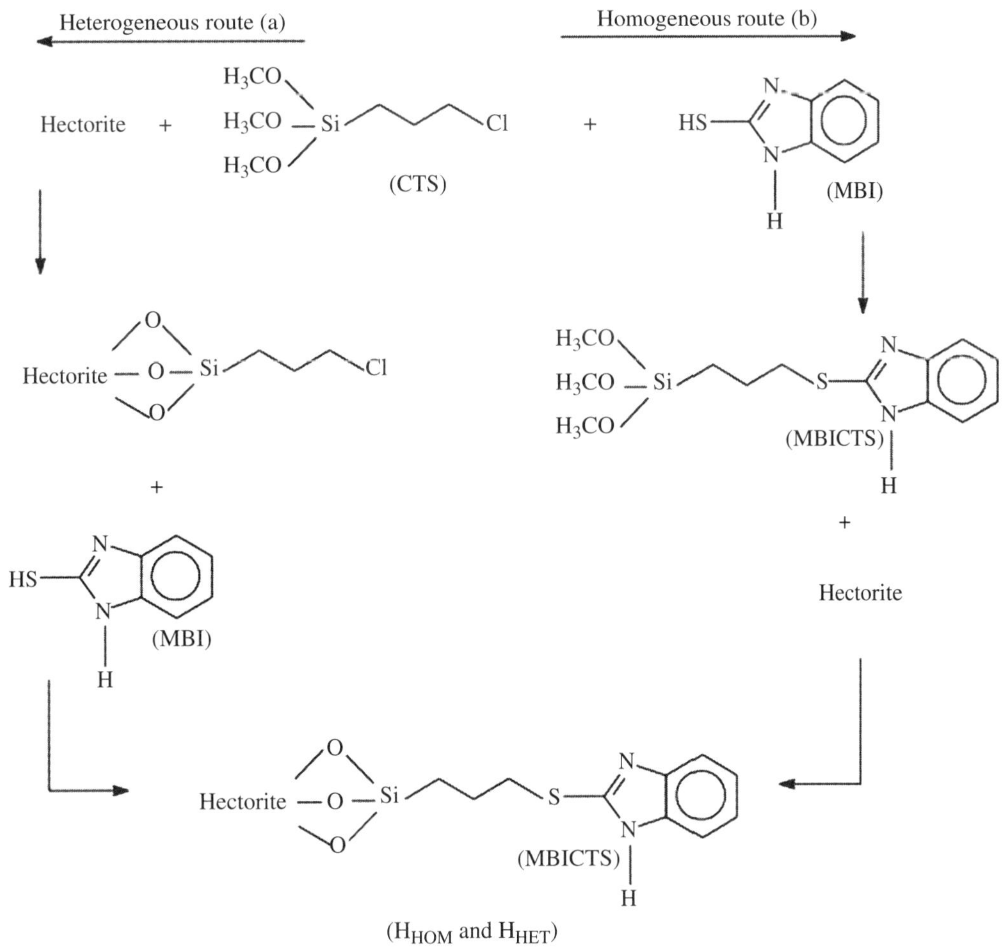

FIGURE 16.8 Incorporation of MBI onto natural hectorite (Hect.) surface by heterogeneous (a) and homogeneous (b) processes.

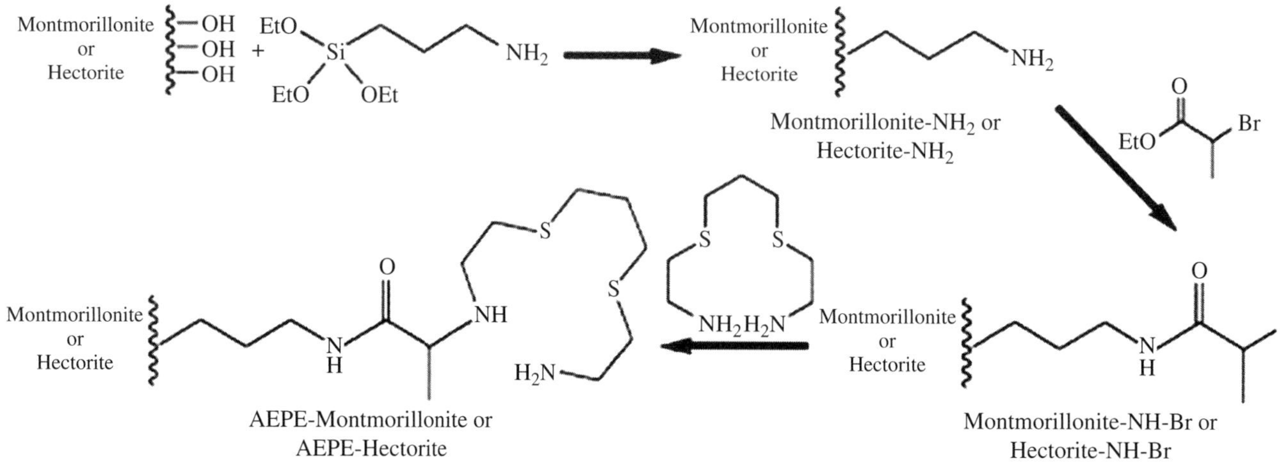

FIGURE 16.9 Grafting of montmorillonite (MMT) and hectorite (Hect.) with 2-(3-(2-aminoethylthio)propylthio)ethanamine (AEPE).

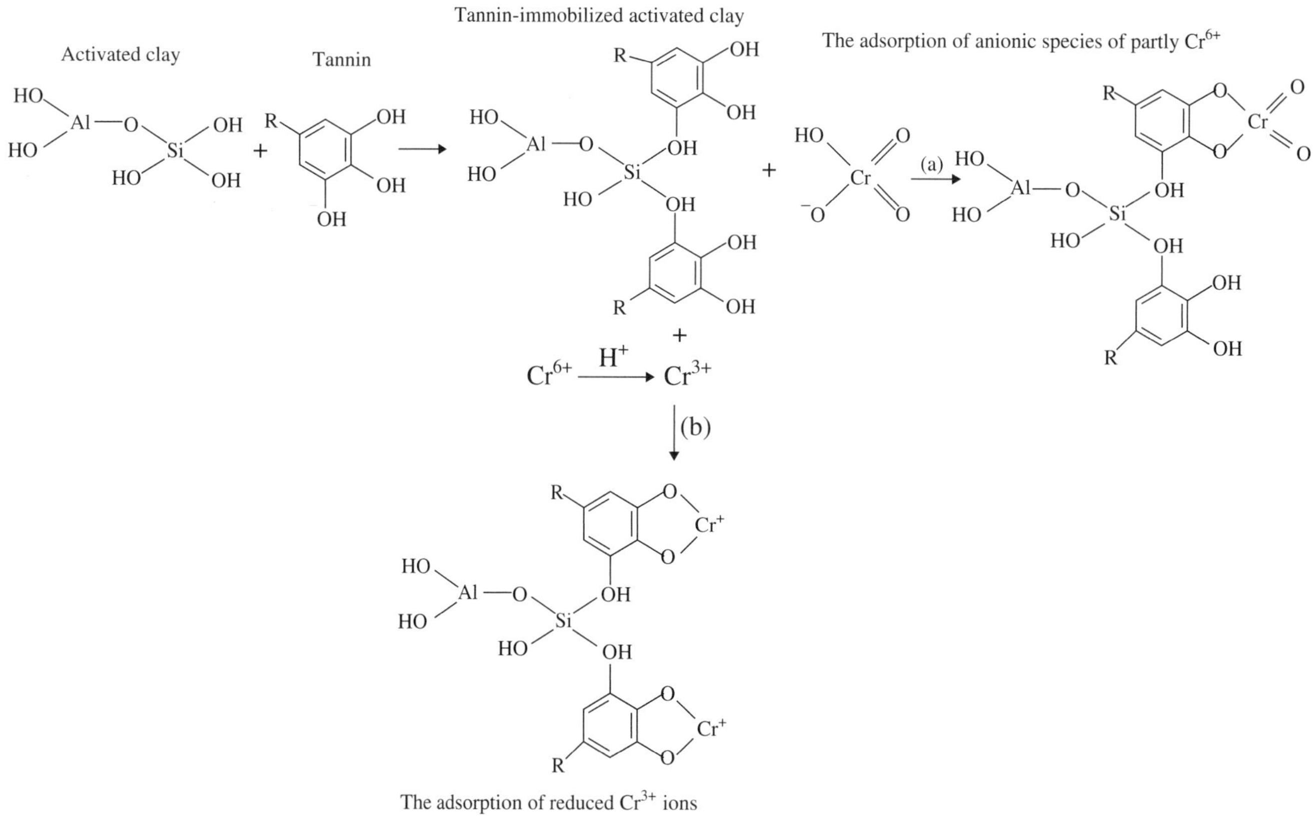

FIGURE 16.10 Adsorption mechanism of total Cr (TCr) onto tannin-immobilized activated clay (TA–AC).

Novel organo-clay minerals AEPE–MMT and AEPE–Hect. were synthesized by grafting AEPE onto the clay minerals (Fig. 16.9) [80]. The XRD patterns indicated that the chelating agents (AEPE) were mainly grafted on the external surface of MMT, while AEPE was grafted on both the external and interlayer surfaces of Hect. The AEPE-modified clay minerals were good chelating materials for Hg(II) ions, compared to the unmodified clay minerals. The adsorption capacity for Hg(II) of AEPE–MMT and AEPE–Hect. was 46.1 and 54.7 mg/g, respectively, for solution containing 140 mg/l Hg(II) ions (pH 4).

Tannin-immobilized activated clay (TA–AC) as a spent adsorbent was used to remove Cr(VI) ions from the aqueous solution (Fig. 16.10) [81]. The maximum adsorption capacity was 24.09 mg/g from the Langmuir isotherm model at 330 K with initial pH = 2.5. Oxidation–reduction was found to occur during the adsorption process. It enhanced the removal capability by partly reducing Cr(VI) ions to Cr(III) ions. So TA–AC demonstrated a high adsorption capacity for Cr(VI) comparable to other adsorbents.

The organo-modified clay minerals were also used as soil-flushing agents for heavy metal–contaminated soil. Natural beidellite, Hect., and vermiculite were transformed into SH-grafted pillared clays, but difficulties were encountered to keep a regular and significant interlayer distance until the last step [82]. The well-crystalline high-charged vermiculite of Santa Olalla (Spain), after exchanging with alkyl ammonium cations and grafting of interlayer Si pillars, showed the highest capacity of Cd^{2+} and Pb^{2+} among the three functionalized tested clay minerals. The cadmium uptake by ray-grass grown on polluted soil was quite depressed by amendments with 0.17% or 0.5% of functionalized synthetic beidellite, but these amendments were inefficient for lead.

16.3 POLYMER MODIFICATION

The water-soluble polymers exhibit several advantages, such as high solubility in water, easy and cheap route of synthesis, an adequate molecular weight and molecular weight distribution, chemical stability, high affinity for one or more metal ions, and selectivity for the metal ion of interest [83]. However, they are difficult to be reused after the adsorption of HMI. To overcome some of the problems associated with the reusability of water-soluble polymers as adsorbents for HMI, they have been grafted onto water-insoluble materials such as clay minerals as supports or used in the form of hydrogels or hybrid hydrogels [84]. In this chapter, the preparation and application of adsorbents for the removal of HMI are covered only with the functional water soluble polymers–grafted nano-clay minerals.

16.3.1 Ionic Interactions

Huang et al. synthesized a novel adsorbent of BNT modified with N-2-hydroxypropyl trimethyl ammonium chloride chitosan (HACC) [85]. The adsorption of Cd(II) onto the HACC–BNT was examined in aqueous solution with respect to the pH, adsorbent dosage, contact time, temperature, and initial concentration. The maximum amount of Cd(II) adsorbed (q_m), as evaluated by the Langmuir isotherm, was 22.23 mg/g of HACC–BNT at pH 7.0 or so and 20°C, and the best desorption performance was obtained with HNO_3, HCl, and EDTA as eluents. The change of adsorption enthalpy indicated that the adsorption of Cd(II) onto the HACC–BNT may be carried out via cation exchange.

Humic acid (HA) was immobilized onto the amine-modified polyacrylamide/BNT composite (Am-PAA-B), which was prepared by direct intercalation polymerization technique, and the product (HA–Am-PAA-B) was used as an adsorbent in the removal of copper(II) ions from aqueous solutions (Fig. 16.11) [86]. The adsorbent behaved like a cation exchanger, and more than 99.0% Cu(II) ion removal was observed in the pH range 5.0–6.0. The desorption of adsorbed Cu(II) ions was achieved by 0.1 M HCl, and four adsorption/desorption cycles were performed without a significant decrease in the adsorption capacity.

Kumar et al. reported an effective methodology for the detoxification of chromium using cellulose–MMT composite material as the adsorbent (Fig. 16.12) [87]. The interaction of surfactant-modified Na–MMT with cellulose biopolymer was followed by the subsequent adsorption of Cr(VI) from aqueous solution as bichromate anion onto the surface of the biocomposite material. The material exhibited a maximum adsorption capacity of 22.2 mg/g in accordance with the Langmuir isotherm model. The composite material could be regenerated using sodium hydroxide as the eluent. The adsorbent could be reused with quantitative recovery for 10 adsorption–desorption cycles. The applicability of the method was demonstrated in the quantitative removal of total chromium from a chrome tannery effluent sample.

16.3.2 Covalent Grafting

Polyacrylamide was successfully grafted from the surfaces of the organo-modified fibrillar clay ATP via the copper-mediated surface-initiated atom transfer radical polymerization (SI-ATRP) technique in water (Fig. 16.13a) [88]. The product, polyacrylamide-grafted ATP (PAM—ATP), was applied preliminarily as an adsorbent in the removal of heavy metal ion (Hg(II)) and two dyes (cationic dye: MB and anionic dye: methyl orange (MO)) from aqueous solutions. Compared with bare ATP, PAM–ATP had much higher adsorption capacities toward Hg(II) ion because of the chelation with Hg(II) ion by formation of either monoamido- or diamido-Hg structures (Fig. 16.13b), which provides a means of capturing Hg(II) from aqueous solution. They also had much higher adsorption capacities toward the cationic dye MB likely due to the flotation with cationic dye of the grafted polyacrylamide chains. However, the adsorption capacities toward the anionic dye MO did not improve from the surface graft polymerization of acrylamide (Table 16.1).

Zhao and Chen et al. synthesized the polyacrylamide/ATP (PAM/ATP) adsorbent by grafting acrylamide (AM) onto the silane coupling reagent–modified ATP (OATP) using solution polymerization (Fig. 16.14) [89–91]. Equilibrium and kinetic adsorption data showed that PAM/ATP displays a high selectivity toward one metal in a two-component or a three-component

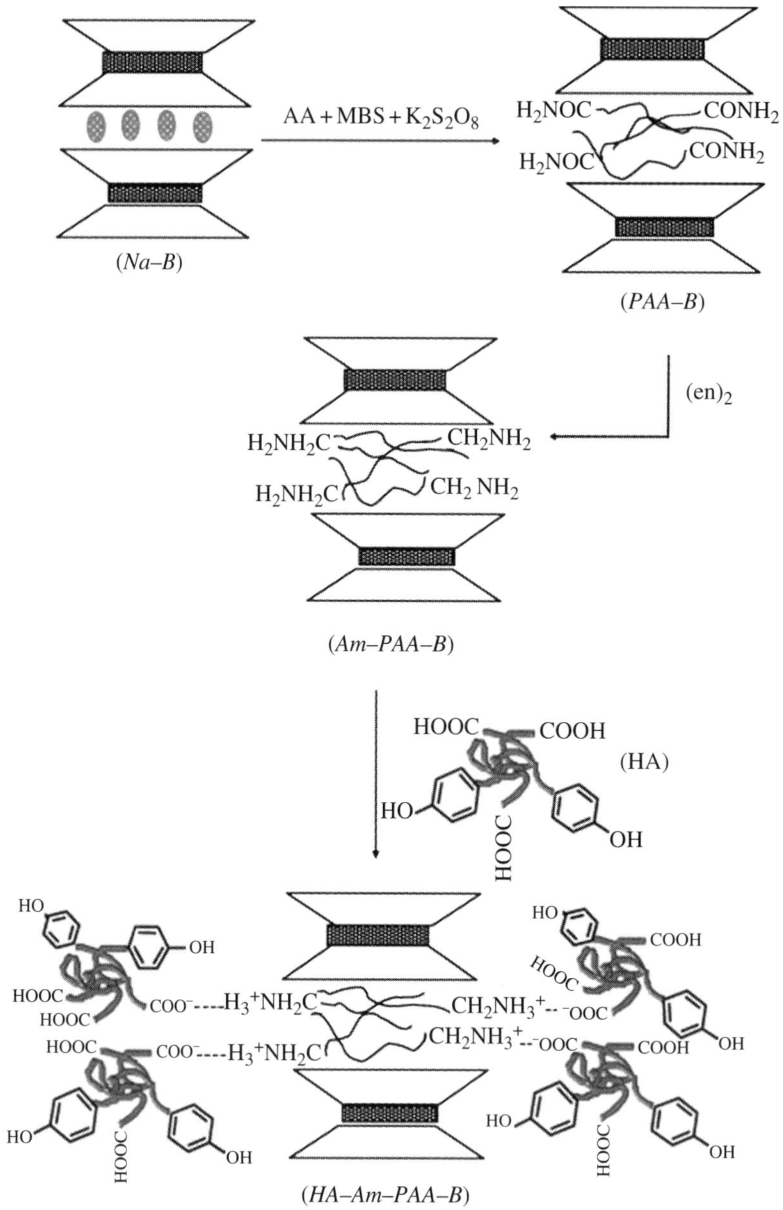

FIGURE 16.11 Preparation route of HA–Am–PAA–B.

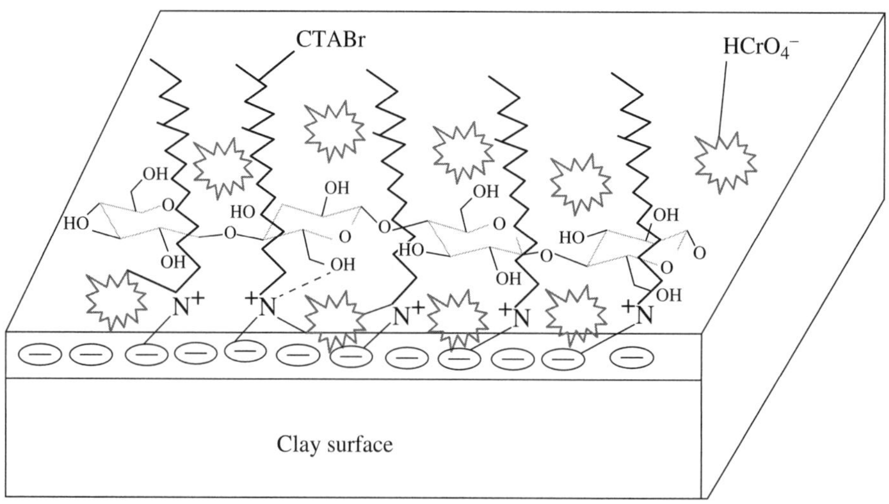

FIGURE 16.12 Conceptual illustration showing the interaction of bichromate ion with the biopolymerclay adsorbent surface.

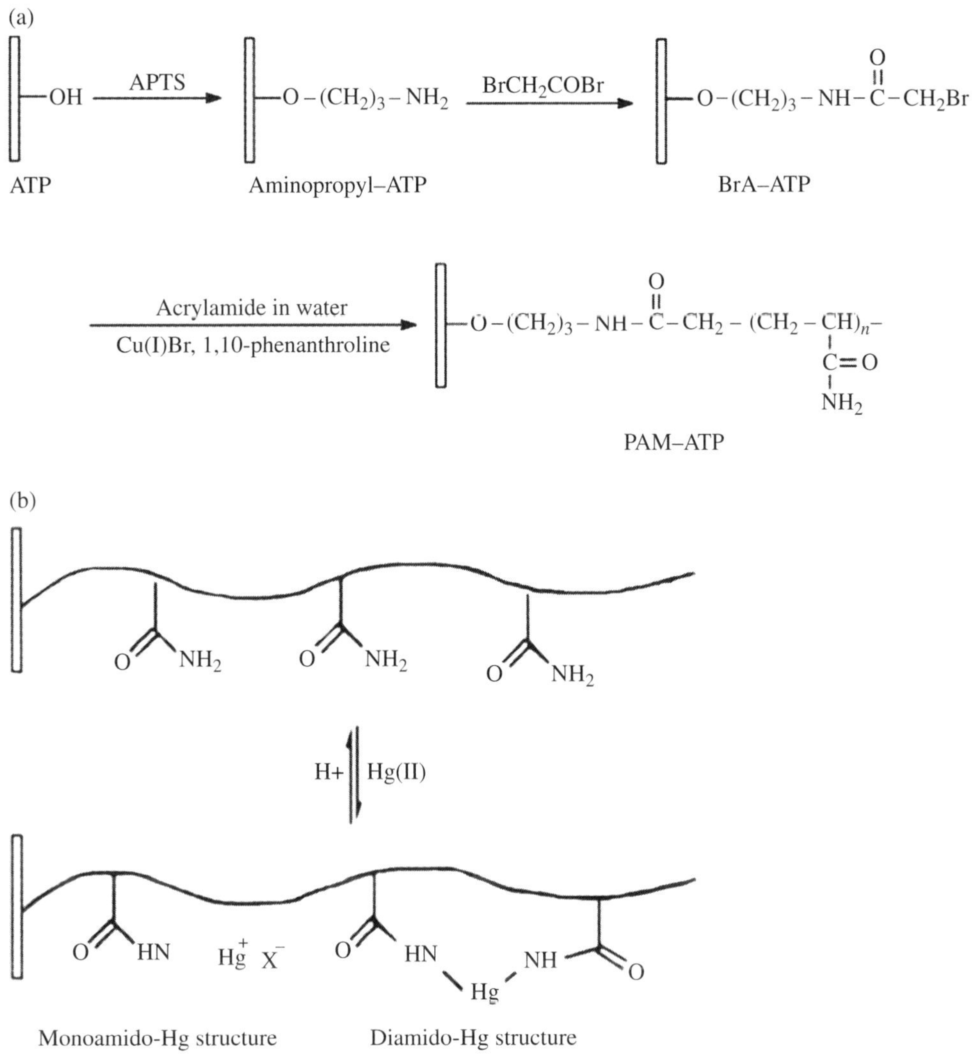

FIGURE 16.i3 Preparation procedure of PAM–ATP (a) and the formation of the amido-Hg complexes (b).

TABLE 16.1 Adsorption capacities of ATP nanofibrils

	Adsorption capacities (mmol/g)		
Samples	Hg(II)	MB	MO
Bare ATP	0.033	0.028	0.023
PAM–ATP	1.12	0.073	0.024

system with an affinity order of Hg^{2+} > Pb^{2+} > Co^{2+} [92]. It was found that the Hg^{2+} adsorption capacity of the PAM/ATP prepared under these conditions is more than sixfold compared with those of ATP and OATP. The adsorption process was rapid; 88% of adsorption occurred within 5 min and equilibrium was achieved at around 40 min. The equilibrium data fitted well with the Langmuir sorption isotherms, and the maximum adsorption capacity of Hg^{2+} onto PAM/ATP was found to be 192.5 mg/g. The Hg^{2+} ions adsorbed onto PAM/ATP could be effectively desorbed in hot acetic acid solution, and the adsorption capacity of the regenerated adsorbents could still be maintained at 95% by the sixth cycle.

Hyperbranched aliphatic polyester (HAPE) had been grafted from the surfaces of the ATP via melt polycondensation of an AB_2-type monomer, 2,2-bis (hydroxymethyl) propionic acid (bis-MPA), with *p*-toluenesulfonic acid (*p*-TSA) as catalyst

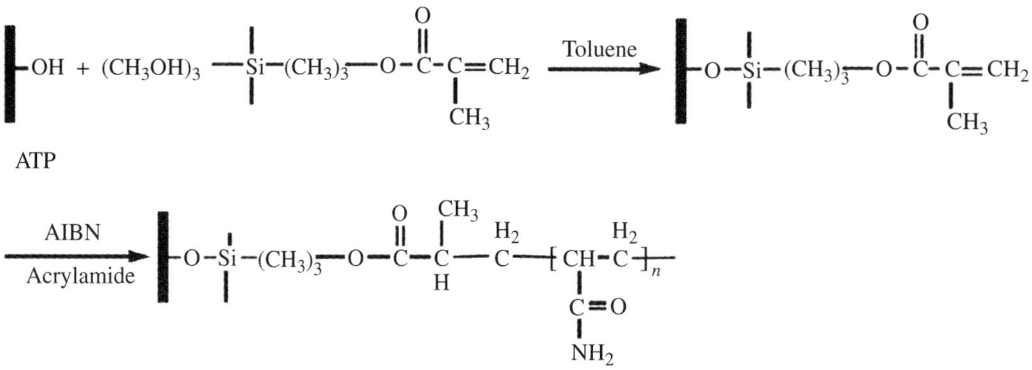

FIGURE 16.14 Preparation procedure of PAM/ATP.

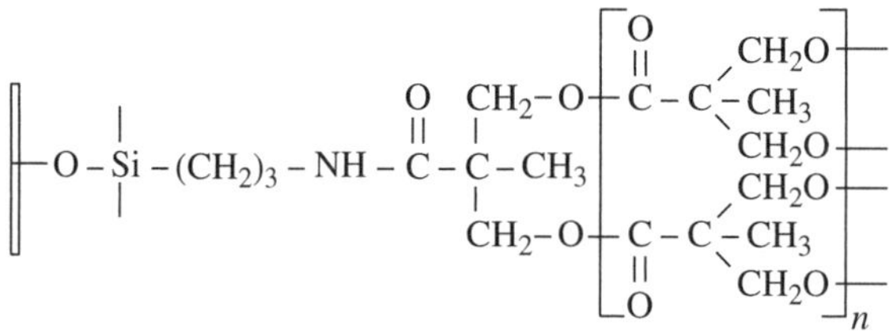

FIGURE 16.15 Ideal structure of the HAPE–ATP.

TABLE 16.2 Competitive adsorption data of ATP adsorbents

	Adsorption capacities (mg/g)			
Samples	Cu(II)	Hg(II)	Zn(II)	Cd(II)
Bare ATP	0.41	0.32	0.97	0.37
A-ATP	2.45	5.29	0.18	0.34
HAPE–ATP	1.76	3.52	0.45	0.29

(Fig. 16.15) [93]. The competitive adsorption capacities of the three clay adsorbents, bare ATP, aminopropyl ATP (A-ATP), and HAPE-grafted ATP (HAPE–ATP), toward the four HMI (Cu(II), Hg(II), Zn(II), and Cd(II)), were also compared (Table 16.2). It was found that bare ATP had higher adsorption capacities for Zn(II) and Cd(II) ions. However, the A-ATP and the HAPE–ATP adsorbents had higher adsorption capacities for Cu(II) and Hg(II) ions. Furthermore, the adsorption capacities for Cu(II) and Hg(II) ions of the HAPE–ATP clay adsorbent were slightly lower than those of the A-ATP clay adsorbent because the groups containing the N element were enshrouded with HAPE, which is not soluble in water, and the ester and hydroxyl groups of HAPE had lower complexation abilities toward these HMI. However, HAPE–ATP had a faster sedimentating rate than A-ATP. This is an advantage in practical applications.

The organic–inorganic hybrid of poly(acrylic acid–acrylonitrile)/ATP, P(A-N)/AT nanocomposites, was prepared by in situ polymerization and composition of acrylic acid (AA) and acrylonitrile (AN) onto modified ATP nanoparticles (Fig. 16.16) [94]. The resultant P(A-N)/AT nanocomposites were transformed into a novel nanoadsorbent of poly(AA-acryloamidoxime)/ATP by further functionalization, that is, P(A-O)/AT nanoadsorbent. The adsorption properties of P(A-O)/AT toward metal ions were determined, and the results indicated that adsorbents with nanocomposite structure had good selectivity toward Pb^{2+} among the numerous metal ions. The maximum removal capacity of Pb^{2+} ions was up to 109.9 mg/g, and it is notable to see that the adsorption removal of the P(A-O)/AT nanoadsorbent for Pb^{2+} ions could achieve more than 96.6% when the initial concentration of Pb^{2+} ions was 120.0 mg/l. The kinetics, isotherm models, and conductivity indicated that it could be a chemisorption process and the best coordination took place when AO:AA:Pb^{2+} = 1:1:1. In addition, after treatment with cetyltrimethylammonium bromide (CTAB), the P(A-O)/AT nanoadsorbent showed better adsorption properties for phenol than similar kinds of materials.

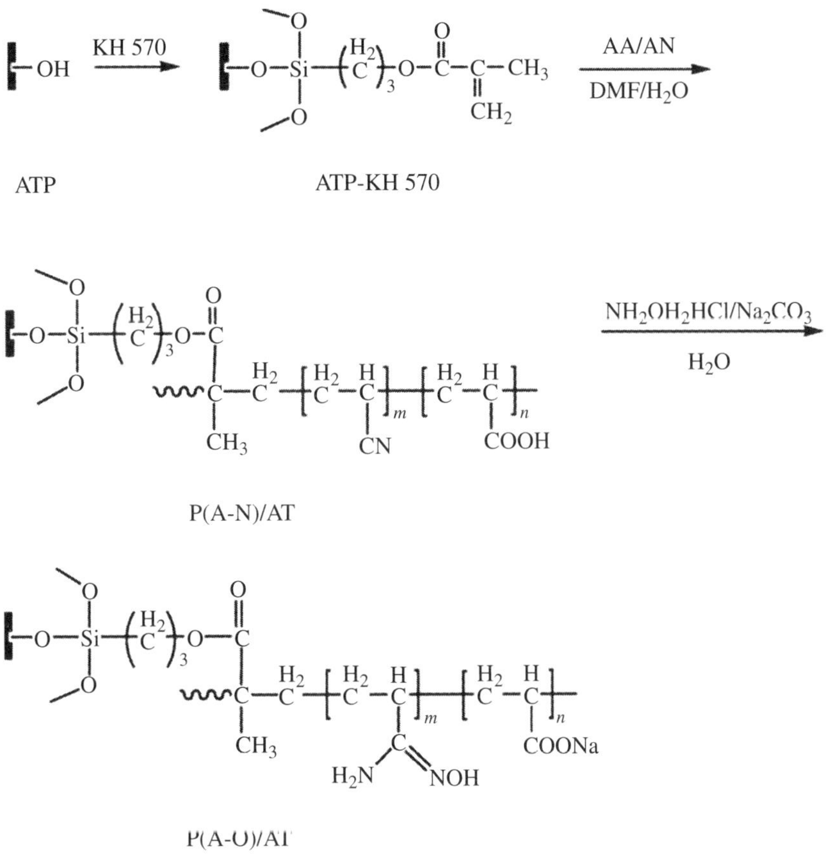

FIGURE 16.16 Scheme for the preparation procedure to P(A-O)/AT.

A novel amphiphilic hybrid material ATP-P(S-b-DVB-g-AO) (ASDO) was synthesized via the combination of SI–ATRP and radical grafting polymerization, after transforming AN units into acrylamide oxime (AO) as a hydrophilic segment (Fig. 16.17) [95]. The amphiphilic hybrid material was employed in the absorption of heavy metal and organic pollutants. The adsorption capacity of ASDO for Pb(II) was 131.6 mg/g, and the maximum removal capacity of ASDO toward phenol was found to be 18.18 mg/g in the case of monolayer adsorption at 30°C. The optimum pH was 5 for both lead and phenol adsorption. Desorption of the adsorbed contaminants from ASDO can take place in alkali solution (1 mol/l NaOH). Repeating the test for six times revealed that the synthesized hybrid materials maintained the stable high adsorption capacity during the adsorption/desorption.

Previous research works suggest that chemically modified clay minerals with small organic molecules or polymers present a new and promising class of adsorbents for water purification and industrial wastewater treatment, especially for the adsorption of metal ions from aqueous solutions [96]. It has been found that these clay minerals would have as great an efficiency as activated carbon if the difficulty experienced in the recovery of the adsorbent from filters after use could be overcome. This also makes the regeneration and possible reuse of a clay adsorbent very difficult.

Unuabonah et al. synthesized a novel water-stable ion exchange resin for the adsorption of metal ions with polyvinyl alcohol (PVA) kaolinite clay as a PVA–nanoclay adsorbent, in order to address the problem of the recovery of adsorbent/resin from filters after use [97]. This adsorbent was found to have an adsorption capacity of 56.18 mg/g for Pb^{2+} ions and 41.67 mg/g for Cd^{2+} ions. The adsorption data obtained was well explained by the Diffuse Layer Model (DLM), which implies that the adsorption of both metal ions onto the modified adsorbent was via an inner-sphere surface complexation mechanism. Virtually complete desorption (ca. 99%) of both metal ions occurred from the PVA–nanoclay within 3 min.

16.4 CONCLUSIONS AND PROSPECTS

The organo-clay nanohybrid adsorbents have been investigated intensively particularly in the removal of a variety of toxic HMI from water. The clay minerals functionalized with organo-compounds become more hydrophobic with enhanced interspace basal spacing, enabling enhanced and selective affinity toward inorganic oxyanions/cations. Although these materials possess

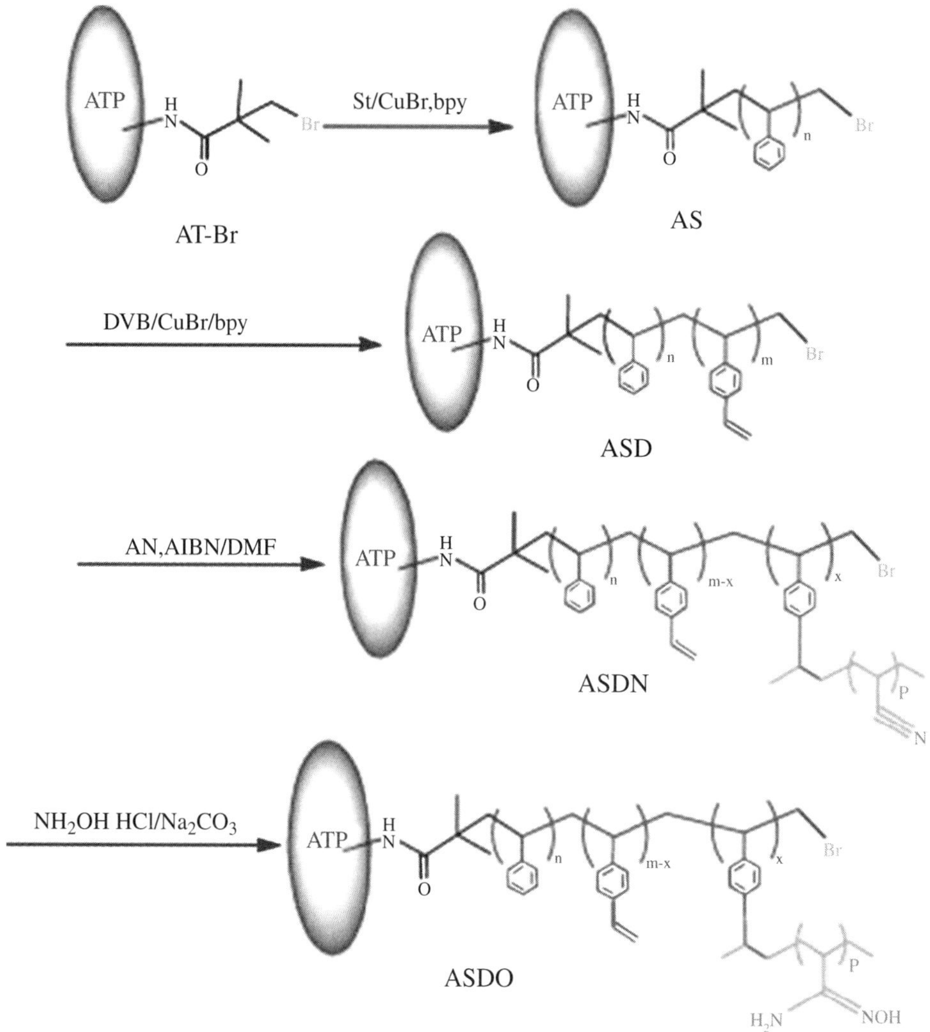

FIGURE 16.17 Scheme of the preparation procedure of ASDO.

several advantages, several difficulties or drawbacks are still underlying, which have to be addressed in future for wider/practical applications.

The adsorption capacity of clay minerals without being exfoliated into the nanoscale is limited due to their small surface area. The development of innovative organo-nanoparticles of layered silicates and similar nanocomposites with large specific surface area merits attention for environmental remediation studies; it needs to be studied in-depth. However, nanoscaled adsorbents are difficult to be separated from the water treating system, although they usually exhibit much higher adsorption capacity. So the adsorption capacity and the practical maneuverability should be comprehensively considered for future work.

The organo-modification of clays via noncovalent interactions seems easier than those with covalent interactions. However, the stability and reusability of the organo-modification of clays via noncovalent interactions might be of concern. The modified organic molecules might fall off from the clays, decreasing their adsorption capacity. Moreover, the dropped organic molecules might become a secondary pollutant.

The studies conducted till now were mostly based on laboratory experiments. Wastewaters containing several cations/anions as well as organic molecules may likely hamper the removal efficiency. Hence, more comprehensive works concerning selectivity are likely to be conducted.

Before organo-modification of clay minerals, they are usually activated with acids. Handling acidic solutions containing a large amount of metal ions produced in the acid-activated process may be another problem to solve.

REFERENCES

[1] Danil de Namor AF, Gamouz AE, Frangie S, Martinez V, Valiente L, Webb OA. Turning the volume down on heavy metals using tuned diatomite.A review of diatomite and modified diatomite for the extraction of heavy metals from water. J Hazard Mater 2012;241–242:14–31.

[2] Khan MA, Rao RAK, Ajmal M. Heavy metal pollution and its control through nonconventional adsorbents (1998–2007): a review. J Int Environ Appl Sci 2008;3:104–141.

[3] Kobya M, Demirbas E, Senturk E, Ince M. Adsorption of heavy metal ions from aqueous solutions by activated carbon prepared from apricot stone. Bioresour Technol 2005;96:1581–1521.

[4] Holan ZR, Volesky B. Biosorption of lead and nickel by biomass of marine aglae. Biotechnol Bioeng 1994;43:1001–1009.

[5] Schiewer S, Fourest E, Chong KH, Volesky B. Description of two-metal biosorption equi-libria by Langmuir-type models. Biotechnol Bioeng 1995;47:451–160.

[6] Fu FL, Wang Q. Removal of heavy metal ions from waste waters: a review. J Environ Manage 2011;92:407–418.

[7] Pan BJ, Pan BC, Zhang WM, Lv L, Zhang QX, Zheng SR. Development of polymeric and polymer-based hybrid adsorbents for pollutants removal from waters. Chem Eng J 2009;151:19–39.

[8] Babel S, Kurniawan TA. Low-cost adsorbents for heavy metals uptake from contaminated water: a review. J Hazard Mater 2003;97:219–243.

[9] Demirbas A. Heavy metal adsorption onto agro-based waste materials: a review. J Hazard Mater 2008;157:220–229.

[10] Alinnor IJ. Adsorption of heavy metal ions from aqueous solution by fly ash. Fuel 2007;86:853–857.

[11] Awan MA, Qazi IA, Khalid I. Removal of heavy metals through adsorption using sand. J Environ Sci China 2003;15:413–416.

[12] Pinnavaia TJ. Intercalated clay catalysts. Science 1983;220:365–371.

[13] Bhattacharyya KG, Gupta SS. Adsorption of a few heavy metals on natural and modified kaolinite and montmorillonite: a review. Adv Colloid Interface Sci 2008;140:114–131.

[14] Liu P, Zhang LX. Adsorption of dyes from aqueous solutions or suspensions with clay nano-adsorbents. Separ Purif Technol 2007;58:32–39.

[15] Cornejo J, Celis R, Pavlovic I, Ulibarri MA. Interactions of pesticides with clays and layered double hydroxides: a review. Clay Miner 2008;43:155–175.

[16] Ahmaruzzaman M. Adsorption of Phenolic compounds on low-cost adsorbents: a review. Adv Colloid Interface Sci 2008;143:48–67.

[17] Bergaya F, Theng BKG, Lagaly G. *Handbook of Clay Science*. Volume 1, Amsterdam: Elsevier Science B.V.; 2006. p 263.

[18] Liu YS, Liu P, Su ZX, Li F, Wen F. Attapulgite–Fe3O4 magnetic nanoparticles via co-precipitation technique. Appl Surf Sci 2008;255:2020–2025.

[19] Zhao GH, Wang JZ, Li YF, Chen X, Liu YP. Enzymes immobilized on superparamagnetic Fe3O4@clays nanocomposites: preparation, characterization, and a new strategy for the regeneration of supports (SCI). J Phys Chem C 2011;115:6350–6359.

[20] Zhang Y, Yang HM. Halloysite nanotubes coated with magnetic nanoparticles. Appl Clay Sci 2012;56:97–102.

[21] Kathleen AC. Synthetic organo- and polymer-clays: preparation, characterization, and materials applications. Appl Clay Sci 2000;17:1–23.

[22] van Oss CJ, Giese RF. Surface modification of clays and related materials. J Dispers Sci Technol 2003;24:363–376.

[23] Wypych F. *Clay Surfaces: Fundamentals and Applications*. Volume 1, Amsterdam: Elsevier Science B.V.; 2004.

[24] Beall GW, Goss M. Self-assembly of organic molecules on montmorillonite. Appl Clay Sci 2004;27:179–186.

[25] Liu P. Polymer modified clay minerals: a review. Appl Clay Sci 2007;38:64–76.

[26] de Paiva LB, Morales AR, Valenzuela Diaz FR. Organoclays: properties, preparation and applications. Appl Clay Sci 2008;42:8–24.

[27] Pavlidou S, Papaspyrides CD. A review on polymer–layered silicate nanocomposites. Progr Polym Sci 2008;33:1119–1198.

[28] Malakul P, Srinivasan KR, Wang HY. Metal adsorption and desorption characteristics of surfactant modified clay complexes. Ind Eng Chem Res 1998;37:4296–4301.

[29] Gupta SS, Bhattacharyya KG. Removal of Cd(II) from aqueous solution by kaolinite, montmorillonite and their poly (oxo zirconium) and tetrabutylammonium derivatives. J Hazard Mater 2006;128:247–257.

[30] Bhattacharyya KG, Gupta SS. Kaolinite, montmorillonite, and their modified derivatives as adsorbents for removal of Cu(II) from aqueous solution. Separ Purif Technol 2006;50:388–397.

[31] Bhattacharyya KG, Gupta SS. Calcined tetrabutylammonium kaolinite and montmorillonite and adsorption of Fe(II), Co(II) and Ni(II) from solution. Appl Clay Sci 2009;46:216–221.

[32] Abou-El-Sherbini KS, Hassainien MM. Study of organically-modified montmorillonite clay for the removal of copper(II). J Hazard Mater 2010;184:654–661.

[33] Hassanien MM, Abou-El-Sherbini KS, Al-Muaikel NS. Immobilization of methylene blue onto bentonite and its application in the extraction of mercury (II). J Hazard Mater 2010;178:94–100.

[34] Krishna BS, Murty DSR, Jai Prakash BS. Surfactant-modified clay as adsorbent for chromate. Appl Clay Sci 2001;20:65–71.

[35] Thanos AG, Katsou E, Malamis S, Psarras K, Pavlatou EA, Haralambous KJ. Evaluation of modified mineral performance for chromate sorption from aqueous solutions. Chem Eng J 2012;211–212:77–88.

[36] Akar ST, Yetimoglu Y, Gedikbey T. Removal of chromium (VI) ions from aqueous solutions by using Turkish montmorillonite clay: effect of activation and modification. Desalination 2009;244:97–108.

[37] Brum MC, Capitaneo JL, Oliveira JF. Removal of hexavalent chromium from water by adsorption onto surfactant modified montmorillonite. Miner Eng 2010;23:270–272.

[38] Huang Y, Ma XY, Liang GZ, Yan YX, Wang SH. Adsorption behavior of Cr(VI) on organic-modified rectorite. Chem Eng J 2008;138:187–193.

[39] Hong HL, Jiang W-T, Zhang XL, Tie LY, Li ZH. Adsorption of Cr(VI) on STAC-modified rectorite. Appl Clay Sci 2008;42:292–299.

[40] Wang JH, Zhang X, Zhang B, Zhao YF, Zhai R, Liu JD, Chen RF. Rapid adsorption of Cr (VI) on modi?ed halloysite nanotubes. Desalination 2010;259:22–28.

[41] Sarkar B, Xi YF, Megharaj M, Krishnamurti GSR, Rajarathnam D, Baidu R. Remediation of hexavalent chromium through adsorption by bentonite based Arquad (R) 2HT-75 organoclays. J Hazard Mater 2010;183:87–97.

[42] Diltz A, An J-H, Riebe B. Appl Clay Sci 2012;67-68:125–133.

[43] Atia AA. Adsorption of chromate and molybdate by cetylpyridinium bentonite. Appl Clay Sci 2008;41:73–84.

[44] Su J, Huang H-G, Jin X-Y, Lu X-Q, Chen Z-L. Synthesis, characterization and kinetic of a surfactant-modified bentonite used to remove As(III) and As(V) from aqueous solution. J Hazard Mater 2011;185:63–70.

[45] Majdan M, Pikus S, Gajowiak A, Sternik D, Zieba E. Uranium sorption on bentonite modified by octadecyltrimethylammonium bromide. J Hazard Mater 2010;184:662–670.

[46] Guerra DL, Silva EM, Lara W, Batista AC. Removal of Hg(II) from an aqueous médium by adsorption onto natural and alkyl-amine modified Brazilian bentonite. Clay Clay Miner 2011;59:568–580.

[47] Kumar ASK, Ramachandran R, Kalidhasan S, Rajesh V, Rejash N. Potential application of dodecylamine modified sodium montmorillonite as an effective adsorbent for hexavalent chromium. Chem Eng J 2012;211–212:396–405.

[48] Kozak O, Praus P, Machovic V, Klika Z. Adsorption of zinc and copper ions on natural and ethylenediamine modified montmorillonite. Ceramics-Silikaty 2010;54:78–84.

[49] Sheng G, Xu S, Boyd SA. A dual function organoclay sorbent for lead and chlorobenzene. Soil Sci Soc Am J 1999;63:73–78.

[50] Faghihian H, Nejati-Yazdinejad M. Sorption performance of cysteine-modified bentonite in heavy metals uptake. Journal of the Serbian Chemical Society 2009;74:833–843.

[51] Phothitontimongkol T, Siebers N, Sukpirom N, Unob F. Preparation and characterization of novel organo-clay minerals for Hg(II) ions adsorption from aqueous solution. Applied Clay Science 2009;43:343–349.

[52] De Leon AT, Nunes DG, Rubio J. Adsorption of Cu ions onto a 1.10 phenanthroline-grafted Brazilian bentonite. Clays and Clay Minerals 2003;51:58–64.

[53] Erdem B, Ozcan A, Gok O, Ozcan AS. Immobilization of 2,2-dipyridyl onto bentonite and its adsorption behavior of copper(II) ions. Journal of Hazardous Materials 2009;163:418–426.

[54] Ozcan AS, Gok O, Ozcan A. Adsorption of lead(II) ions onto 8-hydroxyquinoline-immobilized bentonite. Journal of Hazardous Materials 2009;161:499–509.

[55] Bentouami A, Oulai MS. Cadmium removal from aqueous solutions by hydroxy-8 quinoleine intercalated bentonite. Journal of Colloid and Interface Science 2006;293:270–277.

[56] Gok O, Ozcan A, Erdam B, Ozcan AS. Prediction of the kinetics, equilibrium and thermodynamic parameters of adsorption of copper(II) ions onto 8-hydroxy quinoline immobilized bentonite. Colloids and Surfaces A: Physicochemical and Engineering Aspects 2008;317:174–185.

[57] Filho NLD, Polito WL, Gushikem Y. Adsorption and preconcentration of some heavy metals by 2-mercaptobenzothiazole-clay. Talanta 1995;42:1031–1036.

[58] Manohar DM, Krishnan KA, Anirudhan TS. Removal of mercury(II) from aqueous solutions and chlor-alkali industry wastewater using 2-mercaptobenzimidazole-clay. Water Research 2002;36:1609–1619.

[59] Sing RP, Rawat JP, Kumar R. Effect of cationic, non-ionic and anionic surfactants on the adsorption of carbofuran on three different types of Indian soil. Adsorption Science and Technology 2000;18:333–346.

[60] Zhang ZZ, Sparks DL, Scrivner NS. Sorption and desorption of quaternary amine570cations on clays. Environmental Science and Technology 1993;27:1625–1633.

[61] Lin S-H, Juang R-S. Heavy metal removal from water by sorption using surfactant-modified montmorillonite. Journal of Hazardous Materials 2002;92:315–326.

[62] Wu PX, Zhang Q, Dai YP, Zhu NW, Dang Z, Li P, Wu JH. Adsorption of Cu(II), Cd(II) and Cr(III) ions from aqueous solutions on humic acid modified Ca-montmorillonite. Geoderma 2011;164:215–219.

[63] Sivaiah MV, Venkatesan KA, Sasidhar P, Krishna RM, Murthy GS. Unusal extraction behaviour of crown ether when intercalated in bentonite. New Journal of Chemistry 2005;29:564–569.

[64] Liu Y, Shen X, Xian QM, Chn HD, Zou HX, Gao SX. Adsorption of copper and lead in aqueous solution onto bentonite modified by 4 '-methylbenzo-15-crown-5. Journal of Hazardous Materials 2006;137:1149–1155.

[65] Celis R, Hermosin MC, Cornejo J. Heavy metal adsorption by functionalized clays. Environmental Science and Technology 2000;34:4593–4599.

[66] Guimaraes ADF, Ciminelli VST, Vasconcelos WL. Smectite organofunctionalized with thiol groups for adsorption of heavy metal ions. Applied Clay Science 2009;42:410–414.

[67] Marjanovic V, Lazarevic S, Jankovic-Castvan I, Potkonjak B, Janackovic D, Petrovic R. Chromium (VI) removal from aqueous solutions using mercaptosilane functionalized sepiolites. Chemical Engineering Journal 2011;166:198–206.

[68] Demirbas O, Alkan M, Dogan M, Turhan Y, Namli H, Turan P. Electrokinetic and adsorption properties of sepiolite modified by 3-aminopropyltriethoxysilane. Journal of Hazardous Materials 2007;149:650–656.

[69] Guerra DL, Lemos VP, Angelica RS, Airoldi C. Metal ions adsorption from ethanol solutions on ethylenediamine-modified poly(dimethylsiloxane) elastomeric network. Colloids and Surfaces A: Physicochemical and Engineering Aspects 2008;322:79–86.

[70] Guerra DL, Viana RR, Airoldi C. Adsorption of mercury cation on chemically modified clay. Mater Res Bull 2009;44:485–491.

[71] Wu PX, Dai YP, Long H, Zhu NW, Li P, Wu JH, Dang Z. Characterization of organo-montmorillonites and comparison for Sr(II) removal: equilibrium and kinetic studies. Chem Eng J 2012;191:288–296.

[72] Cui H, Qian Y, Li Q, Wei ZB, Zhai JP. Fast removal of Hg(II) ions from aqueous solution by amine-modified attapulgite. Appl Clay Sci 2013;72:84–90.

[73] Erdemoglu M, Erdemoglu S, Sayilkan F, Akarsu M, Sener S, Sayilkan H. Organo-functional modified pyrophyllite: preparation, characterisation and Pb (II) ion adsorption property. Applied Clay Science 2004;27:41–52.

[74] Dogan M, Turhan Y, Alkan M, Namli H, Turan P, Demirbas O. Functionalized sepiolite for heavy metal ions adsorption. Desalination 2008;230:248–268.

[75] Turhan Y, Turan P, Dogan M, Alkan M, Namli H, Demirbas O. Characterization and adsorption properties of chemically modified sepiolite. Industrial and Engineering Chemistry Research 2008;47:1883–1895.

[76] Guerra DL, Leidens VL, Viana RR, Airoldi C. Journal of Hazardous Materials 2010;180:683–692.

[77] Li DD, Chang XJ, Hu Z, Wang QH, Li RJ, Chai XL. Samarium (III) adsorption on bentonite modified with N-(2-hydroxyethyl) ethylenediamine. Talanta 2011;83:1742–1747.

[78] Chang Y, Liu HW, Zha F, Chen HK, Ren XN, Lei ZQ. Adsorption of Pb(II) by N-methylimidazole modified palygorskite. Chem Eng J 2011;167:183–189.

[79] Guerra DL, Viana RR, Airoldi C. Use of raw and chemically modified hectorites as adsorbents for Th(IV), U(VI) and Eu(III) uptake from aqueous solutions. Desalination 2010;260:161–171.

[80] Phothitonmongkol T, Siebers N, Sukpirom N, Unob F. Applied Clay Science 2009;43:343–349.

[81] Li W, Tang YK, Zeng YT, Tong ZF, Liang DW, Cui WW. Adsorption behavior of Cr(VI) ions on tannin-immobilized activated clay. Chem Eng J 2012;193–194:88–95.

[82] Diaz M, Cambier P, Brendle J, Prost R. Functionalized clay heterostructures for reducing cadmium and lead uptake by plants in contaminated soils. Appl Clay Sci 2007;37:12–22.

[83] Rivas B, Pereira ED, Moreno-Villoslada I. Water-soluble polymer–metal ion interactions. Progr Polymer Sci 2003;28:173–208.

[84] Jiang GH, Wang L, Yu HJ, Amer WA, Zhang L. Recent progress on study of hybrid hydrogels for water treatment. Colloid Surf A Physicochem Eng Asp 2013;416:86–94.

[85] Huang RH, Wang B, Yang BC, Zheng DS, Zhang ZQ. Equilibrium, kinetic and thermodynamic studies of adsorption of Cd(II) from aqueous solution onto HACC-bentonite. Desalination 2011;280:297–304.

[86] Anirudhan TS, Suchithra PS. Humic acid-immobilized polymer/bentonite composite as an adsorbent for the removal of copper(II) ions from aqueous solutions and electroplating industry wastewater. J Ind Eng Chem 2010;16:130–139.

[87] Kumar ASK, Kalidhasan S, Rajesh V, Rajesh N. Application of cellulose-clay composite biosorbent toward the effective adsorption and removal of chromium from industrial wastewater. Ind Eng Chem Res 2012;51:58–69.

[88] Liu P, Guo JS. Polyacrylamide grafted attapulgite (PAM-ATP) via surface-initiated atom transfer radical polymerization (SI-ATRP) for removal of Hg(II) ion and dyes. Colloid Surf A Physicochem Eng Asp 2006;282–283:498–503.

[89] Chen Y, Zhao YJ, Zhou SY, Chu XZ, Xing WH. J Chem Eng Chin Univ 2010;24:274–279.

[90] Zhao YJ, Chen Y, Li MS, Zhou SY, Xue AL, Xing WH. Adsorption of Hg2+ from aqueous solution onto polyacrylamide/attapulgite. J Hazard Mater 2009;171:640–646.

[91] Chen Y, Zhao YJ, Zhou SY, Chu XZ, Yang LL, Xing WH. Preparation and characterization of polyacrylamide/palygorskite. Appl Clay Sci 2009;46:148–152.

[92] Zhou SY, Xue AL, Zhao YJ, Wang QW, Chen Y, Li MS, Xing WH. Competitive adsorption of Hg2+, Pb2+ and Co2+ ions on polyacrylamide/attapulg. Desalination 2011;270:269–274.

[93] Liu P, Wang TM. Adsorption properties of hyperbranched aliphatic polyester grafted attapulgite towards heavy metal ions. J Hazard Mater 2007;149:75–79.

[94] Jin XL, Yu C, Li YF, Qi YX, Yang LQ, Zhao GH, Hu HY. Preparation of novel nano-adsorbent based on organic-inorganic hybrid and their adsorption for heavy metals and organic pollutants presented in water environment. J Hazard Mater 2011;186:1672–1680.

[95] Jin XL, Li YF, Yu C, Ma YX, Yang LQ, Hu HY. Synthesis of novel inorganic-organic hybrid materials for simultaneous adsorption of metal ions and organic molecules in aqueous solution. J Hazard Mater 2011;198:247–256.

[96] Lee SM, Tiwari D. Organo and inorgano-organo-modified clays in the remediation of aqueous solutions: an overview. Appl Clay Sci 2012;59-60:84–102.

[97] Unuabonah EI, Olu-Owolabi BI, Adebowale KO, Yang LZ. Removal of lead and cadmium ions from aqueous solution by polyvinyl alcohol-modified Kaolinite clay: a novel nano-clay Adsorbent. Adsorpt Sci Technol 2008;26:383–406.

PART V

MEMBRANES ON NANOMATERIALS BASIS

17

WATER REMEDIATION USING NANOPARTICLE AND NANOCOMPOSITE MEMBRANES

Kian Fei Yee, Qian Wen Yeang, Yit Thai Ong, Vel Murugan Vadivelu, and Soon Huat Tan
School of Chemical Engineering, Engineering Campus, Universiti Sains Malaysia, Pulau Pinang, Malaysia

17.1 INTRODUCTION

Globally, the demand for clean water has increased due to the increasing world population. The rapid development of the industrial and agricultural sectors that consume large amounts of hazardous chemicals, toxic materials, and pesticides has resulted in the contamination of groundwater and the generation of contaminated wastewater [1]. This phenomenon further increased the demand for clean water, particularly drinking water. The World Health Organization (WHO) has set a guideline on the allowable concentration of chemicals in drinking water, without causing health issues [2]. Some of the major chemicals that often exist in contaminated water have been selected, and the guideline values are given in Table 17.1. Thus, there is an urgent need for new technologies that reduce the contaminants in water.

Advances in nanotechnology suggest that water quality problems could be solved through the utilization of nanomaterials in water remediation because nanomaterials possess properties suitable for water remediation, such as large surface area, high sorbent capacities, and the ability to be functionalized to enhance affinity and selectivity [3]. There are several methods to remove contaminants from water using nanoparticles, including adsorption, photocatalytic degradation, and disinfection. Such methods can be performed using the following: nanoadsorbents, such as nanoscale iron particles (Fe^0) and nanosized metal oxides; nanophotocatalysts, such as titanium dioxide (TiO_2) and zinc oxide (ZnO) nanoparticles; and disinfection agents, such as biogenic silver nanoparticles (bio-Ag^0). However, modifications are still required to improve the efficiency of nanoparticles in the remediation of contaminants [4–7]. These modifications will eventually lead to the development of new and novel nanomaterials with specific properties and functions, including bimetal nanoadsorbents, rigidly supported nanoadsorbents, and core–shell systems of nanosized photocatalysts. Meanwhile, nanoparticles with magnetic properties have always been preferred in water remediation, for the easy separation of contaminant-loaded nanoparticles from treated water, by applying a magnetic field. In the water disinfection process, metallic Ag and TiO_2 nanoparticles are the frequently used disinfection agents because of their bactericidal effects and the ability to inactivate a wide range of harmful microorganisms.

Membrane technologies have also been gaining significant interest in the field of water and wastewater treatment because of their efficiency in contaminant removal. Due to the recent advances in nanotechnology, various nanoparticles have been incorporated into membranes to produce synergistic effects on the treatment processes. Astounding effects of the introduction of nanoparticles for the development of membranes for the said purposes have been widely reported, including fouling mitigation, improved permeation quality, and flux enhancement. TiO_2, Ag, aluminum oxide (Al_2O_3), and silicon dioxide (SiO_2) nanoparticles are among the many nanomaterials being employed. This chapter focuses on the role of various nanoparticles in water remediation, as well as the role of nanomaterials in membrane technology for water and wastewater treatment processes.

Nanomaterials for Environmental Protection, First Edition. Edited by Boris I. Kharisov, Oxana V. Kharissova, and H. V. Rasika Dias.

TABLE 17.1 Guideline values for chemicals that have health significance in drinking water [1]

Chemical	Guideline value (mg/l)
1,1,1-trichloro-2,2′bis(p-chlorophenyl)ethane (DDT)	0.001
2,4,6-trichlorophenol (TCP)	0.2
Arsenic (As)	0.01
Cadmium (Cd)	0.003
Chromium (Cr)	0.05
Copper (Cu)	2
Lead (Pb)	0.01
Mercury (Hg)	0.006
Nickel (Ni)	0.07
Nitrate (as NO_3^-)	50
Tetrachloroethene (C_2Cl_4)	0.04

17.2 NANOPARTICLES IN WATER REMEDIATION

Nanoparticles have shown great potential in water remediation because their small sizes (1–100 nm) allow them to cover larger surface areas with no resistance on internal diffusion [8]. They can also be anchored onto a solid matrix, such as activated carbon (AC) or zeolite, to enhance their effectiveness in water remediation [9], thereby presenting opportunities to resolve or greatly ameliorate water remediation problems [3, 10]. Water remediation can be achieved through adsorption [11–13], photocatalytic degradation [14–17], and disinfection [18–20] using nanoadsorbents, nanophotocatalysts, and bioactive nanoparticles.

17.2.1 Water Remediation by Nanoadsorbents via the Adsorption Technique

The adsorption technique using nanoadsorbents is one of the most common and widely used methods in removing certain classes of water contaminants. In principle, the adsorption technique is beneficial for the removal of water contaminants as well as for recovery purposes [21]. The adsorbed toxic materials are recovered in a concentrated form for disposal, and the valuable adsorbed materials are recycled for industrial purposes [22]. The adsorption technique is a simple, low-cost, and effective technique for removing organic and inorganic contaminants. In addition, dyes [23] and phenolic contaminants [24, 25] with low biodegradability, as well as heavy metal ions, such as As [12, 23, 26–29], Cu [11], Hg [11], Cd [11], Pb [11, 30], Cr [11, 23, 30], and cobalt (Co) [24], can also be effectively removed by nanoadsorbents.

Nanoadsorbents with high surface area to volume ratios are required to achieve high adsorption capacity for water contaminants [10]. Nanosized particles are more efficient than larger-sized particles as adsorbents in water remediation because the former are thermodynamically metastable and tend to transform into more stable forms by adsorbing substances, such as molecules, onto their surfaces to reduce the total free energy and approach the equilibrium state. There are several types of nanoparticles that are suitable and can be used as nanoadsorbents for the removal of water contaminants, such as Fe^0 [9, 25–27, 29–32] and nanosized metal oxides [8, 11, 12, 21, 23, 33]. Other than the selection of suitable adsorbents, the adsorption mechanisms of nanoadsorbents should be studied to improve adsorption efficiency during water remediation [27].

17.2.1.1 Nanoscale Particles Fe^0 is a nontoxic, inexpensive, highly stable, and durable substance. Hence, it is suitable for water remediation and can be sustained for a long period of time without any loss in activity [25]. Furthermore, nanoscale Fe^0 is an effective reductant in treating a wide variety of water contaminants, such as chlorinated benzenes, chlorinated ethenes, pesticides, organic dyes (Orange II, acid orange, acid red, chrysoidine, and tropaeolin O), heavy metal ions (Hg, Cd, Ni, and silver (Ag)), organic contaminants (*N*-nitrosodimethylamine and trinitrotoluene(TNT)), and inorganic anions (As, dichromate ($Cr_2O_7^{2-}$), perchlorate (ClO_4^-), and NO_3^-) [9]. The mechanism of nanoscale Fe^0 in water remediation often involves the iron (Fe) oxidation process. The oxidation stoichiometry of nanoscale Fe^0 is given by Equations 17.1 and 17.2. In the presence of oxygen and water, nanoscale Fe^0 is rapidly oxidized to form free Fe ions through the release of electrons [9]. Therefore, highly reactive nanoscale Fe^0 should be stored in ethanol, not in water, because water contains higher levels of dissolved oxygen than ethanol. Zhang et al. [9] reported that the electron released from the oxidation of nanoscale Fe^0 is accepted by C_2Cl_4 in water, and then reduced to ethane (C_2H_4) in accordance with Equation 17.3, as given in Table 17.2.

TABLE 17.2 General reactions for the remediation of several types of contaminants in water by nanoscale iron particles

General reactions	Contaminants	Equations	References
$C_2Cl_4 + 4Fe^0 + 4H^+ \rightarrow C_2H_4 + 4Fe^{2+} + 4Cl^-$	Tetrachloroethene	17.3	[9]
$RCl + Fe^0 + H^+ \rightarrow RH + Fe^{2+} + Cl^-$	p-Chlorophenol	17.4	[25]
$NO_3^- + 4Fe^0 + 10H^+ \rightarrow NH_4^+ + 4Fe^{2+} + 3H_2O$	Nitrate	17.5	[32]
$2Fe^0 + 2H_2CrO_4 + 3H_2O \rightarrow 3(Cr_{.67}Fe_{.33})(OH)_3 + FeOOH$	Chromium	17.6	[30]
$2Fe^0 + 3Pb(C_2H_3O_2)_2 + 4H_2O \rightarrow 3Pb^0 + 2FeOOH + 4(HC_2H_3O_2) + 2H^+$	Lead	17.7	[30]
$C_2Cl_4 + 5Fe^0 + 6H^+ \rightarrow C_2H_6 + 5Fe^{2+} + 4Cl^-$	Tetrachloroethene	17.8	[36]
$C_2HCl_3 + 4Fe^0 + 5H_2O \rightarrow C_2H_6 + 4Fe^{2+} + 5OH^- + 3Cl^-$	Trichloroethene	17.9	[38]

Cheng et al. [25] revealed that nanoscale Fe^0 successfully reduced chlorophenols (RCl) to phenol (RH), while some of the chlorophenols were transformed to oxidation compounds, as given by Equation 17.4.

$$2Fe^0 + 4H^+ + O_2 \rightarrow 2Fe^{2+} + 2H_2O \tag{17.1}$$

$$Fe^0 + 2H_2O \rightarrow Fe^{2+} + H_2 + 2OH^- \tag{17.2}$$

In the past few years, nanoscale zero-valent iron (nZVI), consisting of an Fe^0 core and a predominantly magnetite shell [31], has become one of the most common adsorbents for the rapid removal of arsenic, including both arsenite (As(III)) and arsenate (As(V)), [26, 27, 29] as well as halogenated hydrocarbons, such as trichloroethylene (TCE) [31], NO_3^{32-}, Cr, and Pb [30]. Kanel et al. [27] revealed that the removal of As(III) is due to the spontaneous adsorption and coprecipitation of As(III), with the reactive surface sites of nZVI as oxidation products. X-ray diffractograms demonstrated that nZVI was converted to amorphous iron (III) oxide (Fe_2O_3), iron (III) hydroxide, magnetite (Fe_3O_4), or maghemite (γ-Fe_2O_3) corrosion products, mixed with lepidocrocite (γ-FeOOH). Hence, As(III) adsorption was enhanced with the increase of corrosion products.

To achieve a higher adsorption rate and removal efficiency, Hwang et al. [32] investigated the optimum conditions for the reduction of NO_3^- using nZVI. Their results showed that the supply of excessive thermal energy to the system increased the rate of nitrate reduction to nitrite and ammonia, where the activation energy barrier was overcome. In addition, a lower aqueous pH favors nitrate removal, in which ferrous hydroxide ($Fe(OH)_2$) and other protective layers deposited on the nZVI surface were dissolved, thereby yielding fresh reactive sites for the reduction process. The general reduction process of nitrate is given by Equation 17.5. Suzuki et al. [34] reported that the reduction of nitrate can be further enhanced by augmenting aqueous iron ions (Fe^{2+}), where the role of Fe^{2+} has been revealed to facilitate the electron transfer from the nZVI core to nitrate. Furthermore, a higher adsorbent dose can increase the removal efficiency of arsenate, and complete arsenic removal is achievable in acidic and neutral pH ranging from pH 3 to pH 7, as reported by Kanel et al. [29].

Although the addition of nZVI to the aquifer showed no effect on the geochemistry and indigenous microbial communities during TCE removal, as reported by Kirschling et al. [31], materials such as AC are recommended as support to nZVI to avoid loss and prevent the agglomeration of nZVI nanoparticles in the treated water. The use of support materials can also enhance the specific surface area of nZVI, thereby leading to higher adsorption capacity [30]. Zhu et al. [26] reported that AC-supported nZVI was easily regenerated by elution with 0.1 M sodium hydroxide (NaOH) for 12 h. Great reusability was shown because there was no observed significant reduction in As removal efficiency after eight cycles of adsorption and desorption. Meanwhile, Ponder et al. [30] revealed that supported nZVI showed great performance in the remediation of Cr(VI) and Pb(II) from aqueous solutions. Fe^0 was rapidly oxidized to goethite (α-FeOOH), while Cr(VI) and Pb(II) were reduced to Cr(III) and zero-valent lead (Pb^0), respectively, as given by Equations 17.6 and 17.7.

Fe becomes oxidized when it is in contact with a less active metal, such as Ag and palladium (Pd). Hence, the said concept is applied in water remediation to enhance the reactivity of monometallic nZVI [35]. Bimetallic nZVI, such as Ag/Fe, Pd/Fe, and Ni/Fe, have been shown to be effective in the transformation of chlorinated compounds [4, 36–38]. Xu et al. [4] evaluated the effectiveness of Ag/Fe bimetallic nZVI in the transformation of hexachlorobenzenes to tetra-, tri-, and dichlorobenzenes in an aqueous solution at room temperature. The results showed that the transformation rate was positively correlated to the metal loading, as well as the location of chlorine on the benzene ring. Lien et al. [36] and Tee et al. [38] demonstrated the utilization of Pd/Fe and Ni/Fe bimetallic nZVI on the degradation of chlorinated ethenes. Trichloroethene (C_2HCl_3) and C_2Cl_4 were reduced completely, with C_2H_4 as the primary reduction product, in accordance with Equations 17.8 and 17.9, as given in Table 17.2. The presence of metal on the nZVI surface significantly increased its reactivity due to the increase in the specific surface area for reactivity. The degradation of TCE in an aqueous solution is shown in Figure 17.1. Moreover, Tian et al. [37] demonstrated that

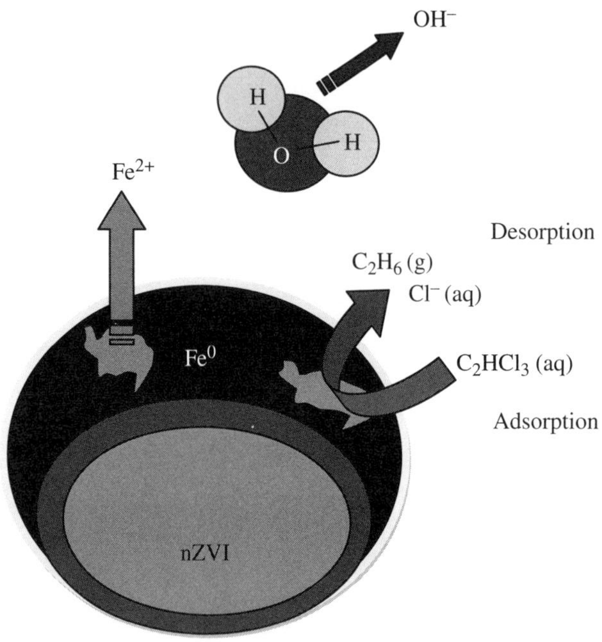

FIGURE 17.1 Degradation of TCE by nanoscale iron nanoparticles.

the degradation rate of dichloro-diphenyl-trichloroethane (DDT) was faster with bimetallic Ni/Fe nanoparticles than with monometallic nZVI. Their results also revealed that weak acidic or alkaline conditions favor the degradation of DDT.

In real scenarios, however, wastewater contains more than one contaminant. Thus, the effects of competing anions are significant. Anions, such as bicarbonate (HCO_3^-), $H_4SiO_4^0$, and $H_2PO_4^{2-}$, have potential competing agents in As(V) adsorption [27, 29]. The common divalent metal cations, such as magnesium ions (Mg^{2+}) and calcium ions (Ca^{2+}), present in contaminated water increase the affinity of nZVI toward As(V) anions. This is attributed to the fact that metal cations cause the nZVI surface to be more positively charged [26]. Therefore, more experiments are required to investigate the correlation among common contaminants in wastewater with nZVI adsorption capability.

17.2.1.2 Nanosized Oxides Nanoscale metal oxides rather than microscale metal oxides are preferred as adsorbents due to the larger specific surface area they cover, which favors adsorption. However, the separation of these nanosized materials from treated water has also raised some issues. Therefore, nanoadsorbents with good adsorption and magnetic properties have often been chosen for the said purpose, where separation can be easily achievable using an external magnetic field right after water remediation. Among nanosized metal oxides, nanosized Fe_2O_3, such as hematite (α-Fe_2O_3), γ-Fe_2O_3, and Fe_3O_4 nanoparticles, exhibited magnetic properties.

The ability of α-Fe_2O_3, γ-Fe_2O_3, and Fe_3O_4 nanoparticles to remove Orange II, a common azo dye in the textile industry, was investigated by Zhong et al. [23]. Their results showed that all three adsorbents successfully removed Orange II, with similar adsorption capabilities. Fe_2O_3 containing Orange II was also regenerated by catalytic combustion at 300°C in air for 3 h, and the readsorption performance remained almost unchanged. Furthermore, Hu et al. [21] demonstrated the potential of γ-Fe_2O_3 nanoparticles as a magnetic nanoadsorbent for the removal of Cr(VI) in wastewater. The adsorption of Cr(VI) by γ-Fe_2O_3 nanoparticles proceeded rapidly and reached equilibrium within 15 min, independent of adsorbate concentration. The existence of competitive ions, such as sodium ions (Na^+), copper ions (Cu^{2+}), nickel ions (Ni^{2+}), Ca^{2+}, Mg^{2+}, NO_3^-, and chloride ions (Cl^-), in wastewater has no significant effect on the adsorption of Cr(VI) by γ-Fe_2O_3 nanoparticles. Furthermore, Cr(VI)-loaded γ-Fe_2O_3 nanoparticles were recovered successfully, and were reused in the readsorption of Cr(VI) ions.

To increase the adsorption rate of contaminants from water, Chang et al. [8, 33] reported the effectiveness of chitosan-bound magnetic nanoadsorbents using Fe_3O_4 nanoparticles as cores and chitosan as ionic exchange groups, in the removal of heavy metal ions, such as Cu(II) and cobalt ion (Co(II)). Chitosan bound covalently on the surface of magnetite was capable of adsorbing metal ions due to the amino groups that serve as chelation sites. The time required to achieve adsorption equilibrium was 1 min for the removal of Cu(II) and Co(II) ions. The rapid adsorption of heavy metal ions using chitosan-bound magnetite nanoparticles from aqueous solutions was due to the high specific surface area and the absence of internal diffusion resistance.

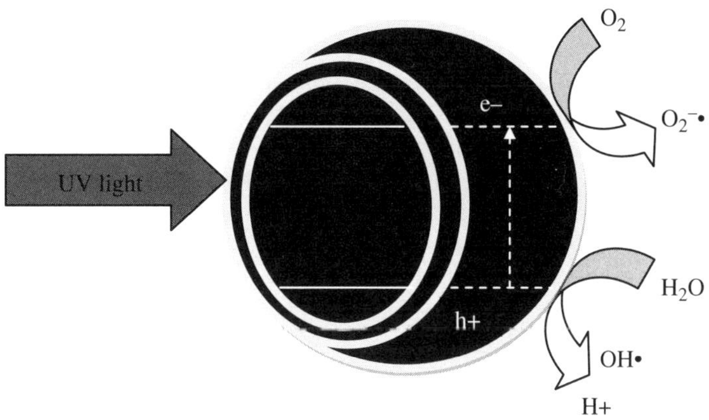

FIGURE 17.2 Schematic of general photocatalysis activity under UV irradiation.

Nanosized Al_2O_3 is one of the most widely studied nanosized metal oxides, aside from nanosized Fe_2O_3 [39]. Pacheco et al. [11] showed that nanosized Al_2O_3 was capable of removing five different metal ions, including Pb, Cr, Cu, Cd, and Hg, from aqueous solutions. However, the adsorption of Pb was quite low (50%), whereas the adsorption of other cations exceeded 90%. Meanwhile, Jang et al. [12] studied As(V) adsorption using highly ordered nanostructured SiO_2 media impregnated with metal oxides. SBA-15 was impregnated with three types of metal oxides, namely, Al_2O_3, zinc oxide (ZnO), and Fe_2O_3. Aluminum-impregnated SBA-15 showed higher As(V) adsorption capacity and faster adsorption rate than activated alumina. This result was due to the large number of active sites created by the dispersion of Al_2O_3 in the mesopore SiO_2 structures. The homogeneous and open pore structure of aluminum-impregnated SBA-15 also contributed to the transport of arsenate species, thereby increasing adsorption performance. Önnby et al. [40] demonstrated that nanosized Al_2O_3 was successfully incorporated into crygels, resulting in higher adsorption capacity and faster kinetics for the treatment of arsenic-contaminated waters, compared with nanosized Al_2O_3. However, the As removal efficiency of the aforementioned nanoadsorbent was reduced due to the existence of co-ions in water. The adsorption of phosphate ions (PO_4^{3-}), NO_3^-, and sulfate ions (SO_4^{2-}) by the nanoadsorbent was 69, 9, and 6%, respectively.

17.2.2 Water Remediation by Nanophotocatalysts via Photocatalytic Degradation

Photocatalytic degradation is one of the advanced oxidation processes (AOPs) using ultraviolet (UV) light as an energy source with a suspended powder or an immobilized form of semiconductor. AOPs are chemical remediation methods that involve the use of semiconductors [41], where active hydroxyl radicals (OH•) species are generated by a high-energy source and serve as the primary oxidant to induce chemical reduction or oxidation reaction [42]. Photocatalytic degradation possesses several advantages compared to other AOPs such as Fenton and photo-Fenton catalytic reactions and hydrogen peroxide (H_2O_2)/UV processes. The advantages included the ability to operate under ambient conditions, the use of inexpensive and nontoxic photocatalysts, as well as atmospheric air as oxidant [43].

Photocatalytic degradation can be categorized as either homogeneous or heterogeneous [44]. Homogeneous photocatalytic degradation involves the dissolution of catalysts and oxidants, such as H_2O_2 or ozone, by direct UV photolysis of the solution. Meanwhile, heterogeneous photocatalytic degradation, also called semiconductor photocatalytic degradation, involves the use of a heterogeneous semiconductor using oxygen (O_2) as an electron scavenger and UV light to remove organic impurities [45]. The basic mechanism of heterogeneous photocatalytic degradation involves the migration of electrons (e–) and holes (h+) to the surface of the photocatalyst through UV light irradiation. The photogenerated electrons and holes react with water (H_2O) and O_2, which is adsorbed on the surface of the photocatalyst, to produce reactive oxygen species, such as oxygen radicals (O_2^-·) and OH· [46]. Figure 17.2 shows the schematic of the general photocatalytic activities for photocatalysis under UV irradiation.

17.2.2.1 Titanium Dioxide Nanoparticles TiO_2 or titania nanoparticles are the most frequently used semiconductor for photocatalytic degradation of contaminated water, with good UV scattering effect [14]. The anatase crystalline phase of TiO_2 has been assumed to be more efficient than rutile and brookite TiO_2 because anatase TiO_2 have smaller particles (<50 nm), a larger surface area, and a higher degree of crystallinity, which favors photocatalytic degradation [47]. In addition, anatase TiO_2 also contains high concentrations of surface hydroxyl groups (OH) [48]. TiO_2 nanoparticles are inexpensive, nonselective, nontoxic, insoluble in water, photostable, and highly photoactive with a wide band-gap energy (3.2 eV), and can completely mineralize toxic and nonbiodegradable organics to carbon dioxide (CO_2), water, and inorganic constituents [6, 44, 45, 49–52]. However, there are some drawbacks in using TiO_2 nanoparticles, including the following: the use of a UV light–limited irradiation source because only light below 388 nm can produce electron and hole pairs; recombination of electron and holes; and separation and agglomeration problems [44, 49, 53]. These limitations lead to a decrease in the photocatalytic efficiency of TiO_2 nanoparticles and eventually reduce the photodegradation of contaminants in wastewater. Hence, numerous studies have been undertaken to utilize TiO_2 nanoparticles more efficiently, such as doping noble metals (Ag, platinum (Pt), gold (Au)), metals, metal ions, or metal oxides, on TiO_2 [5, 42, 44, 45, 48, 50, 54–59], and/or the immobilization of TiO_2 on a rigid support, such as AC and silicon carbide (SiC) [6, 7, 49, 51, 60–62]. The deposited metals and doped ions onto TiO_2 can extend the light absorption capacity into the visible range and suppress the recombination of electrons and holes by acting as electron traps in photocatalysts, leading to better availability of the holes and consequently to better degradation efficiency [45, 55].

Sobana et al. [45] demonstrated that the photocatalytic degradation of dyes was improved by doping noble metal Ag nanoparticles on TiO_2 as electron traps. Results showed that Ag nanoparticles could enhance electron–hole separation by transferring the trapped electrons to the adsorbed O_2, acting as an electron acceptor to produce reactive $O_2^{-}\cdot$ OH· was also generated during the process. In this study, the observed optimum loading ratio to achieve the highest photocatalytic degradation efficiency of dyes was approximately 1.5% Ag concentration in relation to TiO_2 concentration. However, due to the high prices of noble metals, most of the researchers focused on the use of other metals as dopants, such as transition metals and alkaline metals.

Park et al. [5] demonstrated the enhancement in photocatalytic degradation of 4-chlorophenol (4-CP) by doping Cu into TiO_2 nanoparticles. The increase in photocatalytic activity was due to the successful entry of Cu atoms into the TiO_2 lattice, which created charge compensating anion vacancies in the lattice points of TiO_2. Hydrochloric acid (HCl), CO_2, and water were the products formed after the photocatalysis of 4-CP. In the photocatalysis of 2-propanol, bimetal-incorporated TiO_2 photocatalysts (FeZn–TiO_2) were prepared, and the photocatalytic activity was investigated by Park et al. [48]. The highest activity was achieved at the element ratio of 1.5 between Fe and Zn. In this Fe/Zn ratio, the charge recombination of photoexcited electrons and holes was circumvented, thereby enhancing the degradation efficiency. Penpolcharoen et al. [44] showed the positive effect of nanohematite, embedded onto TiO_2, in the photocatalysis of sucrose and NO_3^-, wherein the photogenerated electrons were trapped by iron (III) ion (Fe^{3+}), which led to an improvement in the electron–hole separation and increased photocatalytic activity. Jeon et al. [42] investigated the photonic efficiency of nanosized molybdenum-doped TiO_2-mixed (Mo/Ti) oxide photocatalysts in the degradation of dichloroacetate (DCA). The optimum Mo content was 0.5 mol% to obtain the highest degradation efficiency of 28%.

The degradation efficiency of TiO_2 could be improved by doping with alkaline metals, such as lithium (Li), sodium (Na), and potassium (K). According to Bessekhouadet et al. [56], these alkaline metals lead to the creation of negative charges on the surface of the photocatalyst, facilitating adsorption of cationic molecules of malachite green oxalate (MG) contaminants through electrostatic interactions. The photoformed electrons then move from the lowest unoccupied molecular orbital (LUMO) band of MG to the conduction band of TiO_2. This process enhances electron production by the photocatalyst under illumination, thereby increasing the generation of radicals for the degradation of organic contaminants. An optimum content of dopant on TiO_2 can increase the ability of the doped photocatalyst in absorbing the light spectrum near the visible region. This phenomenon was revealed by Fan et al. [54], where TiO_2 was doped with 0.4% cerium and an enhancement in the photocatalytic degradation of phenol was observed when compared with that of pure TiO_2. The absorption spectrum of TiO_2 increased from 390 to 450 nm after doping, with cerium causing more light to be absorbed for photocatalytic activities.

Aside from noble, transition, and alkaline metals as dopants on TiO_2, metal oxide–modified heterogeneous nanoparticles are also important for maximizing the efficiency of photocatalytic degradation of wastewater contaminants. Dong et al. [59] indicated that cationic and anionic dyes in mesoporous TiO_2–SiO_2 nanocomposites exhibit high decolorization and degradation efficiency because of the synergistic effect of coupled adsorption and photocatalytic oxidation provided by the nanocomposites. The cationic and anionic dyes were absorbed by surface Si–OH and Ti–OH groups, respectively. Meanwhile, Han et al. [55] confirmed that the photocatalytic degradation of phenol begins with the breakup of carbon–carbon bonds by TiO_2–SrO photocatalysts, during which butane (C_4H_{10}) and propanal (C_3H_6O) ions are detected in the aqueous solution. The fastest photocatalytic degradation rate of phenol (14 h) has been achieved using 60 wt% of SrO in TiO_2 photocatalysts under visible light, compared with the 20 and 40 wt% of SrO. Meanwhile, doping $ZnFe_2O_4$ onto TiO_2 successfully enhanced the photocatalytic degradation of rhodamine B, as reported by Liu et al. [58] Moreover, O_2 is an important oxidant in photooxidation, acting as an electron acceptor to decrease the recombination rate between photogenerated electrons and holes. O_2 also promotes the generation of OH· radicals for the degradation of organic contaminants.

The photocatalytic activity of TiO_2 significantly affects the degradation of wastewater contaminants. Thus, aside from metals or metal oxides, ions can be doped onto TiO_2 to enhance its photocatalytic activity. Venkatachalam et al. [50] indicated that doping zirconium ions (Zr^{4+}) onto TiO_2 nanoparticles enhanced the photocatalytic degradation of 4-CP. Zr^{4+} successfully enters the TiO_2 lattice and increases photocatalytic activity. Charge-compensating anion vacancy was created in the TiO_2 lattice points, thereby improving the adsorption of 4-CP. Electron–hole recombination was also minimized under a pH of 5, as suggested by the increased degradation rate of 4-CP. However, according to Liu et al. [57], high levels (i.e., above the optimum level) of dopant ions such as Zn^{2+} limit the diffusion of Zn^{2+} on the TiO_2 lattice. Therefore, many ions are deposited on the particle surface, thereby inhibiting photocatalytic activity. Thus, the optimum Zn^{2+} ion content was set at 0.5 mol% in the photocatalytic degradation of rhodamine B.

Immobilizing TiO_2 on a rigid support, such as AC [6, 49, 62], perlite granules [60], SiC [51], SiO_2 spheres [61], and magnetic cores, [7] enhances photocatalytic degradation of organic contaminants and the separation properties of suspended TiO_2 photocatalysts from the solution. Li et al. [6] doped TiO_2 on AC grains to hybridize the photocatalytic activity of TiO_2 with adsorptivity of AC for photocatalytic degradation of methyl orange. AC enhanced the photoefficiency of TiO_2 photocatalysts by preventing the recombination of electron–hole pairs and extending its adsorption of methyl orange. Methyl orange was then transferred to TiO_2 and photocatalytically degraded. He et al. [62] investigated the use of TiO_2 on AC in removing rhodamine B under microwave irradiation. Rhodamine B is stable in aqueous solutions under microwave irradiation. However, the degradation efficiency of rhodamine B improved significantly to 96% 20 min after AC-supported TiO_2 was introduced into the solution, in which rhodamine B was mineralized to CO_2 and water after the conjugated structure of rhodamine B was destroyed by photocatalysts. Perlite and SiC are also effective supports for TiO_2 in the photocatalytic degradation of phenol [60] and 2-propanol [51]. Hosseini et al. [60] revealed that the rate of phenol degradation is positively affected by UV light intensity. Yamashita et al. [51] indicated that for photocatalytic degradation, the hydrophobic surface of TiO_2–SiC photocatalysts is preferred over that of 2-propanol diluted in water, which produces CO_2 and water.

Core–shell systems, composed of nanosized photocatalysts as the shell and a core with support, are required for easily reclaiming photocatalysts after photocatalytic degradation by either sedimentation or filtration of the aqueous solution. The properties of core–shell photocatalysts can be varied depending on the application, including the size, composition, and thickness of both the core and the shell. Using an inexpensive core in these systems reduces the preparation cost compared with non–core–shell systems, which use solely pure photocatalysts. Fu et al. [7] reported on using nanoscale hexaferrites ($BaFe_{12}O_{19}$) as a magnetic core to enhance the separation properties of photocatalysts from the aqueous solution. They also indicated that the TiO_2 shell is responsible for the photocatalytic degradation of organic contaminants. Wilhelm et al. [61] coated TiO_2 on SiO_2 spheres to photodegrade rhodamine B and found that rhodamine B was decomposed completely to colorless products, such as CO_2, mineral acids, and water, after illumination. Sun et al. [49] found that nanosized Sn(IV)/TiO_2/AC photocatalysts enhance the photodegradation activity of orange G and exhibit effective separation from the aqueous solution. Tin, at a dosage of approximately 2.5 % and pH of about 2, exhibited the highest photocatalytic activity. However, the degradation efficiency dropped with SO_4^{2-} and $H_2PO_4^-$ in the solution because these ions are deposited on the surface of photocatalysts and reduce vacancies for orange G adsorption, thereby reducing photodegradation activity.

17.2.2.2 Zinc Oxide Nanoparticles ZnO is a cheap and nontoxic photocatalyst used in degrading and mineralizing environmental contaminants to CO_2, water, and mineral acids [43]. ZnO has approximately the same band-gap energy as TiO_2 (3.2 eV) and is thus expected to exhibit photocatalytic degradation activity similar to that of TiO_2 [16, 17]. However, ZnO nanoparticles outperform TiO_2 nanoparticles in photocatalytically degrading wastewater contaminants. Sobana et al. [43] and Chen et al. [63] found that ZnO is more reactive than TiO_2 in the photocatalytic degradation of dyes under UV light irradiation. Moreover, the photodegradation rate is faster in alkaline conditions. The surface of ZnO in such conditions is negatively charged, favoring the absorption of dye with a positive charge. Thus, more active OH• radicals are generated, thereby enhancing activity. Furthermore, the efficiency of degrading estrone in water using ZnO is higher than that of degradation using TiO_2 [64]. The efficiency is higher because ZnO exhibits higher UV absorbance than TiO_2 within the wavelength range of 320–385 nm. The excellent activity of ZnO is also due to its intrinsic properties and unique open structure, which acts as a harvesting site for photogenerated electrons during optical excitation, thereby enhancing charge separation. Therefore, ZnO is a suitable alternative to TiO_2 because its photocatalytic degradation outperforms that of TiO_2 and both have a similar photocatalytic degradation mechanism [65]. Jang et al. [66] compared the photocatalytic degradation capacity of methylene blue between ZnO and ZnO nanocrystalline particles. ZnO nanoparticles exhibited greater photocatalytic degradation capacity than ZnO nanocrystalline particles because of the larger surface area provided by the former. The photodegradation efficiency of the dye increases by increasing photocatalyst loading (i.e., expanding the total active surface area) or decreasing the initial concentration of methylene blue solution. However, increasing photocatalyst loading decreases degradation activity because UV light penetration decreases upon loading beyond the optimum limit.

Researchers have made some modifications to increase the efficiency of ZnO in photocatalytically degrading contaminants. Wu et al. [67] found that ZnO coated with TiO_2 degraded phenol better than pure ZnO. Moreover, the light absorption scope of composite nanoparticles is wider than that of pure TiO_2. Yu et al. [68] investigated the performance of pure and Ag-modified ZnO as photocatalysts in the photocatalytic degradation of rhodamine B dye. Activity increased after incorporation of Ag into

ZnO. This result is consistent with that obtained by Anandan et al. [69], who found that the degradation rate of TCP over lanthanum (La)-doped ZnO is much higher than that of pure ZnO and TiO_2. The photocatalytic degradation rate of methyl orange dyes on nanosized coupled oxides (ZnO/SnO_2) is also faster than that on ZnO by about 66% [70]. The former is higher because modified ZnO photocatalysts have higher charge separation efficiency. Sobana et al. [41] modified ZnO with AC to enhance the photocatalytic activity of pure ZnO. Direct blue 53 dye molecules were initially adsorbed on AC, and then diffused to ZnO, which was deposited within the AC pores and degraded. As mentioned earlier, alkaline pH enhances photocatalytic degradation compared with acidic pH. Therefore, ZnO can be modified with metals or metal oxides to enhance its efficiency. Modification reduces the band-gap energy and improves charge separation between photogenerated electrons and holes, thereby enhancing the efficiency of ZnO in photocatalytically degrading contaminants.

17.2.3 Water Remediation by Nanodisinfection Agents via Disinfection

Conventional chlorination, ozonation, advanced filtration, and UV radiation processes are effective water disinfection techniques. However, these techniques have several limitations, including the formation of toxic by-products and high energy consumption [71]. Therefore, new methods for water disinfection, such as nanotechnology, that can overcome such limitations are being explored. The photocatalysis of heterogeneous nanosized particles, such as TiO_2 and metallic Ag nanoparticles, degrades hazardous contaminants in water, as discussed in previous sections. Moreover, these nanoparticles kill bacteria and inactivate various harmful microorganisms [18–20, 72]. These nanoparticles can do so because their small size and high surface to volume ratio enable them to interact closely and directly with pathogens [73].

Ag nanoparticles are effective bactericides, exhibiting strong toxicity against various microorganisms [20]. Morones et al. [73] proposed the general route of bio-Ag^0 against *Escherichia coli*, *Pseudomonas aeruginosa*, *Vibrio cholerae*, and *Salmonella typhi*. The bactericidal process begins by disrupting the functions of the bacteria, such as respiration, after the nanoparticles attach to the bacteria cell membrane. The penetration of nanoparticles into the bacteria damages the DNA, a major sulfur- and phosphorus-containing compound. Ag ions (Ag^+) also contribute to such bactericidal effect. The bactericidal properties of the nanoparticles are size- and surface area–dependent, and only nanoparticles of about 1–10 nm interact well with Gram-negative bacteria. Figure 17.3 illustrates the pathway of Ag^+ ions against bacteria.

According to Lalueza et al. [74], the bactericidal effect of Ag nanoparticles is weak because of the slow generation rate of Ag^+ ions resulting from the small number of Ag nanoparticles. Thus, the number of Ag^+ ions should be sufficient to ensure full interaction with thiol groups in proteins and interfere with DNA replication in bacteria as one disinfection pathway [75]. The bactericidal effects with the same total silver content are ranked as follows: $AgNO_3$> Ag-ZSM-5 > Ag_2O > pellets > granular > Ag nanoparticles. Thus, bactericidal efficiency largely depends on the easy release of Ag^+ ions from the material to yield high silver concentration. De Gusseme et al. [20] found that biologically produced bio-Ag^0 exhibits a faster inactivation rate of *Enterobacteraerogenes*-infecting bacteriophage (UZ1) than $AgNO_3$ and chemically produced bio-Ag^0 with the same Ag

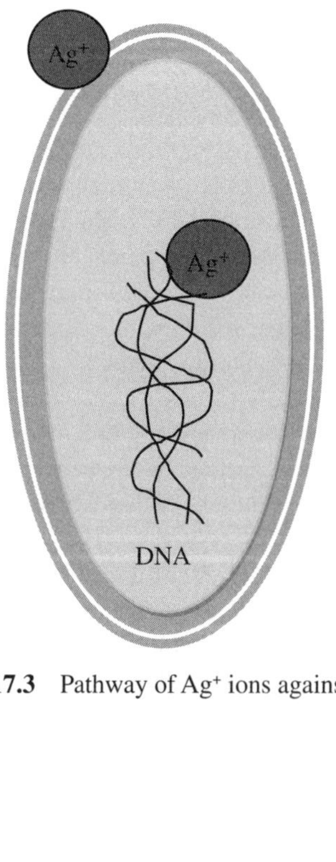

FIGURE 17.3 Pathway of Ag^+ ions against bacteria.

concentration. The advance properties shown by biogenic Ag^0 was due to the role of the dead lactic acid bacterial cell wall within the nanoparticles as microscale carrier matrix for the nanoparticles and prevent the aggregation problem.

Nangmenyi et al. [19] reported that Ag nanoparticles impregnated on an inorganic fiberglass surface exhibit both bacteriostatic and bactericidal properties for disinfecting *E. coli* because of its inorganic fibrous network that offers high contact efficiency and faster kinetics. Gangadharan et al. [18] used modified $AgNO_3$ methacrylic acid copolymer beads in disinfecting Gram-negative and -positive bacteria. *E. coli*, *P. aeruginosa*, and *Staphylococcus aureus* were completely inhibited, but not for spore-forming *Bacillus subtilis*. The inhibited bacteria had little resistance toward the Ag nanoparticle–bound copolymer beads, with 99.9% reduction within the same time frame. In addition, no bacteria adhered to or were adsorbed onto the copolymer beads containing Ag nanoparticles. Thus, these beads have potential applications in water disinfection. Lin et al. immobilized Ag nanoparticles on TiO_2 (TiO_2/Ag) to form hybrid nanocomposite particles [76]. The antibacterial activity of TiO_2/Ag hybrid particles against *E. coli* is similar, with or without UV irradiation. However, the bactericidal behavior of hybrid particles is significantly affected by Ag concentration in the nanocomposite, which is favorable for deactivating *E. coli*.

17.3 NANOCOMPOSITE MEMBRANES IN WATER REMEDIATION

Membrane separation technology is of interest in water remediation because of its high separation efficiency and ease of operation. This technology uses polymeric and ceramic membranes composed of organic and inorganic materials, respectively. Polymeric membranes have become the focus of membrane research because of better control of the pore-forming mechanism, higher flexibility, and lower costs as compared to ceramic membranes. Polymeric membranes have been widely studied including polyethersulfone (PES), poly(vinylidene) difluoride (PVDF), polyamide (PAm), polysulfone (PSf), and polyacrylonitrile (PAN). However, polymeric membranes exhibit high hydrophobicity, fouling and biofouling, low flux, and low mechanical strength. Meanwhile, ceramic membranes made of refractory oxides, such Al_2O_3, TiO_2, and zirconium oxide (ZrO_2), have excellent chemical, thermal, and pH stability compared with polymeric membranes. Moreover, given their long lifetime and resistance to high temperature, pressure, and corrosive solutions, ceramic membranes are very cost-effective for water treatment.

With the objective to improve the properties and functionalities of polymeric and ceramic membranes in water remediation, nanoparticles have been introduced into polymeric and ceramic membranes. Generally, the main goals behind the incorporation of nanoparticles into membranes include membrane fouling and biofouling mitigation, disinfection, and photocatalytic degradation of pollutants.

17.3.1 Development of Membranes with Antifouling Properties

17.3.1.1 Fouling Mitigation in Polymeric Membranes Nanoparticles are added to membranes to minimize fouling and enhance separation properties. Membrane fouling caused by organic solutes and biofouling are the most common types of fouling. The addition of nanoparticles to the membranes increases skin layer thickness and skin surface porosity, suppresses macrovoid formation, increases membrane permeability, and increases or decreases rejection [77]. TiO_2 nanoparticles are the most widely used in remedying membrane fouling caused by organic solutes.

17.3.1.1.1 Titanium Dioxide Nanoparticles One problem in the fabrication of TiO_2/polymer nanocomposites is the dispersion of TiO_2 nanoparticles into the polymer matrix because these nanoparticles weaken the performance of nanocomposites by agglomeration. Thus, Wu et al. [78] modified the TiO_2 surface with γ-aminopropyltriethoxysilane before introducing TiO_2 to the PES matrix to overcome agglomeration. TiO_2 content was optimal at 0.5 wt%. Antifouling properties were improved based on the filtration of bovine serum albumin (BSA) solution. Yang et al. [79] employed sodium dodecyl sulfate to modify the surface of TiO_2 nanoparticles before incorporating them into the PSf matrix. The membranes exhibited outstanding water permeability, hydrophilicity, mechanical strength, and good antifouling properties at 2 wt% TiO_2 content. The membranes also effectively treated emulsified oil wastewater.

Aside from homogeneously dispersing TiO_2 into the polymer matrix, other preparation methods, including membrane surfaces coated with TiO_2 nanoparticles, have been explored to fabricate nanocomposite membranes. TiO_2 nanoparticle self-assembly membranes were prepared by immersing a poly(styrene-alt-maleic anhydride) and PVDF blend membrane in a TiO_2 nanoparticle solution [80]. TiO_2 nanoparticles were assembled on the membrane surface via electrostatic interaction established between TiO_2 nanoparticles and the -COOH groups on the membrane surface. Bae et al. [81] successfully fabricated sulfonated poly(ether) sulfone (SPES) membranes surface-coated with TiO_2 via the electrostatic interaction between TiO_2 nanoparticles and anionic groups on the surface of SPES membranes. Permeability and antifouling ability were enhanced by the TiO_2 layer on the membrane surface [80, 81]. Rahimpour et al. [82] discovered that the performance and antifouling properties of TiO_2-coated membranes are better than those of TiO_2-entrapped membranes.

The effect of UV irradiation on the performance of TiO_2 nanocomposites has also been examined [82–84]. Membrane performance and antifouling properties significantly improved with the increased hydrophilicity of the membrane and the photocatalysis reaction initiated by the UV irradiation, thereby decomposing and removing the foulants, especially organic compounds.

The size of TiO_2 nanoparticles also significantly affects the performance of TiO_2 nanocomposite membranes. Vatanpour et al. [85] and Cao et al. [86] studied the effect of TiO_2 nanoparticle size on the performance of PES and PVDF membranes, respectively. The antifouling property of the TiO_2 nanocomposite membrane improved with decreasing TiO_2 nanoparticle size because of the large surface area and high adsorption of water.

17.3.1.1.2 Silicon Dioxide Nanoparticles Aside from TiO_2, SiO_2 is used to mitigate fouling in polymeric membranes. SiO_2 has been widely used in fabricating thin-film nanocomposites for various applications because of its availability and low price compared with TiO_2. Jadav and Singh [87] incorporated self-synthesized SiO_2 nanoparticles via the hydrolysis of tetraethyl orthosilicate in an acidic medium into PAm films coated over a porous PSf support through interfacial polymerization to produce a thin-film nanocomposite membrane. The flux increased as a function of SiO_2 loading because of increasing pore size. Excellent separation efficiency and productivity flux are obtainable with nanocomposite membranes containing about 1–2 wt% SiO_2.

Similarly, commercialized SiO_2 nanoparticles incorporated into thin-film nanocomposite membranes were synthesized by adopting poly(amidoamine) dendrimers as aqueous monomers, trimesoyl chloride as organic monomer, and PSf as substrate [88]. The addition of 1.0 wt% SiO_2 nanoparticles increased the permeation performance of the PAm–SiO_2 membrane by nearly 50%, without a decrease in the salt rejection rate. When subjected to one cycle of filtration with oily wastewater, the PAm–SiO_2 membrane demonstrated higher stable flux and nearly 50% salt removal. Jin et al. [89] reported that using PAm–SiO_2 thin-film nanocomposite membranes prepared by a similar method in treating raw surface water enhanced membrane separation and antifouling properties.

Yin et al. [90] synthesized MCM-41 (~100 nm) and nonporous spherical silica nanoparticles (~100 nm) as nanofillers to study the influence of the internal pore structures of SiO_2 nanoparticles on the performance of thin-film nanocomposite membranes. The MCM-41 nanoparticles incorporated in the membrane showed significant improvement in permeating water flux, while maintaining high rejection of salts. This result suggests that the internal pores of MCM-41 nanoparticles significantly improved water permeability compared with those of nonporous silica nanoparticles.

Apart from the incorporation of SiO_2 nanoparticles into thin-film nanocomposite membranes by in situ polymerization, the hydrophilicity of PSf ultrafiltration membranes functionalized with SiO_2 nanoparticles via solution blending was also enhanced [91]. The functionalized PSf membrane exhibited good antifouling properties and an increased permeability that is 16 times higher than that of the neat PSf membrane in the filtration of oil-in-water emulsion, which is attributed to the larger pore size and higher hydrophilicity.

17.3.1.1.3 Aluminum Oxide Nanoparticles Like other metal oxide nanoparticles, aluminum-based nanoparticles possess properties that improve the performance of nanocomposite membranes. Yan et al. [92] dispersed commercialized Al_2O_3 nanoparticles (10 nm) into PVDF membranes homogeneously via ultrasonication for the treatment of oily wastewater. Antifouling ability was enhanced as a result of decreased contact angle at constant surface porosity. The antifouling ability of the membranes was also proven by filtration of α-amylase solutions [93] and oil field wastewater [94]. Yang et al. [95] also found that the integration of nanosized Al_2O_3 into the PVDF membrane structure significantly enhanced the antifouling ability of the PVDF membrane.

Aside from homogeneously dispersing the nanoparticles into a polymer matrix, as discussed previously, Saleh and Gupta [96] fabricated PAm nanocomposite membranes containing Al_2O_3 nanoparticles by in situ interfacial polymerization. They used Al_2O_3 nanoparticles with an average size of 14 nm, self-generated by an aqueous sol–gel method using precursors of aluminum nitrate and citric acid mixed solution. The permeate flux was enhanced, while maintaining the salt rejection with the introduction of nanoparticles to the PAm membranes.

Apart from Al_2O_3 nanoparticles, boehmite, an aluminum oxide hydroxide (γ-AlOOH) particle containing extra hydroxyl groups on its surface, was used by Vatanpour et al. [97] in preparing PES-blended membranes. Incorporating boehmite nanoparticles improved the hydrophilicity and pure water flux of the membranes. Boehmite nanoparticle–embedded membranes also exhibited superior characteristics and antifouling ability compared with γ-Al_2O_3/PES membranes. Nevertheless, pure water flux decreased at high nanoparticle concentrations because of agglomeration.

17.3.1.1.4 Zinc Oxide Nanoparticles ZnO nanoparticles, one of the multifunctional inorganic nanoparticles, have gained increasing attention because of their remarkable physical and chemical properties, including high catalytic activity and effective antibacterial and bactericidal abilities. In addition, ZnO nanoparticles are economically more practical than TiO_2 and Al_2O_3 nanoparticles [98]. Furthermore, incorporating ZnO nanoparticles enhanced the hydrophilicity and mechanical properties of the polymer matrix.

ZnO nanocomposite membranes are developed by dispersing different dosages of ZnO nanoparticles homogenously into the polymer matrix, followed by phase inversion [98–100]. In treating reclaimed water, Hong and He [99] reported maximum pure water flux, minimum surface roughness, and minimum membrane resistance at an optimum dosage of 0.005 wt% of ZnO nanoparticles in the PVDF membranes because of enhanced hydrophilicity. Enhanced membrane separation properties were also observed by Liang et al. [98] and Balta et al. [100] Given the successful implantation of ZnO nanoparticles into the membrane inner surface or pore wall, the membrane exhibited remarkable anti-irreversible fouling property and doubled water permeability at the optimum dosage of 6.7 wt% ZnO nanoparticles when the PVDF membranes modified with ZnO were used in multicycle filtration experiments [98]. Aside from lower flux and improved permeability, Balta et al. [100] observed a prominent improvement in dye rejection potential induced by the addition of ZnO nanoparticles to PES membranes even at unusually low concentrations. Given that ZnO nanoparticles exhibit properties similar to those of TiO_2, but are comparatively cheaper, the former have been predicted as an excellent alternative to TiO_2 for use as an antifouling material.

17.3.1.2 Fouling Mitigation for Ceramic Membranes Metal oxides, such as Al_2O_3, TiO_2 and Fe_2O_3 are the most commonly used nanoparticles for the fabrication of nanostructured ceramic membranes for applications in water treatment. The integration of metal oxide nanoparticles with ceramic membranes provides additional functionalities to catalyze and degrade foulants under oxidizing conditions.

17.3.1.2.1 Titanium Dioxide Nanoparticles TiO_2 nanoparticles have drawn significant research attention in the fabrication of ceramic membranes because of their excellent chemical resistance, high water flux, and photochemical and catalytic properties [101]. Previous studies fabricated ceramic membranes by incorporating highly hydrophilic TiO_2 nanoparticles to create preferential passages for the transport of water. For example, Zhang et al. [102] prepared Al_2O_3 ceramic membranes doped with 5% TiO_2 by the solid-state sintering technique. Compared with pure Al_2O_3 membranes, the Al_2O_3–TiO_2 nanocomposite membranes exhibited up to 30% improvement in permeate flux in the filtration of oily wastewater under the same operating conditions because of their enhanced membrane hydrophilicity and antifouling properties resulting from the doping of TiO_2. Monash and Pugazhenthi [103] prepared low-cost porous ceramic membranes with different loadings of TiO_2 nanoparticles. The prepared nanocomposite membranes with optimum TiO_2 loading exhibited a maximum rejection of 99% for an oil concentration of 200 ppm in the oil–water emulsion separation.

Yang and Li [104] fabricated inside-out tubular TiO_2/Al_2O_3 composite membranes for electrofiltration of chemical–mechanical polishing wastewater. The composite membranes were fabricated by coating the TiO_2 coating layer inside the tubular substrate of Al_2O_3 using sol–gel/slip-casting technique and then followed by firing process. The incorporation of 5% TiO_2 to Al_2O_3 increased the permeate flux up to 20%, indicating enhanced hydrophilicity of TiO_2–Al_2O_3 composite membranes compared with pure Al_2O_3 membranes. Zhang et al. [105] fabricated a hierarchical TiO_2 nanowire ultrafiltration membrane made of two dimensions of TiO_2 nanowires, 20–100 nm TNW20 and 10 nm TNW10, by hydrothermal synthesis, followed by a filtration and hot-press process. TNW20 acts as the supporting layer of the membrane, providing mechanical strength, while TNW10 serves as the functional layer, providing outstanding permeability and separation properties. TiO_2 nanowire ultrafiltration membranes exhibited exceptional performance in photodegradation and antifouling tests because of the degradation and inactivation of organic and biological pollutants under UV irradiation.

17.3.1.2.2 Aluminum-Based Nanoparticles In addition to the application of aluminum-based nanoparticles as nanofillers in nanocomposite membranes, Al_2O_3 nanoparticles can also be used as ceramic precursors. Jones et al. [106] fabricated asymmetric Al_2O_3 ultrafiltration membranes with acetic acid surface-stabilized Al_2O_3 nanoparticles (A-alumoxanes). A defect-free Al_2O_3 membrane was obtained by contacting α-Al_2O_3 supports with an aqueous solution of A-alumoxanes followed by firing. The alumoxane-derived membranes possessed high porosity and similar or improved permeability as compared with commercially available alumina membranes. The smooth surface of alumoxane-derived membranes was confirmed by scanning electron microscopy (SEM), and their high hydrophilicity was verified by atomic force microscopy (AFM), indicating great potential in fouling remediation of membranes. Bailey et al. [107] also found that alumoxane-coated ceramic membranes are more hydrophilic than uncoated membranes.

γ-Al_2O_3, another type of aluminum-based nanoparticles, has been extensively studied because of their unique surface charge characteristics defined by the amphoteric behavior of their surface sites (hydroxyl groups). Majhi et al. [108] prepared a low-cost γ-Al_2O_3 membrane on a macroporous clay support using the dip-coating method. The production cost was reduced by fabricating the γ-Al_2O_3 top layer from a boehmite sol prepared via a sol–gel route employing aluminum chloride ($AlCl_3$) salt as a starting material. The maximum rejection values for $MgCl_2$ and $AlCl_3$ were up to 72 and 88%, respectively, at a salt concentration of 3000 ppm.

Instead of employing self-generated γ-Al_2O_3, Madaeni et al. [109] employed a commercialized γ-Al_2O_3-based ceramic microfiltration membrane to eliminate coke particles from petrochemical wastewaters prior to the introduction of coalescers.

The γ-Al_2O_3-based ceramic microfiltration membrane achieved 100% coke rejection at harsh operating conditions of 70°C and 15 bar pressure. DeFriend et al. [101] doped Al_2O_3 membranes fabricated using A-alumoxanes with transit metals, such as Fe, La, and manganese (Mn), to further improve the separation performance of Al_2O_3-based ceramic membranes. $FeAlO_3$ and $LaAlO_3$ membranes showed permeate flux increment of 50%, while the $MnAlO_3$ membrane showed decreased permeate flux compared with the Al_2O_3 homogeneous membrane when used for filtration of dye solutions. Therefore, permeate flux improvement was influenced by chemical functionalization of the Al_2O_3 surface.

17.3.1.2.3 Silicon Dioxide Nanoparticles A separation skin layer with a uniform pore size distribution and a smaller pore size than an inorganic membrane support is required to improve the separation properties of the ceramic membrane [110]. Mesoporous SiO_2 with uniform monodispersed pore sizes and ordered pore structure, which provide large surface area and high pore volume, is one of the promising candidates for the fabrication of the skin layer of a membrane to provide high separation efficiency.

Modified ceramic membranes coated with a layer of mesoporous SiO_2 have been synthesized for different applications to investigate their separation performance. Fakhfakh et al. [111] prepared ceramic membranes derived from mesoporous SiO_2 coated on a clay tubular support and studied their separation properties. The performances of the membranes M1 and M2, with 12.5 and 27-μm thickness, respectively, were compared. The water permeability of M1 was higher at 77.3 l/h·m^2·bar, while that of M2 was 34.56 l/h·m^2·bar. On the contrary, the BSA retention of M2 was higher at 98.64% than that of M1 at 78.33%.

Oh et al. [110] further grafted the SiO_2 coating layer with 3-aminopropyltriethoxysilane (APTS) to facilitate the removal of Cu^{2+} from the aqueous solution. The resulting membrane showed improved rejection and selectivity compared with unmodified ceramic membranes, suggesting that the introduction of mesoporous SiO_2 and amine functionality exhibited great potential in the removal of heavy metal ions from wastewater streams.

In addition to mesoporous SiO_2, Duke et al. [112] fabricated Al_2O_3 ceramic membranes coated with an amorphous molecular sieve of SiO_2. The fabricated membranes exhibited excellent separation properties, with a flux of approximately 1.8 kg/m^2·h and sodium chloride (NaCl) rejection up to 98% with 3.5 wt% (seawater-like) feed. However, knowledge of the effects of employing real seawater feed, long-term operation, and scale-up on membrane performance are essential for future studies.

17.3.1.2.4 Iron-Based Nanoparticles Fe-based nanoparticles have been employed in the modification of ceramic membranes because of their relatively high specific surface area, low cost, availability, easy production, and catalytic activities when combined with strong oxidants and/or UV/visible light. Cortalezzi et al. [113] fabricated ceramic membranes deposited with Fe-based nanoparticles, ferroxane, to examine the effect of the number of coating layers on membrane permeability by measuring clean water fluxes. The results revealed that the membrane permeability varied with the number of coating layers. The rejection tests using dextran molecules showed that the ceramic membrane with two layers of ferroxane nanoparticle coating demonstrated the highest rejection.

A previous study applied Fe_2O_3-coated tubular AZT (a mixture of Al_2O_3, ZrO_2, and TiO_2) ceramic membranes in a combined ozonation–ultrafiltration process through a layer-by-layer technique to eliminate the disinfection by-products and their precursors [114]. More than 85% reduction in the concentration of dissolved organic carbon was obtained using the resulted membrane. In addition, up to 90% decrease in the concentrations of simulated distribution system total trihalomethanes and 85% decrease in simulated distribution system halo acetic acids were observed. Compared with uncoated membranes, the concentrations of aldehydes, ketones, and keto acids in the permeate were decreased by >50% with the Fe_2O_3-coated ceramic membranes. Harman et al. [115] also reported better natural organic matter (NOM) rejection with the coating of Fe_2O_3 on γ-Al_2O_3 ceramic membrane in water with a high specific UV absorbance (SUVA) value, resulting from NOM sorption onto the Fe_2O_3 surfaces.

17.3.2 Development of Membranes with Antibiofouling Properties

17.3.2.1 Antibiofouling in Polymeric Membranes The incorporation of antimicrobial nanomaterials into membranes has been a potential alternative to the conventional forms of biofouling control as polymeric membranes cannot withstand the corrosiveness of chemical cleaners during pretreatment or chemical cleaning. Ag nanoparticles are widely applied in biofouling mitigation of polymeric membranes because of their antibacterial property. A few types of Ag nanoparticles are incorporated into polymeric membranes including commercialized Ag, self-derived Ag from $AgNO_3$, and bio-Ag^0.

Lee et al. [116] employed commercialized Ag nanoparticles in the fabrication of PAm membranes with 10 wt% Ag nanoparticles by in situ polymerization. The nanocomposite membranes exhibited dramatic antibiofouling effects on *Pseudomonas* with little influence on water flux and salt rejection. In addition, SEM analysis also proved that all *Pseudomonas* were made inactive on the nanocomposite membrane.

In another study carried out by Zodrow et al. [117], commercial Ag nanoparticles integrated PSf ultrafiltration membranes synthesized by the wet-phase inversion process showed antimicrobial properties toward different types of bacteria, including

E. coli K12, *Pseudomonas mendocina* KR1, and MS2 *bacteriophage*. The results showed that the membrane hydrophilicity was improved with the addition of Ag nanoparticles. Nevertheless, a loss of Ag from the membrane surface caused the loss of antibacterial and antiviral activities.

Meanwhile, Zhang et al. [118] coated amidoxime surface functionalized PAN (ASFPAN) nanofibers with silver ions (Ag^+) generated from $AgNO_3$ by coordination with the amidoxime functional groups followed by reduction into Ag nanoparticles. The resulting membranes demonstrated the capability of killing the tested microorganisms of *S. aureus* and *E. coli* within 30 min. Furthermore, the resulting membranes showed good transport properties in water permeability tests, suggesting their suitability for filtration applications.

Sawada et al. [119] developed a hydrophilic acrylamide-grafted PES hollow fiber membrane containing Ag nanoparticles with both organic antifouling and antibacterial properties using the method used by Zhang et al. [118] A reduction in membrane fouling by BSA was observed with the nanocomposite membrane because of the improved hydrophilicity resulting from the grafting with acrylamide. High antibacterial activity was exhibited by the Ag-loaded membrane when evaluated against *E. coli* by the halo zone test with an agar culture medium. Li et al. [120] also reported an improvement in antibiofouling performance and surface hydrophilicity because of the immobilization of Ag nanoparticles on the PVDF membrane surface grafted with poly(acrylic acid) (PAA).

Ag nanoparticles, commonly synthesized by the chemical reduction method, are often associated with problems, such as particle stability and the tendency of agglomeration at high concentrations or when the average particle size is less than 40 nm [121]. Therefore, Zhang et al. [122] employed a biological route to synthesize bio-Ag^0 nanoparticles using *Lactobacillus fermentum* LMG 8900 and embedded them into PES membranes. The results showed that the bio-Ag^0/PES composite membranes had excellent antibacterial activity and prevented the attachment of bacteria to the membrane surface as well as reduced biofilm formation in the 9-week test.

17.3.2.2 Antibiofouling in Ceramic Membranes Ceramic membranes are widely used in drinking water treatment because of their long lifetime and resistance to high temperature, pressure, and corrosive solutions. However, biofilms are formed on the surface of ceramic membranes because of the attachment of microorganisms. Hence, Ag nanoparticles are coated onto ceramic membranes because of their antibacterial properties.

Lv et al. [123] coated Ag nanoparticles on porous ceramic modified by an aminosilane coupling agent, APTS, via the coordination bonds formed between the -NH_2 group of the APTS molecule and the Ag atoms on the surface of the nanoparticles. The results showed that the resulting composite can be stored for long periods of time and is resilient under washing in spite of ultrasonic irradiation without loss of nanoparticles. The output count of *E. coli* was zero when tested in filtered water with a bacterial load of approximately 105 colony-forming units (CFU) per milliliter. Seo et al. [124] deposited large amounts of Ag nanoparticles on the surface of mesoporous $Al(OH)_3$ film using the polyol process. The results showed that the Al foam filter coated with Ag/$Al(OH)_3$ mesoporous nanocomposite film showed enhanced antimicrobial filtration properties compared with the bare Al foam filter.

Ma et al. [125] employed TiO_2 modified with Ag nanoparticles in the fabrication of Ag–TiO_2/hydroxyapatite (HAP, $Ca_{10}(PO_4)_6(OH)_2$)/Al_2O_3 bioceramic composite membranes through a simple two-step approach, involving the sol–gel method followed by calcinations, to integrate membrane separation and photocatalytic bacterial inactivation. The resulting membranes exhibited remarkable bactericidal activity because of the improved photobiocide activity of the Ag–TiO_2 nanocomposite and the bacterial adherence of metallic Ag, with the presence of weak UV illumination.

17.3.3 Development of Membranes with Photocatalytic Functionalities

In addition to antifouling properties and flux improvement, the introduction of nanoparticles in membranes also provides catalytic functionalities in the development of photocatalytic membranes for application in photocatalytic reactors for water treatment. Aside from TiO_2 nanoparticles, other metal oxides, such as ZnO, Fe_2O_3, and manganese oxide (MnO), have also been utilized and investigated for this purpose [126, 127].

Photocatalytic reactors for water treatment can be categorized into two main configurations, namely, reactors with suspended photocatalyst particles and reactors with the photocatalyst immobilized on the membrane [128]. The difference between these two configurations is that the former requires an additional downstream separation unit to recover the photocatalyst particles, whereas the latter enables continuous operation. Recently, the development of photocatalytic membranes has gained increasing attention because of their ability to induce photocatalytic reactions on the membrane surface or inside the pores of the membrane. Furthermore, water treatment can be continuously discharged without risking the loss of photocatalyst particles. Significant results were obtained when the catalytic process on the membrane surface was combined with the oxidation process, such as ozonation and UV illumination.

As a strong oxidizing agent, ozone possesses high reactivity with organic compounds responsible for the fouling of membranes in water treatment [129]. Furthermore, the integration of ozonation with other oxidation processes, such as photocatalysis, in the degradation of organic substances was successful [130–132]. Polymeric membranes are not feasible in the combination of ozonation and membrane processes because of their poor stability when exposed to ozone [133]. Meanwhile, ceramic membranes with superior thermal properties and high resistance against corrosion and chemicals are suitable in combined ozonation and membrane processes for water or wastewater treatment [133–135].

17.3.3.1 Polymeric Membranes with Photocatalytic Functionalities Under UV irradiation, polymeric membranes incorporated with TiO_2 nanoparticles exhibit excellent photocatalytic and separation properties in various applications [136–138]. In the removal of RB5 dye under UV via photolysis and photocatalysis mechanisms, composite PVDF/TiO_2 demonstrated a higher rate of RB5 removal than a neat PVDF membrane [138].

Meanwhile, You et al. [137] investigated the photocatalytic and separation performances of a PAA plasma-grafted PVDF membrane with self-assembled TiO_2 nanoparticles in the removal of RB5 dye. The results showed that the flux behavior improved significantly after UV irradiation, which activated the photodegradation of strongly attached foulants. The resulting membrane also demonstrated the capability for photocatalytic removal of RB5 dye.

Song et al. [136] reported that polyethylene glycol (PEG)–PVDF membranes doped with TiO_2 nanoparticles exhibited improved removal of NOM and reduced membrane fouling. In addition, the PEG–TiO_2–doped PVDF membranes also exhibited excellent self-cleaning ability because of their photocatalytic properties.

17.3.3.2 Ceramic Membranes with Photocatalytic Functionalities In addition to polymeric membranes, ceramic membranes have been employed in the fabrication of catalytic membranes for the integrated oxidation and separation processes under UV irradiation. Ma et al. [139] fabricated an asymmetrical inorganic membrane by coating silicon (Si)-doped TiO_2 layers on a commercial Al_2O_3 membrane by a sol–gel technique and found that the removal efficiency of the dye reactive red ED-2B was enhanced significantly under UV irradiation than with photocatalysis or membrane separation alone because of simultaneous separation and degradation of the pollutant.

Similar results were reported by Ma et al. [140] using a composite membrane incorporated with Ag-doped TiO_2 nanoparticles (Ag–TiO_2/HAP/Al_2O_3) for humic acid (HA) removal by microfiltration-coupled photocatalysis. Owing to the synergistic effect of the photocatalytic degradation of foulants and filtration, HA removal and the antifouling property of the fabricated membrane was enhanced under UV irradiation.

In addition to employing doped TiO_2 nanoparticles in the integrated oxidation and separation processes, Liu et al. [141] deposited Ag nanoparticles on electrospun TiO_2 nanofibers by in situ polyol synthesis in the fabrication of Ag/TiO_2 nanofiber membranes. Remarkably high permeation flux was exhibited by the Ag/TiO_2 nanofiber membrane. The Ag/TiO_2 nanofiber membranes exhibited bacterial inactivation up to 99.9% and dye degradation up to 80% under solar irradiation of 30 min.

Hong et al. [142] demonstrated a different approach in the degradation of organic solvents using needle-shape TiO_2 nanofibers inside a micro-channeled Al_2O_3–ZrO_2 composite porous membrane system through a sol–gel technique for the integrated photocatalytic filtration process. In this study, TiO_2 nanofibers were grown on the surface frame of the micro-channeled Al_2O_3–ZrO_2 composite membrane, and the interior of the micro-channeled pores were fully covered. Notably, the specific surface area of the TiO_2-coated membrane system was significantly increased by over 100-fold by following this procedure compared with that of the noncoated membrane.

In addition to TiO_2 nanoparticles, Ke et al. [143] studied the performance of Ag nanoparticles doped with Al_2O_3 nanofiber filtration membranes for the removal of organic dyes. Excellent results were obtained in removing organic dyes under the combination of photocatalysis and filtration processes because of the ability of Ag nanoparticles in stimulating photocatalytic degradation under visible light irradiation.

Membrane processes combined with ozonation are only feasible with ceramic membranes because of their excellent physical and chemical properties. Zhu et al. [144] combined filtration with ozonation to treat wastewater from a major wastewater treatment plant to eliminate organic compounds using a tubular ceramic membrane (α-Al_2O_3) deposited with TiO_2 nanoparticles via a dip-coating technique. The permeate flux increased significantly because of the reduced membrane fouling contributed by the ozonation process. The resulting membrane not only functions as a physical barrier to separate the particulates, but also simultaneously breaks down the organic matter catalytically.

Aside from TiO_2 nanoparticles, Byun et al. [127] investigated the performance of catalytic membranes coated with different metal oxides (MnO or Fe_2O_3 nanoparticles) in a hybrid ozonation–ceramic membrane filtration system. The MnO-coated membrane exhibited the best performance in the filtration of water obtained from a borderline eutrophic lake. The membrane exhibited the fastest recovery in permeate flux when ozone was applied and the highest reduction in the total organic carbon

(TOC) in the permeate. Moreover, Zhu et al. [145] improved membrane antifouling, separation, and degradation of organic pollutants using a TiO_2-nanorod-assembled membrane coated with a mesoporous titanium (Ti)–Mn catalyst layer in the treatment of simulated wastewater (aqueous solution of dye Red-3BS and aniline) because of its high catalytic ozonation capability.

17.4 CONCLUSIONS

Environmental problems, especially water contamination, are becoming very serious and gaining significant attention worldwide in the search for effective remediation methods to meet the continuously growing demand for clean water. Advances in nanotechnology suggest that water quality problems can be solved. This chapter discussed the application of nanoparticles in water remediation through adsorption, photocatalytic degradation, and disinfection. Fe^0 and nanosized metal oxides have potential as efficient nanoadsorbents of toxic materials because of their high surface area to volume ratio, high stability, and high durability. Hybrid nanomaterials, such as AC/chitosan/SiC–supported and bimetal nanoadsorbents, can be fabricated to enhance the specific surface area for higher adsorption capacity and increase the removal of contaminants from water. TiO_2 and ZnO are inexpensive, nontoxic photocatalysts with great capability in degrading contaminants in water. The deposition of metals or ions onto photocatalysts and the immobilization of photocatalysts on a rigid support are necessary to enhance degradation efficiency. Fe^0, nanosized Fe_2O_3, and core–shell systems of nanosized photocatalysts with super paramagnetic properties are preferable for easy reclaiming using a magnetic field. Meanwhile, TiO_2 and Ag nanoparticles showed great bactericidal effects and inactivation of various harmful microorganisms during water disinfection. The small size and high surface to volume ratio of these nanoparticles allows them to interact closely and directly with pathogens. This chapter also reviews the development of membranes incorporated with various nanoparticles including TiO_2, SiO_2, Al_2O_3, ZnO, Fe_2O_3, and Ag with remarkably enhanced performances in membrane fouling and biofouling mitigation, disinfection, and photocatalytic functionalities. Various strategies have been employed for the fabrication of nanoparticle-incorporated membranes. Prior to the introduction of nanoparticles into the membranes, modification of the nanoparticles, such as doping or surface functionalization, may be performed to improve the functionalities and enhance the dispersion or attachment of nanoparticles. Eventually, the nanoparticles are integrated into membranes via different routes, including homogeneous dispersion, surface coating, or as ceramic membrane precursors. The dosage of the nanoparticles should be optimized to produce more cost-competitive and better-performance membranes in the future. Thus, the properties of nanoparticles and the ecotoxicity effects of the release of nanoparticles to the environment need to be investigated further for drinking water applications.

ACKNOWLEDGMENTS

The authors acknowledge USM fellowship supports from Universiti Sains Malaysia (USM) and MyPhD program of the Ministry of Higher Education, Malaysia. This research was also financially supported by the Universiti Sains Malaysia Research University (RU) grant, USM Membrane Cluster Grant, Fundamental of Research Grant Scheme (FRGS), and Postgraduate Research Grant Scheme (PRGS).

REFERENCES

[1] Pandiyan T, Martinez Rivas O, Orozco Martinez J, Burillo Amezcua G, Martinez-Carrillo MA. Comparison of methods for the photochemical degradation of chlorophenols. J Photochem Photobiol A Chem 2002;146:149–155.

[2] WHO. *Guidelines for drinking-water quality*. fourth ed. Geneva: World Health Organization; 2011. p 1–541.

[3] Goyal AK, Johal ES, Rath G. Nanotechnology for water treatment. Curr Nanosci 2011;7:640–654.

[4] Xu Y, Zhang W-X. Subcolloidal Fe/Ag particles for reductive dehalogenation of chlorinated benzenes. Ind Eng Chem Res 2000;39:2238–2244.

[5] Park HS, Kim DH, Kim SJ, Lee KS. The photocatalytic activity of 2.5 wt% Cu-doped TiO_2 nano powders synthesized by mechanical alloying. J Alloys Compd 2006;415:51–55.

[6] Li Y, Li X, Li J, Yin J. Photocatalytic degradation of methyl orange by TiO_2-coated activated carbon and kinetic study. Water Res 2006;40:1119–1126.

[7] Fu W, Yang H, Li M, Chang L, Yu Q, Xu J, Zou G. Preparation and photocatalytic characteristics of core-shell structure $TiO_2/BaFe_{12}O_{19}$ nanoparticles. Mater Lett 2006;60:2723–2727.

[8] Chang YC, Chang SW, Chen DH. Magnetic chitosan nanoparticles: studies on chitosan binding and adsorption of Co(II) ions. React Funct Polym 2006;66:335–341.

[9] Zhang WX. Nanoscale iron particles for environmental remediation: an overview. J Nanopart Res 2003;5:323–332.

[10] Shan G, Yan S, Tyagi R, Surampalli R, Zhang T. Applications of Nanomaterials in Environmental Science and Engineering: review. Pract Period Hazard Toxic Radioactive Waste Manage 2009;13:110–119.

[11] Pacheco S, RodrÃguez R. Adsorption properties of metal ions using alumina nano-particles in aqueous and alcoholic solutions. J Solgel Sci Technol 2001;20:263–273.

[12] Jang M, Shin EW, Park JK, Choi SI. Mechanisms of arsenate adsorption by highly-ordered nano-structured silicate media impregnated with metal oxides. Environ Sci Technol 2003;37:5062–5070.

[13] Damjanovic L, Rakic V, Rac V, Stosic D, Auroux A. The investigation of phenol removal from aqueous solutions by zeolites as solid adsorbents. J Hazard Mater 2010;184:477–484.

[14] Chen H, Lee SW, Kim TH, Hur BY. Photocatalytic decomposition of benzene with plasma sprayed TiO_2-based coatings on foamed aluminum. J Eur Ceram Soc 2006;26:2231–2239.

[15] Lee K, Lee NH, Shin SH, Lee HG, Kim SJ. Hydrothermal synthesis and photocatalytic characterizations of transition metals doped nano TiO_2 sols. Mater Sci Eng B Solid State Mater Adv Technol 2006;129:109–115.

[16] Kamat PV, Huehn R, Nicolaescu R. A "sense and shoot" approach for photocatalytic degradation of organic contaminants in water. J Phys Chem B 2002;106:788–794.

[17] Peng F, Wang H, Yu H, Chen S. Preparation of aluminum foil-supported nano-sized ZnO thin films and its photocatalytic degradation to phenol under visible light irradiation. Mater Res Bull 2006;41:2123–2129.

[18] Gangadharan D, Harshvardan K, Gnanasekar G, Dixit D, Popat KM, Anand PS. Polymeric microspheres containing silver nanoparticles as a bactericidal agent for water disinfection. Water Res 2010;44:5481–5487.

[19] Nangmenyi G, Yue Z, Mehrabi S, Mintz E, Economy J. Synthesis and characterization of silver-nanoparticle-impregnated fiberglass and utility in water disinfection. Nanotechnology 2009;20:495705.

[20] De Gusseme B, Sintubin L, Baert L, Thibo E, Hennebel T, Vermeulen G, Uyttendaele M, Verstraete W, Boon N. Biogenic silver for disinfection of water contaminated with viruses. Appl Environ Microbiol 2010;76:1082–1087.

[21] Hu J, Chen G, Lo IMC. Removal and recovery of Cr(VI) from wastewater by maghemite nanoparticles. Water Res 2005;39: 4528–4536.

[22] Singh VK, Tiwari PN. Removal and recovery of chromium(VI) from industrial waste water. J Chem Technol Biotechnol 1997;69: 376–382.

[23] Zhong LS, Hu JS, Liang HP, Cao AM, Song WG, Wan LJ. Self-assembled 3D flowerlike iron oxide nanostructures and their application in water treatment. Adv Mater 2006;18:2426–2431.

[24] Wang Y-H, Lin S-H, Juang R-S. Removal of heavy metal ions from aqueous solutions using various low-cost adsorbents. J Hazard Mater 2003;102:291–302.

[25] Cheng R, Wang J.l., Zhang, W. x., Comparison of reductive dechlorination of p-chlorophenol using Fe^0 and nanosized Fe^0. J Hazard Mater 2007;144:334–339.

[26] Zhu H, Jia Y, Wu X, Wang H. Removal of arsenic from water by supported nano zero-valent iron on activated carbon. J Hazard Mater 2009;172:1591–1596.

[27] Kanel SR, Manning B, Charlet L, Choi H. Removal of arsenic(III) from groundwater by nanoscale zero-valent iron. Environ Sci Technol 2005;39:1291–1298.

[28] Su C, Puls RW. Arsenate and arsenite removal by zerovalent iron: effects of phosphate, silicate, carbonate, borate, sulfate, chromate, molybdate, and nitrate, relative to chloride. Environ Sci Tech 2001;35:4562–4568.

[29] Kanel SR, Greneche JM, Choi H. Arsenic(V) removal from groundwater using nano scale zero-valent iron as a colloidal reactive barrier material. Environ Sci Tech 2006;40:2045–2050.

[30] Ponder SM, Darab JG, Mallouk TE. Remediation of Cr(VI) and Pb(II) aqueous solutions using supported, nanoscale zero-valent iron. Environ Sci Technol 2000;34:2564–2569.

[31] Kirschling TL, Gregory KB, Minkley EG Jr, Lowry GV, Tilton RD. Impact of nanoscale zero valent iron on geochemistry and microbial populations in trichloroethylene contaminated aquifer materials. Environ Sci Tech 2010;44:3474–3480.

[32] Hwang YH, Kim DG, Ahn YT, Moon CM, Shin HS. Fate of nitrogen species in nitrate reduction by nanoscale zero valent iron and characterization of the reaction kinetics. Water Sci Technol 2010;61:705–712.

[33] Chang YC, Chen DH. Preparation and adsorption properties of monodisperse chitosan-bound Fe_3O_4 magnetic nanoparticles for removal of Cu(II) ions. J Colloid Interface Sci 2005;283:446–451.

[34] Suzuki T, Moribe M, Oyama Y, Niinae M. Mechanism of nitrate reduction by zero-valent iron: equilibrium and kinetics studies. Chem Eng J 2012;183:271–277.

[35] Crane RA, Scott TB. Nanoscale zero-valent iron: future prospects for an emerging water treatment technology. J Hazard Mater 2012;211–212:112–125.

[36] Lien H-L, Zhang W-X. Nanoscale iron particles for complete reduction of chlorinated ethenes. Colloids Surf A Physicochem Eng Asp 2001;191:97–105.

[37] Tian H, Li J, Mu Z, Li L, Hao Z. Effect of pH on DDT degradation in aqueous solution using bimetallic Ni/Fe nanoparticles. Sep Purif Technol 2009;66:84–89.

[38] Tee Y-H, Bachas L, Bhattacharyya D. Degradation of trichloroethylene by iron-based bimetallic nanoparticles. J Phys Chem C 2009;113:9454–9464.

[39] Hua M, Zhang S, Pan B, Zhang W, Lv L, Zhang Q. Heavy metal removal from water/wastewater by nanosized metal oxides: a review. J Hazard Mater 2012;211–212:317–331.

[40] Önnby L, Pakade V, Mattiasson B, Kirsebom H. Polymer composite adsorbents using particles of molecularly imprinted polymers or aluminium oxide nanoparticles for treatment of arsenic contaminated waters. Water Res 2012;46:4111–4120.

[41] Sobana N, Swaminathan M. Combination effect of ZnO and activated carbon for solar assisted photocatalytic degradation of direct blue 53. Solar Energy Mater Solar Cell 2007;91:727–734.

[42] Jeon MS, Yoon WS, Joo H, Lee TK, Lee H. Preparation and characterization of a nano-sized Mo/Ti mixed photocatalyst. Appl Surf Sci 2000;165:209–216.

[43] Sobana N, Swaminathan M. The effect of operational parameters on the photocatalytic degradation of acid red 18 by ZnO. Sep Purif Technol 2007;56:101–107.

[44] Penpolcharoen M, Amal R, Brungs M. Degradation of sucrose and nitrate over titania coated nano-hematite photocatalysts. J Nanopart Res 2001;3:289–302.

[45] Sobana N, Muruganadham M, Swaminathan M. Nano-Ag particles doped TiO_2 for efficient photodegradation of direct azo dyes. J Mol Catal A Chem 2006;258:124–132.

[46] Kato S, Hirano Y, Iwata M, Sano T, Takeuchi K, Matsuzawa S. Photocatalytic degradation of gaseous sulfur compounds by silver-deposited titanium dioxide. Appl Catal Environ 2005;57:109–115.

[47] Li Y, Lee NH, Hwang DS, Song JS, Lee EG, Kim SJ. Synthesis and characterization of nano titania powder with high photoactivity for gas-phase photo-oxidation of benzene from $TiOCl_2$ aqueous solution at low temperatures. Langmuir 2004;20:10838–10844.

[48] Park H, Neppolian B, Jie HS, Ahn JP, Park JK, Anpo M, Lee DY. Preparation of bimetal incorporated TiO_2 photocatalytic nano-powders by flame method and their photocatalytic reactivity for the degradation of diluted 2-propanol. Curr Appl Phys 2007;7:118–123.

[49] Sun J, Wang X, Sun J, Sun R, Sun S, Qiao L. Photocatalytic degradation and kinetics of Orange G using nano-sized Sn(IV)/TiO_2/AC photocatalyst. J Mol Catal A Chem 2006;260:241–246.

[50] Venkatachalam N, Palanichamy M, Arabindoo B, Murugesan V. Enhanced photocatalytic degradation of 4-chlorophenol by Zr^{4+} doped nano TiO_2. J Mol Catal A Chem 2007;266:158–165.

[51] Yamashita H, Nishida Y, Yuan S, Mori K, Narisawa M, Matsumura Y, Ohmichi T, Katayama I. Design of TiO_2-SiC photocatalyst using TiC-SiC nano-particles for degradation of 2-propanol diluted in water. Catal Today 2007;120:163–167.

[52] Doh SJ, Kim C, Lee SG, Lee SJ, Kim H. Development of photocatalytic TiO_2 nanofibers by electrospinning and its application to degradation of dye pollutants. J Hazard Mater 2008;154:118–127.

[53] Di Paola A, García-López E, Marcì G, Palmisano L. A survey of photocatalytic materials for environmental remediation. J Hazard Mater 2012;211–212:3–29.

[54] Fan C, Xue P, Sun Y. Preparation of Nano-TiO_2 doped with cerium and its photocatalytic activity. J Rare Earth 2006;24:309–313.

[55] Han JK, Choi SM, Lee GH. Synthesis and photocatalytic activity of nanocrystalline TiO_2-SrO composite powders under visible light irradiation. Mater Lett 2007;61:3798–3801.

[56] Bessekhouad Y, Robert D, Weber JV, Chaoui N. Effect of alkaline-doped TiO_2 on photocatalytic efficiency. J Photochem Photobiol A Chem 2004;167:49–57.

[57] Liu G, Zhang X, Xu Y, Niu X, Zheng L, Ding X. The preparation of Zn^{2+}-doped TiO_2 nanoparticles by sol-gel and solid phase reaction methods respectively and their photocatalytic activities. Chemosphere 2005;59:1367–1371.

[58] Liu GG, Zhang XZ, Xu YJ, Niu XS, Zheng LQ, Ding XJ. Effect of $ZnFe_2O_4$ doping on the photocatalytic activity of TiO_2. Chemosphere 2004;55:1287–1291.

[59] Dong W, Sun Y, Ma Q, Zhu L, Hua W, Lu X, Zhuang G, Zhang S, Guo Z, Zhao D. Excellent photocatalytic degradation activities of ordered mesoporous anatase TiO_2-SiO_2 nanocomposites to various organic contaminants. J Hazard Mater 2012;229–230:307–320.

[60] Hosseini SN, Borghei SM, Vossoughi M, Taghavinia N. Immobilization of TiO_2 on perlite granules for photocatalytic degradation of phenol. Appl Catal Environ 2007;74:53–62.

[61] Wilhelm P, Stephan D. Photodegradation of rhodamine B in aqueous solution via SiO_2@TiO_2 nano-spheres. J Photochem Photobiol A Chem 2007;185:19–25.

[62] He Z, Yang S, Ju Y, Sun C. Microwave photocatalytic degradation of Rhodamine B using TiO_2 supported on activated carbon: mechanism implication. J Environ Sci 2009;21:268–272.

[63] Chen CC. Degradation pathways of ethyl violet by photocatalytic reaction with ZnO dispersions. J Mol Catal A Chem 2007;264: 82–92.

[64] Han J, Liu Y, Singhal N, Wang L, Gao W. Comparative photocatalytic degradation of estrone in water by ZnO and TiO_2 under artificial UVA and solar irradiation. Chem Eng J 2012;213:150–162.

[65] Pirkanniemi K, Sillanpaa M. Heterogeneous water phase catalysis as an environmental application: a review. Chemosphere 2002;48: 1047–1060.

[66] Jang YJ, Simer C, Ohm T. Comparison of zinc oxide nanoparticles and its nano-crystalline particles on the photocatalytic degradation of methylene blue. Mater Res Bull 2006;41:67–77.

[67] Wu W, Cai YW, Chen JF, Shen SL, Martin A, Wen LX. Preparation and properties of composite particles made by nano zinc oxide coated with titanium dioxide. J Mater Sci 2006;41:5845–5850.

[68] Yu D, Cai R, Liu Z. Studies on the photodegradation of Rhodamine dyes on nanometer-sized zinc oxide. Spectrochim Acta A Mol Biomol Spectrosc 2004;60:1617–1624.

[69] Anandan S, Vinu A, Mori T, Gokulakrishnan N, Srinivasu P, Murugesan V, Ariga K. Photocatalytic degradation of 2,4,6-trichlorophenol using lanthanum doped ZnO in aqueous suspension. Catal Commun 2007;8:1377–1382.

[70] Cun W, Jincai Z, Xinming W, Bixian M, Guoying S, Ping'an P, Jiamo F. Preparation, characterization and photocatalytic activity of nano-sized ZnO/SnO_2 coupled photocatalysts. Appl Catal Environ 2002;39:269–279.

[71] Ortega-Gómez E, Fernández-Ibáñez P, Ballesteros Martín MM, Polo-López MI, Esteban García B, Sánchez Pérez JA. Water disinfection using photo-Fenton: effect of temperature on *Enterococcus faecalis* survival. Water Res 2012;46:6154–6162.

[72] Sadiq IM, Chandrasekaran N, Mukherjee A. Studies on effect of TiO_2 nanoparticles on growth and membrane permeability of *Escherichia coli*, Pseudomonas aeruginosa, and Bacillus subtilis. Curr Nanosci 2010;6:381–387.

[73] Morones JR, Elechiguerra JL, Camacho A, Holt K, Kouri JB, Ramírez JT. The bactericidal effect of silver nanoparticles. Nanotechnology 2005;16:2346–2353.

[74] Lalueza P, Monzón M, Arruebo M, Santamaría J. Bactericidal effects of different silver-containing materials. Mater Res Bull 2011;46:2070–2076.

[75] Matsumura Y, Yoshikata K, Kunisaki S-I, Tsuchido T. Mode of bactericidal action of silver zeolite and its comparison with that of silver nitrate. Appl Environ Microbiol 2003;69:4278–4281.

[76] Lin W-C, Chen C-N, Tseng T-T, Wei M-H, Hsieh JH, Tseng WJ. Micellar layer-by-layer synthesis of TiO_2/Ag hybrid particles for bactericidal and photocatalytic activities. J Eur Ceram Soc 2010;30:2849–2857.

[77] Taurozzi JS, Arul H, Bosak VZ, Burban AF, Voice TC, Bruening ML, Tarabara VV. Effect of filler incorporation route on the properties of polysulfone–silver nanocomposite membranes of different porosities. J Membr Sci 2008;325:58–68.

[78] Wu G, Gan S, Cui L, Xu Y. Preparation and characterization of PES/TiO_2 composite membranes. Appl Surf Sci 2008;254:7080–7086.

[79] Yang Y, Zhang H, Wang P, Zheng Q, Li J. The influence of nano-sized TiO_2 fillers on the morphologies and properties of PSF UF membrane. J Membr Sci 2007;288:231–238.

[80] Li JH, Xu YY, Zhu LP, Wang JH, Du CH. Fabrication and characterization of a novel TiO_2 nanoparticle self-assembly membrane with improved fouling resistance. J Membr Sci 2009;326:659–666.

[81] Bae T-H, Kim I-C, Tak T-M. Preparation and characterization of fouling-resistant TiO_2 self-assembled nanocomposite membranes. J Membr Sci 2006;275:1–5.

[82] Rahimpour A, Madaeni SS, Taheri AH, Mansourpanah Y. Coupling TiO_2 nanoparticles with UV irradiation for modification of polyethersulfone ultrafiltration membranes. J Membr Sci 2008;313:158–169.

[83] Rahimpour A, Jahanshahi M, Mollahosseini A, Rajaeian B. Structural and performance properties of UV-assisted TiO_2 deposited nanocomposite PVDF/SPES membranes. Desalination 2012;285:31–38.

[84] Mansourpanah Y, Madaeni SS, Rahimpour A, Farhadian A, Taheri AH. Formation of appropriate sites on nanofiltration membrane surface for binding TiO_2 photo-catalyst: performance, characterization, and fouling-resistant capability. J Membr Sci 2009; 330:297–306.

[85] Vatanpour V, Madaeni SS, Khataee AR, Salehi E, Zinadini S, Monfared HA. TiO_2 embedded mixed matrix PES nanocomposite membranes: influence of different sizes and types of nanoparticles on antifouling and performance. Desalination 2012;292:19–29.

[86] Cao X, Ma J, Shi X, Ren Z. Effect of TiO_2 nanoparticle size on the performance of PVDF membrane. Appl Surf Sci 2006;253: 2003–2010.

[87] Jadav GL, Singh PS. Synthesis of novel silica-polyamide nanocomposite membrane with enhanced properties. J Membr Sci 2009;328: 257–267.

[88] Jin LM, Yu SL, Shi WX, Yi XS, Sun N, Ge YL, Ma C. Synthesis of a novel composite nanofiltration membrane incorporated SiO_2 nanoparticles for oily wastewater desalination. Polymer 2012;53:5295–5303.

[89] Jin L, Shi W, Yu S, Yi X, Sun N, Ma C, Liu Y. Preparation and characterization of a novel PAm-SiO_2 nanofiltration membrane for raw water treatment. Desalination 2012;298:34–41.

[90] Yin J, Kim E-S, Yang J, Deng B. Fabrication of a novel thin-film nanocomposite (TFN) membrane containing MCM-41 silica nanoparticles (NPs) for water purification. J Membr Sci 2012;423–424:238–246.

[91] Ahmad AL, Majid MA, Ooi BS. Functionalized PSf/SiO_2 nanocomposite membrane for oil-in-water emulsion separation. Desalination 2011;268:266–269.

[92] Yan L, Li YS, Xiang CB. Preparation of poly(vinylidene fluoride) (PVDF) ultrafiltration membrane modified by nano-sized alumina (Al2O3) and its antifouling research. Polymer 2005;46:7701–7706.

[93] Yan L, Li YS, Xiang CB, Xianda S. Effect of nano-sized Al_2O_3-particle addition on PVDF ultrafiltration membrane performance. J Membr Sci 2006;276:162–167.

[94] Yan L, Hong S, Li ML, Li YS. Application of the Al_2O_3-PVDF nanocomposite tubular ultrafiltration (UF) membrane for oily wastewater treatment and its antifouling research. Sep Purif Technol 2009;66:347–352.

[95] Yang JX, Shi WX, Yu SL, Lu Y. Influence of DOC on fouling of a PVDF ultrafiltration membrane modified by nanosized alumina. Desalination 2009;239:29–37.

[96] Saleh TA, Gupta VK. Synthesis and characterization of alumina nano-particles polyamide membrane with enhanced flux rejection performance. Sep Purif Technol 2012;89:245–251.

[97] Vatanpour V, Madaeni SS, Rajabi L, Zinadini S, Derakhshan AA. Boehmite nanoparticles as a new nanofiller for preparation of antifouling mixed matrix membranes. J Membr Sci 2012;401–402:132–143.

[98] Liang S, Xiao K, Mo Y, Huang X. A novel ZnO nanoparticle blended polyvinylidene fluoride membrane for anti-irreversible fouling. Journal of Membrane Science 2012;394–395:184–192.

[99] Hong J, He Y. Effects of nano sized zinc oxide on the performance of PVDF microfiltration membranes. Desalination 2012;302:71–79.

[100] Balta S, Sotto A, Luis P, Benea L, Van der Bruggen B, Kim J. A new outlook on membrane enhancement with nanoparticles: the alternative of ZnO. J Membr Sci 2012;389:155–161.

[101] DeFriend KA, Wiesner MR, Barron AR. Alumina and aluminate ultrafiltration membranes derived from alumina nanoparticles. J Membr Sci 2003;224:11–28.

[102] Zhang Q, Fan Y, Xu N. Effect of the surface properties on filtration performance of Al_2O_3–TiO_2 composite membrane. Sep Purif Technol 2009;66:306–312.

[103] Monash P, Pugazhenthi G. Effect of TiO_2 addition on the fabrication of ceramic membrane supports: a study on the separation of oil droplets and bovine serum albumin (BSA) from its solution. Desalination 2011;279:104–114.

[104] Yang GCC, Li CJ. Tubular TiO_2/Al_2O_3 composite membranes: preparation, characterization, and performance in electrofiltration of oxide-CMP wastewater. Desalination 2008;234:354–361.

[105] Zhang X, Zhang T, Ng J, Sun DD. High-performance multifunctional TiO_2 nanowire ultrafiltration membrane with a hierarchical layer structure for water treatment. Adv Funct Mater 2009;19:3731–3736.

[106] Jones CD, Fidalgo M, Wiesner MR, Barron AR. Alumina ultrafiltration membranes derived from carboxylate–alumoxane nanoparticles. J Membr Sci 2001;193:175–184.

[107] Bailey DA, Jones CD, Barron AR, Wiesner MR. Characterization of alumoxane-derived ceramic membranes. J Membr Sci 2000;176:1–9.

[108] Majhi A, Monash P, Pugazhenthi G. Fabrication and characterization of γ-Al_2O_3–clay composite ultrafiltration membrane for the separation of electrolytes from its aqueous solution. J Membr Sci 2009;340:181–191.

[109] Madaeni SS, Ahmadi Monfared H, Vatanpour V, Arabi Shamsabadi A, Salehi E, Daraei P, Laki S, Khatami SM. Coke removal from petrochemical oily wastewater using γ-Al_2O_3 based ceramic microfiltration membrane. Desalination 2012;293:87–93.

[110] Oh S, Kang T, Kim H, Moon J, Hong S, Yi J. Preparation of novel ceramic membranes modified by mesoporous silica with 3-aminopropyltriethoxysilane (APTS) and its application to Cu^{2+} separation in the aqueous phase. J Membr Sci 2007;301:118–125.

[111] Fakhfakh S, Baklouti S, Baklouti S, Bouaziz J. Preparation, characterization and application in BSA solution of silica ceramic membranes. Desalination 2010;262:188–195.

[112] Duke MC, Mee S, da Costa JCD. Performance of porous inorganic membranes in non-osmotic desalination. Water Res 2007;41:3998–4004.

[113] Cortalezzi MM, Rose J, Wells GF, Bottero J-Y, Barron AR, Wiesner MR. Ceramic membranes derived from ferroxane nanoparticles: a new route for the fabrication of iron oxide ultrafiltration membranes. J Membr Sci 2003;227:207–217.

[114] Karnik BS, Davies SH, Baumann MJ, Masten SJ. Fabrication of catalytic membranes for the treatment of drinking water using combined ozonation and ultrafiltration. Environ Sci Technol 2005;39:7656–7661.

[115] Harman BI, Koseoglu H, Yigit NO, Beyhan M, Kitis M. The use of iron oxide-coated ceramic membranes in removing natural organic matter and phenol from waters. Desalination 2010;261:27–33.

[116] Lee SY, Kim HJ, Patel R, Im SJ, Kim JH, Ryul Min B. Silver nanoparticles immobilized on thin film composite polyamide membrane: characterization, nanofiltration, antifouling properties. Polym Adv Technol 2007;18:562–568.

[117] Zodrow K, Brunet L, Mahendra S, Li D, Zhang A, Li Q, Alvarez PJJ. Polysulfone ultrafiltration membranes impregnated with silver nanoparticles show improved biofouling resistance and virus removal. Water Res 2009;43:715–723.

[118] Zhang L, Luo J, Menkhaus TJ, Varadaraju H, Sun Y, Fong H. Antimicrobial nano-fibrous membranes developed from electrospun polyacrylonitrile nanofibers. J Membr Sci 2011;369:499–505.

[119] Sawada I, Fachrul R, Ito T, Ohmukai Y, Maruyama T, Matsuyama H. Development of a hydrophilic polymer membrane containing silver nanoparticles with both organic antifouling and antibacterial properties. J Membr Sci 2012;387–388:1–6.

[120] Li J-H, Shao X-S, Zhou Q, Li M-Z, Zhang Q-Q. The double effects of silver nanoparticles on the PVDF membrane: surface hydrophilicity and antifouling performance. Appl Surf Sci 2013;265:663–670.

[121] Mafune F, Kohno J, Takeda Y, Kondow T, Sawabe H. Structure and stability of silver nanoparticles in aqueous solution produced by laser ablation. J Phys Chem B 2000;104:8333–8337.

[122] Zhang M, Zhang K, De Gusseme B, Verstraete W. Biogenic silver nanoparticles (bio-Ag^0) decrease biofouling of bio-Ag^0/PES nanocomposite membranes. Water Res 2012;46:2077–2087.

[123] Lv Y, Liu H, Wang Z, Liu S, Hao L, Sang Y, Liu D, Wang J, Boughton RI. Silver nanoparticle-decorated porous ceramic composite for water treatment. J Membr Sci 2009;331:50–56.

[124] Seo YI, Hong KH, Kim SH, Chang D, Lee KH, Kim YD. Removal of bacterial pathogen from wastewater using Al filter with Ag-containing nanocomposite film by in situ dispersion involving polyol process. J Hazard Mater 2012;227–228:469–473.

[125] Ma N, Fan X, Quan X, Zhang Y. Ag–TiO_2/HAP/Al_2O_3 bioceramic composite membrane: fabrication, characterization and bactericidal activity. J Membr Sci 2009;336:109–117.

[126] Meshram S, Limaye R, Ghodke S, Nigam S, Sonawane S, Chikate R. Continuous flow photocatalytic reactor using ZnO–bentonite nanocomposite for degradation of phenol. Chem Eng J 2011;172:1008–1015.

[127] Byun S, Davies SH, Alpatova AL, Corneal LM, Baumann MJ, Tarabara VV, Masten SJ. Mn oxide coated catalytic membranes for a hybrid ozonation–membrane filtration: comparison of Ti, Fe and Mn oxide coated membranes for water quality. Water Res 2011;45:163–170.

[128] Pozzo RL, Giombi JL, Baltanás MA, Cassano AE. The performance in a fluidized bed reactor of photocatalysts immobilized onto inert supports. Catal Today 2000;62:175–187.

[129] Yavich AA, Lee K-H, Chen K-C, Pape L, Masten SJ. Evaluation of biodegradability of NOM after ozonation. Water Res 2004;38: 2839–2846.

[130] Zou L, Zhu B. The synergistic effect of ozonation and photocatalysis on color removal from reused water. J Photochem Photobiol A Chem 2008;196:24–32.

[131] Wang S, Shiraishi F, Nakano K. A synergistic effect of photocatalysis and ozonation on decomposition of formic acid in an aqueous solution. Chem Eng J 2002;87:261–271.

[132] Sánchez L, Peral J, Domènech X. Aniline degradation by combined photocatalysis and ozonation. Appl Catal Environ 1998;19: 59–65.

[133] Karnik BS, Davies SHR, Chen KC, Jaglowski DR, Baumann MJ, Masten SJ. Effects of ozonation on the permeate flux of nanocrystalline ceramic membranes. Water Res 2005;39:728–734.

[134] Schlichter B, Mavrov V, Chmiel H. Study of a hybrid process combining ozonation and microfiltration/ultrafiltration for drinking water production from surface water. Desalination 2004;168:307–317.

[135] Lehman SG, Liu L, Adham S. Application of new generation ceramic membranes for challenging waters. Proceedings of Singapore International Water Week Convention; June 23–27. Singapore: Singapore International Water Week Pte Ltd; 2008.

[136] Song H, Shao J, He Y, Liu B, Zhong X. Natural organic matter removal and flux decline with PEG–TiO_2-doped PVDF membranes by integration of ultrafiltration with photocatalysis. J Membr Sci 2012;405–406:48–56.

[137] You S-J, Semblante GU, Lu S-C, Damodar RA, Wei T-C. Evaluation of the antifouling and photocatalytic properties of poly(vinylidene fluoride) plasma-grafted poly(acrylic acid) membrane with self-assembled TiO_2. J Hazard Mater 2012;237–238:10–19.

[138] Damodar RA, You S-J, Chou H-H. Study the self cleaning, antibacterial and photocatalytic properties of TiO_2 entrapped PVDF membranes. J Hazard Mater 2009;172:1321–1328.

[139] Ma N, Quan X, Zhang Y, Chen S, Zhao H. Integration of separation and photocatalysis using an inorganic membrane modified with Si-doped TiO_2 for water purification. J Membr Sci 2009;335:58–67.

[140] Ma N, Zhang Y, Quan X, Fan X, Zhao H. Performing a microfiltration integrated with photocatalysis using an Ag-TiO_2/HAP/Al_2O_3 composite membrane for water treatment: evaluating effectiveness for humic acid removal and anti-fouling properties. Water Res 2010;44:6104–6114.

[141] Liu L, Liu Z, Bai H, Sun DD. Concurrent filtration and solar photocatalytic disinfection/degradation using high-performance Ag/TiO_2 nanofiber membrane. Water Res 2012;46:1101–1112.

[142] Hong H-J, Sarkar SK, Lee B-T. Formation of TiO_2 nano fibers on a micro-channeled Al_2O_3–ZrO_2/TiO_2 porous composite membrane for photocatalytic filtration. J Eur Ceram Soc 2012;32:657–663.

[143] Ke X, Ribbens S, Fan Y, Liu H, Cool P, Yang D, Zhu H. Integrating efficient filtration and visible-light photocatalysis by loading Ag-doped zeolite Y particles on filtration membrane of alumina nanofibers. J Membr Sci 2011;375:69–74.

[144] Zhu B, Hu Y, Kennedy S, Milne N, Morris G, Jin W, Gray S, Duke M. Dual function filtration and catalytic breakdown of organic pollutants in wastewater using ozonation with titania and alumina membranes. J Membr Sci 2011;378:61–72.

[145] Zhu Y, Chen S, Quan X, Zhang Y, Gao C, Feng Y. Hierarchical porous ceramic membrane with energetic ozonation capability for enhancing water treatment. J Membr Sci 2013;431:197–204.

PART VI

GREEN METHODS IN NANOMATERIALS SYNTHESIS

18

GREEN METHODOLOGIES IN THE SYNTHESIS OF METAL AND METAL OXIDE NANOPARTICLES

ANIRUDDHA B. PATIL AND BHALCHANDRA M. BHANAGE

Department of Chemistry, Institute of Chemical Technology, Mumbai, India

18.1 INTRODUCTION

As early as 1959 the renowned physicist Richard Feynman proposed that the concept of "nanotechnology," in its traditional sense, means building things from the bottom up, with atomic precision. The prefix "nano" is derived from Greek word *nanos* meaning "dwarf" that denotes things one billionth (10^9 m) in size. Nowadays nanotechnology research is developing as a cutting-edge technology interdisciplinary with chemistry, physics, material science, biology, and medicine. In the field of material science, a material is called a "nanoparticle" when "a particle is having one or more dimensions of the order of 100 nm or less." There is a note associated with this definition: "Novel properties that differentiate nanoparticles from the bulk material typically develop at a critical length scale of under 100 nm."

The "novel properties" stated are completely dependent on the fact that at the nanoscale, their properties are unlike the properties of the bulk material (Fig. 18.1). Nanoparticles shows distinctive chemical, physical, electrical, electronic, mechanical, magnetic, dielectric, thermal, optical, and biological properties different than that of the bulk materials [1, 2]. The dimension of nanoparticles has a distinct effect on the physical properties that is significantly dissimilar from those of the bulk material, caused by their large surface atoms, large surface energy, spatial confinement, and reduced imperfections. Nanoparticles show characteristic properties in the field of physicochemistry, optoelectronics, and electronics that mostly depend on their size, shape, and crystallinity. Because of such wide applicability and distinct characteristic properties, nanoparticles are considered as key in the next generation of electronics, optoelectronics, and various chemical and biochemical sensors [3, 4]. Therefore, the recent research activities focus on the synthesis of monodispersed nanoparticles with various shapes and sizes.

Though several physical and chemical methods are comprehensively applied for the synthesis of monodispersed nanoparticles, the stability and the use of toxic chemicals is the topic of predominant concern. The use of nonpolar solvents and toxic chemicals on the surface of nanoparticles in the synthesis procedure limits their applications in the clinical field. Therefore, the development of nontoxic, clean, biocompatible, and ecofriendly methods for nanoparticle synthesis deserves merit. In this regard, recent research is focused on the novel protocol that links it to the green aspects. The green aspects in nanoparticle preparation bypass the conventional techniques and toxic organic solvents. In the literature it has been observed that the current research focuses on the novel, nonconventional, and green methods for metal and metal oxide nanoparticle synthesis. In addition to the benign environmental properties, the new green protocol is inexpensive and time-saving.

This chapter deals with green techniques involving biological methods, supercritical solvents, microwave, sonochemistry, electrochemistry, sonoelectrochemistry, and hydrothermal energy for the synthesis of metal and metal oxide nanoparticles. The microbial synthesis of nanoparticles is one of the most promising fields of research in nanobiotechnology interconnecting biotechnology

Nanomaterials for Environmental Protection, First Edition. Edited by Boris I. Kharisov, Oxana V. Kharissova, and H. V. Rasika Dias.

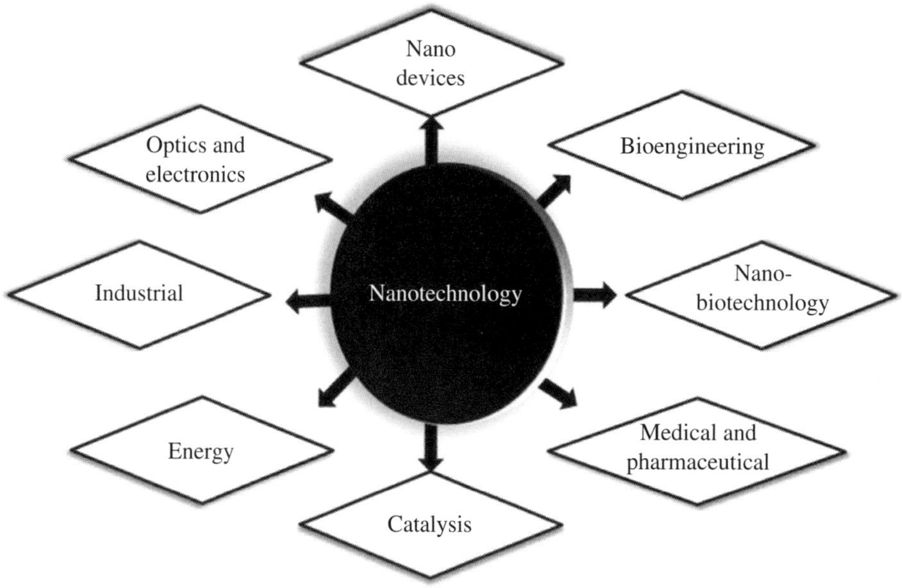

FIGURE 18.1 Applications of nanoscale particles in various fields.

and nanotechnology. In addition to the green techniques mentioned earlier, some of the techniques utilize the vast and easily available solar energy for the synthesis of metal nanoparticles. In recent investigations Bhanage and coworkers developed the concept of concentrated solar energy for the synthesis of metal and metal oxide nanoparticle and proved their applicability in catalysis.

18.2 BIOSYNTHESIS METHODS

Though biological processes are associated with a variety of problems like lengthy time periods (due to culturing of microbes), size distribution and crystallinity, difficulty in shape control synthesis, slow reduction rate, these methods are safe, inexpensive, sustainable, and environmentally friendly. The related limitations are overcome by apt strain selection, incubation temperature, and time, concentration of metal ions, optimal pH conditions, and the quantity of biological material that produce them on a large scale and for commercial applications. Microbial synthesis of nanoparticles is a green chemistry approach that combines microbial biotechnology and nanotechnology. Till date biosynthesis of various nanoparticles, such as gold, silver, gold–silver, alloy, selenium, tellurium, platinum, palladium, silica, titania, zirconia, quantum dots, magnetite and uraninite (UO) nanoparticles, by bacteria, actinomycetes, fungi, and yeasts has been reported [5].

18.2.1 Nanoparticles by Bacteria

Bacteria produce inorganic materials in nanoscale dimensions with attractive morphology either by extra- or by intracellular mechanism. Microbial systems can detoxify the metal ions by reduction and/or precipitation from soluble toxic inorganic ions to insoluble nontoxic metal nanoclusters. This detoxification can be made by intracellular bioaccumulation or extracellular biosorption, biomineralization, complexation, or precipitation. In intracellular production, the accumulated particles are of particular dimension and with less polydispersity; therefore, metal nanoparticles produced extracellularly have more commercial applications in various fields.

18.2.1.1 Bacteria-Assisted Intracellular Nanoparticle Synthesis The various classes of bacteria used for intracellular synthesis of metal and metal oxide nanoparticles such as *Bacillus subtilis* 168 are used to reduce water-soluble Au ions to metallic Au having octahedral morphology inside the cell walls in the range of 5–25 nm [6, 7]. However, in recent investigations *Escherichia coli* DH5α-mediated bioreduction process has been reported for Au(0) nanoparticles production from chloroauric acid; the collected particles on the cell surface were mostly spherical with some having quasi-hexagonal and triangular morphology [8].

In case of silver nanoparticle synthesis, an airborne *Bacillus* sp. isolated from the atmosphere has been reported recently to reduce Ag ions to metallic Ag in the size range of 5–15 nm [9]. Metal ion-reducing bacteria, *Shewanella algae*, was incubated anaerobically at room temperature and neutral pH in an aqueous solution of H_2PtCl_4, to reduce ionic platinum in the existence

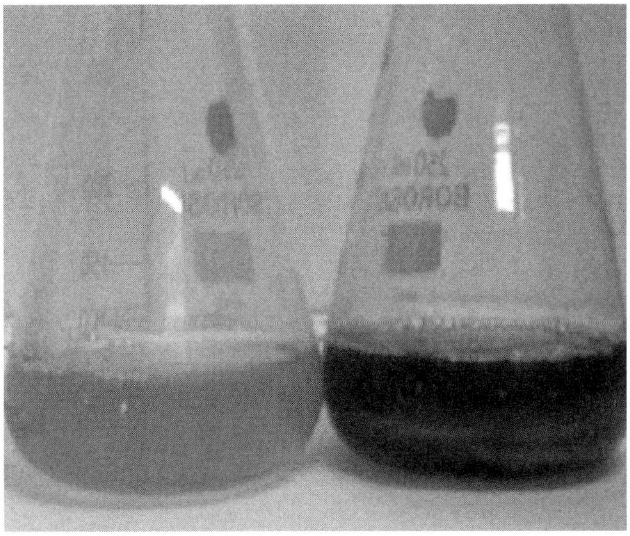

FIGURE 18.2 Conical flasks with silver nitrate (1 mmol l^{-1}) before (left) and after (right) exposure to the culture supernatant of *Phoma glomerata*. Reprinted with permission from Ref. [27]. © 2008, The Society for Applied Microbiology.

of lactate as the electron donor to metallic platinum obtaining black color from pale yellow. The distinct approximately 5-nm-sized platinum nanoparticles were found to be deposited in periplasmic space between the inner and outer membrane of *S. algae* [10]. Yong et al. have synthesized palladium nanoparticles anaerobically by using the sulfate-reducing bacterium *Desulfovibrio desulfuricans* NCIMB 8307 through bacterium bioreduction and biocrystallization. The ionic palladium (II) converts to palladium nanoparticles on the cells surface at neutral pH in the presence of exogenous electron donor [11]. Furthermore, Mann et al. used sediments-isolated microaerophilic bacterium *Aquaspirillum magnetotacticum* for the synthesis of magnetite (Fe_3O_4) nanoparticles [12]. *Desulfosporosinus* sp., a Gram-positive sulfate-reducing microbe, was used for the reduction of hexavalent uranium U (VI) to tetravalent uranium U (IV), which precipitated UO. These UO crystals coated on the cell surface have a size range of 1.5–2.5 nm [13].

18.2.1.2 Bacteria-Assisted Extracellular Nanoparticle Synthesis Metal nanoparticle synthesis by microbes depends on the localization of the reductive components of the cell. When the cell wall-soluble secreted enzymes or reductive enzymes are involved in the metal ion reduction process then the metal nanoparticles are obtained by the extracellular mechanism. In comparison with intracellular accumulation, extracellular production of nanoparticles has broader applications in electronics, optoelectronics, bioimaging, and sensor technology. So far there have been several reports on bacteria obtained for the production of nanoparticles such as *Rhodopseudomonas capsulata*, a prokaryotic bacterium, which was used for the synthesis of spherical Au nanoparticles in the size range of 10–20 nm by reduction of ionic Au at room temperature under neutral condition [14]. However, variation in shapes and sizes was obtained by change in pH; at pH 4.0 triangular nanoparticles of size 50–400 nm and spherical nanoparticles of size 10–50 nm appeared. Remarkably, extracellular synthesis of silver nanoparticles with other bacteria such as *Aeromonas* sp. SH10 [15], *Aeromonas xylinum* [16], and *Morganella* sp. was reported [17] Prasad et al. prepared titanium nanoparticles with spherical aggregates of 40–60 nm extracellularly using the culture filtrate of *Lactobacillus* sp. at room temperature [18]. Unlike in abiotic conditions, an ancient Gram-negative cyanobacterium, *Pediastrum boryanum* UTEX 485, produced Pt (II) organics and metallic platinum nanoparticles extracellularly with different morphologies such as spherical, dendritic, and bead-like chains in the size range of 30 nm to 0.3 μm [19]. In addition to this, *Micrococcus lactilyticus* cell-free extracts were used for production of U(IV) by the reduction of U(VI) [20].

18.2.2 Nanoparticle Synthesis by Fungi

In recent decades, the synthesis of silver and gold nanoparticles using fungi such as *Fusarium oxysporum* [21], *Colletotrichum* sp. [22], *Fusarium semitectum* [23], *Aspergillus fumigatus* [24], *Coriolus versicolor* [25], *Aspergillus niger* [26] has been reported. Birla et al. reported silver nanoparticles by *Phoma glomerata* [27]. They observed black color formation indicates reduction of silver nanoparticles (Fig. 18.2). Fungi are more beneficial than other microorganisms in various ways; for example, fungal mycelia mesh can resist flow pressure and agitation in bioreactors compared to plant materials and bacteria. These are fastidious to grow and easy to handle and fabricate. Also, since the nanoparticles precipitated outside the cell are devoid of unnecessary cellular components, they can be directly used for various applications.

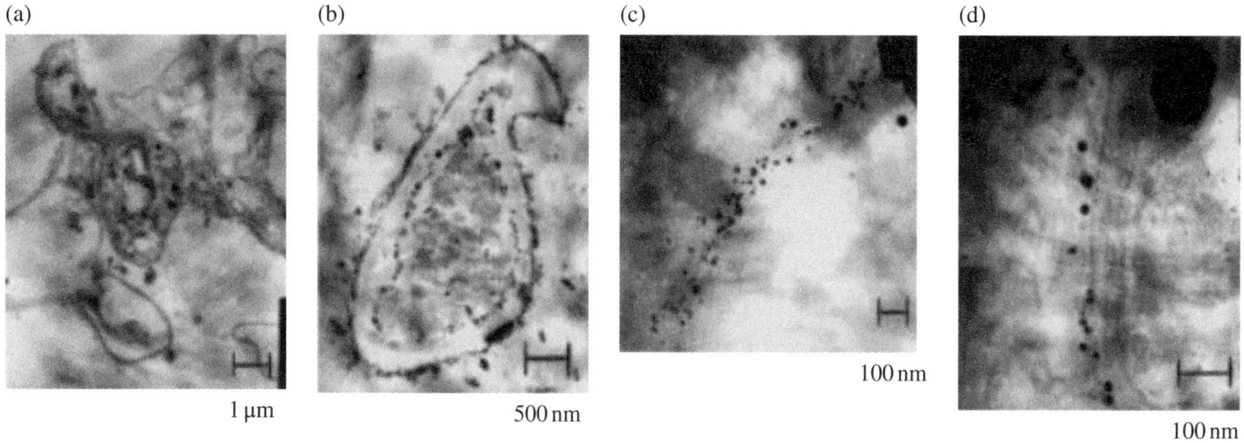

FIGURE 18.3 TEM images at different magnifications (a–d) of thin sections of stained Verticillium cells after reaction with $AuCl_4$ ions for 72 h. Reprinted with permission from Ref. [28]. © 2001, Wiley-VCH Verlag GmbH, Weinheim, Fed. Rep. of Germany.

18.2.2.1 Fungi-Assisted Intracellular Nanoparticle Synthesis In continuation with the earlier discussion, various research groups reported different fungi for nanoparticle synthesis; for example, Mukherjee et al. reported the eukaryotic microorganism *Verticillium* sp. (AAT-TS-4) for the biological synthesis of gold nanoparticles (Fig. 18.3). Gold particles of size 20 nm were prepared on the surface and on the cytoplasmic membrane of fungal mycelium [28].

Recently, Vigneshwaran et al. prepared silver nanoparticles up to 9 nm in size using silver nitrate in the presence of *Aspergillus flavus* on the surface of its cell wall and incubated for 72 h [29].

18.2.2.2 Fungi-Assisted Extracellular Nanoparticle Synthesis Extracellular synthesis of nanoparticles has many applications as unnecessary adjoining cellular components are absent from the cell. Because of the enormous secretory components that are involved in the reduction and capping of nanoparticles, fungi produce nanoparticles extracellularly. In this regard Shankar et al. found an endophytic fungus, *Colletotrichum* sp., isolated from the leaves of the geranium plant, which is helpful in the synthesis of spherical gold nanoparticles [22]. Bhainsa and D'Souza reported the synthesis of silver nanoparticles by use of *Aspergillus fumigatus* with silver nitrate. The spherical and triangular shape nanoparticles were observed in the range of 5–25 nm [24]. This is the fastest reduction protocol by a biological method and is faster than even physical and chemical methods. Cationic proteins secreted by *F. oxysporum* were used for the synthesis of zirconia nanoparticles when incubated with zirconium fluoride [30].

It was also found that *F. oxysporum* produced optoelectronic nanocrystalline material Bi_4 in the size range of 5–8 nm with quasi-spherical morphology by extracellular mechanism, when bismuth nitrate was added as precursor [31].

18.2.3 Actinomycete-Mediated Synthesis of Nanoparticles

Actinomycetes have popularly been known as ray fungi. A novel extremophilic actinomycete, the *Thermomonospora* sp. was found to synthesize extracellular spherical gold nanoparticles with size up to 8 nm [32]. The obtained nanoparticles were stable for more than 6 months. In contrast, the alkalotolerant actinomycete *Rhodococcus* sp. intracellularly accumulated gold nanoparticles with a dimension of 5–15 nm. The availability of reductases on the cell wall reduced ionic Au and accumulated as Au on the cell wall and on the cytoplasmic membrane [33].

18.2.4 Yeast-Mediated Synthesis of Nanoparticles

Lin et al. reported baker's yeast, *Saccharomyces cerevisiae*, for the reduction of Au^{3+} to metallic gold in the peptidoglycan layer of the cell wall by the aldehyde group present in reducing sugars [34]. Extracellular hexagonal silver nanoparticles in the size range of 2–5 nm were reported by using yeast MKY3 [35]. Yeast *Pichia jadinii* (*Candida utilis*) has been used for intracellular synthesis of gold nanoparticles up to 100 nm size with spherical, triangular, and hexagonal morphologies [36, 37]. However, Jha et al. reported *S. cerevisiae* for the synthesis of spherical-shaped face-centered cubic unit cell antimony oxide (Sb) nanoparticles in the size of 2–10 nm at room temperature [38].

18.3 MICROWAVE-ASSISTED NANOPARTICLE SYNTHESIS

Microwaves are radio waves with wavelengths ranging from as long as 1 m to as short as 1 mm, or equivalently, with frequencies between 300 MHz (0.3 GHz) and 300 GHz. In addition to applications in communication, radar, radio astronomy, and navigation, microwave-assisted heating has attracted significant attention in organic synthesis and inorganic material preparation. The fast reaction times, high-throughput capabilities, and beneficial crystallization effects induced by "hotspot" heating helps in the preparation of nanoparticles. Specifically, microwave chemistry has been employed recently in the synthesis of semiconducting nanoparticles, f-block oxide nanostructures, and some interesting hetero-bimetallic d-metal nanoparticles. In this regard, Peng et al. recently reported synthesis of silver nanoparticles using bamboo hemicelluloses as stabilizer and glucose as reducing agent under microwave heating in aqueous medium. The results specified the synthesis of spherical, nanometer-sized particles in the range of 8.3–14.8 nm [39]. Using microwave irradiation Raghunandan et al. also reported silver nanoparticles of 26 ± 5 nm from guava (*Psidium guajava*) leaf extract. The reaction occurs very rapidly within 90 s and results in spherical nanoparticle formation [40]. Recently, Eluri and Paul applied the microwave irradiation method for the synthesis of nickel nanoparticles (Ni NP) from the nickel acetate tetrahydrate [$Ni(CH_3CO_2)_2.4H_2O$] precursor, using sodium hypophosphite monohydrate ($NaPH_2O_2.H_2O$) reagent. The effect of additive concentration on particle size and morphology was studied under microwave heating. The isolated spherical particles (7.2 ± 1.5 nm) changed to agglomerated nano flowers (72 ± 14 nm) by increasing NaOH concentration, whereas addition of small amounts of cetyltrimethyl ammonium bromide (CTAB) or polyvinylpyrolidone (PVP) resulted in slight increase in particle size [41].

Galletti et al. reported an efficient and highly reproducible process for palladium nanoparticle synthesis using the microwave-assisted solvothermal technique. Nanoparticle synthesis has been carried using ethanol solution of $Pd(OAc)_2$ in the presence of PVP as capping agent and by adopting mild reaction conditions and very short irradiation times. The preparation of γ-Al_2O_3-supported catalysts has been suitably carried out in absence of PVP, leading to supported palladium nanoparticles with an average particle size of 5–8 nm [42].

Yu et al. reported shape-controlled synthesis of palladium nanoparticles using the microwave irradiation technique. Palladium nanocubes and nanobars with a mean size of approximately 23.8 nm were readily prepared with H_2PdCl_4 as a precursor and tetraethylene glycol (TEG) as both a solvent and a reducing agent in the presence of PVP and CTAB under microwave irradiation for 80 s. The $PdBr_4^{2-}$ formation due to the coordination replacement of the ligand Cl^- ions by Br^- ions in the presence of bromide was responsible for the synthesis of Pd nanocubes and nanobars. A milder reducing power and higher viscosity with stronger affinity of TEG were helpful in producing larger-sized Pd nanocubes and nanobars [43].

Dahal et al. [44] conducted a widespread comparative study on the effects of microwave versus conventional heating on the nucleation and growth of monodispersed Rh, Pd, and Pt nanoparticles. The obtained results reveal that microwave-assisted heating allows the convenient preparation of nanoparticles with improved morphological control, monodispersity, and higher crystallinity, compared with the conventional heating method under identical conditions. This work indicates a variety of beneficial effects of the microwave heating method such as (i) under identical reaction conditions, morphological control and crystallinity of NPs are significantly improved due to microwave heating; (ii) the microwave-induced nucleation process is faster and more reproducible compared to conventional heating methods; (iii) microwave synthesis results in cubic RhNPs that are more highly active catalysts for hydrogenation reactions.

In addition to metal nanoparticle synthesis, this technique was also explored for the synthesis of various nanocrystalline metal oxides; for example, Bhosale et al. reported synthesis of Cu_2O nanoparticles using benzyl alcohol under microwave irradiation. Benzyl alcohol as solvent, stabilizer, and reagent helps in the synthesis of spherical nanoparticles up to size of 15 nm [45].

However, Bhatte et al. reported additive-free synthesis of nanocrystalline magnesium hydroxide (MH) and magnesium oxide (MgO) using the microwave technique. Use of 1,3-propanediol as a solvent played multiple roles in circumventing the additional requirement of any extraneous species such as base and other capping agents [46].

In continuation with this work, the same group reported an additive-free synthesis of nanocrystalline zinc oxide using the microwave technique. The microwave irradiation technique was found to be faster, cleaner, and more cost-effective for the synthesis of zinc oxide nanocrystalline materials than the conventional method. The results prove that microwave heating can produce polygonal zinc oxide within a short span of time [47].

Idalia et al. showed a fast synthesis route to a variety of binary and ternary metal oxide nanoparticles with high crystallinity and proposed a way to control the reaction rate by applying microwave irradiation to accelerate the organic reaction pathway occurring in a parallel way to nanoparticle formation. In contrast to the parallel synthesis procedures performed in the autoclave, the reactions were much faster and the crystallite size was tuneable [48]. The X-ray diffraction (XRD) pattern of the synthesized metal oxide nanoparticles helped confirm the crystalline nature of the synthesized nanoparticles (Fig. 18.4); however, transmission electron microscopy (TEM) images provide a morphological image of the nanomaterials (Fig. 18.5).

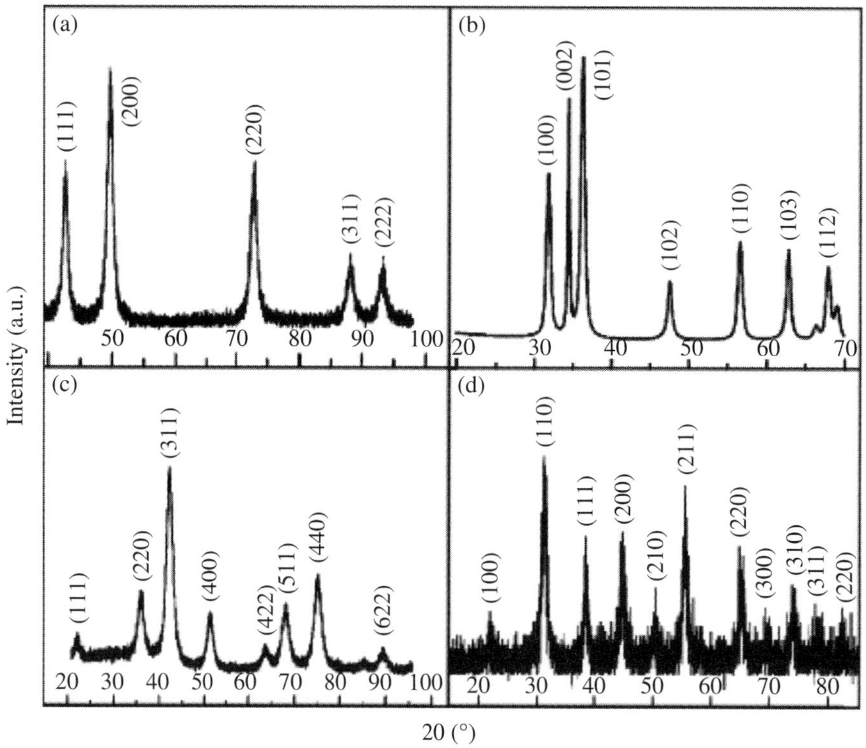

FIGURE 18.4 X-ray diffraction patterns of (a) CoO, (b) ZnO, (c) Fe_3O_4, and (d) $BaTiO_3$. Reprinted with permission from Ref. [47]. © 2008, Royal Society of Chemistry.

18.4 SONOCHEMICAL SYNTHESIS OF NANOPARTICLES

Sonochemical study is concerned with understanding the effect of sonic waves and wave properties on chemical systems. The chemical effects of ultrasound do not come from a direct interaction with molecular species. Instead, sonochemistry arises from acoustic cavitation: the formation, growth, and implosive collapse of bubbles in a liquid, which creates high pressure and temperature followed by a high rate of cooling [49]. These acoustic cavitations are the driving force in a variety of reactions and in materials synthesis. The effectiveness of cavitations depends on the contents of the collapsing bubble and hence on the nature of the solvent. High-boiling hydrocarbons are highly effective in ultrasound methodology because of their low vapor pressure and extremely high temperature within bubbles during cavitations. Nowadays, ultrasound-assisted synthesis is gaining attention for the synthesis of nanocrystalline materials. The methodology works due to acoustic cavitations, that is, formation, growth, and immediate collapse of bubbles in the solvent. Normally an ultrasonic horn is used for the synthesis of various metal and metal oxide nanoparticles (Fig. 18.6).

18.4.1 Metal Nanoparticle Synthesis by Sonochemical Route

Dhas and Gedanken reported palladium nanoclusters at room temperature by sonochemical reduction of palladium acetate, $Pd(CH_3CO_2)_2$, using myristyltrimethylammonium bromide, $[CH_3(CH_2)13N(CH_3)_3Br]$ (NR_4X), in THF or methanol. Apart from having a stabilizing effect, NR_4X acts as a reducing agent, probably due to the decomposition that occurs at the liquid-phase region immediately surrounding the collapsing cavity and provides reducing radicals. The obtained Pd nanoclusters are catalytically active toward carbon–carbon coupling, in the absence of phosphine ligands, to a moderate extent of 30 conversions [50].

Fujimoto et al. reported novel metal Pd and Pt nanoparticles by sonochemical reduction of H_2PtCl_6 or K_2PdCl_4 solutions, wherein atmospheric gas effect on the particle size distribution was investigated. The particle size of Pd and N_2 (Pd/N_2) were found to be 3.6 ± 0.7 nm under Ar (Pd/Ar) and 2.0 ± 0.3 nm in (Pd/N_2). In the case of Pt, a smaller and sharper distribution of the particle size was observed under a Xe atmosphere. These can be explained in terms of a hot-spot temperature created by acoustic cavitation [51].

Bimetallic nanoparticles composed of gold and palladium were synthesized by Mizukoshi et al. using the sonochemical method. Under ultrasound irradiation in an aqueous medium, Au (III) and Pd (II) ions of sodium tetrachloroaurate (III) dihydrate and sodium tetrachloropalladate (II) were reduced by sodium dodecyl sulfate (SDS). In addition to having a stabilizing effect,

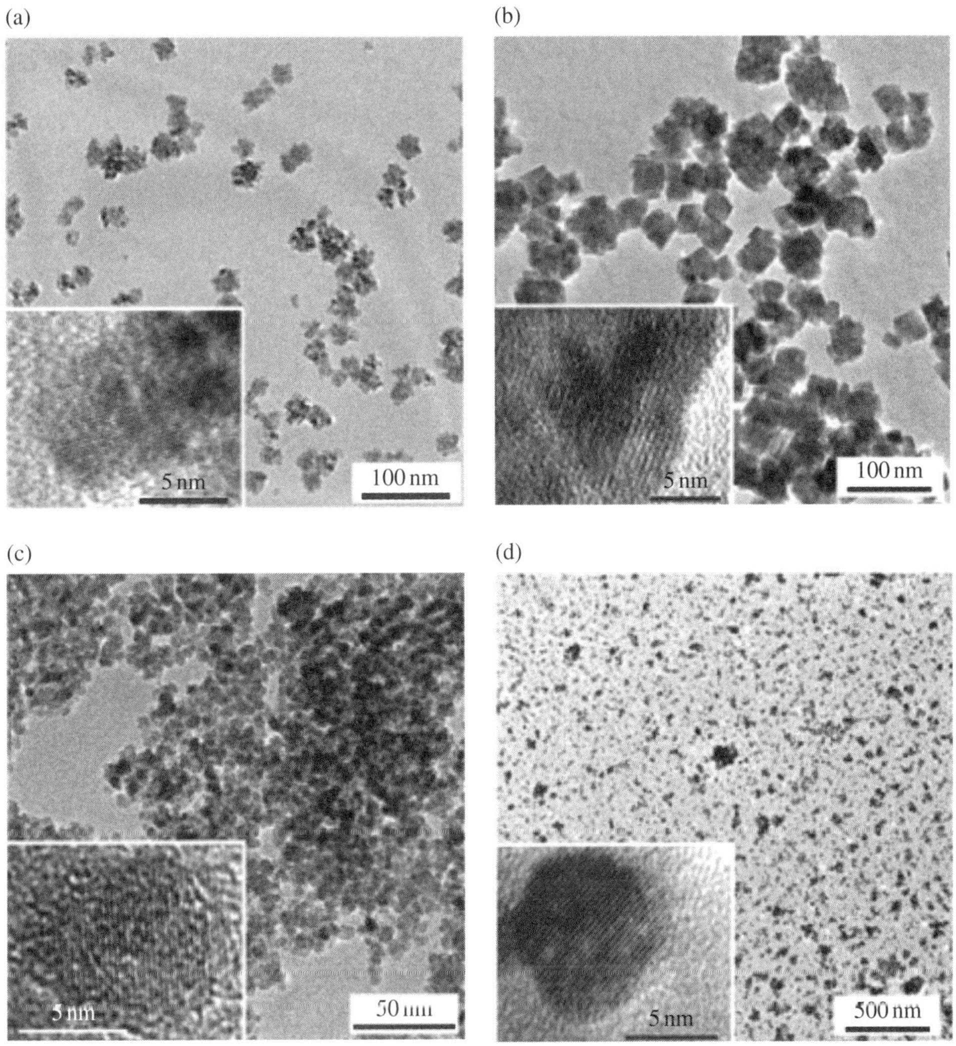

FIGURE 18.5 TEM and high-resolution transmission electron microscopy (HRTEM) images (insets) of (a) CoO, (b) MnO, (c) Fe_3O_4, and (d) $BaTiO_3$. Reprinted with permission from Ref. [48]. © Royal Society of Chemistry.

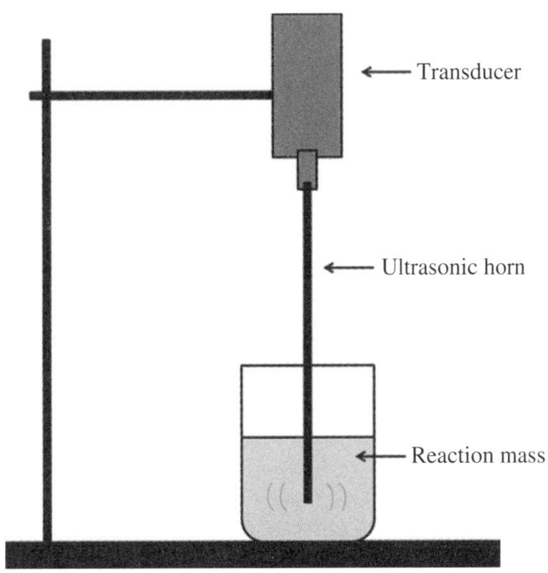

FIGURE 18.6 Schematic representation of the ultrasonic horn.

SDS remarkably enhanced the reduction rate, maybe due to the thermal decomposition that occurs at the interfacial region between bulk solution and cavitation bubbles. A spherical particle with a geometric mean diameter of about 8 nm was obtained [52].

18.4.2 Metal Oxide Nanoparticles by the Sonochemical Route

In addition to being used for metallic and bimetallic nanoparticle synthesis, sonication was also explored for the synthesis of metal oxide nanoparticles. In this regard, Bhatte et al. reported additive-free nanocrystalline zinc oxide synthesis using zinc acetate and 1,4-butanediol through sonication. The solvent 1,4-butanediol played a dual role of fuel as well as capping agent, eliminating the need for addition of any extraneous species [53].

18.5 ELECTROCHEMICAL SYNTHESIS OF NANOPARTICLES

Electrochemistry is a branch of chemistry that deals with chemical reactions that occur in a solution at the interface of an electron conductor (the electrode: a metal or a semiconductor) and an ionic conductor (the electrolyte) and that need electron transfer between the electrode and the electrolyte or species in solution.

Chemical reactions where electrons are transferred between molecules are called redox (oxidation/reduction) reactions. In general, electrochemistry addresses situations where oxidation and reduction reactions are separated in space or time, connected by an external electric circuit. Electrochemistry has been proven to have various applications in the field of physics and chemistry. In recent investigations, electrochemistry has been successfully applied for nanoparticle synthesis. Electrochemical methods have several advantages in terms of quickness, ease of operation, cost, and energy efficiency.

Reetz and Helbig described an electrochemical procedure for the formation of palladium and nickel nanoparticles [54]. The formed intermediate metal salt is reduced at the cathode, to provide metallic particles that are stabilized by tetra-alkyl ammonium salts along with coelectrolyte acetonitrile/tetrahydrofuran.

Recently, Deshmukh et al. reported a novel, one-stroke potential controllable, electrochemical method for PdNP synthesis in the electrolyte as well as on the working electrode. In this case, ionic liquid (IL) plays the role of an electrolyte cum stabilizer. In the typical synthesis, electrodeposition of palladium metal is carried out from a 1-butyl-3-methylimidazolium acetate IL ([BMim][OAc]), which is a common anion with a metal precursor $Pd(OAc)_2$ responsible for the dissolution of metal. The presence of a common acetate anion leads to an increase in the solubility of the metal salt in IL. At a certain potential, deposition of metallic Pd NPs took place at the working electrode; also, the formation of Pd NPs was observed in the bulk electrolyte. The effect of various process parameters on the morphology of synthesized nanoparticles was further investigated [55].

Khaydarov et al. reported a novel electrochemical method for the synthesis of long-lived silver nanoparticles suspended in an aqueous solution as well as silver powders. The method did not include any chemical stabilizing agents. Ag nanoparticles suspended in water solution that were produced by the present protocol are almost spherical in the size range of 2–20 nm. Ag nanoparticles synthesized by the proposed method were sufficiently stable for more than 7 years even under ambient conditions [56].

Li et al. defined a simple, green, and controllable approach for electrochemical synthesis of nanocomposites made from electrochemically reduced graphene oxide (ERGO) and gold nanoparticles. This material retains the specific features of both graphene and gold nanoparticles. The obtained result reveals a homogeneous distribution of gold nanoparticles on the graphene sheets [57].

Taleb et al. describe the electrochemical preparation of dendritic silver films with unusual wetting properties coming from the use of a self-assembled gold nanoparticle (Au NP) template. It shows that the Au NP self-assembled monolayer on the highly ordered pyrolytic graphite (HOPG) surface is responsible for the formation of the dendritic morphology, which is not observed for the same deposition conditions on a bare HOPG substrate. The dendritic structure of the deposited silver film was revealed at a later stage of the electrodeposition process. Surprising wetting properties in terms of hydrophobic surface were also revealed [58]. Mohanty reported the synthesis of various nanoscale materials, such as nanoparticles and nanowires of Au, Pt, Ni Co, Fe, Ag, and so on, by electrodeposition techniques in his article. Both potentiostatic and galvanostatic methods were employed to carry out the electrodeposition process under different potential ranges, time durations, and current densities [59].

18.6 SONOELECTROCHEMICAL REDUCTION

Introducing ultrasonic waves into homogeneous solutions has a considerable effect upon mass transport processes because of macroscopic streaming and microscopic interfacial cavitation events. In electrochemical systems, cavitational erosion permits electrode activation by continuously removing the material present at the electrode surface. In literature studies, nanoporous

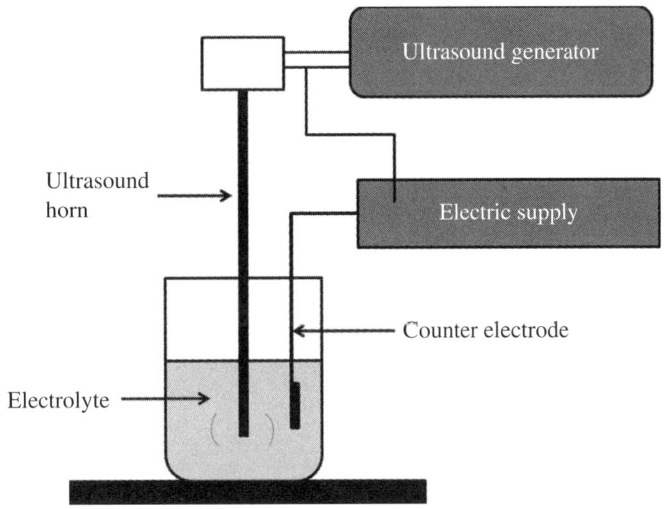

FIGURE 18.7 Schematic representation of sonoelectrochemical setup used for nanoparticle synthesis.

metals were usually prepared by electrodeposition into the template preformed on the surface of the electrode, but it was difficult to release the products from the template to obtain nanoporous metal powders. Therefore, those methods could not meet the requirement for the production of nanoporous metal powders with high yields. Among the electrochemical methods for nanoporous metal synthesis, sonoelectrochemical reduction is effective since the product can be separated from the cathode and collected by centrifugation. In this protocol, metal ions were reduced on the cathode surface and dispersed into the solution by the cavitation effect of the ultrasound. The typical setup for the sonoelectrochemical protocol consists an ultrasonic horn with electrochemical assembly (Fig. 18.7). In this regard, Reisse et al. first came up with a device for the production of metal powders by sonoelectrochemical reduction [60, 61]. Since then, several kinds of metal and semiconductor nanoparticles have been prepared using this method [62–64]. Qui et al. [63] discusses PbTe nanoparticles having a rod shape and spherical morphology (Fig. 18.8). For instance, Durant et al. prepared Zn nanoparticles in an aqueous system and studied catalytic activity [65].

In their review article, Sáez and Mason reported the nanomaterials that were prepared by pulsed sonoelectrochemistry. The majority of nanomaterials produced by this technique are pure metals such as silver, palladium, platinum, zinc, nickel, and gold. The preparation of nanosized metallic alloys and metal oxide semiconductors was also included. A main advantage of this tech nique is that the shape and size of the nanoparticles can be adjusted by changing the operating parameters, which include ultrasonic power, current density, deposition potential, and the ultrasonic versus electrochemical pulse times. In addition, the effect of pH, temperature and composition of the electrolyte in the sonoelectrochemistry cell is also possible [66].

18.6.1 Metallic Nanopowders

Haas et al. prepared copper nanoparticles from an aqueous acidic solution of $CuSO_4$ using PVP as a stabilizer [67]. Spherical copper nanoparticles with a diameter range of 25–60 nm were observed by applying a range of current densities between 55 and 100 mA cm. The first step was the formation of a coordinative bond between PVP and copper ions, forming a Cu^{2+}–PVP complex. When the current pulse was applied to the formed complex in the solution, the Cu was reduced to Cu(0) when polyvinyl alcohol (PVA) was used as a stabilizing agent, and copper with dendritic morphologies was obtained [68].

Zin et al. reported synthesis of platinum nanoparticles from an aqueous chloroplatinic solution [69]. The obtained platinum nanoparticles were spherical with an average size ranging from 10 to 20 nm. The obtained particles get aggregated into secondary structures with a mean size ranging between 100 and 200 nm. Tridimensional dendritic Pt nanostructures were prepared when PVP was used as stabilizer [70].

Aqil et al. prepared gold nanoparticles using pulsed sonoelectrochemistry in an aqueous medium [71]. The aggregation of nanoparticle was prevented by the addition of electrolyte. The electrodeposition of gold was carried out by applying a potential in the range of −850 to −1300 mV versus the normal hydrogen electrode (NHE), and in the presence of α-methoxy-ω-hydroxyl polyethylene (MPEO) the nanoparticles aggregated and settled down in the electrochemical cell. However, using a MPEO/PVP polymer mixture, a stable violet suspension was obtained without sedimentation. Most of the nanoparticles are up to the size of 12 nm, together with a few larger particles (30 nm). In the presence of polyethylene oxide (PEO) disulfide polymer, a very stable suspension of gold nanoparticles with an average diameter of 35 nm was obtained. Jiang et al. synthesized silver nanoparticles

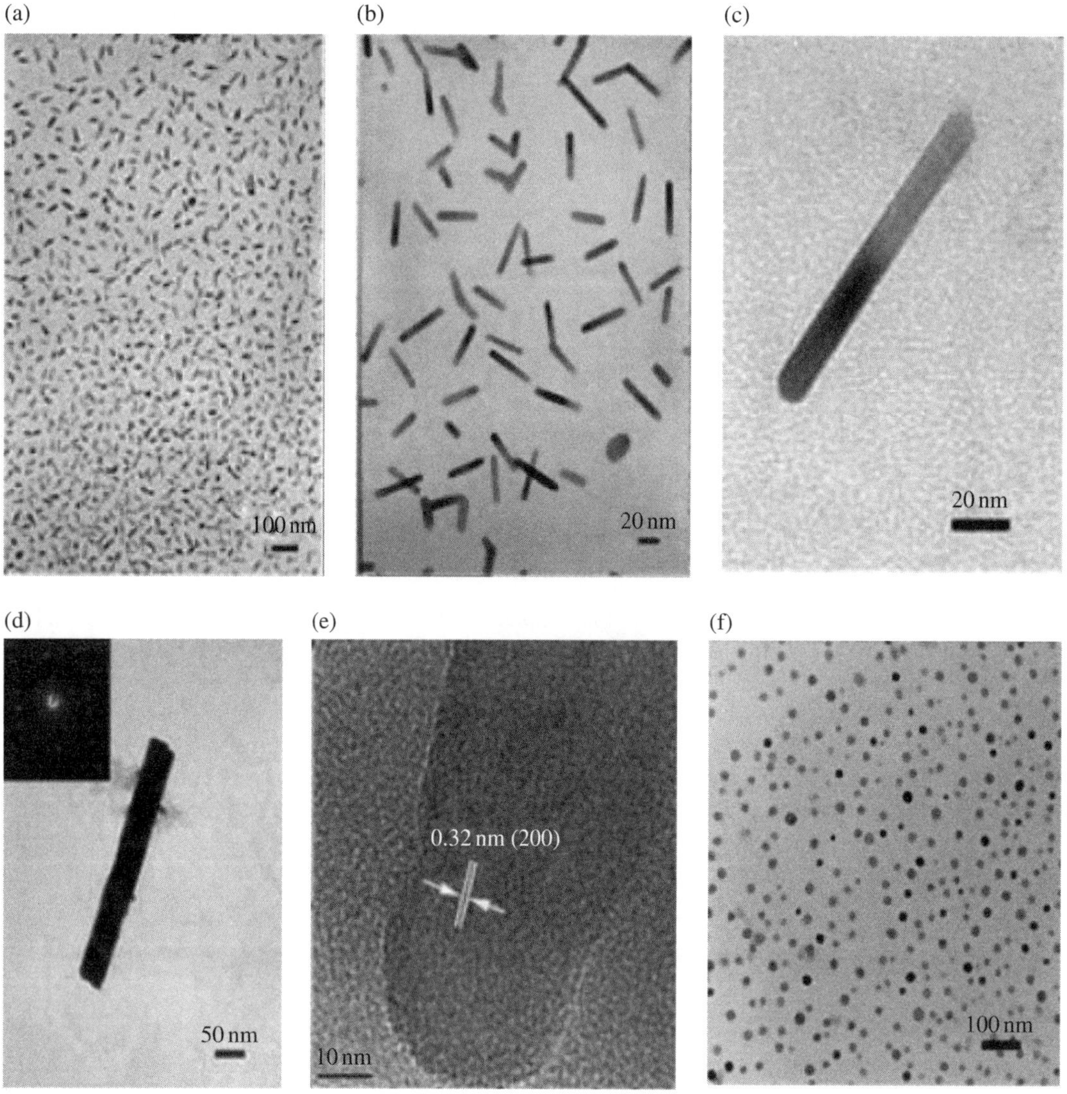

FIGURE 18.8 (a) TEM image of PbTe nanorods prepared by sonoelectrochemistry at room temperature for 30 min. (b) Enlarged TEM image of PbTe nanorods prepared under identical conditions. (c) TEM image of a nanorod obtained after 45 min of reaction. (d) TEM image of a nanorod obtained after 1.5 h of reaction; the inserted selected area electron diffraction pattern indicates the single-crystal nature of the shown nanorod. (e) HRTEM image of PbTe, a nanorod indicating the (200) lattice plane spacing. (f) Spherical PbTe nanoparticles synthesized under a low concentration of Pb^{2+} ions (2 mm). Reprinted with permission from Ref [63]. © 2005, Wiley-VCH Verlag GmbH & Co. KGaA, Weinheim.

with a face-centered cubic structure from a saturated solution of silver citrate in the presence of PVP [64]. Under the experimental conditions, spherical Ag nanoparticles with an average size of 20–25 nm were prepared. In contrast, amorphous silver nanoparticles of 20-nm size were prepared from an aqueous solution of AgBr in the presence of gelatine [72].

However, Lei et al. prepared spherical nanoparticles of tungsten by pulsed sonoelectrochemistry [73]. The mean diameter of these nanoparticles was 30 nm, and some aggregated particles were observed.

Qui et al. synthesized highly dispersed palladium nanoparticles by pulsed sonoelectrochemistry methods with diverse sizes and shapes using an ionic palladium solution in the presence of cetyltrimethylammonium bromide [74]. Mostly, spherical nanoparticles with an average size of 4–5 nm were obtained. However, a longer reaction time than 2.5 h resulted in dendritic-structured Pd. The dendritic palladium was made up of numerous spherical Pd nanoparticles with a diameter of approximately 10 nm. Zinc [61, 65], nickel [61, 75], and cobalt [61] nanoparticles have also been successfully synthesized by pulsed sonoelectrochemical methods.

Nanoparticles of very reactive metals with a high negative reduction potential, for example, magnesium and aluminum, can also be synthesized using pulsed sonoelectrochemistry.

In the case of aluminum nanoparticles a solution of $LiAlH_4$ and $AlCl_3$ in THF was used for the synthesis, and the resulting nanoparticles were found to be in the size range of 10–20 nm [76]. The technique has been also used for the synthesis of binary and ternary alloyed nanopowders containing iron, cobalt, and nickel. These alloys were synthesized using electrolyte bases on Aotani's formulation. Binary and ternary alloys were deposited galvanostatically at 8000 A m^{-2}, and particles with a mean diameter of 100 nm were produced [77].

18.6.2 Semiconductor Nanopowders

Semiconducting materials are the foundation of modern electronics and have applications in photochemistry, in dye-sensitized solar cells, and in the photocatalytic treatment of chemical waste. Nanocrystalline CuO has been prepared in the potentiostatic mode [78]. The work was based on a previous voltametric study that revealed that at an applied potential range between −0.65 and −1 V/SSE it was possible to avoid the formation of a mixture of Cu_2O and Cu. TEM micrographs showed agglomerates of nanoparticles of various sizes; in addition, isolated particles were also found with diameters ranging between 7 and 20 nm.

In literature, Lei et al. reported this technique for the synthesis of tungsten nanoparticles. The electrolyte contained sodium tungstate, trisodium citrate, ferrous sulfate, and citric acid. A titanium alloy horn was used as the cathode and a platinum slice as the anode. A 20-kHz ultrasound generated by an ultrasound generator was used for the synthesis. Because of the combined effect of the electrochemical reaction and the cavitation effect of the ultrasound, iron–tungsten aggregates at the cathode were dispersed into the electrolyte and iron atoms were dissolved in the acidic environment. In this manner tungsten nanoparticles having body-centered-cubic (bcc) structure were obtained by controlling the density of the electric current, the ultrasound pulse period, and the amplitude [73].

18.7 SUPERCRITICAL SOLVENTS FOR NANOPARTICLE SYNTHESIS

A supercritical fluid (SCF) is a substance at a temperature and pressure above its critical point, where distinct liquid and gas phases do not exist. It can flow out through solids like a gas and dissolve materials like a liquid. SCFs are found to be a suitable substitute for organic solvents in various laboratory and industrial processes. Carbon dioxide and water are the most commonly used SCFs, being used for decaffeination and power generation, respectively. On the basis of the characteristic properties and applications discussed, various research groups have applied SCFs for nanoparticle synthesis.

A general classification of SCF-based nanoparticle synthesis techniques can be proposed based on the role of SCF in the process. Certainly, SCFs have been proposed as solutes, solvents, antisolvents, and reaction media. E. Reverchon has put up a detailed review on SCFs for nanoparticle synthesis [79].

18.7.1 Rapid Expansion of Supercritical Solutions

The rapid expansion of supercritical solutions (RESS) is associated with the saturation of the SCF with a solid substrate; then depressurization into a low-pressure chamber through a heated nozzle produces a rapid nucleation of the substrate in the form of very small particles that are collected from the gaseous stream. The resulting morphology depends on the chemical structure of the material and on the RESS parameters such as temperature, pressure drop, impact distance of the jet against a surface, nozzle geometry [80]. The fast release of the solute in the gaseous medium results in the production of very small particles. This process is particularly attractive since it eludes organic solvents. In this regard, Sun et al. reported Ag nanoparticle synthesis wherein water was used in supercritical CO_2 (w/c) as a microemulsion, and the modified supercritical solvent was used to dissolve $AgNO_3$ [81]. A w/c microemulsion containing $AgNO_3$ was rapidly expanded into a room-temperature solution of sodium borohydride for the synthesis of Ag nanoparticles with an average particle size of 7.8 nm. In a subsequent work Meziani et al. reported nanocrystalline Ag particles with a bimodal distribution, with the smaller ones centered around 3.1 nm and larger ones around 10 nm [82].

18.7.2 SCF as Reductant

A variety of reports have documented the synthesis of nanoparticles using SCF as a reductant; for example, Korgel et al. reported Ag nanoparticle synthesis by reduction of silver acetylacetonates in SC–CO_2 in the presence of organic ligands that were used as stabilizers for the nanoparticles [83]. Using the same technique, silicon nanoparticles in the range of 2 and 20 nm were also synthesized in SC–hexane. Kameo et al. produced Pd and Ag nanoparticles by reduction of palladium acetate and silver acetylacetonate in supercritical CO_2 [84].

18.7.3 SCF for Hydrolysis

In supercritical water (SCW) Ziegler et al. synthesized copper oxide (Cu_2O) nanoparticles from the copper nitrate precursor by hydrolysis [85]. The reaction was performed with and without ligands. In the absence of alkanethiol ligands, polydisperse nanoparticles of Cu_2O with size range of 10–35 nm were obtained by hydrolysis; however, addition of 1-hexanethiol results in the formation of Cu nanocrystals of about 7 nm. The alkanethiol ligand helps to stabilize the synthesis of nanocrystals and control their oxidation by reduction to Cu nanoparticles. Ligands that bind on the nanoparticle surface can block the growth of nanoparticles, with a stabilization process [86].

Han and coworkers synthesized Ag nanoparticles in water-in-isooctane continuous phase using w/o reverse micelles and sodium bis(2-ethylhexyl)succinate as surfactant [87]. Water-in-CO_2 tetraethylene glycol dodecyl ether was added as the cosurfactant. The reactants $AgNO_3$ and KBH_4 were separately loaded in two micellar solutions and then mixed. Supercritical CO_2 was used to eliminate the organic phase. The obtained Ag nanoparticles were in the size range between 2 and 5 nm (minimum size) and 6 and 20 nm (maximum size) and obtained by varying the pressure and the w/o ratio.

Hong et al. synthesized TiO_2 nanoparticles by controlled hydrolysis in water-in-CO_2 reverse micelles [88]. The average particle size increases from 8 to 18 nm by increasing the water to surfactant ratio. Cason and Roberts reported Cu nanoparticle preparation in w/c reverse micelles using SC–ethane [89]. The authors compared the rate of particle growth in liquid solvents, and the process is faster in SCF due to improved transport properties. The same research group also used reverse micelles (w/c) in compressed propane or supercritical ethane to produce Ag and Cu nanoparticles by reduction of the corresponding precursors [90].

18.7.4 SCF for Thermal Deposition

Cansell et al. reported the thermal degradation of metallic precursors dissolved in a supercritical fluid (ammonia) [91]. The precursors of Cu and Fe acetylacetonates that show good solubility in ammonia were used for nanoparticle synthesis; the temperature of the reactor was then increased, thus inducing decomposition of the precursors and the precipitation of the corresponding oxides and nitrites. Aggregates of about 50 nm were obtained. The same group subsequently proposed the protocol for nanoparticle synthesis of several other compounds including Cr, Co, Cu, Ni, Al, Ti, and Ga using supercritical ammonia–methanol mixtures.

18.7.5 Supercritical H_2O

Over the past 10 years, synthesis of micro- and nanosized metal oxides has been reported by using the supercritical water method (FT-SCW) by continuous hydrothermal synthesis [92]. In this protocol, an aqueous solution of starting materials is pressurized and fed into a mixing tee that combines reactants with preheated water. Rapid heating results from the mixing and allows the hydrothermal reactions. The hydrothermal reaction rate and metal oxide solubility can be greatly varied when water in its near-critical or supercritical state is used, since the reaction solvent properties are strongly dependent on thermodynamic conditions in these regions. Therefore, the technique can be used to change size, morphology, and crystal structure of many types of particles [93].

SCW is an excellent reaction medium for hydrothermal synthesis since it can vary the rate of reaction and equilibrium by a shift of the dielectric constant and solvent density with temperature and pressure. In this regard, Cote et al. produced nanocrystals of α-Fe_2O_3 and Co_3O_4 using two disparities of the continuous hydrothermal technique: cold mixing and hot mixing of the reactants [94]. In the experiments, compressed water (subcritical) was used. In a subsequent work they also obtained nanocrystalline $CoFe_2O_4$ using the same process variations [95]. In case of metal oxide nanoparticles, Viswanathan and Gupta reported ZnO nanoparticle synthesis [96]. The reaction was carried out in a continuous tubular reactor starting from zinc acetate and using various flow rates and feed concentrations to obtain spherical nanoparticles with the average particle size of 39 nm.

18.8 SOLAR ENERGY

Due to global warming, special attention has been paid to the development of green techniques with available renewable energy sources as a replacement for conventional energy sources. Till date various conventional and nonconventional methods for nanoparticle synthesis have been reported, but these reported methods have some limitations such as harsh reaction conditions, expensive energy source, sophisticated equipment source, the need for a skilful operator. Solar energy is one of the carbon-free, easily available, and nonpolluting energy sources. There have been very few reports on the application of solar energy for the synthesis of nanoparticles.

In this context, Luo reported the size-controlled preparation of dendrimer-protected gold (Au) nanoparticles by application of sunlight [97]. Subsequent to this work, Luo reported one-step dendrimer-protected gold nanoparticles using sunlight [98]. Recently, Chien et al. reported the synthesis of gold nanoparticles under normal sunlight for 5 h wherein the reaction temperature was 34°C [99]. They showed that the synthesis of gold nanoparticles takes place in the temperature range of 30–50°C.

18.9 CONCENTRATED SOLAR ENERGY

In these cases, researchers achieved success in the synthesis of nanoparticles using solar as a green and inexpensive energy source, but the same protocol fails for the synthesis of various other metal and metal oxide nanoparticle as it requires high-intensity energy and this is not possible from natural solar energy sources. To overcome this problem and fulfill the necessity of a higher driving force for nanoparticle synthesis, Bhanage and coworkers introduced the concept of concentrated solar energy. Using this protocol, the authors reported a temperature rise to up to 95°C by using a Fresnel lens as a solar concentrator. This method helps in the faster synthesis of nanoparticles because of the combination of radiation and thermal effects; this dual-energy effect overcomes the limitation of insufficient driving force for nanoparticle synthesis. This novel technique has proven its application for metal and metal oxide nanoparticle synthesis.

This concept was used for the first time for the synthesis of palladium nanoparticles [100, 101]. The reaction mixture was irradiated under concentrated solar energy for palladium nanoparticle synthesis. The nanoparticles in the size range of 30–45 nm have been reported with mixed morphology. The obtained nanoparticles come in various shapes like, triangular, octahedral, decahedral, and icosahedral.

In continuation with this work, the same author applied this concept for shape-selective nanoparticle synthesis, wherein 70% decahedral nanoparticle selectivity was obtained. The synthesized recyclable catalyst was then used for catalysis [102].

Subsequent to this work, Patil et al. used this technique for the synthesis of metal oxides such as zinc oxide nanoparticles. After 6 h of irradiation of $Zn(CH_3COO)_2$ and 1,4-butanediol mixture a milky white ZnO nanocrystalline material was obtained consisting of nanoparticles in the size range of 10–15 nm [103]. Recently, the same author reported MgO nanoparticle synthesis by concentrated solar energy. The obtained nanoparticles show excellent catalytic applications for the Claisen–Schmidt condensation reaction [104].

18.10 CONCLUSION

This chapter provides a comprehensive review of current research activities that are dedicated to green synthesis methods for metal and metal oxide nanoparticle preparation. In this chapter, typical green methods that involve the use of nonconventional energy sources like ultrasonication, microwaves, hydrothermal energy, supercritical CO_2, biosynthesis, and solar energy and avoid toxic reagents are discussed. It also covers the various examples for the shape-selective synthesis of metal and metal oxide nanoparticles using green methods. The nanostructured materials that can be synthesized by these techniques include nanoparticles, nanowires, nanofibers, and nanocomposites. Additionally, special weightage has been given to the solar energy concept for metal and metal oxide nanoparticle synthesis, which also covers the use of concentrated solar energy, which was recently introduced in this field.

REFERENCES

[1] Schmid G. Large clusters and colloids. Metals in the embryonic state. Chem Rev 1992;92:1709–1727.

[2] Daniel MC, Astruc D. Gold nanoparticles: assembly, supramolecular chemistry, quantum-size-related properties, and applications toward biology, catalysis, and nanotechnology. Chem Soc Rev 2004;104:293–346.

[3] Wong TS, Schwaneberg U. Protein engineering in bioelectrocatalysis. Curr Opin Biotechnol 2003;14:590–596.

[4] Ramanaviciusa A, Kausaite A, Kausaite A, Ramanaviciene A. Biofuel cell based on direct bioelectrocatalysis. Biosens Bioelec 2005;20:1962–1967.

[5] Narayanan KB, Sakthivel N. Biological synthesis of metal nanoparticles by microbes. Adv Colloid Interf 2010;156:1–13.

[6] Beveridge TJ, Murray R. Sites of metal deposition in the cell wall of *Bacillus subtilis*. J Bacteriol 1980;141:876–887.

[7] Southam G, Beveridge T. The in vitro formation of placer gold by bacteria. Geochim Cosmochim Acta 1994;58:4527–4530.

[8] Du L, Jiang H, Xiaohua H, Wang E. Biosynthesis of gold nanoparticles assisted by *Escherichia coli* DH5α and its application on direct electrochemistry of hemoglobin. Electrochem Commun 2007;9:1165–1170.

[9] Pugazhenthiran N, Anandan S, Kathiravan G, Prakash N, Crawford S, Ashokkumar M. Microbial synthesis of silver nanoparticles by *Bacillus* sp. J Nanopart Res 2009;11:1811–1815.

[10] Konishi Y, Ohno K, Saitoh N, Nomura T, Nagamine S, Hishida H, Takahashi Y, Uruga T. Bioreductive deposition of platinum nanoparticles on the bacterium Shewanella algae. J Biotechnol 2007;128:648–653.

[11] Yong P, Rowsen NA, Farr J, Harris IR, Macaskie LE. Bioreduction and biocrystallization of palladium by *Desulfovibrio desulfuricans* NCIMB 8307. Biotechnol Bioeng 2002;80:369–379.

[12] Mann S, Frankel RB, Blakemore RP. Structure, morphology and crystal growth of bacterial magnetite. Nature 1984;310:405–407.

[13] Suzuki Y, Kelly SD, Kemner KM, Banfield JF. Radionuclide contamination: nanometre-size products of uranium bioreduction. Nature 2002;419:134–134.

[14] He S, Guo Z, Zhang Y, Zhang S, Wang J, Gu N. Biosynthesis of gold nanoparticles using the bacteria *Rhodopseudomonas capsulata*. Mater Lett 2007;61:3984–3987.

[15] Mouxing F, Qingbiao L, Daohua S, Yinghua L, Ning H, Xu D, Wang H, Huang J. Rapid preparation process of silver nanoparticles by bioreduction and their characterizations. Chin J Chem Eng 2006;14:114–117.

[16] Barud HS, Barrios C, Regiani T, Marques R, Verelst M, Dexpert-Ghys J, Messaddeqa Y, Ribeiroa S. Self-supported silver nanoparticles containing bacterial cellulose membranes. Mater Sci Eng C 2008;28:515–518.

[17] Parikh RY, Singh S, Prasad B, Patole M, Sastry M, Shouche Y. Extracellular synthesis of crystalline silver nanoparticles and molecular evidence of silver residence from Morganella sp. towards understanding biochemical synthesis mechanism. Chem Biol Chem 2008;9:1415–1422.

[18] Prasad K, Jha AK, Kulkarni AR. Lactobacillus assisted synthesis of titanium nanoparticles. Nanoscale Res Lett 2007;2:248–250.

[19] Lengke MF, Fleet ME, Southam G. Synthesis of platinum nanoparticles by reaction of filamentous cyanobacteria with platinum (IV)–chloride complex. Langmuir 2006;22:7318–7323.

[20] Woolfolk CA, Whiteley HR. Reduction of inorganic compounds with molecular hydrogen by *Micrococcus lactilyticus*. I. Stoichiometry with compounds of arsenic, selenium, tellurium, transition and other elements. J Bacteriol 1962;84:647–658.

[21] Mukherjee P, Senapati S, Mandal D. Extracellular synthesis of gold nanoparticles by the fungus *Fusarium oxysporum*. Chem Biol Chem 2002;3:461–463.

[22] Shankar SS, Ahmad A, Pasricha R, Sastry M. Bioreduction of chloroaurate ions by geranium leaves and its endophytic fungus yields gold nanoparticles of different shapes. J Mater Chem 2003;13:1822–1826.

[23] Basavaraja S, Balaji SD, Lagashetty A, Rajasab AH, Venkataraman A. Extracellular biosynthesis of silver nanoparticles using the fungus *Fusarium semitectum*. Mater Res Bull 2008;43:1164–1170.

[24] Bhainsa KC, D'Souza SF. Extracellular biosynthesis of silver nanoparticles using the fungus *Aspergillus fumigatus*. Coll Surf B: Interf 2006;47:160–164.

[25] Sanghi R, Verma P. Biomimetic synthesis and characterisation of protein capped silver nanoparticles. Bioresour Technol 2009;100:501–504.

[26] Gade AK, Bonde PP, Ingle AP, Marcato P, Duran N, Rai MK. Exploitation of *Aspergillus niger* for fabrication of silver nanoparticles. J Biobased Mater Bioenerg 2008;2:243–247.

[27] Birla SS, Tiwari VV, Gade AK, Ingle AP, Yadav AP, Rai MK. Fabrication of silver nanoparticles by *Phoma glomerata* and its combined effect against *Escherichia coli*, *Pseudomonas aeruginosa* and *Staphylococcus aureus*. Lett Appl Microbiol 2009;48:173–179.

[28] Mukherjee P, Ahmad A, Mandal D, Senapati S, Sainkar SR, Khan MI, Ramani R, Parischa R, Ajayakumar PV, Alam M, Sastry M, Kumar R. Bioreduction of $AuCl_4^-$ ions by the fungus, *Verticillium* sp. and surface trapping of the gold nanoparticles formed. Angew Chem Int Ed 2001;40:3585–3588.

[29] Vigneshwaran N, Ashtaputre NM, Varadarajan PV, Nachane RP, Paralikar KM, Balasubramanya RH. Biological synthesis of silver nanoparticles using the fungus Aspergillus flavus Mater Lett 2007;61:1413–1418.

[30] Bansal V, Rautaray D, Ahmad A, Sastry M. Biosynthesis of zirconia nanoparticles using the fungus *Fusarium oxysporum*. J Mater Chem 2004;14:3303–3305.

[31] Uddin I, Adyanthaya S, Syed A, Selvaraj K, Ahmad A, Poddar P. Structure and microbial synthesis of sub-10 nm Bi_2O_3 nanocrystals. J Nanosci Nanotechnol 2008;8:3909–3913.

[32] Ahmad A, Senapati S, Khan MI, Kumar R, Sastry M. Extracellular biosynthesis of monodisperse gold nanoparticles by a novel extremophilic actinomycete. Thermomonospora sp. Langmuir 2003;19:3550–3553.

[33] Ahmad A, Senapati S, Khan MI, Kumar R, Ramani R, Srinivas V, Sastry M. Intracellular synthesis of gold nanoparticles by a novel alkalotolerant actinomycete, *Rhodococcus* species. Nanotechnology 2003;14:824–828.

[34] Lin Z, Wu J, Xue R, Yang Y. Spectroscopic characterization of Au^{3+} biosorption by waste biomass of *Saccharomyces cerevisiae*. Spectrochim Acta 2005;61:761–765.

[35] Kowshik M, Ashtaputre S, Kharrazi S, Vogel W, Urban J, Kulkarni SK, Paknikar KM. Extracellular synthesis of silver nanoparticles by a silver-tolerant yeast strain MKY3. Nanotechnology 2003;14:95–100.

[36] Gericke M, Pinches A. Biological synthesis of metal nanoparticles. Hydrometallurgy 2006;83:132–140.

[37] Gericke M, Pinches A. Microbial production of gold nanoparticles. Gold Bull 2006;39:22–28.

[38] Jha AK, Prasad K, Prasad K. A green low-cost biosynthesis of Sb_2O_3 nanoparticles. Biochem Eng J 2009;43:303–306.

[39] Peng H, Yang A, Xiong J. Green, microwave-assisted synthesis of silver nanoparticles using bamboo hemicelluloses and glucose in an aqueous medium. Carbohydr Polym 2013;91:348–355.

[40] Raghunandan D, Mahesh BD, Basavaraja S, Balaji SD, Manjunath SY, Venkataraman A. Microwave-assisted rapid extracellular synthesis of stable bio-functionalized silver nanoparticles from guava (*Psidium guajava*) leaf extract. J Nanopart Res 2011;13:2021–2028.

[41] Eluri R, Paul B. Microwave assisted green synthesis of nickel nanoparticles using sodium hypophosphite. Mater Lett 2012;76:36–39.

[42] Galletti A, Antonetti C, Venezia AM, Giambastiani G. An easy microwave-assisted process for the synthesis of nanostructured palladium catalysts and their use in the selective hydrogenation of cinnamaldehyde. Appl Catal A 2010;386:124–131.

[43] Yu Y, Zhao Y, Huang T, Liu H. Microwave-assisted synthesis of palladium nanocubes and nanobars. Mater Res Bull 2010;45:159–164.

[44] Dahal N, García S, Zhou J, Humphrey SM. Beneficial effects of microwave-assisted heating versus conventional heating in noble metal nanoparticle synthesis. ACS Nano 2012;6:9433–9446.

[45] Bhosale MA, Bhatte KD, Bhanage BM. A rapid, one pot microwave assisted synthesis of nanosize cuprous oxide. Powder Technol 2013;235:516–519.

[46] Bhatte K, Sawant D, Deshmukh K, Bhanage BM. Additive free microwave assisted synthesis of nanocrystalline $Mg(OH)_2$ and MgO. Particuology 2012;10:384–387.

[47] Bhatte K, Tambade P, Fujita S, Arai M, Bhanage B. Microwave-assisted additive free synthesis of nanocrystalline zinc oxide. Powder Technol 2010;203:415–418.

[48] Idalia B, Igor D, Markus N. One-minute synthesis of crystalline binary and ternary metal oxide nanoparticles. Chem Commun 2008:886–888.

[49] Suslick KS. The chemical effects of ultrasound. Sci Am 1989 (2);260:80–86.

[50] Dhas NA, Gedanken AA. Sonochemical preparation and properties of nanostructured palladium metallic clusters. J Mater Chem 1998;8:445–450.

[51] Fujimoto T, Terauchi S, Umehara H, Kojima I, Henderson W. Sonochemical preparation of single-dispersion metal nanoparticles from metal salts. Chem Mater 2001;13:1057–1060.

[52] Mizukoshi Y, Okitsu K, Maeda Y, Yamamoto TA, Oshima R, Nagata Y. Sonochemical preparation of bimetallic nanoparticles of gold/palladium in aqueous solution. J Phys Chem B 1997;101:7033–7037.

[53] Bhatte K, Fujita S, Arai M, Pandit A, Bhanage B. Ultrasound assisted additive free synthesis of nanocrystalline zinc oxide. Ultrason Sonochem 2011;18:54–58.

[54] Reetz MT, Helbig W. Size-selective synthesis of nanostructured transition metal clusters. J Am Chem Soc 1994;116:1401–1402.

[55] Deshmukh KM, Qureshi ZS, Bhatte KD, Venkatesan KA, Srinivasan TG, VasudevaRao PR, Bhanage BM. One-pot electrochemical synthesis of palladium nanoparticles and their application in the Suzuki reaction. New J Chem 2011;35:2747–2751.

[56] Khaydarov RA, Khaydarov RR, Gapurova O, Estrin Y, Scheper T. Electrochemical method for the synthesis of silver nanoparticles. J Nanopart Res 2009;11:1193–1200.

[57] Li S, Deng D, Shi Q Liu S. Electrochemical synthesis of a graphene sheet and gold nanoparticle-based nanocomposite, and its application to amperometric sensing of dopamine. Microchim Acta 2012;177:325–331.

[58] Taleb A, Mangeney C, Ivanova V. Electrochemical synthesis using a self-assembled Au nanoparticle template of dendritic films with unusual wetting properties. Nanotechnology 2011;22:205301 (9 pp.).

[59] Mohanty US. Electrodeposition: a versatile and inexpensive tool for the synthesis of nanoparticles, nanorods, nanowires, and nanoclusters of metals. J Appl Electrochem 2011;41:257–270.

[60] Reisse J, Francois H, Vandercammen J, Fabre O, Kirsch-De Mesmaeker A, Maerschalk C, Delplancke JL. Sonoelectrochemistry in aqueous electrolyte: a new type of sonoelectroreactor. Electrochim Acta 1994;39:37–39.

[61] Reisse J, Caulier T, Deckerkheer C, Fabre O, Vandercammen J, Delplancke JL, Winand R. Quantitative sonochemistry. Ultrason Sonochem 1996;3:S147–S151.

[62] Mastai Y, Polsky R, Koltypin Y, Gedanken A, Hodes G. Pulsed sonoelectrochemical synthesis of cadmium selenide nanoparticles. J Am Chem Soc 1999;121:10047–10052.

[63] Qiu XF, Lou YB, Anna CS, Anando D, James D, Burgess SD, Clemens B. PbTe nanorods by sonoelectrochemistry. Angew Chem Int Ed Engl 2005;44:5855–5857.

[64] Jiang LP, Wang AN, Zhao Y, Zhang JR, Zhu JJ. A novel route for the preparation of monodisperse silver nanoparticles via a pulsed sonoelectrochemical technique. Inorg Chem Commun 2004;7:506–509.

[65] Durant A, Delplancke JL, Winand R, Reisse J. A new procedure for the production of highly reactive metal powders by pulsed sonoelectrochemical reduction. Tetrahedron Lett. 1995;36:4257–4260.

[66] Sáez V, Mason TJ. Sonoelectrochemical synthesis of nanoparticles. Molecules 2009;14:4284–4299.

[67] Haas I, Shanmugam S, Gedanken A. Pulsed sonoelectrochemical synthesis of size-controlled copper nanoparticles stabilized by poly(*N*-vinylpyrrolidone). J Phys Chem B 2006;110:16947–16952.

[68] Haas I, Shanmugam S, Gedanken A. Synthesis of copper dendrite nanostructures by a sonoelectrochemical method. Chem Eur J 2008;14:4696–4703.

[69] Zin V, Pollet BG, Dabala M. Sonoelectrochemical (20 kHz) production of platinum nanoparticles from aqueous solutions. Electrochim Acta 2009;54:7201–7206.

[70] Shen Q, Jiang L, Zhang H, Min Q, Hou W, Zhu JJ. Three-dimensional dendritic Pt nanostructures: sonoelectrochemical synthesis and electrochemical applications. J Phys Chem C 2008;112:16385–16392.

[71] Aqil A, Serwas H, Delplancke JL, Jerome R, Jerome C, Canet L. Preparation of stable suspensions of gold nanoparticles in water by sonoelectrochemistry. Ultrason Sonochem 2008;15:1055–1061.

[72] Liu S, Huang W, Chen S, Avivi S, Gedanken A. Synthesis of X-ray amorphous silver nanoparticles by the pulse sonoelectrochemical method. J Non-Cryst Solids 2001;283:231–236.

[73] Lei H, Tang YJ, Wei JJ, Li J, Li XB, Shi HL. Synthesis of tungsten nanoparticles by sonoelectrochemistry. Ultrason Sonochem 2007;14:81–83.

[74] Qiu XF, Xu JZ, Zhu JM, Zhu JJ, Xu S, Chen HY. Controllable synthesis of palladium nanoparticles via a simple sonoelectrochemical method. J Mater Res 2003;18:1399–1404.

[75] Jia F, Hu Y, Tang Y, Zhang L. A general nonaqueous sonoelectrochemical approach to nanoporous Zn and Ni particles. Powder Technol 2007;176:130–136.

[76] Mahendiran C, Ganesan R, Gedanken A. Sonoelectrochemical synthesis of metallic aluminum nanoparticles. Eur J Inorg Chem 2009;14:2050–2053.

[77] Delplancke JL, Dille J, Reisse J, Long GJ, Mohan A, Grandjean F. Magnetic nanopowders: ultrasound-assisted electrochemical preparation and properties. Chem Mater 2000;12:946–955.

[78] Mancier V, Daltin AL, Leclercq D. Synthesis and characterization of copper oxide (I) nanoparticles produced by pulsed sonoelectrochemistry. Ultrason Sonochem 2008;15:157–163.

[79] Reverchon E, Adami R. Nanomaterials and supercritical fluids. J Supercrit Fluids 2006;37:1–22.

[80] Jung J, Perrut M. Particle design using supercritical fluids: literature and patent survey. J Supercrit Fluids 2001;20:179–219.

[81] Sun Y, Atorngitjawat P, Meziani MJ. Preparation of silver nanoparticles via rapid expansion of water in carbon dioxide microemulsion into reductant solution. Langmuir 2001;17:5707–5710.

[82] Meziani MJ, Pathak P, Beacham F, Allard LF, Sun YP. Nanoparticle formation in rapid expansion of water-in-supercritical carbon dioxide microemulsion into liquid solution. J Supercrit Fluids 2005;34:91–97.

[83] Korgel BA, Shah PS, Pell LE, Johnston KP. Nanocrystal stabilization and synthesis in supercritical solvents. Proceedings of the Fourth International Symposium on High Pressure Technology and Chemical Engineering; September 22–25; Venice, Italy. Milan: Associazione Italians Di Ingegneria Chimica; 2002, 2, 633.

[84] Kameo A, Yoshimura T, Esumi K. Preparation of noble metal nanoparticles in supercritical carbon dioxide. Colloids Surf A 2003;215:181–189.

[85] Ziegler KJ, Doty RC, Johnston KP, Korgel BA. Synthesis of organic monolayer-stabilized copper nanocrystals in supercritical water. J Am Chem Soc 2001;123:7797–7803.

[86] Shah PS, Husain S, Johnston KP, Korgel BA. Nanocrystal arrested precipitation in supercritical carbon dioxide. J Phys Chem B 2001;105:9433–9440.

[87] Zhang J, Han B, Liu J, Zhang X, He J, Liu Z, Jiang T, Yang G. Recovery of silver nanoparticles synthesized in AOT/$C_{12}E_4$ mixed reverse micelles by antisolvent CO_2. Chem Eur J 2002;8:3879–3833.

[88] Hong S, Lee MS, Hwang H, Lim K, Park S, Ju C, Lee G. Preparation of titanium dioxides in the W/C microemulsions and their photocatalytic activity. Sol Energy Mater 2003;80:273–282.

[89] Cason JP, Roberts CB. Metallic copper nanoparticle synthesis in AOT reverse micelles in compressed propane and supercritical ethane solutions. J Phys Chem B 2000;104:1217–1221.

[90] Cason JP, Khambaswadkar K, Roberts CB. Supercritical fluid and compressed solvent effects on metallic nanoparticle synthesis in reverse micelles. Ind Eng Chem Res 2000;39:4749–4755.

[91] Cansell F, Chevalier B, Demourgues A, Etourneau J, Even C, Pessey V, Petit S, Tressaud A, Weill F. Supercritical fluid processing: a new route for materials synthesis. J Mater Chem 1999;9:67–75.

[92] Sue K, Suzuki M, Arai K, Ohashi T, Ura H, Matsui K, Hakuta Y, Hayashi H, Watanabe M, Hiaki T. Size-controlled synthesis of metal oxide nanoparticles with a flow-through supercritical water method. Green Chem 2006;8:634–638.

[93] Desmoulins-Krawiec S, Aymonier C, Loppinet-Serani A, Weill F, Gorsse S, Etourneau J, Cansell F. Synthesis of nanostructured materials in supercritical ammonia: nitrides, metals and oxides. J Mater Chem 2004;14:228–232.

[94] Cote LJ, Teja AS, Wilkinson AP, Zhang ZJ. Continuous hydrothermal synthesis and crystallization of magnetic oxide nanoparticles. J Mater Res 2002;17:2410–2416.

[95] Cote LJ, Teja AS, Wilkinson AP, Zhang ZJ. Continuous hydrothermal synthesis of $CoFe_2O_4$ nanoparticles. Fluid Phase Equilib 2003;210:307–317.

[96] Viswanathan R, Gupta RB. Formation of zinc oxide nanoparticles in supercritical water. J Supercrit Fluids 2003;27:187–193.

[97] Luo Y. Size-controlled preparation of dendrimer-protected gold nanoparticles: a sunlight irradiation-based strategy. Mater Lett 2008;62:3770–3772.

[98] Luo X. One-step synthesis and characterization of dendrimer-protected gold nanoparticles. Colloid J 2009;71:281–284.

[99] Chien Y, Huang C, Wang S, Yeh C. Synthesis of nanoparticles: sunlight formation of gold nanodecahedra for ultra-sensitive lead-ion detection. Green Chem 2011;13:1162–1166.

[100] Patil AB, Lanke SR, Deshmukh KM, Pandit AB, Bhanage BM. Solar energy assisted palladium nanoparticles synthesis in aqueous medium. Mater Lett 2012;79:1–3.

[101] Patil AB, Bhanage BM. Solar energy assisted starch-stabilized palladium nanoparticles and their application in C–C coupling reactions. J Nanosci Nanotechnol 2013;13:1–8.

[102] Patil AB, Patil DS, Bhanage BM. Selective and efficient synthesis of decahedral palladium nanoparticles and its catalytic performance for Suzuki coupling reaction. J Mol Catal A 2012;365:146–153.

[103] Patil AB, Patil DS, Bhanage BM. ZnO nanoparticle by solar energy and their catalytic application for α-amino phosphonates synthesis. Mater Lett 2012;86:50–53.

[104] Patil AB, Bhanage BM. Novel and green approach for the nanocrystalline magnesium oxide synthesis and its catalytic performance in Claisen–Schmidt condensation. Catal Commun 2013;36:79–83.

19

AN ENVIRONMENTALLY FRIENDLY AND GREEN APPROACH FOR SYNTHESIS AND APPLICATIONS OF SILVER NANOPARTICLES

Muniyandi Jeyaraj, Muralidharan Murugan, Kevin John Pulikotil Anthony, and Sangiliyandi Gurunathan

GS Institute of Bio and Nanotechnology, Coimbatore, India

19.1 INTRODUCTION

This new millennium has begun with a lot of technological advancements in every sector. As a result, basic living standards have seen many improvements in the recent past. Many scientific discoveries have opened the way for new innovations in the health sector, food processing, and product recovery. However, there are considerable negative consequences, and they have a great impact on the environment. Increasing industrial activities eventually leads to environmental pollution. One of the most promising fields that has the potential to manage our environment could be "nanotechnology." Nanotechnology in the broadest and the simplest sense refers to all the technological developments at the nano scale, exactly one billionth of a meter. Nanoparticles (NPs) are the particles whose size ranges from 1 to 100 nm. Their increased surface to volume ratio makes them more catalytic [1]. NPs of metals have different physical and chemical properties than that of bulk metals. Even though NPs have a lot of applications, the way in which they are synthesized is a matter of concern. Chemical methods are available for NP synthesis; this adds to pollution because of the use of toxic chemicals and the generation of hazardous by-products. Physical methods are limited to low yield [2]. Biogenesis of NPs is a possible solution, which uses only biological agents for the development of metal NPs. Among the biological components, bacteria are used the most as they are easy to handle and product recovery is easy with possible genetic manipulations [3, 4].

19.2 THE STORY OF SILVER NPs

The earliest usage of NPs was reported during Roman times, as they used NPs for staining glasswares [5]. Faraday's findings for colloidal gold provided a better understanding of NPs [6]. M. C. Lea in 1889 introduced a method to synthesize citrate-stabilized silver colloid with particle size averaging 7–9 nm [7, 8]. Silver NPs were available commercially for medical applications even as early as 1897 under the commercial name "collargol" [9] whose mean size was later determined to be 10 nm [10, 11]. Soon after, a lot more techniques for silver NP synthesis were developed, which involved both physical and chemical methods.

The history of biogenesis of NPs has its origin in bioremediation studies of microbes on toxic metals. Many microbes were found to have bioremedial action over heavy metals, which eventually eliminated heavy metal contamination. Close observations showed that these metal ions were adsorbed by the microbes as aggregates over their cell wall or even into

Nanomaterials for Environmental Protection, First Edition. Edited by Boris I. Kharisov, Oxana V. Kharissova, and H. V. Rasika Dias.

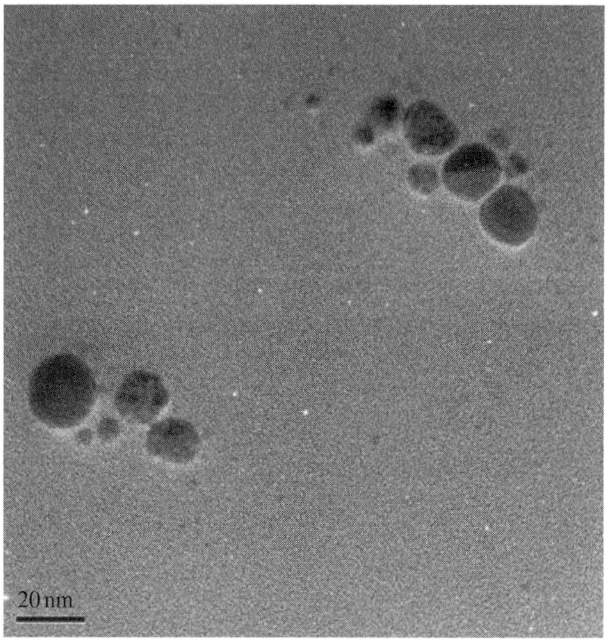

FIGURE 19.1 Biologically synthesized silver nanoparticles (NPs) using *B. flexus* GS5.

their cytoplasm. Nevertheless, the metal aggregates were found only in those organisms that were capable of withstanding concentrated heavy metals. These were termed as competent organisms; they possessed the ability to bind with metal ions forming aggregates. This phenomenon was observed in both Gram-positive and Gram-negative bacteria [12–14]. However, the metal aggregates were not identified as NPs at that time [13]. Gram-positive bacteria like *Bacillus subtilis* showed higher adsorption capacity for various metals than Gram-negative bacteria like *Escherichia coli* [14]. The biological synthesis of silver NPs came into the limelight in 1984 when Haefeli and his colleagues first reported synthesis of silver NPs from a silver mine bacterial strain *Pseudomonas stutzeri* AG259 [15–17]. Several *Bacillus* species also have been reported for the synthesis of silver NPs. We found that *Bacillus flexus* GS, a novel isolate from silver mines, produces silver NPs (Fig. 19.1).

19.3 WHY MICROBIAL SYNTHESIS?

Recently biogenesis of silver NPs has gained attention because of its green approach. A lot more physical and chemical methods are also available for the synthesis of silver NPs. Physical methods like spark discharge, pyrolysis, and attrition are available for NP synthesis. The chemical methods involve reduction of metal ions into NPs and prevention of their aggregation. A lot more reducing agents are available, including sodium borohydride, methoxypolyethylene glycol, potassium bitartarate, and hydrazine [2, 18, 19, 20]. Stabilizers are used to prevent the aggregation of NPs. Some of the stabilizers are sodium dodecyl benzyl sulfate and poly vinyl pyridone [19, 20].

Physical and chemical methods are subject to some drawbacks. High yield is not possible in the physical method, and the chemical method involves usage of toxic solvents like thioglycerol and 2-mercaptoethanol and the production of hazardous by-products [2]. Since silver NPs are used for medical purposes, the toxic components that are involved in their synthesis cause problems. It was found that chemically synthesized NPs have chemicals over their surfaces [21]. This limits the application of silver NPs in the medical field.

In order to overcome the problems faced in physical and chemical methods, biosynthesis of NPs came into focus, since there is no usage of toxic chemicals and the scaling-up process is affordable. Moreover, the method is green, and it uses only biological components for NP synthesis. It is simple and less time-consuming when compared with other techniques. Hence, NPs can be produced at lower costs with reduced toxicity [22]. There is no need for any stabilizers as that in chemical methods because the proteins in the microorganisms act as stabilizers [23]. The biological approach adds to further extensions like size control over the NPs synthesized. It was proven that the size of the NPs synthesized varies according to the pH and temperature of the reaction conditions [24].

19.4 BIOGENERATORS OF SILVER NPs

Soon after the biosynthesis of silver NPs came into the spotlight, a lot of works were conducted for a wide range of organisms encompassing both prokaryotes and eukaryotes. Moreover, because this method of synthesis is pollution free, it adds to the value of the biological reduction method. A wide range of biogenerators have been shown to synthesize silver NPs including a variety of bacteria, fungi, algae, and plant species. Among them bacteria are shown to be the best platform for synthesis as they are easy to handle and are easy subjects for genetic modifications [3, 4].

Bacteria like *Bacillus licheniformis*, *E. coli*, and *Brevibacterium casei* have been shown to synthesize silver NPs mostly around the range of 50 nm [22, 24, 23]. Some of the fungal species also synthesized silver NPs, notably *Fusarium oxysporum*, *Aspergillus niger*, *Fusarium semitectum*, and *Trichoderma asperellum* [25–28]. Silver NPs have also been synthesized in plants due to the presence of phytochemicals that reduce silver ions. For instance, anthraquinone emodin leads to the synthesis of silver NPs in *Bryophyllum* species [29].

19.5 PRINCIPLE OF NP BIOSYNTHESIS

Metal aggregates are reported to be formed over the cellular surfaces of the microorganisms [30]. Metal aggregates are formed only if metal ions are reduced, and the process of this reduction is undertaken by cellular components. Such reducing components are proteins, amino acids, polysaccharides, and even vitamins, which are the host components of the microorganism itself. This reduction is possible only if the microorganisms have the resistant machinery against metal ions. At higher concentrations, silver ions are found to inhibit the growth of microorganisms by destroying them. This could be the key reason for silver's antimicrobial activity. Therefore, only those microorganisms possessing the resistant machinery can be used for metal NP synthesis. Moreover, the effectiveness of the microorganisms is decided by parameters like reduction potential and the capacity of the system. Enzymes are found to be the key source of reduction of metal ions. The role of enzymes in the reduction of metal ions was studied by Anil Kumar and his colleagues [31]. They carried out the experiment by adding purified enzyme nitrate reductase from the fungi *F. oxysporum* with silver nitrate and nicotinamide adenine dinucleotide phosphate-oxidase (NADPH). The change of color to brown indicated the formation of silver NPs. This experiment provided direct evidence that enzymes produced by the microorganisms are responsible for reducing the metal ions to their NPs.

19.6 MECHANISM OF SILVER NP SYNTHESIS

Even though a lot of work has been done in the biosynthesis of silver NPs, the exact mechanism by which NPs are synthesized is not clear. Many mechanisms have been hypothesized for the synthesis of NPs involving biological components, including those that are used for the reduction of metal ions, and also some electro shuttles found in the microorganisms.

19.7 PEPTIDES

Naik et al [32] introduced the idea that peptides can be used as stabilizers for the synthesis of silver NPs. They showed that the peptides would bind over nuclei and enhance crystal growth. Thus, the peptides create a reducing environment that enables easy reduction of silver ions to form silver NPs. This model was confirmed by adding peptides into a solution of silver ions. Upon addition, these peptides interacted with the nanoclusters present in the solution, and it produced a reducing environment that reduced the silver ions at the peptide–metal interface. Thus, the peptides provide a reducing environment over the nanoclusters, enabling crystal formation.

Certain amino acid moieties are found to be involved in the recognition and reduction of silver ions. Mostly amino acids like arginine, cysteine, lysine, and methionine are found to be involved in the formation of silver crystals from silver ions, which gives a better understanding of silver with peptides [32]. In addition, tyrosine is found to reduce silver ions in the alkaline condition. Thus in an alkaline environment, tyrosine acts as a reducing agent where reduction occurs by ionization of the phenolic group

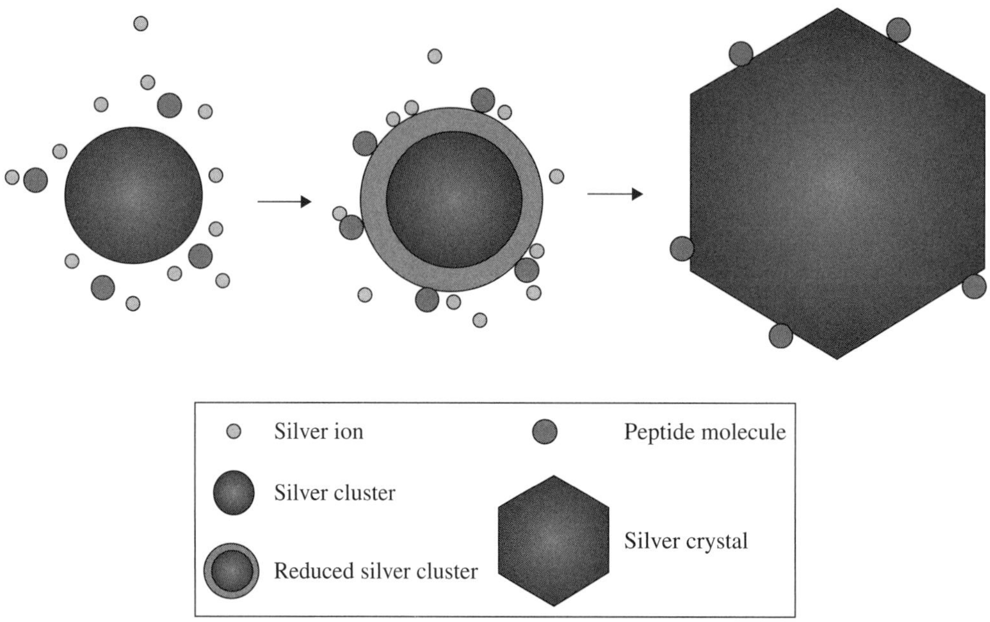

FIGURE 19.2 Peptide aided the growth of the silver crystal. Modified from Naik et al. [32].

of the tyrosine. Finally, the phenolic group turns out to become a semi-quinone structure. Then, the formed silver NPs are separated [33]. It is also found to have reducing action for gold NP synthesis [34] (Fig. 19.2).

19.8 ENZYMES

As stated earlier, the role of enzymes in the biosynthesis of silver NPs was well established by the work of Anil Kumar et al. [31]. They had taken purified nitrate reductase from *F. oxysporum* along with NADPH to synthesize silver NPs. The brown color indicated the formation of silver NPs [31]. Another experiment with *B. subtilis* was performed by Saifuddin et al. [35]. They performed extracellular synthesis of silver NPs in the culture supernatant of *B. subtilis*. The silver ion (Ag^+) when subjected to microwave irradiation was shown to synthesize silver NPs. The synthesized NPs were found to be stable without aggregation mainly because of the protein capping over the silver NPs. Moreover, the culture supernatant also had a considerable amount of nitrate reductase activity. They proposed that the reductase enzyme along with electron shuttling compounds and other peptides could be the reason for silver ion reduction, leading to the formation of silver NPs, as in the case of fungi [36, 37].

An extended view on the function of nitrate reductase enzyme for the synthesis of silver NPs using bacteria was proposed by Kalimuthu et al. [22] who studied silver NP synthesis in *B. licheniformis*. Nitrate reductase helps in the conversion of nitrate to nitrite, and it is mainly involved in the nitrogen cycle [37]. They proposed a mechanism in which electron shuttle enzymatic metal reduction occurs. Earlier, Ahmed et al. [25] had highlighted the importance of NADH- and NADH-dependent nitrate reductase in the synthesis of metal NPs. *In vitro* analysis provided the need for nitrate reductase in the synthesis of silver NPs [31]. *B. licheniformis* is also found to express NADH- and NADH-dependent reductases as an electron carrier. Thus silver ion (Ag^+) is reduced to free silver metal (Ag^0) [22] (Fig. 19.3).

Primarily metal NPs are produced only if the microorganisms are resistant against metal ions. Parikh et al. [4] demonstrated that extracellular synthesis of silver NPs using silver-resistant Morganella species and suggested that the microorganisms would have separate unique mechanisms for synthesis. In that case, the silver crystals were formed in the extracellular matrix. Sintubin et al. [38] suggested a mechanism for synthesizing metal NPs in lactic acid bacteria in which the effect of pH was studied. At higher levels of pH, monosaccharides like glucose are converted to open chain aldehydes. These aldehydes are responsible for reducing metal ions as they are oxidized to their corresponding carboxylic acid.

The influence of visible light on NP synthesis was studied in the culture supernatant of *Klebsiella pneumoniae*. Silver NPs were formed, and the size of the silver NPs was found to be around 3 nm. The reduction of silver ions in the presence of light was made by reducing agents/electron shuttles of Enterobacteriaceae. Even if the bacterial cells were absent, the silver ions were reduced, which confirmed the release of reducing agents from the Entero bacteria. This suggested that reduction of silver ions occurred after conjugation of the photosensitive electron shuttles with nitrate reductase [39].

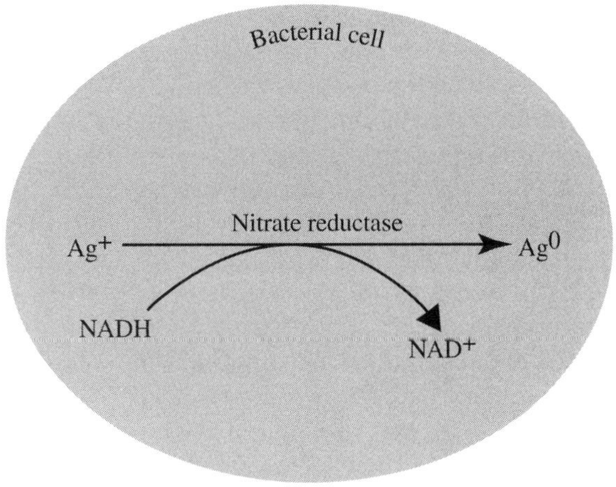

FIGURE 19.3 The possible mechanism by which nanoparticles (NPs) are produced in a bacterial cell. The enzyme nitrate reductase present in the bacterial cell mediates the reduction of silver ions to silver NPs. Modified from Kalimuthu et al. [22].

Fungi are also found to show a similar kind of reductase-mediated mechanism found in bacteria for silver NP synthesis. The involvement of NADH-dependent reductases for silver NP synthesis was assayed by Ahmed et al. [25] in the fungi *F. oxysporum*. They suggested that fungi that lack NADH-dependent reductases might not be involved in silver NP production. Duran et al. [37] verified that with nitrate-dependent reductase and quinine, an extracellular electron shuttle is responsible for the reduction of silver ions in the fungi *F. oxysporum*. The NPs were synthesized extracellularly, and their size ranged from 20 to 50 nm [37]. It was also found that the NPs produced were stabilized by fungal proteins [40]. Ingle et al. [41] confirmed the existence of nitrate reductase by nitrate reductase disks in the fungal filtrate, thereby confirming that NADH-dependent nitrate reductases are important in the synthesis of NPs as they reduce silver ions to silver NPs.

19.9 EXTERNAL FACTORS AFFECTING NPs PRODUCTION

It has been reported that factors like pH, temperature, and even silver nitrate concentration have an effect on the production of silver NPs [24]. The effect of silver nitrate concentration on the synthesis of silver NPs was studied, and the results showed that the synthesis of NPs increased with the increase in silver nitrate concentration. The synthesis of NPs was confirmed by measuring the absorbance at 420 nm. The maximum optimum concentration at which the synthesis of silver NPs occurred the maximum was found to be 5 nM in case of *E. coli* [24]. By increasing the concentration over 5 nM, absorbance was reduced, which indicated the reduction in production of silver NPs. Also, the size of NPs decreased with increase in concentration up to 5 nM after which the size increased. There is no clear explanation as to why the size of NPs varied with varying concentrations of silver nitrate.

Increase in temperature caused an increase in the rate of synthesis of silver NPs up to a certain threshold temperature after which the rate reduced. In *E.* that threshold temperature is found to be 60°C after which the rate of synthesis of silver NPs is reduced. Even as the temperature was increased, NPs were produced in smaller sizes after that threshold temperature the size of the synthesized NPs was found to be increasing. The higher rate of reduction of silver ions might be because enzymes are induced more at that temperature.

The pH is found to have a profound impact on the synthesis of silver NPs. Usually NP production increases with an increase in pH. This is mainly because at higher pH, the hydroxide ions increase the reduction in capacity of the enzymes involved in the synthesis of silver NPs, leading to increased silver NP production. For example, in *E. coli*, maximum production occurred at pH 10.0, and above this pH the reduction decreased. This was observed by a decrease in the absorbance value at 420 nm [24].

19.10 PURIFICATION OF NPs

NPs have amazing properties, most of which are due to their unique size and shape. Synthesized NPs are mostly of different sizes; hence it is necessary to separate and select them according to their size and shape, so that certain functions of the particles can be determined. The size and shape of NPs have an effect on the catalytic activity and cytotoxicity [42, 43]. They are also used in many nanodevices where the size of the particle plays a major role [44–47]. There are various techniques used to purify NPs; a few of them are described below:

1. *Centrifugation*: This is the most widely adopted technique used for the separation and purification of NPs. This involves the separation of particles based on their size and shape by the action of a centrifugal force. A particle in a centrifugal field experiences three main types of forces: centrifugal force, buoyant force, and frictional force. The centrifugal force and buoyant force act in one direction, while the frictional force acts in the opposite direction. The particles are accelerated under the centrifugal field, until all three forces are balanced, after which they settle under constant velocity. The main basis for the separation of NPs by centrifugation is that particles of different sizes and shapes have different settling velocities. For examples, gold NPs and gold nanorods have been separated by centrifugation at 5600 rpm for half an hour [48]. A drawback that persists in the normal centrifugation process is that particles that are similar would have almost similar settling velocity, and hence result in reduced quality of separation. This drawback can be overcome by the use of superior centrifugation techniques like density gradient centrifugation, isopycnic centrifugation, rate zonal centrifugation. Lighter NPs like carbon nanotubes have been efficiently purified using isopycnic centrifugation [49, 50]. Heavier NPs like metallic and inorganic NPs have been effectively separated using rate zonal centrifugation [51].
2. *Electrophoresis*: This technique implies the separation of charge particles using an electric field. When these particles are placed in an electric field, they drift toward the oppositely charged electrode, making this an effective tool for the separation of NPs based on their charge, size, and shape. The most frequently employed electrophoresis techniques for separation of charged NPs include gel electrophoresis [52] and free-flow electrophoresis [53]. Isoelectric focusing is also an important electrophoresis technique used to separate proteins based on their isoelectric points. Gel electrophoresis has been used effectively to separate both gold and silver NPs [54]. Ho and his coworkers showed that free-flow electrophoresis can be used as an effective tool to separate semiconductor NPs. [55]
3. *Magnetic field*: This can be an effective tool to separate the NPs that have magnetic receptiveness, based on their sizes. For example, iron oxide NPs can be separated by this method. In this case it has been stated theoretically that particles of sizes up to 50 nm can be efficiently separated and that smaller particles show hindrance due to thermal diffusion and Brownian movement [56]. Practically, the separation of the smaller NPs was possible due to certain unique properties exhibited by the NPs compared to the bulk material and also due to the overlook of dipole–dipole interactions among the magnetic moment of particles, which can be present even in the absence of an external magnetic field. Moeser and coworkers visualized aggregates of magnetic NPs after high gradient magnetic separation (HGMS) [57]; these were mainly due to the dipole–dipole interaction, which causes aggregation of NPs. This explained the ability of a magnetic field to separate particles smaller than predicted by theory [57, 58]. The magnetic NPs can also be separated based on their material composition with the help of magnetic field flow fractionation (MFFF) This separation takes place mainly due to the competition between magnetic forces and hydrodynamic forces [59]. Thus the magnetic field is one of the major purification techniques that is used to separate and purify magnetic NPs; practically these NPs can also be purified by other techniques like chromatography, centrifugation [60]
4. *Chromatography*: Chromatography is one of the techniques used to separate nonmagnetic NPs. The driving force is mainly the variation in their partition coefficient. The NPs that need to be purified are taken in the mobile phase and passed through the stationary phase. Among the various chromatography techniques available, size exclusion chromatography (SEC) is the most preferred and commonly used technique for purification of NPs. For example, SEC is used to separate gold [61] and silica [62] NPs. A drawback associated with SEC is the irreversible binding of the NPs to the stationary phase, and this difficulty was overcome with the addition of sodium dodecyl sulfate (SDS) to the mobile phase. The idea behind this was the simple concept that like charges repel each other and the negatively charged SDS surfactant would electrostatically repel the negatively charged particles [63, 64]. Other chromatographic techniques like High-performance liquid chromatography (HPLC) have also been reported. Other chromatographic techniques like HPLC have also been reported to purify nanoparticles [65, 66].
5. *Precipitation*: Selective precipitation is a technique that separates the NPs based mainly on their size by precipitating them on the basis of their physical and chemical properties. Gold NPs have been successfully purified by this method [67].
6. *Filtration*: Membrane filtration remains one of the most widely preferred methods for the separation of NPs. This technique separates the particles based mainly upon the pore size of the membrane. Polymeric membrane filters have been shown to be an efficient means for the purification of gold NPs [68]. Sweeney and his fellow workers showed that this filtration technique could be an effective tool for the separation of soluble NPs. Their experiment involved the purification of soluble gold NPs. Their experimental setup showed a continuous filtration system that contained a reservoir, a filter membrane, and a pump. The sample addition rate and the elution rate were kept constant. Based on the size of the filter membrane, the smallest NPs were elected while the larger particles were retained [69] (Fig. 19.4).

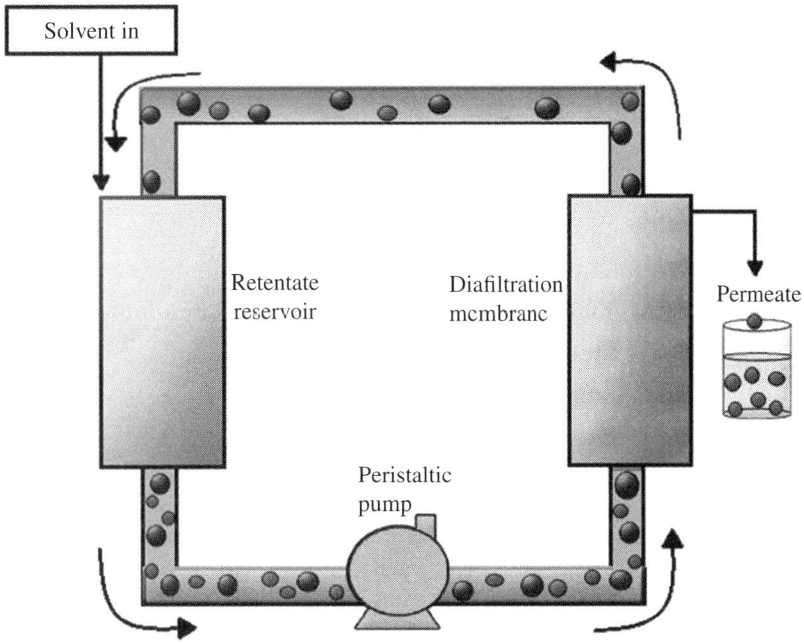

FIGURE 19.4 Schematic representation of the experimental setup maintained by Sweeney and coworkers. In this experiment, the sample containing soluble nanoparticles (NPs) was added to the reservoir; this was pumped through the filter membrane. The filter membrane filters the smaller particles while retaining the larger particles. The filtration process that takes place in the diafiltration membrane is also shown in an expanded manner. This figure has been modified from Sweeney et al. [69].

7. *Extraction*: The process of extraction mainly involves the separation of compounds on the basis of their relative solubilities in two different immiscible liquid phases, the most common being water and an organic solvent. Both organic and inorganic compounds have been separated by this method. For example, gold and silver dendrimer-encapsulated NPs have been separated by this method [70].
8. *Other purification techniques*. Many other techniques have also been used for the purification of NPs. One, not so popular, technique is the use of supercritical ethane for the purification of NPs, which has been used to purify alkanethiol-stabilized gold NPs [71]. Another technique used to purify NPs based on the shape involves the use of surfactants like cetyl trimethyl ammonium bromide (CTAB) because these surfactants have the ability to precipitate rods and platelets in a concentrated solution while not affecting spherical NPs [72]. Other techniques include micelle induction [73, 74], the use of ultrasound-assisted phase transfer for shape-dependent separation [75], and periodic precipitation [76].

19.11 APPLICATIONS OF NPs

The application of NPs has been extended to multiple fields. Especially in the fields of medicine and the environment it holds distinct advantages and is being applied extensively. From the early days, silver salts have been used in many novel therapies such as for the treatment of nicotine addiction and mental illness. It has also been used in the treatment of diseases like syphilis and gonorrhea [77–80]. The main advantage of NPs is their higher surface area due to which they have higher reactivity [81], and these unique properties are not exhibited by them in their bulk nature. The antibacterial property of nanosilver has been widely used in various applications such as in bandages, wound dressing, and ointments [82]. According to some theories, silver ions released from NPs have the ability to inactivate essential bacterial enzymes, drain intracellular Adenosine triphosphate (ATP) levels, hinder DNA replication in bacteria, damage the cytoplasmic membrane of bacteria, and in the end cause cell lysis [83]. A coat of Pd on nanosilver enables them to be used as a catalyst, and it can also be used to treat pollutants in groundwater efficiently [84]. Immobilization of certain enzymes on NPs is found to have a direct impact on their stability. Upon immobilization, controlled release of the enzyme is possible thereby improving the reactivity of the enzyme, and it also has an additional advantage in that it shields the enzyme from degradation [85]. This antibacterial property of silver can be augmented when it is persists for a longer time in a specified location, and by controlled release of silver NP this is possible. Dlugosz M and his coworkers developed a hybrid material made of calcium carbonate and silver NPs. The calcium component in this hybrid released the silver NPs in a controlled manner, which enhanced the antibacterial activity. An additional advantage is that it can be stored as dry powder and reused without any decrease in the bactericidal effect of the NP [86].

19.12 MEDICINAL APPLICATIONS

One of the major applications of NPs in medicine is the treatment of diabetic retinopathy. Diabetic retinopathy is associated with the loss of vision mainly due to angiogenesis. Nanosilver comprises the activity to inhibit angiogenesis. The silver ions released from silver NPs were at sufficiently lower concentration as they did not show any detrimental effects on the development of embryos [87].

19.12.1 Cancer Therapy

Most of the techniques used in cancer therapy have a major drawback in that they tend to damage normal cells in addition to cancer cells. This is seen mainly in the cases of chemotherapy and radiotherapy. NPs have been found to have both direct and indirect effects on tumor cells, and hence they can be one of the game changers in cancer therapy. The *in vivo* and *in vitro* studies in Dalton's lymphoma tumor–induced mice and normal mice showed that nanosilver has the ability to specifically target the tumor cells with negligible effect on normal cells [88]. Cancer therapeutic drugs face a problem of effective delivery to the specific cells, and they have chances of being degraded in this process. In such cases, NPs are used as vehicles for the movement of these drugs to specific cancer cells. This process directly minimizes the dose [89] and quantity required [90], reduces the chances of degradation of the drug [91], and also improves its stability [92].

19.12.2 Molecular Imaging

One of the most important techniques employed in detecting the exact location of a cancer cell is molecular imaging. Hybrid NPs are being generated by a combination of either organic or inorganic agents. They can be used as probes for imaging cancer cells [93]. Gold NPs are one of the many NPs to be effectively used for this imaging technique [94]. Wang et al. developed an immunosensor by hybridizing silver on porous silica NPs. This sensor had the ability to detect the tumor marker, the prostate specific antigen (PSA), on prostate cancer cells. Silica NPs enhanced the binding of the primary antibody, and the electron transfer rates were increased with the help of silver NPs in the component. This technique helped to image the prostate cancer cells without the help of a labeling agent [95].

19.12.3 Drug Delivery Across the Blood–Brain Barrier

The blood–brain barrier is a layer of endothelial cells that regulates the influx of most of the blood components to the brain. Apart from endothelial cells, it consists of two other cellular elements: astrocyte end-feet and pericytes. The tight junctions (TJs) found between the endothelial cells of the brain are responsible for the formation of a selective diffusion barrier, and they prevent most of the bloodborne compounds from entering the brain. This forms a protective barrier for the brain and maintains a different environment for brain cells excluding many bloodborne substances [96]. But it still remains as a barrier for the administration of many drugs to the brain [97–99].

NPs can be used as a better platform for delivering drugs across the blood–brain barrier. Earlier studies revealed that NPs of polymers like poly (butyl cyanoacrylate) that are coated with polysorbate 80 can cross the blood–brain barrier and be used to deliver drugs that are not able to enter the brain across the blood–brain barrier. Hexapeptide dalagrin has been successfully transported across the blood–brain barrier [100–106]. This opened up a new way for treating brain diseases before which surgery followed by radiotherapy, chemotherapy and photo dynamic therapy (PDT) are made. However, they have many limitations [107]. But the researches with NPs have shown them to be an effective vehicle for delivering drugs and in therapeutic functions of the brain.

However, the mechanism by which the NPs cross across the blood–brain barrier is not clearly known. Poly(butyl cyanoacrylate) NPs coated with polysorbate 80 show increased interactions with endothelial cells of the blood–brain barrier when coupled with apolipo proteins [108].

19.13 ENVIRONMENTAL APPLICATIONS

When it comes to the environment, NPs are used in water treatment and desalination. The ability of NPs to desalinate stands out to be one of the most potential applications for the future [109].

19.13.1 Wastewater Treatment

The decrease in fresh water due to global warming, drought, and various other factors has increased the need for wastewater treatment especially in most of the underdeveloped countries. Most of the diseases that occur in children are mainly due to the uptake of contaminated water having many harmful pathogenic microbes [110–114]. Wastewater treatment follows a series of steps, among which one of the major steps is the destruction of bacteria and other pathogenic organisms found in wastewater [115]. The commonly used technique for this step is the use of ultraviolet (UV), light chlorine, and other halogens [116]. There is a rising concern regarding toxicity associated with the use of chlorine in water treatment. The ammonium ion is another major contaminant in drinking water. The presence of this ion decreases the dissolved oxygen in water and reduces its toxicity as the ammonium ion is toxic to most fish species [117]. Water treatment also involves the substitution of these ions by biologically suitable ions, for example, Na^+, K^+. Certain studies involve evaluating the property of synthetically and naturally produced zeolite, metal ions like Ag^+, and polymer films as effective antibacterial agents [83, 118–124]. The larger surface area of NPs enables them to be more reactive, and hence they have the ability to directly react with contaminants more effectively [125, 126]. Around the twentieth century nanofilters were developed, which were very useful in eradicating the hardness and organic solutes that were in the range of 1000–3000 Da [127]. Synthesized MgO and magnesium NPs proved to be effective bactericidal agents against both Gram-negative and Gram-positive bacteria like *E. coli* and *Bacillus megaterium*, respectively [126]. Jain and Pradeep extensively studied the use of silver NPs for water treatment in Gram-negative, rod-shaped coliform bacteria due to its heavy presence in wastewater [128]. Later, various other studies to detect the bactericidal property of silver NPs were undertaken for many other Gram-positive and Gram-negative microbes like *E. coli*, *Staphylococcus aureus*, *K. pneumoniae*, and *Pseudomonas aeruginosa* [128, 129]. Silver NPs have the ability to destroy bacterial cells by binding to their cell wall and proteins at multiple sites and inhibiting their function. They also infiltrate the bacterial DNA and RNA at the precise site and cause cell lysis [130, 131]. The coating of silver NPs on various materials began mainly in 1993 when silver ions were coated on sand filters [132]. This was further expanded to many other materials like zeolites [133] and more recently on fiberglass [134]. Thus the future of nanotechnology for water treatment would mainly be the use of combinations of NPs on filters like ceramic and activated charcoal; this would result in enhanced bactericidal activity and efficient removal of other insoluble impurities (Fig. 19.5).

19.13.2 Nanobiotics

This term basically refers to the use of antibiotics conjugated to NPs. This complex enhanced the bactericidal activity of NPs. Ping li and his fellow workers considered the effect of silver NP on *E. coli* using amoxicillin as a conjugate. This complex was found to have a higher bactericidal activity than silver NPs when they were used alone. The chances for the development of resistance were decreased considerably as either of the two was sufficient for destroying the bacteria. Other such antibiotics used in combination were beta-lactam antibiotics [135] (Fig. 19.6).

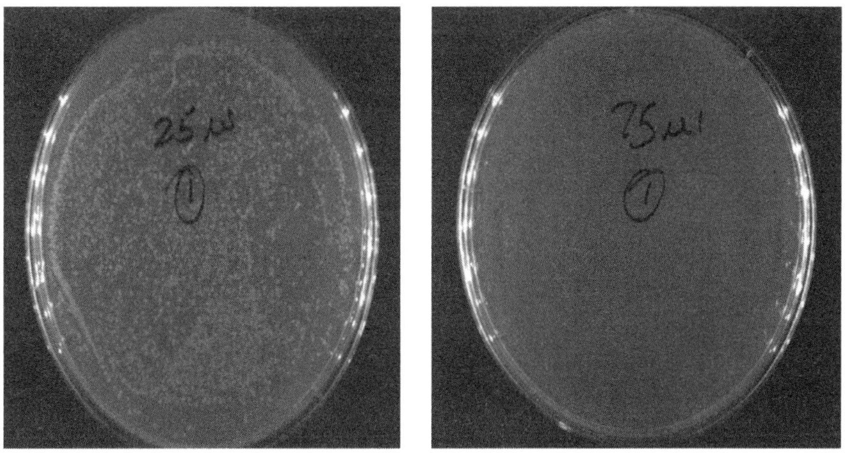

FIGURE 19.5 Sewage water treatment using Ag NPs.

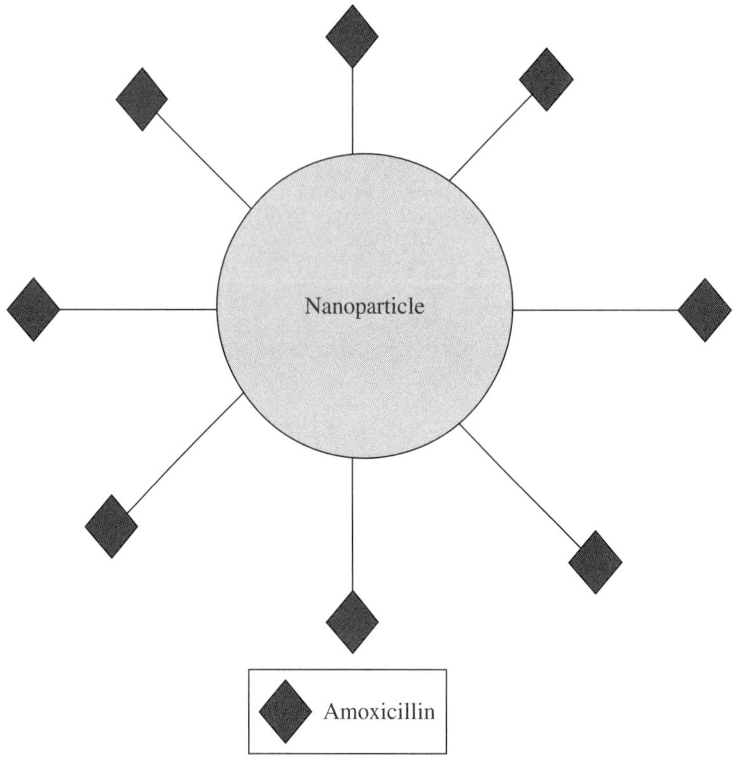

FIGURE 19.6 Binding of amoxicillin with silver nanoparticle (NP). Modified from Fing et al. [135].

19.14 RISING CONCERN OVER NANOSILVER TOXICITY

At lower concentrations, silver NPs exhibit strong antimicrobial activity. Silver NPs release silver ions, which have the ability to destroy bacterial cells by lowering the transmembrane pH gradient and terminating its proton motive force. Although this antimicrobial activity is highly potent, it has serious negative consequences at higher concentrations. As the silver ions accumulate, their concentration increases and at higher concentrations the silver ions have a serious impact on health and on the environment [136]. Certain skin diseases like argyrosis and argyria mainly result due to accumulation of silver in the skin [137]. In the environment, silver NPs inhibit some helpful microbes like nitrogen-fixing and ammonifying bacteria in the soil [138, 139]. They are also found to have an impact on aquatic life. The interaction of silver ions in the gills of fishes affects Na^+ and K^+ uptake and indirectly affects osmo regulation in fishes [140]. These are the main concerns in the widespread use of silver NPs.

19.15 CONCLUSION

Green synthesis of NPs holds the future for nanotechnology. NPs have widespread applications due to their low cost, environmentally friendly nature, ease of synthesis, and greater yield. When compared with other available methods for synthesis, like chemical and physical, they have more potential. The chemically synthesized NPs are found to require an additional stabilizer to prevent aggregation of the particles. Moreover, the chemical process used to synthesize NPs uses chemicals that are toxic to the environment. The physical method of synthesis is comparatively tedious and often produces lower yields. Hence, all these factors support biogenesis for the synthesis of NPs. Among the various biological routes available for synthesis, the bacterial method is the most widely preferred, as genetic manipulation is possible and handling is relatively easy. NPs have properties that differ greatly from that of bulk particles. The unique properties of silver NPs make them a vital tool for many applications. As their synthesis does not involve the use of any toxic substance, they have a wide array of medical and environmental applications. The biogenesis of NPs results in bulk production of NPs. If this potential is rightfully utilized, they can have numerous industrial applications. One of the rising concerns in the use of silver NPs is its toxicity at higher concentrations. If this problem can be overcome, silver NPs would become one of the most sought-after resource in the future with biogenesis being the major route for synthesis.

REFERENCES

[1] Haruta M. When gold is not noble: catalysis by nanoparticles. Chem Rec 2003;3:75–87.

[2] Mallick K, Witcomb MJ, Scurrell MS. Polymer stabilized silver nanoparticles: a photochemical synthesis route. J Mater Sci 2004; 39:4459–4463.

[3] Pugazhenthiran N, Anandan S, Kathiravan G, Kannaian N, Prakash U, Craw Ford S, Ashokkumar M. Microbial synthesis of silver nanoparticles by *Bacillus sp*. J Nanopart Res 2009;11:1811–1815.

[4] Parikh RY, Singh S, Prasad BLV, Patole MS, Sastry M, Shouche YS. Extracellular synthesis of crystalline silver nanoparticles and molecular evidence of silver resistance from *Morganella Sp*.: towards understanding biochemical synthesis mechanism. Chembiochem 2008;9:1415–1422.

[5] Giljohann DA, Seferos DS, Daniel WL, Massich MD, Patel PC, Mirkin CA. Gold nanoparticles for biology and medicine. Angew Chem Int Ed Engl 2010;49:3280–94.

[6] Faraday M. The Bakerian lecture: Experimental relations of gold (and other metals) to light. Philsos Trans Soc Lond 1857;147:145.

[7] Lea MC. On allotropic forms of silver. Am J Sci 1889;37:476–491.

[8] Frens G, Overbeek JT. Carey Leas colloidal silver. Kolloid Z.Z Polym 1969;233 (1–2):922.

[9] Boese K. Uber Collargol, seine Anwendung und seine Erfolge in der Chirurgie und Gyn€akologie. Dtsch Z Chir 1921;163 (1–2):62–84.

[10] Bechhold H. Die Gallertfiltration. Z. Chem. Ind. Kolloide. 1907;2 (3–9):33–41.

[11] Bogdanchikova NE, Kurbatov AV, Tret'yakov VV, Rodionov PP. Activity of colloidal silver preparations towards smallpox virus. Pharm Chem J 1992;26 (9–10):778–779.

[12] Beveridge TJ, Murray RGE. Uptake and retention of metals by cell walls of *Bacillus subtilis*. J Bacteriol 1976;127:1502–1518.

[13] Mullen MD, Wolf DC, Ferris FG, Beveridge TJ, Flemming CA, Bailey GW. Bacterial sorption of heavy metals. Appl Environ Microbiol 1989;55:3143–3149.

[14] Beveridge TJ, Fyfe WS. Metal fixation by bacterial cell walls. Can J Earth Sci 1985;22:1893–1898.

[15] Haefeli C, Franklin C, Hardy K. Plasmid-determined silver resistance in *Pseudomonas stutzeri* isolated from silver mine. J Bacteriol 1984;158:389 392.

[16] Zhang H, Li Q, Lu Y, Sun D, Lin X, Deng X. Biosorption and bioreduction of diamine silver complex by *Corynebacterium*. J Chem Technol Biotechnol 2005;80:285–290.

[17] Nair B, Pradeep T. Coalescence of nanoclusters and formation of submicron crystallites assisted by *Lactobacillus* strains. Cryst Growth Des 2002;2:293–298.

[18] Kim JS, Kuk E, Yu KN, Kim JH, Park SJ, Lee HJ, Kim SH, Park YK, Park YH. Antimicrobial effects of silver nanoparticles. Nanomed Nanotechnol Biol Med 2007;3:95–101.

[19] Tan Y, Dai Y, Li Y, Zhua DJ. Preparation of gold, platinum, palladium and silver nanoparticles by the reduction of their salts with a weak reductant-potassium bitartrate. Mater Chem 2003;13:1069–1075.

[20] Li Y, Duan X, Qian Y, Li Y, Liao H. Nanocrystalline silver particles: synthesis, agglomeration, and sputtering induced by electron beam. J Colloid Interface Sci 1999;209:347–349.

[21] Singh A, Jain D, Upadhyay MK, Khandelwal N, Verma HN, Dig J. Green synthesis of silver nanoparticles using *Argimone mexicana* leaf extract and Evaluation of their antimicrobial activities. Digest J Nanomater Biostruct 2010;5 (2):483.

[22] Kalimuthu K, Babu RS, Venkataraman D, Mohd B, Gurunathan S. Biosynthesis of silver nanocrystals by *Bacillus licheniformis*. Colloids Surf B 2008;65:150–153.

[23] Kalishwaralal K, Deepak V, Pandian SRK, Kottaisamy M, BarathManiKanth S, Kartikeyan B, Gurunathan S. Biosynthesis of silver and gold nanoparticles using *Brevibacterium casei*. Colloids Surf B 2010;77 (2):257–262.

[24] Gurunathan S, Kalishwaralal K, Vaidyanathan R, Venkataraman D, Pandian SRK, Muniyandi J, Hariharan N, Eom SH. Biosynthesis, purification and characterization of silver nanoparticles using *Escherichia coli*. Colloids Surf B 2009;74 (1):328–335.

[25] Ahmad A, Mukherjee P, Senapati P, Mandal D, Khan MI, Kumar R, Santry M. Extracellular biosynthesis of silver nanoparticles using the fungus Fusarium oxysporum. Colloids Surf B Biointerfaces 2003c;28:313–318.

[26] Basavaraja S, Balaji SD, Lagashetty A, Rajasab AH, Venkataraman A. Extracellular biosynthesis of silver nanoparticles using the fungus *Fusarium semitectum*. Mater Res Bull 2008;43 (5):1164–1170.

[27] Gade AK, Bonde P, Ingle AP, Marcato PD, Dura´n N, Rai MK. Exploitation of *Aspergillus niger* for synthesis of silver nanoparticles. J Biobased Mater Bioenergy 2008;3:123–129.

[28] Mukherjee P, Roy M, Mandal BP, Dey GK, Mukherjee PK, Ghatak J, Tyagi AK, Kale SP. Green synthesis of highly stabilized nanocrystalline silver particles by a non-pathogenic and agriculturally important fungus *T. asperellum*. Nanotechnology 2008;19:075103.1–075103.7.

[29] Jha AK, Prasad K, Kulkarni AR. Plant system: nature's nanofactory. Colloids Surf. B Biointerfaces 2009;73 (2):219–223.

[30] Pandian SRK, Deepak V, Kalishwaralal K, Viswanathan P, Gurunathan S. Mechanism of bactericidal activity of silver nitrate—a concentration dependent bi-functional molecule. Braz J Microbiol 2010;41:805–809.

[31] Anil Kumar S, Abyaneh MK, Gosavi Sulabha SW, Ahmad A, Khan MI. Nitrate reductase mediated synthesis of silver nanoparticles from $AgNO_3$. Biotechnol Lett 2007;29:439–445.

[32] Naik RR, Stringer SJ, Agarwal G, Jones SE, Stone MO. Biomimetic synthesis and patterning of silver nanoparticles. Nat Mater 2002;1:169–172.

[33] Selvakannan PR, Swami A, Srisathiyanarayanan D, Shirude PS, Pasricha R, Mandale AB, Sastry M. Synthesis of aqueous Au core–Ag shell nanoparticles using tyrosine as a pH-dependent reducing agent and assembling phase-transferred silver nanoparticles at the air–water interface. Langmuir 2004a;20:7825–7836.

[34] Selvakannan PR, Mandal S, Phadtare S, Gole A, Pasricha R, Adyanthaya SD, Sastry M. Water-dispersible tryptophan-protected gold nanoparticles prepared by the spontaneous reduction of aqueous chloroaurate ions by the amino acid. J Colloid Interface Sci 2004b;269:97–102.

[35] Saifuddin N, Wong CW, Yasumira AAN. Rapid biosynthesis of silver nanoparticles using culture supernatant of bacteria with microwave irradiation. E J Chem 2009;6:1–70.

[36] Newman DK, Kolter R. A role for excreted quinones in extracellular electron transfer. Nature 2000;405:94–96.

[37] Duran N, Marcato PD, Alves OL, Souza G, Souza GI, Esposito E. Mechanistic aspects of biosynthesis of silver nanoparticles by several *Fusarium oxysporum* strains. J Nanobiotechnol 2005;3:8.

[38] Sintubin L, Windt WE, Dick J, Mast J, Ha DV, Verstarete W, Boon N. Lactic acid bacteria as reducing and capping agent for the fast and efficient production of silver nanoparticles. Appl Microbiol Biotechnol 2009;84:741–761.

[39] Mokhtari N, Daneshpajouh S, Seyedbagheri S, Atashdehghan R, Abdi K, Sarkar S, Minaian S, Shahverdi HR, Shahverdi AR. Biological synthesis of very small silver nanoparticles by culture supernatant of Klebsiella pneumonia: the effects of visible-light irradiation and the liquid mixing process. Mater Res Bull 2009;44:1415–1421.

[40] Durán N, Marcato PD, De Souza GIH, Alves OL, Esposito E. Antibacterial effect of silver nanoparticles produced by fungal process on textile fabrics and their effluent treatment. J Biomed Nanotechnol 2007;3:203–208.

[41] Ingle A, Gade A, Pierrat S, Sonnichsen C, Rai MK. Mycosynthesis of silver nanoparticles using the fungus Fusarium acuminatum and its activity against some human pathogenic bacteria. Curr Nanosci 2008;4:141–144.

[42] Wei YH, Han SB, Kim J, Soh SL, Grzybowski BA. Photoswitchable catalysis mediated by dynamic aggregation of nanoparticles. J Am Chem Soc 2010;132:11018–20.

[43] Pan Y, Neuss S, Leifert A, Fischler M, Wen F, Simon U, Schmid G, Brandau W, Dechent WJ. Size-dependent cytotoxicity of gold nanoparticles. Small 2007;3:1941–1949.

[44] Narayanan R, El-Sayed MA. Shape-dependent catalytic activity of platinum nanoparticles in colloidal solution. Nano Lett 2004;4:1343–1348.

[45] Euliss LE, DuPont JA, Gratton S, DeSimone J. Imparting size, shape, and composition control of materials for nanomedicine. Chem Soc Rev 2006;35:1095–1104.

[46] Wang W, Asher SA. Photochemical incorporation of silver quantum dots in monodisperse silica colloids for photonic crystal applications. J Am Chem Soc 2001;123:12528–12535.

[47] Nakanishi H, Bishop KJM, Kowalczyk B, Nitzan A, Weiss EA, Tretiakov KV, Apodaca MM, Klajn R, Stoddart JF, Grzybowski BA. Photoconductance and inverse photoconductance in films of functionalized metal nanoparticles. Nature 2009;460:371–375.

[48] Sharma V, Park K, Srinivasarao M. Shape separation of gold nanorods using centrifugation. Proc Natl Acad Sci USA 2009;106:4981–4985.

[49] Arnold MS, Stupp SI, Hersam MC. Enrichment of single-walled carbon nanotubes by diameter in density gradients. Nano Lett 2005;5:713–718.

[50] Arnold MS, Green AA, Hulvat JF. Sorting carbon nanotubes by electronic structure using density differentiation. Nat Nanotechnol 2006;1:60–65.

[51] Sorensen SB, Pedersen TG, Ottesen M. Fractionation of protein-components from beer by density gradient centrifugation. Carlsberg Res Commun 1982;47:227–231.

[52] Vestal CR, Zhang ZJ. Synthesis of CoCrFeO4 nanoparticles using microemulsion methods and size-dependent studies of their magnetic properties. Chem Mater 2002;14:3817–3822.

[53] Wagner H. Free-flow electrophoresis. Nature 1989;341:669–670.

[54] Hanauer M, Pierrat S, Zins I, Lotz A, Sönnichsen C. Separation of nanoparticles by gel electrophoresis according to size-and shape. Nano Lett 2007;7:2881–2885.

[55] Ho SS, Critchley K, Lilly GD, Shim B, Kotov NA. Free flow electrophoresis for the separation of CdTe nanoparticles. J Mater Chem 2009;19:1390–1394.

[56] Fletcher D. Fine particle high-gradient magnetic entrapment. IEEE Trans Magn 1991;27:3655–3677.

[57] Moeser GD, Roach KA, Green WH, Hatton TA, Laibinis PE. High-gradient magnetic separation of coated magnetic nanoparticles. AlchE J 2004;50:2835–2848.

[58] Ditsch A, Lindenmann S, Laibinis PE, Wang DIC, Hatton TA. High-gradient magnetic separation of magnetic nanoclusters. Ind Eng Chem Res 2005;44:6824–6836.

[59] Latham AH, Freitas RS, Schiffer P, Williams ME. Capillary magnetic field flow fractionation and analysis of magnetic nanoparticles. Anal Chem 2005;77:5055–5062.

[60] Rheinlander T, Kotitz R, Weitschies W, Semmler W. Different methods for the fractionation of magnetic fluids. Colloid Polym Sci 2000;278:259–263.

[61] Siebrands T, Giersig M, Mulvaney P, Fischer CH. Steric exclusion chromatography of nanometer-sized gold particles. Langmuir 1993;9:2297–2300.

[62] Kirkland JJ. High-performance size-exclusion liquid-chromatography of inorganic colloids. J Chromatogr 1979;185:273–288.

[63] Fischer CH, Lilie J, Weller H, Katsikas L, Henglein A. Photochemistry of colloidal semiconductors 29. Fractionation of Cds sols of small particles by exclusion chromatography. Phys Chem Chem Phys 1989;93:61–64.

[64] Wei GT, Liu FK. Separation of nanometer gold particles by size exclusion chromatography. J Chromatogr A 1999;836:253–60.

[65] Jimenez VL, Leopold MC, Mazzitelli C, Jorgenson JW, Murray RW. HPLC of monolayer-protected gold nanoclusters. Anal Chem 2003;75:199–206.

[66] Wilcoxon JP, Martin JE, Provencio P. Size distributions of gold nanoclusters studied by liquid chromatography. Langmuir 2000;16:9912–9920.

[67] Lee JS, Stoeva SI, Mirkin CA. DNA-induced size-selective separation of mixtures of gold nanoparticles. J Am Chem Soc 2006;128:8899–8903.

[68] Akthakul A, Hochbaum AI, Stellacci F, Mayes AM. Size fractionation of metal nanoparticles by membrane filtration. Adv Mater 2005;17:532–5.

[69] Sweeney SF, Woehrle GH, Hutchison JE. Rapid purification and size separation of gold nanoparticles via diafiltration. J Am Chem Soc 2006;128:3190–3197.

[70] Wilson OM, Scott RWJ, Garcia-Martinez JC, Crooks RM. Separation of dendrimer-encapsulated Au and Ag nanoparticles by selective extraction. Chem Mater 2004;16:4202–4204.

[71] Williams DP, Satherley J. Size-selective separation of polydisperse gold nanoparticles in supercritical ethane. Langmuir 2009;25:3743–3747.

[72] Jana NR. Nanorod shape separation using surfactant assisted self-assembly. Chem Commun. 2003:1950–1951.

[73] Park K, Koerner H, Vaia RA. Depletion-induced shape and size selection of gold nanoparticles. Nano Lett 2010;10:1433–1439.

[74] Hollamby MJ, Eastoe J, Chemelli A, Glatter O, Rogers S, Heenan RK, Grillo I. Separation and purification of nanoparticles in a single step. Langmuir 2010;26:6989–6994.

[75] Kang ZH, Tsang CHA, Ma DDD, Wong NB, Lee ST. Nanomaterials separation by an ultrasonic assisted phase transfer method. Chem Phys Lett 2008;455:252–255.

[76] Lagzi I, Kowalczyk B, Grzybowski BA. Liesegang rings engineered from charged nanoparticles. J Am Chem Soc 2010;132:58–60.

[77] Marshall JP, Schneider RP. Systemic argyria secondary to topical silver nitrate. Arch Dermatol 1977;113:1077–1079.

[78] Shelley WB, Shelley ED, Burmeister V. Argyria: the intradermal "photograph," a manifestation of passive photosensitivity. J Am Acad Dermatol 1987;16:211–217.

[79] Gulbranson SH, Hud JA, Hansen RC. Argyria following the use of dietary supplements containing colloidal silver protein. Cutis 2000;66:373–376.

[80] Drake PL, Hazelwood KJ. Exposure-related health effects of silver and silver compounds: a review. Ann Occup Hyg 2005;49:575–585.

[81] Duran N, Marcato PD, De Conti R, Alves OL, Costa FTM, Brocchi M. Potential use of silver nanoparticles on pathogenic bacteria, their toxicity and possible mechanisms of action. J Brazil Chem Soc 2010;21:949–959.

[82] Gurunathan S, Lee KJ, Kalishwaralal K, Sheikpranbabu S, Vaidyanathan R, Eom SH. Antiangiogenic properties of silver nanoparticles. Biomaterials 2009a;30:6341–6350.

[83] Feng QL, Wu J, Chen GQ, Cui FZ, Kim TN, Kim JO. A mechanistic study of the antibacterial effect of silver ions on *Escherichia coli* and *Staphylococcus aureus*. J Biomed Mater Res 2000;52:662–668.

[84] Nutt OM, Hughes BJ, Wong SM. Designing Pd-on-Au bimetallic nanoparticles catalysts for trichloethene hydrodechlorination. Environ Sci Technol 2005;39:1346–1353.

[85] Tamber H, Johansen P, Merkle HP, Gander B. Formulation aspects of biodegrad- able polymeric microspheres for antigen delivery. Adv Drug Deliver Rev 2005;57:357–376.

[86] Dlugosz M, Bulwan M, Kania G, Nowakowska M, Zapotoczny S. Hybrid calcium carbonate/polymer microparticles containing silver nanoparticles as antibacterial agents. J Nanopart Res 2012;14 (12):1313.

[87] Taylor U, Garrels W, Petersen S, Kues W, Lucas-Hahn A, Barcikowski S, Rath D. 149 unimpaired development of murine embryos after injection of silver nanoparticles. Reprod Fertil Dev 2012;25 (1):222–223.

[88] Sriram MI, BarathManiKanth S, Kalishwaralal K, Gurunathan S. Antitumor activity of silver nanoparticles in Dalton's lymphoma ascites tumor model. Int J Nanomedicine 2010;3:1–10.

[89] Sahoo SK, Labhasetwar V. Nanotech approaches to drug delivery and imaging. Drug Discov Today 2003;8:1112–1120.

[90] Fahmy TM, Fong PM, Goyal A, Saltzman WM. Targeted for drug delivery. Nanotoday 2005:18–26.

[91] Fahmy TM, Samstein RM, Harness CC, Saltzman WM. Surface modification of biodegradable polyesters with fatty acid conjugates for improved drug targeting. Biomaterials 2005;26:5727–5736.

[92] Couvreur P, Barratt G, Fattal E, Legrand P, Vauthier C. Nanocapsule technology: a review. Crit Rev Ther Drug Carrier Syst 2002;19:99–134.

[93] Åkerman ME, Chan WCW, Laakkonen P, Bhatia SN, Ruoslahti E. Nanocrystal targeting in vivo. Proc Natl Acad Sci USA 2002;99:12617–12624.

[94] Makarova OV, Ostafin AE, Miyoshi H, Norris JR, Meisel D. Adsorption and encapsulation of fluorescent probes in nanoparticles. J Phys Chem B 1999;103:9080–9084.

[95] Wang H, Zhang Y, Yu H, Wu D, Ma H, Li H, Du B, Wei Q. Label-free electrochemical immunosensor for prostate specific antigen based on silver hybridized mesoporous silica nanoparticles. Anal Biochem 2013;434 (1):123–127.

[96] Ballabh P, Braun A, Nedergaard M. The blood–brain barrier: an overview Structure, regulation, and clinical implications. Neurobiol Dis 2004;16:1–13.

[97] Brightman M. Ultrastructure of the brain endothelium. In: Bradbury MW, editor. *Physiology and Pharmacology of the Blood–Brain Barrier Handbook of Experimental Pharmacology*. Volume 103, Berlin: Springer; 1992. p 1–22.

[98] Begley D. The blood–brain barrier: principles for targeting peptides and drugs to the central nervous system. J Pharm Pharmacol 1996;48:136–146.

[99] Davson H, Segal M. *Physiology of the CSF and Blood–Brain Barrier*. Boca Raton: CRC Press; 1996. p 1–192.

[100] Alyautdin R, Gothier D, Petrov V, Kharkevich D, Kreuter J. Analgesic activity of the hexapeptide dalargin adsorbed on the surface of polysorbate 80-coated poly(butyl cyanoacrylate) nanoparticles. Eur J Pharm Biopharm 1995;41:44–48.

[101] Kreuter J, Alyautdin RN, Kharkevich DA, Ivanov AA. Passage of peptides through the blood–brain barrier with colloidal polymer particles (nanoparticles). Brain Res 1995;674:171–174.

[102] Kreuter J, Petrov VE, Kharkevich DA, Alyautdin RN. Influence of the type of surfactant on the analgesic effects induced by the peptide dalargin after its delivery across the blood–brain barrier using surfactant-coated nanoparticles. J Control Release 1997;49:81–87.

[103] Schro¨der U, Sabel BA. Nanoparticles, a drug carrier system to pass the blood–brain barrier, permit central analgesic effects of i.v. dalargin injections. Brain Res 1996;710:121–124.

[104] Schroeder U, Sommerfeld P, Sabel BA. Efficacy of oral dalargin-loaded nanoparticle delivery across the blood–brain barrier. Peptides 1998b;19:777–780.

[105] Schroeder U, Sommerfeld P, Ulrich S, Sabel BA. Nanoparticle technology for delivery of drugs across the blood–brain barrier. J Pharm Sci 1998a;87:1305–1307.

[106] Ramge P, Kreuter J, Lemmer B. Circadian phasedependent antinociceptive reaction in mice after i.v. injection of dalargin-loaded nanoparticles determined by the hot-plate test and the tail-flick test. Chronobiol Int 1999;17:767–777.

[107] Ullrich NJ, Pomeroy SL. Pediatric brain tumors. Neurol Clin 2003;21:897–913.

[108] Kreuter J, Dmitry S, Valery P, Peter R, Klaus C, Claudia KB, Renad A. Apolipoprotein-mediated transport of nanoparticle-bound drugs across the blood–brain barrier. J Drug Target 2002;10 (4):317–325.

[109] Diallo MS, Christie S, Swaminathan P, Johnson JH, Goddard WA. Dendrimer enhanced ultra-filtration recovery of Cu (II) fromaqueous solutions using Gx-NH2-PAMAM dendrimers with ethylene diamine core. Environ Sci Technol 2005;39:1366–1377.

[110] Momba MNB, Kaleni P. Regrowth and survival of indicator microorganisms on the surfaces of household containers used for the storage of drinking water in rural communities of South Africa. Water Res 2002;36:3023–3028.

[111] Edberg SC, Rice EW, Karlin RJ, Allen MJ. *Escherichia coli*: the best biological drinking water indicator for public health protection. Symp Ser Soc Appl Microbiol 2000;88:S106–S116.

[112] Enriquez C, Nwachuku N, Gerba CP. Direct exposure to animal enteric pathogens. Rev Environ Health 2001;16:117–131.

[113] Momba MNB, Notshe TL. The microbiological quality of groundwater-derived drinking water after long storage in household containers in a rural community of South Africa. J Water Supply Res Technol 2003;52:67–77.

[114] Momba MNB, Malakate VK, Theron J. Abundance of phathogenic Escherichia coli, Salmonella typhimurium and Vibrio cholerae in Nkonkobe drinking water sources. J Water Health 2006;4:289–296.

[115] US Environmental Protection Agency. Microbial and disinfection by-product rules. Federal Register 1998b;63:69389–69476.

[116] Droste RL. *Theory and practice of water and wastewater treatment*. New York: John Wiley and Sons; 1997.

[117] Jung JY, Chung YC, Shin HS, Son DH. Enhanced ammonia nitrogen removal using consistent biological regeneration and ammonium exchange of zeolite in modified SBR process. Water Res 2004;38:347–354.

[118] Shearer AEH, Paik JS, Hoover DG, Haynie SL, Kelley MJ. Potential of an antibacterial ultraviolet-irradiated nylon film. Biotech Bioeng 2000;67:141–146.

[119] Mclean RJC, Hussain AA, Sayer M, Vincent PJ, Hughes DJ, Smith TJN. Antibacterial activity of multilayer silver copper surface-films on catheter material. Can J Microbiol 1993;39:895–899.

[120] Chohan ZH, Suparna CT, Scozzafava A. Metalloantibiotics: synthesis and antibacterial activity of cobalt (II), nickel (II) and zinc (II) complex of kefzol. J Enzym Inhib Med Chem 2004;19:79–84.

[121] Ulkuseven B, Tavman A, Otuk G, Birteksoz S. Antimicrobial activity of Fe-III, Cu-II, Ag-I, Zn-II and Hg-II complexes of 2-(2-hydroxy-5-bromo/nitro-phenyl)-1H and 2-(2-hydroxyphenyl)-5-methyl/chloro/nitro-1H benzimidazoles. Folia Microbiol 2002;47:481–487.

[122] Chen Y, Wang L, Jiang S, Yu HJ. Study on novel antibacterial polymer materials preparation of zeolite antibacterial agents and antibacterial polymer composite and their antibacterial properties. J Polymer Mater 2003;20:279–284.

[123] Cik G, Bujdakova H, Sersen F. Study of fungicidal and antibacterial effect of the Cu (II)- complexes of thiophene oligomers synthesized in ZSM-5 zeolite channels. Chemosphere 2001;44:313–319.

[124] Islam MS, Motahar HM, Banu LA, Sultana C, Quadir MA. Antibacterial and antifungal activity of mixed ligand complexes of oxovanadium (IV), titanium (III) and cadmium (II) metal ions. Oriental J. Chem. 2003;19:547–554.

[125] Ichinose N, Ozaki Y, Kashu S. *Superfine particle technology*. London: Springer; 1992.

[126] Stoimenov PK, Klinger RL, Marchin GL, Klabunde KJ. Metal oxide nanoparticles as bactericidal agents. Langmuir 2002;18:6679–6686.

[127] Zeman LJ, Zydney AL. *Microfiltration and Ultra-filtration: Principles and Applications*. New York: Marcel Dekker; 1996.

[128] Jain P, Pradeep T. Potential of silver nanoparticle-coated polyurethane foam as an antibacterial water filter. Biotech Bioeng 2005;90:59–63.

[129] Sons WK, Youk JH, Lee TS, Park WH. Preparation of antimicrobial ultrafine cellulose acetate fibers with silver nanoparticles. Macromol Rapid Commun 2004;25:1632–1637.

[130] Ovington LG. The truth about silver. Ostomy Wound Manage 2004;50:S1–S10.

[131] Rai M, Yadav A, Gade A. Silver nanoparticles as a new generation of antimicrobials. Biotech Adv 2009;27:76–83.

[132] Mahmood SN, Naeem S, Basit N, Usmani TH. Microbial evaluation of silver coated/impregnated sand for purification of contaminated water. Environ Technol 1993;14:151–157.

[133] Matsumura Y, Yoshikata K, Kunisaki SI, Tsuchido T. Mode of bactericidal action of silver zeolite and its comparison with that of silver nitrate. Appl Environ Microbiol 2003;69:4278–4281.

[134] Nangmenyi G, Yue Z, Mehrabi S, Mintz E, Economy J. Synthesis and characterization of Ag nanoparticle impregnated fiberglass and utility in water disinfection. Nanotechnology 2009;20:495–505.

[135] Ping L, Li J, Wu C, Wu Q, Li J. Synergistic antibacterial effect of ß-lactum antibiotic combined with silver nanoparticle. Nanotechnology 2005;16:1912–1917.

[136] Dibrov P, Dzioba J, Gosnik KK, Häse CC. Chemiosmotic mechanism of antimicrobial activity of Ag^+ in *Vibrio cholerae*. Antimicrob Agents Chemother 2002;46:2668–2670.

[137] Chen X, Schluesener HJ. Nanosilver: a nanoproduct in medical application. Toxicol Lett 2007;176:1–12.

[138] Murata T, Kanao-koshikawa M, Takamatsu T. Effects of Pb, Cu, Sb, In and Ag contamination on the proliferation of soil bacterial colonies, soil dehydrogenase activity, and phospholipid fatty acid profiles of soil microbial communities. Water Air Soil Pollut 2005;164:103–118.

[139] Throback IN, Johansson M, Rosenquist M, Pell M, Hansson M, Hallin S. Silver (Ag+) reduces denitrification and induces enrichment of novel nirK genotypes in soil. FEMS microbial Lett 2007;270:189–194.

[140] Wood CM, Playle RC, Hogstrand C. Physiology and modeling of mechanisms of silver uptake and toxicity in fish. Environ Toxicol Chem 1999;18:71–83.

20

GREEN SYNTHESIS OF NANOMATERIALS USING BIOLOGICAL ROUTES

RAJESH RAMANATHAN, RAVI SHUKLA, SURESH K. BHARGAVA, AND VIPUL BANSAL

NanoBiotechnology Research Laboratory, Centre for Advanced Materials and Industrial Chemistry, School of Applied Sciences, RMIT University, Melbourne, Australia

20.1 INTRODUCTION

In today's modern day science, we can relate future discoveries to one of the most famous and influential talks given by Richard Feynman "There is plenty of room at the bottom," wherein the foundation for one of the most important fields known to mankind, that is, nanotechnology, was laid [1]. The past few decades have seen the premises of nanotechnology revolutionizing modern technology with the dawn of several applications including small and flexible devices that can be used for sensing and electronics to generate a wide array of new generation tools capable of identifying and treating infectious agents [2–8].

To realize the true potential of nanotechnology, it was essential to develop efficient fabrication protocols of nanomaterial synthesis and testing their potential applications. This saw the introduction of physical methods such as mechanical grinding toward nanomaterial synthesis, especially in the early years of the field [9]. With gaining importance, solution-based nanomaterial synthesis routes were developed (more commonly known as chemical route) that now enjoy a long history and have typically dominated the nanosphere with well-established protocols for synthesizing particles with excellent control over shape, size, and properties [9]. With the development of physical and chemical methods, the paramount concern for a negative impact on the environment was also heightened due to the use of toxic chemicals, reaction conditions involving extremes of temperature, pressure and pH, as well as the release of harmful by-products. Hence, scientists started investigating alternative strategies toward new "green" environment friendly fabrication routes for nanomaterial synthesis [9–13].

Unsurprisingly, Mother Nature has developed a large repertoire of functional assemblages of proteins, nucleic acids, and other macromolecules to perform complicated tasks that are still daunting for us to emulate in our laboratories [9, 10]. One such task that is an integral part of biological systems is the assimilation of inorganic materials in hard tissues, which are biocomposites containing structural biomolecules in addition to some 60 different kinds of minerals that perform a myriad of vital biological functions [13, 14]. Furthermore, unicellular organisms have also been observed for their ability to synthesize inorganic materials, both intra- and extra-cellularly [15]. Some examples exhibiting bioinorganic materials include magnetotactic bacteria that synthesize magnetite [16, 17] diatoms that synthesize silica [18, 19] and S-layer bacteria that synthesize gypsum and calcium carbonate as surface layers [20]. Although it is known that these inorganic materials serve important biological functions, it was also realized that the study of these minerals from the material science perspective may have significant potential to transform the fields of physical, biological, and material sciences [9]. The complementary knowledge gained from these areas of sciences by incorporating the concepts and ideas derived from the biological world to the inorganic material synthesis has led to a merger between different fields, giving rise to green nanosciences, green nanotechnology and bionanotechnology.

Nanomaterials for Environmental Protection, First Edition. Edited by Boris I. Kharisov, Oxana V. Kharissova, and H. V. Rasika Dias.

20.2 BIOREMEDIATION

During the detailed study of nature's strategies involved in fabricating materials, one of the pivotal discoveries was the important role of microbial activity in transforming elements in the periodic table, which is a result of assimilatory, dissimilatory, or detoxification processes [21]. This process, also referred to as bioremediation, forms a cornerstone of many biogeochemical cycles and typically involves a change in the oxidation state of at least one-third of the elements in the periodic table by biological entities, and includes highly toxic elements such as Hg, Cr, Se, Te, and As [22]. Microorganisms such as bacteria, fungi, algae, and plants are known for their inherent ability to withstand high concentrations of toxic metal ions through specific resistance mechanisms and therefore these are commonly employed toward the remediation of different metal ions [23–25]. The detoxification by these resistance mechanisms typically occurs either by reduction/oxidation of metal ions or by the formation of insoluble complexes that results in the intracellular accumulation of metal ion complexes/nanoparticles [23–25]. Some of the earliest reports that outline the accumulation of inorganic materials date back to the 1960s when elemental gold was reported in Precambrian algal blooms [26], algal cells [27] and bacteria [28]; CdS in bacteria [29] and yeast [30], and magnetite in magnetotactic bacteria [31]. Although a microbiologist's interest purely lies in understanding the biochemical processes involved in the resistance machinery, materials scientists are conversely fascinated by the ability of these biological entities to make technologically important inorganic materials. This has opened up a new avenue of research wherein biological organisms known for withstanding high concentrations of metal ions are deliberately explored as "nanofactories" to achieve potential large-scale "green" synthesis of nanomaterials. This is similar to the commercially successful fermentation technology, which is regularly employed for the large-scale synthesis of enzymes, drugs, etc., and has potential to be employed for large-scale nanoparticle synthesis. Although a great deal of work has been performed in this emerging area of bionanotechnology with organisms including bacteria, fungi, and plants being regularly employed for the synthesis of a wide range of nanomaterials ranging from metals and metal oxides to metal sulfides, the biological synthesis (or biosynthesis) of inorganic nanomaterials is still in its infancy phase, as the actual biochemical processes involved in the formation of inorganic materials largely remain poorly understood [13]. The following sections will outline some of the biological entities employed for nanomaterial biosynthesis.

20.3 METAL NANOPARTICLES

The synthesis of metal nanoparticles especially that of noble metals is of enormous importance due to their interesting optical, electronic, catalytic, and sensing properties [5, 32–39]. Although solution-based chemical routes have dominated the field of nanomaterial synthesis, the recent past has seen an increasing trend in employing biological entities toward the synthesis of metal nanoparticles. These biological entities includes bacteria, fungi, algae, plants, and plant extracts, each of which have their own potential advantages and associated drawbacks in the context of the synthesis of nanoparticles [9–13]. The different biological entities employed for the reduction of metal ions to their nanoparticulate counterparts are summarized in Table 20.1.

Some of the earliest studies in the field date back to the late 1970s, wherein gold was reported to interact with Precambrian algal blooms and bacteria [26]. Following this report, a wide range of bacterial species were reported for their ability to form nanoparticles. Beveridge and Murray showed the ability of *Bacillus subtilis* to accumulate gold nanoparticles of 5–25 nm in the cell wall when incubated with precursor gold salt [28]. The organophosphate compounds secreted by the bacterium were believed to play an important role in the reduction of gold ions [40]. Additionally, Beveridge and co-workers also showed the unique ability of bacterial (*Pseudomonas aeruginosa*) biofilms toward intracellular precipitation of gold nanoparticles with particle diameter of 20 nm [41]. Several groups have further shown the ability of *Shewanella* spp. (*S. algae*, *S. oneidensis*), a marine bacteria belonging to *Proteobacteria* for the synthesis of gold and platinum nanoparticles [42–44]. Bacteria that belong to this genus have been reported for their unique ability to reduce/oxidize metal ions efficiently as these organisms are found under harsh environmental conditions [45]. *S. algae*, an iron-reducing bacterium, showed the ability to reduce gold ions under anaerobic conditions that typically yields gold nanoparticles of 10–20 nm in diameter. This metal salt reduction ability of these metal-reducing organisms through dissimilatory pathways was attributed to the presence of c-type cytochromes. A few other alternative strategies were also hypothesized wherein the involvement of a species-specific hydrogenase when hydrogen was used as an electron donor was speculated [42]. Additionally, another fascinating hypothesis proposed the expression of a gold reductase enzyme in bacteria in response to gold ion stress that might be involved in the direct reduction of gold ions extracellularly [42, 46]. However, these hypothesized biochemical mechanisms for the reduction of gold ions remains invalidated till date.

In a series of reports, Lengke and co-workers investigated the ability of filamentous cyanobacterial strain *Plectonema boryanum* UTEX 485 for its ability to reduce metal ions (gold, silver, platinum, and palladium) [47–50]. Although this cyanobacterial strain had an interesting ability to reduce all these metal ions to their metallic counterparts, the nanoparticles synthesized were

TABLE 20.1 List of biological entities employed for metal nanoparticle synthesis

Microorganism	Metals	References
Bacteria		
Bacillus subtilis	Au	[28]
Pseudomonas aeruginosa	Au	[41]
Ralstonia metallidurans	Au	[51]
Shewanella algae	Au	[42]
Shewanella oneidensis	Au	[44]
Morganella morganii Subgroups (A, B, C, D, E, F, G1, G2)	Ag	[83]
Morganella psychrotolerans	Ag	[80]
Pseudomonas stutzeri AG259	Ag, Ag_2S	[79]
Morganella morganii RP42	Ag, Cu	[84, 85]
Shewanella algae	Pt	[43]
Desulfovibrio desulfuricans	Pd	[88, 89]
Desulfovibrio Oz7	Pd	[88]
Shewanella oneidensis	Pd	[90]
Bacillus selenitireducens	Te, Se	[93, 94]
Selenihalanaerobacter shriftii	Se	[93]
Sulfurospirillum barnesii	Te, Se	[93, 94]
Lactobacilli strains	Au, Ag, Au-Ag	[62]
Cyanobacteria		
Plectonema boryanum UTEX 485	Au, Ag, Pt, Pd	[47–50]
Actinomycetes		
Actinobacter sp.	Au	[55]
Rhodococcus sp.	Au	[54]
Thermomonospora sp.	Au	[53]
Fungi		
Colletotrichum sp.	Au	[58]
Fusarium oxysporum	Au, Ag, Au-Ag	[59–61]
Verticillium sp.	Au, Ag	[56, 57]
Plant Extracts		
Azadirachta indica	Au, Ag, Au-Ag	[66]
Cymbopogon flexuosus	Au	[65]
Cuminum cyminum	Au	[70]
Cinnamomum verum	Au	[73]
Darjeeling Tea	Au	[71]
Glycine max	Au	[69]
Pelargonium graveolens	Au	[58, 64]
Iris pseudocorus	Cu	[66]

of different size and shape. The reduction of gold thiolate and gold chloride complexes resulted in 10–25 nm diameter gold nanoparticles that were found deposited on the outer surface of the cell membrane, while particles smaller than 10 nm were also formed intracellularly [47]. The prolonged incubation of gold ions resulted in the formation of thin triangular and hexagonal sheets with edge lengths ranging up to several microns in length. It was postulated that gold nanoparticles were synthesized and deposited within the membrane vesicles, which resulted from the interaction of gold ions with phosphorous-, sulfur-, and nitrogen ligand-rich vesicle components. Similarly, the same algal biomass reduced Ag (I) ions at 25°C to form 1–15 nm Ag nanoparticles intracellularly as well as extracellularly in the solution. In addition to performing experiments at 25°C, when these experiments were performed at elevated temperatures, the particles were observed to increase in size with 1–40 nm diameter spherical nanoparticles observed at 60°C, while a mixture of spherical and octahedral crystal platelets of 5–200 nm dimensions were observed when the reaction was carried out at 100°C [50]. *P. boryanum* also showed the ability to reduce the platinum group of salts when reacted with aqueous Pt (IV) and Pd (II) ions for 28 days, respectively. The precipitation of an amorphous Pt (II)-organic spherical complex was promoted initially, which was characteristically connected into long bead-like chains by the continuous coating of organic material. On increasing the reaction temperature and reaction time, these bead-like structures were converted to crystalline Pt nanoparticles of 30–300 nm diameter [48]. In contrast, crystalline spherical and elongated Pd nanoparticles of 30 nm diameter were directly precipitated at lower temperatures. Interestingly, the reactions at higher temperatures only yielded palladium hydride with a small amount of Pd metal nanoparticles [49]. Although this organism was found to have the ability to reduce a range of noble metal ions, the exact biochemical mechanism involved in the formation of metal nanoparticles is yet to be elucidated and still remains a mystery.

A recent study also highlighted the ability of bacterium *Ralstonia metallidurans* to precipitate colloidal gold nanoparticles and the ability of this bacterium to contribute toward the mineralization of metallic gold was highlighted [51]. However, the exact mechanism involved in the biomineralization of metallic gold by *R. metallidurans* is also not yet clear. On the other hand, the presence of several mechanisms for heavy metal resistance in this bacterium including cation efflux, cation reduction, cytoplasmic accumulation, and organic compound formation leads us to believe the importance and possible involvement of these biochemical detoxifying mechanisms toward nanoparticle synthesis [52].

In a series of reports, Sastry's group explored different biological entities with more emphasis on fungus, actinomycetes, and plant extracts for the biosynthesis of metal nanoparticles with significant control over morphologies and size distribution. Less control over the size and shape of nanoparticles and the relatively large synthesis time (from several hours to days) were some of the well-known challenges in employing a biological approach for nanoparticle synthesis. To address some of these challenges, an alkali thermophilic actinomycete *Thermomonospora* sp. was shown for the first time for its ability to not only synthesize gold nanoparticles but also obtain a narrow size distribution (Fig. 20.1) [53]. The extracellular manifestation of these particles further overcomes the limitations associated with harvesting of intracellular nanoparticles from microbial cells, as observed in previous reports, thereby reducing yet another step involved in the post-synthesis processing and harvesting of nanoparticles. Further enhancement in the monodispersity was obtained by employing *Rhodococcus* sp., wherein 10 nm Au nanoparticles were observed, however only intracellularly [54]. Further control over size and shape was observed by tuning reaction parameters, wherein Sastry et al. showed the ability of an *Actinobacter* sp. to synthesize triangular and hexagonal Au particles with an edge length of 30–50 nm when the reaction was carried out in the absence of molecular oxygen [55]. Interestingly, at lower temperatures, the reaction was comparatively slow and resulted in smaller triangular Au nanoparticles with an edge length of approximately 10 nm, while the reaction in the presence of atmospheric oxygen was found slower and yielded distinct triangular and hexagonal gold particles with edge lengths of 50–500 nm. The biochemical analysis recognized the important role of the protease enzyme secreted by *Actinobacter* sp. toward the efficient reduction of Au (III) ions. Further research suggested that molecular oxygen slows down the reduction of gold possibly by inhibiting the protease activity [55]. These studies point toward the important role of controlling simple reaction parameters for controlling the size and shape of nanoparticles during biosynthesis. Additional experiments further revealed the probable involvement of cytochrome oxidase as the primarily responsible component for the reduction of aqueous Au (III) ions.

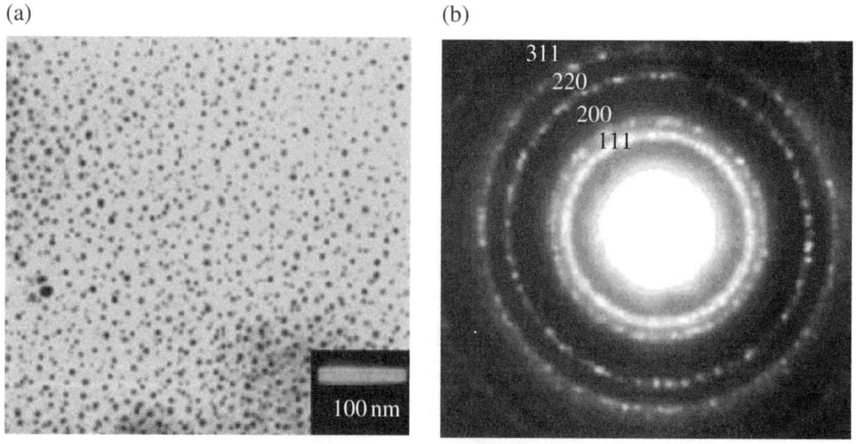

FIGURE 20.1 (a) TEM image and (b) electron diffraction of Au nanoparticles synthesized using *Thermomonospora* sp. Images reprinted with permission from Ref. [53]. © 2003, American Chemical Society.

Furthermore, Bharde et al. identified a group of three proteins that were found to guide the assembly of smaller spherical Au nanoparticles toward triangular nanoparticles. Such investigations started revealing some of the potential biochemical pathways involved in nanoparticle biosynthesis, further leading to the isolation of the metal ion reducing proteins [9]. This knowledge about the involved biomolecular mechanisms may help promote eco-friendly fabrication strategies for large-scale synthesis of technologically important nanomaterials.

In addition to employing actinomycete, Sastry's group systematically investigated a range of fungi for the synthesis of metal nanoparticles including gold, silver, and gold-silver bimetallic nanoparticles. Their initial research demonstrated the ability of fungus *Verticillium* sp. to accumulate Au nanoparticles within the fungal biomass (Fig. 20.2a–d). The accumulation of Au nanoparticles in the fungal biomass was evident due to the conversion of white fungal biomass to a dark purple color and the presence of a characteristic Au surface plasmon resonance (SPR) feature at 550 nm in the biomass [56]. The same fungus, on treatment with Ag (I) ions, also resulted in the intracellular manifestation of Ag nanoparticles (Fig. 20.2e–g) [56, 57]. Although the mechanism of reduction is not clear, it was postulated that the first step might involve the trapping of metal ions on the surface of cells via electrostatic interaction due to the presence of charged amine and carboxylic groups. These ions are further reduced to form nuclei that grow to form nanoparticles as more ions are reduced. Further, preliminary experiments showed the ability of endophytic fungus *Colletotrichum* sp. to successfully yield anisotropic gold nanoparticles of rod-like and prismatic morphology [58]. Similar to *Verticillium* sp., plant pathogenic fungus *Fusarium oxysporum* was also capable of reducing gold ions to gold nanoparticles [59] and silver ions to form silver nanoparticles [60]. These biogenic silver and gold nanoparticles were mostly similar in size typically ranging from 20 to 50 nm in diameter. In addition for its ability to reduce individual metal ions, when *F. oxysporum* was exposed to equimolar solutions of $HAuCl_4$ and $AgNO_3$ simultaneously, it resulted in highly stable Au–Ag alloy nanoparticles of varying mole fractions [61]. The presence of a single SPR feature gradually shifting from nanogold to nanosilver plasmon peaks clearly suggested the formation of gold-silver alloy nanoparticles instead of a segregated metal or core/shell-type structure. Similarly, Nair and Pradeep also showed the ability of different *Lactobacilli* strains to reduce

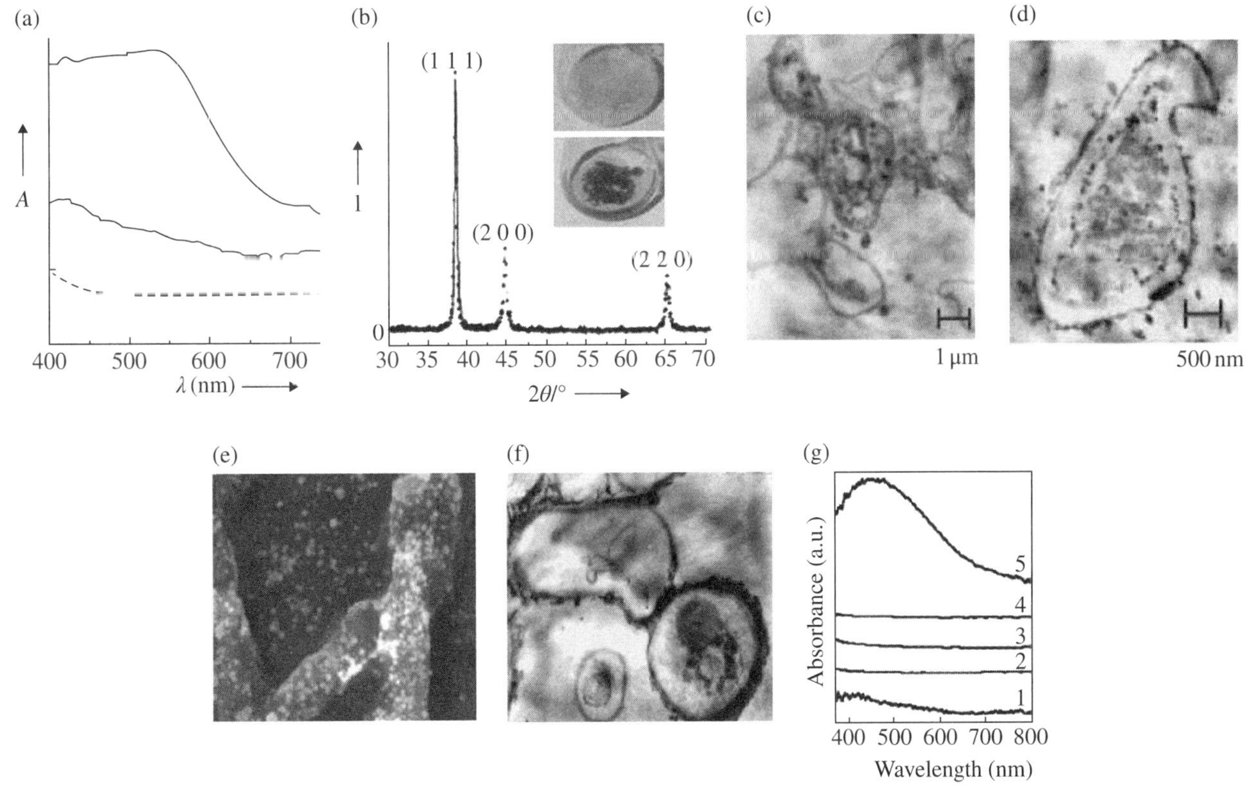

FIGURE 20.2 (a) UV-vis spectra obtained from dried biomass of *Verticillium* sp. before (dashed line) and after (solid line) exposure to gold ions demonstrating the synthesis of Au nanoparticles. (b) XRD pattern of Au nanoparticles from dried biomass of *Verticillium* sp. Inset shows fungal biomass before (upper panel) and after (lower panel) exposure to gold ions indicating Au nanoparticles within fungal biomass. (c-d) TEM images of a thin section of fungal biomass showing the presence of intracellular Au nanoparticles. (e) SEM image and (f) TEM images of *Verticillium* sp. mycelium showing Ag nanoparticles. (g) UV-vis spectrum obtained from *Verticillium* biomass indicating the presence of Ag nanoparticles. Images reprinted with permission from Refs. [56, 57]. © 2001, Wiley-VCH Verlag GmbH & Co. KGaA and © 2001, American Chemical Society.

noble metal ions [62]. Typically, these *Lactobacilli* strains were able to form gold, silver, and gold-silver alloy nanoparticles intracellularly. Comparable to the earlier case, the UV-visible spectra obtained from the nanoparticles synthesized by simultaneous exposure of Au and Ag ions showed a plasmon band at 537 nm, which is centered between the plasmon bands observed for pure Ag and Au nanoparticles, respectively, indicating the formation of an Au-Ag alloy and not a core shell structure. Interestingly, the intracellular manifestation of these nanoparticles created a metal environment within the bacterial cell, which was exploited by Nair and Pradeep toward mapping the bacterial cell using surface-enhanced Raman scattering (SERS). Considering the sensitivity of SERS-based sensors, it may be of utmost importance to explore this avenue of research to further enhance the Raman signals from within the bacterial cell. This could lead to the identification of infectious agents in humans and animals by looking for fingerprint molecules in these microbes [62].

In addition to employing biological entities like bacteria, fungi, and cyanobacterial species that offer a unique ability to transform metal ions to nanoparticles, plant and plant extracts offer an exciting and unique fabrication strategy, as this technique is fast, generally extracellular and with faster rate of reduction. Moreover, several plant species have been reported to have resistance machineries to manage the metal ion stress similar to that found in the bacterial system. For instance, it was recently reported that *Iris pseudocorus* (yellow iris) could tackle copper stress environment by transforming ionic copper to metallic copper nanoparticles in and near its roots with some degree of assistance from the endomycorrhizal fungi present in association with these plants [63]. Although this ability of plants to synthesize nanoparticles in a high stress environment is fascinating, the ability to use leaf extracts of different plants for rapid synthesis of nanoparticles appears more exciting from a practical perspective. Sastry et al. used a range of plant extracts including germanium (*Pelargonium graveolens*) [58, 64], lemongrass (*Cymbopogon flexuosus*) [65], and neem (*Azadirachta indica*) [66] for the synthesis of gold nanoparticles. Interestingly, these particles showed high degree of morphological control by displaying spherical, triangular, decahedral, and icosahedral shapes. The presence of alcohol, terpenoids, flavonoids, and aldehyde/ketone groups present in these extracts were believed to facilitate the reduction of chloroaurate ions.

Furthermore, as an application of biosynthesized metal nanoparticles, the ability of gold nanotriangles biosynthesized by lemongrass extract was elucidated toward the detection of aqueous mercuric ions at ultra-low femtomolar concentrations [67]. This is an important application of biosynthesized nanomaterials, as mercury is one of the most serious environmental pollutants that are known to have serious detrimental effects on living organisms. Particularly, in humans, mercury causes severe aliments to the kidney, heart, nervous system, respiratory system, and muscles. In addition, mercury is known to amalgamate with gold thereby enabling the potential use of gold nanoparticles for environmental sensing applications [38, 68]. The ability of these nanotriangles (thin triangular nanoplates) to detect ultra-low concentrations of mercury was attributed to their thin nature and the presence of thermodynamically unstable high energy sites at the nanotriangle edges, which facilitates a high level of amalgamation of these nanotriangles with mercury. It was observed that during the process, mercury did not get incorporated into the lattice structure of Au nanotriangles; however, mercury caused deconstruction of the high energy tips and edges, resulting in damaging nanostructure morphology. Further experiments revealed that these nanotriangles showed high specificity toward the detection of mercuric ions [67]. In addition, Katti and co-workers reasoned that phytochemicals present within plants possess a plethora of antioxidant properties that could serve as chemical reducing agents for the reduction of metal salts into their corresponding nanoparticulate forms. They hypothesized that the polyphenols, phytoestrogens, and related chemicals in plants could provide a robust coating of tumor-specific phytochemicals on the surface of gold nanoparticles that provides a basis for green nanotechnological processes for the production of cancer-specific gold nanoparticles. The results reported by them provided experimental evidence toward the validation of these hypotheses [69–74]. This group further exploited the reduction capabilities of antioxidant phytochemicals present in green tea, soybean, cumin, and cinnamon for reducing gold salts into nanoparticles with consequent coatings of these polyphenols, phytoestrogens, and a host of other phytochemicals present in these phytoextracts [69, 71, 73]. Interestingly, due to robust phytochemicals surface modification, these nanoparticles showed remarkable stability under a range of pH conditions and did not aggregate in high salt concentrations and other biologically relevant media [69, 71, 73]. Some of these biosynthesized gold nanoparticles were also employed for electrochemistry applications including the study of electron transfer rates in molecules [75, 76], biomedical applications including drug delivery [71, 77], imaging [72, 73], targeted therapy [74], and for catalytic reactions to reduce common environmental pollutants [78].

Unlike gold, ionic forms of silver, copper, selenium, and tellurium are known to be highly toxic to biological cells [79, 80]. Nonetheless, several strains of biological organisms have mechanisms in place for efficient remediation of these toxic metal ions. Incidentally, several silver resistant bacterial species are reported to accumulate metallic silver as much as 25% of the dry weight biomass [79, 81]. The first report on the biosynthesis of silver nanoparticles was on using a silver resistant bacterium *Pseudomonas stutzeri* AG259 isolated from silver mines. In a pioneering study, Klaus et al. exposed *P. stutzeri* AG259 to high concentrations of silver ions that resulted in the intracellular (periplasmic space) accumulation of silver and silver sulfide nanoparticles. These particles typically ranged in diameter from a few nanometers to 200 nm with different morphologies such as triangles, hexagons, and spheres [79]. Furthermore, as an extension of this work, Joerger and co-workers showed the ability to directly synthesize cermet materials by heat treating the Ag nanoparticles containing bacterial biomass [81]. The hard coatings of cermet materials were found to be resistant to mechanical scratching and their optical properties could be tailored by varying the silver loading factor. This ability to synthesize biofilms of metal nanoparticles embedded in a biological matrix has important applications in the synthesis of cermet materials for optically functional thin film coatings. Although the ability of *P. stutzeri* to survive in high silver ion concentration regions was poorly understood at that time, with the advent of sophisticated techniques several groups have now postulated the presence of a silver ion binding protein and other associated proteins on the cell surface as potentially responsible factors for the reduction of silver ions to form silver nanoparticles [80, 82–84].

A more recent study established a direct correlation between the silver resistance machinery (Sil gene cluster) of bacteria *Morganella* spp. and their Ag nanoparticles biosynthesis ability [80, 83, 84]. Additional studies further defined the possible presence of a silver ion reductase enzyme in biological systems, especially in silver resistant species. Furthermore, a *M. psychrotolerans*, psychrotolerant (cold-tolerant) species from the same genus, was demonstrated for shape controlled biosynthesis of Ag nanoplates by controlling the bacterial growth kinetics by growing them at different temperatures (Fig. 20.3) [80]. *Morganella* spp. is the only example so far, wherein all the biogroups belonging to a particular genus have been shown for their ability toward Ag ion reduction, thus establishing this nanoparticle synthesis ability as a genus-wide phenotypic attribute of genus *Morganella* [83]. It was also established that this phenotypic feature of *Morganella* is independent of environmental variable and is strongly associated with the silver resistance machinery of these silver resistant bacterial. Due to the well-known similarities between the silver and copper resistance machineries in different bacterial systems, *M. morganii* was further reported for its ability to synthesize pure metallic copper nanoparticles (without any passivating oxide layer) in an aqueous phase (Fig. 20.4) [85]. This was the first time that the important role of the similarities in the bacterial heavy metal resistance machinery was outlined toward the synthesis of

FIGURE 20.3 TEM images of Ag nanoparticles biosynthesized using Morganella psychrotolerans at (a) 25°C, (b) 20°C, (c) 15°C, and (d) 4°C. Images reprinted with permission from Ref. [80]. © 2011, American Chemical Society.

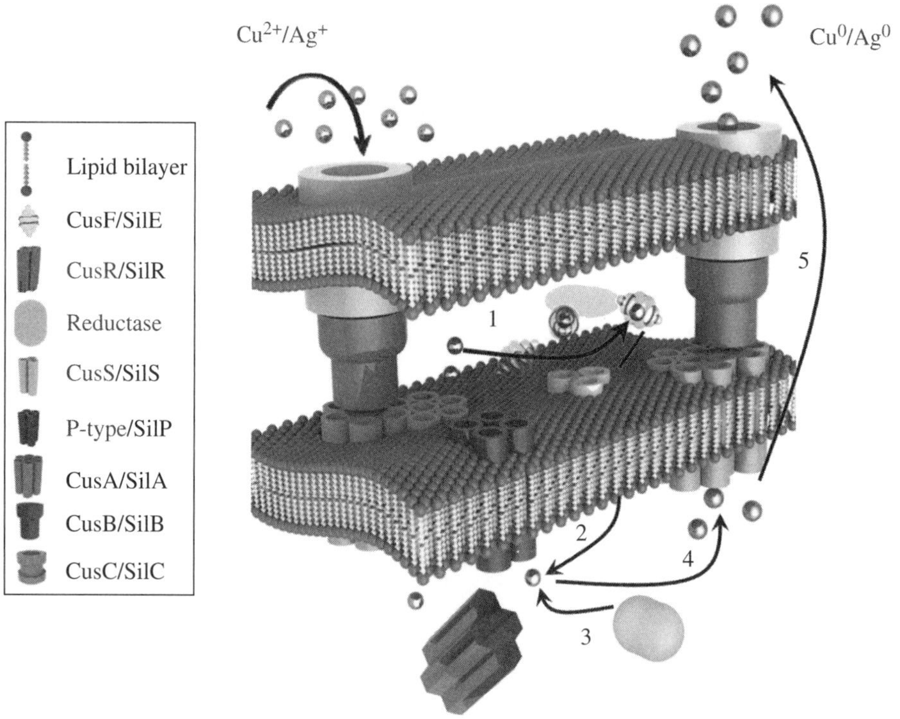

FIGURE 20.4 Schematic representation of the potential similarities between the copper and silver resistance systems that enables *Morganella morganii*, a silver resistant organism, to synthesize CuNPs. The top and the bottom sides of the schematic show the outer surface and the inner cytoplasmic end of the bacterium, respectively. Typically, the process involves the uptake of Cu^{2+} ions by the bacterium (step 1) followed by the presentation of Cu^{2+} ions to the silver resistance machinery (step 2). This initiates a number of processes whereby either a metal ion reductase or similar proteins bind to Cu^{2+} ions (step 3), thereby reducing them to Cu^0 nanoparticles (step 4). The biosynthesized nanoparticles are thereafter released from the cell using a cellular efflux system (step 5). Images reprinted from Ref. [85]. Reproduced by permission of The Royal Society of Chemistry (RSC).

technologically important nanomaterials. The biosynthesis of pure copper nanoparticles is particularly important as chemical synthesis approaches or previously employed biosynthesis approaches led to the stabilization of copper particles in its oxide form due to the propensity of surface oxidation of metallic copper. Chemical synthesis approaches, through which pure metallic copper nanoparticles can be produced, require strong capping agents and laborious synthesis processes to prevent surface oxidation [86, 87].

Similar to gold and silver nanoparticles, palladium nanoparticles biosynthesis is also of great interest with reports appearing in the early 2000s using sulfate reducing bacterium *Desulfovibrio desulfuricans* in the presence of molecular hydrogen or formate as electron donor [88]. Similar to the case in *S. algae* for platinum biosynthesis, the activity of enzymes hydrogenase and/or cytochrome c3 were postulated to be responsible for the reduction of palladium precursors [43]. Interestingly, the absence of anaerobic conditions did not alter the production of palladium nanoparticles, as aerobic conditions are known to be toxic to the bacterium and hydrogenase enzyme. Similarly, other bacterial species belonging to genus *Desulfovibrio* and *Shewanella* also showed ability to reduce palladium precursors [89–91]. These palladium nanoparticles were shown to be efficient catalysts toward dehalogenation reactions with some biologically synthesized Pd catalysts showing better response than chemically synthesized Pd nanoparticles [88–91].

The use of bacteria toward the synthesis of selenium and tellurium nanoparticles is fascinating due to its significant importance in the semiconductor industry [92]. Unlike gold, silver, and copper, selenium and tellurium ions are toxic to bacterial cells at ultra-low concentrations. However, the Mother Nature has devised resistance systems that are unique and intelligent to counter these toxic metalloid ions. Oremland and co-workers have isolated several bacterial species from metalloid-rich environments that have the ability to convert these toxic ions to nanoparticles. These organisms include *Sulfurospirillum barnesii*, *Bacillus selenitireducens*, and *Selenihalanaerobacter shriftii*, wherein the first two species were able to reduce selenate and tellurate oxyanions to selenium and tellurium nanoparticles, respectively, while the latter could only reduce selenate to selenium nanoparticles [93, 94]. Although some bacterial species showed the ability to reduce both these metalloid ions, the morphology of these nanoparticles were unique in each case that span from spherical 300 nm particles to nanorods of approximately 10 nm in diameter by 200 nm length, which clustered together in time to form large rosettes (~1000 nm) containing individual shards of 100 nm width by 1000 nm length. Persistent research by these research groups have unraveled some of the mysteries (metal reductases) and elucidated the biochemical pathways involved in the reduction, but a large void in the available literature continues to persist, particularly in the context of exact biochemical mechanism involved in metal nanoparticle biosynthesis.

In addition to the synthesis of metal nanoparticles by bacteria and fungi, plants and plant extracts have also shown the ability to reduce and synthesize metal nanoparticles. A few examples of plant extract-based metal nanoparticle synthesis like gold and copper have been discussed earlier. Additionally, plant-based routes have led to the formation of silver and gold-silver alloy nanoparticles. For example, *Azadirachta indica* plant extract was earlier reported for its ability to reduce gold ions [66], and it also showed the ability to reduce silver ions [66]. This plant species was further employed to reduce gold and silver together to form a bi-metallic nanoparticulate system thus enabling this route to compete with chemical synthesis fabrication protocols.

20.4 METAL OXIDE NANOPARTICLES

Similar to metal nanoparticles, metal oxides are an important class of inorganic materials that are extensively used in a wide range of applications including catalyst support, semiconductors, separation media, and biology [4, 95–97]. In biological systems, several organisms have already been reported for their ability to process metal ions to their oxide form (Table 20.2). One such important process that has been widely studied is "biosilicification" wherein marine organisms like diatoms, sponges, and radiolarians take up and process soluble silicon to form ornate hierarchical patterns of biosilica [18, 98–101]. Although several biological and bio-inspired methods for the synthesis of silica provide an environmentally benign and energy-conserving process [18, 98–106], there have hitherto only been limited efforts toward the exploration of microorganisms like bacteria and fungi for the biosynthesis of oxide nanoparticles.

Bansal and co-workers explored the possibility of employing eukaryotic organisms such as fungi for the synthesis of metal oxide nanoparticles. To achieve metal oxide nanoparticle biosynthesis, these researchers chose plant pathogenic fungi such as *Fusarium oxysporum* and *Verticillium* sp. as potential candidates based on the rationale that plant pathogenic fungi produce a large amount of extracellular hydrolases. These hydrolytic enzymes might also be able to hydrolyze metal oxide precursors into their respective metal oxide nanoparticle form, when these precursors are exposed to plant pathogenic fungi. Based on this rationale, the biological synthesis of a range of simple binary and complex ternary oxide nanoparticles including silica (Fig. 20.5a) [107], titania (Fig. 20.5b) [107], zirconia (Fig. 20.5c) [108], magnetite [109], and barium titanate [110] could be achieved. Detailed investigations revealed that the fungus *F. oxysporum* releases at least two low molecular weight cationic proteins that were capable of hydrolyzing the aqueous anionic complex precursors

TABLE 20.2 List of biological entities employed for metal oxide nanoparticle synthesis

Microorganism	Oxides	References
Fungi		
Fusarium oxysporum	SiO_2, TiO_2, ZrO_2, $BaTiO_3$, Bi_2O_3, Fe_3O_4	[107–110, 115]
Verticillium sp.	Fe_3O_4	[109]
Humicola sp.	$CuAlO_2$	[119]
Bacteria		
Geobacter metallireducens	Fe_3O_4	[114]
Magnetotactic bacteria	Fe_3O_4	[111]
Actinomycete		
Actinobacter sp.	Fe_3O_4, γ-Fe_2O_3	[112, 113]
Plant and plant extract		
Glycine Max (Soy bean)	Fe_3O_4	[116, 117]
Salix viminalis	ZnO and $LaMnO_3$	[118]

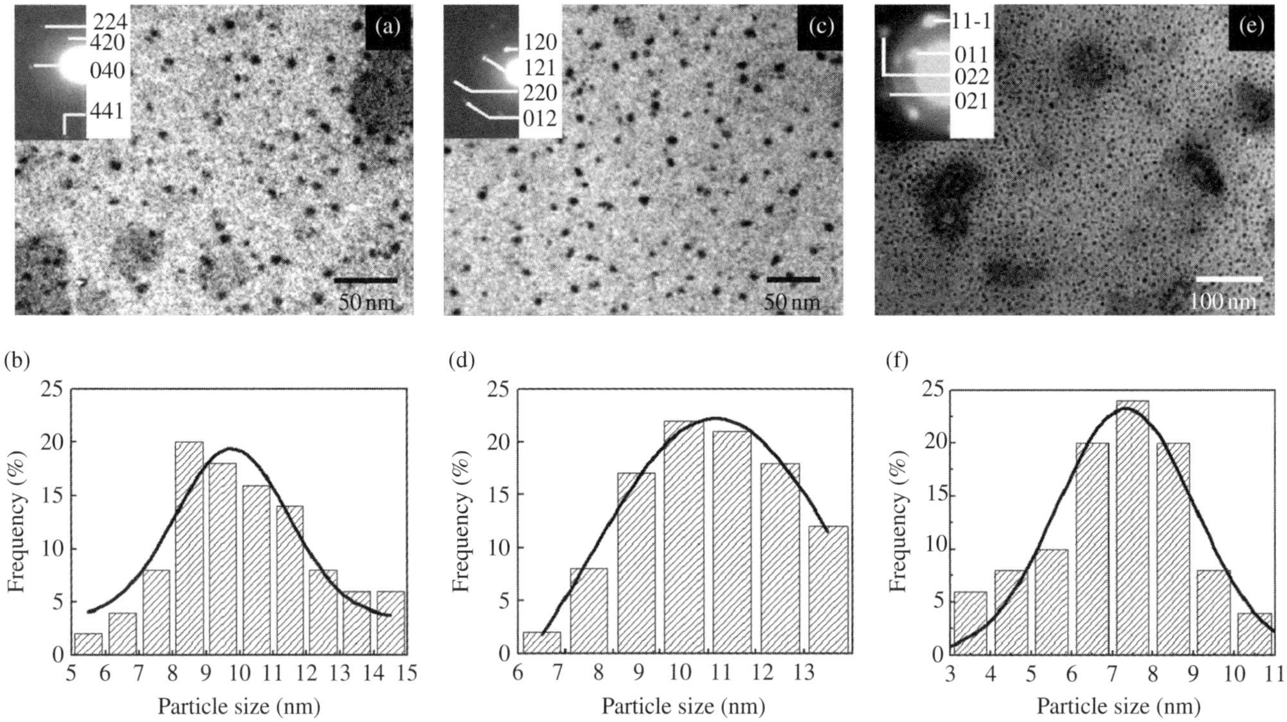

FIGURE 20.5 TEM images of silica (a), titania (c), and zirconia (e) nanoparticles biosynthesized using the fungus *F. oxysporum*. Insets in (a), (c), and (e) correspond to the selected area electron diffraction patterns recorded from the particles shown in the respective main figure. The particle size histograms of the silica, titania, and zirconia particles presented in images (a), (c), and (e) are shown in (b), (d), and (f), respectively. Images reprinted from Refs. [107, 108]. Reproduced by permission of The Royal Society of Chemistry (RSC).

(SiF_6^{2-}, TiF_6^{2-}, and ZrF_6^{2-} ions) employed in these studies [107, 108]. However, the evolutionary role of these hydrolytic enzymes during the life cycle of the fungus is still unclear. Although the morphology of these oxides are similar, magnetite nanoparticles synthesized by *F. oxysporum* and *Verticillium* sp. were significantly different as the former fungus synthesized particles of 20–40 nm in diameter, while the latter yielded a super-assembly of smaller magnetite nanoparticles into cubical structures, with an edge length of 100–400 nm [109]. In addition to fungi, several bacterial species, plants, and actinomycete have also been reported for their ability to synthesize magnetite nanoparticles [111–114] that are outlined in Table 20.2. Recent studies have shown the ability of *F. oxysporum* for the synthesis of extremely small (5–8 nm) bismuth oxide nanoparticles at room temperature [115]. This material typically showed a mixture of monoclinic and tetragonal phase, which is quite remarkable given that these materials exhibit a tetragonal phase when heated above 700°C. Incidentally, several plant species have also been known to accumulate and synthesize simple (Fe_3O_4 and ZnO) [116, 117] and complex ($LaMnO_3$) [118] metal oxide nanomaterials, which are either in response to high metal stress environment or synthesized by deliberately exposing them to metal ions.

In addition to the ability of fungal species to synthesize simple metal oxides, some species of fungi were also able to synthesize complex metal oxides such as barium titanate and chemically "difficult-to-synthesize" multifunctional $CuAlO_2$ nanoparticles [119]. The ability of fungus *F. oxysporum* to synthesize ultra-small sub-10 nm tetragonal barium titanate nanoparticles in the presence of barium acetate and potassium hexafluorotitanate precursors is quite fascinating as this fungus is not known to encounter these metal ions in their natural habitat [110]. It is important to highlight that $BaTiO_3$ is a technologically important material, however this material loses its ferroelectric properties in its nano-form, as the $BaTiO_3$ crystals transform from tetragonal (bulk) to cubic (nano) phase [120]. However, the ultra-small $BaTiO_3$ nanocrystals formed by fungus *F. oxysporum* could retain their ferroelectric properties even below sub-10 nm size regime, as was demonstrated through room-temperature ferroelectricity and ferroelectric-relaxor behavior studies. Additionally, Kelvin probe microscopy further enabled the electrical writing and thereafter reading the information on individual $BaTiO_3$ nanoparticles synthesized by this fungus. This opens up exciting opportunities in terms of synthesizing organic-inorganic hybrid biogenic materials on a large scale for applications in the field of electronics.

20.5 METAL SULFIDE NANOPARTICLES

There is significant interest in developing protocols for efficient and environmental friendly synthesis of semiconductor materials such as CdS for cell labeling applications due to their unique size-dependent chemical and physical properties. The different biological entities employed for the synthesis of metal sulfide nanoparticles are summarized in Table 20.3. Of the different biological entities, bacteria have shown considerable success in the synthesis of CdS nanoparticles with several species including *Escherichia coli* [121], *Klebsiella* sp. [122, 123], and *Clostridium thermoaceticum* [124]. Some of these bacterial strains belonging to *Klebsiella pneumoniae* were specific toward the synthesis of CdS nanoparticles as other metal sulfides were not detected on exposure to Pb and Zn ions [122]. The sulfide source in the case of *Klebsiella* sp. and *C. thermoaceticum* was primarily from the amino acid cysteine, which is either added to the bacterial growth medium or via the activity of cysteine desulfhydrase enzyme. Of the aforementioned bacteria, when *E. coli* was incubated with cadmium chloride and sodium sulfide, it intracellularly synthesized 2–5 nm CdS nanocrystals of wurtzite crystal phase [121]. In contrast to bacterial systems wherein the quantum dot nanocrystals are found intracellularly, the fungus *F. oxysporum* was shown for its ability to synthesize CdS quantum dots extracellularly [125]. Further analysis using polyacrylamide gel electrophoresis (SDS-PAGE) indicated the presence of at least four protein bands in the aqueous extract of the fungal biomass that were proposed to be potentially responsible for the formation of CdS nanocrystals.

The employment of bacteria for the synthesis of CdS nanoparticles was fascinating in itself, but remarkably Dameron et al. showed the unique ability of yeast toward intracellular synthesis of quantum semiconductor crystallites. In these studies, two different yeast species, namely, *Candida glabrata* and *Schizosaccharomyces pombe*, were exposed to aqueous cadmium ions following which 1–3 nm nanocrystallites of CdS nanoparticles were found deposited intracellularly [30, 126]. Phytochelatins with the general structure $(\gamma\text{-Glu-Cys})_n$-Gly, where "*n*" ranged from 2 to 6, were found to control the nucleation and growth of CdS nanoparticles. Furthermore, the synthesis efficiency for CdS nanocrystals was improved by exposing *S. pombe* cells to aqueous cadmium ions in their mid-log phase of growth, suggesting the importance of the growth phase on the synthesis of nanomaterials. In addition to improving the efficiency of nanoparticle synthesis, the authors also demonstrated the ability of *Torulopsis* sp. (yeast) for the synthesis of spherical PbS nanocrystals within the quantum confinement regime [127]. The property of yeast for the synthesis of metal sulfides was postulated to be the result of metal-chelating peptides that nullify the stress generated by these metal ions. A classic bioremediation pathway is followed by yeast cells wherein the exposure of metal ions initially results in a metal ion-γ glutamyl complex with an increase in the intracellular sulfide ion level. This is followed by the complexation of sulfide ions with Cd or Pb ions to form CdS or PbS nanocrystals that further accumulate in the vacuoles within yeast cells. In an interesting study, Banfield and co-workers employed natural biofilms of sulfate-reducing bacteria from the

TABLE 20.3 List of biological entities employed for metal sulfide nanoparticle synthesis

Microorganism	Sulfides	References
Bacteria		
Escherichia coli	CdS	[121]
Clostridium thermoaceticum	CdS	[124]
Klebsiella pneumoniae	CdS	[122]
Klebsiella planticola	CdS	[123]
Sulfate reducing bacteria	ZnS	[128]
Magnetotactic bacteria	Fe_3S_4	[129, 130]
Fungus		
Fusarium oxysporum	CdS	[125]
Actinomycete		
Actinobacter sp.	Fe_3S_4, FeS_2	[113]
Yeasts		
Candida glabrata	CdS	[30]
Schizosaccharomyces pombe	CdS	[30, 126]
Torulopsis sp.	PbS	[127]

family *Desulfobacteriaceae* for the synthesis of sphalerite (ZnS) particles [128]. This complex ZnS biomineralization was found to significantly reduce the concentration of Zn below acceptable levels for drinking water suggesting a cheaper and eco-friendly alternative for water purification. In addition to Cd, Zn, and Pb sulfides, iron sulfide nanoparticles belong to an important class of materials known for their magnetic properties. Incidentally, in the biological world, iron sulfide is naturally found in strains of magnetotactic bacteria [129, 130]. There have been reports using other bacterial systems for the synthesis of iron sulfide nanoparticles that are within the permanent single magnetic domain size. Very recently, Bharde et al. have shown the ability of actinomycete *Actinobacter* sp. for its ability to synthesize magnetic iron sulfide nanoparticles extracellularly under aerobic conditions [113]. These nanoparticles were synthesized with ferric ions in the presence of exogenous sulfate source and typically encompassed Fe_3S_4 and FeS_2 of 20 nm diameter. The important role of bacterial sulfate reductases for the conversion of sulfate into sulfide was postulated with several low molecular weight proteins found to be involved in the stabilization of these particles.

20.6 METAL CARBONATE NANOPARTICLES

Earlier strategies of metal carbonates synthesis focused on using a biomimetic approach, wherein an external source of CO_2 along with metal ions was employed in the presence of different biomacromolecules, which react to form carbonate nanoparticles [9]. Although biomimetic processes have resulted in different mineral polymorph and morphological control, a biosynthesis approach was proposed to overcome the necessity for an external CO_2 source. Sastry's group, in addition to engaging fungus and actinomycete for the synthesis of metal, metal oxide, and metal sulfide nanoparticles, also tailored these organisms for the extracellular synthesis of a range of carbonate materials including $CaCO_3$, $BaCO_3$, $SrCO_3$, $CdCO_3$, and $PbCO_3$. This was achieved as these organisms are known to produce CO_2 as a by-product during the organisms' active aerobic metabolism process. The versatility of these organisms in synthesizing different kinds of nanoparticles is quite fascinating and intriguing. Fungi including *F. oxysporum* [131–134], *Verticillium* sp. [135], *Trichothecium* sp. [131], and actinomycete including *Rhodococcus* sp [132] and *Thermomonospora* sp [135] were employed for the synthesis of the aforementioned carbonate species. Interestingly, these carbonates displayed different crystal phases with different morphologies ranging from flat circular to cubic to star-shaped depending on the fungus/actinomycete employed for the synthesis. The variability in morphology and crystal phases could be attributed to the production of species-specific proteins that might be responsible for the synthesis of these materials. The important fact that the synthesis of these carbonates minerals proceeds via endogenous production of CO_2 makes this biosynthesis approach a truly biogenic method for synthesizing biominerals. The different biological entities employed for the biosynthesis of different biominerals are summarized in Table 20.4.

20.7 BIOLEACHING: A TRULY "GREEN" BIOLOGICAL APPROACH

The aforementioned synthesis routes for the biosynthesis of technologically important nanomaterials are fascinating, but in most cases an external metal ion chemical source is essential. In contrast, bioleaching is a tool promoted by algae, mosses, lichens, plants, animals, actinomycetes, a variety of bacteria, and a few fungi in their natural habitats for the low-cost extraction of various metals [136–138]. Bioleaching employing microorganisms have been used at a commercial level for the recovery of metals like gold, copper, and iron. Although bioleaching approaches are employed for metal recovery, this approach was not originally investigated in the context of nanoparticle formation.

TABLE 20.4 List of biological entities employed for biominerals synthesis

Microorganism	Biominerals	References
Fungi		
Verticillium sp.	$BaCO_3$, $CaCO_3$	[135]
Trichothecium sp.	$CaCO_3$	[131]
Fusarium oxysporum	$CaCO_3$, $CdCO_3$, $PbCO_3$, $SrCO_3$	[131–134]
Actinomycetes		
Rhodococcus sp.	$CaCO_3$	[132]
Thermomonospora sp.	$BaCO_3$ and $CaCO_3$	[135]

For instance, silica-based materials are commonly found in nature, which are formed via biological processes undertaken by biological entities including plants, diatoms, sponges, fungi, and bacteria [18, 85–88]. Several groups have developed biomimetic routes for the synthesis of silica with excellent control over morphology [18, 85–93]. However, a unique bioleaching approach for the efficient extraction of siliceous materials from natural sources was developed by Bansal and co-workers. This involved a biology-based model system wherein silica was extracted from naturally available raw materials like white sand and zircon sand as well as from agro-industrial by-products like rice husk, thereby negating the use of chemical precursors [139]. The versatile fungus *F. oxysporum* was employed for bioleaching of silica from complex silicates present in the white sand that was proposed to proceed via a two-step process. This involved the initial hydrolysis and leaching of silicate complexes in the form of silicic acid from the sand, followed by its condensation to form silica nanoparticles capped by stabilizing proteins from the fungus. The possibility for the extraction of silica from white sand was further extended with investigating raw materials that are available naturally for the synthesis of nanomaterials using a fungus-mediated bioleaching approach. Interestingly, bioleaching of silica from rice husk using *F. oxysporum* could not only leach the amorphous silica present in rice husk, but could also transform this amorphous silica to crystalline silica at room temperature [140]. The critical role of the fungal metabolism in the phase transformation of amorphous to crystalline silica was confirmed by employing extracellular cationic proteins from this fungus for silica bioleaching, which could only leach the amorphous siliceous material from rice husk in the form of silica nanoparticles, without transforming them to a crystalline phase. This ability to use naturally available raw materials for nanoparticle biosynthesis was an interesting concept and therefore the same group further investigated whether the fungus *F. oxysporum* offers selectivity in terms of nanomaterial synthesis. This is because biological catalysts (e.g., enzymes) are well known for their high selectivity toward certain reactions. To check this possibility, Bansal and co-workers exposed *F. oxysporum* to zircon sand ($Zr_xSi_{1-x}O_{4x}$), which is a mixture of zirconia (ZrO_2) and silica (SiO_2). Interestingly, this fungus was able to selectively leach out silica from zircon sand in the form of nanoparticles. Notably, zircon sand is a technologically important refractory material and its commercial value is dependent on the amount of zirconia component present in this material. This selective leaching of silica from zircon sand offered dual advantages, namely, while the leaching of silica from zircon sand yielded highly crystalline silica nanoparticles c as a product, the process added value to the original raw material (zircon sand) by enrichment of zirconia in zircon sand. This unique and efficient bioleaching approach may be extended toward extracting important nanomaterials from other minerals present in large quantity in natural environments.

20.8 NOVEL APPROACH TOWARD BIOSYNTHESIS

The preceding sections have shown the potential of naturally existing biological entities toward the synthesis of nanoparticles. The quest to synthesize new materials with interesting properties has pushed the boundaries to fresher limits and we are developing new fabrication strategies to cope with this demand. One such interesting strategy is to tailor organisms by the incorporation of genes responsible for the production of different biochemical molecules that bind and reduce heavy metal ions. For instance, the biochemical mechanism in certain plant-based synthesis have been elucidated wherein phytochelatins (PCh) and metallothionein (MT) have been identified as potential targets due to the strong binding nature of these proteins to complex heavy metal ions including copper, silver, cadmium, lead, mercury, and zinc [141–143]. In a recent study, a range of fairly monodisperse metal (Au, Ag), semiconductor (CdZn, CdSe, CdTn, SeZn, CdSeZn, Te, CdSeZnTe), magnetic (Fe, FeAg, FeCo, FeMn, FeCoNi, FeCoMn), mono- (CdCs) and di-electric earth, and rare fluoride (PrGd, SrGd, SePr) nanoparticles were synthesized by expressing PCh from *Arabidopsis thaliana* and/or MT from *Pseudomonas putida* in recombinant *E. coli* DH5α cells (Fig. 20.6) [144]. Interestingly, some of the bi- and tri-metallic nanoparticles synthesized using this technique are yet to be achieved via a chemical synthesis route. Moreover, varying the metal ion concentration allowed the particle size to be retained below 10 nm diameter, thereby displaying interesting properties of quantum dots and florescent semiconducting materials. These properties were further validated using experimental data wherein these nanoparticles were employed for in vitro conjugation of biomaterials and cellular-imaging analysis.

20.9 OUTLOOK

Several investigations on the biosynthesis of inorganic nanomaterials by bacteria, fungi, algae, yeasts, and plant and plant extracts have been made. Although this field of nanoparticle biosynthesis has been extensively studied, there are a number of questions that require serious attention before this approach can compete with the existing physical and chemical approaches. One major question that requires considerable focus is to elucidate the mechanistic aspect of nanoparticle biosynthesis by these

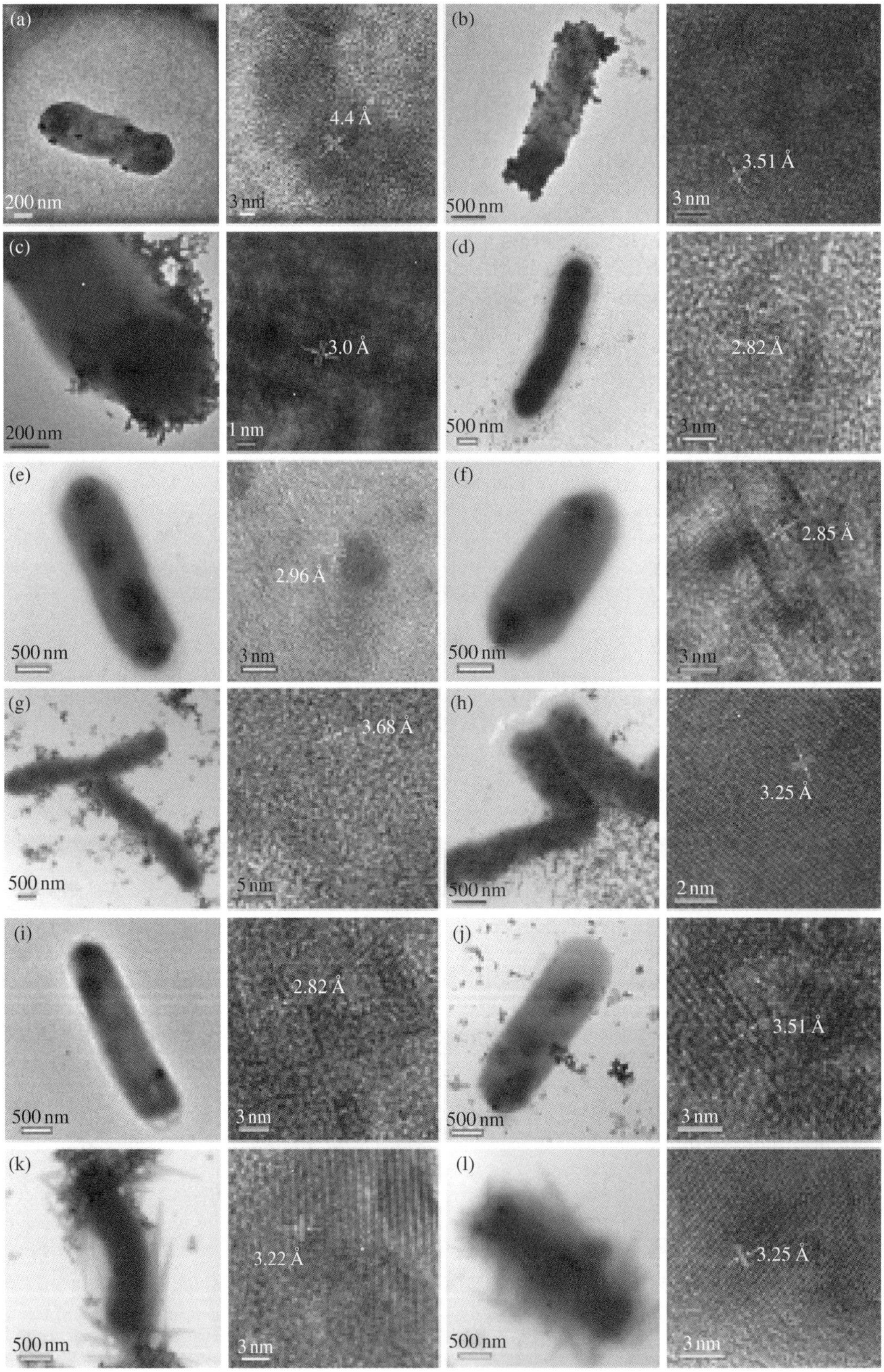

FIGURE 20.6 TEM images of semiconducting nanoparticles synthesized using recombinant *E. coli* cells—expressing phytochelatin synthase from *Arabidopsis thaliana* (a) CdZn, (b) CdSe, (c) CdTe, and (d) SeZn; expressing both *A. thaliana* phytochelatin synthase and *Pseudomonas putida* metallothionein (e) CdZn, (f) CdSe, (g) CdTe, and (h) SeZn. The HRTEM images show the interplanar spacing in the lattice. Images reprinted with permission from Ref. [144]. © 2010, Wiley-VCH Verlag GmbH & Co. KGaA.

organisms. Likewise, the surface chemistry of biosynthesized nanoparticles still requires substantial understanding. As shown in a few investigations [80, 85], the important role of metal resistance in an organism's ability to synthesize nanomaterials would have significant potential to employ a rational approach for nanoparticle biosynthesis. Additionally, only recently engineered organisms modified to over-express phytochelatins and metallothionein (that are known to bind to metal ions) have been employed for the synthesis of a range of important nanomaterials [144], a strategy that requires serious investigation. Lastly, the commercial applicability for the large-scale synthesis of inorganic nanomaterials using a biosynthesis approach is yet to be seen. An understanding of the important questions of the mechanism and surface chemistry will enable this approach to be used on an industrial scale for the synthesis of inorganic nanomaterials.

REFERENCES

[1] Feynman RP. There's plenty of room at the bottom. J Microelectromech Syst 1992;1 (1):60–66.

[2] Cao YC, Jin R, Mirkin CA. Nanoparticles with Raman spectroscopic fingerprints for DNA and RNA detection. Science 2002;297 (5586):1536–1540.

[3] Kang B, Mackey MA, El-Sayed MA. Nuclear targeting of gold nanoparticles in cancer cells induces DNA damage, causing Cytokinesis arrest and apoptosis. J Am Chem Soc 2010;132 (5):1517–1519.

[4] Nassif N, Roux C, Coradin T, Bouvet OMM, Livage J. Bacteria quorum sensing in silica matrices. J Mater Chem 2004;14 (14):2264–2268.

[5] Plowman B, Ippolito SJ, Bansal V, Sabri YM, O'Mullane AP, Bhargava SK. Gold nanospikes formed through a simple electrochemical route with high electrocatalytic and surface enhanced Raman scattering activity. Chem Commun 2009;33:5039–5041.

[6] Sawant PD, Sabri YM, Ippolito SJ, Bansal V, Bhargava SK. In-depth nano-scale analysis of complex interactions of Hg with gold nanostructures using AFM-based power spectrum density method. Phys Chem Chem Phys 2009;11 (14):2374–2378.

[7] Tian L, Yam L, Wang J, Tat H, Uhrich KE. Core crosslinkable polymeric micelles from PEG-lipid amphiphiles as drug carriers. J Mater Chem 2004;14 (14):2317–2324.

[8] Schmid G, Corain B. Nanoparticulated gold: syntheses, structures, electronics, and reactivities. Eur J Inorg Chem 2003;2003 (17):3081–3098.

[9] Bansal V, Bharde A, Ramanathan R, Bhargava SK. Inorganic materials using 'unusual' microorganisms. Adv Colloid Interface Sci 2012;179–182:150–168.

[10] Bansal V, Ramanathan R, Bhargava SK. Fungus-mediated biological approaches towards 'green' synthesis of oxide nanomaterials. Aust J Chem 2011;64 (3):279–293.

[11] Dhillon GS, Brar SK, Kaur S, Verma M. Green approach for nanoparticle biosynthesis by fungi: current trends and applications. Crit Rev Biotechnol 2012;32 (1):49–73.

[12] Thakkar KN, Mhatre SS, Parikh RY. Biological synthesis of metallic nanoparticles. Nanomed Nanotechnol Biol Med 2010;6 (2):257–262.

[13] Sastry M, Ahmad A, Khan MI, Kumar R. Microbial nanoparticle production. In: Niemeyer CM, Mirkin CA, editors. *Nanobiotechnology*. Weinheim: Wiley-VCH Verlag GmbH & Co. KGaA; 2004. p 126–135.

[14] Lowenstam H. Minerals formed by organisms. Science 1981;211 (4487):1126–1131.

[15] Simkiss K, Wilbur KM. *Biomineralization*. New York: Academic Press; 1991.

[16] Frankel RB, Blakemore RP. *Iron Biominerals*. Plenum Press: Springer; 1991.

[17] Philipse AP, Maas D. Magnetic colloids from magnetotactic bacteria: chain formation and colloidal stability. Langmuir 2002;18 (25):9977–9984.

[18] Kröger N, Deutzmann R, Sumper M. Polycationic peptides from diatom biosilica that direct silica nanosphere formation. Science 1999;286 (5442):1129–1132.

[19] Parkinson J, Gordon R. Beyond micromachining: the potential of diatoms. Trends Biotechnol 1999;17 (5):190–196.

[20] Schultze-Lam S, Harauz G, Beveridge T. Participation of a cyanobacterial S-layer in fine-grain mineral formation. J Bacteriol 1992;174 (24):7971–7981.

[21] Stolz JS, Basu PB, Oremland RO. Microbial transformation of elements: the case of arsenic and selenium. Int Microbiol 2002;5 (4):201–207.

[22] Bruins MR, Kapil S, Oehme FW. Microbial resistance to metals in the environment. Ecotoxicol Environ Saf 2000;45 (3):198–207.

[23] Trevors JT, Oddie KM, Belliveau BH. Metal resistance in bacteria. FEMS Microbiol. Lett 1985;32 (1):39–54.

[24] Mehra RK, Winge DR. Metal ion resistance in fungi—molecularmechanisms and their regulated expression. J Cell Biochem 1991;45 (1):30–40.

[25] Nies DH. Microbial heavy metal resistance. Appl Microbiol Biotechnol 1999;51 (6):730–750.

[26] Zumberge JE, Sigleo AC, Nagy B. Molecular and elemental analyses of the carbonaceous matter in the gold-and uranium-bearing Vaal Reef carbon seams, Witwatersrand sequence. Miner Sci Eng 1978;10:223.

[27] Hosea M, Greene B, McPherson R, Henzl M, Dale Alexander M, Darnall DW. Accumulation of elemental gold on the alga Chlorella vulgaris. Inorg Chim Acta 1986;123 (3):161–165.

[28] Beveridge TJ, Murray RG. Sites of metal deposition in the cell wall of Bacillus subtilis. J Bacteriol 1980;141 (2):876–887.

[29] Reese RN, Winge DR. Sulfide stabilization of the cadmium-γ-glutamyl peptide complex of *Schizosaccharomyces pombe*. J Biol Chem 1988;263 (26):12832–12835.

[30] Dameron CT, Reese RN, Mehra RK, Kortan AR, Carroll PJ, Steigerwald ML, Brus LE, Winge DR. Biosynthesis of cadmium sulfide quantum semiconductor crystallites. Nature 1989;338 (6216):596–597.

[31] Blakemore R. Magnetotactic bacteria. Science 1975;190 (4212):377–379.

[32] Bansal V, Jani H, Du Plessis J, Coloe PJ, Bhargava SK. Galvanic replacement reaction on metal films: a one-step approach to create nanoporous surfaces for catalysis. Adv Mater 2008;20 (4):717–723.

[33] Bansal V, Li V, O'Mullane AP, Bhargava SK. Shape dependent electrocatalytic behaviour of silver nanoparticles. CrystEngComm 2010;12 (12):4280–4286.

[34] Bansal V, O'Mullane AP, Bhargava SK. Galvanic replacement mediated synthesis of hollow Pt nanocatalysts: significance of residual Ag in H_2 evolution reactions. Electrochem Commun 2009;11 (8):1639–1642.

[35] Pearson A, Bhargava SK, Bansal V. UV-switchable polyoxometalate sandwiched between TiO_2 and metal nanoparticles for enhanced visible and solar light photococatalysis. Langmuir 2011;27 (15):9245–9252.

[36] Pearson A, Jani H, Kalantar-Zadeh K, Bhargava SK, Bansal V. Gold nanoparticle-decorated keggin ions/TiO_2 photococatalyst for improved solar light photocatalysis. Langmuir 2011;27 (11):6661–6667.

[37] Pearson A, O'Mullane AP, Bansal V, Bhargava SK. Galvanic replacement mediated transformation of Ag nanospheres into dendritic Au–Ag nanostructures in the ionic liquid [BMIM][BF_4]. Chem Commun 2010;46 (5):731–733.

[38] Sabri YM, Ippolito SJ, O'Mullane AP, Tardio J, Bansal V, Bhargava SK. Creating gold nanoprisms directly on quartz crystal microbalance electrodes for mercury vapor sensing. Nanotechnology 2011;22 (30):305501.

[39] Sadek AZ, Bansal V, McCulloch DG, Spizzirri PG, Latham K, Lau DWM, Hu Z, Kalantar-Zadeh K. Facile, size-controlled deposition of highly dispersed gold nanoparticles on nitrogen carbon nanotubes for hydrogen sensing. Sens Actuators B 2011;160 (1):1034–1042.

[40] Southam G, Beveridge TJ. The occurrence of sulfur and phosphorus within bacterially derived crystalline and pseudocrystalline octahedral gold formed in vitro. Geochim Cosmochim Acta 1996;60 (22):4369–4376.

[41] Karthikeyan S, Beveridge TJ. *Pseudomonas aeruginosa* biofilms react with and precipitate toxic soluble gold. Environ Microbiol 2002;4 (11):667–675.

[42] Kashefi K, Tor JM, Nevin KP, Lovley DR. Reductive precipitation of gold by dissimilatory Fe(III)-reducing Bacteria and Archaea. Appl Environ Microbiol 2001;67 (7):3275–3279.

[43] Konishi Y, Ohno K, Saitoh N, Nomura T, Nagamine S, Hishida H, Takahashi Y, Uruga T. Bioreductive deposition of platinum nanoparticles on the bacterium *Shewanella algae*. J Biotechnol 2007;128 (3):648–653.

[44] Suresh AK, Pelletier DA, Wang W, Broich ML, Moon JW, Gu B, Allison DP, Joy DC, Phelps TJ, Doktycz MJ. Biofabrication of discrete spherical gold nanoparticles using the metal-reducing bacterium *Shewanella oneidensis*. Acta Biomater 2011;7 (5):2148–2152.

[45] Hau HH, Gralnick JA. Ecology and biotechnology of the genus Shewanella. Annu Rev Microbiol 2007;61:237–258.

[46] Lovley DR. Anaerobes into heavy metal: dissimilatory metal reduction inanoxic environments. Trends Ecol Evol 1993;8 (6):213–217.

[47] Lengke MF, Fleet ME, Southam G. Morphology of gold nanoparticles synthesized by filamentous cyanobacteria from gold(I)-Thiosulfate and gold(III)-chloride complexes. Langmuir 2006;22 (6):2780–2787.

[48] Lengke MF, Fleet ME, Southam G. Biological synthesis of platinum nanoparticles by filamentous cyanobacteria from platinum(IV)-chloride complex. Langmuir 2006;22 (17):7318–7323.

[49] Lengke MF, Fleet ME, Southam G. Synthesis of palladium nanoparticles by reaction of filamentous cyanobacteria with a palladium(II)-chloride complex. Langmuir 2007;23 (17):8982–8987.

[50] Lengke MF, Fleet ME, Southam G. Biosynthesis of silver nanoparticles by filamentous cyanobacteria from silver(I)-nitrate complex. Langmuir 2007;23 (5):2694–2699.

[51] Reith F, Rogers SL, McPhail DC, Webb D. Biomineralisation of gold: biofilms on bacterioform gold. Science 2006;313 (5784):233–236.

[52] Mergeay M, Monchy S, Vallaeys T, Auquier V, Benotmane A, Bertin P, Taghavi S, Dunn J, van der Lelie D, Wattiez R. *Ralstonia metallidurans*, a bacterium specifically adapted to toxic metals: towards a tentative catalogue of metal-responsive genes. FEMS Microbiol Rev 2003;27 (2–3):385–410.

[53] Ahmad A, Senapati S, Khan MI, Kumar R, Sastry M. Extracellular biosynthesis of monodisperse gold nanoparticles by a novel extremophilic actinomycete, *Thermomonospora sp*. Langmuir 2003;19 (8):3550–3553.

[54] Ahmad A, Senapati S, Khan MI, Kumar R, Ramani R, Srinivas V, Sastry M. Intracellular synthesis of gold nanoparticles by a novel alkalotolerant actinomycete, *Rhodococcus* species. Nanotechnology 2003;14 (7):824.

[55] Bharde A, Kulkarni A, Rao M, Prabhune A, Sastry M. Bacterial enzyme mediated biosynthesis of gold nanoparticles. J Nanosci Nanotechnol 2007;7 (12):4369–4377.

[56] Mukherjee P, Ahmad A, Mandal D, Senapati S, Sainkar SR, Khan MI, Ramani R, Parischa R, Ajaykumar PV, Alam M, Sastry M, Kumar R. Bioreduction of $AuCl^{4-}$ ions by the fungus, Verticillium sp. and surface trapping of the gold nanoparticles formed. Angew Chem Int Ed 2001;40 (19):3585–3588.

[57] Mukherjee P, Ahmad A, Mandal D, Senapati S, Sainkar SR, Khan MI, Parischa R, Ajaykumar PV, Alam M, Kumar R, Sastry M. Fungus-mediated synthesis of silver nanoparticles and their immobilization in the mycelial matrix: a novel biological approach to nanoparticle synthesis. Nano Lett 2001;1 (10):515–519.

[58] Shankar SS, Ahmad A, Pasricha R, Sastry M. Bioreduction of chloroaurate ions by geranium leaves and its endophytic fungus yields gold nanoparticles of different shapes. J Mater Chem 2003;13 (7):1822–1826.

[59] Mukherjee P, Senapati S, Mandal D, Ahmad A, Khan MI, Kumar R, Sastry M. Extracellular synthesis of gold nanoparticles by the fungus *Fusarium oxysporum*. ChemBioChem 2002;3 (5):461–463.

[60] Ahmad A, Mukherjee P, Senapati S, Mandal D, Khan MI, Kumar R, Sastry M. Extracellular synthesis of silver nanoparticles using the fungus *Fusarium oxysporum*. Colloids Surface B 2003;28 (4):313–318.

[61] Senapati S, Ahmad A, Khan MI, Sastry M, Kumar R. Extracellular biosynthesis of bimetallic Au–Ag alloy nanoparticles. Small 2005;1 (5):517–520.

[62] Nair B, Pradeep T. Coalescence of nanoclusters and formation of submicron crystallites assisted by *Lactobacillus* strains. Cryst Growth Des 2002;2 (4):293–298.

[63] Manceau A, Nagy KL, Marcus MA, Lanson M, Geoffroy N, Jacquet T, Kirpichtchikova T. Formation of metallic copper nanoparticles at the soil–root interface. Environ Sci Technol 2008;42 (5):1766–1772.

[64] Shankar SS, Ahmad A, Pasricha R, Khan MI, Kumar R, Sastry M. Immobilization of biogenic gold nanoparticles in thermally evaporated fatty acid and amine thin films. J Colloid Interface Sci 2004;274 (1):69–75.

[65] Shankar SS, Rai A, Ankamwar B, Singh A, Ahmad A, Sastry M. Biological synthesis of triangular gold nanoprisms. Nat Mater 2004;3 (7):482–488.

[66] Shankar SS, Rai A, Ahmad A, Sastry M. Rapid synthesis of Au, Ag, and bimetallic Au core–Ag shell nanoparticles using Neem (Azadirachta indica) leaf broth. J Colloid Interface Sci 2004;275 (2):496–502.

[67] Singh A, Pasricha R, Sastry M. Ultra-low level optical detection of mercuric ions using biogenic gold nanotriangles. Analyst 2012;137 (13):3083–3090.

[68] Kim Y, Johnson RC, Hupp JT. Gold nanoparticle based sensing of "spectroscopically silent" heavy metal ions. Nano Lett 2001;1 (4):165–167.

[69] Shukla R, Nune SK, Chanda N, Katti K, Mekapothula S, Kulkarni RR, Welshons WV, Kannan R, Katti KV. Soybeans as a phytochemical reservoir for the production and stabilization of biocompatible gold nanoparticles. Small 2008;4 (9):1425–1436.

[70] Katti K, Chanda N, Shukla R, Zambre A, Suibramanian T, Kulkarni RR, Kannan R, Katti KV. Nanotechnology from cumin phyto chemicals: Generation of biocompatible gold nanoparticles. Int J Nanotechnol Biomed 2009;1 (1):B39–B52.

[71] Nune SK, Chanda N, Shukla R, Katti K, Kulkarni RR, Thilakavathy S, Mekapothula S, Kannan R, Katti KV. Green nanotechnology from tea: phytochemicals in tea as building blocks for production of biocompatible gold nanoparticles. J Mater Chem 2009;19 (19):2912–2920.

[72] Viator JA, Gupta S, Goldschmidt BS, Bhattacharyya K, Kannan R, Shukla R, Dale PS, Boote E, Katti K. Gold nanoparticle mediated detection of prostate cancer cells using photoacoustic flowmetry with optical reflectance. J Biomed Nanotechnol 2010;6 (2):187–191.

[73] Chanda N, Shukla R, Zambre A, Mekapothula S, Kulkarni RR, Katti K, Bhattacharyya K, Fent GM, Casteel SW, Boote EJ. An effective strategy for the synthesis of biocompatible gold nanoparticles using cinnamon phytochemicals for phantom CT imaging and photoacoustic detection of cancerous cells. Pharm Res 2011;28 (2):279–291.

[74] Shukla R, Chanda N, Zambre A, Upendran A, Katti K, Kulkarni RR, Nune SK, Casteel SW, Smith CJ, Vimal J. Laminin receptor specific therapeutic gold nanoparticles (^{198}AuNP-EGCG) show efficacy in treating prostate cancer. Proc Natl Acad Sci USA 2012;109 (31):12426–12431.

[75] Du L, Jiang H, Liu X, Wang E. Biosynthesis of gold nanoparticles assisted by Escherichia coli DH5α and its application on direct electrochemistry of haemoglobin. Electrochem Commun 2007;9 (5):1165–1170.

[76] Wang Y, He X, Wang K, Zhang X, Tan W. Colloids Surf. Bull Am Meteorol Soc 2009;73 (1):75–79.

[77] Dhar S, Reddy EM, Shiras A, Pokharkar V, Prasad BLV. Natural gum reduced/stabilized gold nanoparticles for drug delivery formulations. Chem Eur J 2008;14 (33):10244–10250.

[78] Sharma NC, Sahi SV, Nath S, Parsons JG, Gardea- Torresde JL, Pal T. Synthesis of plant-mediated gold nanoparticles and catalytic role of biomatrix-embedded nanomaterials. Environ Sci Technol 2007;41 (14):5137–5142.

[79] Klaus-Joerger T, Joerger R, Olsson E, Granqvist C-G. Silver-based crystalline nanoparticles, microbially fabricated. Proc Natl Acad Sci USA 1999;96 (24):13611–13614.

[80] Ramanathan R, O'Mullane AP, Parikh RY, Smooker PM, Bhargava SK, Bansal V. Bacterial kinetics-controlled shape-directed biosynthesis of silver nanoplates using Morganella psychrotolerans. Langmuir 2010;27 (2):714–719.

[81] Klaus-Joerger T, Joerger R, Olsson E, Granqvist C-G. Bacteria as workers in the living factory: metal-accumulating bacteria and their potential for materials science. Trends Biotechnol 2001;19 (1):15–20.

[82] Gupta A, Matsui K, Lo JF, Silver S. Molecular basis for resistance to silver cations in Salmonella. Nat Med 1999;5 (2):183–188.

[83] Parikh RY, Ramanathan R, Coloe PJ, Bhargava SK, Patole MS, Shouche YS, Bansal V. Genus-wide physicochemical evidence of extracellular crystalline silver nanoparticles biosynthesis by Morganella spp. PLoS One 2011;6 (6):e21401.

[84] Parikh RY, Singh S, Prasad BLV, Patole MS, Sastry M, Shouche YS. Extracellular synthesis of crystalline silver nanoparticles and molecular evidence of silver resistance from Morganella sp.: towards understanding biochemical synthesis mechanism. ChemBioChem 2008;9 (9):1415–1422.

[85] Ramanathan R, Field MR, O'Mullane AP, Smooker PM, Bhargava SK, Bansal V. Aqueous phase synthesis of copper nanoparticles: a link between heavy metal resistance and nanoparticle synthesis ability in bacterial systems. Nanoscale 2013;5 (6):2300–2306.

[86] Mott D, Galkowski J, Wang L, Luo J, Zhong C-J. Synthesis of size-controlled and shaped copper nanoparticles. Langmuir 2007;23 (10):5740–5745.

[87] Park BK, Jeong S, Kim D, Moon J, Lim S, Kim JS. Synthesis and size control of monodisperse copper nanoparticles by polyol method. J Colloid Interface Sci 2007;311 (2):417–424.

[88] Yong P, Rowson NA, Farr JPG, Harris IR, Macaskie LE. Bioreduction and biocrystallization of palladium by Desulfovibrio desulfuricans NCIMB 8307. Biotechnol Bioeng 2002;80 (4):369–379.

[89] Baxter-Plant VS, Mikheenko IP, Macaskie LE. Sulphate-reducing bacteria, palladium and the reductive dehalogenation of chlorinated aromatic compounds. Biodegradation 2003;14 (2):83–90.

[90] Lloyd JR, Yong P, Macaskie LE. Enzymatic recovery of elemental palladium by using sulfate-reducing bacteria. Appl Environ Microbiol 1998;64 (11):4607–4609.

[91] Konishi Y, Ogi T, Saito N. Room temperature synthesis and their applications of noble metal nanoparticles by metal ion-reducing bacteria. J Japan Soc Powder Powder Metall 2010;57 (7):508–513.

[92] Murray CB, Norris DJ, Bawendi MG. Synthesis and characterization of nearly monodisperse CdE (E = sulfur, selenium, tellurium) semiconductor nanocrystallites. J Am Chem Soc 1993;115 (19):8706–8715.

[93] Oremland RS, Herbel MJ, Blum JS, Langley S, Beveridge TJ, Ajayan PM, Sutto T, Ellis AV, Curran S. Structural and spectral features of selenium nanospheres produced by Se-respiring bacteria. Appl Environ Microbiol 2004;70 (1):52–60.

[94] Baesman SM, Bullen TD, Dewald J, Zhang D, Curran S, Islam FS, Beveridge TJ, Oremland RS. Formation of tellurium nanocrystals during anaerobic growth of bacteria that use Te oxyanions as respiratory electron acceptors. Appl Environ Microbiol 2007;73 (7):2135–2143.

[95] Soni SK, Ramanathan R, Coloe PJ, Bansal V, Bhargava SK. Self-assembled enzyme capsules in ionic liquid [BMIM][BF_4] as templating nanoreactors for hollow silica nanocontainers. Langmuir 2010;26 (20):16020–16024.

[96] Wang Y, Bansal V, Zelikin AN, Caruso F. Templated synthesis of single-component polymer capsules and their application in drug delivery. Nano Lett 2008;8 (6):1741–1745.

[97] Nicollian EH, Brews JR. *MOS Physics and Technology*. New York: Wiley-Interscience; 1982. p 920.

[98] Kröger N, Deutzmann R, Bergsdorf C, Sumper M. Species specific polyamines from diatoms control silica morphology. Proc Natl Acad Sci USA 2000;97 (26):14133–14138.

[99] Perry C, Keeling-Tucker T. Biosilicification: the role of the organic matrix in structure control. J Biol Inorg Chem 2000;5 (5):537–550.

[100] Shimizu K, Cha J, Stucky GD, Morse DE. Silicatein a: cathepsin L-like protein in sponge biosilica. Proc Natl Acad Sci USA 1998;95 (11):6234–6238.

[101] Swift DM, Wheeler AP. Evidence of an organic matrix from diatom biosilica. J Phycol 1992;28 (2):202–209.

[102] Milligan AJ, Morel FMM. A proton buffering role for silica in diatoms. Science 2002;297 (5588):1848–1850.

[103] Morse DE. Silicon biotechnology: harnessing biological silica production to construct new materials. Trends Biotechnol 1999;17 (6):230–232.

[104] Perry CC, Keeling-Tucker T. Model studies of colloidal silica precipitation using biosilica extracts from Equisetum telmateia. Colloid Polym Sci 2003;281 (7):652–664.

[105] Ramanathan R, Campbell JL, Soni SK, Bhargava SK, Bansal V. Cationic amino acids specific biomimetic silicification in ionic liquid: a quest to understand the formation of 3-D structures in diatoms. PLoS One 2011;6 (3):e17707.

[106] Schultz TF, Egerton-Warburton L, Crawford SA, Wtherbee R. Identification of a 41 kDa protein embedded in the biosilica of scales and bristles isolated from Mallomonas splendens (Synurophyceae, Ochrophyta). Protist 2001;152 (4):315–327.

[107] Bansal V, Rautaray D, Bharde A, Ahire K, Sanyal A, Ahmad A, Sastry M. Fungus-mediated biosynthesis of silica and titania particles. J Mater Chem 2005;15 (26):2583–2589.

[108] Bansal V, Rautaray D, Ahmad A, Sastry M. Biosynthesis of zirconia nanoparticles using the fungus Fusarium oxysporum. J Mater Chem 2004;14 (22):3303–3305.

[109] Bharde A, Rautaray D, Bansal V, Ahmad A, Sarkar I, Yusuf SM, Sanyal M, Sastry M. Extracellular biosynthesis of magnetite using fungi. Small 2006;2 (1):135–141.

[110] Bansal V, Poddar P, Ahmad A, Sastry M. Room-temperature biosynthesis of ferroelectric barium titanate nanoparticles. J Am Chem Soc 2006;128 (36):11958–11963.

[111] Bazylinski DA, Frankel RB, Jannasch HW. Anaerobic magnetite production by a marine, magnetotactic bacterium. Nature 1988;334 (6182):518–519.

[112] Bharde A, Wani A, Shouche Y, Joy PA, Prasad BLV, Sastry M. Bacterial aerobic synthesis of nanocrystalline magnetite. J Am Chem Soc 2005;127 (26):9326–9327.

[113] Bharde AA, Parikh RY, Baidakova M, Jouen S, Hannoyer B, Enoki T, Prasad BLV, Shouche YS, Ogale S, Sastry M. Bacteria mediated precursor-dependent biosynthesis of superparamagnetic iron oxide and iron sulfide nanoparticles. Langmuir 2008;24 (11):5787–5794.

[114] Vali H, Weiss B, Li Y-L, Sears SK, Kim SS, Kirschvink JL, Zhang CL. Formation of tabular single-domain magnetite induced by Geobacter metallireducens GS-15. Proc Natl Acad Sci U S A 2004;101 (46):16121–16126.

[115] Uddin I, Adyanthaya S, Syed A, Selvaraj K, Ahmad A, Poddar P. Structure and microbial synthesis of sub-10 nm Bi_2O_3 nanocrystals. J Nanosci Nanotechnol 2008;8 (8):3909–3913.

[116] Cai Y, Shen Y, Xie A, Li S, Wang X. Green synthesis of soya bean sprouts-mediated superparamagnetic Fe_3O_4 nanoparticles. J Magn Magn Mater 2010;322 (19):2938–2943.

[117] Łukaszewicz J, Wesołowski R, Cyganiuk A. Enrichment of salix viminalis wood in metal ions by phytoextraction. Polish J Environ Stud 2009;18 (3):507–511.

[118] Cyganiuk A, Klimkiewicz R, Olejniczak A, Lukaszewicz JP. Biotechnological fabrication of $LaMnO_3$-carbon catalyst for n-butanol conversion to ketones. Carbon 2010;48 (1):99–106.

[119] Ahmad A, Jagadale T, Dhas V, Khan S, Patil S, Pasricha R, Ravi V, Ogale S. Fungus-based synthesis of chemically difficult-to-synthesize multifunctional nanoparticles of $CuAlO_2$. Adv Mater 2007;19 (20):3295–3299.

[120] Gupta S, Bhattacharjee S, Pandey D, Bansal V, Bhargava S, Peng J, Garg A. Absence of morphotropic phase boundary effects in BiFeO(3)-PbTiO(3) thin films grown via a chemical multilayer deposition method. Appl Phys A 2011;104 (1):395–400.

[121] Sweeney RY, Mao C, Gao X, Burt JL, Belcher AM, Georgiou G, Iverson BL. Bacterial biosynthesis of cadmium sulfide nanocrystals. Chem Biol 2004;11 (11):1553–1559.

[122] Holmes JD, Richardson DJ, Saed S, Evans-Gowing R, Russell DA, Sodeau JR. Cadmium-specific formation of metal sulfide "Q-particles" by Klebsiella pneumoniae. Microbiology 1997,143 (8):2521–2530.

[123] Sharma PK, Balkwill DL, Frenkel A, Vairavamurthy MA. A new Klebsiella planticola strain (Cd-1) grows anaerobically at high cadmium concentrations and precipitates cadmium sulfide. Appl Environ Microbiol 2000;66 (7):3083–3087.

[124] Cunningham DP, Lundie LL Jr. Precipitation of cadmium by Clostridium thermoaceticum. Appl Environ Microbiol 1993;59 (1):7–14.

[125] Ahmad A, Mukherjee P, Mandal D, Senapati S, Khan MI, Kumar R, Sastry M. Enzyme mediated extracellular synthesis of CdS nanoparticles by the fungus, Fusarium oxysporum. J Am Chem Soc 2002;124 (41):12108–12109.

[126] Kowshik M, Deshmukh N, Vogel W, Urban J, Kulkarni SK, Paknikar KM. Microbial synthesis of semiconductor CdS nanoparticles; their characterization, and their use in the fabrication of an ideal diode. Biotechnol Bioeng 2002;78 (5):583–588.

[127] Kowshik M, Vogel W, Urban J, Kulkarni SK, Paknikar KM. Microbial synthesis of semiconductor PbS nanocrystallites. Adv Mater 2002;14 (11):815–818.

[128] Labrenz M, Druschel GK, Thomsen-Ebert T, Gilbert B, Welch SA, Kemner KM, Logan GA, Summons RE, Stasio GD, Bond PL, Lai B, Kelly SD, Banfield JF. Formation of sphalerite (ZnS) deposits in natural biofilms of sulfate-reducing bacteria. Science 2000;290 (5497):1744–1747.

[129] Mann S, Frankel RB, Blakemore RP. Structure, morphology, and crystal growth of bacterial magnetite. Nature 1984;310 (5976):405–407.

[130] Mann S, Sparks NHC, Frankel RB, Bazylinski DA, Jannasch HW. Biomineralization of ferrimagnetic greigite (Fe_3S_4) and iron pyrite (FeS_2) in a magnetotactic bacterium. Nature 1990;343 (6255):258–261.

[131] Ahmad A, Rautaray D, Sastry M. Biogenic calcium carbonate: calcite crystals of variable morphology by the reaction of aqueous Ca^{2+} ions with fungi. Adv Funct Mater 2004;14 (11):1075–1080.

[132] Rautaray D, Ahmad A, Sastry M. Biosynthesis of $CaCO_3$ crystals of complex morphology using a fungus and an actinomycete. J Am Chem Soc 2003;125 (48):14656–14657.

[133] Rautaray D, Sanyal A, Adyanthaya SD, Ahmad A, Sastry M. Biological synthesis of strontium carbonate crystals using the fungus Fusarium oxysporum. Langmuir 2004;20 (16):6827–6833.

[134] Sanyal A, Rautaray D, Bansal V, Ahmad A, Sastry M. Heavy-metal remediation by a fungus as a means of production of lead and cadmium carbonate crystals. Langmuir 2005;21 (16):7220–7224.

[135] Rautaray D, Ahmad A, Sastry M. Biological synthesis of metal carbonate minerals using fungi and actinomycetes. J Mater Chem 2004;14 (14):2333–2340.

[136] Ehrlich HL. How microbes influence mineral growth and dissolution. Chem Geol 1996;132 (1–4):5–9.

[137] Munch JC, Ottow JCG. Preferential reduction of amorphous to crystalline iron oxides by bacterial activity. Soil Sci 1980;129 (1):15–21.

[138] Natarajan KA. *Biogeochemistry of Rivers in Tropical South and South East Asia*. Hamburg: Heft 82, Geologisch-Paläontologisches Inst. und Inst. für Biogeochemie und Meereschemie der Univ. Hamburg; 1999.

[139] Bansal V, Sanyal A, Rautaray D, Ahmad A, Sastry M. Bioleaching of sand by the fungus, Fusarium oxysporum as a means of producing extracellular silica nanoparticles. Adv Mater 2005;17 (7):889–892.

[140] Bansal V, Ahmad A, Sastry M. Fungus-mediated biotransformation of amorphous silica in rice husk to nanocrystalline silica. J Am Chem Soc 2006;128 (43):14059–14066.

[141] Cobbett C, Goldsborough P. Phytochelatins and metallothioneins: roles in heavy metal detoxification and homeostasis. Annu Rev Plant Biol 2002;53 (1):159–182.

[142] Hirata K, Tsuji N, Miyamoto K. Biosynthetic regulation of phytochelatins, heavy metal-binding peptides. J Biosci Bioeng 2005;100 (6):593–599.

[143] Robinson NJ, Whitehall SK, Cavet JS. Microbial metallothioneins. In: *Advances in Microbial Physiology*. Volume 44, London: Academic Press; 2001. p 183–213.

[144] Park TJ, Lee SY, Heo NS, Seo TS. In vivo synthesis of diverse metal nanoparticles by recombinant Escherichia coli. Angew Chem Int Ed 2010;49 (39):7019–7024.

PART VII

CO_2 ADSORPTION

21

NANOMATERIALS FOR CARBON DIOXIDE ADSORPTION

Luis Ángel Garza Rodríguez[1] and Elsa Nadia Aguilera González[2]

[1] *Facultad de Ingeniería y Arquitectura, Universidad Regiomontana A.C., Monterrey, Mexico*

[2] *Corporación Mexicana de Investigación en Materiales S.A. de C.V., Calle Ciencia y Tecnología No. 790, Saltillo 400, Saltillo, Coahuila, Mexico*

21.1 INTRODUCTION

21.1.1 Global Warming

Over millennia, the temperature of the earth has evolved and has experienced warming and cooling cycles. With the passage of time, the earth has reached a dynamic equilibrium between the amount of heat received from the sun and the amount lost by the planet. It is estimated that the amount of energy that reaches the top of the atmosphere during the day is 1.360 W m^{-2} [1], and 30% is reflected back into space. Two-thirds of these reflections are due to clouds and aerosols. The other third is due to the light areas on the earth's surface such as ice, deserts, and snow. The remainder of the energy not reflected is absorbed by the atmosphere and the surface of the earth. Theoretically, the average temperature of the earth should be −18°C in order to be able to support life in hostile conditions. To maintain the thermal balance, the earth in its natural rotation cycle exposes a portion of this energy to sunlight (the adsorbed energy), while the dark side begins a cooling process. During this cooling process, a portion of the energy is radiated from the earth's surface into space (in the form of heat waves) and absorbed by some constituents in the atmosphere, mainly water vapor, carbon dioxide, ozone, oxide monoxide, and methane (known as greenhouse gases), by increasing the average temperature of the earth to 15°C, to keep the water in liquid form and sustain life on the planet. These greenhouse gases naturally regulate the average temperature of the earth to sustain life.

In the last 100 years, the average global temperature has increased by 0.8°C, of which the 0.6°C increase has occurred in the last three decades [2]. This change is largely due to the increase in the concentration of water vapor, methane, nitrogen oxides, ozone, and carbon dioxide, most of which are a result of human activities such as burning fossil fuels, industry, manufacturing of fertilizers, and deforestation, among others.

The most common among these is carbon dioxide (CO_2), which is added every day in large quantities due to the burning of fossil fuels. Global awareness has turned its attention to this gas as the main cause of global warming. This fact can be seen when analyzing the data collected and published by the Division of National Oceanic and Atmospheric Administration (NOAA) of the Department of Commerce of the United States, which noted carbon dioxide concentration in the atmosphere of 396.08 ppm in its report of February 2013 (Fig. 21.1).

The data on CO_2 concentration in the atmosphere obtained by Charles David Keeling at the south pole and in Hawaii since 1958 generated a graph known as the Keeling curve (Fig. 21.2). Obviously, CO_2 concentration has increased by just over 25% over the past 54 years, reaching a level of 398 ppm in 2013, and growing at a rate of about 1.5 ppm each year.

Nanomaterials for Environmental Protection, First Edition. Edited by Boris I. Kharisov, Oxana V. Kharissova, and H. V. Rasika Dias.

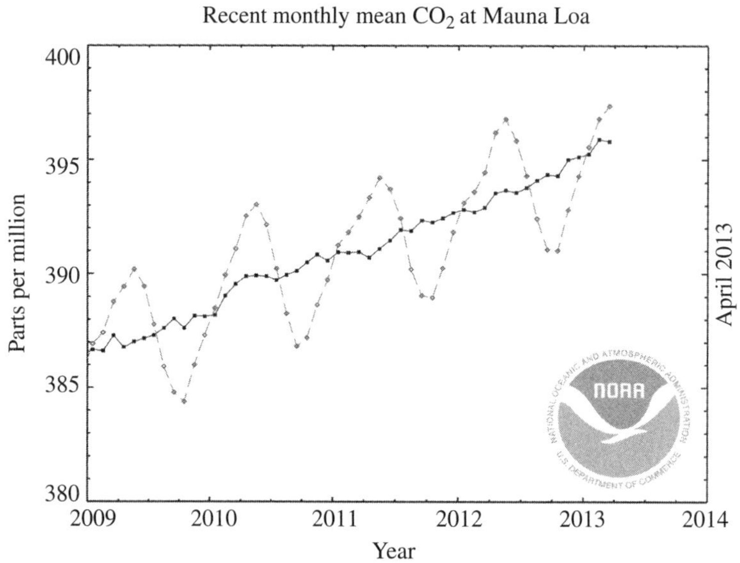

FIGURE 21.1 CO_2 concentration in the atmosphere (ppm) measured at Mauna Loa Observatory Hawaii [3]. The dashed line with diamond symbols represents the monthly mean values, centered on the middle of each month. The line with the square symbols represents the same, after correction, for the average seasonal cycle.

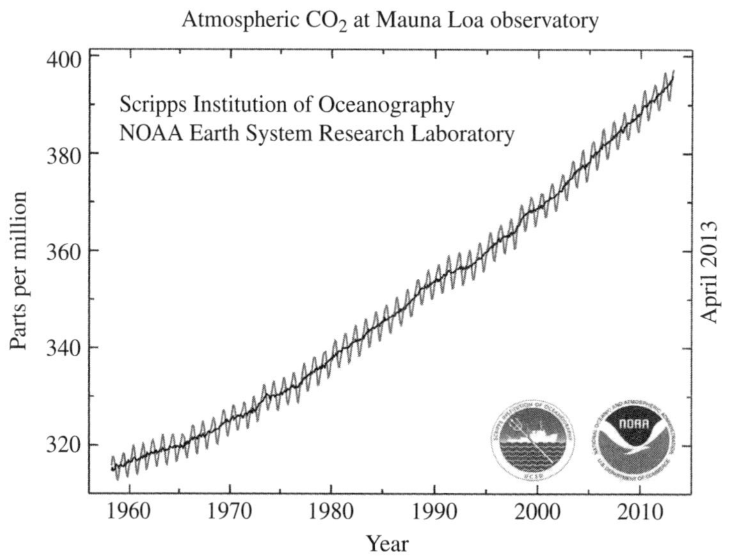

FIGURE 21.2 Monthly mean atmospheric CO_2 at Mauna Loa Observatory, Hawaii [3]. The CO_2 data (sinusoidal curve) is measured as the mole fraction in dry air; the other curve represents the seasonally corrected data.

There is a clear relationship between the increase in global average temperature and the concentration of CO_2 in the atmosphere. Countries like the United States, China, and India with populations of 2.953 million to 2011 million (41.6% of world population) generate just over 52% of CO_2 emissions [4]. Figure 21.3 shows the relationship between the population and CO_2 emissions in 2011.

Figure 21.4a shows greenhouse gas emissions by sector in the United States. Using this categorization, emissions from electricity generation accounted for the largest portion (33%), transportation activities accounted for the second largest portion (27%), while emissions from industry accounted for the third largest portion (20%) of U.S. greenhouse gas emissions in 2010. The global picture is not far behind (Fig. 21.4b). The generation of electricity and heat was by far the largest producer of CO_2 emissions and was responsible for 41% of world CO_2 emissions in 2010.

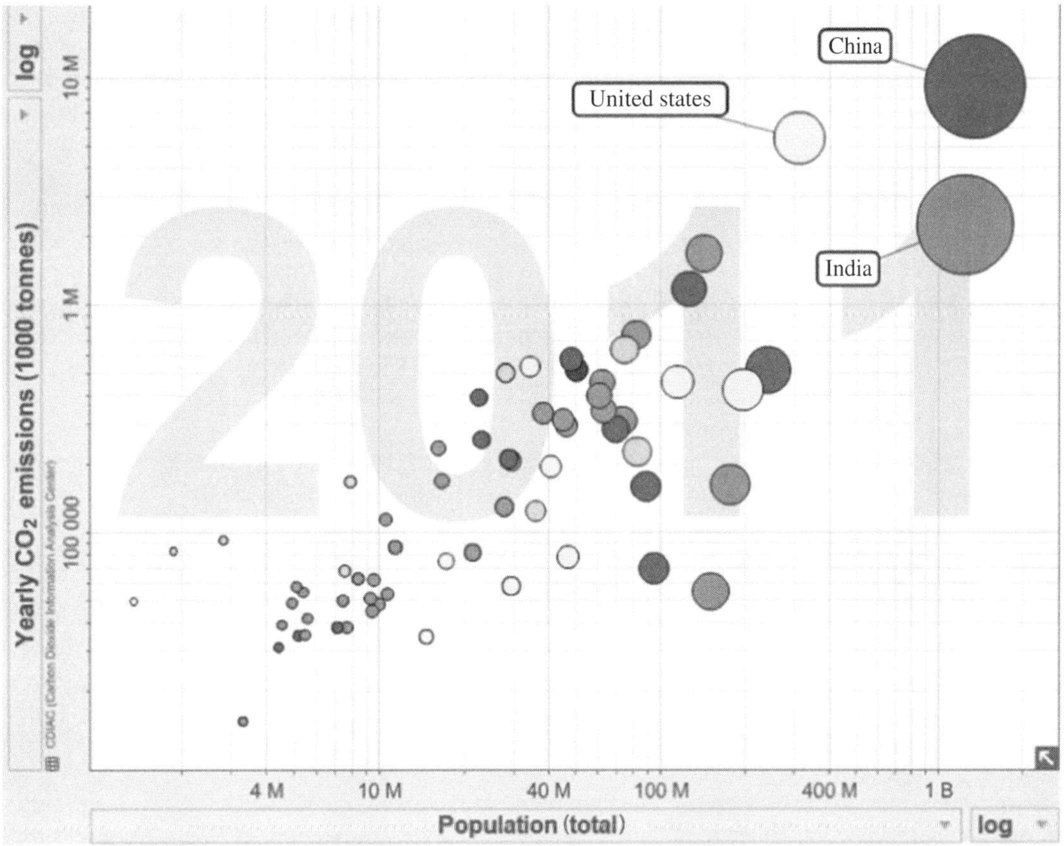

FIGURE 21.3 Relationship between CO_2 emissions in 2011 and the population by country. *Source*: http://www.gapminder.org/world-offline/ [4].

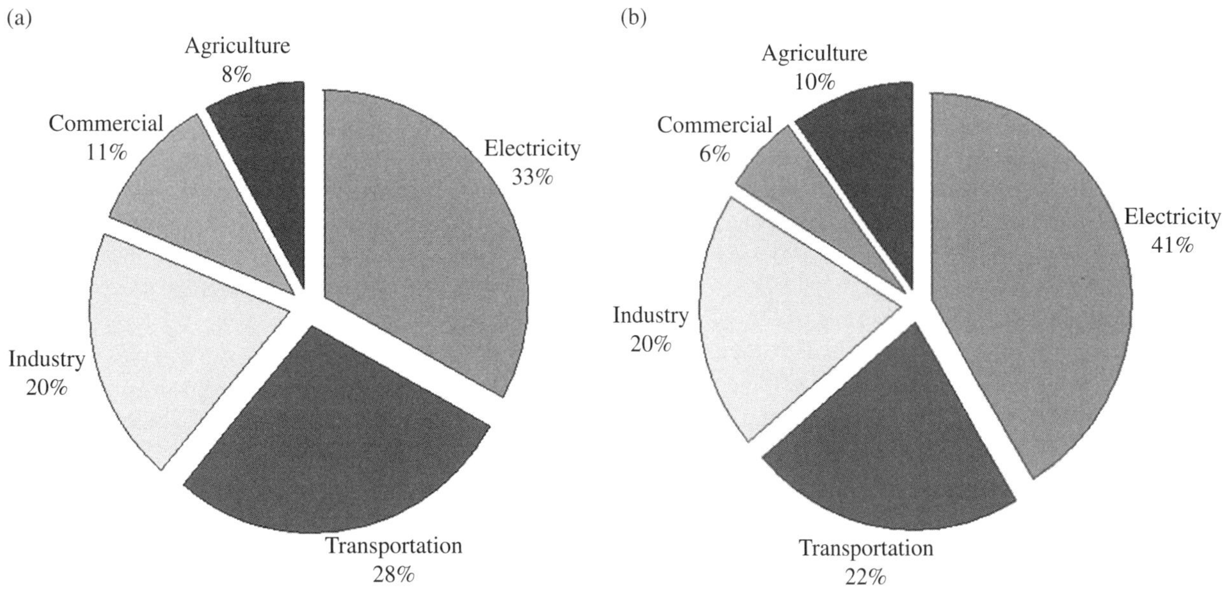

FIGURE 21.4 (a) U.S. greenhouse gas emissions by economic sector in 2010. Adapted from Ref. [5]. (b) World greenhouse gas emissions by economic sector in 2010. Adapted from Ref. [6].

Although global energy is currently used more efficiently, the demand for oil will continue to increase with the increase of the population and the economy. By 2030, it is estimated that oil will contribute to 29% of global energy consumption and solar power to 1%. It is estimated that for the same period, the transport sector will have the largest increase in oil demand [7].

In the industry, CO_2 is used in processes such as the synthesis of urea, methanol, carbonated beverages, in greenhouses [8], supercritical extraction [9], and in extinguishers, among others. The volumes generated worldwide far exceed the requirements of this gas in these processes.

The urgent need for accessible and affordable energy, based on the burning of fossil fuels and consumption projections in the coming decades, requires the development of various technologies to mitigate or minimize CO_2 emissions to the atmosphere.

However, these technologies cannot be used in all sectors. For example, emissions from houses and transportation vehicles, which are important due to the volume they generate, can present significant economic challenges for successful implementation, because these effluents enter the air directly and it is not practical to separate greenhouse gas streams containing a few parts per million. This does not happen in the industrial sector, where it is more feasible to capture CO_2 because of the huge volumes and high concentrations that are generated.

The questions that arise are (i) How can we remove or capture CO_2? and (ii) What do we do with the large amounts of CO_2, once captured? To answer the first question, there are several proven technologies for removing CO_2; examples of such processes in industrial application are precombustion, postcombustion, and oxycombustion. Once removed, greenhouse gases can be disposed of through various proposals, the most studied being the geological sequestration and storage options at the bottom of the sea and mineral carbonation or sequestration.

21.2 CO_2 STORAGE AND SEQUESTRATION

21.2.1 Geological Sequestration

The most readily available method of sequestration is underground containment. Capturing CO_2 from major stationary sources, transporting it usually by pipeline, and injecting it into suitable deep rock formations for geological storage provide a way of avoiding the emission of CO_2 into the atmosphere. There are a number of oil and gas wells, coal mines, and abandoned salt domes or low production yields. These sites have huge open spaces where it is feasible to store CO_2 at high pressure (Fig. 21.5). The engineered injection of CO_2 into subsurface geological formations was first undertaken in Texas in the United States in the early 1970s, as part of enhanced oil recovery (EOR) projects [10]. In 1996, the world's first large-scale storage project was initiated by Statoil and its partners at the Sleipner Gas Field in the North Sea. The world's first large-scale CO_2 storage project in a gas reservoir was conducted in Salah, Algeria [11].

To geologically store CO_2, it must first be compressed, usually to a dense fluid state known as "supercritical." Depending on the rate at which temperature increases with depth (the geothermal gradient), the density of CO_2 will increase with depth, until at about 800 m or greater, the injected CO_2 will be in a dense supercritical state (see Fig. 21.6).

The most effective storage sites are those where CO_2 is immobile because it is trapped permanently under a thick, low-permeability seal or is converted into solid minerals (e.g., carbonates) or adsorbed onto surfaces of coal micropores or through a combination of physical and chemical trapping mechanisms.

CO_2 has different water solubilities depending on the temperature, pressure, and dissolved salts. Once dissolved in water, it can form carbonic acid or other insoluble carbonates depending on the rock mineralogy. Underground coal is also an adsorbent of many substances, including CO_2, H_2S, and SO_2, among others. The carbon affinity for CO_2 is 2–8 times greater than for methane. By contrast, the H_2S found in gas and sulfur oxides (SO_x), which are found in the flue gases, have higher affinity for carbon as CO_2.

The geological disposal of CO_2 therefore needs to meet three requirements:

1. Capacity: the disposal unit has to have sufficient capacity to receive and retain the intended volume of CO_2.
2. Injectivity: the ability to inject CO_2 deep into the ground at the rate that it is supplied from the CO_2 source.
3. Confinement: if CO_2 is not confined, then, due to its buoyancy (being lighter than water), it will flow upward, ultimately entering the shallow hydrosphere (including potable groundwater), the biosphere, and the atmosphere.

CO_2 disposal in geological media has not yet been implemented as a mitigation measure for climate change, although CO_2 injection and disposal have occurred for different reasons in the last three decades. There are other challenges facing the large-scale development of the geological disposal of CO_2, but they are of an economic, financial, legal, and regulatory nature and are also likely to be linked to public attitude toward such developments [14].

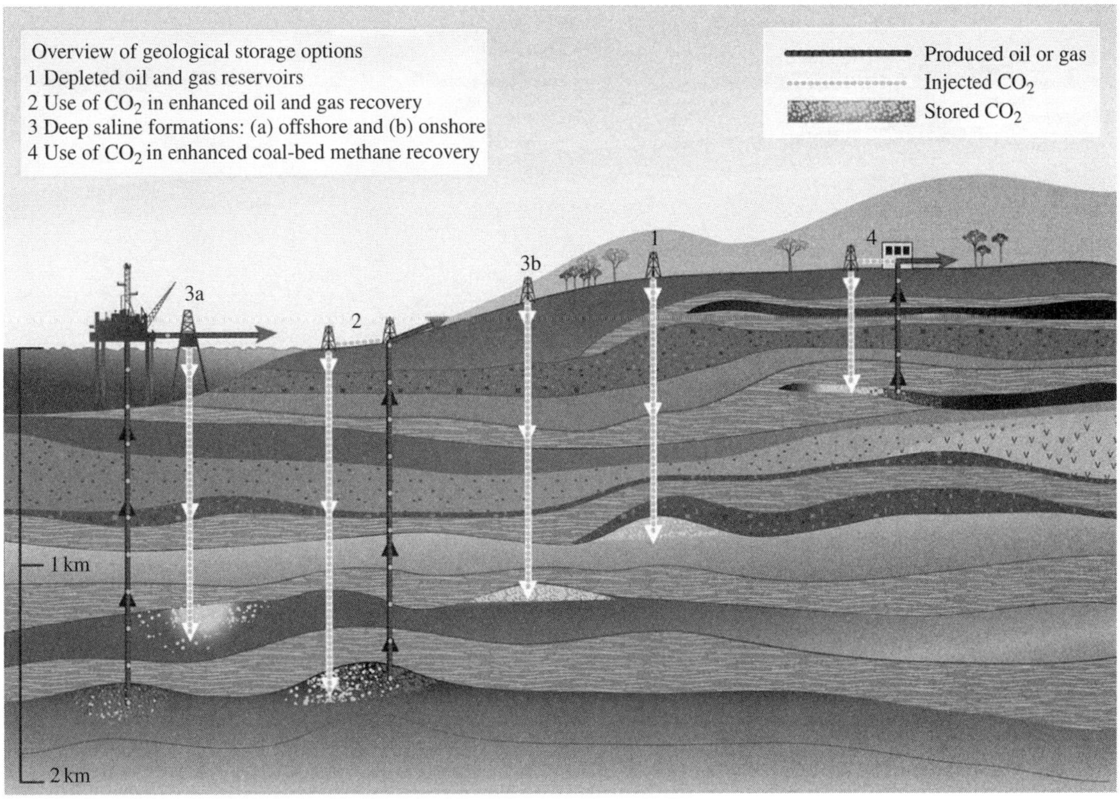

FIGURE 21.5 Methods for storing CO_2 in deep underground geological formations. Courtesy from Ref. [13].

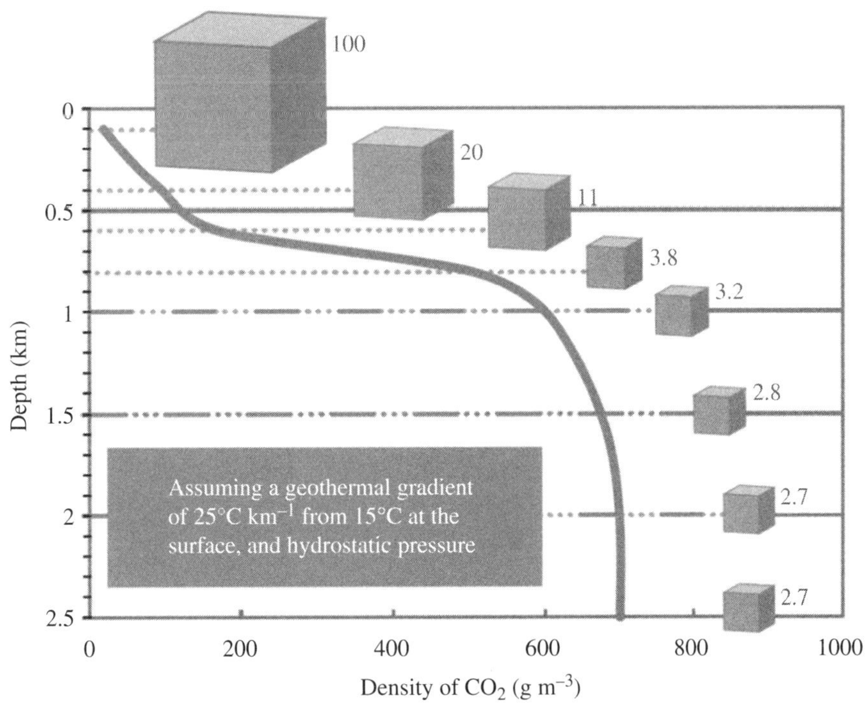

FIGURE 21.6 Variation of CO_2 density with depth, assuming hydrostatic pressure and a geothermal gradient of 25°C km^{-1} from 15°C at the surface [12]. CO_2 density increases rapidly at a depth of approximately 800 m, when the CO_2 reaches a supercritical state. Cubes represent the relative volume occupied by the CO_2 and down to 800 m; this volume can be seen to dramatically decrease with depth. At depths below 1.5 km, the density and specific volume become nearly constant. Courtesy from Ref. [13].

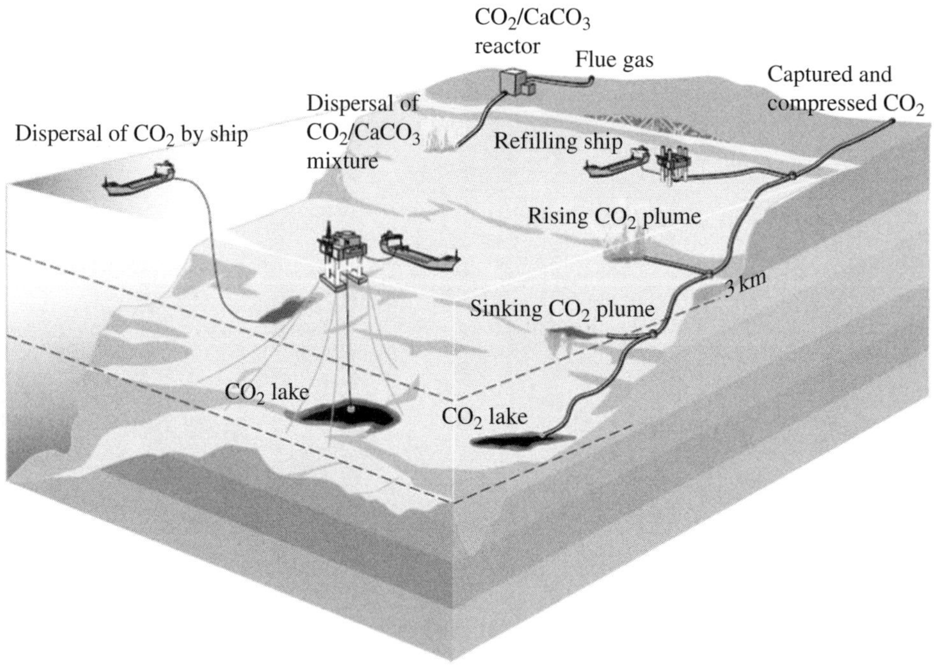

FIGURE 21.7 Some of the ocean storage strategies for CO_2. Courtesy from Ref. [13].

21.2.2 Ocean Storage

Captured CO_2 could be deliberately injected into the ocean at great depth, where most of it would remain isolated from the atmosphere for centuries. CO_2 can be transported via pipeline or ship for release into the ocean or onto the sea floor.

The concept involves the slow exchange of CO_2 between the bottom water and the sea surface, allowing gas retention for millennia.

Analyses of ocean observations and models agree that injected CO_2 will remain isolated from the atmosphere for several hundred years and that the fraction retained will tend to be larger with deeper injection. Additional concepts to prolong CO_2 retention include forming solid CO_2 hydrates and liquid CO_2 lakes on the sea floor, as well as increasing CO_2 solubility by, for example, dissolving mineral carbonates (Fig. 21.7).

21.2.2.1 Impacts on the Marine Environment Overall, there is limited knowledge of the population in deep water, the community structure, and the ecological interactions in deep water. The sensitivity of deep ocean ecosystems to store CO_2 remains largely unknown. Most ocean storage proposals try to minimize the volume of water with high concentrations of CO_2, either by diluting the CO_2 in large volumes of water or by isolating the CO_2 in a small volume (e.g., lakes).

Ocean storage of CO_2 could occur deep in the ocean where there is virtually no light and a lack of photosynthetic organisms; hence, those that could be affected are heterotrophic the organisms, that is, most animals. Therefore, the effects of CO_2 must be identified at individual levels (physiological) of the ecosystem.

21.2.2.2 Mineral Carbonation Mineral carbonation results from the reaction of CO_2 with metal oxide–bearing materials, the most attractive metals being calcium and magnesium. In nature, such a reaction is called silicate weathering and it takes place on a geological timescale. However, in an industrial scenario, high concentrations of captured CO_2 are put into contact with metal oxide–bearing materials, thus forming insoluble carbonates [15, 16].

The kinetics of natural mineral carbonation is slow; hence, all currently implemented processes require energy-intensive preparation of solid reactants to achieve affordable conversion rates and/or additives that must be regenerated and recycled using external energy sources. The resulting carbonated solids must be stored at an environmentally suitable location. The technology is still in the development stage and is not yet ready for implementation. The best case studied so far is the wet carbonation of the natural silicate olivine, which costs between 50 and 100 US\$/tCO_2 stored and translates into a 30–50% energy penalty on the original power plant.

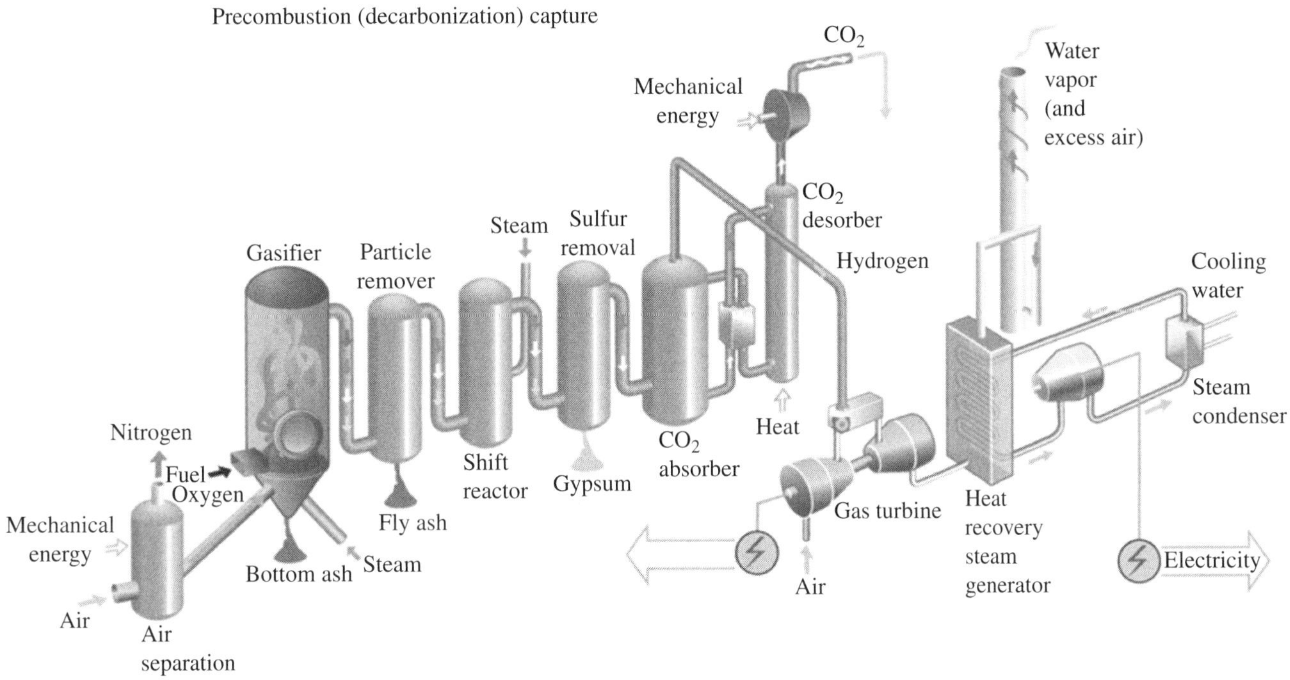

FIGURE 21.8 Precombustion schematic diagram flow. Courtesy from Ref. [17].

21.2.3 Processes for Carbon Dioxide Capture

21.2.3.1 Precombustion Capture Figure 21.8 shows the diagram of CO_2 removal using the precombustion process. The system involves capturing CO_2 before burning the fuel. The fuel gas, liquid, or solid is first transformed into synthesis gas, mainly CO and H_2O, by reforming processes (for liquid and gaseous fuels) or gasification reactors (solid fuel). The latter reaction is commonly called "partial oxidation" and proceeds according to reaction 21.1. The synthesis gas is cleaned to remove particles that could damage the turbine or cause problems in subsequent steps of the process.

$$C_xH_y + xH_2O \rightleftharpoons xCO_2 + \left(\frac{x+y}{2}\right)H_2 \tag{21.1}$$

Then CO is converted to CO_2 and H_2 by reacting CO with water vapor through a shift reaction (water–gas shift reaction):

$$CO + H_2O \rightleftharpoons CO_2 + H_2 \tag{21.2}$$

At this stage, the gas is composed mainly of CO_2 and H_2. The concentration of CO_2 is about 15–60% (dry basis) and the total pressure is typically between 2 and 7 MPa. CO_2 is removed by an absorption process. The typical solvents used for precombustion capture are either physical (i.e., Selexol [18] and Rectisol) or chemical (i.e., methyldiethanolamine [19]). Unlike the solvents for postcombustion capture, these technologies are mature and proven on a large commercial scale [20].

Although the initial fuel conversion steps of precombustion are costlier and more elaborate, the higher concentrations and higher pressure of CO_2 in the gas stream make the separation and the subsequent compression of CO_2 easier than in postcombustion. The removal rate of CO_2 is over 90%. In precombustion technology, the energy required for water–gas shift reaction, solvent regeneration, and compression of CO_2 could represent anywhere from 14 to 25% of the output in a natural gas combined cycle (NGCC) [21].

21.2.3.2 Postcombustion Capture In this method, CO_2 is captured in a scrubber from flue gases that are close to atmospheric pressure using an absorption process based on chemical solvents, such as amines (e.g., monoethanolamine). This is because chemical solvents are less dependent on partial pressure than physical solvents, and the partial pressure of CO_2 in the flue gas is low, typically 4–14% by volume. The gases pass through the absorption column where the solvent reacts with the CO_2, chemically

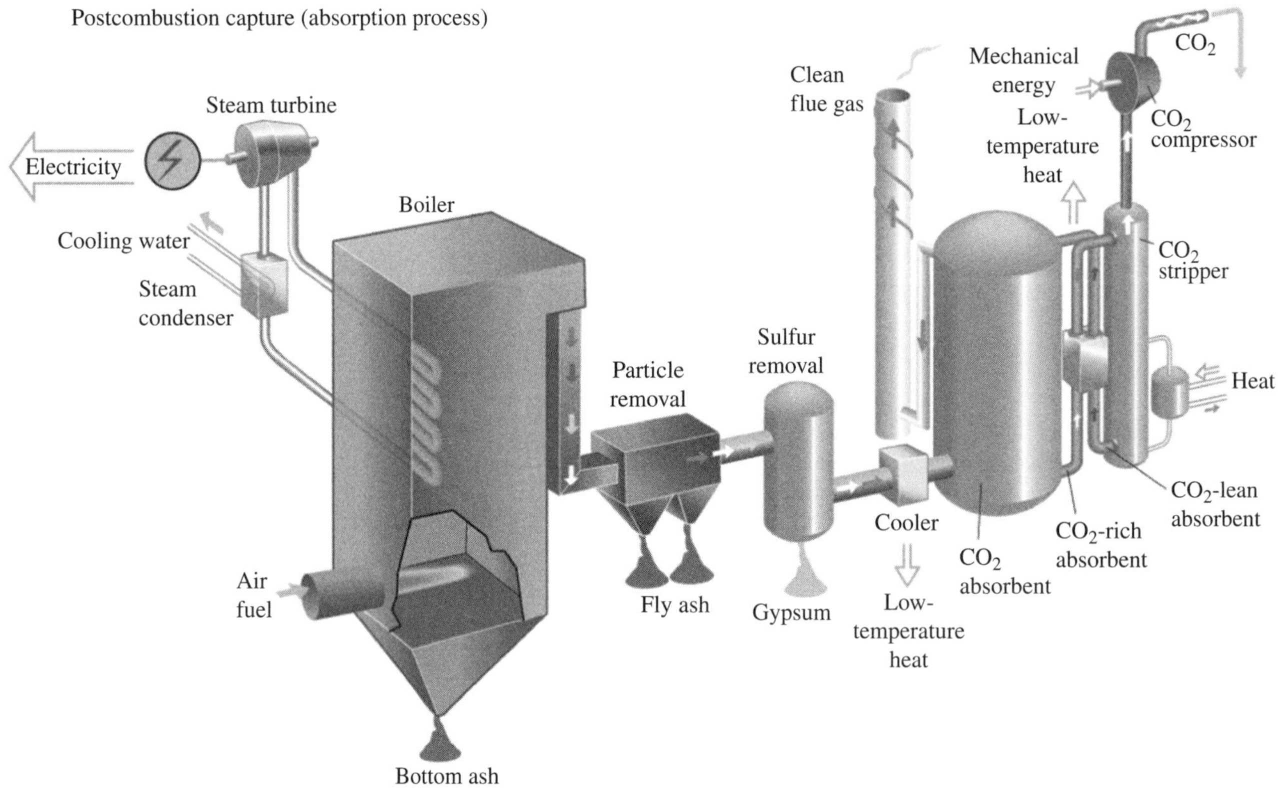

FIGURE 21.9 Postcombustion schematic diagram flow. Courtesy from Ref. [22].

binding it and removing it from the gas stream. After absorption, the cleaned flue gases primarily contain water vapor and climatic inactive nitrogen (see Fig. 21.9). On leaving the scrubber, the solvent is heated to release the nearly pure CO_2 and the solvent can then be reused (Fig. 21.10). Heat from the power plant's steam turbines is used to raise the temperature to the desired level. The captured CO_2 can be transported to a storage site.

In the case of a power plant using postcombustion capture, the energy required for solvent regeneration and CO_2 compression could represent anywhere from 25 to 35% of its output.

21.2.3.3 Oxyfuel Combustion Capture Oxyfuel combustion is still in the demonstration phase and uses high-purity oxygen. This results in relatively higher CO_2 concentrations in the gas stream and, hence, in easier separation of CO_2 but at the cost of increased energy requirements for the separation of oxygen from air. A large amount of oxygen is required for combustion, which is obtained from an air separation unit. The flue gas from oxycombustion is compressed and chilled to separate out nitrogen, oxygen, and other impurities. The resulting CO_2 concentration is typically 95 mol% or more (Fig. 21.11).

21.3 NOVEL TECHNOLOGIES

A major challenge in the implementation of carbon capture and storage (CCS) is that current industrial carbon capture technologies are energy-intensive and not cost-effective [24]. To be viable, a capture technology must achieve 90% CO_2 capture with a maximum energy penalty of 10% [25]. The conventional technologies have many disadvantages: the liquid–gas interaction, that is, the aqueous amine solutions, requires huge capital and operational costs because of the energy needed for regeneration. Recently, gas–solid adsorption processes have been proposed as promising technologies for carbon capture [26]. Solid sorbents have the potential to reduce the energy demand of capture processes because of their potentially higher loading capacities, the absence of solvent heating and vaporization during regeneration, lower material heat capacities, and lower heat of sorption [27–29]. These materials require a large surface area-to-mass ratio

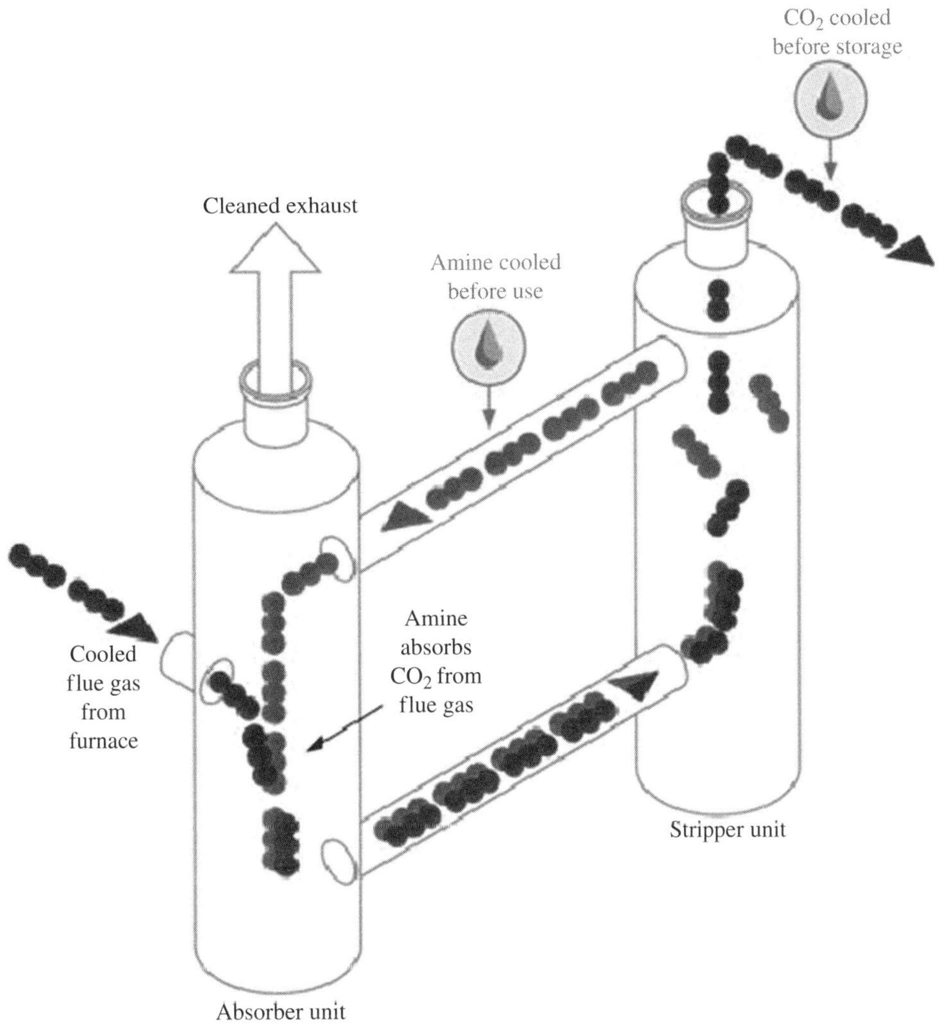

FIGURE 21.10 Simplified process flow of the absorption on CO_2 from flue gas with an amine-based absorber. Adapted from http://spectrum.ieee.org/energy/environment/the-water-cost-of-carbon-capture.

and a preferential interaction with CO_2 to be efficient and effective. Activated carbon, zeolites, metal–organic frameworks (MOF) [30, 31], and amine-modified carbon nanotubes (amine-CNTs) have recently been identified as promising sorbents for CO_2 capture [32].

21.3.1 Nanocapture of Carbon Dioxide

Environmental nanotechnology is considered to play a key role in shaping current environmental engineering and science. The nanoscale has stimulated the development and use of novel and cost-effective technologies for pollution detection, monitoring, and remediation [33]. The use of nanoparticles may have an advantage over conventional methods due to the much larger surface area of nanoparticles on a mass basis. The unique structure and electronic properties of some nanoparticles makes them adsorbent to pollutants. Many nanomaterials have adsorbent properties, depending on their size. Chemically modified nanomaterials have also attracted a lot of attention, especially nanoporous materials due to their exceptionally high surface area [34].

CCS produced in fossil-fuelled power plants has received significant attention since the Kyoto Protocol came into force on February 16, 2005. Various CO_2 capture technologies including absorption, adsorption, cryogenic, membrane, and others have been investigated [35, 36]. The Intergovernmental Panel on Climate Change (IPCC) concluded that the design of a large-scale

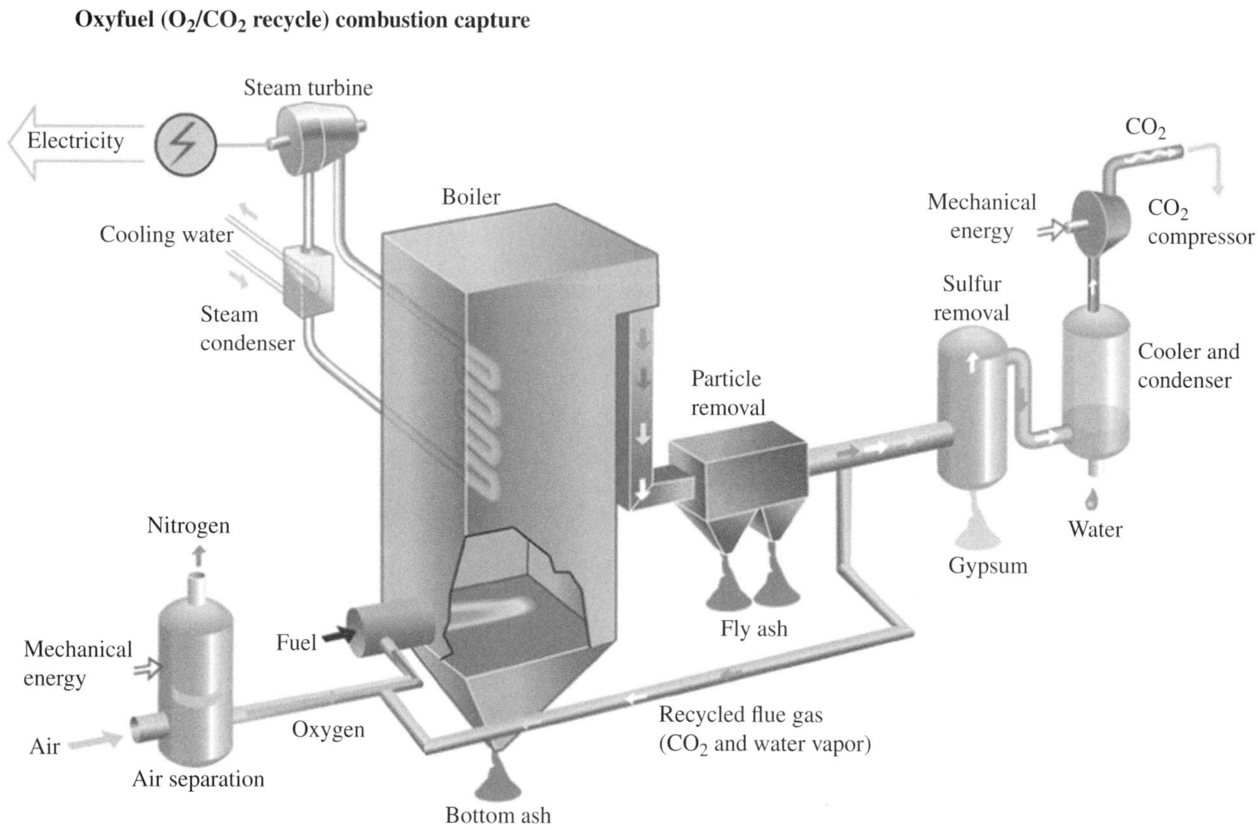

FIGURE 21.11 Oxyfuel schematic diagram flow. Courtesy from Ref. [23].

adsorption process might be feasible and the development of a new generation of material that is capable of adsorbing CO_2 efficiently will undoubtedly enhance the competitiveness of adsorptive separation in a flue gas application.

21.3.1.1 Carbon Nanotubes CNTs are most the most well known among nanohollow structured materials with their dimensions ranging from 1 to 10 nm in diameter and 200 to 500 nm in length. These new materials have unique properties such as uniform porosity, high pore volume, high specific surface area, and low mass density [37]. CNTs have been proven to possess good potential for CO_2 capture from flue gas due to their unique physicochemical properties as well as their high thermal and chemical stability [38]. CNTs are attractive for adsorbing gases because they have high surface area, pore structure, wide spectrum of surface functional groups, and relatively controllable porosity. The adsorption of CO_2 in carbonaceous materials such as single-walled carbon nanotubes (SWCNTs) [39] and multiwalled carbon nanotubes (MWCNTs) corresponds to the amount of CO_2 adsorption which takes place near the carbon surface solid only due to the physical forces (van der Waals interactions, physisorption) that carbon atoms exert on CO_2 molecules. Physisorption occurs due to van der Waals forces between adsorbate molecules and adsorbents while chemisorption takes place due to chemical interactions between the adsorbate molecules and the surface functional groups of adsorbents. These nanomaterials have been proven to possess good potential as superior adsorbents for removing many kinds of organic and inorganic pollutants in air streams [40] or from aqueous environments [41]. Both SWCNTs and MWCNTs have been tested as adsorbents of CO_2 [42]. For example, Lee et al. [42] compare MWCNTs and granular activated carbon (GAC) for CO_2 adsorption. Under the same conditions, the capacity of MWCNTs and GAC was 1.57 and 1.65 mmol $CO_2\,g^{-1}$ sorbent, respectively.

21.3.1.2 Amine Types At the industrial level, the most common amine compounds used are monoethanolamine (MEA) and diethanolamine (DEA) [43]. Recently, Chowdhury et al. [44] investigated CO_2 sorption rate, loading capacity, and heat of reaction measurements of 25 amine-based absorbents. Correlating their findings with the differences in chemical structure of the amines, they suggested that the specific amine moieties and structures used may be selected or modified to tune the CO_2 capture process. Tertiary amine–CO_2 reaction chemistry suggests that the tertiary amine–CNTs may have high CO_2 capacity with a relatively low heat of sorption.

The amines ethylene diamine (EDA), tetraethylenepentamine (TEPA), *N*-dimethylaminopropyltrimethoxysilane (DMAPS), 3-aminopropyltriethoxysilane (APTS), and polyethyleneimine (PEI) are the most common compounds used in CNT modification reported in the literature.

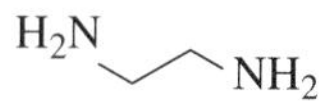

EDA

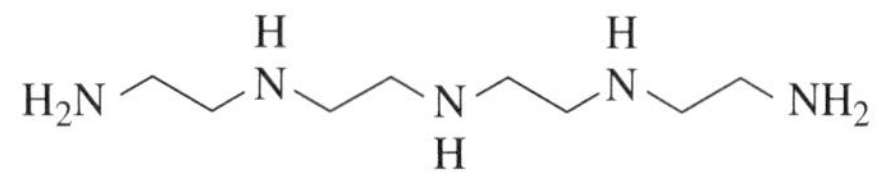

TEPA

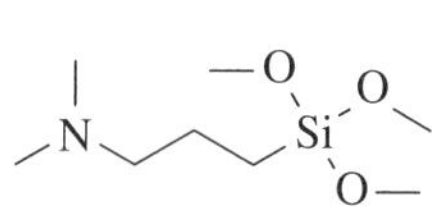

DAMPS

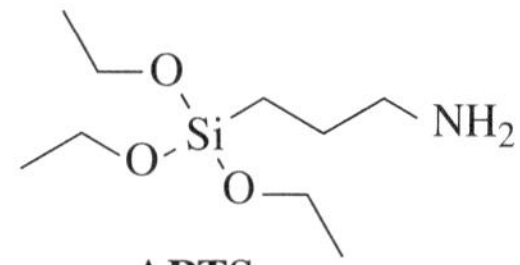

APTS

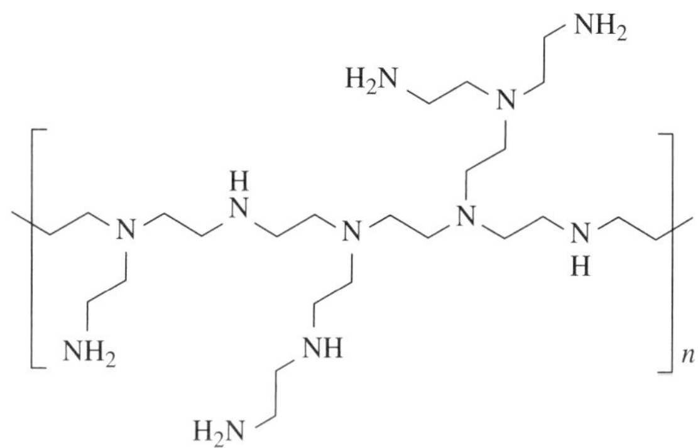

PEI

21.3.1.2.1 Amine-Modified Carbon Nanotubes The chemical modification of the CNTs with functional groups containing amine groups improved the capacity to capture CO_2. CNT interactions with gases may be tuned for chemisorption interactions through the functionalization of the nanotube surface with various functional groups. Covalent functionalization of SWCNTs has also been well studied [45] and can be performed by facile acid treatment, which creates hydroxyl (–OH) and carboxylic acid (–COOH) groups on the ends and walls of the SWCNTs [46]. The enhancement occurs because of the acidity of CO_2, a Lewis acid molecule, which prefers to adsorb on a nitrogen-rich basic surface.

Some conditions affect the efficiency of adsorption of CNTs. In general, in comparison to typical commercial adsorbents, such as zeolites [47], the performance of CO_2 adsorption on amine-modified CNTs increases with the presence of moisture till a certain limit, and then decreases. In most cases, the adsorption decreases with the increase in temperature due to changes in the mechanisms of interaction and the initial CO_2 concentration also displays an interesting behavior.

The general proceedings for testing CNT adsorption capacity in the laboratory recommend a stream of CO_2 with a concentration between 2 and 50%, which was selected as representative of various CO_2 emission levels from combustion in industrial activities, fossil fuel power plants, coal gasification systems, and confined spaces like submarines and space capsules. A range of temperatures between 20 and 100°C, preferably 50°C, is used to simulate a typical flue gas from an industrial operation. The selection of water vapor typically ranges between 8 and 12% in a flue gas [48].

21.3.1.2.2 Effect of Temperature Su et al. observed the adsorption isotherms of CO_2 via MWCNTs impregnated with APTS at temperatures ranging from 20 to 100°C. Comparing the differences in the capacity of CNTs with and without amine modifications (Fig. 21.12a), the results showed an increase of 1.5–2.0 times in the q_e for the latter. Another interesting work from Ye et al. [49] reports the adsorption of low concentrations of CO_2 with carbon nanotubes (MWCNTs) impregnated with TEPA, a long aliphatic amine with primary and secondary amine groups, at temperatures ranging from 282 to 313°C. They evaluated the adsorption capacity of CNT sorbents with four different TEPA loadings: 10 (CT-10), 20, 30, and 40%. Figure 21.12 shows the adsorption capacity curves. Contrary to the expected results, the adsorption capacity of the raw CNTs experienced a modest increase to 0.49 mmol g^{-1} from 283 to 298°C and then a gradual decrease to 0.33 mmol g^{-1} with

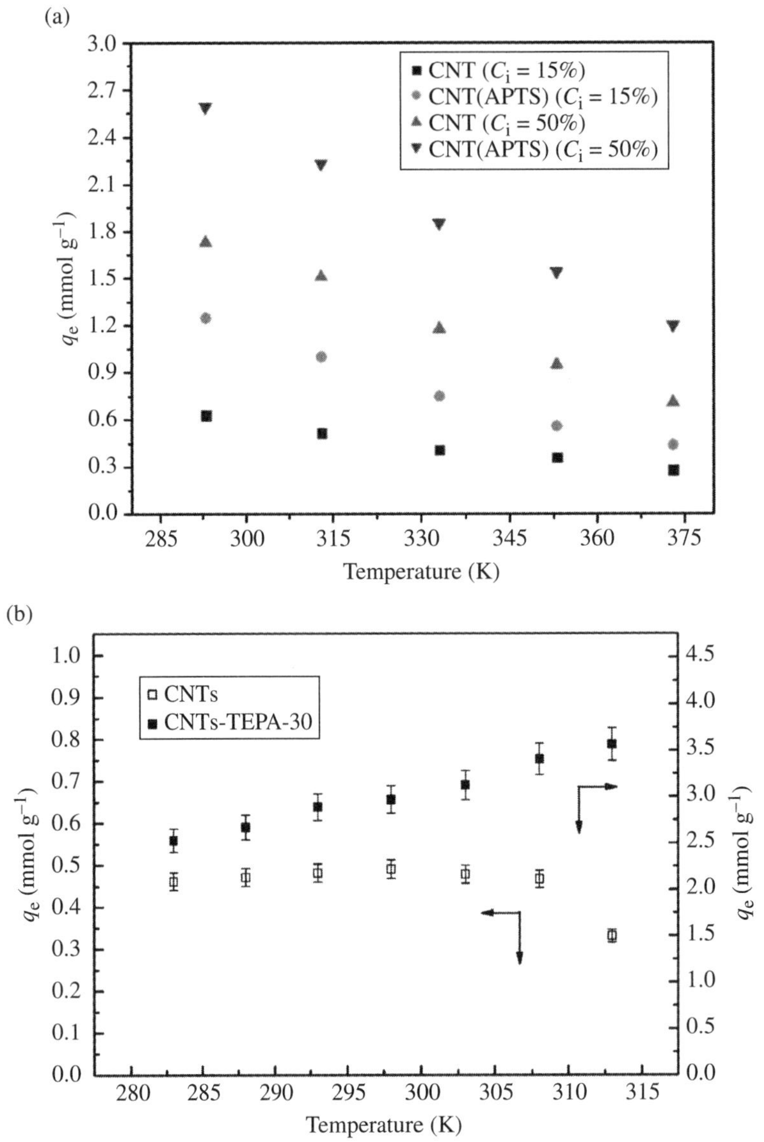

FIGURE 21.12 (a) Effect of temperature and CO_2 concentration on adsorption capacity via CNTs and CNT(APTS). Adapted from Ref. [38]. (b) Effect of temperature on CO_2 adsorption capacity for CNTs and CT-30 (gas flow rate, 50 ml min^{-1}; CO_2 concentration, 2.0 vol%). Reprinted with permission from Ref. [49]. © 2012, American Chemical Society.

a further increase of temperature to 313°C (physisorption). However, the CNT(TEPA) were capable of both chemical and physical adsorption [50]. Ye et al. [49] suggested that the increased molecular flexibility of TEPA loaded in the CNT channels at higher temperatures enables more amino groups to be exposed and be accessible, causing a higher CO_2 adsorption capacity.

Dillon et al. [51] conducted experiments for the adsorption of CO_2 in a covalently attached PEI-functionalized SWCNT having a different molecular weight than the polymer (600–25,000 Da). The q_e increased with the molecular weight of the polymer (Table 21.1) and decreased with the increase in temperature (Fig. 21.13). Even though the thermal effect in CNT(TEPA) was not observed by Ye et al. in the presence of the aliphatic polyamine [49], it is suggested that the polymer chain limits the flexibility and hence the exposure of the amino groups.

The decrease in adsorption with higher temperatures may be attributed to the weakening of the van der Waals force between CO_2 and the adsorbent surface with a rise in temperature, also suggesting that the adsorption/desorption of CO_2 can be conducted via a temperature swing operation. In fact, the CNTs and CNT(APTS) are shown to be thermally stable in air up to 400°C, which is much higher than the reported desorption temperature of CO_2 from many CNT surfaces (~120°C). For example, Wilcox et al. conducted the desorption of CNT(APTS) by heating up to 120°C and passing a flow of pure N_2.

TABLE 21.1 Selected physical and spectroscopic data for branched PEI-SWNTs

PEI (M_W) (Da)	PEI:C_{SWNT}	Raman D:G	CO_2 absorption at 75°C (%)	CO_2 efficiency, mol of CO_2 mol^{-1} of PEI
600	1:155	0.448	4.1	0.0217
1,800	1:405	0.291	4.8	0.0827
10,000	1:1910	0.273	5.2	0.143
25,000	1:2065	0.273	7.2	0.159

Reprinted with permission from Ref. [51]. © 2008, American Chemical Society.

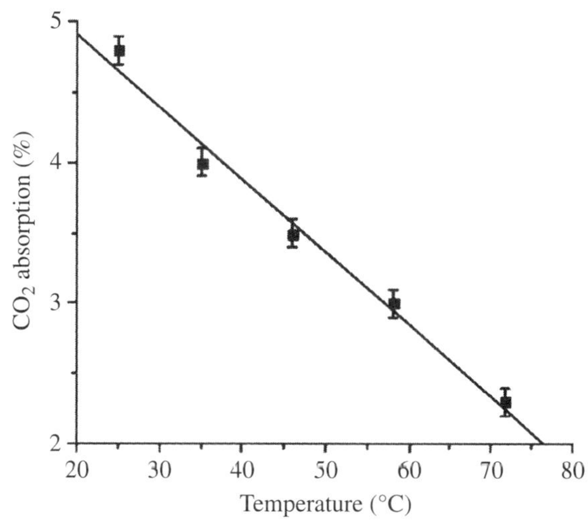

FIGURE 21.13 A plot of CO_2 adsorption efficiency as a function of temperature for PEI(10000)-SWNT. Reprinted with permission from Ref. [51]. © 2008, American Chemical Society.

21.3.1.2.3 Effect of Moisture Due to the presence of moisture in the flue gas streams, it is important to evaluate its effect on the adsorption capacity. Water vapor plays a significant role in the CO_2 adsorption reaction. It may inhibit the CO_2 adsorption capacity of hydrophilic materials [52] while increasing that of hydrophobic materials, for example, CNTs. Su et al. reported that the adsorption capacity (q_e) of CNT(APTS) increased from 85.7 to 108 mg g^{-1} sorbent as the water vapor increased from 0 to 2.2%, but decreased from 108 to 87.5 mg g^{-1} sorbent as the water vapor further increased from 2.2 to 8.9% (Fig. 21.14). Similar results were reported by Ye et al. [49] (Fig. 21.15). This could be explained as a further increase in water vapor led to competitive adsorption between CO_2 and H_2O at the same sorption sites.

This improvement in the adsorption of CO_2 in the presence of moisture may be due to several reasons. First, the surface of CNT(APTS) contains a primary amine (RNH_2), which can react with CO_2 and, in the absence of water, lead to the formation of a carbamate ion by reactions 21.3 and 21.4. The presence of water then regenerates amine molecules by reactions 21.5 and 21.6 [53].

$$CO_2 + 2RNH_2 \rightleftharpoons RNH_3^+ + RNHCOO^- \quad (21.3)$$

$$CO_2 + 2R_2NH \rightleftharpoons R_2NH_2^+ + R_2NCOO^- \quad (21.4)$$

$$RNHCOO^- + H_2O \rightleftharpoons RNH_2 + HCO_3^- \quad (21.5)$$

$$R_2NCOO^- + H_2O \rightleftharpoons R_2NH_2 + HCO_3^- \quad (21.6)$$

Second, the amine groups can also directly react with CO_2 and water to form a bicarbonate ion (HCO_3^-), as shown in reactions 21.7 and 21.8.

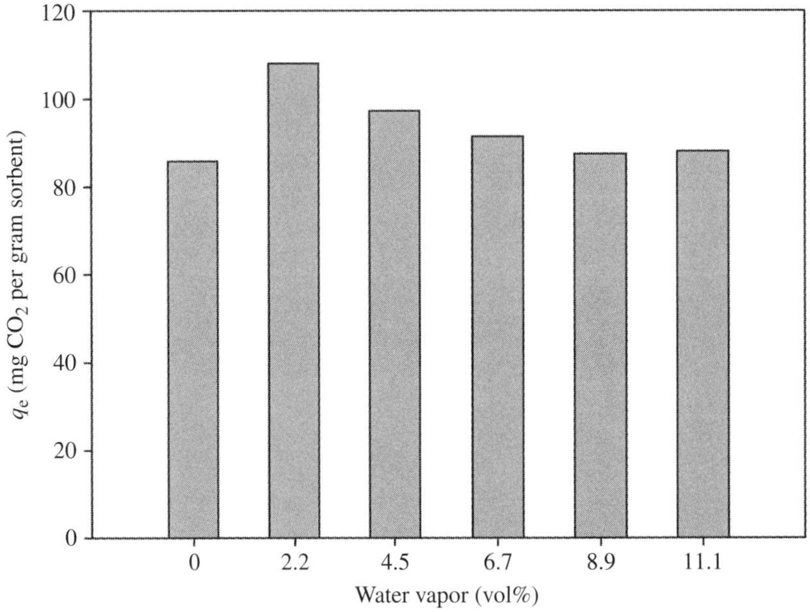

FIGURE 21.14 Effects of water vapor on 15% CO_2 adsorption on CNT(APTS) at 50°C. Courtesy from Environmental Protection Administration, ROC (Taiwan).

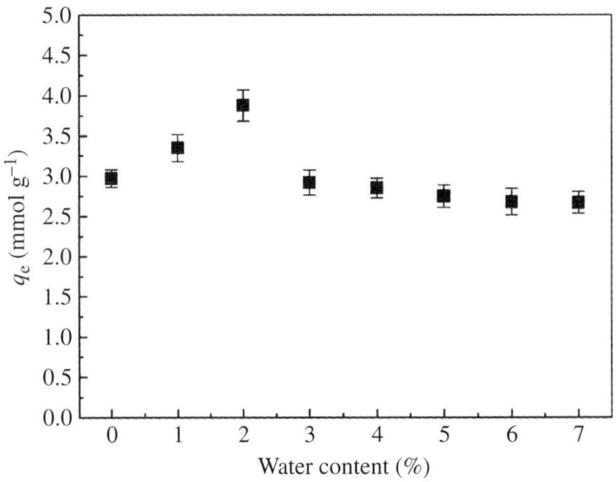

FIGURE 21.15 Effect of moisture on CO_2 adsorption of CT-30 at 298°C (gas flow rate, 50 ml min^{-1}; CO_2 concentration, 2.0 vol%). Reprinted with permission from Ref. [49]. © 2012, American Chemical Society.

$$CO_2 + RNH_2 + H_2O \rightleftharpoons RNH_3^+ + HCO_3^- \tag{21.7}$$

$$CO_2 + R_2NH + H_2O \rightleftharpoons R_2NH_2^+ + HCO_3^- \tag{21.8}$$

The decrease in q_e with the increase in water vapor percentage could be explained by the competitive adsorption between CO_2 molecules and water molecules at the same adsorption sites.

21.3.1.2.4 Effect of CO_2 Concentration As mentioned earlier, the typical concentration of CO_2 in a flue gas discharge is on the order of 5–50%. It is reported that with higher CO_2 concentrations, the adsorption capacity also increases (Fig. 21.16).

The CNT impregnated with TEPA, as reported by Ye et al., shows a CO_2 adsorption capacity of about 130.6 mg g^{-1} sorbent at a temperature and CO_2 concentration of 25°C and 2.0%, respectively (Fig. 21.17 and Table 21.2). Their work focuses on the requirement

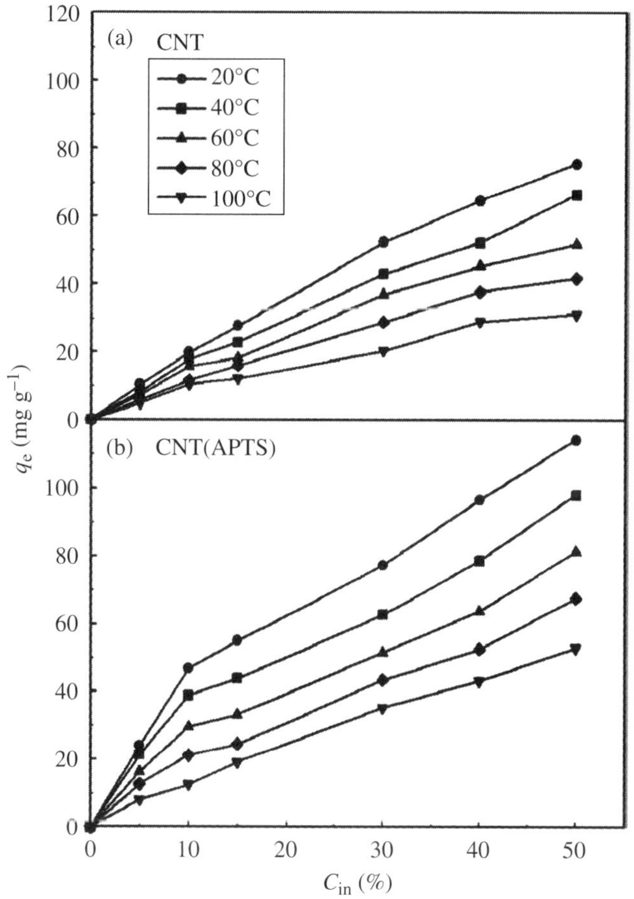

FIGURE 21.16 Adsorption isotherms of CO_2 via CNTs and CNT(APTS) at multiple temperatures. Reprinted with permission from Ref. [38]. © 2009, Elsevier.

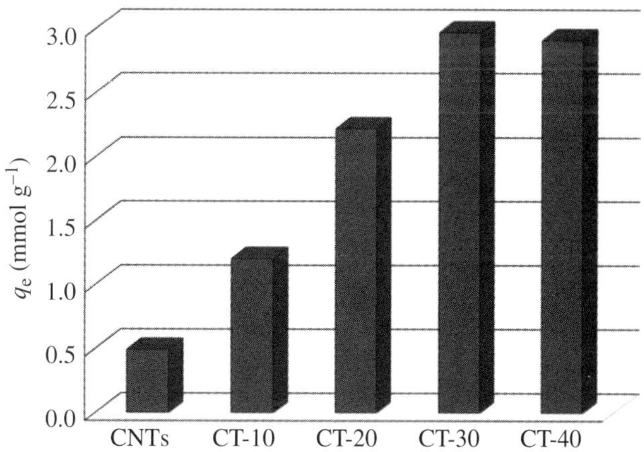

FIGURE 21.17 CO_2 adsorption capacity of CNTs and CNTs–TEPA-n at 25°C (gas flow rate, 50 ml min^{-1}; CO_2 concentration, 2.0 vol%) Reprinted with permission from Ref. [49]. © 2012, American Chemical Society.

for long durations of human operation in a confined space, which has made low CO_2 concentration capture a critical technology. They evaluated the adsorption capacity of CNT sorbents with four different TEPA loadings: 10 (CT-10), 20, 30, and 40%.

Considering the low concentration of CO_2 in the feed (2%), the adsorption capacity of CNT-30 of 2.97 mmol g^{-1} sorbent (equivalent to 130.68 mg CO_2 g^{-1} sorbent) is much higher than that of other amine-functionalized adsorbents reported in the literature under the same operational conditions [55–58].

TABLE 21.2 Comparisons of q_e via various raw and modified CNTs

Adsorbent	Modification	q_e (mg g^{-1})	Conditions	References
CNT	APTS	43.3	C_{in}:15% T: 20°C	[38]
		114.0	C_{in}: 50% T: 20°C	[38]
SWCNT	—	87.0	C_{in}: 99% T: 35°C	[39]
SWCNT	PEI	72.0	C_{in}: 100% T: 75°C	[51]
CNT	TEPA	130.6	C_{in}: 02% T: 25°C	[49]
MWCNT		514.8	C_{in}: 100% T: 25°C P: 11 bar	[54]

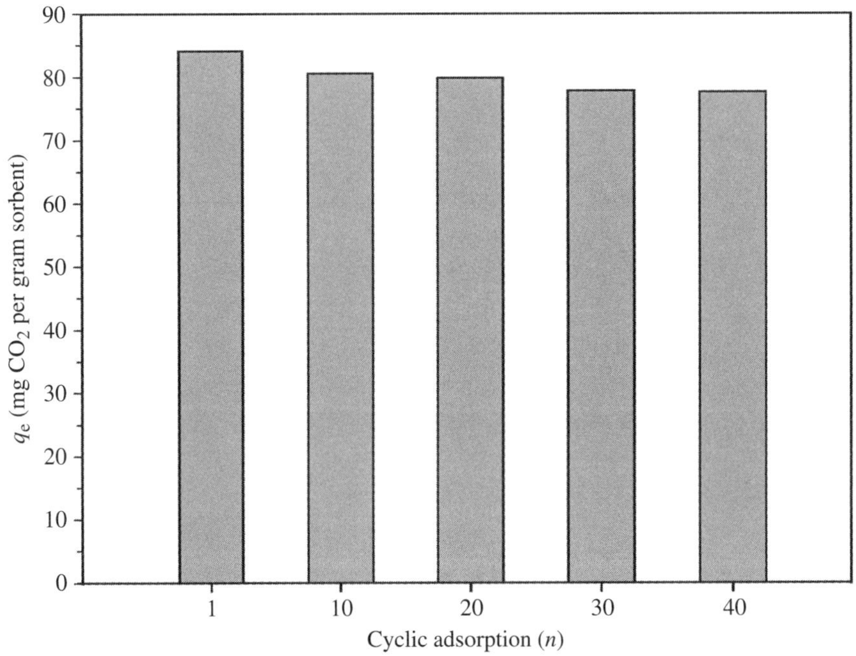

FIGURE 21.18 Cyclic CO_2 adsorption on CNT(APTS) via a thermal/vacuum desorption (C_{in}=15%, T_{ad}=50°C, $T_{desorption}$=130°C, P=0.03 atm). Adapted from http://ehs.epa.gov.tw/DocList/DownloadPublication/2722.

21.3.1.3 Adsorbent Regenerability The regenerability of reactivation of adsorbents is another important parameter for lengthy operations. If the regeneration of CNTs is not carried out, the practical use of CNTs in the field is not possible. If the regeneration of CNTs is performed, they should be successfully regenerated and reused through a number of cycles and then they can be possibly accepted as cost-effective adsorbents for capturing CO_2 from flue gas. A comparison of the energy consumption between CNT(APTS) and the traditional 30% MEA adsorption mixture reveals a remarkably lower energy input for the regeneration process, with 111.1 kJ mol^{-1} and 2.010 kJ mol^{-1} for CNT(APTS) and 30% MEA, respectively.

Some authors have reported results from the cyclic adsorption/desorption process and found an excellent behavior. The regeneration of CNT(APTS) via vacuum temperature swing adsorption (VTSA) achieves only 7.7% attrition after 40 cycles (Fig. 21.18).

Although CNT(APTS) shows good performance in CO_2 adsorption, the unit cost of the employed CNTs is approximately US\$5 g^{-1}, which is much higher than the typical cost of GAC at US\$1 kg^{-1}. Although the use of CNTs for large-scale CO_2 sequestration is impractical at this moment, a lightweight reusable absorption system offers a potential solution in closed systems such as spacecraft and orbiting space stations as well as in submarines.

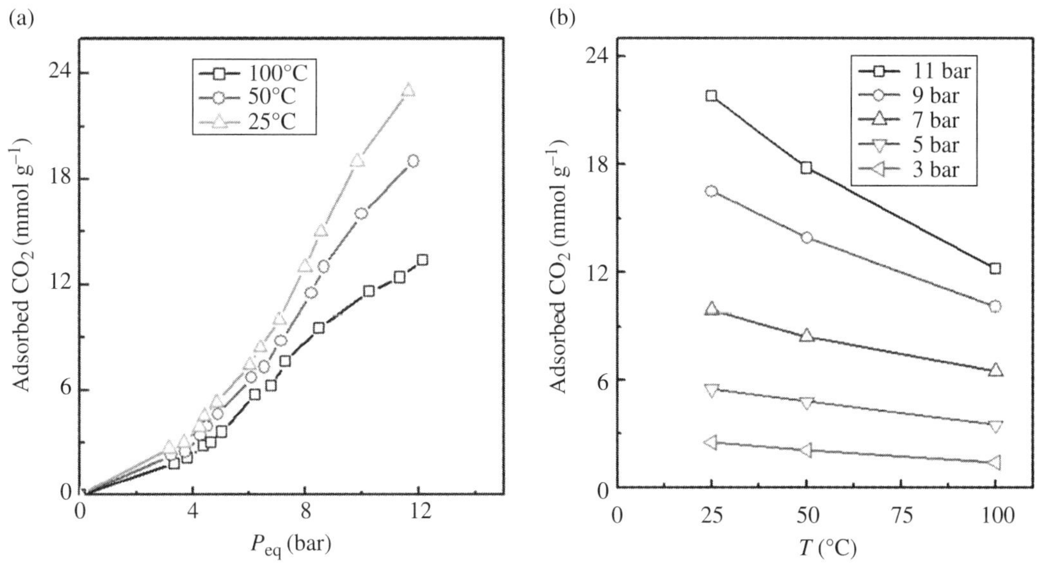

FIGURE 21.19 (a) CO_2 adsorption isotherms and (b) temperature variation of adsorption capacity for HEG. Reprinted with permission from Ref. [68]. © 2011, American Institute of Physics.

21.3.1.4 Adsorption of CO_2 in Graphene Graphene, one of the allotropes (such as CNTs, fullerene, and diamond) of elemental carbon, is a planar monolayer of carbon atoms arranged in a two-dimensional (2D) honeycomb lattice. It could be tailored chemically and structurally in numerous ways: by depositing metal atoms [59] or molecules [60] on top; by incorporating nitrogen or boron in its structure [61], or by using different substrates to modify the electronic structure [62]. Furthermore, the main resource for various electrical and optical applications is stemmed due to the interactions between graphene and various chemical molecules [63]. Graphene, which possesses exceptional mechanical, electrical, thermal, and optical properties, has numerous potential applications in areas like sensors, solar cells, transistors, and hydrogen storage [64, 65]. As mentioned in the previous section, CNTs have also been investigated as alternatives for CO_2 adsorption owing to their large surface area and high porosity [54]. But the cost of CNTs is still prohibitive. Graphene, as a new class of carbon nanomaterials, is found to be economical and has novel properties similar to CNTs.

The storage capacity of graphene for different gases has been suggested in theoretical studies and CO_2 adsorption capacity has been demonstrated at very low temperature (195°C), which does not have much practical implication [66, 67]. Mishra and Ramaprabhu [68] suggested the use of graphene oxide for CO_2 adsorption, which can be achieved by hydrogen-induced thermal exfoliation of graphene oxide at 200°C for possible large-scale production of graphene. They confirmed the physical adsorption of CO_2 in graphene oxide using Fourier transform infrared spectroscopy. These authors also reported that maximum adsorption capacities of 21.6, 18, and 12 mmol g^{-1} sorbent of hydrogen-exfoliated graphene (HEG) were observed at 11 bar pressure and temperatures of 25, 50, and 100°C, respectively (see Fig. 21.19). These values are higher than that of other adsorbents. The adsorption capacity of HEG for CO_2 was found to be higher than other carbon nanostructures (e.g., activated carbons (ACs), CNTs) and zeolites at the same pressure and temperature. Siriwardane et al. [69] have reported approximately 7 mmol g^{-1} sorbent of CO_2 adsorption in AC. Gensterblum et al. [70] have reported nearly 6 mmol g^{-1} sorbent of CO_2 adsorption in AC at 45°C and 11 bar pressure. Zhang et al. [71] have reported around 20% enhancement in CO_2 uptake by modifying the AC with nitrogen at room temperature and have shown around 16 mmol g^{-1} sorbent adsorption capacity at 11 bar. Cavenati et al. [72] have reported approximately 3.2 mmol g^{-1} of CO_2 adsorption in 13X zeolite at 12 bar pressure and room temperature. A high-pressure CO_2 adsorption study on different MOFs by Millward and Yaghi exhibits CO_2 adsorption capacity ranging from 2 to 8 mmol g^{-1} under similar conditions [30].

Inorganic nanoparticles, that is, metal [73] and metal oxides [74], have been successfully incorporated between graphene sheets, and their performance has been investigated in a wide variety of applications in the fields of catalysis [75], gas sorption [76], and electrode materials [77]. The incorporation of metal atoms and oxides over the surface area of graphene shows high CO_2 uptake volume. Meanwhile, nanoparticles incorporated between graphene sheets effectively prevent the aggregation of graphene–nanoparticle composites in the hybrid system [78, 79], and their high porosity could increase their performance as an adsorbent and in other applications.

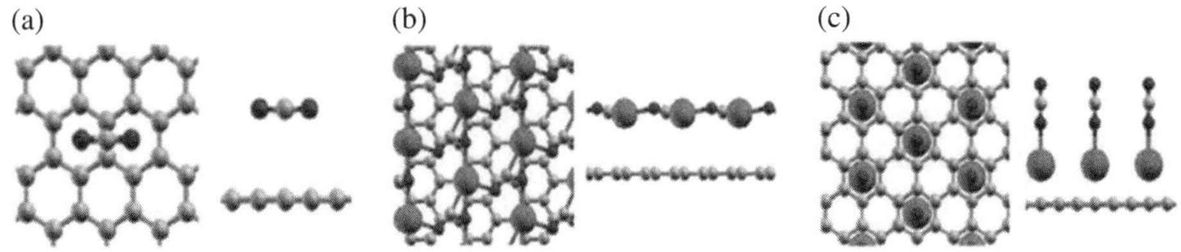

FIGURE 21.20 Carbon capture geometry optimized structure: (a) pristine graphene, (b) Ca/graphene at 12.5%, and (c) Ca/graphene at 16.67%. Reprinted with permission from Ref. [80]. © 2011, American Chemical Society.

TABLE 21.3 Gas-adsorption capacities of Ca/graphene as a function of Ca content

Ca content (%)	CO_2 uptake (mmol $CO_2\,g^{-1}$ sorbent)
0	3.48
0–5	3.05
5–15	7.34
15–20	8.91

Adapted from Ref. [80].

Cazorla et al. [80] found that Ca-decorated graphene (Fig. 21.20) possesses unusually large CO_2 uptake capacities in the range of 2.27–9.09 mmol $CO_2\,g^{-1}$ sorbent under a low-gas-pressure regime as a result of its unique topology and a strong interaction between the metal nanoparticles and CO_2 molecules. It is also reported that the interactive strength of the Ca metal and CO_2 molecule can be efficiently tuned as a function of the Ca loading content (Table 21.3).

Guo et al. [80] confidently proposed that these Ca metal/carbon-based nanomaterial composites exhibit characteristics of suitable adsorbents for CO_2 capture and sequestration applications.

Moreover, Carrillo et al. [81] studied the adsorption of CO_2 on a Ti–graphene system with high metal coverage using the density functional theory and molecular dynamics simulation. Positively charged Ti atoms on graphene surfaces attract negatively charged oxygen atoms toward the graphene surfaces. This force is stronger than the initial repulsion on the C atom by the Ti atoms. When a CO_2 molecule approaches the graphene surface, it cannot be linear any longer. The O atoms in a bent CO_2 molecule are under different force fields. Thus, one O atom traps an electronic charge from the Ti atoms of the upper plane and ends up bonded to four Ti atoms, resulting in the CO_2 molecule interacting very strongly with the Ti atoms of the upper plane.

Zhou et al. [82] prepared graphene–Mn_3O_4 (GMNO) materials with high porosity through hydrothermal reaction. They tested them to the adsorption of CO_2. Owing to the different contents of Mn_3O_4, the surface area values of the GMNO samples vary from about 140 to 685 $m^2\,g^{-1}$. It is suggested that the adsorption capacity for the GMNO samples (Fig. 21.21) comes from the surface area and the basic sites of the Mn_3O_4 nanoparticles.

21.3.1.4.1 Graphene Membranes Inorganic membranes are much more stable at high temperatures and in the presence of chemicals than polymeric membranes [83]. Zeolite and silica porous materials can act as molecular sieves, separating gas (e.g., CO_2, N_2, H_2) molecules by their respective sizes [84]. However, because the sizes of CO_2, N_2, H_2, and other combustion-specific gases are very similar, membrane porocity must be controlled on a scale comparable to the size differences among these gas molecules. The separation of gas mixtures using nanoporous materials is an emerging field of research with many potential applications including fuel cells, batteries, gas sensors, and gas purification. The materials that are currently being investigated for these applications include organic polymer-based membranes, porous carbon, and inorganic membranes made of ceramics, metals, or glass, and graphene. Graphene is impermeable to all gases, even to helium, because of the electron density of its aromatic rings. In order to create a membrane, pores must be created synthetically. Nano-sized pores have been artificially produced to obtain functional gas membranes using graphene. Post-treatment techniques like electron–hole drilling are suitable for pores in the nanometer range, and smaller pores can be created by catalyzed polymerization at high temperatures.

Unfortunately, the gas-separation nanomembrane technology is still not at this stage, but active research continues in this direction.

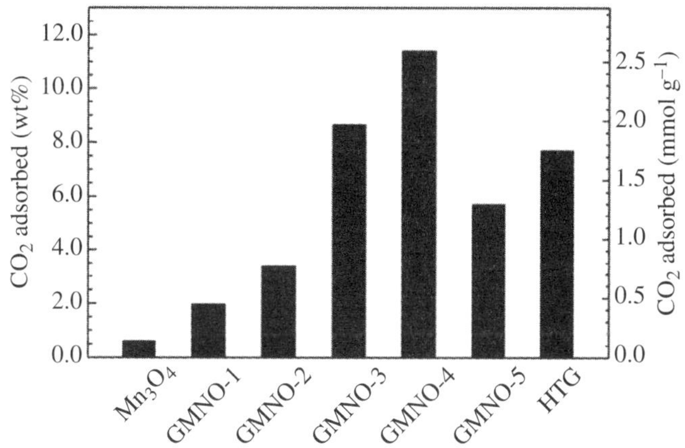

FIGURE 21.21 CO_2 adsorption capacity comparison of hydrothermal-reduced graphene (HTG), Mn_3O_4, and GMNO samples. Reprinted with permission from Ref. [82]. © 2012, Springer.

FIGURE 21.22 Illustration of a single molecular-sized pore in a graphene membrane separating CO_2 from nitrogen. Image credit: Zhangmin Huang, University of Colorado, Boulder.

It is more difficult to separate CO_2 from a mixture of CO_2 and N_2 (two principal components in a flue gas) because they have similar kinetic diameters and neither has an electric dipole moment (see Fig. 21.22). It was demonstrated that the chemical functionalization of graphene can also affect the separation performance of porous graphene membranes. Shan et al. [85] modeled the effect of the chemical functionalization of the nanopores in a graphene sheet on the separation of N_2 and CO_2. The pore rim was simulated with modified structures of all carbons (unmodified pore), all-H passivated, N–H modified, and all-N modified. The chemical functionalization of the pore rim plays an important role in the selectivity of CO_2 owing to an enhanced electrostatic interaction. The all-N functionalized pore shows the best CO_2 selectivity ($sCO_2/N_2 = 11$).

REFERENCES

[1] Thekaekara MP. Solar irradiance: total and spectral and its possible variations. Appl Optics 1976;15:915–920.

[2] America's Climate Choices. Washington, DC: Committee on America's Climate Choices; National Research Council; National Academies Press; 2011.

[3] Trans P, Keeling R. *Trends in Atmospheric Carbon Dioxide*. Earth System Research Laboratory, Global Monitoring Division. Available at http://www.esrl.noaa.gov/gmd/ccgg/trends/mlo.html. Accessed March 12, 2013.

[4] Data in Gapminder World. Available at http://www.gapminder.org/data/. Accessed April 16, 2013.

[5] Inventory of U.S. greenhouse gas emissions and sinks: 1990–2010. EPA 430-R-12-001 [Online]. Available at http://www.iea.org/co2highlights/co2highlights.pdf. Accessed April 22, 2013.

[6] CO_2 emissions from fuel combustion 2012 edition highlights. Available at www.iea.org/co2highlights. Accessed June 3, 2014.

[7] Longoria P. *Alternative Fuels*. NorthWest Missouri State University. Missouri Center of Excellence for Plants Biologics; February 2006.

[8] Ryle GJA, Powell CE, Tewson V. Effect of elevated CO_2 on the photosynthesis, respiration and growth of perennial ryegrass. J Exp Bot 1992;43:811–818.

[9] Stewart G. Dry cleaning with liquid carbon dioxide. In: DeSimone JM, Tumas W, editors. *Green Chemistry Using Liquid and Supercritical Carbon Dioxide*. New York: Oxford University Press; 2003. p 215–227.

[10] Walsh MP, Lake LW, editors. *Generalized Approach to Primary Hydrocarbon Recovery*. Amsterdam: Elsevier; 2003. Handbook of petroleum exploration and production, 4.

[11] Riddiford FA, Tourqui A, Bishop CD, Smith TB. A cleaner development: the In Salah Gas Project, Algeria. In: Gale J, Kaya Y, editors. Proceedings of the 6th International Conference on Greenhouse Gas Control Technologies (GHGT-6); October 1–4, 2002; Kyoto, Japan; Vol. I 2003. p 601–606.

[12] Angus S, Armstrong B, de Reuck KM. International Thermodynamic Tables of the Fluid State. Volume 3, Carbon dioxide. London: Pergamon Press; 1973. p 266–359. IUPAC chemical data series.

[13] Metz BO, Davidson HC, de Coninck M, Loos M, Meyer LA. 2005. IPCC special report on carbon dioxide capture and storage [Online]. Available at http://www.ipcc.ch/pdf/special-reports/srccs/srccs_wholereport.pdf/. Accessed April 20, 2013.

[14] Bachu S. CO_2 storage in geological media: role, means, status and barriers to deployment. Prog Energ Combust 2008;34:254–273.

[15] Lackner KS, Wendt CH, Butt DP, Joyce EL, Sharp DH. Carbon dioxide disposal in carbonate minerals. Energy 1995;20:1153–1170.

[16] Dunsmore HE. A geological perspective on global warming and the possibility of carbon dioxide removal as calcium carbonate mineral. Energ Convers Manage 1992;33:565–572.

[17] Precombustion capture. Available at http://www.vattenfall.com/en/ccs/precombustion.htm?WT.ac=search_success. Accessed April 23, 2013.

[18] Kohl A, Nielsen R. *Gas Purification*. 5th ed. Houston, TX: Gulf Publishing Company; 1997.

[19] Korens N, Simbeck DR, Wilhelm DJ. Report prepared for U.S. Department of Energy. Process screening analysis of alternatives gas treating and sulfur removal for gasification. Pittsburg, PA: National Energy Technology Laboratory; 2002.

[20] Strait R, Nagvekar M. Carbon dioxide capture and storage in the nitrogen and syngas industries [Online]. 2010, KBR Technology Internal Newspaper, 303. Available at http://www.kbr.com/Newsroom/Publications/Articles/Carbon-Dioxide-Capture-and-Storage-in-the-Nitrogen-Syngas-Industries.pdf. Accessed June 2, 2014.

[21] Rubin ES, Keith DW, Gilboy CF. 2005. Comparative assessments of fossil fuel power plants with CO_2 capture and storage [Online]. Available at http://repository.cmu.edu/cgi/viewcontent.cgi?article=1082&context=epp. Accessed April 30, 2013.

[22] Postcombustion capture picture. Available at http://www.vattenfall.com/en/ccs/postcombustion.htm. Accessed April 23, 2013.

[23] Oxyfuel combustion capture. Available at http://www.vattenfall.com/en/ccs/oxyfuel-combustion.htm. Accessed April 23, 2013.

[24] Haszeldine RS. Carbon capture and storage: how green can black be? Science 2009;325:1647–1652.

[25] Ramezan M, Skone TJ, Nsakala N, Liljedahl, GN. 2007. Carbon dioxide capture from existing coal-fired power plants [Online]. U. S. Department of Energy, National Energy Technology Laboratory. DOE/NETL-401/110907. Available at http://www.netl.doe.gov/File%20Library/Research/Energy%20Analysis/Coal/CO2-Retrofit-From-Existing-Plants-Revised-November-2007.pdf. Accessed June 2, 2014.

[26] Choi S, Drese JH, Jones CW. Adsorbent materials for carbon dioxide capture from large anthropogenic point sources. ChemSusChem 2009;2:796–854.

[27] Sjostrom S, Krutka H. Evaluation of solid sorbents as a retrofit technology for CO_2 capture. Fuel 2010;89:1298–1306.

[28] Hicks JC, Drese JH, Fauth DJ, Gray ML, Qi G. Designing adsorbents for CO_2 capture from flue gas-hyperbranched aminosilicas capable of capturing CO_2 reversibly. J Am Chem Soc 2008;130:2902–2903.

[29] Hedin N, Chen LJ, Laaksonen A. Sorbents for CO_2 capture from flue gas-aspects from materials and theoretical chemistry. Nanoscale 2010;2:1819–1841.

[30] Millward AR, Yaghi OM. Metal–organic frameworks with exceptionally high capacity for storage of carbon dioxide at room temperature. J Am Chem Soc 2005;127:17998–17999.

[31] Kishan MR, Tian J, Thallapally PK, Fernandez CA, Dalgarno SJ, Warren JE. Flexible metal–organic supramolecular isomers for gas separation. Chem Commun 2010;46:538–540.

[32] Hsu SC, Lu C, Su F, Zeng W, Chen W. Thermodynamics and regeneration studies of CO_2 adsorption on multiwalled carbon nanotubes. Chem Eng Sci 2010;65:1354–1361.

[33] Environmental Protection Agency (EPA). *Nanotechnology white paper*. EPA 100/B-07/001 [Online]. Washington, DC: EPA; 2007. Available at http://nepis.epa.gov/Exe/ZyPDF.cgi/60000EHU.PDF?Dockey=60000EHU.PDF Accessed June 2, 2014.

[34] Feng X, Fryxell GE, Wang Q, Kim AY, Liu J, Kemner KM. Functionalized monolayers on ordered mesoporous supports. Science 1997;276:923.

[35] White CM, Strazisar BR, Granite EJ, Hoffman JS, Pennline HW. Separation and capture of CO_2 from large stationary sources and sequestration in geological formations-coalbeds and deep saline aquifers. J Air Waste Manag Assoc 2003;53:645–715.

[36] Aaron D, Tsouris D. Separation of CO_2 from flue gas: a review. Separ Sci Technol 2005;40:321–348.

[37] Leela Mohana Reddy A, Ramaprabhu S. Hydrogen adsorption properties of single-walled carbon nanotube—nanocrystalline platinum composites. Int J Hydrogen Energ 2008;33:1028–1034.

[38] Su F, Lu C, Cnen W, Bai H, Hwang JF. Capture of CO_2 from flue gas via multiwalled carbon nanotubes. Sci Total Environ 2009;407:3017–3023.

[39] Cinke M, Li J, Bauschlicher CW Jr, Ricca A, Meyyappan M. CO_2 adsorption in single walled carbon nanotubes. Chem Phys Lett 2003;376:761–766.

[40] Agnihotri S, Rood MJ, Rostam-Abadi M. Adsorption equilibrium of organic vapors on single-walled carbon nanotubes. Carbon 2005;43:2379–2388.

[41] Su F, Lu C. Adsorption kinetic, thermodynamic and desorption studies of natural organic matters by thermally treated multiwalled carbon nanotubes. J Environ Sci Health A 2007;42:1543–1552.

[42] Lee ZH, Lee KT, Bhatia S, Mohamed AR. Post-combustion carbon dioxide capture: evolution towards utilization of nanomaterials. Renew Sust Energ Rev 2012;16:2599–2609.

[43] Orr FM Jr. CO_2 capture and storage: are we ready? Energy Environ Sci 2009;2:449–458.

[44] Chowdhury FA, Okabe H, Shimizu S, Onoda M, Fujioka Y. Development of novel tertiary amine absorbents for CO_2 capture. Energy Procedia 2009;1:1241–1248.

[45] Hirsch A. Functionalization of single-walled carbon nanotubes. Angew Chem Int Ed 2002;41:1853–1859.

[46] Peng HQ, Alemany LB, Margrave JL, Khabashesku VN. Sidewall carboxylic acid functionalization of single-walled carbon nanotubes. J Am Chem Soc 2003;125:15174–15182.

[47] Lu C, Bai H, Wu B, Su F, Hwang JF. Comparative study of CO_2 capture by carbon nanotubes, activated carbons, and zeolites. Energ Fuel 2008;22:3050–3056.

[48] Chaffee AL, Knowles GP, Liang Z, Zhang J, Xiao P, Webley PA. CO_2 capture by adsorption: materials and process development. Int J Greenhouse Gas Control 2007;1 (1):11–18.

[49] Ye Q, Jiang J, Wang C, Liu Y, Pan H, Shi Y. Adsorption of low-concentration carbon dioxide on amine-modified carbon nanotubes at ambient temperature. Energ Fuel 2012;26:2497–2504.

[50] Hsu S, Lu C. Modification of single-walled carbon nanotubes for enhancing isopropyl alcohol vapor adsorption from air streams. Separ Sci Technol 2007;42:2751–2766.

[51] Dillon EP, Crouse CA, Barron AR. Synthesis, characterization, and carbon dioxide adsorption of covalently attached polyethyleneimine-functionalized single-wall carbon nanotubes. ACS Nano 2008;2:156–164.

[52] Jadhav P, Chatti R, Biniwale R, Labhsetwar N, Devotta S, Rayalu S. Monoethanol amine modified zeolite 13X for CO_2 adsorption at different temperatures. Energ Fuel 2007;21:3555–3559.

[53] Xu XC, Song CS, Miller BG, Scaroni AW. Influence of moisture on CO_2 separation from gas mixture by a nanoporous adsorbent based on polyethylenimine-modified molecular sieve MCM-41. Ind Eng Chem Res 2005;44:8113–8119.

[54] Mishra AK, Ramaprabhu S. Nano magnetite decorated multiwalled carbon nanotubes: a robust nanomaterial for enhanced carbon dioxide adsorption. Energy Environ Sci 2011;4:889.

[55] Chang ACC, Chuang SSC, Gray MM, Soong Y. In-situ infrared study of CO_2 adsorption on SBA-15 grafted with γ-(aminopropyl) triethoxysilane. Energ Fuel 2003;17:468–473.

[56] Huang HY, Yang RT, Chinn D, Munson CL. Amine-grafted MCM-48 and silica xerogel as superior sorbents for acidic gas removal from natural gas. Ind Eng Chem Res 2003;42:2427–2433.

[57] Serna-Guerrero R, Da'na E, Sayari A. New insights into the interactions of CO_2 with amine-functionalized silica. Ind Eng Chem Res 2008;47:9406–9412.

[58] Belmabkhout Y, Serna-Guerrero R, Sayari A. Adsorption of CO_2 containing gas mixtures over amine-bearing pore-expanded MCM-41 silica: application for gas purification. Ind Eng Chem Res 2009;49:359–365.

[59] Uchoa B, Lin CY, Castro Neto AH. Tailoring graphene with metals on top. Phys Rev B 2008;77:035420.

[60] Wehling TO, Novoselov KS, Morozov SV, Vdovin EE, Katsnelson MI, Geim AK, Lichtenstein AI. Molecular doping of graphene. Nano Lett 2008;8:173.

[61] Martins TB, Miwa RH, da Silva AJR, Fazzio A. Electronic and transport properties of boron-doped graphene nanoribbons. Phys Rev Lett 2007;98:196803.

[62] Das A, Chakraborth B, Sood AK. Raman spectroscopy of graphene on different substrates and influence of defects. Mater Sci B 2008;31:579.

[63] AlZahrani AZ. First-principles study on the structural and electronic properties of graphene upon benzene and naphthalene adsorption. Appl Surf Sci 2010;257:807.

[64] Liu ZK, Li JH, Sun ZH, Tai GA, Lau SP, Yan F. The application of highly doped single-layer graphene as the top electrodes of semitransparent organic solar cells. ACS Nano 2012;6:810–818.

[65] Hui YY, Tai GA, Sun ZH, Xu ZH, Wang N, Yan F, Lau SP. n- and p-Type modulation of ZnO nanomesh coated graphene field effect transistors. Nanoscale 2012;4:3118–3122.

[66] Ghosh A, Subrahmanyam KS, Krishna KS, Datta S, Govindaraj A, Pati SK, Rao CNR. Uptake of H_2 and CO_2 by graphene. J Phys Chem C 2008;112:15704.

[67] Wang L, Lee K, Sun YY, Lucking M, Chen Z, Zhao JJ, Zhang SB. Graphene oxide as an ideal substrate for hydrogen storage. ACS Nano 2009;3:2995.

[68] Mishra AK, Ramaprabhu S. Carbon dioxide adsorption in graphene sheets. AIP Adv 2011;1:032152.

[69] Siriwardane RV, Shen M, Fisher EP, Poston J. Adsorption of CO_2 on molecular sieves and activated carbon. Energ Fuel 2001;15:279.

[70] Gensterblum Y, van Hemert P, Billemont P, Busch A, Charrière D, Li D, Krooss BM, deWeireld G, Prinza D, Wolf KHAA. European inter-laboratory comparison of high pressure CO_2 sorption isotherms. I: activated carbon. Carbon 2009;47:2958.

[71] Zhang Z, Xu M, Wang H, Li Z. Enhancement of CO_2 adsorption on high surface area activated carbon modified by N_2, H_2 and ammonia. Chem Eng J 2010;160:571.

[72] Cavenati S, Grande CA, Rodrigues AE. Adsorption equilibrium of methane, carbon dioxide, and nitrogen on zeolite 13X at high pressures. Chem Eng J 2004;49:1095.

[73] Si Y, Samulski E. Exfoliated graphene separated by platinum nanoparticles. Chem Mater 2008;20:6792–6797.

[74] Williams G, Seger B, Kamat P. TiO_2 UV-assisted photocatalytic reduction of graphene oxide. ACS Nano 2008;2:1487–1491.

[75] Scheuermann G, Rumi L, Steurer P, Bannwarth W, Mülhaupt R. Palladium nanoparticles on graphite oxide and its functionalized graphene derivatives as highly active catalysts for the Suzuki-Miyaura coupling reaction. J Am Chem Soc 2009;131:8262–8270.

[76] Zhou D, Han BH. Graphene-based nanoporous materials assembled by mediation of polyoxometalate nanoparticles. Adv Funct Mater 2010;20:2717–2722.

[77] Paek S, Yoo E, Honma I. Enhanced cyclic performance and lithium storage capacity of SnO_2/graphene nanoporous electrodes with three-dimensionally delaminated flexible structure. Nano Lett 2009;9:72–75.

[78] Zhu Y, Murali S, Cai W, Li X, Suk JW, Potts JR, Ruoff RS. Graphene and graphene oxide: synthesis, properties, and applications. Adv Mater 2010;22:3906–3924.

[79] Wei D, Liu Y. Controllable synthesis of graphene and its applications. Adv Mater 2010;22:3225–3241.

[80] Cazorla C, Shevlin SA, Guo ZX. Calcium-based functionalization of carbon materials for CO_2 capture: a first-principles computational study. J Phys Chem C 2011;115:10990.

[81] Carrillo I, Rangel E, Magana LF. Adsorption of carbon dioxide and methane on graphene with a high titanium coverage. Carbon 2009;47:2758.

[82] Zhou D, Liu Q, Cheng QY, Cheng QY, Zhao YC, Cui Y, Wang T, Han B. Graphene-manganese oxide hybrid porous material and its application in carbon dioxide adsorption. Chin Sci Bull 2012;57:3059–3064.

[83] Kai T, Kazama S, Fujioka Y. Development of cesium-incorporated carbon membranes for CO_2 separation under humid conditions. J Memb Sci 2009;342:14–21.

[84] Fujioka Y, Yamada K, Kazama S, Yogo K, Kai T, Uoe K. Development of innovative gas separation membranes through sub-nanoscale materials control. Research Institute of Innovative Technology for the Earth (RITE) [Online]: Tokyo, Japan; 2009. Available at http://gcep.stanford.edu/pdfs/E2aAjl5DtYBmHJN7TWaiBg/2.4.5.Fujioka_07_web.pdf Accessed June 2, 2014.

[85] Shan M, Xue Q, Jing N, Ling C, Zhang T, Yan Z, Zheng J. Influence of chemical functionalization on the CO_2/N_2 separation performance of porous graphene membranes. Nanoscale 2012;4:5477–5482.

PART VIII

INTELLIGENT NANOMATERIALS

22

DEVELOPMENT OF INTELLIGENT NANOMATERIALS AS A STRATEGY TO SOLVE ENVIRONMENTAL PROBLEMS

JOSE RUBEN MORONES-RAMÍREZ

Chemical Engineering Department, School of Chemistry, Universidad Autónoma de Nuevo León, Pedro de Alba, S/N. Ciudad Universitaria, San Nicolás de los Garza, Mexico

In the twenty-first century, our global society will face an important challenge: it will have to generate enough resources for the increasing world population to subsist, while maintaining homeostasis in our earth's numerous ecosystems. In order to face such a complex challenge, there will have to be significant breakthroughs and advances in the fields of environmental pollution, as well as sustainable production of energy, water, and food. Nanotechnology, specifically in the areas of material synthesis and development, will play an essential role in allowing us to accomplish such difficult tasks. This chapter will address the synthesis, characterization, and environmental applications of "intelligent" materials, which have earned their name due to their capability of dynamically responding to either external or internal stimuli. This characteristic allows them to be used in synthetic systems that pretend to improve or at least mimic the properties within a biological system. The materials we will discuss here are metallic and nonmetallic nanostructures and hybrid polymer–metal nanomaterials. Research in the engineering and design of all the classified materials mentioned is imperative to stimulate the creation of new technologies and devices that will raise the standard of living of the world population while being environment-friendly at the same time.

22.1 ENVIRONMENTALLY RESPONSIVE MATERIALS AS MECHANICAL ACTUATORS

The fabrication of environmentally responsive mechanical actuators—valves being among the ones most studied—has been possible with the use of environmentally responsive polymers [1–10]. These materials are capable of undergoing conformational changes and phase transitions upon local or external stimuli, and are therefore termed "smart" or "intelligent" materials [11, 12]. As a result of their exhibited mechanical responses, they are excellent candidates for incorporation into microfluidic systems. These materials have therefore successfully controlled flows [2–5, 13, 14] and delivered chemical entities [2, 15, 16] in response to different environmental stimuli that turn actuators from "on" to "off." They have potential applications in generating chemical gradients, which will translate into achieving a better spatiotemporal control of delivered reagents. In the following sections, we proceed to describe some of the different responsive materials that have been used in the literature to create such mechanical actuators.

Figure 22.1 shows how responsive polymers are generally categorized as locally triggered or bulk-triggered. However, among the most commonly studied responsive materials, there are four major types [11]: (1) bulk-triggered temperature-sensitive polymers, such as the widely used poly (*N*-isopropyl acrylamide) (PNIPAAm) [17] and elastin-like polypeptides [18, 19]; (2) bulk-triggered pH-sensitive polymers [2, 20, 21], such as the commonly used acrylic acid [22], which contains carboxylate groups; (3) other, not so frequently used polymers with bulk-triggered sensitivities to a wide range of physical stimuli such as pressure

Nanomaterials for Environmental Protection, First Edition. Edited by Boris I. Kharisov, Oxana V. Kharissova, and H. V. Rasika Dias.

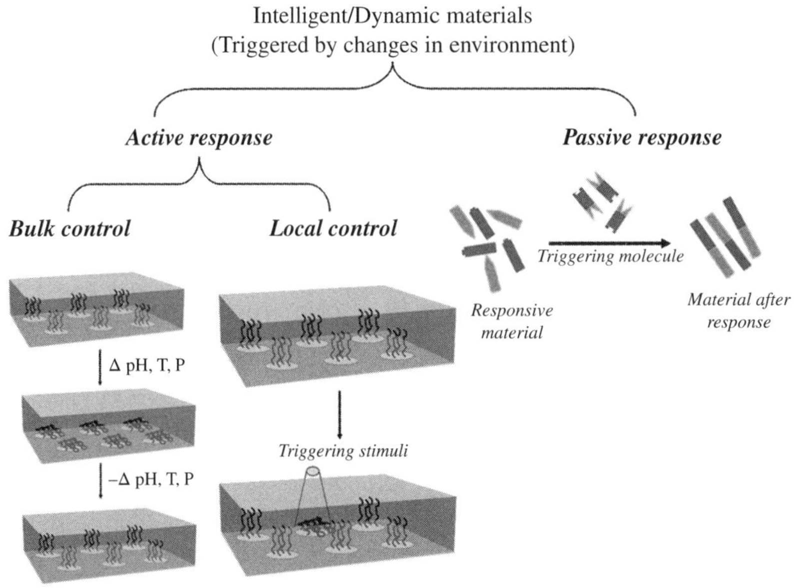

FIGURE 22.1 Environmentally responsive materials can be categorized into three different types: passive responsive materials, which respond to the presence or absence of a specific molecule; active responsive materials, which can be reversibly triggered externally in response to a stimulus, and among these active materials there are two categories. These are materials that respond to bulk changes in their surroundings, such as pH and temperature, and materials that respond to local changes, such as light.

and applied voltage [23, 24]; and (4) locally triggered materials responsive to light [25–31] or the presence of a specific molecule [32, 33]. In addition, several of these response polymers can be coupled or modified with specific functional chemical groups to exhibit combinatory responses, such as designing a copolymeric material with both pH and temperature sensitivity.

22.1.1 Bulk-Triggered Responsive Materials (pH and Temperature)

Bulk-triggered responsive materials have found a series of applications in the delivery and precise release of chemical reagents, such as drugs and growth factors, both *in vivo* and *in vitro*. Some examples include hydrogels capable of responding to different chemical and physical external stimuli (such as pH and temperature) that release a drug (protein or peptide) at specific sites in the body [34]. In addition, the synthesis of systems with higher complexity has been designed to respond to specific enzymes [35, 36] and chemical entities. This allows sophisticated cellular targeting and delivery mechanisms, such as the hydrogels employed as insulin delivery systems, which are capable of responding to the presence of glucose [37]. These and other cleverly designed delivery devices have improved therapies for a wide variety of diseases [37–39].

Additional applications, which this work is most interested in, are those in which these materials are used to deliver reagents into *in vitro* microfluidic systems. The most successful attempts to fabricate valves have been accomplished using these bulk-triggered responsive materials. These systems present several advantages, such as their capacity to externally control flows (redirecting) and to introduce chemical reagents with fast response times by simply changing the bulk conditions of the system, such as pH [40], pneumatic pressure [41], temperature [42], or chemical entities like glucose [16]. However, their design poses a main problem for their use in cell culture systems, that is, their driven mechanical action is triggered by changing the bulk conditions of the environment. Changing the bulk conditions of the system causes cells to be exposed to hostile environments not apt for their survival. Such challenges have been recently addressed by developing materials that present localized responses to specific light wavelengths, [13, 43] or by implementing nanofabricated chemical reagent sources containing chemical reagents that are dispensed to the system through pressure [44] or simple diffusion [45].

22.1.2 Locally Responsive Materials (Light)

Optically responsive materials offer great advantages and present appropriate characteristics that enable them to be used in mechanical actuators for controlling the flow and delivery of chemical reagents in cell culture systems. Although other cell culture designs have been proposed for the delivery of chemical reagents without changing the bulk conditions of the system, such as pressurized reagent sources [44], they require complicated circuitry and do not offer the capability of stimulating highly specific sites within the system as easily as it is done using an external light source (laser) [14].

Light-responsive materials are typically triggered by conformational changes of certain dye molecules in the polymer backbones or as pendent groups [29, 46–50], or by the local generation of heat due to dyes linked either to a thermally responsive polymer or to a hydrogel [27, 30]. However, dyes are usually chemically reactive, are susceptible to bleaching, and their responses are generally very slow [14, 25, 35, 51]. Therefore, research has been focused on developing alternative materials, such as nanocomposites that couple two or more different materials and synergize their individual physical properties.

22.1.3 Nanocomposite Materials (Thermally Responsive Polymers and Metallic Nanoparticles)

Developing hybrid or nanocomposite materials that couple metallic nanoparticles to responsive polymers is a field of research that has received a lot of attention due to the extensive research done in the synthesis and design of each of the individual components (responsive polymers and metallic nanoparticles). Specifically, coupling thermally responsive polymers to metallic nanoparticles has allowed the development of materials that can be triggered both optically and thermally since the incorporated metallic nanoparticles act as amplifiers and converters of light of a specific wavelength into heat [52]. There are two major advantages of these materials over the traditional optical responsive polymers (using dyes). The first is the higher chemical stability of both the nanoparticles and the polymer, and the second is the fast optical response of the nanocomposite, since it is triggered by a coupled mechanism between the fast thermal response of the polymer and the capability of the metallic nanoparticles to convert absorbed light into heat, taking advantage of the rapid heat conduction at small scales. These optothermally responsive nanocomposites have been synthesized as temperature-responsive PNIPAM hydrogels that have entrapped metallic particles within their structure [13, 53–55] and have found applications to control flow in microfluidic devices. Synthesis of these systems presents an advantage since it is simple to physically entrap nanoparticles within the hydrogel; however, the downside is the long time taken to trigger the response of these systems and their requirement for high-power lasers. It is therefore the aim of this work to develop a synthetic route that couples polymer chains to metallic nanoparticles, and thus obtain stable nanocomposites. These can be further used as "on–off" valves to control flow in microfluidic systems using low-power lasers and faster switching times.

As a result, in the remainder of this chapter, we will focus on the characteristic of each of the components that play a role in the synthesis and response of these optothermally responsive materials.

22.2 THERMALLY RESPONSIVE POLYMERS (PNIPAM)

PNIPAM is one of the most well-studied thermally responsive materials in the literature. PNIPAM is a polymer that contains both hydrophobic and hydrophilic chemical groups as illustrated in Figure 22.2. Due to these chemical characteristics, it has been determined to be a temperature-responsive polymer with a lower critical solubility temperature (LCST) of around 32°C in water. Such behavior of solubility in water can be explained thermodynamically with the Gibbs free energy equation of the PNIPAM/water system: $\Delta G_{mix} = \Delta H_{mix} - T\Delta S_{mix}$. Since a negative Gibbs free energy determines spontaneous processes, and PNIPAM has been determined to have an LCST, we must assume that both ΔH_{mix} and ΔS_{mix} must be negative. In this case, as the T increases, the entropy term becomes more positive and eventually reaches the enthalpy term to switch the Gibbs energy from a negative value that determines a miscible system to a positive value that determines an immiscible system.

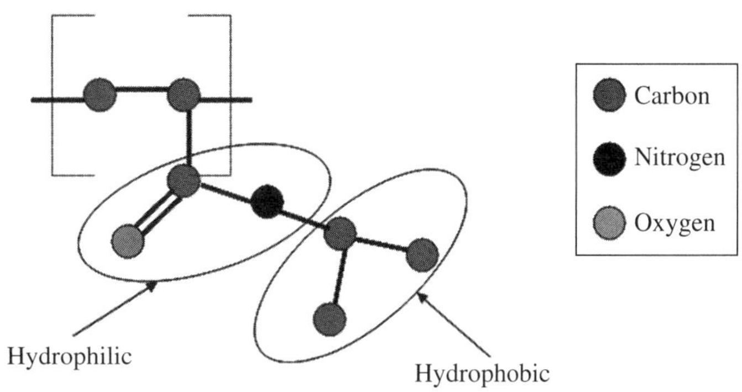

FIGURE 22.2 Chemical structure of poly-*N*-isopropylacrylamide. The hydrophilic amide groups are capable of forming hydrogen bonds at temperatures below 32°C. The hydrophobic isopropyl group becomes dominant at 32°C and causes the polymer to phase separately from man's aqueous environment. This gives the polymer its responsive characteristics.

This phenomenon can also be explained from the hydrophilic/hydrophobic chemical nature of PNIPAM. At lower temperatures, it is water-soluble and hydrophilic. When temperature increases, PNIPAM is more hydrophobic due to the breaking of the hydrogen bonds between water and the NH and CO groups, and so there is a tendency for the polymer to associate with itself rather than with the solvent. This leads to a phase separation, forming two pure phases. This translates as a decrease in entropy and a positive increase in the Gibbs energy.

The conformation of the polymer chains above the LCST is a hydrated and expanded state, and below the LCST, the chains are dehydrated and aggregated. This behavior exhibited by PNIPAM has been used for several applications such as sensors, externally triggered drug delivery systems, and biological separations [56, 57].

22.3 NOBLE METAL NANOPARTICLES

Metallic nanostructures have received a lot of attention since they exhibit very different optical, electrical, thermodynamic, and chemical properties from the bulk material. These novel properties found in these materials have earned them a place in applications of sensors [58], molecular labels [59], and even as biocides and antimicrobials [60, 61]. Such properties exhibited at the nanoscale are exemplified in the way gold interacts with light at different scales. Gold in bulk looks shiny yellow in reflected light, but when very thin gold films are prepared, they look blue in transmitted light. Additionally, the colors change to orange and several tones of purple and red as the film thickness decreases up to a few nanometers. This phenomenon is due to the collective oscillation of conducting electrons at the interface between the metal film and the surrounding dielectric [62, 63]. The conducting electrons at the interface of metallic nanostructures respond to electromagnetic fields with electronic resonant absorption in the visible wavelength range, called localized surface plasmon resonance (LSPR), and give rise to very intense colors.

The LSPR is highly sensitive to size and shape of the nanostructures as well as to other parameters such as the refractive index of the surrounding medium and the distance between neighboring particles. Methods of calculating theoretical extinction spectra of metallic nanoparticles will be described in the next section. When the nanostructures are nonspherical and exhibit higher growth in one dimension, to form wires or rods, the plasmon band splits into two bands corresponding to oscillation of the conducting electrons along (longitudinal) and perpendicular to (transverse) the long axis [64]. The transverse mode resonance resembles the observed peak for spheres, but the longitudinal mode is considerably shifted toward red and depends strongly on the difference in length compared to width. The absorption spectrum is able to shift from the blue to the red by forming elongated wires, tubes, belts, and other more sophisticated structures (stars, rings, cubes, nanoshells, and onion-like structures) that incorporate different metals [65–69]. All these shapes are synthesized to move the absorption peaks along the absorption spectrum and design sensors and molecular labels. The high sensitivity to size and shape that the absorption peaks present need to be synthesized in order to produce highly monodisperse metallic structures.

22.4 CALCULUS OF EXTINCTION SPECTRA IN METALLIC NANOPARTICLES USING MIE THEORY

Optical properties of metal nanostructures have very strong dependence on their specific size and shape as well as on several other parameters, such as the refractive index of the surrounding medium and the distance between neighboring particles. Specifically, the optical properties of nanoparticles can be characterized and calculated theoretically with accuracy using Mie theory.

Mie theoretical calculations are based on the rigorous solutions of Maxwell's equations with spherical boundary conditions at the sphere. When the particles have a surface stabilizing agent such as a polymer shell or thick film surrounding them, Mie theory requires to take into account the dielectric function of both the metallic particle and the surrounding medium. Therefore, in the case of metallic particles, the optical properties can be described by simply determining the nanoparticle radii and the bulk frequency–dependent dielectric constant (in the case of aqueous solution, the water dielectric constant is used). When a different medium is taken into account, such as a thin film on the surface of the nanoparticle or a dielectric shell, the dielectric constant used has a dependence of position and varies along the radius [70, 71]. Examples of these structures are illustrated in Figure 22.3.

22.5 ENERGY CONVERSION OF METALLIC NANOPARTICLES

Metallic nanoparticles interact with light at different wavelengths through scattering and absorption. The scattering component of light is enhanced by the increase in particle size after approximately 100 nm; however, the absorption stays mostly constant and slowly increases with size. The nanoparticles show an extinction spectrum, with a higher intensity at the surface

Metallic nanoparticle	Metallic nanoparticle with a stabilizing polymer surface or core
	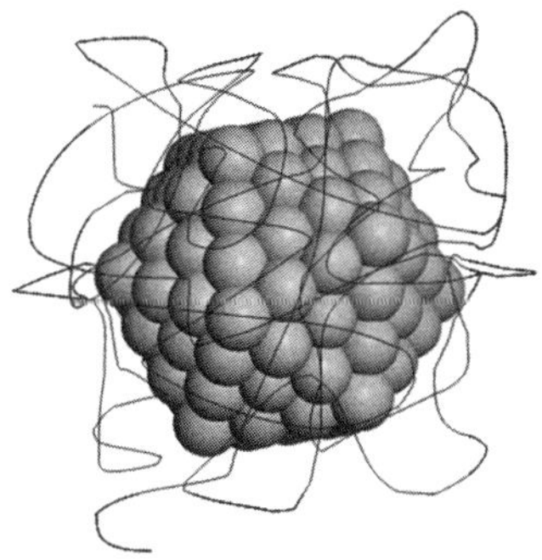

FIGURE 22.3 Schematic of a metallic nanoparticle in the left column of the table. In the right column of the table, a schematic of a metallic nanoparticle with a shell or film covering its surface is shown.

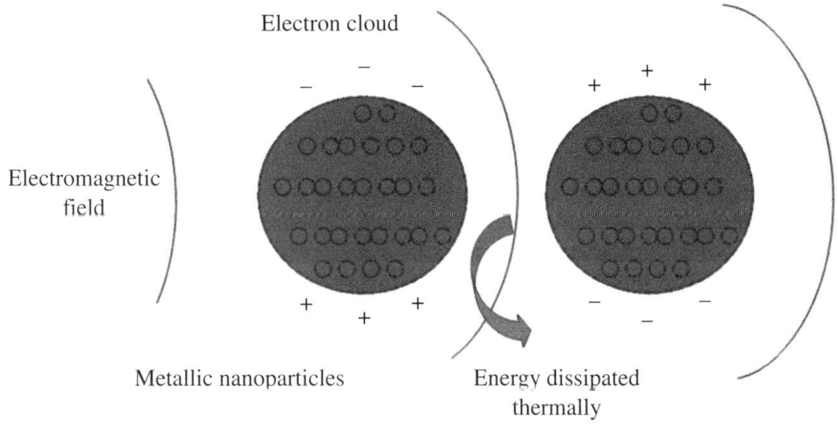

FIGURE 22.4 Illustration of the mechanism of heat dissipation by the surface electrons present in metallic nanoparticles when they are exposed to a specific wavelength and excited by an electric field.

plasmon–resonance wavelength, and follow a photothermal phenomenon where, due to energy relaxation of the surface electrons to the phonon bath, the resonant energy is dissipated as heat to the local surroundings (Fig. 22.4). This raises the temperature of the particles' surroundings when illuminated by a light source, such as a laser. Smaller particles and structures such as nanorods, nanorings, and nanoshells convert most of the absorbed energy into heat since they present larger surface-to-volume ratios and less scattering, which therefore enhance the optothermal conversion efficiency, thus producing larger increments of temperature in the surrounding environment [43, 68].

Temperature increments of gold and silver nanoparticles have been measured using polarization interference, which allows detection of slight phase changes induced by heating. Efficient energy conversion increases the local temperature in the immediate vicinity of the particle up to 15 K [72], and work on different structures such as nanocrescent has achieved increments of up to 60 K in the nanostructure vicinity [43].

22.6 SYNTHESIS ROUTES OF ENVIRONMENTALLY SENSITIVE POLYMER–METAL NANOPARTICLE HYBRIDS

One popular synthesis method to produce metallic nanostructures is wet chemistry reactions since they are relatively cheap and capable of producing colloidal suspensions of homogeneous particles in large quantities. The ingredients in the liquid phase are the metal salt precursor, which is reduced to its metallic form by a reducing agent such as ethylene glycol [63, 73, 74], sodium citrate [75, 76], or sodium borohydrate [77, 78]. Previous review papers have discussed several mechanisms in the formation of metallic particles by wet chemistry methods. The steps described include the creation of metal seeds, followed by the growth of metallic clusters, and finally agglomeration and further growth of the metallic clusters to form nanoparticles that reach sizes of a few nanometers in diameter (illustrated in Fig. 22.5). The control of the nucleation, aggregation, and growth steps of the reaction is key in the production of metallic structures with homogeneous sizes and shapes [73, 74, 79]. However, this is not a thermodynamically stable process since particles in the colloid will tend to minimize their surface energy by agglomerating with other particles. This leads to the formation of large agglomerates with polydisperse sizes and shapes. Size and shape can be controlled by the addition of a surface stabilizer present in the surrounding solvent during the formation of the metallic structures. These surface stabilizers are molecules, usually organic polymers that prevent high affinity to the metal surface and the metallic ions [80, 81]. The final material is then a hybrid polymer–metal nanocomposite where the polymers act as capping agents; they control the growth of the particles in the synthesis process and prevent surface contact of the particles, which prevents agglomeration during synthesis and storage [63, 81–83].

So far, the most effective and well-studied stabilizing molecules for silver and gold are poly-(vinylpyrrolidone) (PVP) and sodium citrate [63, 80, 81, 84], respectively. Their interaction with the metals allows fabrication of highly stabilized nanostructures with controlled shapes and sizes. However, in the production of hybrid materials where coupling of the metallic nanostructure with a responsive polymer is required, PVP or sodium citrate adhered to the surface must be removed before further surface conjugation can be achieved. This is done by a ligand exchange process, previously done with thiols [85], using molecules with stronger bonds to the gold surface than PVP.

Various studies have focused on the coupling of polymers and metal nanoparticles to form hybrids. Some of the methods developed (Fig. 22.6) include surface grafting by functionalizing the responsive polymers with thiol groups to promote adsorption on the surface, and surface-initiated polymerization by prior treatment of the metallic nanoparticle surface with radical polymerization initiators [75, 76]. Previous reports have observed a strong interaction of molecules that contain chemical groups with N and O, such as PVP and the amide group in PNIPAM and PAM, with the noble metals, platinum [86–88], gold [89], and silver [73, 74, 90]. The interaction with the surface of the metallic particles has been attributed to hydrophobic forces

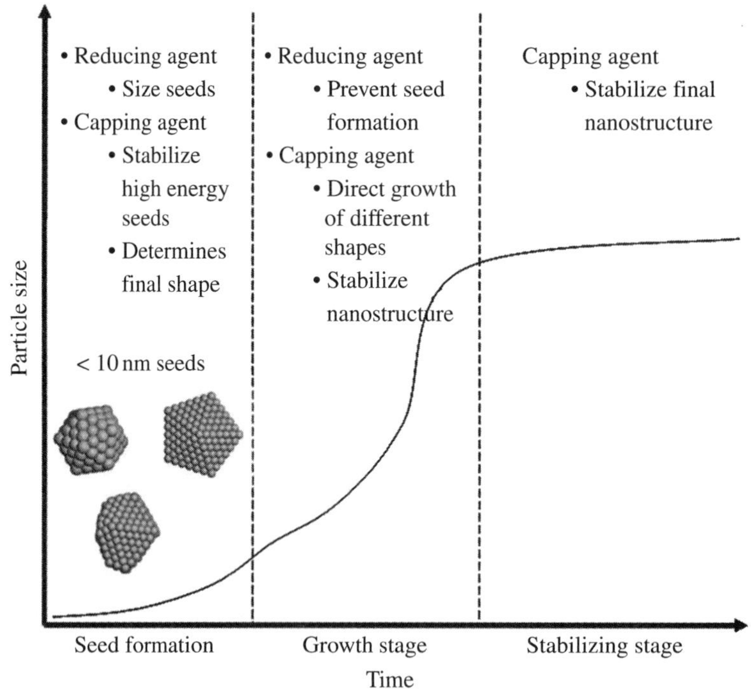

FIGURE 22.5 Graphs show the different stages observed for metallic nanoparticle synthesis. The seeding stage consists of the formation of particles with diameters less than 10 nm. In this stage, the reducing agent plays a major role in the size of the particles depending on how fast it reduces the metallic salts. The stabilizing agent therefore plays a role in stabilizing different particle shapes of different surface energies, and the appropriate stabilization leads to the formation of various shapes (spheres, wires, tripods, cubes, stars, etc.). The growth stage involves agglomeration and growth of small particles. The capping and reducing agent must inhibit seed formation and promote growth, which leads to homogeneous size and shape distributions. The stabilizing stage only involves the capping agent that is able to stabilize the resulting metallic nanostructures.

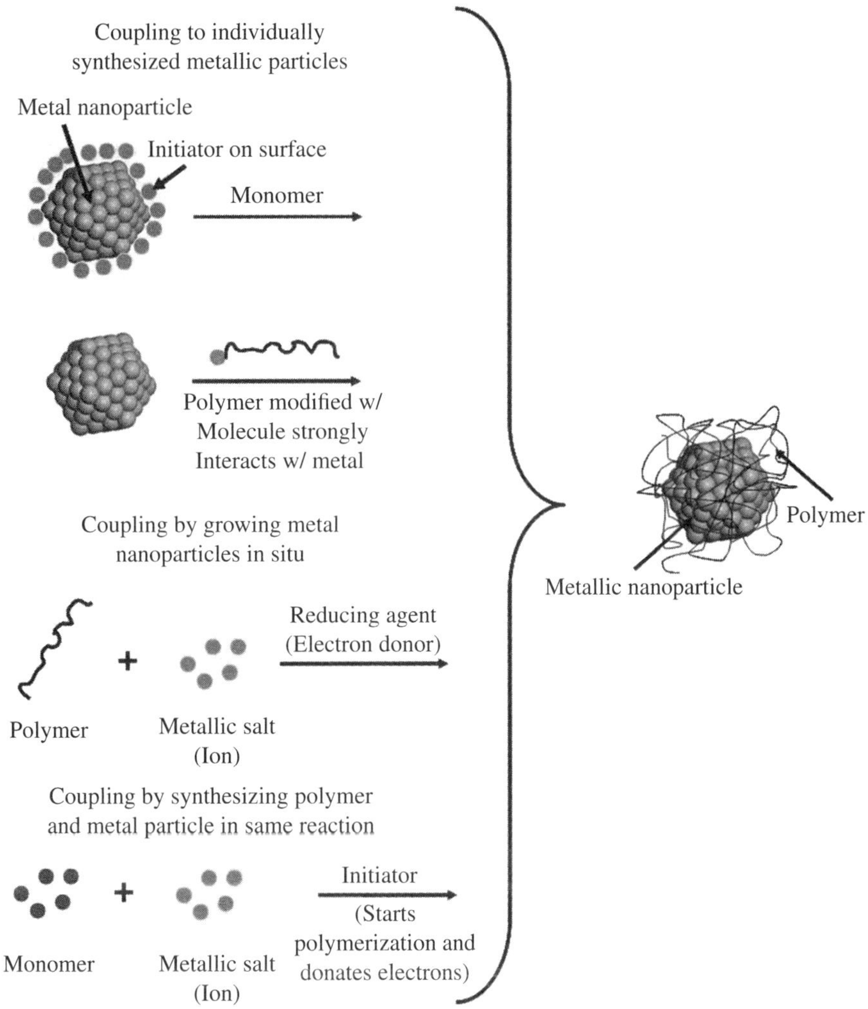

FIGURE 22.6 Different mechanisms to couple metallic nanoparticles and polymeric materials are illustrated and labeled respectively. The first process involves making the metal particles separately and then treating their surface to start polymerization on them or treating a polymer with functional groups that attach to the metal. The second process involves *in situ* reduction of metal salts in the presence of the polymer. The last process involves using the initiator as an electron donor and initiator of polymerization; therefore, both polymerization of the monomer and formation of the particles are performed at the same time. All these lead to the synthesis of coupled polymer–metal nanocomposites with different characteristics.

as well as possible chemical coordination bonds between the polymer and the metal surface. PNIPAM has been previously used as a capping and stabilizing agent in the synthesis of platinum particles reduced with ethanol [86–88].

Additionally, synthesis of silver [91, 92] and bimetallic platinum–gold [93] colloids has been approached by using simultaneous dispersion polymerization of polystyrene and NIPAM in the presence of metallic precursors. The products yield metallic particles within the PNIPAM grafts on the surface of polystyrene spheres, allowing enhanced catalytic properties and effective ways of recuperating precious metals. These reports have used X-ray photoelectron spectroscopy (XPS) and Fourier transform infrared (FTIR) to study the interactions with PNIPAM and silver in these syntheses. They found that the peaks for the carboxyl and nitrogen hydrogen bonds have shifted, indicating that there is an interaction of silver with these specific groups. This interaction provides a potential to develop novel synthesis routes to synthesize colloidal hybrid nanomaterials.

22.7 CONCLUSIONS

Current cell culture systems present a challenge of externally controlling the conditions of local microenvironments in order to emulate *in vivo*–like conditions to study cellular responses to chemical and physical changes in the environment. The current most successful alternatives to control chemical conditions in microenvironments is the incorporation of responsive mechanical actuators, such as "on–off" valves, to control the flow and delivery of chemical reagents. The main component of mechanical actuators that gives it the switching capability is the responsive material that is capable of undergoing conformational changes in response to different stimuli. Responsive materials can respond to bulk and/or localized stimuli, where, for applications controlling cellular microenvironment conditions, it is necessary to use responsive materials to localized triggers. Being optically responsive materials, which is the most used and appropriate option for this application, thermally responsive polymer–metal nanocomposites offer the best option for use as the optical switching component of "on–off" valves. The challenges that arise are the synthesis pathways to couple both metal particles and the responsive polymer. A solution proposed is to take advantage of the previously reported interaction of N- and O-containing compounds, such as PNIPAM, with different metals, such as silver and gold, to perform a wet chemistry *in situ* reduction of metallic salts and achieve stable, well-controlled particles coupled with the responsive polymer. Lastly, different methodologies to incorporate the developed nanocomposites into nanoporous membranes can be addressed which in future work will be used as optothermally responsive nanovalves in *in vitro* cell culture systems to control the delivery of reagents and *in vivo*–like conditions of the microenvironments.

Successful development of these cell culture systems will have a profound impact on the development of pathways to drug discovery and novel therapeutics of diseases that develop due to cellular malfunction such as cancer and diabetes.

REFERENCES

[1] Beebe DJ, Moore JS, Bauer JM, Yu Q, Liu RH, Devadoss C, Jo BH. Functional hydrogel structures for autonomous flow control inside microfluidic channels. Nature 2000;404 (6778):588–590.

[2] Eddington DT, Beebe DJ. Flow control with hydrogels. Adv Drug Deliv Rev 2004;56 (2):199–210.

[3] Lue SJ, Hsu J-J, Chen C-H, Chen B-C. Thermally on–off switching membranes of poly(N-isopropylacrylamide) immobilized in track-etched polycarbonate films. J Membr Sci 2007;301 (1–2):142–150.

[4] Peng T, Cheng Y-L. Temperature-responsive permeability of porous PNIPAAm-g-PE membranes. J Appl Polym Sci 1998;70: 2133–2142.

[5] Peng T, Cheng Y-L. pH-responsive permeability of PE-g-PMAA membranes. J Appl Polym Sci 1999;76:778–786.

[6] Peng T, Cheng Y-L. PNIPAAm and PMAA co-grafted porous PE membranes: living radical co-grafting mechanism and multi-stimuli responsive permeability. Polymer 2001;42:2091–2100.

[7] Chu L-Y, Niitsuma T, Yamaguchi T, Nakao S-I. Thermoresponsive transport through porous membranes with grafted PNIPAM gates. AIChE J 2003;49 (4):896–909.

[8] Huang J, Wang X, Chen X, Yu X. Temperature-sensitive membranes prepared by the plasma-induced graft polymerization of N-isopropylacrylamide into porous polyethylene membranes. J Appl Polym Sci 2003;89:3180–3187.

[9] Lee YM, Ihm YS, Shim JK, Kim JH. Preparation of surface-modified stimuli-responsive polymeric membranes by plasma and ultraviolet grafting methods and their riboflavin permeation. Polymer 1995;36:81–85.

[10] Xie R, Chu L-Y, Chen W-M, Xiao W, Wang H-D, Qu J-B. Characterization of microstructure of poly(N-isopropylacrylamide)-grafted polycarbonate track-etched membranes prepared by plasma-graft pore-filling polymerization. J Membr Sci 2005;258 (1–2):157–166.

[11] Gil ES, Hudson SM. Stimuli-responsive polymers and their biconjugates. Prog Polym Sci 2004;29 (12):1173–1222.

[12] Langer R, Tirrell DA. Designing materials for biology and medicine. Nature 2004;428:487–492.

[13] Sershen SR, Mensing GA, Ng M, Halas NJ, Beebe DJ, West JL. Independent optical control of microfluidic valves formed from optomechanically responsive nanocomposite hydrogels. Adv Mater 2005;17:1366–1368.

[14] Brinker CJ. Evaporation-induced self assembly: functional nanostructures made easy. MRS Bull September 2004;2004:631–640.

[15] Podual K, Doyle F III, Peppas NA. Modeling of water transport in and release from glucose-sensitive swelling-controlled release systems based on poly(diethylaminoethyl methacrylate-g-ethylene glycol). Ind Eng Chem Res 2004;43:7500–7512.

[16] Cartier S, Horbett TA, Ratner DB. Glucose-sensitive membrane coated porous filters for control of hydraulic permeability and insulin delivery from a pressurized reservoir. J Membr Sci 1995;106:17–24.

[17] Schild HG. Poly(N-isopropylacrylamide): experiment, theory and application. Prog Polym Sci 1992;17 (2):163–249.

[18] Frey W, Meyer DE, Chilkoti A. Dynamic addressing of a surface pattern by a stimuli-responsive fusion protein. Adv Mater 2003;15 (3): 248–251.

[19] Frey W, Meyer DE, Chilkoti A. Thermodynamically reversible addressing of a stimuli responsive fusion protein onto a patterned surface template. Langmuir 2003;19 (5):1641–1653.

[20] Peppas NA, Khare AR. Preparation, structure and diffusional behavior of hydrogels incontrolled-release. Adv Drug Deliv Rev 1993; 11 (1–2):1–35.

[21] Dai S, Ravi P, Tam KC. pH-Responsive polymers: synthesis, properties and applications. Soft Matter 2008;4 (3):435–449.

[22] Hilt JZ, Gupta AK, Bashir R, Peppas NA. A microsensor based on a microcantilever patterned with an environmentally sensitive hydrogel. In: Manginell RP, Borenstein JT, Lee LP, Hesketh PJ, editors. *BioMEMS and Bionanotechnology*, Volume 729, MRS, Pittsburgh, PA; . 2002. p 173–178.

[23] Shoenfeld NA, Grodzinsky AJ. Contribution of electrodiffusion to the dynamics of electrically stimulated changes in mechanical- properties of collagen membranes. Biopolymers 1980;19 (2):241–262.

[24] Zheng J, Xu C, Hirai T. Electromechanical nonionic gels. New J Phys 2008;10:023016.

[25] Kungwatchakun D, Irie M. Photoresponsive polymers. Photocontrol of the phase separation temperature of aqueous solutions of poly Nisopropylacrylamide- co-N-(4-phenylazophenyl) acrylamide. Die Makromolekulare Chemie, Rapid Commun 1988;9 (4):243–246.

[26] Irie M, Kungwatchakun D. Stimuli-responsive polymers—photostimulated reversible phase-separation of aqueous-solutions of poly(N-isopropylacrylamide) with pendant azobenzene groups. Proc Jpn Acad Ser B Phys Biol Sci 1992;68 (8):127–132.

[27] Nayak S, Lyon LA. Photoinduced phase transitions in poly(N-isopropylacrylamide) microgels. Chem Mater 2004;16 (13):2623–2627.

[28] Radt B, Smith TA, Caruso F. Optically addressable nanostructured capsules. Adv Mater 2004;16 (23–24):2184–2189.

[29] Jiang H, Kelch S, Lendlein A. Polymers move in response to light. Adv Mater 2006;18 (11):1471–1475.

[30] Slocik JM, Tam F, Halas NJ, Naik RR. Peptide-assembled optically responsive nanoparticle complexes. Nano Lett 2007;7 (4): 1054–1058.

[31] Szilagyi A, Sumaru K, Sugiura S, Takagi T, Shinbo T, Zrinyi M, Kanamori T. Rewritable microrelief formation on photoresponsive hydrogel layers. Chem Mater 2007;19 (11):2730–2732.

[32] Byrne ME, Park K, Peppas NA. Molecular imprinting within hydrogels. Adv Drug Deliv Rev 2002;54 (1):149–161.

[33] Liu J, Lu Y. Smart nanomaterials responsive to multiple chemical stimuli with controllable cooperativity. Adv Mater 2006;18 (13): 1667–1671.

[34] Peppas NA, Hilt JZ, Khademhosseini A, Langer R. Hydrogels in biology and medicine: from molecular principles to bionanotechnology. Adv Mater 2006;18 (11):1345–1360.

[35] Shimoboji T, Larenas E, Fowler T, Kulkarni S, Hoffman AS, Stayton PS. Photoresponsive polymer-enzyme switches. Proc Natl Acad Sci U S A 2002;99 (26):16592–16596.

[36] Shimoboji T, Larenas E, Fowler T, Hoffman AS, Stayton PS. Temperature-induced switching of enzyme activity with smart polymerenzyme conjugates. Bioconjug Chem 2003;14 (3):517–525.

[37] Qiu Y, Park K. Environment-sensitive hydrogels for drug delivery. Adv Drug Deliv Rev 2001;53:321–339.

[38] Peppas NA, Kim B. Stimuli-sensitive protein delivery systems. J Drug Deliv Sci Technol 2006;16:11–18.

[39] Bashir R, Hilt JZ, Elibol O, Gupta A, Peppas NA. Micromechani calcantilever as an ultrasensitive pH microsensor. Appl Phys Lett 2002;81:3091–3093.

[40] Park JY, Oh HJ, Kim DJ, Baek JY, Lee SH. A polymeric microfluidic valve employing a pH-responsive hydrogel microsphere as an actuating source. J Micromech Microeng 2005;16:656–663.

[41] Grover WH, Skelley AM, Liu CN, Lagally ET, Mathies RA. Monolithic membrane valves and diaphragm pumps for practical large-scale integration into glass microfluidic devices. Sens Actuators B 2003;89:315–323.

[42] Stoeber B, Zhihao Y, Liepmann D, Muller SJ. Flow control in microdevices using thermally responsive triblock copolymers. J Microelectromech Syst 2005;14:207–213.

[43] Liu GL, Kim J, Lu Y, Lee LP. Optofluidic control using photothermal nanoparticles. Nat Mater 2005;10:1–6.

[44] Keenan TM, Hsu C-H, Folch A. Microfluidic "jets" for generating steady-state gradients of soluble molecules on open surfaces. Appl Phys Lett 2006;89:114103-1–114103-3.

[45] Abhyankar VV, Lokuta MA, Huttenlocher A, Beebe DJ. Characterization of a membrane-based gradient generator for use in cell-signaling studies. Lab Chip 2006;6:389–393.

[46] Higuchi A, Hamamura A, Shindo Y, Kitamura H, Yoon BO, Mori T, Uyama T, Umezawa A. Photon-modulated changes of cell attachments on poly(spiropyran-co-methyl methacrylate) membranes. Biomacromolecules 2004;5 (5):1770–1774.

[47] Edahiro J, Sumaru K, Tada Y, Ohi K, Takagi T, Kameda M, Shinbo T, Kanamori T, Yoshimi Y. In situ control of cell adhesion using photoresponsive culture surface. Biomacromolecules 2005;6 (2):970–974.

[48] Nayak A, Liu H, Belfort G. An optically reversible switching membrane surface. Angew Chem Int Ed 2006;45 (25):4094–4098.

[49] Zhu LY, Wu WW, Zhu MQ, Han JJ, Hurst JK, Li ADQ. Reversibly photoswitchable dual-color fluorescent nanoparticles as new tools for live-cell imaging. J Am Chem Soc 2007;129 (12):3524–3526.

[50] Garcia A, Marquez M, Cai T, Rosario R, Hu ZB, Gust D, Hayes M, Vail SA, Park CD. Photo-, thermally, and pH-responsive microgels. Langmuir 2007;23 (1):224–229.

[51] Hugel T, Holland NB, Cattani A, Moroder L, Seitz M, Gaub HE. Single-molecule optomechanical cycle. Science 2002;296 (5570):1103–1106.

[52] Chou C-H, Chen C-D, Wang CCR. Highly efficient, wavelength-tunable, gold nanoparticle based optothermal nanoconvertors. J Phys Chem B 2005;109:11135–11138.

[53] Sershen SR, Westcott SI, West JL, Halas NJ. An opto-mechanical nanoshell-polymer composite. Appl Phys B 2001;73:379–381.

[54] Sershen SR, Westcott SL, Halas NJ, West JL. Independent optically addressable nanoparticle-polymer optomechanical composites. Appl Phys Lett 2002;80 (24):4609.

[55] Jones CD, Lyon AL. Photothermal patterning of microgel/gold nanoparticle composite colloidal crystals. J Am Chem Soc 2003; 125:460–465.

[56] Akiyama H, Tamaoki N. Polymers derived from N-isopropylacrylamide and azobenzenecontaining acrylamides: photoresponsive affinity to water. J Polym Sci Part A Polym Chem 2004;42:5200–5214.

[57] Bergbreiter DE, Case BL, Liu Y-S, Caraway JW. Poly(n-isopropylacrylamide) soluble polymer supports in catalysis andsynthesis. Macromolecules 1998;31:6053–6062.

[58] Nath N, Chilkoti A. Label free colorimetric biosensing using nanoparticles. J Fluoresc 2004;14 (4):377–389.

[59] Bruchez MJ, Moronne M, Gin P, Weiss S, Alivisatos AP. Semiconductor nanocrystals as fluorescent biological labels. Science 1998; 281 (5385):2013–2016.

[60] Morones JR, Elechiguerra JL, Camacho-Bragado A, Holt K, Kouri JB, Tapia Ramirez J, Jose Yacaman M. The bactericidal effect of silver nanoparticles. Nanotechnology 2005;16:2346–2353.

[61] Elechiguerra JL, Burt JL, Morones JR, Camacho-Bragado A, Gao X, Lara H, Jose Yacaman M. Interaction of silver nanoparticles with HIV-I. J Nanobiotechnol 2005;3:6.

[62] Liz-Marzan LM. Nanometals: formation and color. Mater Today 2004;7:26–31.

[63] Carotenuto G, Pepe GP, Nicolais L. Preparation and characterization of nano-sized Ag/PVP composites for optical applications. Euro Phys J B 2000;16:11–17.

[64] Xia Y, Halas N. Shape-controlled synthesis and surface plasmonic properties of metallic nanostructures. MRS Bull 2005;30:338–348.

[65] Halas N. Playing with plasmons: tuning the optical resonant properties of metallic nanoshells. MRS Bull 2005;30:362–367.

[66] Chen J, Wiley B, Li Z-Y, Campbell D, Saeki F, Cang H, Au L, Lee J, Li X, Xia Y. Gold nanocages: engineering the structure for biomedical applications. Adv Mater 2005;17:2255–2261.

[67] Hirsch LR, Gobin AM, Lowery AR, Tam F, Drezek RA, Halas NJ, West JL. Metal nanoshells. Ann Biomed Eng 2006;34:15–22.

[68] Lu Y, Liu GL, Kim J, Mejia YX, Lee LP. Nanophotonic crescent moon structures with sharp edge for ultrasensitive biomolecular detection by local electromagnetic field enhancement effect. Nano Lett 2004;5:119–124.

[69] Aizpurua J, Hanarp P, Sutherland DS, Kall M, Bryant GW, Garcia de Abajo FJ. Optical properties of gold nanorings. Phys Rev Lett 2003;90:57401-1–57401-4.

[70] Bohren CF, Huffman DR. Absorption and Scattering of Light by Small Particles. 1st ed. Weinheim: Wiley-VCH; 1998.

[71] Jain PK, Lee KS, El-Sayed IH, El-Sayed MA. Calculated absorption and scattering properties of gold nanoparticles of different size, shape, and composition: applications in biological imaging and biomedicine. J Phys Chem B 2006;110 (14):7238–7248.

[72] Boyer D, Tamarat P, Maali A, Lounis B, Orrit M. Photothermal imaging of nanometer-sized metal particles among scatterer. Science 2002;297:1160–1163.

[73] Silvert P-Y, Herrera-Urbina R, Duvauchelle N, Vijayakrishnan V, Tekaia-Elhsissen K. Preparation of colloidal silver dispersions by the polyol process. Part 1—Synthesis and characterization. J Mater Chem 1996;6:573–577.

[74] Silvert P-Y, Herrera-Urbina R, Tekaia-Elhsissen K. Preparation of colloidal silver dispersions by the polyol process. J Mater Chem 1997;7:293–299.

[75] Raula J, Shan J, Nuopponen M, Niskanen A, Jiang H, Kauppinen EI, Tenhu H. Synthesis of gold nanoparticles grafted with a thermoresponsive polymer by surface-induced reversible- addition-fragmentation chain-transfer polymerization. Langmuir 2003;19:3499–3504.

[76] Kim DJ, Heo J-Y, Kim KS, Choi IS. Formation of thermoresponsive poly (nisopropylacrylamide)/dextran particles by atom transfer radical polymerization. Macromol Rapid Commun 2003;24:517–521.

[77] Hyning DLV, Klemperer WG, Zukoski CF. Silver nanoparticle formation: predictions and verification of the aggregative growth model. Langmuir 2001;17:3128–3135.

[78] Hyning DLV, Zukoski CF. Formation mechanisms and aggregation behavior of borohydride reduced silver particles. Langmuir 1998; 14:7034–7046.

[79] Xiong Y, Xia Y. Shape-controlled synthesis of metal nanostructures: the case of palladium. Adv Mater 2007;19 (20):3385–3391.

[80] He B, Juei Juei T, Liew KY, Liu H. Synthesis of size controlled Ag nanoparticles. J Mol Catal A Chem 2004;221:121–126.

[81] Zhang Z, Zhao B, Hu L. PVP protective mechanism of ultrafine silver power synthesized by chemical reduction processes. J Solid State Chem 1996;121:105–110.

[82] Wiley B, Sun Y, Mayers B, Xia Y. Shape-controlled synthesis of metal nanostructures: the case of silver. Chem A Euro J 2005;11 (2): 454–463.

[83] Shan J, Nuopponen M, Jiang H, Viitala T, Kauppinen E, Kontturi K, Tenhu H. Amphiphilic gold nanoparticles grafted with poly(N-isopropylacrylamide) and polystyrene. Macromolecules 2005;38 (7):2918–2926.

[84] Wang H, Qiao X, Chen J, Ding S. Preparation of silver nanoparticles by chemical reduction method. Colloid Surf A Physicochem Eng Asp 2005;256:111–115.

[85] Carotenuto G, Nicolais L. Size-controlled synthesis of thiol-derivatized gold clusters. J Mater Chem 2003;13:1038–1041.

[86] Chen C-W, Akashi M. Synthesis, characterization, and catalytic properties of colloidal platinum nanoparticles protected by poly(N-isopropylacrylamide). Langmuir 1997;13:6465–6472.

[87] Chen C-W, Akashi M. Preparation of ultrafine platinum particles protected by poly(Nisopropylacrylamide) and their catalytic activity in the hydrogenation of Allyl alcohol. J Polym Sci Part A Polym Chem 1997;35:1329–1332.

[88] Miyazaki A, Nakano Y. Morphology of platinum nanoparticles protected by poly(Nisopropylacrylamide). Langmuir 2000;16:7109–7111.

[89] Sun Y, Xia Y. Shape-controlled synthesis of gold and silver nano-particles. Science 2002;298:2176–2179.

[90] Kim D, Jeong S, Moon J. Synthesis of silver nanoparticles using the polyol process and the influence of precursor injection. Nanotechnology 2006;17:4019–4024.

[91] Chen C-W, Chen M-Q, Serizawa T, Akashi M. In-situ formation of silver nanoparticles on poly(N-isopropylacrylamide)-coated polystyrene microspheres. Adv Mater 1998;14:1122–1126.

[92] Chen C-W, Serizawa T, Akashi M. Synthesis and characterization of poly(N-isopropylacrylamide)- coated polystyrene microspheres with silver nanoparticles on their surfaces. Langmuir 1999;15:7998–8006.

[93] Chen C-W, Serizawa T, Akashi M. In situ formation of Au/Pt bimetallic colloids on polystyrene microspheres: control of particle growth and morphology. Chem Mater 2002;14:2232–2239.

PART IX

DESALINATION

23

ENGINEERED NANOMATERIALS FOR PURIFICATION AND DESALINATION OF PALATABLE WATER

VIJAY C. VERMA[1], SWECHHA ANAND[2], MAYANK GANGWAR[1,3], AND SANTOSH K. SINGH[2]

[1]*Department of Microbiology and Infectious Diseases, Institute of Medical Sciences, Banaras Hindu University, Varanasi, India*

[2]*Centre of Experimental Medicine and Surgery, Institute of Medical Sciences, Banaras Hindu University, Varanasi, India*

[3]*Department of Pharmacology, Institute of Medical Sciences, Banaras Hindu University, Varanasi, India*

23.1 INTRODUCTION

Water is a limited natural resource that is vital for life on earth, and the depletion and pollution of its sources further complicate the seriousness of this problem. Despite the fact that more than 70% of the earth is covered by water, most of it is useless because of its unpalatability. However, contamination of groundwater, which is a major source of safe drinking water, is widespsread in many parts of the world. Clean water is a basic requirement for any society, and in recent years many parts of the world have faced substantial increases in the demand-to-supply ratio of drinking water. The rate of global water consumption is almost more than twice the rate of growth of the human population. According to the United Nations, approximately 1 billion people on the earth are facing severe scarcity of fresh drinking water, and if this situation continues, the demand for freshwater will reach an approximately 56-fold increase by 2025 as compared to the present. Many strategic consultancies are speculating concern over "water" becoming the cause of conflict and possible war in the future. To prevent these unpleasant consequences, it is worthwhile using water wisely and promoting awareness for saving water in our society, apart from expediting technological advancement for rendering surface water appropriate for drinking. Recent statistics show that in developing countries a number of people die annually due to consumption of unhygienic water and this number is even greater than the lives lost in wars and violence combined. In the African continent, more than 5.7 million people lack access to safe and hygienic water (Fig. 23.1), and about 20 million lack sanitation services [1]. As per the United Nations Millennium Development Goals (UN-MDGs), access to safe drinking water, free from disease-causing bacteria and viruses, is a basic human right, and essential to maintain a healthy society. The international community has committed to provide safe drinking water and accesses to sanitation services by 2015. To achieve these MDGs, scientists are looking toward nanotechnology-enabled solutions. By harnessing the unique and novel properties of nanoscale materials, they are trying to adapt them for water treatment purposes. Nanotechnology offers great opportunity to refine and optimize current and future techniques of purifying water. Nanotechnology-based solutions in the water sector will find wide applications if they are cost-effective, efficient, and able to provide clean drinking water even in very remote parts of the world.

In this chapter, we try to explore the possibilities of sustainable access to safe drinking water by adopting cutting-edge tools of nanotechnology. We discuss how nano-enabled technologies can efficiently process surface and groundwater for drinking. In recent years, there has been significant advancement in nanotechnology in virtually every domain of science [2, 3]. So optimists

Nanomaterials for Environmental Protection, First Edition. Edited by Boris I. Kharisov, Oxana V. Kharissova, and H. V. Rasika Dias.

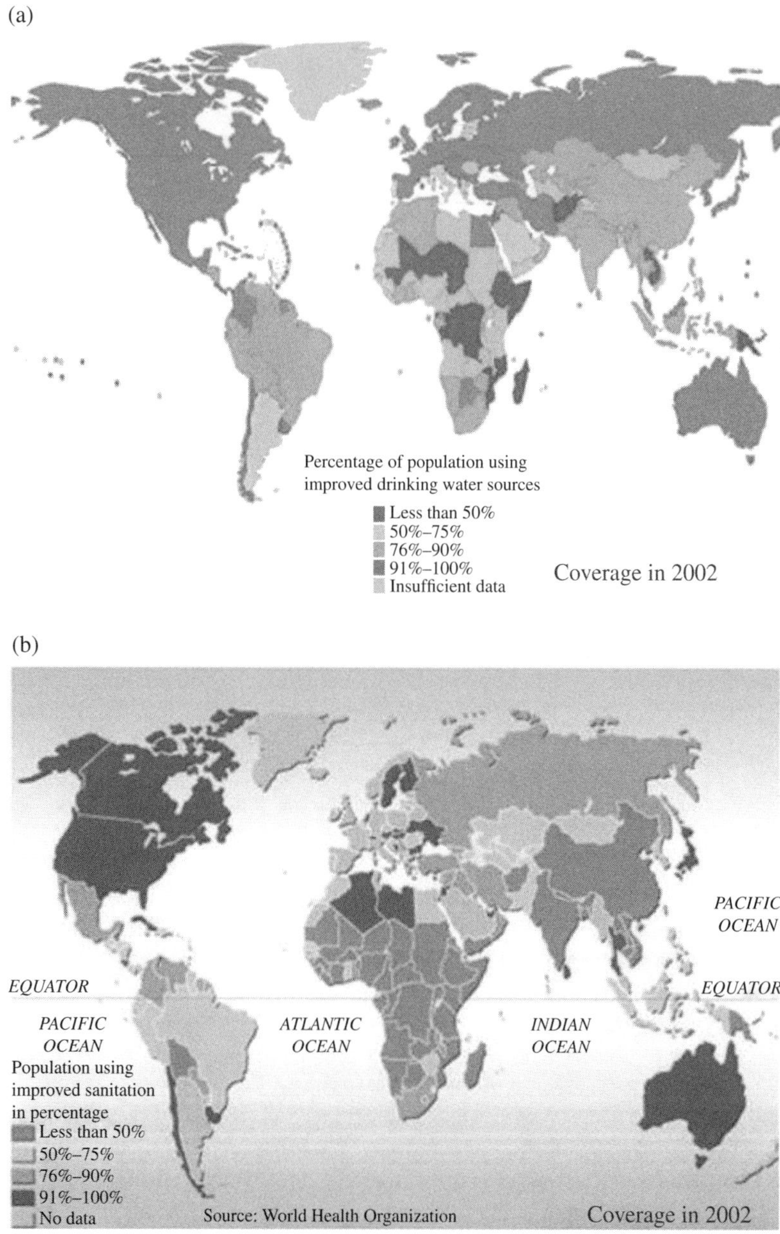

FIGURE 23.1 (a) Reduction by half of the proportion of people without sustainable access to safe drinking water and basic sanitation between 1990 and 2015. (b) Reduction in the proportion of households without access to hygienic sanitation facilities and affordable and safe drinking water by at least one-third.

are viewing nanotechnology as a revolutionary technology that might be able to solve many problems that have become a concern for human society.

23.2 DESALINATION

Nanotechnology could be used in water treatment for monitoring, desalinization, purification, and waste water treatment [4]. These might be some vital sectors having a significant role in coping with future water shortage [5, 6]. However, one must understand that nanotechnology itself has its limitations of accessibility and affordability, and is not a "magic stick" to solve all problems. Many nanotechnology-based applications are still undergoing research and development and none of them has yet been scaled up to industry level [7]. Over the last few decades, desalination technologies have been used increasingly throughout

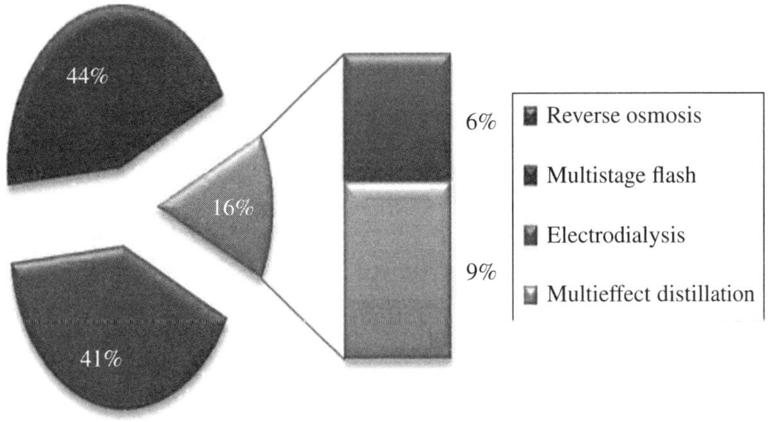

FIGURE 23.2 Major desalination technologies and their relative contributions to worldwide installed capacity for seawater and brackish water desalination. MSF accounts for 44%, RO 41%, and 15% is shared by other thermal methods like MED (9%) and ED (6%) [7, 9].

the world to produce drinking water from brackish groundwater and seawater, and also to improve the quality of existing supplies of freshwater for drinking and industrial purposes, apart from the treatment of industrial and municipal wastewater prior to discharge or reuse. In the early 1950s, about 225 land-based desalination plants were installed worldwide with a combined capacity of about 27 million gallons per day (mgd) of waste water treatment. Today there are about 3500 plants worldwide with a production capacity of about 3000 mgd. As the demand for freshwater increases and the quality of existing supplies deteriorates, the use of desalination technologies will increase [8] significantly.

Membrane-based RO or thermal-based multistage flash (MSF) and multieffect distillation (MED) contribute to over 90% of the global desalination capacity. RO plants, with typical capacities of approximately 20,000 m^3/day, account for approximately 41% of the total desalination capacity, while MSF plants, with typical capacities of approximately 76,000 m^3/day, account for approximately 44%. Other technologies, such as electrodialysis, account for only a small fraction (~5.6%) of the desalination capacity and are more suited for applications in brackish and groundwater treatment. The system-level schematics of the different types of desalination plants and their relative installed capacities are shown in Figure 23.2.

23.3 AQUAPORINS

Aquaporin is a well-known membrane-bound water channel found in all known living systems [10]. It allows water to pass from it as a single molecule; actually, aquaporins are so selective that they will exclude the passage of any sort of contamination including bacteria, virus, protein, DNA, salts, reagents, and even dissolved gases [11]. The technology is based on a discovery that was awarded the Nobel Prize in chemistry in 2003. Profesor Peter Agre is a molecular biologist from the Johns Hopkins University in Baltimore, the United States, who discovered during his studies that there is a special protein that is responsible for the rapid permeation of water in cells. Agre and his team named these proteins "aquaporins," as they function as water pores on the nanoscale [12]. In plants, they work like the plumbing system of cells and ensure a highly efficient but selective transport of water; only H_2O molecules can pass through these channels [13]. For plants, these filters are a lifesaver, as they ensure that the cell does not lose any minerals [14]. This specificity of aquaporin can be harnessed to prepare selective filter membranes for water purification. Some bacterial aquaporins, like aquaporin Z (ApqZ), could be candidates to design and develop nanomembranes because of their ruggedness and stability at high voltage, heat, detergent, and pH. Under natural systems, aquaporins move water from areas of low dissolved solids to areas of high dissolved solids; however, when a pressure is applied, the process can be reversed, similar to RO. Some commercial settings of ApqZ are now available [15]: they are inserted into 5 nm thick and 200 nm diameter polymer vesicles (Fig. 23.3). The highly concentrated proteins are deposited on inexpensive nitrocellulose filters and cross-linked to one another with ultraviolet (UV) light.

A European research project named MEMBAQ, led by Professor Claus Helix Nielsen and his team at the Technical University of Denmark, has undertaken further challenges to develop an aquaporin-based commercial technology for water purification [17]. According to them, the stability of the protein and the construction of a durable filter membrane are still very difficult to achieve. A thin film of perforated Teflon is used to further support the protein-bound membrane. Later, another layer made of porus hydrogel is constructed, which further stabilizes the membrane.

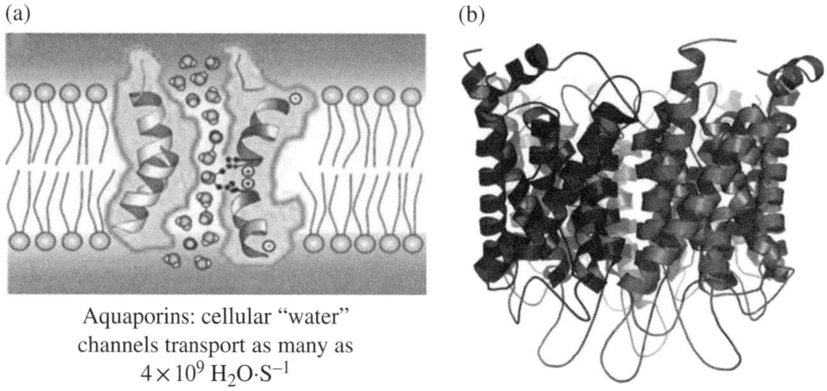

FIGURE 23.3 Pure water is forced through the aquaporin Z (AqpZ) network under pressure [16]. (a) A depiction of interaction of water molecules with AqpZ network and (b) the 3D view of the aqpZ.

23.4 FERRITINS

The biological systems in recent years have attracted much attention for inspiring research and development in water purification. Ferritin is a cage-like iron-storing protein found in animals, plants, and microbial systems and is involved in the sequestration and storage of protein [18]. Ferritin is composed of self-assembled 24 polypeptide subunits; apo-ferritin, an iron-free assemblage, has an inner cavity of approximately 8 nm diameter. When iron molecules are diffused into this cavity, they mineralize the apo-ferritin into nanoparticles of ferrihydrite, which is actually a ferric oxyhydroxide. Ferritin then becomes a potential candidate for the remediation, and particularly the photoreduction, of contaminants such as toxic metals like chlorocarbons [19]. As we know, when iron oxide undergoes photoreduction from Fe (III) to Fe (II), it makes the catalyst inactive, and since ferritin makes this conversion naturally, the encaged iron oxide is prevented from undergoing photoreduction, thus maintaining its stability. In addition to the conventional use of ferritin for the remediation and removal of contamination, a recent development is to use it to synthesize nanoparticles. It is evident that both metallic and nonmetallic hydroxide particles can be synthesized using ferritin. Recently, both iron and cobalt nanoparticles were synthesized using ferritin, which facilitates the assemblage of particles in a solution. After assembling, ferritin was dried on a solid support and cleaned with ozone to remove the protein cage. This leaves well-dispersed nano oxide particles with a relatively high degree of control for particle sizes between 2 and 8 nm. Further exposure of the particles to hydrogen and high temperature can be used to convert the metal oxides to metallic particles [20]. Water contamination with oxyanions (arsenate and phosphate) and metal ions over wide ranges of concentrations is a problem for water and related systems that clean water. Eutrophication of surface water, caused by a high level of phosphate, is an environmental issue. Further, arsenate ($HAsO_4/H_2AsO_4$) is a health- and life-threatening contaminant of drinking water in many areas of the world, notably in Asia and also in parts of the United States and Europe. There are numerous techniques available for the removal of phosphate and arsenate from water. In general, all the available techniques suffer from a low-affinity problem and, therefore, are inefficient in the low-concentration range (below 5 ppb). In addition, current water purification installations suffer from biofouling due to the accumulation of phosphate in the RO systems at concentrations favorable for microbial growth. Recently, this problem was addressed by using ferritin (hyperthermophilic protein nanocage), which can form and hold an iron-based nanoparticle inside the protein. The nanoparticle thus formed is capable of adsorbing oxyanions with higher affinity even below 1 ppb. The experimental program used a wide range of aqueous phosphate concentrations to monitor the phosphate removal by ferritin. Concentrations ranged from 5 to 200 ppb. The specific ferritin concentrations were also varied. The approach for the adsorption experiment was a standard batch liquid-phase equilibration. Radioactively labeled phosphate solution was equilibrated with ferric iron–loaded ferritin solution. The system was left overnight for equilibration. The aqueous phosphate solution and ferritin were separated using a column filter with a cutoff filter of 3 kDa. The ^{32}P concentration in the permeate was measured using a liquid scintillation counter, thus producing the final aqueous phase equilibrium concentration [21, 22].

23.5 SINGLE ENZYME NANOPARTICLES

Enzymes are the most versatile candidates for biosensing and bioremediation in areas of chemical conversions. Only certain limitations like short life span and lack of stability underscore their ability to provide cost-effective options. However, nowadays, advancements in technology like enzyme immobilization and genetic modification make them more suitable candidates for bioremediation purposes. A recent interesting development is single enzyme nanoparticles (SENs). A SEN is nothing but a single molecule of an enzyme, surrounded by a protective silicate cage of a few nanometers, but the active sites are chemically accessible to the ligands [23]. The use of individual enzymes for remediation processes has recently become more common since it offers a variety of advantages as compared to traditional microbial methods. Major advantages are their stability under extreme conditions such as high and low pH, high salinity, and temperature.

23.6 CARBON NANOTUBES

Significant advancements in nanotube research have been witnessed in recent years, especially in CNT. A wide range of commercial applications have been identified for CNTs [24]. In recent years, several new reports pertaining to the role of CNTs in pollution control, waste water purification [25], and desalination [26] have created a new surge of interest in nanotube research (Table 23.1).

CNTs are the most versatile adsorbents for an array of pollutants like heavy metals, environmental residues, and organic pollutants such as chloroform, benzene, dichlorobenzene, trihalomethanes, and polyaromatic hydrocarbon [27]. CNTs also have a phenomenal sorbent potential for inorganic pollutants such as fluoride, along with several divalent metal ions (Table 23.2). Interestingly, it was observed that crude and poor-quality CNTs having a high surface area have shown better adsorption capacity as compare to aligned CNTs. This suggests that modification and surface activation of functional groups of CNTs causes the removal of amorphous carbon [28]. Activation of CNTs under oxidizing conditions with chemicals such as HNO_3, $KMnO_4$, H_2O_2, NaOCl, H_2SO_4, KOH, and NaOH has been widely reported. During activation, the impurities and catalyst-supporting materials that alter the surface characteristics by introducing new functional groups are dissolved.

Recently it was reported that ceramic pore channels inducted with CNTs can efficiently remove oils from water [29]. This is a significant development for the control of oil spillage in water bodies. CNT-embedded magnesium oxide composite is reported to efficiently remove lead (II) from water [30].

TABLE 23.1 Basic advantages and functions of CNT-based water filtration systems

Functions	Applications
Removal of organic molecules and metal ions	Water filtration and remediation
Removal of bacteria and viruses	Point-of-use treatment systems
Advance performances	**Applications**
1. High and fast retention of pollutants	Improved adsorption efficiency and capacity
2. Fast mass transport	Gravity driven or low-pressure operation
3. High mechanical properties and light weight	High compacted and portable filters
4. Possible regeneration by heat or chemical treatment	Reusable and economical systems

TABLE 23.2 Various types of CNT-based nanomaterials used in water purification specifically for pollutants

Nanomaterials	Pollutants	Sorption	References
SWCNTs	Chloroform ($CHCl_3$)	3.158 mg/g	Lu et al. (2005)
CNT polymer composite	Trichloroethylene	Nondetectable <0.01 PPB	Salipira et al. (2007)
CNT polymer composite	*P*-Nitrophenol	99% removal from a 10 mg/l spiked water	Salipira et al. (2007)
CeO_2-CNT	As (V)	82 mg/g	Peng (2005)
CeO_2-aligned CNT	Cr (V)	30.2 mg/g	Di (2006)
Acid-treated MWCNTs	Pb (II)	97.08 mg/g	Li et al. (2003)
	Cu (II)	24.49 mg/g	
	Cd (II)	10.86 mg/g	
CNTs (diameter 2 nm)	Microcystin toxin (MC) Herbicides	14.8 mg/g	Yan et al. (2006)

23.7 POROUS MEDIA AND CERAMICS

Porous ceramics are another good candidate for the development of water filters because of their increased surface area during nanoparticle growth [31]. Some ceramics like maganese oxide can effectively change the valence state of metal ion in water, which can easily removed by reaction with iron oxide to form a stable material [32]. Very often these reactions can be enhanced by adding other nanomaterials such as copper and copper oxide and this combination can effectively remove phosphate, heavy metals, lead, arsenic, and other pollutants. According to established norms, the permissible arsenic level in safe drinking water is about 50–10 µg/l; however, many regions of the world have very high arsenic levels in available groundwater. It has been shown that a combination of nano manganese oxide fibers combined with nano iron oxide is effective in removing arsenic from groundwater [33]. The nano MnO_2 and Fe_2O_3 base medium has higher breakthrough bed volume than other media; additionally, it is equally effective for As (III) and As (V). As we know, As (III) is more toxic than As (V). This medium is also effective for high Pb (II) adsorption. Porous ceramics activated with iron oxide nanoparticles can also provide effective remediation of pollutants like trichloroethylene (TCE) and tetrachloroethylene.

23.8 NANOFILTRATION

Nanofiltration can remove dissolved solids, but is often used to soften water by removing dissolved organic carbon. The nanofiltration membrane works in a similar manner to RO, except that a relatively low pressure is required because of the large pore size (0.05–0.005 µm). Membrane technologies including ultrafiltration, nanofiltration, and RO are now emerging as key components in the domain of advanced water purification and desalination systems (Fig. 23.4). Some recent reviews in the literature have focused on the importance of nanofiltration for removing cations, natural organic matter, biological contaminants, organic pollutants, nitrate, and arsenic from ground- and surface water [34]. It was also reported that nanofiltration can efficiently remove traces of uranium from seawater. Nanofiltration has also been evaluated for desalination of seawater, and it was observed that in combination with RO it could effectively convert saline water to potable water [35]. Nanofiltration can be a good solution for improving water quality even in large water distribution systems by substantially reducing organic and biological contaminants [36, 37].

Engineered nanoparticles nowadays provide unprecedented opportunities for efficient water purification catalysts and redox active media by their optical electronic and catalytic virtues. It is important to understand that chemical groups are key biological constituents that make these nanomaterials functionalize. Recently it has been reported that CNTs can be successfully used to fabricate carbon nanofilters. These CNT nanofilters are hollow cylinders with radially aligned CNT walls. It was observed that these nanofilters can efficiently reduce bacterial contaminations (Fig. 23.5) such as *Salmonella typhi*, *Escherichia coli*, and *Staphylococcus aureus*, as well as polio virus sabin 1 from contaminated water [38].

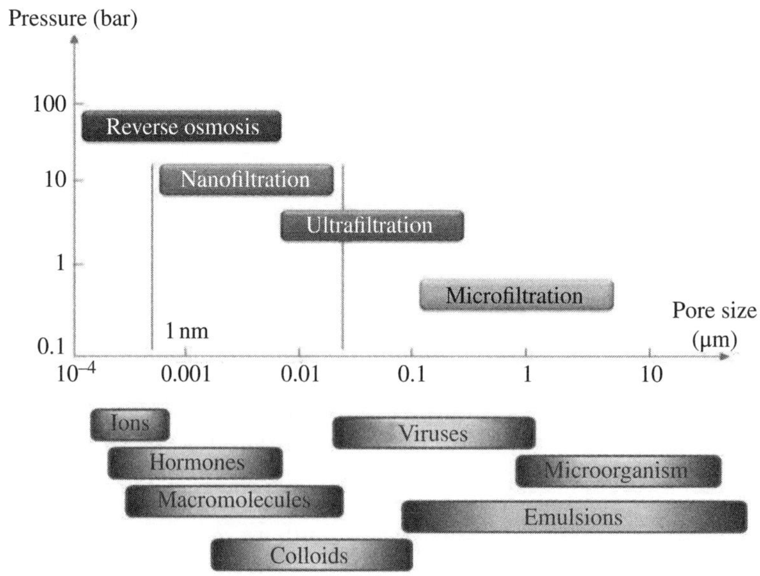

FIGURE 23.4 Ranges for nanofiltration, ultrafiltration, RO, and microfiltration, and indicative molecules passing through these membranes.

(a) (b) (c) (d) (e)

FIGURE 23.5 (a) Unfiltered water containing *E. coli*. (b) *E. coli* culture from polluted water. (c) Filtration by using a CNT filter. (d) Water filtered through a nanotube filter. (e) Culture of filtered water showing the absence of bacteria.

Alumina nanoparticles (7–25 nm) were successfully used for ultrafiltration membrane; the pore size and molecular cutoff weight of this membrane depends on the uniformity of the nanoparticles [39]. It was also observed that doping of alumina nanoparticles with Fe, Mn, and La can increase the selectivity and permeate flux. A novel nanofilter membrane has been fabricated by depositing 4.5–5 layer pairs of styrene sulfonate/poly (allylaminehydrochloride) onto porous alumina [40]. This new nanofilter membrane shows high water flux and retention of divalent cations like chlorides and sulfate. Another new approach was the deposition of multilayers of charged polypeptide (poly-9-glutamic acid or ploy-L-lysine) inside the pores of a highly functionalized polycarbonate membrane.

23.9 DENDRIMERS

Dendrite polymers are another good nanomaterial for the use of water purification; these are basically highly branched macromolecules with a core, interior branched cells, and terminal branched cells. They are usually symmetrical and spherical macromolecules with a potential to bind with an array of receptors on cell membranes or other biological surfaces. Many nanocomposites based on dendrimers have the capacity to enhance ultrafiltration. The first complete dendrimer family is PolyAMidoAMine (PAMAM), which was commercialized in 1990. Dendritic polymers exhibit very unique features like small size (1–20 nm) and high stability, which make them an attractive functional material for water purification. These nanocomposites can also be used in recyclable unimolecular micelles for recovering organic solutes from water, and to prepare scaffold and template for redox and catalytically active nanoparticles.

23.10 METAL NANOPARTICLES

Metallic nanoparticles exhibit a very strong UV-VIS absorption band that is not present in the spectrum of the bulk metal. These absorption bands are due to the collective excitation of the conduction electrons when the size of the particles is less than the mean free path of the electron in the metal. This is known as the localized surface plasmon resonance (LSPR). LSPR can be supported by a wide variety of structures like spheres, rods, and triangles. Particle size and shape are even more important in determining the position of the resonance. As the size of the particle increases, there is an increase in the amount of absorption and scattering, both of which contribute to the optical extinction of a metallic nanoparticle. Moreover, as the size increases, scattering takes over from absorption as the dominant contributor to the extinction, and there is a change in the position and width of the LSPR. Metal nanoparticles have been used as attractive adsorbents because they have a much larger surface area than bulk particles, in addition to their easy functionalization. It was observed that MgO nanoparticles are a very effective biocide both against gram-positive and gram-negative bacteria. These magnesium nanoparticles absorb large amounts of halogen molecules, sometimes up to 20% by weight. Gold nanoparticles duly coated with palladium can effectively remove TCE from groundwater. Zinc oxide nanoparticles have been used to remove arsenic from water, while the bulk zinc oxide cannot even adsorb arsenic. Ferrites and other iron-containing minerals such as akaganeite, feroxyhyte, ferrihydrite, goethte, hematite, lepidocrocite, maghemite, and magnetite have been used for waste water treatment. Magnetite was used for the removal of actinide and heavy metals. Besides iron, other metal nanoparticles like Co, Ni, and Pd are also widely used as magnetic nanoparticles for the removal of contaminants from water. Titanium dioxide (TiO_2) is a naturally occurring oxide of titanium and is found in nature as the well-known minerals rutile, anatase, and brookite. Titanium dioxide nanoparticles are commonly used as a photocatalyst. Magnetite (Fe_3O_4) is a mineral with the chemical formula Fe_3O_4, one of several iron oxides [41].

23.11 GRAPHENE

A graphene sheet is an infinite two-dimensional (2D) layer consisting of sp-hybridized carbon atoms that belongs to one of the five 2D Bravais lattices called the hexagonal (triangular) lattice. It is noteworthy that by piling up graphene layers, in an orderly way, one can form 3D graphite. Graphene was initially considered as a theoretical building block used to describe the graphite crystal, and to study the formation of CNTs (rolled graphene sheets).

Gold nanoparticle–graphene nanocomposites have been used for desalination purposes and it was observed that this nanocomposite system can enhance the palatability of water fulfilling several criteria. This nanocomposite was prepared by the reduction of gold ions using sodium citrate as the reducing and capping material. Both graphene–gold and graphene–silver nanocomposites are efficient photothermal materials used for water desalination. In one method, a mixture of 0.1 g graphene and 50 μg gold particles was added to 1 l of seawater, and exposed to sunlight. The rate of temperature rise upon exposure to light from a halogen lamp after adding different nanomaterials was measured. The mixture of gold and graphene was much more efficient than plasmonic nanomaterials alone. The obtained water via desalination using the graphene nanocomposite was tested chemically and biologically. All chemical testing was carried out according to the standard international protocols in order to determine water quality, namely, salinity, pH, dissolved oxygen test, hardness, heavy metal content, and the total amount of phosphate, sulfate, and carbonate ions (Table 23.3). The result indicated that water obtained from this nanocomposite was totally pure and free from all salts, metal ions, and heavy metals.

Many protozoa, bacteria, viruses, algae, and fungi are found in natural water systems. Some are pathogenic (typhoid, cholera, and amebic dysentry can result from waterborne pathogens). The obtained distilled water using this nanocomposite was examined and indicated that the water was totally clean and did not contain any microorganism. This might be because of the photothermal effect of the nanocomposite, which raises the temperature and causes bacterial death.

23.12 WATER SITE REMEDIATION/NANOREMEDIATION

In recent years, nanoremediation has become the main focus of research and development. There is great potential for the use of this technology to clean up contaminated sites and protect the environment from pollution. This eco-friendly technology is considered to be an effective alternative to the current practices of site remediation.

Nanoremediation methods involve the application of reactive materials for the detoxification and transformation of pollutants. These materials initiate both chemical reduction and catalysis of the pollutants of concern. The unique properties of nanomaterials make them best suited for *in situ* applications. Their small size and novel surface coatings enable them to achieve farther and wider distribution when compared to large-sized particles.

Over the years, the field of remediation has grown and evolved significantly into new technologies in attempts to improve the remediation process. One of the most established systems is "pump-and-treat" [42]. Pump-and-treat systems operate on the basis of removing contaminated groundwater from the ground, downstream of the contamination site, and then treating it

TABLE 23.3 Comparative analysis of gold–graphene nanocomposite for its efficacy in the removal of major ionic contaminants and also on various other parameters of water palatability

SN	Water quality	Graphene-gold nanocomposite	Common drinking water
1	pH	7	6.8
2	Calcium content	0	50 ppm
3	Chlorine content	0	100 ppm
4	Oxygen	1.5%	2.2%
5	Sulfate ions	0	20 ppm
6	Mg ions	0	30 ppm
7	Salinity	0	0.03%
8	Hardness	0	2%

before returning it to the ground. The only limitation of this technology is that it takes a long time to achieve cleanup goals [42, 43]. Pump-and-treat systems at some sites with dense nonaqueous phase liquids (DNAPLs) may need to operate for considerably longer periods [43]. Despite the number of limitations exhibited by these systems, pump-and-treat remedies still account for 67% of groundwater remedies proposed or in progress [42, 43].

In the early 1990s, the reducing capabilities of metallic substances, such as zero-valent iron (ZVI), began to be examined for their ability to treat a wide range of contaminants in wastewater. The most common deployment of ZVI has been in the form of permeable reactive barriers (PRBs) designed to remediate them [42]. PRBs, first installed at the field-scale in 1994, offer a substitute for the more established pump-and-treat systems. The first full-scale commercial PRB was approved for use in the state of California by the San Francisco Regional Water Quality Control Board (RWQCB) in 1994 [42]. This passive treatment system has been used to treat pollutants, including chlorinated hydrocarbons, nitro aromatics, polychlorinated biphenyls (PCBS), pesticides, and even chromate. The reducing capabilities of ZVI, can dechlorinate chemicals such as TCE and PCBs [44]. ZVI can reduce hexavalent chromium Cr (VI) to trivalent chromium Cr (III), and precipitate it out of the solution, immobilizing it as Cr (III) hydroxides or chromium–iron hydroxide solid solutions [42, 45].

23.13 NANOSCALE ZERO-VALENT IRON

Iron nanoparticles are an attractive component for nanoremediation. Iron at the nanoscale was synthesized from Fe (II) and Fe (III), using borohydride as the reductant. Nanoscale zero-valent iron (nZVI) particles range from 10 to 100 nm in diameter. They exhibit a typical core shell structure. The core consists primarily of zero-valent or metallic iron whereas the mixed valent, that is, Fe (II) and Fe (III) oxide, shell is formed as a result of oxidation of the metallic iron. Iron typically exists in the environment as iron (II) and iron (III) oxides. nZVI is generally preferred for nanoremediation because of the large surface area of the nanoparticles and the larger number of reactive sites [46], and also because it possesses dual properties of adsorption and reduction, as shown in Figure 23.6.

This enables it to be used for the remediation of a wide range of contaminants present *in situ*. Moreover, when ZVI was allowed greater access to the contamination site, it was found to give out less hazardous waste during the treatment process. ZVI can also be modified based on the contaminants present. It could be modified to include catalysts like palladium, coatings such as polyelectrolyte or triblock polymers [47], or be enclosed in emulsified oil micelles [48]. In 2003, nanoscale iron particles were investigated for their effect on a number of common pollutants in groundwater and contaminated soil. The results showed that the nanoscale iron particles were highly effective for the transformation and detoxification of a number of pollutants (Table 23.4). To remove the contaminants, the super paramagnetic property of iron nanoparticles was manipulated and retrieved using a magnetic field without ignoring its release into the environment. Laboratory tests have indicated that more than 99% of arsenic in water samples can be removed using 12 nm diameter iron oxide nanoparticles [46].

TCE, a hazardous organic contaminant present in water, can be removed by ZVI nanoparticles modified to contain an oil–liquid membrane. This oil–liquid membrane is generally composed of food-grade surfactants; the biodegradable oil and water are hydrophobic and form an emulsion with ZVI. This is termed as emulsified zero-valent iron (EZVI) [50]. Since all DNAPLs, such as TCE, are hydrophobic, the emulsion is miscible with the contaminant, allowing an increased contact between TCE DNAPL and the ZVI present within the oil emulsion droplet [51]. While the ZVI in the emulsion remains reactive, the chlorinated compounds are continuously dechlorinated within the aqueous emulsion droplet, producing a concentration gradient within the

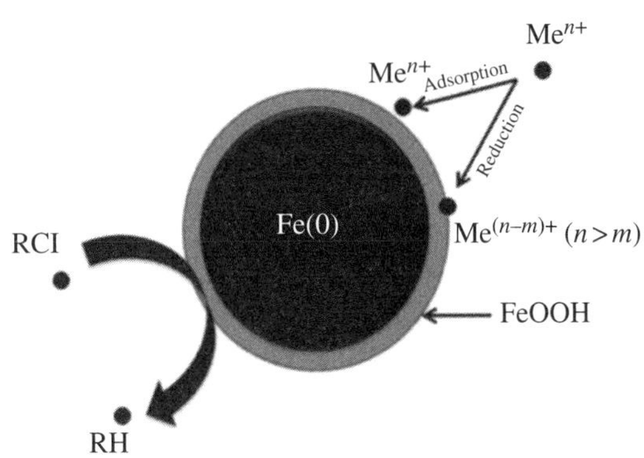

FIGURE 23.6 Schematic diagram of nanoscale zero-valant iron (nZVI).

TABLE 23.4 Contaminants remediated by nanoscale zero-valent iron (nZVI) [49]

Carbon tetrachloride	Chrysoidine	*cis*-Dichloroethene	Dichlorobenzenes
Chloroform	Tropaeolin	*trans*-Dichloroethene	Bromoform
Dichloromethane	Acid Orange	1,1-Dichloroethene	TNT
Chloromethane	Acid Red	Vinyl chloride	Chlorobenzene
Hexachlorobenzene	Mercury	PCBs	Dibromochloromethane
Pentachlorobenzene	Nickel	Dioxins	Dichromate
Tetrachlorobenzenes	Silver	Pentachlorophenol	DDT
Trichlorobenzenes	Cadmium	NDMA	Dichlorobromomethane
Arsenic	Lindane	Tetrachloroethene	Perchlorate
Orange II	Trichloroethene	Nitrate	

oil membrane, which in turn acts as a driving force to allow additional TCE migration into the membrane and additional degradation is carried out. A potential benefit of EZVI over nZVI for environmental applications is that the hydrophobic membrane surrounding the nZVI protects it from other groundwater constituents, such as some inorganic compounds, which might otherwise react with the nZVI, reducing its capacity or passivating the iron [52].

Another type of nanoparticle used for environmental applications is the bi-metallic nanoparticle (BNP). Bi-metallic nanoparticles consist of elemental iron or other metals in conjunction with a metal catalyst, such as platinum (Pt), gold (Au), nickel (Ni), or palladium [51]. The combination of metals to form a nanoparticle increases the kinetics of redox reaction, therefore catalyzing the reaction. The most commonly used and commercially available BNPs are the palladium and iron BNPs (Pd/Fe). The surface area normalized rate constant of iron BNPs combined with palladium (nZVI/Pd) was two orders of magnitude higher than that of nZVI [53]. Palladium and iron BNPs are generally used for the removal of TCE. In one of the studies, palladium was used to convert TCE into ethane with minimal formation of vinyl chloride and other chlorinated intermediates that often occur with anaerobic bioremediation and with iron metal [54].

nZVI and reactive nanoscale iron product (RNIP) comprise the most basic form of the nano iron technology [49, 55]. Particles of nZVI, typically about 100–200 nm in diameter, consist solely of ZVI (Fe^0). The most common approach to nZVI synthesis employs sodium borohydride as the key reductant [49]. In 1997, Wang et al. first produced the nanoscale iron particles in the laboratory using the method of sodium borohydride ($NaBH_4$) reduction [56]. By mixing $NaBH_4$ with $FeCl_3{\cdot}6H_2O$, Fe^{3+} is reduced according to the following reaction:

$$Fe\left(H_2O\right)_6^{3+} + 3BH_4 + 3H_2O \rightarrow Fe^0 + 3B\left(OH\right)_3 + 10.5H_2$$

In a laboratory-scale production of nZVI, Wang et al. achieved a particle size distribution of less than 100 nm for 90% of the particles produced. The BET surface area (Brunauer–Emmett–Teller (BET) theory that explains adsorption of gas molecules on a solid surface and is a important analysis technique for the measurement of the specific surface area of a material) for the particles was determined to be 33.5 m^2/g [56]. Following the reaction, the reduced particles of iron (Fe^0) created could be directly used for contaminant destruction. The stoichiometry of the reduction of TCE to ethane, a typical decontamination reaction, would proceed as follows:

$$C_2HCl_3 + 4Fe^0 + 5H^+ \rightarrow C_2H_6 + 4Fe^{2+} + 3Cl^-$$

RNIP particles vary slightly from nZVI particles, in that RNIP particles consist of approximately a 50:50 wt.% mixture of iron and magnetite (Fe_3O_4). The core of the particles consists of the elemental iron (α-Fe), while the Fe_3O_4 surrounds the Fe, forming an outer shell [55].

23.14 CONCLUSIONS

The aim of this chapter is to give an overall perspective of the use of nanoparticles to solve potential issues such as the more effective treatment of contaminated water for the purposes of drinking and reuse than through conventional means. Nanoremediation has the potential to clean up large contaminated sites *in situ*, reduce cleanup time, and eliminate the need for the removal of contaminants, hence reducing the contaminant concentration to near zero. A great deal of care needs to be taken if it has to be implemented in real life to avoid deleterious effects of unhygienic water. The success of these techniques in field conditions is a factor for interdisciplinary collaboration of chemistry, material science, and geology to cope with the challenges of this research.

REFERENCES

[1] WHO. 2004. Household water treatment and safe storage. Available at www.who.int/household_water/en/. Accessed April 13, 2014.

[2] McIntyre RA. Common nano-materials and their use in real world applications. Sci Prog 2012;95:1–22.

[3] Malsh I. The importance of interdisciplinary approaches: the case of nanotechnology. IPTS Report. Seville, Spain: IPTS; 1997. Report nr 13.

[4] Shannon MA. Science and technology for water purification in the coming decades. Nature 2008;452:301–310.

[5] Cygan RT, Brinker CJ. A molecular basis for advanced materials in water treatment. MRS Bull 2008;33:42–47.

[6] Savage N, Diallo MS. Nanomaterials and water purification: opportunities and challenges. J Nano Res 2005;7:331–342.

[7] Miller JE. Review of water resources and desalination technologies. Sandia National Laboratories report. Albuquerque (NM): Sandia National Laboratories; 2003. Report nr SAND 2003–0800.

[8] Wagner L. Water desalination—tap into the liquid gold. Research report. London: Mora Associates; 2007.

[9] Fritzmann C, Lowenberg J, Wintgens T, Melin T. State-of-the-art of reverse osmosis desalination. Desalination 2007;216:1–76.

[10] Benga G. Birth of water channel proteins-the aquaporins. Cell Biol Int 2003;27:701–709.

[11] Kaufman Y, Berman A, Freger V. Supported lipid bilayer membranes for water purification by reverse osmosis. Langmuir 2010;26: 7388–7395.

[12] Agre P, Mathai JC, Smith BL, Preston GM. Functional analyses of aquaporin water channel proteins. Methods Enzymol 1999; 294:550–572.

[13] Yakata K, Tani K, Fujiyoshi Y. Water permeability and characterization of aquaporin-11. J Struct Biol 2011;174:315–320.

[14] Chrispeels MJ, Agre P. Aquaporins: water channel proteins of plant and animal cells. Trends Biochem Sci 1994;19:421–425.

[15] Li X, Wang R, Tang C, Vararattanavech A, Zhao Y, Torres J, Fane T. Preparation of supported lipid membranes for aquaporin Z incorporation. Colloids Surf B Biointerfaces 2012;94:333–340.

[16] Murata K, Mitsuoka K, Hirai T, Walz T, Agre P, Heymann JB, Engel A, Fujiyoshi Y. Structural determinants of water permeation through aquaporin-1. Nature 2000;407:599–605.

[17] Bomholt J, Hélix-Nielsen C, Scharff-Poulsen P, Pedersen PA. Recombinant production of human aquaporin-1 to an exceptional high membrane density in Saccharomyces cerevisiae. PLoS One 2013;8:e56431.

[18] Douglas T, Ripoll DR. Calculated electrostatic gradients in recombinant human H-chain ferritin. Protein Sci 1998;7:1083–1091.

[19] Kim I, Hosein H, Strongin DR, Douglas T. Photochemical reactivity of ferritin for Cr (VI) reduction. Chem Mater 2002;14:4874–4879.

[20] Hosein HA, Strongin DR, Allen M, Douglas T. Iron and cobalt oxide and metallic nanoparticles prepared from ferritin. Langmuir 2004;20:10283–10287.

[21] Kampschreur MJ, Temmink H, Kleerebezem R, Jetten MS, van Loosdrecht MC. Nitrous oxide emission during wastewater treatment. Water Res 2009;43:4093–4103.

[22] Martins AM, Pagilla K, Heijnen JJ, van Loosdrecht MC. Filamentous bulking sludge—a critical review. Water Res 2004;38:793–817.

[23] Kim J, Grate JW. Single-enzyme nanoparticles armored by a nanometer-scale organic/inorganic network. Nano Lett 2003;3:1219–1222.

[24] De Volder MF, Tawfick SH, Baughman RH, Hart AJ. Carbon nanotubes: present and future commercial applications. Science 2013;339: 535–539.

[25] Maggini L, Raquez JM, Marega R, Jensen Ahrens J, Pineux F, Meyer F, Dubois P, Bonifazi D. Magnetic poly(vinylpyridine)-coated carbon nanotubes: an efficient supramolecular tool for wastewater purification. ChemSusChem 2013;6:367–373.

[26] Gethard K, Sae-Khow O, Mitra S. Water desalination using carbon-nanotube-enhanced membrane distillation. ACS Appl Mater Interfaces 2011;3:110–114.

[27] Brooks AJ, Lim HN, Kilduff JE. Adsorption uptake of synthetic organic chemicals by carbon nanotubes and activated carbons. Nanotechnology 2012;23:294008.

[28] Rybak-Smith MJ, Sim RB. Complement activation by carbon nanotubes. Adv Drug Deliv Rev 2011;63:1031–1041.

[29] Chen X, Hong L, Xu Y, Ong ZW. Ceramic pore channels with inducted carbon nanotubes for removing oil from water. ACS Appl Mater Interfaces 2012;4:1909–1918.

[30] Saleh TA, Gupta VK. Column with CNT/magnesium oxide composite for lead (II) removal from water. Environ Sci Pollut Res Int 2012;19:1224–1228.

[31] Lin YH. Nitrification/denitrification in swine wastewater using porous ceramic sticks with plastic rings as supporting media in two-stage fixed-biofilm reactors. Water Sci Technol 2010;62:985–994.

[32] Brown J, Sobsey MD. Ceramic media amended with metal oxide for the capture of viruses in drinking water. Environ Technol 2009; 30:379–391.

[33] Najafpour MM, Rahimi F, Aro EM, Lee CH, Allakhverdiev SI. Nano-sized manganese oxides as biomimetic catalysts for water oxidation in artificial photosynthesis: a review. J R Soc Interface 2012;9:2383–2395.

[34] Van der Bruggen B, Lejon L, Vandecasteele C. Reuse, treatment, and discharge of the concentrate of pressure-driven membrane processes. Environ Sci Technol 2003;37:3733–3738.

[35] Mohsen MS, Jaber JO, Afonso MD. Desalination of brakish water by nanofiltration and reverse osmosis. Desalination 2003;157:167.

[36] Peltier S, Cotte E, Gatel D, Herremans L, Cavard J. Nanofiltration improvements of water quality in a large distribution system. Water Sci Technol 2003;3:193–200.

[37] Modise CM, Bendick JA, Miller CJ, Neufeld RD, Vidic RD. Use of hydrophilic and hydrophobic microfiltration membranes to remove microorganisms and organic pollutants from primary effluents. Water Environ Res 2006;78:557–564.

[38] Srivastava A, Srivastava ON, Talapatra S, Vajtai R, Ajayan PM. Carbon nanotube filters. Nature Mater 2004;3:610–614.

[39] DeFriend KA, Wiesner MR, Barron AR. Alumina and aluminate ultrafiltration membranes derived from alumina nanoparticles. J Membr Sci 2003;224:11–28.

[40] Stanton BW, Harris JJ, Miller MD, Bruening ML. Ultrathin, multilayered polyelectrolyte films as nanofiltration membranes. Langmuir 2003;19:7038–7042.

[41] Mallampati R, Valiyaveettil S. Biomimetic metal oxides for the extraction of nanoparticles from water. Nanoscale 2013;5:3395–3399.

[42] U.S. EPA. Permeable reactive barriers technologies for contaminant remediation. Remedial Technology Development Forum report. Washington, DC: U.S. Environmental Protection Agency; 1998. Report nr EPA/600/R-98/125.

[43] U.S. EPA. Cleaning up the nation's waste sites: markets and technology trends. Washington, DC: U.S. Environmental Protection Agency; 2004. Report nr EPA 542-R-04-015.

[44] Lin Y, Fryxell GE, Wu H, Engelhard M. Selective sorption of cesium using self-assembled monolyaers on mesoporous supports. Environ Sci Technol 2001;35:3962–3966.

[45] U.S. EPA. Use of monitored natural attenuation at superfund, RCRA corrective action, and underground storage tank sites. OSWER Directive 9200.4–17. Washington, DC: U.S. Environmental Protection Agency; 1997.

[46] Rickerby DG, Morrison M. Nanotechnology and the environment: a European perspective. Sci Technol Adv Mater 2007;8:19–24.

[47] Saleh N, Sirk K, Liu Y, Phenrat T. Surface modifications enhance nano iron transport and NAPL targeting in saturated porous media. Environ Eng Sci 2007;24:45–57.

[48] Hydutsky BW, Mack EJ, Beckerman BB, Skluzacek JM, Mallouk TE. Optimization of nano- and micro iron transport through sand columns using polyelectrolyte mixtures. Environ Sci Technol 2007;41:6418–6424.

[49] Zhang WX. Nanoscale iron particles for environmental remediation: an overview. J Nanopart Res 2003;5:323–332.

[50] Quinn J, Geiger C, Clausen C, Brooks C, Coon K. Field demonstration of DNAPL dehalogenation using emulsified zero-valent iron. Environ Sci Technol 2005;39:1309–1318.

[51] Otto M, Floyd M, Bajpai S. Nanotechnology for site remediation. Remediation 2008;19:99–108.

[52] Hara O, Krug S, Quinn TJ, Clausen C, Geiger C. Field and laboratory evaluation of the treatment of DNAPL source zones using emulsified zero-valent iron. Remediation 2006;16:35–56.

[53] Zhang WX, Elliot DW. Applications of iron nanoparticle in ground water remediation. Remediation 2006;16:7–21.

[54] Nutt MO, Heck KN, Alvarez P, Wong MS. Improved Pd-on-Au bimetallic nanoparticle catalysts for aqueous phase trichloroethane hydrodechlorination. Appl Catal B 2006;69:115–125.

[55] Okinaka K, Jasdanian A, Shimizu H, Okita T, Kakuya K. Treatment of 1,1,1-trichloroethane with reactive nanoscale iron products in simulated groundwater. Proceedings of the Fourth International Conference on Remediation of Chlorinated and Recalcitrant Compounds; Paper 2E-01. Columbus: Battelle Memorial Institute; 2004.

[56] Wang CB, Zhang WX. Synthesizing nano scale iron particles for rapid and complete dechlorination of TCE and PCBs. Environ Sci Technol 1997;31:2154–2156.

PART X

NANOCATALYSIS

24

NANOCATALYTIC WASTEWATER TREATMENT SYSTEM FOR THE REMOVAL OF TOXIC ORGANIC COMPOUNDS

Sodeh Sadjadi
Nuclear Science and Technology Research Institute, Tehran, Iran

24.1 INTRODUCTION

Water is a limited resource; the world is facing formidable challenges in meeting the rising demand for clean water as the available supplies of freshwater are decreasing due to contamination of this natural resource.

Heterogeneous catalysis has received a tremendous amount of interest, both from a scientific and an industrial perspective. Catalysts are also essential in converting hazardous waste into less harmful products. One area of catalysis that is developing at a rapid pace is nanocatalysis. Nanoparticles have great potential as water-purifying catalysts and redox-active media due to their large surface areas and their size- and shape-dependent optical, electronic, and catalytic properties [1]. To avoid the dangerous accumulation of organic pollutants in the aquatic environment, powerful oxidation methods are developed for their complete destruction from natural waters and wastewaters. Although different isolation, physical separation, biological, and chemical methods can be utilized, the most promising techniques are the so-called advanced oxidation processes based on the in situ generation of the hydroxyl radical (·OH) as an oxidant of organic matter. These processes involve catalyzed chemical, photochemical, and electrochemical techniques to bring about chemical degradation of organic pollutants [2].

This chapter is divided into six sections. Following the introductory section, Sections 24.2–24.6 highlight the results of selected studies on the use of nanomaterials as catalysts for different advanced oxidation processes.

24.2 PHOTOCATALYTIC OXIDATION

24.2.1 Introduction

Photocatalysis may be described as a photoinduced reaction that is accelerated by the presence of a catalyst [3]. When a semiconductor photocatalyst is irradiated with photons of energy equal to or greater than its band gap, electrons are excited from the valence band to the conduction band, leaving positively charged holes in the valence band, thus leading to a charge separation [4]. The photogenerated electrons can react with electron acceptors such as O_2 adsorbed on the catalyst surface or dissolve in water to give O_2^{-} radicals [5]. The photogenerated holes react with OH^- or H_2O, oxidizing them into ·OH radicals [6]. These active radicals are responsible for the decomposition of organic compounds (Fig. 24.1a). Photocatalysts also promote a photocatalytic reaction by acting as mediators for the charge transfer between two adsorbed molecules (Fig. 24.1b). Photocatalysts quench the excited state either by accepting an electron or transferring the charge to another substrate [7]. The efficiency of photocatalysis depends on how well one can prevent this charge recombination [8]. There has been considerable interest in the use photocatalysis

Nanomaterials for Environmental Protection, First Edition. Edited by Boris I. Kharisov, Oxana V. Kharissova, and H. V. Rasika Dias.

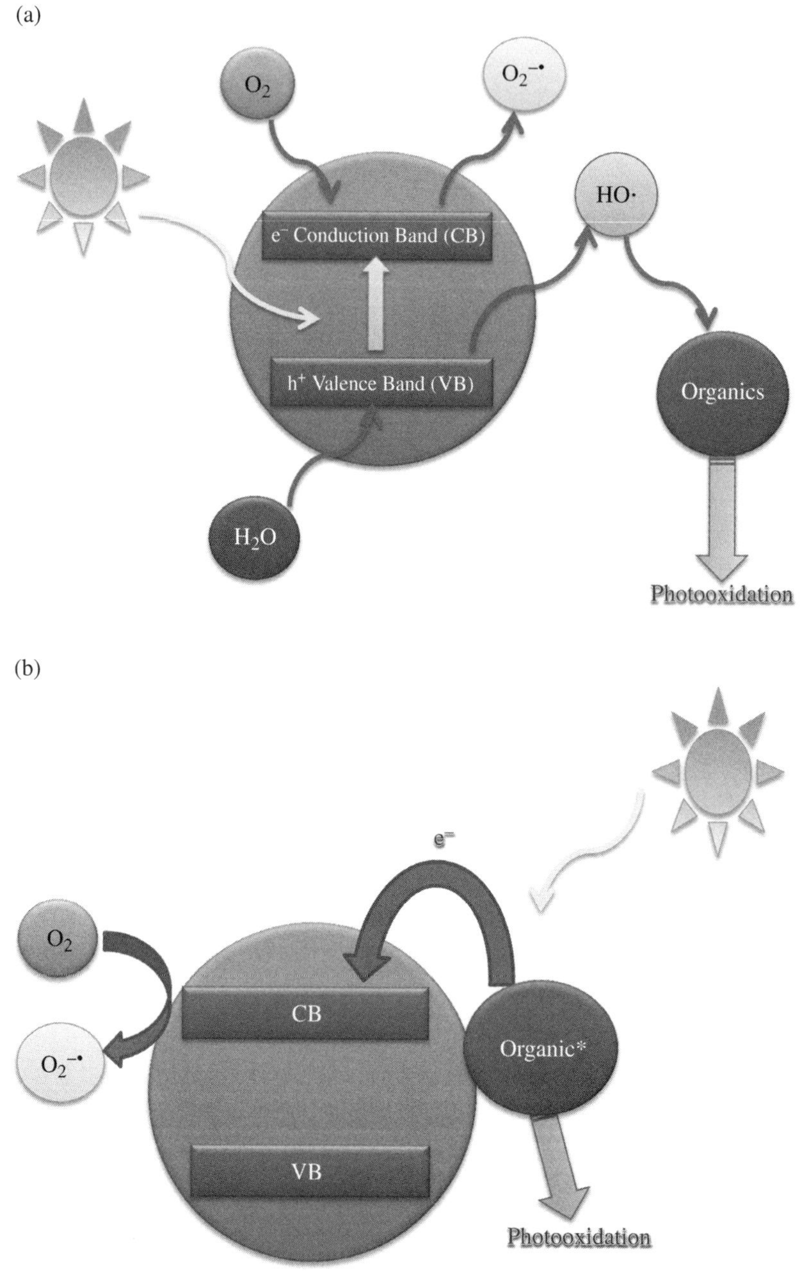

FIGURE 24.1 Photoinduced charge-transfer processes in semiconductor nanoparticles (a) upon bandgap excitation. (b) Nanoparticles act as mediators for the charge transfer between two adsorbed molecules.

for environmental remediation. Currently, much literature exists on the development of new photocatalytic nanomaterials with regard to this aspect. In this section, we review the basic concepts of nanocrystalline photocatalysts and ways to enhance their photocatalytic activities. Some examples of nanocrystalline semiconductors, used in the photocatalysis of organic compounds, are mentioned here (Table 24.1; [9]) but the reader may refer to reference articles for a detailed discussion on basic synthesis strategies and photocatalytic reaction conditions.

24.2.2 Nanocrystalline Photocatalysts

As the size of the semiconductor particle is reduced below a critical diameter, splitting of the energy bands into discrete, quantized levels occurs. Particles that exhibit these quantization effects are often called "Q-particles." The optical, electronic, and catalytic properties of Q-particles drastically differ from those of the corresponding macrocrystalline substance [10]. The band

TABLE 24.1 Some examples of nanocrystalline semiconductors used in photocatalysis of organic compounds

Nanocatalyst	Organic pollutants	References	Nanocatalyst	Organic pollutants	References
$ZnFe_2O_4$	Phenol	[9a]	Polyoxometalates supported on yttrium-doped TiO_2 ($H_3PW_{12}O_{40}$–Y–TiO_2)	Methyl orange	[9u]
TiO_2 films	Methyl orange	[9b]	$ZnAl_2O_4$ nanoparticles	Gaseous toluene	[9v]
ZnO and Pt–ZnO films	Phenol	[9c]	TiO_2 nanoparticles supported on porous glass beads	Methyl orange	[9w]
Core/shell-structured ZnO/SiO_2 nanoparticles	Rhodamine B	[9d]	ZnO–bentonite nanocomposite	Phenol	[9x]
Hydrophilic ZnS nanocrystals	Basic violet 5BN	[9e]	Yttrium Orthovanadate (YVO_4) nanoparticles	Direct blue 53	[9y]
Ag-loaded TiO_2 nanotube arrays	Methylene blue	[9f]	Nb-loaded ZnO Nanoparticles	Phenol	[9z]
MWCNTs loaded with Ag nanoparticles	Rhodamine B	[9g]	Pr-doped TiO_2	Phenol	[9aa]
Hydroxyapatite-supported Ag_3PO_4 nanoparticles	Methylene blue	[9h]	Layered $La_2Ti_2O_7$ nanosheets	Methyl orange	[9ab]
Bi_3NbO_7 nanoparticles	Rhodamine B	[9i]	SnS_2	Methyl orange	[9ac]
Co-doped TiO_2 nanoparticles	2-Chlorophenol	[9j]	MoS_2/TiO_2	Phenol	[9ad]
Nano-aluminum oxide	Pyridine	[9k]	$BiVO_4/Bi_2O_2CO_3$ nanocomposites	Rhodamine B	[9p]
Lanthanide(La^{3+}, Nd^{3+}, or Sm^{3+})-doped ZnO nanoparticles	4-Nitrophenol	[9l]	CNTs/P–TiO_2	Methyl orange	[9ae]
$CdS/La_2Ti_2O_7$	Methyl orange	[9m]	M@TiO_2 (M = Au, Pt, Ag)	Benzene	[9af]
$TiO_2/SiO_2/NiFe_2O_4$	Violet 5B	[9n]	$CoFe_2O_4$–Cr_2O_3–SiO_2	Methylene blue	[9ag]
C-doped $Zn_3(OH)_2V_2O_7$ nanorods	Methylene blue	[9o]	$MgFe_2O_4/TiO_2$ composite	Rhodamine B	[9ah]
$BiVO_4/Bi_2O_2CO_3$	Rhodamine B	[9p]	Fe_3O_4/ZnO	Methyl orange	[9ai]
Ag–AgI/Fe_3O_4@SiO_2	Rhodamine B and 4-chlorophenol	[9q]	NiO/$AgNbO_3$	Methylene blue	[9aj]
CuxS/TiO_2 copper sulphide/titanium oxide	Methylene blue, methyl orange	[9r]	Fe–Co–TiO_2	Rhodamine B	[9ak]
$Ln_2Ti_2O_7$	Methylene blue	[9s]	Chitosan/CdS	Cango red	[9al]
V-doped TiO_2 nanoparticles	Methylene blue and 2,4-dichlorophenol	[9t]	Fe^{3+}–TiO_2–zeolite	Methyl orange	[9am]

gap energy of Q-particles is increased by reducing the particle size, which is a great practical advantage. One disadvantage of nanosized particles is the need for light with a shorter wavelength for photocatalyst activation than for bulk materials [11].

Crystal size and surface area can play an important role in affecting the photocatalytic activity of photocatalysts. Wang et al. conducted experiments on the decomposition of chloroform and observed an optimal TiO_2 nanoparticle size for maximum photocatalytic efficiency. The particle's photocatalytic activity increased as the particle size was decreased from 21 to 11 nm, but when the size was reduced further to 6 nm, the activity decreased [12]. As an explanation to these observations, Zhang et al. suggested that the high surface area of the nanoparticles increases the photonic efficiency due to the increased interfacial charge–carrier transfer rates. However, below a certain particle size, charge recombination is faster than interfacial charge–carrier transfer processes [13]. Thus, there exists an optimum particle size for maximum photocatalytic efficiency. In another investigation, Liao et al. manipulated TiO_2 nanoparticles into spheres, cubes, ellipsoids, and nanorods using various surfactants. Cube-shaped and smaller-sized TiO_2 nanoparticles showed increased red shift in the UV–vis light region. This reduces the energy needed for photocatalysis [7].

24.2.3 Nanocatalyst Enhancements

Among all semiconductors, TiO_2 is the most widely used photocatalyst. It has two stable crystalline phases—anatase and rutile—which are achieved at synthesis temperatures of typically around 350 and 800°C, respectively. Nanocatalyst systems that include iron oxides, sulphides, and selenides are considered unstable for catalysis because they readily undergo photoanodic corrosion [10a, 14]. ZnO is also unstable, as it readily hydrolyzes to form $Zn(OH)_2$ on the particle surface, which deactivates the catalyst [14b]. Therefore, it is natural that most of the studies reviewed here are centered around TiO_2 nanocluster properties. There are several ways to improve the photocatalytic activity of TiO_2-based catalysts. The sensitization of TiO_2 with a second component to enhance activity and shift the wavelength of irradiation into the visible region is a goal that has been eagerly pursued [11]. There are several ways to achieve this goal such as doping TiO_2 nanomaterials with other elements, coupling two semiconductor systems, organic–inorganic nanostructured composites, and dye sensitization of nanocrystalline films.

24.2.3.1 Doping Doping is defined as the inclusion or substitution of a foreign atom into the TiO_2 lattice to induce new optical transitions that are not available in pure TiO_2.

24.2.3.1.1 Self-Doped TiO_2 (Reduced) Self-doping is caused by a stoichiometric imbalance in which oxygen is lost and reduced cationic sites are incorporated into the lattice. This imbalance can be found in many forms, for example, formation of isolated point defects such as oxygen vacancies or titanium interstitials, or formation of networks of such point defects. New optical excitation events and subsequently new photoactivity (such as visible light photoactivity) for self-doped TiO_2 is due to the presence of reduced cationic sites on the surface of TiO_2 [15].

24.2.3.1.2 Cation-Doped TiO_2 Ion doping into titanium dioxide can influence the performance of these photocatalysts. Several research groups have observed visible light photoactivity for cation-doped TiO_2. In a study conducted by Lee et al. on the photocatalytic decomposition of methanol, anatase doped with Cr^{3+}, Ni^{2+}, and Fe^{3+} showed greater activity in the visible light region than pure TiO_2 [16]. Studies by different research groups can lead to different, sometimes contradictory results. For example, Cr- and Fe-doped anatase showed longer-lived trapped holes than undoped anatase for the photodecomposition of acetaldehyde and toluene, but it did not result in increased UV photoactivity [17]. Some cationic dopants (such as Fe^{3+} and TV^{4+}) exhibited much longer-lived carrier lifetimes than undoped TiO_2, whereas other cations (e.g., Al^{3+}) resulted in shorter lifetimes [15]. The dopant ions may occupy either lattice or interstitial sites and, depending on their location, they can function either as hole and electron traps, or mediate interfacial charge transfer [14b]. According to Choi and coworkers [18], the optimum dopant concentration balances the ability of these centers to prevent charge recombination. Zhang et al. found that this optimum concentration is particle size–dependent and decreases with an increase in size [13].

24.2.3.1.3 Anion-Doped TiO_2 The most extensively studied anion dopant is nitrogen (N), but other anionic dopants (e.g., C, halides, P, and B) have also been examined. Modification of TiO_2 with N results in increased UV photoactivity [15]. The key question that is open to dispute is the chemical nature and the location of the species that shift the TiO_2 absorption properties into the visible region. Species such as NO_x, NH_x, and N^{2-} have been proposed, not to mention the NO^-, NO_2^-, and NO_3^- species that have been confirmed [19]. As N-doped TiO_2 samples are prepared using different methods that result in different N-doping concentrations, crystalline types, and surface areas, the behavior of the various samples cannot be compared accurately [20]. The visible light photoactivity of N-doped TiO_2 is often much less than under UV. These photocatalysts are also often less active when compared with TiO_2 under UV irradiation [19b].

24.2.3.2 Coupling Two Semiconductor Systems Recently, many researchers have regarded coupled and capped semiconductor systems as promising photocatalysts (such as CdS–TiO_2, CdS–ZnO, CdS–Ag_2S, ZnS–CdS, CdS–ZnS, ZnS–CdSe, ZnO–ZnS, ZnO–ZnSe, and CdS–HgS [7]). In dual semiconductor systems, the photogenerated electrons in one semiconductor are injected into the lower-lying conduction band of the second semiconductor (Fig. 24.2). As illustrated in parts (a) and (b) of Figure 24.2, the charge separation mechanism governing the two capped and coupled semiconductor systems is similar. However, the interfacial charge transfer in this multicomponent semiconductor system is significantly different [21]. While in coupled systems the two semiconductors are in contact with each other and both holes and electrons are accessible for selective redox processes on different particle surfaces, the capped nanoclusters essentially have a core–shell geometry. Only one of the charge carriers is accessible at the surface. Hence, the electron gets trapped within the core particle, and is not readily accessible for the reduction reaction [11]. The control of the shell thickness or the particle radius of the core is a delicate point in the fabrication of core–shell nanostructures and deserves special attention. These factors control the charge separation and photocatalytic properties of core–shell nanocatalysts. In capped systems, if the shell is too thick, only the holes are accessible at the surface of the photocatalyst, resulting in selective interfacial charge transfer and the photooxidation reaction efficiencies of the resulting photocatalyst generally improve. However, the reducing charge carriers are not utilized during the photocatalyic reaction, since these nanoparticles do not often exhibit photoreduction activity [21].

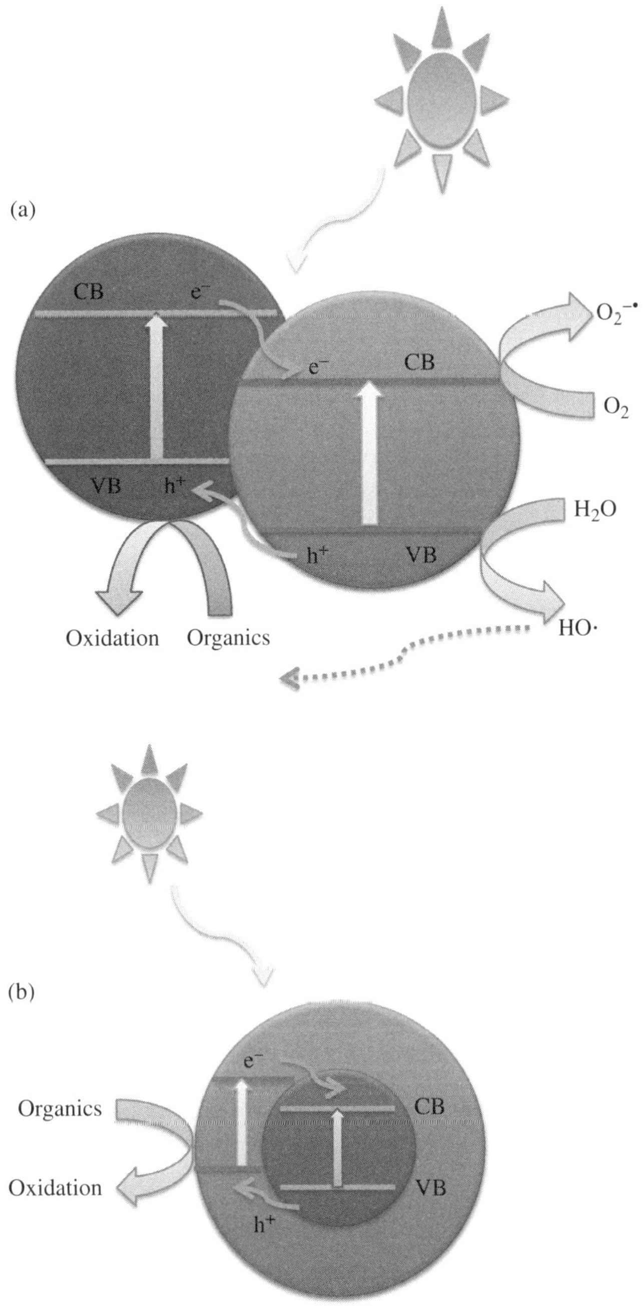

FIGURE 24.2 Charge transfer in (a) a coupled and (b) a capped semiconductor system.

Three-layered colloidal particles (such as CdS–HgS–CdS [22], Ag–TiO_2–SiO_2, Pt–TiO_2–SiO_2, or CdS–ZnO–CdO [7]) are another development in the field of surface-modified semiconductor nanoparticles.

24.2.3.3 Organic–Inorganic Nanostructured Composites In recent years, organic–inorganic materials have attracted increasing research attention because of their potential usage as catalytic nanomaterials. Considerable effort has been undertaken to incorporate organic components within an inorganic lattice to achieve symbiosis of the properties of both components. For example, Braun et al. reported the synthesis of organic superlattices based on cadmium sulphide and cadmium selenide [23]. Tenne et al. also synthesized inorganic WSe_2 and PtS_2 compounds, with a graphite-like crystal structure. These compounds can be used to construct fullerene-like structures and nanotubes, which are promising candidates for applications in catalysis [24].

24.2.3.4 Nanocrystalline Films and Dye Sensitization Thin metal oxide films exhibit interesting photochromic, electrochromic, photocatalytic, and photoelectrochemical (PECh) properties. Their nanoparticles are in electronic contact, allowing for electric charge percolation. This charge transport is highly efficient, and the quantum yield is practically unity [21]. TiO_2 nanotube array films and TiO_2 nanoparticulate films were investigated for the photodegradation of organic compounds [25]. The degradation efficiency of TiO_2 nanotube array films was much higher as compared with the TiO_2 nanoparticulate films with similar thickness and geometric area due to the higher internal surface area of the nanotube structure. The photoactivity of the TiO_2 nanotube films was strongly influenced by the thickness and very slightly by the tube diameter [25b, f, 26]. The effects of the nanotube structure including tube length, tube wall thickness, and crystallinity on the photocatalytic activity for the degradation of organic compounds were investigated in detail by several researchers [27]. Generally, the reaction rate increased gradually by increasing TiO_2 film thickness till a maximum due to the increment of the TiO_2 photocatalyst that may participate in the photodegradation. However, a further increase in TiO_2 film thickness resulted in somewhat decreasing the photoreaction rate. When the film is thicker than the light penetration depth, the active thickness is practically constant and the bottom film serves only as a support. The photoreactants have a longer diffusion path in the longer nanotubes; therefore, the photoreaction rate decreases [19b]. Nanostructured films made up of wide bandgap semiconductors, such as TiO_2, respond only in the UV region. In this film, the effective surface area can be enhanced 1000-fold; thus, light absorption can be modified with only a dye or a short bandgap semiconductor monolayer adsorbed on each particle [28]. For example, by successively dipping the ZnO film in Cd^{2+} and S^{2-} solutions, one can cast a thin film of CdS nanocrystallites. These heterostructured semiconductor films are photoelectrochemically active and generate photocurrent under visible light excitation.

24.2.3.5 Plasmon-Sensitized TiO_2 Nanoparticles Plasmonic nanoparticles such as silver and gold nanoparticles have been employed as an attractive approach to promote the performance of photocatalysts due to their surface plasmon resonance, which can enhance the visible light absorption and consequently increase the overall efficiency of the photocatalysts [9af, 29]. Core–shell nanoparticles have been applied to enhance photocatalytic activity by preventing the electron–hole recombination. It is expected that, due to the coupling of plasmon resonances in the core with the electron–hole pair generation in the shell, these hybrid Ag/Au TiO_2 nanoparticles will exhibit photocatalytic activity in the visible spectral range, thereby making more efficient use of solar energy. Zheng et al. [9af] recently synthesized plasmonic photocatalysts M–TiO_2 (M=Au, Pt, Ag) and investigated their visible light photoactivities for the oxidation of benzene. Ag/AgCl [30], Ag/AgBr/$WO_3{\cdot}H_2O$ [9ak], and Ag and Au nanoparticles supported on ZrO_2 and SiO_2 are also efficient visible light active photocatalysts for both oxidative and reductive reactions [31].

24.2.3.6 Immobilization of TiO_2 In many of the reported photocatalytic applications, nanocatalysts were used as nanoparticle slurry. But using nanoparticles in slurry form has several practical constraints. Particle–particle aggregation, increased process cost due to additional solid/liquid separation, reduced penetration depth of incident light because of the solution turbidity, and health hazard due to fugitive emissions of nanoparticles during slurry preparation are some of the practical constraints [32]. Immobilization of the photocatalysts onto a support medium can eliminate most of the limitations linked to the use of nanoparticles in slurry form. However, this can lead to a decrease in the surface area available for reactions compared to suspended systems. These problems can be resolved by supporting nanoparticles on large–surface area materials like quartz, silica gel, alumina, zeolites, clays, activated carbon (AC), or glass spheres [33], which could avoid the formation of macroscopic aggregates of photoactive particles [34].

Khatamian et al. have reported the preparation of TiO_2 nanoparticles and metal ion–doped TiO_2 nanoparticles supported on ZSM-5 zeolite [35]. The degradation efficiency of dye was much higher as compared with bare TiO_2 due to the high adsorption capacity of the supported TiO_2 that results in the higher concentration of dye molecules around the TiO_2 particles. Furthermore, the strong electrostatic field present in the zeolite framework can effectively reduce the recombination rate of electron–hole pairs, thus increasing the photocatalytic activity [35].

Another possible solution to facilitate the separation of catalysts is the use of solid catalysts with magnetic properties, which can be efficiently separated from the reaction medium applying an external magnetic field. Most of the magnetic photocatalysts developed so far contain a magnetic core (magnetite (Fe_3O_4), maghemite (γ-Fe_2O_3), or ferrite (e.g., $NiFe_2O_4$)), and a catalyst shell (TiO_2) [36]. An intermediate and inert layer is necessary between the TiO_2 shell and magnetic core in order to avoid photodissolution of iron and to prevent the magnetic core from acting as an electron–hole recombination center. Silica is chemically inert and does not affect any redox reactions at the core surface. Silica is also optically transparent, permitting light energy to penetrate.

Recently, there have been great advances in the development of TiO_2 photocatalytic membranes, since their simultaneous physical separation and chemical decomposition of water contaminants are promising for practical and full-scale applications [37].

Carbon-based TiO_2 composites have attracted much attention and become a very active field of research due to their unique properties. A combination of TiO_2 and AC could create many active sites for photocatalytic degradation. AC not only increases the surface area but also modifies the acid–base properties and the UV spectrum of TiO_2 [19b]. Promising results were also obtained with carbon nanotubes (CNTs)–TiO_2 hybrids, tested for the photodegradation of some organic compounds [38]. The photoactivity of the CNTs–TiO_2 is generally higher than that of TiO_2 and TiO_2–AC samples. It is considered that photo-induced charge transfer occurs in the electronic interaction between the carbon layers or walls of the multiwalled carbon nanotube (MWCNT) and TiO_2. The electrons excited in the conduction band of TiO_2 may migrate into the nanocylinder of the CNTs and lead to a reduced electron–hole pair recombination and thus to an increase in photon efficiency. O_2 adsorbed on the surface of the CNTs may accept the electron and form the $^{\cdot}$OH radical, which oxidizes the adsorbed dye directly on the surface [19b].

24.3 CATALYTIC OZONATION

24.3.1 Introduction

Ozonation is an attractive and increasingly important method for the degradation of water and wastewater organic pollutants. Unfortunately, total mineralization of organic pollutants cannot be achieved by this process. Ozonation of organic compounds usually leads to the formation of aldehydes and carboxylic acids, which cannot be degraded with ozone and may produce more toxic and resistant by-products [39]. Additionally, ozonation rate is relatively slow and this reaction is strongly pH-dependent and occurs faster with an increase of pH [40]. It has been demonstrated that ozonation of organic compounds can effectively be improved in the presence of a catalyst. In particular, solid metal oxides are more useful in catalytic ozonation than ionized metals, as ionized metals are considered toxic substances in water [41]. These processes are also very often characterized by a lower usage of ozone when compared with ozonation alone. In contrast to ozonation, catalytic ozonation allows for the effective formation of hydroxyl radicals even at a low pH. The ozone efficiency in heterogeneous catalytic ozonation is higher than that found in homogeneous catalytic ozonation. The catalytic ozonation can proceed by direct molecular ozone reactions and by an indirect pathway forming OH radicals after ozone decomposition.

24.3.2 Heterogeneous Catalytic Ozonation

A short overview of the literature concerning catalytic ozonation indicates that different or contradictory results are published in the literature. The controversies in this field are due to the unknown mechanisms of catalytic processes. Some authors suggest that hydroxyl radicals are the active species responsible for oxidation reactions, while others indicate that catalytic oxidation proceeds without hydroxyl radical formation. In the first case, it is proposed that catalysts cause ozone decomposition leading to hydroxyl radical formation. Another important issue is the adsorption of ozone and/or organics on the surface of the catalyst, as the usage of catalysts in aqueous solutions will lead to competition between water, ozone, and organic compounds for catalytic active sites [40].

The catalytic ozonation process is significantly influenced by key factors such as temperature and pH. The pH is the main factor for both ozone stability and catalyst surface properties such as the surface charge of the particles, which influences the adsorption capacity of the catalyst. TiO_2 P25 is the most commonly tested catalyst for the ozonation of organic pollutants in water. The main characteristics influencing the TiO_2 activity are surface area, crystalline phase, particle size, and the aggregate size in suspension [3]. Owing to the very strong Van der Waals interactions, TiO_2 nanoparticles are almost inclined to form aggregates. The pH and surface charges of the solution mainly determine the stability of TiO_2 nanoparticles in aqueous solution [42]. While agglomeration has been shown to be an important factor for the catalyst's activity, its measurement is difficult. An activity decrease could be expected by bringing the surface close to neutrality.

In an investigation on catalytic ozonation of nitrobenzene in the presence of nano-TiO_2, it was found that only the nanometer rutile TiO_2 is catalytically active Adsorption of nitrobenzene was found to play an important role and catalytic oxidation proceeds via hydroxyl radical formation. High mineralization of ozonation by-products takes place in the presence of the nanometer rutile TiO_2 [43].

Nanosized TiO_2 supported on Zeolite was prepared by Wang et al. and good catalytic performance of TiO_2/Zeolite was achieved in catalytic ozonation [9m].

Rodríguez et al. synthesized TiO_2-supported nickel nanoparticles. The presence of two phases (NiO/Ni) in the Ni/TiO_2 slightly improved the conversion of 2,4-dichlorophenoxyacetic acid compared with TiO_2 [44].

Recently, photocatalytic ozonation with TiO_2 (O_3/UV/TiO_2) has received enormous attention for the oxidation of organic pollutants. Jing et al. prepared an anatase TiO_2 photocatalyst by hydrothermal method. TiO_2 displayed more photocatalytic activity for dimethyl phthalate degradation than that of TiO_2 produced by a sol–gel process. The effectiveness of $TiO_2/UV/O_3$ system was more than that of the $TiO_2/UV/O_2$ (UV/O_3) system, implying a synergistic effect between photocatalysis and ozonation [45].

The efficiency of nano-ZnO particles has been studied for two different sizes: nano- and microsized particles, to study the effect of catalyst size on catalytic ozonation of 4-nitrochlorobenzene (4NCB). The effect of pH was also examined. According to the results, with increasing particle size from nanosize to microsize range and pH from 3.0 to 9.0, a higher degradation of 4NCB was obtained. Based on these results, it is suggested that the hydroxyl radical does not affect the degradation of 4NCB, but the surface of the catalyst plays an important role [46]. In contrast, a high degree of degradation during catalytic ozonation of dye and formaldehyde was achieved using MgO nanoparticles. In both cases, the catalyst has the ability to enhance the decomposition of ozone and promotes the formation of $OH^{\cdot}$ radicals [47].

Catalytic ozonation was carried out in the presence of nanosized magnetite and biogenic magnetite. Both of them were found to enhance the degradation of *para*-chlorobenzoic acid. The surface functional groups are responsible for the catalytic activity. Despite having more surface functional groups, the catalytic efficiency of biogenic magnetite is less than that of nanosized magnetite, as a result of the formation of bigger aggregates of biogenic magnetite than of nanosized magnetite [41].

To determine the effect of morphology on catalytic performance, CeO_2 nanoparticles with varying morphology (aggregates of irregularly shaped particles, crystalline, irregular aggregate morphology) were prepared to study the catalytic ozonation of phenol. The differences in the catalytic activity demonstrated by these three oxides were attributed to amounts of Ce^{3+} on the CeO_2 surface and, consequently, to the demand for oxygen to burn each precursor [48].

The combined use of ozone and AC has been identified as an interesting alternative to destroy toxic organic compounds. However, AC is easily oxidized in the process of ozonation. Ozone attacks AC, leading to soluble organic matter production, which increases the dissolved total organic carbon concentration in the absence of organic compounds [49]. CNTs may be more suitable for catalyst support than AC in liquid phase reactions. Liu et al. [50] showed that a higher catalytic activity was obtained for Pt deposited on CNTs than for Pt deposited on AC. The mechanism assumes radical chain reactions. MWCNTs with different surface chemical properties were prepared by Gonçalves et al. The correlation between surface properties of the MWCNTs and their performances as catalysts for the degradation of oxalic and oxamic acids through ozonation were evaluated. Generally, a low acid character of the catalysts enhances the efficiency of this process. MWCNTs present a higher catalytic activity compared to AC, which was justified by the differences in their surface chemistry [51].

Depositing noble metals (e.g., Au, Ag, and Pt) on the semiconductors is one of the best-known methods used to enhance catalytic activity. It is reported that noble metals can act as electron sinks to enhance the interfacial charge transfer processes [52]. However, it requires a solid support or stabilizing agent to interact with the various substrates. Some attempts have been made recently to study the ozonation reactions in the presence of gold nanoparticles [9t, 53]. For example, the catalytic ozonation of acid orange 10 in the presence of $Au\text{-}Bi_2O_3/Bi_2O_3$ nanocatalysts was investigated by Pugazhenthiran et al. [2]. $\beta\text{-}MnO_2$ nanowires have potential utility as catalysts for catalytic ozonation of organic compounds. For example, Dong et al. [54] revealed high efficiency of $\beta\text{-}MnO_2$ nanowires as a catalyst of phenol ozonation. A free radical mechanism was proposed which involves ozone decomposition and hydrogen peroxide formation. Furthermore, it was observed that Mn ions dissolving into the solution negatively impact the efficiency of the catalytic process [40].

Membrane application in surface water treatment provides many advantages over conventional treatments. However, this effort is hampered by the fouling issue. Ozone has been used to resolve the fouling issue; however, only ceramic membranes can be used in combination with ozone, as they are resistant to ozone. Coating the ceramic membranes with catalytic materials such as iron oxide [55] or manganese oxide has been shown to reduce their fouling [56] as they catalyze the oxidation of the organic foulants that deposit on the membrane surface. Corneal et al. discussed the effect of the number of coating layers on the structure of the resulting catalytic coating, and the filtration performance of the manganese oxide–coated membranes. The results showed that intermittent catalytic ozonation effectively maintained high permeate fluxes and prevented membrane fouling, as compared to that observed with the uncoated membrane [57].

24.4 CATALYTIC ELECTROCHEMICAL PROCESSES

Electrochemical methods for environmental applications are not a new technology. Anodic oxidation of wastewater was first mentioned as early as 1890. However, regarding cost-effectiveness, electrochemical techniques are still unfeasible in comparison with traditional biological techniques. In recent years, many investigations have been especially focused on the improvement of the electrocatalytic activity of electrode materials to facilitate the degradation efficiency of pollutants [58]. Attempts for an electrochemical oxidation treatment for organic pollutants can be divided into two categories: direct oxidation at the anode and indirect oxidation using appropriate, anodically formed oxidants. The direct electro-oxidation rate of organic pollutants is dependent on the diffusion rates of organic compounds in the active points of the anode, which results in low degradation efficiency. In addition, direct electro-oxidation occurs via the production of hydroxyl radicals, which are produced by the oxidation of water at the anode; water electrolysis often consumes much energy. The decomposition of the water increases with increasing voltage; however, the degradation efficiency of pollutants may not increase proportionally. Additionally, the pH of wastewater, which is an important parameter in the electrochemical treatment, could be changed by water electrolysis. For these reasons, the development of efficient catalytic electrochemical methods for water treatment based on the indirect electro-oxidation of pollutants involving electrogeneration of strong oxidants (such as hydroxyl radicals) is now in progress. These catalytic electrodes can produce oxidants and the addition of other chemicals is not required. In this process, the waste is oxidized in the bulk solution by these oxidants. In recent years, the electrode surface modification process via advances in nanofabrication to improve electrocatalytic properties has been one of the most active research areas. The improvement in performance appears therefore to be primarily caused by the increase in surface area, which results in high contacting rates and pollutant adsorption. The addition of catalytically active nanoparticles is shown to be an effective method to increase the transfer speed of electrons, reactivity, and diffusion in reaction [59]. In recent years, several electrodes have been evaluated to rank anode materials in terms of high stability, high activity toward organic oxidation, and cost. The types of electrodes tested included TiO_2, ZnO, Nb_2O_5, WO_3, SnO_2, ZrO_2, CdS, and ZnS, among others.

The PECh degradation of organic pollutants is an extension of the heterogeneous photocatalytic process, in which the catalyst is placed on an electrode that is controlled potentiostatically. In a PECh system, the photogenerated electron–hole pairs are separated by means of an externally applied electric field [60]. A particulate TiO_2 film electrode is usually used as a photoanode by coating TiO_2 nanoparticles on the supporting media such as a conducting glass or a metallic material. In these systems, UV light must be used in order to excite TiO_2 [61]. To efficiently utilize the solar light and improve the photooxidation technique for wastewater purification, TiO_2 film electrodes were modified by coupling with other semiconductors (such as CdS, ZnO, and SnO_2), and doping with metal or nonmetal ions. Dopants can shift the light absorption toward visible light (such as Fe-doped TiO_2 or N-doped TiO_2 electrodes) or exhibit a significant blue shift in their UV absorption spectrums (such as Si-doped TiO_2 nanofilm electrodes) [62]. TiO_2 film thickness can affect the efficiency of both light energy conversion and electron transfer. Yan et al. demonstrated that there exists an optimal thickness of the TiO_2 film for the maximum degradation rate of phenol. In their experiments, they observed an improvement in activity when the thickness was increased, but the activity decreased when the thickness was increased further to an 11-layer film. They attributed this to its relatively high resistance to electron transfer. In addition, the thick film can shield the catalysts from light [62, 63].

Highly oriented titania nanotube arrays with {101} crystal face were prepared by Hou et al. The PECh catalytic activity toward the oxidation of acid orange 7 of nanotube arrays was particularly higher for P-25 TiO_2 than for TiO_2 films [64]. The degradation rate of organic pollutants was enhanced using Zn-doped TiO_2 nanotube [65] and TiO_2 nanotube/Sb-doped SnO_2 electrodes [66]. The PECh degradation of dyes has also been performed using other nanostructured films like indium tin oxide film [67] and Bi_2WO_6 nanoflake film electrodes [61]. A higher PECh degradation of naphthol blue-black has been observed for WO_3 film electrodes than for TiO_2 nanoparticulate film electrodes [68].

It has been demonstrated that CNTs functionalized with ultrathin foreign species coatings are an ideal electrode material in the field of electrocatalytic degradation. Recently, SnO_2/CNT composite electrodes, and Eu_2O_3/CNT and CeO_2/CNT core–shell structures have been synthesized. The preparation method has a significant effect on the photoelectrocatalytic degradation of the dye. For example, Zhang et al. synthesized SnO_2/CNT composite electrodes by a liquid deposition method and a solvothermal method, which show a higher electrocatalytic performance than that of the liquid deposition method [69].

24.5 CATALYTIC FENTON PROCESS

24.5.1 Introduction

Fenton chemistry involves the generation of a hydroxyl radical from hydrogen peroxide (H_2O_2) in the presence of Fe^{2+} or Fe^{3+} ion (Eq. 24.1). The generated hydroxyl radicals can decompose a wide range of organic compounds [70].

$$Fe^{2+} + H_2O_2 \rightarrow Fe^{3+} + HO^{\bullet} + OH^- \quad (24.1)$$

It is well known that the homogeneous Fenton process has many disadvantages; it requires to be operated at pH < 3.0 for preventing the precipitation of Fe^{2+} and Fe^{3+} ions [71], and therefore this system suffers from the draining of the ferrous catalyst and the hydrolysis of iron ions (limited pH range) [72]. The resulting sludge of ferric hydroxide requires a further separation and disposal, which makes it a costly process [73].

24.5.2 Heterogeneous Fenton-Like Systems

Heterogeneous Fenton-like systems using iron-supported catalysts, for example, zero-valent iron ((ZVI) Fe^0) [74], goethite (α-FeOOH) [75], Fe_3O_4 [76], and Fe^0/Fe_3O_4 [77], have recently been developed [78]. However, the reaction rate of heterogeneous Fenton catalysts is generally slower than that of homogeneous catalysts due to the agglomeration of solid particles, decreasing reactive sites and catalytic activity [71, 79]. In order to improve the rate of the catalytic degradation reaction, most heterogeneous Fenton systems require an external energy input, such as irradiation with UV light or sunlight [80]. Chu et al. developed a hybrid system of Fenton and photo–Fenton processes to increase the degradation rate as well as power consumption [81]. The deposition of iron species supported over different materials (zeolites, clays, silicas, CNTs, etc.) has been widely reported as photo–Fenton-like heterogeneous catalysts to overcome the recovery of the homogeneous iron ions [82]. However, these matrices introduce mass transfer resistance between catalytic sites and organic pollutants and reduce efficiency; they also reduce light penetration and inhibit photodegradation [72, 79]. This disadvantage can be overcome with the use of nanomagnetic Fenton's catalysts. These catalysts increase the rate of mass transfer near the catalyst surface and enhance light absorption. They also combine the advantage of the high surface area of nanoparticles with the settling properties of much larger particles by the application of an external magnetic field [82a]. Several studies on the degradation of organic pollutants investigate the potential use of magnetic heterogeneous Fenton catalysts. For example, Zhu and Tang [83] synthesized a magnetic nano-$BiFeO_3$ catalyst for the photo–Fenton decomposition of organic pollutants in the presence of H_2O_2. In another study, Zhang et al. investigated a heterogeneous Fenton system based on Fe–Fe_2O_3 core–shell nanowires [84].

The application of nanoparticles as catalysts of the Fenton-like and photo–Fenton reactions has been described by several investigators (Table 24.2 [72, 79, 83, 85]). In comparison with their microsized counterparts, nanoparticles show a higher catalytic activity. The advantage of using nanoparticles as catalysts for Fenton-like reactions would more than offset the disadvantage of requiring UV radiation to accelerate the reaction [86]. For example, Valdés-Solís et al. [87] have developed manganese ferrite nanoparticles with a high surface area that can accelerate the Fenton-like reaction without requiring UV radiation. Zelmanov and Semiat [88] investigated the catalytic degradation of ethylene glycol and phenol by iron(III) oxide nanoparticles in the absence of UV radiation.

Recently, Fenton processes without needing the addition of toxic chemical reagents based on H_2O_2 electrogeneration have gained a lot of interest. H_2O_2 is formed through an electrochemical reaction of oxygen (cathodic half-reaction (Eq. 24.2) [89]:

$$O_2 + 2H^+ + 2e^- \rightarrow H_2O_2 \tag{24.2}$$

In the electro–Fenton process, hydroxyl radicals are produced from a reaction between electrogenerated H_2O_2 and added Fe^{2+} through Fenton's reaction (Eq. 24.1) in the acidic solution [90]. Fe^{2+} can also be regenerated by the reduction of Fe^{3+} at the cathode via reaction 24.3 [91]:

$$Fe^{3+} + e^- \rightarrow Fe^{2+} \tag{24.3}$$

Since the optimal pH for the photo–Fenton process ($pH<3$) is not agreeable with the optimal pH values for the electrochemical generation of H_2O_2 ($pH>7$), the degradation rate in the electro–Fenton oxidation is not fast. Fe@Fe_2O_3 core–shell nanowires as an iron reagent offer one quantitative solution to this problem. A neutral electro–Fenton system is easily obtainable with their combination with CNTs and AC fibers as cathodes [84, 92].

An alternative technology to overcome this problem was developed by Ding et al., who designed a PECh/electro–Fenton system by coupling visible light–driven PECh oxidation and electro–Fenton oxidation in an undivided cell. Bi_2WO_6 nanoplates deposited on fluorine-doped tin oxide (FTO) glass and Fe@Fe_2O_3 core–shell nanowires supported on AC fiber were used as the anode and the cathode in this system, respectively [93].

Photoelectro–Fenton (PEF) is another technology that has been proposed in recent years in which the solution treated under electro–Fenton conditions is simultaneously irradiated with UV light. The irradiation causes the photolysis of $Fe(OH)^{2+}$, the predominant species of Fe^{3+} in the pH range 2.5–4.0, yielding more Fe^{2+} and $^{\cdot}OH$ and accelerating the photolysis of complexes of Fe^{3+} with oxidation by-products such as carboxylic acids [70a, 90b]. One of the new approaches in developing an efficient advanced oxidation processes is to use PEF combined with heterogeneous photocatalysis [85h, 94]. Recent work has reported an enhancement of PEF treatment when it is combined with TiO_2 [85a] or ZnO photocatalysis (Table 24.2). As mentioned earlier, the modification of TiO_2 with a short bandgap semiconductor such as Fe_2O_3 can enhance activity and shift the wavelength of irradiation into the visible region [95]. The negative effect of Fe_2O_3 modification on TiO_2 was also reported. Jeon et al. modified TiO_2 with Fe_2O_3. In comparison to the pure TiO_2, the modified TiO_2 (Fe_2O_3/TiO_2) nanoparticles displayed a lower PECh activity that was attributed to the hematite-induced charge recombination due to an energy level mismatch between TiO_2 and

TABLE 24.2 Catalytic Fenton and coupled systems

No.	Organic pollutants	Method	Catalyst	Optimum reaction conditions	Important findings	References
1.	Rhodamine B (RB)	Photo–Fenton	Magnetic $NiFe_2O_4$	RhB 10 mg/l, 0.20 g $NiFe_2O_4$	$NiFe_2O_4$ exhibits photo–Fenton catalytic activity in the presence of oxalic acid	[72]
				1.0 mM oxalic acid, pH 3.0	Catalyst is very stable, recoverable, and easy to separate using an external magnetic field	
2.	Orange 4 (RO4)	Photo–Fenton	TiO_2–P25 nanoparticles	Ferrous photo–Fenton: Fe^{2+} 0.1 mmol, H_2O_2 10 mmol, 100 mg catalyst, pH 3	Increase of UV light intensity increases the degradation rate linearly	[85a]
				Ferrioxalate photo–Fenton: Fe^{3+} 0.15 mmol, H_2O_2 15 mmol, 200 mg catalyst, pH 3	UV/ferrous/H_2O_2/TiO_2 and UV/ferrioxalate/H_2O_2/TiO_2 processes are found to be more efficient than the individual photo–Fenton and UV/TiO_2–P25 processes	
3.	Bisphenol A (BPA)	Sono–Fenton	Fe_3O_4 magnetic nanoparticles	40 kHz, 100 W, catalyst 585 mg/l, H_2O_2 160 mmol/l, BPA 20 mg/l	The catalyst performed similar activity within a wide pH range from 3 to 9 and showed good stability and activity even after five recycles	[85b]
					The degradation efficiencies were remarkably enhanced by ultrasound	
4.	2,4-dichloro phenol (2,4-DCP)	Fenton-like degradation	Fe_3O_4 magnetic nanoparticles	H_2O_2 12 mM, catalyst 1.0 g/l, 2,4-DCP 100 mg/l, pH 3.0, 30°C	Catalysts that can remove 2,4-DCP efficiently	[85c]
					Two-stage first-order kinetics of 2,4-DCP degradation was observed	
5.	Reactive black 5 (RB5) and orange II	Heterogeneous Fenton	Fe_3O_4–poly(3,4-ethylenedioxythiophene) (PEDOT)	H_2O_2 10 mM, 100 mg/l catalysts, 10 mM RB5 or orange II	Compared with Fe_3O_4 nanoparticles, Fe_3O_4–PEDOT core–shell nanoparticles demonstrated outstanding catalytic performance	[79]
6.	2-Chlorophenol (2-CP)	Photo–Fenton	Zero-valent iron (ZVI) nanoparticles	2-CP 0.39 mM (50 ppm), pH 2.8 Fe_{Tot} = 18 ppm	The reaction rate improved when the pH was decreased	[85d]
					The reaction conversion by applying radiation	

(*Continued*)

TABLE 24.2 ***(Continued)***

No.	Organic pollutants	Method	Catalyst	Optimum reaction conditions	Important findings	References
7.	Phenol	Photoelectro–Fenton-like	Nanoscale zero valent iron (nZVI)	30 min, 0.5 g/l of nZVI, 500 mg/l of H_2O_2, phenol 200 mg/l, pH 6.2, 12 Ma/cm^2 of current density	The removal efficiency was increased with an increase in nZVI dosage and decreased with an increase in initial phenol concentration and initial pH Pseudo-first-order kinetics	[85e]
8.	Basic red 46 (BR46)	Photoelectro–Fenton (PEF)	TiO_2 nanoparticles immobilized on glass plates	Fe(III) 0.1 mM, BR46 15 mg/l, 35 min, applied current 300 mA	PEF/UV/TiO_2 was the most efficient processes in comparison with UV/TiO_2, EF, PEF, and PEF/UV/TiO_2	[85f]
9.	Phenol	Photoelectro–Fenton	α-Fe_2O_3/TiO_2 nanotubes (TiO_2 NTs)	Catalyst 2 mg, 10 mg/l phenol, cathode nickel, TiO_2 NTs anode, 5 V, 1 ml of the H_2O_2 solution (3%)	α-Fe_2O_3/TiO_2 nanotubes electrodes synthesized by electrochemical deposition method This electrode showed an enhanced absorbance in the visible light region and had good stability	[85g]
10.	C.I. Basic yellow 28 (BY28)	Photoelectro–Fenton (PEF)	ZnO nanoparticles immobilized on glass plates	20 mg/l of BY28, pH 3.0, Na_2SO_4 0.05 M, I = 100 mA, Fe^{3+} 0.1 mM	A comparison of PEF/ZnO processes with different systems follows a decreasing order: PEF/ZnO > PEF > UV/ZnO > EF > UV-C	[85h]
11.	Rhodamine B (RB)	Fenton-like	Nanoscaled $BiFeO_3$	RB 10 μmol/l, H_2O_2 10 mmol/l, catalyst 0.5 g/l, pH 5	Nanoscaled $BiFeO_3$ effectively catalyzed the decomposition of H_2O_2 into OH radicals The catalyst was demonstrated to have stability and reusability	[83]

hematite [96]. The preparation process probably plays an important role in increasing or decreasing the PECh activity of Fe_2O_3/TiO_2 composite electrodes [85g]. Researchers have shown that nanoscale zero-valent iron (nZVI) can be implicated as a heterogeneous PEF-like catalyst for the removal of environmental contaminants [85e].

In Table 24.2, we outline additional examples of reported environmental pollutants and their respective nanocatalysts in heterogeneous Fenton-like systems.

24.6 SONOCATALYTIC PROCESSES

24.6.1 Introduction

Since the past few decades, there has been considerable interest in the use of ultrasound for the treatment of water and wastewater [97]. Nevertheless, sonolysis alone is not generally considered to be attractive for application in large-scale treatment processes because for many organic pollutants the sonochemical degradation rate is very low, and for some stable and complicated organic pollutants, incomplete degradation may occur. Moreover, it requires costly equipment and consumes a high amount of energy. In order to solve these problems, a method using ultrasonic irradiation coupled with Fe^{2+}, H_2O_2, and Fenton reagent has been attempted. Much effort has also been devoted to heterogeneous sonochemical degradation, namely, sonocatalytic degradation, in the presence of various types of catalysts such as metal oxides.

It is presumed that the efficiency achieved using ultrasound irradiation is due to the cavitation phenomenon. An ultrasonic wave is a pressure wave with alternate compression and rarefaction, which is able to break the intermolecular forces maintaining the cohesion of the liquid and produce a cavity in the rarefaction section of the wave. The chemical and physical effects of ultrasound derive primarily from acoustic cavitation, which includes formation, growth, and collapse of the cavity. Bubble collapse in liquids results in an enormous concentration of energy from the conversion of kinetic energy of liquid motion into heating of the contents of the bubble [98]. Water sonolysis will induce the splitting of water molecules into free radical species ($H^{\cdot}$, $HO^{\cdot}$) in a cavitation bubble. In the absence of any solutes, these primary radicals could recombine to form H_2O or H_2O_2 and then be released into the bulk solution (Eqs. 24.4–24.7).

$$H_2O \xrightarrow{\text{ultrasonic waves}} {}^{\bullet}OH + {}^{\bullet}H \quad (24.4)$$

$${}^{\bullet}OH + {}^{\bullet}H \rightarrow H_2O \quad (24.5)$$

$${}^{\bullet}OH + {}^{\bullet}OH \rightarrow H_2O_2 \quad (24.6)$$

$$2\,{}^{\bullet}OH \rightarrow H_2O + {}^{\bullet}O \quad (24.7)$$

The presence of dissolved oxygen is reported to improve sonochemical reactions. If the solution is saturated with oxygen, additional radicals are produced in the gas phase (Eqs. 24.8–24.11) [99]:

$$O_2 \xrightarrow{\text{ultrasonic waves}} 2\,{}^{\bullet}O \quad (24.8)$$

$${}^{\bullet}O + H_2O \rightarrow 2\,{}^{\bullet}OH \quad (24.9)$$

$$O_2 + {}^{\bullet}H \rightarrow {}^{\bullet}OOH \quad (24.10)$$

$$2\,{}^{\bullet}OOH \rightarrow H_2O_2 + O_2 \quad (24.11)$$

The two proposed sonodegradation mechanisms are pyrolysis and hydroxyl radical generation in the cavitations bubble, which subsequently oxidize the organic compounds [100].

24.6.2 Heterogeneous Sonocatalytic Degradation

24.6.2.1 Sonochemical Degradation of Organic Compounds by TiO_2-Based Catalysts Sonodegradation efficiency can be improved by adding a catalyst. The amount of heterogeneous catalyst plays a crucial role in the rates of sonochemical degradation reactions. The presence of an excessive amount of catalyst could result in the scattering of ultrasound waves that may decrease the rates of reactions. However, an appropriate amount of catalyst promotes the formation of free radicals, thereby increasing the rate of degradation of the organic compound. The most common catalyst under ultrasonic irradiation of an organic compound is TiO_2. It is now well known that the particle size and crystalline phase of TiO_2 powders determine the ultimate catalytic activity. For example, it has been demonstrated that the activity of nanometer anatase TiO_2 powder is better than that of ordinary anatase TiO_2 powder in the sonocatalytic degradation of methyl parathion [101]. In addition, the degradation of the methyl orange in the presence of nanometer anatase and rutile TiO_2 powders was compared. According to Wang et al., the sonocatalytic activity of the nanometer anatase TiO_2 powder is obviously better than that of the nanometer rutile TiO_2 powder [102], while previous research works showed that the sonocatalytic activity of ordinary rutile TiO_2 powder was remarkably higher than that of the ordinary anatase TiO_2 powder [103]. The sonocatalytic degradation mechanisms were found to be different in nanometer anatase TiO_2 and nanometer rutile TiO_2. Organic compounds undergo direct oxidation by holes on the surface of nanometer anatase TiO_2 particles. Thus, the surface adsorption of the organic compounds increase as the size of anatase TiO_2 particles decreases [104]. However, the holes on the surface of nanometer rutile TiO_2 particles first react with water molecules and then produce the ·OH radicals. That is, the holes indirectly degrade the organic pollutants through ·OH radicals [105]. The effect of ultrasonic power on the degradation of organic compounds using anatase and rutile nanometer TiO_2 catalysts is also different because of these different degradation mechanisms. In general, in the case where onefold ultrasonic irradiation is used in the presence of a rutile TiO_2 catalyst, the degradation of organic compounds gradually increases as the output power increases due to the increased ·OH generation. Contrarily, the degradation ratio is in inverse proportion to the output power of ultrasound in the presence of an anatase TiO_2 catalyst, which is a consequence of destruction of anatase TiO_2 structure and desorption of organic compounds on the surface of nanoparticles [102].

The differences in the crystal structures between nanometer anatase and rutile TiO_2 particles caused the different reusability behaviors of the catalysts. It was found that the degradation of organic compounds in the presence of reused anatase TiO_2 nanoparticles declines gradually, but the sonocatalytic activity of reused rutile TiO_2 nanoparticles is hardly reduced compared with the fresh one. As mentioned earlier, the metastable structure of anatase TiO_2 particles changes under ultrasonic irradiation, while the nanometer rutile TiO_2 particles possess a stable crystal structure [102].

Recent research attention is focused on the improvement of the catalytic activity of TiO_2 catalysts. In the past few decades, doping with various transitions, anchoring TiO_2 particles onto materials with a large surface area, such as carbon-based materials [106], and using TiO_2 nanotubes [107] or metal-doped TiO_2 nanotubes [108] have all been used to improve the sonocatalytic activity of TiO_2 (Table 24.3).

24.6.2.2 Sonochemical Degradation of Organic Compounds by Other Nanocatalysts Among nanoparticles, nanosized ZnO as a sonocatalyst has been widely used for its high efficiency, nontoxic nature, and low cost. Previously, the sonocatalytic degradation of some organic pollutants in aqueous solution using nanosized ZnO powder has been reported [109] (Table 24.3). It has been found that some inorganic oxidants such as $KClO_4$, $KClO_3$, and $Ca(ClO)_2$ can effectively assist sonocatalytic degradation in the presence of nanosized ZnO powder [109b].

Sonocatalysis using magnetic nanoparticles, such as ZVI, also enhances the overall degradation of the organic compounds [110]. The increase in efficiency is related to factors such as enhanced mass transfer rates, enlarged surface area of Fe^0 due to the shockwave created by cavitational collapse, which may cause direct erosion on the particle's surface and deaggregation of particles to hinder agglomeration, and ultrasonic cleaning and decontamination of surfaces by the shockwave that remove the precipitation of iron oxides/hydroxides on the iron surface. Additionally, the degradation also rises via Fenton's reaction between Fe(II) that forms on the iron and ultrasonically produced H_2O_2 [111].

Table 24.3 [100, 102, 105–110, 112] summarizes reported findings on ultrasonic conditions, important conclusions, and the performance of catalysts in the enhancement of the degradation of organic compounds.

24.6.3 Combination of Sonocatalytic Degradation with Other Technologies

It is very difficult to treat non- or low-transparent wastewaters using the photocatalytic degradation method due to the low penetrating ability of any kind of light source, whereas the penetrating ability of ultrasound is very strong through any water medium [113]. The simultaneous use of sonocatalytic and photocatalytic degradation is called sonophotocatalytic degradation. In comparison to photolysis, sonolysis, and sono-photolysis, sonophotocatalytic degradation generally occurs faster than during the respective individual processes. Sonolysis might increase the photocatalytic reaction rate by increasing the amounts of hydroxyl radicals, enhancing the mass transfer of organics between the liquid phase and the catalyst surface, and increasing the surface area of photocatalyst particles and thus the catalytic activity of the semiconductor catalyst [114].

Ultrasound is also used to enhance the efficiency of photocatalytic-assisted electrochemical oxidation and the sonophotoelectrocatalysis process [115].

The combination of ultrasound with the adsorption process was found to be more promising in the elimination of organic pollutants. Ultrasound not only promoted desorption but also enhanced the rate of mass transfer in the sorption process and increased/facilitated pore diffusion [116].

TABLE 24.3 Sonochemical degradation of organic compounds by nanocatalysts

No.	Organic pollutants	Catalyst	Optimum reaction conditions	Important findings	References
1.	Parathion (P)	Nanometer rutile TiO_2	50 mg/l P, 0.05 g TiO_2, 40 kHz, 50 W	The degradation rate was enhanced in the presence of TiO_2 Pseudo-first-order reaction kinetics	[112a]
2.	Basic blue 41 (BB41)	Nanosized TiO_2	15 mg/l BB41, 0.01 g TiO_2, 35 kHz, 160 W	H_2O_2 accelerated the reaction Pseudo-first-order reaction kinetics	[100]
3.	Acid red B (ARB)	TiO_2 ZnO TiO_2/ZnO	10 mg/l ARB, 0.01 g catalyst, 40 kHz, 50 W	Sonocatalytic activity of TiO_2/ZnO (4:1) was the highest Pseudo-first-order reaction kinetics	[109]
4.	Methyl parathion (MP)	Microsized rutile TiO_2 Nanosized anatase TiO_2	50 mg/l MP, 0.06 g TiO_2, 40 kHz, 50 W	TiO_2 and H_2O_2 increased the degradation rate Pseudo-first-order reaction kinetics	[112b]
5.	Methyl orange (MO)	Nanometer anatase and rutile TiO_2	10–20 mg/l MO, 500–750 mg/l TiO_2, 40–60 kHz, 50 W, 40°C, 100–120 min	Sonocatalytic activity of nanometer anatase TiO_2 powder is obviously better than that of nanometer rutile TiO_2 powder Pseudo-first-order reaction kinetics	[102]
6.	Acid red B (ARB)	Nanometer anatase and rutile TiO_2	1.5 g/l TiO_2, 10 mg/l ARB, 30 W, pH 3.0, 30°C	The sonocatalytic performance of nanometer anatase TiO_2 is obviously better than that of the nanometer rutile TiO_2 powder	[105]
7.	Rhodamine B (RB)	Fe-doped TiO_2 nanotubes	2 g/l of catalyst, RB 50 mg/l, 35 kHz and 50 W	The optimum Fe doping at Fe:Ti = 0.005 showed the highest sonocatalytic activity and exceeded that of undoped TiO_2 nanotubes Leached Fe^{3+} could catalyze Fenton-like reaction Fe-doped TiO_2 nanotubes could retain high degradation efficiency even after being reused for four cycles	[108]
8.	Rhodamine B (RB)	TiO_2 nanotube (TiO_2 NTs)	20–100 W, 35 and 130 kHz, 8 mM H_2O_2, 1 l/min of air flow, neutral pH, 50 mg/l RB, 120 min, 2 g/l TiO_2 NTs.	High sonocatalytic activity of TiO_2 NTs was achieved due to their high specific surface area and unique pore system Pseudo-first-order reaction kinetics	[107]
9.	Black 5 (RB5)	Nanosized TiO_2 Nanosized CdS CdS/TiO_2 core–shell nanocomposite	RB5 100 mg/l, nanocatalyst 0.05 g, 20 kHz, 41 W, 10 min, pH 6.5, 38°C.	Nanocomposite of TiO_2/CdS exhibited a high sonocatalytic activity in comparison with pure nanosized TiO_2 and CdS Complete mineralization of the organic compound into SO_4^{2-} and NO_3^- ions	[112c]
10.	Methylene blue (MB)	C–Cr codoping in TiO_2 nanoparticles (C–Cr–TiO_2)	MB 10 mg/l, 100 mg, C–Cr–TiO_2, pH 7.0, 40 kHz, 80 W	The C–Cr–TiO_2 sonocatalysts show a higher sonocatalytic activity than that of undoped TiO_2 samples When the amount of C or Cr dopant is less than 1 wt.%, the sonocatalytic activity is increased, while more than this amount decreases the sonocatalytic activity	[112d]

(Continued)

TABLE 24.3 ***(Continued)***

No.	Organic pollutants	Catalyst	Optimum reaction conditions	Important findings	References
11.	Acid red B (ARB)	Er^{3+}:$YAlO_3/TiO_2$–ZnO nanocomposite	ARB 10 mg/l, 1.0 g/l sonocatalyst, 150 min, pH 5.5, 40 kHz, 50 W, 25°C	The sonocatalyst with 1:1 Ti/Zn molar ratio heat-treated at 550°C for 60 min showed the highest sonocatalytic activity Reactions follow the pseudo-first-order kinetics	[112e]
12.	Methyl orange (MO)	$InVO_4/TiO_2$ nanojunction composites	28 kHz, $InVO_4/TiO_2$ 0.03 g, MO ≈ 1.0 × 10 − 5 M	(1:50) $InVO_4/TiO_2$ nanojunction composite has exhibited the highest sonocatalytic activity	[112f]
13.	Methylene blue (MB)	Fe–fullerene/TiO_2 nanocomposite	20 kHz, MB 1×10^{-5} mol/l, Fe–fullerene/TiO_2 0.05 g	With the increase in the amount of Fe, the degradation rate of methylene blue solution increased	[112g]
14.	Rhodamine B (RhB)	CdS/CNT–TiO_2 nanocomposite	20 kHz, catalysts 0.01 g, RhB 1.0×10^{-5} M	Under ultrasonic irradiation, RhB was efficiently degraded by the CdS/CNT–TiO_2 composites Compared with the pristine MWCNTs, the surface areas of CNT–TiO_2 and CdS/CNT–TiO_2 were greatly decreased	[112h]
15.	Acid red B (ARB)	CeO_2/TiO_2 SnO_2/TiO_2 ZrO_2/TiO_2 composites	10 mg/l catalyst, 1000 mg/l Acid red B, pH = 7.0, 25°C, 80 min, 40 kHz, 50 W	Sonocatalytic degradation decreases in order: $CeO_2/TiO_2 > SnO_2/TiO_2 > TiO_2 > ZrO_2/TiO_2 > SnO_2 > CeO_2 > ZrO_2$ Pseudo-first-order kinetics reactions	[112i]
16.	Rhodamine B (RhB)	Graphene–TiO_2 composites (TiO_2 nanoparticles on the graphene nanosheets)	RhB 2.00×10^{-5} mol/l, catalyst 0.02 g/100 ml, 20 kHz, 750 W	The graphene–TiO_2 composites possessed a high specific surface area, which increased the decolorization rate for RhB solution First-order reaction kinetics	[106]
17.	Acid red B (ARB)	US/ZnO/$Ca(ClO)_2$ US/ZnO/$KClO_3$ US/ZnO/$KClO_4$	Oxidant 10.0 mmol/l, 40°C pH 7.0, ARB 10 mg/l, catalyst 1.0 g/l, 30 min, 40 kHz, 50 W	Oxidants can assist the sonocatalytic degradation and the effect of oxidants changes to the order $KClO_4 > KClO_3 > Ca(ClO)_2$ Pseudo-first-order kinetics reactions	
18.	(RhB, MB, Congo red, Bromothymol Blue, Alizarin)	$H_3PW_{12}O_{40}/ZrO_2$	30 ppm RB, 20 mg catalyst, pH 1.27, oxygen flow rate 5 ml/min 20 kHz, 20–80 W/cm^2, 30 min	The $H_3PW_{12}O_{40}/ZrO_2$ nanocomposite showed higher activity than pure polyoxometalate or pure ZrO_2 The reaction rate constants of sonocatalytic degradation of dyes are higher than that of photocatalytic degradation	[112j]
19.	Diclofenac (DCF)	Nanoparticles of zero-valent iron (ZVI)	DCF 30 μM, pH 3.0, ZVI = 0.2 g/l 861 kHz	Sonolysis with ZVI nanoparticles improves the efficiency of sonochemical degradation of DCF High frequency was more effective than power ultrasound (20 kHz), and the efficiency was maximum at 861 kHz	[110]

TABLE 24.4 Combination of sonocatalytic degradation with other technologies

No.	Organic pollutants	Method	Catalyst	Optimum reaction conditions	Important findings	References
1.	Malachite green (MG)	Sonosorption	Polyacrylic acid–nanoclay nanocomposite (PAA–nanoclay)	22.5 kHz, 120 W, 0.5 g PAA–nanoclay, 500 mg/l MG	The adsorption of dye increases with increase in pH The experimental data fit very well with Freundlich adsorption isotherm, indicating multilayer adsorption	[116]
2.	Methyl orange (MO)	Sonophotoelectrocatalytic degradation	TiO_2 nanotube	40 kHz, 150 W, 253.7 nm UV lamp with 11 W; TiO_2 nanotube as electrode, MO 5×10^{-5} M, 40 min; 25°C; 0.5 M Na_2SO_4 electrolyte	The hybrid processes could efficiently enhance the degradation efficiency of MeO Pseudo-first-order kinetics	[117a]
3.	Phenol	Combination of ultrasound with adsorption	Nanoclay	22.5 kHz, 120 W, phenol 250–5000 mg/l, 0.5 g clay	Adsorption phenomenon is improved due to sonication Freundlich adsorption isotherm favors nanoclay	[117b]
4.	Methylene blue (MB)	Sonophotocatalytic degradation	Fe/TiO_2 Fe/TiO_2–MWCNT	23 kHz, Fe/TiO_2–MWCNT 0.05 g, MB 1.0×10^{-5} M, irradiation system with visible light	The MWCNT deposited with Fe/TiO_2 catalyst showed much higher sonocatalytic activity than Fe/TiO_2 The MB can be effectively degraded under visible light, ultrasonic irradiation, and coupling the two processes with higher efficiency Pseudo-first-order kinetic	[114]
5.	Rhodamine B (RhB)	Sono–Fenton process	$Fe@Fe_2O_3$ core–shell nanowires	25 kHz, 100 W, 0.9 mmol $Fe@Fe_2O_3$ nanowires, 5 mg/l^{-1} RhB, air was bubbled (0.1 m^3/h^{-1})	This system worked well at neutral pH It was found that $Fe@Fe_2O_3$ core–shell nanowires showed significantly higher catalytic activity than Fe^0, Fe^{2+}, and Fe^{2+}/Fe_2O_3 for the degradation of RhB	[84b]
6.	Methyl orange (MO)	Sonophotocatalytic degradation	Nanosized Ag/TiO_2	Ag/TiO_2 60 mg/l, MeO 37 mg/l, 40 kHz, 200 W, illuminated by light generated by a xenon lamp	The activity of nanosized Ag/TiO_2 was obviously better than that of both Degussa P25 and Yili TiO_2 Pseudo-first-order kinetics	[117c]

Fenton reactions have been widely utilized to degrade organic compounds. However, due to some drawbacks (Section 24.5) of this process, a new technology in treating organic compounds is needed. In order to overcome this drawback, more recently, a novel sono-Fenton system was developed that could offer advantages including working at neutral pH, no sludge generation, and the possibility of recycling the iron reagent [84b].

Some new studies on the use of ultrasonic irradiation in photocatalysis, electro-oxidation, and Fenton processes are summarized in Table 24.4 [84b, 114, 116, 117].

24.7 CONCLUSION

Since the definition of the concept of a catalyst given by Berzelius in 1836, the field of catalysis has experienced an astonishing transformation. The success of basic research in the field of catalysis has had a direct effect on solving many fundamental technological and environmental problems. Regardless of the intended application, the common challenge of achieving long-lasting, highly active, selective, and environmental-friendly catalytic materials and processes still remains. Nanocatalysis is one of the most exciting subfields to have emerged from nanoscience. Nanomaterials also provide unprecedented opportunities to develop more efficient water-purification catalysts and redox-active media. This chapter has given further evidence to this issue and has tried to address what all the potential environmental impacts of the nanocatalyst might be. Despite recent positive progress, additional investigations of the risks of nanomaterials to the environment as well as their toxicity and persistence are still needed, as this type of understanding is a requirement for the rational design of efficient catalysts.

REFERENCES

[1] Savage N, Diallo MS. Nanomaterials and water purification: opportunities and challenges. J Nanopart Res 2005;7:331–342.

[2] Pugazhenthiran N, Sathishkumar P, Murugesan S, Anandan S. Effective degradation of acid orange 10 by catalytic ozonation in the presence of Au-Bi_2O_3 nanoparticles. Chem Eng J 2011;168:1227–1233.

[3] Mills A, Le Hunte S. An overview of semiconductor photocatalysis. J Photochem Photobiol A Chem 1997;108:1–35.

[4] Carp O, Huisman CL, Reller A. Photoinduced reactivity of titanium oxide photoinduced reactivity of titanium oxide. Solid State Chem 2004;32:33–177.

[5] Konstantinou IK, Albanis TA. TiO_2-assisted photocatalytic degradation of azo dyes in aqueous solution: kinetic and mechanistic investigations—a review. Appl Catal B Environ 2004;49:1–14.

[6] Akpan UG, Hameed BH. Parameters affecting the photocatalytic degradation of dyes using TiO_2-based photocatalysts: a review. J Hazard Mater 2009;170:520–529.

[7] Mohamed RM, McKinney DL, Sigmund WM. Enhanced nanocatalysts. Mater Sci Eng R 2012;73:1–13.

[8] Gerven TV, Mul G, Moulijn J, Stankiewicz A. A review of intensification of photocatalytic processes. Chem Eng Proc 2007; 46:781–789.

[9] (a) Cao S-W, Zhu Y-J, Cheng G-F, Huang Y-H. $ZnFe_2O_4$ nanoparticles: microwave-hydrothermal ionic liquid synthesis and photocatalytic property over phenol. J Hazard Mater 2009;171:431–435.
(b) Zhang Y, Wan J, Ke Y. A novel approach of preparing TiO_2 films at low temperature and its application in photocatalytic degradation of methyl orange. J Hazard Mater 2010;177:750–754.
(c) Morales-Flores N, Pal U, Sánchez Mora E. Photocatalytic behavior of ZnO and Pt-incorporated ZnO nanoparticles in phenol degradation. Appl Catal A Gen 2011;394:269–275.
(d) Zhai J, Tao X, Pu Y, Zeng X-F, Chen J-F. Core/shell structured ZnO/SiO_2 nanoparticles: preparation, characterization and photocatalytic property. Appl Surf Sci 2010;257:393–397.
(e) Li J, Xu Y, Liu Y, Wu D, Sun Y. Synthesis of hydrophilic ZnS nanocrystals and their application in photocatalytic degradation of dye pollutants. China Particuology 2004;2:266–269.
(f) Sun L, Li J, Wang C, Li S, Lai Y, Chen H, Lin C. Ultrasound aided photochemical synthesis of Ag loaded TiO_2 nanotube arrays to enhance photocatalytic activity. J Hazard Mater 2009;171:1045–1050.
(g) Yan Y, Sun H, Yao P, Kang S-Z, Mu J. Effect of multi-walled carbon nanotubes loaded with Ag nanoparticles on the photocatalytic degradation of Rhodamine B under visible light irradiation. Appl Surf Sci 2011;257:3620–3626.
(h) Hong X, Wu X, Zhang Q, Xiao M, Yang G, Qiu M, Han G. Hydroxyapatite supportedAg_3PO_4 nanoparticles withhighervisiblelight photocatalytic activity. Appl Surf Sci 2012;258:4801–4805.
(i) Fang J, Ma J, Sun Y, Liu Z, Gao C. Synthesis of Bi_3NbO_7 nanoparticles with a hollow structure and their photocatalytic activity under visible light. Solid State Sci 2011;13:1649–1653.

(j) Barakat MA, Schaeffer H, Hayes G, Ismat-Shah S. Photocatalytic degradation of 2-chlorophenol by Co-doped TiO_2 nanoparticles. Appl Catal B Environ 2004;57:23–30.
(k) Zhao H, Liang L, Liu H. Fast photo-catalytic degradation of pyridine in nano aluminum oxide suspension systems. J Environ Sci 2011;23:S156–S158.
(l) Khatamian M, Khandar AA, Divband B, Haghighi M, Ebrahimiasl S. Heterogeneous photocatalytic degradation of 4-nitrophenol in aqueous suspension by Ln (La^{3+}, Nd^{3+} or Sm^{3+}) doped ZnO nanoparticles. J Mol Catal A Chem 2012;365:120–127.
(m) Wang R, Xu D, Liu JB, Li KW, Wang H. Preparation and photocatalytic properties of CdS/$La_2Ti_2O_7$ nanocomposites under visible light. Chem Eng J 2011;168 (1):455–460.
(n) Jin Y, Yong-kang L, Yu L, Jun-ping L. Synthesis and characterization of magnetic TiO_2/SiO_2/$NiFe_2O_4$ composite photocatalysts. Chem Res Chinese U 2010;26 (2):278–282.
(o) Guo H, Chen J-F, Weng W, Li S. Hydrothermal synthesis of C-doped $Zn_3(OH)_2V_2O_7$ nanorods and their photocatalytic properties under visible light illumination. Appl Surf Sci 2011;257 (9):3920–3923.
(p) Madhusudan P, Ran J, Zhang J, Yu J, Liu G. Novel urea assisted hydrothermal synthesis of hierarchical $BiVO_4$/$Bi_2O_2CO_3$ nanocomposites with enhanced visible-light photocatalytic activity. Appl Catal B Environ 2011;110:286–295.
(q) Guo JF, Ma B, Yin A, Fan K, Dai W-L. Photodegradation of Rhodamine B and 4-chlorophenol using plasmonic photocatalyst of Ag–AgI/Fe_3O_4@ SiO_2 magnetic nanoparticle under visible light irradiation. Appl Catal B Environ 2011;101 (3–4):580–586.
(r) Andronic L, Isac L, Duta A. Photochemical synthesis of copper sulphide/titanium oxide photocatalyst. J Photochem Photobiol A Chem 2011;221 (1):30–37.
(s) Uzunova BM, Todorovska R, Dimitrov D, Todorovskya D. Lanthanide-doped titanium dioxide layers as photocatalysts. Appl Surf Sci 2008;254 (22):7296–7302.
(t) Tian B, Li C, Gu F, Jiang H, Hu Y, Zhang J. Flame sprayed V-doped TiO_2 nanoparticles with enhanced photocatalytic activity under visible light irradiation. Chem Eng J 2009;151:220–227.
(u) Yajun W, Kecheng L, Changgen F. Photocatalytic degradation of methyl orange by polyoxometalates supported on yttrium-doped TiO_2. J Rare Earth 2011;29:866–871.
(v) Li X, Zhu Z, Zhao Q, Wang L. Photocatalytic degradation of gaseous toluene over $ZnAl_2O_4$ prepared by different methods: a comparative study. J Hazard Mater 2011;186:2089–2096.
(w) Shen C, Wang YJ, Xu JH, Luo GS. Facile synthesis and photocatalytic properties of TiO_2 nanoparticles supported on porous glass beads. Chem Eng J 2012;209:478–485.
(x) Meshram S, Limaye R, Ghodke S, Nigam S, Sonawane S, Chikate R. Continuous flow photocatalytic reactor using ZnO–bentonite nanocomposite for degradation of phenol. Chem Eng J 2011;172:1008–1015.
(y) Mohamed RM, Harraz FA, Mkhalid IA. Hydrothermal synthesis of size-controllable Yttrium Orthovanadate (YVO_4) nanoparticles and its application in photocatalytic degradation of direct blue dye. J Alloy Compd 2012;532:55–60.
(z) Kruefu V, Ninsonti H, Wetchakun N, Inceesungvorn B, Pookmanee P, Phanichphant S. Photocatalytic degradation of phenol using Nb-loaded ZnO nanoparticles. Eng J Canada 2012;16:91–99.
(aa) Chiou C-H, Juang R-S. Photocatalytic degradation of phenol in aqueous solutions by Pr-doped TiO_2 nanoparticles. J Hazard Mater 2007;149:1–7.
(ab) Li KW, Wang Y, Wang H, Zhu M, Yan H. Hydrothermal synthesis andphotocatalytic properties of layered $La_2Ti_2O_7$ nanosheets. Nanotechnology 2006;17:4863–4867.
(ac) Zhang YC, Du ZN, Li KW, Zhang M. Size-controlled hydrothermal synthesis of SnS_2 nanoparticles with high performance in visible light-driven photocatalytic degradation of aqueous methyl orange. Sep Sci Technol 2011;81 (1):101–107.
(ad) Pourabbas B, Jamshidi B. Preparation of MoS_2 nanoparticles by a modified hydrothermal method and the photo-catalytic activity of MoS_2/TiO_2 hybrids in photo-oxidation of phenol. Chem Eng J 2008;138 (1–3):55–62.
(ae) Wang S, Zhou S. Photodegradation of methyl orange by photocatalyst of CNTs/P-TiO_2 under UV and visible-light irradiation. J Hazard Mater 2011;185 (1):77–85.
(af) Zheng Z, Huang B, Qin X, Zhang X, Dai Y, Whangbo MH. Facile in situ synthesis of visible-light plasmonic photocatalysts M@TiO_2 (M = Au, Pt, Ag) and evaluation of their photocatalytic oxidation of benzene to phenol. J Mater Chem 2011;21:9079–9087.
(ag) Senapati KK, Borgohain C, Sarma KC, Phukan P. Photocatalytic degradation of methylene blue in water using $CoFe_2O_4$–Cr_2O_3–SiO_2 fluorescent magnetic nanocomposite. J Mol Catal A Chem 2011;346 (1–2):111–116.
(ah) Zhang L, He Y, Wu Y, Wu T. Photocatalytic degradation of RhB over $MgFe_2O_4$/TiO_2 composite materials. Mater Sci Eng B 2011;176 (18):1497–1504.
(ai) Xia J, Wang A, Liu X, Su Z. Preparation and characterization of bifunctional, Fe_3O_4/ZnO nanocomposites and their use as photocatalysts. Appl Surf Sci 2011;257 (23):9724–9732.
(aj) Shu H, Xie J, Xu H, Li H, Gu Z, Sun G, Xu Y. Structural characterization and photocatalytic activity of NiO/$AgNbO_3$. J Alloy Compd 2010;496 (1–2):633–637.
(ak) Wang P, Huang B, Qin X, Zhang X, Dai Y, Whangbo MH. Ag/AgBr/WO_3·H_2O: visible-light photocatalyst for bacteria destruction. Inorg Chem 2009;48:10697–10702.
(al) Zhu H, Chen X, Zheng Z, Ke X, Jaatinen E, Zhao J, Guo C, Xie T, Wang D. Mechanism of supported gold nanoparticles as photocatalysts under ultraviolet and visible light irradiation. Chem Commun 2009 (48):7524–7526.
(am)Wang C, Shi H, Li Y. Synthesis and characteristics of natural zeolite supported Fe^{3+}-TiO_2 photocatalysts. Appl Surf Sci 2011;257 (15):6873–6877.

[10] (a) Hoffmann MR, Martin ST, Choi W, Bahnemann D. Environmental applications of semiconductor photocatalysis. Chem Rev 1995;95:69–96.
(b) Kamat PV. Tailoring nanostructured thin films. Chemtech 1995:22–28.

[11] Beydoun D, Amal R, Low G, McEvoy S. Role of nanoparticles in photocatalysis. J Nanopart Res 1999;1:439–458.

[12] Wang C-C, Zhang Z, Ying JY. Photocatalytic decomposition of halogenated organics over nanocrystalline titania. Nanostruct Mater 1997;9:583–586.

[13] Zhang Z, Wang C-C, Zakaria R, Ying JY. Role of particle size in nanocrystalline TiO_2-based photocatalysts. J Phys Chem B 1998;102:10871–10878.

[14] (a) Fischer C-H, Lillie J, Weller H, Katsikas L, Henglein A. Photochemistry of colloidal semiconductors. 29. Fractionation of CdS sols of small particles by exclusion chromatography. Ber Bunsenges Phys Chem 1989;93:61–64.
(b) Howe RF. Recent developments in photocatalysis. Dev Chem Eng Mineral Process 1998;6 (1):55–84.

[15] Henderson MA. A surface science perspective on TiO_2 photocatalysis. Surf Sci Rep 2011;66:185–297.

[16] Lee K, Lee NH, Shin SH, Lee HG, Kim SJ. Hydrothermal synthesis and photocatalytic characterizations of transition metals doped nano TiO_2 sols. Mater Sci Eng B 2006;129:109–115.

[17] Salmi M, Tkachenko N, Lamminmaki RJ, Karvinen S, Vehmanen V, Lemmetyinen H. Femtosecond to nanosecond spectroscopy of transition metal-doped TiO_2 particles. J Photochem Photobiol A Chem 2005;175:8–14.

[18] Choi W, Termin A, Hoffmann MR. The role of metal ion dopants in quantum-sized TiO_2: correlation between photoreactivity and charge carrier recombination dynamics. J Phys Chem B 1994;98:13669–13679.

[19] (a) Emeline AV, Kuznetsov VN, Rybchuk VK, Serpone N. Visible-light-active titania photocatalysts: the case of N-doped TiO_2s—properties and some fundamental issues. Int J Photoenergy 2008;2008:1–19.
(b) Di Paola A, García-López E, Marcì G, Palmisano L. A survey of photocatalytic materials for environmental remediation. J Hazard Mater 2012;211–212:3–29.

[20] Yates HM, Nolan MG, Sheel DW, Pemble ME. The role of nitrogen doping on the development of visible light-induced photocatalytic activity in thin TiO_2 films grown on glass by chemical vapour deposition. J Photochem Photobiol A Chem 2006;179:213–223.

[21] Bedja I, Kamat PV. Capped semiconductor colloids. Synthesis and photoelectrochemical behaviour of TiO_2-capped SnO_2 nanocrystallites. J Phys Chem 1995;99:9182–9188.

[22] Mews A, Eychmuller A, Giersig M, Schooss D, Weller H. Preparation, characterization, and photophysics of the quantum dot quantum well system CdS/HgS/CdS. J Phys Chem 1994;98:934–941.

[23] Braun PV, Osenar P, Stup S-I. Semiconducting superlattices templated by molecular assemblies. Nature 1996;380 (6572):325–328.

[24] Tenne R. Fullerene-like structures and nanotubes from inorganic compounds. Endeavour 1996;20 (3):97–104.

[25] (a) Liu Z, Zhang X, Nishimoto S, Jin M, Tryk DA, Murakami T, Fujishima A. Highly ordered TiO_2 nanotube arrays with controllable length for photoelectrocatalytic degradation of phenol. J Phys Chem C 2008;112:253–259.
(b) Zhuang H, Lin C, Lai Y, Sun L, Li J. Some critical structure factors of titanium oxide nanotube array in its photocatalytic activity. Environ Sci Technol 2007;41:4735–4740.
(c) Xu H, Vanamu G, Nie Z, Konishi H, Yeredla R, Phillips J, Wang Y. Photocatalytic oxidation of a volatile organic component of acetaldehyde using titanium oxide nanotubes. J Nanomater 2006;2006:1–8.
(d) Macak JM, Zlamal M, Krysa J, Schmuki P. Self-organized TiO_2 nanotube layers as highly efficient photocatalysts. Small 2007;3:300–304.
(e) Lai Y, Sun L, Chen Y, Zhuang H, Lin C, Chin JW. Effects of the structure of TiO_2 nanotube array on Ti substrate on its photocatalytic activity. J Electrochem Soc 2006;153:D123–D128.
(f) Tan LK, Kumar MK, An WW, Gao H. Transparent, well-aligned TiO_2 nanotube arrays with controllable dimensions on glass substrates for photocatalytic applications. ACS Appl Mater Interfaces 2010;2:498–503.

[26] Liu Z, Zhang X, Nishimoto S, Murakami T, Fujishima A. Efficient photocatalytic degradation of gaseous acetaldehyde by highly ordered TiO_2 nanotube arrays. Environ Sci Technol 2008;42:8547–8551.

[27] (a) Fu P, Zhang P. Enhanced photoelectrochemical properties and photocatalytic activity of porous TiO_2 films with highly dispersed small Au nanoparticles. Thin Solid Films 2011;519:3480–3486.
(b) Liang H-C, Li X-Z. Effects of structure of anodic TiO_2 nanotube arrays on photocatalytic activity for the degradation of 2,3-dichlorophenol in aqueous solution. J Hazard Mater 2009; 162:1415–1422.

[28] Hagfeldt A, Gratzel M. Light-induced redox reactions in nanocrystalline systems. Chem Rev 1995;95:49–68.

[29] (a) Kamat PV. Photophysical, photochemical and photocatalytic aspects of metal nanoparticles. J Phys Chem B 2002; 106: 7729–7744.
(b) Eustis S, El-Sayed MA. Why gold nanoparticles are more precious than pretty gold: noble metal surface plasmon resonance and its enhancement of the radiative and nonradiative properties of nanocrystals of different shapes. Chem Soc Rev 2006;35:209–217.

[30] Wang P, Huang B, Qin X. Ag@AgCl: a highly efficient and stable photocatalyst active under visible light. Angew Chem Int Ed Engl 2008;47:7931–7933.

[31] Chen X, Zheng Z, Ke X. Supported silver nanoparticles as photocatalysts under ultraviolet and visible light irradiation. Green Chem 2010;12:414–419.

[32] (a) Murray CA, Parsons SA. Preliminary laboratory investigation of disinfection by-product precursor removal using an advanced oxidation process. Water Environ 2006;20:123–129.

(b) Klavarioti M, Mantzavinos D, Kassinos D. Removal of residual pharmaceuticals from aqueous systems by advanced oxidation processes. Environ Int 2009;35:402–417.

[33] Khataee AR, Zarei M, Fathinia M, Khobnasab Jafari M. Photocatalytic degradation of an anthraquinone dye on immobilized TiO_2 nanoparticles in a rectangular reactor: destruction pathway and response surface approach. Desalination 2011;268:126–133.

[34] Kibanova D, Cervini-Silva J, Destaillats H. Efficiency of clay–TiO_2 nanocomposites on the photocatalytic elimination of a model hydrophobic air pollutant. Environ Sci Technol 2009;43:1500–1506.

[35] Khatamian M, Hashemian S, Yavari A, Saket M. Preparation of metal ion (Fe^{3+} and Ni^{2+}) doped TiO_2 nanoparticles supported on ZSM-5 zeolite and investigation of its photocatalytic activity. Mater Sci Eng B 2012;177 (18):1623–1627.

[36] Álvarez PM, Jaramillo J, López-Piñero F, Plucinski PK. Preparation and characterization of magnetic TiO_2 nanoparticles and their utilization for the degradation of emerging pollutants in water. Appl Catal B Environ 2010;100:338–345.

[37] Choi H, Stathatos E, Dionysiou DD. Sol-gel preparation of mesoporous photocatalytic TiO_2, films and TiO_2,/Al_2O_3 composite membranes for environmental applications. Appl Catal B Environ 2006;63:60–67.

[38] (a) Yu Y, Yu JC, Yu J-G, Kwok Y-C, Che Y-K, Zhao J-C, Ding L, Ge W-K, Wong P-K. Enhancement of photocatalytic activity of mesoporous TiO2 by using carbon nanotubes. Appl Catal A Gen 2005;289:186–196.

(b) Bouazza N, Ouzine M, Lillo-Rodenas MA, Eder D, Linares-Solano A. TiO_2 nanotubes and CNT-TiO_2 hybrid materials for the photocatalytic oxidation of propene at low concentration. Appl Catal B Environ 2009;92:377–383.

(c) Wang W, Serp P, Kalck P, Faria JL. Photocatalytic degradation of phenol on MWNT and titania composite catalysts prepared by a modified sol–gel method. Appl Catal B Environ 2005;56:305–312.

(d) Yu H, Quan X, Chen S, Zhao H, Zhang Y. TiO_2–carbon nanotube heterojunction arrays with a controllable thickness of TiO_2 layer and their first application in photocatalysis. J Photochem Photobiol A Chem 2008;200:301–306.

[39] Zhai X, Chen Z, Zhao S, Wang H, Yang L. Enhanced ozonation of dichloroacetic acid in aqueous solution using nanometer ZnO powders. J Environ Sci 2010;22 (10):1527–1533.

[40] Nawrocki J, Kasprzyk-Hordernb B. The efficiency and mechanisms of catalytic ozonation. Appl Catal B Environ 2010;99:27–42.

[41] Jung H, Kim J-W, Choi H, Lee J-H, Hur H-G. Synthesis of nanosized biogenic magnetite and comparison of its catalytic activity in ozonation. Appl Catal B Environ 2008;83:208–213.

[42] Safaei-Naeini Y, Aminzare M, Golestani-Fard F, Khorasanizadeh F, Salahi E. Suspension stability of titania nanoparticles studied by UV-VIS spectroscopy method. Iran J Mater Sci Eng 2012;9:62–68.

[43] Yang Y, Ma J, Qin Q, Zhai X. Degradation of nitrobenzene by nano-TiO_2 catalyzed ozonation. J Mol Catal A Chem 2007;267:41–48.

[44] Rodríguez JL, Valenzuela MA, Pola F, Tiznado H, Poznyak T. Photodeposition of Ni nanoparticles on TiO_2 and their application in the catalytic ozonation of 2,4-dichlorophenoxyacetic acid. J Mol Catal A Chem 2012;353–354:29–36.

[45] Jing Y, Li L, Zhang Q, Lu P, Liu P, Lü X. Photocatalytic ozonation of dimethyl phthalate with TiO_2 prepared by a hydrothermal method. J Hazard Mater 2011;189:40–47.

[46] Tabatabael SM, Mehrizad A, Gharbani P. Nano-catalytic ozonation of 4-nitrochlorobenzene in aqueous solutions. E J Chem 2012;9 (4):1968–1975.

[47] (a) Moussavi G, Mahmoudi M. Degradation and biodegradability improvement of the reactive red 198 azo dye using catalytic ozonation with MgO nanocrystals. Chem Eng J 2009;152:1–7.

(b) Moussavi G, Khavanin A, Alizadeh R. The integration of ozonation catalyzed with MgO nanocrystals and the biodegradation for the removal of phenol from saline wastewater. Appl Catal B Environ 2010;97:160–167.

[48] Pinheiro da Silva MF, Soeira LS, Daghastanli KRP, Martins TS, Cuccovia IM, Freire RS, Isolani PC. CeO_2-catalyzed ozonation of phenol. J Therm Anal Calorim 2010;102:907–913.

[49] Rivera-Utrilla J, Sánchez-Polo M. Ozonation of 1,3,6-naphthalenesulphonic acid catalysed by activated carbon in aqueous phase. Appl Catal B Environ 2002;39 (4):319–329.

[50] Liu Z-Q, Ma J, Cui Y-H. Carbon nanotube supported platinum catalysts for the ozonation of oxalic acid in aqueous solutions. Carbon 2008;46:890–897.

[51] Gonçalves AG, Figueiredo JL, Órfão JJM, Pereira MFR. Influence of the surface chemistry of multi-walled carbon nanotubes on their activity as ozonation catalysts. Carbon 2010;48:4369–4381.

[52] Kamat PV, Meisel D. Nanoparticles in advanced oxidation processes. Curr Opin Colloid Interface Sci 2002;7:282–287.

[53] (a) Puckett SD, Heuser JA, Keith JD, Spendel WU, Pacey GE. Interaction of ozone with gold nanoparticles. Talanta 2005; 66: 1242–1246.

(b) Hao Z, Cheng D, Guo Y, Liang Y. Supported gold catalysts used for ozone decomposition and simultaneous elimination of ozone and carbon monoxide at ambient temperature. Appl Catal B Environ 2001;33:217–222.

[54] Dong Y, Yang H, He K, Song S, Zhang A. β-MnO_2 nanowires: a novel ozonation catalyst for water treatment. Appl Catal B Environ 2009;85:155–161.

[55] Karnik BS, Baumann MJ, Masten SJ, Davies SH. AFM and SEM characterization of iron oxide coated ceramic membranes. J Mater Sci 2006;41:6861–6870.

[56] Corneal LM, Masten SJ, Davies SHR, Tarabara VV, Byun S, Baumann MJ. AFM, SEM and EDS characterization of manganese oxide coated ceramic water filtration membranes. J Memb Sci 2010;360:292–302.

[57] Corneal LM, Baumann MJ, Masten SJ, Davies SHR, Tarabara VV, Byun S. Mn oxide coated catalytic membranes for hybrid ozonation-membrane filtration: membrane microstructural characterization. J Memb Sci 2011;369:182–187.

[58] (a) Comninellis C, Nerini A. Anodic oxidation of phenol in the presence of NaCl for wastewater treatment. J Appl Electrochem 1995;25:23–28.
(b) Houk LL, Johnson SK, Feng J, Houk RS, Johnson DC. Electrochemical incineration of benzoquinone in aqueous media using a quaternary metal oxideelectrode in the absence of a soluble supporting electrolyte. J Appl Electrochem 1998;28: 1167–1177.
(c) Bock C, MacDougall B. Anodic oxidation of p-benzoquinone and maleic acid. J Electrochem 1999;146:2925–2932.

[59] (a) Yuan S, Hu S. Characterization and electrochemical studies of Nafion/nano- TiO_2 film modifies electrodes. Electrochim Acta 2004;49:4287–4293.
(b) Liu Z, Hong L, Guo B. Physicochemical and electrochemical characterization of anatase titanium dioxide nanoparticles. J Power Sources 2005;143:231–235.
(c) Rhun VL, Garnier E, Pronier S, Alonso-Vante N. Electrocatalysis on nanoscale ruthenium-based material manufactured by carbonyl decomposition. Electrochem Commun 2000;2:475–479.

[60] Luo J, Hepel M. Photoelectrochemical degradation of naphthol blue black diazo dye on WO_3 film electrode. Electrochim Acta 2001;46:2913–2922.

[61] Li J, Zhang X, Ai Z, Jia F, Zhang L, Lin J. Efficient visible light degradation of Rhodamine B by a photo-electrochemical process based on a Bi_2WO_6 nanoplate film electrode. J Phys Chem C 2007;111:6832–6836.

[62] Zhang Y, Xiong X, Han Y, Zhang X, Shen F, Deng S, Xiao H, Yang X, Yang G, Peng H. Photoelectrocatalytic degradation of recalcitrant organic pollutants using TiO_2 film electrodes: an overview. Chemosphere 2012;88:145–154.

[63] Yan X, Shi H, Wang D. Photoelectrocatalytic degradation of phenol using a TiO_2/Ni thin-film electrode. Korean J Chem Eng 2003;20:679–684.

[64] Hou Y, Li X, Liu P, Zou X, Chen G, Yue P-L. Fabrication and photo-electrocatalytic properties of highly oriented titania nanotube arrays with {101} crystal face. Sep Purif Technol 2009;67:135–140.

[65] Zhao H, Chen Y, Quan X, Ruan X. Preparation of Zn-doped TiO_2 nanotubes electrode and its application in pentachlorophenol photo-electrocatalytic degradation. Chin Sci Bull 2007;52:1456–1461.

[66] Li PQ, Zhao GH, Cui X, Zhang YG, Tang YT. Constructing stake structured TiO_2-NTs/Sb-doped SnO_2 electrode simultaneously with high electrocatalytic and photocatalytic performance for complete mineralization of refractory aromatic acid. J Phys Chem C 2009;113:2375–2383.

[67] Habibi MH, Talebian N, Choi J-H. Characterization and photocatalytic activity of nanostructured indium tin oxide thin-film electrode for azo-dye degradation. Thin Solid Films 2006;515:1461–1469.

[68] Hepel M, Luo J. Photoelectrochemical mineralization of textile diazo dye pollutants using nanocrystalline WO_3 electrodes. Electrochim Acta 2001;47:729–740.

[69] Zhang D, Pan C, Shi L, Mai H, Gao X. Controllable synthesis and highly efficient electrocatalytic oxidation performance of SnO_2/CNT core-shell structures. Appl Surf Sci 2009;255:4907–4912.

[70] (a) Ting WP, Lu MC, Huang YH. The reactor design and comparison of Fenton, electro-Fenton and photoelectro-Fenton processes for mineralization of benzene sulfonic acid (BSA). J Hazard Mater 2008;156:421–427.
(b) dos Santos ACV, Masini JC. Applying sequential injection analysis (SIA) and response surface methodology for optimization of Fenton-based processes. Talanta 2009; 77:1081–1086.

[71] Yan J, Tang H, Lin Z, Anjum MN, Zhu L. Efficient degradation of organic pollutants with ferrous hydroxide colloids as heterogeneous Fenton-like activator of hydrogen peroxide. Chemosphere 2012;87:111–117.

[72] Liu S-Q, Feng L-R, Xu N, Chen Z-G, Wang X-M. Magnetic nickel ferrite as a heterogeneous photo-Fenton catalyst for the degradation of Rhodamine B in the presence of oxalic acid. Chem Eng J 2012;203:432–439.

[73] Gajović A, Silva AMT, Segundo RA, Šturm S, Jančar B, Čeh M. Tailoring the phase composition and morphology of Bi-doped goethite–hematite nanostructures and their catalytic activity in the degradation of an actual pesticide using a photo-Fenton-like process. Appl Catal B Environ 2011;103:351–361.

[74] Zhou T, Li YZ, Ji J, Wong FS, Lu XH. Oxidation of 4-chlorophenol in a heterogeneous zero valent iron/H_2O_2 Fenton-like system: kinetic, pathway and effect factors. Sep Purif Technol 2008;62:551–558.

[75] Ortiz de la Plata GB, Alfano OM, Cassano AE. Decomposition of 2-chlorophenol employing goethite as Fenton catalyst. I. Proposal of a feasible, combined reaction scheme of heterogeneous and homogeneous reactions. Appl Catal B Environ 2010;95:1–13.

[76] Xue XF, Hanna K, Abdelmoula M, Deng NS. Adsorption and oxidation of PCP on the surface of magnetite: kinetic experiments and spectroscopic investigations. Appl Catal B Environ 2009;89:432–440.

[77] Costa RCC, Moura FCC, Ardisson JD, Fabris JD, Lago RM. Highly active heterogeneous Fenton-like systems based on Fe^0/Fe_3O_4 composites prepared by controlled reduction of iron oxides. Appl Catal B Environ 2008;83:131–139.

[78] Xu L, Wang J. A heterogeneous Fenton-like system with nanoparticulate zero-valent iron for removal of 4-chloro-3-methyl phenol. J Hazard Mater 2011;186:256–264.

[79] Shin S, Yoon H, Jang J. Polymer-encapsulated iron oxide nanoparticles as highly efficient Fenton catalysts. Catal Commun 2008;10:178–182.

[80] (a) Lam FLY, Hu X. A high performance bimetallic catalyst for photo-Fenton oxidation of Orange II over a wide pH range. Catal Commun 2007;8:2125–2129.
(b) Flox C, Garrido JA, Rodríguez RM, Cabot PL, Centellas F, Arias C, Brillas E. Mineralization of herbicide mecoprop by photo-electro-Fenton with UVA and solar light. Catal Today 2007;129:29–36.
(c) Muthuvel I, Swaminathan M. Photoassisted Fenton mineralisation of Acid Violet 7 by heterogeneous Fe(III)–Al_2O_3 catalyst. Catal Commun 2007;8:981–986.

[81] Chu W, Chan KH. Hybrid system to upgrade conventional Fenton's process by incorporating photo-Fenton as a successive treatment process: degradation of monuron. Ind Eng Chem Res 2007;46:1505–1510.

[82] (a) Molina R, Segura Y, Martínez F, Melero JA. Immobilization of active and stable goethite coated-films by a dip-coating process and its application for photo-Fenton systems. Chem Eng J 2012;203:212–222.
(b) He F, Shen X-Y, Lei L-C. Photochemically enhanced degradation of phenol using heterogeneous Fenton-type catalysts. J Environ Sci 2003;15:351–355.
(c) Iurascu B, Siminiceanu I, Vione D, Vicente MA, Gil A. Phenol degradation in water through a heterogeneous photo-Fenton process catalyzed by Fe-treated laponite. Water Res 2009;43:1313–1322.
(d) Pariente MI, Martinez F, Melero JA, Botas JA, Velegraki T, Xekoukoulotakis NP, Mantzavinos D. Heterogeneous photo-Fenton oxidation of benzoic acid in water: effect of operating conditions, reaction by-products and coupling with biological treatment. Appl Catal B Environ 2008;85:24–32.
(e) Yuranova T, Enea O, Mielczarski E, Mielczarski J, Albers P, Kiwi J. Fenton immobilized photo-assisted catálisis through a Fe/C structured fabric. Appl Catal B Environ 2004;49:39–50.

[83] Luo W, Zhu L, Wang N, Tang H, Cao M, She Y. Efficient removal of organic pollutants with magnetic nanoscaled $BiFeO_3$ as a reusable heterogeneous Fenton-like catalyst. Environ Sci Technol 2010;44:1786–1791.

[84] (a) Ai Z, Lu L, Li J, Zhang L, Qiu J, Wu M. Fe@Fe_2O_3 core–shell nanowires as iron reagent. 1. Efficient degradation of Rhodamine B by a novel sono-Fenton process. J Phys Chem C 2007;111:4087–4093.
(b) Ai Z, Lu L, Li J, Zhang L, Qiu J, Wu M. Fe@Fe_2O_3 core–shell nanowires as iron reagent. 2. An efficient and reusable sono-Fenton system working at neutral pH. J Phys Chem C 2007;111:7430–7436.
(c) Ai Z, Mei T, Li J, Li J, Jia F, Zhang L, Qiu J. Fe@Fe_2O_3 core–shell nanowires as iron reagent. 3. Their combination with CNTs as an effective oxygen-fed gas diffusion electrode in a neutral electro-Fenton system. J Phys Chem C 2007;111:14799–14803.

[85] (a) Selvam K, Muruganandham M, Sobana N, Swaminathan M. Enhancement of UV-assisted photo-Fenton degradation of reactive orange 4 using TiO_2-P25 nanoparticles. Sep Purif Technol 2007;54:241–247.
(b) Huang R, Fang Z, Yan X, Cheng W. Heterogeneous sono-Fenton catalytic degradation of bisphenol A by Fe_3O_4 magnetic nanoparticles under neutral condition. Chem Eng J 2012;197:242–249.
(c) Xu L, Wang J. Fenton-like degradation of 2,4-dichlorophenol using Fe_3O_4 magnetic nanoparticles. Appl Catal B Environ 2012;123–124:117–126.
(d) Alfano OM, Cassano AE, Ortiz de la Plata GB. 2-chlorophenol degradation via photo Fenton reaction employing zero valent iron nanoparticles. J Photochem Photobiol A Chem 2012;233:53–59.
(e) Babuponnusami A, Muthukumar K. Removal of phenol by heterogenous photo electro Fenton-like process using nano-zero valent iron. Sep Purif Technol 2012;98:130–135.
(f) Khataee AR, Zarei M, Ordikhani-Seyedlar R. Heterogeneous photocatalysis of a dye solution using supported TiO_2 nanoparticles combined with homogeneous photoelectrochemical process: molecular degradation products. J Mol Catal A Chem 2011; 338:84–91.
(g) Cong Y, Li Z, Zhang Y, Wang Q, Xu Q. Synthesis of α-Fe_2O_3/TiO_2 nanotube arrays for photoelectro-Fenton degradation of phenol. Chem Eng J 2012;191:356–363.
(h) Iranifam M, Zarei M, Khataee AR. Decolorization of C.I. Basic Yellow 28 solution using supported ZnO nanoparticles coupled with photoelectro-Fenton process. J Electroanal Chem 2011;659:107–112.

[86] Garrido-Ramírez EG, Theng BKG, Mora ML. Clays and oxide minerals as catalysts and nanocatalysts in Fenton-like reactions—a review. Appl Clay Sci 2010;47:182–192.

[87] (a) Valdés-Solís T, Valle-Vigón P, Álvarez S, Marbán G, Fuertes AB. Manganese ferrite nanoparticles synthesized through a nanocasting route as a highly active Fenton catalyst. Catal Commun 2007;8:2037–2042.
(b) Valdés-Solís TP, Valle-Vigón P, Sevilla M, Fuertes AB. Encapsulation of nanosized catalysts in the hollow core of a mesoporous carbon capsule. J Catal 2007;251:239–243.

[88] Zelmanov G, Semiat R. Iron(3) oxide-based nanoparticles as catalysts in advanced organic aqueous oxidation. Water Res 2008;42:492–498.

[89] Brillas E, Calpe JC, Casado J. Mineralization of 2,4-D by advanced electrochemical oxidation processes. Water Res 2000; 34:2253–2262.

[90] (a) Isarain-Chávez E, Arias C, Cabot PL, Centellas F, Rodríguez RM, Garrido JA, Brillas E. Mineralization of the drug b-blocker atenolol by electro-Fenton and photoelectro-Fenton using an air-diffusion cathode for H_2O_2 electrogeneration combined with a carbon-felt cathode for Fe^{2+} regeneration. Appl Catal B Environ 2010;96:361–369.
(b) Khataee AR, Safarpour M, Naseri A, Zarei M. Photoelectro-Fenton/nanophotocatalysis decolorization of three textile dyes mixture: response surface modeling and multivariate calibration procedure for simultaneous determination. J Electroanal Chem 2012;672:53–62.

[91] Brillas E, Sires I, Oturan MA. Electro-Fenton process and related electrochemical technologies based on Fenton's reaction chemistry. Chem Rev 2009;109:6570–6631.

[92] (a) Lu L, Ai Z, Li J, Zheng Z, Li Q, Zhang L. Synthesis and characterization of $Fe@Fe_2O_3$ core–shell nanowires and nanonecklaces. Cryst Growth Des 2007;7:459–464.
(b) Li J, Ai Z, Zhang L. Design of a neutral electro-Fenton system with $Fe@Fe_2O_3$/ACF composite cathode for wastewater treatment. J Hazard Mater 2009;164:18–25.

[93] Ding X, Ai Z, Zhang L. Design of a visible light driven photo-electrochemical/electro-Fenton coupling oxidation system for wastewater treatment. J Hazard Mater 2012;239–240:233–240.

[94] Zarei M, Khataee AR, Ordikhani-Seyedlar R, Fathinia M. Photoelectro-Fenton combined with photocatalytic process for degradation of an azo dye using supported TiO_2 nanoparticles and carbon nanotube cathode: neural network modeling. Electrochim Acta 2010;55:7259–7265.

[95] (a) Baker DR, Kamat PV. Photosensitization of TiO_2 nanostructures with CdS quantum dots: particulate versus tubular support architectures. Adv Funct Mater 2009;19:805–811.
(b) Hou Y, Li XY, Zou XJ, Quan X, Chen GC. Photoeletrocatalytic activity of a Cu_2O-loaded self-organized highly oriented TiO_2 nanotube array electrode for 4-chlorophenol degradation. Environ Sci Technol 2009;43 (3):858–863.

[96] Jeon TH, Choi WY, Park HW. Photoelectrochemical and photocatalytic behaviors of hematite-decorated titania nanotube arrays: energy level mismatch versus surface specific reactivity. J Phys Chem C 2011;115:7134–7142.

[97] (a) Suslick KS, Doktycz SJ, Flint EB. On the origin of sonoluminescence and sonochemistry. Ultrasonics 1990;28 (5):280–290.
(b) Petrier C, Jiang Y, Lamy MF. Ultrasound and environment: sonochemical destruction of chloroaromatic derivatives. Environ Sci Technol 1998;32 (9):1316–1318.
(c) Wang L, Zhu L, Luo W, Wu Y, Tang H. Drastically enhanced ultrasonic decolorization of methyl orange by adding CCl4. Ultrason Sonochem 2007;14 (2):253–258.
(d) Wang J, Guo Y, Liu B, Jin X, Liu L, Xu R, Kong Y, Wang B. Detection and analysis of reactive oxygen species (ROS) generated by nano-sized TiO_2 powder under ultrasonic irradiation and application in sonocatalytic degradation of organic dyes. Ultrason Sonochem 2011;18:177–183.

[98] Sadjadi S, Sadjadi S, Hekmatshoar R. Ultrasound-promoted greener synthesis of benzoheterocycle derivatives catalyzed by nanocrystalline copper(II) oxide. Ultrason Sonochem 2010;17:764–767.

[99] Tezcanli-Guyer G, Ince NH. Individual and combined effects of ultrasound, ozone and UV irradiation: a case study with textile dyes. Ultrasonics 2004;42:603–609.

[100] Abbasi M, Razzaghi Asl N. Sonochemical degradation of Basic Blue 41 dye assisted by nanoTiO_2 and H_2O_2. J Hazard Mater 2008;153 (3):942–947.

[101] Wang J, Pan Z, Zhang Z, Zhang X, Wen F, Ma T, Jiang Y, Wang L, Xu L, Kang P. Sonocatalytic degradation of methyl parathion in the presence of nanometer and ordinary anatase titanium dioxide catalysts and comparison of their sonocatalytic abilities. Ultrason Sonochem 2006;13:493–500.

[102] Wang J, Ma T, Zhang Z, Zhang X, Jiang Y, Pan Z, Wen F, Kang P, Zhang P. Investigation on the sonocatalytic degradation of methyl orange in the presence of nanometer anatase and rutile TiO_2 powders and comparison of their sonocatalytic activities. Desalination 2006;195:294–305.

[103] Wang J, Guo B, Zhang X, Zhang Z, Han J, Wu J. Sonocatalytic degradation of methyl orange in the presence of TiO_2 catalysts and catalytic activity comparison of rutile and anatase. Ultrason Sonochem 2005;12:331–337.

[104] Pang YL, Abdullah AZ, Bhatia S. Review on sonochemical methods in the presence of catalysts and chemical additives for treatment of organic pollutants in wastewater. Desalination 2011;277:1–14.

[105] Wang J, Jiang Y, Zhang Z, Zhang X, Ma T, Zhang G, Zhao G, Zhang P, Li Y. Investigation on the sonocatalytic degradation of Acid Red B in the presence of nanometer TiO_2 catalysts and comparison of catalytic activities of anatase and rutile TiO_2 powders. Ultrason Sonochem 2007;14:545–551.

[106] Lei Z, Ghosh T, Park C-Y, Ze-Da M, Won-Chun O. Enhanced sonocatalytic degradation of Rhodamine B by graphene-TiO_2 composites synthesized by an ultrasonic-assisted method. Chin J Catal 2012;33:1276–1283.

[107] (a) Pang YL, Bhatia S, Abdullah AZ. Process behaviorofTiO_2 nanotube-enhanced sonocatalyticdegradationof Rhodamine Binaqueoussolution. Sep Purif Technol 2011;77:331–338.

(b) Pang YL, Abdullah AZ. Comparative study on the process behavior and reaction kinetics in sonocatalytic degradation of organic dyes by powder and nanotubes TiO_2. Ultrason Sonochem 2012;19:642–651.

[108] Pang YL, Abdullah AZ. Effect of low Fe^{3+} doping on characteristics, sonocatalytic activity and reusability of TiO_2 nanotubes catalysts for removal of Rhodamine B from water. J Hazard Mater 2012;235–236:326–335.

[109] (a) Wang J, Jiang Z, Zhang L, Kang P, Xie Y, Lv Y, Xu R, Zhang X. Sonocatalytic degradation of some dyestuffs and comparison of catalytic activities of nano-sized TiO_2, nano-sized ZnO and composite TiO_2/ZnO powders under ultrasonic irradiation. Ultrason Sonochem 2009;16:225–231.

(b) Wang J, Jiang Z, Zhang Z, Xie Y, Lv Y, Li J, Deng Y, Zhang X. Study on inorganic oxidants assisted sonocatalytic degradation of Acid Red B in presence of nano-sized ZnO powder. Sep Purif Technol 2009;67:38–43.

(c) Wang J, Jiang Z, Zhang Z, Xie Y, Wang X, Xing Z, Xu R, Zhang X. Sonocatalytic degradation of Acid Red B and Rhodamine B catalyzed by nano-sized ZnO powder under ultrasonic irradiation. Ultrason Sonochem 2008;15:768–774.

[110] Ziylan A, Koltypin Y, Gedanken A, Ince NH. More on sonolytic and sonocatalytic decomposition of Diclofenac using zero-valent iron. Ultrason Sonochem 2013;20:580–586.

[111] (a) Shen ZZ, Shen JZ. The use of ultrasound to enhance the degradation of the Basic Green by cast iron. Ultrasonics 2006; 44:353–356.

(b) Zhang H, Duan LJ, Zhang Y, Wu F. The use of ultrasound to enhance the decolorization of the CI Acid Orange 7 by zero-valent iron. Dyes Pigments 2005;65 (1):39–43.

(c) Zhang H, Jiang M, Wang ZQ, Wu F. Decolorisation of CI Reactive Black 8 by zero-valent iron powder with/without ultrasonic irradiation. Color Technol 2007;123 (3):203–208.

[112] (a) Wang J, Ma T, Zhang Z, Zhang X, Jiang Y, Dong D, Zhang P, Li Y. Investigation on the sonocatalytic degradation of parathion in the presence of nanometer rutile titanium dioxide (TiO_2) catalyst. J Hazard Mater 2006;137 (2):972–980.

(b) Wang J, Sun W, Zhang Z, Zhang X, Li R, Ma T, Zhang P, Li Y. Sonocatalytic degradation of methyl parathion in the presence of micron-sized and nano-sized rutile titanium dioxide catalysts and comparison of their sonocatalytic abilities. J Mol Catal A Chem 2007;272 (1–2):84–90.

(c) Ghows N, Entezari MH. Exceptional catalytic efficiency in mineralization of the reactive textile azo dye (RB5) by a combination of ultrasound and core–shell nanoparticles (CdS/TiO_2). J Hazard Mater 2011;195:132–138.

(d) Zhang S. Synergistic effects of C–Cr codoping in TiO_2 and enhanced sonocatalytic activity under ultrasonic irradiation. Ultrason Sonochem 2012;19:767–771.

(e) Gao J, Jiang R, Wang J, Kang P, Wang B, Li Y, Li K, Zhang X. The investigation of sonocatalytic activity of Er^{3+}:$YAlO_3$/TiO_2-ZnO composite in azo dyes degradation. Ultrason Sonochem 2011;18:541–548.

(f) Min YL, Zhang K, Chen YC, Zhang YG. Sonodegradation and photodegradation of methyl orange by $InVO_4$/TiO_2 nanojunction composites under ultrasonic and visible light irradiation. Ultrason Sonochem 2012;19:883–889.

(g) Meng Z-D, Oh W-C. Sonocatalytic degradation and catalytic activities for MB solution of Fe treated fullerene/TiO_2 composite with different ultrasonic intensity. Ultrason Sonochem 2011;18:757–764.

(h) Zhu L, Meng Z-D, Park C-Y, Ghosh T, Oh W-C. Characterization and relative sonocatalytic efficiencies of a new MWCNT and CdS modified TiO_2 catalysts and their application in the sonocatalytic degradation of Rhodamine B. Ultrason Sonochem 2013; 20:478–484.

(i) Wang J, Lv Y, Zhang L, Liu B, Jiang R, Han G, Xu R, Zhang X. Sonocatalytic degradation of organic dyes and comparison of catalytic activities of CeO_2/TiO_2, SnO_2/TiO_2 and ZrO_2/TiO_2 composites under ultrasonic irradiation. Ultrason Sonochem 2010;17:642–648.

(j) Salavati H, Tavakkoli N, Hosseinpoor M. Preparation and characterization of polyphosphotungstate/ZrO_2 nanocomposite and their sonocatalytic and photocatalytic activity under UV light illumination. Ultrason Sonochem 2012;19:546–553.

[113] Wang J, Ma T, Zhang Z, Zhang X, Jiang Y, Zhang G, Zhao G, Zhao H, Zhang P. Investigation on transition crystal of ordinary rutile TiO_2 powder and its sonocatalytic activity. Ultrason Sonochem 2007;14:246–252.

[114] Zhang K, Oh WC. Kinetic study of the visible light-induced sonophotocatalytic degradation of MB solution in the presence of Fe/TiO_2-MWCNT catalyst. Bull Korean Chem Soc 2010;31:1589–1595.

[115] (a) Zhang H, Lv YJ, Liu F, Zhang DB. Degradation of CI Acid Orange 7 by ultrasound enhanced ozonation in a rectangular air-lift reactor. Chem Eng J 2008;138 (1–3):231–238.

(b) Eren Z. Ultrasound as a basic and auxiliary process for dye remediation: a review. J Environ Manage 2012;104:127–141.

[116] Sonawane SH, Chaudhari PL, Ghodke SA, Parande MG, Bhandari VM, Mishra S, Kulkarni RD. Ultrasound assisted synthesis of polyacrylic acid–nanoclay nanocomposite and its application in sonosorption studies of malachite green dye. Ultrason Sonochem 2009;16:351–355.

[117] (a) Zhang Z, Yuan Y, Liang L, Fang Y, Cheng Y, Ding H, Shi G, Jin L. Sonophotoelectrocatalytic degradation of azo dye on TiO_2 nanotube electrode. Ultrason Sonochem 2008;15:370–375.

(b) Sonawane S, Chaudhari P, Chodke S, Ambade S, Gulig S, Mirikar A, Bane A. Combined effect of ultrasound and nanoclay on adsorption of phenol. Ultrason Sonochem 2001;15:1033–1037.

(c) Wang H, Niu J, Long X, He Y. Sonophotocatalytic degradation of methyl orange by nano-sized Ag/TiO_2 particles in aqueous solutions. Ultrason Sonochem 2008;15:386–392.

25

CATALYST DESIGN BASED ON NANO-SIZED INORGANIC CORE OF ENZYMES: DESIGN OF ENVIRONMENTALLY FRIENDLY CATALYSTS

Mohammad Mahdi Najafpour

Department of Chemistry, Center for Research in Climate Change and Global Warming, and Institute for Advanced Studies in Basic Sciences (IASBS), Zanjan, Iran

Catalysts are very important compounds in our life [1–3]. Catalysts with new pathways enable the highly selective formation of desired products at rates that are commercially viable [1–3]. Some important reactions in human life are water oxidation, water reduction, nitrogen reduction, and carbon dioxide reduction, which function very efficiently at moderate temperature in organisms. We need efficient, stable, environmentally friendly, low-cost, and multielectron transfer catalysts to perform tasks at an industrial scale. Multielectron transfer catalysts have the ability to accommodate, accumulate, and transfer multiple electrons to reaction substrates at the same time. The compounds that catalyze multielectron reactions are prone to structural rearrangement and instability during turnover.

While heavy metals are widely used as catalysts, organisms manage to use abundant, low-cost, and environmentally friendly transition metals for the same purpose. Biomimetics is biologically inspired design that could be defined as the study of the structure and function of biological systems as models for the design and engineering of materials and machines [4]. For biologically inspired catalyst designs, we could look at enzymes as they are the most efficien catalysts.

Probably, the first enzymes had broad substrate specificity and were not efficient [5]. It is believed that a relatively small number of enzymes might have sufficed to support the metabolic network of the first proto-organisms. These first enzymes were known as generalists, which changed to *modern enzymes* [5]. Metal clusters are used as the active sites of some *modern enzymes*. Over past years, the clusters were studied by many methods, most importantly X-ray diffraction methods. The structures may serve as a helpful and fundamental basis for a better understanding of the mechanism and provide a framework for generating hypotheses for catalysis design.

From a certain viewpoint, we may consider enzymes to be nano-sized inorganic cores in a protein matrix. To understand the subject we should find the chemistry of interactions between proteins and solid surfaces.

Interactions between proteins and solid surfaces were studied using many methods by several groups [6–11]. Interactions of proteins with clays or minerals have been reviewed by a few authors [12, 13]. Many reports indicated that enzyme molecules were not intercalated in the mineral structure but were immobilized at the external surfaces and at the edges of the mineral oxide [14]. This result also showed the protein's molecular conformation after binding to the mineral colloid surfaces.

The solid surfaces might have an important role in the origin of life and the formation of biomolecules [15]. Wächtershäuser's group proposed that the earliest form of life originated in a volcanic hydrothermal flow at high pressure and high temperature (~100°C). In this condition, a composite of mineral compounds with catalytic transition metal centers could play important roles in forming the first biological molecules [16, 17]. Cairns-Smith's group also proposed that imperfect clay crystals were the first genes [18]. It was also known that minerals and clays could induce selection, concentration, and organization of

Nanomaterials for Environmental Protection, First Edition. Edited by Boris I. Kharisov, Oxana V. Kharissova, and H. V. Rasika Dias.

prebiotic and simple molecules into biomolecules. For example, Kauffman proposed that the mineral protometabolic surface could recruit random polypeptides from their environment and improve the efficiency of catalytic activities [19, 20].

Russell's group proposed that minerals could attach to proteins to form enzymes [21]. In other words, minerals might be used initially by polypeptides or carboxylate groups to become active centers of the enzyme precursors. Then, they might initially catalyze the primary and important reactions of energy conversion and nutrient cycling; for example, pyrophosphate in successive main chain NH peptide, protoferredoxins as thiolated iron sulfide in peptide, precursors to carbon monoxide dehydrogenase/acetyl CoA synthetase as nickel sulfide in a peptide, and manganese oxide minerals to the active center of the water-oxidizing complex (WOC) in Photosystem II [21, 22]. It was observed that there are specific size-dependent interactions that drive the binding of proteins to nanoparticles [23].

Instead of the "ready-made mineral incorporated into an existing protein theory" proposed by Russell's group, the author proposes that the cofactor formation in particular enzymes is biomineralization in the presence of the protein. According to this hypothesis, mineral-like clusters may form in the presence of a protein and be used as the first inorganic core of enzymes [24]. First, I consider the hypothesis and then I try to explain how the hypothesis could be used to design new environmentally friendly catalysts.

The first and simplest reaction of a protein with nanominerals is the extraction of ions from the minerals by the protein (Fig. 25.1a). In these reactions, minerals decompose and produce ions that coordinate with amino acid side chains.

The collecting of proteins around (nano)minerals is the second reaction (Fig. 25.1b and c) [26]. An important factor in these reactions is the surface charge of both protein and particles [26]. For example, Vallee reported that on the surfaces of titania and silica, carboxylate groups are the preferential peptide-binding groups because of their strong electrostatic interactions with ions on surfaces [26]. The surface charges of both proteins and oxide particles are largely dependent on the pH of the solution, and these interactions are governed by physical or chemical forces [27, 28]. It is important to note that adsorption of a protein on a surface may induce conformation changes in both the biomolecule and on the surface of the particles [26].

Russell and other groups suggested minerals as a "ready-made catalyst" [22, 29]. For example, iron–sulfur centers are potentially important for catalysis via oxidation–reduction reactions since they can exist as Fe^{2+} or Fe^{3+}, and the ability to stabilize

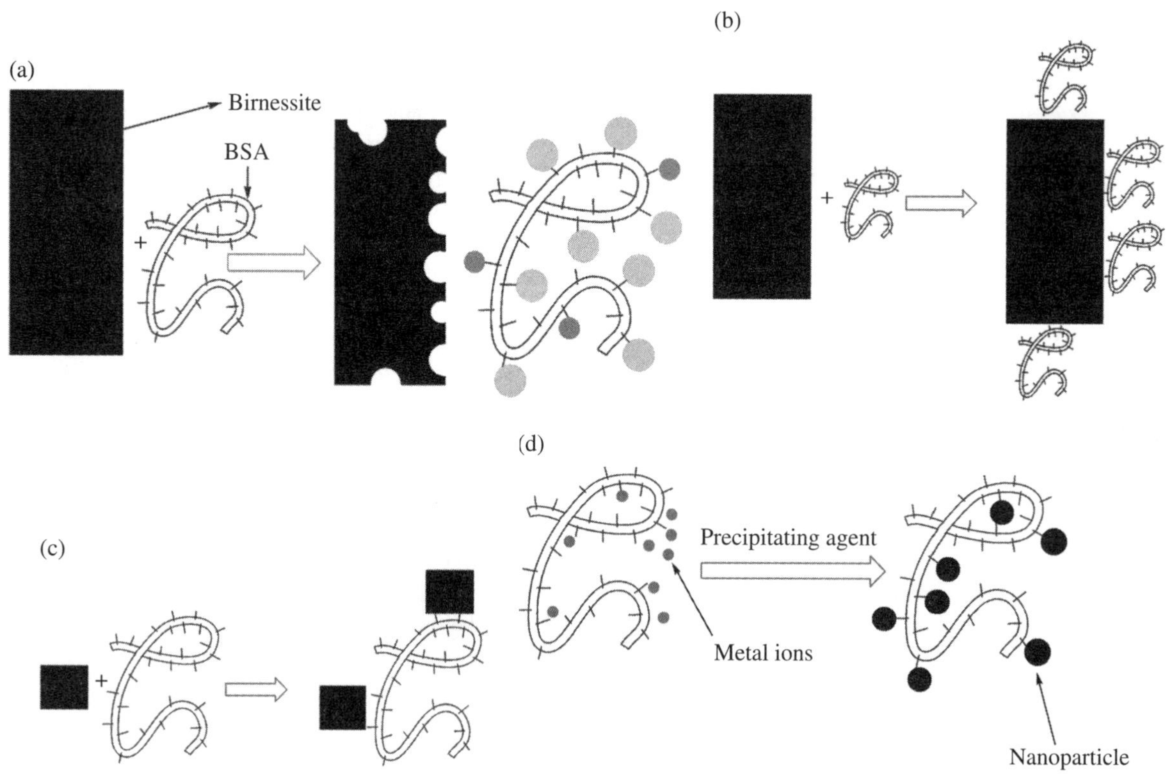

FIGURE 25.1 (a) An important reaction between mineral and a protein could be the extraction of ions of minerals. (b, c) Arrangements of nanoparticles and proteins depend on their size. (d) When nanoparticles are formed in the presence of a protein, the protein not only induces nucleation, but could also inhibit the further growth of nanoparticles. Reprinted with permission from Ref. [25]. © 2013, Springer.

and sequester these centers as peptide nests must have an advantage for organisms [21]. Another example is that the $CaMn_4O_5$ cluster in Photosystem II was derived from a component of a natural early marine manganese precipitate that contained a $CaMn_4O_5$ cluster [21]. It is to be noted that the bond types and lengths of the manganese ions in the tunnel mineral hollandite ($Ba_{0.2}Ca_{0.15}K_{0.3}Mn_{6.9}Al_{0.2}Si_{0.3}O_{16}$) are directly comparable to those in the $CaMn_4O_5$ cluster in Photosystem II and can be considered as both structural [30] and functional [31] models for the $CaMn_4O_5$ cluster in Photosystem II; it is also proposed that these minerals may be assimilated by early cyanobacteria during the Archaean period, approximately 3200–2800 million years ago, to form the cluster in the biological WOC [22, 30].

The third reaction of (nano)minerals with proteins is observed when nanoparticles are formed in the presence of proteins (Fig. 25.1c) [32–35]. Precipitating agents could be reductants, oxidants, anions, cations, wetting/drying cycles [11], or light that precipitate metal ions to form particles in the presence of proteins. In this condition, the results show that the protein not only induces the nucleation, but could also inhibit the further growth of nanominerals, as, without the protein, larger particles are formed [32–35]. Numerous and very small nanoparticles could form in this condition [32–35].

The dispersivity and stability of nanoparticles in this condition are usually unique. The cofactor formation in particular enzymes could be considered to be mineralization in the presence of the protein. In other words, one of the numerous and very small nanoparticles (Fig. 25.1c) produced in the presence of protein could be formed in an appropriate location and be used as a primitive inorganic core (cofactor). The structure of the primitive cofactor could be changed and modified by amino acid side groups.

WOC in Photosystem II, nitrogenase, nitrous oxide reductase or oxidase, hydrogenase, and dehydrogenase are enzymes with an inorganic core, and mineralization could be important in their formation (Scheme 25.1).

It should be considered that there is a tendency to form special structures similar to inorganic cofactors in enzymes from simple ions (manganese, iron, calcium, and so on) in water even in the absence of any amino acids (Scheme 25.2) [37–39].

To show the interactions between proteins and minerals, the interactions of birnessite, a common manganese oxide in the environment, and bovine serum albumin (BSA), a protein that has a strong affinity for a variety of inorganic molecules, are shown as examples in Figure 25.2.

When birnessite is formed in the presence of BSA, the results show that the BSA not only induces the nucleation, but could also inhibit the further growth of manganese oxide, as, without BSA, larger particles are formed (Fig. 25.3). BSA also helps to disperse these nanoparticles in solution and the dispersivity of particles (or minerals) in water can increase the rate of reaction of particles with other substrates.

The dispersivity of manganese oxide particles in this condition is unique and may be considered a soluble form of nano-sized colloidal manganese (III, IV) oxide [40].

All these observations show similarities between inorganic cofactors in enzymes and nanoparticles formed in the presence of proteins. For example, in Photosystem II, Mn_4CaO_5 is a nano-sized inorganic cofactor [41, 42], and the authors have proposed that the Archaean ocean may have sufficient manganese and calcium ions that, in alkaline conditions, may have enabled protocyanobacteria to assemble oxide mineral as functional compliments of cofactors in Photosystem II [43].

The hypothesis may be extended to other enzymes with inorganic cofactors (Scheme 25.1) and we may assume the cofactor formation reaction, in many enzymes, might have been mineralization in the presence of proteins instead of direct adsorption of minerals onto protein. Many of the inorganic (nano)particles, which could be considered as inorganic cores or cofactors in enzymes, could have catalyzed related reactions (Table 25.1).

(a) (b) (c)

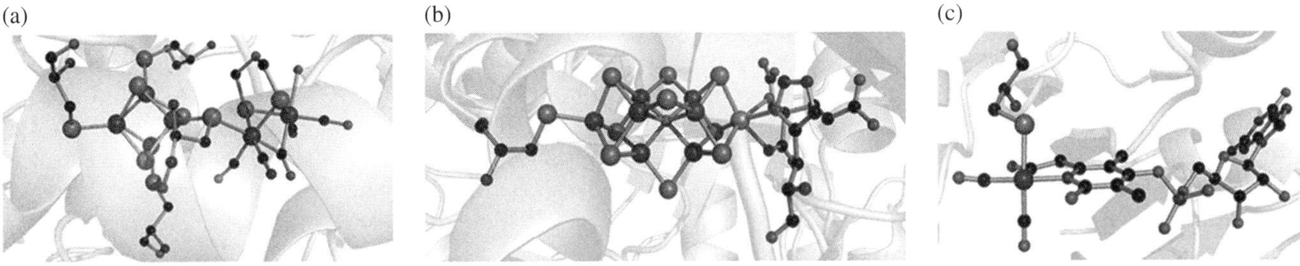

SCHEME 25.1 Examples of complex inorganic cofactors in biology. Crystal structures of (a) the [FeFe]-hydrogenase H-cluster, (b) the Mo-nitrogenase FeMo cofactor, and (c) the [Fe]-hydrogenase Fe-guanylyl cofactor (FeGP). Reprinted with permission from Ref. [36]. © 2012, Elsevier.

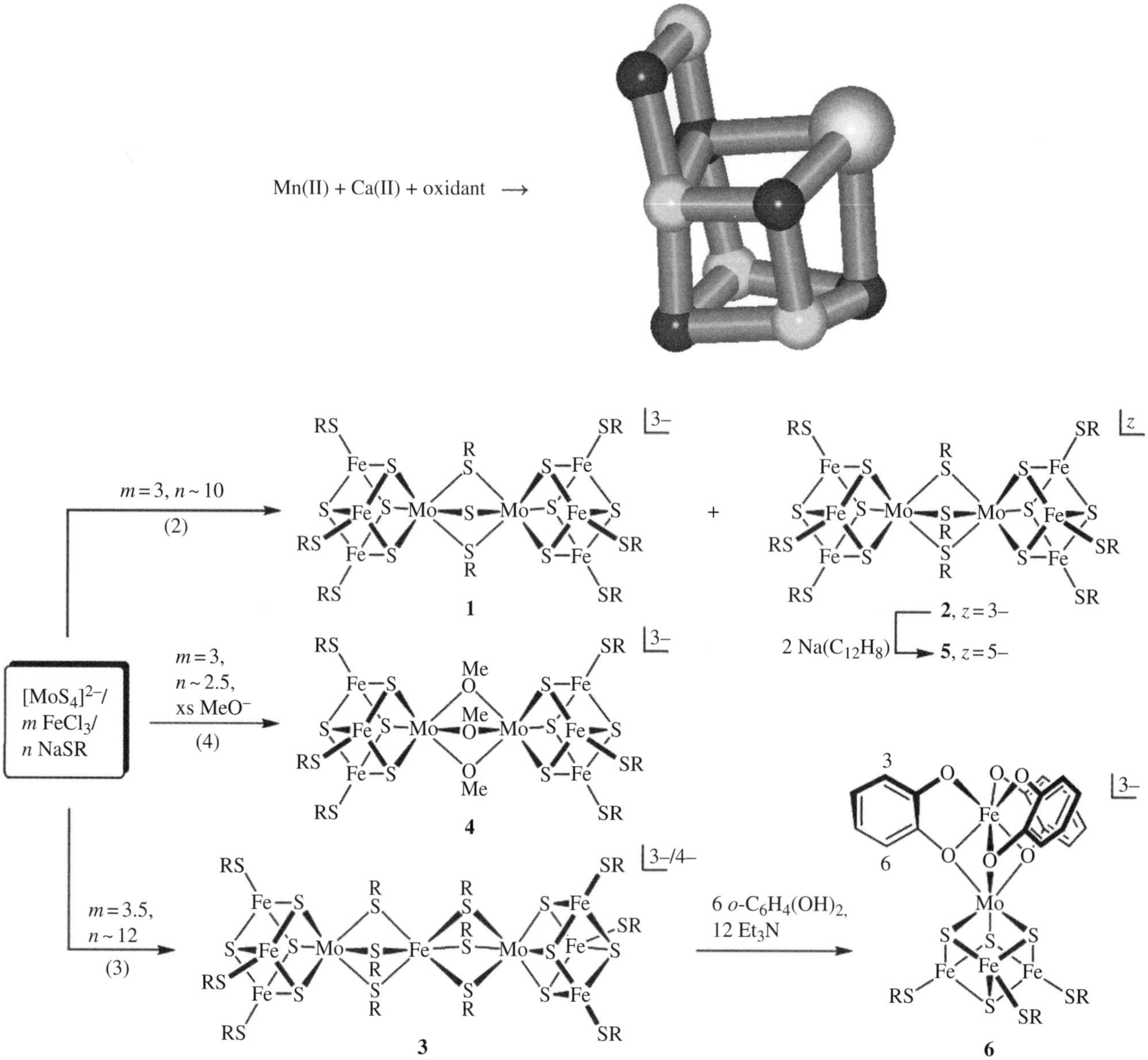

SCHEME 25.2 Self-assembly of inorganic core models of WOC [37, 38] and nitrogenase [39] from simple ions in water even in the absence of any amino acids. Image from Ref. [39].

For example, manganese oxides have been reported as heterogeneous catalysts for water oxidation [41]. Interestingly, Najafpour et al. have shown that the incorporation of calcium ions into manganese oxides can improve the catalytic activity of synthetic manganese oxides [37, 38]. In addition to these calculations, Siegbahn showed that there is great similarity between manganese–calcium oxides and manganese–calcium clusters in water oxidation in Photosystem II [57]. Siegbahn used a hypothetical Mn-O crystal constructed with all Mn–Mn distances set to 2.70 Å and all Mn–O distances set to 1.90 Å. In the next step, one calcium atom was inserted into the surface of this crystal, with all Ca–Mn distances equal to 3.4 Å and all Ca–O distances set to 2.4 Å. These distances are very similar to the ones in the manganese calcium cluster in Photosystem II (Fig. 25.4). This comparison also shows similarity between the inorganic cofactor of enzymes and metal oxides [57].

In another report, Dörr et al. described a method for the synthesis of ammonia from dinitrogen using H_2S as a reductant and iron sulfide, which could be considered an inorganic core for the nitrogenase [44]. The reductant, as well as the reaction conditions (atmospheric nitrogen pressure and temperatures on the order of 70–80°C), is rather mild and comparable to biological processes [44]. Other models are shown in Table 25.1.

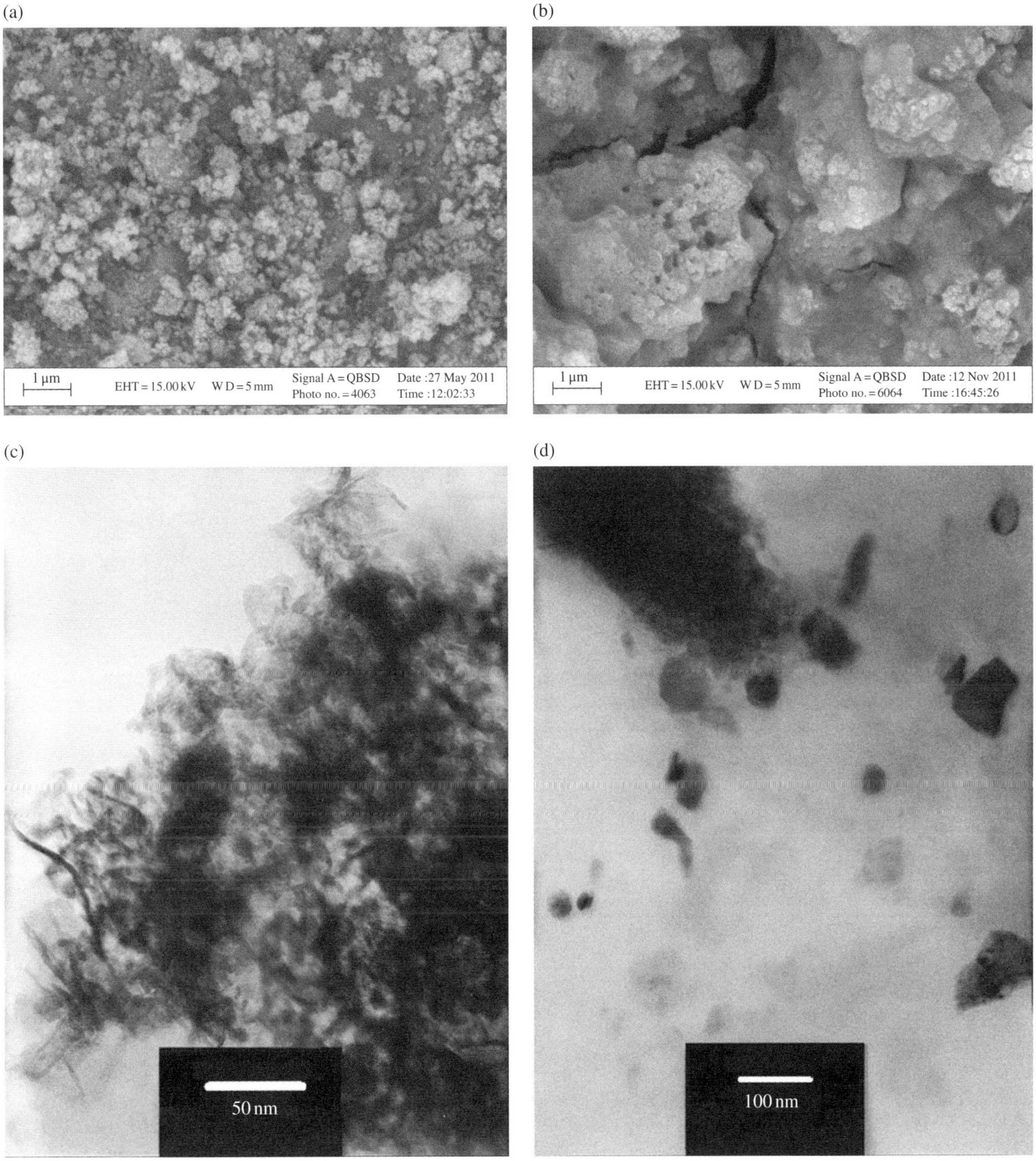

FIGURE 25.2 SEM images show nano-sized birnessite (a) before and (b) after treating with BSA solution. TEM images show nano-sized birnessite (c) before and (d) after treating with BSA solution. For experimental details, see experimental methods. Reprinted with permission from Ref. [24]. © 2012, Springer.

Recently, a biomimetic chalcogel was reported that is capable of photocatalytic hydrogen evolution under simulated solar illumination [47]. The new chalcogels are prepared from double-cubane $Mo_2Fe_6S_8$ cluster units linked by Sn_2S_6 ligands, forming a random, amorphous network with strong optical absorption. The switch from Fe_4S_4 cubane units to $Mo_2Fe_6S_8$ double-cubane units illustrates the flexibility and ease with which we can synthesize chalcogels with a variety of

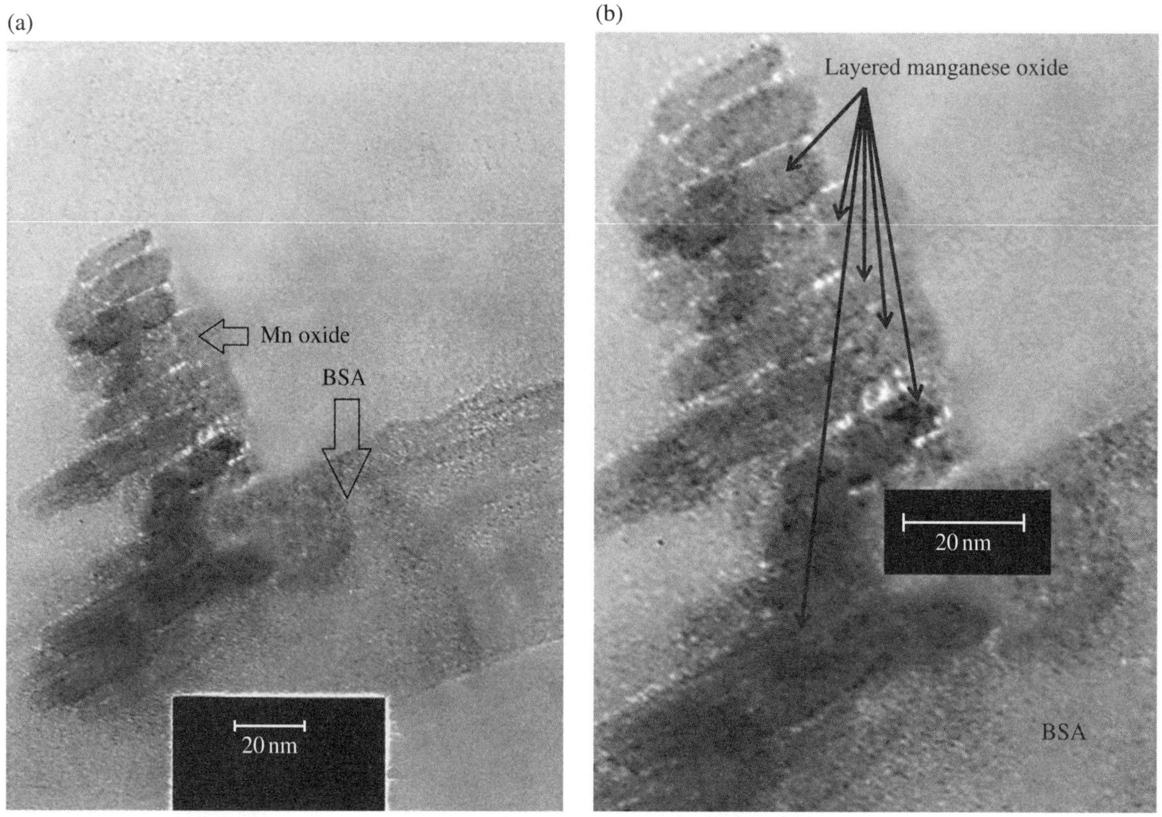

FIGURE 25.3 (a and b) The manganese oxide prepared in the presence of BSA. Scale bars for TEM images are 20 nm. Images (a) and (b) are reproduced with permission from Ref. [24]. © Springer.

redox-active biomimetic cofactors. The chalcogels can be easily functionalized with cationic chromophores, such as $[Ru(bpy)_3]^{2+}$, enabling photoexcited electron transfer from dye to cluster. The biomimetic chalcogels are fully stable in aqueous solutions under illumination for at least 3 weeks, which contrasts greatly with the water-unstable molecular $[Mo_2Fe_6S_8]$-based clusters [47].

More interestingly, in many cases, mechanisms of these related reactions (Table 25.1) with only inorganic (nano)particles, such as metal oxides or sulfides, and enzymes could be very similar (Scheme 25.3) [58]. In conclusion, it is worth pointing out that in some cases the metal oxides or sulfides, which are considered as cofactors in enzymes, are efficient and unique catalysts for related reactions in vitro. These synthetic or natural simple compounds, as minerals, are formed in no special or preorganized precursors [37–39], and in purely aqueous solution usually show structures similar to inorganic cores of enzymes. Such structures could be formed, changed, and modified by environmental groups (amino acids) in the first enzymes.

It is worth noting that the assembly of inorganic cofactors in these enzymes could also be considered stepwise mineralization [59]. As an example, a proposed assembly of the WOC in Photosystem II is shown in Figure 25.5 [59]. The biosynthesis of the FeMo cofactor could also be considered complex biomineralization (Fig. 25.5) [60].

Both enzymes and related metal oxides and sulfides that we discussed catalyze multielectron transfer reactions and, therefore, it is necessary to find theoretical concepts for multielectron transfer reactions. For these reactions, the overall electron transfer is essentially self-organized or concerted, and individual electron transfer steps are related to each other; therefore, the set of corresponding equations can be reduced to a single one, which then reflects improved, multistep, electron transfer. The theoretical formalism for such an approach has been discussed by Tributsch and Pohlmann [61]. In multielectron reactions, high-energy intermediate states disappear and the entire multistep electron transfer reaction can be described with one single kinetic equation [61]; in addition, an important property of catalysts is that the first electron transfer for the catalysts facilitates the transfer of the second electron. Then, this transfer facilitates the transfer of the third, and the transfer of the fourth electron, or even further electrons usually by triggering a molecular change.

TABLE 25.1 Relation between catalytic reactions of inorganic cores of enzymes in vivo or related inorganic compounds in vitro

Elements	Oxide or sulfide	Enzyme	Related minerals	Enzymatic reaction(s)	Reaction(s)	Reference
Ca, Mn, and O	Calcium and manganese oxide	Water-oxidizing complex in Photosystem II	Birnessite, Hollandite	$2H_2O \rightarrow 4H^+ + O_2 + 4e^-$	$2H_2O \rightarrow 4H^+ + O_2 + 4e^-$	[40–43]
Fe and S	Iron sulfide	Nitrogenase	Pyrite	$N_2 + 6e^- + 6H^+ \rightarrow 2NH_3$	$N_2 + 6e^- + 6H^+ \rightarrow 2NH_3$	[44]
Fe and O	Iron oxide	Hydrolyzing phosphoester	Hematite	Hydrolyzing phosphoester	Hydrolyzing phosphoester	[45, 46]
Fe and S	Iron sulfide	Hydrogenase	Pyrite	$2H^+ + 2e^- \rightarrow H_2$	$2H^+ + 2e^- \rightarrow H_2$	[47]
Cu and S	Nickel and iron and sulfide	Nitrous oxide reductase or oxidase	Chalcocite (Cu_2S); covellite CuS	$N_2O + 2H^+ + 2e^- \rightarrow N_2 + H_2O$	Reduction of nitrogen oxides compounds	[48]
Mo and S	Molybdenum sulfide	Hydrogenase	Molybdenite	$2H^+ + 2e^- \leftrightarrow H_2$	$2H^+ + 2e^- \leftrightarrow H_2$	[49]
Mo and S	Molybdenum sulfide	DMSO reductase	Molybdenite	$DMSO + 2H^+ \rightarrow DMS + H_2O$	$SO_2 \rightarrow S$	[50]
W and S	Tangestan sulfide	DMSO reductase	Tungstenite	$DMSO + 2H^+ \rightarrow DMS + H_2O$	$SO_2 \rightarrow S$	[50]
Ni, Fe, and S	Nickel and iron and sulfide	Hydrogenase and dehydrogenase	Pentlandite	$2H^+ + 2e^- \leftrightarrow H_2$	$2H^+ + 2e^- \leftrightarrow H_2$	[51, 52]
V, S	Vanadium and sulfide	Nitrogenase/hydrogenase	Colimaite	$N_2 + 6e^- + 6H^+ \rightarrow 2NH_3$ $2H^+ + 2e^- \rightarrow H_2$	$2H^+ + 2e^- \rightarrow H_2$	[53, 54]
Mo, Fe, and S	Molybdenum iron sulfide	Nitrogenase/hydrogenase	Femolite	$N_2 + 6e^- + 6H^+ \rightarrow 2NH_3$ $2H^+ + 2e^- \rightarrow H_2$	$2H^+ + 2e^- \rightarrow H_2$	[55, 56]

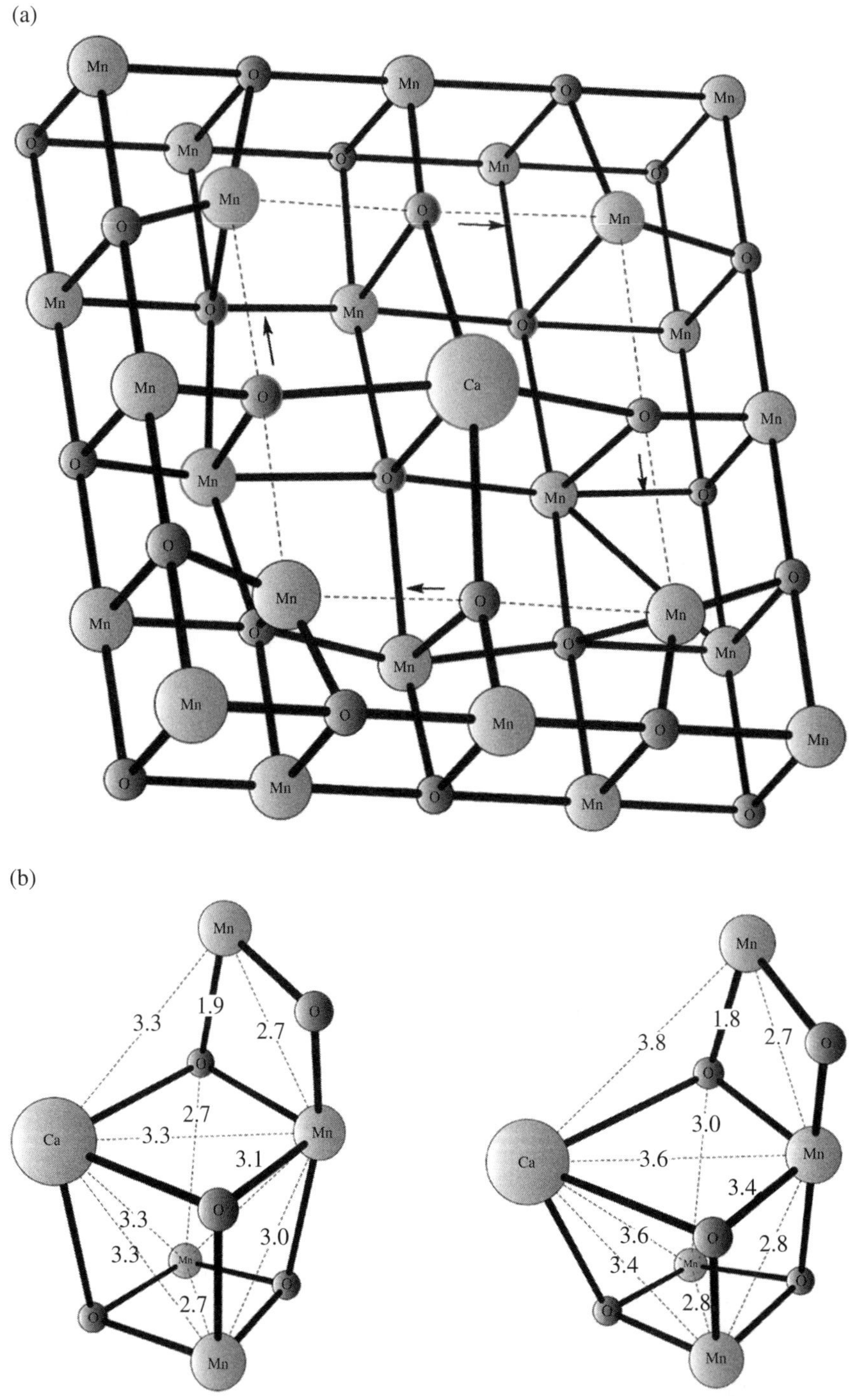

FIGURE 25.4 (a) Siegbahn [57] inserted a calcium ion into the manganese oxide crystal. After optimization, the atoms circled were selected for the Mn_4CaO_5 cluster to compare with the WOC in Photosystem II. (b) The cluster to the left is taken out of a mineral after some relaxation and the right is taken out of the optimized WOC in Photosystem II. Reprinted with permission from Ref. [57]. © 2011, John Wiley & Sons.

Multicenters of transition metals such as metal oxides or sulfides can do that, but mono-, and even dinuclear, centers cannot get involved in such a reaction, because they cannot accommodate and accumulate multielectrons without huge changes in their structures. Thus, both inorganic cores of enzymes and related inorganic compounds could catalyze multielectron transfer reactions.

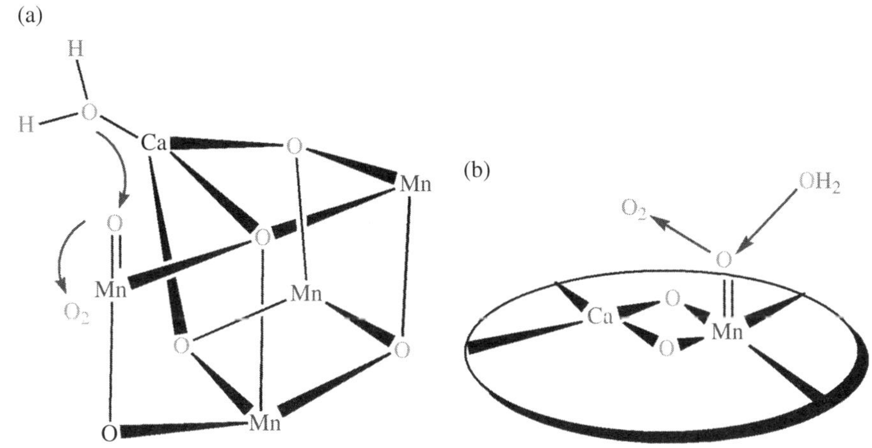

SCHEME 25.3 Proposed mechanisms of oxygen evolution by the WOC in (a) Photosystem II and (b) calcium manganese oxides.

Using density functional calculations, Nørskov's group for the biological catalysts and active catalysts for hydrogen reduction showed that in terms of "being able to stabilize intermediates involving atomic hydrogen they have very similar properties" [62]. In other words, they considered the following mechanism for hydrogen evolution:

$$H^{+} + e^{-} + * \rightarrow H* \tag{25.1}$$

$$2H* \rightarrow H_2 + 2* \tag{25.2a}$$

$$H^{+} + e^{-} + H* \rightarrow H_2 + * \tag{25.2b}$$

By calculating the free energy of atomic hydrogen bonding to the catalyst, the group compared different metal surfaces as catalysts. They found that for a fast hydrogen evolution, reaction steps cannot be associated with large changes in the free energy. Thus, the compound that forms strong bonds to atomic hydrogen is not a good catalyst because the hydrogen release step will be slow (Eqs. 25.2a and 25.2b) [62]. The compound that does not bind to atomic hydrogen is also not a good catalyst because the proton/electron transfer step will be slow. This approach is a quantifier to the Sabatier principle [63]: the interaction between the catalysts and the adsorbate should be neither too strong nor too weak (Fig. 25.4). Regarding biological hydroge nase with FeMo cofactor, Nørskov's group found that hydrogen atoms can only bind exothermically to the three equatorial sulfur ligands (μ_2S ligands) on the FeMo cofactor (Fig. 25.6). In this condition, it results in a binding energy close to that of Pt. Interestingly, MoS_2 showed a similar diagram (Fig. 25.6).

The calculations show that the metal ions and the first atoms coordinated to them (usually S and O) are important as reactions. Thus, in addition to enzymes, related metal sulfides or oxides with similar arrangements could be efficient catalysts for the related reactions (Table 25.1).

However, the 3 billion years of evolutionary experiments by *Nature* have provided better catalysts with regard to appropriate residues around inorganic cores. In other words, it is important to note that, in addition to cofactors, enzymes have hundreds of amino acids; however, only a small fraction of the residues come into contact with the inorganic cores, and an even smaller fraction, three to four residues on an average, are directly involved in metal ions. The roles of the residues that come into direct contact with the inorganic cores could be the regulation of charges and electrochemistry of the inorganic cores, helping in coordinating water molecules at appropriate metal sites, and proton transfer in the stability of these inorganic cores. Other residues are not in direct contact with the inorganic cores but play very important roles in enzymes, and their deletion from enzymes causes a dramatic decline in the rate of reactions. Many of these amino acid residues are important in the substrates for proton transfer. To design a biomimetic *catalyst*, a deep understanding of the roles of these amino acid residues in related enzymes is necessary, but these are not discussed here. The amino acid side groups could improve catalytic reactions efficiently. Thus, the addition of these groups to metal oxide or sulfide models could increase the efficiency of catalytic activity (Table 25.1). Recently, we considered the ability of manganese oxide monosheets to self-assemble and synthetize layered structures of manganese oxide including guanidinium and imidazolium groups. The compounds can be considered new structural models for the WOC of Photosystem II [64].

On the other hand, Nakmora et al. used in situ spectroelectrochemical techniques to show that the stabilization of surface-associated intermediate Mn(III) species, on the surface of manganese oxide, relative to charge disproportionation is an effective strategy to lower the overpotential for water oxidation by MnO_2 [65]. The formation of N–Mn bonds via the coordination of

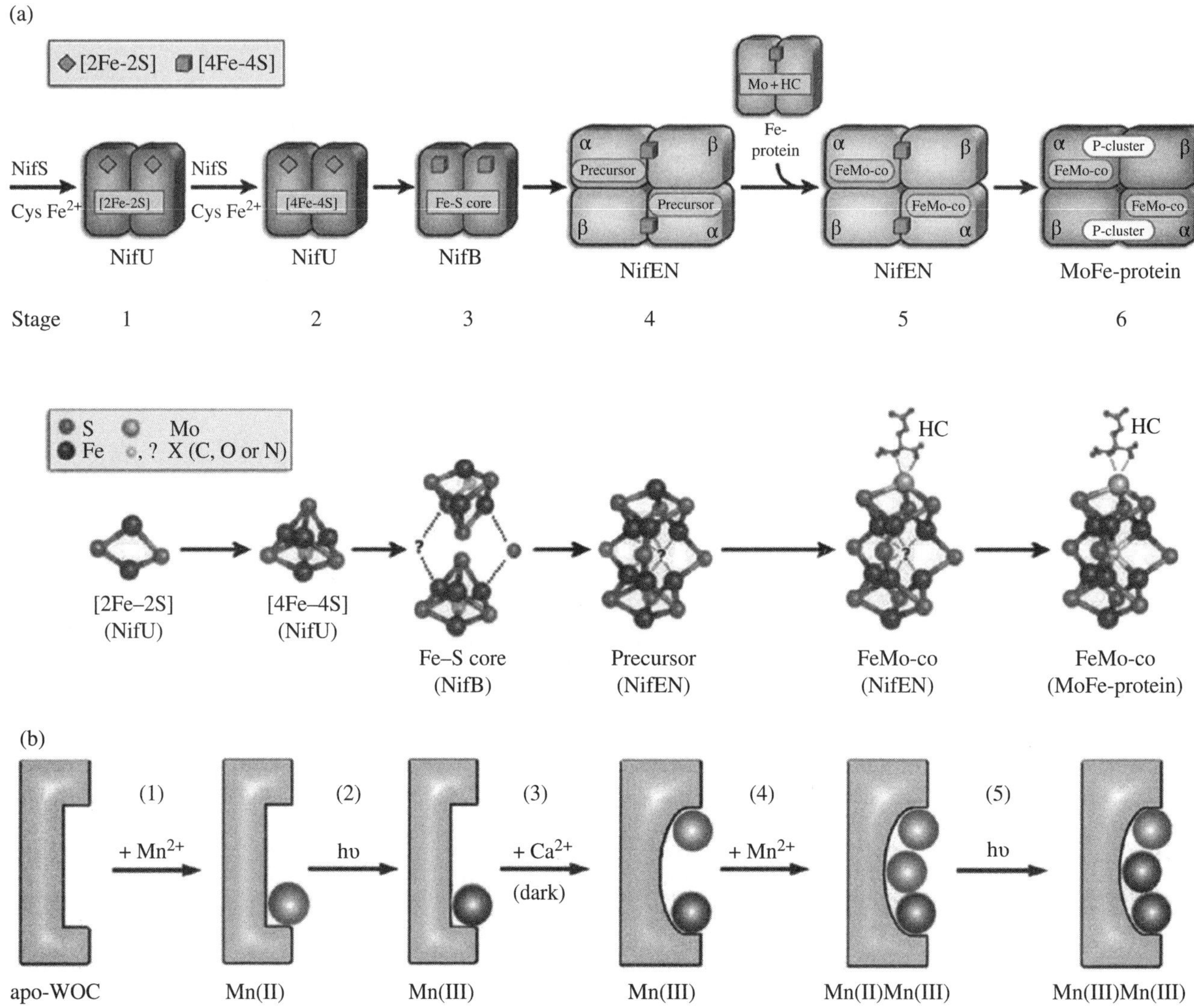

FIGURE 25.5 Sequence of events during FeMo cofactor assembly. (a) The biosynthetic flow of FeMo cofactor is NifU → NifS → NifB → NifEN → MoFe-protein. HC, homocitrate. Structures of intermediates during FeMo cofactor assembly. Shown are the cluster types that have been identified (on NifU, NifEN, and MoFe-protein) or proposed (for NifB). Hypothetically, NifB could bridge two [4Fe-4S] clusters by inserting a sulfur atom along with the central atom, X, thereby generating a FeeS scaffold that could be rearranged into a precursor closely resembling the core structure of the mature FeMo cofactor. In the case of the NifEN-associated precursor, only the 8Fe model is shown. The potential presence of X in the intermediates of FeMo cofactor biosynthesis is indicated by a question mark. Reprinted with permission from Ref. [60]. © 2009, Nature Publishing Group. (b) Assembly of the WOC in Photosystem II. The proposed stepwise binding of Mn(II) (1), Ca(II) to the apo-WOC (2), as well as the photo-oxidation of Mn(II) to Mn(III) are shown in this figure. Initially, a single Mn(II) is bound to the apo-WOC and oxidized by a first quantum of light. After binding of Ca(II) and a subsequent light-independent structural rearrangement (3) a second Mn(II) is bound (4) and photooxidized (5). The binding of the two missing manganese ions and chloride is not depicted in this scheme as it could not be kinetically resolved. Reprinted with permission from Ref. [59]. © 2011, Elsevier.

amine groups of poly(allylamine hydrochloride) to the surface of Mn sites of MnO_2 electrodes effectively stabilized the Mn(III) species, resulting in approximately 500 mV negative shift of the onset potential for water oxidation at neutral pH [65]. It showed that additional organic groups of manganese, which exist in Photosystem II, may increase the catalytic activity of these clusters.

In conclusion, from a certain viewpoint, an enzyme is considered a nano-sized inorganic core in a protein matrix. A model system by BSA and manganese oxide was selected and it was observed that when nanoparticles form in the presence of proteins, the protein not only induces nucleation, but can also inhibit the further growth of nanominerals, and numerous and very

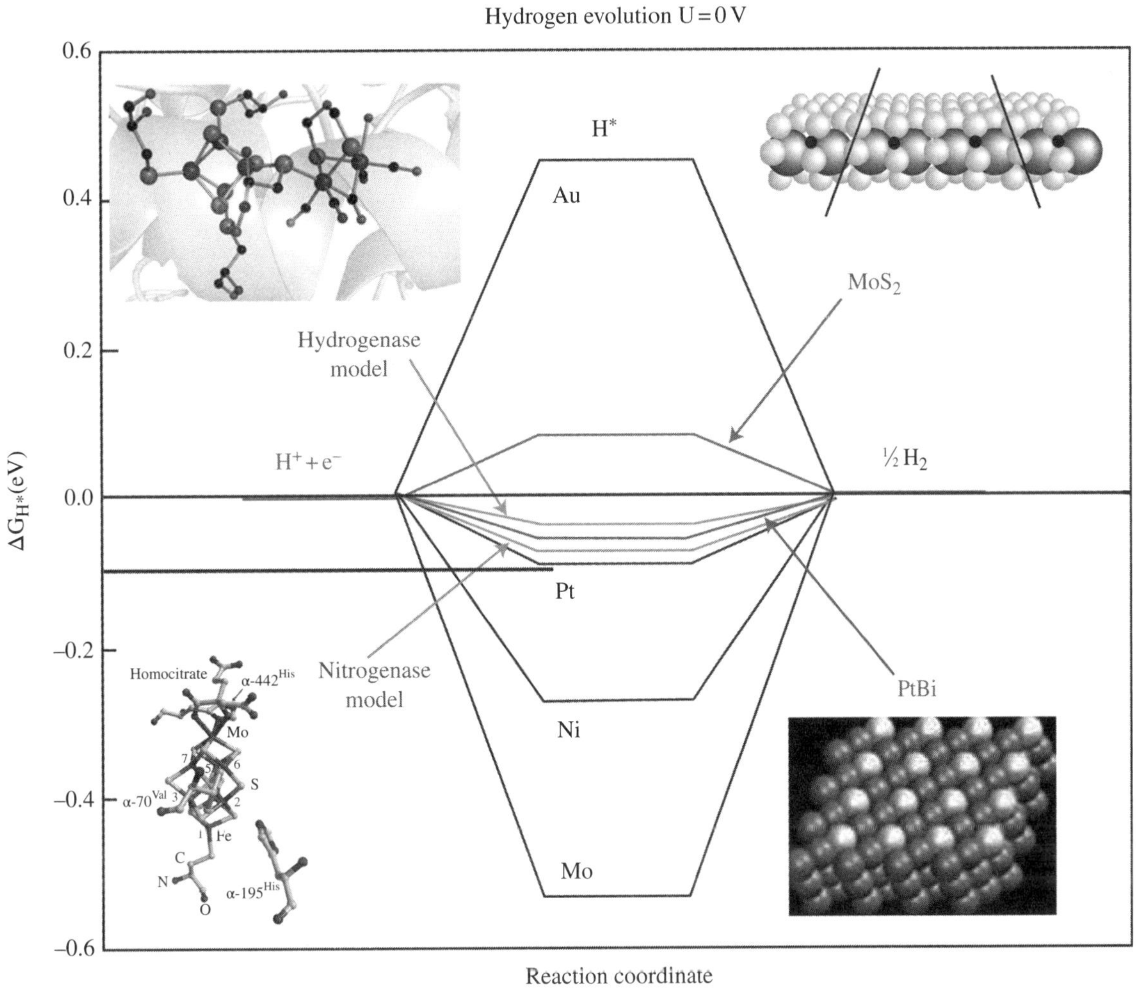

FIGURE 25.6 Calculated free energy for hydrogen evolution at a potential U = 0 relative to the standard hydrogen electrode at pH = 0. In this diagram, Au is shown as a compound that does not bind to atomic hydrogen and Mo is shown as a compound that forms strong bonds with atomic hydrogen. For both cases, hydrogen evolution is slow. Reprinted with permission from Ref. [62]. © 2005, American Chemical Society.

small nanoparticles could form in this condition. Without the protein, larger particles are formed. BSA can also help to disperse these nanoparticles in a solution and the dispersity of particles (or minerals) in water could increase the rate of reaction of particles with other substrates. Other groups reported similar results. One of the numerous and very small nanoparticles produced in the presence of protein could be formed in an appropriate location in proteins and be used as a primitive inorganic core (cofactor) of an enzyme. In other words, the author proposes that the cofactor formation in particular enzymes could be biomineralization in the presence of a protein.

25.1 METHODS

25.1.1 Material

All reagents and solvents were purchased from commercial sources and were used without further purification. Mid-infrared (MIR) spectra of KBr pellets of compounds were recorded on a Bruker vector 22 in the range between 400 and 4000 cm^{-1}. Transmission electron microscopy (TEM) and scanning electron microscopy (SEM) were carried out with Philips CM120 and LEO 1430VP, respectively. The X-ray powder patterns were recorded with a Bruker, D8 ADVANCE (Germany) diffractometer (Cu-Kα radiation). Manganese atomic absorption spectroscopy (AAS) was performed on an Atomic Absorbtion Spectrometer Varian Spectr AA 110.

25.1.2 Experimental Methods

Nano-sized birnessite was prepared by a simple method:

Solution 1: $CaCl_2.4H_2O$ (4.0 mmol, 732.2 mg) and $Mn(CH_3COO)_2.4H_2O$ (5.6 mmol, 1372 mg) were dissolved in water (5 ml). The mixture was stirred for about 10 min at room temperature.

Solution 2: KOH was added to a solution of $KMnO_4$ (2.4 mmol, 379 mg) in 60 ml water to obtain a hot saturated KOH solution.

Addition of solution 1 to solution 2 under vigorous stirring resulted in a dark precipitate. Then the mixture was allowed to cool with continued stirring for 2 h.

The obtained suspension was filtered and washed using distilled water (3 l) before being allowed to dry for 12 h at 60°C in an oven. Then the solid was heated to 400°C for 10 h in air to obtain a black powder. The yield was >%95.

Extraction of ions by water: Nano-sized birnessite (10 mg) was added to water (20 ml). After 24 h, the filtration solution was diluted to 25 ml and analyzed by AAS to calculate the amount of calcium and manganese extracted by water.

Extraction of ions by BSA: Nano-sized birnessite (10 mg) was added to a BSA solution (0.5 g BSA in 20 ml water). After 1 month, the filtration solution was diluted to 25 ml and analyzed by AAS to calculate the amount of calcium and manganese extracted by BSA (the first reaction of proteins and nanominerals). The solid was studied by SEM and TEM to observe the morphology of birnessite after reaction by BSA (the second reaction of proteins and nanominerals).

To study the third reaction of nano-sized birnessite and BSA, manganese (II) chloride (161 mg, 1.24 mmol) was added to a solution of BSA (1 g) in water (80 ml) and the solution was stirred for 4 h. Then $KMnO_4$ (88 mg in 16 ml water) was added to the solution under constant stirring. The solution was studied by SEM and TEM to observe the morphology of manganese oxide.

ACKNOWLEDGMENT

The author is grateful to the Institute for Advanced Studies in Basic Sciences for providing financial support.

REFERENCES

[1] Masel RI. *Chemical Kinetics and Catalysis*. New York: Wiley-Interscience; 2001.

[2] Knözinger H, Kochloefl K. *Heterogeneous Catalysis and Solid Catalysts; In Ullmann's Encyclopedia of Industrial Chemistry*. Weinheim: Wiley-VCH; 2002.

[3] Jencks WP. *Catalysis in Chemistry and Enzymology*. New York: McGraw-Hill; 1969.

[4] Vincent JF, Bogatyreva OA, Bogatyrev NR, Bowyer A, Pahl AK. Biomimetics: its practice and theory. J R Soc Interface 2006;3:471–482.

[5] Copley SD. Evolution and the enzyme. In: Mander L, Liu HW, editors. *Comprehensive Natural Products II: Chemistry and Biology*. New York: Elsevier; 2010. p 9–46.

[6] Rode BM, Son HL, Suwannachot Y, Bujdák J. The combination of salt-induced peptide formation reaction and clay catalysis: a way to higher peptides under primitive earth conditions. Orig Life Evol Biosph 1997;27:325–337.

[7] Liu R, Orgel LE. Polymerization on the rocks: β-amino acids and arginine. Orig Life Evol Biosph 1998;28:245–257.

[8] Zaia DA, Zaia CT, Santana H. Which amino acids should be used in prebiotic chemistry studies? Orig Life Evol Biosph 2008;38:469–488.

[9] Poller R, Boehme C, Marx D. Ab initio simulations of desorption and reactivity of glycine at a water–pyrite interface at "iron–sulfur world" prebiotic conditions. Orig Life Evol Biosph 2006;36:363–379.

[10] Vieira AP, Berndt G, de Souza Junior IG, Di Mauro E, Paesano A Jr, de Santana H, da Costa AC, Zaia CT, Zaia DA. Adsorption of cysteine on hematite, magnetite and ferrihydrite: FT-IR, Mossbauer, EPR spectroscopy and X-ray diffractometry studies. Amino Acids 2011;40:205–214.

[11] Lambert J-F. Adsorption and polymerization of amino acids on mineral surfaces: a review. Orig Life Evol Biosph 2008;38:211–242.

[12] Naidja A, Huang PM, Bollag JM. Enzyme-clay interactions and their impact on transformations of natural and anthropogenic organic compounds in soil. J Environ Qual 2000;29:677–691.

[13] Boyd SA, Mortland MM. Enzyme interactions with clays and clay-organic matter complexes. In: Bollag J-M, Stotzky G, editors. *Soil Biochemistry*. Volume 6, New York: Marcel Dekker; 1990. p 1–20.

[14] Naidja A, Liu C, Huang PM. Formation of protein-birnessite complex, XRD, FTIR, and AFM analysis. J Colloid Interface Sci 2002;251:46–56.

[15] Bernal JD. *The Physical Basis of Life*. London: Routledge & Kegan Paul Ltd.; 1951.

[16] Wächtershäuser G. Before enzymes and templates: theory of surface metabolism. Microbiol Mol Biol Rev 1988;52 (4):452–484.

[17] Wächtershäuser G. On the chemistry and evolution of the pioneer organism. Chem Biodivers 2007;4 (4):584–602.

[18] Cairns-Smith AG. *Seven Clues to the Origin of Life*. Cambridge: Cambridge University Press; 1985.

[19] Kauffman SA. *The Origins of Order: Self-Organization and Selection in Evolution*. New York: Oxford University Press; 1993.

[20] Kaufman ED, Belyea J, Johnson MC, Nicholson ZM, Ricks JL, Shah PK, Bayless M, Pettersson T, Feldotö Z, Blomberg E, Claesson P, Franzen S. Probing protein adsorption onto mercaptoundecanoic acid stabilized gold nanoparticles and surfaces by quartz crystal microbalance and zeta-potential measurements. Langmuir 2007;23:6053–6062.

[21] Russell MJ, Allen JF, Milner-White EJ. Inorganic complexes enabled the onset of life and oxygenic photosynthesis. In: Allen JE, Gantt E, Golbeck JH, Osmond B, editors. *Energy from the Sun: 14th International Congress on Photosynthesis*. Dordrecht: Springer; 2008. p 1193–1198.

[22] Milner-White EJ, Russell MJ. Functional capabilities of the earliest peptides and the emergence of life. Genes 2011;2:671–688.

[23] Cedervall T, Lynch I, Foy M, Berggård T, Donnelly SC, Cagney G, Linse S, Dawson KA. Detailed identification of plasma proteins adsorbed on copolymer nanoparticles. Angew Chem Int Ed 2007;46 (30):5754–5756.

[24] Najafpour MM. Biomineralization: a proposed evolutionary origin for inorganic cofactors of enzymes. Theory Biosci 2012; 131:265–272.

[25] Najafpour MM. An approach for catalyst design in artificial photosynthetic systems: focus on nanosized inorganic cores within proteins. Photosynth Res 2013;117:197–205.

[26] Vallee A, Vincent H, Pradier C. Peptide interactions with metal and oxide surfaces. Accounts Chem Res 2010;43:1297–1306.

[27] Haynes CA, Norde W. Globular proteins at solid/liquid interfaces. Colloids Surf B 1994;2 (6):517–566.

[28] Arai T, Norde W. The behaviour of some model proteins at solid-liquid interfaces. 1. Adsorption from single protein solutions. Colloids Surf B 1990;51:1 15.

[29] Bianchi A, Giorgi C, Ruzza P, Toniolo C, Milner-White J. A synthetic hexapeptide designed to resemble a proteinaceous p-loop nest is shown to bind inorganic phosphate. Proteins 2012;80:1418–1424.

[30] Sauer Y, Yachandra V. A possible evolutionary origin for the Mn_4 cluster of the photosynthetic water oxidation complex from natural MnO_2 precipitates in the early ocean. Proc Natl Acad Sci USA 2002;99 (13):8631–8636.

[31] Najafpour MM. Hollandite as a functional and structural model for the biological water oxidizing complex: manganese-calcium oxide minerals as a possible evolutionary origin for the $CaMn^4$ cluster of the biological water oxidizing complex. Geomicrobiol J 2011;28 (8):714–718.

[32] Yang L, Shen QM, Zhou JG, Jiang K. Biomimetic synthesis of CdS nanocrystals in the pepsin solution. Mater Lett 2005; 59:2889–2892.

[33] Yang L, Xing RM, Shen QM, Jiang K, Ye F, Wang JY, Ren QS. Fabrication of protein-conjugated silver sulfide nanorods in the bovine serum albumin solution. J Phys Chem B 2006;110:10534–10539.

[34] Meziani MJ, Sun YP. Protein-conjugated nanoparticles from rapid expansion of supercritical fluid solution into aqueous solution. J Am Chem Soc 2003;125:8015–8018.

[35] Mamedova NN, Kotov NA, Rogach AL, Studer J. Albumin-CdTe nanoparticle bioconjugates: preparation, structure, and interunit energy transfer with antenna effect. Nano Lett 2001;1:281–286.

[36] Duffus BR, Hamilton TL, Shepard EM, Boyd ES, Peters JW, Broderick JB. Radical AdoMet enzymes in complex metal cluster biosynthesis. Biochim Biophys Acta 2012;1824 (11):1254–1263.

[37] Zaharieva I, Najafpour MM, Wiechen M, Haumann M, Kurz P, Dau H. Synthetic manganese–calcium oxides mimic the water-oxidizing complex of photosynthesis functionally and structurally. Energy Environ Sci 2011;4:2400–2408.

[38] Najafpour MM, Ehrenberg T, Wiechen M, Kurz P. Calcium manganese (III) oxides ($CaMn_2O_4.xH_2O$) as biomimetic oxygen-evolving catalysts. Angew Chem Int Ed 2010;49:2233–2237.

[39] Lee SC, Holm RH. The clusters of nitrogenase: synthetic methodology in the construction of weak field clusters. Chem Rev 2004;104:1135–1157.

[40] Najafpour MM, Jafarian Sedigh D, King'ondu CK, Suib SL. Nano-sized manganese oxide–bovine serum albumin was synthesized and characterized. It is promising and biomimetic catalyst for water oxidation. RSC Adv 2012;2:11253–11257.

[41] Najafpour MM, Rahimi F, Aro EM, Lee CH, Allakhverdiev SI. Nano-sized manganese oxides as biomimetic catalysts for water oxidation in artificial photosynthesis: a review. J R Soc Interface 2012;9 (75):2383–2395.

[42] Najafpour MM, Moghaddam AN, Yang YN, Aro E, Carpentier R, Eaton-Rye JJ, Lee C, Allakhverdiev SI. Biological water-oxidizing complex: a nano-sized manganese-calcium oxide in a protein environment. Photosynth Res 2012;114:1–13.

[43] Najafpour MM. Amorphous manganese-calcium oxides as a possible evolutionary origin for the CaMn4 cluster in photosystem II. Orig Life Evol Biosph 2011;41:237–247.

[44] Dörr M, Kässbohrer J, Grunert R, Kreisel G, Brand WA, Werner RA, Geilmann H, Apfel C, Robl C, Weigand W. A possible prebiotic formation of ammonia from dinitrogen on iron sulfide surfaces. Angew Chem Int Ed 2003;42:1540–1543.

[45] Zheng Y, Duanmu C, Gao Y. A magnetic biomimetic nanocatalyst for cleaving phosphoester and carboxylic ester bonds under mild conditions. Org Lett 2006;8 (15):3215–3217.

[46] Zhu H, Alexeev D, Hunter DJ, Campopiano DJ, Sadler PJ. Oxo-iron clusters in a bacterial iron-trafficking protein: new rolesfor a conserved motif. Biochem J 2003;376:35–41.

[47] Yuhas BD, Smeigh AL, Douvalis AP, Wasielewski MR, Kanatzidis MG. Photocatalytic hydrogen evolution from FeMoS-based biomimetic chalcogels. J Am Chem Soc 2012;134:10353–10356.

[48] McCandless FP, Hodgson KM, White RH, Bowman JD. Reduction of nitric oxide with metal sulfides. Ind Eng Chern Process Des Dev 1980;19:108–113.

[49] Laursen AB, Kegnæs S, Dahl S, Chorkendorff I. Molybdenum sulfides—efficient and viable materials for electro—and photoelectrocatalytic hydrogen evolution. Energy Environ Sci 2012;5:5577–5591.

[50] Mulligan DJ, Berk D. Reduction of SO_2 with CH_4 over selected transition metal sulfides. Ind Eng Chem Res 1989;28:926–931.

[51] Watanabe S, Arai T, Matsumi R, Atomi H, Imanaka T, Miki K. Crystal structure of HypA, a nickel-binding metallochaperone for [NiFe] hydrogenase maturation. J Mol Biol 2009;394:448–459.

[52] Vandenborre H, Vermeiren P, Leysen R. Hydrogen evolution at nickel sulphide cathodes in alkaline medium. Electrochim Acta 1984;29 (3):297–301.

[53] Eady RR. Current status of structure function relationships of vanadium nitrogenase. Coord Chem Rev 2003;237:23–30.

[54] Naman SA. Photoproduction of hydrogen from hydrogen sulfide in vanadium sulfide colloidal suspension—effect of temperature and pH. Int J Hydrogen Energy 1997;22 (8):783–789.

[55] Mayer SM, Lawson DM, Gormal CA, Roe SM, Smith BE. New insights into structure-function relationships in nitrogenase: A 1.6 A resolution X-ray crystallographic study of Klebsiella pneumoniae MoFe-protein. J Mol Biol 1999;292 (4):871–891.

[56] Merki D, Vrubel H, Rovelli L, Fierro S, Hu X. Fe, Co, and Ni ions promote the catalytic activity of amorphous molybdenum sulfide films for hydrogen evolution. Chem Sci 2012;3:2515–2525.

[57] Siegbahn PEM. The effect of backbone constraints: the case of water oxidation by the oxygen-evolving complex in PSII. ChemPhysChem 2011;12:3274–3280.

[58] Shevela D, Koroidov S, Najafpour MM, Messinger J, Kurz P. Calcium manganese oxides as oxygen evolution catalysts: O_2 formation pathways indicated by 18O-labelling studies. Chemistry 2011;17 (19):5415–5423.

[59] Becker K, Cormann KU, Nowaczyk MM. Assembly of the water-oxidizing complex in photosystem II. J Photochem Photobiol B 2011;104:204–211.

[60] Schwarz G, Mendel RR, Ribbe MW. Molybdenum cofactors, enzymes and pathways. Nature 2009;460:839–847.

[61] Tributsch H, Pohlmann L. Electron transfer: classical approaches and new frontiers. Science 1998;279:1891–1895.

[62] Hinnemann B, Moses PG, Bonde J, Jørgensen KP, Nielsen JH, Horch S, Chorkendorff I, Nørskov JK. Biomimetic hydrogen evolution: MoS_2 nanoparticles as catalyst for hydrogen evolution. J Am Chem Soc 2005;127:5308–5309.

[63] Sabatier P. Ber Deutsch Chem Ges 1911;44:1984–2001.

[64] Najafpour MM, Allakhverdiev SI. Manganese compounds as water oxidizing catalysts for hydrogen production via water splitting: from manganese complexes to nano-sized manganese oxides. Int J Hydrogen Energy 2012;37:8753–8765.

[65] Takashima T, Hashimoto K, Nakamura R. Mechanisms of pH-dependent activity for water oxidation to molecular oxygen by MnO_2 electrocatalysts. J Am Chem Soc 2012;134:1519–1527.

PART XI

NANOSENSORS

26

NEUTRON-FLUENCE NANOSENSORS BASED ON BORON-CONTAINING MATERIALS

LEVAN CHKHARTISHVILI

Department of Engineering Physics, Georgian Technical University, Tbilisi, Georgia

Laboratory for Boron and Powdered Composite Materials, F. Tavadze Institute of Metallurgy and Materials Science, Tbilisi, Georgia

Neutron-fluence nanosensors are nanotech-integrated monitoring tools that are helpful in reducing environmental risks. Concern over the consequences of harsh radionuclide pollution has resulted in an increasing demand for suitable means of monitoring neutron radiation sources: radioactive pollutants generated directly or indirectly, as a result of nuclear waste, influence different substances used in nuclear power stations (water, gases, metals, etc.) or places existing near radioactive materials. The development of a range of neutron-sensor materials has provided devices with enhanced selectivity and sensitivity for neutron radiation by a number of substances from harsh sites.

On the one hand, boron-containing materials possess extraordinary potential for solid-state neutron detector applications because boron ^{10}B isotope has a capture cross section of ~3835 barns for thermal neutrons, which is several orders of magnitude larger than isotopes of other chemical elements. On the other hand, state-of-the-art nanotechnology offers multiple benefits for neutron radiation–sensing applications: the ability to incorporate nano-sized radiation indicators into widely used materials such as paints, coatings, and ceramics to create nanocomposite materials; the development of ultralow-power, flexible detection systems that can be portable or embedded in clothing or other systems; and so on.

Boron belongs to the group of least abundant chemical elements: B content in the earth's crust makes up no more than about 0.005 wt.%. However, its role in forming the various molecular and solid-state structures is incommensurably great. Understanding the structural diversity of boron-containing solid phases comes down to the B atom's electronic structure: it is a strongly distinct acceptor and elemental boron structures should be electron-deficient. This is why all-boron crystalline forms, as well as amorphous ones, exhibit very complex, clustered structures, in which icosahedron B_{12} serves as a main building block. Besides, elemental boron forms diatomic molecule B_2, small molecular clusters B_n, and some nanophases that are also constructed from the interconnected icosahedra. Acceptor behavior of B atoms and their clusters favors the formation of a huge number of borides, that is, compounds of boron with metals, which are usually characterized by the donor behavior. Only binary B/Me crystalline compounds with chemical formulas from Me_5B to MeB_{66} can be counted as approximately 250. Higher borides are characterized by the clustered structures based on icosahedron and/or other 3D boron cages, while among the lower borides one can find a wide variety of 1D and 2D structural motifs—various chains and plane networks of boron atoms. Boron compounds with nonmetals that are characterized by a higher electron affinity, that is, boron carbides, nitrides, oxides, etc., show less complicated structures. They also form nanosystems like nanotubes and fullerenes. Strong B–B bonds makes all the boron-rich solids refractory and resistant against aggressive environments, while the diversity of geometric and electronic structures and, consequently, the diversity of sets of their physical properties, enable their usage in large spheres of technical and technological applications.

Nanomaterials for Environmental Protection, First Edition. Edited by Boris I. Kharisov, Oxana V. Kharissova, and H. V. Rasika Dias.

The chemical element boron has two stable isotopes: ^{10}B and ^{11}B. The natural abundance of boron stable isotopes ^{nat}B is approximately 19.6% ^{10}B and approximately 80.4% ^{11}B. These isotopes differ by a single neutron in the $1p^{3/2}$ shell in the nucleus. However, this difference makes the neutron capture cross section of ^{10}B significantly exceed that of ^{11}B. The total absorption cross section of ^{10}B, ^{11}B, and ^{nat}B for "room-temperature" neutrons (with velocity ~2200 m s^{-1}) equals 3835, 0.0055, and 767 barns, respectively. Thus, boron has two principal isotopes, which are chemically almost identical, but have quite different neutron-absorbing characteristics. The effectiveness of boron as a neutron absorber is due to the cross section of the ^{10}B isotope, while ^{11}B is essentially nonreactive with neutrons. This provides a great deal of flexibility in the use of boron-containing materials in nuclear systems, since the mixture of boron isotopes can be adjusted. The value of an absorbing component can be increased by a factor of approximately 4.5 by substituting with enriching natural materials without changing the sample's dimensions, which is not possible with other candidate materials. Especially, boron enriched in the ^{10}B isotope to greater than its natural isotopic abundance is useful for neutron irradiation detection equipment, as well as neutron fluence–measuring instruments.

It is worth mentioning that neutrons were discovered by producing them from boron nuclei by bombardment with α-particles: $B + He \rightarrow N + n$ [1]. Since neutrons are not ionizing particles, their detection depends on the ionizing properties of the products of neutron capture reactions, the most important of which is thermal neutron capture by ^{10}B nucleus [2]:

$$6\% : {}^{10}_{5}B + {}^{1}_{0}n \rightarrow {}^{11}_{5}B \rightarrow {}^{7}_{3}Li\left(1.01\ \text{MeV}\right) + {}^{4}_{2}He(1.78\ \text{MeV}),$$
$$94\% : {}^{10}_{5}B + {}^{1}_{0}n \rightarrow {}^{11}_{5}B \rightarrow {}^{7}_{3}Li\left(0.84\ \text{MeV}\right) + {}^{4}_{2}He(1.47\ \text{MeV}) + {}^{0}_{0}\gamma(0.48\ \text{MeV}).$$

As for the general method of measuring intermediate and fast neutrons, one must first moderate their energies to thermal values so that they can be detected with a thermal neutron detector. One of the principal measurement systems that moderate intermediate and fast neutrons is so-called long counter, containing a boron fluoride (BF_3) tube, surrounded by an inner paraffin moderator. It responds uniformly to neutron energies from approximately 10 keV to approximately 5 MeV. Incident neutrons cause a direct response after being thermalized in the inner paraffin layer. Those from other directions are either reflected or thermalized by the outer paraffin jacket and then absorbed in a layer of boron oxide (B_2O_3). The disadvantage of this arrangement is that the probability that a moderated neutron will enter the BF_3 tube and be counted is not dependent on the initial energy and, therefore, no information on the spectral distribution of neutron energies is obtained.

In the early Review [3], a brief description of the applications of the nuclear properties of boron isotopes was provided. These included counters of neutrons and energy sensors for neutron detectors. Some neutron-capture applications of boron cluster compounds were also mentioned [4, 5]. Neutron-sensor device structures based on boron-containing materials were also described in the newer Review [6] devoted to the isotopic effects of boron in solid materials. Recently, in the special issue of the SPIE Proceedings [7], devoted to the achievements in radiation detectors physics, boron-containing epitaxial layers proposed for neutron-sensor application have been described. It should be emphasized that, in all the solid-state neutron sensors, the working body is a nanolayer because ^{10}B is an excellent neutron absorber and neutrons are rapidly stopped in a ^{10}B-enriched material.

Crystalline semiconducting boron was used in neutron thermometers [8]. Such neutron sensors offered advantages over the conventional (e.g., gas-filled) ones. Neutron detectors constructed from the boron-rich semiconductors could be particularly effective due to the easily detectable products of the neutron reaction with the ^{10}B nucleus. A boron-based neutron detector was first described by Gaulé et al. [9]. A pair of thermoresistors were matched with respect to their semiconductor properties, but with different nuclear properties: one thermistor was made of ^{10}B, and the other of ^{11}B. Both isotopes were stable, but, when exposed to neutrons with thermal energies, only ^{10}B nuclei underwent transformation according to reaction ${}^{10}_{5}B + {}^{1}_{0}n \rightarrow {}^{7}_{3}Li + {}^{4}_{2}He + 2.79\ \text{MeV}$. The significant amount of the averaged energy (2.79 MeV) released caused the ^{10}B-thermistor to become warmer than the ^{11}B-thermistor. This temperature difference can be converted into an electrical signal. The formation of the rectifying contact on the ^{10}B sample allowed its application in a different type of solid-state neutron detector [10]. Its principle of operation is based on the generation charge carriers due to absorbing ${}^{4}_{2}He$, that is, α-particles, ejected in the process of neutron interaction with boron. In the neutron detector [11] based on semi-insulating boron-containing semiconductor crystal with an external bias voltage, a neutron interaction with ^{10}B nucleus produces ${}^{4}_{2}He$ and ${}^{7}_{3}Li$ nuclei. Both are rapidly stopped (within ~10 μm) in the semiconductor, exciting thousands of electron–hole pairs. These free charges then drift in the electric field on the crystal, generating a measurable current pulse.

In a study by Emin and Aselage [12], a boron carbide–based thermoelectric device for the detection of a thermal-neutron flux was proposed. It exploited, on the one hand, very high melting temperature and the radiation tolerance of boron carbides, making them suitable for use within hostile environments, for example, within nuclear reactors, and, on the other hand, boron carbides' anomalously large Seebeck coefficient, by proposing a relatively sensitive detector of the local heating that follows the absorption of a neutron by a ^{10}B nucleus. The possibility of elaboration of the boron carbide–based neutron detectors was also considered by Emin [13]. As ^{10}B is an excellent absorber of thermal neutrons, they cannot penetrate (for much more than ~1 mm) into a ^{10}B-enriched material before being absorbed. Upon absorbing a neutron, $^{10}_{5}B$ decays into $^{7}_{3}Li$ and $^{4}_{2}He$, thereby heating the material in the immediate vicinity of the neutron absorption. A temperature difference is thereby established between the faces of a boron carbide sample. It can be monitored using boron carbide's large Seebeck effect. All told, boron carbides may serve as simple, small, and robust (as the neutron-radiation damage arising from the decay of ^{10}B isotope in the icosahedral boron-rich solids is minimized by the self-healing that characterizes such structures) neutron detectors.

Real-time solid-state neutron detectors were also fabricated [14] from different semiconducting boron–carbon alloys prepared by the plasma-enhanced chemical vapor deposition (PECVD) method. Single neutrons were detected and signals induced by gamma-rays were determined to be insignificant. The source gas closo-1,2-dicarbadodecaborane (ortho-carborane) was used to fabricate the boron–carbon alloys with only the natural isotopic abundance of ^{10}B. Devices made of thicker boron–carbon alloy layers enriched in ^{10}B led to increased detection efficiency. Active diodes could use the inherent micron-scale spatial resolution, increasing the range of applications.

Radioisotopes and particle accelerators are widely utilized as neutron sources: neutron therapy, radiography, and damage evaluation of fusion reactor materials are typical examples of these up-to-date applications. With such an increase in neutron facilities, personal dosimetry, environmental monitoring, and evaluation of leakage neutrons become more important. In a study by Oda et al. [15], a plastic track detector was applied to thermal neutron dosimetry by combining it with a ceramic boron nitride converter using (n, α) reactions. BN seems to be a promising converter because of its high boron concentration and smooth surface. The efficiency was 1.0×10^{-3} pits n^{-1}, corresponding to the sensitivity of 9.6×10^{2} pits mm^{-2} mSv; the linear response was between 0.035 and 0.7 mSv; and the minimum detectable dose equivalent was estimated to be approximately 0.005 mSv. Boron neutron capture therapy uses high intensity neutron radiation (up to $10^{9}\,cm^{-2}s^{-1}$). To evaluate the prescribed dosage, it is important to separately detect the thermal neutron flux and the gamma dose. Tanaka et al. [16] proposed a method using a glass-rod dosimeter (GRD) with thermal neutron shielding. In order to evaluate the thermal neutron fluence, the combination with GRD and BN as a thermal neutron converter was also developed.

Recently, epitaxial layers of hexagonal boron nitride (h-BN) synthesized by metal organic chemical vapor deposition (MOCVD) have been proposed [17] for neutron-sensor applications. Measurements indicated that the thermal neutron absorption coefficient and absorption length of h-BN epilayers with natural boron-isotopic composition are approximately 0.0036 μm^{-1} and 277 μm, respectively. To partially address the key requirement of long carrier lifetime and diffusion length for a solid-state neutron detector, microstrip metal–semiconductor–metal (MSM) detectors were fabricated on the same basis and were tested. A good current response was generated in these detectors using continuous irradiation with a thermal neutron beam corresponding to an effective conversion efficiency of approximately 80% for absorbed neutrons. Graphene-oxide—a derivative of the single-layer graphite—exhibits an ability to provide sufficient change in luminescent properties when exposed to neutron radiation. Utilizing this property, Robinson et al. [18] investigated the integration of hexagonal boron nitride (h-BN) with some graphene-based structures to evaluate the radiation-induced conductivity in nanoscale devices and discussed the successful integration of h-BN with large-area graphene electrodes as a means to provide the foundation for large-area nanoscale radiation sensors.

Boron phosphide (BP) single crystals were also studied [19] to develop refractory electric devices, such as solid-state neutron detectors utilizing a large cross section of the ^{10}B isotope. The isotopic composition of the wafer used was $(^{10}BP)_{0.95}(^{11}BP)_{0.05}$.

Thermo-luminescent dosimeters (TLD), which are one of the few known types of thermal-neutron dosimeters commonly used for radiation-protection purposes of personnel, utilize (n, α) reaction of ^{10}B nucleus [20]. It should be emphasized that these dosimeters are sensitive to γ-rays as well. This decreases the accuracy of neutron dosimetry to some extent, because at many working places thermal neutrons coexist with γ-rays. This type of neutron dosimetry can be performed, for example, by the spark counting of tracks in the boron-doped film [21]. For this purpose, thin cellulose nitrate films are doped with a boron compound (0.1%). After thermal neutron irradiation, the tens of millimemeter-thick films are etched in an aqueous solution of NaOH and then the etch-pits caused by ^{10}B (n, α) ^{7}Li reactions are punched and counted at high voltages. The ratio of the spark density to the thermal neutron fluence was found to be 1.0×10^{-4} for a boron concentration of 1%. A thermal neutron dose up to 3×10^{-6} Sv can be measured with this system. As a highly sensitive, simple, and nonradioactive neutron dosimeter, the plastic plates (CR–39) doped with another boron compound—ortho-carborane—were prepared [22]. After thermal neutron irradiation, the plates were etched in an aqueous solution of KOH. The etch-pits generated by ^{10}B (n, α) ^{7}Li reactions were then counted using an optical microscope or an automatic track-counting system. When the etching time is kept constant, the etch-pit density is proportional to the irradiated thermal neutron fluence. The proportional constant is termed "sensitivity," which was found to be 4.2×10^{-4} for a plate containing ortho-carborane at a concentration of 0.5 wt.% for an etching time of 16 h. By considering background counts, it was established that a thermal neutron dose of 0.025 mSv can be measured with this plate. These plates are insensitive to visible, UV, X-, β-, and γ-rays and are easy to handle because here the detector and converter are incorporated together. Alternative boron compound, lithium tetraborate $Li_2B_4O_7$, was also considered [23] as promising converter for neutron dosimeters. The poor reliability of sintered crystals as dosemeter was overcome by making it forms of glass ceramics $Li_2B_4O_7$ + 5% SiO_2 + 0.01% Pt or thin evaporated layers on a LiF single crystal. From the information on the thermally-stimulated-electron-emission glow pattern of the duplicated structure sample it may be possible to measure the dose of each kind of radiation separately in a mixed radiation field.

Nanostructured films of germanium (Ge) and silicon (Si) doped with boron isotope ^{10}B are known to be the best materials for the preparation of novel, highly effective, and sustainable neutron-radiation sensors [24]. Boron atoms, originally shallow acceptors for Ge and Si, change their charge because of (n, α) nuclear reaction on ^{10}B stimulated by neutron irradiation bearing Li, which is a shallow donor for these semiconductors. Charge carrier concentration changes (a single captured neutron reverses a single charge) are easy to fix with precision with the help of an electrical measuring instrument. This provides an opportunity to measure neutron fluence up to neutrons of very high density. Novel sensors should rely on one or more sensing mechanisms and produce a signal that indicates the nuclear/ionizing radiation value. For them it is also convenient to operate by so-called smart or intelligent sensory systems. Ion-implanted semiconductor neutron-sensor was designed by Guldamashvili et al. [25, 26]. It contains a p–n junction with an inversion layer of p-type made by doping n-Si with ^{10}B-ions. The sensor can work both in counter and dosimeter modes.

Finally, we have evaluated [27] the key physical-technical characteristics of neutron detectors made from ^{10}B-enriched semiconducting materials: thickness of the effective working layer ~10 mm, releasing rate of the ^{10}B–n interaction products ~10^{15} cm^{-3} s, electron–hole pairs generating rate in process of neutron absorption ~10^{22} cm^{-3} s, rate of rise in the temperature ~10K s^{-1}, and device mean operating time ~10^{-4}s.

In conclusion, nanolayers made up of ^{10}B isotope-enriched elemental boron crystalline modifications, semiconducting boron compounds (boron carbides, nitrides, phosphides, etc.), and boron-doped common semiconductor materials can serve as working bodies for effective neutron-fluence electronic sensors of various types.

REFERENCES

[1] Shadduck HA. The neutron. J Chem Educ 1936;13:303–308.

[2] Martin JE. Physics for Radiation Protection. A Handbook. Weinheim: Wiley-VCH; 2006. p 666–668.

[3] Potapov SP. On application of stable isotopes of boron. Atom Energy 1961;10:244–252.

[4] Plešek J. Potential applications of the boron cluster compounds. Chem Rev 1992;92:269–273.

[5] Grimes RN. Boron clusters come of age. J Chem Educ 2004;81:658–672.

[6] Chkhartishvili LS. Isotopic effects of boron (Review). Trends Inorg Chem 2009;11:105–167.

[7] James RB, Burger A, Franks LA, Fiederle M. Hard X-Ray, Gamma-Ray, and Neutron Detector Physics XIV (SPIE Proceedings). Volume 8507, San Diego: SPIE; 2012. p 237.

[8] Cooper HS. Boron. In: Hampel CA, editor. Rare Metals Handbook. New York/London: Reinhold Publishing Corporation/Chapman & Hall Ltd.; 1961. p 69–81.

[9] Gaulé GK, Ross RL, Bloom JL. $^{10}B/^{11}B$ thermistors pairs and their applications. In: Gaulé GK, editor. Boron: Preparation, Properties and Applications. New York: Plenum Press; 1965. p 317–338.

[10] Costato M, Fontanesi S. Studio delle propretiá fisiche del Boro. Atti Semin Mat Fis Univ Modena 1969;18:231–281.

[11] Lund JC, Olscher F, Shah KS. Solid state neutron detectors from boron-rich semiconducting compounds. In: Abstracts of the 10th international symposium on boron, borides and related compounds. Albuquerque: New Mexico University; 1990. p 100.

[12] Emin D, Aselage TL. A proposed boron-carbide-based solid-state neutron detector. J Appl Phys 2005;97:013529.

[13] Emin D. Unusual properties of icosahedral boron-rich solids. J Solid State Chem 2006;179:2791–2798.

[14] Robertson BW, Adenwalla S, Harken A, Welsch P, Brand JI, Dowben PA, Claassen JP. A class of boron-rich solid-state neutron detectors. Appl Phys Lett 2002;80:3644–3646.

[15] Oda K, Miyake H, Michijima M. CR39–BN detector for thermal neutron dosimetry. J Nucl Sci Technol 1987;24:129–134.

[16] Tanaka H, Sakurai Y, Suzuki M, Masunaga S, Kinashi Y, Marunashi A, Ono K. Study on the dose evaluation using glass rod dosimeter for boron neutron capture therapy. IFMBE Proc 2013;39:1142–1144.

[17] Majety S, Li J, Cao XK, Dahal R, Lin JY, Jiang HX. Metal–semiconductor–metal neutron detectors based on hexagonal boron nitride epitaxial layers. SPIE Proc 2012;85070R:1–9.

[18] Robinson JA, Wetherington M, Hughes Z, La Bella M III, Bresnehan M. Investigation of graphene-based nanoscale radiation sensitive materials. SPIE Proc 2012;8373:1–9.

[19] Kumashiro Y, Kudo K, Matsumoto K, Okada Y, Koshira T. Thermal neutron irradiation experiments on ^{10}BP single crystal wafers. In: Proceedings of the 9th international symposium on boron, borides and related compounds. Duisburg: Universität Duisburg Gesamthochschule; 1987. p 371–372.

[20] Griffith RV, Hankins DE, Gammage RB, Wheeler RV. Recent developments in personnel neutron dosimeters-alpha (Review). Health Phys 1979;36:253–260.

[21] Tsuruta T, Takagaki M. Neutron dosimetry by the spark counting of tracks in boron-doped film. Health Phys 1982;43:705–710.

[22] Tsuruta T, Juto N. Neutron dosimetry with boron-doped CR–39 plastic. J Nucl Sci Technol 1984;21:871–876.

[23] Kikuchi R, Kawamoto T, Lee CH, Kawanishi M. The TSEE from the sintered $Li_2B_4O_7$ crystals, the glass ceramics of $Li_2B_4O_7$ + SiO_2 and the thin evaporated layer of $Li_2B_4O_7$ on the LiF single crystal. Radiat Prot Dosimetry 1983;4:196–200.

[24] Kervalishvili PJ. Boron isotopes doped germanium and silicon based gamma and neutron radiation nanosensors. In: Contents of the 2nd International Conference "Nanotechnologies". Tbilisi: Publ. House Nekeri; 2012. p 120–121.

[25] Guldamashvili AI, Bokuchava GV, Kutelia RN, Nebieridze TsM. High-effective ion-implanted semiconductor neutron detector. Bull Georgian Acad Sci 2005;172:250–233.

[26] Guldamashvili AI, Bokuchava GV, Nardaya YuI, Nebieridze TsM, Sichinava AV. Ion-implanted semiconductor neutron sensor (IISNS). In: Kervalishvili PJ, editor. Nuclear Radiation Nanosensors and Nanosensory Systems. Tbilisi: Publ. House Tech. Univ.; 2014. p 122–124.

[27] Chkhartishvili L, Tsagareishvili O, Tavadze G. Neutron detectors based on ^{10}B-containing nanomaterials. In: Kervalishvili PJ, editor, Nuclear Radiation Nanosensors and Nanosensory Systems. Tbilisi: Publ. House Tech. Univ.; 2014. p 81–83.

PART XII

NANORESERVOIRS FOR HYDROGEN STORAGE

27

HYDROGEN NANORESERVOIRS MADE OF BORON NITRIDE

LEVAN CHKHARTISHVILI

Department of Engineering Physics, Georgian Technical University, Tbilisi, Georgia

Laboratory for Boron and Powdered Composite Materials, F. Tavadze Institute of Metallurgy and Materials Science, Tbilisi, Georgia

Hydrogen nanoreservoirs are nanotech tools that are useful for green energy production. Hydrogen exhibits the highest heating value per mass of all chemical fuels. Furthermore, it is regenerative and environmental-friendly. As is known, reducing the world's dependence on fossil fuels is one of the global research priorities. One strategy calls for powering vehicles with fuel cells that use hydrogen extracted from a carrier-material to generate electricity. But usually the extraction requires high temperatures, and that limits efficiency. Fortunately, these limits may be loosened by using catalysts to drastically lower the temperature to that required to liberate hydrogen. Attractive energy carriers are rich in hydrogen and do not produce greenhouse gases upon oxidation. In addition, they remain liquid or solid over a wide temperature range, allowing it to be handled. According to a recent reference book [1] on energy change, two principal questions related to utilizing hydrogen fuel need to be answered: "Is the hydrogen economy around the corner?" and "Is hydrogen an alternative to fossil energy carriers?"

Current interest in the use of hydrogen as a transportation fuel has driven extensive research into novel gas storage materials. The current materials lack the ability to store the necessary amounts of hydrogen under technologically useful conditions. Hence, there is a need for new materials to solve the hydrogen storage problem. Many recent efforts have been focused on boron-containing hydrogenous materials primarily with regard to the light constitutional elements and the resulting high hydrogen storage capacities. In particular, *Handbook of Nanophysics* [2] discusses important possibilities in the use of boron clusters (with proper addition of metal atoms) for efficient hydrogen storage. According to a review [3] on the utilization of boron and its compounds, boron is a promising element for hydrogen storage with its hydrides and nanostructural forms.

The hydrogen storage capacity of boron-rich nanostructured materials was estimated in a lecture [4]. Physisorption of hydrogen needs lower energy, but produces lower capacities. As for chemisorption (usually in the form of metal hydrides, CH_4, etc.), this corresponds to higher hydrogen capacities, but requires higher energy as well. Between these limits, there exists a window of 0.2–0.6 eV/H_2 necessary for efficient reversible storage at room temperature and moderate pressure for onboard automotive applications. Materials needed for effective hydrogen storage should be able to bind hydrogen molecules chemically, without nondissociation. The collective monograph [5] also illustrates the practical versatility of boron, including the application of boron chemistry (e.g., designing large molecules containing icosahedral boron clusters) for hydrogen storage.

Analytical geometric models suggested for boron nitride [6] and all-boron [7] nanosystems seem to be helpful in calculating the available volumes of corresponding nanoreservoirs for hydrogen storage.

Nanomaterials for Environmental Protection, First Edition. Edited by Boris I. Kharisov, Oxana V. Kharissova, and H. V. Rasika Dias.

27.1 BN SURFACES

Fourier-transform infrared (FT–IR) spectrometry analysis of the turbostratic boron nitride (t-BN) powder surface activity revealed [8] BHN_2, B_2NH, and some other hydrogen-containing species. Shrestha et al. [9] also reported the results of BN-substrate adsorption studies performed on powder samples subjected various cleaning treatments. The adsorption of H species on B- versus N-atoms on the hexagonal boron nitride (h-BN) (001) surface was investigated [10] theoretically within the density functional theory (DFT) using a cluster approach. Only the B atoms were subjected to a local transformation from the hexagonal (h-BN) to cubic (c-BN) phase. Upon adsorption of the H species, a neighboring surface B atom was substituted by a C atom (hence, the number of electrons in the system increased): in the presence of surface C impurities on the electron-deficient B surface, the adsorption of H on these C impurities will yield an embryonic cubic nucleus. The adsorption of H species resulted in a local transformation from sp^2- to sp^3-hybridization with a hereby connected negative adsorption energy (−55 kJ/mol). This adsorption process is therefore highly unlikely to occur in a chemical vapor deposition (CVD) synthesis, that is, at temperatures in the 600–1000°C range. Reactivities of hydrogen atom/ions with different boron nitride, as well as carbon, phases were studied in detail [11] by means of frontier orbital theory based on Hartree–Fock (HF) calculations. The results showed that there is a significant difference in the reactivity of atomic hydrogen with the sp^2- and the sp^3-carbon phases such that phase selectivity is facilitated during the CVD of diamond. In contrast, these reactivities of atomic hydrogen with BN phases are similar, indicating the difficulty in obtaining a pure c-BN phase via CVD. In addition, hydrogen ions show higher reactivity than their neutrals, whereas hydrogen anions show similar reactivity with the two carbon or the two BN phases. CH_3 species were found to be promoters for the preferential etching selectivity of hydrogen in BN growth. The joining of methyl species in the etching process of hydrogen over BN phases would alter the etching preference of the hydrogen.

The electronic properties of 2D hydrogenated and semihydrogenated h-BN sheets were investigated [12] using first-principles calculations. It was found that the hydrogenation effects in the BN sheets are quite different from those in the graphene sheets. Hydrogenation changes the band character of BN sheets, which causes the hydrogenated BN sheet to have a smaller band gap than the pristine one. While for the semihydrogenated sheet, the stable B-semihydrogenated BN sheet is a ferromagnetic metal due to the unpaired $2p_z$ electrons of N atoms. These studies demonstrated that the electronic properties of BN sheets can be well tuned by hydrogenation. A control on the H functionalizations of H–BN structures by carrier doping was revealed [13] using DFT calculations. When the system is electron-doped, H-adatoms will exclusively bond with B-atoms, resulting in possible magnetization of the system, whereas hole-doping favors the adatoms to form insulating orthodimer structures on the BN structures. This behavior is caused by a peculiar chemical bond between the N- and H-adatoms, whose strength significantly depends on the carrier type and level. Moreover, the adatoms' diffusion on these BN structures can be steered along a designable path by the carrier doping still attributed to the carrier-dependent bond stability. This carrier control of functionalizations is robust via H-adatom concentration and the physical conditions of BN structures, thus offering an easy route to controllably anchor the properties of functionalized BN systems for desired applications.

The structural and electronic characteristics of fully hydrogenated BN layers and zigzag-edged nanoribbons were investigated [14] using dispersion-corrected DFT calculations. In the fully hydrogenated BN structure, the hydrogen atoms adsorb on top of the B and N sites, alternating on both sides of the h-BN plane in a specific periodic manner. Among various low-energy hydrogenated membranes referred to as chair, boat, twist-boat, and stirrup, the stirrup conformation is the most energetically favorable one. The zigzag-edged BN nanoribbon, prominently fabricated in experiments, possesses intrinsic semimetallicity with full hydrogenation. The semimetallicity can be tuned by applying a transverse electric bias, thereby providing a promising route for spintronics device applications. Using DFT, a series of calculations of structural and electronic properties of hydrogen vacancies in a fully hydrogenated BN layer were conducted [15]. By dehydrogenating the H–BN structure, B-terminated vacancies can be created which induce complete spin polarization around the Fermi level, irrespective of the vacancy size. On the contrary, the H–BN structure with N-terminated vacancies can be a small-gap semiconductor, a typical spin gapless semiconductor, or a metal depending on the vacancy size. Utilizing such vacancy-induced band gap and magnetism changes, possible applications in spintronics can be proposed, and a special H–BN-based quantum dot device designed.

Using first-principles calculations, Ding et al. [16] investigated the structural and electronic properties of monolayer porous BN, and also of graphene (C) and BC_2N sheets, with different hydrogen passivations. All these porous sheets with one-hydrogen passivation exhibit direct band gap semiconducting behaviors. The porous BN sheet has a larger band gap than the porous C sheet, whereas the porous BC_2N sheets have variable band gaps depending on the atomic arrangements of B, C, and N atoms. The stablest conformation of porous BC_2N sheets is composed of C and BN hexagons, whereas with two-hydrogen passivation, it becomes a structure containing continuous BN and interrupted C zigzag lines. Furthermore, due to the sp^3-hybridization of the edge atoms, the two-hydrogen passivation induces the changes of band gaps as well as direct-to-indirect band gap transitions in all the porous sheets. These studies demonstrated that the porous C, BN, and BC_2N sheets have semiconducting behaviors with practical band engineering by different values of hydrogen passivation. The attention paid to potential applications of boron nitrides in hydrogen gas sorption was insufficient. In a study by Weng et al. [17], a novel BN material, that is, porous microbelts, with the highest specific surface area up to 1488 m^2/g, was obtained through one-step template-free reaction of a boron acid-melamine precursor with ammonia. The obtained BN phase was partially disordered and belonged to an intermediate state between h-BN and amorphous a-BN phases. By changing the synthesis temperatures, the textures of obtained porous microbelts are adjustable. H_2 sorption evaluations demonstrated that the materials exhibit high and reversible H_2 uptake from 1.6 to 2.3 wt.% at 77 K and at a relatively low pressure of 1 MPa.

Zhao et al. performed [18] a first-principles theoretical investigation of the adsorption of hydrogen molecules between bilayers of solid matrix layers—BN sheets (BBN) and graphene/BN heterobilayers (GBN)—with variable interlayer distance. It was found that the H_2 adsorption energy has a minimum by expanding the interlayer spacing, along with further interlayer expansion, arising from many H_2 binding states and electrostatic interaction induced by the polar nature of B–N bonds. To determine whether successive addition of H_2 molecules is indeed possible using the minimal H_2 adsorption energy as the reference state, the hydrogen storage capacity of BBN and GBN was simulated with different stacking types, and it was found that the GBN with Bernal stacking is superior for reversible hydrogen storage. Up to eight molecules of H_2 can be adsorbed with an average adsorption energy of approximately 0.20 eV/H_2, corresponding to approximately 7.69 wt.% hydrogen uptake.

Gradient-corrected DFT computations were performed [19] to probe the local chemical reactivity of Stone–Wales defects and edge sites in zigzag- and armchair-edge BN nanoribbons with the CH_2 cycloaddition. Independent of the nanoribbon types and the defect orientations, the reactions at Stone–Wales defect sites were found to be more exothermic than those at the center of perfect BN nanoribbons. The intriguing electronic and magnetic properties of fully and partially hydrogenated BN nanoribbons were investigated [20] by means of first-principles computations. They showed that independent of ribbon width, fully hydrogenated armchair nanoribbons are nonmagnetic semiconductors, while the zigzag counterparts are magnetic and metallic. The partially hydrogenated zigzag nanoribbons (using hydrogenated and pristine ones as building units) exhibit diverse electronic and magnetic properties: they are nonmagnetic semiconductors when the percentage of hydrogenated nanoribbons blocks is minor, while a semiconductor → semimetal → metal transition occurs, accompanied by a nonmagnetic → magnetic transfer, when the hydrogenated part is dominant. Although the semimetallic property is not robust when the hydrogenation ratio is large, this behavior is sustained for partially hydrogenated zigzag BN nanoribbons with a smaller degree of hydrogenation. Thus, controlling the hydrogenation ratio can precisely modulate the electronic and magnetic properties, which endows BN nanomaterials many potential applications in novel integrated functional nanodevices. A quantum mechanical description was reported [21] based on the DFT of the structures and electronic properties of armchair BN nanoribbons edge-terminated with O atoms and OH groups. The O-edge termination was found to give a peroxide-like structure that is nonmagnetic and semiconducting. The O-terminated BN nanoribbon was stabilized by the reduction of the peroxide groups with H atoms leading to a polyol-like structure. The two chains of hydrogen bonds created along the edges led to alternating five- and seven-membered rings and caused the ribbon to become nonplanar with rippled edges. Three configurations of different ripple periods and amplitudes were found with energy differences up to 2 eV per unit cell but with virtually the same band gap of 4.2 eV.

Bhattacharya et al. performed [22] DFT calculations to explore the possibility of using a metal-functionalized hydrogenated BN sheet for the storage of molecular hydrogen. The chair BHNH conformer is ideally suited for the adsorption of metal adatoms on the surface of the sheet. The Li metal, in particular, binds to the sheet with a binding energy of ~0.88 eV/Li atom and becomes cationic, which thereby attracts hydrogen molecules. However, the interaction of the BHNH sheet and the absorbed H_2 molecules with Li^+ is different from the conventionally known Dewar coordination or Kubas-type interaction for hydrogen storage. Each Li^+ can adsorb up to four H_2 molecules, and the hydrogen binding energy is in the desired energy window for effective storage of molecular hydrogen. The fully Li-functionalized BHNH sheet yields a reasonably high gravimetric density, which is >7 wt.%. By DFT calculations, Chen et al. investigated [23] the adsorption of transition-metal (TM) atoms (TM = Sc, Ti, V, Cr, Mn, Fe, Co, and Ni) on a carbon-doped h-BN sheet and the corresponding cage $B_{12}N_{12}$. The carbon-doped BN nanostructures with dispersed Sc could store up to five and six molecules of H_2, respectively, with the average binding energy of 0.3–0.4 eV, indicating the possibility of fabricating hydrogen storage media with high capacity. It was also demonstrated that the geometrical effect is important for hydrogen storage, leading to a modulation of the charge distributions of *d*-levels, which dominate the binding between H_2 and TM atoms. In a study by Shahgaldi et al. [24], titanium-coated boron nitride nanofibers were produced by the electrospinning method, and the effect of heat treatment on the nanofibers was studied. TiO_2-coated BN nanofibers, with a diameter of 100 nm, were obtained after heat treatment and nitridation. The X-ray diffraction (XRD) and FT–IR spectroscopy depicted hexagonal structures of BN with sharp peaks related to titanium. The hydrogen uptake capacities of the nanofibers were investigated by pressure composition isotherms in the range of 1–70 bars at room temperature. Different heat treatment temperatures resulted in different morphologies and specific surface areas for the nanofibers, which had direct effects on the hydrogen absorption of the samples. Hydrogen absorption measurements revealed that coatings on BN nanofilms that were synthesized at a lower temperature during the nitridation process had the highest hydrogen absorption at room temperature. This result can be explained by the presence of titanium and the specific morphology and higher surface area of the nanofibers. A series of MgO/h-BN composites with different mass ratios were synthesized by the impregnation method and used as supports for Ru catalysts in ammonia synthesis reaction [25]. The catalysts were characterized, in particular, by N_2 physical adsorption and temperature-programmed reduction of H_2. The activity measurements of ammonia synthesis were carried out in a reactor with a mixture of N_2 and H_2 atmosphere under a steady-state condition. The results showed that the rate of ammonia formation was strongly influenced by the h-BN content used in the catalysts' preparation process. At N_2:H_2 = 1:3 atmosphere, the optimum activity was achieved when MgO:h-BN = 8:2 was used as the catalytic support.

As is known, the c-BN phase is an extremely promising multifunctional material. However, to exploit all possible applications, large-area CVD of c-BN films is required. For a successful CVD growth of high-quality c-BN films, one must obtain a deeper understanding of the structural and electronic properties of the dominant c-BN growth surfaces under CVD conditions, that is, the (100), (110), and (111) surfaces, and their modification in the presence of surface-stabilizing atomic hydrogen (H). In a study by Karlsson and Larsson [26], the surface-stabilizing effect of H on the B- and N-terminated surfaces of c-BN (100) was investigated using DFT calculations. It was found that a 100% surface coverage of on-top H on some B-terminated c-BN (100) surfaces is not able to uphold an ideal bulk-like structure. For the N-terminated c-BN (100) surfaces, opposite observations were made. The process of H abstraction, with gaseous atomic H, was found to be significantly more favorable for a B-terminated c-BN (100) surface than for an N-terminated c-BN (100) surface. It was also found that N radical sites are more stable toward radical surface site collapse than B radical sites. In the work of Karlsson and Larsson [27], the ability of BH_x and NH_x species (x=0, 1, 2, 3) to act as growth species for the CVD of c-BN, in an H-saturated gas phase, was investigated using DFT calculations. It was found that the optimal growth species for CVD growth of c-BN are B, BH, BH_2, N, NH, and NH_2, that is, decomposition of the incoming BH_3 and NH_3 growth species is very crucial for CVD growth of c-BN. It was also found that it would be most preferable to use a CVD method where the incoming BH_3 and NH_3 growth species are separately introduced into the reactor, for example, by using an atomic layer deposition type of method.

A systematic and comparative investigation of the number of different ^{IV}B–^{VI}B transition metal carbides, nitrides, sulfides, silicides, and borides as well as main group element ceramics including h-BN for their electrocatalytic performance toward the hydrogen evolution reaction was presented by Wirth et al. [28]. The performances of the electrocatalysts were compared with Pt and Ni benchmarks.

27.2 BN NANOTUBES

Like the hydrogenated BN surfaces, BN nanotubes have potential applications in hydrogen storage and were suggested as better hydrogen storage media than C nanotubes. Physisorption of a hydrogen molecule together with chemisorption of atomic hydrogen on BN nanotubes was studied in detail. According to the theoretical results, chemisorption of hydrogen to boron is exothermic. Chemisorption of hydrogen alters the electronic properties, indicating that the hydrogenated BN nanotubes have additional potential applications, such as nanoscale electronic devices.

Hydrogen uptake capacities of 1.8 and 2.6 wt.% were obtained [29] on BN multiwalled and bamboo-like nanotubes, respectively, under 10 MPa at room temperature. For BN nanotubes synthesized by CVD over a wafer made by an $LaNi_5$/B mixture and Ni powder at 1473 K, which were straight with a diameter of 30–50 nm and a length of up to several microns, it was first verified [30] that the BN nanotubes can store hydrogen by means of an electrochemical method. The mechanism of the electrochemical hydrogen storage process can be summarized as follows:

$$BN + H_2O + e^- \leftrightarrow BN - H_{ads} + OH^-,$$

$$BN - H_{ads} \leftrightarrow BN - H_{abs},$$

$$2BN - H_{ads} \leftrightarrow 2BN + H_2.$$

$BN-H_{ads}$ and $BN-H_{abs}$ are denoted as adsorbed and absorbed hydrogen on BN nanotubes, respectively. The hydrogen desorption of nonelectrochemical recombination in cyclic voltammograms, which can be considered as the slow reaction in BN nanotubes, suggested the possible existence of strong chemisorption of hydrogen. The cathodic adsorption peak of hydrogen was found in the cyclic voltammogram. The hydrogen desorption peak of nonelectrochemical recombination appeared at the range between −0.78 and −0.85 V. It was concluded that the improvement of the electrocatalytic activity by surface modification with metal or alloy would enhance the electrochemical hydrogen storage capacity of BN nanotubes. The binding energy of hydrogen atoms to a (10, 0) single-walled BN nanotube was calculated [31] at 25, 50, 75, and 100% coverage using the DFT.

The average binding energy is the highest at 50% coverage when the H atoms are adsorbed on the adjacent B and N atoms along the tube axis, and the value is −53.93 kcal/mol, which is similar to half of the H–H binding energy. In addition, the band gap (−4.29 eV) of the pristine (10, 0) nanotube was decreased up to −2.01 eV for the 50% hydrogen coverage. Zhang et al. [32] performed DFT calculations to investigate the adsorption of atomic hydrogen H on the wrapping axis of nonpolar armchair (5, 5) and chiral (8, 4) BN nanotubes with a view toward understanding the chemisorption-induced polarization field in BN nanotubes. The adsorption of H along the zigzag B–N bonds that lie on the wrapping axis of the BN nanotubes enhances the macroscopic polarization field. Depending on whether the B or N site near the edge of the nanotube is adsorbed with H, the direction of the polarization field, as well as the work function of the tube ends, can be changed significantly. The relationships between the chemical effect as well as the geometric distortion of the tube caused by H-chemisorptions and the induced polarization were investigated, respectively. These results have implications for the application of BN nanotubes as electron emitters. Zhao and Ding systematically studied [33] the effects of several gaseous adsorbates, including H_2, on the electronic properties of open edges of BN nanotubes by using DFT calculations. The results indicated that H_2 molecules dissociate and chemisorb on open BN nanotube edges with large adsorption energy because the tube edge has either an open or capped structure and thus has dangling bonds or pentagonal defects. The high reactivity of an open-ended BN nanotube can be comparable with that of its carbon counterpart, although the wall of the BN nanotube is chemically more stable than a single-walled carbon nanotube's wall. Moreover, it was noted that adsorption of the gas at the tips of open BN nanotubes can modify their electronic properties. A considerable amount of charge transferred for the adsorption of gaseous hydrogen on the open BN nanotubes may account for the changes of the electronic properties. Interestingly, the open (5, 5) BN nanotube exhibits the properties of wide–band gap materials when gas is adsorbed at top sites, while a smaller band gap is observed when this gas is adsorbed on seat sites. These results might be helpful in the design of BN nanotube–based nanomaterials such as field emitters or nanojunctions. Ab initio local spin density approximation calculations were performed [34] to study the magnetic properties of hydrogenated BN nanotubes (H–BN). It was found that the adsorption of a single H atom on the external surface of a BN nanotube can induce spontaneous magnetization in the H–BN nanotubes, whereas no magnetism was observed when two H atoms were adsorbed on two neighboring N atoms or on two neighboring B and N atoms. However, spontaneous magnetization was also found in the H–BN nanotubes with two H atoms on two B atoms not next to each other. This may be experimentally accessible when the coverage of H atoms adsorbed on the external surface of BN nanotubes is low. Defects produce spontaneous magnetization on BN nanotubes. When one or three H atoms are adsorbed on vacancy defects, the H–BN nanotubes are nonmagnetic, while magnetic H–BN nanotubes can be obtained by two H atoms adsorbed on vacancy defects, which may be difficult to control experimentally. The key issue for magnetism is the existence of unpaired electrons, which can be realized either by low coverage of hydrogen atoms or by making defects on perfect BN nanotubes. This indicates that it is possible to tune the magnetic properties of BN nanotubes by hydrogenation or defects, thus providing a new synthetic route toward metal-free magnetic materials. Further theoretical investigations of chemical hydrogen storage showed [35] that the dehydrogenation of chemisorbed hydrogen atoms on BN nanotubes could be triggered by appropriate reagents through simultaneous proton and hydride transfer. The computed free energy of the activation barrier for the reduction of formaldehyde to methanol by chemisorbed hydrogen atoms on a zigzag BN nanotube was predicted to be 12.7 kcal/mol. The thermodynamic and kinetic feasibilities of H_2 dissociation on some zigzag nanotubes including the BN ones have been investigated [36] theoretically by calculating the dissociation and activation energies. The tubes were determined to be inert toward H_2 dissociation, both thermodynamically and kinetically. The reaction is endothermic by 5.8 kcal/mol, exhibiting high activation energy of 38.8 kcal/mol.

The structural and electronic characteristics of fully hydrogenated BN nanotubes were determined [37] by quantum chemical methods. Single-walled nanotubes up to and over 10 nm in diameter were fully optimized by periodic calculations, made possible by the utilization of line group symmetries. The preferred fully *exo*-hydrogenated BN nanotubes have diameters below 1 nm. Partial *endo*-hydrogenation was shown to stabilize large BN nanotubes, producing energetically favored tubes with diameters of 3.5 nm. The calculated band gaps suggest that perhydrogenated boron nitride nanotubes could be insulators, the band gaps being practically equal for zigzag and armchair tubes.

Adsorption of some chemical species, including H, H_2, and NH_3, on the sidewall of zigzag (8, 0) and armchair (5, 5) BN nanotubes was studied [38] using DFT. Particular attention was paid to searching for the most stable configuration of the adsorbates and the surface reactivity at a perfect site and near a Stone–Wales defect. Reactivity near the Stone–Wales defect is generally higher than that at the perfect site because of the formation of frustrated B–B and N–N bonds and the local strain caused by pentagonal and heptagonal pairs. The adsorption of NH_3 on the sidewall of BN nanotubes can be described as molecular chemisorption due to modest interaction between highest-occupied molecular orbital (HOMO) of NH_3 with lowest-unoccupied molecular orbital (LUMO) of BN. Adsorption of NH_3 can affect electronic properties of BN nanotubes by raising the Fermi level. As such, NH_3 can be viewed as an n-type impurity. As for H_2, it can only be physisorbed on the sidewall of BN nanotubes through van der Waals interaction. The geometries, formation energies, electronic properties, and reactivities of Stone–Wales defects in single-walled (8, 0) BN nanotubes were investigated [39] by means of gradient-corrected DFT computations. Stone–Wales defects deform BN nanotubes severely and result in local curvature changes at defect sites. The energies of defect formation increase with increasing tube diameters and depend on orientation. The reaction energies of model H_2 addition are mostly endothermic for defective as well as pristine BN nanotubes, but reactions at the most favorable sites near the Stone–Wales defect (homoelement N–N bonds followed by the B–B bond sites) are exothermic. This and the fact that the band structures of BN nanotubes are only slightly changed by Stone–Wales and vacancy-type defects as well as by chemical additions at low modification ratios endow BN nanotubes with great application potential.

The physisorption and chemisorption of hydrogen in BN nanotubes, investigated by DFT, were compared [40] with carbon nanotubes. The physisorption of H_2 on BN nanotubes was found to be less favorable energetically than on carbon nanotubes. The chemisorption of H_2 molecules on pristine BN nanotubes is endothermic. The reaction energy data for BN nanotubes were summarized as follows:

$$BN + H^{\bullet} \rightarrow BN - H^{\bullet} - 6.54\ \text{kcal/mol},$$

$$BN - H^{\bullet} + H^{\bullet} \rightarrow BN - 2H - 78.72\ \text{kcal/mol},$$

$$BN + 2H^{\bullet} \rightarrow BN - 2H - 85.26\ \text{kcal/mol},$$

$$BN + H_2 \rightarrow BN - 2H \ + 19.64\ \text{kcal/mol}.$$

Consequently, perfect BN nanotubes are not good candidates for hydrogen storage by either mechanism. Other strategies must be utilized if BN nanotubes are to be employed as hydrogen storage media such as utilizing them as supporting media for hydrogen-absorbing metal nanoclusters. The adsorption of atomic and molecular hydrogen on carbon-doped BN nanotubes was investigated [41] with DFT calculations. The binding energy was found to be substantially increased when compared with hydrogen on nondoped BN nanotubes. These results are in agreement with experimental results for BN nanotubes where dangling bonds are present. The atomic hydrogen makes a chemical covalent bond with carbon substitution, while physisorption occurs for molecular hydrogen. For the H_2 molecule adsorbed on top of a carbon atom in a boron site ($BN+C_B-H_2$), a donor defect level is present, while for the H_2 molecule adsorbed on top of a carbon atom in a nitrogen site ($BN+C_N-H_2$), an acceptor defect level is present. The binding energies of H_2 molecules absorbed on C-doped BN nanotubes are in the optimal range in order to work as a hydrogen storage medium. Although the properties of the C-doped BN nanotubes depend on their composition, these nanotubes are promising candidates for storing hydrogen. Durgun et al. investigated [42] the hydrogen absorption capacity of Ti-covered single-walled BN nanotubes using the first-principles plane-wave method. The interaction of H_2 molecules with the outer surface of the bare nanotube, which is normally very weak, can be significantly enhanced upon functionalization by Ti atoms. Each Ti atom adsorbed on the BN nanotube can bind up to four H_2 molecules with an average binding energy suitable for room temperature storage. BN nanotubes were synthesized [43] over both Fe^{3+}-impregnated mobile-composition-of-matter (MCM) and Fe_2O_3/MCM complex catalyst systems at relatively low temperatures for 1 h by the CVD technique in large quantities. The formation of BN nanotubes was tailored at different reaction temperatures by changing the catalyst type. The diameters were in the range of 2.5–4.0 nm for thin tubes and 20–60 nm for thick ones. The thin tube formation originated due to an average pore size of 4 nm. Higher reaction temperatures caused side product formations. The gas uptake capacity measurements at room temperature showed that BN nanotubes could adsorb 0.85 wt.% hydrogen, which is two times larger than that for carbon nanotubes. The adsorption of H_2 on Ce-doped BN nanotubes was investigated [44] by the means of DFT calculations. For the Ce/BN nanotube system, it is found that Ce preferentially occupies the hollow site over the BN hexagon. The results indicate that seven molecules of H_2/Ce can be adsorbed and 5.68 wt.% H_2 can be stored in a Ce_3/BN nanotube system. Among nanotubes doped with metals, Ce exhibits the most favorable hydrogen adsorption characteristics in terms of adsorption energy and uptake capacity. Both hybridization of the Ce 5*d* orbital with the H 1*s* orbital and the polarization of the H_2 molecules contribute to the hydrogen adsorption. Ce clustering can be suppressed by preferential binding of Ce atoms on BN nanotubes, which denotes that a BN nanotube as a hydrogen storage substrate is better than a carbon nanotube due to its heteropolar binding nature.

Wu et al. [45] investigated the properties of chemically modified BN nanotubes with NH_3 and four other amino functional groups on the basis of DFT calculations. Unlike the case of carbon nanotubes, NH_3 can be chemically adsorbed on top of the B atom, with a charge transfer from NH_3 to the BN nanotube. The minimum-energy path calculation shows that a small energy barrier is encountered during the adsorption. Similarly, a small energy barrier (~0.42 eV) is also involved in the desorption, suggesting that both adsorption and desorption can be realized even at room temperature. For chemically modified BN nanotubes with various amino functional groups, the adsorption energies are typically less than that of NH_3. The trend of adsorption energy change can be correlated with the trend of relative electron-withdrawing or electron-donating capability of the amino functional groups. Overall, the chemical modification of BN nanotubes with the amino groups results in few changes in the electronic properties. To help understand the mechanism of chemical peeling of BN nanotubes, single-walled model systems were studied [46] using DFT. Dimethyl sulfoxide acts as a water transporter across the interphase by the formation of a water-dimethyl sulfoxide complex, which links to an active site on the wall of the tube. Afterward, a unimolecular hydrolysis reaction takes place. In contrast, when dimethyl sulfoxide is absent, the hydrolysis reaction is bimolecular, which implies that the activation energy is higher owing to the larger entropic factor in comparison with the unimolecular mechanism. The effect of the surface curvature was analyzed by studying the hydrolysis reaction on tubes of increasing diameters. It was found that for the larger studied system, that is, the $B_{48}N_{48}H_{24}$ (12, 0) zigzag model, the reaction with the water-dimethyl sulfoxide complex is bimolecular and does not yield hydrolysis but rather hydration, which indicates that planarity of the surface must be strongly perturbed in order to achieve the peeling.

Quasi-1D cylindrical pores of single-walled BN and C nanotubes efficiently differentiate [47] adsorbed hydrogen isotopes at 33 K. Extensive path integral Monte Carlo simulations revealed that the mechanisms of quantum sieving for both types of nanotubes are quantitatively similar; however, the stronger and heterogeneous external solid–fluid potential generated from single-walled BN nanotubes enhanced the selectivity of deuterium over hydrogen both at zero coverage and at finite pressures. It was shown that this enhancement of the D_2/H_2 equilibrium selectivity results from larger localization of hydrogen isotopes in the interior space of single-walled BN nanotubes in comparison to that of equivalent single-walled C nanotubes. The operating pressures for efficient quantum sieving of hydrogen isotopes are strongly dependent on both the type and the size of the nanotube.

The electromagnetic nonbounded interactions of the $NH_2BHNBHNH_2$ molecule inside the $B_{18}N_{18}$ ring were investigated [48] with hybrid DFT for a physicochemical explanation of electromagnetic nonbounded interactions within these nanosystems.

27.3 BN FULLERENES

A medial-neglecting-by-differential-overlapping (MNDO) study of enthalpy of formation of boron–nitrogen analogues of buckminsterfullerene, including (carbo)hydrogenated ones $B_3N_3H_6$, $B_5N_5H_8$ and $B_4N_4C_2H_8$, was presented by Xis et al. [49]. The possibility of H_2 gas storage in fullerene-like materials such as boron nitride, as well as carbon, clusters was investigated [50] by molecular dynamics (MDin) calculations. These authors showed that these clusters have energy barriers for H_2 molecules to pass through atomic rings. H_2 molecules enter into $B_{36}N_{36}$ from hexagonal rings more easily than tetragonal rings. H_2 molecules would pass from hexagonal rings of $B_{36}N_{36}$ more easily than hexagonal rings of C_{60} due to the larger curvature because of tetragonal rings in the BN cage structure. BN and C fullerenes are sublimed at 1800°C and 600°C, respectively. Consequently, BN fullerenes have greater stability at high temperature. Thus, BN fullerene materials would be better candidates for H_2 storage. Using a first principles theory based on DFT formulation, Sun et al. [51] studied the energetics and thermal stability of storing hydrogen in B–N-based nanostructures. The hydrogen molecule was found to enter through the hexagonal face of the $B_{36}N_{36}$ cage, preferring to remain inside the cage in molecular form. The energy barriers for the hydrogen molecule to enter into or escape from the cage are, respectively, 1.406 and 1.516 eV. As the concentration of hydrogen inside the cage increases, the cage expands and the bond length of the hydrogen molecule contracts, resulting in significant energy cost. At zero temperature, up to 18 hydrogen molecules can be stored inside a $B_{36}N_{36}$ cage corresponding to a gravimetric density of 4 wt.%. However, MDin simulation at room temperature (300 K) indicated that high weight percentage hydrogen storage cannot be achieved in B–N cage structures. The interaction between nanoclusters of $B_{12}N_{12}$ and some small molecules including H_2 was investigated [52] by using DFT computations, exploiting the structural and electronic properties of the adsorbate/cluster complex. The calculated adsorption energy of the most stable state was −1.35 kJ/mol. It was revealed that the adsorption of H_2 molecules has no significant effect on the electronic properties of the cluster.

The structural principles of perhydrogenated BN fullerenes were determined [53] by quantum chemical calculations. The octahedral BN fullerenes were classified into structural families, which were systematically studied. Each family was developed into a series of cages, up to $B_{444}N_{444}H_{888}$, subjecting them to perhydrogenation. Chiral, T-symmetric nanostructures are favored in energy, possessing an optimal diameter of around 2.8 nm, corresponding to a molecular formula of $B_{228}N_{228}H_{456}$. The facets of the octahedral nanostructures are clearly curved, placing B–H bonds on the convex side and N–H bonds on the concave side. The curvature is due to the partially ionic character of the B–N bond, which polarizes the hydrogen atoms. The described structural principles of perhydrogenated BN fullerenes are expected to provide new insight into structural characteristics and properties of nanomaterials based on BN.

By the incorporation of C into $B_{12}N_{12}$ fullerenes, the theoretical investigation [54] showed that the hydrogenation reaction on C-doped $B_{11}N_{12}C$ clusters is both thermodynamically favored and kinetically feasible under ambient conditions. The C atom can work as an activation center to dissociate the H_2 molecule and provide the free H atom for further hydrogenation on the $B_{11}N_{12}C$ fullerene, which saves the materials cost in practical applications for hydrogen storage. The curvature also plays an important role in reducing the activation barrier for the hydrogen dissociation on the BN fullerenes.

27.4 BN NANOCOMPOUNDS

A DFT study of small base molecules and tetrahedral and cubane-like group V clusters encapsulated in B_{80} showed [55] that the boron buckyball is a hard acid and prefers hard bases like NH_3 or N_2H_4 to form stable off-centered complexes. The boron cap atoms are electrophilic centers, and prefer mainly to react with electron-rich nucleophilic sites. The stability of the complexes will be governed by the size and electron-donating character of the encapsulated clusters.

The adsorption of atomic and molecular hydrogen on armchair and zigzag boron carbonitride nanotubes was investigated [56] with DFT calculations. The adsorption of atomic H on the BC_2N nanotubes presents properties that are promising for nanoelectronic applications. Depending on the adsorption site for H, the Fermi energy moves toward the bottom of the conduction band or toward the top of the valence band, leading the system to exhibit donor or acceptor characteristics, respectively. The H_2 molecules are physisorbed on the BC_2N surface for both chiralities. The binding energies for the H_2 molecules are slightly dependent on the adsorption site, and they are close to the range that can work as a hydrogen storage medium. First-principles calculations were carried out [57] to investigate Ti (Sc)-decorated 2D boron–carbon-nitride BC_2N sheets for their application as hydrogen storage materials. The results showed that with four H_2 molecules attached to each metal atom, the Ti (Sc)-decorated BC_2N can store up to 7.6 (7.8) wt.% of hydrogen in molecular form. The adsorption energy is within the range of 0.40–0.56 (0.13–0.27) eV/H_2, which is suitable for ambient temperature hydrogen storage. An extensive first-principles investigation of both exohedral and endohedral Ti-decorated BC_4N nanotubes for hydrogen storage also was reported by Bhattacharya et al. [58]. The results revealed that an endohedral capping of Ti is energetically favorable compared to an exohedral capping, albeit marginally by approximately 0.1–0.4 eV/Ti atom. However, this endohedral insertion process is difficult since it requires overcoming of a rather high energy barrier of approximately 4 eV/Ti atom, as obtained from nudge elastic band calculation of the minimum-energy path. It was observed that each Ti^+ ion sitting on the hexagonal face of the exohedral Ti–BC_4N can bind up to four hydrogen molecules with successive energies of adsorption lying in the range of approximately 0.4–0.7 eV. Further, it was predicted that at high Ti coverage, the system can absorb up to 5.6 wt.% of hydrogen. After establishing the adsorption of hydrogen molecules on the Ti–BC_4N nanotube, MDin simulation was performed to understand the desorption behavior. It was observed that at 300 K the system remains stable with all four H_2 molecules attached to Ti, while at 500 K hydrogen gets released in molecular form from the Ti–BC_4N nanotube without breaking the cage. Thus, the desorption temperature and kinetics are quite favorable. This investigation underscores the potential of Ti-decorated BC_4N nanotubes as a promising nanostructure candidate material for H_2 storage. When a planar sheet of carbon (graphene) or isoelectronic boron nitride is decorated by Ti atoms for a high-capacity hydrogen storage material, due to charge transfer, the Ti becomes cationic and helps to adsorb hydrogen in the molecular form. The principal bottleneck for using such a system for efficient H storage is the problem of metal clustering on the surface. Using the first-principles DFT calculations, it was shown [59] that clustering of Ti atoms can be avoided by using a composite BC_4N planar sheet that is obtained through chemical modification of the graphene surface by systematically replacing C atoms by B and N atoms. This Ti–BC_4N system shows hydrogen storage capacity with a reasonably high gravimetric efficiency (~8.4 wt.%). Further, using ab initio MDin simulation, it was shown that this system is stable at 300 K, while desorption starts at approximately 500–600 K, which is lower compared to that of conventional graphene or BN planar structures.

Motivated by successful fabrication of monolayer materials consisting of hybrid graphene and boron nitride domains, Liu et al. [60] performed a first-principles study of hybrid graphene/boron nitride (C–BN) nanoribbons with dihydrogenated edge(s). The study suggested that hybrid C–BN nanoribbons may possess semimetallicity with a certain range of widths for the graphene and BN sections. In general, the hybrid C–BN nanoribbons can undergo the semiconductor-to-semimetal-to-metal transitions as the width of both graphene and BN nanoribbons increases. The calculated electronic structures of the hybrid C–BN nanoribbons suggest that dihydrogenation of the boron edge can induce localized edge states around the Fermi level, and the interaction among the localized edge states can lead to the semiconductor-to-semimetal-to-metal transitions.

Boron nitride materials in the form of borazine polymers with high specific surface of approximately 50–700 m^2/g (highly macroporous BN aerogels) were produced [61] and shown to adsorb H_2 and other gases. A series of porous BN materials with specific surface of approximately 440–710 m^2/g were also produced [62] using polymeric precursors. As they can adsorb H_2 and some other gases, these materials can serve as selective gas adsorbents utilizing the local polar characteristics of the individual B–N bonds not present in the carbon structures. Ammonia borane (NH_3–BH_3, AB) is known as a lightweight material containing a high density of hydrogen H_2 that can be readily liberated for use in fuel cell–powered applications. However, in the absence of a straightforward and efficient method for regenerating AB compound from dehydrogenated polymeric spent fuel, its full potential as a viable H_2 storage material will not be realized. It was demonstrated [63] that the spent fuel type derived from the removal of more than two equivalents of H_2 per molecule of AB, that is, polyborazylene, can be converted back to AB nearly quantitatively by 24 h treatment with hydrazine N_2H_4 in liquid ammonia NH_3 at 40°C in a sealed pressure vessel. Nanoscale h-BN additive for ammonia borane (AB) was shown [64] to decrease the onset temperature for hydrogen release. Both the nano-BN and the AB:nano-BN samples were prepared by ball milling. The hydrogen release was measured by a volumetric gas burette system.

The dehydrogenation of AB was found to occur in three steps that are generally represented by the transformation of AB to polyaminoborane (PAB, $(H_2NBH_2)_x$) and H_2, PAB to polyiminoborane (PIB, $(HNBH)_x$) and H_2, and PIB to BN and H_2. As previously mentioned, ammonia borane is under significant investigation as a possible hydrogen storage material. While chemical additives have been shown to lower the temperature for hydrogen release from ammonia borane, this may result in additional complications in the regeneration cycle. Mechanically alloyed h-BN (nano-BN) was shown [65] to facilitate the release of hydrogen from ammonia borane at lower temperature, with minimal induction time and less exothermicity, and inert nano-BN may be easily removed during any regeneration of the spent ammonia borane. The samples were prepared by mechanically alloying ammonia borane with nano-BN (i.e., physical mixtures). The ^{11}B magic angle spinning (MAS) solid state nuclear magnetic resonance (NMR) spectrum of the decomposition products showed that diammoniate of diborane is present in the mechanically alloyed mixture, which drastically shortens the induction period for hydrogen release from ammonia borane. Analysis also showed that all the borazine produced in the reaction comes from ammonia borane and that increasing the nano-BN surface area results in increased amounts of borazine. However, under high temperature, for example, approximately 150°C, isothermal conditions, the amount of borazine released significantly decreases. The electron microscopy of the initial and final nano-BN additive provided evidence for loss of crystallinity but not significant chemical changes. The higher concentration of borazine observed for low-temperature dehydrogenation of AB/nano-BN mixtures versus neat ammonia borane was attributed to a surface interaction that favors the formation of precursors, which ultimately result in borazine. This pathway can be avoided through isothermal heating at temperatures lower than 150°C. Several effects of the mixtures of AB:nano-BN were shown to be beneficial in comparison with neat ammonia borane: decrease of the dehydrogenation temperature, decrease in NH_3 formation, as well as decrease of the exothermicity of hydrogen release with increasing nano-BN concentration. Hydrogen is produced via the following reactions:

$$H_3NBH_3 \rightarrow (H_2NBH_2)_x + H_2 \quad 90-120°C,$$

$$(H_2NBH_2)_x \rightarrow (HNBH)_x + xH_2 \quad 120-160°C,$$

$$(HNBH)_x \rightarrow BN + xH_2 \quad \text{well above } 500°C.$$

The state-of-the-art for hydrogen storage is compressed H_2 at 700 bar and the development of a liquid-phase hydrogen storage material has the potential to take advantage of the existing liquid-based distribution infrastructure. Luo et al. [66] described a liquid phase hydrogen storage material, BN methylcyclopentane, that is a liquid under ambient conditions (i.e., at 20°C and 1 atm), air- and moisture-stable, and recyclable; releases H_2 controllably and cleanly at temperatures below or at the proton exchange membrane fuel cell waste-heat temperature of 80°C; utilizes catalysts that are cheap and abundant for H_2 desorption; features reasonable gravimetric and volumetric storage capacity; and does not undergo a phase change upon H_2 desorption. Among many promising hydrogen-rich boron-containing materials, ammonia borane (NH_3BH_3, AB) received much attention because of its satisfactory air stability, relatively low molecular mass, and remarkably high energy storage densities (gravimetric and volumetric hydrogen capacities are 19.6 wt.% and 140 g/l, respectively). However, the direct use of pristine ammonia borane as a hydrogen energy carrier in onboard/fuel-cell applications is prevented beacuse of its very slow dehydrogenation kinetics below 100°C and the concurrent release of detrimental volatile by-products such as ammonia, borazine, and diborane. Yıldırım [67] discussed different approaches to understand and control the properties of ammonia borane and other boron-based materials to be practical in terms of reduced dehydrogenation temperatures, accelerated H_2 release kinetics, and/or minimized borazine release: (i) developing of a new family of metal borohydride ammonia borane complexes (mixed metal amidoboranes $Na_2Mg(NH_2BH_3)_4$ and $Li_2(BH_4)_2NH_3BH_3$, and $Ca(BH_4)_2(NH_3BH_3)_2$ are some examples), from which >11 wt.% hydrogen can be released; (ii) exploring the effect of nano-confinement and catalytic activity of various metal-organic frameworks on the hydrogen release of ammonia borane; and (iii) designing and synthesizing a new nanoporous material, the so-called graphene oxide framework, as a potential storage medium for hydrogen gas.

27.5 OTHER BORON-RICH NANOMATERIALS

Hydrogen adsorption on the pristine boron sheets and nanotubes was investigated [68] by DFT calculations. Both molecular physisorption and dissociative atomic chemisorption were considered. Molecular hydrogen physisorption energies were found to be approximately 30–60 meV/molecule, actually lower than in graphene and in carbon nanotubes and far from the energies of 300–400 meV/molecule necessary for efficient hydrogen storage at room temperature and moderate pressures for onboard automotive applications. Chemisorption binding energies on B nanotubes should be approximately 2.4–2.9 eV/H atom, similar to the values obtained in C nanotubes. Finally, the energy barrier from molecular physisorption to dissociative chemisorption of hydrogen is about 1.0 eV/molecule. Therefore, the calculations predict physisorption as the leading adsorption mechanism of hydrogen at moderate temperatures and pressures. Wu et al. [69] investigated the feasibility of bare and metal-coated boron buckyball B_{80} with metals Li, Na, K, Be, Mg, Ca, Sc, Ti, and V for hydrogen storage using the DFT approach. Ca and Sc were found to be the best candidates for hydrogen storage with moderate adsorption energy of H_2 avoiding clustering of Sc and Ca on the B_{80} surface. Further, it was observed that an isolated cluster $Ca_{12}B_{80}$ ($Sc_{12}B_{80}$) can bind up to 66 (60) H_2 molecules with an average binding energy of 0.096 (0.346) eV/H_2, leading to a hydrogen storage capacity of 9.0 wt.% (7.9 wt.%). Two adsorption mechanisms, charge-induced dipole interaction and the Dewar–Kubas interaction, were demonstrated and shown to be responsible for high hydrogen storage capacity of $Ca_{12}B_{80}$ and $Sc_{12}B_{80}$. Most interestingly, the hydrogen-loaded $B_{80}Sc_{12}$–48 H_2 complex can further adsorb 12 H_2 through charge-induced dipole interaction. In other words, these two mechanisms dominate the adsorption of different parts of H_2 in the same cluster of $B_{80}Sc_{12}$–60 H_2. A comprehensive study was performed [70] on hydrogen adsorption and storage in Ca-coated boron fullerenes and nanotubes by means of DFT computations. Ca strongly binds to boron fullerene and nanotube surfaces due to the charge transfer between Ca and the B substrate. Accordingly, Ca atoms do not cluster on the surface of the boron substrate, while transition metals (such as Ti and Sc) persist in clustering on the B_{80} surface. B_{80} fullerene coated with 12 Ca atoms can store up to 60 H_2 molecules with a binding energy of 0.12–0.40 eV/H_2, corresponding to a gravimetric density of 8.2 wt.%, while the hydrogen storage capacity in a (9, 0) B nanotube is 7.6 wt.% with a binding energy of 0.10–0.30 eV/H_2. The Ca-coated boron fullerenes and nanotubes proposed are favorable for reversible adsorption and desorption of hydrogen at ambient conditions.

First-principles quantum chemical methods were used [71] to study ground-state energies and geometrical configurations of boron–hydrogen chains. The ground-state energies were found to be comparable with those of pristine boron clusters. The ground-state energies of the dimerized $(BH)_n$ chain were fitted into a model, in order to determine the corresponding parameters. This dimerization induces a band gap of approximately 0.6 eV. Tam et al. [72] reported spectroscopic observations on B atoms isolated in cryogenic parahydrogen p-H_2, normal deuterium n-D_2, and some noble-gas matrices. The $2s^23s(^2S)$–$2s^22p(^2P)$ B atom Rydberg absorption suffers large gas-to-matrix blue shifts, increasing in the sequence with n-D_2 < p-H_2. Much smaller shifts are observed for the $2s2p^2(^2D)$–$2s^22p(^2P)$ B atom core-to-valence transition. Ultraviolet (UV) absorption spectra of B/p-H_2 solids showed two strong peaks at 216.6 and 208.9 nm, which were assigned to the matrix-perturbed $2s^23s(^2S)$–$2s^22p(^2P)$ and $2s^2p^2(^2D)$–$2s^22p(^2P)$ B atom absorptions, respectively. This assignment is supported by the previous quantum path integral simulations. Laser-induced fluorescence emission spectra of B/p-H_2 solids showed a single line at 249.6 nm, coincident with the gas-phase wavelength of the $2s^23s(^2S)$–$2s^22p(^2P)$ B atom emission. The UV laser irradiation results in photobleaching of the B atom emission and absorptions, accompanied by the formation of B_2H_6. The review by Grimes [73] cited U.S. Patents on a widely used air bag propellant system for automobiles employing the dicesium salt of the $B_{12}H_{12}^{2-}$ ion as a burning accelerant to ensure rapid but controlled inflation of the bag. On the one hand, hydrogen-terminated icosahedral $B_{12}H_{12}^{2-}$, which has the same structure as the unit in solids, is the most stable molecule among the various polyhedral boranes synthesized. On the other hand, small boron clusters favor planar or nearly planar structures. The fact that the stable structure of boron clusters depends on the hydrogen contents means that the structure is tunable bycontrolling the number of hydrogen atoms. Ohishi et al. [74–76] reported the formation of icosahedral $B_{12}H_8^+$ through ion–molecule reactions of the decaborane ion $B_{10}H_n^+$ (n = 6–14) with decaborane $B_{10}H_{14}$ and diborane B_2H_6 molecules in an external quadrupole-static attraction ion trap. In the process of ionization, a certain number of hydrogen and boron atoms are detached from decaborane ions by the energy caused by the charge transfer from ambient gas ion to decaborane molecule. The hydrogen content n of $B_{12}H_n^+$ was determined by the analysis of the mass spectrum. The result reveals that $B_{12}H_8^+$ is the main product. First-principles calculations indicated that $B_{12}H_8^+$ preferentially forms an icosahedral structure rather than a quasiplanar structure. The energies of the formation reactions of $B_{12}H_{14}^+$ and $B_{12}H_{12}^+$ between $B_{10}H_x^+$ (x = 6 and 8) ions, which were considered to be involved in the formation of $B_{12}H_n^+$, and a B_2H_6 molecule were calculated. The calculations of the detachment pathway of H_2 molecules and H atoms from the product ions $B_{12}H_{14}^+$ and $B_{12}H_{12}^+$ indicate that the intermediate state has a relatively low energy, enabling the detachment reaction to proceed owing to the sufficient reaction energy. This autodetachment of H_2 accounts for the experimental result that $B_{12}H_8^+$ is the most abundant product, even though it does not have the lowest energy among $B_{12}H_n^+$ molecules. The hydrogen and boron contents of the $B_{10-y}H_x^+$ cluster were controlled by charge transfer from ambient gas (Ne or He) ions. This gas leads to the generation of $B_{10-y}H_x^+$ clusters with x = 4–10 and y = 0–1 (with x = 2–10 and y = 0–2). The introduction of ambient gas also increases the production of clusters. The DFT calculations were conducted to investigate the structure and the mechanism of formation of $B_{10-y}H_x^+$ and $B_{12}H_n^+$ clusters.

Among complex hydrides for hydrogen storage reviewed by Orimo et al. [77] are the boron-containing ones $Li_4BN_3H_{10}$ amide, and $LiBH_4$, $NaBH_4$, and some other tetrahydroborates. Hydrogen storage in metal hydrides is considered to be one of the most attractive methods. Ozturk and Demirbas [78] compared the hydrogen absorption–desorption behavior of the boron compounds with that of metal hydrides, which are considered as one of the most attractive hydrogen reservoirs. Boron compounds have a very high energy density, much better than that of liquid hydrogen, and also a lot safer. $LiBH_4$ is a complex hydride that consists of 18 wt.% of hydrogen. It has stability compared with other chemical hydrides and an easy conversion to H_2. Thus, there are good reasons that hydrogen storage materials for $LiBH_4$ will be used for power sources. Metallaboranes constitute an attractive class of compounds intermediate between borane cages and transition metal clusters. While carbon substitution into a borane cage is rather common and gives rise to an entire class of metallaheteroborane compounds, other main-group atom substitution is rather rare. It was shown that the reaction excess of $LiBH_4$, followed by thermolysis with excess of $BH_3 \cdot THF$, leads to the formation of oxamolybdaborane clusters. These are notable examples of oxametallaborane compounds where oxygen is contiguously bound to both cluster metals and boron atoms. Similarly, the reaction with $LiBH_4$ is followed by thermolysis with chalcogen powders of S or Se. The insertion of O, S, and Se atoms into the parent clusters shows a shortening of metal–metal bond distances. Theoretical calculations of DFT type were carried out [79] to study the geometries, energetics, and bonding properties in these metallaheteroborane compounds, aiming at completing their characterization and, in particular, establishing the exact number of hydrogens in the structures.

Mechanically milling ammonia borane and lithium borohydride in equivalent molar ratio results [80] in the formation of a complex $LiBH_4 \cdot NH_3BH_3$. This complex was studied in terms of its decomposition behavior and reversible dehydrogenation property. This study found that $LiBH_4 \cdot NH_3BH_3$ first disproportionates into $(LiBH_4)_2 \cdot NH_3BH_3$ and NH_3BH_3, and the resulting mixture exhibits a three-step decomposition behavior upon heating to 450°C, totally yielding approximately 15.7 wt.% hydrogen. Metal borohydrides are of interest as hydrogen storage materials as BH_4^- are 27% hydrogen by weight. When the borohydride anion is paired with a light-weight cation such as Li^+ or Na^+, the compound has a hydrogen weight percentage that is sufficiently high to be of potential practical importance for hydrogen storage. To optimize hydrogen release from materials containing the BH_4^- anion requires an understanding of the thermal decomposition mechanism. In particular, it is important to identify any stable intermediates with a lower hydrogen-to-boron ratio that may form in the course of the decomposition of borohydrides. One such intermediate is the $B_{12}H_{12}^{2-}$ anion, which was indentified in studies of borohydride decomposition. Transmission IR spectroscopy was used [81] to characterize the temperature dependence of the vibrational spectra of $LiBH_4$, $NaBH_4$, KBH_4, and $K_2B_{12}H_{12}$. The tetrahedral BH_4^- species has two IR active fundamentals, the asymmetric B–H stretch and the asymmetric BH_4 deformation, which are observed at 2292 and 1179 cm^{-1}, respectively, for KBH_4. In addition, two other peaks are observed in the B–H stretch region at 2225 and 2387 cm^{-1} in KBH_4, due to the overtone of the asymmetric bend and the combination band of the symmetric and asymmetric bending modes. Peak positions are at similar values for $NaBH_4$ and $LiBH_4$. The high symmetry of the icosahedral $B_{12}H_{12}^{2-}$ anion leads to only three IR active fundamentals, which were observed at 716, 1076, and 2485 cm^{-1} for $K_2B_{12}H_{12}$. Upon heating to 550 K, the IR spectrum of KBH_4 shows transformation into a new species with peaks at 716, 1074, and 2150 cm^{-1}. The changes in the spectrum provide good evidence for the formation of $B_{12}H_{12}^{2-}$ as an intermediate in the decomposition of the BH_4^- anion. For the first time, it was demonstrated [82] that hydrogen can be released and reabsorbed from a promising storage material, overcoming a major hurdle to its use as an alternative fuel source. Nanoparticles of sodium borohydride were synthesized and encased inside nickel shells. Their unique "core–shell" nanostructure showed remarkable hydrogen storage properties, including the release of energy at much lower temperatures than previously observed: one can expect initial energy release at 50°C, and significant release at 350°C.

DFT calculation was performed [83] to investigate the electronic structures of cage $B_{12}H_n$ for up to $n \leq 12$ and $AlB_{12}H_n$ for up to $n \leq 13$. Moreover, the computations were extended to the charged clusters of $(B_{12}H_{12})_q$, $(AlB_{12}H_{12})_q$, and $(AlB_{12}H_{13})_q$, where $q = \pm 1$ and ± 2. Their energies were calculated and structural analysis was carried out. The cage form of B_{12} remains stable against hydrogen adsorptions. The binding energies of $B_{12}H_n$ and $AlB_{12}H_n$ are in a decreasing trend with n. The HOMO–LUMO energy gaps show that $B_{12}H_{11}$ has relatively higher chemical hardness. $B_{12}H_2$, $B_{12}H_4$, $B_{12}H_7$, and $B_{12}H_{10}$ are energetically more stable clusters. The $AlB_{12}H_3$, $AlB_{12}H_8$, $AlB_{12}H_{10}$, and $AlB_{12}H_{12}$ clusters are also obtained as relatively more stable. In the charged $AlB_{12}H_n$ clusters, structural orientations are observed for $n = 12$ and 13. There is considerable interest in bare boron and metal–boron clusters as they offer potential for materials suited for hydrogen storage. Böyükata and Güvenç [84] presented studies on Al-doped and hydrogenated B_n cage $AlB_{12}H_n$ clusters ($n = 1$–14), as well as clusters and hydrogenated cage structures of $MB_{12}H_n$ complexes (for M = Ti, Cr, Fe, Co and $n \leq 13$). For the computations of these systems DFT was utilized. Interaction energies of H, 2H, and H_2 were analyzed. One of the findings showed that the metal-coated cage structure of B_{12} is stable against H adsorption.

TM atoms bound to B-doped fullerenes were proposed [85] as adsorbents for high-density, room-temperature, ambient-pressure storage of hydrogen. $C_{48}B_{12}$ disperses TMs by charge transfer interactions to produce stable organometallic buckyballs. A particular Sc can bind as many as 11 hydrogen atoms per TM, 10 of which are in the form of dihydrogen that can be adsorbed and desorbed reversibly. In this case, the calculated binding energy is about 0.3 eV/H_2, which is ideal for use onboard vehicles. The theoretical maximum retrievable H_2 storage density is approximately 9 wt.%. In the work of Zhao et al. [86], TM boride and carboride nanostructures were studied as model organometallic materials for hydrogen storage. The dispersed TM atoms' function is H_2 sorption centered on the surface of the boron or carbon–boron substrate. The flexibility offered in the variety of possible structures permits the study of the effect of the TM–TM distance on the storage capacity. When the TMs are too close to one another, TM–TM bonding reduces the capacity. Even when separated by distances larger than the normal TM–TM bond length, delocalization of TM valence electrons can still lower the hydrogen capacity. An optimal TM–TM distance for the structural motifs studied was approximately 6Å. The study also permitted the evaluation of new TM boride nanostructures. It was predicted that a low-energy single-walled scandium triboride ScB_3 nanostructures can bind approximately 6.1 wt.% hydrogen with the energy of 22–26 kJ/mol. A new family of porous boron-substituted carbon BC_x materials with controlled structure was investigated [87]. The chemistry involves a B-precursor polymer containing templates in the form of inorganic additives. Amorphous carbon-like BC_x materials containing up to 12% B were prepared, which show an extended fused hexagonal ring structure with B-puckered curvature. This maintains its electron deficiency out of planar B moiety, due to limited *p*-electron delocalization, and exhibits superactivated properties to enhance H_2 binding energy (20–10 kJ/mol) and adsorption capacity. After removing the inorganic additives by water-washing, the resulting porous BC_x shows a surface area of 500–800 m²/g. A porous BC_6 material exhibits a reversible hydrogen physisorption capacity of 0.5 and 3.5 wt.% H_2 per 500 m²/g surface area of the material at 293 and 77 K, respectively, under moderate hydrogen pressure (<100 bars). Both values are more than three times higher than H_2 absorption capacities in the corresponding carbonaceous materials. The physisorption results were further warranted by absorption isotherms, indicating a binding energy of hydrogen molecules between 10 and 20 kJ/mol, significantly higher than the 4 kJ/mol reported on most graphitic C surfaces.

The investigation of the interaction of a hydrogen atom with a $B_{12}P_{12}$ nanocluster based on DFT calculations indicated [88] that this process is energetically more favorable than that with $B_{12}N_{12}$ cluster.

Meng et al. [89] investigated the potential for hydrogen storage of a new class of nanomaterials, metal–diboride nanotubes. These materials have the advantages of a high density of binding sites on the tubular surfaces without the adverse effects of metal clustering. Using the TiB_2 (8, 0) and (5, 5) nanotubes as prototype examples, it was shown through first-principles calculations that each Ti atom can host two intact H_2 units, leading to a retrievable hydrogen storage capacity of 5.5 wt.%. Most strikingly, the binding energies fall in the desirable range of 0.2–0.6 eV/H_2 molecule, endowing these structures with the potential for room-temperature, near–ambient pressure applications.

REFERENCES

[1] Wengenmayr R, Bührke T, editors. *Renewable Energy: Sustainable Concepts for the Energy Change*. Berlin: Wiley – VCH; 2012.

[2] Sadrzadeh A, Yakobson BI. Boron fullerenes. In: Sattler KD, editor. *Handbook of Nanophysics: Clusters and Fullerenes*. Boca Raton: CRC Press; 2010. p 1–9.

[3] Fakioğlu E, Yürüm Y, Veziroğlu TN. A review of hydrogen storage systems based on boron and its compounds. Int J Hydrogen Energy 2004;29:1371–1376.

[4] Boustani I. *Towards Novel Boron Nanostructured Materials: Theoretical Predictions and Experimental Confirmations*. Dresden: Technische Universität Dresden; 2009. Available at http://nano.tu-dresden.de/pubs/slides_others/2009_04_30_Ihsan_Boustani.pdf. Accessed April 15, 2014.

[5] Hosmane NS, editor. *Boron Science: New Technologies and Applications*. Boca Raton: CRC Press; 2011.

[6] Chkhartishvili L. Boron nitride nanosystems of regular geometry. J Phys Conf Ser 2009;176:1–17.

[7] Chkhartishvili L. Geometrical models for bare boron nanotubes. In: Borisenko VE, Gaponenko SV, Gurin VS, Kam CH, editors. *Physics, Chemistry and Applications of Nanostructures*. Singapore: World Scientific; 2011. p 118–121.

[8] Baraton MI, Merle T, Quintard P, Lorenzelli V. Surface activity of a boron nitride powder: a vibrational study. Langmuir 1993;9:1486–1491.

[9] Shrestha P, Alkhafaji MT, Lukowitz MM, Yang G, Migone AD. Adsorption studies on boron nitride substrates. Langmuir 1994;10:3244–3249.

[10] Mårlid B, Larsson K, Carlsson J-O. Hydrogen and fluorine adsorption on the h-BN (001) plane. J Phys Chem B 1999;103:7637–7642.

[11] Zhang RQ, Chu TS, Lee CS, Lee ST. A theoretical study on the interactions of hydrogen species with various carbon and boron nitride phases. J Phys Chem B 2000;104:6761–6766.

[12] Wang Y. Electronic properties of two-dimensional hydrogenated and semihydrogenated hexagonal boron nitride sheets. Phys Stat Sol RRL 2010;4:34–36.

[13] Zhang Z, Guo W. Controlling the functionalizations of hexagonal boron nitride structures by carrier doping. J Phys Chem Lett 2011;2:2168–2173.

[14] Samarakoon DK, Wang X-Q. Intrinsic half-metallicity in hydrogenated boron-nitride nanoribbons. Appl Phys Lett 2012;100:1–4.

[15] Zhou YG, Wang ZG, Nie JL, Yang P, Sun X, Khaleel MA, Zu XT, Gao F. Vacancies in fully hydrogenated boron nitride layer: implications for functional nanodevices. Phys Stat Sol RRL 2012;6:105–107.

[16] Ding Y, Wang Y, Shi S, Tang W. Electronic structures of porous graphene, BN, and BC_2N sheets with one- and two-hydrogen passivations from first principles. J Phys Chem C 2011;115:5334–5343.

[17] Weng Q, Wang X, Zhi C, Bando Y, Golberg D. Boron nitride porous microbelts for hydrogen storage. Nano Lett 2013;7:1558–1565.

[18] Zhao G, Li Y, Liu C, Wang Y, Sun J, Gu Y, Wanga Y, Zeng Z. Boron nitride substrate-induced reversible hydrogen storage in bilayer solid matrix via interlayer spacing. Int J Hydrogen Energy 2012;37:9677–9687.

[19] Chen W, Li Y, Yu G, Zhou Z, Chen Z. Electronic structure and reactivity of boron nitride nanoribbons with Stone–Wales defects. J Chem Theory Comput 2009;5:3088–3095.

[20] Chen W, Li Y, Yu G, Li C-Z, Zhang SB, Zhou Z, Chen Z. Hydrogenation: a simple approach to realize semiconductor–half-metal–metal transition in boron nitride nanoribbons. J Am Chem Soc 2010;132:1699–1705.

[21] Lopez–Bezanilla A, Huang J, Terrones H, Sumpter BG. Structure and electronic properties of edge-functionalized armchair boron nitride nanoribbons. J Phys Chem C 2012;116:15675–15681.

[22] Bhattacharya S, Bhattacharya A, Das GP. Anti-Kubas type interaction in hydrogen storage on a Li decorated BHNH sheet: a first-principles based study. J Phys Chem C 2012;116:3840–3844.

[23] Chen M, Zhao Y-J, Liao J-H, Yang X-B. Transition-metal dispersion on carbon-doped boron nitride nanostructures: applications for high-capacity hydrogen storage. Phys Rev B 2012;86:1–7.

[24] Shahgaldi S, Yaakob Z, Khadem DJ, Wan Daud WR. Characterization and the hydrogen storage capacity of titania-coated electrospun boron nitride nanofibers. Int J Hydrogen Energy 2012;37:11237–11243.

[25] Yang X, Tang L, Xia C, Xiong X, Mu X, Hu B. Effect of MgO/h-BN composite support on catalytic activity of Ba – Ru/MgO/h-BN for ammonia synthesis. Chin J Catal 2012;33:447–453.

[26] Karlsson J, Larsson K. Hydrogen-enduced de/reconstruction of the c-BN (100) surface. J Phys Chem C 2010;114:3516–3521.

[27] Karlsson J, Larsson K. Adsorption of growth species on the c-BN (100) surface. J Phys Chem C 2011;115:16977–16983.

[28] Wirth S, Harnisch F, Weinmann M, Schröder U. Comparative study of IVB–VIB transition metal compound electrocatalysts for the hydrogen evolution reaction. Appl Catal B Environ 2012;126:225–230.

[29] Ma R, Bando Y, Zhu H, Sato T, Xu C, Wu D. Hydrogen uptake in boron nitride nanotubes at room temperature. J Am Chem Soc 2002;124:7672–7673.

[30] Chen X, Gao XP, Zhang H, Zhou Z, Hu WK, Pan GL, Zhu HY, Yan TY, Song DY. Preparation and electrochemical hydrogen storage of boron nitride nanotubes. J Phys Chem B 2005;109:11525–11529.

[31] Han SS, Lee SH, Kang JK, Lee HM. High coverage of hydrogen on a (10,0) single-walled boron nitride nanotube. Phys Rev B 2005;72:1–4.

[32] Zhang J, Loh KP, Wu P, Sullivan MB, Zheng J. Chemisorption-induced polarization of boron nitride nanotube. J Phys Chem C 2008;112:10279–10286.

[33] Zhao J-X, Ding Y-H. Theoretical studies of the interaction of an open-ended boron nitride nanotube (BNNT) with gas molecules. J Phys Chem C 2008;112:20206–20211.

[34] Li F, Zhu Z, Zhao M, Xia Y. Ab initio calculations on the magnetic properties of hydrogenated boron nitride nanotubes. J Phys Chem C 2008;112:16231–16235.

[35] Roy L, Mittal S, Paul A. Breaking the myth of the recalcitrant chemisorbed hydrogen on boron nitride nanotubes: a theoretical perspective. Angew Chem Int Ed Engl 2012;51:4152–4156.

[36] Beheshtian J, Soleymanabadi H, Kamfiroozi M, Ahmadi A. The H_2 dissociation on the BN, AlN, BP and AlP nanotubes: a comparative study. J Mol Model 2012;18:2343–2348.

[37] Tanskanen JT, Linnolahti M, Karttunen AJ, Pakkanen TA. Structural and electronic characteristics of perhydrogenated boron nitride nanotubes. J Phys Chem C 2008;112:2418–2422.

[38] An W, Wu X, Yang JL, Zeng XC. Adsorption and surface reactivity on single-walled boron nitride nanotubes containing Stone–Wales defects. J Phys Chem C 2007;111:14105–14112.

[39] Li Y, Zhou Z, Golberg D, Bando Y, von Ragué Schleyer P, Chen Z. Stone–Wales defects in single-walled boron nitride nanotubes: formation energies, electronic structures, and reactivity. J Phys Chem C 2008;112:1365–1370.

[40] Zhou Z, Zhao J, Chen Z, Gao X, Yan T, Wen B, von Ragué Schleyer P. Comparative study of hydrogen adsorption on carbon and BN nanotubes. J Phys Chem B 2006;110:13363–13369.

[41] Baierle RJ, Piquini P, Schmidt TM, Fazzio A. Hydrogen adsorption on carbon-doped boron nitride nanotube. J Phys Chem B 2006;110:21184–21188.

[42] Durgun E, Jang Y-R, Ciraci S. Hydrogen storage capacity of Ti-doped boron-nitride and B/Be-substituted carbon nanotubes. Phys Rev B 2007;76:1–4.

[43] Okan BS, Kocabaş ZO, Ergün AN, Baysal M, Letofsky–Papst I, Yürüm Y. Effect of reaction temperature and catalyst type on the formation of boron nitride nanotubes by chemical vapor deposition and measurement of their hydrogen storage capacity. Ind Eng Chem Res 2012;51:11341–11347.

[44] Zhang Z-W, Zheng W-T, Jiang Q. Hydrogen adsorption on Ce/BNNT systems: a DFT study. Int J Hydrogen Energy 2012;37:5090–5099.

[45] Wu X, An W, Zeng XC. Chemical functionalization of boron-nitride nanotubes with NH_3 and amino functional groups. J Am Chem Soc 2006;128:12001–12006.

[46] Avila F, Arenas JF, Peláez D, Otero JC, Bando Y, Golberg D, Huang Q, Soto J. Role of dimethyl sulfoxide in the hydrolytic peeling of boron nitride nanotubes. J Phys Chem C 2009;113:15565–15568.

[47] Kowalczyk P, Gauden PA, Terzyk AP. Cryogenic separation of hydrogen isotopes in single-walled carbon and boron-nitride nanotubes: insight into the mechanism of equilibrium quantum sieving in quasi-one-dimensional pores. J Phys Chem B 2008;112:8275–8284.

[48] Monajjemi M, Lee VS, Khaleghian M, Honarparvar B, Mollaamin F. Theoretical description of electromagnetic nonbonded interactions of radical, cationic, and anionic $NH_2BHNBHNH_2$ inside of the $B_{18}N_{18}$ nanoring. J Phys Chem C 2010;114:15315–15330.

[49] Xis X, Jelski DA, Bowser JR, George TF. MNDO study of boron–nitrogen analogues of buckminsterfullerene. J Am Chem Soc 1992;114:6493–6496.

[50] Narita I, Oku T. Molecular dynamics calculation of H_2 gas storage in C_{60} and $B_{36}N_{36}$ clusters. Diam Relat Mater 2002;11:945–948.

[51] Sun Q, Wang Q, Jena P. Storage of molecular hydrogen in B–N cage: energetics and thermal stability. Nano Lett 2005;5:1273–1277.

[52] Beheshtian J, Peyghan AA, Bagheri Z, Kamfiroozi M. Interaction of small molecules (NO, H_2, N_2, and CH_4) with BN nanocluster surface. Struct Chem 2012;23:1567–1572.

[53] Karttunen AJ, Linnolahti M, Pakkanen TA. Structural characteristics of perhydrogenated boron nitride fullerenes. J Phys Chem C 2008;112:10032–10037.

[54] Wua H, Fan X, Kuo J-L. Metal free hydrogenation reaction on carbon doped boron nitride fullerene: a DFT study on the kinetic issue. Int J Hydrogen Energy 2012;37:14336–14342.

[55] Muya JT, Lijnen E, Nguyen MT, Ceulemans A. Encapsulation of small base molecules and tetrahedral/cubane-like clusters of group V atoms in the boron buckyball: a density functional theory study. J Phys Chem A 2011;115:2268–2280.

[56] Rossato J, Baierle RJ, Schmidt TM, Fazzio A. First-principles study of the adsorption of atomic and molecular hydrogen on BC_2N nanotubes. Phys Rev B 2008;77:1–6.

[57] Song N, Wang Y, Sun Q, Jia Y. First-principles study of hydrogen storage on Ti (Sc)-decorated boron–carbon-nitride sheet. Appl Surf Sci 2012;263:182–186.

[58] Bhattacharya S, Majumder C, Das GP. Hydrogen storage in Ti-decorated BC_4N nanotube. J Phys Chem C 2008;112:17487–17491.

[59] Bhattacharya S, Majumder C, Das GP. Ti-decorated BC_4N sheet: a planar nanostructure for high-capacity hydrogen storage. J Phys Chem C 2009;113:15783–15787.

[60] Liu Y, Wu X, Zhao Y, Zeng XC, Yang J. Half-metallicity in hybrid graphene/boron nitride nanoribbons with dihydrogenated edges. J Phys Chem C 2011;115:9442–9450.

[61] Borek TT, Ackerman W, Hua DW, Paine RT, Smith DM. Highly microporous boron nitride for gas adsorption. Langmuir 1991;7:2844–2846.

[62] Janik JF, Ackerman WC, Paine RT, Hua D-W, Maskara A, Smith DM. Boron nitride as a selective gas adsorbent. Langmuir 1994;10:514–518.

[63] Sutton AD, Burrell AK, Dixon DA, Garner EB III, Gordon JC, Nakagawa T, Ott KC, Robinson JP, Vasiliu M. Regeneration of ammonia borane spent fuel by direct reaction with hydrazine and liquid ammonia. Science 2011;331:1426–1429.

[64] Neiner D, Karkamkar A, Linehan JC, Arey B, Autrey T, Kauzlarich SM. Promotion of hydrogen release from ammonia borane with mechanically activated hexagonal boron nitride. J Phys Chem C 2009;113:1098–1103.

[65] Neiner D, Luedtke A, Karkamkar A, Shaw W, Wang J, Browning ND, Autrey T, Kauzlarich SM. Decomposition pathway of ammonia borane on the surface of nano-BN. J Phys Chem C 2010;114:13935–13941.

[66] Luo W, Campbell PG, Zakharov LN, Liu S-Y. A single-component liquid-phase hydrogen storage material. J Am Chem Soc 2011;133:19326–19329.

[67] Yıldırım T. Boron based materials for hydrogen storage. In: Contents of the 17th symposium on boron, borides and related materials. Ankara: BKM; 2011. p 41.

[68] Cabria I, López MJ, Alonso JA. Density functional calculations of hydrogen adsorption on boron nanotubes and boron sheets. Nanotechnology 2006;17:778–785.

[69] Wu G, Wang J, Zhang X, Zhu L. Hydrogen storage on metal-coated B_{80} buckyballs with density functional theory. J Phys Chem C 2009;113:7052–7057.

[70] Li M, Li Y, Zhou Z, Shen P, Chen Z. Ca-coated boron fullerenes and nanotubes as superior hydrogen storage materials. Nano Lett 2009;9:1944–1948.

[71] Sabra MK, Boustani I. New chains of boron and boron hydrogen. Europhys Lett 1998;42:611–618.

[72] Tam S, Macler M, de Rose ME, Fajardo ME. Electronic spectroscopy of B atoms and B_2 molecules isolated in para-H_2, normal-D_2, Ne, Ar, Kr, and Xe matrices. J Chem Phys 2000;113:9067–9078.

[73] Grimes RN. Boron clusters come of age. J Chem Educ 2004;81:658–672.

[74] Ohishi Y, Kimura K, Yamaguchi M, Uchida N, Kanayama T. Formation of hydrogenated boron clusters in an external quadrupole static attraction ion trap. J Chem Phys 2008;128:1–7.

[75] Ohishi Y, Kimura K, Yamaguchi M, Uchida N, Kanayama T. Synthesis and formation mechanism of hydrogenated boron clusters $B_{12}H_n$ with controlled hydrogen content. J Chem Phys 2010;133:1–7.

[76] Ohishi Y, Kimura K, Yamaguchi M, Uchida N, Kanayama T. Synthesis and formation mechanism of hydrogenated boron clusters $B_{12}H_n^+$ through charge transfer from ambient gas ion. In: Contents of the 17th symposium on boron, borides and related materials. Ankara: BKM; 2011. p 215.

[77] Orimo S-I, Nakamori Y, Eliseo JR, Züttel A, Jensen CM. Complex hydrides for hydrogen storage. Chem Rev 2007;107:4111–4132.

[78] Ozturk T, Demirbas A. Boron compounds as hydrogen storage materials. Energy Sources A 2007;29:1415–1423.

[79] Krishnamoorthy BS, Kahlal S, Ghosh S, Halet J-F. A theoretical investigation of the structure, bonding and reactivity of metallaborane cluster compounds. In: Contents of the 17th symposium on boron, borides and related materials. Ankara: BKM; 2011. p 269.

[80] Luo J, Wu H, Zhou W, Kang X, Fang Z, Wang P. $LiBH_4 \cdot NH_3BH_3$: a new lithium borohydride ammonia borane compound with a novel structure and favorable hydrogen storage properties. Int J Hydrogen Energy 2012;37:10750–10757.

[81] Jash P, Meaux K, Trenary M. Transmission infrared spectroscopy of hydrogen storage materials. In: Contents of the 17th symposium on boron, borides and related materials. Ankara: BKM; 2011. p 234.

[82] UNSW. 2012. Nano-structures to release hydrogen's energy potential. Available at http://www.materialstodey.com. Accessed August 17, 2012.

[83] Böyükata M, Güvenç ZB. DFT study of Al doped cage $B_{12}H_n$ clusters. Int J Hydrogen Energy 2011;36:8392–8402.

[84] Böyükata M, Güvenç ZB. Investigation of metal-doped boron clusters and their hydrogenated cage. In: Contents of the 17th symposium on boron, borides and related materials. Ankara: BKM; 2011. p 87.

[85] Zhao Y, Kim Y-H, Dillon AC, Heben MJ, Zhang SB. Hydrogen storage in novel organometallic buckyballs. Phys Rev Lett 2005;94:1–4.

[86] Zhao Y, Lusk MT, Dillon AC, Heben MJ, Zhang SB. Boron-based organometallic nanostructures: hydrogen storage properties and structure stability. Nano Lett 2008;8:157–161.

[87] Jeong Y, Chung TCM. The synthesis and characterization of a super-activated carbon containing substitutional boron (BC_x) and its applications in hydrogen storage. Carbon 2010;48:2526–2537.

[88] Beheshtian J, Kamfiroozi M, Bagheri Z, Ahmadi A. Theoretical study of hydrogen adsorption on the $B_{12}P_{12}$ fullerene-like nanocluster. Comput Mater Sci 2012;54:115–118.

[89] Meng S, Kaxiras E, Zhang Z. Metal–diboride nanotubes as high-capacity hydrogen storage media. Nano Lett 2007;7:663–667.

PART XIII

FUEL CELLS ON NANOMATERIALS BASIS

28

FUEL CELLS WITH NANOMATERIALS FOR ECOLOGICALLY PURE TRANSPORT

GENNADY GERASIMOV

Institute of Mechanics, Moscow State University, Moscow, Russia

28.1 INTRODUCTION

Among the key technologies for the transition from environmental-unfriendly fossil fuel use to the hydrogen-based economy, fuel cells are promising devices for direct conversion of chemical energy into electricity [1]. Fuel cell vehicles are developed to replace conventional internal combustion engines. The aim is to produce lower pollutant emissions, reduce fuel consumption, and allow long-range transportation, contrary to electric vehicles with batteries [2]. At present, fuel cell technology is shifting very fast from fundamental research to real design [3]. The Californian market is the leader in the development of passenger vehicle strategy today. Using the automaker survey results, the California Fuel Cell Partnership described a way to build retail hydrogen stations to meet the demand for hydrogen in 2014 and to prepare for the 50,000 vehicles coming by 2017 [4].

Nanotechnology offers different possibilities to increase the energy conversion efficiency of fuel cells within catalysts, membranes, and hydrogen storage. The most frequently used catalyst for the electrochemical oxidation of hydrogen in fuel cells is platinum or its alloys [5]. Usually, a platinum layer is deposed on the surface of commercial carbon. With the aim to reduce the platinum quantity and to increase its efficiency, the possibility of using Pt nanoclasters and thin layers on fuel cell electrodes has been under study since years [6]. On the other hand, the application of carbon nanostructures, particularly the latest, that is, carbon nanotubes (CNTs), as a support for highly dispersed Pt nanoparticles leads to higher activity of the oxygen reduction reaction and better performance of the fuel cell, as compared to a catalyst supported on commercial carbon [7, 8].

A large portion of the current work in the fuel cell technology is devoted to proton exchange fuel cells (PEFC), as they are the most suitable fuel cells for vehicular applications [9]. The most highly developed membranes for PEFCs are polymer membranes [10]. Therefore, these fuel cells are sometimes referred to as polymer electrolyte fuel cells. Some recent investigations show the advantage of track-etched membrane applications in fuel cells [11]. Track-etched membranes may be used as a host matrix for the preparation of polymer–polymer nanocomposite membranes with polyelectrolyte nanodomains oriented normal to the plane of the membrane [12]. The membranes synthesized in this way demonstrate the ability to enhance transport in the desired direction. Nanomaterials also possess the potential to be lightweight and highly efficient storage media for hydrogen as applied to fuel cell vehicles [13]. Among them, carbon nanomaterials (nanotubes, fullerenes, and graphene-like materials) attract considerable attention as reversible hydrogen storage media under ambient conditions [14].

This chapter gives a brief review of the recent progress in the application of nanomaterials for the improvement of fuel cells that will facilitate the transition of fuel cell vehicles to the commercial market. The review includes the examination of available experimental data as well as main technical aspects of fuel cell technology. Since the number of publications on this subject is too large, the review quotes the most typical works on each of the considered items.

Nanomaterials for Environmental Protection, First Edition. Edited by Boris I. Kharisov, Oxana V. Kharissova, and H. V. Rasika Dias.

28.2 FUEL CELL VEHICLES

PEFC is the device that generates a current in a set of electrochemical reactions. The working principle of this device is shown in Figure 28.1. On the anode side of the fuel cell, molecular hydrogen diffuses to the anode catalyst, where it dissociates into protons and electrons. The protons are conducted through the electrolyte membrane to the cathode whereas the electrons flow through the external electrical circuit because the membrane is electrically insulated. On the cathode side of the fuel cell, oxygen molecules react with the electrons and protons on the cathode catalyst, with the formation of water molecules. The corresponding electrochemical reactions within the fuel cell can be written as follows:

Anode side: $H_2 \rightarrow 2H^+ + 2e^-$
Cathode side: $2H^+ + 2e^- + 1/2O_2 \rightarrow H_2O$
Overall reaction: $H_2 + 1/2O_2 \rightarrow H_2O$

Proton exchange membranes are the key components in the fuel cell system. Researchers strive to reach the proton exchange membrane with high proton conductivity, low electronic conductivity and low permeability to fuel, low electroosmotic drag coefficient, good chemical and thermal stability, good mechanical properties, and low cost [15]. Attempts to understand the complex and interrelated transport and electrochemical phenomena inside the membrane have activated the appearance of numerous theoretical and experimental investigations. Numerical simulations have been used to analyze and compare the effects of fuel cell orientation, operating conditions, and geometry parameters on the cell performance [16–18].

A subcategory of PEFCs is direct-methanol fuel cells (DMFCs), in which methanol is used as the fuel. These devices can be recognized as an appropriate solution for mobile power sources since they operate at low temperatures with a high-energy density of methanol that is stable liquid at all environmental conditions [19]. The reactions at the electrodes and overall cell reaction in this case are as follows:

Anode side: $CH_3OH + H_2O \rightarrow CO_2 + 6H^+ + 6e^-$
Cathode side: $3/2O_2 + 6H^+ + 6e^- + 1/2O_2 \rightarrow 3H_2O$
Overall reaction: $CH_3OH + 3/2O_2 \rightarrow CO_2 + 2H_2O$

Methanol on the anode side of the fuel cell is usually found in a weak solution, because in high concentrations it has the tendency to diffuse through the polymer membrane and react with molecular oxygen from the cathode side (the so-called crossover effect). A comprehensive review of the state-of-the-art studies of mass transport phenomena in DMFCs is presented by Zhao et al. [20].

Any hydrogen-fueled PEFC system has four major subsystems, namely, the fuel cell stack, air supply, hydrogen supply, and thermal and water management system. Figure 28.2 shows the schematic diagram of such a system for vehicle applications [21].

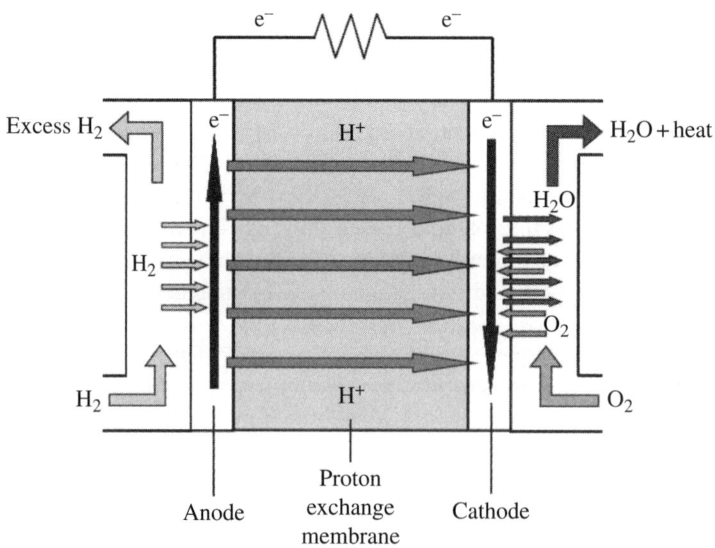

FIGURE 28.1 Working principle of a fuel cell. Reprinted with permission from Ref. [10]. © Elsevier.

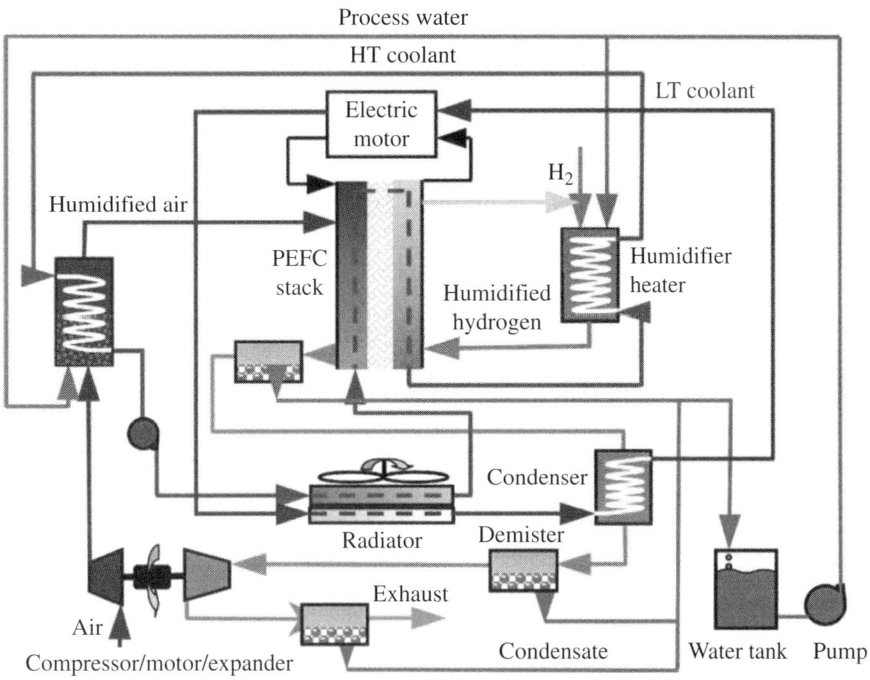

FIGURE 28.2 Schematic diagram of a hydrogen-fueled, polymer electrolyte fuel cell system for automotive applications. Reprinted with permission from Ref. [21]. © Elsevier.

The fuel cell stack in Figure 28.2 is an assembly of elementary cells (anode, polymer membrane, cathode) interconnected by bipolar plates in a planar technology configuration. The thermal and water management system consists of two coolant circuits and a process water circuit. The high-temperature circuit delivers the coolant (mixture of ethylene glycol and water) to the stack and rejects the stack waste heat in an air-cooled radiator. The high-temperature coolant also preheats the cathode and anode streams and provides the latent heat of humidification. The low-temperature circuit delivers the coolant to the shell-and-tube condenser and also cools the electric motor. The condenser of the vehicle's air-conditioning system can be cooled by means of the low-temperature coolant as well. The process water circuit uses deionized water to humidify the anode and cathode streams.

A general overview of the current research on fuel cells for vehicular applications is given by Veziroglu and Macario [22]. It includes technical aspects of fuel cell vehicle development, environmental impact, economic analysis, and comparison of different hydrogen vehicle technologies. The technical aspects are connected with such subjects as degradation and durability of the proton exchange membrane [23], potential hydrogen production methods [24, 25], and onboard hydrogen storage [26]. The subject of degradation and durability has attracted attention over the last years [27]. Various factors affect the fuel cell performance, inducing irreversible changes in the kinetic and transport properties of the cell. The main factors are fuel and oxidant impurities [28], temperature and relative humidity conditions [29], cyclic current-loading conditions [30], as well as problems of start-up and shutdown [31], and fuel and air starvation [32]. A comprehensive review of the current progress in the study of fuel cell degradation was performed by Borup et al. [33].

The problem of hydrogen production, storage, distribution, and dispensing is critical in the development of hydrogen fuel cell vehicles. Hydrogen can be produced on-site at the filling station or in larger production plants, from where it is distributed to the filling stations in pipelines or by trucks. In the latter case, hydrogen can be carried in either compressed or liquid form [34]. The main methods of hydrogen industrial production now are steam methane reforming (close to 50% of the global demand for hydrogen), oil or naphtha reforming from refinery or chemical industrial off-gases (about 30%), coal and biomass gasification (about 18%), water electrolysis (3.9%), and others (0.1%) [35]. Hydrogen onboard storage technologies are an important factor in the overall performance of hydrogen fuel vehicles [36]. Storage options include metal hydrides [37], carbon nanostructures [38], compressed gas [39], and liquid organic carriers [40].

The potential change in emissions and energy use by the replacement of fossil-fuel on-road vehicles with hydrogen fuel cell vehicles was examined by Paster et al. [41]. As estimations show, the net quantity of most types of emissions associated with air pollution would decrease considerably, including nitrogen oxides, volatile organic compounds, particulate matter, ammonia, and carbon monoxide. On the other hand, as noted by Huo et al. [42], pollutant emissions from vehicles must be separated into total and urban emissions to differentiate the locations of emissions, because the main emissions are related to fuel production, which is typically located in rural areas. In this connection, coal-based hydrogen would likely lead to increased emissions compared to conventional vehicles using petroleum-based fuels [43].

28.3 NANOMATERIALS AS CATALYSTS

One of the most important challenges for the ultimate commercialization of fuel cells is the preparation of active, robust, and low-cost catalysts [44]. The catalyst most frequently used for the electrochemical oxidation of hydrogen and methanol in the PEFC system consists of carbon-supported Pt or multimetallic nanoparticles (Pt, Co, Ni, V, Fe, Cu, Pd, W, Ag, Au, etc.) [5]. Figure 28.3 illustrates possible processes of surface adsorption and reactivity for the methanol oxidation reaction in the case of a gold–platinum alloy catalyst in DMFCs [45, 46]. The catalytic reaction of Pt in the alloy occurs via a combination of reaction steps, such as the adsorption of CH_3OH on Pt followed by dehydrogenation, the formation of intermediate CO_{ad}/Pt, the transfer of CO_{ad}/Pt to neighboring Au-atop sites forming CO_{ad}/Au, the formation of OH_{ad}/Au or surface oxides on gold, and the reactions of Pt–CO_{ad} + Au–OH_{ad} and Au–CO_{ad} + Au–OH_{ad} toward the final product (CO^{-3}) [45].

Several techniques are used to depose uniform catalyst spherical nanoparticles on the high-surface carbon support. These techniques can be subdivided into chemical, electrochemical, and physical ones. Chemical methods use colloidal, impregnation-reduction, chemical vapor deposition, and reverse micelles processes. A simple, environment-friendly colloidal-precipitation method was successfully employed by Yao et al. [47] to prepare carbon-supported platinum nanoparticles with a narrow size distribution. In this method, $(NH_4)_2WO_4$ was used to react with H_2PtCl_6, which forms $(NH_4)_2PtCl_6$ and H_2WO_4 simultaneously. The precipitation of $(NH_4)_2PtCl_6$ and the colloidal of H_2WO_4 can protect the formation of the Pt nanoparticles. In the impregnation method, the metal precursors impregnate the carbon support in aqueous and/or organic media followed by reduction using a reducing agent such as organic alcohols [48]. Fuel cell electrode catalysts with improved electrochemical properties were prepared by Kim and Moon [49] by dispersing Pt nanoparticles onto CNTs using a chemical vapor deposition method. Pt particles synthesized by this method have a relatively uniform size of approximately 1 nm, which is substantially smaller than in the case of a commercial Pt/carbon black catalyst (≤4.5 nm) prepared by wet impregnation. A reverse micelle synthesis was used by Cheney et al. [50] to improve the nanoparticle size uniformity of bimetallic Pt/Ni nanoparticles supported on γ-Al_2O_3.

The electrochemical method for Pt nanoparticle deposition on porous and high-surface carbon substrates such as carbon black and CNTs is an alternative way to prepare gas diffusion electrodes for PEFCs that has the ability to localize the metal particles on the surface of the electrode and reduce the thickness of the catalysis layer [51]. The deposition procedure can be carried out by using either a potentiostatic or a galvanostatic technique, which involves direct and pulse techniques. Electrodeposition via the galvanostatic pulse technique is considered convenient to improve the current distribution. Therefore, it is easy to control the particle size and composition of the alloy simply by varying experimental parameters such as on/off time and peak current density [52]. Figure 28.4 illustrates the schematic diagram of the pulse electrodeposition method.

The physical sputter deposition technique allows obtaining ultralow levels of catalyst loading (on the order of 5 μgPt cm^{-2} contrary to a commercial catalyst having a Pt loading of 0.3 mgPt cm^{-2}), localization of the catalyst on the uppermost surface of the carbon support, and increasing the activity for methanol oxidation compared to commercial catalysts [53]. Advantages of the sputter deposition compared with chemical deposition are easy preparation of the catalyst (absence of reducing and deflocculating agents, no heat treatment in hydrogen) and low Pt loading, allowing a cost reduction in the product together with easy industrial transfer of the process. It should be noted that the sputter deposition method is already a commercial technique for thin film deposition for other industrial applications [54].

Recently, numerous efforts were made to replace Pt-based catalysts in PEFCs due to the high cost of the metal [5]. The progress in nanotechnology and material science has raised high expectations in the design and synthesis of alternative new

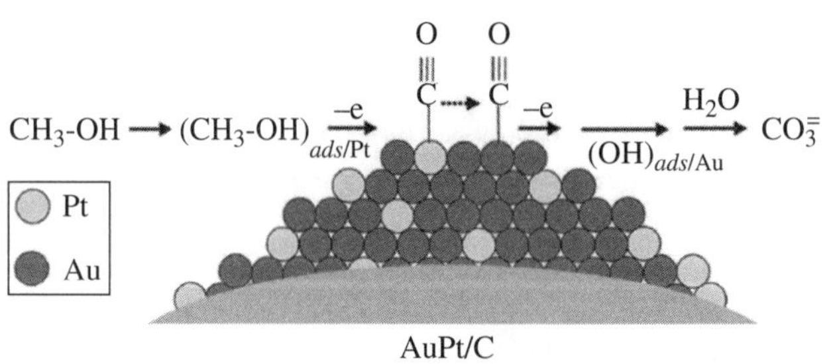

FIGURE 28.3 Schematic illustration of the electrocatalytic oxidation of methanol on an AuPt/C catalyst in alkaline electrolyte. Reprinted with permission from Ref. [45]. © Elsevier.

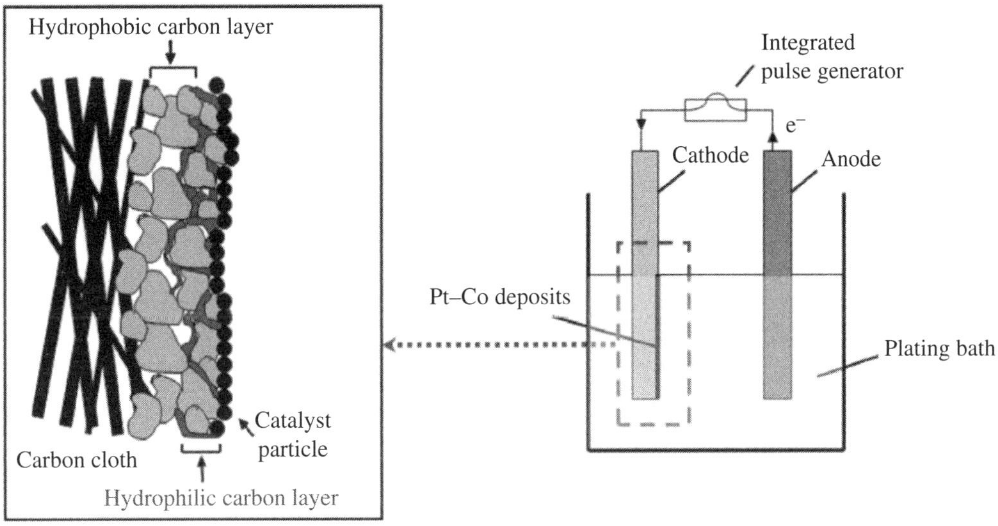

FIGURE 28.4 Schematic diagram of pulse electrodeposition method. Reprinted with permission from Ref. [52]. © Elsevier.

nonprecious-metal catalysts with high activity and practical durability for the methanol oxidation and oxygen reduction reactions in PEFCs. Attention was particularly given to the oxygen reduction reaction since this appears to be responsible for major voltage losses within the cell. For this reaction, some promising Pt-free catalsysts were studied intensively: nonprecious-metal chalcogenides [55], transition metal carbides [56], nitrides [57], and organometallic components [58]. The review by Othman and Dicks [59] outlines classes of nonprecious-metal systems that have been investigated over the past 10 years.

Although CNTs are used as supports for a nanoparticle metal catalyst, it was recently demonstrated that surface-modified CNTs and graphene show electrocatalytic activity for the oxygen reduction reaction in alkaline and acidic media [5]. Particularly, nitrogen-doped graphene that was synthesized by Qu et al. [60] by chemical vapor deposition of methane in the presence of ammonia acts as a metal-free electrode with a much better electrocatalytic activity, long-term operation stability, and tolerance to crossover effect than platinum catalysts. As demonstrated by Wang et al. [61], some polyelectrolytes have a strong electron-accepting ability to withdraw electrons from carbon atoms in the CNT plane to induce the net positive charge, facilitating the catalytic activity of the nitrogen-free CNTs.

28.4 NANOMATERIALS AS CATALYST SUPPORT

The performance and durability of fuel cells seriously depend on catalyst support materials [5]. The nature of the support is most important, since it determines the dispersion and stability of the metal crystallites, the electronic properties of the metal, and mass transfer and ohmic resistances of the catalyst layer [62]. Current catalysts for PEFCs are Pt or Pt alloys supported on carbon black in the form of highly dispersed nanoparticles. Recent studies, however, show that Pt-based nanoparticles supported on other forms of carbon, such as carbon nanofibers, CNTs, and graphene, display higher electrocatalytic activities [63]. Although the underlying mechanisms for activity enhancement are still not well understood, it was suspected that the surface structures and electronic properties of CNTs may well be responsible. Figure 28.5 shows the transmission electron microscopy (TEM) images of highly dispersed Pt nanoparticles with very small and uniform size distribution (2–3 nm) that were obtained on carbon nanofibers and CNT supports [64].

CNTs have high specific surface area, corrosion resistance, good electronic conductivity, and high stability [65]. The main methods for the preparation of Pt-based catalysts supported on CNTs involve the reduction of Pt(II) salts by microemulsion, chemical vapor deposition, and electrochemical deposition, among others. A series of green techniques for synthesizing CNT-supported platinum nanoparticles for methanol fuel cell applications have been reported by Shimizu et al. [66]. These techniques utilize either the supercritical fluid carbon dioxide or water as a medium for depositing platinum nanoparticles on surfaces of multiwalled or single-walled CNTs. An improved catalyst support for DMFCs on the base of polyphosphazene-coated CNTs was prepared by Qian et al. [67].

Graphene has a unique advantage as a catalyst support material compared to other carbon nanomaterials such as carbon black and CNTs owing to much larger surface-to-mass ratio. An overview of graphene nanosheets used as supports for fuel cell catalysts is presented by Antolini [68]. A systematic and comparative study of the effects of structure, composition, and carbon support properties on the electrocatalytic activity and stability of graphene-supported Pt–Ni alloy nanocatalysts for methanol

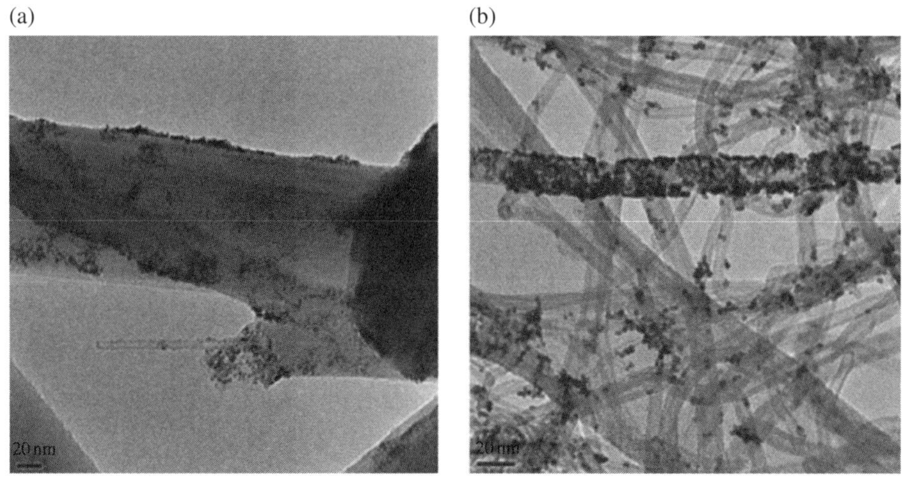

FIGURE 28.5 TEM images of Pt nanoparticles deposited on (a) carbon nanofibres and (b) carbon nanotubes. Reprinted with permission from Ref. [64]. © Elsevier.

oxidation was performed by Hu et al. [69]. Uniform and porous graphene nanoflake structures were prepared by Shang et al. [7] as an excellent support for Pt-based fuel cells. Pt nanoclusters with different thicknesses are deposited on the graphene surface by the combination of a simple chemical vapor deposition and magnetron sputtering techniques. The hybrid films are found to exhibit distinctly superior electrocatalytic activities toward methanol oxidation with an ultralow metal loading. Pt nanoparticles supported on nitrogen-doped reduced graphene oxide were investigated by He et al. [70]. As the measurements show, the catalytic activity of this nanomaterial in the oxygen reduction reaction is practically unchanged, indicating its excellent stability.

Nanostructured thin film (NSTF) catalysts are the leading electrode technology at present for polymer electrolyte fuel cells because of their significant advantages in durability and specific activity. This type of catalyst consists of a single layer of crystalline organic whiskers of about 1 μm in length, about 50 nm in diameter, and with a number density of $5 \times 10^9 \, cm^{-2}$ [71]. These whiskers are obtained by temperature annealing of a vacuum-deposited organic red pigment. The deposition of Pt or Pt alloy thin films on these support structures are performed by the physical vapor deposition or sputter deposition techniques. These crystalline whiskers have high thermal, chemical, and electrochemical stability under fuel cell operation conditions [72]. The roughness of the catalyst coating on the whiskers can vary from very smooth to very rough, and it depends on the deposition conditions and the metal elements.

28.5 NANOSTRUCTURED MEMBRANES

Nanostructured materials are also used in the development of cheaper and more efficient membranes for fuel cells. Most commercial membranes are based on the state-of-the-art hydrated perfluorosulfonic acid material under the commercial name of Dupont Nafion [73], which combines high ionic conductivity and low gas permeability with good chemical, mechanical, and thermal strength. However, these fluorine-containing membranes are expensive and don't operate efficiently at temperatures below 0°C or above 80°C due to the evaporation of water from the membrane structure. For practical applications, the operation of fuel cells at temperatures above 100°C is desired, both for hydrogen and methanol-fuelled cells, which requires the development of new classes of membranes. An overview of the research and development of nanostructured membranes for different fuel cell applications is given by Thiam et al. [74].

The first approach to improve the membrane properties is the use of nanocomposite hybrids. The addition of nanocrystalline ceramic oxides to Nafion [75] enhances the membrane characteristics, allowing its operation up to 130°C, when the cell is fuelled with hydrogen, and up to 145°C in the DMFCs. A new ceramic-based reinforcing porous substrate for fuel cell applications was developed by Seol et al. [76]. It consists of hygroscopic silica (SiO_2) nanoparticles interconnected by 3-glycidoxypropyltrimethoxysilane (GPTMS)-based silicate binders and a poly(paraphenylene terephthalamide) (PPTA) nonwoven support. This unusual ceramic substrate is featured with strong mechanical strength, well-developed nanoporous structure, high hydrophilicity, and good water retention capability.

Carbon nanostructures are widely studied at present for the purpose of their use in membrane material preparation. The effect of sulfonic acid–functionalized graphene oxide nanosheets was investigated by Zarrin et al. [77] as inorganic fillers in a

Nafion composite membrane, fabricated by a simplistic solution casting method. This novel nanocomposite material demonstrates a significant proton conductivity improvement (four times) over the unmodified one and can be considered as a potential proton exchange membrane replacement for applications having low humidity of 30% and high temperature of 120°C. The nanocomposite prepared by the covalent functionalization of multiwalled CNTs by grafting sulfanilic acid and their dispersion into sulfonated poly(ether ether ketone) was investigated by Tripathi et al. [78] as a novel membrane material for applications in alcohol or hydrogen fuel cells.

Some recent investigations show advantages of track-etched membranes in fuel cell applications such as no need to wet the membrane, larger proton conductivity, resistance to extreme conditions, etc. [11]. The ion-track technology is based on the irradiation of thin films of various materials with accelerated heavy ions [79]. Heavy ions produce so-called latent tracks with high density of matter radiation damage along their path in the film with a diameter between 1 and 10 nm depending on the ion type and its kinetic energy. After the etching procedure, the latent tracks may be used for the production of track-etched membranes [80]. The grafting functionalization of the ion tracks in polymer thin films leads to the formation of highly proton-conductive electrolyte membranes for fuel cell applications [81]. Such ion track–grafted membranes have lower crossover of methanol, same proton conductibility, lower ion-exchange capacity, and superior mechanical properties compared to Nafion membranes, which make them promising materials for widespread application in direct methanol fuel cells [82]. Track-etched membranes may be also used as a host matrix for the preparation of polymer–polymer nanocomposite membranes with polyelectrolyte nanodomains oriented normal to the plane of the membrane [12]. The membranes synthesized in this way demonstrate the ability to enhance transport in the desired direction.

28.6 NANOMATERIALS FOR HYDROGEN STORAGE

Although hydrogen is widely recognized as a promising energy carrier for the transportation sector, widespread adoption of hydrogen and fuel cell technologies depends critically on the ability to store hydrogen at adequate densities. At present, no known material or storage means exists that satisfies all requirements to enable high-volume automotive application [83]. Hydrogen can be stored as pressurized gas, cryogenic liquid, or as a suitable solid-state material [84]. High-pressure gaseous hydrogen storage offers the simplest solution in terms of infrastructure requirement and has become the most popular and highly developed method [85]. Recent progress in cryogenic storage of liquid hydrogen has been reviewed by Ho and Rahman [86]. Solid-state storage is potentially the most convenient and safest method from a technological point of view, because this storage technology supposes the presence of near-ambient temperatures and pressures [87]. Moreover, storage of hydrogen in liquid or gaseous form requires a large container and poses important safety problems for onboard transport applications. An attempt to receive some insight into almost all classes of known hydrogen storage materials was made by Pukazhselvan et al. [88].

The hydrogen storage materials in solid state can be divided into several categories in terms of the strength of hydrogen bonding: chemisorption, physisorption, and quasi-molecular bonding, which is the intermediate one between chemisorption and physisorption [14]. In the case of chemisorption, the H_2 molecule dissociates into individual atoms, migrates into the material, and binds chemically with a binding energy lying in the 2–4 eV range. The bonding is strong, and desorption takes place at high temperatures. This category of hydrogen storage materials consists of simple metal and intermetallic hydrides with maximum emphasis on magnesium-based hydrides, which can be treated as promising candidates for competitive hydrogen storage with reversible hydrogen capacity up to 7.6 wt% for automotive applications. A recent review [89] reports current developments of metal hydrides in properties including hydrogen storage capacity, kinetics, cyclic behavior, toxicity, pressure, and thermal response.

A group of Mg-based nanostructured hydrides is under intensive investigation at present. The hydrogen sorption property of magnesium MgH_2 in the form of sandwiched Pd/Mg/Pd films of nanoparticles with sizes on the order of 50 nm was studied by Barcelo et al. [90]. A nanostructured MgH_2/TiH_2 composite was synthesized [91] directly from Mg and Ti metals by ball milling under an initial hydrogen pressure of 30 MPa. The desorption temperature of this composite is more than 100°C lower compared to commercial MgH_2. The improved properties are due to the catalyst and nanostructure introduced during high-pressure ball milling. The synthesis, hydrolysis, structure, dynamics, intermediate compound, and improvement of de/rehydrogenation properties of $LiBH_4$ as hydrogen storage material with high gravimetric hydrogen capacity of 18.4 wt% were investigated by Li et al. [92]. The hydrogen desorption properties of the nanostructured MgH_2–Ti alloy composite for hydrogen storage produced via combined vacuum arc remelting and mechanical alloying were studied and compared with pure magnesium hydride [93].

The physisorption of hydrogen in porous materials is a viable mechanism for hydrogen storage in automotive applications. This storage mechanism has the advantage of possessing fast kinetics, low heat of adsorption, and being completely reversible. The investigation of hydrogen adsorption capability of different nanostructured carbon materials is one of the ongoing strategic research areas in science and technology [94]. The challenges, distinguishing traits, and apparent contradictions of carbon-based

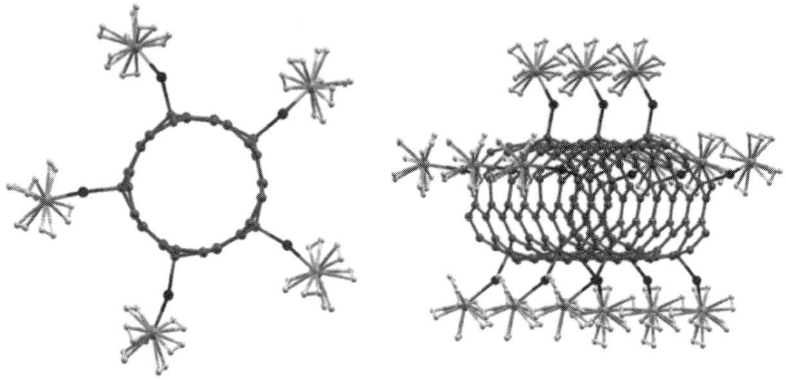

FIGURE 28.6 Atomic structures of Ti–hydroxyl complexes attached to CNTs. The left figure is the cross-sectional view and the right figure is the oblique view. Reprinted with permission from Ref. [101]. © Elsevier.

hydrogen storage technologies have been reported by Yürüm et al. [95]. Among all porous materials, metal-organic frameworks are the best candidates for H_2 adsorption, since they consist of light atoms, they are highly porous, and their pore dimensions can be tailored by chemical engineering [96]. The most recent achievements in developing novel microporous silicon-based structures, aluminosilicates, and related materials have been described by Azzouz [97].

The transformation from the physisorption state to the chemisorption state of H_2 molecules in nanostructured carbon materials as applied to a fullerene C_{20} and a B-doped fullerene $C_{19}B$ system was investigated by Tian et al. [98] using different density function methods. The convenient carbon-based adsorbent material could form the basis of technologically viable hydrogen storage systems is CNTs. Recent applications of CNTs in hydrogen production and storage were examined by Oriňáková and Oriňák [38]. The investigation of the structure and hydrogen storage behavior of Ca-decorated graphene shows that Ca dimers act as nucleation positions of hydrogen adsorption [99]. The hydrogen storage capacity via the spillover mechanism in Ca-adsorbed graphene depends on the Ca content and could approach 7.7 wt%. Theoretical analysis [100] shows that the magnesium-decorated boron fullerene B_{80} has a high hydrogen storage capacity storing up to 96 H_2 molecules with an ideal hydrogen uptake of 14.2%. This suggests a possible method of engineering new structures for high-capacity hydrogen storage materials with the reversible adsorption and desorption of hydrogen molecules. Various forms of storage materials, which are obtained by modifying well-known nanomaterials using Ti-functional group complexes, were investigated by Lee et al. [101]. As calculations show, Ti-decorated 2-mercaptoethyl sulfide, hydroxylfunctionalized polyethylene, and CNTs can store H_2 molecules with a gravimetric density of 8.8, 11.5, and 5.5 wt%, respectively. The atomic structure of Ti–hydroxyl complexes attached to CNTs is shown in Figure 28.6.

28.7 CONCLUSION

Nanotechnology occupies an important place in the production and investigation of new functional materials for applications in fuel cell vehicles. One of the most important challenges for ultimate commercialization of the fuel cell technology is the preparation of active, robust, and low-cost catalysts on the base of nanomaterials, such as Pt or multimetallic nanoparticles, nonprecious-metal chalcogenides, organometallic compounds, surface-modified CNTs, and graphene. The performance and durability of fuel cells seriously depend on catalyst support materials. Recent studies show that catalyst nanoparticles supported on nanostructured carbon materials (carbon nanofibers, nanotubes, and graphene) display higher electrocatalytic activity than those supported on traditional ones.

Nanostructured materials are also used in the development of cheaper and more efficient membranes for fuel cells (nanocomposite membranes with the addition of nanocrystalline ceramic oxides or functionalized graphene oxide nanosheets to the commercial Nafion material, nanocomposites prepared on the base of multiwalled CNTs, etc.). Some recent investigations show advantages of track-etched membranes in fuel cell applications such as no need to wet the membrane, larger proton conductivity, and resistance to extreme conditions.

Although hydrogen is widely recognized as a promising energy carrier for the transportation sector, widespread adoption of hydrogen and fuel cell technologies depends critically on the ability to store hydrogen at adequate densities. Solid-state storage is potentially the most convenient and safest method from a technological point of view, because this storage technology supposes the presence of near-ambient temperatures and pressures. This category of storage media includes a group of Mg-based nanostructured hydrides, novel microporous silicon-based structures, and different nanostructured carbon materials.

REFERENCES

[1] Colella WG, Jacobson MZ, Golden DM. Switching to a U.S. hydrogen fuel cell vehicle fleet: the resultant change in emissions, energy use and greenhouse gases. J Power Sources 2005;150:150–181.

[2] Bernay C, Marchand M, Cassir M. Prospects of different fuel cell technologies for vehicle application. J Power Sources 2002;108:139–152.

[3] Helmolt R, Eberle U. Fuel cell vehicles: status 2007. J Power Sources 2007;165:833–843.

[4] Dunwoody C. Transition to a commercial fuel cell vehicle market in California. Fuel Cell Bull 2009;11:12–14.

[5] Rabis A, Rodriguez P, Schmidt TJ. Electrocatalysis for polymer electrolyte fuel cells: recent achievements and future challenges. ACS Catal 2012;2:864–890.

[6] Lee W-D, Lim D-H, Chun H-J, Lee H-I. Preparation of Pt nanoparticles on carbon support using modified polyol reduction for low-temperature fuel cells. Int J Hydrogen Energy 2012;37:12629–12638.

[7] Shang N, Papakonstantinou P, Wang P, Silva SRP. Platinum integrated graphene for methanol fuel cells. J Phys Chem C 2010;114:15837–15841.

[8] Jeng K-T, Hsu N-Y, Chien C-C. Synthesis and evaluation of carbon nanotube-supported RuSe catalyst for direct methanolfuel cell cathode. Int J Hydrogen Energy 2011;36:3997–4006.

[9] Zhang H, Shen PK. Recent development of polymer electrolyte membranes for fuel cells. Chem Rev 2012;112:2780–2832.

[10] Dupuis A-C. Proton exchange membranes for fuel cells operated at medium temperatures materials and experimental techniques. Prog Mater Sci 2011;56:289–327.

[11] Hanot H, Ferain E. Industrial applications of ion track technology. Nucl Instrum Meth Phys Res B 2009;267:1019–1022.

[12] Chen H, Palmese GR, Elabd YA. Membranes with oriented polyelectrolyte nanodomains. Chem Mater 2006;18:4875–4881.

[13] Varin, R. A.; Czujko, T.; Wronski, Z. S. *Nanomaterials for Solid State Hydrogen Storage*; Springer, New York : 2009.

[14] Jena P. Materials for hydrogen storage: past, present, and future. J Phys Chem Lett 2011;2:206–211.

[15] Peighambardoust SJ, Rowshanzamir S, Amjadi M. Review of the proton exchange membranes for fuel cell applications. Int J Hydrogen Energy 2010;35:9349–9384.

[16] Zhang Y, Pitchumani R. Numerical studies on an air-breathing proton exchange membrane (PEM) fuel cell. Int J Heat Mass Transfer 2007;50:4698–4712.

[17] Beicha A. Modeling and simulation of proton exchange membrane fuel cell systems. J Power Sources 2012;205:335–339.

[18] Hwang J-J, Chen Y-J, Kuo J-K. The study on the power management system in a fuel cell hybrid vehicle. Int J Hydrogen Energy 2012;37:4476–4489.

[19] Park J-Y, Seo Y, Kang S, You D, Cho H, Na Y. Operational characteristics of the direct methanol fuel cell stack on fuel and energy efficiency with performance and stability. Int J Hydrogen Energy 2012;37:5946–5957.

[20] Zhao TS, Xu C, Chen R, Wang WW. Mass transport phenomena in direct methanol fuel cells. Prog Energy Combust Sci 2009;35:275–292.

[21] Ahluwalia RK, Wang X. Direct hydrogen fuel cell systems for hybrid vehicles. J Power Sources 2005;139:152–164.

[22] Veziroglu A, Macario R. Fuel cell vehicles: State of the art with economic and environmental concerns. Int J Hydrogen Energy 2011;36:25–43.

[23] Silva RA, Hashimoto T, Thompson GE, Rangel CM. Characterization of MEA degradation for an open air cathode PEM fuel cell. Int J Hydrogen Energy 2012;37:7299–7308.

[24] Cetinkaya E, Dincer I, Naterer GF. Life cycle assessment of various hydrogen production methods. Int J Hydrogen Energy 2012;37:2071–2080.

[25] Pilavachi PA, Chatzipanagi AI, Spyropoulou AI. Evaluation of hydrogen production methods using the analytic hierarchy process. Int J Hydrogen Energy 2009;34:5294–5303.

[26] De Wit MP, Faaij APC. Impact of hydrogen onboard storage technologies on the performance of hydrogen fuelled vehicles: A techno-economic well-to-wheel assessment. Int J Hydrogen Energy 2007;32:4859–4870.

[27] Yuan X-Z, Li H, Zhang S, Martin J, Wang H. A review of polymer electrolyte membrane fuel cell durability test protocols. J Power Sources 2011;196:9107–9116.

[28] Zamel N, Li X. Effect of contaminants on polymer electrolyte membrane fuel cells. Prog Energy Combust Sci 2011;37:292–329.

[29] Choi K-S, Kim H-M, Yoon HC, Forrest ME, Erickson PA. Effects of ambient temperature and relative humidity on the performance of Nexa fuel cell. Energy Convers Manage 2008;49:3505–3511.

[30] Liu D, Case S. Durability study of proton exchange membrane fuel cells under dynamic testing conditions with cyclic current profile. J Power Sources 2006;162:521–531.

[31] Shen Q, Hou M, Liang D, Zhou Z, Li X, Shao Z, Yi B. Study on the processes of start-up and shutdown in proton exchange membrane fuel cells. J Power Sources 2009;189:1114–1119.

[32] Jang I, Hwang I, Tak Y. Attenuated degradation of a PEMFC cathode during fuel starvation by using carbon-supported IrO_2. Electrochim Acta 2013;90:148–156.

[33] Borup R, Meyers J, Pivovar B, Kim YS, Mukundan R, Garland N, Myers D, Wilson M, Garzon F, Wood D, Zelenay P, More K, Stroh K, Zawodzinski T, Boncella J, McGrath JE, Inaba M, Miyatake K, Hori M, Ota K, Ogumi Z, Miyata S, Nishikata A, Siroma Z. Scientific aspects of polymer electrolyte fuel cell durability and degradation. Chem Rev 2007;107:3904–3951.

[34] Markert F, Nielsen SK, Paulsen JL, Andersen V. Safety aspects of future infrastructure scenarios with hydrogen refuelling stations. Int J Hydrogen Energy 2007;32:2227–2234.

[35] Muradov NZ, Veziroglu TN. From hydrocarbon to hydrogen-carbon to hydrogen economy. Int J Hydrogen Energy 2005;30:225–237.

[36] Ananthachar V, Duffy JJ. Efficiencies of hydrogen storage systems onboard fuel cell vehicles. Solar Energy 2005;78:687–694.

[37] Li C, Peng P, Zhou DW, Wan L. Research progress in $LiBH_4$ for hydrogen storage: a review. Int J Hydrogen Energy 2011;36:14512–14526.

[38] Oriňáková R, Oriňák A. Recent applications of carbon nanotubes in hydrogen production and storage. Fuel 2011;90:3123–3140.

[39] Galassi MC, Acosta-Iborra B, Baraldi D, Bonato C, Harskamp F, Frischauf N, Moretto P. Onboard compressed hydrogen storage: fast filling experiments and simulations. Energy Procedia 2012;29:192–200.

[40] Ahluwalia RK, Hua TQ, Peng JK. On-board and Off-board performance of hydrogen storage options for light-duty vehicles. Int J Hydrogen Energy 2012;37:2891–2910.

[41] Paster MD, Ahluwalia RK, Berry G, Elgowainy A, Lasher S, McKenney K, Gardiner M. Hydrogen storage technology options for fuel cell vehicles: well-to-wheel costs, energy efficiencies,and greenhouse gas emissions. Int J Hydrogen Energy 2011;36:14534–14551.

[42] Huo H, Wu Y, Wang M. Well-to-wheels assessment of criteria pollutantemissions from various vehicle/fuel systems. Atmos Environ 2009;43:1796–1804.

[43] Jaramillo P, Samaras C, Wakeley H, Meisterling K. Greenhouse gas implications ofusing coal for transportation: life cycle assessment of coal-to-liquids, plug-in hybrids,and hydrogen pathways. Energy Policy 2009;37:2689–2695.

[44] Zhong C-J, Luo J, Fang B, Wanjala BN, Njoki PN, Loukrakpam R, Yin J. Nanostructured catalysts in fuel cells. Nanotechnology 2010;21:062001.

[45] Mott D, Luo J, Njoki PN, Lin Y, Wang L, Zhong C-J. Synergistic activity of gold-platinum alloy nanoparticle catalysts. Catal Today 2007;122:378–385.

[46] Luo J, Njoki PN, Lin Y, Wang L, Zhong C-J. Activity-composition correlation of AuPt alloy nanoparticle catalysts in electrocatalytic reduction of oxygen. Electrochem Comm 2006;8:581–587.

[47] Yao S, Feng L, Zhao X, Liu C, Xing W. Pt/C catalysts with narrow size distribution prepared by colloidal-precipitation method for methanol electrooxidation. J Power Sourc 2012;217:280–286.

[48] Chen L, Guo M, Zhang H-F, Wang X-D. Characterization and electrocatalytic properties of PtRu/C catalysts prepared by impregnation-reduction method using Nd_2O_3 as dispersing reagent. Electrochim Acta 2006;52:1191–1198.

[49] Kim H, Moon SH. Chemical vapor deposition of highly dispersed Pt nanoparticles on multi-walled carbon nanotubes for use as fuel-cell electrodes. Carbon 2011;49:1491–1501.

[50] Cheney BA, Lauterbach JA, Chen JG. Reverse micelle synthesis and characterization of supported Pt/Ni bimetallic catalysts on gamma-Al2O3. Appl Catal Gen 2011;394:41–47.

[51] Paoletti C, Cemmi A, Giorgi L, Giorgi R, Pilloni L, Serra E, Pasquali M. Electro-deposition on carbon black and carbon nanotubes of Pt nanostructured catalysts for methanol oxidation. J Power Sourc 2008;183:84–91.

[52] Woo S, Kim I, Lee JK, Bong S, Lee J, Kim H. Preparation of cost-effective Pt-Co electrodes by pulse electrodeposition for PEMFC electrocatalysts. Electrochim Acta 2011;56:3036–3041.

[53] Gruber D, Ponath N, Müller J, Lindstaedt F. Sputter-deposited ultra-low catalyst loading for PEM fuel cells. J Power Sourc 2005;150:67–72.

[54] Alvisi M, Galtieri G, Giorgi L, Giorgi R, Serra E, Signore MA. Sputter deposition of Pt nanoclusters and thin films on PEM fuel cell electrodes. Surf Coat Technol 2005;200:1325–1329.

[55] Tritsaris GA, Nørskov JK, Rossmeisl J. Trends in oxygen reduction and methanol activation on transition metal chalcogenides. Electrochim Acta 2011;56:9783–9788.

[56] Stottlemyer AL, Kelly TG, Meng Q, Chen JG. Reactions of oxygen-containing molecules on transition metal carbides: surface science insight into potential applications in catalysis and electrocatalysis. Surf Sci Rep 2012;67:201–232.

[57] Di Noto V, Negro E, Polizzi S, Riello P, Atanassov P. Preparation, characterization and single-cell performance of a new class of Pd–carbon nitride electrocatalysts for oxygen reduction reaction in PEMFCs. Appl Catal Environ 2012;111–112:185–199.

[58] Zagal JH, Griveau S, Silva JF, Nyokong T, Bedioui F. Metallophthalocy-anine-based molecular materials as catalysts for electrochemical reac-tions. Coord Chem Rev 2010;254:2755–2791.

[59] Othman R, Dicks AL, Zhu Z. Non precious metal catalysts for the PEM fuel cell cathode. Int J Hydrogen Energy 2012;37:357–372.

[60] Qu L, Liu Y, Baek J-B, Dai L. Nitrogen-doped graphene as efficient metal-free electrocatalyst for oxygen reduction in fuel cells. ACS Nano 2010;4:1321–1326.

[61] Wang S, Yu D, Dai L. Polyelectrolyte functionalized carbon nanotubes as efficient metal-free electrocata-lysts for oxygen reduction. J Am Chem Soc 2011;133:5182–5185.

[62] Figueiredo JL, Pereira MFR, Serp P, Kalck P, Samant PV, Fernandes JB. Development of carbon nanotube and carbon xerogel supported catalysts for the electro-oxidation of methanol in fuel cells. Carbon 2006;44:2516–2522.

[63] Li L, Xing Y. Pt-Ru nanoparticles supported on carbon nanotubes as methanol fuel cell catalysts. J Phys Chem C 2007;111: 2803–2808.

[64] Andersen SM, Borghei M, Lund P, Elina Y-R, Pasanen A, Kauppinen E, Ruiz V, Kauranen P, Skou EM. Durability of Carbon Nanofiber (CNF) & Carbon Nanotube (CNT) as catalyst support for proton exchange membrane fuel cells. Solid State Ion 2013;231:94–101.

[65] Karousis N, Tagmatarchis N, Tasis D. Current progress on the chemical modification of carbon nanotubes. Chem Rev 2010;110:5366–5397.

[66] Shimizu K, Wang JS, Wai CM. Application of green chemistry techniques to prepare electrocatalysts for direct methanol fuel cells. J Phys Chem A 2010;114:3956–3961.

[67] Qian J, Wei W, Huang H, Tao Y, Chen K, Tang X. A study of different polyphosphazene-coated carbon nanotubes as a Pt–Co catalyst support for methanol oxidation fuel cell. J Power Sourc 2012;210:345–349.

[68] Antolini E. Graphene as a new carbon support for low-temperaturefuel cell catalysts. Appl Catal Environ 2012;123–124:52–68.

[69] Hu Y, Wu P, Yin Y, Zhang H, Cai C. Effects of structure, composition, and carbon support properties on the electrocatalytic activity of Pt-Ni-graphene nanocatalysts for the methanol oxidation. Appl Catal B: Environ 2012;111–112:208–217.

[70] He D, Jiang Y, Lv H, Pan M, Mu S. Nitrogen-doped reduced graphene oxide supports for noble metal catalysts with greatly enhanced activity and stability. Appl Catal Environ 2013;132–133:379–388.

[71] Debe MK, Poirier RJ. Postdeposition growth of a uniquely nanostructured organic film by vacuum annealing. J Vac Sci Technol 1994;A12:2017–2022.

[72] Gancs L, Kobayashi T, Debe MK, Atanasoski R, Wieckowski A. Crystallographic characteristics of nanostructured thin film fuel cellelectrocatalysts—a HRTEM study. Chem Mater 2008;20:2444–2454.

[73] Mauritz KA, Moore RB. State of understanding of Nafion. Chem Rev 2004;104:4535–4586.

[74] Thiam HS, Daud WRW, Kamarudin SK, Mohammad AB, Kadhum AAH, Loh KS, Majlan EH. Overview on nanostructured membrane in fuel cell applications. Int J Hydrogen Energy 2011;36:3187–3205.

[75] Licoccia S, Traversa E. Increasing the operation temperature of polymer electrolyte membranes for fuel cells: from nanocomposites to hybrids. J Power Sources 2006;159:12–20.

[76] Seol J-H, Won J-H, Yoon K-S, Hong YT, Lee S-Y. SiO_2 ceramic nanoporous substrate-reinforced sulfonated poly(arylene ether sulfone) composite membranes for proton exchange membrane fuel cells. Int J Hydrogen Energy 2012;37:6189–6198.

[77] Zarrin H, Higgins D, Jun Y, Chen Z, Fowler M. Functionalized graphene oxide nanocomposite membrane for low humidity and high temperature proton exchange membrane fuel cells. J Phys Chem C 2011;115:20774–20781.

[78] Tripathi BP, Schieda M, Shahi VK, Nunes SP. Nanostructured membranes and electrodes with sulfonic acid functionalized carbon nanotubes. J Power Sources 2011;196:911–919.

[79] Waheed A, Forsyth D, Watts A, Saad AF, Mitchell GR, Farmer M, Harris PJF. The track nanotechnology. Radiat Meas 2009;44:1109–1113.

[80] Chakarvarti SK. Track-etch membranes enabled nano-/microtechnology: a review. Radiat Meas 2009;44:1085–1092.

[81] Clochard M-C, Berthelot T, Baudin C, Betz N, Balanzat E, Gébel G, Morind A. J Power Sourc 2010;195:223–231.

[82] Yoshida M, Kimura Y, Chen J, Asano M, Maekawa Y. Preparation of PTFE-based fuel cell membranes by combining latent track formation technology with graft polymerization. Radiat Phys Chem 2009;78:1060–1066.

[83] McWhorter S, Read C, Ordaz G, Stetson N. Materials-based hydrogenstorage: attributes for near-term, early market PEM fuel cells. Curr Opin Solid State Mater Sci 2011;15:29–38.

[84] Mao SS, Shen S, Guo L. Nanomaterials for renewable hydrogen production, storage and utilization. Progr Nat Sci Mater Int 2012;22:522–534.

[85] Zheng J, Liu X, Xu P, Liu P, Zhao Y, Yang J. Development of high pressure gaseous hydrogen storage technologies. Int J Hydrogen Energy 2012;37:1048–1057.

[86] Ho SH, Rahman MM. Forced convective mixing in a zero boil-off cryogenic storage tank. Int J Hydrogen Energy 2012;37:10196–10209.

[87] Conte M, Prosini PP, Passerini S. Overview of energy/hydrogen storage: state-of-the-art of the technologies and prospects for nanomaterials. Mater Sci Eng B 2004;108:2–8.

[88] Pukazhselvan D, Kumar V, Singh SK. High capacity hydrogen storage: basic aspects, new developments and milestones. Nano Energy 2012;1:566–589.

[89] Sakintuna B, Lamari-Darkrim F, Hirscher M. Metal hydride materials for solidhydrogen storage: a review. Int J Hydrogen Energy 2007;32:1121–1140.

[90] Barcelo S, Rogers M, Grigoropoulos CP, Mao SS. Hydrogen storage property of sandwiched magnesium hydride nanoparticle thin film. Int J Hydrogen Energy 2010;35:7232–7235.

[91] Shao H, Felderhoff M, Schüth F. Hydrogen storage properties of nanostructured MgH_2/TiH_2 composite prepared by ballmilling under high hydrogen pressure. Int J Hydrogen Energy 2011;36:10828–10833.

[92] Li C, Peng P, Zhou DW, Wan L. Research progress in $LiBH_4$ for hydrogen storage: a review. Int J Hydrogen Energy 2011;36:14512–14526.

[93] Mahmoudi N, Kaflou A, Simchi A. Synthesis of a nanostructured MgH_2–Ti alloy composite for hydrogen storage via combined vacuum arc remelting and mechanical alloying. Mater Lett 2011;65:1120–1122.

[94] Armandi M, Bonelli B, Cho K, Ryoo R, Garrone E. Study of hydrogen physisorption on nanoporous carbon materials of different origin. Int J Hydrogen Energy 2011;36:7937–7943.

[95] Yürüm Y, Taralp A, Veziroglu TN. Storage of hydrogen in nanostructured carbon materials. Int J Hydrogen Energy 2009;34:3784–3798.

[96] Hirscher M, Panella B, Schmitz B. Metal-organic frameworks for hydrogen storage. Micropor Mesopor Mater 2010;129:335–339.

[97] Azzouz A. Achievement in hydrogen storage on adsorbents with high surface-to-bulk ratio – Prospects for Si-containing matrices. Int J Hydrogen Energy 2012;37:5032–5049.

[98] Tian C, Wang Z, Jin M, Zhao W, Meng Y. Transformation mechanism of a H2 molecule from physisorption to chemisorption in pristine and B-doped C20 fullerenes. Chem Phys Lett 2011;511:393–398.

[99] Gao Y, Zhao N, Li J, Liu E, He C, Shi C. Hydrogen spillover storage on Ca-decorated grapheme. Int J Hydrogen Energy 2012;37:11835–11841.

[100] Li JL, Hu ZS, Yang GW. High-capacity hydrogen storage of magnesium-decorated boron fullerene. Chem Phys 2012;392:16–20.

[101] Lee H, Nguyen MC, Ihm J. Titanium-functional group complexes for high-capacity hydrogen storage materials. Solid State Comm 2008;146:431–434.

PART XIV

REMEDIATION OF RADIONUCLIDES

29

HUMIC FUNCTIONAL DERIVATIVES AND NANOCOATINGS FOR REMEDIATION OF ACTINIDE-CONTAMINATED ENVIRONMENTS

Irina V. Perminova[1], Stepan N. Kalmykov[1], Natalia S. Shcherbina[1,2], Sergey A. Ponomarenko[1,3], Vladimir A. Kholodov[1,4], Alexander P. Novikov[5], Richard G. Haire[6], and Kirk Hatfield[7]

[1] *Chemistry Department, Lomonosov Moscow State University, Moscow, Russia*

[2] *Paul Scherrer Institute (PSI), Villigen, Switzerland*

[3] *Enikolopov Institute of Synthetic Polymeric Materials of RAS, Moscow, Russia*

[4] *Dokuchaev Soil Science Institute of RAAS, Moscow, Russia*

[5] *Vernadsky Institute of Geochemistry and Analytical Chemistry of RAS, Moscow, Russia*

[6] *Oak Ridge National Laboratory, Oak Ridge, TN, USA*

[7] *Engineering School of Sustainable Infrastructure & Environment, University of Florida, Gainesville, FL, USA*

29.1 INTRODUCTION

Actinides in higher-valence states (e.g., V and VI) represent one of the most difficult classes of contaminants due to both their high radiotoxicity and mobility. The interaction of actinides with other materials is an important topic affecting migration, release, and natural and designed controls within the environment. Actinides interact strongly with humic substances (HS), many of which occur naturally in both mobile and immobile forms [1–8]. HS can enhance actinide solubility through complexation and by forming stable colloids. If HS exist as coatings on mineral surfaces, they can aid actinide retention on these surfaces. Both methods of control of actinide migration in the environment are shown in Figure 29.1.

HS are natural hyperbranched polyelectrolytes with a vast functional periphery dominated by carboxyl and hydroxyl groups. HS account for 50–80% of the organic carbon in soils, natural waters, and bottom sediments [9]. They are characterized with substantial molecular heterogeneity and disordered structures, which contribute to the longevity of HS spanning from hundreds to thousands of years. The recalcitrant nature of HS is of particular relevance for soil/aquifer remediation technologies predicated on a reactive matrix that is not consumed by microorganisms during remediation. On the other hand, the structural complexity inherent in HS (illustrated by Fig. 29.2) creates opportunities for a broad range of chemical interactions.

These HS can be oxidized by strong oxidants; act as reducing agents; take part in protolytic, ion exchange, and complexation reactions; participate in donor–acceptor interactions; engage in hydrogen bonding; and take part in van-der-Waals interactions [10]. Hence, the HS can interact practically with all chemicals released to the environment. This unique constellation of reactive features strongly suggests that HS have the potential to address a broad spectrum of needs within the focal area of environmental remediation [11], which is confirmed by multiple examples of actual remedial applications [12–14]. However, despite multiple

Nanomaterials for Environmental Protection, First Edition. Edited by Boris I. Kharisov, Oxana V. Kharissova, and H. V. Rasika Dias.

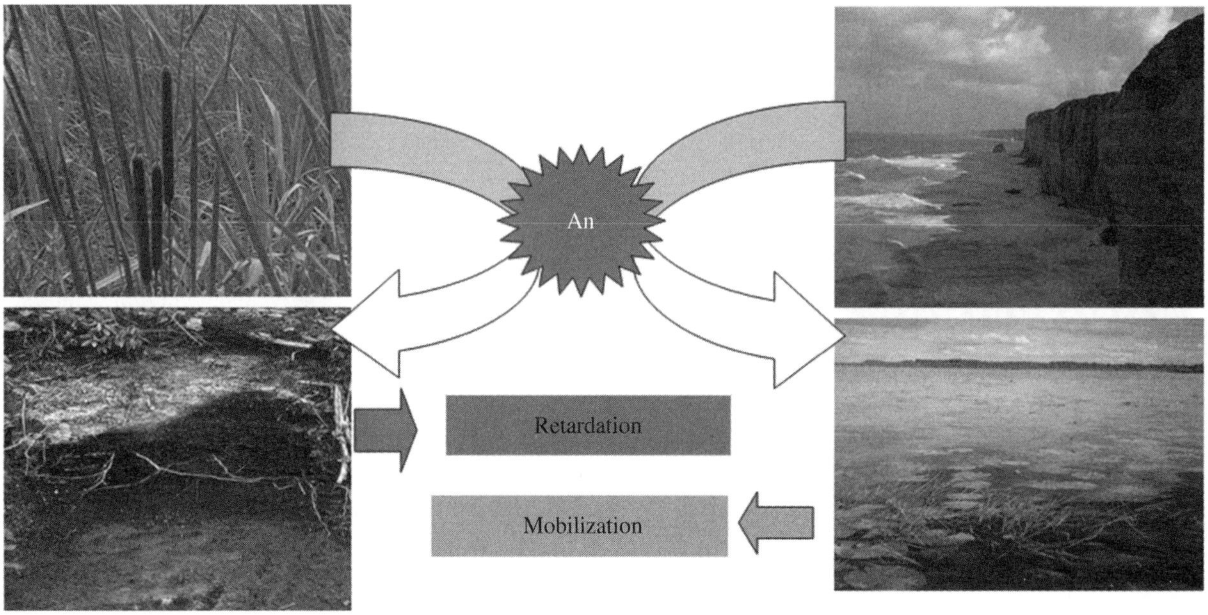

FIGURE 29.1 Humic substances (HS) influence the migration of actinides from two perspectives: (a) as mobile organomineral colloids and complexes in aquatic environments (right panel); and (b) as immobile HS (e.g., humic coatings on mineral phases in soils—left panel) that can retard actinide migration.

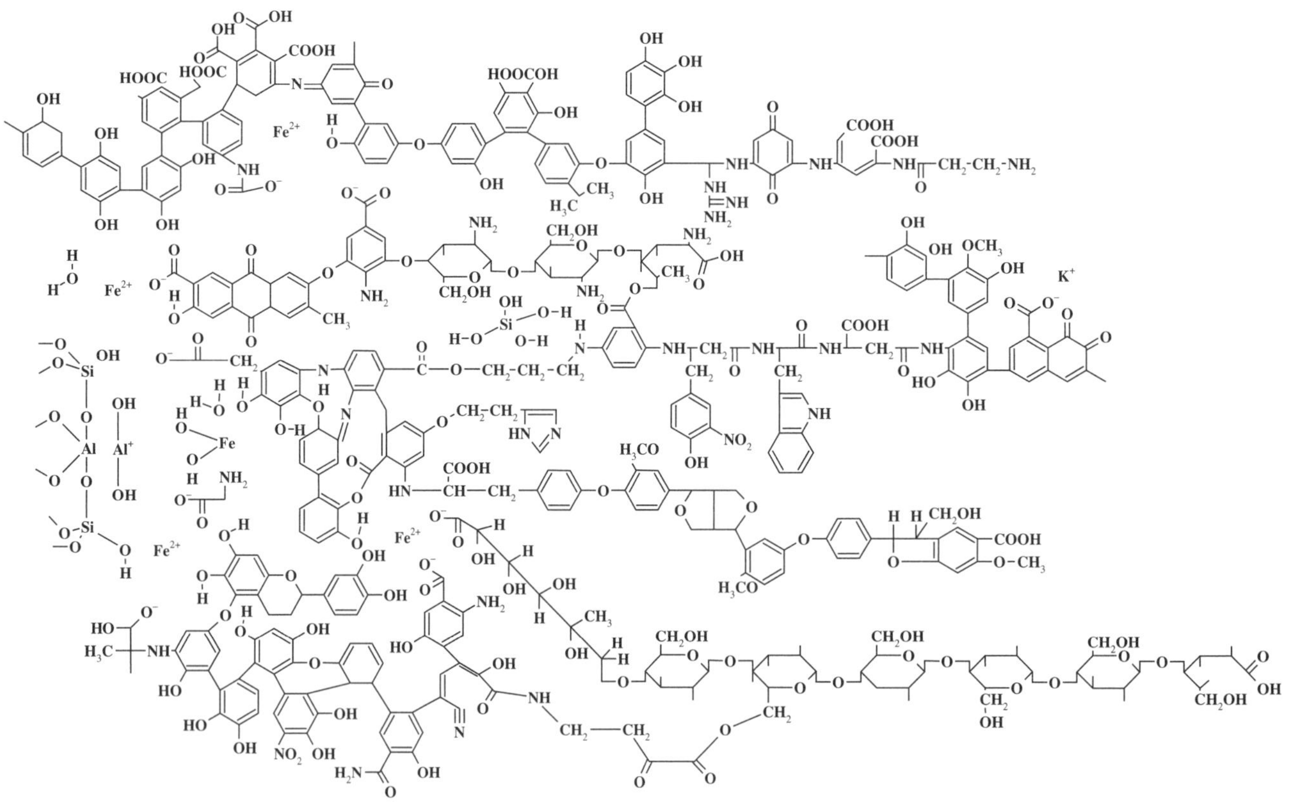

FIGURE 29.2 Hypothetical structural fragment of soil humic substances (Kleinhempel, 1970). The fragment illustrates the extreme structural heterogeneity of natural HS, which contain various oxygen-containing functional groups. The latter enables HS to engage redox, complexation, and sorption interactions.

reactivities and huge reserves of humic raw materials such as lignite, peat, and lignin, they are not frequently tapped for remedial needs. This is due to the immense structural heterogeneity that greatly hinders their use in practical applications.

To overcome the issues of structural heterogeneity that define humic materials and also undermine their practical utility for environmental remediation, an original approach was developed [11] to design humic materials with customized properties. Based on the previously described roles of HS affecting actinide migration, this new approach must achieve three goals. First, it must enhance the redox capacity of HS and consequently enable reduction and immobilization of the most mobile actinide species. The second goal requires that the method enhance the surface activity of HS for creating humic coatings on solid supports for interception and long-term stewardship of water-borne actinides. This requires a comprehensive study of interactions between designed humic materials and actinides in the higher-valence states. The third goal assumes that novel environmental protection technologies will be suggested based on the use of these designed humic materials and their nanocoatings.

29.2 REDOX-ENHANCED HUMIC MATERIALS AND THEIR INTERACTIONS WITH ACTINIDES

Reported values of formal electrode potentials for HS from different sources vary from +0.15 to +0.79 V versus standard hydrogen electrodes [15–20]. This is due in part to their complex makeup. From this range and the reversibility of such redox transformations, it may be surmised that the redox properties of HS are attributable to the quinonoid moieties present in the aromatic backbone [21]. Moreover, direct electrochemical evidence exists for the quinonoid nature of the redox-active units. It shows that natural organic matter (NOM) (particularly, the polyphenol fraction) gives an electrode response similar to that of model quinones such as juglone, lowsone, and anthraquinone disulphonate (AQDS). Hence, similar to quinones, the HS can participate in both abiotic and biotic redox transformations of contaminants in polluted environments. Several studies can be cited where HS were shown to participate in abiotic redox transformations of selected actinides. For example, the reduction of highly oxidized species of Pu and Np (Pu(VI, V) and Np(VI)) has been demonstrated [22–24]. In addition, direct abiotic reduction of Cr(VI) by HS was also reported [25–27]. However, U(VI) and Np(V) reduction was not observed in the presence of natural HS [7].

Given the discussed importance of quinonoid units for the redox behavior of HS with respect to actinides, we hypothesized that humic materials with enhanced redox properties can be manufactured by incorporating quinonoid units with known properties into humic backbones.

29.2.1 Directed Synthesis of Quinonoid-Enriched Humic Materials and Assessment of Their Redox Properties

To realize an approach of enhancing the redox properties of humic materials, the corresponding customized humic materials were synthesized by incorporating different quinonoid units as described by Perminova et al. [28] and schematically shown in Figure 29.3 for the example of hydroquinone.

All modification reactions were run on humic acids isolated from leonardite (oxidized coal). The latter is widely used for the manufacture of commercial humates. The quinonoid monomers used for the modification of humic materials included hydroquinone (HQ), catechol (CT), and 1,4-benzoquinone (BQ). The monomers were chosen as ubiquitously being present in the biopolymers. They were incorporated into the humic backbones at different monomer-to-HS ratios: that is, 100, 250, and

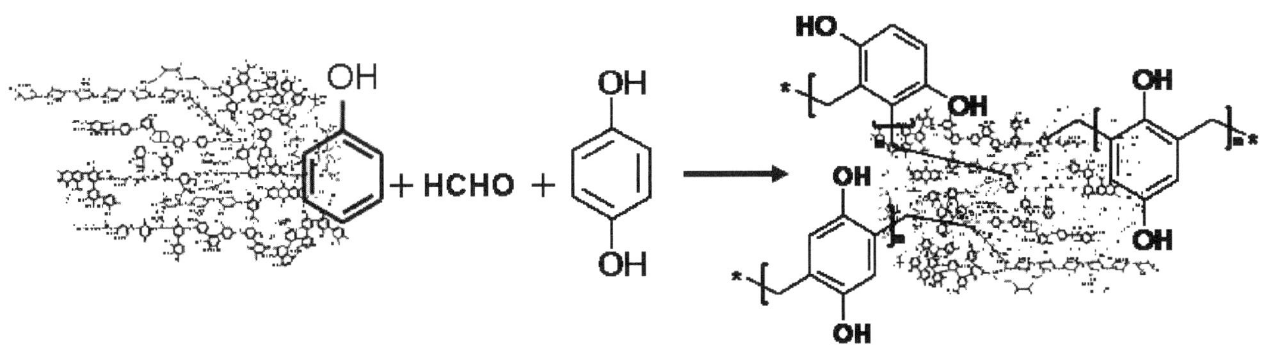

FIGURE 29.3 Synthetic pathway for manufacturing quinonoid-enriched humic derivatives using phenolformaldehyde polycondensation (with the example of hydroquinone).

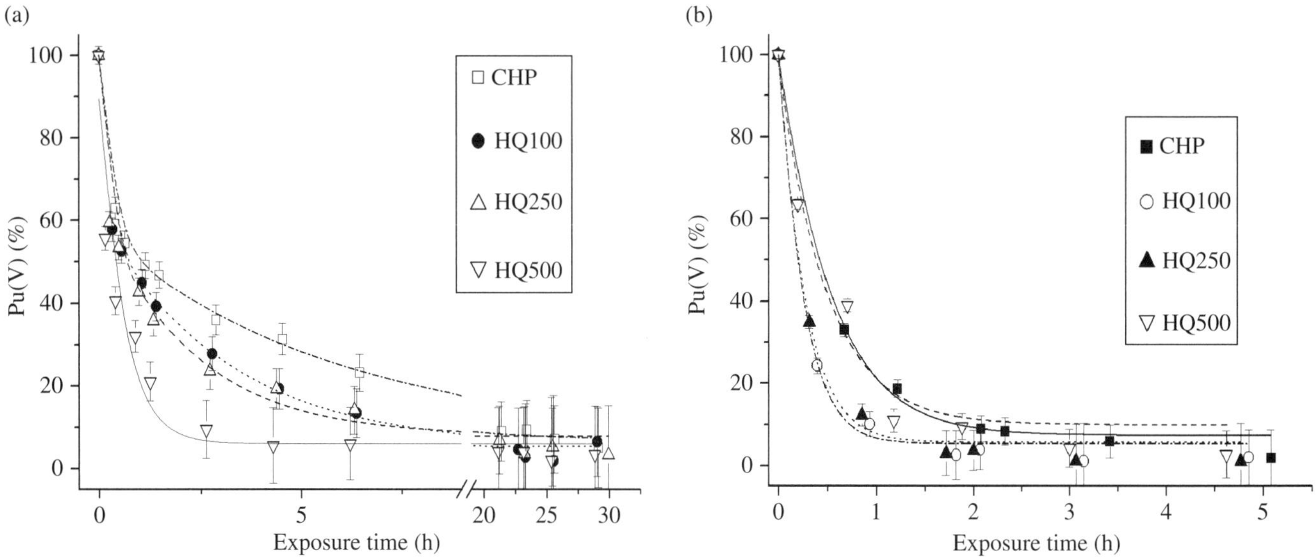

FIGURE 29.4 Reduction of Pu(V) by hydroquinone-enriched humic derivatives. Conditions: Ar atmosphere, $C_{tot}(Pu)=2.3\cdot10^{-9}$ M, $C_{tot}(HS)=10\,mg\cdot l^{-1}$, $I=0$ M. (a) pH 7.5; (b) pH 4.5. Reproduced with permission from Ref. [30]. © Elsevier B.V.

500 mg of the monomer per 1 g of the parent HS using phenolformaldehyde polycondensation for HQ and CT, and oxidative polymerization for BQ. It should be noted that contrary to HQ and CT monomers that were joined to aromatic rings of HS via CH2 bridges, BQ produced biphenyl-type structures with C–C bonds between humic and hydroquinone rings. The redox capacities of these derivatives measured as described by Matthiessen [29] were substantially higher (up to 4 mmol/g) as compared to the parent humic material (0.6 mmol/g). The largest increase was observed within the hydroquinone-enriched derivatives as opposed to catechol or benzoquinone products. Of particular importance is that the redox capacity was proportional to the incorporated amount of quinonoid centers, which enables the manufacturing of quinonoid-enriched derivatives with predicted redox capacities.

29.2.2 Reduction of Actinides in the Higher-Valence State by the Quinonoid-Enriched Humic Materials

The performance of quinonoid-enriched humic derivatives was tested with regard to the reduction of Pu(V) and Np(V). Corresponding data were reported by Shcherbina et al. [30, 31]. Since qiononoid monomers are sensitive to the presence of oxygen, it was important to study the redox properties of the quinonoid-enriched humic derivatives both under oxic and anoxic conditions and at different pH values. The actinide concentrations were set at 10^{-9} M for Pu(V) and 10^{-5} M for Np(V). Such a substantial difference in actinide concentrations was predicated on different sensitivities of analytical techniques used for monitoring Pu(V) and Np(V) speciation in the solution: liquid–liquid extraction followed by liquid scintillation counting in the case of Pu(V) [23, 32], and spectrophotometry in the ultraviolet–visible–near-infrared (UV/vis/NIR) range for Np(V) [33]. Consequently, the concentrations of HS used in the experiments with Pu(V) were set to a lower value (10 mg/l) compared to Np(V) (250 and 500 mg/l).

It was found that, in general, all humic derivatives studied (including the parental HS) were more effective at reducing Pu(V) than Np(V). As shown in Figure 29.4a, under anoxic conditions, the complete reduction of Pu(V) to Pu (IV) was observed within 4–5 h. Liquid extraction did not reveal Pu(III) in the solution once equilibrium was achieved given its stabilty in such solutions. In the presence of oxygen, Pu(V) reduction was still detected, although with much slower kinetics: complete Pu(V) reduction occurred within 145 h (not shown).

For the quinonoid-enriched derivatives, a significant relationship was observed between the degree of modification and the reduction performance. This can be clearly seen for the case of plutonium from Figure 29.4b. The derivatives could be placed in the following ascending order according to their reducing efficiency: CHP < HQ100 ≤ HQ250 < HQ500.

As the reduction of An(V) is pH-dependent, the kinetics of Pu(V) reduction under varying pH conditions was also studied. Faster reduction rates occurred in acidic (4.5) versus neutral pH (7.5).

For Np(V), the reduction was much slower, reaching 30% (by HQ500) within 120 h [30]. The reduction of Np(V) was monitored as described by Sachs and Bernhard, and Keller [33, 34], by measuring the absorbance of the aqueous NpO_2^+ ion and the

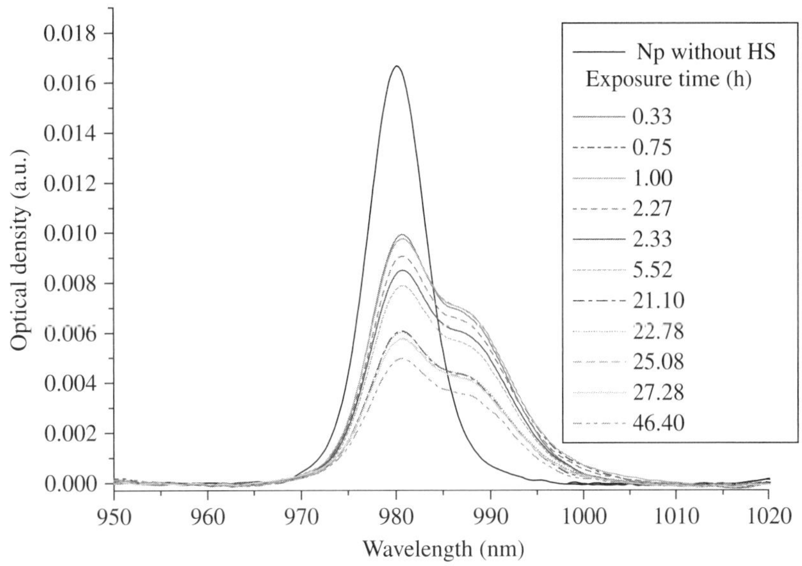

FIGURE 29.5 Typical spectra of NpO_2^+ and a Np(V)–humic complex (NpO_2^+HS); $C(Np)_{tot} = 3.8\ 10^{-5}$ M, C(HS) = 250 mg/l. The spectrum of Np(V)–humic complex is shown for the dynamics during 46,4 h of exposure. It can be seen that within the time the absorption intensity of both NpO_2^+ and the HS complex decreased as a result of Np(V) reduction and complexation by the quinonoid-enriched humic derivatives.

Np(V)–HS complex at 981.5 and 987 nm, respectively (Fig. 29.5). The molar absorbance of the NpO_2^+ ion was determined to be 377.71 mol^{-1} cm^{-1}, which is close to the 395 l·mol^{-1} cm^{-1} value reported by Sachs and Bernhard [33], and Keller [34].

The Np(V) reduction by humic derivatives containing different quinonoid moieties is shown in Figure 29.6. The most effective reduction of Np(V) was observed for a hydroquinone-enriched humic substance followed by catechol and benzoquinone (Fig. 29.6a). As in the case of Pu(V), a direct relationship was observed between the reducing performance of the derivative and its degree of modification (Fig. 29.6b).

The reduction rate for a hydroquinone derivative with a moderate degree of modification (HQ250) was on the same level as with the samples enriched with catechol and benzoquinone units with higher degrees of modifications (CT 500 and BQ 500). In addition, the hydroquinone derivatives brought about the most complete reduction of Np(V).

The results show better Np(V) reduction with hydroquinone humic derivatives, which suggests that they may function better as sequestering agents for waterborne actinides in their higher-valence states. With regard to a contaminated site remediation, a humic-based technology would function best under conditions that were both anoxic and acidic (e.g., at pH < 5). These conditions are met, for example, at the Field Research Center of the Oak Ridge National Laboratory, U.S. DOE [35], where high concentrations of uranium (mg/l) and technetium are also found. These conditions would be considered acceptable for the effective use of modified humic derivatives for the purpose of actinide reduction, thereby reducing their mobility in the subsurface environment.

For actinide contaminated sites having oxygenated groundwaters and pH values close to neutral, such as the Siberian Chemical Industrial Complex, Russia (SCIC) [36], and another highly contaminated Russian site—the Production Association "Mayak" in Western Ural [37]—the proposed humic-based technology might not provide for desired lowering of neptunium mobility, but will be still efficient for immobilizing Pu(V).

29.2.3 Interactions of Np with Hydroquinone-Enriched Humic Materials on the Surface of Colloidal Goethite Particles

To assess redox-mediating activities of the hydroquinone-enriched humic materials under environmental conditions, they were introduced to neptunium(V)-goethite (α-FeOOH) colloid systems. The latter were to mimic omnipresent mineral colloids [38]. For this investigation, goethite suspensions were spiked with ^{237}Np(V) to create a concentration of $(6.0 \pm 0.6) \cdot 10^{-7}$ M. The parental and modified humic materials (CHP and CHP–HQ100, respectively) were added separately to the pre-equilibrated Np–goethite suspensions at concentrations of 43 mg/l and continuously stirred for a month. The experiments were run at two different pH values: 3.5 and 7.5. It was found that in the presence of hydroquinone derivatives of HS, neptunium–HS complexes formed on the surface of the goethite, whereas they were not observed with only the parent humic materials. The Np–HS

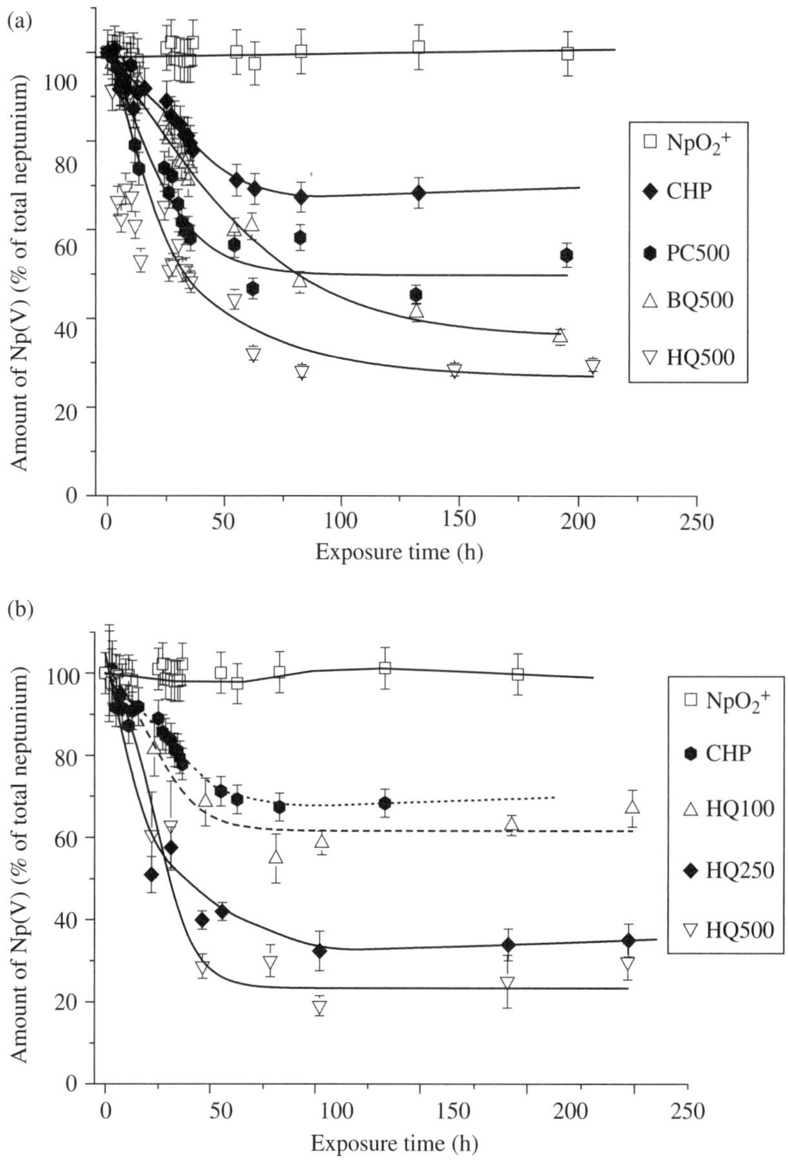

FIGURE 29.6 Reduction and removal of Np(V) by the humic derivatives under anoxic conditions, pH 4.7, $C_0(Np) = 5.4 \cdot 10^{-5}$ M, $C_0(HS) = 500\,mg \cdot l^{-1}$, Np(V):HS ratio = 1:40. (a) Effect of quinonoid monomer nature incorporated into humic structure (HQ, hydroquinone; CT, catechol; BQ, benzoquinone); (b) Effect of reagent ratio (parent HA-to-quinonoid monomer) for HQ-enriched derivatives. HQ100, HQ250, and HQ500 stand, respectively, for derivatives with reagent ratios of 100, 250, and 500 mg of HQ per 1000 mg of CHP. Reproduced with permission from Ref. [31]. © American Chemical Society.

complexes were investigated using X-ray photoelectron spectroscopy (XPS) as described by Kalmykov et al. [39]. XPS is a powerful synchrotron technique that measures the elemental compositions, as well as chemical and electronic states of the elements that exist within a material. The XPS spectra of Np equilibrated with goethite and CHP–HQ100 at pH 3.5 and pH 7.5 are presented in Figure 29.7. Complete reduction of Np(V) to Np(IV) occurred at pH 3.5, but it remained in pentavalent state at pH 7.5. The higher reducing power of hydroquinone-enriched humic derivatives is in good agreement with what is predicted by the Nernst equation.

The functional distribution of humic materials sorbed onto a goethite surface at different pH values was also studied using scanning transmission X-Ray microscopy (STXM), which accommodates both a good 2D spatial resolution (up to 10 nm) and the capability of producing high-quality spectra (NEXAFS). The purpose of this study was to investigate the structure of the humic coating on the goethite surface under different pH conditions. The sample equilibrated at pH 3.5 showed distinct structures in the STXM C1s ratio image analysis of the selected sample regions (Fig. 29.8).

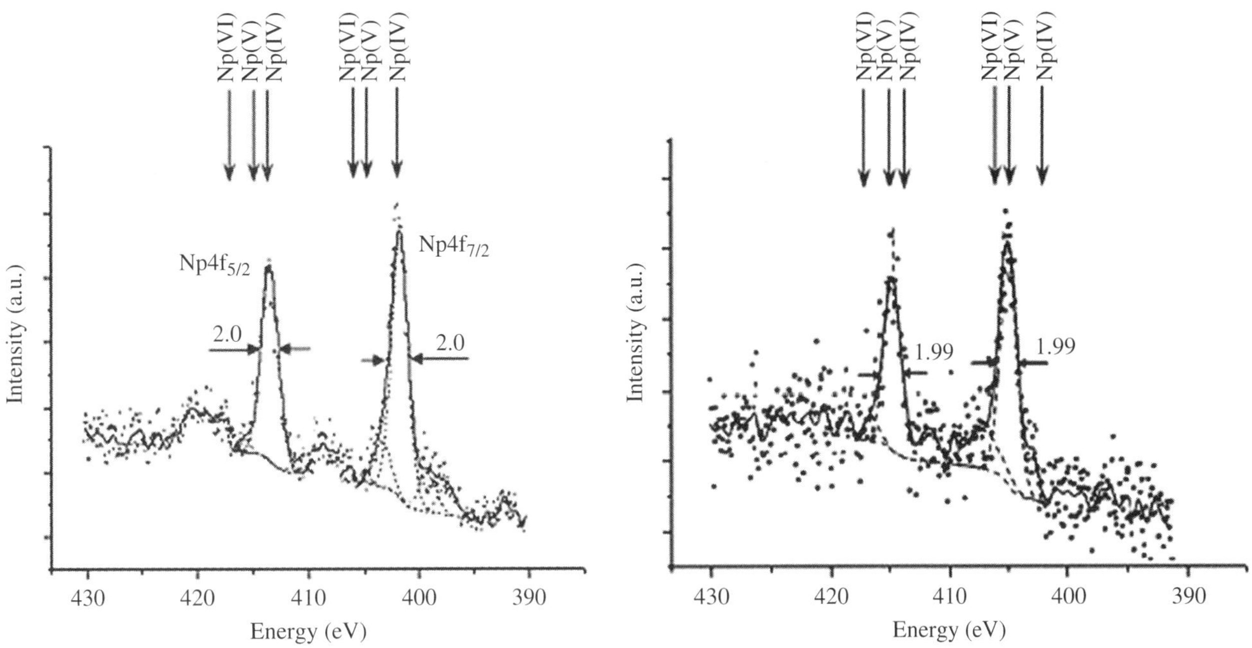

FIGURE 29.7 Np 4f electron spectra taken for ternary Np–CHP–HQ100 goethite system: left – pH = 3.5, right – pH = 7.5, I($NaClO_4$) = 0.1 M, $c(Np(V))_T = 5.8 \cdot 10^{-7}$ M, c(α-FeOOH) = 0.22 g/l, c(CHP–HQ100) = 44 mg/l.

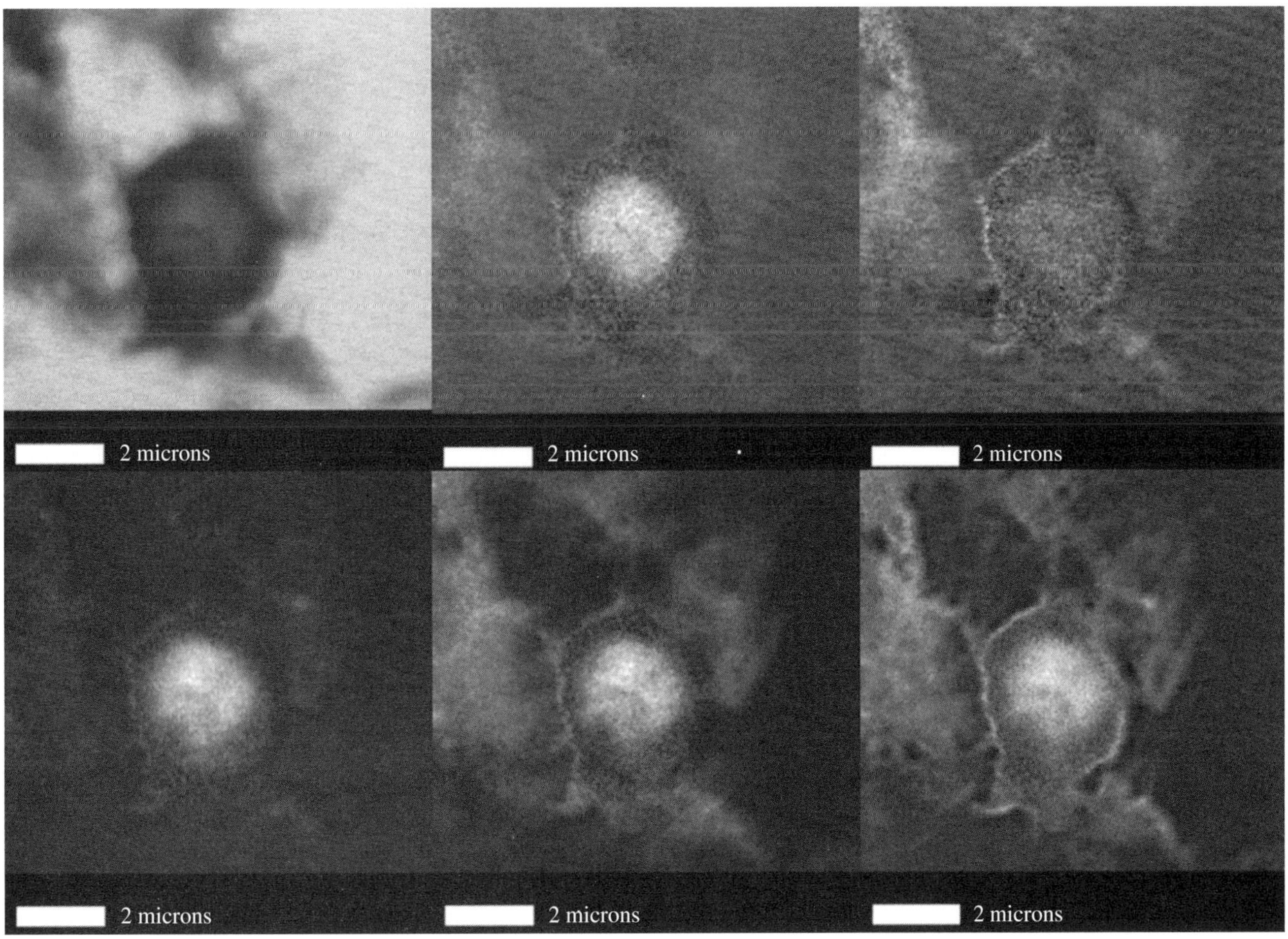

FIGURE 29.8 The STXM C1s ratio image analysis of the goethite sample equilibrated with CHP–HQ100 at pH 3.5. Upper row (from left to right): absorption image at 280 eV below the C1s edge; ratio images showing the distribution of aromatics, phenol-type groups and lower row aliphatics, carboxyl-type groups and total organics. Bright gray values indicate high concentrations of organic functionality.

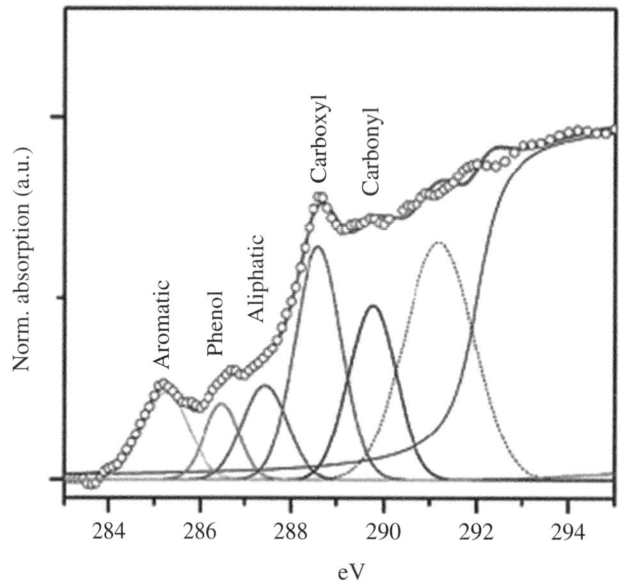

FIGURE 29.9 Deconvolution fit of CHP–HQ100 with different carbon functionalities as indicated by the labels. The open dots represent the smoothed measured spectra obtained by cluster analysis and the solid line indicates the fitted spectra.

The "hot spot" produced by the Np–organic matter was observed in the central region of the goethite particle and contained high amounts of aromatic and aliphatic structures, whereas edge regions of the particle contained lower amounts of these functionalities. The particle's edge structures and the surroundings of the particle itself seemed to be enriched with oxygen-containing functional groups as deduced from the ratio images of phenol-type and carboxyl-type structures. Similar observations of the spatial heterogeneity of organic matter functionalities sorbed onto inorganic particles have been reported in the literature for clay minerals and soil aggregates [40, 41].

Principle component analysis (PCA) and cluster analysis revealed one dominant cluster, whereas the spectral quality of additional clusters was very poor (not shown here). The cluster spectrum was deconvoluted and quantified, as has been described elsewhere [42–44]. The spectrum shown in Figure 29.9 is consistent with that reported for fulvic acids from groundwater of the Gorleben site [45]. At the same time, no organic carbon was detected using carbon K-edge measurements for the sample equilibrated at pH 7.5: a flat line was detected in the energy region 280–310 eV, indicating the optical density (OD) of inorganic phases (goethite). These findings can be interpreted as representing an absence of adsorption of organic matter.

The conclusion was that the reduction of Np(V) at low pH values in the ternary Np–goethite–HS system was initiated when hydroquinone-enriched humic derivatives were used. This was not the case with the parent (leonardite) HS. According to STXM analysis, the modified humics formed surface coatings on goethite colloids at low pH as well as the surrounding cloud that was rich in organic matter. Such goethite–humic aggregates could serve as effective scavengers for actinides (e.g., Np(V)) from aqueous solutions.

In general, the experimental results discussed here provide positive evidence in support of using quinonoid-enriched humic materials as agents to facilitate the immobilization of hazardous actinides by inducing their reduction to less mobile forms. Other potent applications of humic materials could be derived if they could be immobilized as organic coatings onto other solid supports. These coatings could intercept actinides in their higher-valence state and provide a mechanism for their long-term stewardship. To solve this problem, another type of modification of humic materials was proposed based on the incorporation of surface-active silicon-containing groups into the humic backbone.

29.3 HUMIC NANOCOATINGS AND THEIR USE FOR SEQUESTRATION OF MOBILE ACTINIDES IN THEIR HIGHER-VALENCE STATE

29.3.1 Mineral-Adhesive Silanized Humic Derivatives and Their Use for In Situ Preparations of Humic Nanocoatings

The original approach to the in situ preparation of humic nanocoatings under aquatic conditions was developed and described by Perminova et al. [46] and Karpiouk et al. [47]. This approach was based on the use of "mineral-adhesive," silanized humic derivatives, which are soluble in water, but undergo a phase switch upon contact with the hydroxyl-carrying surface of the solid

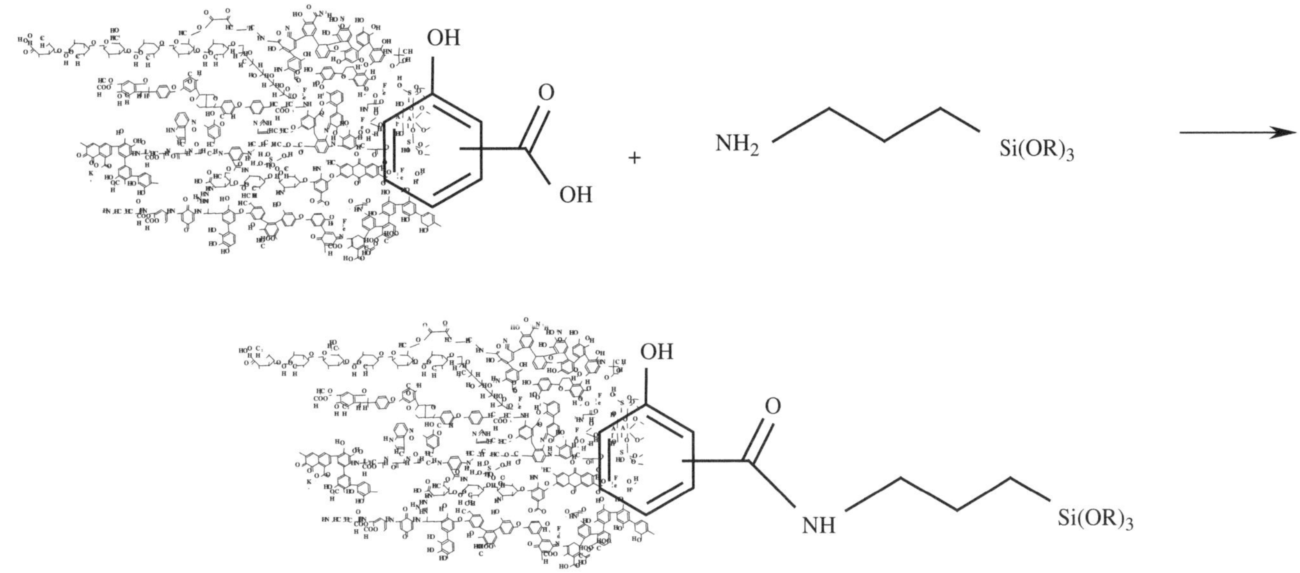

FIGURE 29.10 Schematic reaction pathway for synthesis of alkoxysilyl humic derivatives with mineral-adhesive properties not inherent within the parent humic material.

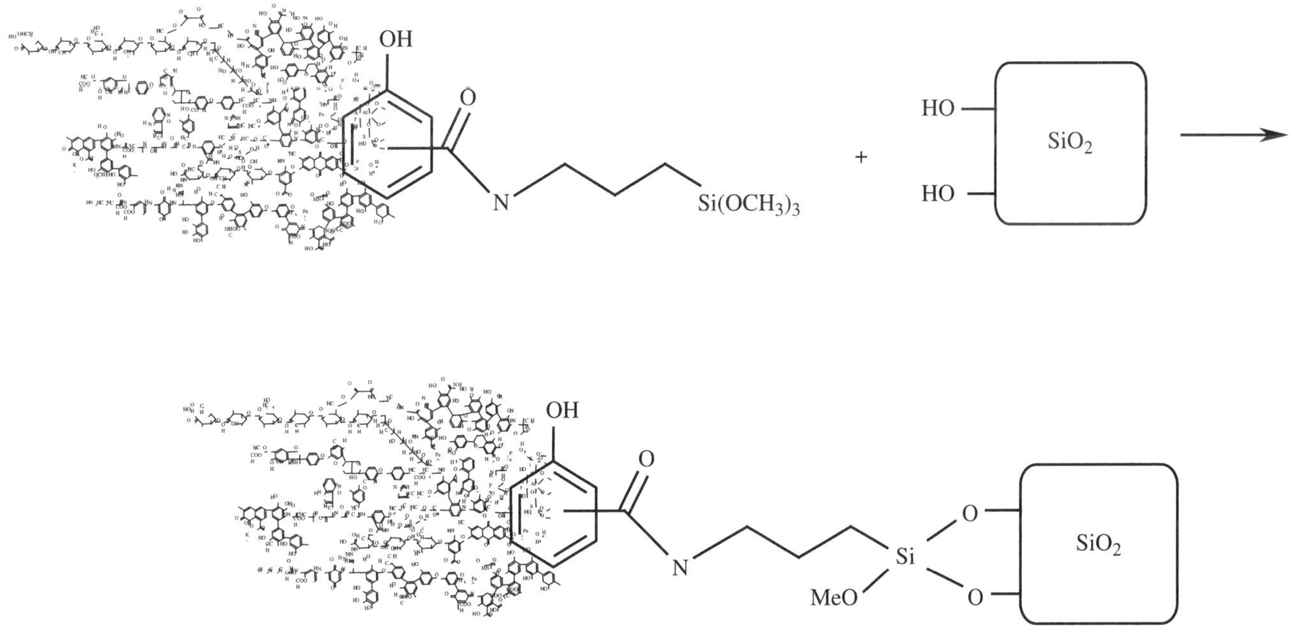

FIGURE 29.11 Self-assembly of silanized humic derivatives into adlayers at the silica gel surface under ambient conditions in aquatic environments.

support. The pertinent derivatives were produced via the incorporation of alkoxysilyl groups into the humic backbone by treating the parent humic material with 3-aminopropyltrialkoxysilane. The synthetic pathway appears in Figure 29.10.

The alkoxysilyl groups (Si–OAlk, where Alk is a hydrocarbon radical, e.g., $-CH_3$) undergo hydrolysis upon contact with water-producing silanol groups (Si–OH) capable of covalent bonding to mineral surfaces carrying hydroxyls. The formation of Si–O–Si or Si–O–M (where M is a metal, e.g., Fe) bonds provides a phase switch of the humic derivatives and their immobilization on the surface of mineral support. The corresponding reaction scheme is shown in Figure 29.11.

The covalent bonding is viewed as being of particular advantage in the proposed approach, since it precludes the facile remobilization of the humic coating and any contaminants bound to that coating. The spatial homogeneity and morphology

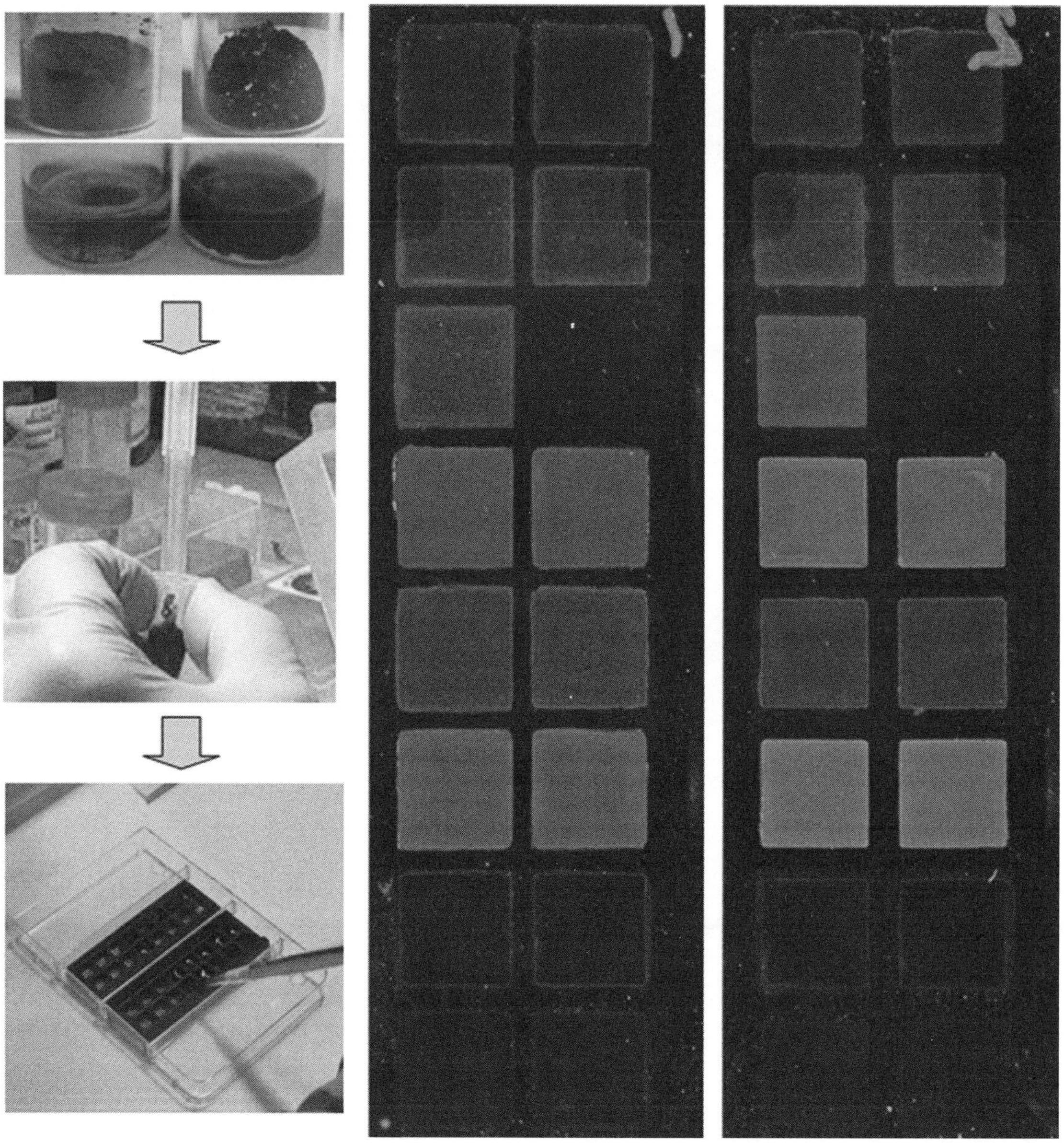

FIGURE 29.12 General view of the alkoxysilanized humic probes in the solid and dissolved states applied on the surface of DNA array glass slides that are segregated into multiple patches by the plastic chamber (left part) and the "artificial" color images of the humic adlayers. The latter are self-assembled in situ at the water/solid interface (right part). The materials in four glass beakers show humic probes made from coal materials (darkly colored) and peat materials (lighter color). The slides were scanned in a laser microarray scanner at two channels: Ex 532 nm, em 570–590 nm (green); and Ex 635 nm, em 680–700 nm (red). The samples are listed from the bottom and are as follows: the bottom (first) row is a solution of a nonmodified sample of coal humics (CHP); the second row is CHP–GPTS-100; the third row is AHS–APTS-100; the fourth row is CHP–APTS-100; the fifth row is PHA–APTS-100; the sixth row is a buffer solution (right side, the background control); and CHP–APTS-20 nonheated (left side); the seventh row is CHP–APTS-20 nonheated; the eighth row is CHP–APTS-20 (heated). Reproduced with permission from Ref. [47]. © Royal Society of Chemistry.

of the humic coatings formed as a result of the self-assembly of the silanized humic derivatives into adlayers at the water–solid interface were studied using confocal laser microscopy (CLM) and atomic force microscopy (AFM) [47]. This was achieved by preparing humic coatings suitable for AFM studies using DNA array slides as solid supports. The immobilization procedure for the alkoxysilyl humic derivatives and the CLM images of the resultant coatings are shown in Figure 29.12.

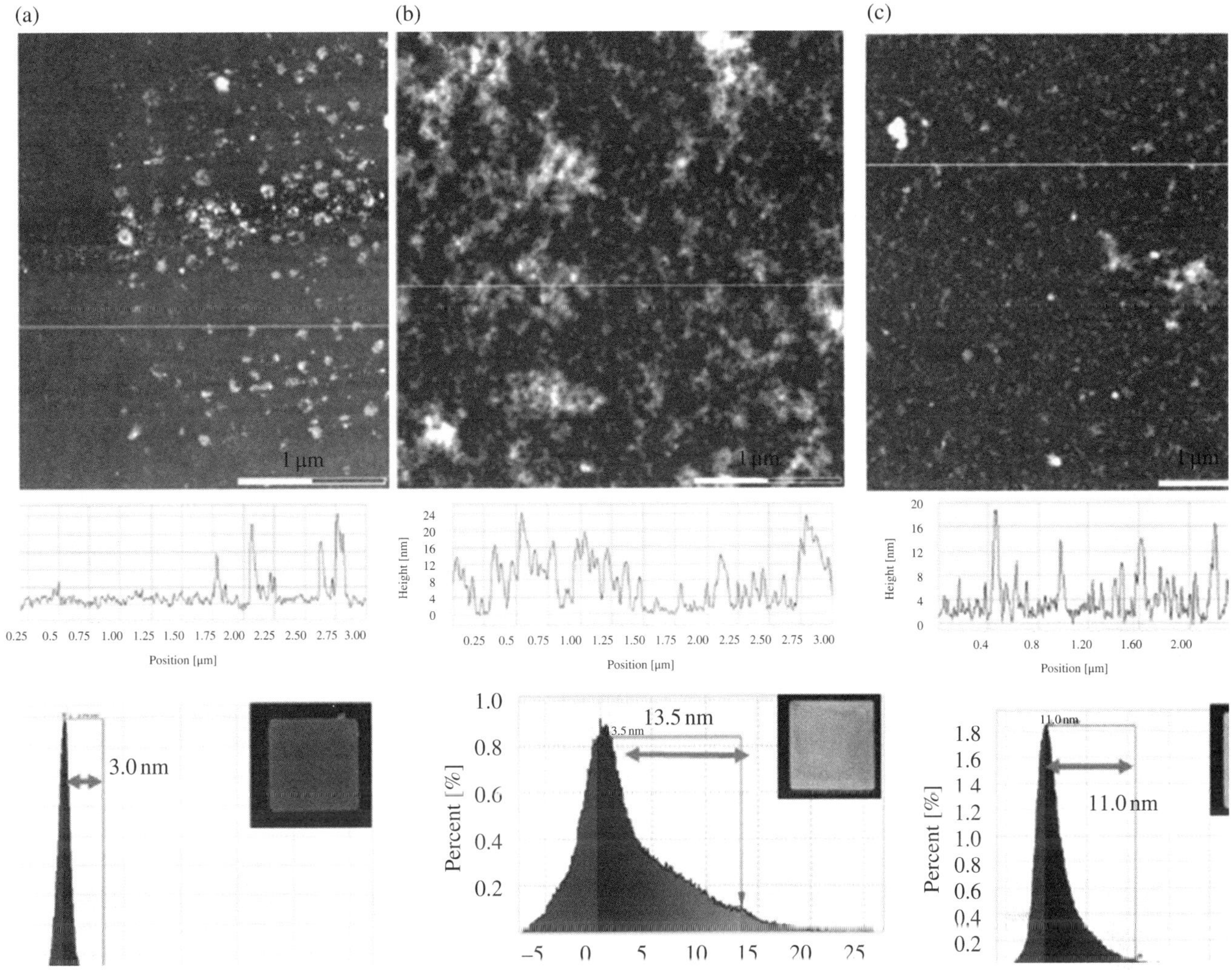

FIGURE 29.13 Height scanning force microscopy (SFM) images of the glass surfaces modified with the alkoxysilanized humic substances from different sources. (a) CHP–APTS-100-treated surface with an RMS roughness of about 1.5 nm; (b) PHS–APTS-100-treated surface with an RMS roughness of 5.9 nm; and (c) AHS–APTS-100-treated surface with an RMS roughness of 3.8 nm. The roughness was determined for a $3 \times 3\,\mu m^2$ surface area. For claritym we also include the fluorescence image of the same sample. The graphs below each image represent a cross section profile along the horizontal line as indicated, and the corresponding height distributions. Reproduced with permission from Ref. [47]. © Royal Society of Chemistry.

At the micron scale (resolution of CLM), results show silanized HS producing somewhat homogeneous coatings on DNA array glass. At the same time, substantial differences could be seen from the fluorescence of HS coatings depending on the origin of parent materials (peat, coal, and water) modified with 3-aminopropyltrimethoxysilane (APTS). The highest fluorescence intensities were observed for peat HS (PHS–APTS-100) and aquatic HS (AHS–APTS-100), which may vary with the thickness of the self-assembled layer or the amount of attached humic material. The results indicate that peat humic materials produced the thickest layer. Of particular note is the lack of fluorescence over surfaces treated with parental humic materials. Nonmodified humic materials, as expected, did not attach to the glass surface due to a lack of positively charged vacancies at the water–solid interface, which drives their interaction with alumosilicates in nature. The function of positive vacancies was successfully replaced in this study by incorporating alkoxysilyl groups as anchoring moieties into the humic structure.

The finer-scale structure of these humic coatings was studied on the nanometer scale using the AFM technique. The results are given in Figure 29.13. AFM images show the surface topography of all three samples—coal, peat, and aquatic humic materials, immobilized on a DNA glass slide under aqueous conditions. All images are accompanied by the corresponding height distributions.

The AFM results show that all humic adlayers were separated into multiple domains of nano- and submicrosize. Again substantial differences can be seen in the morphology of coatings produced by the humic materials of different origins. The roughest surfaces were observed for PHS–APTS and AHS–APTS at 5.9 and 3.8 nm, respectively, which is in line with indications of fluorescent images showing the largest portion of humic matter attached to the surfaces in the cases of PHS and AHS. For coal-based HS, a much smoother surface was observed with an rms roughness of 1.5 nm. The coal HS adlayer consisted of flat circular particles (2–15 nm in diameter) homogeneously distributed over the glass surface. The particles did not exhibit intimate contact with one another. This might be indicative of repulsive interactions between particles produced by the negatively charged humic polyanions, which preclude coalescence and induce the appearance of the diffuse structure in the adlayer. The domains observed in the adlayer assembled by peat humic acids are very different from those of coal and present much larger particles of 10–50 nm in diameter, aggregates of these particles (20–100 nm), and ridge-like assemblies (up to 200 nm in length and up to 30 nm in height). Aquatic humic materials produced an intermediate picture consisting of domains ranging from 5 to 30 nm and chain-like assemblies (up to 100 nm). The particles had "geometric" shapes dissimilar to those of both peat and coal materials. The aquatic adlayer was characterized with the highest heterogeneity provided by the irregular location of large particle chains reaching 100 nm in length and up to 40 nm in height.

The nanostructures described earlier for the adlayers of silanized humic materials are in agreement with the AFM images observed for the natural humic colloids and for the humic adlayers immobilized onto different mineral surrogates (glass, oxidized silicon wafer, mica, carbonaceous or goethite surfaces) [48–52]. The coatings composed of separated islands of aquatic humic materials were observed for immobilized SRFA and SRHA (IHSS standard humic materials from the Suwannee River) [49]. Close estimates of particle shapes and sizes are reported for these in situ studied humic colloids: (i) flat particles (8–13 nm in diameter); (ii) aggregates of particles (20–100 nm); and (iii) chain-like assemblies, networks, and torus-like structures [53]. The results provide an insight into the molecular features of the surface morphology of humic coatings, which can be immobilized onto the hydroxyl-carrying solid supports using the guided self-assembly of alkoxysilanized humic materials.

Of particular interest was the comparative assessment of the approach with other immobilization techniques. For this purpose, the trimethoxysilyl derivatives of coal and peat humic acids were also immobilized onto silica gels and the corresponding quantities of HS were determined [46]. The amounts found were 210 and 232 mg of HS/g SiO_2, respectively. A comparison of these results with those reported elsewhere [54] for immobilized humic materials using alternative techniques showed that the new immobilization method generates 2–10 times more humic material on the silica gel. Immobilization of HS onto the APTS-modified silica gel in the aqueous phase by the adsorption mechanism without further heating yielded 65 mg HS/g SiO_2. The immobilization method based on HS binding to SiO_2 modified with APTS in the presence of N-(3-dimethylaminopropyl)-N'-ethylcarbodiimide hydrochloride (EDS) had a higher yield (107 mg HS/g SiO_2), and the binding of the HS to APTS-modified silica gel at high temperature in anhydrous solvents (e.g., dimethylformamide (DMF)) yielded 124 mg HS/g SiO_2. When binding HS to the APTS-modified silica via glutaric aldehyde, a value of 60 mg HS/g SiO_2 was obtained. An additional advantage of the proposed approach is that the binding of HS was practically irreversible, while alternative techniques released up to 50% of the immobilized HS into the solution upon washing with 0.1 M NaCl at pH 10 [54].

The conclusion reached is that the unique properties of these developed silanized humic derivatives include (i) their solubility in water; (ii) the capability to adhere to different minerals; and (iii) their environmental compatibility. A progressive combination of these properties allows us to consider these derivatives as important reagents for the treatment of both actinide- and metal-contaminated aquatic environments. They can serve either as liquid scavengers—phase-switchers—or as reactive agents for in situ installation of permeable reactive barriers (PRBs) aimed at sequestering mobile higher-valence actinides from contaminated groundwaters.

29.3.2 Sequestration of Mobile Actinides in Their Higher-Valence States onto Humic Nanocoatings

To further develop the concept of a remedial technology based on the use of "mineral-adhesive" humic agents, consideration was given to the in situ installation of PRBs, which are capable of sequestering actinides in the higher-valence states, and the execution of experiments for demonstrating the immobilization of waterborne Np(V) and Pu(V) [55]. Humic coatings were produced by self-assembly of the alkoxysilanized humic derivatives onto silica gel as described in the previous paragraph. Leonardite humic acid (HA) and its hydroquinone derivatives were used as parent humic materials for the incorporation of alkoxysilyl groups. The amount of organic carbon immobilized onto silica gel in case of the leonardite HA was 9.2% (240 mg of HA per 1 g of silica gel), whereas for the hydroquinone derivative it was much less accounting for 3.3% (100 mg HA per 1 g of SiO_2). Concentrations of Np(V) and Pu(V) in the working solutions were 4.68 10^{-6} M and $4.90 \cdot 10^{-9}$ M, respectively. Experiments with Np(V) were executed at pH 4.5. Experiments with Pu(V) were conducted at three pH values: 3.5, 4.6, and 7.7. All experiments were executed in the absence of oxygen under batch conditions. The sequestration of Np(V) and Pu(V) by the humic-coated silica gels is shown in Figure 29.14 and Figure 29.15, respectively.

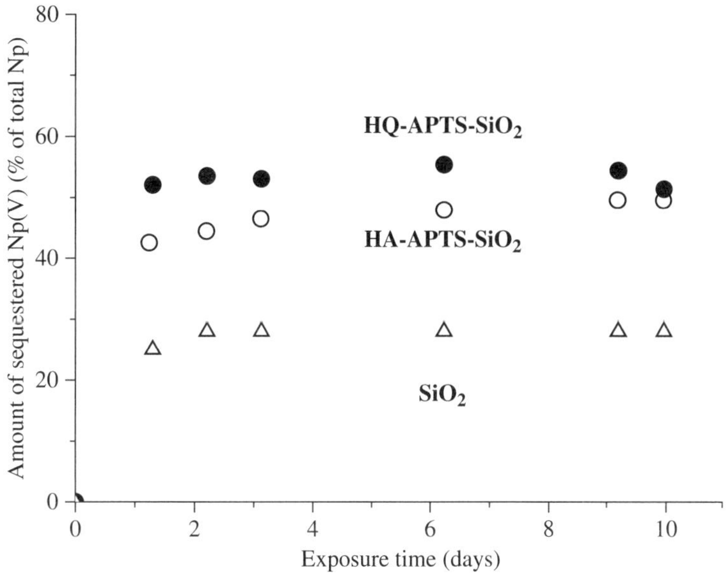

FIGURE 29.14 Sequestration kinetics of Np(V) in the presence of humic-coated silica gels: leonardite humic acid (HA–APTS–SiO_2) and its hydroquinone-enriched derivative (HQ–APTS–SiO_2); $C_0(Np) = 4.68 \times 10^{-6}$ M, pH 4.5, mass/volume ratios for HA–APTS–SiO_2 and HQ–APTS–SiO_2 are 3.5 and 2.0 g/l, respectively. Reproduced with permission from Ref. [55]. © Elsevier B.V.

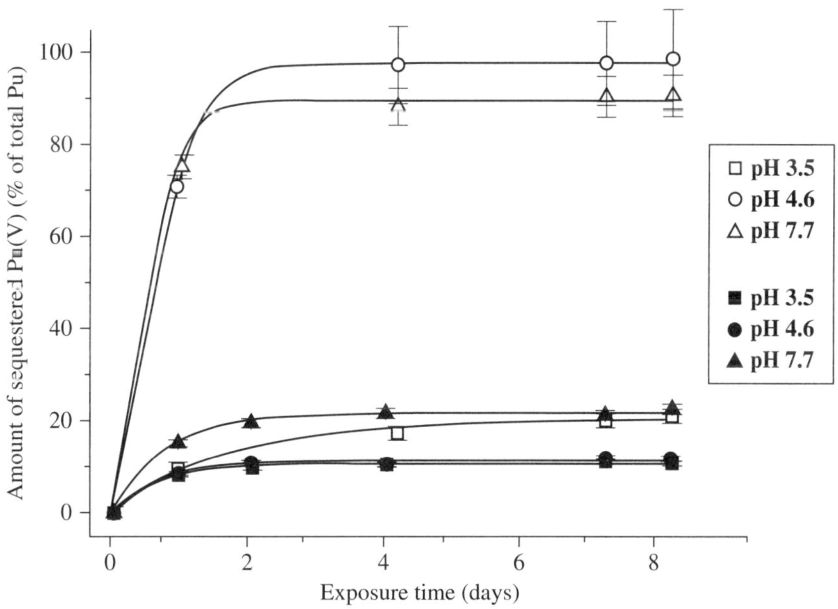

FIGURE 29.15 Sequestration kinetics of Pu(V) by humic-coated silica gel (alkoxysilyl derivative of leonardite HA (HA–APTS–SiO_2)) at different pH values; $C_0(Pu) = 4.9 \times 10^{-9}$ M, mass/volume ratio for HA–APTS–SiO_2 is 0.345 g/l. Reproduced with permission from Ref. [55]. © Elsevier B.V.

Both humic-coated silica gels effectively sorbed Np(V). There was no substantial difference observed in the sequestration performance of silica gel coated with the parental humic acid versus its hydroquinone-enriched derivative. The amount of sequestered neptunium was 55% for HQ–APTS–SiO_2 versus 45% for HA–APTS–SiO_2.

Much higher sorption of Pu(V) was observed onto the humic-coated silica gels as compared to Np(V). System equilibrium was reached after 4 days of exposure. At this time, almost 90% sorption of initial Pu was observed in systems containing humic-coated silica gels at pH 4.6 and 7.7. Pu sorption onto bare silica gels under the same conditions never exceeded 10%. At the same time, less Pu sorption was observed at pH 3.5. This result was in line with previous findings of lower Pu sorption under acidic conditions [56]. One explanation might be that at low pH values the binding affinity for actinides decreases due to strong competitive interactions between protons and actinide ions.

From Figure 29.14, Np partition coefficients (K_d) for the humic-coated silica gels were estimated after the system reached equilibrium. They ranged between 3×10^2 and 5×10^2 ml/g for both sorbents studied, showing no substantial benefit from the incorporation of hydroquinone units into the humic material. The measured K_d values were an order of magnitude higher than those of pure silica gel (30 ml/g). It can be expected that K_d values for Np will increase with pH.

The Pu partition coefficients were calculated from the data after 4 days of exposure. They were 8×10^3, 3×10^4, and 1.9×10^5 ml·g^{-1} for pH values of 3.5, 7.7, and 4.6, respectively. It should be noted that the calculated values of K_d for humic-coated silica gels were about two orders of magnitude larger than those for bare SiO_2 ($6.30 \cdot 10^2$ l·kg^{-1}). This is in line with the significant enhancement of the sequestration capacity of HS-coated silica gel. In general, determined plutonium K_d values for HS-modified silica gels corroborated with estimates reported from batch experiments with soil and sediment particles, which were on the order of 10^4 l·kg^{-1} [57, 58]. Corresponding K_{OC} values ranged from 1.2 10^5 to 4.4 10^5 l·kg^{-1}, which were well within the range of partition coefficients reported for organic-rich geosorbents [58, 59].

Experimental results given for steady-state conditions demonstrate that the affinity of humic derivatives for solid support is of key importance for remedial purposes, since it determines the amount of humic material that can be immobilized onto the solid support.

The experiment here demonstrates the unique property of alkoxysilyl HS derivatives to bind to OH-carrying surfaces of silica under ambient conditions—in an aqueous phase and at the room temperatures. Resultant humic coatings on minerals facilitate the retention and sequestration of waterborne contaminants from the aquifer. These results may have important implications for an inexpensive alternative to installing PRBs by excavation.

29.4 CASE STUDY ON HUMIC-BASED NANOTECHNOLOGY SOLUTIONS TO REMEDIATION

The experimental findings reported in the previous sections have demonstrated that humic derivatives modified to acquire the needed remedial functions such as reducing capacity or surface adhesion activity can be successfully used for developing "nature-like" nanotechnology solutions for environmental remediation. For example, "mineral-adhesive" humic derivatives capable of forming covalent bonds with mineral surfaces in water can be used for the in situ installation of humic PRB. This can be accomplished by injecting solutions of alkoxysilyl derivatives directly into contaminated aquifers using a "fence row" of wells as proposed in a patent application [60]. The injected humics would form an organic coating on mineral surfaces of aquifer materials present and create a "humic curtain" capable of sequestering actinides from the contaminated groundwater. The scheme of humic PRB installed in the contaminated aquifer is shown in Figure 29.16. This method of installation may be particularly advantageous in the remediation of deep diluted plumes of contamination.

Further work, however, is needed to develop and test humic PRBs under environmental conditions before this technology can be brought into practice. Detailed kinetic, batch, and flow-through studies need to be conducted to define physical–chemical and technological parameters of the processes involved including humic reagent loadings, hydraulic residence times, and other parameters are needed for the design and optimization of the humic PRB performance under natural subsurface conditions.

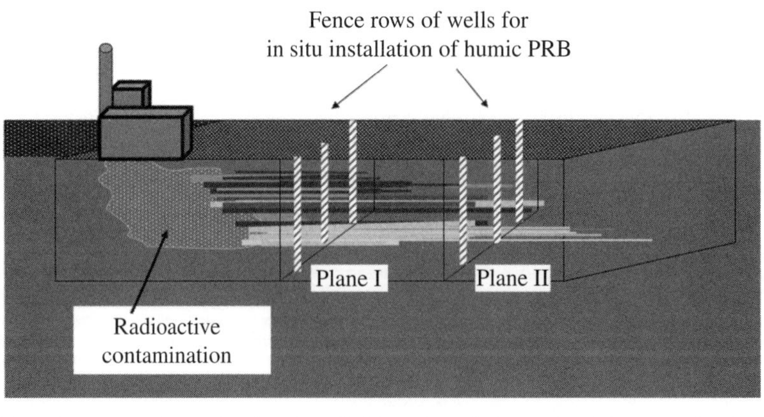

FIGURE 29.16 Scheme of humic permeable reactive barriers (HPRBs) installation using "fence row" of wells for injecting a solution of "mineral-adhesive" humic derivates.

29.5 CONCLUSIONS

The results summarized in this chapter provide positive evidence for the potential use of hybrid humic nanomaterials and nanocoatings as safe and environmentally benevolent agents to build a suite of new environmental remediation technologies. This is because the properties of chemically modified humic materials can be manipulated efficiently and in a reproducible manner.

Beyond PRB applications, HS can be used as stabilizing agents to prevent the aggregation of magnetic nanoparticles of zero-valent iron and iron oxides in aqueous environments and engineered systems [61–63]. HS have been shown to stabilize "nano-iron," preventing it from aggregation and enhancing its mobility in aquifers [64]. The benefits of using HS, as compared to synthetic modifiers (e.g., synthetic polymers), include their biocompatibility and remarkable capacity to detoxify and mitigate the risk of potential hazards associated with the release of nanoparticles into the environment. Additionally and contrary to biomacromolecules, HS are typically resistant to biodegradation under both anoxic and oxic conditions in soils and the subsurface. In addition, engineered nanoparticles stabilized with HS can be synthesized in situ or a priori in HS solutions to reduce the costs of remediation [65]. These properties create new opportunities for preparing agents of innovative remedial technologies that combine the benevolent green properties of HS and their ability to self-assemble into hyperbranched functionalized macromolecules with high performance and reactivity of engineered nanoparticles. This will bring about the development of new treatment strategies based on nature-like nanotechnology solutions for remediation in general and for actinide-contaminated environments in particular.

ACKNOWLEDGMENTS

We dedicate this effort to Richard G. Haire in honor of his receiving the ACS 2013 Glenn T. Seaborg Award in nuclear chemistry. We acknowledge the support of the funding agencies that facilitated the reported research including the joint research program of US DOE and Russian Academy of Sciences (project RUC2-20006 MO-04), Russian Foundation for Basic Research (11-03-12177-OFI-M-2011), Russian Ministry for Science and Education (State Contract No. 16.740.11.0183), and NATO-CLG (grant ESP.EAP.CLG 983197).

REFERENCES

[1] Czerwinski KR, Buckau G, Scherbaum F, Kim JI. Complexation of the uranyl ion with aquatic humic acid. Radiochim Acta 1994;65:111–119.

[2] Kim JI, Delakowitz B, Zeh P, Klotz D, Lazik D. A column experiment for the study of colloidal radionuclide migration in Gorleben Aquifer Systems. Radiochim Acta 1994;66/67:165–167.

[3] Kim JI, Sekine T. Complexation of Np(V) with humic acid. Radiochim Acta 1991;55:187–192.

[4] Kribeck B, Podlaha J. The stability constant of the UO_2^{2+} humic acid complex. Organic Geochem 1980;2:93–97.

[5] Marquardt CM, Kim JI. Complexation of Np(V) with humic acid: intercomparison of results from different laboratories. Radiochim Acta 1998;80:129–137.

[6] Nash KL, Choppin GR. Interaction of humic and fulvic acids with Th(IV). J Inorg Nucl Chem 1980;42:1045–1050.

[7] Rao L, Choppin GR. Thermodynamic study of the complexation of neptunium(V) with humic acids. Radiochim Acta 1995;69:87–95.

[8] Torres R, Choppin GR. Europium (III) and Americium (III) stability constants with humic acid. Radiochim Acta 1984;35:143–148.

[9] Thurman EM. *Organic geochemistry of natural waters*. Dordrecht: Martinus Nijhof & Dr. W. Junk Publishers; 1985.

[10] Hayes MH, McCarthy P, Malcolm RL, Swift RS. *Humic substance II: In search of structure: Setting the scene*. New York: John Wiley & Sons Inc.; 1989. p 3.

[11] Perminova IV, Hatfield K. Use of humic substances to remediate polluted environments: from theory to practice. In: Perminova IV, Hatfield K, Hertkorn N, editors. *NATO Science Series: IV: Earth and Environmental Sciences*. Volume 52, Dordrecht: Springer; 2005. p 3–36.

[12] Lipczynska-Kochany E, Kochany J. Effect of humic substances on the Fenton treatment of wastewater at acidic and neutral pH. Chemosphere 2008;73 (5):745–750.

[13] Van Stempvoort DR, Lesage S, Molson J. Use of humic substances to remediate polluted environments: from theory to practice. In: Perminova IV, Hatfield K, Hertkorn N, editors. *NATO Science Series: IV: Earth and Environmental Sciences*. Volume 52, Dordrecht: Springer; 2005. p 233–256.

[14] Wan J, Dong W, Tokunaga TK. Method to attenuate U(VI) mobility in acidic wasteplumes using humic acids. Environ Sci Technol 2011;45 (6):2331–2337.

[15] Ziechmann W. *Huminstoffe*. Weinheim: Verlag Chemie; 1980.

[16] Visser SA. Oxidation-reduction potentials and capillary activities of humic acids. Nature 1964;204:581.

[17] Helburn RS, MacCarthy P. Determination of some redox properties of'humic acid byalkaline ferricyanide titration. Anal Chim Acta 1994;295:263–272.

[18] Skogerboe RK, Wilson SA. Reduction of ionic species by fulvic acid. Anal Chem 1981;53:228–232.

[19] Oesterberg R, Shirshova L. Non-equilibrium oscillating redox properties of humic acids. Geochim Cosmochim Acta 1997;61:4599–4604.

[20] Struyk Z, Sposito G. Redox properties of standard humic acids. Geoderma 2001;102:329–346.

[21] Scott DT, McKnight DM, Blunt-Harris EL, Kolesar SE, Lovley DR. Quinone moieties act as electron acceptors in the reduction of humic substances by humics-reducing microorganisms. Environ Sci Technol 1998;32:2984–2989.

[22] Bondietti EA, Reynolds SA, Shanks MN. *Transuranic Nuclides in the Environment*. Vienna: IAEA; 1976.

[23] Andre C, Choppin GR. Reduction of Pu(V) by humic acid. Radiochim Acta 2000;88:613–616.

[24] Zhang YJ, Bryan ND, Livens FR, Jones MN. Selectivity in the complexation of actinides by humic substances. Environ Pollut 1997;96:361–367.

[25] Fukusima M, Nakayasu K, Tanaka S, Nakamara H. Speciation analysis of chromium after reduction of cromium (VI) by humic acid. Toxicol Environ Chem 1997;62:207–215.

[26] Wittbrodt PR, Palmer CD. Effect of temperature, ionic strength, background electrolytes and Fe (III) on the reduction of hexavalent chromium by soil humic substances. Environ Sci Technol 1996;30:2470–2477.

[27] Zhilin DM, Schmitt-Kopplin P, Perminova IV. Reduction of Cr(VI) by peat and coal humic substances. Environ Chem Lett 2004;2:141–145.

[28] Perminova IV, Kovalenko AN, Schmitt-Kopplin P, Hatfield K, Hertkorn N, Belyaeva EY, Petrosyan VS. Design of quinonoid-enriched humic materials with enhanced redox properties. Environ Sci Technol 2005;39 (21):8518–8524.

[29] Matthiessen A. Determining the redox capacity of humic substances. Vom Wasser 1995;84:229–235.

[30] Shcherbina NS, Kalmykov SN, Perminova IV, Kovalenko AN. Reduction of actinides in higher oxidation states by hydroquinone-enriched humic derivatives. J Alloy Comp 2007;444–445:518–521.

[31] Shcherbina NS, Perminova IV, Kalmykov SN, Kovalenko AN, Haire RG, Novikov AP. Redox and Complexation Interactions of Neptunium(V) with Quinonoid-Enriched Humic Derivatives. Environ Sci Technol 2007;41 (20):7010–7015.

[32] Morgenstern C, Choppin GR. Kinetics of the oxidation of Pu(IV) by manganese oxide. Radiochim Acta 2002;90:69–74.

[33] Sachs S, Bernhard G. NIR spectroscopic study of the complexation of neptunium(V) with humic acids: influence of phenolic OH groups on the complex formation. Radiochim Acta 2005;93:141–145.

[34] Keller C. *The Chemistry of the Transuranium Elements*. New York: Verlag Chemie; 1971.

[35] Watson DB, Doll WE, Gamey TJ, Jardine PM. *Technical Report*. Oak Ridge, TN: Oak Ridge Field Research Center; 2001.

[36] Rybaltchenko, A. I.; Pimenov, M. K; Kostin, P. P. *The Deep Injection of LRW*; IzdAT: Moscow, 1994, (in Russian).

[37] Novikov AP, Kalmykov SN, Utsunomiya S, Ewing R, Horreard F, Merkulov A, Clark SB, Tkachev VV, Myasoedov BF. Colloid transport of plutonium in the far-field of the Mayak Production Association, Russia. Science 2006;314:638–641.

[38] Khasanova AB, Kalmykov SN, Shcherbina NS, Kovalenko AN, Perminova IV, Clark S. *Recent Advances in Actinide Science*. In Proceedings of the Eighth Actinide Conference, Actinide 2005 Alvarez, R.; Bryan ND, May I, editors. 2005 July 4–8; Manchester: RSC Publishing; 2006. p 86–88.

[39] Kalmykov SN, Schaefer T, Claret F, Perminova IV, Petrova (Khasanova) AB, Shcherbina NS, Teterin YA. Sorption of neptunium onto goethite in the presence of humic acids with different hydroquinone group content. Radiochim Acta 2008;96 (9–11):685–690.

[40] Claret F, Bauer A, Schaefer T, Griffault L, Lanson B. Experimental investigation of the interaction of clays with high-pH solutions: case study from the Callovo–Oxfordian formation, Meuse-Haute Marne underground laboratory (France). Clay Clay Miner 2002;50 (5):633–646.

[41] Kretzschmar R, Schafer T. Metal retention and transport on particles in the environment. Elements 2005;1 (4):205–210.

[42] Lerotic M, Jacobsen C, Schäfer T, Vogt S. Cluster analysis of soft x-ray spectromicroscopy data. Ultramicroscopy 2004;100 (1–2):35–57.

[43] Schafer T, Hertkorn N, Artinger R, Claret F, Bauer A. Functional group analysis of natural organic colloids and clay association kinetics using C(1s) spectromicroscopy. J Phys 2003;IV (104):409–412.

[44] Claret F, Schafer T, Rabung T, Wolf M, Bauer A, Buckau G. Differences in properties and Cm(III) complexationbehavior of isolated humic and fulvic acid derived from Opalinus clay and Callovo-Oxfordian argillite. Appl Geochem 2005;20:1158–1168.

[45] Schaefer T, Buckau G, Artinger R, Kim JI, Geyer S, Wolf M, Bleam WF, Wirick S, Jacobsen C. Origin and mobility of fulvic acids in the Gorlebenaquifer system: implications from isotopic data andcarbon/sulfur XANES. Org Geochem 2005;36 (4):567–582.

[46] Perminova IV, Karpiouk LA, Ponomarenko SA, Konstantinov AI, Hertkorn N, Hatfield K, Muzafarov AM. Controlling aqueous sorption of humic substances on silica gel by directed alkoxysilyl-derivatization of their functionalities. Colloid Surf A 2012;396:224–232.

[47] Karpiouk LA, Ponomarenko SA, Mourran A, Bochkariov D, Muzafarov AM, Hatfield K, Perminova IV. Self-assembly of alkoxysilanized humic substances into multidomain adlayers at the water-solid interface: linking surface morphology to molecular structures of adsorbate. Soft Matter 2012;8 (8):2452–2459.

[48] Abu-Lail L, Liu Y, Atabek A, Camesano TA. Quantifying the adhesion and interaction forces between *Pseudomonas aeruginosa* and natural organic matter. Environ Sci Technol 2007;41:8031–8037.

[49] Liu AG, Wu RC, Eschenazi E, Papadopoulos K. AFM on humic acid adsorption on mica. Colloid Surf A 2000;174:245–252.

[50] Sparks DL, Grundl TJ, editors. Mineral-Water Interfacial Reactions. Volume 715, 1998. ACS Symposium Series; p 57.

[51] Namjesnik-Dejanovic K, Maurice PA. Conformations and aggregate structures of sorbed natural organic matter on muscovite and hematite. Geochim Cosmochim Acta 2001;65:1047–1057.

[52] Plaschke M, Romer J, Klenze R, Kim JI. In situ AFM study of sorbed humic acid colloids at different pH. Colloid Surf A 1999;160:269–279.

[53] Wilkinson KJ, Balnois E, Leppard GG, Buffle J. Characteristic features of the major components of freshwater colloidal organic matter revealed by transmission electron and atomic force microscopy. Colloid Surf A 1999;155:287–310.

[54] Koopal LK, Yang Y, Minnaard AJ, Theunissen PLM. Chemical immobilization of humic acid on silica. Colloid Surf 1998;141:385–395.

[55] Perminova IV, Karpiouk LA, Shcherbina NS, Ponomarenko SA, Kalmykov SN, Hatfield K. Preparation and use of humic coatings covalently bound to silica gel for Np(V) and Pu(V) sequestration. J Alloys Comp 2007;444-445:512–517.

[56] Prout WE. Adsorption of radioactive waste by Savannah River plant soil. Soil Sci 1958;86:13–17.

[57] Mahara Y, Matsuzuru H. Mobile and immobile plutonium in a groundwater environment. Water Res 1989;23 (1):43–50.

[58] U.S. Environmental Protection Agency. Understanding Variation in Partition Coefficient, Kd, Values: Volume II; Office of Radiation and Indoor Air, Office of Solid Waste and Emergency Response, EPA [and] Office of Environmental Restoration, U.S. DOE, Washington, DC, EPA 402-R-99-004B: 1999.

[59] Ames LL, editor. *Proceedings of Actinide-Sediment Reactions Working Meeting, Seattle, Washington*. Battelle Pacific Northwest Laboratories; 1976. p 225.

[60] Perminova IV, Ponomarenko SA, Karpiouk LA, Hatfield K. WO2007102750 (A1) 2007) September 13, Department of Chemistry of the Lomonosov Moscow State University, Russia, and University of Florida, USA.

[61] Tombácz E, Bica D, Hajdú A, Illés E, Majzik A, Vékás L. Surfactant double layer stabilized magnetic nanofluids for biomedical application. J Phys Condens Matter 2008;20 (20):204103–204109.

[62] Illés E, Tombácz E. The role of variable surface charge and surface complexation in the adsorption of humic acid on magnetite. Colloid Surf A. 2004;230:99–109.

[63] Ghosh S, Jiang W, McClements JD, Xing B. Colloidal stability of magnetic iron oxide nanoparticles: influence of natural organic matter and synthetic polyelectrolytes. Langmuir 2011;27 (13):8036–8043.

[64] Johnson RL, Johnson GOB, Nurmi JT, Tratnyek PG. Natural organic matter enhanced mobility of nanozerovalent iron. Environ Sci Technol 2009;43:5455–5460.

[65] Polyakov AY, Goldt AE, Sorkina TA, Perminova IV, Pankratov DA, Goodilin EA, Tretyakov YD. Constrained growth of anisotropic magnetic delta-FeOOH nanoparticles in the presence of humic substances. Cryst Eng Comm 2012;14:8097–8102.

PART XV

ENVIRONMENTAL RISKS AND TOXICITY

30

ENVIRONMENTAL RISKS OF NANOTECHNOLOGY: EVALUATING THE ECOTOXICITY OF NANOMATERIALS

MIGUEL A. MÉNDEZ-ROJAS[1], JOSÉ LUIS SÁNCHEZ-SALAS[1], ARACELY ANGULO-MOLINA[2], AND TERESA DE JESÚS PALACIOS-HERNÁNDEZ[3]

[1]*Departamento de Ciencias Químico-Biológicas, Universidad de las Américas Puebla, Puebla, Mexico*

[2]*Departamento de Ciencias de la Salud, Universidad de las Américas Puebla, Puebla, México*

[2]*Departamento de Ciencias Biológicas, Universidad Popular Autónoma del Estado de Puebla, Puebla, México*

30.1 INTRODUCTION

Ecotoxicology is a young, multidisciplinary field of science concerned with the study of contaminants (chemicals or biological organisms) in the biosphere and the understanding of their effects on the population, community, and ecosystem level, including humans [1, 2]. It is a multidisciplinary field that uses tools and concepts from biology, chemistry, medicine, toxicology, and ecology. In that sense, nano-ecotoxicology is an emergent branch of ecotoxicology specifically dedicated to engineered and natural nanomaterials (NMs). For such a specialized field of interest, it is important to make a complete assessment of the life cycle of the product, in order to understand the potential environmental and health hazards of such materials.

Ultrafine particles (UFPs) suspended in the air have been traditionally included as a topic of interest in toxicology, and their sizes range in the nanometer scale; when present in water or soil, they are referred to as colloids. In the normal toxicological terminology, particles with diameters less than 0.1 μm (100 nm, 0.1 μm) are called UFPs, although they can be grouped into three general categories: (1) those with diameters less that 100 nm; (2) those with sizes between 100 and 2500 nm (resulting from aggregation of UFPs); and (3) coarse-mode particles larger than 2500 nm. On the other hand, the term colloid is applied to particles with sizes in the 1–1000 nm range (0.001–1 μm). This kind of extremely fine and small materials may be the product of natural processes, they may be produced as by-products of anthropogenic activities, or inclusively they may be specifically manufactured for very specific applications.

Natural nanostructured materials have been around us for a long time. Humans and living beings have been exposed to naturally produced NMs since the beginning of life sources (soil erosion, ocean water evaporation, forest fires, photochemical reactions, volcanic eruptions, viruses, biogenic magnetite biosynthesized by magnetotactic bacteria, mollusks, arthropods, fish, birds, or from disintegration of iron meteorites when entering into the atmosphere) [3, 4]. They are also artificially produced, both intentionally for very specific applications (pigments, quantum dots, magnetic nanoparticles, catalysts, coatings, cosmetics, among several more examples) and as by-products of several manufacturing and industrial processes (fuel and charcoal combustion, mineral processing, cooking, welding, smoking, building demolition, consumer products containing NM degradation, etc.), many of them potentially toxic (Fig. 30.1).

Engineered nanomaterials (ENMs), on the other hand, are sources of concern as they have not been around us for a long time and living beings may have not developed appropriate biological barriers or trapping systems to avoid undesirable interactions that may harm individuals. There are several reports of toxicological studies of NMs which suggest that several of them may be dangerous, although the results are sometimes not conclusive and even contradictory [1] (Fig. 30.2).

Nanomaterials for Environmental Protection, First Edition. Edited by Boris I. Kharisov, Oxana V. Kharissova, and H. V. Rasika Dias.

FIGURE 30.1 Some natural sources of nanomaterials. Clockwise: a forest fire, dust storms, seawater evaporation, volcanic eruptions.

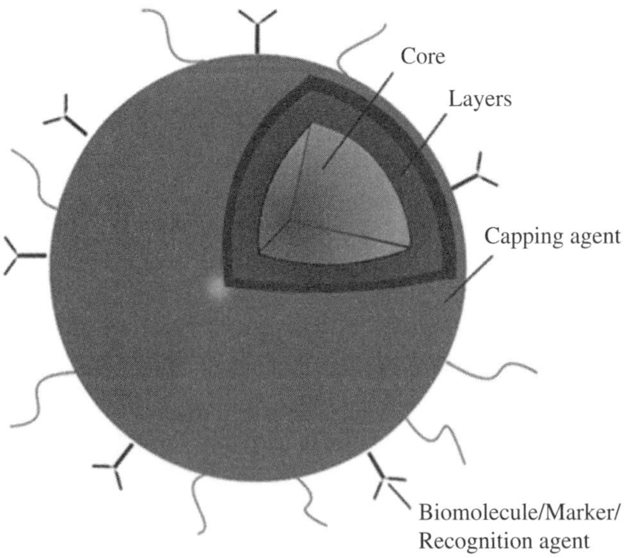

FIGURE 30.2 Schematic representation of an engineered nanomaterial (ENM). The core determines some of the physical properties of the material; the layer consists of molecules acting as stabilizers or modifying agents; the capping acts as an agent to increase biocompatibility, change charge, and solubility; the biomarker/fluorophore/recognition agent generates specifity for recognizing a substrate or detecting the nanomaterial.

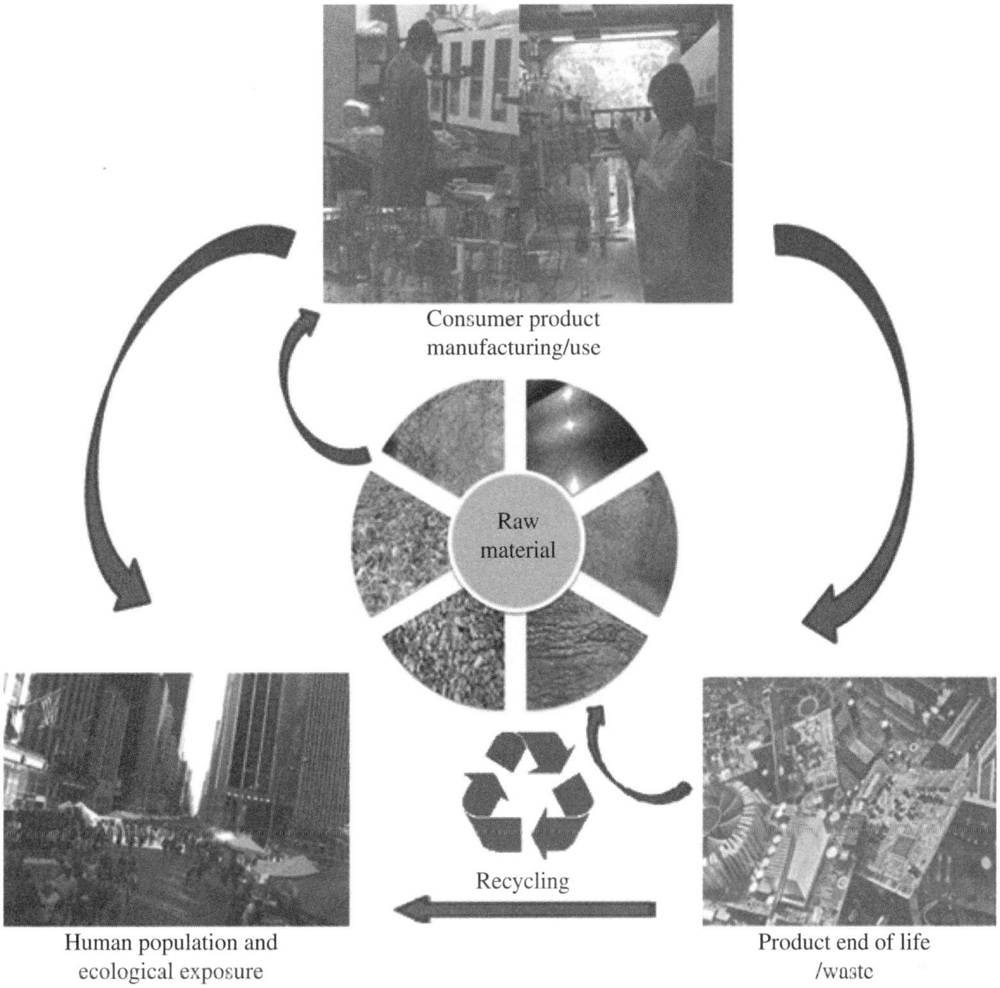

FIGURE 30.3 Life cycle of a consumer product containing nanomaterials.

For example, nano TiO_2 and nano ZnO have been reported as toxic to soil bacterial communities that may alter environmentally important soil processes [5]. However, other reports found that toxicity may arise not directly from the nanosized particles acting on bacteria but rather from metal ions known to be toxic for the bacteria and also from the chemical and biological dissolution of metal oxides and sulphates in the environment [6]. So one needs to establish the mechanisms underlying the real source of toxicity before jumping to a conclusion. It is therefore highly recommended that the toxicological effect of NMs be clarified before their commercial or practical applications or, on the other hand, to halt or modify their toxicity.

The concern on how some engineered or natural NMs may become hazardous pollutants posing a serious threat to public and environmental health is alive and growing, as careful studies to understand and model their complex interrelations with life systems are moving slower than the rate at which they are being introduced into new consumer products. Toxicological studies related to NMs started two decades ago, but most of the published papers and reports are still limited to *in vitro* studies or *in vivo* analyses of laboratory animal models mainly concerning human health impact. Although the first reports on environmental impacts of NMs are more recent [7–9], the terms "ecotoxicity" and "econanotoxicity" were not yet used (Fig. 30.3).

According with Kahru and Duborguier, nanostructured TiO_2, ZnO, CuO, Ag, single-walled carbon nanotubes (SWCNTs), mutiwalled carbon nanotubes (MWCNTs), and fullerenes, C_{60}, are among the NMs more likely to have environmental and health impact, due to their high volumes of production or extended use in consumer products [1]. For example, the large-scale manufacturing of SWCNTs was estimated to reach 1500 tons per year in 2011, while the total production of nanostructured metal oxide for cosmetic use was 1000 tons per year from 2005 to 2010 [10, 11]. The list may be extended to some other commercially important materials such as nano Au, nano zero-valent iron (Fe), quantum dots (CdS, CdSe, CdTe, ZnSe), nano iron oxides (Fe_2O_3, Fe_3O_4), nano CeO_2, nano SiO_2, graphene, as well as a long list of nanocomposites. All the later materials are

FIGURE 30.4 Consumer products containing nanomaterials that are already in the market.

included in a list of representative manufactured NMs published by the Working Party on Manufactured Nanomaterials (WPMN), a committee formed in 2006 by the Organization for Economic Co-operation and Development (OECD) to address the safety challenges of NMs.

By 2010, between 880 and 1000 different consumer products containing ENMs were identified in the market, a number that has been steadily increasing every year [12, 13]. Just in the span of 3 years (2007–2010), the number increased sixfold, recording the largest increment for personal care and coating products, including cosmetics, textiles, and antiwetting products. If we add to that the number of natural NMs, the exposure to nanostructured materials is an actual, real, and complex problem that needs to be carefully analyzed to understand the potential risks and to determine the right protective measures that should be implemented for our own safety (Fig. 30.4).

In order to determine whether an NM is toxic (or not) for the environment or human health, it will need a very rigorous characterization to know its precise physical and chemical characteristics and to understand the biological consequences of its use. Some relevant properties of NMs need to be considered in order to assess their potential toxicity: how they react (chemically), the sorption of chemicals on their surface or their own sorption onto a biological surface, the size/shape relationship, whether they are soluble or not in some specific solvent, their pH range or physical state, whether they are susceptible to form aggregates or to agglomerate, and, finally, whether a coating is present or not among some others. Some relevant effects to monitor are the generation of reactive oxygen species (ROS), whether they are able to act as carriers of toxic substances, their changes in oxidation state, their bioaccumulation, the molecular interactions that they are able to generate, and other indirect effects. Solubility is important, as it affects the bioavailability of a material. Further transformation of an NM prior to and after interacting with a biological system has to be considered, because an innocuous material may become toxic and vice versa [14].

30.2 THE BIOSYSTEM—NANOMATERIAL INTERACTION

The chemical and physical properties of bulk materials can vary greatly with respect to their nanostructured forms. They may become toxic and harmful, in contrast to being inert in their macroscopic form. The potential toxicity of NMs has been recognized by several authors [3, 10, 15–22]. A better understanding of the risks associated with specific NMs may reduce environmental damage or adverse health effects to the living beings in an ecosystem [23, 24]. Interpretation of toxicity may be a complex issue, as sometimes the synthetic methodology may affect the results because the processing of the material may incorporate additives, surfactants, and solvents that are not completely removed from the final products, especially if their physical, chemical, and biological interactions are not known in detail. For example, C_{60} was initially considered to be toxic, but later studies indicate that such toxicity was related to residual tetrahydrofuran (THF) used in the processing of the material [25]. Then, biological activity may depend on other components present in the chemical formulation of the material. Commercial sources of NMs do not often provide information regarding the synthesis or the use of stabilizing/capping agents, so a careful characterization *a priori* is highly recommended.

Interactions between nanostructured materials and biological systems may occur in several ways, being simple or very complex. As the scale of biologically relevant objects such as membrane structures, biomolecules (enzymes, proteins, DNA, RNA, antibodies), virus, bacteria, or eukaryotic cells is comparable in several kinds of NMs, there exist different unknown potential levels of complex interactions. Toxicity is a complex event *in vivo* and currently it is difficult to monitor systemic and physiological effects *in vitro*, so most assays determine effects at the cellular level. Most assays oversimplify the events they measure and are selected because they are cheap, easy to quantify, and reproducible (Fig. 30.5).

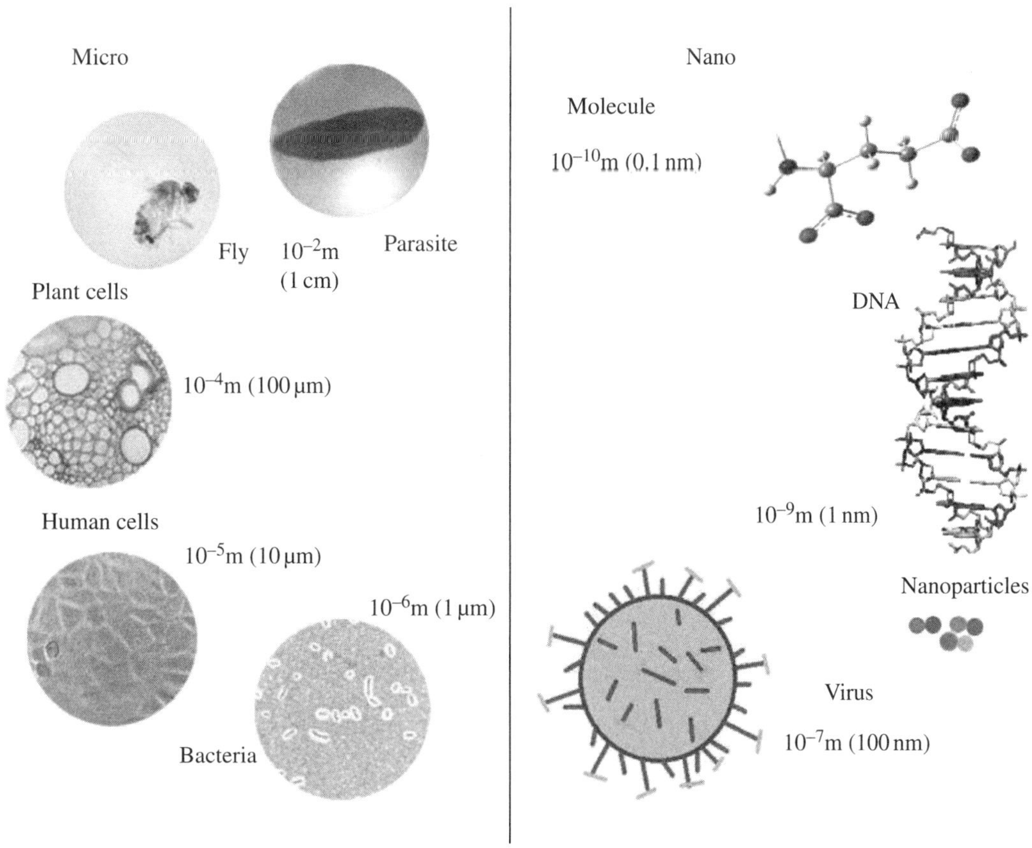

FIGURE 30.5 Comparative size scales of nanomaterials and biological systems.

Due to their reduced size, NMs may pass through several important biological barriers. An average cell membrane is able to avoid internalization of nanoparticles larger than 6 nm, although by endocytosis, materials up to 100 nm may enter the intracellular space. The nuclear membrane can stop particles smaller than 40 nm. The blood–brain barrier (BBB) filters particles up to 35 nm, while the alveolar–capillary barrier, up to 10–24 nm. In the kidney, the renal system is able to resist particles in the range of 8–12 nm, while the skin has a dermal barrier efficient in the range from 20 to 30 nm. The gastric mucosae are not very selective, allowing particles less than 500 nm in size to move across [26].

Although apparently we may have a good knowledge of an NM's chemical and physical characteristics, there is a lack of understanding of the intracellular activity and impact of engineered NMs on cell function. They may interact with a single cell in different ways than a tissue or whole organism, determining that simple *in vivo* models may not be suitable for complete interpretation. They may also coat their surfaces, such as, proteins, antibodies, and small biomolecules, depending on the type of biological fluid they are in contact with (blood, plasma, interstitial fluid), avoiding the immune system. They may even affect intracellular responses, inducing damage or beneficial responses. Nanostructured metal oxides, for example, may generate reactive oxygen species such as singlet oxygen, superoxide, and peroxide, as well as participating in oxidation–reduction processes on the cell surface, which may degrade cell membranes, proteins, and even DNA. Interaction with biomolecules may also induce changes in their functional structures or block the active sites of enzymes, which in turn will not always have good metabolic consequences [3] (Fig. 30.6).

In econanotoxicology, it is important to understand how NMs will interact with a living organism from the moment they are exposed to it till their degradation or elimination, as well as whether these materials (or their by-products) are bioacummulated within cells, tissues, and organs, thus inducing intracellular changes, inflammatory responses, or undesirable effects culminating in metabolic illness. Because nanotoxicology is a new research topic of interest, there are many contributions as attempts to standardize the evaluation of NM toxicity, considering that the interaction of these materials with dying agents, DNA, and cellular structures could cause some variability in data interpretation and must be validated carefully [27, 28].

Ecotoxicity tests are tools used within environmental hazard assessment frameworks to answer questions about the intrinsic dangers of chemical substances that may be released into the environment [29]. These tools can be applied to NMs when they are evaluated, and the exposure scenarios should be replicated using *in vitro* and *in vivo* toxicity assays to know the potential

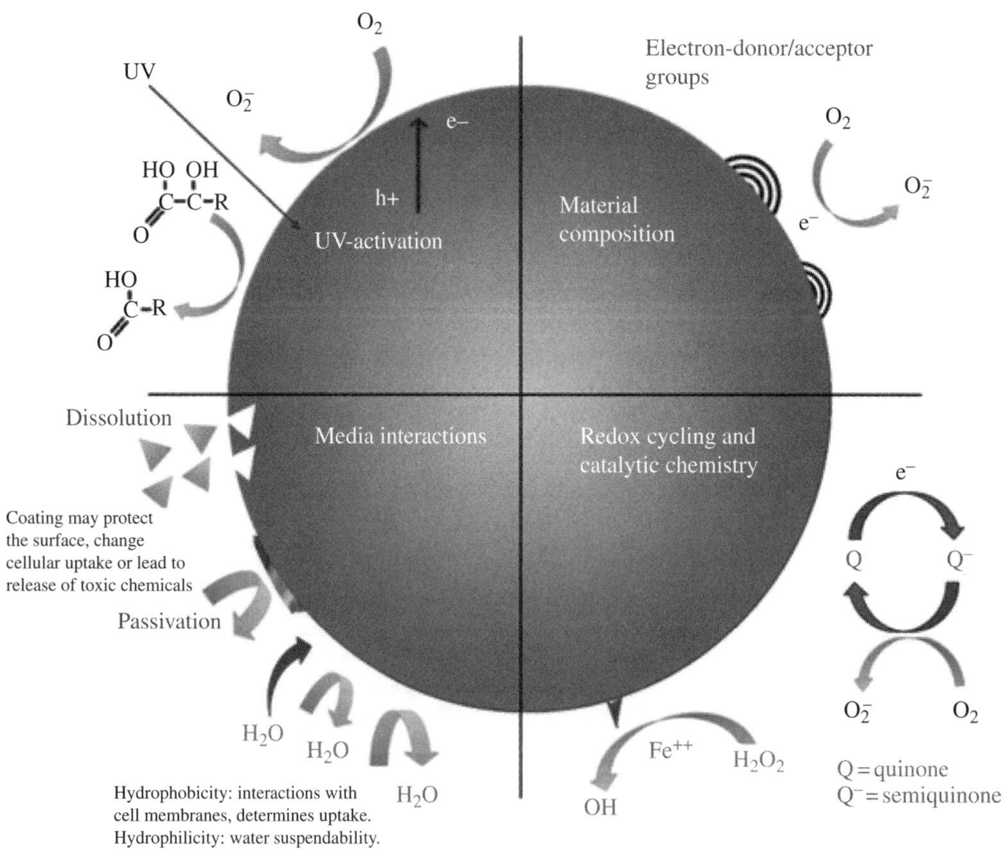

FIGURE 30.6 Possible mechanisms by which nanoparticles interact with biological tissue. Adapted from Ref. [3].

health risks. One problem with *in vitro* assays is that the results that are obtained in this study cannot guarantee biocompatibility *in vivo*, and therefore data from *in vitro* studies may be misleading and will require verification through animal evaluations [30].

In order to understand the impact of NMs on the environment and living systems, several specific methods have been developed. They can be grouped into four categories: (1) chemical and physical characterization; (2) microbiological assays; (3) *in vitro* assays; and (4) *in vivo* assays.

30.2.1 Chemical and Physical Characterization

Very sophisticated and specialized analytical instrumentation has been developed to obtain some fundamental physical information about NMs that we desire to study and is already available in major facilities around the globe [31]. Several well-established techniques are also available such as scanning or transmission electron microscopy (SEM, TEM) to obtain precise information about the size, morphology, and chemical composition (when energy-dispersive X-ray (EDX) detectors are available) of NMs. With dynamic light scattering (DLS) instrumentation it is possible to determine the hydrodynamic radii of the nanoparticles when dispersed in a liquid, and it is possible to study the influence of pH on the surficial charge (Zeta potential), the nature of the solvent, its temperature, the effect of capping agents and detergents in the stabilization of nanoparticles, the stability against time, pH, and temperature, and to understand the kinetics of aggregation in a solution. The specific area of powdered materials can be obtained by using Brunauer–Emmett–Teller (BET) analyzers; thermal stability and transformation may be determined by thermal gravimetric analyzers (TGAn); and chemical composition and presence of contaminants can be determined by atomic absorption spectrophotometry (AASp) or inductively coupled plasma mass spectrometry (ICP-MS). Other spectroscopies such as visible-ultraviolet, infrared, or Raman may also be useful in defining the existence of organic or inorganic coatings, chemical modifications in the surface, and chemical identity, among other characteristics.

Of course, once the chemical and physical analyses show some relevant data, *in vitro* and *in vivo* studies may give us complementary information. Biotests using bacterial, cells or tissue cultures, animal models (mice, rats, rabbits, dogs, fishes, etc.), and ecotoxicity models (*Micropterus salmoides*, *Caenorhabditis elegans*, *Daphnia magna*, *Ceriodaphnia dubia*, common fruit fly, *Drosophila melanogaster*, and some invertebrates and small vertebrates) are currently among the most used in research laboratories around the world.

30.2.2 Microbiological Assays

Different approaches can be used to assess bacterial toxicity using well-characterized materials and standard bacterial assay systems. It is possible to examine the effects of nanoparticle concentration, particle size, exposure time, growth medium, and pH on the growth and viability of bacterial cells like *Escherichia coli*, *Bacillus subtilis*, and *Shewanella oneidensis*. Among other methods to assess NM bacterial toxicity, we can mention the following:

Disk diffusion tests. Bacterial sensitivity to different-sized NMs is tested by disk diffusion test as described by Ruparelia [32]. Small filter paper disks of uniform size (i.e., 6 mm diameter) are placed separately in each of the different nanoparticle suspensions for 5 min; then the disks are carefully removed using sterile forceps. After the bacterial suspension (100 µl of 10^4–10^5 CFU ml^{-1}) is uniformly plated on Luria–Bertani (LB) agar plates or other rich media, a disk containing nanoparticles is placed at the center of each plate and the plate is incubated at 37°C for 18 h. The average diameter of the inhibition zone (DIZ) surrounding the disks is measured to determine inhibition. This simple method gives us an idea as to whether some NMs have any activity; however, sometimes one might need to use minimum media in place of rich media (i.e., M9 media) to see if any effect is present.

Determination of minimum inhibitory concentrations (MIC). The MIC is defined as the lowest concentration of a compound that inhibits the growth of an organism [33]. The MIC test can be determined for *E. coli* in LB medium at pH 7.2 and/or in M9 minimal medium at pH 6.4 [34], 7.2, and 7.8. For *B. subtilis*, MIC can be tested in LB and minimal media at pH 7.2 [35], and for *S. oneidensis* in LB and horse blood agar (HBA) minimal media at pH 7.2 [36]. Reactions are carried out in test tubes containing 5 ml of the logarithmic-phase (OD_{600} ~0.098) bacterial cultures and different-sized nanoparticles at various concentrations (i.e., 50, 100, and 150 mg/l). Tubes with sterile media containing no nanoparticles and nanoparticles only served as controls. Samples are incubated on a shaker (200 rpm) at 37°C (*E. coli* and *B. subtilis*) or 30°C (*S. oneidensis*), with growth monitored by obtaining measurements of the optical density at 600 nm (OD_{600}) every 30 min for 8 h. In the end, the last tube with no growth corresponds to the MIC of that compound.

Colony-forming unit (CFU) measurements. Studies of *E. coli* and *B. subtilis* viability are performed in liquid cultures at a nanoparticle concentration of 100 mg/l (or the proper concentration according to the NM). Aliquots are taken at 0, 1, 5, and 24 h and serially diluted in the appropriate minimal medium, and the dilutions are seeded on LB agar plates. After overnight incubation at 37°C, the numbers of CFU are counted manually.

Live/dead viability assays. E. coli and *B. subtilis* cultures grown to logarithmic phase in M9 medium and *B. subtilis* in minimal medium, respectively, are treated with different concentrations (i.e., 50, 100, and 150 mg/l) of nanoparticles. Following exposure, the impact on bacterial membrane integrity is assessed using a live/dead BacLight bacterial viability kit. To quantify the relative numbers of live and dead cells, the relative fluorescence intensities are measured using a fluorescence plate reader (excitation at 485 nm, emission at 525 and 625 nm).

Monitoring superoxide production. Superoxide production upon exposure of bacterial suspensions to various concentrations of nanoparticles is monitored by following the absorbance at 470 nm due to the reduction of 100 µM 2,3-bis(2-methoxy-4-nitro-5-sulfophenyl)-2*H*-tetrazolium-5-carboxanilide (XTT) to XTT-formazan by superoxide (O_2^-) [37, 38].

Microarray hybridization and analysis. For microarray experiments, an overnight *E. coli* culture is used to inoculate 250 ml flasks containing 100 ml of prewarmed M9 medium to an OD600 of approximately 0.1 and incubated at 37°C with shaking at 200 rpm until the mid-log phase (OD600, ~0.5). Cultures are treated with either prewarmed nanoparticles (100 mg/l) or Milli-Q water. After 1 h, cells are harvested by rapid centrifugation (5000 × *g*, 2 min at 4°C) and snap-freezing in liquid N_2. Three separate controls and three experimental cultures are examined for each condition. Total cellular ribonucleic acid (RNA) is isolated as described by Brown and Pelletier: the cells are first resuspended in TE (Tris 10 mM-EDTA 1 mM, pH 7.6) buffer and incubated with 1 mg/ml of lysozyme to lyse the cells [39, 40]. Purified, fluorescently labeled cDNA is hybridized to *E. coli* K-12 gene expression 4-by-72 K arrays (or other microarrays) using a Nimblegen hybridization system. Microarrays are washed according to the array manufacturer's procedure. Briefly, microarray mixers are removed in 42°C Nimblegen wash buffer I and then washed manually in room-temperature buffers: wash buffer I for 2 min, wash buffer II for 1 min, and wash buffer III for 15 s. Microarrays are dried for 80 s using a Maui wash system and then scanned and the images quantified. Microarray data are normalized using the Lowess normalization algorithm, and an analysis of variance (ANOVA) is performed to determine significant differences in gene expression levels between conditions and time points using the false discovery rate testing method ($P < 0.01$).

Cyanobacteria and green algae models. These microorganisms have been also used to determine the toxicity of an NM due to their ecological position at the base of the aquatic food chain and their essential role in nutrient cycling and oxygen production. Cyanobacteria constitute a phylum of bacteria that obtain their energy through plant-like photosynthesis. They are the most widespread primary producers in the marine food chain and are crucial in many other habitats including freshwater bodies, saline lakes, and biological soil crust.

For example, the toxicity of nano-CeO_2 suspension was determined by monitoring the growth inhibition of the green alga *Pseudokirchneriella subcapitata* and by determining the constitutive luminescence inhibition of the recombinant bioluminescent cyanobacterium *Anabaena* CPB4337. The bioassays using the bioluminescent cyanobacterium *Anabaena* CPB4337 are based on the inhibition of constitutive luminescence caused by the presence of toxins [41].

30.2.3 *In Vitro* Assays

In order to evaluate the biological activity and/or toxicity of NMs, some alternatives have been explored to determine the effect of a particle upon a living organism. Conventional *in vitro* analyses and cell-based assays were performed to obtain an estimate that could mimic the *in vivo* physiologic environment of a living being, and thus determine their possible biological risk in case the material is toxic. To determine the metabolic state of a group of cells we must consider a concept known as cell viability, which indicates the potential of this group of cells to proliferate and grow. A normal cell population must be metabolically active in culture, which must indicate that all their functions are normal. In toxicology, there are many ways to determine cell viability, from simple dye exclusions to the use of sophisticated instruments. In nanotoxicology, the same techniques have been explored that have been used in toxicology to evaluate the effect of an NM when a cell population is exposed; nevertheless, these studies have not resulted in the creation of standards that could be useful to most of the new NMs released into the environment [28]. Besides, *in vitro* analyses are very popular because of their established methodologies, low costs, broad number of replicates, small setups, safety and efficacy, and a few ethical issues. The important advantage of *in vitro* testing of NMs is that it is a solution for the replacement or reduction of laboratory animals, reducing at the same time the uncertainty caused due to the variability between individuals [42]. As with other man-made materials such as cosmetics and drugs, *in vitro* evaluation of NMs needs to be performed because of the increase in nanostructured materials and nanoparticles that are released into the environment. Table 30.1 summarizes the most popular *in vitro* analyses employed in nanotoxicology, some of them validated by the OECD, the European Union Reference Laboratory for Alternatives to Animal Testing (EURL ECVAM), and the National Institutes of Health via the Nanotechnology Characterization Laboratory (NCL-NIH). Many of them include common dye exclusions, indirect determinations of metabolic disruptions, microscopy analyses, and cell viability determinations by impedance.

TABLE 30.1 Summary of common toxicity assays for different nanomaterials

Assay	Purposes	Applications in nanotoxicology	Cell line	References
Neutral red uptake	Cell viability	Carbon nanotubes	A549 pulmonary cell line	OECD Draft Guidance 129 [43]
	Cell death	C_{60}	THP-I pulmonary cell line	ECVAM validated [43]
	Phototoxicity	Fe_3O_4	Human monocyte macrophages	[44]
Trypan blue	Proliferative ability	Colloidal Ag	Bronchial epithelial cell	[45]
	Cell viability	MoO_3	BEAS-2B	[46]
	Cell growth	Fe_3O_4	Rat liver cells BRL 3A	[47]
		TiO_2	Human epidermal cells A431	[48]
		Carbon-based nanoparticles		
		Gold nanoparticles		
Fluorescein diacetate derivatives	Reactive oxygen species	Copper oxide	Human epithelial cells HEp-2	[49]
		TiO_2	SD Primary hepatocytes	[50]
Lactate dehydrogenase assay	Disruptions in cell membrane	Fe_3O_4	Hepatocarcinoma cells Hep-G2	[51]
		Metallic Cu	Kidney cells LLC-PK1	[52]
		CuO	Rat liver cells BRL 3A	[46]
		TiO_2		
		Metallic Ag		
		Titania stoichiometric		
		Metallic cobalt		
		Copper–zinc mixed oxide variants		
		Nickel		
		Nickel oxide		
		Zirconia		
		Alumina		
		Tungsten carbide		
		CdO		
		MoO_3		
		MnO_2		
Tetrazolium salts (MTT, MTS)	Letal dose 50%	Fe_3O_4	Hepatocarcinoma cells Hep-G2	[53]
	Mitochondrial activity	Metallic Cu	A549 pulmonary cell line	[51]
		CuO	THP-I pulmonary cell line	[52]
		TiO_2	Kidney cells LLC-PK1	
		Metallic Ag	Rat liver cells BRL 3A	
		Titania stoichiometric		
		Metallic cobalt		
		Copper–zinc mixed oxide variants		
		Nickel		
		Nickel oxide		
		Zirconia		
		Alumina		
		Tungsten carbide		
		CdO		
		MoO_3		
		MnO_2		

(*Continued*)

TABLE 30.1 (*Continued*)

Assay	Purposes	Applications in nanotoxicology	Cell line	References
Glutathione reduced assay	Cell signaling activity	CdO TiO_2 MoO_3 Metallic Al Colloidal Ag	Rat liver cells BRL 3A Human epidermal cells A431	[46] [47] [54]
Caspase activation kits	Cell apoptosis	Fe_3O_4 TiO_2	Human fibroblasts hTERT-BJ1 Kidney cells LLC-PK1 Hepatocarcinoma cells Hep-G2	[55] [56] [57] [58]
Real-time cell electronic sensing system	Cytotoxicity	TiO_2 Gold nanoparticles	NK 92 Cells Breast cancer cells MDA-231-B and MCF-7 Human alveolar epithelial cells L-132 Human glioblastoma cells T98G Human primary fibroblast cells AGO-1522B	[59] [60]
Electron microscopy	Nanoparticle shape and intracellular localization	CNT TiO_2 CeO Fe_3O_4	Human epidermal keratinocytes (HEK) Neural stem cells Bronchial epithelial cells	[26] [61] [62] [55] [63]

[a]OECD and EVCAM guidelines are summarized in Ref. [43].

In this way, cytotoxicity tests can be used to predict the acute toxicity of NMs, and 2D and 3D assays are used to address their specific localization. It is important to mention that in all colorimetric assays—in addition to counting the cell number—it is very important to measure any specific damage, and not necessarily only total death, because if there is any inhibition of growth during the experiment that is not caused directly by the application of nanoparticles, this decrease in cell growth could be estimated as a false positive. Besides, all colorimetric assays have demonstrated that they are liable to the interaction between dye agents and NMs, and it is very difficult to wash out the remaining dye [28, 43]. At this moment, 2D and 3D systems such as light and electron microscopy, and real-time analyses by impedance or platting efficiency assays appear to be the most reliable systems because of the elimination of additional chemical treatments and the constant monitoring of the culture [43, 50, 64, 60, 65].

30.2.4 *In Vivo* Assays

Living systems are potentially exposed to NMs through ingestion, ocular, dermal, or inhalation pathways. This exposure can occur when environmental pollutans are present in air, water, or soil. We do not know the effects of many of these materials on our health, and if the *in vitro* assays do not satisfactorily demonstrate the potential effects, then testing in an *in vivo* model is needed. In order to select an appropriate and representative *in vivo* model it is important to consider the NMs' route of exposure. For example, for the oral route, although several mammalian test species may be used, the rat is the preferred species. In the case of dermal exposure, the most common animal models are the rat, the rabbit, and the guinea pig, but the albino rabbit is preferred because of its size, ease of handling, skin permeability, and extensive database. Commonly used laboratory strains must be employed. If a species other than rats, rabbits, or guinea pigs is used, the tester must provide justification and reasoning for its selection [66].

Using a variety of techniques, typical *in vivo* assessments include the determination of physiological localization and the concentration of material in specific tissues, rate of excretion, macroscopic tissue analysis, and organism toxicity [67].

Short-term ("acute") tests are generally used first, with observations of organism survival being the most common measurement of effect. Longer-term ("chronic") tests (with observation of sublethal effects on organism growth or reproduction being the most common measurement of effect) are then used when results from short-term tests combined with large safety factors suggest that there may be risks to the environment [29].

Generally, the initial step for the assessment and evaluation of the toxic characteristics of a substance is to determine the acute oral toxicity. This provides information on health hazards likely to arise from short-term exposure by the oral route [66]. The term acute toxicity is used to describe the adverse effects of a substance that may result from a single dose of a substance or multiple doses given within a 24 h period. The studies are carried out via oral, dermal, and inhalation routes of exposure for the purpose of estimating doses that cause lethality. Acute effects may be local and/or systemic [66, 68].

Acute toxicity tests are performed to obtain information on the biological activity of a chemical and its mechanism of action at different levels including cell components. This permits us to obtain information for the identification and risk management in the related context of production, handling, and use of the chemical [69]. The LD_{50} (median lethal dose) value is currently the basis for toxicologic classification of chemicals and is defined as the statistically derived dose that, when administered in an acute toxicity test, is expected to cause death in 50% of the treated animals in a given period. The LD_{50} value is expressed in terms of weight of test substance per unit weight of test animal (milligrams per kilogram). In the last years, acute systemic toxicity studies have been among the most criticized of all toxicology tests on both scientific and ethical grounds. Newer, preferred methods are now trying to employ dose selection lethality limits instead of LD_{50}, applying the 3Rs principle (refinement, reduction, and replacement of animal use) [66, 68].

In case there is a need to use animals, the requirement should be reviewed and approved by the institutional animal ethics committee. The National Toxicology Program (NTP) regularly evaluates substances for a variety of health-related effects. Rodents are the most common animal models used by the NTP [70]. Studies for general toxicology using rodents include single-dose acute studies, repeat-dose studies at 2, 4, 13, 26, or 52 week's duration, carcinogenic studies with or without genetically modified animal models, as well as sensitization and irritation studies.

Generally, the testing laboratories adhere to the principles enunciated in the *Guide for the Care and Use of Laboratory Animals* [71]. The regular NTP *in vivo* procedures are as follows:

Perinatal exposure. This range-finding study determines whether there is maternal toxicity and/or toxicity to the pups in order to provide a basis for determining the doses for the subsequent toxicity study (13-week or 2-year study). The animals are exposed to the substance during in utero development, through their mother's milk, and via dosed feed, dosed water, or gavage administration.

14-Day Toxicity Protocol. The goal of this is to provide a basis for identifying potential target organs and toxicities and to assist in setting doses for the 13-week exposure study. After a 10–14-day quarantine period, animals are assigned randomly to treatment groups. The study includes five treatment groups, each administered a different concentration of test article per sex per species plus a control group. Each group per sex per species contains five animals. The animals receive the test article through a designated route of exposure and the control animals receive the vehicle alone.

13-Week Toxicity Study. In addition to obtaining toxicological data, the purpose of this study is to determine the treatments for each strain and species to be used in the 2-year toxicology/carcinogenesis study. Basically, after a 10–14-day quarantine period, animals are assigned randomly to treatment groups. The study includes five treatment groups, each administered a different concentration of the problem material plus a control group. Each group contains 10 animals per sex per species. The animals receive the test chemical by a designated route of exposure. Controls receive untreated water or feed or vehicle alone in gavage and dermal studies. For dosed-feed and dosed-water studies, animals are exposed for 90 days, after which they are sacrificed with no recovery period. For inhalation, gavage, and dermal studies, animals are exposed five times per week, on weekdays only, until the day prior to necropsy.

2-Year Study Protocol. The purpose of this study is to determine the toxicological and/or carcinogenic effects of long-term exposure on rats and mice. Typically, after a 10–14-day quarantine period, animals are assigned randomly to treatment groups. Rats and mice receive the test agent for 104 weeks via a defined route of exposure at three treatment concentrations plus controls. For inhalation, gavage, and dermal studies, animals are treated five times per week, on weekdays only [70].

In *in vivo* assays, the evaluation includes identifying the treatment-related lesions in target organs. In mammals, the organ weights of at least liver, thymus, right kidney, right testis, heart, and lung are recorded from all animals surviving until the end of the study. A complete necropsy is performed on all treated and control animals that either die or are sacrificed. All tissues required for complete histopathology are prepared and stained with hematoxylin and eosin for histopathology evaluation.

In the NTP's *in vivo* procedures, all the studied animals are weighed individually on day one of the test, after 7 days, and at weekly periods thereafter. Animals are observed twice daily, at least 6 h apart, including on holidays and weekends, for morbidity and death. Animals found moribund or showing clinical signs of pain or distress are humanely euthanized. Formal clinical observations are performed and recorded weekly. For dosed-feed or dosed-water studies, food consumption/water

TABLE 30.2 Blood clinical measurements

Hematology	Clinical chemistry
Red blood cell count (RBC)	Sorbitol dehydrogenase (SDH)
Mean corpuscular volume (MCV)	Alkaline phosphatase (ALP)
Hemoglobin (Hb)	Creatine kinase (CK)
Hematocrit	Creatinine
Mean corpuscular hemoglobin (MHC)	Total protein
Mean corpuscular hemoglobin concentration (MCHC)	Albumin
Erythrocyte morphologic assessment	Blood urea nitrogen (BUN)
Leukocyte count (WBC)	Total bile acids
Leukocyte differential	Alanine aminotransferase (ALT)
Reticulocyte count	Glucose
Platelet count and morphologic assessment	Cholesterol
	Triglycerides

consumption is measured and recorded weekly. In 2-year study procedures, individual animal body weights for test and control group animals are recorded on day one of the test and at 4-week intervals thereafter except for dosed-feed and dosed-water studies, in which these are recorded weekly for the first 13 weeks and monthly thereafter [70].

Additionally, specific toxicological parameters can be evaluated and processed for hematology and clinical chemistry determinations. Blood is collected from core study mice at the end of the study for hematology determinations (Table 30.2). Other studies such as micronuclei determinations in blood cells, genotoxicity, sperm morphology, and vaginal cytology evaluations are also used.

Another type of studies includes immunotoxicology probes. Assessment of the adverse effects on the immune system is an important component for evaluating the overall health and safety of NMs. The immune system is constantly functioning to maintain homeostasis, eliminating pathogens and removing cancerous cells. Small modifications to the immune system, which may occur following NM exposure, could lead to impaired protection or inappropriate immune response resulting in autoimmunity and damage to the host [72]. The most common effects include an increased susceptibility to infections or cancer, autoimmune diseases, chronic inflammation, or allergies. There are large spectrums of *in vitro* and *in vivo* immunological assays in comprehensive immunotoxicity studies. These include assays of immunochemistry (quantification of cytokines), immunogenicity (antibodies), immunopathology (relative weight and histopathology of lymphatic organs), immunophenotyping (analysis of cell origin), functional test (analysis of macrophage and granulocyte functions), hypersensivity testing, infection models (bacterial, viral, fungal models), and asthma models.

For example, Lee evaluated the immunotoxicity of silica nanoparticles *in vivo* [30]. These nanoparticles have been used in chemical mechanical polishing, varnishes, cosmetics, food, and biomedical devices. Although silica is generally considered to be noncytotoxic, designing silica as NMs may change its biocompatibility because of the changes in its physicochemical properties. In an *in vivo* assay, animals received silica nanoparticles suspended in distilled water for 4 weeks (5 days/week). The results indicated that *in vivo* exposure to silica nanoparticles caused damage to systemic immunity through the dysregulation of the spleen, but the *in vivo* data were inconsistent with *in vitro* data, which showed lower cytotoxicity for silica nanoparticles. This is an example of the importance of verifying biocompatibility both *in vitro* and *in vivo* during the design of new NMs and therefore data from *in vitro* studies require verification through animal evaluations [30].

On the other hand, in humans, the most critical exposure routes for NMs are inhalation and skin contact, although the adverse effects are mainly expected to occur in the lungs [73]. *In vivo*, there are combinations of particle delivery techniques such as intratracheal instillation/aspiration/inhalation or nose-only/whole-body inhalation that could be used as a means to study the pulmonary and systemic effects of nanoparticles. The evaluation of respiratory tract toxicity from airborne materials frequently involves exposure of animals via inhalation. This provides a natural route of entry into the host and, as such, is the preferred method for the introduction of toxicants into the lungs. However, for various reasons, this technique cannot always be used, and the direct instillation of a test material into the lungs via the trachea has been employed in many studies as an alternative exposure procedure.

For example, Horie and coworkers [74] evaluated the pulmonary toxicity of MWCNTs by intratracheal instillation in rat. The MWCNT dispersion was administered to rat lung by single intratracheal instillation at doses of 0.2 mg and 0.6 mg/rat. Bronchoalveolar lavage fluid (BALF) was collected at 3 days, 1 week, 1 month, 3 months, and 6 months after instillation. They found that the intratracheal instillation of MWCNTs induced persistent inflammation in rat lung not only in the high-dose group but also in the low-dose group [74].

More recently, efforts have been focused on the development and validation of new alternative test systems (sensitive, specific, rapid) for toxicological research that will reduce, replace, or refine animal use. Model systems under development include nonmammalian species, transgenic species, genetically engineered *in vitro* cell systems, microchip array technology, and computer-based predictive toxicology models [70]. Fish and amphibian embryo models are gaining increasing popularity in the area of toxicology, both in research and potential regulatory application. The fish and amphibian embryo models provide an ethically acceptable small-scale analysis system with the complexity of a complete organism.

The OECD is an intergovernmental organization in which representatives from 34 industrialized countries of North and South America, Europe, Asia, and the Pacific region, as well as the European Commission, meet to coordinate and harmonize policies, discuss issues of mutual concern, and work together to respond to international problems such as NM ecotoxicity. The OECD's WPMN was established in 2006 to promote international cooperation in human health and environmental safety aspects of manufactured NMs. The OECD program has focused on generating appropriate methods and strategies to ensure potential safety issues through the following approaches:

- Establishing an OECD database on manufactured NMs to inform and analyze research activities and strategies on environmental, human health, and safety issues
- Testing specific NMs for human health and safety evaluation, while ensuring appropriate testing methods (*in vivo* and *in vitro*).

Some standards of OECD ecotoxicity *in vivo* tests include water flea acute (*Daphnia magna*), where the dosing method is natural water by 48 h and the test end point is the half maximal effective concentration (EC_{50}) [75]. *D. magna* is an organism widely used as an indicator in aquatic environmental risk assessment because *Daphnia* filter large volumes of water and water-suspended particles. It also plays an important role in freshwater food chains [76, 77]. These features make *D. magna* a particularly useful test animal for assessing the accumulation of NMs, because their uptake in this organism could result in transfer throughout the food chain. *D. magna* may be grown in artificial freshwater (Ca + Mg hardness 2.5 mM, pH 6.5–7.1) with a photoperiod of 16:8 light:dark at 20 ± 2°C. The population is fed three times a week with a green algae culture of *Scenedusmus sp.* (dominant species), *Monoraphidium contortum*, and *Selenastrum capricornutum*. Organisms used in tests must be 5–7 days old at the beginning of the experiments.

For fish acute (*Zebrafish*), the test medium is natural water by 96 h and the test end point is 50% of the maximum lethal concentration (LC_{50}). In the case of prolonged fish toxicity (*Zebrafish*), the study is monitored for 14 days following EC/LC_{50} until the test end point. There are *in vivo* assays using birds, such as bird dietary toxicity, where the doses are applied in the basal diet for 8 days and the test end point is LC_{50}. Other *in vivo* assays include other fish species, honey bee, earthworms, and plants. These are some examples of *in vivo* assays approved by the OECD [68].

Aditionally, studies to evaluate how NMs may affect the different development stages of plants are also an easy alternative to assess their potential environmental effects. For example, the germination and growing of seeds of *Lactuca sativa* were tested by Yang and Watts [78]. *L. sativa* is one species that is used and recommended by the Environmental Protection Agency (EPA) regularly for measuring pesticide and toxic substances in the environment. The germination average rate is usually 85%, and the seeds have to be stored in dry and dark places at room temperature. Initially, the seeds are wetted in a bleach solution (10% from commercial product) for 10 min to eliminate biological contaminants. The seeds are then rinsed three times and set up for germination immediately. In a plastic tray (transparent), squares of 2.25 cm^2 are drawn to accommodate the seeds in each intersection. The number of seeds will depend on how many substances should be tested and must be by triplicate. The system will include the NM to be tested in solution at different concentrations (i.e., 0.1, 0.5, 0.75 mg/ml) in sterile deionized water. All positive and negative controls must be considered.

The seeds are incubated by 168 h using a photoperiod of 12 h (light and dark) and temperature at 25°C ± 0.5°C. Observation and counting registration of germination is recorded each 24 h. The root size (mm) is also registered using a Vernier scale and the exposed seeds are compared with the nonexposed ones. The elongation root during the exposure is calculated using the formula

$$\mathrm{ER} = \mathrm{L}_{\text{treated}} - \mathrm{L}_{\text{non-treated}}$$

$$\mathrm{RRG} = \mathrm{ER}_{\text{sample}} - \mathrm{ER}_{\text{control}}$$

where L_{treated} and $L_{\text{non-treated}}$ are the length of roots with or without treatment, respectively. The relative root growth (RRG) is calculated according to Schildknecht and de Campos-Vidal (2002, cited in Ref. [70]).

30.3 CONCLUSIONS

Although no large-scale spills have been reported and documented to evaluate the real ecological impacts of NMs in the environment, there is a genuine concern by several groups claiming the implementation of international standard methods and procedures for environmental, health, and safety testing, in order to establish solid arguments to confirm or deny the potential and highly polemic hazards of NMs in actual or future use. There are several opportunities to develop new methods for experimentally testing the potential impact of NMs on the environment, in particular, simple, cheap, and fast methods that may correlate specific physical properties with biological activity. Due to the lack of definitive information for most of the actual (and future) available NMs, nanoecotoxicology seems to be a field of opportunities of research for scientists in materials and environmental sciences. Gross tests of cytotoxicity are still required to screen many effects, but there is a growing need to supplement them with more subtle tests of metabolic pathway regulation and signaling and biological models, as the responses will vary with different compounds. It is important to interpret *in vitro* results in terms of *in vivo* responses on the same or similar cells. However, the *in vitro* system regularly lacks many factors that the *in vivo* system posesses, such as blood stream, blood pressure, O_2/CO_2 pressure and concentration, hormonal changes, and osmolality, among several others. The nature of the response has to be considered carefully. A toxic response *in vitro* may be the result of changes in cell survival metabolism, whereas the major problem *in vivo* may be a tissue response (e.g., inflammatory reaction, fibrosis, organ failure) or a systemic response (e.g., pyrexia, vascular dilatation). For *in vitro* testing to become effective, models of these responses must be developed and simulated *in vitro*.

Several environmental groups, nongovernmental organizations, and academic organizations have been involved in public discussions about the fears surrounding the production, commercial use, and disposal of NMs, but we have no conclusive information to definitively answer the central questions regarding the environmental impacts of ENMs.

In order to avoid a public rejection and misinformation of the topic, it is important to have a continuous and responsible exchange of information between the society and scientists, discussing real scientific facts and not only fears instilled by sci-fi books, partial interpretation of facts, or pseudo-scientific ideas. In a way, it seems like the most real—and immediate—danger for humanity, involving nanotechnology, has resulted from the misinterpretation of its real benefits and hazards. Some civil, nongovernmental groups such as the ETC an international organization dedicated to the conservation and sustainable advancement of cultural and ecological diversity and human rights, among several others follow very closely the development of new technologies (including genomics, biotechnology, and nanotechnology), but sometimes share very limited and biased documents with their followers, exposing polemic points of view regarding problems related to such technologies [79]. A coalition of this organization and other consumer safety and environmental groups (CTA International Center for Technology Assessment, Center for Environmental Health, Food, and Water Watch, Friends of the Earth, and the Institute for Agriculture and Trade Policy) even filed a lawsuit against the Food and Drug Administration (FDA) over the health and environmental risks of nanotechnology and NMs [80]. It is likely that nanotechnology may become a double-edged sword if it falls in wrong hands. Misconceptions have to be cleared to avoid public misinterpretation of their real utility—or danger [81].

ACKNOWLEDGMENTS

Our thanks to Xiomara G. Fernandez, Violeta Fernández, Fernando Arteaga, Lizette A. Minjarez, Enrique Gonzalez, and Astrid Espinoza (UDLAP) for helping with the design and the making of graphic art for this manuscript.

REFERENCES

[1] Kahru A, Duborguier H–C. From ecotoxicology to nanoecotoxicology. Toxicology 2010;269:105–119.

[2] Newman MC, Zhao Y. Ecotoxicology nomenclature: LC, LD, LOC, LOEC, MAC. In: Jorgensen SE, Fath B, editors. *Encyclopedia of Ecology*. Amsterdam, Netherlands: Elsevier; 2008. p 1187–1193.

[3] Nel A, Xia T, Mädler L, Li N. Toxic potential of materials at nanolevel. Science 2006;311:622–627.

[4] Buzea C, Pacheco II, Robbie K. Nanomaterials and nanoparticles: sources and toxicity. Biointerphases 2007;2:MR17–MR71.

[5] Ge Y, Schimel JP, Holden PA. Evidence for negative effects of TiO2 and ZnO nanoparticles on soil bacterial communities. Environ Sci Technol 2011;45:1659–1664.

[6] Rousk J, Ackermann K, Curling SF, Jones DL. Comparative toxicity of nanoparticulate CuO and ZnO to soil bacterial communities. PLOS ONE 2012;7:e34197. Available at http://www.plosone.org/article/info%3Adoi%2F10.1371%2Fjournal.pone.0034197. Accessed March 19, 2013.

[7] Colvin VL. The potential environmental impact of engineered nanomaterials. Nat Biotechnol 2003;21:1166–1170.

[8] Oberdöster E, Larkin P, Rogers J. Rapid environmental impact screening for engineered nanomaterials: a case study using microarray technology. In Project on Emerging Nanotechnologies. Wilson Center and the Pew Charitable Trust; 2005. Available at http://eprints.internano.org/77/1/Environmental_Impact_Screening.pdf. Accesed March 13, 2013.

[9] Lewinski N. Nanomaterials: what are the environmental and health impacts? Chem Eng Prog 2008;104:37–40.

[10] Klaine SJ, Alvarez PJJ, Batley GE, Fernandes TF, Handy RD, Lyon DY, Mahendra S, McLaughlin MJ, Lead JR. Nano-materials in the environment: behavior, fate, bioavailability, and effects. Environ Toxicol Chem 2008;27:1825–1851.

[11] Alvarez PJJ, Colvin VL, Lead J, Stone V. Research priorities to advance eco-responsible nanotechnology. ACS Nano. 2009;3: 1616–1619.

[12] (a) Wijnhoven, SWP, Dekkers S, Kooi M, Jongeneel WP, de Jong WH. 2011. Nanomaterials in consumer products, RIVM Report 34370003/2010. National Institute for Public Health and the Environment, Netherlands. Available at http://www.rivm.nl/bibliotheek/rapporten/320029001.pdf. Accessed March 19, 2013.
(b). Wijnhoven SWP., Dekkers S, Hagens WI, de Jong WH. 2009. Exposure to nanomaterials in consumer products, RIVM Letter Report 340370001/2009, National Institute for Public Health and the Environment, Netherlands. Available at http://www.rivm.nl/bibliotheek/rapporten/340370001.pdf. Accessed March 19, 2013.

[13] Project of Emerging Nanotechnologies. 2009. Consumer products inventory. Available at http://www.nanotechproject.org/inventories/consumer/. Accessed March 19, 2013.

[14] Som C, Berges M, Chaudhry Q, Dusinska M, Fernandes TF, Olsen SI, Nowack B. The importance of live cycle concepts for the developments of safe nanoproducts. Toxicology 2010;269:160–169.

[15] Barnard AS. Nanohazards: knowledge is our first defense. Nat Mater 2006;5:245–248.

[16] Brayner R, Ferrari-Iliou R, Brivois N, Djediat S, Benedetti MF, Fievet F. Toxicological impact studies based on Escherichia coli bacteria in ultrafine ZnO nanoparticles colloidal medium. Nano Lett 2006;6:866–870.

[17] Ju-Nam Y, Lead JR. Manufactured nanoparticles: an overview of their chemistry, interactions and potential environmental implications. Sci Total Environ 2008;400:396–414.

[18] Suresh AK, Pelletier DA, Wang W, Moon J-W, Gu B, Mortensen NP, Allison DP, Joy DC, Phelps TJ, Doktycz MJ. Silver nanocrystallites: biofabrication using Shewanella oneidensis, and an evaluation of their comparative toxicity on Gram-negative and Gram-positive bacteria. Environ Sci Technol 2010;44:5210–5215.

[19] Thill A, Zeyons O, Spalla O, Chauvat F, Rose J, Auffan M, Flank AM. Cytotoxicity of CeO_2 nanoparticles for Escherichia coli. Physico-chemical insight of the cytotoxicity mechanism. Environ Sci Technol 2006;40:6151–6156.

[20] Auffan M, Rose J, Bottero JY, Lowry GV, Jolivet JP, Wiesner MR. Towards a definition of inorganic nanoparticles from an environmental, health and safety perspective. Nat Nanotechnol 2009;4:634–641.

[21] Handy RD, Owen R, Valsami-Jones E. The ecotoxicology of nanoparticles and nanomaterials: current status, knowledge gaps, challenges, and future needs. Ecotoxicology 2008;17:315–325.

[22] Nowack B, Bucheli TD. Occurrence, behavior and effects of nanoparticles in the environment. Environ Pollut 2007;150:5–22.

[23] Guzman KAD, Taylor MR, Banfield JF. Environmental risks of nanotechnology: national nanotechnology initiative funding, 2000–2004. Environ Sci Technol 2006;40:1401–1407.

[24] Kirchner C, Liedl T, Kudera S, Pellegrino T, Javier AM, Gaub HE, Stolzle S, Fertig N, Parak WJ. Cytotoxicity of colloidal CdSe and CdSe/ZnS nanoparticles. Nano Lett 2005;5:331–338.

[25] Henry TB, Menn FM, Fleming JT, Wilgus J, Compton RN, Sayler GS. Attributing effects of aqueous C60 nano-aggregates to tetrahydrofuran decomposition products in larval zebrafish by assessment of gene expression. Environ Health Perspect 2007;115: 1059–1065.

[26] Oberdörster G, Oberdörster E, Oberdörster J. Nanotoxicology: an emergent discipline evolving from studies of ultrafine particles. Environ Health Perspect 2005;113:823–839.

[27] Singh N, Manshian B, Jenkins GJS, Griffiths SM, Williams PM, Maffeis TGG, Wright CJ, Doak SH. Nanogenotoxicology: the DNA damaging potential of engineered nanomaterials. Biomaterials 2009;30:3891–3914.

[28] Laurent S, Burtea C, Thirifays C, Häfeli UO, Mahmoudi M. Crucial ignored parameters on nanotoxicology: The importance of toxicity assay modifications and "cel vision". Plos One 2012;7:e29997. Available at http://www.plosone.org/article/info:doi/10.1371/journal.pone.0029997. Accessed March 19, 2013.

[29] Crane M, Handy RD, Garrod J, Owen R. Ecotoxicity test methods and environmental hazard assessment for engineered nanoparticles. Ecotoxicology 2008;17:421–437.

[30] Lee S, Kim MS, Lee D, Kwon TK, Khang D, Yun HS, Kim SH. The comparative immunotoxicity of mesoporous silica nanoparticles and colloidal silica nanoparticles in mice. Int J Nanomedicine 2013;8:147–158.

[31] Sayre P, Klaper R, editors. 2011. Nanomaterials and the environment & instrumentation, metrology, and analytical methods. Project Report 2009–2010, National Nanotechnology Initiative Series of EHS Workshops and Reports. Available at http://www.nano.gov/sites/default/files/pub_resource/nanoandenvironmentandinstrumentation.pdf. Accessed March 19, 2013.

[32] Ruparelia JP, Chatterjee AK, Duttagupta SP. Strain specificity in a antimicrobial activity of silver and copper nanoparticles. Acta Biomater 2008;4:707–716.

[33] Qi L, Xu Z, Jiang X, Hu C, Zou X. Preparation and antibacterial activity of chitosan nanoparticles. Carbohydr Res 2004;339: 2693–2700.

[34] Kim YS, Seo JH, Cha HJ. Enhancement of heterologous protein expression in Escherichia coli by co-expression of nonspecific DNA binding stress protein. Dps. Enzyme Microb Technol 2003;33:460–465.

[35] Stein DC, Kopec LK, Yasbin RE, Young FE. Characterization of Bacillus subtilis DSM704 and its production of 1-deoxynojirimycin. Appl Environ Microbiol 1984;48:280–284.

[36] Gorby YA, Yanina S, McLean JS, Rosso KM, Moyles D, Dohnalkova A, Beveridge TJ, Chang IS, Kim BH, Kim KS, Culley DE, Reed SB, Romine MF, Saffarini DA, Hill EA, Shi L, Elias DA, Kennedy DW, Pinchuk G, Watanabe K, Ishii SI, Logan B, Nealson KH, Fredrickson JK. Electrically conductive bacterial nanowires produced by Shewanella oneidensis strain MR-1 and other microorganisms. Proc Natl Acad Sci USA 2006;103:11358–11363.

[37] Dumas EM, Ozenne V, Mielke RE, Nadeau JL. Toxicity of CdTe quantum dots in bacterial strains. IEEE Trans Nanobiosci 2009;8:58–64.

[38] Lyon DY, Brunet L, Hinkal GW, Wiesner MR, Alvarez PJJ. Antibacterial activity of fullerene water suspensions (nC(60)) is not due to ROS-mediated damage. Nano Lett 2008;8:1539–1543.

[39] Brown SD, Thompson MR, VerBerkmoes NC, Chourey K, Shah M, Zhou J, Hettich RL, Thompson DK. Molecular dynamics of the Shewanella oneidensis response to chromate stress. Mol Cell Proteomics 2006;5:1054–1071.

[40] Pelletier DA, Suresh AK, Holton GA, McKeown CK, Wang W, Gu B, Mortensen NP, Allison DP, Joy DC, Allison MR, Brown SD, Phelps TJ, Doktycz MJ. Effects of engineered cerium oxide nanoparticles on bacterial growth and viability. Appl Environ Microbiol 2010;6:7981–7989.

[41] Rodea-Palomares I, Gonzalez-Garcia C, Leganes F, Fernandez-Piñas F. Effect of pH, EDTA, and anions on heavy metal toxicity toward a bioluminescent cyanobacterial bioreporter. Arch Environ Contam Toxicol 2009;57:477–487.

[42] Valant J, Iavicoli I, Drovne D. The importance of a validated standard methodology to define in vitro toxicity of nano-TiO2. Protoplasma 2012;249:493–502.

[43] Dusinka M, Runden-Pran E, Carreira SC, Saunders M. Critical evaluation of toxicity tests. In: Fadeel B, Pietroiusti A, Shvedova AA, editors. Adverse Effects of Engineered Nanomaterials. Exposure, Toxicology, and Impact on Human Health. New York, NY: Academic Press; 2012. p 63–83.

[44] Lewinski N, Colvin N, Drezer V. Cytotoxicity of nanoparticles. Small 2008;4:26–49.

[45] Goodman CM, McCusker CD, Yilmaz T, Rotello V. Toxicity of gold nanoparticles functionalized with cationic and anionic side chains. Bioconj Chem 2004;15:897–900.

[46] Hussain SM, Hess KL, Gearhart JM, Geiss KT, Schlager JJ. In vitro toxicity of nanoparticles in BRL 3A rat liver cells. Toxicol Vitro 2005;19:975–983.

[47] Shukla RK, Sharma V, Pandey AK, Singh S, Sultana S, Dhawan A. ROS-mediated genotoxicity induced by titanium dioxide nanoparticles in human epidermal cells. Toxicol Vitro 2011;25:231–241.

[48] Herzog E, Casey A, Lyng FM, Chambers G, Byrne HJ, Davoren M. A new approach to the toxicity testing of carbon-based nanomaterials—the clonogenic assay. Toxicol Lett 2007;174:49–60.

[49] Fahmy B, Cormier SA. Copper oxide nanoparticles induce oxidative stress and cytotoxicity in airway epithelial cells. Toxicol Vitro 2009;23:1365–1371.

[50] Stern SE, Zolnik BS. 2010. NCL Method GTA-7. Hepatocyte primary ROS assay. Nanotechnology Characterization Laboratory. Available at http://ncl.cancer.gov/NCL_Method_GTA-7.pdf. Accessed March 13, 2013.

[51] Stern SE, Potter T. 2010. NCL Method GTA-2. Hep-G2 Hepatocarcinoma Cytotoxicity Assay. Nanotechnology Characterization Laboratory. Available at http://ncl.cancer.gov/NCL_method_GTA-2.pdf. Accessed March 13, 2013.

[52] Stern SE, Potter T. 2010. NCL Method GTA-1. LLC-PK1 Kidney Cytotoxicity Assay. Nanotechnology Characterization Laboratory. Available at http://ncl.cancer.gov/NCL_Method_GTA-1.pdf. Accessed March 13, 2013.

[53] Lanone S, Rogerieux F, Geys J, Dupont A, Maillot-Marechal E, Boczkowski J, Lacroix, G, Hoet P. Comparative toxicity of 24 manufactured nanoparticles in human alveolar epithelial and macrophage cell lines. Part Fibre Toxicol. 2009;6:4. Available at http://www.particleandfibretoxicology.com/content/6/1/14. Accessed March 19, 2013.

[54] Stern SE, Potter T, Neun BW. 2010. NCL Method GTA-3. Hep-G2 Hepatocyte Glutathione Assay. Nanotechnology Characterization Laboratory. Available at http://ncl.cancer.gov/NCL_Method_GTA-3.pdf. Accessed March 13, 2013.

[55] Berry CC, Wells S, Charles S, Curtis ASG. Cell response to dextran-derivatized iron oxide nanoparticles post-internalization. Biomaterials 2003;24:4551–5413.

[56] Stern SE, Potter T. 2010. NCL Method GTA-5. LLC-PK1 Kidney Cell Apoptosis Assay. Nanotechnology Characterization Laboratory. Available at http://ncl.cancer.gov/ncl_method_gta-5.pdf. Accessed March 13, 2013.

[57] Stern SE, Potter T. 2010. NCL Method GTA-6. HepG2 Hepatocarcinoma Apoptosis Assay. Nanotechnology Characterization Laboratory. Available at http://ncl.cancer.gov/NCL_Method_GTA-6.pdf. Accessed March 13, 2013.

[58] Stern SE, Neun BW. 2010. NCL Method GTA-14. Hep G2 Hepatocarcinoma Homogeneous Apoptosis Assay. Nanotechnology Characterization Laboratory. Available at http://ncl.cancer.gov/NCL_Method_GTA-14.pdf. Accessed March 13, 2013.

[59] Dobrovolskaia M, Potter T. 2009. NCL Method ITA-11. Measurement of Nanoparticle Effects on Cytotoxic Activity of NK Cells by Label-Free RT-CES System. Nanotechnology Characterization Laboratory. Available at http://ncl.cancer.gov/NCL_Method_ITA-11.pdf. Accessed March 13, 2013.

[60] Butterworth KT, Coulter JA, Jain S, Forker J, McMahon SJ, Schettino G, Prise KM, Currell FJ, Hirst DG. Evaluation of cytotoxicity and radiation enhancement using 1.9 nm gold particles: potential application for cancer therapy. Nanotechnology 2010;21:295101.

[61] Monteiro-Riviere NA, Nemanich RJ, Inman AO, Wang YY, Riviere JE. Multi-walled carbon nanotube interactions with human epidermal keratinocytes. Toxicol Lett 2010;2:2004–2010.

[62] Wang Y, Wang J, Wu M, Deng X, Wen T, Chen C, Zhang H, Liu Y, Jiao Z. Internalization, translocation and biotransformation of silica-coated titanium dioxide nanoparticles in neural stem cells. J Nanosci Nanotechnol 2010;10:7121–7125.

[63] Raemy DO, Limbach LK, Rothen-Rutishauser B, Grass RN, Gehr P, Birbaum K, Brandenberger C, Günther D, Stark WJ. Cerium oxide nanoparticle uptake kinetics from the gas-phase into lung cells in vitro is transport limited. Eur J Pharm Biopharm 2011;77:368–375.

[64] Ma, Y. *In vitro* models for nanotoxicity testing. In: Nanotoxicity: From In vivo and In vitro Models to Health Risks; Sahu, S.; Casciano, D., editors; John Wiley and Sons Ltd.: Hoboken, NJ; 2009; pp. 349–377.

[65] Arora S, Rajwade JM, Paknikar KM. Nanotoxicology and in vitro studies: the need of the hour. Tox App Pharm 2012;258:151–165.

[66] Code of Federal Regulation. 799.9110—TSCA acute oral toxicity. Protection of Environment. 2010; Available at http://cfr.vlex.cm/vid/799-tsca-acute-oral-toxicity-19834216. Accessed February 17, 2013.

[67] Maurer-Jones MA, Hayne CL. In Vivo and In Vitro nanotoxicology studies. SYMPOSIUM 1: nanodiagnostics and nanotherapeutics: building research ethics and oversight. J Law Med Ethics 2012;40:795–801.

[68] Seidle T, Robinson S, Holmes T, Creton S, Prieto P, Scheel J, Chlebus M. Cross-sector review of drivers and available 3Rs approaches for acute systemic toxicity testing. Toxicol Sci 2010;116:382–396.

[69] Walum E. Acute oral toxicity. Environ Health Perspect 1998;106 (Supplement 2):497–503.

[70] National Toxicological Program. 2010. Department of Health and Human Services, National Institute of Environmental Health Sciences. Available at http://ntp.niehs.nih.gov/. Accessed February 23, 2013.

[71] National Research Council (NRC). 2011. Guide for the care and use of Laboratory Animals, 8th ed. Available at http://grants.nih.gov/grants/olaw/Guide-for-the-Care-and-Use-of-Laboratory-Animals.pdf. Accessed April 12, 2013.

[72] Wang X, Reece SP, Brown JM. Immunotoxicological impact of engineered nanomaterial exposure: mechanisms of immune cell modulation. Toxicol Mech Meth 2013;23:168–177.

[73] Groso A, Fink AP, Magrez A, Riediker M, Meyer T. Management of nanomaterials safety in research environment. Part Fibre Toxicol. 2010;7:40. Available at http://www.particleandfibretoxicology.com/content/7/1/40. Accessed March 13, 2013.

[74] Horie M, Stowe M, Kambara T, Lee BW, Endoh S, Maru J, Oyabu T, Myojo T, Ogami A, Uchida K, Yamamoto K, Kobayashi N, Kuroda E, Nakazato T, Morimoto Y. Pulmonary inflammation of well-dispersed multi-wall carbon nanotubes following intratracheal instillation: toxicity by fiber of 1–5 μm in Length. Materials 2012;5:2833–2849.

[75] Tervonen K, Waissi G, Petersen EJ, Akkanen J, Kukkonen VK. Analysis of fullerene-C60 and kinetic measurements for its accumulation and depuration in Daphnia magna. Environ Toxicol Chem 2010;29:1072–1078.

[76] Geller W, Müller H. The filtration apparatus of cladocera; Filter mesh-sizes and their implications on food selectivity. Oecologia 1981;49:316–321.

[77] Ebert D. Ecology, Epidemiology and Evolution of Parasitism in Daphnia. Bethesda, MD, USA: *National Library of Medicine, National Center for Biotechnology Information*; 2005.

[78] Yang L, Watts DJ. Particle surface characteristics may play an important role in phytotoxicity of alumina nanoparticles. Toxicol Lett 2005;158:122–132.

[79] ETC Group. 2003. May 2. Nanotech and the Precautionary Prince. Genotype.Available at http://www.etcgroup.org/sites/www.etcgroup.org/files/publication/158/01/precautionprince.pdf. Accessed December 21, 2012.

[80] International Center for Technology Assessment, Citizen Petition to FDA. 2006. Petition requesting FDA amend its regulations for products composed of engineered nanoparticles generally and sunscreen drug products composed of engineered nanoparticles specifically, Docket 2006P-0210. Available at http://www.icta.org/doc/Nano%20FDA%20petition%20final.pdf. Accessed March 19, 2013.

[81] Warheit B. Debunking some misconceptions about nanotoxicology. Nano Lett 2010;10:4777–4782.

31

ENVIRONMENTAL RISK, HUMAN HEALTH, AND TOXIC EFFECTS OF NANOPARTICLES

Jamuna Bai A. and Ravishankar Rai V.
Department of Studies in Microbiology, University of Mysore, Mysore, India

31.1 INTRODUCTION

Nanotechnology is the science and technology of developing novel materials known as nanoparticles ranging in sizes of 100 nm or less. Because of their intrinsic properties, nanoparticles have a wide range of applications. The reduction in size from their bulk counterparts increases their surface-to-volume ratio and consequently improves their electromagnetic, catalytic, pharmacokinetic, and targeting properties [1]. The unique physicochemical characterisitics of nanoparticles have led to its exploitation in the field of medicine and industry. They have found numerous applications in electronics, photovoltaic, catalysis, environmental, and space engineering, cosmetic industry, and as therapeutics and diagnostics in medicine and pharmacy. Nanotechnology is at the forefront of technology and has revolutionized our lives with its use in commercial, technological, and therapeutic applications. Nonetheless, recent investigation on the extensive application of nanoparticles implies that these have potential toxic effects on environmental ecosystems and on living organisms [2]. Due to the ever-increasing applications of nanoparticles, the artificial synthesis of the nanoparticles has escalated. Furthermore, the rapid production and application of engineered nanoparticles will lead to their release in the environment, causing their increased exposure and interaction with the biotic and abiotic components of the ecosystems. Although the application of nanomaterials is noteworthy, their presence and long-term exposure in the ecosystem is not bereft of hazardous and toxic biological effects. Hence, it is necessary to understand the interaction of released nanoparticles with the environment, possible ecotoxicological effects, consequences of long-term exposure of engineered nanoparticles to human health, as well as hazard identification and risk assessment of nanotechnology for the safe application of engineered nanoparticles [3].

31.2 APPLICATION OF NANOPARTICLES

31.2.1 Biology and Medicine

The size-dependent physical properties of nanoparticles make them useful in a wide range of applications. Apart from these, the optical and magnetic effects are the ones most used for biological applications. Some of the potential applications of nanomaterials in biology and medicine are in fluorescent biological labels for diagnostics, in therapeutics for drug and gene delivery, pathogen detection, detection of proteins, tissue engineering, destruction of tumor cells by heating (hyperthermia), separation and purification of biological molecules, cell sorting, magnetic resonance imaging, contrast enhancement, and phagokinetic studies [4].

Nanomaterials for Environmental Protection, First Edition. Edited by Boris I. Kharisov, Oxana V. Kharissova, and H. V. Rasika Dias.

31.2.2 Nanoparticles as Antimicrobials

Some nanoparticles such as silver, titanium dioxide, and zinc oxide are finding numerous applications as antimicrobials and additives in consumer, health-related, and industrial products. Silver nanoparticles have a broad-spectrum antimicrobial activity against several pathogens and are used as additives in health-related products such as bandages, catheters, and other materials to prevent infection, especially in healing of wounds and burns. Nanoparticles of titanium dioxide are used in cosmetics, filters that exhibit strong germicidal properties, and in conjunction with silver as an antimicrobial agent. Moreover, due to its photocatalytic activity, it has been used in wastewater treatment. Zinc oxide (ZnO) and copper oxide (CuO) nanomaterials are being incorporated into a variety of medical and skin coatings due to their antimicrobial property. ZnO nanoparticles are used in wallpapers in hospitals as antimicrobials. ZnO powder is an active ingredient for dermatological applications in creams, lotions, and ointments on account of its antibacterial properties [5].

31.2.3 Nanoparticles in Textiles and Coatings

Engineered nanoparticles can find good application in textile designing. They can be used to improve the quality of textiles used by health and defence personnel as they have diverse properties such as self-cleaning; water- and dirt-repellent; antimicrobial, conductive, and antistatic; solvent, UV, and abrasion resistance; and decreased gas permeability and flammability. For example, carbon nanotubes have been used to improve electrical, antistatic, thermal-conductive, and flame-retardant properties of textile composite fibers. Metal and carbonaceous nanoparticles are usually used in the synthesis of nanotextiles. Nanoparticles of layered silicate such as montmorillonite will find applications as carriers of agents and to improve the dyeing properties of fibers [6].

31.2.4 Nanotechnologies in the Food Industry

In the food industry, nanoparticles such as micelles, liposomes, nanoemulsions, biopolymeric nanoparticles and cubosomes, and nanosensors are finding increased application as they ensure long shelf life and safety of food. They are also known to enhance the nutritive value and quality of food. For example, nanocapsules containing nutraceuticals are incorporated in cooking oil, flavor enhancers are nanoencapsulated, and nanoparticles having the ability to selectively bind and remove chemicals are used in foods. Nanoparticles have also been used for food packaging. Bionanocomposites are hybrid nanostructured materials with improved mechanical, thermal, and gas barrier properties that are used in food packaging. Apart from protecting food and increasing shelf life, these bionanocomposites are considered to be environmental-friendly as they minimize the use of plastics as packaging materials [7].

31.3 TOXICITY OF INORGANIC NANOPARTICLES

Metal oxide nanoparticles are often used as industrial catalysts and in therapeutics, and elevated levels of these particles have been clearly demonstrated in the ecosystem. In recent years, there has been an increase in studies focusing on the toxicity of nanoparticles (Table 31.1).

31.3.1 Silver Nanoparticles

Silver nanoparticles (Ag NPs) have been the most commercialized nanomaterial used in diverse applications. The antimicrobial property of Ag NPs has led to their use in surgical and wound dressings and coatings in medical devices, but potential adverse effects have also been reported in the literature. Although they are one of the most commercially viable nanotechnological products, safety issues have been raised regarding the use of such nanoparticles due to unintentional health and environmental impacts. There are many studies focusing on the mechanisms by which Ag NP are toxic as their increasing application could lead to their release into the environment and eventually result in environmental toxicity or ecotoxicity. To understand the effect of exposure to Ag NP in human cells, a study was conducted to evaluate Ag NP–induced DNA damage, cell death, and functional impairment in human mesenchymal stem cells (hMSCs). Cytotoxic effects were seen at a concentration of 10 μg/ml on exposing Ag-NPs to hMSCs for 1, 3 and 24 h. Transmission electron microscopy (TEM) studies revealed that AgNPs were distributed to the cytoplasm and nucleus. Comet assay and chromosomal aberration test revealed that DNA damage occurred at a concentration of 0.1 μg/ml. Therefore, AgNPs could cause potential cyto- and genotoxic effects in hMSCs [8]. In another study, the contribution of silver ion to the toxicity of Ag NPs in A549 lung cells was investigated. Cell viability measured by the for 3-(4,5-dimethythiazol-2-yl)-2,5-diphenyl tetrazolium bromide (MTT) assay revealed that Ag NP suspensions were more toxic when the initial

TABLE 31.1 *In vitro* and *in vivo* toxicity of metallic nanoparticles

Nanoparticle (size, dosage, and exposure time)	Experimental model/ biological system	Biological effects	Reference
Silver nanoparticle			
10 μg/ml	Human mesenchymal stem cells	DNA damage, functional impairment, and cell death	[8]
	A549 lung cells	Apoptosis and cytotoxic	[9]
		ROS-dependent cytotoxic effect and ROS-independent cell cycle assortment	[10]
40–80 nm		Impairment of mitochondrial function	[11]
	Caenorhabditis elegans	Internalization of nanoparticle and cellular toxicity	[12]
20 nm	L929 fibroblast RAW 264.7 macrophage	Cytotoxic in L929 fibroblast	[13]
	Murine macrophages	Internalization and localization in cytoplasm Mitochondrial damage, apoptosis, and cell death	[14]
5–100 μg/ml	Skin epithelial A431, lung epithelial A549, and murine macrophage Raw 2647	ROS generation, cellular disruption, expression of stress markers pp38, TNF2, and hsp 70	[15]
Gold nanoparticle			
40, 200, and 400 μg/kg/day	Mice intraperitoneal dosage	Bioaccumulation but no subacute physiological damage	[16]
5–15 nm ≥50 μM	Balb/3T3 mice fibroblast	Disruption of actin cytoskeleton Degradation of clatharin heavy chain	[17]
8 nm 50 nM	Human liver cell line HL7702	Interaction with glutathione (GSH) resulting in intracellular GSH depletion, mitochondrial membrane depolarization, and apoptosis	[18]
1.5 nm Negatively charged (50 μg/ml) and positively charged (10 μg/ml)	Zebrafish embryo	Abnormal startle and adult behavior Low survivorship in adults	[19]
Zinc oxide nanoparticle			
300 mg/kg	Mice	Oxidative stress, DNA damage, and apoptosis Pathological lesions in liver	[20]
10–50 μg/ml	Human nasal mucosal cells	DNA damage and cytotoxic effect	[21]
Titanium dioxide nanoparticle			
	Rat and human glial cells C6 and U373	Vesicle formation after internalization, apoptosis Cytotoxic	[22]
	Intratracheal instillation in mice	Inflammation and bleeding in lung Accumulation of ROS and increased lipid peroxidation Decreased antioxidant capacity	[23]
	Zebrafish (*Danio rerio*)	Accumulation in gill, liver, heart, and brain	[24]
	Mice	Accumulation in kidney—nephric inflammation, necrosis, and dysfunction	[25]
Silicon dioxide nanoparticle 20–100 nm	Renal proximal tubular cell lines Human HK-2 and porcine LLC-PK1	ROS generation, lipid peroxidation—oxidative stress	[26]
Copper oxide nanoparticle			
100 nm	Human pulmonary epithelial cell A549	Dose-dependent oxidative stress Genotoxic effects	[27]
23.5 nm	Mice model	Kidney, liver, and spleen are target organs	[28]
Nanocopper nanoparticle 50–200 mg/kg	Transcriptome profiling of rat kidney	Renal proximal tubule necrosis in rat kidney	[29]
SiO_2, Fe_2O_3, and CuO nanoparticle	Airway epithelial cells HEp-2 cells	Oxidative stress due to redox coupling Cytotoxicity is dose-dependent and more in CuO nanoparticles	[30]

silver ion fraction was higher. Flow-cytometric analyses of cell cycle and apoptosis showed that there was no significant difference between the treatment with Ag NP suspension and Ag NP supernatant. At high silver ion fractions (≥5.5%), the Ag NP did not add measurable toxicity to the Ag NP suspension, whereas at low silver ion fractions (≤2.6%), Ag NP suspensions were more toxic than their supernatant [9]. Some of the manufactured nanomaterials are redox-active and are capable of passing through cell membranes into mitochondria. Ag NPs (40 and 80 nm) can cause impairment of mitochondrial function by alterations of mitochondrial membrane permeability, resulting in an uncoupling effect on the oxidative phosphorylation system. Thus, the exposure to Ag NPs causes mitochondrial toxicity, leading to cellular toxicity [11]. The physicochemical behavior, uptake, toxicity (growth inhibition), and mechanism of toxicity of three Ag NPs with different sizes and polyvinylpyrrolidone (PVP) or citrate coatings were characterized in the animal model *Caenorhabditis elegans*. Significant aggregation and extraorganismal dissolution of silver, organismal uptake, and transgenerational transfer of Ag NPs were observed. Growth inhibition by all tested Ag NPs at concentrations in the low mg/l levels was seen. Although all tested Ag NPs were internalized (passed cell membranes) in *C. elegans*, at least part of the toxicity observed was mediated by ionic silver [12]. The mechanism of action of Ag NP toxicity has also been studied in relationship to the generation of reactive oxygen species (ROS) in A549 cells. Ag NPs caused ROS formation in the cells, a reduction in their cell viability and mitochondrial membrane potential (MMP), an increase in the proportion of cells in the sub-G1 (apoptosis) population, S phase arrest, and downregulation of the cell cycle–associated proliferating cell nuclear antigen (PCNA) protein, in a concentration- and time-dependent manner. Pretreatment of the A549 cells with the antioxidant *N*-acetyl-cysteine decreased the effects of Ag NPs on the reduced cell viability, caused changes in the MMP and proportion of cells in the sub-G1population, but had no effect on the Ag NP–mediated S phase arrest or downregulation of PCNA. Thus, it could be concluded that the *in vitro* toxic effects of Ag NPs on A549 cells were mediated by both ROS-dependent (cytotoxicity) and ROS-independent (cell cycle arrest) pathways [10]. The effect of particle size on Ag NPs' cytotoxicity, inflammation, genotoxicity, and developmental toxicity revealed that Ag NPs of 20 nm were more toxic than the larger nanoparticles. However, 20 nm sized Ag NPs were more cytotoxic in comparison to silver ions in L929 fibroblasts than in but not in RAW 264.7 macrophages. This implies that the potency of silver to induce cell damage depends on both the cell type and and its size [13].

Ag NPs can also induce stress responses along with cytotoxicity in mammalian cells. Ag NPs are efficiently internalized via scavenger receptor–mediated phagocytosis in murine macrophages, get localized in the cytoplasm, cause mitochondrial damage, and induce apoptosis and cell death. In the presence of Ag ion–reactive thiol-containing compounds, these effects are more, which implies that Ag ions play a major role in Ag NP toxicity. Ag NPs are internalized by scavenger receptors, trafficked to the cytoplasm, and induce toxicity by releasing Ag ions [14]. In another study, Ag NPs synthesized by reduction with tannic acid (TSNPs) and sodium borohydride were assessed for toxicity in a cellular environment using skin epithelial A431, lung epithelial A549, and murine macrophage RAW264.7 cells over a range of doses (5–100 µg/ml). TSNPs exhibited a higher negative zeta-potential and also higher toxicity and dose-dependent increase in ROS generation and cellular disruption. TSNPs induced a dose-dependent increase in the expression of stress markers pp38, TNF-α, and HSP-70. The toxicity of nanoparticles was found to vary according to their surface potential. This proves that the cytotoxicity of Ag NPs changes with the surface potential of nanoparticles and cell type. Thus, physicochemical properties of Ag NPs direct their toxicity [15].

31.3.2 Gold Nanoparticles

Gold nanoparticles (Au NPs) offer great promise in biomedical applications. Au NPs have shown promising biological and military applications due to their unique electronic and optical properties. Au NPs are usually considered to be bioinert, but questions have been raised regarding their safety. However, little is known about their cytotoxicity when they come into contact with a biological system. Hence, it is essential to study the cytotoxicity of Au NPs. Au NPs are able to cross the blood–brain barrier and accumulate in the neural tissue. However, no evidence of toxicity was observed on evaluating the bioaccumulation and toxic effects of different doses (40, 200, and 400 µg/kg/day) of 12.5 nm Au NPs upon intraperitoneal administration in mice every day for 8 days. The gold levels in blood did not increase with the dose administered, whereas in all the organs examined there was a proportional increase in gold, indicating efficient tissue uptake. Thus, tissue accumulation patterns of Au NPs depend on the doses administered and the accumulation of the particles does not produce subacute physiological damage [16]. In another study, Balb/3T3 mouse fibroblasts were exposed to Au NPs (5 and 15 nm) citrate stabilized for 72 h. Only Au NPs 5 nm in size at concentrations ≥50 µM exhibited cytotoxic effects. Disruption of the actin cytoskeleton was evident and reduction of the expression and degradation of the clathrin heavy chain were observed in cells exposed for 72 h to Au NPs [17]. It has been shown that small Au NPs can be endocytosed by cells and form aggregates inside the cell, resulting in cytotoxicity [31]. The interaction of Au NPs with glutathione (GSH) and their role in the sequence of apoptotic signaling events that occur after modulation of the cellular redox state were studied in HL7702 cells (human liver cell line). After incubation with 8 nm AuNPs at 50 nM, there was an early decline in cytosolic GSH, which initiated mitochondrial transmembrane potential depolarization and apoptosis. Au NPs (8 nm) possess strong Au–S bonding interactions with GSH resulting in intracellular GSH depletion.

A decrease in GSH alone can act as a potent early activator of apoptotic signaling. Increased H_2O_2 production following mitochondrial GSH depletion represents a crucial event, which commits HL7702 cells to apoptosis through the mitochondrial pathway [18]. To understand the long-term consequences of developmental nanoparticle exposure, zebrafish embryos were acutely exposed to three Au NPs that possess functional groups with differing surface charges. Embryos were exposed to 50 µg/ml of 1.5 nm Au NPs possessing negatively charged 2-mercaptoethanesulfonic acid (MES), neutral 2-(2-(2-mercaptoethoxy) ethoxy)ethanol (MEEE) ligands, and positively charged 10 µg/ml of the Au NPs containing trimethylammoniumethanethiol (TMAT). Both MES–Au NP- and TMAT–Au NP-exposed embryos exhibited hypolocomotor activity, while those exposed to MEEE–Au NPs did not. A subset of embryos that were exposed to 1.5 nm MES–Au NPs and TMAT–Au NPs during development from 6 to 120 h post fertilization were raised to adulthood. Behavioral abnormalities and the number of survivors into adulthood were evaluated at 122 days post fertilization. It was found that both treatments induced abnormal startle behavior following a tap stimulus. However, the MES–Au NPs-exposed group also exhibited abnormal adult behavior in light and had a lower survivorship into adulthood. Thus, acute, developmental exposure to 1.5 nm MES–Au NPs and TMAT–Au NPs, two nanoparticles differing only in the functional group, affects larval behavior, with effects of behavior persisting into adulthood [19].

Thus, it is clear that long-term retention of Au NPs in the body increases the potential for toxicity. However, it was shown in a recent study that nanoclustering of gold and iron as a nanoparticle could be easily biodegraded in lysosomes of macrophages. On administering nanorose (nanoclustered gold and iron nanoparticle) to C57BL/6 mice, the gold concentration was significantly reduced in 11 murine tissues in as few as 31 days ($P<0.01$). Moreover, the hematology studies showed no toxicity of nanorose injected into mice for up to 14 days after administration. This implies that degradation of nanoparticles into smaller subunits can enhance and facilitate their rapid excretion. Therefore, such strategies to decrease the particle size for excretion can reduce the potential toxicity of Au NPs [32].

31.3.3 Zinc Oxide Nanoparticles

Zinc oxide nanoparticles (ZnO NPs) are finding applications in a wide range of products including cosmetics, food packaging, and antimicrobial coatings. Therefore, there are high chances of human exposure to these nanoparticles through dermal, inhalation, and oral routes. Toxicity studies on the effect of ZnO NPs on ecological receptors across different taxa such as bacteria, algae and plants, as well as aquatic and terrestrial invertebrates and vertebrates have shown that relatively high to acute toxicity of ZnO NPs (in the low mg/l levels) to environmental species is highly dependent on test species, physicochemical properties of the material, and test methods. Particle dissolution to ionic zinc and particle-induced generation of ROS represented the primary modes of action for ZnO NP toxicity across all species tested, and photoinduced toxicity associated with its photocatalytic property was the other important mechanism of toxicity under environmentally relevant UV radiation [33]. Studies on oral toxicity of ZnO NPs in mice revealed that a significant accumulation of nanoparticles in the liver leads to cellular injury after subacute oral exposure of ZnO NPs (300 mg/kg) for 14 consecutive days. This was evident by the elevated alanine aminotransferase (ALT) and alkaline phosphatase (ALP) serum levels and pathological lesions in the liver. ZnO NPs were also found to induce oxidative stress indicated by an increase in lipid peroxidation. Fpg-specific DNA lesions were observed in the liver, indicating that oxidative stress was the cause of DNA damage. Subacute oral exposure to ZnO NPs in mice leads to an accumulation of nanoparticles in the liver, causing oxidative stress–mediated DNA damage and apoptosis. These results also suggest the need for a complete risk assessment of any new engineered nanoparticle before its arrival into the consumer market [20].

Similarly, in a study focusing on the cyto- and genotoxicity of ZnO NPs and ZnO powder in primary human nasal mucosa cells cultured in the air–liquid interface, cytotoxic effects and DNA damage were shown at a ZnO NP concentration of 50 ($P<0.01$) and 10 µg/ml ($P<0.05$), respectively. This indicates cyto- and genotoxic properties as well as a proinflammatory potential of ZnO NPs in nasal mucosa cells. Therefore, precautionary measures should be taken before wide-scale industrial and dermatological application of these nanoparticles. However, further investigations on long-term exposure to ZnO NPs are required to understand their impact on the environment and human health [21].

31.3.4 Titanium Dioxide Nanoparticles

Titanium dioxide nanoparticles (TiO_2 NPs) are an important material used as an additive in pharmaceutical and cosmetic products. Due to their high surface-to-mass index, TiO_2 NPs show unique physical and chemical characteristics compared to the bulk substance [34]. TiO_2 NPs are widely used in the chemical, electrical, and electronic industries. They can enter directly into the brain through the olfactory bulb and get deposited in the hippocampus region. The effect of TiO_2 NPs on rat and human glial cells, C6 and U373, has been studied to understand their mechanism of action in the nervous system. TiO_2 NPs inhibited proliferation and induced morphological changes that were related with a decrease in immunolocation of F-actin fibers. TiO_2 NPs were internalized and the formation of vesicles was observed. TiO_2 NPs induced apoptosis after

96 h of treatment. Hence, TiO_2 NPs have a cytotoxic effect on glial cells, suggesting that exposure to TiO_2 NPs could cause brain injury and be hazardous to health [22]. Exposure to TiO_2 NPs has been demonstrated to result in pulmonary inflammation in animals. The oxidative stress and molecular mechanism associated with pulmonary inflammation in chronic lung toxicity caused by the intratracheal instillation of TiO_2 NPs for 90 consecutive days in mice were evaluated. TiO_2 NPs significantly accumulated in the lung, leading to an obvious increase in lung indices, inflammation, and bleeding in the lung. Exposure to TiO_2 NPs significantly increased the accumulation of ROS and the level of lipid peroxidation, and decreased antioxidant capacity in the lung. Furthermore, TiO_2 NPs exposure activated nuclear factor-κB (NF-κB), increased the levels of tumor necrosis factor-α, cyclooxygenase-2, heme oxygenase-1, interleukin-2, interleukin-4, interleukin-6, interleukin-8, interleukin-10, interleukin-18, interleukin-1β, and CYP1A1 expression. However, TiO_2 NP exposure decreased NF-κB–inhibiting factor and heat shock protein 70 expression. Thus, the generation of pulmonary inflammation caused by TiO_2 NPs in mice is closely related to oxidative stress and the expression of inflammatory cytokines [23]. The long-term risk of TiO_2 NP exposure was studied by using zebrafish (*Danio rerio*) as *in vivo* model to assess the chronic toxicity of TiO_2 NPs. The adverse effect to zebrafish was seen as concentration-dependent and time-dependent inhibition of growth and decrease in the liver weight ratio of zebrafish. TiO_2 NPs could translocate among organs and pass through the blood–brain and the blood–heart barrier after long-term exposure and accumulate and get distributed in gill, liver, heart, and brain [24]. TiO_2 NPs are also known to damage the kidney of mice on exposure. The molecular mechanism of TiO_2 NPs–induced nephric injury has been investigated. TiO_2 NPs accumulate in the kidney, resulting in nephric inflammation, cell necrosis, and dysfunction. NF-κB gets activated by TiO_2 NP exposure, promoting the expression levels of tumor necrosis factor-α, macrophage migration inhibitory factor, interleukin-2, interleukin-4, interleukin-6, interleukin-8, interleukin-10, interleukin-18, interleukin-1β, cross-reaction protein, transforming growth factor-β, interferon-γ, and CYP1A1. Thus, TiO_2 NP–induced nephric injury of mice is associated with alteration of inflammatory cytokine expression and reduction of detoxification of TiO_2 NPs [25].

31.3.5 Silicon Dioxide Nanoparticles

Silica nanoparticles (nano-SiO_2) are one of the most popular nanomaterials used in industrial manufacturing and medicine. While inhalation of nanoparticles causes pulmonary damage, nano-SiO_2 can be transported into the blood and get deposited in target organs where they exert potential toxic effects. The kidney is one such secondary target organ. However, toxicological information of their effect on renal cells and the mechanisms involved remain sparse. On investigating the cytotoxicity of nano-SiO_2 of different sizes on two renal proximal tubular cell lines (human HK-2 and porcine LLC-PK_1), it was shown that internalization process occurs by macropinocytosis and clathrin-mediated endocytosis for 100 nm nano-SiO_2. These nanoparticles were localized in vesicles. Toxicity was size- and time-dependent (24, 48, 72 h) and increased as nanoparticles became smaller. Second, analysis of oxidative stress based on the assessment of ROS production (dihydroethidium) or lipid peroxidation (malondialdehyde) clearly demonstrated the involvement of oxidative stress in the toxicity of 20 nm nano-SiO_2. The induction of antioxidant enzymes (catalase, GSTpi, thioredoxin reductase) could explain their lesser toxicity with 100 nm nano-SiO_2 [26].

31.3.6 Copper Oxide Nanoparticles

Copper oxide nanoparticles (CuO NPs) are increasingly used in various applications. Oxidative stress is considered to be the main cause of the cytotoxicity of CuO NPs in mammalian cells. However, little is known about the mechanism of genotoxicity of CuO NPs following exposure to human cells. CuO NPs were known to induce genotoxic response through p53 pathway in human pulmonary epithelial cells (A549). CuO NPs reduced cell viability and the degree of reduction was dose-dependent. CuO NPs were also found to induce oxidative stress in a dose-dependent manner indicated by depletion of glutathione and induction of lipid peroxidation, catalase, and superoxide dismutase. The expression of Hsp70, the first-tier biomarker of cellular damage, was induced by CuO NPs. Further, CuO NPs upregulated the cell cycle checkpoint protein p53 and the expression of DNA damage repair proteins Rad51 and MSH2. These results demonstrate that CuO NPs possess a genotoxic potential in A549 cells that may be mediated through oxidative stress [27]. To assess the toxicity of CuO NPs (23.5 nm) *in vivo*, LD_{50}, morphological changes, pathological examinations, and blood biochemical indexes of experimental mice were studied comparatively with microcopper particles (17 μm) and cupric ions ($CuCl_2 \cdot 2H_2O$). The LD_{50} for the CuO NPs, microcopper particles, and cupric ions exposed to mice via oral gavage were 413, >5000, and 110 mg/kg body weight, respectively. The toxicity classes of nano and ionic copper particles are both class 3 (moderately toxic), and microcopper is class 5 (practically nontoxic) on the Hodge and Sterner Scale. The kidney, liver, and spleen were found to be target organs of CuO NPs. Nanoparticles induced toxicological effects and heavy injuries on the kidney, liver, and spleen of experimental mice, but microcopper particles did not, on a mass basis [28]. The mechanisms of nanocopper-induced nephrotoxicity were studied by analyzing renal gene expression profiles phenotypically anchored to conventional toxicological outcomes. Male Wistar rats were given nanocopper (50, 100, 200 mg/kg) and microcopper (200 mg/kg) at different doses for 5 days. Nanocopper induced widespread renal proximal tubule necrosis in rat kidneys with increased levels of blood urea nitrogen and creatinine. Whole-genome transcriptome profiling of rat kidneys revealed significant alterations in the expression of many genes involved in valine, leucine, and isoleucine degradation, complement and coagulation cascades, oxidative phosphorylation, cell cycle, mitogen-activated protein kinase signaling pathway, and glutathione metabolism. Systems used in nanotoxicology studies have proved to provide valuable insight and toxicogenomic approaches are presenting an unprecedented amount of mechanistic information on molecular responses to nanocopper, thus helping hazard identification and risk assessment [29].

To understand the impact of airborne nanoparticles on the respiratory system, airway epithelial (HEp-2) cells were exposed to increasing doses of silicon oxide (SiO_2), ferric oxide (Fe_2O_3), and copper oxide (CuO) nanoparticles, the leading metal oxides found in ambient air surrounding factories. CuO induced the greatest amount of cytotoxicity in a dose-dependent manner, while even high doses (400 μg/cm^2) of SiO_2 and Fe_2O_3 were nontoxic to HEp-2 cells. Although all metal oxide nanoparticles were able to generate ROS in HEp-2 cells, CuO could overcome the cellular antioxidant defenses (e.g., catalase and glutathione reductase). A significant increase in the level of eight-isoprostanes and in the ratio of GSSG to total glutathione in cells exposed to CuO suggested that ROS generated by CuO induced oxidative stress in HEp-2 cells. Cotreatment of cells with CuO and the antioxidant resveratrol increased cell viability, suggesting that oxidative stress may be the cause of the cytotoxic effect of CuO. These studies demonstrated that there is a high degree of variability in the cytotoxic effects of metal oxides and that this variability is not due to the solubility of the transition metal but due to the sustained oxidative stress possibly due to redox cycling [30]. CuO had already been reported to be highly cytotoxic; however, carbon-coated copper nanoparticles are much less cytotoxic and more tolerated. Measuring the two materials' intra- and extracellular solubility in model buffers explained this difference on the basis of altered copper release when supplying copper metal or the corresponding oxide particles to the cells. These observations are in line with a Trojan horse–type mechanism and illustrate the dominating influence of physicochemical parameters on the cytotoxicity of a given metal [35].

31.4 MECHANISM OF CYTOTOXICITY AND GENOTOXICTIY OF NANOPARTICLES

Nanoparticles can gain access to the human body through ingestion, injection, transdermal delivery, and inhalation. Further, they can translocate to secondary organs [36]. Airborne nanoparticles enter through the respiratory system. Endocytosis of nanoparticles through alveolar epithelial cells plays an important role in its translocation to other organs. Nanoparticles are able to translocate from the lung to the liver, spleen, heart, and other organs [37]. If the inhaled nanoparticles are able to gain entry into other organs via the olfactory bulb, it could cause potential hazard as they would have direct access to the central nervous system [38]. Nanoparticles can enter the body via the skin through injection during drug delivery and imaging studies or by application of cosmetics and antimicrobials for wound healing. Nanoparticles injected into the skin can elicit photocatalytic activity in the dermal layers, causing the formation of free radicals in skin cells, damaging DNA, and disrupting normal cell functions and cell viability [39]. Nanoparticles administered in the dermis can migrate to regional lymph nodes, potentially via skin macrophages and Langerhans cells, raising potential concern for immunomodulation [40]. Nanoparticles used in food packaging and processing and as additives can be ingested. Following ingestion, translocation of particles into and across the gastrointestinal mucosa can occur but uptake via gut and translocation is probably size-dependent. Further, nanoparticles can also enter the bloodstream via gastrointestinal assimilation [41].

The biological effects of nanoparticles are due to both its physical and chemical mechanisms of action [42]. The production of ROS, dissolution and release of toxic ions, disturbance of electron/ion cell membrane transport activity, oxidative damage through catalysis, lipid peroxidation, and surfactant properties are considered to be chemical mechanisms. The generation of ROS is implied to be the main underlying chemical process in nanotoxicology leading to secondary processes that cause cell damage and eventually cell death. ROS are also involved in inflammatory processes. Free radical formation also has direct impact on cell integrity. The physical mechanisms known to cause biological effects are due to the nanoparticle size and surface properties. They involve membrane damage and disruption of membrane activity, and effect transport processes, protein conformation/folding, and protein aggregation/fibrillation [43].

The biological response to nanoparticles is a result of the various processes occurring as a result of the chemical and physical interactions in the cell. These cellular responses can occur before or after internalization of particles, or as a response to the uptake mechanism itself (Fig. 31.1). Membrane stability can be affected by nanoparticles either directly (e.g., physical damage) or indirectly (e.g., oxidation) and can lead to cell death. The cell membranes are able to control intracellular homeostasis through selective permeability and transport mechanisms, and this makes them a vulnerable target for possible damaging

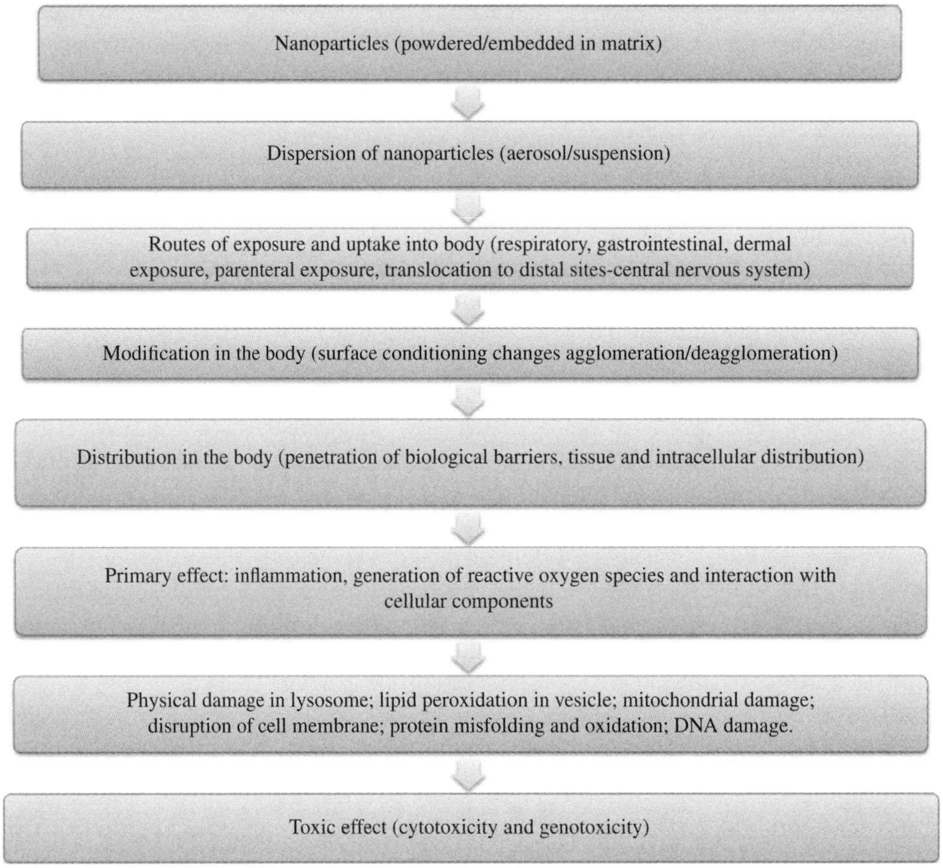

FIGURE 31.1 Biological effects of inorganic or metal-based nanoparticles.

effects of nanoparticles. Interactions of nanoparticles with membranes depend largely on the nanoparticles' surface properties and size. The nanoparticles' size is known to influence surface pressure and adhesion forces during interactions [44].

Once internalized, nanoparticles interact with various cellular organelles and macromolecules present in the cell, resulting in cytotoxic and genotoxic effects. Cytoxicity leads to cell death as a result of membrane lipid peroxidation, membrane rupture, energy depletion, or organelle destruction. Genotoxicity occurs when adverse effects result at a genetic level and this can cause cell death due to apoptosis and changes in interconnected signaling pathways [45].

Nanoparticles having the same size and magnitude as that of the protein molecules can interfere with cell signaling processes. Nanoparticles interact with proteins, either by chaperone-like activity or by changing the configuration of proteins, leading to protein misfolding [46].

Nanoparticle–DNA interaction has attracted special attention when assessing potential toxicological risks caused by nanomaterials. Nanoparticles can affect the genes either directly or indirectly, which is termed primary or secondary damage, respectively. The mechanism of potential DNA damage by nanoparticles is still being understood. The main mechanism inducing genotoxicity is the ability of the nanoparticles to produce oxidative stress. Oxidative stress results due to the imbalance between ROS and antioxidant conditions in the cell. Nanoparticles can intercalate or directly interact with DNA and cause mutagenicity [47]. Apart from direct intercalation or physical and/or electrochemical interaction with nanoparticles, ROS production by nanoparticles is again believed to play a major role in DNA damage [48]. ROS is responsible for oxidation of DNA bases, strand breakage in DNA, and lipid peroxidation–mediated DNA adducts. In the primary mechanism, metal nanoparticles, depending on their physicochemical properties, can directly produce oxidants, such as highly reactive hydroxyl radicals ($OH^{\cdot}$). Insolubility and surface properties of nanoparticles also directly affect cell genotoxicity. It has been shown that small nanoparticles can directly affect DNA by permeating into the cell nucleus. For example, the nanoparticles of titanium dioxide and silica are known to penetrate the nucleus and cause intranuclear protein aggregates that can block vital cellular mechanisms such as replication, transcription, and cell proliferation [49, 50]. The free metal ions generated from nanoparticles can also induce permeability of nucleus barriers and indirectly damage DNA molecules. Nanoparticles can also stimulate mitochondria in cells to produce ROS, which can destabilize the genetic material of the cells. In the secondary mechanism, nanoparticles can induce inflammatory cells, such as macrophages, to produce and deliver oxidants to cells, thus causing genotoxicity from inflammation [51]. The direct interaction of nanoparticles and the DNA molecule or DNA-related proteins can lead to physical damage to the genetic material. Nanoparticles can also cause DNA damage by interacting with other cellular proteins such as those involved in the cell division process. Thus, nanoparticles can induce various cellular responses that can lead to genotoxicity, such as oxidative stress, inflammation, and aberrant signaling responses [52].

31.5 *IN VITRO* AND *IN VIVO* TOXICITY ASSAYS

The toxicity of nanoparticles can be assessed by a number of *in vitro* and *in vivo* studies. The reduction of tetrazolium salts such as MTT (3-(4,5-dimethylthiazol-2-yl)-2,5-diphenyltetrazolium bromide) and XTT (2,3-bis-(2-methoxy-4-nitro-5-sulfophenyl)-2*H*-tetrazolium-5-carboxanilide) can be used to estimate cell viability and proliferation. However, tetrazolium salts get reduced by superoxides and produce absorbant formazan end products. For example, nano-TiO_2 induces superoxide formation in different mammalian cells and the use of MTT/XTT to measure viability or proliferation in such cytological investigations can lead to false positive results. Therefore, MTT and XTT assays may inaccurately predict cell toxicity or overestimate cell viability, respectively [53].

In vivo studies are necessary to elucidate mechanisms, pathways, and entry routes of nanoparticles in a complex multicellular organism. This is required not only for nanomaterials used in industrial processes, where human exposure could occur via the environment, but also for nanomaterial use where human exposure is part of the design, for example, nanomedicines. For nanoparticles produced on an industrial scale, the extent to which factory workers, specific population subgroups, or the public in general are exposed needs to be established. It has been demonstrated that nanoparticles gain access to the body mainly via the airways, the skin, or by ingestion. They are also able to translocate to secondary organs; however, this has only been demonstrated in small quantities. For example, observations on the toxic effects of nanoparticles can be carried out using zebrafish as a model due to its fast embryonic development and transparent body structure. The other advantages of using zebrafish as an organism for studying the potential toxic effects of nanoparticles are that it has a high degree of homology and similar physiological reactions as in mammals. This allows for real-time, noninvasive visualization of the uptake and translocation of nanoparticles, and their effects on organogenesis and morphological development [54]. Cell culture–based assays are used as a prescreening tool to understand the biological effects of nanoparticles. However, along with the *in vitro* assays, it is necessary to confirm the *in vivo* biological activities of nanoparticles in animal models to study the suitability of their application [55].

Apart from *in vivo* and *in vitro* studies, the use of genomic tools such as gene expression profiling is being considered as a potentially rapid and cost-effective approach for identifying and assessing prospective hazard, characterizing chemical (or particle) mode of action, and assessing human relevance in support of human health risk assessment. Once it is proved that mRNA/protein expression profiles can effectively predict the modes of action and biological outcomes of exposure at relevant doses, gene expression data will be more reliable for use in risk assessment studies. Gene expression profiling can be extremely useful in identifying effects at low doses and be used to distinguish between doses that elicit an adaptive response and those that result in adverse toxic effects. So far, the application of gene expression profiling in regulatory toxicology has focused more on the qualitative identification of the chemical modes of action and transcription biomarkers that can predict specific toxicities than on the quantitative determination of threshold values, such as benchmark doses [56]. The advantages of using toxicogenomic profiling is that it can be used as a screening tool to prioritize the specific assays that should be conducted from the standard battery of tests, and this will minimize the use of animals, and be more cost-effective and time-saving [57]. The other benefits are that the global analyses of transcriptional changes provide a wealth of information that will be useful in identifying the probable modes of action and to query their relevance to human adverse health outcomes [58].

Proteomics involves the study of the complete profile of proteins in a given cell, tissue, or biological system at a given time. It uses a set of high-throughput methodologies with a wide dynamic range that enables the discovery of novel biomarkers. The application of proteomic tools such as two-dimensional electrophoresis and mass spectrometry methods could also help address the growing environmental threat posed by nanoparticles and provide useful information while increasing knowledge about nano–bio interactions. However, the limitations and challenges of this approach are the shortage of genome sequence data necessary for protein identification and the growing requirement for more stringent study designs [59].

However, before the full-fledged applicability of toxigenomic and proteomic approaches in human health risk assessment, it is essential to know whether gene expression profiling and other such genomic and proteomic tools can identify hazards, assess risk of exposure via quantitative dose–response analysis, and identify adverse effects associated with specific modes of action of nano–bio interactions [60].

31.6 HAZARD IDENTIFICATION AND RISK ASSESSMENT OF NANOPARTICLES

Prior to studying the impact of nanoparticles on the environment and health, minimal characterization of the engineered nanoparticles has to be carried out. The physicochemical properties of the nanoparticles must be understood in detail for analyzing nanoparticle-induced toxicity. As the field of nanotoxicity is emerging and the specific properties of nanoparticles influencing cellular toxicity are still unclear, it is essential to completely characterize nanoparticles [61]. Nanoparticle characterization involves knowing its chemical composition and purity; particle size and particle size distribution; shape, structure, and specific surface morphology (crystal phase, form); surface chemistry/coating/charge/area; agglomeration or aggregation state, or particle size under experimental conditions as appropriate; stability over time; dissolution, water solubility, dosimetry, and uptake. For assessing the effects of nanoparticles on the environment, the indication of their hazardous effects, the effect of dissolution in water on thier toxicity, their tendency for agglomeration or sedimentation, their fate during wastewater treatment, and their stability during incineration need to be understood. Acute toxicity, chronic toxicity, impairment of DNA, crossing and damaging of tissue barriers, brain damage, and translocation and effects of engineered nanoparticles in the skin, gastrointestinal, or respiratory tract are some of the effects that have to be studied to assess the safety of nanoparticles for human health [6].

Hazard identification and risk assessment of engineered nanoparticles can be briefly summarized as follows. The initial step is the physicochemical characterization of the nanoparticles. It will involve nanoparticle structure alerts and their behavior in aerosols and suspensions. This is followed by the identification of structure or composition of the nanoparticle as the main hazard. The next step is to test the nanoparticles for the production of ROS in a cellular environment. Further, the nanoparticles have to be tested for genotoxicity, immunotoxicity, skin toxicity, ocular, liver, and kidney toxicity under *in vitro* conditions. After *in vitro* experiments, *in vivo* tests are carried out for immunotoxicity, organ toxicity, genotoxicity, and reproductive toxicity. The nanoparticles showing positive genotoxicity and mutagenicity will be analyzed for carcinogenicity and adverse effects on reproductive behavior. The risk assessment of nanoparticles involves the evaluation of the magnitude of risk at different exposure levels, setting of the exposure levels, and other regulatory limits, that is, hazard identification, hazard characterization, exposure assessment, and risk characterization. Based on the hazard assessment of nanoparticles, the knowledge on experimental levels of exposure to nanoparticles and toxic effects induced by nanoparticles will be combined and used to enable effective management and safer application of nanoparticles [62].

31.7 CONCLUSION

The toxicological effect of nanoparticles on the ecosystem and human health has been of increasing interest. The interaction between the highly increased reactive surfaces of nanoparticles due to the increased surface-to-volume ratio and the biotic and abiotic components of the ecosystem constitutes the basis of nanotoxicological studies. The interaction mechanism between nanoparticles and living organisms is not yet completely understood. Apart from the nanoparticle material, size, shape, surface, charge, coating, dispersion, agglomeration, aggregation, concentration, and matrix, dosimetrics should also be considered for toxicological studies. Hazard identification at the *in vivo* level with respect to entry routes and putative targets has been identified. But there are few epidemiological studies on the long-term exposure to engineered nanoparticles. Therefore, there is a need for long-term evaluation of the risks associated with nanoparticles before their large-scale production and application for commercial and medical purposes.

REFERENCES

[1] Linkov I, Steevens J, Adlakha-Hutcheon G, Bennett E, Chappell M, Colvin V, Davis JM, Davis T, Elder A, Hansen S, Hakkinen PB, Hussain SM, Karkan D, Korenstein R, Lynch I, Metcalfe C, Ramadan AB, Satterstrom FK. Emerging methods and tools for environmental risk assessment, decision-making, and policy for nanomaterials: summary of NATO Advanced Research Workshop. Journal of Nanoparticle Research 2009;11:513–527.

[2] Gajewicz A, Rasulev B, Dinadayalane TC, Urbaszek P, Puzyn T, Leszczynska D, Leszczynski J. Advancing risk assessment of engineered nanomaterials: application of computational approaches. Advanced Drug Delivery Reviews 2012;64:663–693.

[3] Bhatt I, Tripathi BN. Interaction of engineered nanoparticles with various components of the environment and possible strategies for their risk assessment. Chemosphere 2011;82:308–317.

[4] Salata OV. Applications of nanoparticles in biology and medicine. Journal of Nanobiotechnology 2004;2:1–6.

[5] Rai RV, Bai JA. Nanoparticles and their potential application as antimicrobials. In: Mendez-Vilas A, editor. *Science Against Microbial Pathogens: Communicating Current Research and Technological Advances*. Spain: Formatex Research Center; 2011. p 197–209.

[6] Som C, Wick P, Krug H, Nowack B. Environmental and health effects of nanomaterials in nanotextiles and facade coatings. Environment International 2011;37:1131–1142.

[7] Sozer N, Kokini JL. Nanotechnology and its applications in the food sector. Trends in Biotechnology 2009;27:82–89.

[8] Hackenberg S, Scherzed A, Kessler M, Hummel S, Technau A, Froelich K, Ginzkey C, Koehler C, Hagen R, Kleinsasser N. Silver nanoparticles: evaluation of DNA damage, toxicity and functional impairment in human mesenchymal stem cells. Toxicology Letters 2011;201:27–33.

[9] Beer C, Foldjberg R, Hayashi Y, Sutherland DS, Autruo H. Toxicity of silver nanoparticles—nanoparticle or silver ion? Toxicology Letters 2012;208:286–292.

[10] Chairuangkitti P, Lawanprasert S, Roytrakul S, Aueviriyavit S, Phummiratch D, Kulthong K, Chanvorachote P, Maniratanachote R. Silver nanoparticles induce toxicity in A549 cells via ROS-dependent and ROS-independent pathways. Toxicology in Vitro 2013;27:330–338.

[11] Teodoro JS, Simoes MA, Duarte FV, Rolo PA, Murdoch CR, Hussain MS, Palmeira MC. Assessment of the toxicity of silver nanoparticles in vitro: a mitochondrial perspective. Toxicology in Vitro 2011;25:664–670.

[12] Meyer JN, Lord AC, Yang XY, Turner EA, Badireddy AR, Marinakos SM, Chilkoti A, Weisner MR, Auffan M. Intracellular uptake and associated toxicity of silver nanoparticles in Caenorhabditis elegans. Aquatic Toxicology 2010;100:140–150.

[13] Park MVDZ, Neigh MA, Vermeulen JP, de la Fonteyne LJJ, Verharen HW, Briede JJ, van Loveren H, de Jong WH. The effect of particle size on the cytotoxicity, inflammation, developmental toxicity and genotoxicity of silver nanoparticles. Biomaterials 2011;32: 9810–9817.

[14] Singh PS, Ramarao P. Cellular uptake, intracellular trafficking and cytotoxicity of silver nanoparticles. Toxicology Letters 2012;213:249–259.

[15] Kaur J, Tikoo K. Evaluating cell specific cytotoxicity of differentially charged silver nanoparticles. Food and Chemical Toxicology 2013;51:1–14.

[16] Lasagna-Reeves C, Gonzalez-Romero D, Barria MA, Olmedo I, Clos A, Sadagopa Ramanujam VM, Urayama A, Vergara L, Kogan MJ, Soto C. Bioaccumulation and toxicity of gold nanoparticles after repeated administration in mice. Biochemical and Biophysical Research Communications 2010;393:649–655.

[17] Coradeghini R, Gioria S, Garcia CP, Natvia P, Franchini F, Gilliland D, Ponti J, Rossi F. Size-dependent toxicity and cell interaction mechanisms of gold nanoparticles on mouse fibroblasts. Toxicology Letters 2013;217:205–216.

[18] Gao W, Xu K, Ji L, Tang B. Effect of gold nanoparticles on glutathione depletion-induced hydrogen peroxide generation and apoptosis in HL7702 cells. Toxicology Letters 2011;205:86–95.

[19] Truong L, Saili SK, Miller JM, Hutchinson JE, Tanguay RL. Persistent adult zebrafish behavioral deficits results from acute embryonic exposure to gold nanoparticles. Comparative Biochemistry and Physiology Part C: Toxicology & Pharmacology 2012;155:269–274.

[20] Sharma V, Singh P, Pandey AK, Dhawan A. Induction of oxidative stress, DNA damage and apoptosis in mouse liver after sub-acute oral exposure to zinc oxide nanoparticles. Mutation Research 2012;745:84–91.

[21] Hackenberg S, Scherzed A, Technau A, Kessler M, Froelich K, Ginzkey C, Koehler C, Burghartz M, Hagen R, Kleinsasser N. Cytotoxic, genotoxic and pro-inflammatory effects of zinc oxide nanoparticles in human nasal mucosa cells in vitro. Toxicology in Vitro 2011;25:657–663.

[22] Márquez-Ramírez SG, Delgado-Buenrostro NL, Chirino YI, Iglesias GG, López-Marure R. Titanium dioxide nanoparticles inhibit proliferation and induce morphological changes and apoptosis in glial cells. Toxicology 2012;302:146–156.

[23] Sun Q, Tan D, Ze Y, Sang X, Liu X, Gui S, Cheng Z, Cheng J, Hu R, Gao G, Liu G, Zhu M, Zhao X, Sheng L, Wang L, Tang M, Hong F. Pulmotoxicological effects caused by long-term titanium dioxide nanoparticles exposure in mice. Journal of Hazardous Materials 2012;235–236:47–53.

[24] Chen J, Dong X, Xin Y, Zhao M. Effects of titanium dioxide nano-particles on growth and some histological parameters of zebrafish (Danio rerio) after a long-term exposure. Aquatic Toxicology 2011;101:493–499.

[25] Gui S, Zhang Z, Zheng L, Cui Y, Liu X, Li N, Sang X, Sun Q, Gao G, Cheng Z, Cheng J, Wang L, Tang M, Hong F. Molecular mechanism of kidney injury of mice caused by exposure to titanium dioxide nanoparticles. Journal of Hazardous Materials 2011;195: 365–370.

[26] Passagne I, Morille M, Rousset M, Pujalte I, L'Azou B. Implication of oxidative stress in size-dependent toxicity of silica nanoparticles in kidney cells. Toxicology 2012;299:112–124.

[27] Ahamed M, Siddiqui MA, Akhtar MJ, Ahmad I, Pant BA, Alhadaq AH. Genotoxic potential of copper oxide nanoparticles in human lung epithelial cells. Biochemical and Biophysical Research Communications 2010;396:578–583.

[28] Chen Z, Meng H, Xing G, Chen C, Zhao Y, Jia G, Wang T, Yuan H, Ye C, Zhao F, Chai Z, Zhu C, Fang X, Ma B, Wan L. Acute toxicological effects of copper nanoparticles in vivo. Toxicology Letters 2006;163:109–120.

[29] Liao MY, Liu HG. Gene expression profiling of nephrotoxicity from copper nanoparticles in rats after repeated oral administration. Environmental Toxicology and Pharmacology 2012;34:67–80.

[30] Fahmy B, Cormier AS. Copper oxide nanoparticles induce oxidative stress and cytotoxicity in airway epithelial cells. Toxicology in Vitro 2009;23:1365–1371.

[31] Cui W, Li J, Zhang Y, Rong H, Lu W, Jiang L. Effects of aggregation and the surface properties of gold nanoparticles on cytotoxicity and cell growth. Nanomedicine 2012;8:46–53.

[32] Jenkins JT, Halaney DL, Soklov KV, Ma LL, Shipley HJ, Mahajan S, Louden CL, Asmis R, Milner TE, Johnston KP, Feldman MD. Excretion and toxicity of gold–iron nanoparticles. Nanomedicine 2013;9:356–365.

[33] Ma H, Williams PL, Stephen A. Diamond ecotoxicity of manufactured ZnO nanoparticles—a review. Environmental Pollution 2013;172:76–85.

[34] Hackenberg S, Friehs G, Froelich K, Ginzkey C, Koehler C, Scherzed A, Burghartz M, Hagen R, Kleinasser N. Intracellular distribution, geno- and cytotoxic effects of nanosized titanium dioxide particles in the anatase crystal phase on human nasal mucosa cells. Toxicology Letters 2010;195:9–14.

[35] Studer MA, Limbach LK, Duc LV, Krumeich F, Athanassiou EK, Gerber LC, Moch H, Stark WJ. Nanoparticle cytotoxicity depends on intracellular solubility: comparison of stabilized copper metal and degradable copper oxide nanoparticles. Toxicology Letters 2010;197: 169–174.

[36] Stern ST, McNeil SE. Nanotechnology safety concerns revisited. Toxicological Sciences 2008;101:4–21.

[37] Nakajima H, Ozaki K, Hongyo T, Narama I, Todo T. A rapid and easy method for the qualitative detection of intracellular deposition of inhaled nanoparticles. Nanomedicine 2011;7:881–888.

[38] Liu Y, Gao Y, Zhang L, Wang T, Wang J, Jiao F, Li W, Li Y, Li B, Chai Z, Wu G, Chen C. Potential health impact on mice after nasal instillation of nano-sized copper particles and their translocation in mice. Journal of Nanoscience and Nanotechnology 2009;9: 6335–6343.

[39] Crosera M, Bovenzi M, Maina G, Adami G, Zanette C, Florio C, Larese FF. Nanoparticle dermal absorption and toxicity: a review of literature. International Archives of Occupational and Environmental Health 2009;82:1043–1055.

[40] Kim S, Lim YT, Soltesz EG, De Grand AM, Lee J, Nakayama A, Parker JA, Mihaljevic T, Laurence RG, Dor DM, Cohn LH, Bawendi MG, Frangioni JV. Near-infrared fluorescent type II quantum dots for sentinel lymph node mapping. Nature Biotechnology 2004;22:93–97.

[41] Powell JJ, Faria N, Thomas-McKay E, Pele LC. Origin and fate of dietary nanoparticles and microparticles in the gastrointestinal tract. Journal of Autoimmunity 2010;34:226–233.

[42] Nel AE, Madler L, Velegol D, Xia T, Hoek EM, Somasundaran P, Klaessig F, Castranova V, Thompson M. Understanding biophysicochemical interactions at the nano-bio interface. Nature Materials 2009;8:543–557.

[43] Elsaesser A, Howard CV. Toxicology of nanoparticles. Advanced Drug Delivery Reviews 2012;64:129–137.

[44] Vasir JK, Labhasetwar V. Quantification of the force of nanoparticle–cell membrane interactions and its influence on intracellular trafficking of nanoparticles. Biomaterials 2008;29:4244–4252.

[45] Jin Y, Kannan S, Wu M, Zhao JX. Toxicity of luminescent silica nanoparticles to living cells. Chemical Research in Toxicology 2007;20:1126–1133.

[46] Wagner SC, Roskamp M, Pallerla M, Araghi RR, Schlecht S, Koksch B. Nanoparticle-induced folding and fibril formation of coiled-coil-based model peptides. Small 2010;6:1321–1328.

[47] Pereira R, Rocha-Santos TAP, Antunes FE, Rasteiro MG, Ribeiro R, Goncalves F, Soares AMVM, Lopes I. Screening evaluation of the ecotoxicity and genotoxicity of soils contaminated with organic and inorganic nanoparticles: the role of ageing. Journal of Hazardous Materials 2011;194:345–354.

[48] Bhabra G, Sood A, Fisher B, Cartwright L, Saunders M, Evans WH, Surprenant A, Lopez-Castejon G, Mann S, Davis SA, Hails LA, Ingham E, Verkade P, Lane J, Heesom K, Newson R, Case CP. Nanoparticles can cause DNA damage across a cellular barrier. Nature Nanotechnology 2009;4:876–883.

[49] Chen M, von Mikecz A. Formation of nucleoplasmic protein aggregates impairs nuclear function in response to SiO2 nanoparticles. Experimental Cell Research 2005;305:51–62.

[50] Liu L, Takenaka T, Zinchenko AA, Chen N, Inagaki S, Asada H. Cationic silica nanoparticles are efficiently transferred into mammalian cells. International Symposium on Micro-NanoMechatronics and Human Science 2007;1–2:281–285.

[51] Jaganathan H, Godin B. Biocompatibility assessment of Si-based nano- and micro-particles. Advanced Drug Delivery Reviews 2012; 64:1800–1819.

[52] Singh N, Manshain B, Jenkins GJS, Griffith MS, Williams PW, Maffeis TGG, Wright CJ, Doak SH. NanoGenotoxicology: the DNA damaging potential of engineered nanomaterials. Biomaterials 2009;30:3891–3914.

[53] Wang S, Yu H, Wickliffe JK. Limitation of the MTT and XTT assays for measuring cell viability due to superoxide formation induced by nano-scale TiO2. Toxicology in Vitro 2011;25:2147–2151.

[54] Fent K, Weisbord CJ, Wirth-Heller A, Pieles U. Assessment of uptake and toxicity of fluorescent silica nanoparticles in zebrafish (Danio reria) early life stages. Aquatic Toxicology 2010;100:218–228.

[55] Schrand AM, Rahman MF, Hussain SM, Schlager JJ, Smith DA, Syed AF. Metal-based nanoparticles and their toxicity assessment. Nanomedicine and Nanobiotechnology 2010;2:544–568.

[56] Thomas RS, Allen BC, Nong A, Yang L, Bermudez E, Clewell HJ III, Andersen ME. A method to integrate benchmark dose estimates with genomic data to assess the functional effects of chemical exposure. Toxicological Sciences 2007;98:240–248.

[57] Dix D, Martin H, Richard A, Setzer R, Kavlock R. The ToxCast program for prioritizing toxicity testing of environmental chemicals. Toxicological Sciences 2007;95:5–12.

[58] Currie R. Toxicogenomics: the challenges and opportunities to identify biomarkers, signatures and thresholds to support mode-of-action. Mutation Research 2012;746:97–103.

[59] Sheehan D. The potential of proteomics for providing new insights into environmental impacts on human health. Reviews on Environmental Health 2007;22:175–194.

[60] Bourdon JA, Williams A, Kuo B, Moffat I, White PA, Halappanavar S, Vogel U, Wallin H, Yauk CL. Gene expression profiling to identify potentially relevant disease outcomes and support human health risk assessment for carbon black nanoparticle exposure. Toxicology 2013;303:83–93.

[61] Love SA, Maurer-Jones MA, Thompson JW, Lin Y-S, Haynes C-L. Assessing nanoparticle toxicity. Annual Review of Analytical Chemistry 2012;5:181–205.

[62] Savolainen K, Alenius H, Norppa H, Pylkkanen L, Tuomi T, Kasper G. Risk assessment of engineered nanomaterials and nanotechnologies—a review. Toxicology 2010;269:92–104.

32

IMPLICATIONS OF THE USE OF NANOMATERIALS FOR ENVIRONMENTAL PROTECTION: CHALLENGES IN DESIGNING ENVIRONMENTALLY RELEVANT TOXICOLOGICAL EXPERIMENTS

Rute F. Domingos[1] and José P. Pinheiro[2]

[1]*Centro de Química Estrutural, Instituto Superior Técnico, Universidade de Lisboa, Lisbon, Portugal*

[2]*Centro de Biomedicina Molecular Estrutural (CBME)/IBB, LA, DQF/FCT, Universidade do Algarve, Faro, Portugal*

32.1 INTRODUCTION

Deliberate application of nanomaterials (NMs) for environmental protection could result in intentionally diffuse inputs of these engineered materials into the environment. The anticipated new or enhanced activity of these NMs will inevitably result in both new risks and new benefits to human and environmental health. *The highly dynamic behavior of NMs in environmental systems results in a set of chemical and physical transformations* such as dissolution, and/or size and shape changes, including homo- (between particles of the same type) and heteroaggregation (between particles of different types), that will greatly affect their fate.

Due to NMs' inherent reactivity, stabilizers are usually added to improve their mobility by protecting the NMs' surface from unwanted reactions, preventing the interparticle aggregation by charge repulsion and/or steric effects, and also serving as a diffusion-controlled release carrier. The stabilizer is thus a barrier between the NM and the matrix where it is immersed. Previously, the stabilizers used were low-molecular organic compounds such as carbonic acids, alcohols, and amides, and natural polymers like gelatin, gum Arabic, agar-agar, starch, and cellulose [1]. Currently, synthetic polymers are more frequently employed [2]. Charged polymers are more commonly used since the stabilization they provide is quite effective and yields relatively stable core–shell entities that work well for their designed purpose. Due to their chemical stability, the commonest acid groups used in stabilizers are sulfonic ($-SO_3^-$) (polystyrene sulfonate (PSS)), carboxylic (–COOH) (like polyacrilic acid (PAA)), and numerous combinations and variations such as block copolymers, and cross-linked configurations. Uncharged stabilizers are less used and less relevant for our analysis since they tend to bind metal ions poorly or not at all. These include polymers like polyvinil pirrolidone (PVP) and polyethylene glycol (PEG) that stabilize the particles mostly by steric effects. The possible effects of stabilizers on the behavior of NMs showed that complex interactions are possible between several different processes, affecting their physicochemical properties and consequently their fate. However, most of these interactions have not yet been examined systematically.

It is vital to predict and validate the different forms of occurrence of nanoscale materials in the environmental compartments, given that it is most likely that these forms have different mobility, bioavailability, and, consequently, different toxicological effects. The rigorous quantification of the dynamic contribution of various possible NM-containing species is extremely necessary to accomplish this aim. In this chapter, several key questions that underlie the effort to understand how NMs behave in natural systems and how they may impact the environment will be addressed: (i) *How do the NMs' properties and their*

Nanomaterials for Environmental Protection, First Edition. Edited by Boris I. Kharisov, Oxana V. Kharissova, and H. V. Rasika Dias.

nanoscale features affect their behavior and interactions with other environmentally relevant parameters? (ii) *What transformations are likely to occur in natural systems?* (iii) *How do the transformations affect the NMs' properties, behaviors, and bioavailability?* These questions will be addressed by discussing two interdependent essential issues: physicochemical transformations of the NMs and their possible interactions when dispersed on a matrix. The discussion of the environmental fate of a NM, which will determine the NM species to which ecosystems and the general population will be exposed to, will ultimately facilitate ecotoxicologists in their choice of appropriated toxicological experiments (including the organisms and the exposure route).

The discussion presented in this chapter will be focused on metal-containing NMs for two main reasons: (1) for comparison with the well-established methodology existing for toxicological experiments with metal ions, and (2) for changes needed in the methodology as the metal-containing NMs are toxic *per se* and simultaneously a source of metal ions. Carbon-based NMs will not be addressed since the quantification methodologies for these NMs are still limited and much less sensitive when compared to those available for metal-containing NMs.

32.2 FATE OF NMs IN SOLUTION

32.2.1 Physicochemical Transformations

The reliable detection, quantification, and characterization of NMs is of crucial importance for the exposure and risk assessment, which in turn provides a justifiable and robust regulatory framework that will allow the continued innovation and development of the nanotechnology industry. State-of-the-art analytical techniques and complementary analytical methodologies for the characterization of NMs are under development [3]. An additional set of challenges emerges when the objective is to quantify and characterize NMs in complex matrices, including natural waters, soils, sediments, and biological tissues because of the environmentally and biologically relevant concentrations involved. This is an area of intensive research where the development of new and/or more robust and sensitive techniques and methodologies is urgently necessary.

Properties such as composition, shape, and structure vary greatly according to the type of NM and are also expected to vary along the NM's life cycle. These properties are expected to directly determine the NM's environmental fate and transformations that occur during the NM's life cycle. Other factors affecting the stability of an NM include matrix effects, changes in pH or ionic strength, impurities effects, degradation of coatings, and functionalization. Representative chemical and physical transformations of metal-containing NMs include (Fig. 32.1) (i) dissolution, (ii) homoaggregation, and (iii) heteroaggregation.

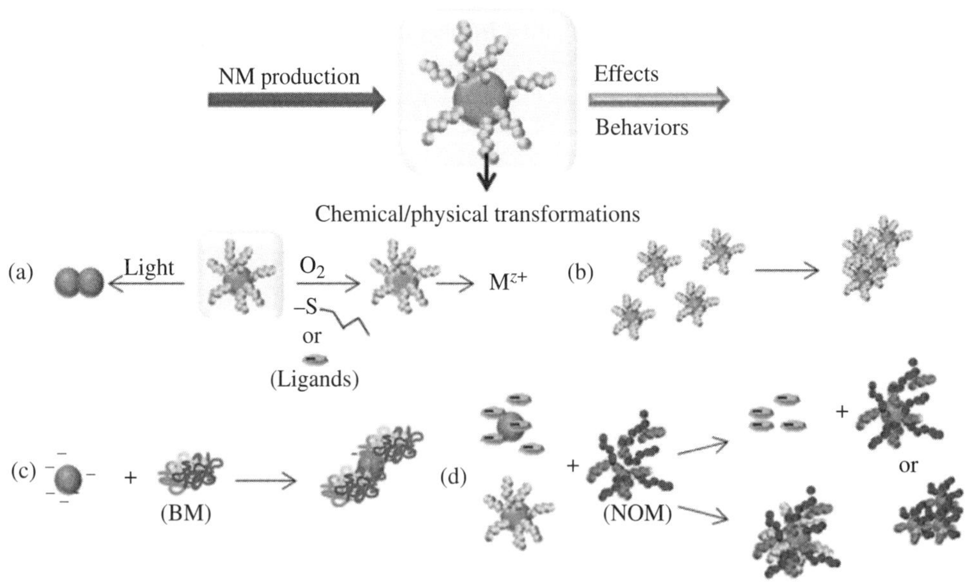

FIGURE 32.1 Representative chemical and physical transformations of metal-containing NMs: (a) photodegradation of coatings that can lead to aggregation, and oxidation, thiol ligands and other ligands that can promote dissolution, (b) homoaggregation, (c) adsorption of biomacromolecules (BM) that can affect the aggregation, (d) adsorption of NOM that can displace smaller ligands and promote stabilization or flocculation, or that can lead to poorly characterized heterogeneous mixed polymer–NOM layers.

32.2.1.1 Dissolution A large number of NMs may be composed of or can contain constituents that can undergo chemical phenomena such as reduction and/or oxidation in aquatic environments resulting in their dissolution. The dissolution process of an NM is of great importance since it strongly affects the stability of the NM and, therefore, directly influences its long-term use in all solution-based applications. An understanding of the NM's dissolution mechanism will lead to a much more rational safer-to-design efficient synthesis route for high-quality and safer NMs. Despite the importance of a NM's dissolution, very little is known about its underlying mechanism and reaction pathways under environmentally relevant conditions.

In Ag NMs, the oxidation of Ag^0 to Ag^I results in the NMs' dissolution and subsequent release of bactericidal Ag^+ [4]. Sulfur and selenium, major components of quantum dots (QD), can also undergo oxidation resulting in the release of soluble toxic metal ions such as Cd or Pb [5]. Moreover, the use of stabilizers that have a binding capacity through metal ions can also promote the dissolution of these photoluminescent particles [6]. Photoreactions can be very important transformation processes because they can affect the NM's coatings, oxidation state, generation of reactive oxygen species (ROS), and persistence. NMs such as TiO_2 and ZnO are innately photoactive, potentially producing ROS when exposed to sunlight [7], and promoting dissolution mainly in the case where NMs contain ZnO particles. NMs that constitute Ag, Zn, and Cu are mainly affected by dissolution and sulfidation because they form partially soluble metal oxides and they have a strong affinity for inorganic and organic sulfide ligands. The dissolution of these NMs results in the release of toxic cations [8] such that their persistence is reduced but their toxicity is increased. Evidently, complete dissolution of metal-containing NMs allows the prediction of their impact using existing models for metal ion speciation and their effects.

The quantification of the dissolved ions from a NM dispersed in an environmental compartment is a complicated issue because the dissolution process is most often not in equilibrium. The real-time kinetic measurement of the dissolution process limits the storage of whole unfractionated samples for subsequent ion analysis using appropriate analytical techniques, because the dissolution rate may be fast or not attain the equilibrium during the experimental time.

32.2.1.2 Aggregation Once in the environment, NMs will interact with naturally occurring bio- or geomacromolecules, including proteins, polysaccharides, and natural organic matter (NOM), which will significantly affect their surface chemistry. Homo- and heteroaggregation are dynamic processes that can lead to a decrease in the available surface area of the materials, thereby decreasing their reactivity. However, this decrease is dependent on the particle number, size distribution, and the fractal dimensions of the aggregate [9].

The diffusion of the particles in dispersion is driven by the Brownian motion, where the temperature and particle number concentration determine the particle–particle collision frequency. The particles affect each other by attractive and repulsive forces such as Borne repulsion, diffuse double-layer potential, and van der Waals attraction that act on different length scales. The extended Derjaguin–Landau–Verwey–Overbeek (DLVO) theory [10] weights the attractive and repulsive forces acting on two closely adjacent particles. However, this theory is only applicable if there is no interference with these diffusive or attractive forces, including the effects of particle shape, charge heterogeneity, and surface roughness that may influence the collision efficiency [11]. The addition of salt ions to the medium results in an important effect on the homoaggregation of particles: some of the salt ions will accumulate in the electrostatic diffuse double layer (which consists of a layer of charge at the surface of a particle and the electric field generated by the charged surface), screening some of the surface charge of the particles. The result is a decrease in the electrostatic diffuse double layer and, consequently, a decrease in the repulsion forces and an increase in the aggregation between the particles. Multiple charged cations and anions such as Ca^{2+} and PO_4^{3-} may also promote the aggregation of NMs [12]. The presence of highly negatively charged natural colloids in natural environmental systems is also known to have a key effect on the aggregation of positively charged NMs. The effect exerted depends on the concentration of the NOM: a low concentration of NOM may lead to a reduction in the surface charge of the particles, an intermediate NOM concentration to a net neutral surface charge, while elevated concentrations can provide the particles with a net negative surface charge by charge reversal, stabilizing the colloidal dispersion. However, the outcome of such colloidal dispersions composed by NOM and NMs is not easily predicted, and needs to be analyzed experimentally by performing acid–base titrations and electrophoretic mobility measurements. In fact, it is known that NOM may provide both charge and steric stabilization [13] of the NMs, although they may also result in bridging flocculation [12]. Steric stabilization can occur even at higher levels of ionic strength [13a], where the DLVO theory would predict strong aggregation. For NMs stabilized with higher-molecular-weight (MW) polymeric coatings, mixed polymer–NOM layers may form on NMs, but depending on the nature of the coating polymer, interactions with NOM may be minimal [14]. NOM effects are complex and difficult to predict; however, it is extremely important to explore these interactions since NOM concentrations are typically orders of magnitude higher in concentration than NMs, and so are likely to substantially modify their properties and behaviors.

As for the quantification of the dissolution process of a NM, the determination of the aggregate sizes in an environmental compartment may also require real-time kinetic measurements, because the aggregation rate can be fast or the aggregate's size distribution may not reach equilibrium within the experimental time window. The size of a NM will affect its bioavailability to organisms: when aggregates or heteroaggregates become too large for direct transport across the cell wall and/or membrane, uptake may be prevented.

32.2.1.3 Experimental Techniques Nanoparticulate systems are not uniquely composed by single discrete molecular species, resulting in the appearance of specific artifacts that must mandatorily be taken into account. Consequently, the identification and quantification of these systems need rigorous requirements relying on numerous analytical techniques and/or methodologies in order to validate the obtained results [15]. Furthermore, appropriate traceable standards, quality control procedures, and, when available, standard reference materials must be used in order to validate the analytical techniques and/or methodologies. Unfortunately, very few standard reference materials are available for NM analysis, and currently no standard that consists of complex matrices with certified concentrations of NMs is available. This is in fact an area of great research but remains a major challenge due to the inherent reactivity and instability of the NMs.

The quantification of the dissolution of NMs is not trivial. Besides the difficulties in adapting the existing analytical techniques and/or methodologies, another important factor in the assessment of NM dissolution is the separation of particles from the dissolved components. The flawed separation of the particles from the dissolved components results in over/underquantification of the particles' dissolution. (Ultra)filtration [16], dialysis [17], or (ultra)centrifugation [18] methods are usually used to separate the particles from the dissolved species. Dialysis and (ultra)filtration methods are membrane separation methods based on physical separation that are dependent on the size of the membrane pores, and differ only in the driving force, which for (ultra)filtration is the applied pressure and for dialysis the concentration gradient. A drawback is the size of the filters currently available; the lowest are 20 nm pore size filters. The use of the combination of centrifugation and filtration (centrifugal ultrafiltration) has been recently seen as an effective way to overcome this problem due to the low pore sizes of the membranes available (cutoff value down to 3 kDa). Once the separation of particles from the dissolved components is ensured, a range of analytical techniques can be applied to quantify the dissolution (inductively coupled plasma mass spectrometry (ICP-MS); inductively coupled plasma atomic emission spectroscopy (ICP-AES), atomic absorption spectrometry (AAS), voltammetric techniques). The choice of the analytical technique is based on the sensitivity and detection limit intended. The operation of the ICP-MS instrument in the single-particle mode (SP-ICP-MS) [19] allows the quantification of the dissolved fraction, but with the advantage of not requiring a separation process between particulates and dissolved species (discussed later). The optimal recovery of the dissolved fraction is the main challenge for the effective separation of particles and dissolved species. This is a still more critical challenge when stabilizers are used or when the adsorption of ligands/colloids in NMs occurs that have a binding capacity through the dissolved species (discussed in the following section).

Many different state-of-the-art analytical techniques are available for size measurements including microscopy techniques, dynamic light scattering (DLS), fluorescence correlation spectroscopy (FCS), nanoparticle tracking analysis (NTA), SP-ICP-MS, and separation-based techniques such as field flow fractionation (FFF) or hydrodynamic chromatography (HDC). Each technique has its strengths and weaknesses and results can vary greatly according to the technique selected, the way it is used, and the types of particles studied [15].

Microscopy techniques such as transmission electron microscopy (TEM) [20] and atomic force microscopy (AFM) [21] can achieve sufficient spatial resolution, allowing the distinction of even the smallest single particles (1 nm resolution, which in AFM is limited to the z dimension). For a large number of observations, TEM provides a number average diameter, and when coupled with energy-dispersive spectrometry (EDSp) and electron diffraction, it can additionally provide information on the elemental composition and crystal phase, respectively. For particles that are of similarly size or smaller than the radius of curvature of the AFM tip, the measurement of the lateral distances is usually biased, and, therefore, the height measurements are considered to be more accurate. The height quantifications will correspond to number average diameters when spherical particles are measured. Both microscopy techniques can provide reasonably accurate number average dimensions that have little experimental bias when the sample preparation is free from artifacts, which can be a major challenge when nanoparticulate systems are studied [15]. However, these techniques often suffer from tedious sample preparation, and low sample numbers coupled with high capital and operating costs, especially when attempting to obtain quantitative data for a representative sample.

In DLS [22], one of the most used techniques for NM characterization, the light from a coherent source is scattered when it is directed at the particle suspension. The scattering fluctuates with time due to the random Brownian motion of the particles, allowing the determination of a z-averaged translational diffusion coefficient (D) from the autocorrelation of the Doppler shifts. DLS has the advantage of being a nondestructive technique with a minimum of sample processing. An important drawback is its strong dependence on the Rayleigh scattering on the particle radius (sixth power dependency). Consequently, the presence of aggregates or even contaminating particles such as dust can mask the NPs' signal. The increase in particle scattering by increasing their concentration, most often beyond those found in the environment [23], does not result in an advantage since the consequence is often greater particle aggregation. However, DLS can be very useful in following the kinetics of NM aggregation since it measures the gradual evolution of the z-average hydrodynamic diameter with time.

FCS [24] can be used to measure the D of fluorescent macromolecules or particles under a wide range of solution conditions, including low concentrations ($\approx 0.5\,mg\,L^{-1}$). In this technique, a laser light is focused into the sample of interest using confocal optics, creating an illuminated confocal volume with approximately 0.5–1.0 μm^3. Temporal fluctuations in the measured fluorescence intensity are used to derive an autocorrelation curve, which, in the absence of other processes that affect sample fluorescence (e.g., chemical reactions), is related to the translational diffusion of the fluorescent molecule or particle. The advantage of this technique is that it works at or near single-molecule detection, providing a number average D for singly labeled particles or a weight average D for conditions in which several fluorescent labels are bound to each particle. Two important limitations of FCS are that the particles must be fluorescent, which can be achieved by using labeling techniques [25], and the upper size limit of the aggregates is restricted by the size of the confocal volume (~1 μm^3).

NTA [15, 26], based on a laser-illuminated microscopical technique, also relies, like DLS, on the scattering generated from particles undergoing Brownian motion. However, the Brownian motion of the particles is analyzed in real time by a charged coupled device (CCD) camera, and the mean squared distances that are traveled by each particle in two dimensions are simultaneously visualized and tracked by a particle-tracking image analysis program, in order to determine the number-based D. Since the scattering of small particles varies significantly with particle radius, larger particles can mask the signal of the smaller ones.

SP-ICP-MS [19, 27] requires an adequate time resolution and low particle number concentrations in order to ensure that each pulse corresponds to one particle only. Thousands of individual intensity readings are acquired, each with a very short dwell time (~10 ms), as a function of time, where the pulses above the background represent the measurement of an individual particle. The number of counts of the pulse is related to the number of analyte atoms in the particle, and the frequency of the pulses is proportional to the number concentration of particles. This technique requires little sample preparation and uncomplicated additional method development for a given matrix and/or analyte. Moreover, beside the determination of the particle concentration and size (as low as 20 nm), it also allows the quantification of dissolved metal concentrations (on the ng L^{-1} range). However, this technique is highly dependent on the signal-to-noise ratio of the ICP-MS in use, which may significantly hinder analysis of smaller-sized NPs.

FFF [28] is a chromatography-like elution technique based on the physical interactions of the sample with a field applied at right angles to a thin, open channel. This field allows the separation of particles based on their D. The separation occurs because the flow through the channel is laminar with a parabolic cross-sectional velocity profile. Mass-based D values are determined when the FFF is coupled with an absorbance detector. This is a promising technique when combined with a more sensitive element-specific detector such as ICP-MS.

In HDC [29], particle separation is solely based on particle size independently of particle type and density. It can separate particles in the size range from 5 to 100 nm, but can go well above this limit, depending on the size column used. The column is packed with nonporous beads building up flow channels or capillaries, resulting in the separation of particles by flow velocity and the velocity gradient across the particle. In the narrow conduits, the larger particles cannot fully access slow-flow regions, resulting in faster elution of these larger particles from the column, and higher retention times for smaller particles. HDC is sufficiently robust to require no sample pretreatment, even for samples as complex as lake water and sewage sludge supernatant [29]. When coupled with ICP-MS it allows the simultaneous analysis of most of the commonly used NMs in a single run, having the great advantage over FFF of a fast sample analysis time, which is less than 10 min per sample. A drawback is the quantitative aspect, which is still to be fully addressed, as quantitative ionic standards are not suitable for use in HDC columns. On-column and post-column calibration using a range of standards with different sizes and concentrations for the NM of interest could be used in order to surpass this limitation, but no such standards are still available for NMs.

In addition to size and size distribution, surface chemistry and functionalization should also be taken into account when considering the transport of NMs. An overview of the numerous spectroscopic techniques available for characterizing the surfaces of particles has been provided by Handy et al. [30].

Once NMs are released, their environmental fate determines how they will be biogeochemically processed, or weathered, and which receptors will be exposed (organisms and exposure route). These transformations add up on other modifications of the NMs that take place during their life cycle in an NM-enabled product, ultimately generating the NM form to which ecosystems and the general population will be exposed to. Despite the critical impact that such transformations may have on the chemical and biological activity of NMs, the knowledge available is scarce.

32.2.2 Matrix Interactions

When considering the release of a stabilized NM into the environment, two key aspects must be analyzed: (1) the input of metal ions into the matrix by dissolution of the NM, and (2) the interaction of the stabilizer soft shell with the released metal ions, as well as with other metal ions and/or other macromolecules present in the matrix. To achieve this aim, it is not only necessary to understand how the different metal ion species in a solution behave but also to quantify them. Thus, the theory of metal binding in colloidal systems and the available instrumental techniques capable of quantifying the speciation will be reviewed.

32.2.2.1 Understanding Trace Metal Ion Speciation in Colloidal Dispersions Understanding dynamic speciation requires not only the knowledge of metal-binding equilibrium parameters but also the diffusion and/or kinetic fluxes of the various metal species in a solution, both depending on the timescale considered. Lability is a concept used to define the input of metal species into an overall flux toward a consuming interface constructed upon the magnitudes of the diffusive mass transport and metal complex dissociation fluxes [31]. Two limiting cases apply: (1) complex species do not have the time to dissociate/associate in the diffusion layer and will not contribute to the total flux (static system, inert complexes); and (2) the rates of metal complex association/dissociation are high enough so that the kinetic flux arising from the dissociation of the complex in the diffusion layer is greater than the diffusion-limited flux (dynamic system). For a static system, only the free metal contributes to the overall flux, while for a dynamic system, two situations are possible: (1) the kinetic flux is much larger than the diffusive one so that the free metal ion will be in equilibrium with its complex forms all along the diffusion layer, and all the metal present will contribute (labile complexes); and (2) the kinetic and diffusive fluxes are of the same order of magnitude, thus the free metal and part of the bound metal will contribute to the overall flux (nonlabile complexes).

Besides the kinetic rate parameter, the D of the complex species in the diffusion layer also affects the flux in dynamic systems. The presence of NMs or macromolecular ligands such as humic matter slows down the flux of dynamic metal ions due to lower D of those ligands, resulting in a significant impact on the metal availability toward the consuming interface. For labile complexes, a mean D, taking into account all the diffusing species, should be considered [32].

These are the essential features of dynamic trace metal speciation, and understanding them is important in order to understand why the biouptake equilibrium models (like the free ion activity model (FIAM) [33] and the biotic ligand model (BLM) [34]) might fail to describe the metal toxicity [35]. Nevertheless, we will focus mostly on the equilibrium situation for the sake of simplicity. More information on the dynamic aspects of trace metal speciation can be found by Mota et al. [36].

The equilibrium-binding modeling in charged colloidal systems needs to consider two contributions: (1) the electrostatic binding due to the charges and (2) the chemical intrinsic binding due to the specific interactions between the metal ions and the binding groups in the colloidal particle. The electrostatic contribution is usually described using a Boltzmann equation or more commonly in colloidal or biocolloidal systems using Donnan potentials [37]. For example, Donnan potentials have been used to model the binding of metal ions to the cell walls of bacteria [38] and metal ion interaction with different types of biomass in biosorption studies [39]. For colloidal and/or particulate systems, it is common that more than one chemical ligand is present in the system (chemical heterogeneity). In this case, a discrete-site model with a large number of binding sites (e.g., Windermere Humic Aqueous Model computer code (WHAMC) [40] or a continuous model described by a continuous probability distribution of binding energies such as the nonideal competitive adsorption ((NICA)–Donnan [41]) model is better at modeling the metal binding.

32.2.2.2 The Interaction of Metal Ions with the Stabilizing Soft Shell Layer The charged polymer layer that stabilizes the NM is effectively a soft shell that possesses a different physicochemical environment compared to the bulk, namely, an overall negative potential. Depending on the polymer's nature and cross-linking, this layer can behave more like a gel or even a polymer brush; either way, it will react strongly to ionic strength changes by shrinking or swelling. Cations can be bound by this layer by electrostatic interactions or even covalent binding, depending on the nature of the charged group and the metal in question. The ratio between bound and total metal in the stabilizing layer and bulk phase will depend on the respective layer and matrix composition, especially the amount of metal-binding ligands present. Nonetheless, the free metal ion concentration will be larger in the soft shell layer than in the bulk. This effect is well known in gels where the enrichment in free metal ion concentration in the gel regarding the bulk solution is given by a Boltzmann equation of the following type:

$$\frac{\left(\mathrm{M}^{z+}\right)_{\mathrm{gel}}}{\left(\mathrm{M}^{z+}\right)_{\mathrm{sol}}} = \exp\left(-\frac{zF}{RT}\psi_{\mathrm{D}}\right) \tag{32.1}$$

where M^{z+} is a metal ion with charge z, ψ_D is the Donnan potential resulting from the fixed charged groups in the gel phase, F is the Faraday constant, R is the gas constant, and T is the absolute temperature. Since the potential in the gel is directly related with the ionic strength in the solution, this effect is correspondingly larger for lower ionic strengths and negligible at high ionic strengths. Due to the ion charge dependence, this effect is significantly larger for divalent and trivalent ions than for monovalent ones. As an example, Davis et al. [42] reported an increase of free cadmium of 3.2× and 5.8× in 0.2 and 0.5% acrylamide gels for an ionic strength of 1.0 mM, but only an increase of 1.5× and 2.2× for an ionic strength of 5.0 mM, respectively, and barely no increase within the experimental error for an ionic strength of 10 mM. Acrylamide gels are a good example of a modestly charged gel with no significant intrinsic chemical binding. The same authors [43] also used alginate gels and found an increase of free cadmium in the gel (3×) at an ionic strength of 10 mM, but, in contrast to acrylamide gels, the total level of free cadmium was enhanced by a factor of approximately 60, resulting predominantly from the specific binding of the cadmium by the uronic acids of the alginate gel. Alginate gels are not only more charged than acrylamide gels but also have a significant ability to chemically bind cadmium, thus further enriching the gel phase in cadmium ions. These examples illustrate not only the tremendous potential that the polymer layer may have to bind metal ions but also their variability depending on their respective charge densities and the presence or absence of specific binding sites for the metal ions.

32.2.2.3 Self-Interaction An interesting and often neglected aspect of the NM's fate and toxicity evaluation is that one of the prime binding ligands for the dissolved metal ions from the NM is very probably the charged polymer in the stabilizing layer. For a medium with low ionic strength and relatively high pH (higher than the pK_a of the acid groups of the shell), a reasonably charged soft shell will have a major impact on the stability of the particle. In this situation, the free metal ion concentration in the soft shell layer will be much larger than in the bulk, and, consequently, the NM metal core will be much less likely to dissolve, since it will be in the vicinity of a much higher free metal concentration, closer to the saturation value. This is a feature that is commonly used by manufacturers to effectively prevent their NMs from dissolving. Evidently, when exposed to changes in the bulk, especially increases of ionic strength and lowest pH environments, the potential in the layer might change significantly and the NMs will dissolve, becoming a source of metal ions to the bulk.

32.2.2.4 Stabilizer–Matrix Interactions A stabilized NM is likely to suffer slower physical changes than a nonstabilized NM, thus being available to interact with the components of the matrix. This leads to another often ignored aspect related to a NM's ability to bind the metal ions previously present in the matrix, thus changing their speciation [6, 44]. The ability of the core–shell NMs to interact with other matrix components, like humic and fulvic acid fractions [45], the rich variety of particulate matter (among those several positively charged metal oxide particles), and living organisms (bacterial cell walls, algae membranes, phytoplankton shells, etc.), is also well documented. On the face of it, it is important to remember that in toxicological experiments, upon introducing the NM into the exposure media, the trace metal speciation and possibly the speciation of other components of the matrix will change. Therefore, in order to understand the biouptake of toxic species, we first must understand the nature of the changes yielded by NM–matrix interactions.

32.2.2.5 Speciation Techniques The trace metal speciation techniques of interest can be separated in two groups: equilibrium and dynamic techniques. Equilibrium techniques usually report the free metal ion concentration, while dynamic techniques report the free metal ion plus the labile fraction of the complexed metal.

Although we will not describe it, we would like to mention the competitive ligand exchange adsorptive cathodic stripping voltammetry (CLE-AdSV) [46]. The reason we do not a priori recommend this technique is the need to add a competitive ligand to the solution, which is not desirable in many toxicological experiments.

32.2.2.5.1 Free Metal Ion Determination Among the techniques that are able to determine the free metal ion concentration, the most useful are the potentiometry with ion-selective electrodes (ISEs), the Donnan membrane technique (DMT), and the absence of gradients and Nernstian equilibrium stripping technique (AGNES).

Potentiometry using ISEs is a well-established and known technique, and is probably the ideal method for free metal ion determination since the electrodes are easy to operate, and directly measure the free metal ion without disturbing the equilibrium or the sample composition [47]. The drawback is that the ISEs' use is restricted to the quantification of proton and those elements present in total concentrations above 10^{-6} mol L^{-1} due to the solubilization and contamination/adsorption of the electrode membrane [48]. Although in the last decade the development of potentiometric sensors using polymeric membranes has increased the ISEs' detection limit [49], their operation is still difficult and most are not commercially available. When the ISE is placed in a solution containing the ion, a potential (E) is established between the membrane interface and the solution. This E is measured against a reference electrode, and its magnitude is proportionally dependent on the free metal ion concentration in the bulk, c^*_M:

$$E = k + \frac{S}{n} \log c^*_M \qquad (32.2)$$

where S/n represents the slope (theoretically $2.303RT/F$) and k is a constant dependent on electrode construction.

The DMT is a speciation technique based on the transport of the metal ions from the sample solution, through a negatively charged semipermeable cation exchange membrane, to an acceptor solution with the same ionic strength as the sample solution. This technique was first proposed by Lampert [50] but gained wider acceptance only since its development by Temminghoff et al. [51]. The membrane is initially in equilibrium with the chosen salt solution on the acceptor side. When the sample solution is added, the cations will permeate the membrane until they attain equilibrium, whereas the rate of transport of anionic species is impeded by the negative charge of the membrane. If the salt concentrations of the donor and acceptor solutions are equal, they will result in equal concentrations of the free metal ions on both sides of the membrane, reaching what is called Donnan equilibrium. If the free ion concentrations are equal, then a simple measurement of the total metal in the acceptor will yield the free metal ion concentration in the sample. This technique has the advantage of being able to perform simultaneous measurements of several elements; nevertheless, the equilibration time is very long (often 24 or 48 h), and the common membranes used are calcium-based. This implies that the background electrolyte will necessarily have calcium ions, with the disadvantage of modification of the ionic strength of the sample solution and the introduction of a significant amount of a divalent cation that competes with our metal ions of interest for the binding ligands present in the solution. DMT is probably the best technique to determine free metal ions in a solution as long as the system under study already has a medium to large concentration of calcium and is stable over longer time windows.

AGNES is an electrochemical stripping technique designed to quantify low free metal ion concentrations (in particular conditions, as low as 10^{-10} mol L^{-1}). This is a two-step technique: a deposition step where the metal ion is amalgamated in the working electrode and a stripping step where the concentration of amalgamated metal is quantified. The essential concept in this particular technique is that during the deposition step, the amalgamation of the metal in the working electrode is allowed to proceed up until the equilibrium value between oxidized and reduced forms on both sides of the interface, which is set by the applied potential and Nernst equation. When this equilibrium state is reached, there will be no concentration gradients, either in solution or in the amalgam. The constant ratio is maintained between the concentrations of the electroactive couple due to the Nernstian equilibrium (determined by the applied potential E_1) and is given by [52]

$$Y = \frac{c^*_{\mathrm{M}^0}}{c^*_{\mathrm{M}}} = \exp\left[\frac{-nF\left(E_1 - E^{0'}\right)}{RT}\right] \tag{32.3}$$

where Y is the concentration gain (or preconcentration factor), $c^*_{M^0}$ is the metal concentration inside the mercury electrode, and $E^{0'}$ stands for the standard formal potential of the redox couple of the metal. The stripping signal is the characteristic measured signal that depends on the electrochemical technique used in the stripping mode, and can be, for instance, the charge in a stripping chronopotenciometry measurement [53]. One clear advantage of AGNES is its simple Nernstian response, similar to an ISE. Effectively, the amalgamated metal is linearly related with the free metal ion concentration [52] with a proportionality factor h (which can be obtained from the calibration), since, according to the following equation, $c^*_{\mathrm{M}^0}$ is Yc^*_{M}:

$$\text{Stripping\_signal} = hc^*_{\mathrm{M}} \tag{32.4}$$

Although AGNES has very good selectivity and low detection limit, two main disadvantages need to be point out: (1) the limitation to metal ions that can amalgamate reversibly on the working electrode, usually a mercury electrode (cadmium, lead, zinc, indium, thalium, bismuth, tin, and antimony, although the last one has very low solubility in mercury); and (2) the time involved in the preconcentration step to reach equilibrium (minutes to hours), which increases with the decrease of bulk concentration.

32.2.2.5.2 Dynamic Speciation Techniques Although in this chapter we focus on the free metal ion concentration, it is worthwhile remembering that the physicochemical systems under study are seldom at equilibrium. Therefore, it is interesting to briefly mention the main trace metal dynamic speciation techniques used: (i) stripping electrochemical techniques, among which anodic stripping voltammetry (ASV) and scanned stripping chronopotentiometry (SSCP) are the most commonly used, and (ii) diffusive gradients in thin films (DGT). The interpretation of the DGT signal for an NM system is quite complicated and is currently under revision, so it will not be detailed here [54].

Among the electrochemical techniques, SSCP has many significant advantages over ASV, and, therefore, we will focus on it. SSCP curves are constructed from a series of individual measurements made over an interval of deposition potentials, E_d, covering the situations where no metal is reduced at the electrode, passing through the Nernstian regime to the region were all the metal reaching the electrode is reduced (diffusion-limited situation). Each individual measurement consists of a two-step process involving (1) the deposition (accumulation) step, where metal ions are reduced at the chosen E_d for a fixed period of time, followed by (2) the stripping step, where the accumulated metal is reoxidized through the application of a low constant oxidizing current (stripping current, I_s). The analytical signal is the electrolysis time, τ (transition time), and therefore an SSCP curve is a representation of τ over E_d, where the analytical signal τ always reflects the magnitude of the original deposition flux, irrespective of its nature (i.e., diffusion controlled or kinetically controlled). The thermodynamic complex stability constant can be calculated from the shift in the half-wave deposition potential, $\Delta E_{d,1/2}$ (analogous to the DeFord–Hume expression), irrespective of the degree of lability of the system [55]:

$$\Delta E_{d,1/2} = -\frac{RT}{nF}\left[\ln\left(1+K'\right)+\ln\left(\frac{\tau^*_{M+L}}{\tau^*_M}\right)\right] \tag{32.5}$$

where τ^*_{M+L} and τ^*_M are the limiting wave heights in the presence and absence of ligands, respectively, n is the number of electrons involved in the faradaic process, and K' is the conditional equilibrium parameter in conditions of excess ligand ($=Kc_{L,t}=c_{ML}/c_M$).

SSCP is especially advantageous when using a thin mercury film electrode (TMFE) since the rapid metal transport inside the thin film during the stripping step implies that the measurements are always performed under conditions of complete depletion. This allows the use of much higher stripping currents with the great advantage of the oxygen not having the time to chemically oxidize the amalgamated metal [56]. The comparison of the stability constants obtained from the two SSCP signals (the shift of the half-wave deposition potential and the decrease of transition time) provides information about the dynamic nature of the metal complexes formed with the ligands present, in reasonable agreement with the predictions from the dynamic theory for colloidal systems [55, 57].

The simultaneous use of AGNES and SSCP to study metal binding with heterogeneous ligands results in a great advantage as they provide complementary data [6, 58].

32.2.2.6 ***Bioavailability and Toxicity of NMs*** Metal speciation is a crucial feature for the comprehension of the transport, bioavailability, and toxicity of metal ions. FIAM and BLM, based on the concept that the formation of any complex in solution will reduce trace metal uptake, successfully describe more than 90% of the metal ion bioavailability studied cases. Exceptions to these models can be explained taking into account the dynamic aspects of interconversion of species and the possibility of some inert complexes to be transported directly through the biological cell membrane (Fig. 32.2).

Bioavailability of metal-containing NMs differs from the free metal ion case in three aspects: (1) NMs may induce cell wall and cell membrane damage; (2) NMs can be internalized; and (3) NMs will provide a point source of metal ions when dissolving.

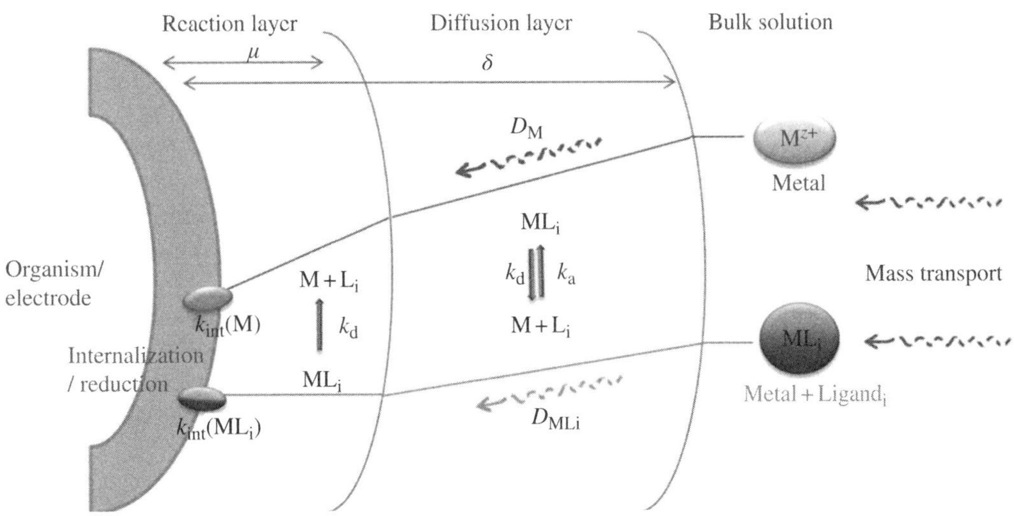

FIGURE 32.2 Schematic representation of the concentration profiles for the metal (M) and the metal–ligand species (ML_i) as a function of distance from the organism/electrode surface. The figure illustrates the biological, chemical, and physical parameters that govern the biouptake/reduction of trace compounds: k_{int}, internalization rate constant; k_a, association rate constant of ML_i; k_d, dissociation rate constant of ML_i; D_M, diffusion coefficient of M; D_{ML}, diffusion coefficient of the ML_i species; μ, thickness of the reaction layer; and δ, thickness of the diffusion layer.

Regarding this last point, an often ignored aspect is that when an NM sticks to an organism interface, it starts dissolving, yielding a tremendously high local metal ion concentration. This is ascribed not only for cell walls or membranes but also for fish gills, mollusk filtration mechanisms, and digestive tubes. The potential damage of highly concentrated point sources cannot be neglected, especially when considering chronic effects on the organism's health.

32.3 RECOMMENDATIONS/GUIDELINES

In this section, it is assumed that the metal-containing NMs and their uses are known, thus defining the environmental compartment where they are ultimately released. A proper literature survey should then be performed to choose the relevant test organism(s). The guidelines to be followed for an environmentally relevant toxicological experimental protocol for metal-containing NMs are based on a four-tier approach (Fig. 32.3):

1. *Exposure solution.* The physicochemical conditions of the exposure solution should mimic as far as possible those of the environmental compartment. Care should be taken to avoid the starvation of the test organism. In the speciation and toxicological assays, the following parameters need to be either fixed or varied in a controlled way: temperature, pH, ionic strength, light, other metals, and organic ligands including NOM. In certain specific cases, other parameters might also be important, for example, dissolved oxygen.
2. *Speciation.* These studies should be used to define the experimental conditions for toxicological assays, whether for short-term (acute effects) or long-term (chronic effects) experiments. The most important parameter that needs to be defined is the NM concentration. The lowest NM concentration should always coincide with the detection limit of the available techniques since they are likely larger than their environmental concentration. The upper limit should be chosen by analogy to the metal ion concentrations used in studies of acute effects. To accomplish this aim, it is necessary to monitor the following aspects in the exposure media during the relevant time period: (i) NM dissolution and (dis)aggregation, (ii) free metal concentration, and (iii) NM interaction with other components of the matrix.

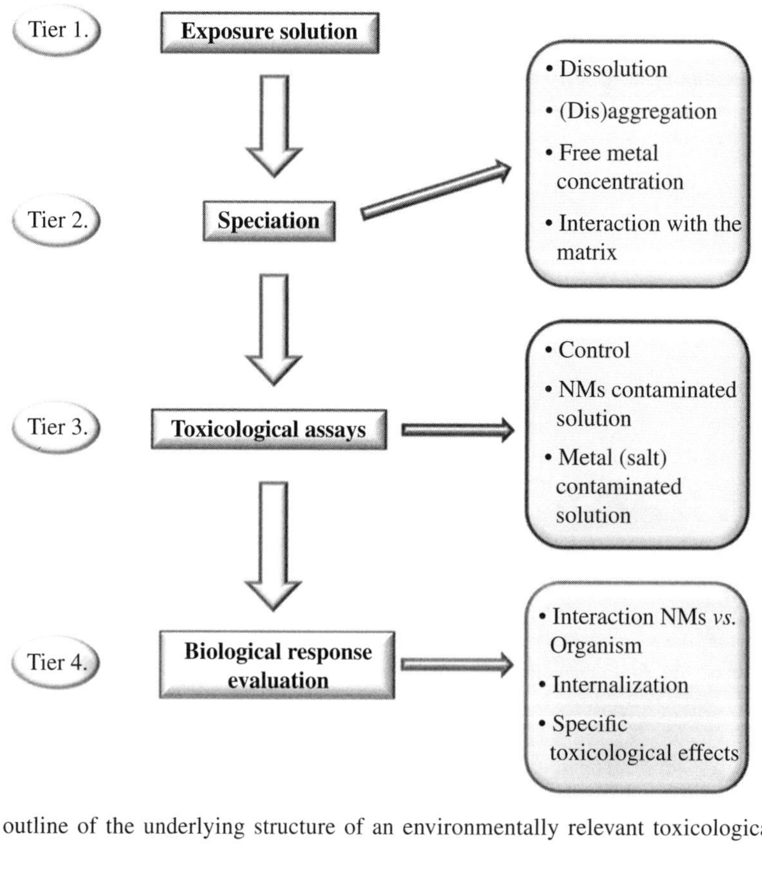

FIGURE 32.3 Schematic outline of the underlying structure of an environmentally relevant toxicological experimental protocol for metal-containing NMs.

3. *Toxicological assays.* The speciation data will be used to select the most adequate test organism available for the environmental compartment of interest; for example, in the case of NM sedimentation, the choice of the test organism will be limited to sediment dwellers. A typical toxicological assay includes a control and the contaminated solution. In the case of metal-containing NMs, an additional control containing the same total metal concentration (added as a salt) is necessary. Care should be taken to ensure that the results are statistically significant.
4. *Biological response evaluation.* The achievement of this objective includes (i) visualization of the interaction between the NM and the organism (such as internalization and adsorption at the interface); (ii) quantification of the metal ion and, when possible, of the NM at the biological interface and internalized in the organism (or in the relevant parts of the organism), and (iii) comparison of the specific toxicological effects induced by the metal ions and the metal-containing NMs inferred from genomics and/or proteomics.

32.4 FUTURE WORK AND PRESSING PROBLEMS

Once in natural systems, NMs will be undoubtedly subjected to a dynamic physical and chemical environment that will consequently result in different and unknown end points and products that are far from their pristine or manufactured state. Unknown factors include (i) the NM form that will be present in commercial products; (ii) the potential for the material to be released into the environment; and (iii) the transformations of the material that may affect exposure. Another aspect is that the detection limit of the available analytical techniques for the quantification and characterization of NMs is still high as compared with their predicted levels in the environment.

Currently, the major pressing problem is our present level of understanding of the NMs' fate and the consequent lack of modeling tools available for prediction. A great deal of research effort will thus be required over the coming years to investigate the mobility of NMs through the environmental compartments in order to enhance our understanding of the underlying mechanisms and identify the key influencing factors. Further research regarding the environmental fate and how it may evolve with time under realistic environmental conditions will also be required in order to achieve more robust risk assessments of NMs.

ACKNOWLEDGMENTS

Funding for this work was provided by the Fundação para a Ciência e Tecnologia (FCT), Portugal: Science 2008 IST-CQE3 "Environmental Chemistry" Assistant Researcher position to RFD and Project PTDC/AAC-AMB/111998/2009. JPP acknowledges the FCT funding support: Project Pest-OE/EQB/LA0023/2013.

REFERENCES

[1] Pomogailo AD, Kestelman VN, editors. Principles and mechanisms of nanoparticle stabilization by polymers. In: *Metallopolymer Nanocomposites*. Volume 81, Berlin Heidelberg: Springer; 2005. p 65–113.

[2] (a) Green JM, Beestman GB. Recently patented and commercialized formulation and adjuvant technology. Crop Protection 2007;26:320–327.

(b) Nair R, Varghese SH, Nair BG, Maekawa T, Yoshida Y, Kumar DS. Nanoparticulate material delivery to plants. Plant Science 2010;179:154–163.

[3] (a) van der Kammer F, Ferguson PL, Holden PA, Masion A, Rogers KR, Klaine SJ, Koelmans AA, Horne N, Unrine JM. Analysis of engineered nanomaterials in complex matrices (environment and biota): general considerations and conceptual case studies. Environmental Toxicology and Chemistry 2012;31:32–49.

(b) Pettitt ME, Lead JR. Minimum physicochemical characterization requirements for nanomaterial regulation. Environment International 2013;52:41–50.

[4] Lok CN, Ho CM, Chen R, He QY, Yu WY, Sun H, Tam PKH, Chiu JF, Che CM. Silver nanoparticles: partial oxidation and antibacterial activities. Journal of Biological Inorganic Chemistry 2007;12:527–534.

[5] Derfus AM, Chan WCW, Bhatia SN. Probing the cytotoxicity of semiconductor quantum dots. Nano Letters 2004;4:11–18.

[6] Domingos RF, Franco C, Pinheiro JP. Stability of core/shell quantum dots—role of pH and small organic ligands. Environmental Science and Pollution Research 2013;20 (7):4872–4880.

[7] (a) Hund-Rinke K, Simon M. Ecotoxic effect of photocatalytic active nanoparticles (TiO_2) on algae and daphnids. Environmental Science and Pollution Research International 2006;13 (4):225–232.

(b) Zhang L, Jiang Y, Povery Y, York D. Investigation into the antibacterial behavior of suspensions of ZnO nanoparticles (ZnO nanofluids). Journal of Nanoparticle Research 2007;9:479–489.

[8] (a) Heinlaan M, Ivask A, Blinova I, Dubourguier H-C, Kahru A. Toxicity of nanosized and bulk ZnO, CuO and TiO_2 to bacteria Vibrio fischeri and crustaceans Daphnia magna and Thamnocephalus platyurus. Chemosphere 2008;71:1308–1316.
(b) Mudunkotuwa IA, Rupasinghe T, Wu C-M, Grassian VH. Dissolution of ZnO nanoparticles at circumneutral pH: a study of size effects in the presence and absence of citric acid. Langmuir 2012;28:396–403.

[9] Hotze EM, Bottero JY, Wiesner MR. Theoretical framework for nanoparticle reactivity as a function of aggregation state. Langmuir 2010;26:11170–11175.

[10] (a) Derjaguin BV, Landau LD. Theory of the stability of strongly charged lyophobic sols and of the adhesion of strongly charged particles in solutions of electrolytes. Acta Physicochimica 1941;14:733–762.
(b) Verwey EJW, Overbeek JTG. *Theory of the Stability of Lyophobic Colloids: The Interaction of Sol Particles Having an Electric Double Layer*. New York: Elsevier; 1948.

[11] (a) Elimelech M, Gregory J, Jia X, Williams RI. *Particle Deposition and Aggregation: Measurement, Modelling and Simulation*. Woburn: Butterworth-Heinemann; 1995.
(b) Grasso D, Subramaniam K, Butkus M, Strevett K, Bergendahl J. A review of non-DLVO interactions in environmental colloidal systems. Reviews in Environmental Science and Biotechnology 2002;1:17–38.

[12] Domingos RF, Peyrot C, Wilkinson KJ. Aggregation of titanium dioxide nanoparticles: role of calcium and phosphate. Environmental Chemistry 2010;7:61–66.

[13] (a) Domingos RF, Tufenkji N, Wilkinson KJ. Aggregation of titanium dioxide nanoparticles: role of a fulvic acid. Environmental Science and Technology 2009;43 (5):1282–1286.
(b) Baalousha M, Manciulea A, Cumberland S, Kendall K, Lead JR. Aggregation and surface properties of iron oxide nanoparticles: influence of pH and natural organic matter. Environmental Toxicology and Chemistry 2008;27:1875–1882.
(c) Bian S-W, Mudunkotuwa IA, Rupasinghe T, Grassian VH. Aggregation and dissolution of 4 nm ZnO nanoparticles in aqueous environments: influence of pH, ionic strength, size, and adsorption of humic acid. Langmuir 2011;27:6059–6068.

[14] Stankus DP, Lohse SE, Hutchison JE, Nason JA. Interactions between natural organic matter and gold nanoparticles stabilized with different organic capping agents. Environmental Science and Technology 2011;45:3238–3244.

[15] Domingos RF, Baalousha MA, Ju-Nam Y, Reid MM, Tufenkji N, Lead JR, Leppard GG, Wilkinson KJ. Characterizing manufactured nanoparticles in the environment: multimethod determination of particle sizes. Environmental Science and Technology 2009; 43:7277–7284.

[16] (a) Schmidt J, Vogelsberger W. Dissolution kinetics of titanium dioxide nanoparticles: the observation of an unusual kinetic size effect. Journal of Physical Chemistry B 2006;110:3955–3963.
(b) Li M, Zhu L, Lin D. Toxicity of ZnO nanoparticles to Escherichia coli: mechanism and the influence of medium components. Environmental Science and Technology 2011;45:1977–1983.

[17] (a) Kittler S, Greulich C, Diendorf J, Koller M, Epple M. Toxicity of silver nanoparticles increases during storage because of slow dissolution under release of silver ions. Chemistry of Materials 2010;22:4548–4554.
(b) Misra SK, Dybowska A, Berhanu D, Croteau MN, Luoma SN, Boccaccini AR, Valsami-Jones E. Isotopically modified nanoparticles for enhanced detection in bioaccumulation studies. Environmental Science and Technology 2012;46:1216–1222.

[18] (a) Li X, Lenhart JJ, Walker HW. Aggregation kinetics and dissolution of coated silver nanoparticles. Langmuir 2012;28:1095–1104.
(b) Ma R, Levard C, Marinakos SM, Chen Y, Liu J, Michel FM, Brown GE, Lowry GV. Size-controlled dissolution of organic-coated silver nanoparticles. Environmental Science and Technology 2012;46:752–759.

[19] Laborda F, Jiménez-Lamana J, Bolea E, Castillo JR. Selective identification, characterization and determination of dissolved silver(I) and silver nanoparticles based on single particle detection by inductively coupled plasma mass spectrometry. Journal of Analytical Atomic Spectrometry 2011;26:1362–1371.

[20] Leppard GG. Nanoparticles in the environment as revealed by transmission electron microscopy: detection, characterisation and activities. Current Nanoscience 2008;4:278–301.

[21] Balnois E, Papastavrou G, Wilkinson KJ. Force microscopy and force measurements of environmental colloids. In: Wilkinson KJ, Lead J, editors. *Environmental colloids: behaviour, structure and characterisation*. Volume 10, Chichester: John Wiley & Sons, Ltd; 2007. p 405–467.

[22] (a) Filella M, Zhang J, Newman ME, Buffle J. Analytical applications of photon correlation spectroscopy for size distribution measurements of natural colloidal suspensions: capabilities and limitations. Colloids and Surfaces A: Physicochemical and Engineering Aspects 1997;120:27–46.
(b) Schurtenberger P, Newman ME. Characterization of biological and environmental particles using static and dynamic light scattering. In: Buffle J, van Leeuwen HP, editors. *Environmental Particles*. Volume II, Boca Raton, FL: Lewis; 1993. p 37–115.

[23] (a) Gottschalk F, Sonderer T, Scholz RW, Nowack B. Modeled environmental concentrations of engineered nanomaterials (TiO_2, ZnO, Ag, CNT, fullerenes) for different regions. Environmental Science and Technology 2009;43:9216–9222.
(b) Gottschalk F, Ort C, Scholz RW, Nowack B. Engineered nanomaterials in rivers—exposure scenarios for Switzerland at high spatial and temporal resolution. Environmental Pollution 2011;159:3439–3445.
(c) Gottschalk F, Nowack B. The release of engineered nanomaterials to the environment. Journal of Environmental Monitoring 2011;13:1145–1155.

[24] (a) Thompson NL. Fluorescence correlation spectroscopy. In: Lakowicz JR, editor. *Topics in Fluorescence Spectroscopy, Techniques*. Volume 1, New York: Plenum Press; 1991. p 337–378.
(b) Fatin-Rouge N, Buffle J. *Study of Environmental Systems by Means of Fluorescence Correlation Spectroscopy*. Volume 10, Chichester: John Wiley and Sons, Ltd; 2007.
(c) Starchev K, Wilkinson KJ, Buffle J. *Application of Fluorescence Correlation Spectroscopy to the Study of Environmental Systems In Fluorescence Correlation Spectroscopy: Theory and Applications*. Heidelberg: Springer; 2000.

[25] Starchev K, Zhang J, Buffle J. Applications of fluorescence correlation spectroscopy—particle size effect. Journal of Colloid and Interface Science 1998;203:189–196.

[26] Gallego-Urrea JA, Tuoriniemi J, Hassellov M. Applications of particle-tracking analysis to the determination of size distributions and concentrations of nanoparticles in environmental biological and food samples. Trends in Analytical Chemistry 2011;30:473–483.

[27] (a) Pace HE, Rogers NJ, Jarolimek C, Coleman VA, Higgins CP, Ranville JF. Determining transport efficiency for the purpose of counting and sizing nanoparticles via single particle inductively coupled plasma mass spectrometry. Analytical Chemistry 2011;83:9361–9369.
(b) Mitrano DM, Lesher EK, Bednar A, Monserud J, Higgins CP, Ranville JF. Detecting nanoparticulate silver using single-particle inductively coupled plasma-mass spectrometry. Environmental Toxicology and Chemistry 2012;31:115–121.

[28] Hassellov M, van der Kammer F, Beckett R. Characterization of aquatic colloids and macromolecules by field-flow fractionation. In: Wilkinson KJ, Lead JR, editors. *Environmental Colloids and Particles: Behaviour, Structure and Characterisation*. Volume 10, Chichester: John Wiley and Sons, Ltd; 2007. p 223.

[29] (a) Tiede K, Boxall ABA, Tiede D, Tear SP, David H, Lewis J. A robust size-characterization methodology for studying nanoparticle behaviour in "real" environmental samples, using hydrodynamic chromatography coupled to ICP-MS. Journal of Analytical Atomic Spectrometry 2009;24:964–972.
(b) Tiede K, Boxall ABA, Wang X, Gore D, Tiede D, Baxter M, David H, Tear SP, Lewis J. Application of hydrodynamic chromatography-ICP-MS to investigate the fate of silver nanoparticles in activated sludge. Journal of Analytical Atomic Spectrometry 2010;25:1149–1154.

[30] Handy RD, van der Kammer F, Lead JR, Hassellov M, Owen R, Crane M. The ecotoxicology and chemistry of manufactures nanoparticles. Ecotoxicology 2008;17:287–314.

[31] van Leeuwen HP. Revisited the conception of lability of metal complexes. Electroanalysis 2001;13:826–830.

[32] van Leeuwen HP, Cleven R, Buffle J. Voltammetric techniques for complexation measurements in natural aquatic media. Pure and Applied Chemistry 1989;61:255–274.

[33] (a) Morel FMM, Hering JG. *Principles and Applications of Aquatic Chemistry*. Chichester: John Wiley & Sons, Ltd; 1993.
(b) Campbell PGC. Interactions between trace metals and aquatic organisms: a critique of the free-ion activity model. In: Tessier A, Turner DR, editors. *Metal Speciation and Bioavailability in Aquatic Systems*. Volume 3, Chichester: John Wiley & Sons, Ltd; 1995. p 45–102.

[34] (a) Playle RC. Modeling metal interaction at fish gills. Science of the Total Environment 1998;219:147–163.
(b) Hassler CS, Slaveykova VI, Wilkinson KJ. Some fundamental (and often overlooked) considerations underlying the free ion activity and biotic ligand models. Environmental Toxicology and Chemistry 2004;23:283–291.

[35] Slaveykova VI, Wilkinson KJ. Predicting the bioavailability of metals and metal complexes: critical review of the biotic ligand model. Environmental Chemistry 2005;2:9–24.

[36] Mota AM, Pinheiro JP, Gonçalves MLS. Electrochemical methods for speciation of trace elements in marine waters. dynamic aspects. Journal of Physical Chemistry A 2012;116:6433–6442.

[37] Davies JT, Rideal EK. *Interfacial Phenomena*. London: Academic Press; 1963.

[38] (a) Plette ACC, van Riemsdijk WH, Benedetti MF, van der Wal A. pH dependent charging behavior of isolated cell walls of a gram-positive soil bacterium. Journal of Colloid and Interface Science 1995;173:354–363.
(b) Yee N, Fowle DA, Ferris FG. A Donnan potential model for metal sorption onto Bacillus subtilis. Geochimica et Cosmochimica Acta 2004;68:3657–3664.

[39] (a) Schiewer S, Volesky B. Ionic strength and electrostatic effects in biosorption of divalent metal ions and protons. Environmental Science and Technology 1997;31:2478–2485.
(b) Rey-Castro C, Lodeiro P, Herrero R, de Vicente MES. Acid-base properties of brown seaweed biomass considered as a Donnan gel. A model reflecting electrostatic effects and chemical heterogeneity. Environmental Science and Technology 2003;37:5159–5167.

[40] Tipping E. WHAMC—a chemical equilibrium model and computer code for waters, sediments and soils incorporating a discrete-site/electrostatic model of ion-binding by humic substances. Computers and Geosciences 1994;20:973–1023.

[41] Koopal LK, Saito T, Pinheiro JP, van Riemsdijk WH. Ion binding to natural organic matter: general considerations and the nica-donnan model. Colloids and Surfaces A: Physicochemical and Engineering Aspects 2005;265:40–54.

[42] Davis TA, Yezek LP, Pinheiro JP, van Leeuwen HP. Measurement of Donnan potentials in gels by in-situ microelectrode. Journal of Electroanalytical Chemistry 2005;584:100–109.

[43] Davis TA, Kalis EJJ, Pinheiro JP, Town RM, van Leeuwen HP. Cd(II) speciation in alginate gels. Environmental Science and Technology 2008;42:7242–7247.

[44] (a) Hartmann NB, Legros S, van der Kammer F, Hofmann T, Baun A. The potential of TiO_2 nanoparticles as carriers for cadmium uptake in Lumbriculus variegatus and Daphnia magna. Aquatic Toxicology 2012;118–119:1–8.
(b) Zhang X, Sun H, Zhang Z, Niu Q, Chen Y, Crittenden JC. Enhanced bioaccumulation of cadmium in carp in the presence of titanium dioxide nanoparticles. Chemosphere 2007;67:160–166.

[45] Diegoli S, Manciulea AL, Begum S, Jones IP, Lead JR, Preece JA. Interaction between manufactured gold nanoparticles and naturally occurring organic macromolecules. Science of the Total Environment 2008;402:51–61.

[46] van den Berg CMG. Potentials and potentialities of cathodic stripping voltammetry of trace elements in natural waters. Analytica Chimica Acta 1991;250:265–276.

[47] Pesavento M, Alberti G, Biesuz R. Analytical methods for determination of free metal ion concentration, labile species fraction and metal complexation capacity of environmental waters: a review. Analytica Chimica Acta 2009;631:129–141.

[48] Buffle J. *Study of Complexation Properties by Voltammetric Methods*. Chichester: Ellis Horwood; 1988.

[49] (a) Batley GE, Apte SC, Stauber JL. Speciation and bioavailability of trace metals in water: progress since 1982. Australian Journal of Chemistry 2004;57:903–919.
(b) Pretsch E. The new wave of ion-selective electrodes. Trends in Analytical Chemistry 2007;26:47–51.

[50] Lampert JK. *Measurement of Trace Cation Activities by Donnan Membrane Equilibrium and Atomic Absorption Analysis*. Wisconsin, MI: University of Wisconsin-Madison; 1982.

[51] Temminghoff EJM, Plette ACC, van Eck R, van Riemsdijk WH. Determination of the chemical speciation of trace metals in aqueous systems by the Wageningen Donnan membrane technique. Analytica Chimica Acta 2000;417:149–157.

[52] Galceran J, Companys E, Puy J, Cecília J, Garcés JL. AGNES: a new electroanalytical technique for measuring free metal ion concentration. Journal of Electroanalytical Chemistry 2004;566:95–109.

[53] Parat C, Authier L, Aguilar D, Companys E, Puy J, Galceran J, Potin-Gautier M. Direct determination of free metal concentration by implementing stripping chronopotentiometry as the second stage of AGNES. Analyst 2011;136:4337–4343.

[54] (a) van der Veeken PLR, Pinheiro JP, van Leeuwen HP. Metal speciation by DGT/DET in colloidal complex systems. Environmental Science and Technology 2008;42:8835–8840.
(b) Puy J, Uribe R, Mongin S, Galceran J, Cecilia J, Levy J, Zhang H, Davison W. Lability criteria in diffusive gradients in thin film. Journal of Physical Chemistry A 2012;116:6564–6573.

[55] Pinheiro JP, van Leeuwen HP. Scanned stripping chronopotentiometry of metal complexes: lability diagnosis and stability computation. Journal of Electroanalytical Chemistry 2004;570:69–75.

[56] Parat C, Schneider A, Castetbon A, Potin-Gautier M. Determination of trace metal speciation parameters by using screen-printed electrodes in stripping chronopotentiometry without deaerating. Analytica Chimica Acta 2011;688:156–162.

[57] (a) Pinheiro JP, Minor M, van Leeuwen HP. Metal speciation dynamics in colloidal ligand dispersion. Langmuir 2005;21:8635–8642.
(b) Pinheiro JP, Domingos RF, Minor M, van Leeuwen HP. Metal speciation dynamics in colloidal ligand dispersions. Part 3: lability features of steady-state systems. Journal of Electroanalytical Chemistry 2006;596:57–64.

[58] Domingos RF, Huidobro C, Companys E, Galceran J, Puy J, Pinheiro JP. Comparison of AGNES (absence of gradients and Nernstian equilibrium stripping) and SSCP (scanned stripping chronopotentiometry) for trace metal speciation analysis. Journal of Electroanalytical Chemistry 2008;617:141–148.

CONCLUDING REMARKS

The field of nanoscience and nanotechnology has developed increasingly during the past 20 years and nanomaterials have become very useful and effective in several distinct areas of environmental protection. For example, in water sector, nanomaterials, mainly in the forms of nanomembranes (organic polymers or inorganic ceramics), are utilized in waste water treatment (as commercialized products or in product development/testing stages). Nanofiltration is used for drinking water treatment, seawater desalination, and in food industry. Nanosorbents may be used for the remediation of toxic elements, such as arsenic. Nanoscale titanium dioxide is used as a catalyst in waste water treatment by photocatalysis; this results in the formation of reactive hydroxyl radicals, organic pollutants resistant to biological degradation. Among elemental metal or bimetallic nanomaterials, zero-valent iron nanoparticles (nZVI) (pure form or on activated carbon) are used for in situ groundwater remediation, mainly for the elimination of chlorine-containing volatile organic compounds. Nano-based coatings (nanosealers) are also used, in particular, in ceramic industry. In case of air sector, nanomaterials such as nanoscale noble metal catalysts are utilized in catalytic converters in the automobile industry. Nanomaterials, mostly in the nanofiber form, are frequently used in the filters for passenger cabin air and for dust removal in fabrics. Nanomaterials are used for eliminating CO_2 from power plant flue gases, which is also an important field of application.

As it is evident from the discussions made in the preceding chapters, the bulk of the activities are focused on water purification. Obviously, it is not necessary to again emphasize that clean water (i.e., water that is free of toxic chemicals and pathogens) is essential to human health. Clean water is also a critical feedstock in a variety of key industries including electronics, pharmaceuticals, and food. The world is facing formidable challenges in meeting rising demands of clean water as the available supplies of freshwater are decreasing due to (i) extended droughts, (ii) population growth, (iii) more stringent health-based regulations, and (iv) competing demands from a variety of users [1]. Nanosorbents, nanocatalysts and redox active nanoparticles, nanostructured and reactive membranes, bioactive nanoparticles, dendrimer-enhanced ultrafiltration, and other nanomaterials/nanoprocesses are attractive as separation media for water purification. On a mass basis, they have much large surface areas than bulk particles. Nanomaterials can also be functionalized with various chemical groups to increase their affinity toward a given compound. They can also serve as high capacity/selective and recyclable ligands/adsorbents for toxic metal ions and radionuclides, and organic and inorganic solutes/anions in aqueous solutions. Regarding nanomembranes, it is to be noted that since approximately 97% of water on Earth is salt water, desalination using engineered membranes containing nanoparticles has currently become an attractive option for the production of drinking water.

Rather than the use of metals, metal oxides, and metal complexes and their composites, the use of ZVI NPs, iron oxides, and other Fe-containing nanomaterials for remediation of both organic and inorganic pollutants from groundwater and contaminated soils has become a relatively significant topic in nanotechnology. Due to the relatively low toxic nature of

Nanomaterials for Environmental Protection, First Edition. Edited by Boris I. Kharisov, Oxana V. Kharissova, and H. V. Rasika Dias.

iron-containing nanoparticles, they can be applied in free and supported forms without any significant restrictions to obtain a considerable decontamination effect of the environment. Nevertheless, the evaluation of toxicity of the Fe-containing nanomaterials remains an object of permanent research and polemics. Since real long-time effects of the presence of iron-containing nanoparticles in rivers, groundwater, lakes, and oceans are still unknown, the idea [2] of introducing iron-containing nanoparticles in a large scale into oceans, in order to stimulate the growth of phytoplankton and increase the adsorption of CO_2, was further rejected. Carbon nanotubes (CNTs) offer tremendous opportunity and possibilities for addressing the problems related to water quality. They have high sorption capacity toward heavy metals and microorganisms, can inhibit cell activities, and are viewed as novel membrane materials that induce immense improvement in membrane-separation processes. Titanium dioxide, another "heart" of nanotechnology, possesses highly photoreactive (001) surface, allowing redox reactions with adsorbed water molecules and hydroxide ions, leading to the formation of reactive species that can break down toxic compounds and kill bacteria in water. Also, its use allows removal of arsenic, chromium, uranium, mercury, and lead. Related combination, colloidal nanocarbon–metal nanocomposites with TiO_2 as the metal oxide, can also serve as a nanophotocatalyst for the photodegradation of dichloro-diphenyl-trichloroethane, aldrin ($C_{12}H_8Cl_6$), lindane ($C_6H_6Cl_6$), and polychlorinated biphenyls. Ceramic materials with perovskite-type structure ($NaTaO_3$, La:$NaTaO_3$, and Sm:$NaTaO_3$), pyrochlore-type structure (Sm_2MTaO_7 (M = Fe, In, Ga), Bi_2MTaO_7 (M = Fe, In, Ga)), and rectangular tunnel-type structure ($Na_2Ti_6O_{13}$) have proved to be efficient catalysts for the degradation of organic pollutants present in aqueous solution and for water conversion into hydrogen and oxygen. Several metal complexes have also been utilized for environmental purposes.

Nanoscale adsorbents (nanoscale particles from organic or inorganic materials that have a high affinity to adsorb substances) are being intensively studied in respect to their environmental applications. Because of their high porosity, small size, and active surface, nano-adsorbents not only are capable of sequestering contaminants with varying molecular size, hydrophobicity, and speciation behavior but also they enable the manufacturing processes as raw materials are utilized efficiently without releasing their toxic payload. In particular, clay nanomaterials have been widely used for the removal of toxic metal ions from contaminated water, in which the toxic metal ions were adsorbed onto the clay nanomaterials via electrostatic interactions or intermolecular forces. Numerous other adsorbents, for example, activated carbon and other carbon nanomaterials, silica gel, zeolites, low-cost adsorbents from agro-industrial wastes, biosorbents, mineral-based adsorbents, and layered-double hydroxides, have been examined for their potential toward the removal (adsorption) of diverse type of aquatic pollutants as well as gas phase adsorption of environmental contaminants. Nanomembranes can be incorporated with various nanoparticles including TiO_2, SiO_2, Al_2O_3, ZnO, Fe_2O_3, Fe, and Ag via different routes, including homogeneous dispersion and surface coating. Thus produced nanomembranes can serve for drinking water purposes.

Nanomaterials/nanoscale objects can be fabricated by a variety of techniques, from classic wet chemistry routes to sophisticated methods such as flame, laser, microwave, ultrasonic, radiation, CVD, sputtering methods, wire explosion, and arc discharge. However, much milder "greener" fabrication processes are available. Products from nature or those derived from natural products, such as extracts of various plants or parts of plants, tea, coffee, banana, simple amino acids, as well as wine, table sugar, and glucose, have been used as reductants and as capping agents during the synthesis. Polyphenols found in plant material often play a key role in these processes. The techniques involved are simple, environmentally friendly, and generally one-pot processes. Tea extracts with high polyphenol content act as both chelating/reducing and capping agents for nanoparticles. The majority of greener synthetic efforts reported thus far are dedicated to obtaining silver nanoparticles, perhaps due to their ease of preparation, importance in disinfection science, and utility in several applications. Several other metals (Au, Fe, Pt, Pd, Cu), their alloys, oxides, and salts have also been obtained by "greener" methods. Bacteria and fungi can also lead to similar products.

In addition to the uses noted, nanomaterials are also used for the synthesis of nanosensors, which can serve for neutron detection, for hydrogen storage purposes, fuel cells, drug delivery, cosmetics, and clothes. At the same time, although for decades research has focused on most of the nanomaterials, it is very likely that the long-term effects have been overlooked or even the data presently available on their potential toxicity are not adequate to reach a conclusion. Therefore, the development of eco-toxicology (a multidisciplinary field of study that aims to evaluate and predict the impact of toxic chemicals on biological organisms) in relation to nanosystems is extremely important to evaluate the potential risks and effects of nanomaterials on environment and human health. In this respect, some technologically important nanomaterials (Ag, Au, Fe, CdS, CdSe, CdTe, ZnSe, TiO_2, Fe_2O_3, Fe_3O_4, CeO_2, SiO_2, ZnO, CuO, SWCNT, MWCNT, fullerenes, graphene, as well as several nanocomposites) are discussed in this book, concluding that before large-scale production and application for commercial and medical purposes, nanomaterials should be evaluated for their long-term risk.

In conclusion, we could state that the use of nanomaterials/nanocomposites/nanoparticles for the protection of the environment is primarily at the research and development level, especially when air and soil protection is considered. In case of water sector, there has been a significant progress and several processes are more frequently applied, but they are not competing enough with the existing and classic well-established technologies.

THE EDITORS

REFERENCES

[1] Savage N, Diallo MS. Nanomaterials and water purification: opportunities and challenges. J Nanopart Res 2005;7:331–342.

[2] Boyd PW, Watson AJ, Law CS, Abraham ER, Abraham ER, Trull T, Murdoch R, Bakker DC, Bowie AR, Buesseler KO, Chang H, Charette M, Croot P, Downing K, Frew R, Gall M, Hadfield M, Hall J, Harvey M, Jameson G, LaRoche J, Liddicoat M, Ling R, Maldonado MT, McKay RM, Nodder S, Pickmere S, Pridmore R, Rintoul S, Safi K, Sutton P, Strzepek R, Tanneberger K, Turner S, Waite A, Zeldis J. A mesoscale phytoplankton bloom in the polar Southern Ocean stimulated by iron fertilization. Nature 2000;407: 695–702.

AUTHOR INDEX

Nanomaterials for Environmental Protection, First Edition. Edited by Boris I. Kharisov, Oxana V. Kharissova, and H. V. Rasika Dias.

SUBJECT INDEX

Nanomaterials for Environmental Protection, First Edition. Edited by Boris I. Kharisov, Oxana V. Kharissova, and H. V. Rasika Dias.